NUMERICAL MATHEMATICS AND SCIENTIFIC COMPUTATION

NUMERICAL MATHEMATICS AND SCIENTIFIC COMPUTATION

Books in the series

Monographs marked with an asterisk (*) appeared in the series 'Monographs in Numerical Analysis' which is continued by the current series.

For a full list of titles please visit
http://www.oup.co.uk/academic/science/maths/series/nmsc

* J. H. Wilkinson: *The algebraic eigenvalue problem*
* I. Duff, A. Erisman, and J. Reid: *Direct methods for sparse matrices*
* M. J. Baines: *Moving finite elements*
* J. D. Pryce: *Numerical solution of Sturm-Liouville problems*

C. Schwab: *p- and hp- finite element methods: theory and applications to solid and fluid mechanics*
J. W. Jerome: *Modelling and computation for applications in mathematics, science, and engineering*
A. Quarteroni and A. Valli: *Domain decomposition methods for partial differential equations*
G. Em Karniadakis and S. J. Sherwin: *Spectral/hp element methods for CFD*
I. Babuška and T. Strouboulis: *The finite element method and its reliability*
B. Mohammadi and O. Pironneau: *Applied shape optimization for fluids*
S. Succi: *The lattice Boltzmann equation: for fluid dynamics and beyond*
P. Monk: *Finite element methods for Maxwell's equations*
A. Bellen and M. Zennaro: *Numerical methods for delay differential equations*
J. Modersitzki: *Numerical methods for image registration*
M. Feistauer, J. Felcman, and I. Straškraba: *Mathematical and computational methods for compressible flow*
W. Gautschi: *Orthogonal polynomials: computation and approximation*
M. K. Ng: *Iterative methods for Toeplitz systems*
M. Metcalf, J. Reid, and M. Cohen: *Fortran 95/2003 explained*
G. Em Karniadakis and S. Sherwin: *Spectral/hp element methods for computational fluid dynamics, second edition*
D. A. Bini, G. Latouche, and B. Meini: *Numerical methods for structured Markov chains*
H. Elman, D. Silvester, and A. Wathen: *Finite elements and fast iterative solvers: with applications in incompressible fluid dynamics*
M. Chu and G. Golub: *Inverse eigenvalue problems: theory, algorithms, and applications*
J.-F. Gerbeau, C. Le Bris, and T. Lelièvre: *Mathematical methods for the magnetohydrodynamics of liquid metals*
G. Allaire and A. Craig: *Numerical analysis and optimization: an introduction to mathematical modelling and numerical simulation*
K. Urban: *Wavelet methods for elliptic partial differential equations*
B. Mohammadi and O. Pironneau: *Applied shape optimization for fluids, second edition*
K. Böhmer: *Numerical methods for nonlinear elliptic differential equations: a synopsis*
M. Metcalf, J. Reid, and M. Cohen: *Modern Fortran Explained*

Numerical Methods for Nonlinear Elliptic Differential Equations

A Synopsis

Klaus Böhmer
University of Marburg

OXFORD
UNIVERSITY PRESS

OXFORD
UNIVERSITY PRESS

Great Clarendon Street, Oxford OX2 6DP

Oxford University Press is a department of the University of Oxford.
It furthers the University's objective of excellence in research, scholarship,
and education by publishing worldwide in

Oxford New York

Auckland Cape Town Dar es Salaam Hong Kong Karachi
Kuala Lumpur Madrid Melbourne Mexico City Nairobi
New Delhi Shanghai Taipei Toronto

With offices in

Argentina Austria Brazil Chile Czech Republic France Greece
Guatemala Hungary Italy Japan Poland Portugal Singapore
South Korea Switzerland Thailand Turkey Ukraine Vietnam

Oxford is a registered trade mark of Oxford University Press
in the UK and in certain other countries

Published in the United States
by Oxford University Press Inc., New York

First published 2010

British Library Cataloguing in Publication Data
Data available

Library of Congress Control Number: 2009943744

Typeset by SPI Publisher Services, Pondicherry, India
Printed in Great Britain
on acid-free paper by
CPI Antony Rowe, Chippenham, Wiltshire

ISBN 978-0-19-957704-0

1 3 5 7 9 10 8 6 4 2

With love to my late wife Inge,
my daughters Annette and Christine, and
my grandchildren Joachim, Dorothea, Leon, Luca, and Rebekka

* *To God Alone Be The Glory, Organ St. Michaelis, Lüneburg*

Contents

Preface xiv

PART I ANALYTICAL RESULTS

1 From linear to nonlinear equations, fundamental results 3
1.1 Introduction 3
1.2 Linear versus nonlinear models 3
1.3 Examples for nonlinear partial differential equations 10
1.4 Fundamental results 13
1.4.1 Linear operators and functionals in Banach spaces 13
1.4.2 Inequalities and $L^p(\Omega)$ spaces 18
1.4.3 Hölder and Sobolev spaces and more 20
1.4.4 Derivatives in Banach spaces 27

2 Elements of analysis for linear and nonlinear partial elliptic differential equations and systems 32
2.1 Introduction 32
2.2 Linear elliptic differential operators of second order, bilinear forms and solution concepts 36
2.3 Bilinear forms and induced linear operators 45
2.4 Linear elliptic differential operators, Fredholm alternative and regular solutions 54
2.4.1 Introduction 54
2.4.2 Linear operators of order $2m$ with C^∞ coefficients 58
2.4.3 Linear operators of order 2 under C^k conditions 64
2.4.4 Weak elliptic equation of order $2m$ in Hilbert spaces 69
2.5 Nonlinear elliptic equations 77
2.5.1 Introduction 77
2.5.2 Definitions for nonlinear elliptic operators 79
2.5.3 Special semilinear and quasilinear operators 81
2.5.4 Quasilinear elliptic equations of order 2 88
2.5.5 General nonlinear and Nemyckii operators 96
2.5.6 Divergent quasilinear elliptic equations of order $2m$ 100
2.5.7 Fully nonlinear elliptic equations of orders 2, m and $2m$ 108
2.6 Linear and nonlinear elliptic systems 113
2.6.1 Introduction 113
2.6.2 General systems of elliptic differential equations 114
2.6.3 Linear elliptic systems of order 2 118

2.6.4 Quasilinear elliptic systems of order 2 and variational methods 125
2.6.5 Linear elliptic systems of order $2m, m \geq 1$ 132
2.6.6 Divergent quasilinear elliptic systems of order $2m$ 137
2.6.7 Nemyckii operators and quasilinear divergent systems of order $2m$ 140
2.6.8 Fully nonlinear elliptic systems of orders 2 and $2m$ 146
2.7 Linearization of nonlinear operators 147
2.7.1 Introduction 147
2.7.2 Special semilinear and quasilinear equations 149
2.7.3 Divergent quasilinear and fully nonlinear equations 151
2.7.4 Quasilinear elliptic systems of orders 2 and $2m$ 157
2.7.5 Linearizing general divergent quasilinear and fully nonlinear systems 158
2.8 The Navier–Stokes equation 163
2.8.1 Introduction 163
2.8.2 The Stokes operator and saddle point problems 163
2.8.3 The Navier–Stokes operator and its linearization 167

PART II NUMERICAL METHODS

3 A general discretization theory 173
3.1 Introduction 173
3.2 Petrov–Galerkin and general discretization methods 175
3.3 Variational and classical consistency 185
3.4 Stability and consistency yield convergence 189
3.5 Techniques for proving stability 194
3.6 Stability implies invertibility 203
3.7 Solving nonlinear systems: Continuation and Newton's method based upon the mesh independence principle (MIP) 205
3.7.1 Continuation methods 205
3.7.2 MIP for nonlinear systems 206

4 Conforming finite element methods (FEMs) 209
4.1 Introduction 209
4.2 Approximation theory for finite elements 212
4.2.1 Subdivisions and finite elements 212
4.2.2 Polynomial finite elements, triangular and rectangular K 214
4.2.3 Interpolation in finite element spaces, an example 221
4.2.4 Interpolation errors and inverse estimates 229
4.2.5 Inverse estimates on nonquasiuniform triangulations 233
4.2.6 Smooth FEs on polyhedral domains, **with O. Davydov** 238
4.2.7 Curved boundaries 250
4.3 FEMs for linear problems 257

4.3.1 Finite element methods: a simple example, essential tools 258
4.3.2 Finite element methods for general linear equations and systems of orders 2 and $2m$ 264
4.3.3 General convergence theory for conforming FEMs 266
4.4 Finite element methods for divergent quasilinear elliptic equations and systems 273
4.5 General convergence theory for monotone and quasilinear operators 277
4.6 Mixed FEMs for Navier–Stokes and saddle point equations 281
4.6.1 Navier–Stokes and saddle point equations 281
4.6.2 Mixed FEMs for Stokes and saddle point equations 282
4.6.3 Mixed FEMs for the Navier–Stokes operator 286
4.7 Variational methods for eigenvalue problems 288
4.7.1 Introduction 288
4.7.2 Theory for eigenvalue problems 289
4.7.3 Different variational methods for eigenvalue problems 292

5 Nonconforming finite element methods 296
5.1 Introduction 296
5.2 Finite element methods for fully nonlinear elliptic problems 298
5.2.1 Introduction 298
5.2.2 Main ideas and results for the new FEM: An extended summary 299
5.2.3 Fully nonlinear and general quasilinear elliptic equations 305
5.2.4 Existence and convergence for semiconforming FEMs 308
5.2.5 Definition of nonconforming FEMs 311
5.2.6 Consistency for nonconforming FEMs 317
5.2.7 Stability for the linearized operator and convergence 319
5.2.8 Discretization of equations and systems of order $2m$ 332
5.2.9 Consistency, stability and convergence for $m, q \geq 1$ 336
5.2.10 Numerical solution of the FE equations with Newton's method 341
5.3 FE and other methods for nonlinear boundary conditions 345
5.4 Quadrature approximate FEMs 346
5.4.1 Introduction 346
5.4.2 Quadrature and cubature formulas 348
5.4.3 Quadrature for second order linear problems 350
5.4.4 Quadrature for second order fully nonlinear equations 357
5.4.5 Quadrature FEMs for equations and systems of order $2m$ 361
5.4.6 Two useful propositions 367
5.5 Consistency, stability and convergence for FEMs with variational crimes 368
5.5.1 Introduction 368
5.5.2 Variational crimes for our standard example 370

5.5.3 FEMs with crimes for linear and quasilinear problems 380
5.5.4 Discrete coercivity and consistency 387
5.5.5 High order quadrature on edges 390
5.5.6 Violated boundary conditions 392
5.5.7 Violated continuity 399
5.5.8 Stability for nonconforming FEMs 406
5.5.9 Convergence, quadrature and solution of FEMs with crimes 411
5.5.10 Isoparametric FEMs 414

6 Adaptive finite element methods, by W. Dörfler 420
6.1 Introduction 420
6.1.1 The model problem 421
6.1.2 Singular solutions 421
6.1.3 A priori error bounds 423
6.1.4 Necessity of nonuniform mesh refinement 425
6.1.5 Optimal meshes – A heuristic argument 425
6.1.6 Optimal meshes for 2D corner singularities 427
6.1.7 The finite element method–Notation and requirements 428
6.2 The residual error estimator for the Poisson problem 430
6.2.1 Upper a posteriori bound 430
6.2.2 Lower a posteriori bound 432
6.2.3 The a posteriori error estimate 433
6.2.4 The adaptive finite element method 434
6.2.5 Stable refinement methods for triangulations in $\mathbb{R}^2$ 436
6.2.6 Convergence of the adaptive finite element method 438
6.2.7 Optimality 442
6.2.8 Other types of estimators 447
6.2.9 *hp* finite element method 448
6.3 Estimation of quantities of interest 449
6.3.1 Quantities of interest 449
6.3.2 Error estimates for point errors 449
6.3.3 Optimal meshes–A heuristic argument 451
6.3.4 The general approach 451

7 Discontinuous Galerkin methods (DCGMs), with V. Dolejší 455
7.1 Introduction 455
7.2 The model problem 459
7.3 Discretization of the problem 461
7.3.1 Triangulations 461
7.3.2 Broken Sobolev spaces 462
7.3.3 Extended variational formulation of the problem 463
7.3.4 Discretization 469
7.4 General linear elliptic problems 472

7.5 Semilinear and quasilinear elliptic problems 474
7.5.1 Semilinear elliptic problems 474
7.5.2 Variational formulation and discretization of the problem 475
7.5.3 Quasilinear elliptic systems 477
7.5.4 Discretization of the quasilinear systems 478
7.6 DCGMs are general discretization methods 482
7.7 Geometry of the mesh, error and inverse estimates 486
7.7.1 Geometry of the mesh 487
7.7.2 Inverse and interpolation error estimates 487
7.8 Penalty norms and consistency of the J_h^σ 491
7.9 Coercive linearized principal parts 494
7.9.1 Coercivity of the original linearized principal parts 494
7.9.2 Coercivity and boundedness in $\mathcal{V}^h$ for the Laplacian 495
7.9.3 Coercivity and boundedness in $\mathcal{V}^h$ for the general linear and the semilinear case 499
7.9.4 $\mathcal{V}^h$-coercivity and boundedness for quasilinear problems 502
7.10 Consistency results for the c^h, b^h, ℓ^h 503
7.10.1 Consistency of the c_h and b_h 503
7.10.2 Consistency of the ℓ^h 505
7.11 Consistency properties of the a_h 507
7.11.1 Consistency of the a_h for the Laplacian 507
7.11.2 Consistency of the a_h for general linear problems 511
7.11.3 Consistency of the semilinear a_h 514
7.11.4 Consistency of the quasilinear a_h for systems 518
7.11.5 Consistency of the quasilinear a_h for the equations of Houston, Robson, Süli, and for systems 523
7.12 Convergence for DCGMs 527
7.13 Solving nonlinear equations in DCGMs 532
7.13.1 Introduction 532
7.13.2 Discretized linearized quasilinear system and differentiable consistency 532
7.14 *hp*-variants of DCGM 538
7.14.1 *hp*-finite element spaces 539
7.14.2 *hp*-DCGMs 540
7.14.3 *hp*-inverse and approximation error estimates 540
7.14.4 Consistency and convergence of *hp*-DCGMs 542
7.15 Numerical experiences 546
7.15.1 Scalar quasilinear equation 546
7.15.2 System of the steady compressible Navier–Stokes equations 554

8 Finite difference methods 560
8.1 Introduction 560
8.2 Difference methods for simple examples, notation 562

8.3 Discrete Sobolev spaces 566
8.3.1 Notation and definitions 566
8.3.2 Discrete Sobolev spaces 569
8.4 General elliptic problems with Dirichlet conditions, and their difference methods 572
8.4.1 General elliptic problems 572
8.4.2 Second order linear elliptic difference equations 574
8.4.3 Symmetric difference methods 581
8.4.4 Linear equations of order $2m$ 583
8.4.5 Quasilinear elliptic equations of orders 2, and $2m$ 584
8.4.6 Systems of linear and quasilinear elliptic equations 586
8.4.7 Fully nonlinear elliptic equations and systems 587
8.5 Convergence for difference methods 588
8.5.1 Discretization concepts in discrete Sobolev spaces 589
8.5.2 The operators P^h, Q'^h 591
8.5.3 Consistency for difference equations 594
8.5.4 $\mathcal{V}_b^h$-coercivity for linear(ized) elliptic difference equations 600
8.5.5 Stability and convergence for general elliptic difference equations 604
8.6 Natural boundary value problems of order 2 610
8.6.1 Analysis for natural boundary value problems 611
8.6.2 Difference methods for natural boundary value problems 613
8.7 Other difference methods on curved boundaries 622
8.7.1 The Shortley–Weller–Collatz method for linear equations 623
8.8 Asymptotic expansions, extrapolation, and defect corrections 626
8.8.1 A difference method based on polynomial interpolation for linear, and semilinear equations 627
8.8.2 Asymptotic expansions for other methods 630
8.9 Numerical experiments for the von Kármán equations, **with C.S. Chien** 633

9 Variational methods for wavelets, with S. Dahlke 635
9.1 Introduction 635
9.2 The scope of problems 637
9.3 Wavelet analysis 639
9.3.1 The discrete wavelet transform 640
9.3.2 Biorthogonal bases 644
9.3.3 Wavelets and function spaces 646
9.3.4 Wavelets on domains 647
9.3.5 Evaluation of nonlinear functionals 652
9.4 Stable discretizations and preconditioning 653
9.5 Applications to elliptic equations 659
9.6 Saddle point and (Navier–)Stokes equations 664
9.6.1 Saddle point equations 664
9.6.2 Navier–Stokes equations 666

9.7 Adaptive wavelet methods, **by T. Raasch** 669
9.7.1 Nonlinear approximation with wavelet systems 672
9.7.2 Wavelet matrix compression 675
9.7.3 Adaptive wavelet–Galerkin methods 678
9.7.4 Adaptive descent iterations 680
9.7.5 Nonlinear stationary problems 683

Bibliography 686

Index 733

Preface

Nonlinear problems play an increasingly important role in mathematics, science and engineering. Thus, an exciting interplay originates between the different disciplines, stimulated by the insight that linear problems often provide only poor approximations for the often nonlinear original problems. This applies particularly for all the bifurcation and induced dynamical regimes in nonlinear differential equations. In fact, this book, *A 'Synopsis'*, and the following *Numerical Methods for Bifurcation and Center Manifolds in Nonlinear Elliptic and Parabolic Equations* [120], are motivated by these interrelations and their numerical realization.

A major part of these phenomena is governed by elliptic and the corresponding parabolic partial differential equations and systems of orders 2 and $2m$. For these nonlinear equations, bifurcation and the related, hence local, dynamics are mainly governed by the underlying elliptic partial differential equations and a small dimensional time dependent model, the center manifold. So we can avoid most of the problems of time discretization of parabolic partial differential equations and concentrate on the space discretization of elliptic partial differential equations. The problem of convergence of the solutions of space discretized elliptic problems to those of the original nonlinear problems is the core of this book.

Astonishing is the fact that in the many excellent books on numerical methods for elliptic problems, see, e.g. below, nonlinear phenomena are either totally omitted or only treated by one or two important examples. Most books concentrate on just one of the many available discretization methods. Certainly, one of the reasons is the fact that by linearization many of the nonlinear equations and their numerical methods can be reduced to the linear case. However, for many problems, e.g. all types of bifurcation scenarios, a closer relation between the original nonlinear problem and its nonlinear discretization is mandatory.

For linear problems the existence, uniqueness and regularity of solutions are well known. For nonlinear problems this is no longer correct and the results are widely spread in a huge number of textbooks, monographs and much more original papers. They are obtained by very different methods. So we have to summarize and formulate those analytical results, necessary for the following proofs for and the convergence rates of the numerical methods. In particular, we need a systematic listing of conditions guaranteeing the existence, uniqueness and regularity for the solutions of these nonlinear elliptic problems, and exact conditions for the linearization. We include hints to available textbooks and monographs for this area and to special problems, not covered here. With a few exceptions, we do not refer to nor list the results of original papers.

In contrast to other books, the goal here is a systematic study, in particular of the joint structure, of the discretization methods for linear and *nonlinear elliptic problems.*

Necessarily this has to start with the linear foundations but has to proceed to fully nonlinear problems, where the highest derivatives occur nonlinearly. In particular, this last highlight is possible, since recently the author, probably for the first time, has published a finite element method and proved its convergence for this wide class of general nonlinear problems, cf. Böhmer [118, 119]. So, we prove stability and convergence for the numerical methods below and for all the linear and nonlinear problems studied in Chapter 2.

It cannot be the goal of this book to provide an encyclopedia of all available space discretization methods for elliptic problems, since new methods are emerging all the time. Rather we have chosen the most important examples of different types. New methods will probably fit into our framework, certainly with appropriate generalizations and modifications.

It would be pretty unrealistic to assume that a standard reader would study the whole book. Rather, a scientist would look for the class of elliptic problems and the corresponding numerical method most interesting for him. Therefore this preface gives a much extended survey of the content of each chapter and thus allows an appropriate choice. We formulate two examples at the end of this preface for a goal-oriented selection of material from the book. The book presents the result that these types of short studies are possible.

To give a feeling for the type of our elliptic problems, we formulate a hierarchy of increasingly nonlinear model problems, generalizing the Laplacian $\Delta u = \sum_{i=1}^{n} \partial^i \partial^i \ u = g$, with g independent of the solutions u. Its obvious linear generalization in $\mathbb{R}^n$ is $\sum_{i,j=0}^{n} (-1)_{j>0} \partial^j (a_{ij} \partial^i \ u) = g$. We use the notation $\partial^j \ u := \partial u / \partial x_j$, $\partial^0 \ u := u$, $\nabla u := (\partial^j u)_{j=1}^n$ $\nabla^2 u := (\partial^i \partial^j u)_{i,j=1}^n$, and $(-1)_{j>0} = -1$ for $j > 0$ and else $= 1$. Special semilinear forms are $\Delta u + f(u) = 0$ or, the more general, $\Delta u + f(x, u, \nabla u) = 0$. Quasilinear equations in divergence form are $-\sum_{i=0}^{n} \partial^i (a_i(x, u, \nabla u)) = 0$. Fully nonlinear problems are the most difficult: $a(x, u, \nabla u, \nabla^2 u) = 0$. These second order equations are generalized in Chapter 2 to equations of order $2m$ and to systems of order $2m, m \geq 1$, and the generalized Laplacian. We study them on polygonal or curved domains mainly with generalized Dirichlet boundary conditions. For $m = 1$ we admit (nonlinear) boundary conditions, induced by the (nonlinear) differential operators.

The goal for our discretization approach is the proof of the unique existence of discrete solutions for all these previous problems, their computability and convergence to the exact solutions. We reach this goal for *finite element methods*, FEMs, and adaptive forms, difference, mesh-free and spectral methods. For discontinuous Galerkin methods it allows linear to quasilinear equations and systems of order 2. Due to the open problem of calculating complicated nonlinear operators, wavelet methods are restricted to semilinear equations and systems of order $2m, m \geq 1$.

So this monograph is essentially oriented towards theoretical numerical mathematics. The only problems computed here are quasilinear and steady compressible Navier–Stokes equations in Section 7.15 and the van Kármán equations in Section 8.9. All the other case studies, e.g. the dew drops on a spider's web, the local dynamics in the earth mantel, and weather models, and the beginning chaos in Navier–Stokes equations require the tools of numerical symmetry breaking. They are delayed to [120].

Linearization techniques are the main tools for both books. The trick which makes it work relies upon the bounded invertibility of the derivative of the operator in a neighborhood of a solution. If this is assumed, it may be shown that even an *equibounded* invertibility, hence stability, holds for the chosen discretizations. Most previous approaches for quasilinear problems rely upon monotonicity results for the equation at hand. This limits the nonlinearities and also the discretizations which may be used. The new approach permits a unified convergence theory for the previously mentioned problems and discretizations. Nevertheless, each of these combinations requires a specific theoretical "lifting" to establish convergence and stability. The author is pretty sure that by modifying this lifting the techniques in this book can be applied to the other space discretization methods as well. Indeed, it has worked very well starting with FEMs and proceeding to all our other methods.

The linearization technique for the previous FEMs and all the following methods does not allow convergence results for monotone and quasilinear operators in $W^{m,p}(\Omega)$ for $1 < p < 2$, cf. Subsection 4.3.3 and Lemma 2.77, Section 2.7, and Chapters 4 and 5. However it yields, for $2 \le p < \infty$, convergence of the expected order with respect to the discrete $H^m(\Omega)$ norms for all types of elliptic problems.

On the other hand, the monotone operator techniques yield convergence results with different orders for $1 < p < \infty$ with respect to discrete $W^{m,p}(\Omega)$ norms, cf. Theorem 4.67. Strongly monotone operators or quasilinear operators in $H^2(\Omega)$ allow specific results. In fact, many important applications are modeled by monotone operators. So we embed the monotone operator approach, independent of linearization, into our general discretization theory. We do not aim for a full proof of these results. Rather we combine the excellent presentation in Zeidler [678], with Chapter 3. We formulate the results in Section 4.5 such that the other discretization methods, in particular the nonconforming methods in Chapters 5 and 7, are included. These results seem to extend the state of the art in several directions.

These two independent approaches, here linearization and monotone operator techniques, complement each other very appropriately. Both yield complementing results for the existence, uniqueness and convergence of the discrete solutions for all previous discretizations and for quasilinear to fully nonlinear problems.

Back to the linearized problems. These are split into a coercive part and a compact perturbation, thus allowing the highly successful perturbation techniques. Many methods are nonconforming, so the approximating functions violate the boundary and/or smoothness properties of the exact solutions. Then the relations between the strong and the weak form of the (linearized) problem are essential. A so-called *anticrime transformation* masters these difficulties for the convergence theory. For the adaptive FEMs in Chapter 6, Dörfler uses related techniques.

These techniques allow the above-mentioned formulation of a new class of FEMs, difference, mesh-free and spectral methods for fully nonlinear problems, cf. [118–120] This is different from the earlier handling of quasilinear problems via monotonicity arguments. They do not seem to be applicable to fully nonlinear problems. The next book [120] is the motivation for this book. There all the bifurcation and local dynamical scenarios are again governed by linearization of the corresponding problems. So the possibility of linearization has to be carefully studied. The class of nonlinear problems amenable to these techniques is characterized in Section 2.7. It essentially requires the above-mentioned boundedly invertible linearized operator.

Finally, many methods for solving nonlinear discrete problems are based upon the corresponding linear problems, e.g. all types of Newton and continuation techniques. Furthermore, we obtain the Fredholm alternative and can prove stability and convergence in a unified way for the different types of space discretization methods and the large class of linear and nonlinear problems characterized before. Despite this unified convergence theory, the different methods presented here, cf. the following paragraphs, and in [120], require carefully distinguishing their different approximation characteristics.

So we have many good reasons for linearization as the main tool.

It is a fascinating phenomenon that an appropriate structure allows very similar arguments for a wide range of problems.

The most frequently used method are FEMs, including nonconforming FEMs. Some of the techniques in FEMs can be modified for the other methods as well. Adaptive FEMs employ a posteriori estimates for determining problem-adapted FEMs. *Discontinuous Galerkin methods*, DCGMs, are closely related to and modify FEMs by penalty terms. The oldest, the *finite difference method,* is still a favorite in bifurcation problems. *Wavelets* are one of the quite recent methods with still many open technical problems. Only *spectral methods* allow the numerical realization of infinite symmetry groups. The *mesh-free or radial basis methods* allow a very flexible positioning of their centers, again very advantageous for maintaining the symmetry of problems in their discretization. Therefore both methods are omitted here and used in Böhmer [120] as demonstration examples for the general discretization theory employed in both books, cf. Chapter 3 and the new approach in [120].

We omit several interesting methods, e.g. some generalizations of finite elements, the finite volume, cf. Bey [90] and Knabner and Angermann [445]. For boundary integral methods, several interesting books are available by Steinbach and Rjasanov [556, 595] and Hsiao and Wendland [413]. They have been applied to nonlinear problems by linearizing the original problem and combining Newton–Kantorowich methods and continuation. The corresponding linear equations can then be solved with these methods. The same techniques have recently been used for wavelet and mesh-free methods, now included in our convergence theory. For boundary integral methods we did not see this possibility.

Originally this book was planned as part of [120]. As a basis for numerical methods in bifurcation of nonlinear differential equations, we had to summarize the state of the art of numerical methods here for elliptic problems. But, for most of the methods studied here, these results turned out to be rather incomplete towards nonlinearity, in particular full nonlinearity. So we had to extend many known results for nonlinear problems into different and missing directions. This was possible by systematically using the above general discretization theory. It even allowed the proof of the convergence results for nonlinear boundary operators. These are problems which seem to be not much discussed in the literature. Forced by this wide range of problems, we split the presentation into two essentially independent books.

As mentioned, most of the many excellent books on numerical methods for PDEs are concentrated on linear problems. In particular, they intensively discuss the corresponding linear solvers. Many standard solvers for large nonlinear problems are essentially based upon linear solvers. The discretization methods presented here are so-called "linear" methods. This is the essential condition for numerical methods adapted to bifurcation and for the *mesh independence principle* (MIP). This states that the Newton method applied to the discrete problem converges, for a small enough "step size", h, essentially as fast and independent of h, as that applied to the analytic problem. This implies quadratic and, e.g. linear convergence for the original and modified Newton's method. So the convergence of these iteration methods to discrete solutions is guaranteed for our wide range of discretization methods and nonlinear problems. The other tools are continuation methods. Both methods solve sequences of linear problems. The nonlinearity only enters via the computation of the defects or residuals of the approximate discrete solution in the nonlinear problem. A combination of Newton with continuation techniques provides efficient solvers for nonlinear discrete systems. Usually this is combined with available multilevel methods.

Since these linear solvers are pretty different for the different types of the above methods and problems, a synopsis of the linear solvers does not make too much sense. Hence, we do not discuss linear solvers. Modifications for bifurcation and approximations for the eigenvalues and invariant subspaces for the linearized operators are delayed to [120]. So this synopsis on numerical methods for nonlinear elliptic problems essentially studies stability, consistency, convergence, and computability.

The following references, here and throughout the book, are chosen to support specific points in the text. It would be totally impossible to list all relevant books and papers. The different discretization methods are studied in hundreds or even thousands of papers. So we usually only cite books, unless specific points have to be made. This implies that the contributions of many scientists represent extremely important contributions to the area in this book without being mentioned here. One example are the numerical methods for the Navier–Stokes equations, with 3,731 citations in October 2009 in the Zentralblatt.

According to the previous discussion, this book is organized as follows.

Part I is devoted to *analytical results.* Chapter 1 demonstrates, for the *simple* mechanical *example of a bent rod*, the change in the character from linear to nonlinear regimes. This is followed by several examples for different types of nonlinear elliptic differential equations, e.g. in mathematics – the Monge–Ampère equation – in science – the reaction–diffusion systems – and in engineering – the von Kármán and the Navier–Stokes equations. These problems and methods require the analytical results necessary for our (space-) discretization methods. The necessary tools from functional analysis and calculus in Banach spaces are listed in Section 1.4.

In Chapter 2, we summarize the indicated *analytical results.* We include general linear, special semilinear, semilinear, quasilinear and fully nonlinear elliptic differential equations and systems of orders 2 and $2m$, e.g. the Navier–Stokes equations. An early form of such a collection of results for engineering problems are the two volumes of Ames [26,27]. Numerical methods are studied there as well. Extensions to science and the necessary analytic and group theoretic tools are collected in Rogers and Ames [557]. The essentials are existence, uniqueness and regularity of their solutions and properties of the linearizations. This linearization is applicable to nearly all the nonlinear elliptic problems discussed before (see Section 2.7). Excluded are nonlinear problems that are not elliptic in the standard sense, but monotone. Their FEMs are discussed in Section 4.5. The linearization and its bounded invertibility for the included problems yield the Fredholm alternative and the stability of space discretization methods, respectively.

Linearization is not always appropriate for the analysis of nonlinear problems. Different techniques have to be used for other aspects of nonlinear problems: existence, uniqueness and other properties of solutions are obtained by methods, usually independent of linearization. They are appropriately studied via monotone and Nemyckii operators, systematically described, e.g. in Zeidler's impressive "multigraph" [675]–[678]. Other examples are discussed, e.g. by Haslinger *et al.* [393], for unilateral boundary value problems, by Feistauer and Zenísek [316, 317], and by Zenísek [681], with monotone and quasimonotone methods and variational crimes. Haslinger et al. [393] consider mainly mechanical contact problems via inequalities.

Often maximum principles, and different inequalities are employed, cf. Gilbarg and Trudinger, [346]. In a fascinating series of papers, [215–228] Crandall, Lions and colleagues study *viscosity solutions*, which we do not consider here. In Caffarelli and Glowinski, [154], they are applied to numerical methods for the Pucci equation.

Another example is the huge area of practically relevant problems, e.g. in fluid and solid mechanics and in modeling. In solid mechanics the nonlinearity is complicated by additional corner and edge singularities and different scales for the interesting scenarios. Here and in related problems, adaptive mixed FEMs are mandatory, cf. e.g., Braess, [135–137] and with Carstensen, and Reddy, [138], Johannsen et al. and Kawohl et al. [421, 438] Stein/Sagar, [594]. Essential phenomena, based upon a posteriori error estimates, employ the nonlinearity directly. Some of these problems are indicated in Chapter 6, but we do not study them explicitly. The question how well these problems can be solved with lineraization techniques is still open.

We turn to **Part II** numerical methods. Most standard monographs or textbooks, present the convergence theory of space discretizations only for one specific discretization method and only linear problems. In other books, e.g., Brenner and Scott, [142], FEMs for nonlinear problems are essentially described by a few examples. The books of Glowinski, partially with Lions and Tremolieres, and surveys [350, 351, 353, 354], systematically study numerical methods for variational inequalities and nonlinear variational problems. Again nonlinear problems play an important role in Glowinski's many proceedings edited, e.g. with Absi, Angrand, Bertin, Bristeau, Chen, Dervieux, Desideri, Fortin, Larrouturou, Lascaux, Le Tallec, Lions, Liou, Mantel, Neittanmäki, Pekka, Périaux, Pouletty, Tezduyar, Tong, Tremolieres, Veysseyre, Zhang. We only list the first and last one [355, 356]. Glowinski discusses special fully nonlinear problems with Dean and recently with Caffarelli and Guidoboni, Juarez, Pan [154, 272–276, 352].

The last two papers use viscosity solutions. These allow other problems than our linearization, e.g. the Pucci equation.

Numerical methods for nonlinear problems are intensively studied in China as well, cf. e.g. the proceedings edited by Chen, Chow, Glowinski, Kako, Li, Périaux, Shi, Tong, Ushijima, Widlund, Zhang. e.g. [586,642]. Different methods are discussed, e.g. in Temam, [621–624], Feistauer, Felcman and Straškraba, [315]. [621–624] considers for the Naviér–Stokes problem, finite difference and finite element methods. [315] discuss, for compressible flow problems, in addition finite volume and DCGMs, with proofs in Novotný, Straškraba, [515].

Finite difference and finite element methods for linear problems are discussed by Hackbusch [387]. Different strategies are presented for proving convergence, e.g. M-matrix techniques for finite differences versus the standard bilinear forms for FEMs. Grossmann *et al.* [374–376] mainly study linear problems, approximated by difference, FE and boundary integral methods, cf. Sloan, Süli and Vandewalle [592]. Convergence results for nonlinear boundary operators seem to be nearly totally missing.

In the earlier approaches, studying general convergence theories, e.g. as in [28–30, 32, 371, 441, 528–532, 547, 596, 675–678], spectral or FEMs with variational crimes, DCGMs, wavelet and mesh-free methods are necessarily missing. Sometimes finite difference and FEMs are studied via outer and inner approximation methods, e.g. in [528–532, 678]. Conforming FEMs are considered as inner approximation schemes, and classical difference methods as external approximation schemes, cf. Temam, Vainikko, Schumann and Zeidler [574, 622, 645, 647, 678]. More complicated methods, e.g. nonconforming FEMs, and systems and fully nonlinear equations are not included.

Here we employ a new theory, combining generalized Petrov–Galerkin methods and one of the classical theories, here that by Stetter [596]. This yields, in Chapter 3, the necessary existence, stability, convergence and computing results for all these problems and discretization methods in a unified way. In particular, it avoids distinguishing inner and external approximation schemes. The standard "consistency and stability imply convergence" result is proved.

Chapter 3, with its *general discretization theory*, strongly depends upon linearization and requires discussing its limitations and advantages. As previously mentioned, linearization for quasilinear problems, defined on Sobolev spaces $W^{m,p}(\Omega)$, is limited to $1 \leq p < \infty$. We obtain stability and convergence results for $1 \leq p \leq 2$ discrete $W^{m,p}(\Omega)$ norms, but for $2 \leq p < \infty$ only with respect to discrete $H^m(\Omega)$ norms. This is complemented by the monotone convergence approach mentioned above.

Based upon the previous analytical existence, uniqueness, and regularity results, obtained by any other method, linearization allows proving, in a unified way and for a major part of these problems, the desired stability, convergence, and Fredholm results. In fact, the stability of the nonlinear system is a consequence of the stability of the linearized discrete equation. For general fully nonlinear problems this seems, until now, the only possible way at all. Finally, we have already mentioned the appropriate combination of Newton and continuation techniques in the MIP.

Most problems considered in this book can be studied in their weak form. Then the spaces of ansatz functions, including the exact solution, and of test functions are identical. For the exceptional nondivergent quasilinear and the fully nonlinear problems this is no longer correct, cf. Subsection 2.5.7 and Sections 5.2 and 5.3. So Chapter 3 is formulated to include these weak cases.

Additionally, we assume that the derivative of the nonlinear operator, evaluated in the exact (isolated) solution, is boundedly invertible. We do have to defend this condition: The arguments are all closely related to the numerically necessary condition of a (locally) well conditioned problem. This is often guaranteed by monotonicity arguments, cf. Subsections 2.5.5 and 2.5.6. For an existing locally unique solution, Frechet differentiability is satisfied for many practically relevant problems. Exceptions might be unilateral problems, which seem to have been studied until now mainly for linear problems, see e.g. Haslinger, Hlavácek and Nečas [393].

As a consequence of the Taylor formula and the Fredholm alternative for these operators, the bounded invertibility of the derivative is usually, but not necessarily, satisfied as well. Otherwise a nontrivial kernel indicates bifurcation, see below, or the second derivative has to be positive or negative definite, or other combinations with higher derivatives have to be satisfied to fit the above monotonicity arguments. Thus there will be (exceptional) operators with locally unique solutions violating the condition of bounded invertibility. For our numerical approaches we are luckily able to monitor this danger. If this *exceptional case* is detected, the *convergence* for the following numerical methods simply *is not proved.* Furthermore, except for monotonicity and iteration techniques, cf. Zeidler [678], Section 26.4, or small dimensional systems, there are not too many numerical methods for solving the nonlinear discrete equation without requiring a boundedly invertible derivative in one form or the other. The latter, in particular, has to be correct for linear solvers and the Newton method for the discrete problem. Unless the derivative is nearly singular, the corresponding discrete linear equation is stable, hence boundedly invertible. This fact is often monitored automatically.

There is another important reason for assuming the bounded invertibility of the derivative, strongly related to the Fredholm alternative for the linearized elliptic operator. If it were violated, small perturbations of the original problem, e.g. corresponding to round-off errors, could be embedded into a parameter-dependent problem. This would usually have a bifurcation point or another singularity. The standard numerical methods would have to be replaced by numerical methods for bifurcation, see e.g. Böhmer [113–115, 120] Caloz and Rappaz [156], and Cliffe, Spence and Tavener [180].

Conversely, the bounded invertibility of the linearized original operator as a consequence of the stability of the linearized discrete operator is proved in Section 3.6. This last result is particularly interesting: For the more complicated elliptic problems the available analytical results become more and more lucanary. Then continuation techniques can be combined with the results in Section 3.6, relating the stability of the linearized A^{h} with the invertibility of A, and with methods similar to the classical existence results, obtained e.g. by Courant, Friedrichs and Lewy [213]. This allows research trips into analytically unknown areas, e.g. systems of the fully nonlinear Monge–Ampère equations. All these results apply to the above *wide class of nonlinear problems* and the chosen *important types of space discretization methods.*

So there are enough good and numerically necessary reasons for assuming a boundedly invertible derivative.

There is a totally different other reason for this assumption in several discretization methods, see above. For example, for boundary integral methods a boundary integral formulation of nonlinear problems seems to be difficult. For wavelet and mesh-free methods stability and convergence results for nonlinear problems have not been available. So the different Newton–Kantorowich and continuation techniques are applied to the analytical problem. For these originating linear problems the above-mentioned discretization methods yield convergence. However, most scientists in the numerical community seem to prefer a discretization of the nonlinear problem compared to this linear detour. For wavelet and mesh-free methods we succeeded in proving convergence for a wide range of nonlinear problems in Chapter 9 and in [120].

We want to return to stability and convergence of these discretizations. Stability is reduced in Chapter 3 to a few well-known basic concepts from functional analysis and approximation theory. We combine coercive bilinear forms or monotone operators, their compact perturbations, bordered systems (for bifurcation and center manifolds), exact and approximate projectors with interpolation, best approximation and inverse estimates for approximating spaces. This allows a unified proof for stability for the above discretization methods and operator equations. Stability for the linear operator suffices. Its principal part is essentially coercive, hence its discretization is stable. The stability of the linear operator A^{h} itself, with A, a compact perturbation of its principal part, turns out to be a consequence of the bounded invertibility of A. Examples are the von Kármán or the Navier–Stokes equations. Sections 2.7 and 2.8 show that our elliptic linear and nonlinear equations and systems do satisfy this condition. Consistency is simple for conforming FE, wavelet and spectral methods. For the nonconforming cases we have to employ appropriate tricks. Often the relation between the weak and strong form is essential. The necessary conditions and results, including the mesh independence principle, are formulated.

Following this rather extended discussion of stability and convergence for discretization methods and their differences compared to other approaches, we present short summaries of the results for our different methods in the following chapters. The corresponding introductions formulate extended summaries.

It is certainly justified considering the nowadays most important FEMs more extensively than the others. In Chapter 4 we start the discussion with *conforming FEMs*. Thus the ansatz and test functions satisfy the appropriate regularity and boundary conditions. We start by summarizing the well-known approximation theory of FEs. Linear to quasilinear problems allow strong and weak formulations. The original and discrete forms have to be compared. In fact, the weak, simultaneously variational forms are directly used for these problems to yield the conforming FEMs. Based upon Chapter 3 the proof of consistency, stability and convergence is pretty simple. Consistency is nearly obvious. Stability is proved by compactly perturbing the coercive principal parts, or for the Navier–Stokes problem, the Stokes operator. We do not study super convergence results, cf. Bank, Xu and Griewank, Reddien and Jiang, Shi, Xue, [66, 368, 417].

However, the strong requirements for conforming FEMs with respect to ansatz and test functions, exactly evaluated test integrals, and restrictions to weak forms is a severe drawback. This is solved by considering *nonconforming FEMs* in Chapter 5.

Only to these nonconforming FEMs do we apply and prove convergence for the full spectrum of linear to fully nonlinear equations and systems of orders 2 and $2m$. Weak forms are no longer possible for fully nonlinear problems. So specific techniques are applied. They relate the strong form of the fully nonlinear system to the weak form of the linearized equations with stability properties. We present, with O. Davydov, special differentiable FEs, needed for these problems, cf. Subsection 4.2.6. So new FEMs are formulated for *general, nondivergent* quasilinear and *fully nonlinear equations and systems,* see Böhmer [118, 119]. In particular for nonlinear problems and nonconforming FEMs, the relation between classical and variational consistency has to be discussed. This technique can even be used for proving convergence for nonlinear boundary operators. The necessary quadrature approximations for all FEMs are described. Variational crimes for FEs violating regularity and boundary conditions are studied in Section 5.5 in $\mathbb{R}^2$ for linear and quasilinear problems. This allows high order conforming and nonconforming versions, particularly important for bifurcation, and yields several new results. An essential tool is the *anticrime transformation.*

As a consequence of the dominant role of FEMs, the presentation of numerical solutions of five classes of problems is restricted to FEMs: the *variational methods for eigenvalue problems* are presented in Section 4.7. The *convergence theory for monotone operators as applied to quasilinear problems* is described in Section 4.5. Section 5.2 is devoted to *FEMs for fully nonlinear elliptic problems.* In Section 5.3 we study *FEMs for nonlinear boundary conditions.* Convergence results for these problems are hardly, if at all, discussed in the literature. *Quadrature approximate FEMs* are the topic in Section 5.4. We thus close several gaps in the literature. The reason for this restriction to FEMs is very simple. These *stability, convergence and computing results are valid essentially for all the other methods* as well. In fact, the proofs for all these results remain correct for all the other methods. This could be shown by only slightly and obviously modifying our unified approach on the basis of the specific consistency and stability results available for each method. This claim has to be slightly restricted for DCGMs and wavelet methods. Our DCGM approach is available only for problems of order 2, and is not possible for fully nonlinear elliptic problems. Wavelet methods are limited to semilinear problems.

In the meantime, for many of the difficult problems in applications, e.g. the Navier–Stokes problem in fluid dynamics, or in elastomechanics and electromagnetism, mixed FEMs are combined with nonconforming FEs, cf. Braess [135, 136] and Brezzi and Fortin [144]. Totally new potentials seem to develop within the mesh-free methods, cf. [120].

In Chapter 6, W. Dörfler describes the idea of *a posteriori error estimation* and *adaptive mesh refinement.* The main points are the description of the adaptive finite element algorithm and its convergence and the sketch of the optimal complexity of this algorithm with respect to an approximation class of solutions. The techniques in this chapter, presented essentially for linear elliptic problems, can be combined with those for nonlinear problems in the previous chapters. As indicated in the last subsection, this might allow generalizing many results to nonlinear problems. Interesting results for nonlinear problems are due to Verführt [657].

Discontinuous Galerkin methods (DCGMs) with V. Dolejší in Chapter 7 are another type of nonconforming FEMs. For avoiding rather highly technical conditions and proofs we restricted DCGMs to quasilinear equations and systems of order 2. DCGMs violate boundary conditions and continuity of the piecewise polynomials. These violations are compensated by additional penalty functions in the discrete weak form. The above technique in FEMs applied to fully nonlinear problems seems to be impossible for DCGMs.

In Chapter 8, *difference methods* are treated, as the other methods, by applying our new approach to the difference equations in weak form. This is different from most earlier approaches, either based upon the strong forms or studied as external approximation schemes. There weak forms have been considered in many papers and books, cf. [574, 622, 647, 678], and in lectures, cf. Bank [62]. Our results again are formulated for linear to quasilinear equations and systems of orders $2m, m \geq 1$. In contrast to the earlier M-matrix approach, the discretization errors are here estimated with respect to a discrete Sobolev norm of order m for a problem of order $2m$, similar to our FEMs results. This allows efficient defect correction methods as an appropriate strategy for formulating high order methods as in Böhmer [105–109] and Hackbusch [384]. Particularly interesting are symmetric forms, again for equations and systems of order $2m, m \geq 1$. For cuboidal domains with faces parallel to the axes the boundary condition are exactly reproduced. This implies convergence of order 2 and $2k$ for k defect corrections, compared to order 1 and k for the unsymmetric forms. Only partially do we prove the results for the difference methods for two reasons. The techniques are very similar to FEMs, including the case of violated boundary conditions in Sections 8.2–8.6. The special methods for curved boundaries in Section 8.7 are known only for special cases and require highly technical proofs based upon totally different techniques. They do not seem to be applied too much these days. We finally apply, with C.S. Chien, difference, combined with extrapolation methods, to the von Kármán equation.

Wavelet methods are applied in Chapter 9 with S. Dahlke for the first time in this generality and appropriate for nonlinear problems and bifurcation. This presents an obvious extension of the first papers on wavelets for nonlinear elliptic problems, Böhmer and Dahlke [121, 122]. This contrasts with the recent books on wavelets, combined with elliptic PDEs, by Cohen [195] and Urban [640, 641]. The most far-reaching results for linear problems until now were obtained for self-adjoint and saddle point problems requiring a *nonsingular* linear operator A. For these cases proofs for convergence are available, directly employing the special properties of wavelets. As a consequence of the difficulties with evaluating nonlinear functions and operators with wavelet arguments, cf. Subsection 9.3.5, general quasilinear and fully nonlinear problems have to be excluded. With this exception, the whole spectrum of corresponding wavelet methods is shown to be stable and convergent. Again the corresponding linearized operator has to be boundedly invertible. T. Raasch finishes this chapter with adaptive wavelet methods and presents some numerical results.

Spectral methods in Böhmer [120] inherit, for correct choices of approximating spaces, even Γ-equivariance for infinite symmetry groups, e.g. $O(3) \times D_4$, from the original to its discrete bifurcation problem. This is the key condition for reproducing bifurcation scenarios by discretization. *Mesh-free methods* have been successfully applied to nonlinear problems. Convergence proofs have been totally missing for more than 20 years. So our linearization approach again shows its power by proving, only very recently, the missing convergence for these problems and the standard mesh-free methods in [120]. Due to the rather free placement of their "centers", they are very appropriate for problems with symmetries as well. So we delay the presentation of *spectral* and *mesh-free methods* to [120].

We discuss dicretization methods for problems in $\mathbb{R}^n, n \geq 2$. The analysis for FEMs works well in $\mathbb{R}^n$. The methods in Section 5.5 seem to be the only type of FEMs, analytically restricted to $\mathbb{R}^2$, losing accuracy in possible generalizations to $\mathbb{R}^3$. However, there is a severe restriction of FEMs on triangulations essentially to $\mathbb{R}^n, n \leq 3$, caused by the tremendously increasing complexity of the software, in particular for adaptive FEMs. This situations is maintained for DCGMs. FEMs and difference methods on cuboidal subdivisions allow general $\mathbb{R}^n$, but do not admit adaptivity. The wavelet methods in Chapter 9 have not yet reached the level of FEMs and DCGMs. For the spectral methods in [120] we have excluded domain decomposition techniques, usually not necessary for equivariant problems. So all these methods in our presentation allow convergence in $\mathbb{R}^n$, but the problem of highly technical software is not yet solved.

The situation is totally different for the mesh-free methods in [120]. The analysis is possible for $\mathbb{R}^n, n \geq 1$, but the approximation quality strongly increases with increasing n. The software is nearly independent of n. So for many cases mesh-free methods might be the future methods of choice for $n \geq 3$. In [120] we give a proof for convergence of these methods. This is an approach totally different from Chapter 3 in this book. For the first time it shows convergence of meshfree methods applied to nonlinear problems. This can be applied to the other methods as well and allows modifications and adaptivity for some of them.

Summarizing, we obtain stability and convergence for the different types of space discretization methods and all the elliptic linear and nonlinear equations and systems of orders 2 and $2m$. This synopsis allows the extensions and generalizations described in the previous paragraphs, partially much beyond the present state of the art. Thus it fills a gap in the available literature.

This book is aimed at three groups of people. Firstly, graduate students and scientists who want to study and to numerically analyze nonlinear elliptic problems in mathematics, science and engineering. They will find the necessary tools and convergence results. This simultaneously serves as a solid foundation for many in the huge number of papers in scientific computing in this area. Finally, the bounded invertibility of the linearized original operator as a consequence of the stability of discrete operator in Section 3.6 can be combined with methods similar to the classical existence results as in Courant, Friedrichs and Lewy [213], for indicating new existence results. All these methods apply to the above wide class of nonlinear and even fully nonlinear problems and *the important types of space discretization methods*. In particular, many of the existence, uniqueness and regularity results are only known for the case $H_0^m(\Omega)$. Generalizations to the $W_0^{m,p}(\Omega, \mathbb{R}^q)$ situation would be worthwhile. In fact, results as in Theorem 2.122, form the basis for numerical methods for the

$W_0^{m,p}(\Omega, \mathbb{R}^q)$ setting, opening the possibility for research into new problems outside the previous scope.

Secondly, we have previously indicated, and update that later on, several gaps in our numerical results. These might stimulate future research. For example, FEMs violating continuity and boundary conditions and DCGMs are restricted here to order 2, these FEMs even to $\mathbb{R}^2$. Possible tools might be complementary boundary conditions and higher order cubature formulas. Asymptotic expansions for the discretization error and defect correction methods on this basis can probably be generalized with our techniques to FE and the other methods as well. Extensions of our results to other methods, e.g. the excluded finite volume methods, might be worthwhile. The complicated problems in elastomechanics and electromagnetism, approximated by mixed FEMs combined with nonconforming FEs, might find some new tools in combination with adaptive FEMs. These adaptive FEMs are presented here essentially for the Laplacian with some hints for generalizations. So there are still interesting areas for further research.

Thirdly, material for many different graduate courses advanced seminars or research projects can be chosen. The starting point is the motivation for nonlinear problems in Chapter 1. The next step would lead to the existence, uniqueness and regularity results for the desired class of elliptic problems in Chapter 2. The next Chapter 3 would be mandatory for all elliptic problems and their numerical methods. Then e.g. a DCGM for a quasilinear operator, F, is chosen. If the Frechet derivative, F', of this operator is coercive, then stability and convergence of the DCGM is completely proved in Chapter 7. For a more general, only boundedly invertible F', the proof for stability for the nonconforming FEM has to be updated to the DCGM. This straight forward job is left as a (not quite trivial) task to the interested reader. Or the new numerical methods for fully nonlinear problems allow different new research projects. In geometry a surface with a rather general prescribed Gauss curvature, closely related to the Monge–Ampère equation, could be numerically solved for the first time even in higher dimensions. It would be desirable to extend the deep results for Monge–Ampère equations to systems. Again our numerical methods for fully nonlinear problems can be combined with continuation methods indicating new possibilities. These examples show the wide range of applications from pure mathematics to science and engineering.

So the author hopes that this book will stimulate further research, e.g., elliptic problems on manifolds. Concentrating my work for decades on numerical methods for general defect corrections and bifurcation of PDEs, and only recently on FEMs, certainly some important results for the different discretization methods will have escaped me. In a huge project of ca. 1500 pages in both books, I will have missed a lot of misprints and hopefully only a few minor errors. So I will gratefully accept critics, suggestions, or remarks.

Acknowledgements: I am very grateful to many people who have strongly supported and influenced my books or encouraged me in their genesis. First of all my co-authors of this volume, C.S. Chien, S. Dahlke, O, Davydov, W. Dörfler, V. Dolejší, and T. Raasch have done a great job. It was always fascinating how initial ideas developed to the final forms. Then my good friends and cooperators

E. Allgower and Z. Mei, cf. the Bibliogrphy, were particularly influential for many results in this and the next book. The co-authors of [120], E. Allgower, P. Beltram, P. Chossat, A. Cliffe, G. Dangelmayr, C. Geiger, K. Georg$^{(\dagger)}$, N. Jangle, D. Janovska, V. Janovski, Z. Mei, R. Schaback, K.-H. Schild, B. Schmitt, J. Tausch and my other friends and colleagues P. Antonietti, M. Buhmann, E. Doedel, P. DuChateau, B. Fiedler, J. Gwinner, S. Feistauer, W. Hackbusch, R. OMalley, K. Nickel$^{(\dagger)}$, L. Schumaker, Z.-C. Shi, A. Spence, W. Wendland, G. Wittum, E. Zeidler, stimulated the progress with many discussions. Last but not least most of my Ph.D. students, J. Bošek, U. Garbotz, N. Sassmannshausen, A. Schwarzer, R. Sebastian, and additional guests in Marburg, P. Ashwin, W. Govaerts, A. Hoy, and S. F. McCormick provided essential results.

The Deutsche Forsungsgemeinschaft (DFG), and the Deutscher Akademischer Austauschdienst (DAAD), provided the necessary additional financial support.

The Philipps University Marburg offered a stimulating atmosphere for teaching and research. Essential parts of this book had been used in year long lectures. Without the help of my many masters and Ph.D. students in particular [120] would have been impossible. N. Sassmannshausen contributed good ideas to Chapter 3. Several colleagues in our Department, W. Gromes, V. Mammitzsch, C. Portenier, G. Schumacher, H. Upmeier, were always ready to answer my many questions. Most important was the support by our group numerical mathematics, mainly by S. Dahlke, J. Kappei, T. Raasch and B. Schmitt. F. Muth has done a great job. She has typed approximately half of the early forms of Chapters 1, 2, 4, 5.

Very supporting was Oxford University Press' highly professional, very constructive, kind and personal advice by S. Adlung, E. Süli, A. Warman, M. Johnstone, P. Hendry and many others.

Finally, the love, caring, and encouragement of my late wife through all the many years with successes and intrigues is the basis for this book.

Marburg
May 2010

Klaus Böhmer

Part I

Analytical Results

1

From linear to nonlinear equations, fundamental results

1.1 Introduction

Why do linear models describe nonlinear reality sometimes correctly and sometimes totally insufficiently? To get a feeling and foster the insight for this difference the first two examples are chosen as very simple mechanical problems, deduced by physical reasoning. In this chapter we do not intend to give a rigorous analysis of the problems, but rather motivate, by the following examples, the rigorous analysis in later chapters. So we do not care at the moment to present the mathematical equations in the appropriate function spaces, nor do we refer much to physics, chemistry or biology. Usually varying parameters play a dominant role. The first examples even indicate bifurcation effects in differential equations. These phenomena are the goal of Böhmer [120]. We end this chapter by summarizing the necessary basic results for this book.

1.2 Linear versus nonlinear models

To answer the introductory question we show why the following linear model cannot describe the situation correctly. For more explicit mechanical arguments, see, e.g. Troger and Steindl [637].

Example 1.1. *Bending rod with perpendicular load*

Let a vertically positioned rod *be clamped at the lower end* and be free at the upper end, see Figure 1.1. A load P at the end of the rod, originally perpendicular to the rod, forces it to bend sideways. Its displacement x is proportional to P, for small P, i.e. the following *linear relationship* holds:

$$x = cP \quad \text{for an appropriate } c. \tag{1.1}$$

However, in a simple experiment we observe that the rod will deform nonlinearly or even break when the load P passes a certain critical value. In other words, for large P, the linear law (1.1) does not represent the “reality” appropriately.

So we have to incorporate the “nonlinear” effects of deformation. We want to derive a *nonlinear* mathematical/physical *model* for the stationary position of the bent rod. There is a strong analogy between the relatively unknown strain energy of a rod and

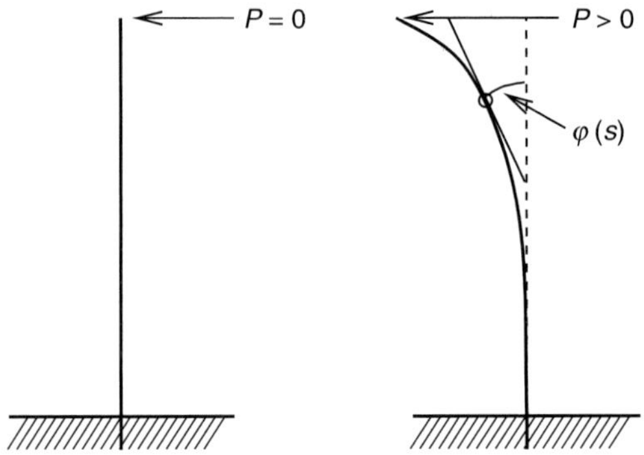

Figure 1.1 Rod with load perpendicular to the rod.

the well-known kinetic energy E of a moved body. This is described by the formula $E = m/2\ (du(t)/dt)^2$ with the mass m, the velocity, $v = du(t)/dt$, the distance traversed, $u(t)$, and the time, t. *Analogously,* we obtain for the (local) strain energy, U, at a point with arc length s on the deformed rod the rule

$$U = \frac{\alpha}{2}\left(\frac{d\varphi}{ds}(s)\right)^2 .$$

Here α is the bending stiffness and $\varphi(s)$ represents the angle between the vertical axis and the tangent in the point with arc length s (see Figure 1.1). The total bending energy, U_{bend}, of the deformed rod is given by

$$U_{\text{bend}} = \frac{\alpha}{2}\int_0^L \left(\frac{d\varphi}{ds}\right)^2 ds \quad \text{for} \quad \varphi \in C^1[0,L],$$

with the total length of the rod, L. The potential energy, due to the moving of the tip of the rod in the direction P, is given by $-Pd(L)$, where $d(L)$ indicates the displacement at the end of the rod. Figure 1.1 shows that

$$d(L) = \int_0^L \sin\varphi(s) ds.$$

Therefore, the total energy, i.e. the so-called potential, is the sum of these two terms, see Troger and Steindl [637]:

$$V(\varphi) := \frac{\alpha}{2}\int_0^L \left(\frac{d\varphi}{ds}\right)^2 ds - P\int_0^L \sin\varphi\, ds \quad \text{for} \quad \varphi \in C^1[0,L]. \tag{1.2}$$

One of the fundamental principles of mechanics states that a position of *equilibrium* of a mechanical system, here the equilibrium of the rod bent by φ, is *characterized by a minimum of its potential,* here V, i.e. for every small function ψ we have necessarily

$$V(\varphi + \psi) \geq V(\varphi). \tag{1.3}$$

We *claim* that this minimizing function φ is characterized by a *nonlinear boundary value problem* with the *differential equation*

$$G(\varphi, \lambda) := \frac{d^2\varphi}{ds^2} + \lambda \cos\varphi = 0, \quad \text{for } \lambda := P/\alpha, \tag{1.4}$$

and the *boundary conditions*, indicated by $_b$ in the corresponding subspace,

$$C_b^2[0,L] := \left\{ \varphi \in C^2[0,L]; \quad \varphi(0) = \frac{d\varphi}{ds}(L) = 0 \right\}. \tag{1.5}$$

This represents a nonlinear model for the equilibrium or stationary position of the bent rod. Now we find, by the standard linearization of (1.2), for small ψ and $\sin\psi = \psi + \mathcal{O}(\|\psi\|^2)$, compare Definition 1.41 and Theorem 1.42,

$$0 \le V(\varphi+\psi) - V(\varphi) = \alpha \int_0^L \left(\frac{d\varphi}{ds}\frac{d\psi}{ds} \right) ds - P \int_0^L \psi \cos\varphi ds + \mathcal{O}(\|\psi\|^2).$$

Here we use the standard notation

$$\begin{aligned} & g(\psi) = \mathcal{O}(\|\psi\|^\alpha) \ \text{ is equivalent to} \\ & \|g(\psi)\|/\|\psi\|^\alpha \quad \text{remains bounded for } \ \|\psi\| \to 0, \ \text{and} \\ & g(\psi) = o(\psi) \iff \lim_{\|\psi\|\to 0} = 0, \end{aligned} \tag{1.6}$$

with $\|\psi\|$ indicating any norm of ψ. By partial integration we obtain

$$0 \le V(\varphi+\psi) - V(\varphi) = \alpha \left(\psi \frac{d\varphi}{ds} \right) \Big|_0^L - \int_0^L \psi \left(\alpha \frac{d^2\varphi}{ds^2} + P\cos\varphi \right) ds + \mathcal{O}(\|\psi\|^2). \tag{1.7}$$

For the situation in Figure 1.1, φ has to satisfy the boundary conditions, see (1.5),

$$\varphi(0) = \frac{d\varphi}{ds}(L) = 0, \tag{1.8}$$

since the rod at the endpoint is not bent. The first term in (1.7) vanishes if we require, according to (1.8), $\psi(0) = (d\varphi/ds)(L) = 0$. This has to be satisfied for every small perturbation, ψ, for a rod, $\varphi + \psi$, still clamped at $s = 0$. Thus

$$0 \le V(\varphi+\psi) - V(\varphi) = - \int_0^L \psi \left(\alpha \frac{d^2\varphi}{ds^2} + P\cos\varphi \right) ds + \mathcal{O}(\|\psi\|^2). \tag{1.9}$$

We started with $\varphi, \psi \in C^2[0,L]$, but find $\psi \in C[0,L]$ with $\psi(0) = 0$ sufficient for (1.9). These changes in differentiability will play an increasingly important role for more complicated existence results, see e.g. Example 1.7 below. We have split the increment $V(\varphi+\psi) - V(\varphi)$ into a linear operator (here a functional), acting upon ψ, and a higher order remainder term $\mathcal{O}(\|\psi\|^2)$, sometimes denoted as **h.o.t.**

We have to show that the minimizing solution φ of (1.3) indeed solves the nonlinear differential equation (1.4). Using the standard argument of the fundamental lemma of the calculus of variations, we assume, contrary to (1.4), that

$$\frac{d^2\varphi}{ds^2} + \lambda \cos\varphi \not\equiv 0, \quad \text{for} \quad \varphi \in C^2[0,L] \text{ and } \lambda := P/\alpha.$$

Choosing $d\psi_0 := -(d^2\varphi/ds^2 + \lambda\cos\varphi) \in C^0[0,L]$ leads to

$$c_0 := -\int_0^L \psi_0 \left(\alpha \frac{d^2\varphi}{ds^2} + P\cos\varphi\right) ds = \alpha \int_0^L (\psi_0)^2 ds > 0.$$

The remainder term in (1.9) satisfies

$$\mathcal{O}(\|\mu\psi_0\|^2) = \mu^2 c_1, \quad \text{with} \quad c_1 := \mathcal{O}(\|\psi_0\|^2).$$

Hence, for small enough μ we have $|\mu c_1| < c_0/2$, hence $c_0 + \mu c_1 > c_0/2$, and

$$V(\varphi + \mu\psi_0) - V(\varphi) > \mu c_0/2 \text{ for } \mu > 0,$$
$$V(\varphi + \mu\psi_0) - V(\varphi) < \mu c_0/2 \text{ for } \mu < 0.$$

This contradicts the fact that φ is a minimum for V, thus showing (1.4), indeed a nonlinear model for the stationary position of the bent rod.

To study the relationship of (1.4) and (1.5) to (1.1), we start the discussion with the case of small angles φ. From $\cos\varphi \approx 1$ and (1.4) we obtain

$$\frac{d^2\varphi}{ds^2} + \lambda\cos\varphi \approx \frac{d^2\varphi}{ds^2} + \lambda = 0.$$

Hence $\varphi \approx -\lambda s^2/2 + \mu s + \nu$ and, by (1.5), we find $\nu = 0,\ \mu = L\lambda = LP/\alpha$, and

$$0 \leq \varphi(s) = \lambda(-s^2/2 + Ls) \leq \varphi(L) = P\frac{L^2}{2\alpha} \ll 1 \text{ for small } P.$$

This implies in particular, that a nontrivial solution exists for every $\lambda \neq 0$.

Similarly we may use $\sin\varphi \approx \varphi$ and find

$$d(L) = \int_0^L \sin\varphi(s)ds \approx \int_0^L \varphi(s)ds = \lambda\left(-\frac{s^3}{6} + L\frac{s^2}{2}\right)\Bigg|_0^L = P\frac{L^3}{3\alpha}; \tag{1.10}$$

thus we have reproduced the form (1.1).

When we proceed to larger values of P, and hence φ, the solution of the boundary value problem (1.4), (1.5) requires the use of numerical methods or elliptic functions.

Physical arguments indicate that the rod will break at a distance s, where, for increasing P, for the first time, the tension $\sigma = \sigma(s)$ on the surface of the rod becomes larger than the critical tension σ_{crit} of the material used. Furthermore, it is well known that the curvature $\kappa = d\varphi/ds$ is proportional to $\sigma = cd\varphi/ds$. If we assume $|\varphi(s)| < \pi/2$, we derive from (1.4) and (1.5) that

$$\frac{d^2\varphi}{ds^2}(s) = -\lambda\cos\varphi < 0.$$

Hence

$$\frac{d\varphi}{ds}(0) > \frac{d\varphi}{ds}(s) > \frac{d\varphi}{ds}(L) = 0 \quad \text{for} \quad 0 < s < L.$$

So the rod will break for $s = 0$ if P is so large that $cd\varphi/ds(0) > \sigma_{\text{crit}}$. Certainly the linear relation (1.1) can no longer describe this situation correctly.

In particular, the linear model (1.1) is not appropriate, since we have assumed at different stages that either φ or ψ is small. Only in this case does neglecting the **h.o.t.s** makes sense, otherwise they will strongly influence the solutions for a nonlinear model. ■

In Example 1.1, showing the need for nonlinear models, the solutions for the linear and nonlinear problem were pretty similar for small values of P. The next Example 1.2 deals with a qualitative difference in the behavior of solutions of linear and nonlinear problems.

Example 1.2. *Rod with axial load*

Again, we consider deformations of a thin inextensible rod of length L. This time, we study an axial load P. The rod is pinned at one end (for a horizontally positioned rod this is denoted as a *simply supported end*); the other end moves freely in the vertical direction. So, the rod tends to deform in a plane. We perform the experiment indicated in Figure 1.2. If we slowly increase P, starting with $P = 0$, the rod stays unchanged. For a further increase of P beyond a critical P_0, deviations of the rod from the vertical position are observable. These deviations increase in size with P.

For modeling the new situation we can use the same bending energy as in Example 1.1. The total length equals the total arc length, $s = L$, of the rod. The potential energy is again given as $-Pd(L)$ with the displacement at the upper end of the rod, $d(L)$, here in the vertical direction. Figure 1.2 shows that, instead of the original L,

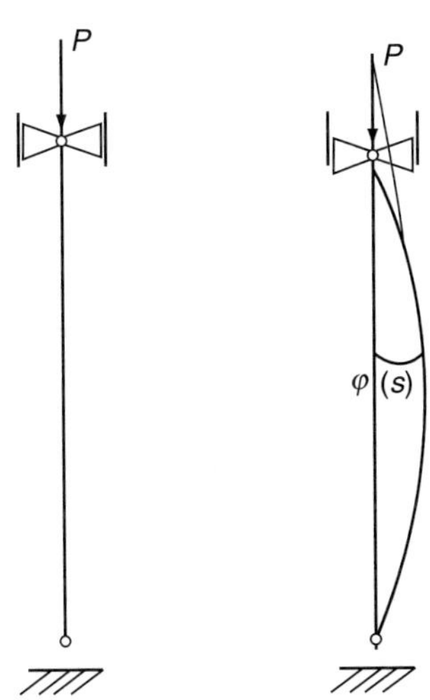

Figure 1.2 Rod with axial load.

contrasting to (1.10), the upper end has come down by

$$d(L) = \int_0^L (1 - \cos\varphi)ds.$$

With this $d(L)$ we obtain the total energy $V(\varphi)$ and the difference $V(\varphi + \psi) - V(\varphi)$, see also (1.2), as

$$V(\varphi) := \alpha \int_0^L \left(\frac{d\varphi}{ds}(s)\right)^2 ds/2 - P \int_0^L (1 - \cos\varphi)ds \text{ for } \varphi \in C^2[0, L], \tag{1.11}$$

$$V(\varphi + \psi) - V(\varphi) = \alpha \left(\psi \frac{d\varphi}{ds}\right)\Big|_0^L - \int_0^L \psi \left(\alpha \frac{d^2\varphi}{ds^2} + P \sin\varphi\right) ds + \mathcal{O}(\|\psi\|^2).$$

An analogous argument as in Example 1.1 shows that the *bent rod in the stationary position necessarily minimizes* $V(\varphi)$ in (1.11). The first term in $V(\varphi + \psi) - V(\varphi) \geq 0$ vanishes, typically, for the following boundary conditions.

$$\frac{d\varphi}{ds}(0) = \frac{d\varphi}{ds}(L) = 0. \tag{1.12}$$

Equation (1.12) represent the so-called *natural* or *Neumann boundary conditions* as in Figure 1.2: the rod is not bent at the endpoints, but the lower endpoint is fixed, and the upper moves along the vertical axis. Imposing (1.12), the minimized $V(\varphi)$ in (1.11) yields

$$0 \leq V(\varphi + \psi) - V(\varphi) = -\int_0^L \psi \left(\alpha \frac{d^2\varphi}{ds^2} + P \sin\varphi\right) ds + \mathcal{O}(\|\psi\|^2).$$

The minimizing function is characterized by a now different *boundary value problem*

$$G(\varphi, \lambda) := \frac{d^2\varphi}{ds^2} + \lambda \sin\varphi = 0 \text{ with } \lambda := P/\alpha, \text{ and } \frac{d\varphi}{ds}(0) = \frac{d\varphi}{ds}(L) = 0. \tag{1.13}$$

Obviously, $\varphi \equiv 0$ represents a trivial solution for (1.13), independent of λ, L. This $G(\varphi, \lambda) = 0$ in (1.13) is a nonlinear equation for φ. The question of whether (1.13) has a (unique) solution or not has to be answered by mathematical results. For the case of elliptic partial differential equations we will present those results in Chapter 2. If a (unique) solution φ of (1.13) exists, similarly for (1.4), (1.5) in Example 1.1, then an analogous argument shows that φ minimizes $V(\varphi)$ and thus solves our problem. ■

These two examples show totally different behavior, as a consequence of the different nonlinearities in the differential equations, different boundary conditions, and loads. In Example 1.1 a solution $\varphi \equiv 0$ is possible only for $\lambda = 0 = P$. Already, a very small P causes displacements $d(L) > 0$. The experiments in Example 1.2 show that up to a certain $P = P_0$ we only have the trivial solution $\varphi(s) \equiv 0$. If P increases beyond P_0, the rod starts to buckle in the form of a, at the beginning, sinusoidal function. If we were to fix the rod at its middle point, a much higher P would be necessary to deform the rod. So why and when are nontrivial solutions generated? To explain these observations, we again study the nontrivial solutions for (1.13), originating from the trivial state $\varphi(s) \equiv 0$. We start with a very small function φ and consider the approximation

$$\sin\varphi = \varphi - \varphi^3/3! + \varphi^5/5! + \cdots = \varphi + \textbf{h.o.t.}$$

The small nontrivial solutions φ of (1.13) have to satisfy the linearized problem

$$\frac{d^2\varphi}{ds^2} + \lambda\varphi = 0, \qquad \frac{d\varphi}{ds}(0) = \frac{d\varphi}{ds}(L) = 0,\ \lambda = P/\alpha, \tag{1.14}$$

see (1.4). This linear boundary eigenvalue problem has nontrivial solutions of the form $\varphi = c\cos\sqrt{\lambda}\,s$ if and only if

$$\frac{d\varphi}{ds}(L) = -c\sqrt{\lambda}\sin\sqrt{\lambda}L = 0 = \frac{d\varphi}{ds}(0), \tag{1.15}$$

hence if and only if $\sqrt{\lambda}L = n\pi$, $\lambda = (n\pi/L)^2$ and $\varphi(s) = c\cos n\pi s/L$. In this case the linearized operator has a nontrivial kernel or null space. This observation shows that only for certain discrete values of $P = \alpha(\pi n/L)^2$ may the linearized problem (1.14) have a nontrivial solution, originating in the trivial solution. This seems to be and in fact is a necessary condition for a nontrivial solution "bifurcating from the trivial solution" of the original problem, see Figure 1.3. We will study "bifurcation problems", their local dynamical behavior, and their numerical approximation for PDEs in Böhmer [120].

These very simple examples indicate two features, essential for further studies. For a nonlinear operator its linearized form provides essential information. If "bifurcation phenomena" of the original nonlinear problem are interesting the null space of its linearized operator plays a major role. This is one of the reasons why in the following chapters the problem of linearization and its corresponding Fredholm alternative, hence results about the null space of its linearized operator, are so important.

As we indicated already in the Preface, linearization is an essential feature of our later convergence proofs for all the numerical methods studied in the two books. Nevertheless, there are different opinions about combining linearization and numerical

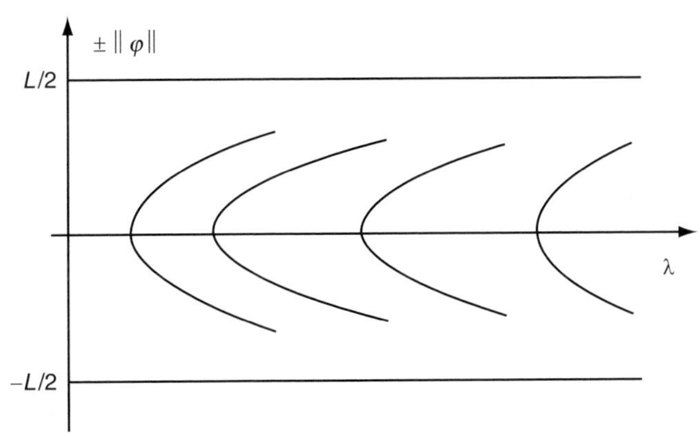

Figure 1.3 Global structure of bifurcating solution branches.

methods. Some use linearizations for the original analytical problems. The corresponding numerical methods are only applied to linear problems. Hence a nonlinear convergence theory is unnecessary. The many papers on nonlinear convergence theory show that most mathematicians disagree with this idea. On the other hand, adaptive mixed FEMs, e.g. for mechanical problems, still seem to require the full nonlinear techniques.

1.3 Examples for nonlinear partial differential equations

We consider mathematical models for important problems in different sciences. We find a spectrum of parameter-dependent semi- to fully nonlinear partial elliptic differential equations. As usual in mathematics and physics, they are appropriately simplified models. We do not go into details, but only give the corresponding equations and the meaning of the occurring functions and usually omit the exact function spaces in this introduction. Most of these examples will be more intensively discussed in the later case studies in this book and in Böhmer [120]. For many other interseting, however mostly linear applications, cf. Dautray and Lions [256, 257].

Example 1.3. *Chladny sound figures*

Assume that we have a thin flexible square plate, horizontally fixed at its center [13–16]. On the surface of the plate we uniformly distribute tiny particles, for example, sand. Above the plate we fix a source of sound waves with variable frequency related to the λ in (1.16). We want to examine the reaction of the thin plate for varying λ. The pressure of the sound waves acts as a load on the surface of the plate. In the case of resonance, a vibration with (unmoved) nodal lines is induced on the plate. There the tiny particles are collected. Thus, the distribution of these particles with respect to varying λ illustrates the vibration of the plate. Let $\Omega := [0, L] \times [0, L]$ represent the plate and $u = u(x, y)$ denote the maximal deviation of the vibration from the trivial flat position of the plate at the point (x, y). Then $u(x, y)$ satisfies the strongly simplified mathematical model equation, compare (1.13) and Theorem 2.55,

$$G(u, \lambda) := \Delta u + \lambda \sin u = 0 \text{ in } \Omega = [0, L] \times [0, L] \text{ defined on} \tag{1.16}$$

$$C_b^2(\Omega) := \left\{ u \in C^2(\Omega); \quad \frac{\partial u}{\partial n} = 0 \text{ on } \partial\Omega \right\}. \tag{1.17}$$

Here $\Delta := \partial^2 \ /\partial x^2 + \partial^2 \ /\partial y^2$ is the *Laplacian operator* and $\partial u/\partial n$ the *normal derivative* in the outer direction of $\partial\Omega$. For arbitrary λ the trivial flat state $u(x, y) \equiv 0$ represents a solution of (1.16), (1.17). As in Example 1.2 we linearize Equation (1.16) with respect to u at $(u, \lambda) = (0, \lambda)$, cf. Definition 1.41 and Theorem 1.42, and obtain, similarly to (1.14), (1.15), with the trivial solution $v = 0$ for arbitrary λ,

$$\partial_u G(0, \lambda)v = \Delta v + \lambda v = 0 \Longleftrightarrow \Delta v = -\lambda v \quad \text{in} \quad \Omega \quad \text{for} \quad v \in C_b^2(\Omega). \tag{1.18}$$

The $-\lambda$, allowing nontrivial solutions in the kernel of $\partial_u G(0,\lambda)$, are the

$$\text{eigenvalues } \lambda_0 = -\lambda = -(m^2+n^2)\pi^2/L^2 \text{ of } \Delta \text{ for all } m,n \in \mathbb{N}_0 \text{ s.t.} \tag{1.19}$$

$$v(x,y) = v_{m,n}(x,y) := \cos\frac{m}{L}\pi\, x\, \cos\frac{n}{L}\pi y, \text{ and } v(x,y) = v_{n,m}(x,y) \text{ are} \tag{1.20}$$

$$\text{the eigenfunctions of } \Delta, \text{ hence, } \Delta v = \Delta\left(\cos\frac{m}{L}\pi\, x\, \cos\frac{n}{L}\pi\, y\right) = -\frac{m^2+n^2}{L^2}\pi^2\, v.$$

Therefore, we obtain, e.g. for $(m,n) = (1,1)$, $(m,n) = (2,3)$ and $(m,n) = (5,5)$, $(m,n) = (1,7)$ the eigenvalues $\lambda_0 = -2\pi^2/L^2, -13\pi^2/L^2$, and $-50\pi^2/L^2$ with one, two and three linearly independent solutions $v_{m,n}, v_{n,m}$. All their linear combinations solve (1.18) as well, for fixed λ_0. Hence, these λ_0 for the sound well are candidates for bifurcations to nontrivial u, and for the generation of various Chladny figures.

Figure 1.4 shows $v_{1,3} - v_{3,1}$, one of the many possible linear combinations with specific symmetry properties. This Chladny problem is a typical example of *multiple bifurcations* in nonlinear problems with symmetries. ■

Example 1.4. *Buckling of a plate*

Consider a plate P under compression of the form $\Omega \in \mathbb{R}^2$. It is defined by the Airy stress function $w(x,y)$ of the plate P at the point (x,y) and the deviation or deflection $u(x,y)$ for P from its flat trivial state. They are modeled by the von Kármán equations.

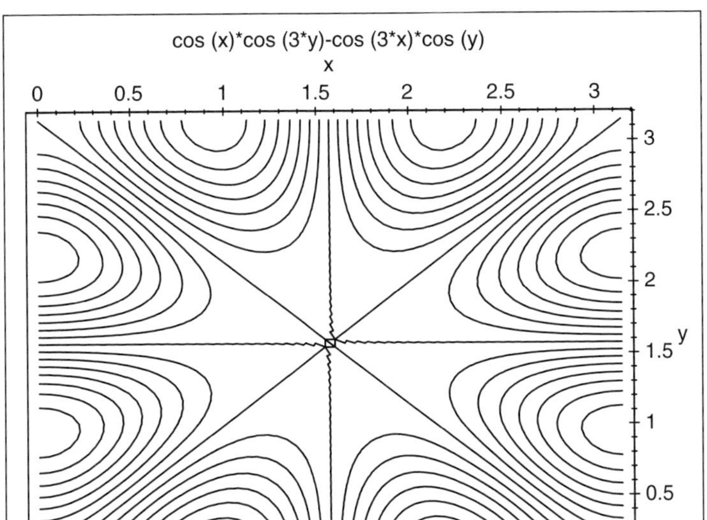

Figure 1.4 Chladny sound figures on $[0,\pi] \times [0,\pi]$. The straight (!) lines through the origin are nodal lines (fixed, elevation 0); the curves show elevation lines. Notice the high symmetry, and neglect the wriggles near the origin.

A mathematical justification for this model is given by Berger and Five [82–84], and, as the first terms in an asymptotic expansion, by Ciarlet [175]:

$$G_s(u,w) := \begin{pmatrix} \Delta^2 u & -[u,w] \\ \Delta^2 w & +\frac{1}{2}[u,u] \end{pmatrix} = \begin{pmatrix} f \\ 0 \end{pmatrix} \text{ in } \Omega \subset \mathbb{R}^2. \tag{1.21}$$

Here λ is the external load, $\Delta^2 = \Delta\Delta$ is the two-dimensional biharmonic operator, Ω the shape of the plate, and the bracket operator $[\cdot,\cdot]$ is defined by

$$[u,v] = u_{xx}v_{yy} - 2u_{xy}v_{xy} + u_{yy}v_{xx}.$$

This semilinear system will be discussed in Subsection 2.6.6, cf. (2.403) ff. Bifurcation regimes should be avoided in realistic constructions. But the mode jumping is perhaps one of the most noteworthy features of experimental and mathematical studies about the buckling of plates. Equation (1.21) is taken up as a case study in Chapter 8. ■

Example 1.5. *Reaction–diffusion equations*

Semilinear systems of reaction–diffusion equations are typical models in nuclear reactor physics, and in chemical and biological reactions. In particular, many chemical reactions, ecological systems and mechanisms of pattern formation are described mathematically as systems of semilinear differential equations. Here we study stationary equations with $\partial\vec{u}/\partial t = 0$ of the form

$$\frac{\partial \vec{u}}{\partial t} = G(\vec{u},\alpha) = D\Delta\vec{u} + \vec{f}(\vec{u},\alpha) = 0 \quad \text{in } \Omega \times (0,T). \tag{1.22}$$

In this process the domain $\Omega \subset \mathbb{R}^n$ represents the area where the chemical reaction occurs or a biological/ecological system lives. $\vec{u} : \mathbb{R}^n \to \mathbb{R}^q$ represents various (intermediate) components in a chemical reaction or species of a biological system. It is a smooth vector function representing the reaction effect of the system. The vector $\alpha \in \mathbb{R}^p$ defines the control parameters in the system. The components of $D \in \mathbb{R}^{q\times q}$, a diagonal matrix, describe the diffusion rate of each species and $\vec{f} : \mathbb{R}^n \times \mathbb{R}^p \to \mathbb{R}^q$ their reactions. Appropriate boundary conditions are combined with (1.22) for each given situation. They can be any of Dirichlet, Neumann, or Robin type, or even in some dynamical forms. The interaction of diffusion and reaction, measured by D and $\vec{f}(\vec{u},\alpha)$, can yield rich and interesting phenomena. This is demonstrated in case studies in Böhmer [120]. An example used in many numerical studies is $D = 1 = q, \vec{u} = u, f(u,\alpha) = \alpha f(u)$, see Theorem 2.58. ■

Example 1.6. *Three quasilinear and fully nonlinear geometric problems*

The solutions of these problems are functions $u : \Omega \to \mathbb{R}$ with prescribed boundary conditions $u = \phi$ on $\partial\Omega$.

The minimal surface equation, see Example 2.72 and Gilbarg and Trudinger [346], p. 339, has the quasilinear form

$$G_s u = Gu = (1 + |Du|^2)\Delta u + \sum_{i,j=1}^{n} \partial^i u \partial^j u \partial^i \partial^j u = 0 \text{ on } \Omega. \tag{1.23}$$

The Monge–Ampère equation is a fully nonlinear problem. For general n and for $n = 2$, see Example 2.78 and [346], pp. 441, 471, it has the form

$$0 = G(u) := \det D^2 u - f(x, u, Du) = u_{xx}u_{yy} - u_{xy}^2 - f(x, u, Du) \text{ for } n = 2. \quad (1.24)$$

These G in (1.24) and in (1.25) are uniformly elliptic in an $u_1 \in C^2(\Omega)$ if $D^2 u_1$ is strictly positive definite and symmetric in Ω.

Equation for a surface with prescribed Gauss curvature: now let $K = K(x)$ at $x \in \Omega$ be the prescribed Gauss curvature. Then (1.24) has the special form, see Example 2.78 and [346], p. 442,

$$G(u) := \det D^2 u - K(1 + |Du|^2)^{(n+2)/2} = 0, \; K(x) > 0. \quad (1.25)$$

■

Example 1.7. *Navier–Stokes equations*

One of the most stimulating equations for research in mathematics and physics is the Navier–Stokes equation. It models, e.g. the flow of water around a moving vessel, and the air around the wing of an airplane. In this context (1.26) with small values of $0 < \nu$ are interesting. Then boundary layers and turbulence develop, modeling the up-wind properties. Mathematically, it was the first extremely important system of nonlinear differential equations. It required an update of the definition of elliptic or parabolic equations, and for small $0 < \nu$ the boundaries between parabolic and hyperbolic problems are no longer well defined. Furthermore, by varying $0 < \nu$ and other parameters, extremely interesting regimes of bifurcation and the corresponding dynamics and chaotic phenomena originate, cf. [120]. Still, for a lot of these problems satisfactory answers are missing.

$$\frac{\partial \vec{u}}{\partial t} = G(\vec{u}, p) := \begin{pmatrix} -\nu\Delta\vec{u} + \sum_{i=1}^{n} u_i \partial_i \vec{u} + \nabla p \\ div\ \vec{u} \end{pmatrix} = \begin{pmatrix} \vec{f} \\ 0 \end{pmatrix} \text{ in } \Omega$$

$$\int_\Omega p dx = 0, \qquad \vec{u} = 0 \text{ on } \Gamma = \partial\Omega. \quad (1.26)$$

Here $\vec{u}, p$ and $\vec{f}$ denote velocity, pressure and forcing term of an incompressible medium, respectively,

$$\vec{u} = (u_1, \ldots, u_n)^\mathsf{T}, \; \vec{f} = (f_1, \ldots, f_n)^\mathsf{T} : \Omega \subset \mathbb{R}^n \to \mathbb{R}^n, \; p : \Omega \to \mathbb{R}, n \le 3,$$

and the divergence and gradient are div $\vec{u} := \partial u_1/\partial x_1 + \cdots + \partial u_n/\partial x_n$ and grad $p = \nabla p = (\partial p/\partial x_1, \ldots, \partial p/\partial x_n)^\mathsf{T}$. The equation $\int_\Omega p dx = 0$ guarantees a unique p. Here we only study the stationary form with $\partial\vec{u}/\partial t \equiv 0$. We will discuss its so-called center manifold approximations as a case study in Böhmer [120]. ■

1.4 Fundamental results

1.4.1 Linear operators and functionals in Banach spaces

Continuous or *bounded* and *compact linear operators* are among the most important concepts in functional analysis. We use them in both books very intensively.

Definition 1.8. Dense, compact sets, linear bounded, compact operators:

1. *A subset* $\mathcal{U}$ *of a* Banach space $\mathcal{X}$ *(a linear space, closed with respect to its norm) is called* dense, *if every* $x \in \mathcal{X}$ *is the limit of a sequence* $\{x_n\}_{n\in\mathbb{N}} \subset \mathcal{U}$. *A* $\mathcal{U}$ *is called* precompact *(or* relatively compact*) and* compact, *if every sequence* $\{x_n\}_{n\in\mathbb{N}} \subset \mathcal{U}$ *contains a convergent subsequence* $\{x_{n_i}\}_{i,n_i\in\mathbb{N}}$ *and additionally satisfies* $\lim_{i\to\infty} x_{n_i} \in \mathcal{U}$, *respectively.*
2. *The linear operator* $L : \mathcal{X} \to \mathcal{Y}$ *is called* bounded *(or equivalently* continuous*) and* compact *if the image of the unit sphere* $\{x \in \mathcal{X} : \|x\| \le 1\}$ *is bounded and precompact, respectively. Hence, for a bounded* L *there exists* $C \in \mathbb{R}_+$ *such that*
$$\|Lx\|_{\mathcal{Y}} \le C\|x\|_{\mathcal{X}} \ \textit{for all}\ \ x \in \mathcal{X}\ \textit{and}\ \|L\| := \|L\|_{\mathcal{Y}\leftarrow\mathcal{X}} := \inf\{C\} \tag{1.27}$$
is called (and is) a norm *for* L. *The set of all these continuous linear operators is denoted as* $\mathcal{L}(\mathcal{X},\mathcal{Y})$. *Furthermore,* $L = 0 \Leftrightarrow \forall x \in \mathcal{X} : Lx = 0$, *and*
$$\begin{aligned} \textit{for}\ L \in \mathcal{L}(\mathcal{X},\mathcal{Y})\quad \textit{we call}\ \ \mathcal{N}(L) &:= \{x \in \mathcal{X} : Lx = 0\}\ \textit{and} \\ \mathcal{R}(L) &:= \{y \in \mathcal{Y} : \exists\ \ x \in \mathcal{X} : y = Lx\} \end{aligned} \tag{1.28}$$
the kernel or null space and range of L, *respectively.*
3. *Analogously we define* bounded linear and bilinear forms, $l : \mathcal{X} \to \mathbb{R}, x \to \langle l, x\rangle := l(x)$ *and* $b : \mathcal{X} \times \mathcal{Y} \to \mathbb{R}$, $l = 0$, $b = 0$, *and their norms. We define the* dual space *as*
$$\mathcal{X}' := \{l : \mathcal{X} \to \mathbb{R} : l \textit{ is a bounded linear form}\}. \tag{1.29}$$

A subset $\mathcal{U} \subset \mathcal{X}$ is precompact, if its closure, $\overline{\mathcal{U}}$, is compact. The compactness of a subset $\mathcal{U}$ is equivalent to the following property: Any *open covering,* $\mathcal{U} \subset \cup_{i\in\mathcal{I}}\mathcal{U}_i$, $\mathcal{U}_i$ open in $\mathcal{X}, \mathcal{I}$ an index set, contains a finite subcovering. For finite-dimensional Banach spaces, $\mathcal{X}$, compactness is equivalent to $\mathcal{U}$ being a closed, bounded set. For infinite-dimensional Banach spaces compactness only implies, but is not equivalent to this property. A change from $\|\cdot\|_{\mathcal{X}}, \|\cdot\|_{\mathcal{Y}}$ to equivalent norms does not change $\mathcal{L}(\mathcal{X},\mathcal{Y})$, but $\|L\|$. Now we have the following

Proposition 1.9. *Simultaneously with* $\mathcal{Y}$, $\mathcal{L}(\mathcal{X},\mathcal{Y})$ *is a Banach space. The functional in (1.27) is indeed a norm for linear operators, and for* $\mathcal{Y} = \mathbb{R}$ *for linear funcionals. It satisfies*
$$\|L\| = \sup_{\|x\|_{\mathcal{X}}=1} \|Lx\|_{\mathcal{Y}}, \|L_2 \circ L_1\| \le \|L_2\|\ \|L_1\|\ \forall\ L, L_1 \in \mathcal{L}(\mathcal{X},\mathcal{Y}), L_2 \in \mathcal{L}(\mathcal{Y},\mathcal{Z}).$$
Analogously for bilinear forms $b : \mathcal{X} \times \mathcal{Y} \to \mathbb{R}$ *we use* $\|b(x,y)\|_{\mathbb{R}} \le C_b\|x\|_{\mathcal{X}} \cdot \|y\|_{\mathcal{Y}}$ *to introduce* $\|b\| := \inf\ \{C_b\}$. *The kernel,* $\mathcal{N}(L)$, *is a closed subspace, the range,* $\mathcal{R}(L)$, *is usually only a subspace, see Theorem 1.19.*

Example 1.10. *Special linear operators*

1. *Linear operators for* $\dim \mathcal{X} < \infty$: Obviously, in this case all linear operators are bounded and have closed kernel and range.

2. *Rod with axial load:* For small φ we have $\sin\varphi \approx \varphi$, hence, see Example 1.2 and its linear counterpart in (1.14), we study

$$L : C^2[0,l] \to C[0,l], \quad L\varphi := \ddot{\varphi} + \lambda\varphi, \tag{1.30}$$

with the norms

$$\|\varphi\|_\infty := \sup_{0\le s\le l} |\varphi(s)| \text{ and } \|\varphi\|_{2,\infty} := \|\varphi\|_\infty + \|\dot{\varphi}\|_\infty + \|\ddot{\varphi}\|_\infty$$

in $C[0,l]$ and $C^2[0,l]$, respectively. Then L is a bounded linear operator, independent of the boundary conditions, since

$$\|L\varphi\| \le |\lambda|\|\varphi\|_\infty + \|\ddot{\varphi}\|_\infty \le \|\varphi\|_{2,\infty}\max\{1,|\lambda|\} \text{ with } \|L\| \le \max\{1,|\lambda|\}.$$

■

Above we have introduced the linear functionals on $\mathcal{X}$ yielding the space $\mathcal{X}'$. In a totally analogous way it is possible to define the linear functionals on $\mathcal{X}'$, thus introducing $\mathcal{X}'' := (\mathcal{X}')'$. Since $\langle l, x\rangle$ is linear in both arguments,

$$\langle \cdot, x\rangle \ \in (\mathcal{X}')' =: \mathcal{X}'' \quad \text{for every} \quad x \in \mathcal{X}$$

it is possible to identify $\mathcal{X}$ with a subset of $\mathcal{X}''$, sometimes even with $\mathcal{X}''$ itself.

Definition 1.11. *The mapping*

$$\langle\cdot,\cdot\rangle =: \langle\cdot,\cdot\rangle_{\mathcal{X}'*\times\mathcal{X}} : \mathcal{X}' \times \mathcal{X} \to \mathbb{R}, \{x', x\} \mapsto \langle x', x\rangle \in \mathbb{R} \ (or \ \in \mathbb{C})$$

is called a pairing (for $\mathcal{X}'$ and $\mathcal{X}$). $\mathcal{X}$ is called reflexive *if there exists a bijection $b : \mathcal{X}'' \to \mathcal{X}$ such that for the pairing $\langle\cdot,\cdot\rangle' : \mathcal{X}'' \times \mathcal{X}' \to \mathbb{R}$,*

$$\begin{aligned} &\langle x'', l\rangle'_{\mathcal{X}''\times\mathcal{X}'} = \langle l, b(x'')\rangle_{\mathcal{X}'\times\mathcal{X}} = \langle l, x\rangle_{\mathcal{X}'\times\mathcal{X}} =: \langle l, x\rangle =: l(x) \\ &\forall\ l \in \mathcal{X}',\ x = b(x'') \ \textit{implying} \ \|x''\| = \|b(x'')\| = \|x\| \quad \forall\ x'' \in \mathcal{X}''. \end{aligned} \tag{1.31}$$

$\|x''\|$ is indeed a norm, and we use the notation $\mathcal{X}'' = \mathcal{X}$ for reflexive $\mathcal{X}$. Sometimes, the analogous

$$\|x\|_d := \sup\{|\langle l, x\rangle| : \ \forall\ l \ \textit{with} \ \|l\|_{\mathcal{X}'} = 1\} \tag{1.32}$$

is called the dual norm *of x and we have $\|x\|_d = \|x\|$.*

The compactness of operators and their products is a powerful property. We introduce

Definition 1.12. Weak convergence: *A sequence $\{x_n\}_{n=1}^\infty \subset \mathcal{X}$, a Banach space, is weakly convergent towards $x_0 \in \mathcal{X}$ iff $\forall l \in \mathcal{X}'$ we have $\lim_{n\to\infty} l(x_n) = l(x_0)$.*

Theorem 1.13. Properties of compact operators:

1. *For Banach spaces $\mathcal{X}, \mathcal{Y}, \mathcal{Z}$ let $L_1 \in \mathcal{L}(\mathcal{X},\mathcal{Y}), L_2 \in \mathcal{L}(\mathcal{Y},\mathcal{Z})$. If one of the L_1, L_2 is compact, then $L_2 \circ L_1$ is compact as well. Furthermore, $L_1 \in \mathcal{L}(\mathcal{X},\mathcal{Y})$ is compact if and only if $(L_1)^d \in \mathcal{L}(\mathcal{Y}',\mathcal{X}')$ is compact; for L^d see (1.33) below.*

2. *For compact* L_1*, and equibounded* $L_2^h \in \mathcal{L}(\mathcal{Y}, \mathcal{Z})$*,* $L_2^h \circ L_1$ *is compact.*[1]
3. *A compact* $L_1 \in \mathcal{L}(\mathcal{X}, \mathcal{Y})$ *maps every weakly convergent convergent sequence into a strongly convergent sequence.*

Among the many important theorems on linear operators we particularly need theorems on extensions, open mappings and closed ranges and some consequences. We summarize the existence of extensions of bounded linear and bilinear forms and operators, extending the *Hahn–Banach extension theorem,* and their inverses:

Theorem 1.14. Extension theorem for bounded forms and linear operators:

1. *Let* $\mathcal{X}_0 \subset \mathcal{X}, \mathcal{Y}_0 \subset \mathcal{Y}$ *be linear subspaces of the Banach spaces* $\mathcal{X}, \mathcal{Y}$*. Let* $l_0 : \mathcal{X}_0 \to \mathbb{R}$*,* $L_0 : \mathcal{X}_0 \to \mathcal{Y}$ *and* $b_0 : \mathcal{X}_0 \times \mathcal{Y}_0 \to \mathbb{R}$ *be bounded linear functionals, operators and bilinear forms with respect to the norms in* $\mathcal{X}, \mathcal{Y}$*, respectively. Then* l_0*,* L_0 *and* b_0 *can be* extended *as* l, L, b *to* $\mathcal{X}$ *and* $\mathcal{X} \times \mathcal{Y}$ *with the same norms. They are well defined bounded linear functionals, operators and bilinear forms on* $\mathcal{X}, \mathcal{Y}$ *with* $\|l\|_{\mathcal{X}} = \|l_0\|_{\mathcal{X}}, \|L\|_{\mathcal{Y}\leftarrow\mathcal{X}} = \|L_0\|_{\mathcal{Y}\leftarrow\mathcal{X}}, \|b\|_{\mathcal{X}\times\mathcal{Y}} = \|b_0\|_{\mathcal{X}\times\mathcal{Y}}$*, respectively.*
2. *If* $\mathcal{X}_0, \mathcal{Y}_0$ *are dense subsets of* $\mathcal{X}, \mathcal{Y}$ *these* l, L, b *are defined by the corresponding limits: For* $x \in \mathcal{X}, y \in \mathcal{Y}$*, choose sequences*

$$\begin{array}{lll} x_n \in \mathcal{X}_0, \quad x = \lim_{n\to\infty} x_n & & lx := \lim_{x_n\to x} l_0 x_n \\ & \text{and define} & Lx := \lim_{x_n\to x} L_0 x_n \\ y_n \in \mathcal{Y}_0, \quad y = \lim_{n\to\infty} y_n & & b(x,y) := \lim_{x_n\to x, y_n\to y} b_0(x_n, y_n). \end{array}$$

Then the l, L, b *are independent of the choice of the convergent sequences* $\{x_n\}_{n\in\mathbb{N}}$*,* $\{y_n\}_{n\in\mathbb{N}}$*. If* $L_0 : \mathcal{X}_0 \subset \mathcal{X} \to \mathcal{Y}$ *is compact, then so is* $L_0 : \mathcal{X} \to \mathcal{Y}$*.*

Theorem 1.15. Banach open mapping theorem: *Let* $L \in \mathcal{L}(\mathcal{X}, \mathcal{Y})$ *be* surjective: $L(\mathcal{X}) = \mathcal{Y}$*. Then the image,* $L(\mathcal{U})$*, of every open subset,* $\mathcal{U} \subset \mathcal{X}$*, is open in* $\mathcal{Y}$*.*

Corollary 1.16. Bijective linear operators are invertible: $L \in \mathcal{L}(\mathcal{X}, \mathcal{Y})$ *is* bijective, *often called an* isomorphism, *iff it is surjective and* injective, *that is* $Lx_1 = Lx_2$ *only for* $x_1 = x_2$*. If* L *is bijective, then* L^{-1} *exists and* $L^{-1} \in \mathcal{L}(\mathcal{Y}, \mathcal{X})$*.*

For the next theorem we need the concepts of *dual operators,* generalizing transpose matrices, and *generalized orthogonal complements.* For $L \in \mathcal{L}(\mathcal{X}, \mathcal{Y})$ the *dual operator* $L^d : \mathcal{Y}' \to \mathcal{X}'$ (to distinguish later derivatives, L', we use L^d for the dual operator), is uniquely defined for every $l \in \mathcal{Y}'$ by

$$\begin{aligned} L^d \in \mathcal{L}(\mathcal{Y}', \mathcal{X}') \text{ with } l \in \mathcal{Y}' &\mapsto L^d l \in \mathcal{X}' : \\ &\langle l, Lx\rangle_{\mathcal{Y}'\times\mathcal{Y}} = \langle L^d l, x\rangle_{\mathcal{X}'\times\mathcal{X}} \ \forall \ x \in \mathcal{X} \end{aligned} \tag{1.33}$$

with the pairings in $\mathcal{Y}', \mathcal{Y}$ and $\mathcal{X}', \mathcal{X}$, often denoted by the same $\langle\cdot,\cdot\rangle$. Only if the specific combination is emphasized do we use the index of the pairing as in

[1] For the later applications to discretizations with parameter h, we allow L_2^h, cf. Hackbusch [387], Lemma 6.4.5, and Wloka [666], Satz 2.

(1.31), (1.33). The above L^d is unique, linear and continuous. Furthermore, for $L, L_1 \in \mathcal{L}(\mathcal{X}, \mathcal{Y}), L_2 \in \mathcal{L}(\mathcal{Y}, \mathcal{Z}), \alpha \in \mathbb{C}$,

$$\|L^d\| = \|L\|, (L_1 \circ L_2)^d = L_2^d \circ L_1^d, (\alpha L)^d = \overline{\alpha} L^d; L^{dd} := (L^d)^d = L; \tag{1.34}$$

the last $L^{dd} = L$ is valid only in reflexive Banach spaces $\mathcal{X} = \mathcal{X}'', \mathcal{Y} = \mathcal{Y}''$.

We denote the dual space of a Hilbert space as $\mathcal{X}^* := \mathcal{X}'$. We obtain the following representation and natural embedding:

Theorem 1.17. Riesz representation theorem: *Let $\mathcal{X}$ be a Hilbert space and $l \in \mathcal{X}' = \mathcal{X}^*$. Then there exists a unique $x_l \in \mathcal{X}$ such that*

$$l(x) = (x,\ x_l)\ \forall\ x \in \mathcal{X} \text{ and } \|x_l\|_{\mathcal{X}} = \|l\|_{\mathcal{X}'}. \tag{1.35}$$

Corollary 1.18. *There exists a unique (Riesz-)isomorphism $J_{\mathcal{X}} \in \mathcal{L}(\mathcal{X}, \mathcal{X}^* = \mathcal{X}')$, see Corollary 1.16, such that*

$$J_{\mathcal{X}} x_l = l,\ J_{\mathcal{X}}^{-1} l = x_l,\ \|J_{\mathcal{X}}\| = \left\|J_{\mathcal{X}}^{-1}\right\| = 1.$$

$\mathcal{X}^* = \mathcal{X}'$ is again a Hilbert space with scalar product $(x', y')_{\mathcal{X}^*} := (J_{\mathcal{X}}^{-1} x', J_{\mathcal{X}}^{-1} y')_{\mathcal{X}}$ and $\|l\|_{\mathcal{X}^*} = \|x_l\|_{\mathcal{X}}$. $\mathcal{X}$ and $(\mathcal{X}^*)^* = \mathcal{X}''$ can be identified by $x_l = J_{\mathcal{X}}^{-1} l$.

For two Hilbert spaces $\mathcal{X}, \mathcal{Y}$ with scalar products $(\cdot,\cdot)_{\mathcal{X}}, (\cdot,\cdot)_{\mathcal{Y}}$, let $L \in \mathcal{L}(\mathcal{X}, \mathcal{Y})$. Besides the dual operator, the *adjoint operator* $L^* : \mathcal{Y} \to \mathcal{X}$ is defined by

$$L^* \in \mathcal{L}(\mathcal{Y}, \mathcal{X}) \text{ and } \forall\ y\ \in \mathcal{Y}:\ (y, Lx)_{\mathcal{Y}} = (L^* y, x)_{\mathcal{X}} \quad \forall\ x \in \mathcal{X}. \tag{1.36}$$

The dual and adjoint operators are related via the Riesz isomorphism as

$$L^* = (J_{\mathcal{X}})^{-1} L^d J_{\mathcal{Y}} \in \mathcal{L}(\mathcal{Y}, \mathcal{X}). \tag{1.37}$$

Only if $\mathcal{X}, \mathcal{Y}$ are identified with $\mathcal{X}', \mathcal{Y}'$, do we find $L^* = L^d$. In particular, we introduce *self-adjoint* or *symmetric* operators and *orthogonal projectors* as

$L \in \mathcal{L}(\mathcal{X}, \mathcal{X})$ is called a self-adjoint or symmetric operator and an orthogonal projector, if, $L = L^*$ and $L = L^2, L = L^*$, respectively.

Pairings may be used to generalize *orthogonal complements*: let subsets or subspaces $\mathcal{U} \subset \mathcal{X}$ and $\mathcal{V}' \subset \mathcal{X}'$ be given. Then

$$^{\perp}\mathcal{U} := \{x' \in \mathcal{X}' :\ \langle x', u\rangle_{\mathcal{X}' \times \mathcal{X}} = 0\ \ \forall\ u \in \mathcal{U}\} \subset \mathcal{X}', \tag{1.38}$$

$$\mathcal{V}'^{\perp} := \{x \in \mathcal{X} :\ \langle v', x\rangle_{\mathcal{X}' \times \mathcal{X}} = 0\ \ \forall\ v' \in \mathcal{V}'\} \subset \mathcal{X},$$

$$^{\perp}(\mathcal{V}') := (\mathcal{V}'^{\perp})'' := \{x'' \in \mathcal{X}'' :\ \langle x'', v'\rangle'_{\mathcal{X}'' \times \mathcal{X}'} = 0\ \ \forall\ v' \in \mathcal{V}'\} \subset \mathcal{X}'',$$

with $\mathcal{X}'' = \mathcal{X}$ for reflexive Banach spaces. The last space $(\mathcal{V}'^{\perp})''$ is used only very rarely. For continuous L (and $\langle\cdot,\cdot\rangle$), the subspaces $^{\perp}\mathcal{U}, \mathcal{V}'^{\perp}$ and $\mathcal{N}(L)$ are closed. For $\mathcal{R}(L)$ we have

Theorem 1.19. Closed range theorem: *For $L \in \mathcal{L}(\mathcal{X}, \mathcal{Y})$ the following four statements are equivalent ($\Longleftrightarrow$):*

$$\begin{aligned}\mathcal{R}(L) \ \text{ is closed } &\Longleftrightarrow \mathcal{R}(L^d) \ \text{ is closed} \Longleftrightarrow \\ \mathcal{R}(L) = \mathcal{N}(L^d)^\perp &\Longleftrightarrow \mathcal{R}(L^d) = {}^\perp\mathcal{N}(L).\end{aligned}$$

Definition 1.20. *For $L \in \mathcal{L}\ (\mathcal{X}, \mathcal{Y})$, let*

1. $\mathcal{N}(L) = \{x \in \mathcal{X} \ : \ Lx = 0\}, \quad \dim \mathcal{N}(L) < \infty,$ *and* (1.39)
2. $\mathcal{R}(L)$ *be closed, hence, a subspace $\overline{\mathcal{N}} \subset \mathcal{Y}$ exists, with*

   $\mathcal{Y} = \overline{\mathcal{N}} \oplus \mathcal{R}(L)$ *and let* $\operatorname{codim} \mathcal{R}(L) := \dim \overline{\mathcal{N}} < \infty.$

Then L is called a Fredholm operator *and the number $i(L) := \dim \mathcal{N}(L) - \operatorname{codim} \mathcal{R}(L)$ its* Fredholm index.

Definition 1.21. Complexification of Banach spaces and operators: *We extend a real $\mathcal{X}$ to a complex Banach space $\mathcal{X}_c$ cf. $\mathbb{R}$ to $\mathbb{C}$, as, cf. [676]*

$$\begin{aligned}&\mathcal{X}_c := \{x + iy : x, y \in \mathcal{X}\}, \quad i^2 = -1, \ (a + ib)(x + iy) := (ax - by) + i(bx + ay),\\ &(x_1 + iy_1)(x_2 + iy_2) = (x_1x_2 - y_1y_2) + i(y_1x_2 + x_2y_1), \ \ \forall a, b \in \mathbb{R}, \ x_1, x_2, y_1, y_2 \in \mathcal{X},\\ &\text{and } \|x + iy\|_{\mathcal{X}_c} := \max_{0 \le \phi \le 2\pi} \|x(\cos\phi) + y(\sin\phi)\|. \qquad (1.40)\end{aligned}$$

Similarly we extend an $A \in \mathcal{L}(\mathcal{X}, \mathcal{Y})$ to

$$A_c(x + iy) := Ax + iAy, \forall x, y \in \mathcal{X}, \ \text{ thus } A_c \in \mathcal{L}(\mathcal{X}_c, \mathcal{Y}_c). \qquad (1.41)$$

Theorem 1.22. Complexified Banach space $\mathcal{L}(\mathcal{X}, \mathcal{Y})$ and linear operator: *Assume Banach spaces $\mathcal{X}$, $\mathcal{Y}$ and $A \in \mathcal{L}(\mathcal{X}, \mathcal{Y})$. Define $\mathcal{X}_c, \|\cdot\|_{\mathcal{X}_c}, A_c$ as in (1.40), (1.41). Then $\mathcal{X}_c$, $\mathcal{Y}_c$ are Banach spaces and $A_c \in \mathcal{L}(\mathcal{X}_c, \mathcal{Y}_c)$ is linear and bounded with $\|A\|_{\mathcal{Y} \leftarrow \mathcal{X}} \le \|A_c\|_{\mathcal{Y}_c \leftarrow \mathcal{X}_c} \le 2\|A\|_{\mathcal{Y} \leftarrow \mathcal{X}}$.*

1.4.2 Inequalities and $L^p(\Omega)$ spaces

We start with some inequalities for finite sums. The *discrete Chebyshev inequality* holds for ordered sums. Then we obtain, e.g. [396], p 99, for the two sets of $f_1 \ge f_2 \ge \ldots \ge f_N \in \mathbb{R}$ and $g_1 \ge g_2 \ge \ldots \ge g_N \in \mathbb{R}$,

$$\sum_{j=1}^N f_j/N \sum_{j=1}^N g_j/N \le \sum_{j=1}^N f_j g_j/N. \qquad (1.42)$$

This implies for arbitrary $f_1, f_2, \ldots, f_N \in \mathbb{R}$ that

$$\sum_{j=1}^N f_j \le \left|\sum_{j=1}^N f_j\right| \le \sqrt{N \sum_{j=1}^N f_j^2}. \qquad (1.43)$$

Often we need the *discrete Hölder inequalities*. We always choose a pair of $1 \le p, q \le \infty$ with $1/p + 1/q = 1$. Then we find, with $\sum_{j=1}^{N} |g_j|^q|^{1/q} := \max_{j=1}^{N} |g_j|$ for $q = \infty$ [396,397] for $1 < p, q < \infty$, and for $1 = p < q = \infty$ as in (1.46),

$$\left|\sum_{j=1}^{N} f_j g_j\right| \le \left(\sum_{j=1}^{N} |f_j|^p\right)^{1/p} \left(\sum_{j=1}^{N} |g_j|^q\right)^{1/q} \quad \text{for } 1 \le p, q \le \infty, 1/p + 1/q = 1. \tag{1.44}$$

For cuboids $Q := \prod_{j=1}^{n} [a_j, b_j] \in \mathbb{R}^n, a_j \le b_j$ we have $\mu(Q) := \prod_{j=1}^{n} (b_j - a_j)$. For the set Ω, we choose two sets A^i and A^o, the union of countably many inner and outer *cuboids* Q_l^i and Q_k^o with disjoint interiors. They have to satisfy $A^o := \cup Q_k^o \supset \Omega$ and $A^i := \cup Q_l^i \subset \Omega$. For $\mu(A^o) := \sum_k \mu(Q_k^o)$ and $\mu(A^i) := \sum_l \mu(Q_l^i)$, let $\inf_{A^o}\{\mu(A^o)\} = \sup_{A^i}\{\mu(A^i)\}$. Then Ω is called μ-measurable, and $\mu(\Omega) := \inf_{A^o}\{\mu(A^o)\}$. For Ω measurable, a function, $f : \Omega \to \mathbb{R} \cup \infty \cup -\infty$, is called measurable, if for every $\alpha \in \mathbb{R}$ the set $\{x \in \Omega : f(x) \ge \alpha\}$ is measurable.

For any (measurable) $\Omega \subset \mathbb{R}^n$ we define and estimate

$$\|f\|_{L^p(\Omega)} := \left(\int_\Omega |f(x)|^p dx\right)^{1/p}, 1 \le p < \infty, \text{ and } \|f\|_{L^\infty(\Omega)} := \text{ess} \sup_{x \in \Omega} |f(x)|$$

$$L^p(\Omega) := \{f : \Omega \to \mathbb{R} : \|f\|_{L^p(\Omega)} < \infty\}, \Omega \subset \mathbb{R}^n, \tag{1.45}$$

$$\begin{aligned} \|fg\|_{L^1(\Omega)} &= \int_\Omega |f(x)g(x)|dx \le \left(\int_\Omega |f(x)|^p dx\right)^{1/p} \left(\int_\Omega |g(x)|^q dx\right)^{1/q} \\ &= \|f\|_{L^p(\Omega)} \|g\|_{L^q(\Omega)} \ \forall\, f \in L^p(\Omega), g \in L^q(\Omega) \\ &\quad \forall\, 1 \le p, q \le \infty \text{ with } 1/p + 1/q = 1, \text{ the } \textit{Hölder inequalities.} \end{aligned}$$

The usually missing case $1 = p, q = \infty$ in (1.44), (1.45) is directly obtainable: $f \in L^1(\Omega), g \in L^\infty(\Omega)$ imply $fg \in L^1(\Omega)$, and similarly for (1.44),

$$\|fg\|_{L^1(\Omega)} = \int_\Omega |f(x)g(x)|dx \tag{1.46}$$

$$\le \|g\|_{L^\infty(\Omega)} \left(\int_\Omega |f(x)|dx\right) = \|f\|_{L^1(\Omega)} \|g\|_{L^\infty(\Omega)}. \tag{1.47}$$

In particular we obtain for the case $p = q = 2$ the *Cauchy–Schwarz inequality*

$$|(u, v)_{L^2(\Omega)}| \le \|u\|_{L^2(\Omega)} \cdot \|v\|_{L^2(\Omega)}. \tag{1.48}$$

For later discussions of nonlinearities we need the extension

$$\left\|\prod_{j=1}^{d} f_j\right\|_{L^1(\Omega)} = \int_{\Omega} \left|\prod_{j=1}^{d} f_j\right| dx \leq \prod_{j=1}^{d} \left(\int_{\Omega} |f_j(x)|^{p_j} dx\right)^{1/p_j} \tag{1.49}$$

$$= \prod_{j=1}^{d} \|f_j\|_{L^{p_j}(\Omega)} \ \forall\ f_j \in L^{p_j}(\Omega) \forall\ 1 \leq p_j \leq \infty \text{ with } \sum_{j=1}^{d} 1/p_j = 1.$$

1.4.3 Hölder and Sobolev spaces and more

For a function $u \in C_0^\infty(\Omega, \mathbb{R})$ the following partial derivatives are well defined, with

$$\alpha := (\alpha_1, \ldots, \alpha_n) \in \mathbb{N}_0^n,\ |\alpha| := \alpha_1 + \cdots + \alpha_n, \alpha\,! := (\alpha_1)! \cdots (\alpha_n)! \tag{1.50}$$

$$\partial^i u := \frac{\partial u}{\partial x_i}, \partial u := (\partial^1 u, \ldots, \partial^n u),\ \partial^\alpha u := \partial_1^{\alpha_1} \ldots \partial_n^{\alpha_n} u := \frac{\partial^{|\alpha|}}{\partial x_1^{\alpha_1} \ldots \partial x_n^{\alpha_n}}\ u.$$

The C^k, $C^{k,1}$ and $C^{k+\mu}$ indicate the usual classical k-times continuously differentiable functions $u : \mathcal{D} \to \mathbb{R}$, with appropriate $\mathcal{D} \subset \mathbb{R}^n$. For $u \in C^{k,1}$ [and $u \in C^{k+\mu}$], respectively, all derivatives $\partial^\alpha u$ for $|\alpha| \leq k$ are *Lipschitz* [and *Hölder*] continuous, hence $|\partial^\alpha u(x) - \partial^\alpha u(y)| \leq C|x-y|^\mu$ for all $x, y \in \mathcal{D}$ for $\mu = 1$ [and $0 < \mu < 1$].

For Sobolev spaces we have to generalize this concept:

Definition 1.23. *For a function $u \in L^p(\Omega), 1 \leq p \leq \infty$, a function $v := \partial^\alpha u \in L^p(\Omega)$ is called a* weak α derivative *of u, iff*

$$(w, v)_{L^p(\Omega)} = (-1)^{|\alpha|}(w^\alpha, u)_{L^p(\Omega)} \textit{ for all } w \in C_0^\infty(\Omega). \tag{1.51}$$

*The **Sobolev space** $W^{k,p}(\Omega), k \in \mathbb{N}_0, 1 \leq p \leq \infty$, is the set of functions with weak derivatives up to order k:*

$$W^{k,p}(\Omega) := \{u \in L^p(\Omega) : \forall \alpha \textit{ with } |\alpha| \leq k : \partial^\alpha u \in L^p(\Omega)\}. \tag{1.52}$$

For $u \in L^p(\Omega)$ let the weak derivative $\partial^\alpha u \in L^p(\Omega)$, and in $\Omega' \subset \Omega$ let the classical derivative $\partial^\alpha u \in L^p(\Omega')$ exist. Then the weak and the classical derivative coincide *almost everywhere (a.e.)* in Ω'.

Theorem 1.24. Sobolev spaces are Banach spaces: *The Sobolev spaces $W^{k,p}(\Omega)$, $k \in \mathbb{N}_0$, $1 \leq p \leq \infty$, are Banach spaces with respect to the norm*

$$\|u\|_{W^{k,p}(\Omega)} := \|u\|_{W^{k,p}} := \|u\|_{k,p} := \left(\sum_{|\alpha| \leq k} \|\partial^\alpha u\|_{L^p(\Omega)}^p\right)^{1/p}. \tag{1.53}$$

For the special case $p = 2$ we have the Hilbert space $H^k(\Omega)$ with

$$(u, v)_k := (u, v)_{H^k(\Omega)} := \sum_{|\alpha| \leq k} (\partial^\alpha u, \partial^\alpha v)_{L^2(\Omega)},\ \|u\|_{H^k(\Omega)} := (u, v)_{H^k(\Omega)}^{1/2}; \tag{1.54}$$

sometimes we use $\|u\|_k := \|u\|_{H^k(\Omega)}$. *The spaces* $W_0^{k,p}(\Omega)$ *and* $H_0^k(\Omega)$, *defined as closure of* $C_0^\infty(\Omega)$ *with respect to the corresponding norm, e.g.* $\|\cdot\|_{H^k(\Omega)}$,

$$W_0^{k,p}(\Omega) := \overline{C_0^\infty(\Omega)}^{\|\cdot\|_{k,p}}, H_0^k(\Omega) := \overline{C_0^\infty(\Omega)}^{\|\cdot\|_{H^k(\Omega)}}, \tag{1.55}$$

are again Banach and Hilbert spaces with the same norm (1.53) and (1.54), respectively. Sometimes, we need the following seminorms

$$|u|_{W^{k,p}(\Omega)} := \left(\sum_{|\alpha|=k} \|\partial^\alpha u\|^p_{L^p(\Omega)} \right)^{1/p}, \; |u|_{H^k(\Omega)} := |u|_{W^{k,p=2}(\Omega)}. \tag{1.56}$$

If Ω *is bounded, then* $\|\cdot\|_{W_0^{k,p}(\Omega)} := |\cdot|_{W^{k,p}(\Omega)}\big|_{W_0^{k,p}(\Omega)}$ *and* $\|\cdot\|_{W^{k,p}(\Omega)}$ *are equivalent norms on* $W_0^{k,p}(\Omega)$*; similarly, for* $p = 2$*, on* $H_0^k(\Omega)$*. In particular, we obtain with* Poincaré's constant $C_P(\Omega)$

$$|u|_{W^{k,p}(\Omega)} \le \|u\|_{W^{k,p}(\Omega)} \le C_P(\Omega)|u|_{W^{k,p}(\Omega)} \forall u \in W_0^{k,p}(\Omega). \tag{1.57}$$

Proposition 1.25. cf. Zeidler, [678], p. 1032*: For* $\Omega \in C^{0,1}$ *the standard norm* $\|u\|_{W^{k,p}(\Omega)}$ *for the Sobolev spaces* $W^{k,p}(\Omega)$ *with* $k \in \mathbb{N}$ *is* equivalent to the norms

$$\left(\int_\Omega \sum_{|\alpha|=k} |\partial^\alpha u|^p dx + \left| \int_\Omega u dx \right|^p \right)^{1/p} \tag{1.58}$$

$$\left(\int_\Omega \sum_{|\alpha|=k} |\partial^\alpha u|^p dx + \left| \int_{\partial\Omega} u \, ds \right|^p \right)^{1/p}$$

$$\left(\int_\Omega \sum_{|\alpha|=k} |\partial^\alpha u|^p dx + \int_{\partial\Omega} |u|^p \, ds \right)^{1/p},$$

where ds indicates the surface element on $\partial\Omega$*. For* $n = 1$ *and* $\Omega = (a, b)$ *we define* $\int_{\partial\Omega} g \, ds := g(b) - g(a)$.

For $\mathcal{X} \subset \mathcal{Y}$ we denote the mapping

$$\vec{I} : \mathcal{X} \to \mathcal{Y}, \; x \mapsto \vec{I}\, x := x \in \mathcal{Y} \text{ for } \mathcal{X} \subset \mathcal{Y}$$

as an *inclusion* or *embedding,* often abbreviated as $\mathcal{X} \hookrightarrow \mathcal{Y}$. If $\vec{I}$ is a continuous and compact operator, then $\mathcal{X}$ is called *continuously* and *compactly embedded into* $\mathcal{Y}$, respectively, so $\|x\|_\mathcal{Y} \le C\|x\|_\mathcal{X} \forall x \in \mathcal{X}$. For the compact embedding, additionally each bounded sequence $\{x_n\}$ in $\mathcal{X}$ has a subsequence $\{x_{n_k}\}$ converging in $\mathcal{Y}$, cf. [677], pp. 1026 ff., [676], p. 248, [387], Example 6.4.7:

Theorem 1.26. Embedding theorem for Sobolev spaces: *Let Ω be a bounded Lipschitz domain, $j \in \mathbb{N}_0$, $k \in \mathbb{N}$, $0 \le j < k$, $1 \le p$, $q < \infty$, $0 < \alpha < 1$ and define $d := 1/p - (k-j)/n$. Then the following embedding results are valid.*[2]

1. $W^{k,p}(\Omega) \hookrightarrow W^{j,q}(\Omega)$ *and* $W_0^{k,p}(\Omega) \hookrightarrow W_0^{j,q}(\Omega)$ *are* continuous *for* $d \le 1/q, j \le k$, *and* compact *for* $d < 1/q, j < k$. *These results remain valid for* $W_0^{k,p}(\Omega)$ *and arbitrary open* Ω.
2. *In particular,* $W^{k,p}(\Omega) \hookrightarrow W^{k,q}(\Omega)$ *is continuous for* $1 \le q \le p \le \infty, 0 \le k$.
3. $W^{k,p}(\Omega) \hookrightarrow C^j(\overline{\Omega})$ *are* compact *for* $d < 0$, *i.e.* $k - j > n/p$, *and* $W^{k,p}(\Omega) \hookrightarrow C^{j,\alpha}(\overline{\Omega}) \hookrightarrow C^j(\overline{\Omega})$ *for* $d + \alpha/n < 0$ *or* $d < 0$. *This inclusion might require changing* $u \in W^{k,p}(\Omega)$ *on a subset of* Ω *of* n*-dimensional measure* 0.
4. *If we replace the above bounded* Ω *by* $\mathbb{R}^n$ *we find:* $W^{k,p}(\mathbb{R}^n) \hookrightarrow C^j(\mathbb{R}^n)$ *for* $d < 0$, *i.e.* $k - j > n/p$, *and* $W^{k,p}(\mathbb{R}^n) \hookrightarrow W^{j,q}(\mathbb{R}^n)$ *for* $d \le 1/q$ *and* $j < k$ *are continuous.*
5. $C^\alpha(\bar{\Omega}) \hookrightarrow L_p(\Omega)$ *and* $C^{0,1}(\bar{\Omega}) \hookrightarrow L_p(\Omega)$ *are* continuous *for* $0 \le \alpha \le 1$ *and* $1 \le p \le \infty$.
6. *For* $0 < \alpha \le 1, j < k$, *the embeddings* $C^{k,\alpha}(\overline{\Omega}) \hookrightarrow C^k(\overline{\Omega}) \hookrightarrow C^j(\overline{\Omega})$ *are compact.*

The previous compact embedding result implies, e.g.

Proposition 1.27. *Under the conditions of Theorem 1.26, for every* $\varepsilon > 0$ *a* $C_\varepsilon > 0$ *exists such that*

$$\|x\|_{W^{j,q}(\Omega)} \le \varepsilon \|x\|_{W^{k,p}(\Omega)} + C_\varepsilon \|x\|_{L^s(\Omega)} \quad \text{for} \quad 0 \le j < k,$$
$$1 \le p, q, s < \infty, \quad 1/p - (k-j)/n < 1/q, \quad 1/q - k/n \le 1/s.$$

In particular, we find for $p = q = s$ *and* $0 < j < k$*:*

$$\|x\|_{W^{j,p}(\Omega)} \le \varepsilon \|x\|_{W^{k,p}(\Omega)} + C_\varepsilon \|x\|_{L^p(\Omega)}.$$

Theorem 1.28. Sobolev–Stein extension theorem [141], p. 31: *Let* $\Omega \subset \mathbb{R}^n$ *have a Lipschitz boundary, let* $k \in \mathbb{N}_0$ *and* $p \in \mathbb{R}$ *with* $1 \le p \le \infty$. *Then there exists a bounded operator* $E : W^{k,p}(\Omega) \to W^{k,p}(\mathbb{R}^n)$ *and a constant,* C, *independent of* v, *such that*

$$Ev|_\Omega = v \text{ and } \|Ev\|_{W^{k,p}(\mathbb{R}^n)} \le C\|v\|_{W^{k,p}(\Omega)} \; \forall v \in W^{k,p}(\Omega). \tag{1.59}$$

Further *bounded extension operators* for an open set with $\overline{\Omega} \subset \Omega'$ of the form $E_D : C_D^{k,\alpha}(\overline{\Omega}) \to C_0^{k,\alpha}(\Omega')$, with $_0$ and $_D$ indicating trivial and nontrivial Dirichlet boundary conditions are discussed in [344], p. 136, [677], pp. 305–306. In Chapter 5, for example, we need Theorem 1.28.

Theorem 1.29. Extension of a boundary function, [344], p 137: *Let* $\Omega \subset \mathbb{R}^n$ *be a* $C^{k,\alpha}$ *domain, let* $k \in \mathbb{N}$ *and let* Ω' *be an open set with* $\overline{\Omega} \subset \Omega'$. *Let* $v \in C^{k,\alpha}(\partial\Omega)$ *Then there exists a function* $w \in C_0^{k,\alpha}(\Omega')$, *such that* $v = w$ *on* $\partial\Omega$.

Trace operators for Sobolev spaces of fractional order, $s \in \mathbb{R}_+$, are necessary in different contexts, in particular for nonconforming, adaptive FE and discontinuous Galerkin methods. We formulate the following results in $H^s(\Omega)$, and $W^{s,p}(\Omega)$ form,

[2] Note that $L^q(\Omega) = W^{0,q}(\Omega)$ and d is independent of n for $k = j$.

cf. Zeidler [678], pp. 1031 ff., Hackbusch [387], Subsections 6.2.4 and 6.2.5. They represent an integral variant of the Hölder spaces $C^{k,\mu}(\Omega)$. We define for these

$$H^s(\Omega), W^{s,p}(\Omega) \text{ for } \partial\Omega \in C^{0,1}, s = k + \mu, k = 0, 1, \ldots, 0 < \mu < 1, 1 < p < \infty, \quad (1.60)$$

the corresponding scalar products and *Sobolev–Slobodeckij norms* as

$$(u, v)_{H^s(\Omega)} := (u, v)_{H^k(\Omega)} + \sum_{|\alpha| \leq k} f(\partial^\alpha u, \partial^\alpha v), \|u\|_{H^s(\Omega)} := \left((u, u)_{H^s(\Omega)}\right)^{1/2}, \quad (1.61)$$

with

$$f(u, v) := \int_{\Omega \times \Omega} \frac{[u(x) - u(y)][v(x) - v(y)]}{|x - y|^{n+2\mu}} dxdy$$

$$\text{and for } 1 \leq p \leq \infty, f_p(u) := \int_{\Omega \times \Omega} \frac{|u(x) - u(y)|^p}{|x - y|^{n+p\mu}} dxdy,$$

$$\|u\|_{W^{s,p}(\Omega)} := (\|u\|^p_{W^{k,p}(\Omega)} + \sum_{|\alpha| \leq k} f_p(\partial^\alpha u))^{1/p}.$$

For $H^s(\partial\Omega), W^{s,p}(\partial\Omega)$ replace in (1.61) the Ω by $\partial\Omega$ and $dxdy$ by $dsds'$, and n by $n - 1$.

Theorem 1.30. Sobolev spaces of fractional order: *The Sobolev spaces $H^s(\Omega)$, $H^s(\partial\Omega)$, and $W^{s,p}(\Omega), W^{s,p}(\partial\Omega)$, $s = k + \mu, k = 0, 1, \ldots, 0 < \mu < 1, 1 < p < \infty$, in (1.60) are Hilbert and reflexive Banach spaces, respectively, with respect to the scalar product and norms in (1.61).*

For considering the boundary and functions defined there, we introduce the concept of an atlas. We achieve that by introducing a parametrization, hence, local bijections between neighborhoods of boundary points and the unit sphere along the boundary. This allows formulating the necessary differentiability conditions for $\partial\Omega$ and a parametrization for the boundary.

Definition 1.31. *Let $k \in \mathbb{N}_0$ [and t with $0 < t \leq \infty, t = k + \lambda, k \in \mathbb{N}_0, 0 \leq \lambda < 1$] be given. [Here and below we indicate by [...] this case $0 < t \leq \infty, \ldots$ and its consequences.] For each $x \in \partial\Omega$ we require a neighborhood $\mathcal{U} = \mathcal{U}(x) \subset \mathbb{R}^n$ and a bijection $\phi : \mathcal{U} \to \mathcal{B}_1(0) := \mathcal{B}_1^n(0) = \{x \in \mathbb{R}^n : |x| \leq 1\}$, the* unit ball *in $\mathbb{R}^n$, such that*

$$\begin{aligned}
&1.\ \phi \in C^{k,1}(\bar{\mathcal{U}}), \phi^{-1} \in C^{k,1}(\overline{\mathcal{B}_1(0)}) \\
&\quad [\text{and } \phi \in C^t(\bar{\mathcal{U}}), \phi^{-1} \in C^t(\overline{\mathcal{B}_1(0)})], \\
&2.\ \phi(\mathcal{U} \cap \partial\Omega) = \{\xi \in \mathcal{B}_1(0) : \xi_n = 0\}, \\
&3.\ \phi(\mathcal{U} \cap \Omega) = \{\xi \in \mathcal{B}_1(0) : \xi_n > 0\}, \\
&4.\ \phi(\mathcal{U} \cap (\mathbb{R}^n \backslash \overline{\Omega})) = \{\xi \in \mathcal{B}_1(0) : \xi_n < 0\}.
\end{aligned} \quad (1.62)$$

Then we denote Ω as $(k, 1)$ [and t] smooth, indicated by $\Omega \in C^{k,1}$ [and $\in C^t$]. $\Omega \in C^{(0,1)}$ is called a Lipschitz domain. *These $(\mathcal{U}, \phi)$ are denoted as an* atlas.

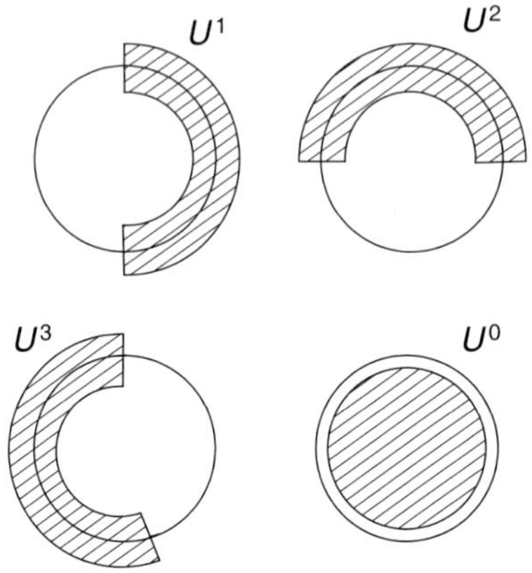

Figure 1.5 Atlas for a circle.

Proposition 1.32. *Let $\Omega \in C^{k,1}, k \in \mathbb{N}_0$ [and $\Omega \in C^t, 0 < t \le \infty$] be open and bounded. Then we need only finitely many neighborhoods, $\mathcal{U}^i$, bijections, $\phi_i, 0 \le i \le N$, and* parametrizations of the boundary, *α_i, $1 \le i \le N$, in (1.62), such that cf. Figures 1.5 and 1.6,*

1. $\mathcal{U}^i$ *is open, bounded,* $0 \le i \le N$, $\bigcup_{i=0}^{N} \mathcal{U}^i \supset \bar{\Omega}$, $\overline{\mathcal{U}^0} \subset \Omega$,
2. $\phi_i : \mathcal{U}^i \to \mathcal{B}_1^n(0)$ *is bijective,* $\phi_i \in C^{k,1}(\overline{\mathcal{U}^i})$ [*and* $\in C^t(\overline{\mathcal{U}^i})$], $0 \le i \le N$,
3. $\mathcal{U}_i := \mathcal{U}^i \cap \partial\Omega$, $\alpha_i := \phi_i|_{\mathcal{U}_i}$, $1 \le i \le N$, $\bigcup_{i=1}^{N} \mathcal{U}_i = \partial\Omega$, (1.63)
4. $\alpha_i : \mathcal{U}_i \to \alpha_i(\mathcal{U}_i) = \mathcal{B}_1^{n-1}(0) \subset \mathbb{R}^{n-1}$ *is bijective,* $1 \le i \le N$,
5. $\alpha_i \circ \alpha_j^{-1} \in C^{k,1}\left(\overline{\alpha_j(\mathcal{U}_i \cap \mathcal{U}_j)} \right)$ [*and* $\alpha_i \circ \alpha_j^{-1} \in C^t\left(\overline{\alpha_j(\mathcal{U}_i \cap \mathcal{U}_j)} \right)$].

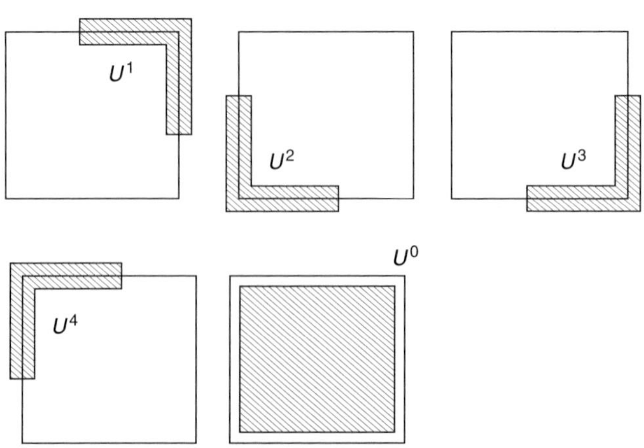

Figure 1.6 Atlas for a square.

These $\mathcal{U}^i, \phi_i, 0 \leq i \leq N$, *define an* atlas, *hence satisfy (1.62). The system* $(\mathcal{U}_i, \alpha_i)$, $1 \leq i \leq N$, *is denoted as a* parametrization or coordinate system *for* $\partial\Omega$. *In particular,* Ω *is sometimes called a* smooth manifold, *if in (1.63) the* ϕ_i, α_j *and their inverses are* $\in C^\infty$.

The above boundary operators have to be applied to functions $u \in W_p^m(\Omega)$ and will yield functions $u|_{\partial\Omega} \in W_p^{m-1/p}(\partial\Omega)$. Using the above ϕ_i, α_i, it is possible to reduce the discussion to functions defined on the $\mathcal{U}^i, \mathcal{U}_i$ or the $\mathcal{B}_1^n(0), \mathcal{B}_1^{n-1}(0)$ or even to $\mathbb{R}^n, \mathbb{R}^{n-1}$. We use the following *partition of unity*:

Proposition 1.33. *Let* $\mathcal{U}^i, 0 \leq i \leq N$, *satisfy (1.63). Then functions* $\sigma_i \in C_0^\infty(\mathbb{R}^n)$, $0 \leq i \leq N$, *exist such that*

$$\begin{aligned} carr\ (\sigma_i) &:= \ closure\ \{x \in \mathbb{R}^n : \sigma_i(x) \neq 0\} \\ &= supp\,(\sigma_i) \subset \mathcal{U}^i, \quad \sum_{i=0}^N \sigma_i^2(x) = 1 \ for\ all\ x \in \bar{\Omega}. \end{aligned} \tag{1.64}$$

This representation is called a partition of unity.

Since, by (1.64), $\sum_{i=0}^N \sigma_i^2(x)u(x) = \sum_{i=0}^N u(x)\sigma_i^2(x) = u(x)$, we may consider each of the $\sigma_i u$ separately. In particular, we will use the above bijections $\alpha_i : \mathcal{U}_i \to \mathcal{B}_1^{n-1}(0)$. Then we study the local parametrizations $(\sigma_i u) \circ (\alpha_i)^{-1} : \mathcal{B}_1^{n-1}(0) \to \mathbb{R}$ and sum over all appropriate terms. With carr $(\sigma_i) \subset \mathcal{U}^i$, we extend $(\sigma_i u) \circ (\alpha_i)^{-1}$ to $(\sigma_i u) \circ (\alpha_i)^{-1} : \mathbb{R}^{n-1} \to \mathbb{R}$.

Definition 1.34. *Let* Ω *be bounded and choose for* $\Omega \in C^{k,1}$ *[and* $\Omega \in C^t$*], an* s *with* $s \leq k+1$ *[and* $s < t$ *for* $t \notin \mathbb{N}$, $t > 1$ *or* $s \leq t \in \mathbb{N}$*]. Furthermore, let the* $\mathcal{U}^i$, ϕ_i, $\mathcal{U}_i$, α_i *and* σ_i, $0 \leq i \leq N$, *satisfy (1.63), (1.64). Then the Sobolev space* $W^{s,p}(\partial\Omega)$ *is defined as*

$$W^{s,p}(\partial\Omega) := \left\{u : \partial\Omega \to \mathbb{R} : (\sigma_i u) \circ \alpha_i^{-1} \in W_{p,0}^s\left(\mathcal{B}_1^{n-1}(0)\right), 1 \leq i \leq N\right\}.$$

By $\overline{\mathcal{U}^0} \subset \Omega$ we may restrict the discussion essentially to $1 \leq i \leq N$.

Theorem 1.35. Sobolev spaces $W^{s,p}(\partial\Omega)$ are Banach spaces: *For bounded* $\Omega \in C^{k,1}$ *[and* C^t*] and* s *as in Definition 1.34, let the* $\mathcal{U}^i$, ϕ_i, $\mathcal{U}_i$, α_i *and* σ_i, $0 \leq i \leq N$, *satisfy (1.63), (1.64). Then the above Sobolev space* $W^{s,p}(\partial\Omega)$ *is a Banach and a Hilbert space, for* $1 \leq p < \infty$, $p \neq 2$ *and* $p = 2$, *respectively, with the obvious modification of the norms in (1.61), and replacing* Ω *by* $\partial\Omega$. *The* $W^{s,p}(\partial\Omega)$ *are reflexive for* $1 < p < \infty$. *For* $p = 2$ *the scalar product is*

$$(u, v)_{H^s(\partial\Omega)} := \sum_{i=1}^N \left((\sigma_i u) \circ \alpha_i^{-1}, (\sigma_i v) \circ \alpha_i^{-1}\right)_{H^s(\mathbb{R}^{n-1})}. \tag{1.65}$$

A different choice of atlas and partition of unity yields the same Sobolev spaces. More precisely, replace the $\mathcal{U}_i$, α_i, σ_i *chosen above by* $\tilde{\mathcal{U}}_i$, $\tilde{\alpha}_i$, $\tilde{\sigma}_i$, $0 \leq i \leq \tilde{N}$, *again satisfying (1.63), (1.64). This defines another* $C^{k,1}$ *[and* C^t*] coordinate system and another partition of unity. The norms, as in Therem 1.24 and the modified (1.61),* *are equivalent and the two Sobolev spaces* $W^{s,p}(\partial\Omega)$ *and* $\tilde{W}^{s,p}(\partial\Omega)$ *contain the same elements, so can be identified.*

Now we are able to define the restriction operators for functions defined on Ω to the boundary $\partial\Omega$. Let $\overline{\Omega} \subset \Omega_0$. Then a restriction to the boundary is uniquely determined for sufficiently smooth functions. So

$$\gamma_0 : C_0^\infty(\Omega_0) \to C^\tau(\partial\Omega), \quad (\gamma_0 u)(x) := u(x) \forall x \in \partial\Omega, \tag{1.66}$$

for appropriate $\tau \in \mathbb{R}$, is well defined.

Definition 1.36. Trace or restriction operator: *Let an extension* γ *of* γ_0 *in (1.66) exist such that* $\gamma_p = \gamma \in \mathcal{L}(W^{s,p}(\Omega), W^{t,p}(\partial\Omega))$ *for appropriate* s, t, *with*

$$\gamma|_{C_0^\infty(\Omega)} = \gamma_0, \quad \textit{and denote it as } u|_{\partial\Omega} := \gamma u.$$

Then this γ *is called a* trace operator *or a* restriction operator.

The following theorem shows that these trace operators exist, if $\Omega \in C^{0,1}$ and the smoothness in $W^{s,p}(\Omega)$ is better than in $W^{t,p}(\partial\Omega)$, cf. [677], p. 1030.

Theorem 1.37. Properties of the trace operator: *Let* Ω *be open and bounded,* $\Omega \in C^{0,1}$, $n \geq 2$, $1 \leq p < \infty$, *and* γ *as in Definition 1.36. Then*

1. *For* $p < \infty$ *the* $\gamma = \gamma_p : W^{1,p}(\Omega) \to W^{1-1/p,p}(\partial\Omega)$ *is linear,* surjective, *and continuous:* $\|\gamma_p u\|_{W^{1-1/p}(\partial\Omega)} \leq C\|u\|_{W^{1,p}(\Omega)}$ $\forall u \in W^{1,p}(\Omega)$.
2. $u \in W_0^{1,p}(\Omega)$ *is equivalent to* $u \in W^{1,p}(\Omega)$ *and* $\gamma\, u = 0$ *on* $\partial\Omega$.
3. *For* $\Omega \in C^{m-1,1}$ *the* $u \in W_0^{m,p}(\Omega)$ *is equivalent to* $u \in W^{m,p}(\Omega)$ *and* $D^\alpha\, u \in W_0^{1,p}(\Omega)$ *or* $\gamma\, D^\alpha\, u = 0 \ \ \forall\ |\alpha| \leq m-1$.
4. *For* $p < \infty$ *there exists a linear continuous extension operator*

$$E_p : W^{1-1/p,p}(\partial\Omega) \to W^{1,p}(\Omega)$$

such that the following diagram is commutative with the identity $I : W^{1,p}(\Omega) \to W^{1,p}(\Omega)$

$$\begin{array}{ccc} W^{1,p}(\Omega) & \xrightarrow{\gamma_p} & W^{1-1/p,p}(\partial\Omega) \\ I \nwarrow & & \swarrow E_p \\ & W^{1,p}(\Omega) & \end{array}$$

Theorem 1.38. Trace theorem: *Let* $\Omega \subset \mathbb{R}^n$ *have a Lipschitz boundary,* $\Omega \in C^{0,1}$, *and let* $1 \leq p \leq \infty$. *Then there exist constants,* C, *such that*

$$\|v\|_{L^p(\partial\Omega)} \leq C\|v\|_{L^p(\Omega)}^{1-1/p} \cdot \|v\|_{W^{1,p}(\Omega)}^{1/p} \quad \forall v \in W^{1,p}(\Omega) \textit{ and} \tag{1.67}$$

$$\|v\|_{L^p(\partial\Omega)} \leq C\|v\|_{L^2(\Omega)}^{1/2}\|v\|_{H^1(\Omega)}^{1/2} \leq C\|v\|_{H^1(\Omega)} \textit{ for } p = 2.$$

The $\|v\|_{W^{k,p}(\Omega)}$ *satisfy a trace estimate as well:*

$$\|v\|_{W^{k,p}(\partial\Omega)} \leq C\|v\|_{W^{k,p}(\Omega)}^{1-1/p} \cdot \|v\|_{W_p^{k+1}(\Omega)}^{1/p} \quad \forall v \in W_p^{k+1}(\Omega). \tag{1.68}$$

Theorem 1.39. Trace operators for Sobolev spaces of fractional order: *For* $\Omega \in C^{k,1}$, *choose* s *such that* $1/2 < s = k+1 \in \mathbb{N}$ *[and for* $\Omega \in C^t$ *choose* s *such that* $1/2 < s \le t \in \mathbb{N}$ *or* $1/2 < s < t \notin \mathbb{N}$*] and* $s > |\alpha| + 1/2, |\alpha| \ge 0$. *We formulate separately for* $|\alpha| = 0$ *and* $|\alpha| > 0$. *Then the following trace and extension statements are correct: the trace operators* $\gamma\partial^\alpha : H^s(\Omega) \to H^{s-|\alpha|-1/2}(\partial\Omega)$ *exist.*

1. $\gamma \in \mathcal{L}(H^s(\Omega), H^{s-1/2}(\partial\Omega))$, *so* $u \in H^s(\Omega)$ *implies* $\gamma u = u|_{\partial\Omega} \in H^{s-1/2}(\partial\Omega)$.
2. $\gamma\partial^\alpha \in \mathcal{L}(H^s(\Omega), H^{s-|\alpha|-1/2}(\partial\Omega))$, *hence, for* $u \in H^s(\Omega)$ *the restriction implies* $\gamma\partial^\alpha u = \partial^\alpha u|_{\partial\Omega} \in H^{s-|\alpha|-1/2}(\partial\Omega)$ *and* $\|\partial^\alpha u|_{\partial\Omega}\|_{H^{s-|\alpha|-1/2}(\partial\Omega)} \le C\|u\|_{H^s(\Omega)}$. *In particular the normal derivatives* $\partial^l u/\partial n^l$ *exist, e.g. for* $t > s > l + 1/2$ *and* $\Omega \in C^t$, *and* $u \in H_0^s(\Omega)$ *with* $s > l + 1/2$ *implies* $\gamma\partial^\alpha u = \partial^\alpha u|_{\partial\Omega} = 0$, *if* $0 \le |\alpha| \le l$.
3. *For each* $w \in H^{s-1/2}(\partial\Omega)$ *there exists* $u \in H^s(\Omega)$ *such that* $w = \gamma u$ *and* $\|u\|_{H^s(\Omega)} \le C_s\|w\|_{H^{s-1/2}(\partial\Omega)}$.

For the Gauss integral theorem, we use the notation $F = (F_1, \cdots, F_n) \in W^{1,1}(\Omega)$

$$\mathrm{div} F = \sum_{i=1}^n \partial F_i/\partial x_i.$$

Theorem 1.40. Gauss integral theorem: *Let* $\Omega \subset \mathbb{R}^n$ *be a* $C^{0,1}$ *compact domain with outer normal vector* $\nu = (\nu_1, \cdots, \nu_n)$ *along* $\partial\Omega$, *and* $F \in W^{1,1}(\Omega, \mathbb{R}^n)$. *Then*

$$\int_\Omega \mathrm{div} F dx = \int_{\partial\Omega} \langle F, \nu\rangle_n ds \text{ and } \int_\Omega \partial F_i/\partial x_i dx = \int_{\partial\Omega} F_i \nu_i ds,\ i = 1, \ldots, n. \quad (1.69)$$

1.4.4 Derivatives in Banach spaces

We introduce two types of derivatives in Banach spaces. We motivate it by the nonlinear operator, cf. (1.13), for the *loaded rod.*

$$G : C^2[0,l] \to C[0,l], G(\varphi) = d^2\varphi/ds^2 + \lambda \sin\varphi. \quad (1.70)$$

(1) Choosing functions $\varphi, \psi \in C^2[0,l]$ with small

$$\|\psi\|_{2,\infty} := \|\psi\|_\infty + \|\dot{\psi}\|_\infty + \|\ddot{\psi}\|_\infty,$$

we obtain the relation, with fixed parameter λ,

$$\begin{aligned} G(\varphi+\psi) - G(\varphi) &= d^2\psi/ds^2 + \lambda(\sin(\varphi+\psi) - \sin\varphi) \\ &= d^2\psi/ds^2 + (\lambda\cos\varphi)\psi + o(\psi) \\ &=: L(\varphi)\psi + o(\psi) \\ &=: G'(\varphi)\psi + o(\psi). \end{aligned}$$

The notation $L(\varphi) = G'(\varphi)$, called the *Frechét derivative*, is formalized in (1.71) below:

$$G'(\varphi) : C^2[0,l] \to C[0,l], G'(\varphi)\psi := d^2\psi/ds^2 + (\lambda\cos\varphi)\psi := \ddot{\psi} + (\lambda\cos\varphi)\psi.$$

This operator $G'(\varphi)$ is obviously linear and it is bounded, see Example 1.2,

$$\|G'(\varphi)\psi\| \le |\lambda\cos\varphi|\|\psi\|_\infty + \|\ddot{\psi}\|_\infty \le \max\{1,|\lambda|\}\|\psi\|_{2,\infty} \text{ or } \|G'(\varphi)\| \le \max\{1,|\lambda|\}.$$

For appropriate φ and ψ we even have equality.

(2) To determine the *directional derivative*, we choose fixed $\varphi, \psi \in C^2[0,l]$, $\tau \in [0,l]$. We introduce a function of a real variable $t \in [0,l]$, as

$$g : \mathbb{R} \to \mathbb{R}, g(t) := G(\varphi(\tau) + t\psi(\tau)) = (d^2(\varphi + t\psi)/ds^2)(\tau) + \lambda\sin(\varphi(\tau) + t\psi(\tau)).$$

We can use the classical differentiation rules, e.g. the chain rule for a function of a real variable to obtain

$$dg/dt\big|_{t=0} = (d^2\psi/ds^2)(\tau) + (\lambda\cos\varphi(\tau))\psi(\tau).$$

By considering this relation for variable $\tau \in [0,l]$, and fixed functions φ, ψ, we could equally well have defined the corresponding

$$\bar{g} : \mathbb{R} \to C[0,l], \bar{g}(t) := d^2(\varphi + t\psi)/ds^2 + \lambda\sin(\varphi + t\psi).$$

Similarly to the above, we can linearize

$$\begin{aligned}\bar{g}(t) - \bar{g}(0) &= td^2\psi/ds^2 + \lambda(\sin(\varphi + t\psi) - \sin\varphi)\\ &= td^2\psi/ds^2 + t(\lambda\cos\varphi)\psi + t\ o(\psi)\\ &= t(L(\varphi)\psi) + o(t\psi)\\ &= t\ G'(\varphi)\psi + o(t\psi).\end{aligned}$$

This $L(\varphi)\psi = G'(\varphi)\psi$ will be called the *Gateaux derivative* in (1.72). Obviously, by fixing the functions and variables, e.g. the φ, ψ, τ above, the computation of these *directional derivatives* only uses classical calculus. The linearization is more complicated, but it has advantages for the bifurcation in [120]. We need a simple criterion which guarantees a directional derivative to be a Frechét derivative, see Definition 1.41 and Theorem 1.42.

Definition 1.41. Frechét (F) and Gateaux (G) derivative:

1. *Let $H : \mathcal{D} \subset \mathcal{X} \to \mathcal{Y}, x_0 \in \mathcal{U} := \mathcal{U}(x_0) \subset \mathcal{D}$ and $\mathcal{U}$ be open. Then F is called* Frechét (F) differentiable *in x_0 if there is an $L_F \in \mathcal{L}(\mathcal{X},\mathcal{Y})$, depending on x_0 but not on h, such that*

$$H(x_0 + h) - H(x_0) = L_F h + o(h) \quad \textit{for} \quad h \to 0, x_0 + h \in \mathcal{U}. \tag{1.71}$$

Then $H'_F(x_0) := H'(x_0) := L_F$ is called the Frechét derivative *of H in x_0.*

2. *H is called* Gateaux (G) differentiable *in $x_0 \in \mathcal{U}$, if for every h (such that $\|h\| = 1$) there is a $H'_G(x_0,h) \in \mathcal{Y}$ such that*

$$H(x_0 + th) - H(x_0) = tH'_G(x_0,h) + o(t) \textit{ for } t \to 0, x_0 + th \in \mathcal{U}; \tag{1.72}$$

this $o(t) = o_h(t)$ depends on h. If (1.72) should be valid only for a fixed h (with $\|h\| = 1$), then H is called Gateaux differentiable in x_0 in the direction of h.

Furthermore, the $H'_G(x_0, h)$ *in (1.72) is called* the Gateaux (G) derivative *of* H *in* x_0 in the direction *of* h.

These Frechét and Gateaux derivatives generalize the total and directional derivatives for $F : \mathcal{D} \subset \mathbb{R}^n \to \mathbb{R}^m$, respectively. We will mainly apply Frechét derivatives. However, by $G(t) := H(x_0 + th)$ we have reduced $H : \mathcal{D} \subset \mathcal{X} \to \mathcal{Y}$ to a $G : I \subset \mathbb{R} \to \mathcal{Y}$ mapping for a real variable t. Thus, all the rules, e.g. sum, product and chain rule, obtained by limits of divided differences, remain correct for Gateaux derivatives as well, if they are interpreted as limits. This requires that x_0 and h are kept fixed during this differentiation. This allows an easier and more straightforward approach. Under appropriate conditions, which essentially cover all our applications, Gateaux derivatives yield a simple tool to determine Frechét derivatives as well.

F and G derivatives are uniquely determined. Sometimes, "one-sided" derivatives are appropriate, in particular, if $\mathcal{U}(x_0)$ is not open and, hence, not all $x_0 + h$ and $x_0 + th$ are in $\mathcal{U}(x_0)$, for sufficiently small h and t. An operator which is F differentiable in x_0 is continuous in x_0. That is not true, if it is only G differentiable. This situation is well known for real valued functions of two variables.

It is obvious that both derivatives are generated by linearizing the difference of operator values: In (1.71) we have linearized with respect to $h \in \mathcal{X}$, in (1.72) with respect to $t \in \mathbb{R}$. In particular we have as a consequence of (1.72)

$$H'_G(x_0, h) = \lim_{t \to 0} \frac{H(x_0 + th) - H(x_0)}{t} = \frac{d}{dt} H(x_0 + th)|_{t=0}.$$

For some examples we have the situation that $L_G \in \mathcal{L}(\mathcal{X}, \mathcal{Y})$ exists such that for all $\|h\| = 1, H'_G(x_0, h) = L_G h$. Sometimes this is even part of the definition of a Gateaux derivative, see [677]. The $o(t)$ term in (1.72) does depends on h. We have

Theorem 1.42. Frechét (F) and Gateaux (G) derivatives and their relations:

1. *The Frechét and Gateaux derivatives are uniquely determined.*
2. *If* H *is Frechét differentiable in* $x_0 \in \mathcal{U}(x_0) \subset \mathcal{D}$*, it is continuous in* x_0*, Gateaux differentiable and* $H'_G(x_0, h) = L_F h$.
3. *Let* H *be Gateaux differentiable in* x_0*, such that*

$$\begin{aligned} L_G = L_G(x_0) = H'_G(x_0) \in \mathcal{L}(\mathcal{X}, \mathcal{Y}) &\textit{ exists, such that} \\ H(x_0 + th) - H(x_0) = tH'_G(x_0)h + o_h(t) &\textit{ and} \\ o_h(t) = o(t)C(h) \quad \textit{with } \|C(h)\| \le C &\textit{ for } \|h\| = 1. \end{aligned} \tag{1.73}$$

Then H *is Frechét differentiable in* x_0 *and for all* h *with* $\|h\| = 1$ *we have* $H'_G(x_0, h) = H'_G(x_0)h = L_G h = L_F h$*, hence* $L_F = H'_G(x_0)$.

4. *Let* H *be Gateaux differentiable in a neighborhood* $\mathcal{U}(x_0)$ *of* x_0 *and such that*

$$\begin{aligned} &H(x + t\,h) - H(x) = t\, H'_G(x)h + o_h(t) \textit{ s.t.} \qquad (1.74) \\ &L_G(x) = H'_G(x) \in \mathcal{L}(\mathcal{X}, \mathcal{Y}) \textit{ is continuous w.r.t } x \textit{ for all } x \in \mathcal{U}(x_0). \end{aligned}$$

Then H *is Frechét differentiable in* $\mathcal{U}(x_0)$ *and* $L_F(x) = L_G(x)$.

Theorem 1.43. Mean value theorem: *Let $F : \mathcal{D} \subset \mathcal{X} \to \mathcal{Y}$ and a, $b \in \mathcal{D}$ given with $\overline{ab} = \{a + t(b-a); 0 \le t \le 1\} \subset \mathcal{D}$, such that F is F differentiable on $\overline{ab}$ (or in $\mathcal{D}$) and*

$$\|F'(a + t(b-a))\| \le g'(t), 0 \le t \le 1,$$

were g' and $f'(t) := F'(a + t(b-a))(b-a)$ are integrable in $[0,1]$. Then the following relations are valid, prefeably with x_0 near $\overline{ab}$,

$$F(b) - F(a) = \int_0^1 F'(a + t(b-a))dt(b-a), \tag{1.75}$$

$$\|F(b) - F(a)\| \le (g(1) - g(0))\|b-a\|, \tag{1.76}$$

$$\|F(b) - F(a)\| \le \|b-a\| \sup_{x \in \overline{ab}} \|F'(x)\|, \tag{1.77}$$

$$\|F(b) - F(a) - F'(x_0)(b-a)\| \le \|b-a\| \sup_{x \in \overline{ab}} \|F'(x) - F'(x_0)\|. \tag{1.78}$$

Modifications for G derivatives into the direction of $b-a$ are straightforward.

Partial derivatives are defined by splitting the variable into components. Let

$$F := (F_1(x), \ldots, F_m(x)) : \mathcal{D} \subset \mathcal{X} = \prod_{i=1}^{n} \mathcal{X}_i \to \mathcal{Y} = \prod_{j=1}^{m} \mathcal{Y}_j \tag{1.79}$$

with different arguments $x = (x, \ldots, x_u) \in \mathcal{X}$ and components, $(F_1(x), \ldots, F_m(x))$. For Banach spaces $\mathcal{X}_i$ and $\mathcal{Y}_j$ and the norm $||x||_{\mathcal{X}} := \max_{i=1}^{n}\{||x_i||_{\mathcal{X}_i}\}$, the $\mathcal{X}, \mathcal{Y}$ are Banach spaces as well. We define *partially differentiable* F:

Definition 1.44. *Choose $x^0 \in \mathcal{U}(x^0) \subset \mathcal{D}, \mathcal{U}(x^0)$ a neighborhood of x_0. Assume there exists a $L_i \in \mathcal{L}(\mathcal{X}_i, \mathcal{Y})$ and a $L_i^j \in \mathcal{L}(\mathcal{X}_i, \mathcal{Y}_j)$, respectively, such that for all small $h_i \in \mathcal{X}_i$,*

$$F(x^0 + h_i) - F(x^0) - L_i h_i = o(h_i) \text{ and } F_j(x^0 + h_i) - F_j(x^0) - L_i^j h_i = o(h_i). \tag{1.80}$$

Then the operator F and its components F_j, are called partially (Frechet) differentiable in x^0 *with respect to x_i and the $(\partial F/\partial x_i)(x_0) := (\partial_{x_i} F)(x_0) := L_i$ and L_i^j are called its* partial derivatives. *By replacing h_i by th_i and $t \to 0$ we obtain the* partial Gateaux differentiability. *For $\mathcal{X}_i = \mathcal{Y}_j = \mathbb{R}$ and $h_i = te^i$ with the i-th* unit vector, e^i, *and small $t \in \mathbb{R}$, we obtain the classical partial derivatives.*

Theorem 1.45. Continuously partially differentiable operators are (Frechet) differentiable:

1. *Let F in (1.79) be continuously partially differentiable in $\mathcal{U}(x^0)$, that is $\partial F/\partial x_i \in C(\mathcal{U}(x^0))$. Then F is (Frechet) differentiable in x^0.*
2. *Let F in (1.79) be differentiable in x^0. Then*

$$F'(x^0) = \left(\frac{\partial F}{\partial x_i}(x^0)\right)_{i=1}^{n} = \left(\frac{\partial F_j}{\partial x_i}(x^0)\right)_{i=1,j=1}^{n \quad m} = \left(L_i^j\right)_{i=1,j=1}^{n \quad m}.$$

With $h = (h_1, \ldots, h_n)$ *this yields the result*

$$F(x^0 + h) - F(x^0) + o(h) = F'(x^0)h = \sum_{i=1}^{n} \frac{\partial F}{\partial x_i}(x_0)h_i = \left(\sum_{i=1}^{n} \frac{\partial F_j}{\partial x_i}(x^0)h_i \right)_{j=1}^{m}.$$

Theorem 1.46. Implicit function theorem: *Let* $\mathcal{U} \subset \mathcal{X}$*, and* $\mathcal{V} \subset \mathcal{Y}$ *be open sets,* $F : \bar{\mathcal{U}} \times \mathcal{V} \to \mathcal{Z}, F \in C^1(\bar{\mathcal{U}} \times \mathcal{V})$*, and let* $(x_0, y_0) \in \mathcal{U} \times \mathcal{V}$ *be a solution of* $F(x_0, y_0) = 0$*. Furthermore, let* $\partial_x F(x_0, y_0) \in \mathcal{L}(\mathcal{X}, \mathcal{Z})$ *be boundedly invertible. Then there exist neighborhoods* $\mathcal{U}(x_0) \subset \overline{\mathcal{U}(x_0)} \subset \mathcal{U}$*,* $\mathcal{V}(y_0) \subset \mathcal{V}$ *for* x_0, y_0*, and an operator,* $g : \mathcal{V}(y_0) \to \overline{\mathcal{U}(x_0)}$ *such that*

$$F(g(y), y) = 0 \ \forall y \in \mathcal{V}(y_0). \tag{1.81}$$

Furthermore, $g \in C^1(\mathcal{V}(y_0))$ *and* g *is unique in the following sense: For every* $(x, y) \in \mathcal{U}(x_0) \times \mathcal{V}(y_0)$ *such that* $F(x, y) = 0$ *this* $x = g(y)$*, and* $\lim_{y \to y_0} g(y) = g(y_0)$.

Definition 1.47. C^1 diffeomorphism: *Let* $\mathcal{M}, \mathcal{N}$ *be arbitrary subsets of Banach spaces* $\mathcal{X}, \mathcal{Y}$ *and let* $F : \mathcal{M} \to \mathcal{N}$.

1. $F : \mathcal{M} \to \mathcal{N}$ *is called a* C^1 diffeomorphism *iff* F *is bijective and* F *and* F^{-1} *are* C^1 *mappings.*
2. F *is called a* C^1 local diffeomorphism *at* $x_0 \in \mathcal{M}$ *iff* $\mathcal{U}(x_0) \subset \mathcal{M}$ *and* $F(\mathcal{U}(x_0)) \subset \mathcal{N}$ *are neighborhoods of* x_0 *and* $F(x_0)$*, respectively, and* $F : \mathcal{U}(x_0) \to F(\mathcal{U}(x_0))$ *is a* C^1 diffeomorphism.

Theorem 1.48. Local inverse function theorem [675], Theorem 4.F: *Let the mapping* $F : \mathcal{U}(x_0) \subset \mathcal{M} \to \mathcal{N}$ *be* C^1 *with Banach spaces* $\mathcal{X}, \mathcal{Y}$*. Then* F *is a local* C^1 *diffeomorphism at* x_0 *iff* $F'(x_0) : \mathcal{U}(x_0) \to F'(x_0)(\mathcal{U}(x_0))$ *is bijective.*

Corollary 1.49. Derivative of inverse functions [676], Corollary 4.37: *Let* F: $\mathcal{U}(x_0) \subset \mathcal{M} \to \mathcal{N}$ *be in* C^1*, and let* $F'(x_0) : \mathcal{M} \to \mathcal{N}$ *be bijective. Then* F *is a local* C^1 *diffeomorphism at* x_0 *and, in a small neighborhood* $\mathcal{U}(x_0)$ *of* x_0:

$$(F^{-1})'(y) = (F'(x))^{-1} \text{ with } y = F(x), x \in \mathcal{U}(x_0). \tag{1.82}$$

Higher order derivatives of order l are obtained by applying the Gateaux definition for l independent $h_1, \ldots, h_l$. Then results are valid analogously to Theorem 1.42. Since they are hardly used in this book, but intensely in [120]. Starting with Theorem 1.46, we could then replace the C^1 by $C^r, r \geq 1$.

2

Elements of analysis for linear and nonlinear partial elliptic differential equations and systems

2.1 Introduction

We attempt to give a necessarily incomplete survey of this huge area. The presentation is restricted to facts which seem to be most important for our context. In particular, it is totally impossible to refer to all the many highly interesting results available in this area. Nonlinear elliptic differential equations or systems and their numerical solutions, including bifurcation and center manifolds in Böhmer [120] are our main goal. Accordingly, we select the references essentially from among textbooks. The study of numerical solutions and bifurcations is based upon linearization and the Fredholm alternative for the *linearized problem*.[3] So we study linear and nonlinear elliptic differential equations and systems and their linearization of order 2 and $2m$. The Fredholm alternative for the linearized problem shows that, in some sense, these systems behave similarly to finite-dimensional systems: the kernel and co-range of the linearized operator, A, have the same finite dimension unless parameters extend the kernel. Solutions for $Au = f$ exist if and only if f is perpendicular to the kernel of the dual operator A^d. We summarize the existence and uniqueness results for solutions, and the Fredholm alternative for the bifurcation context. Regularity results for the solutions are necessary for good convergence of numerical methods. We only formulate proofs for results that are straightforward generalizations of the reviewed books. However, we omit the results related to the different types of maximum principles and oscillations and the concept of viscosity solutions. Monotony and (nonlinear) coercivity techniques yield a lot of results, summarized here for our context. Good references for these important results are the books, by e.g. Gilbarg and Trudinger [346], Dong [299], Chen and Wu [170], Evans [310], Kreiss and Lorenz [454], Showalter [589], Taylor [618–620], and Zeidler [675–678].

The transition from order 2 to $2m$, from one to several space variables, from linear to nonlinear, and from one elliptic differential equation to systems complicates the situation. Correspondingly, the results become less strong. In particular, not many regularity results for quasilinear and nonlinear equations and systems of higher order

[3] Linearized problem always means the derivative of a problem and differentiable is always Frechét differentiable, unless stated otherwise.

seem to be available in the textbooks and monographs used in this survey. For a fascinating survey of the historic development of the analysis of partial differential equations, see Brézis and Browder [143].

As motivation we consider the following simple semilinear model problem

$$A\,u = -\Delta u + f(u) = g \in C(\Omega), \quad u|_{\partial\Omega} = 0, \tag{2.1}$$

with the Laplacian Δ and continuous functions f, g. A straightforward way to solve it would be to determine a "classical solution" $u_0 \in C^2(\Omega)$ satisfying (2.1). In fact this is still one of the (necessary) solution concepts for nonlinear problems, which we will summarize in Section 2.5. For linear problems with $f(u) = cu,\ c \in \mathbb{R}$, and nonlinear problems other important concepts have been developed. In particular, the nonlinear $f(u)$ must not grow too fast for $|u| \to \infty$, so growth conditions have to be imposed for $f(\cdot)$. Often a solution $u_0 \in C^2(\Omega)$ does not exist. So we have to generalize the "classical solution" $u_0 \in C^2(\Omega), Au_0 \in C(\Omega)$ in two steps into $u_0 \in H^2(\Omega), Au_0 \in L^2(\Omega)$, the "strong solution", and $u_0 \in H^1(\Omega), Au_0 \in H^{-1}(\Omega)$, the "weak solution". Accordingly, the boundary condition $u_0|_{\partial\Omega} = 0$ has to be replaced by $u_0 \in H_0^1(\Omega)$, cf. Theorem 1.24 and Remark 2.3. For linear problems the bilinear form $a(\cdot,\cdot)$ induced by A is an essential tool. Sobolev spaces $H^1(\Omega), H_0^1(\Omega), H^{-1}(\Omega)$, and their generalizations $W^{m,p}(\Omega), W_0^{m,p}(\Omega),\ W^{-m,q}(\Omega),\ 1/p + 1/q = 1$, cf. Subsection 1.4.3, the necessary boundary analysis for $\partial\Omega$, and the trace operators are summarized in Section 1.4.

We start in Sections 2.2–2.4 with linear problems. In Section 2.2, motivated by the linear version (2.2) of our above model problem (2.1). We determine appropriate concepts for generalized solutions in Hilbert spaces for a second order elliptic operator and the corresponding bilinear form. In this motivating section, we sometimes refer to definitions and properties introduced later on. Section 2.3 studies, in the Banach space setting, bilinear forms $a(\cdot,\cdot)$ and the induced linear operators A and vice versa. The norms, duals and invertibility of A are characterized by properties of $a(\cdot,\cdot)$. The concepts of Gelfand triples, coercive and elliptic bilinear forms $a(\cdot,\cdot)$ are introduced and relations to extremal problems are discussed. Finally the Riesz–Schauder theory, the Fredholm alternative and conditions for elliptic bilinear forms are studied. These results provide a solid foundation for the later linearization. General $2m$-th order elliptic operators and the corresponding bilinear forms are the topic of Section 2.4. We formulate Fredholm alternatives, estimate the solutions, summarize the essential regularity results in Hölder and Sobolev (Banach) spaces for $2m$-th order elliptic operators with C^∞ coefficients. C^k coefficients are allowed for order 2. Finally, weak solutions in Sobolev (Hilbert) spaces are studied.

In Section 2.5 we introduce nonlinear elliptic problems. Most of them live in Banach spaces, often in Hölder or Sobolev spaces. Usually, growth conditions for the coefficients have to be imposed. We start with the different definitions and types of solutions and study semilinear, quasilinear and fully nonlinear problems. Examples and some existence, uniqueness, regularity and Fredholm-type results are added. In Section 2.6 we turn to linear and nonlinear systems of elliptic equations. Again we distinguish between elliptic and coercive systems, their compact perturbations, and the corresponding bilinear forms.

We do know that the linearized operators play a dominant role for the stability of numerical methods and for bifurcation. So we study in Section 2.7 the linearized forms of the nonlinear operators presented in Sections 2.5 and 2.6. It turns out that, under slightly stronger conditions, Fredholm-type results are available for all of them. One of the most challenging examples of nonlinear systems of elliptic equations are the Navier- Stokes equations, studied in Section 2.8. We start with the Stokes operator. It represents a (noncoercive) saddle point problem. Its extension to the nonlinear version requires trilinear forms and bilinear operators.

Linearization of the nonlinear problems in this book yields linear operators, A, inducing bilinear forms, $a(u,v)$, sums of $\int_\Omega a\partial^\alpha u \partial^\beta v dx, |\alpha|, |\beta| \le m$. For $a \in L^\infty(\Omega)$, these are well defined only for functions $u, v \in W^{m,p}(\Omega)$ with $2 \le p \le \infty$, but not for $1 \le p < 2$, cf. Sections 2.2, 2.3, and Subsection 2.4.4. The A are compact perturbations E.g. the linearized Navier-Stokes operator compactly perturbs the invertible Stokes operator (with stable discretization). Similarly operators B with $H^m(\Omega)$-coercive bilinear forms are induced by their principal parts. For $1 \le p < 2$, we would obtain $W^{m,p}(\Omega)$-coercive bilinear forms; unfortunately they are defined no longer. We will show in Chapters 3 ff. that operators with $H^m(\Omega)$-coercive bilinear forms are stable with respect to discrete $H^m(\Omega)$ norms. Furthermore, linear boundedly invertible operators with stable consistent discretization essentially inherit this stability to their compactly perturbed invertible operators. Consistency results will be available for this large class of linear and nonlinear equations and systems as well. Stability of the linearized problem implies stability of the nonlinear problem as well.

Therefore, linear and nonlinear operators $A : W^{m,p}(\Omega) \to W^{-m,q}(\Omega), 1/p + 1/q = 1$, are particularly important, cf. Corollary 2.44. These results yield the key for a complete convergence theory for the following equations, systems and the main discretization methods for $W^{m,p}(\Omega), 2 \le p \le \infty$ w.r.t the $H^m(\Omega)$ norm for $2 < p$. This allows in many areas strong extensions of the state of the art, e.g. for wavelet and radial basis or mesh-free methods. The excluded case $1 \le p < 2$ allows a complete convergence theory as well, however based upon monotone operators, cf. Section 4.5.

Many problems in science, e.g. in mechanics, are related to variational problems defined via integrals. We want to touch on some aspects. In many cases (systems of) elliptic equations are the so-called *Euler systems* of a variational problem, e.g. in Theorem 2.97. This is a main topic in the calculus of variations, see, e.g. the monographs by Giaquinta and Hildebrandt [341, 342]. The first variation of these integrals applied to a test function yields the weak form, the so-called *Euler system*, and weak solutions of a PDE. A transformation of this equation into the strong form yields a *PDE in divergence form,* sometimes known as a *divergent PDE.* This transition from a variational problem to its first variation or the weak, often divergence form is discussed in Subsection 2.6.4. Many of these problems are related to *monotone operators.* They are systematically discussed by Zeidler [677, 678] and partially summarized in Subsections 2.5.5, 2.5.6 and 2.6.4, Theorems 2.97, 2.98. For linear problems the divergence form and the general form can easily be exchanged, if only appropriate regularity properties for the coefficients are guaranteed. Therefore in most of the literature nonlinear problems are studied in divergence form. We too concentrate our presentation mainly on this form. However, the exceptions in

Subsections 2.5.3, 2.5.7 and 2.6.8 are important. For the general case we refer, e.g. to Ladyženskaja and Uralceva [463, 464, 466]. For fully nonlinear elliptic problems, there is no divergence form corresponding to the strong form, and, hence, no weak form or solution of a PDE in strong form. This is relevant for their dicretizations, formulated for the first time for the general case in this book, mainly as an FEM. In the preface we have already mentioned the viscosity solutions. We do not discuss them here.

Most of the following results are valid only for *uniformly elliptic equations.* The case of *nonuniformly elliptic equations* is indicated in Subsection 2.5.3, and in, e.g. Ivanov [415].

For this chapter we consulted only a few of the original papers, but mainly refer to the textbooks and monographs cited in the text. We begin with a short list, not aiming for completeness. Elliptic PDEs of second order, including quasilinear and fully nonlinear equations are studied by Ladyženskaja and Uralceva [464], Gilbarg and Trudinger [346], and Chen and Wu [170]. More general PDEs essentially for the linear case, are discussed by Garabedian [336], Hörmander [402], Nečas [509, 510] and Wloka [667]. PDEs, now including nonlinear problems, are further studied by Evans [310] and Dong [299]. Evans [310] extensively discusses different representations for solutions for first and second order equations. His, and some of the other, theoretically oriented books extend the theory for linear equations to nonlinear problems. Usually the Hölder, Sobolev and Schauder results are complemented with maximum principles. Dong [299] concentrates on second order equations and presents several important examples mainly of parabolic type. Two impressive "multi-graphs" on linear and nonlinear problems are due to Zeidler [675–678], and Taylor [618–620]. Zeidler's main interest is nonlinear functional analysis with applications to PDEs. He gives a large number of results for linear and nonlinear PDEs. Taylor more directly aims for equations and systems of PDEs. Both combine many of the preceding techniques for elliptic equations and systems, Taylor even on manifolds. Triebel [634, 635] initiated and obtained a new hierarchy of spaces, the so-called Besov–Triebel–Lizorkin spaces and many results for nonlinear equations. Based upon these results, Runst and Sickel [561] develop a very general theory for nonlinear PDEs. Polyanin and Zaitsev [535] collect the exact solutions for many nonlinear elliptic (parabolic and hyperbolic) equations. Necessarily, in our context, the maximum principle, lower and upper solutions and oscillation are neglected compared to the preceding monographs. Morrey [500] and Giaquinta and Hildebrandt [341, 342] study mainly quasilinear differential equations related to variational problems, a main topic in the calculus of variations. For higher order PDEs nonhomogeneous boundary conditions are far more complicated, in contrast to second order; in fact they are a serious problem. They are presented by Lions and Magenes [478–480]. For more literature see the following sections.

In this book, we mainly consider isolated solutions u_0 with boundedly invertible linearized operator $G'(u_0)$. Extensions are described by Böhmer [120]. For isolated solutions, Theorem 1.46 and the results from Section 2.4 usually allow us the restriction to homogeneous boundary conditions. But for fully nonlinear problems, we discuss the separate role of differential and boundary operators. So, in general, it is possible to define an exact extension of the boundary function and use the standard substitution trick. Alternatively, an approximating function, extended to the full domain, yields a small error at the boundary, causing only a small change in the exact solution.

2.2 Linear elliptic differential operators of second order, bilinear forms and solution concepts

In this section we motivate different solution concepts for the easiest case of second order linear equations. The linear form of our above model problem (2.1) is

$$A\,u = -\Delta u + cu = f \in C(\Omega), u|_{\partial\Omega} = 0. \tag{2.2}$$

But even for $c > 0, f \in C(\Omega)$, and a smooth enough $\partial\Omega$, a "classical solution" $u_0 \in C^2(\Omega)$ with $\|u_0\|_{C^2(\Omega)} \leq C\|f\|_{C(\Omega)}$ does not always exist, hence $A : C^2(\Omega) \to C(\Omega)$ is not boundedly invertible. For a counterexample, see Hackbusch [387] and for appropriate modifications see, e.g. our Theorem 2.37. Linear functionals allow us a way out of this dilemma. From Definition 1.8 we do know that for a Hilbert space $\mathcal{V}$ and its dual $\mathcal{V}'$,

$$f \in \mathcal{V}' \quad \text{with} \quad \langle f, v\rangle_{\mathcal{V}'\times\mathcal{V}} = 0 \;\; \forall\, v \in \mathcal{V} \quad \text{is equivalent to} \quad f = 0. \tag{2.3}$$

So we use linear functionals for testing $A\,u = -\Delta u + cu = f$. We relax $A\,u \in C(\Omega)$ in two steps into $A\,u \in L^2(\Omega)$ and then $A\,u \in H^{-1}(\Omega)$.

We have to relate operators, bilinear forms and boundary terms. We use the standard notations for *partial derivatives*

$$\partial^i u := \frac{\partial u}{\partial x_i}, \;\; \partial^i\partial^j u = \frac{\partial^2 u}{\partial x_i \partial x_j}, \;\; \partial^0 u := u, \;\; \Delta u := \sum_{i=1}^{n} (\partial^i)^2 u. \tag{2.4}$$

Unless stated otherwise, we require throughout this chapter:

$$\Omega \subset \mathbb{R}^n \;\text{ is open, bounded and }\; \partial\Omega \text{ is Lipschitz-continuous.} \tag{2.5}$$

For this Ω the corresponding spaces are defined with respect to their *Sobolev (Hilbert) space inner products, norms, and seminorms,* with $H^0(\Omega) = L^2(\Omega)$,

$$(u, v)_{H^1(\Omega)} := \sum_{i=0}^{n} (\partial^i u, \partial^i v)_{L^2(\Omega)}, \;\; (u, v)_{H^2(\Omega)} := \sum_{i,j=0}^{n} (\partial^j\partial^i u, \partial^j\partial^i v)_{L^2(\Omega)}, \tag{2.6}$$

$$\|u\|_{H^i(\Omega)} := (u, u)^{1/2}_{H^i(\Omega)}, \;\; i = 0, 1, 2, \;\; |u|_{H^1(\Omega)} := \Big(\sum_{i=1}^{n} \|\partial^i u\|^2_{L^2(\Omega)}\Big)^{1/2}.$$

As usual, $H^{-i}(\Omega), i = 1, 2$, denote the dual spaces of bounded linear functionals on $H^i(\Omega), i = 1, 2$.

We start testing (2.2) with $v \in L^2(\Omega)$ in the sense of (2.3) and for $u \in H^2(\Omega)$ and introduce *classical, strong and weak operators and bilinear forms.* In contrast to the standard literature we use in this section different notations A_c, for the *strong forms* defined on C^2 and $A_s, a_s(\cdot,\cdot)$, on H^2, sometimes called the *Friedrichs extension,* and

$A, a(\cdot,\cdot)$ for the *weak forms* in (2.11):

$$A_c : C^2(\Omega) \to C(\Omega), \quad A_c\, u = -\Delta u + cu, \tag{2.7}$$

$$A_s : H^2(\Omega) \to L^2(\Omega), \quad A_s\, u = -\Delta u + cu, \text{ and}$$

$$a_s(\cdot,\cdot) : H^2(\Omega) \times L^2(\Omega) \to \mathbb{R} \text{ defined by} \tag{2.8}$$

$$a_s(u,v) := (A_s u, v)_{L^2(\Omega)} = \int_\Omega (-\Delta\, u + cu)\ v dx\ \forall v \in L^2(\Omega).$$

The condition $u \in H^2(\Omega)$ still is too restrictive. We apply (2.9). Zeidler [677], pp. 19 and 20, calls this one of the *cornerstones of modern analysis.* For $v, w \in H^1(\Omega)$ we generalize the classical integration by parts formula, $\int_a^b v'wdx = vw|_a^b - \int_a^b vw'dx$. With the outer normal $\nu = (\nu_1, \ldots, \nu_n)^\mathsf{T}$ and a Lipschitz-continuous manifold $\partial\Omega$, we obtain the *fundamental Green's formula,* cf. Ciarlet [174], p. 14,

$$\int_\Omega w\partial v/\partial x_i dx + \int_\Omega v\partial w/\partial x_i dx = \int_\Omega \partial(vw)/\partial x_i dx = \int_{\partial\Omega} vw\nu_i ds, \tag{2.9}$$

$$\partial\Omega \in C_L, \forall w, v \in H^1(\Omega).$$

We apply this to the last equation in (2.8) to get the *first Green's formula*

$$\int_\Omega (-\Delta\, u + cu)\ vdx = a_s(u,v) = \int_\Omega (\nabla\, u, \nabla\, v)_n + cuvdx - \int_{\partial\Omega} \frac{\partial u}{\partial \nu}\ vds, \tag{2.10}$$

with the *Euclidean product and norm* $(\cdot,\cdot)_n$ and $|\cdot|_n$ in $\mathbb{R}^n$. Note that $\int_{\partial\Omega} \partial u/\partial\nu\ vds = 0$ if $\forall \partial u/\partial\nu \in L^2(\partial\Omega)$ and $\forall v \in H^1(\Omega)$ with $v = 0$ on $\partial\Omega$. By Theorems 1.37, 1.38, we denote this as $\forall v \in H_0^1(\Omega)$ in the trace sense. This allows us to reduce the differentiability conditions for u and the solution u_0. We introduce the *weak bilinear form and operator,* $a(\cdot,\cdot), A$ sometimes called the *energy extension,* as

$$a(\cdot,\cdot) : H^1(\Omega) \times H^1(\Omega) \to \mathbb{R} \text{ and } A : H^1(\Omega) \to H^{-1}(\Omega) \text{ defined by} \tag{2.11}$$

$$a(u,v) := \int_\Omega \nabla\, u \nabla\, v + cuvdx =: \langle Au, v\rangle_{H^{-1}(\Omega)\times H^1(\Omega)}, \text{ thus, for “regular” } u,$$

$$\forall u \in H^2(\Omega), v \in H^1(\Omega) :\ a(u,v) = a_s(u,v) + \int_{\partial\Omega} \frac{\partial u}{\partial\nu}\ vds \text{ and } = a_s(u,v),$$

$$\text{if } \partial\Omega_1 \cup \partial\Omega_2 = \Omega \text{ and } \int_{\partial\Omega} \frac{\partial u}{\partial \nu}\ vds = 0 \text{ for } \forall v|_{\partial\Omega_1} = 0 \text{ and } \forall \left.\frac{\partial u}{\partial\nu}\right|_{\partial\Omega_2} = 0,$$

often with $\partial\Omega_1 = \partial\Omega$ or $\partial\Omega_2 = \partial\Omega$. The restriction $v|_{\partial\Omega_1}$ for $v \in H^1(\Omega)$ yields $v|_{\partial\Omega} \in H^{1/2}(\Omega)$ in the trace sense. For our Ω in (2.5), this is an immediate consequence of Theorems 1.37, 1.39.

Remark 2.1.

1. *For the weak form $a(u,v)$ we need $u, v \in H^1(\Omega)$, and for the strong form $a_s(u,v)$ we need $u \in H^2(\Omega), v \in L^2(\Omega)$. The relation $a(u,v) = a_s(u,v) + \int_{\partial\Omega} \partial u/\partial\nu\ vds$,*

shows that $a(u,v) = a_s(u,v)$ *if* $\int_{\partial\Omega} \partial u/\partial\nu \, v ds = 0$. *In (2.11),* $u \in H^2(\Omega)$ *implies* $\partial u/\partial\nu|_{\partial\Omega_2} \in H^{1/2}(\Omega)$, *and a well-defined* $\int_{\partial\Omega} \partial u/\partial\nu \, v ds$. *Under appropriate conditions even for* $u \in H^1(\Omega)$ *a* $\partial u/\partial\nu|_{\partial\Omega} \in H^{-1/2}(\Omega)$ *is possible, cf. Costabel [211]. With* $v \in H^1(\Omega)$, *hence* $v|_{\partial\Omega} \in H^{1/2}(\Omega)$ *the* $\int_{\partial\Omega} \partial u/\partial\nu \, v ds$ *is still defined. Under those conditions, the previous Neumann boundary value, but not the strong form* $a_s(u,v)$, *in (2.11) is well defined for* $u \in H^1(\Omega)$. *Other results for this problem of boundary traces are discussed, e.g. in Taylor [618], Chapter 4, Proposition 4.5, and Jonsson and Wallin [425], Jerison and Kenig [419], Schwab [575] and Grisvard [373]. We do not consider these extensions here in more detail.*

2. *For the later nonconforming discretization methods in Chapters 5 and 7, and adaptive FEMs in Chapter 6, we will consider (2.11) on subtriangles* $T \subset \Omega$, *indicated as* $a_T(u,v) = a_{s,T}(u,v) + \int_{\partial T} \partial u/\partial\nu \, v ds$. *This will turn out to be one of the deciding tools for estimating the consistency errors in the discrete generalization of the* $a(u,v)$. *The piecewise functions,* $u^h|_T$, *usually polynomials, in FEMs and DCGMs always satisfy* $u^h \in H^k(T), k \geq 2$, *hence* $a_T(u^h,v^h) = a_{s,T}(u^h,v^h) + \int_{\partial T} \partial u^h/\partial\nu \, v^h ds$. *For the corresponding consistency estimates, we will require* u *as close as possible to the* $u \in H^1(\Omega)$ *in the weak form. Therefore motivated by Theorems 1.37, 1.39 and 2.48, we replace the above* $u \in H^1(\Omega)$ *either by* $u \in H^2(\Omega)$ *as in (2.11), or require, in DCGMs,* $u \in H^{3/2+\epsilon}(\Omega)$ *with any* $\epsilon > 0$. *Then* $\partial u/\partial\nu|_{\partial\Omega_2} \in H^\epsilon(\partial\Omega) \subset L^2(\partial\Omega)$ *is valid and* $\int_{\partial\Omega} \partial u/\partial\nu \, v ds$ *is well defined, cf. Remarks 5.51 and 7.1.*
3. *For simplicity, we assume, here and for analogous cases, unless stated otherwise: For natural or Neumann boundary conditions let* $u \in H^2(\Omega)$ *or* $u \in H^{3/2+\epsilon}(\Omega)$, *hence* $\partial u/\partial\nu|_{\partial\Omega_2} \in L^2(\partial\Omega)$.

Obviously the two operators and bilinear forms defined in (2.8) and (2.11) are different. We could even go one step further: By an additional partial integration, again using (2.9), one would end up with the distributional bilinear form

$$a_d(u,v) := \int_\Omega u(-\Delta v + cv)dx \ \forall \ \ v \in C_0^\infty(\Omega), \ \forall \ \ u \in (C_0^\infty(\Omega))'. \tag{2.12}$$

This allows us to formulate solutions $u_0 \in (C_0^\infty(\Omega))'$, the space of distributions. We do not study this possibility in this book; see, e.g. Lions and Magenes [478], Runst and Sickel [561] Showalter [589].

Based upon (2.8) and (2.11), we are now able to introduce the concepts of *classical and strong and weak solutions* $u_0 = u_{c,0}$ and $u_0 = u_{s,0}$ and $u_0 = u_{w,0}$, respectively, of $Au = f$. We use $u_{c,0}$, $u_{s,0}$ or $u_{w,0}$ and $A_c, A_s, A = A_w$ only if the context does not directly show the chosen solution concept. The existence, uniqueness and regularity of the solutions are studied in the following sections. These regularity conditions will show that the type of solution $u_{c,0}, u_{s,0}$ or $u_{w,0}$ is essentially a consequence of the smoothness of the problem. Note that the definition of the linear operators and bilinear forms does not require boundary conditions, but the following definition of the solutions does require boundary conditions, e.g. $u_0 \in H_0^1(\Omega)$. The classical solution $u_0 = u_{c,0}$ and the strong solution $u_{s,0}$ are then defined, e.g. for Dirichlet boundary conditions, by

$$u_0 = u_{c,0} \in C^2(\Omega),\ u_0|_{\partial\Omega} = 0 : A_s\, u_0 = -\Delta u_0 + cu_0 = f \in C(\Omega), \qquad (2.13)$$

$$u_0 = u_{s,0} \in H^2(\Omega) \cap H_0^1(\Omega) : \quad A_s\, u_0 = -\Delta u_0 + cu_0 = f \in L^2(\Omega) \Leftrightarrow \qquad (2.14)$$

$$\int_\Omega f v dx = (A_s u_0, v)_{L^2(\Omega)} = \int_\Omega (-\Delta\, u_0 + cu_0) v dx = a_s(u_0, v)\ \forall\, v \in L^2(\Omega).$$

The weak solution u_0 is defined, with $f \in H^{-1}(\Omega)$ instead of $f \in L^2(\Omega)$, by

$$\begin{aligned} u_0 = u_{w,0} \in H_0^1(\Omega) \Leftrightarrow A\, u_0 = f \in H^{-1}(\Omega) \ \Leftrightarrow \qquad (2.15) \\ \langle Au_0, v\rangle_{H^{-1}(\Omega)\times H_0^1(\Omega)} &= \int_\Omega (\nabla\, u_0\,, \nabla\, v)_n + c\, u_0\, v\, dx = a(u_0, v) \\ &= \int_\Omega f_0 v + \sum_{j=1}^n f_j \partial^j v dx \\ &= \langle f, v\rangle_{H^{-1}(\Omega)\times H_0^1(\Omega)}\ \forall\, v \in H_0^1(\Omega),\ \text{cf. } (2.107). \end{aligned}$$

As a special case of Theorem 2.43 and Corollary 2.44, we obtain

Theorem 2.2. *For $c \geq 0$ and $f \in H^{-1}(\Omega)$ or $f = f_0 \in L^2(\Omega)$ the problem (2.15) has a unique solution $u_0 \in H_0^1(\Omega)$.*

In (2.15) we required $f \in H^{-1}(\Omega)$ and tested with $v \in H_0^1(\Omega)$, indicating the need for $Au_0 = f \in H^{-1}(\Omega)$. In fact, there are reasons for distinguishing and for identifying $H^{-1}(\Omega)$ and $H_0^{-1}(\Omega)$. Since we need this discussion for all the following more general cases, we formulate it here.

Remark 2.3. *We assume a bounded Ω as in (2.5). Then, according to Proposition 2.34, cf. Adams [1], pp. 49 ff. $W_0^{m,p}(\Omega) \subset W^{m,p}(\Omega) \neq W_0^{m,p}(\Omega)$. On the other hand, Taylor [618], Chapter 4, Proposition 5.1, proves, for a smooth Ω, hence $\partial\Omega \in C^\infty$, cf. Proposition 1.32, and $m \in \mathbb{N}_0$, that there is a natural and norm-preserving isomorphism*

$$H_0^{-m}(\Omega) = \text{ dual space of } H_0^m(\Omega) \approx H^{-m}(\Omega).$$

This fact is one of the reasons why sometimes in the literature $H^{-m}(\Omega)$ and $H_0^{-m}(\Omega)$, and $W^{-m,q}(\Omega), 1/p + 1/q = 1$, and $W_0^{-m,q}(\Omega)$ are not distinguished. *On the other hand, often only $f \in W^{-m,q}(\Omega)$ are used for the $f \in W_0^{-m,q}(\Omega)$, e.g. in Proposition 2.34. That $H^{-m}(\Omega)$ and $H_0^{-m}(\Omega)$ are in fact different is demonstrated by (2.107) versus (2.110).*

Now we extend (2.15) to more general boundary and elliptic operators of second order. The strong operator A_s can be directly defined and is related to its *strong bilinear form* $a_s(\cdot,\cdot)$, with real valued coefficients as

$$A_s : H^2(\Omega) \to L^2(\Omega),\ a_s(\cdot,\cdot) : H^2(\Omega) \times L^2(\Omega) \to \mathbb{R} \quad \text{and} \tag{2.16}$$

$$A_s u := \sum_{i,j=0}^{n} (-1)_{j>0} \partial^j (a_{ij} \partial^i\ u) \text{ with } (-1)_{j>0} := 1 \text{ for } j = 0 \text{ else } := -1,$$

$$a_s(u,v) := (A_s u, v)_{L^2(\Omega)}, \forall u \in H^2(\Omega), v \in L^2(\Omega),\ a_{ij} \in W^{1,\infty}(\Omega) \text{ and} \tag{2.17}$$

$$A_c : C^2(\Omega) \to C(\Omega),\ A_c := A_s|_{\ C^2(\Omega)},\ a_{ij} \in C^1(\Omega).$$

Even for $A_c : C^2(\Omega) \to C(\Omega)$ test spaces can be defined as functions of bounded variation defining a bilinear form for $u_{c,0} \in C^2(\Omega)$. We do not consider this here, nor generalize the above distributional bilinear form $a_d(\cdot,\cdot)$ in (2.12).

For the (standard) weak problem we directly define the *weak bilinear form* $a(\cdot,\cdot)$ and the induced A by partial integration of (2.16) and the Green's formula with vanishing boundary terms as

$$a(\cdot,\cdot)\ : H^1(\Omega) \times H^1(\Omega) \to \mathbb{R} \text{ and } A : H_0^1(\Omega) \to H^{-1}(\Omega), \text{ with} \tag{2.18}$$

$$a(u,v) \ := \int_\Omega \sum_{i,j=0}^{n} a_{ij} \partial^i u \partial^j v dx$$

$$=: \langle Au, v\rangle_{H^{-1}(\Omega)\times H_0^1(\Omega)},\ \forall u,v \in H_0^1(\Omega),\ a_{ij} \in L^\infty(\Omega).$$

Definition 2.4. Principal part and ellipticity: *The* principal parts A_p *and* $a_p(u,v)$ *of the operator,* A*, and its bilinear form,* $a(u,v)$*, in (2.18) are*

$$A_p u := \sum_{i,j=1}^{n} (a_{ij} \partial^i\ u) \partial^j \ \in\ H^{-1}(\Omega) \quad \textit{and}$$

$$a_p(u,v) := \int_\Omega \left(\sum_{i,j=1}^{n} a_{ij} \partial^i\ u\ \partial^j\ v \right) dx = \int_\Omega (A_p u) \circ v dx, \tag{2.19}$$

and similarly for A_s, A_c *and* $a_s(\cdot,\cdot)$*. For* $\vartheta = (\vartheta^1, \ldots, \vartheta^n)^\mathsf{T} \in \mathbb{R}^n$ *let the* characteristic polynomial *of* A_p *satisfy;*[4]

$$\sum_{i,j=1}^{n} a_{ij}(x) \vartheta^i \vartheta^j \geq 0 \textit{ and } \geq \epsilon' |\vartheta|_n^2 \text{ , } \textit{for an } \epsilon' > 0,\ \forall x \in \Omega, \vartheta^i \in \mathbb{R},\ \textit{respectively,} \tag{2.20}$$

with the Euclidean norm $|\vartheta| = |\vartheta|_n$ *in* $\mathbb{R}^n$*. Then the operator* A *(or* A_s, A_c*) is called* elliptic *(*≥ 0*), and* strongly elliptic $\left(\geq \epsilon' |\vartheta|_n^2\right)$*. For* $a_{ij} \in L^\infty(\Omega)$ *(or* $a_{ij} \in W^{1,\infty}(\Omega)$*) this condition is only required* almost everywhere (a.e.) *in* Ω*.*

[4] It is always possible to use $\geq$ instead of $\leq$ in (2.20) by multiplying A by -1.

Obviously, the above *bilinear forms* and *linear operators* in (2.16), (2.18) are *continuous or bounded*, that is, positive constants, C, exist, such that

$$\begin{aligned}
|a(u,v)| &< C\|u\|_{H^1(\Omega)}\|v\|_{H^1(\Omega)} \;\forall u,v \in H^1(\Omega), \qquad (2.21)\\
|a_s(u,v)| &< C\|u\|_{H^2(\Omega)}\|v\|_{L^2(\Omega)} \;\; \forall u \in H^2(\Omega), v \in L^2(\Omega),\\
\|Au\|_{H^{-1}(\Omega)} &< C\|u\|_{H^1(\Omega)} \;\; \forall u \in H^1(\Omega),\\
\|A_s u\|_{L^2(\Omega)} &< C\|u\|_{H^2(\Omega)} \;\; \forall u \in H^2(\Omega) \text{ and}\\
\|A_c u\|_{C(\Omega)} &< C\|u\|_{C^2(\Omega)} \;\; \forall u \in C^2(\Omega).
\end{aligned}$$

As an immediate consequence of (2.19), (2.20), we obtain, for a strongly elliptic differential operator of second order, the so-called $H_0^1(\Omega)$ *coercivity* of the principal part:

$$\text{for } a_{ij} \in L^\infty(\Omega), \text{ (2.20) implies} \quad a_p(u,u) \geq \epsilon' |u|^2_{H^1(\Omega)}. \tag{2.22}$$

For $u \in H_0^1(\Omega)$ the seminorm $|u|_{H^1(\Omega)}$ is equivalent to the standard Sobolev norm $\|u\|_{H^1(\Omega)}$, see (2.6). In fact, (2.21), (2.22) show the *equivalence* of $\|\cdot\|_{H^1(\Omega)}$ and $(a_p(\cdot,\cdot))^{1/2}$ in $H_0^1(\Omega)$, for $A_s = -\Delta$, in general, called the *energy norm.*

A generalization of (2.21) to $A : W^{1,p}(\Omega) \to W^{-1,q}(\Omega)$, $1 \leq p \leq \infty$, $1/p + 1/q = 1$, with test spaces $\mathcal{V}_b = W^{1,p}(\Omega)$ and to $A_s : W^{2,p}(\Omega) \to L^q(\Omega)$ with test spaces $L^p(\Omega)$, is possible as well; see Sections 2.4 and 2.5, and, e.g. [561, 618–620, 634, 635].

For higher order problems we will consider $A : W^{m,p}(\Omega) \to W^{-m,q}(\Omega)$. However, generalizations to $A : C^2(\Omega) \to C(\Omega)$, and a corresponding theory is not appropriate, but only for $A : C^{2,\gamma}(\Omega) \to C^\gamma(\Omega), 0 < \gamma < 1$, see Subsections 2.4.2 and 2.4.3.

Again the relation between the weak and strong bilinear forms is based upon the so-called *natural boundary conditions.* With the outer normal $\nu = (\nu_1, \ldots, \nu_n)^\mathsf{T}$ on a C^1 manifold $\partial\Omega$, we obtain from (2.9) or the *Gauss integral theorem,*

$$\int_\Omega v\partial^j(a_{ij}\partial^i u)dx = -\int_\Omega (\partial^j v)a_{ij}\partial^i u + \int_{\partial\Omega} v a_{ij}\partial^i u \nu_j ds. \tag{2.23}$$

This is applied to $a(\cdot,\cdot)$ in (2.18) yielding a *generalized first Green's formula*

$$a(u,v) = \int_\Omega (A_s\, u)v dx + \int_{\partial\Omega} (B_a\, u)v ds = a_s(u,v) + \int_{\partial\Omega} (B_a u)v ds \tag{2.24}$$

with

$$B_a\, u := \sum_{i=0,\ldots,n}^{j=1,\ldots,n} \nu_j a_{ij}\partial^i\, u, \quad \text{the natural boundary operator.} \tag{2.25}$$

Sometimes, instead of the boundary operator, B_a, a more general form is used:

$$Bu := \sum_{i=1}^n b_i\partial^i u + b_0 u \quad \text{with} \quad b = (b_1,\ldots,b_n) \in \mathbb{R}^n, (\nu,b)_n \neq 0. \tag{2.26}$$

$B_a u|_{\partial\Omega}$ and $Bu|_{\partial\Omega} \in H^{-1/2}(\partial\Omega)$ or $Bu|_{\partial\Omega} \in L^2(\partial\Omega)$ are well defined for $u \in H^{3/2+\epsilon}(\Omega)$, cf. Remark 2.1 and Theorem 1.38. In contrast to B we call B_a the *natural*, or *induced boundary operator,* since it is "induced" by $a(\cdot,\cdot)$. For the special case of (2.2) we obtain Neumann boundary conditions

$$A_s u = -\Delta u + cu \quad \text{induces } B_a\ u = \partial u/\partial\nu. \tag{2.27}$$

The last condition, $(\nu, b)_n \neq 0$, in (2.26) *excludes tangential derivatives* $\sum_{i=1}^n b_i \partial^i u$ of u along $\partial\Omega$. By rescaling Bu we can obtain $(\nu, b)_n = \nu^\top b = \nu^\top A\nu \neq 0$ for a strongly elliptic operator A, see Proposition 2.5, Remark 2.6, and Hackbusch [387], Remark 7.4.10.

The differential operator $Au = f$ and on $\partial\Omega$ *Dirichlet* and *natural boundary conditions* $Bu := u = \varphi$ and $B_a u = \varphi$ are tested by $\mathcal{V}_b$, and are realized for $u_0 \in H^1(\Omega)$ with

$$\begin{aligned} &\mathcal{V}_b = H_0^1(\Omega) \text{ for Dirichlet boundary conditions: } H^{1/2}(\Omega) \ni u_0|_{\partial\Omega} = 0 \text{ and} \\ &\mathcal{V}_b = \mathcal{V} = H^1(\Omega) \text{ for natural boundary conditions: we require} \\ &u \in H^2(\Omega), \text{ implying } H^{1/2}(\partial\Omega) \ni B_a u_0|_{\partial\Omega} = \varphi \end{aligned} \tag{2.28}$$

in the trace sense, cf. Remark 2.1 and Theorems 1.38, 1.37 and 2.43. Since $v|_{\partial\Omega} \in H^{1/2}(\partial\Omega)$ for $v \in H^1(\Omega)$, the $\int_{\partial\Omega}(B_a u)v ds$ is again well defined in this case, cf. textbooks on PDEs, e.g. [135, 141, 387].

By the standard trick $u := u - \bar{u}$ with $\bar{u} = \varphi$ on $\partial\Omega$ we obtain homogeneous Dirichlet boundary conditions. In fact we may use Theorem 1.29 for extending the boundary function, φ to Ω, yielding $\bar{\varphi}$ and thus $\bar{u} := \bar{\varphi}$, see Remark 2.7.

An extension to parts of $\partial\Omega$ with Dirichlet and natural boundary conditions as in (2.11) is presented in most textbooks, e.g. [141]. Therefore we only indicate it here; compare Remark 2.7 below. For $u = \varphi$ or $B_a \bar{u} = \varphi$ on an open subset $\partial\Omega_1$ of the boundary $\partial\Omega$ we first extend φ from $\partial\Omega_1$ to $\partial\Omega$ and then to $\bar{u}$ on Ω and again use the above trick, applicable in later sections as well. Then we use as test space

$$\mathcal{V}_b := \{u \in H^2(\Omega) : u|_{\partial\Omega_1} = 0\}. \tag{2.29}$$

We determine the *weak* and *strong solution* $u_0 = u_{w,0}$ and $u_0 = u_{s,0}$ of the *Dirichlet boundary value problem* with the bilinear forms in (2.18) and (2.16) by

$$\begin{aligned} &u_{w,0} = u_0 \in H_0^1(\Omega) : a(u_0, v) = \langle f, v\rangle_{H^{-1}(\Omega)\times H^1(\Omega)}\ \forall v \in H_0^1(\Omega), \text{ and} \\ &u_{s,0} = u_0 \in H_0^1(\Omega) \cap H^2(\Omega) : a_s(u_0, v) = (f, v)_{L^2(\Omega)}\ \forall v \in L^2(\Omega), \\ &\text{hence } Au_0 = Au_{w,0} = f \in H^{-1}(\Omega) \text{ and} \\ &A_s u_0 = A_s u_{s,0} = f \in L^2(\Omega), \text{ respectively} \end{aligned} \tag{2.30}$$

In Section 2.4 we will return to classical solutions, $A_s u_{s,0} = f \in C(\Omega)$.

For a smoother $u_0 \in H^2(\Omega) \cap H_0^1(\Omega)$ and $u_0|_{\partial\Omega} = 0$, (2.24) and (2.25) imply $u_{w,0} = u_0 = u_{s,0}$, since $H^1(\Omega)$ is dense in $L^2(\Omega)$, so

$$u_0 \in H^2 \cap H_0^1(\Omega) : a(u_0, v) = \int_\Omega (A_s u_0) v dx = a_s(u_0, v) = \int_\Omega f v dx\ \forall v \in H_0^1. \tag{2.31}$$

The *natural boundary operator* B_a was "induced" by $a(\cdot,\cdot)$. By partial integration of $a_s(\cdot,\cdot)$ we obtained $a(\cdot,\cdot)$ and a boundary term defined via B_a. But $a(\cdot,\cdot)$ does not necessarily uniquely determine A_s and B_a and vice versa. So the question naturally arises whether a boundary value problem with a general boundary operator B as in (2.26) uniquely determines a bilinear form $a(\cdot,\cdot)$ and vice versa. Let, cf. (2.16), $(,)_n$ *be the Euclidean scalar product in* $\mathbb{R}^n$,

$$A_s u := \sum_{i,j=0}^{n} (-1)_{j>0} \partial^j (a_{ij} \partial^i\, u) \text{ and with } b(x) := (b_1(x), \cdots, b_n(x)) \in \mathbb{R}^n, \quad (2.32)$$

$$B\, u := \sum_{i=0}^{n} b_i \partial^i\, u \text{ assume } (b(x), \nu(x))_n \neq 0 \;\forall x \in \partial\Omega. \quad (2.33)$$

If $(b(x), \nu(x))_n = 0$ for a specific $x \in \partial\Omega$ were allowed, $\sum_{i=1}^n b_i(x) \partial^i\, u$ would be a *tangential derivative* of u in x. This sometimes does not make sense, however see Wloka [667] p. 161 ff., where $b_0 \equiv 0, (b(x), \nu(x)) = 0$ is discussed.

Hackbusch's [387] Theorem 7.4.11 and Remark 7.4.12 show

Proposition 2.5. *Let A_s and B be given as in (2.32), (2.33). Then there exists a modified bilinear form $a_1(\cdot,\cdot) : H^1(\Omega) \times H^1(\Omega)$ as in (2.18), such that the variational formulation as*

$$a_1(u_0, v) = \int_\Omega f v dx \text{ for } u_0 \in H^2(\Omega)\ \forall v \in H^1(\Omega), B_{a_1} u_0 = B u_0|_{\partial\Omega} = \varphi \quad (2.34)$$

corresponds to the classical formulation $A_s u = f$ in Ω and $Bu = \varphi$ on $\partial\Omega$.

If $a(\cdot,\cdot)$ in (2.18), corresponding to A_s in (2.16), is $H^1(\Omega)$ elliptic, cf. Definition 2.17, and the coefficients of A_s and Ω are C^1, this[5] *modified $a_1(\cdot,\cdot)$ is again $H^1(\Omega)$ elliptic.*

Remark 2.6.

1. *This proposition allows us to restrict the discussion for $m = 1$ to the Dirichlet or the natural boundary condition, or a combination, see below.*
2. *It is important to realize the impact of the boundary conditions, cf. Treves [632], Chapter 37. In fact, for $f \in L^2(\Omega)$, we obtain identical strong and weak solutions, $u_0 = u_{s,0} = u_{w,0}$, in (2.31) if one of the following conditions is satisfied, see (2.24). Either $u_0 \in H^2(\Omega)$ satisfies the natural boundary condition $B_a(u_0)|_{\partial\Omega} = 0$, for B_a see (2.25). Then the boundary term drops out*

[5] Hackbusch indicates that it might be advantageous to employ additional boundary integral terms. This would avoid a smooth extension of a function, defined on $\partial\Omega$, to Ω. Theorem 1.29 and its proof in Gilbarg and Trudinger [346] presents a constructive way to solve this problem. Thus $a_1(\cdot,\cdot)$ can be constructed in the form (2.18).

$\forall v \in H^1(\Omega)$. Or u_0 satisfies the Dirichlet boundary condition $u_0|_{\partial\Omega} = 0$. Then the test functions are chosen as $v \in H_0^1(\Omega) = \mathcal{V}_b$. Again the boundary term drops out. Finally, $\mathcal{V}_b$ has to be appropriately modified, cf. (2.143), for a combination as in (2.11).

For natural boundary conditions $B_a u = \varphi$ on $\partial\Omega$ the reduction to $B_a u = 0$ as for Dirichlet conditions, is usually not considered. Instead, we choose

$$u_0 \in \mathcal{U} = \mathcal{V} = H^1(\Omega) \text{ and solve } Au_0 = f \in H^{-1}(\Omega). \tag{2.35}$$

Note that for $u_0 \in H^2(\Omega)$ and (2.5), the strong forms $A_s u_0 \in L^2(\Omega)$ and $B_a u_0 \in H^{1/2}(\partial\Omega)$ are well defined. Even for $u_0 \in H^1(\Omega)$, $B_a u_0 \in H^{-1/2}(\partial\Omega)$, the dual of $H^{1/2}(\partial\Omega)$, might make sense, cf. Remark 2.1 and Theorem 1.37. So for $v \in H^{1/2}(\partial\Omega)$, the $\int_{\partial\Omega} B_a u\ v\ ds$ are well defined.

So the *strong solution* $u_0 = u_{s,0}$ of the *natural boundary value problem* solves

$$u_{s,0} = u_0 \in H^2(\Omega) : A_s u_0 = f \in L^2(\Omega), B_a u_0 = \varphi \in H^{1/2}(\Omega). \tag{2.36}$$

Returning to (2.24), we obtain instead of (2.31), with the smoother $u_0 \in H^2(\Omega)$,

$$\begin{aligned} a(u_0, v) &= \int_\Omega (A_s u_0) v dx + \int_{\partial\Omega} B_a u_0\ v\ ds = a_s(u_0, v) + \int_{\partial\Omega} B_a u_0\ v\ ds \qquad (2.37) \\ &= \int_\Omega f v dx + \int_{\partial\Omega} \varphi\ v\ ds \text{ or, more generally, } = \langle f, v\rangle_{\mathcal{V}_b' \times \mathcal{V}_b} + \int_{\partial\Omega} \varphi\ v\ ds, \end{aligned}$$

with $\mathcal{V}_b = H_0^1(\Omega)$, cf. Proposition 2.34 and Remark 7.1.

Remark 2.7.

1. *Equation (2.37) is used in two steps: Firstly, we insert $v \in H_0^1(\Omega)$ and delete the boundary integral term. This yields $Au_0 - f = 0 \in H^{-1}(\Omega)$ or even $Au_0 - f = 0 \in L^2(\Omega)$. For the previous Dirichlet conditions with $u_0,\ v \in H_0^1(\Omega) = \mathcal{V}_b = \mathcal{U}_b$ we would be done. Natural boundary conditions require a second step: By the previous $a(u_0, v) = \langle f, v\rangle_{\mathcal{V}' \times \mathcal{V}}$ or $= \int_\Omega f v dx$ we are left with $\int_{\partial\Omega} B_a u_0\ v\ ds = \int_{\partial\Omega} \varphi\ v\ ds\ \forall v \in H^1(\Omega)$, yielding $B_a u_0|_{\partial\Omega} = \varphi$.*
2. *An extension to parts of $\partial\Omega$ with Dirichlet and natural boundary conditions as in (2.11) is presented in most textbooks, e.g. Treves [632], Chapter 37, and Hackbusch [386, 387], Section 7.4 and Brenner and Scott [141]. Therefore we only indicate it here. For Dirichlet and natural boundary conditions let $\partial\Omega = \partial\Omega_1 \cup \partial\Omega_2$ be the union of two disjoint "regular" subsets. For $u = \varphi$ on $\partial\Omega_1 \subset\subset \partial\Omega \neq \partial\Omega_1$ we first extend φ from $\partial\Omega_1$ to $\partial\Omega$ and then to $\bar{u}$ on Ω and again use the above trick for enforcing homogeneous Dirichlet conditions for $u_0 - \bar{u}$. Then we use as test and solution space*

$$\mathcal{V}_b := \{v \in H^1(\Omega) : v|_{\partial\Omega_1} = 0\} \tag{2.38}$$

and, cf. (2.11), (2.24), (2.31), define

$$\hat{f}(v) := \int_\Omega f v dx + \int_{\partial\Omega_2} \varphi\ v\ ds \text{ or } = \langle f, v\rangle_{\mathcal{V}_b' \times \mathcal{V}_b} + \int_{\partial\Omega_2} \varphi\ v\ ds.$$

2.3 Bilinear forms and induced linear operators

This presentation is strongly influenced by Hackbusch [386, 387]. His proofs for the Hilbert space situation can usually be simply modified to get the corresponding results for Banach spaces, see e.g. Proposition 2.18. However, motivated by the discussion at the beginning of this chapter, Hilbert spaces are still the most important for this book. We use the notation of many authors in numerical analysis, and exchange, in contrast to [387] ellipticity to coercivity and vice versa. Hence coercive bilinear forms induce (generalized energy) norms and elliptic bilinear forms are induced by elliptic differential operators. We have seen in the last section that the linear differential operators which we want to consider in this context are closely related to bilinear forms. In fact they essentially determine each other uniquely, however see Proposition 2.5. So we start in this section by studying bilinear forms and the properties of linear operators induced by them. We will apply and extend this in the following sections to linear and nonlinear elliptic differential operators of order $2m$ in Sobolev spaces, see Adams, [1]. We assume

$$\mathcal{V} \text{ is a real Banach space,} \tag{2.39}$$

and indicate whenever we need reflexivity. Linear and nonlinear (weak) differential operators related to bilinear or nonlinear forms are mostly studied as $A : \mathcal{V} \to \mathcal{V}'$, tested with $v \in \mathcal{V}$. Often forms $A_s : \mathcal{V} \to \mathcal{Y}$ are used, e.g. for Hölder spaces in the next section. The discussion of bilinear forms in this section allows us to apply these results to linear and nonlinear operators and systems in Sections 2.4–2.6 and their linearization in Section 2.7. In the numerical context, they are often appropriately defined in Banach spaces with different spaces, $\mathcal{U}$, for solutions and, $\mathcal{V}$, for test functions. For our theoretical goals in this section, this is hardly ever relevant. So we mainly discuss the case $\mathcal{U} = \mathcal{V}$.

For many of the following results, e.g. coercivity and the inf–sup condition, boundary conditions are mandatory. If we want to emphasize this we use the notation $\mathcal{V}_b$ for $\mathcal{V}$, where the index $_b$ indicates the boundary conditions. These $\mathcal{V}_b$ are closed subspaces of the original $\mathcal{V}$, hence again are Banach spaces. So we formulate most results in this section with respect to $\mathcal{V}$.

Definition 2.8. Properties of bilinear forms: *Let a* bilinear form $a(\cdot,\cdot)$ *be defined on* $\mathcal{U} \times \mathcal{V} \to \mathbb{R}$, *e.g.* $\mathcal{V} = H^1(\Omega)$ *or* $\mathcal{V} = W_0^{m,p}(\Omega)$. *This* $a(\cdot,\cdot)$ *is called* continuous *or* bounded, *if there exists a constant* $C_b \in \mathbb{R}_+$ *such that*

$$|a(u,v)| \le C_b \|u\|_{\mathcal{U}} \|v\|_{\mathcal{V}} \quad \forall \quad u \in \mathcal{U}, v \in \mathcal{V}. \tag{2.40}$$

If in (2.41)–(2.43), $\mathcal{U} = \mathcal{V}$, *this* $a(\cdot,\cdot)$ *is called* $\mathcal{V}$-coercive, *if it is* bounded *and there exists* $\alpha \in \mathbb{R}_+$ *such that*

$$a(u,u) > \alpha \|u\|_{\mathcal{V}}^2 \; \forall u \; \in \mathcal{V}. \tag{2.41}$$

The dual bilinear form $a^d(\cdot,\cdot) : \mathcal{V} \times \mathcal{V} \to \mathbb{R}$ *for* $a(\cdot,\cdot)$ *is defined as*

$$a^d(u,v) := a(v,u) \quad \forall \quad u, v \in \mathcal{V}. \tag{2.42}$$

The $a(\cdot,\cdot)$ *is called* symmetric *if*

$$a(u,v) = a^d(u,v) = a(v,u) \; \forall \; u,v \in \mathcal{V}. \tag{2.43}$$

(For complex Banach spaces, $\underline{a(u,u)}$ *and* $a(v,u)$ *in (2.41) and (2.42), (2.43) have to be replaced by* $\Re\, a(u,u)$ *and* $\overline{a(v,u)}$*, respectively.)*

Thus a $\mathcal{V}$-coercive bounded bilinear form introduces a norm, $\|\cdot\|_a$. Indeed, see (2.40), (2.41), this $\|\cdot\|_a$ is *equivalent to the norm* $\|\cdot\|_{\mathcal{V}}$ on $\mathcal{V}$. So, positive constants, $\gamma_1 = (\alpha)^{1/2}, \gamma_2 = (C_b)^{1/2}$, exist such that $0 < \gamma_1 < \gamma_2$ and

$$\gamma_1 \|u\|_{\mathcal{V}} < \|u\|_a := (a(u,u))^{1/2} \; < \; \gamma_2 \|u\|_{\mathcal{V}} \forall u \in \mathcal{V}. \tag{2.44}$$

For the special case $A_s u = -\Delta u$ or $a_{ij} = \delta_{ij}, a_{00} = 0$ the corresponding $\|u\|_a = ((\nabla u, \nabla u)_{L^2(\Omega)})^{1/2} = \|\nabla u\|_{L^2(\Omega)}$ is denoted as the *energy norm.*

Remark 2.9. *Often the* bilinear form $a(\cdot,\cdot)$ *is defined, as in Section 2.2 and Subsection 2.4.4, via an elliptic differential operator. Then the standard choice for* $\mathcal{V}$ *is* $H^m(\Omega)$*. An extension to* $\mathcal{V} = W^{m,p}(\Omega)$ *is possible for* $p \geq 1$*. For* Ω *in (2.5),* $W^{m,p}(\Omega) \subset W^{m,q}(\Omega), p \geq q$*, is densely and continuously embedded, cf. Theorem 1.26. For the linearization of quasilinear and fully nonlinear equations, see e.g. Section 2.7.3, Theorem 2.122,* $1 \leq p \leq \infty$*, is required, cf. the beginning of Subsection 2.4.4. Only for* $2 \leq p \leq \infty$ *do we obtain the full convergence results for all discretizations with respect to the discrete* $H^m(\Omega)$*norm.*

We use, for linear operators, $A : \mathcal{U} \to \mathcal{V}'$, with $\mathcal{U}, \mathcal{V}$, Banach spaces, the notation

$$\begin{aligned} \mathcal{L}(\mathcal{U},\mathcal{V}') &:= \{A : \mathcal{U} \to \mathcal{V}'\} \text{ for bounded } A, \\ \mathcal{N}(A) &:= \{u \in \mathcal{U} : Au = 0\} = \ker(A), \\ \mathcal{R}(A) &:= \{v \in \mathcal{V}' : \exists u \in \mathcal{U} : v = Au\} = \text{ range } (A). \end{aligned} \tag{2.45}$$

The notation of a dual operator in (1.33) for a linear operator $A \in \mathcal{L}(\mathcal{U},\mathcal{V}')$ in a reflexive Banach space, $\mathcal{V} = \mathcal{V}''$, is slightly updated here into

$$A^d \in \mathcal{L}(\mathcal{V},\mathcal{U}') : \; \langle u, A^d v\rangle_{\mathcal{U}\times\mathcal{U}'} = \langle Au, v\rangle_{\mathcal{V}'\times\mathcal{V}} \; \forall \; u,v \in \mathcal{V}. \tag{2.46}$$

Proposition 2.10. *Let* $a(\cdot,\cdot)$ *be a continuous bilinear form, see (2.40).*

1. *Then there exists a unique operator* $A \in \mathcal{L}(\mathcal{U},\mathcal{V}')$ *such that*

$$a(u,v) = \langle Au, v\rangle_{\mathcal{V}'\times\mathcal{V}} \quad \forall \; u \in \mathcal{U}, \; v \in \mathcal{V}, \; \|A\|_{\mathcal{V}'\leftarrow\mathcal{U}} \leq C_b. \tag{2.47}$$

A is called the operator induced *by the bilinear form. In particular,*

$$\begin{aligned} \|A\|_{\mathcal{V}'\leftarrow\mathcal{U}} &= \sup_{u\in\mathcal{U},\|u\|_{\mathcal{U}}=1} \Big\{ \sup_{v\in\mathcal{V},\|v\|_{\mathcal{V}}=1} \{|a(u,v)|\} \Big\} \\ &= \sup\{|a(u,v)| : u \in \mathcal{U}, v \in \mathcal{V}, \|u\|_{\mathcal{U}} = \|v\|_{\mathcal{V}} = 1\}. \end{aligned} \tag{2.48}$$

2. *Similarly, for* $\mathcal{U} = \mathcal{V}$, *the* $a^d(u,v) = a(v,u)$ *induces the special dual operator* $A^d \in \mathcal{L}(\mathcal{V}, \mathcal{V}')$ *by*

$$\langle A^d u, v\rangle_{\mathcal{V}' \times \mathcal{V}} = a^d(u,v) = a(v,u) = \langle Av, u\rangle_{\mathcal{V}' \times \mathcal{V}} \tag{2.49}$$
$$\forall\ u, v \in \mathcal{V}\ \Rightarrow\ A^d \in \mathcal{L}(\mathcal{V}, \mathcal{V}').$$

This A^d *is the* dual operator *to* A, *see (1.33), and* $\|A^d\|_{\mathcal{V}' \leftarrow \mathcal{V}} = \|A\|_{\mathcal{V}' \leftarrow \mathcal{V}}$.

3. *Let* $\mathcal{V}_1$ *be a dense subspace of* $\mathcal{V}$ *and let* $a(\cdot,\cdot)$ *be defined on* $\mathcal{V}_1 \times \mathcal{V}_1$, *satisfying (2.40) for* $u, v \in \mathcal{V}_1$ *with the norm* $\|\cdot\|_{\mathcal{V}}$. *Then there is a unique extension* $\vec{a}(\cdot,\cdot)$ *of* $a(\cdot,\cdot)$ *to* $\mathcal{V} \times \mathcal{V}$ *such that (2.40) is still satisfied with the same* C_b *and norms* $\forall\ u, v \in \mathcal{V}$.
4. *If the (real) Hilbert space* $\mathcal{V}$ *is identified with* $\mathcal{V}'$, *then* $A \in \mathcal{L}(\mathcal{V}, \mathcal{V})$ *induces the* adjoint operator $A^* \in \mathcal{L}(\mathcal{V}, \mathcal{V})$, *in the sense of (1.36) by replacing (2.49) by*

$$(A^* u, v)_{\mathcal{V}} = a^d(u,v) = a(v,u) = (u, Av)_{\mathcal{V}}\ \ \forall\ \ v, u \in \mathcal{V}. \tag{2.50}$$

Finally, see (1.37), $\|A^*\|_{\mathcal{V} \leftarrow \mathcal{V}} = \|A\|_{\mathcal{V} \leftarrow \mathcal{V}}$ *and* $A^* = A$ *for a symmetric bilinear form* $a(\cdot,\cdot)$.

5. *An* $a(\cdot,\cdot) : \mathcal{U} \times \mathcal{V} \to \mathbb{R}$, *with* $\mathcal{U} = W^{m,p_1}, \mathcal{V} = W^{m,p_2}$, *with coefficients* $a_{i,j} \in W^{m,p_3}$, *has to satisfy* $1/p_1 + 1/p_2 + 1/p_3 = 1$. *This case plays a certain role in the linearization of nonlinear elliptic equations.*

Remark 2.11. *We have restricted the discussion here to bilinear forms* $a(\cdot,\cdot)$ *and their uniquely induced operator* $A \in \mathcal{L}(\mathcal{U}, \mathcal{V}')$, *and its dual operator* $A^d \in \mathcal{L}(\mathcal{V} = \mathcal{V}'', \mathcal{U}')$. *If, more generally,* $A \in \mathcal{L}(\mathcal{X}, \mathcal{Y})$ *for two normed spaces* $\mathcal{X}, \mathcal{Y}$, *then* $A^d \in \mathcal{L}(\mathcal{Y}', \mathcal{X}')$ *is, for every* $y' \in \mathcal{Y}'$, *defined by the following unique* $x' \in \mathcal{X}'$, *such that*

$$\langle Ax, y'\rangle_{\mathcal{Y} \times \mathcal{Y}'} = \langle x, x'\rangle_{\mathcal{X} \times \mathcal{X}'} \forall\ x \in \mathcal{X}\ \Rightarrow\ A^d \in \mathcal{L}(\mathcal{Y}', \mathcal{X}'),\ A^d y' := x'\ \forall\ y' \in \mathcal{Y}'.$$

In many situations it is important to characterize the invertibility of an induced A by appropriate properties of the bilinear form $a(\cdot,\cdot)$, in particular, the following inf–sup condition. This and the invertibility of A usually require *boundary conditions to be incorporated* into $\mathcal{U}, \mathcal{V}$, see below and [386], p. 127, [135], p. 117. The following theorem is often applied to numerical methods. So it is sometimes important to distinguish the $u \in \mathcal{U} \neq \mathcal{V} \ni v$. It is straightforward to generalize Hackbusch's proof for his Lemma 6.5.3 [387], from his $A \in \mathcal{L}(\mathcal{U}, \mathcal{U}')$ form for Hilbert spaces $\mathcal{U}, \mathcal{U}'$ to our Banach space setting $A \in \mathcal{L}(\mathcal{U}, \mathcal{V}')$.

Theorem 2.12. Brezzi–Babuška condition: *Let* $\mathcal{U}, \mathcal{V}$ *be* reflexive Banach spaces *and* $A \in \mathcal{L}(\mathcal{U}, \mathcal{V}')$ *be induced by the bounded* $a(\cdot,\cdot) : \mathcal{U} \times \mathcal{V} \to \mathbb{R}$, *see (2.47). Then the four* *statements (2.51), (2.52), (2.53) and (2.54) are equivalent:*

$$A^{-1} \in \mathcal{L}(\mathcal{V}',\mathcal{U}) \text{ exists, hence } C \text{ exists, s.t. } \|A^{-1}\|_{\mathcal{U}\leftarrow\mathcal{V}'} \le C, \tag{2.51}$$

$$\exists\ \epsilon, \epsilon' > 0 \text{ s.t. } \sup_{0\ne v\in\mathcal{V}} |a(u,v)|/\|v\|_{\mathcal{V}} \ge \epsilon\|u\|_{\mathcal{U}} \quad \forall u \in \mathcal{U} \tag{2.52}$$

$$\text{and} \quad \sup_{0\ne u\in\mathcal{U}} |a(u,v)|/\|u\|_{\mathcal{U}} \ge \epsilon'\|v\|_{\mathcal{V}} \quad \forall v \in \mathcal{V},$$

$$\exists\ \epsilon > 0 \text{ s.t } \sup_{0\ne v\in\mathcal{V}} |a(u,v)|/\|v\|_{\mathcal{V}} \ge \epsilon\|u\|_{\mathcal{U}} \quad \forall u \in \mathcal{U} \tag{2.53}$$

$$\text{and} \quad \sup_{0\ne u\in\mathcal{U}} |a(u,v)|/\|u\|_{\mathcal{U}} > 0 \quad \forall v \in \mathcal{V},\ 0 \ne v$$

$$\exists\ \epsilon > 0 \text{ s.t } \sup_{0\ne v\in\mathcal{V}} |a(u,v)|/\|v\|_{\mathcal{V}} \ge \epsilon\|u\|_{\mathcal{U}} \quad \forall u \in \mathcal{U} \tag{2.54}$$

$$\text{and} \quad \forall\ v \in \mathcal{V}, v \ne 0\ \exists\ u \in \mathcal{U} : a(u,v) \ne 0.$$

Each of the conditions (2.51)–(2.54) implies $\|A^{-1}\|_{\mathcal{U}\leftarrow\mathcal{V}'} \le 1/\epsilon$. If ϵ, ϵ' are chosen as the suprema in (2.52)–(2.54), then $\|A^{-1}\|_{\mathcal{U}\leftarrow\mathcal{V}'} = 1/\epsilon$.

If one of these (2.51)–(2.54) holds, then $\forall\ f \in \mathcal{V}'\ \exists_1\ u_0 \in \mathcal{U}$ such that

$$Au_0 = f \ \Leftrightarrow a(u_0,v) = \langle Au_0, v\rangle_{\mathcal{V}'\times\mathcal{V}} = \langle f, v\rangle_{\mathcal{V}'\times\mathcal{V}}\ \forall\ v \in \mathcal{V}. \tag{2.55}$$

We do not give a proof, see [386], p. 127, [135], p. 117. But it is straightforward to show that (2.52) or (2.53) are necessary for (2.51): If $0 \ne w \in \mathcal{R}(A)^{\perp}$ exists, choose $0 \ne v \perp w$. Then $a(w,v) = 0$, contradicting (2.52), (2.53).

Remark 2.13. Injective A and the inf–sup condition:

1. *The notation* inf-sup condition *is due to the equivalence of, e.g. the first line of (2.52) with*

$$\inf_{0\ne u\in\mathcal{U}} \left\{ \sup_{0\ne v\in\mathcal{V}} \{|a(u,v)|/\|v\|_{\mathcal{V}}\}/\|u\|_{\mathcal{U}} \right\} \ge \epsilon.$$

   Some authors replace the above $\ge \epsilon$ by $= \epsilon$.
2. *Sometimes we refer to the second line of (2.53) as*

$$\sup\{|a(u,v)| : u \in \mathcal{U}, \|u\|_{\mathcal{U}} = 1\} > 0 \quad \textit{for all} \quad 0 \ne v \in \mathcal{V}. \tag{2.56}$$

3. *The first line of the above condition (2.52) is directly related to the* dual norm of Au *as*

$$\epsilon\|u\|_{\mathcal{U}} \le \sup\{|\langle Au, v\rangle_{\mathcal{V}'\times\mathcal{V}}| : v \in \mathcal{V}, \|v\|_{\mathcal{V}} = 1\} = \|Au\|_{\mathcal{V}'}. \tag{2.57}$$

   This inequality is necessary for an injective A, and for the existence of A^{-1}. A combination with the second condition of the second line of (2.52) yields the bijectivity.

4. *Often Theorem 2.12 is applied to a discrete* $A^h : \mathcal{U}^h \to \mathcal{V}'^h$. *Then the last estimate in (2.51) applied to* $\|(A^h)^{-1}\|_{\mathcal{U}^h \leftarrow \mathcal{V}'^h} \le C$ *yields the stability of* A^h.
5. *Under appropriate conditions, the first and second line of (2.52) are equivalent, see Proposition 2.19.*

Sometimes a corollary of Theorem 2.12 is useful, obtained simply by rescaling.

Corollary 2.14. *Let* $\mathcal{U}, \mathcal{V}$ *be* reflexive Banach spaces *and* $A \in \mathcal{L}(\mathcal{U}, \mathcal{V}')$ *be induced by the bounded* $a(\cdot,\cdot) : \mathcal{U} \times \mathcal{V} \to \mathbb{R}$, *see (2.47). Then the two statements (2.58) and (2.59) are equivalent:*

$$A^{-1} \in \mathcal{L}(\mathcal{V}', \mathcal{U}) \text{ exists, hence } C \text{ exists, s.t. } \|A^{-1}\|_{\mathcal{V}' \leftarrow \mathcal{U}} \le C, \tag{2.58}$$

$$\exists\ \epsilon, \epsilon' > 0 \text{ s.t.} \quad \sup_{v \in \mathcal{V}, \|v\|_{\mathcal{V}}=1} |a(u,v)| \ge \epsilon \ \ \forall u \in \mathcal{U}, \|u\|_{\mathcal{U}} = 1, \tag{2.59}$$
$$\text{and} \quad \sup_{u \in \mathcal{U}, \|u\|_{\mathcal{U}}=1} |a(u,v)| \ge \epsilon' \ \ \forall v \in \mathcal{V}, \|v\|_{\mathcal{V}} = 1.$$

Obviously a $\mathcal{V}$-coercive, hence bounded bilinear form $a(\cdot,\cdot)$ satisfies each of (2.52)–(2.54). Hence a combination of (1.33) and (2.52) shows:

Theorem 2.15. Coercivity implies invertibility; invertible A, A^d:

1. *Let* $\mathcal{V}$ *be reflexive and the bounded* $a(\cdot,\cdot)$ *be either* $\mathcal{V}$*-coercive, see (2.40), (2.41), or satisfy one of (2.52)–(2.54), thus* $A^{-1} \in \mathcal{L}(\mathcal{V}', \mathcal{V})$ *exists, see Theorem 2.12. Then* A *and* A^d *are simultaneously invertible and*

$$A, A^d \in \mathcal{L}(\mathcal{V}, \mathcal{V}'), \|A\|_{\mathcal{V}' \leftarrow \mathcal{V}} = \|A^d\|_{\mathcal{V}' \leftarrow \mathcal{V}} \le C \le C_b \text{ in (2.40)}, \tag{2.60}$$
$$A^{-1}, (A^d)^{-1} \in \mathcal{L}(\mathcal{V}', \mathcal{V}), \|A^{-1}\|_{\mathcal{V} \leftarrow \mathcal{V}'} = \|(A^d)^{-1}\|_{\mathcal{V} \leftarrow \mathcal{V}'} \le C'.$$

2. *If* $a(\cdot,\cdot)$ *is* $\mathcal{V}$*-coercive with* α *as in (2.41), then the previous* $C' \le 1/\alpha$, *or if* $a(\cdot,\cdot)$ *satisfies one of (2.52)–(2.54), then* $C' \le 1/\epsilon$.
3. *In particular,* $Au_0 = f$ *in (2.55) and its dual counterpart are uniquely solvable.*

In the remainder of this section we study $\mathcal{V}$-coercive and $\mathcal{V}$-elliptic bilinear forms. Sometimes, Equation (2.55) is equivalent to a minimum problem:

Theorem 2.16. Extremal problem:

1. *Let* $a(\cdot,\cdot)$ *be* $\mathcal{V}$*-coercive, hence bounded, and symmetric, and* $f \in \mathcal{V}'$ *be given. Then there is a unique minimum* $u_0 \in \mathcal{U} = \mathcal{V}$ *for the* extremal problem

$$J(u_0) = \inf\{J(u) := a(u,u) - 2f(u) : \ \forall \ u \in \mathcal{V}\}. \tag{2.61}$$

2. *This minimum is attained for the (unique) solution* u_0 *of (2.55).*

The elliptic differential operators in Sections 2.2 and 2.4 do not correspond to coercive, but to more general, so-called elliptic bilinear forms. So we need:

Definition 2.17. Embeddings, Gelfand triple, elliptic bilinear forms:

1. *Let* $\mathcal{V} \subset \mathcal{W}$ *be Banach spaces and* $\mathbf{I} : \mathcal{V} \hookrightarrow \mathcal{W}$, $\mathbf{I}v := v \in \mathcal{W}\ \forall v \in \mathcal{V}$. *This embedding* $\mathbf{I} : \mathcal{V} \hookrightarrow \mathcal{W}$ *is called* dense and continuous or compact *if* $\mathcal{V}$ *is dense in* $\mathcal{W}$ *and* $\mathbf{I}$ *is continuous or compact, respectively*
2. *Let* $\mathcal{V}$ *be a Banach and* $\mathcal{W}$ *a Hilbert space, with*[6] $\mathcal{W} = \mathcal{W}'$ *identified. Then* $\mathcal{V} \subset \mathcal{W} \subset \mathcal{V}'$ *is called a* Gelfand triple *if*

$$\mathcal{V} \hookrightarrow \mathcal{W} = \mathcal{W}' \hookrightarrow \mathcal{V}',\ \mathcal{V} \hookrightarrow \mathcal{W} \text{ is dense and continuous.} \tag{2.62}$$

3. *Under this condition, the* bounded *bilinear form* $a(\cdot,\cdot) : \mathcal{V} \times \mathcal{V} \to \mathbb{R}$ *is called* $\mathcal{V}$-elliptic *if there exist constants* $C_c, \alpha \in \mathbb{R}$ *such that*

$$a(u,u) \geq \alpha \|u\|_{\mathcal{V}}^2 - C_c \|u\|_{\mathcal{W}}^2 \quad \forall\ u \in \mathcal{V} \text{ with } \alpha > 0. \tag{2.63}$$

The second inequality is a so-called Gårding inequality. For $C_c = 0$ *we obtain again a* $\mathcal{V}$*-coercive bilinear form* $a(\cdot,\cdot)$.

We will show in Subsection 2.4.4 that elliptic operators induce, under standard conditions, $H^m(\Omega)$-elliptic bilinear forms.

In (2.62) only embedding properties of $\mathcal{V} \subset \mathcal{W}$ are listed. Relation (2.62) implies $\mathcal{W}'$ to be continuously embedded into $\mathcal{V}'$ for Banach spaces $\mathcal{V}, \mathcal{W}$, but $\mathcal{V}, \mathcal{W}$ densely and continuously into $\mathcal{V}'$ for Hilbert spaces, see [387], Corollary 6.3.10. Therefore, see (2.62), with $\mathcal{V} \hookrightarrow \mathcal{W} = \mathcal{W}'$ dense and continuous, we obtain for

$$\text{Hilbert spaces } \mathcal{V}, \mathcal{W} : (2.62) \Rightarrow \mathcal{V} \hookrightarrow \mathcal{W} = \mathcal{W}' \hookrightarrow \mathcal{V}' \text{ dense and continuous.} \tag{2.64}$$

For applications of partial differential equations the most important Gelfand triples are those with compact embeddings. To apply the following Riesz–Schauder theory we need criteria for compact operators, cf. Hackbusch [387], Theorem 6.4.10.

Proposition 2.18. *In the Gelfand triple (2.62) let the embedding* $\mathcal{V} \hookrightarrow \mathcal{W}$ *be compact and* $T \in \mathcal{L}(\mathcal{V}', \mathcal{V})$ *be given. Then the dense embeddings in (2.62) allow us to modify this* $T \in \mathcal{L}(\mathcal{V}', \mathcal{V})$, *e.g.* $A^{-1} = T$, *yielding compact*

$$T_1 \in \mathcal{L}(\mathcal{V}', \mathcal{V}'), T_2 \in \mathcal{L}(\mathcal{V}', \mathcal{W}), T_3 \in \mathcal{L}(\mathcal{V}, \mathcal{V}), T_4 \in \mathcal{L}(\mathcal{W}, \mathcal{V}), T_5 \in \mathcal{L}(\mathcal{W}, \mathcal{W}).$$

Proof. The compact embedding $\mathbf{I} : \mathcal{V} \hookrightarrow \mathcal{W}$ implies the embedding $\mathbf{I_1} : \mathcal{W} = \mathcal{W}' \hookrightarrow \mathcal{V}'$ to be compact as well. If in a product of continuous operators at least one factor is compact, the product is compact as well, see [387], Lemma 6.4.5. Now $T_1 = \mathbf{I_1 I} T : \mathcal{V}' \to \mathcal{V} \to \mathcal{W}' \to \mathcal{V}'$, similarly $T_2 = \mathbf{I}T : \mathcal{V}' \to \mathcal{V} \to \mathcal{W} = \mathcal{W}'$, $T_3 = \mathbf{I}T\mathbf{I_1} : \mathcal{W} = \mathcal{W}' \to \mathcal{V}' \to \mathcal{V} \to \mathcal{W} = \mathcal{W}'$, $T_4 = T\mathbf{I_1} : \mathcal{W} = \mathcal{W}' \to \mathcal{V}' \to \mathcal{V}$, $T_5 = T\mathbf{I_1 I} : \mathcal{V} \to \mathcal{W} = \mathcal{W}' \to \mathcal{V}' \to \mathcal{V}$. □

It is interesting that for the most important Gelfand triples we can reduce the conditions in Theorem 2.12, cf. [387], Lemma 6.5.17:

Proposition 2.19. *In a Gelfand triple, see (2.62), with reflexive Banach space* $\mathcal{V}$ *let the embedding* $\mathcal{V} \hookrightarrow \mathcal{W}$ *be compact. Then in Theorem 2.12 the two conditions in*

[6] A Hilbert space $\mathcal{W}$ can always be identified with its dual $\mathcal{W}'$.

(2.52) are equivalent. In particular, $A^{-1} \in \mathcal{L}(\mathcal{V}', \mathcal{V})$ is guaranteed by one of the two equivalent lines of (2.52).

Later we will apply the Liapunov–Schmidt method. It requires results concerning kernels and ranges of linear(ized) operators. Important candidates are compact perturbations of boundedly invertible operators and operators induced by elliptic bilinear forms. This includes essentially all the following problems, e.g. Navier–Stokes operators. The following *Riesz–Schauder theory* and its modification show that these operators behave in some sense similarly to finite dimensional linear operators. In fact, the *Fredholm alternative* is[7] valid. We formulate it for the two cases indicated above. The form in Theorem 2.20 for the Riesz–Schauder theory is a combination of Hackbusch's Theorem 6.4.12 and the Closed Range Theorem 1.19. For the many cases of operators induced by elliptic bilinear forms we modify this result in the following Theorem 2.21.

Theorem 2.20. Riesz–Schauder theory: *Let $\mathcal{V}$ be a Banach space, $I \in \mathcal{L}(\mathcal{V}, \mathcal{V})$ denote the identity and let $T \in \mathcal{L}(\mathcal{V}, \mathcal{V})$ be compact, for $T^d \in \mathcal{L}(\mathcal{V}', \mathcal{V}')$ cf. Remark 2.11.*

1. *For every $\lambda \in \mathbb{C} \setminus \{0\}$, one of the following alternatives is valid:*
   (i) $(T - \lambda I)^{-1} \in \mathcal{L}(\mathcal{V}, \mathcal{V})$;
   (ii) λ *is an* eigenvalue *for* $T \Leftrightarrow \exists x \neq 0 : Tx = \lambda x$.
   For (i) *the* $(T - \lambda I)x = y$ *is uniquely solvable in* $\mathcal{V}$ *for every* $y \in \mathcal{V}$ *and*
   $$(T - \lambda I)^{-1} \in \mathcal{L}(\mathcal{V}, \mathcal{V}) \Longleftrightarrow (T^d - \bar{\lambda} I)^{-1} \in \mathcal{L}(\mathcal{V}', \mathcal{V}'),$$
   that is, both are simultaneously bijective and boundedly invertible, so
   $$\mathcal{R}(T - \lambda I) = \mathcal{V}, \quad \mathcal{R}(T^d - \bar{\lambda} I) = \mathcal{V}',$$
   $$\mathcal{N}(T - \lambda I) = \{0\} \subset \mathcal{V}, \quad \mathcal{N}(T^d - \bar{\lambda} I) = \{0\} \subset \mathcal{V}'. \tag{2.65}$$
   For (ii) *we find: λ is an eigenvalue for $T \Longleftrightarrow \bar{\lambda}$ is an eigenvalue for T^d. Then there exist finite dimensional* eigenspaces
   $$\mathcal{E}(T, \lambda) = \mathcal{N}(T - \lambda I) \neq \{0\}, \quad \textit{and } \mathcal{E}(T^d, \bar{\lambda}) = \mathcal{N}(T^d - \bar{\lambda} I) \neq \{0\}$$
   with dim $\mathcal{E}(T, \lambda) =$ dim $\mathcal{E}(T^d, \bar{\lambda}) < \infty$ *and* $Tx = \lambda x \quad \forall x \in \mathcal{E}(T, \lambda)$. *Then $Tx - \lambda x = y$ has at least one, and hence infinitely many solutions $x \in \mathcal{V}$ if and only if $\langle y, x' \rangle_{\mathcal{V} \times \mathcal{V}'} = 0 \ \forall \ x' \in \mathcal{E}(T^d, \bar{\lambda})$.*
2. *The* spectrum *of T is defined as the set*
   $$\sigma(T) := \{\lambda \textit{ eigenvalue for } T\} \cup \{0, \textit{ if } T^{-1} \notin \mathcal{L}(\mathcal{V}, \mathcal{V})\}.$$
   This $\sigma(T)$ is at most countable and the only possible limit point in $\mathbb{C}$ is 0. This is the reason for excepting $\lambda = 0$ here and in 1. We find $\sigma(T^d) = \overline{\sigma(T)} := \{\bar{\lambda} : \lambda \in \sigma(T)\}$.

[7] A Fredholm operator, compare Definition 1.20, is a bounded linear operator $A \in \mathcal{L}(\mathcal{V}, \mathcal{Y})$ with finite dimensional kernel and co-range. The following $A - \lambda I$ and $A^d - \bar{\lambda} I$ for $\lambda \in \mathbb{C}$ and its inverses usually requires the complexification of $\mathcal{V}, \mathcal{V}'$ and the operators A, A^d, see Theorem 1.22.

3. *For* (i), (ii) *and* $\lambda \neq 0$, *the closed ranges,* $\mathcal{R}(T-\lambda I), \mathcal{R}(T^d-\bar{\lambda}I)$, *and kernels,* $\mathcal{N}(T-\lambda I)=\mathcal{E}(T,\lambda)$, $\mathcal{N}(T^d-\bar{\lambda}I)=\mathcal{E}(T^d,\bar{\lambda})$, *are subspaces of the same finite co-dimension and dimension, respectively. They are related as*

$$\mathcal{R}(T-\lambda I)=(\mathcal{N}(T^d-\bar{\lambda}I))^{\perp}, \text{ and } \mathcal{R}((T-\lambda I)^d) =^{\perp} (\mathcal{N}(T-\lambda I)) \tag{2.66}$$

equivalent to $\langle y, x'\rangle_{\mathcal{V}\times\mathcal{V}'}=0\ \forall\ y\in\mathcal{R}(T-\lambda I), x'\in\mathcal{E}(T^d,\bar{\lambda})$,

$$\text{and } \langle x, y'\rangle_{\mathcal{V}\times\mathcal{V}'}=0\ \forall\ x\in\mathcal{E}(T,\lambda),\ y'\in\mathcal{R}((T-\lambda I)^d),$$

respectively. Hence, $T-\lambda I$ *are Fredholm operators of* index *0.*

The last part of this theorem is a special case of the Closed Range Theorem 1.19 and usually not included in the Riesz–Schauder theory. The relation between this and the next theorem is mainly based upon the compact embedding $\mathcal{V}\hookrightarrow\mathcal{W}$. This implies a compact embedding $I:\mathcal{V}\hookrightarrow\mathcal{V}'$ as well. Let A be induced by the above bounded elliptic bilinear form $a(\cdot,\cdot)$. Then for large enough $C_K>0$ the $A+C_KI$ is boundedly invertible. So, $T:=(A+C_KI)^{-1}I:\mathcal{V}\to\mathcal{V}$ is compact by Theorem 1.13. Thus Theorem 2.20 can be applied to this $T-\mu I:\mathcal{V}\to\mathcal{V}$ and some parts only have to be slightly changed to yield, cf. [388], Theorem 6.5.15:

Theorem 2.21. Fredholm alternative: *Theorem 2.20 remains correct with the following modifications: Let* $\mathcal{V}\subset\mathcal{W}=\mathcal{W}'\subset\mathcal{V}'$ *be a Gelfand triple, see (2.62), with a Hilbert space* $\mathcal{W}$, *a compact embedding* $\mathcal{V}\hookrightarrow\mathcal{W}$ *(hence* $\mathcal{V}\hookrightarrow\mathcal{V}'$ *is compact as well) and with a* reflexive Banach space $\mathcal{V}=\mathcal{V}''$. *Let* $a(\cdot,\cdot)$ *be bounded elliptic, and let* $A\in\mathcal{L}(\mathcal{V},\mathcal{V}')$ *and* $I\in\mathcal{L}(\mathcal{V},\mathcal{V}')$ *be the induced operator and the embedding (hence* I *is compact), respectively, and* $A^d\in\mathcal{L}(\mathcal{V},\mathcal{V}')$ *be the dual for* A.

1. *For every* $\lambda\in\mathbb{C}$ *(including* $\lambda=0$*!) one of the following alternatives is valid:*
   (i) $(A-\lambda I)^{-1}\in\mathcal{L}(\mathcal{V}',\mathcal{V})$;
   (ii) λ *is an eigenvalue for* A.
   For (i), $(A-\lambda I)u_0=f$ *is uniquely solvable in* $\mathcal{V}$ *for every* $f\in\mathcal{V}'$ *with*
   $$(A-\lambda I)^{-1}\in\mathcal{L}(\mathcal{V}',\mathcal{V}) \Longleftrightarrow (A^d-\bar{\lambda}I)^{-1}\in\mathcal{L}(\mathcal{V}',\mathcal{V}),$$
   that is, both are simultaneously bijective and boundedly invertible, so
   $$\mathcal{R}(A-\lambda I)=\mathcal{R}(A^d-\bar{\lambda}I)=\mathcal{V}', \mathcal{N}(A-\lambda I)=\mathcal{N}(A^d-\bar{\lambda}I)=\{0\}\subset\mathcal{V}. \tag{2.67}$$
   For (ii), $Au_0-\lambda u_0=f$ *has at least one, and in fact* $\dim\mathcal{E}(A^d,\bar{\lambda})$ *many solution* $u_0\in\mathcal{V}$,, *if and only if* $\langle f,u^d\rangle_{\mathcal{V}'\times\mathcal{V}}=0\ \forall\ u^d\in\mathcal{E}(A^d,\bar{\lambda})\subset\mathcal{V}$, *and (2.67) is updated according to (2.66).*
2. *The* spectrum *of* A *is at most countable and admits no limit points in* $\mathbb{C}$, *and is defined, differently from above, as*
   $$\sigma(A):=\{\lambda \text{ eigenvalue for } A\}, \text{ and } \lambda\in\sigma(A)\Leftrightarrow\bar{\lambda}\in\sigma(A^d).$$
3. *For* (i) *and* (ii) *the closed ranges and kernels again satisfy (2.66) and* $A-\lambda I$ *and* $A^d-\bar{\lambda}I$ *are Fredholm operators of* index *0.*

Remark 2.22.

1. *With* $A - \lambda I,\ I \in \mathcal{L}(\mathcal{V}, \mathcal{V}'), I$ *compact and* $(A - \lambda I)^{-1} \in \mathcal{L}(\mathcal{V}', \mathcal{V})$ *the operator* $(A - \lambda I)^{-1} I \in \mathcal{L}(\mathcal{V}, \mathcal{V})$ *is compact by Theorem 1.13.*
2. *Solving* $Au_0 = f$ *and* $(A - \lambda I)u_0 = f$ *the above two cases have to be distinguished. The unique existence of a solution* u_0 *for the problem* $Au_0 = f$ *and* $(A - \lambda I)u_0 = f$ *is correct if and only if* $\lambda = 0$ *and* λ *is not an eigenvalue of the operator* A. *If* $(A - \lambda I)u_0 = f$ *has to be solved for an eigenvalue* λ, *finitely many,* $n' := \dim \mathcal{N}(A - \lambda I)$, *conditions have to be imposed for existence and uniqueness, cf. Theorem 1.19:*

$$f \in \mathcal{R}(A - \lambda I) = (\mathcal{N}(A^d - \bar{\lambda} I))^{\perp} \iff \langle f, y_j \rangle_{\mathcal{V}' \times \mathcal{V}} = 0, j = 1, \ldots, n', \text{ with}$$

$$\text{span } \{y_1, \ldots, y_{n'}\} = \mathcal{N}(A^d - \bar{\lambda} I) \subset \mathcal{V},\ u_0 \in \mathcal{V}^{\perp}, \mathcal{V} = \mathcal{N}(A - \lambda I) \oplus \mathcal{V}^{\perp}.$$

   If a special solution u_s *for* $(A - \lambda I)u_0 = f$ *is known, the set of general solutions* u_0 *is* $u_0 \in \{u_s + \mathcal{N}(A - \lambda I)\}$, *similarly for* $(A^d - \bar{\lambda} I)u_0 = f$ *and* $\mathcal{N}(A^d - \bar{\lambda} I)$.
3. *Not all linearizations of the quasilinear equations, see Section 2.7, satisfy (2.63), particularly in* $W^{m,p}(\Omega)$ *for* $p > 2$. *Then Theorem 2.21 is applicable with appropriate modifications.*

The following two propositions are often useful, cf. [387], Lemmas 6.4.13 and 6.5.18:

Proposition 2.23. *Let* $\mathcal{X} \hookrightarrow \mathcal{Y} \hookrightarrow \mathcal{Z}$ *be continuously embedded Banach spaces, and* $\mathcal{X} \hookrightarrow \mathcal{Y}$ *be compactly embedded. Then for every* $\epsilon > 0$ *there exists* C_ϵ *such that*

$$\|u\|_{\mathcal{Y}} \leq \epsilon \|u\|_{\mathcal{X}} + C_\epsilon \|u\|_{\mathcal{Z}} \quad \forall \quad u \in \mathcal{X}. \tag{2.68}$$

Proposition 2.24. *Let* $a(\cdot,\cdot), b(\cdot,\cdot)$ *be bilinear forms and* $a(\cdot,\cdot)$ *be* $\mathcal{V}$*-elliptic, with the Gelfand triple* $\mathcal{V} \subset \mathcal{W} = \mathcal{W}' \subset \mathcal{V}'$ *and a reflexive Banach space* $\mathcal{V}$, *see (2.63). Then* $a(\cdot,\cdot) + b(\cdot,\cdot)$ *is again* $\mathcal{V}$*-elliptic if* $b(\cdot,\cdot)$ *satisfies one of the following conditions:*

1. *For every* $\epsilon > 0$ *there exists* C_ϵ *such that, compare Proposition 2.23,*

$$|b(u,u)| \leq \epsilon \|u\|_{\mathcal{V}}^2 + C_\epsilon \|u\|_{\mathcal{W}}^2 \quad \forall \quad u \in \mathcal{V}. \tag{2.69}$$

2. *Let the embeddings* $\mathcal{V} \hookrightarrow \mathcal{X}, \mathcal{V} \hookrightarrow \mathcal{Y}$ *be continuous and at least one of them be compact and let*

$$|b(u,u)| \leq C_b \|u\|_{\mathcal{X}} \cdot \|u\|_{\mathcal{Y}} \quad \forall\, u \in \mathcal{V}. \tag{2.70}$$

3. *Let the embeddings* $\mathcal{V} \hookrightarrow \mathcal{X}, \mathcal{V} \hookrightarrow \mathcal{Y}$ *be continuous and (2.70) be satisfied. For* $\|\cdot\|_{\mathcal{X}}$ *or* $\|\cdot\|_{\mathcal{Y}}$, *we assume: For every* $\epsilon > 0$ *a* C_ϵ' *exists such that, cf. (2.68),*

$$\|u\|_{\mathcal{X}} \leq \epsilon \|u\|_{\mathcal{V}} + C_\epsilon' \|u\|_{\mathcal{W}} \text{ or } \|u\|_{\mathcal{Y}} \leq \epsilon \|u\|_{\mathcal{V}} + C_\epsilon' \|u\|_{\mathcal{W}}\ \forall u \in \mathcal{V}. \tag{2.71}$$

   This condition is guaranteed by Proposition 1.27.

We will come back to these bilinear forms in the case of Hilbert spaces, and linear operators A of order $2m$, see the following (2.133) ff., in Section 2.4.4, and to the general case of Banach spaces in Section 2.7, particularly in Subsection 2.7.4, where the linearization of nonlinear differential operators is studied.

2.4 Linear elliptic differential operators, Fredholm alternative and regular solutions

2.4.1 Introduction

The aim of this section is the definition of sufficiently general concepts of linear elliptic differential operators of order $2m$ in the Hölder and Sobolev space settings. We need existence, uniqueness, regularity results and Fredholm alternatives for the solutions, which can be combined with the nonlinear problems in Sections 2.5 ff. and numerical methods in the following chapters. A linear differential operator A of order $2m$, directly generalizing Section 2.2, is

$$A : \mathcal{V} \to \mathcal{V}' \text{ with } \mathcal{V} = H^m(\Omega) \subset \mathcal{W} = L^2(\Omega) = \mathcal{W}' \subset \mathcal{V}' = H^{-m}(\Omega). \tag{2.72}$$

But it makes sense to consider competing very successful approaches in Hölder and general Sobolev spaces. We will present them in this section. In Section 2.2 we introduced the classical, strong and weak operators $A_c : C^2(\Omega) \to C(\Omega)$, $A_s : H^2(\Omega) \to L^2(\Omega)$ and $A : H^1(\Omega) \to H^{-1}(\Omega)$. We will generalize them to Hölder and Sobolev spaces and study elliptic operators $A_c = A_s : C^{2m+k,\gamma}(\Omega) \to C^{k,\gamma}(\Omega)$ and $A_s : W^{2m+k,p}(\Omega) \to W^{k,p}(\Omega)$ in Subsection 2.4.2 under C^∞-conditions for the coefficients of the differential and boundary operators and for Ω. In contrast to Section 2.2 we do no longer use these nonstandard indices in A_c, A_s, A. The regularity results in this section show that the operators are essentially the same; only the spaces and the smoothness conditions vary. We use the A_s, A only if necessary, e.g. when partial integration and the Green's formula require this notation. It is worthwhile realizing that for second order equations the C^∞-conditions can be relaxed to C^k-conditions and still get $A : C^{2+k,\gamma}(\Omega) \to C^{k,\gamma}(\Omega)$ results, summarized in Subsection 2.4.3. Finally, in Subsection 2.4.4 we study the direct generalization of Section 2.2 to elliptic operators of order $2m$ and formulate the necessary results for $A : H^m(\Omega) \to H^{-m}(\Omega)$. In Lions and Magenes [478] and Oden and Reddy [518], further generalizations are discussed. An essential difference between elliptic operators of order 2 and $2m$ is caused by the fact that only for order 2 is it always possible to transform the problem to homogeneous boundary conditions. Lemma 2.26 shows that for order $2m$ problems, $m > 1$, this is only possible for Dirichlet boundary systems. General boundary operators for $2m$ are more complicated and induce surprising effects, see Lions and Magenes [478], Oden and Reddy [518], and Runst and Sickel [561].

For defining linear elliptic operators of order $2m$ we use the standard notation for *multi-indices*, α, *reals or vectors* ϑ^α or ϑ and *(partial) derivatives* $\partial^\alpha u$. The notation for the wide range of problems in the present chapter is a nightmare. Many different possibilities have been suggested in the literature. We choose relatively similar symbols, $\partial^i u, \partial^\alpha u, \partial u$, for partials, and $\vartheta^i, \vartheta^\alpha, \vartheta$, for reals or vectors. Let

$$\alpha := (\alpha_1, \ldots, \alpha_n) \in \mathbf{N}_0^n,\ |\alpha| := \alpha_1 + \cdots + \alpha_n, \alpha\,! := (\alpha_1)! \cdots (\alpha_n)! \quad (2.73)$$

$$\partial^i u := \frac{\partial u}{\partial x_i}, \partial u := (\partial^1 u, \ldots, \partial^n u),\ \partial^\alpha u := \partial_1^{\alpha_1} \ldots \partial_n^{\alpha_n} u := \frac{\partial^{|\alpha|}}{\partial x_1^{\alpha_1} \ldots \partial x_n^{\alpha_n}}\ u,$$

$$\nabla^k u := (\partial^\alpha u)_{|\alpha|=k}, \nabla^{\leq k} u := (\partial^\alpha u)_{|\alpha|\leq k}, \partial^0 u := \nabla^0 u := u, \nabla u := \nabla^1 u, \text{ with}$$

$$u(x), \partial^i u(x), \partial^\alpha u(x) \in \mathbb{R}, \partial u(x) \in \mathbb{R}^n,\ \nabla^k u(x) \in \mathbb{R}^{n_k}, \nabla^{\leq k} u(x) \in \mathbb{R}^{N_k}, \text{ where}$$

$$n_k := \sum_{|\alpha|=k} 1,\ N_k := \sum_{j=0}^{k} n_j, \text{ and } \Theta^k \in \mathbb{R}^{n_k},\ \Theta^{\leq k} \in \mathbb{R}^{N_k}, \Theta^0 \in \mathbb{R}, \Theta := \Theta^1,$$

$$\vartheta := (\vartheta^1, \ldots, \vartheta^n) \in \mathbb{R}^n, \vartheta^\alpha := (\vartheta^1)^{\alpha_1} \cdots (\vartheta^n)^{\alpha_n}, \vartheta^i \in \mathbb{R}, i \geq 0, \vartheta^0 = 1,$$

and sometimes $D^k u := \nabla^k u, D^{\leq k} u := \nabla^{\leq k} u, D^0 u := \nabla^0 u := D^0 u := u, Du := \nabla^1 u.$

Only in Subsections 2.5.4 and 2.5.7, 2.6.4 do we appropriately replace our notation (x, Θ^0, Θ^1) and $(x, \Theta^0, \Theta^1, \Theta^2)$ by (x, z, p) and (x, z, p, r), respectively.

The standard *Sobolev inner products, norms*, and *seminorms* are defined as

$$(u, v)_{H^m(\Omega)} := \sum_{|\alpha|\leq m} (\partial^\alpha u, \partial^\alpha v)_{L^2(\Omega)}, \quad \text{and} \quad \|v\|_{H^m(\Omega)} := ((u, u)_{H^m(\Omega)})^{1/2},$$

$$|v|_{H^m(\Omega)} := \left(\sum_{|\alpha|=m} \|\partial^\alpha v\|^2_{L^2(\Omega)}\right)^{1/2}, \|v\|_{W^{m,p}(\Omega)} := \left(\sum_{|\alpha|\leq m} \|\partial^\alpha v\|^p_{L^p(\Omega)}\right)^{1/p} \quad (2.74)$$

As a consequence of the *polynomial formula* the inequalities[8]

$$1 \leq \binom{m}{\alpha} := \binom{m}{\alpha_1 \cdots \alpha_n} := \frac{m!}{\alpha_1! \cdots \alpha_n!} \leq C_{m,n} \forall m \in \mathbb{N}, \alpha \in \mathbb{N}_0^n, \vartheta \in \mathbb{R}^n,$$

$$\begin{aligned}
\text{imply } |\vartheta|_n^{2m}/C_{m,n} &\leq |\vartheta|_{n,\alpha}^{2m} := \sum_{|\alpha|=m} (\vartheta)^{2\alpha} \\
&= \sum_{|\alpha|=m} (\vartheta^1)^{2\alpha_1} \cdots (\vartheta^n)^{2\alpha_n} \qquad (2.75) \\
&\leq \sum_{|\alpha|=m} \binom{m}{\alpha_1 \cdots \alpha_n} (\vartheta^1)^{2\alpha_1} \cdots (\vartheta^n)^{2\alpha_n} \\
&= \left(\sum_{i=1}^{n} (\vartheta^i)^2\right)^m = |\vartheta|^{2m} \\
&= |\vartheta|_n^{2m} = \left(\sum_{i=1}^{n} (\vartheta^i)^2\right)^m \\
&= \leq C_{m,n} \sum_{|\alpha|=m} (\vartheta)^{2\alpha},\ |\vartheta|_n \text{ Euclidean norm in } \mathbb{R}^n.
\end{aligned}$$

[8] As a consequence of (2.75), we can use, in (2.79), here and throughout, either $|\vartheta|_{n,\alpha}$ or $|\vartheta|_n$ with appropriately modified $\lambda(x), \Lambda(x)$.

Definition 2.25. Linear elliptic differential operators of order $2m$ and different boundary conditions:

1. A linear differential operator of order $2m$ *has the strong form*

$$Au := \sum_{|\alpha|\le 2m} a_\alpha \partial^\alpha u \textit{ with real valued coefficients } a_\alpha; \tag{2.76}$$

for the weak form A we prefer (2.102). The symbol S for A

$$S_A(x,\vartheta) := S(x,\vartheta) := \sum_{|\alpha|\le 2m} a_\alpha(x)\vartheta^\alpha, \tag{2.77}$$

is often called the characteristic polynomial for A. *The* principal part A_p *is*

$$A_p u := \sum_{|\alpha|=2m} a_\alpha(x)\partial^\alpha u. \tag{2.78}$$

2. *Generalizing (2.20), the principal part of the symbol of an elliptic A (or of $-A$) has to satisfy, in $\mathcal{D}\subset\Omega$ and $\forall\vartheta\in\mathbb{R}^n$, $x\in\mathcal{D}$,*

$$\lambda(x)|\vartheta|_n^{2m} \le (-1)^m \sum_{|\alpha|=2m} a_\alpha(x)\vartheta^\alpha \le \Lambda(x)|\vartheta|_n^{2m}\ \forall\vartheta\in\mathbb{R}^n,\ x\in\mathcal{D}. \tag{2.79}$$

3. *This A is called* elliptic *and* strongly elliptic *and* uniformly elliptic, *in $\mathcal{D}$, if $0\le\lambda(x)$ and $0<\epsilon\le\lambda(x)$ and $0<\epsilon\le\lambda(x),\Lambda(x)/\lambda(x)$ bounded $\forall\ x\in\mathcal{D}$, respectively.*
For $a_\alpha\in L^\infty(\mathcal{D})$ or $a_{\alpha,\beta}\in W^{|\beta|,\infty}(\mathcal{D})$ in (2.102), and $a_{\alpha,\beta}\in L^\infty(\mathcal{D})$ in (2.110) this condition is only required almost everywhere in $\mathcal{D}$.
4. *The most important types of boundary operators, in the trace sense, see Remark 2.1 and Theorem 1.37, are imposed $\forall x\in\partial\Omega$ or in complementing subsets $\partial\Omega=\partial\Omega_1\cup\partial\Omega_2$, and are denoted as*

$$\text{Dirichlet operator}\quad B_D u := (\partial^i u/\partial\nu^i)_{i=0}^{m-1} \tag{2.80}$$

$$\text{Neumann operator}\quad \partial^i u/\partial\nu^i\ , m\le i\le 2m-1,\ \textit{and} \tag{2.81}$$

$$\text{General operator}\ B_j u := \sum_{|\alpha|\le m_j} b_{j,\alpha}\partial^\alpha u \textit{ of order } m_j<2m,\ j=1,\cdots,m, \tag{2.82}$$

$$\textit{with real } b_{j,\alpha}, \textit{e.g. the induced } B_a \textit{ or, for } m=1: u, \frac{\partial u}{\partial\nu},\ b_0 u+\frac{\partial u}{\partial\nu}. \tag{2.83}$$

5. *The system (2.82) is called a* Dirichlet system *if $m_j=j-1, j=1,\ldots,m$. It is called a* normal system *if $0\le m_1<\cdots<m_m$ and S_{B_j}, the symbol of B_j, satisfies $S_{B_j}(x,\nu)\ne 0$, $\nu=\nu(x)$ the normal vector $\forall x\in\partial\Omega, j=1,\ldots,m$.* A boundary value problem *is defined by the corresponding pair $(A,B=(B_j)_{j=1}^m)$, with one of the previous $B=(B_j)_{j=1}^m$.*

Here and for our numerical methods we essentially require the case of uniformly elliptic equations, hence

$$0 < \epsilon \leq \lambda(x), \ \Lambda(x)/\lambda(x) \leq \infty \ \forall x \in \Omega. \tag{2.84}$$

If (2.84) is violated, then *Tychonov regularization,* here with $(-1)^m \delta \sum_{|\alpha|=2m} \partial^\alpha u$, $\delta > 0$, may be used for $\delta \to 0$. Inequality (2.84) is violated usually at boundary points. Then for numerical methods, these violating boundary points may be cut off and approximate boundary conditions may be imposed. The case of *nonuniformly elliptic equations* is indicated in Subsection 2.5.3, and systematically studied by, e.g. Ivanov [415].

By a standard trick, inhomogeneous boundary conditions can be transformed into homogeneous boundary conditions. If u_1 and u_2 satisfy the prescribed inhomogeneous linear conditions, then the difference $u_1 - u_2$ satisfies the homogeneous conditions. For a Dirichlet system this transformation is possible for $m \geq 1$ due to the following lemma. In fact, the proof for Lemma 5.1 in Oden and Reddy [518] can be slightly generalized and extended, yielding in particular, the estimate in (2.86):

Lemma 2.26. Transformation into homogeneous boundary conditions: *Let (2.82) be a* Dirichlet system *with* $b_{j,\alpha} \in C^{m-j+1}(\Omega)$, $\partial\Omega \in C^m$ *and* $(g_j)_{j=1}^m$ *with* $g_j \in W^{m-j+1-1/p,p}(\partial\Omega), j = 1, \ldots, m$. *Then there exists* $u_1 \in W^{m,p}(\Omega)$ *such that in the trace sense, see Remark 2.1 and Theorem 1.37 ,*

$$B_j u_1 - g_j = 0 \quad in\ \partial\Omega, \ j = 1, \cdots, m, \ and \tag{2.85}$$

$$\|u_1\|_{W^{m,p}(\Omega)} \leq C \|(g_1, \ldots, g_m)\|_{\Pi_{j=1}^m W^{m-j+1-1/p,p}(\partial\Omega)}. \tag{2.86}$$

With any $u_2 \in W^{m,p}(\Omega)$, *such that* $B_j u_2 = g_j \quad in\ \partial\Omega, \ j = 1, \cdots, m,$ *we obtain* $B_j(u_1 - u_2) = 0 \quad in\ \partial\Omega, \ j = 1, \ldots, m.$

As a consequence of Theorem 1.37, we can extend $g_0 \in W^{k-1/p,p}(\partial\Omega)$ to $g_0 \in W^{k,p}(\Omega)$. This allows us, for the most important special case, $m = 1$, to relax the previous condition $\partial\Omega \in C^1$.

Lemma 2.27. Transformation to homogeneous conditions for $m = 1$: *Choose* $m = 1$, *a Lipschitz boundary* $\partial\Omega \in C^{0,1}$, *and the* Dirichlet condition *with* $u = g_0$ *on* $\partial\Omega$, *with* $p < \infty$. *Then for* $g_0 \in W^{k-1/p,p}(\partial\Omega)$, *a* $u_1 \in W^{1,p}(\Omega)$ *exists, such that in the trace sense,* $u_1 - g_0 = 0$ *in* $\partial\Omega$. *Then again* $(u_1 - u_2) - g_0 = 0$ *in* $\partial\Omega$.

Corresponding lemmas for Dirichlet systems for systems of equations are valid as well, so we study mainly trivial Dirichlet boundary conditions.

For operators A of higher order $m > 1$, the problem of defining induced and well fitting boundary operators B_a is much more complicated than for $m = 1$. We will come back to this in Subsection 2.4.2 (2.92) ff. and Subsection 2.4.4. In particular, for $m > 1$ the transformation of inhomogeneous into homogeneous boundary conditions is only possible in exceptional cases. So nonhomogeneous boundary conditions are studied by Lions and Magenes, [478–480].

2.4.2 Linear operators of order $2m$ with C^∞ coefficients

The topic of this subsection are strong operators of the form (2.76) with C^∞ coefficients and boundary. An adequate existence, uniqueness and regularity theory is not known for $A : C^2(\Omega) \to C(\Omega)$, but only for $A : C^{2+s,\gamma}(\Omega) \to C^{s,\gamma}(\Omega), 0 < \gamma < 1, s \in \mathbb{N}_0$. We essentially formulate the results in *Hölder and Sobolev spaces,* $C^{2m+s,\gamma}(\Omega)$ and $W^{2m+s,p}(\Omega)$. For extensions we refer to Runst and Sickel [561]. Instead of the standard $C^{2m}(\Omega)$ spaces we use the corresponding *Hölder spaces* $C^{k,\gamma}(\Omega)$, where $k = 2m + s$, with the norms

$$\|u\|_{C^k(\Omega)} := \sum_{|\alpha|\le k} \sup_{x\in\Omega} |\partial^\alpha u(x)| \text{ with } k \in \mathbb{N}_0, \gamma \in \mathbb{R}, 0 < \gamma < 1, \text{ and}$$

$$\|u\|_{C^{k,\gamma}(\Omega)} := \|u\|_{C^k(\Omega)} + \sum_{|\alpha|=k} \sup_{y\ne x, y\in\Omega} \frac{|\partial^\alpha u(x) - \partial^\alpha u(y)|}{|x-y|_n^\gamma} \text{ with} \tag{2.87}$$

$$C^k(\Omega) := \{u : \Omega \to \mathbb{R} : \|u\|_{C^k(\Omega)} < \infty\},\ C^{k,\gamma}(\Omega) := \{u \text{ s.t. } \|u\|_{C^{k,\gamma}(\Omega)} < \infty\}.$$

We study BVPs of the form (A, B), see (2.76)–(2.82), with the solution u_0

$$A : C^{2m,\gamma}(\bar\Omega) \to C^\gamma(\bar\Omega) \ni f,\ B_j : C^{2m,\gamma}(\bar\Omega) \to C^{2m-m_j,\gamma}(\partial\Omega) \ni g_j, \tag{2.88}$$

$$Au := \sum_{|\alpha|\le 2m} a_\alpha \partial^\alpha u,\ B := (B_j)_{j=1}^m,\ B_j u := \sum_{|\alpha|\le m_j} b_{j\alpha}\partial^\alpha u,\ \ m_j < 2m,$$

$$Au_0(x) = f(x) \forall x \in \Omega, Bu_0(x) = 0 \text{ or } = g(x) = (g_j(x))_{j=1}^m \forall x \in \partial\Omega, \tag{2.89}$$

for $u, u_0 \in C^{2m,\gamma}(\bar\Omega)$ or $u, u_0 \in C^{2m+k,\gamma}(\bar\Omega)$, compare (2.93).

Instead of this complicated situation one certainly would prefer a simpler setting. For the case $2m = 2$ and Dirichlet boundary conditions we would aim for a solution $u_0 \in C^2(\Omega) \cap C(\bar\Omega), u_0|_{\partial\Omega} = 0$ for, e.g.

$$2m = 2 : \quad A := -\Delta : C^2(\Omega) \to C(\Omega), \quad -\Delta u_0(x) = f(x) \forall x \in \Omega. \tag{2.90}$$

It is known that this solution u_0 is unique and it exists, if the Green function exists for Ω, see Hackbusch [387], Theorems 3.1.2, 3.2.11, 3.2.13. This u_0 does not depend continuously upon f with respect to the sup norms in $C^2(\Omega)$ and $C(\Omega)$. So $\|u_0\|_{C^{2,\gamma}(\Omega)} \le C\|f\|_{C^\gamma(\Omega)}, 0 < \gamma < 1$ is *correct* in the Hölder norms, see e.g. Theorem 2.30 below, but it is *wrong* for the norms $\|u_0\|_{C^2(\Omega)}, \|f\|_{C(\Omega)}$, compare Hackbusch's Theorem 3.2.15. We get for (2.90) an estimate with the "unusual" norm $\|u_0\|_{C(\Omega)}$ instead of $\|u_0\|_{C^2(\Omega)}$. More generally, let A be strongly elliptic on a bounded Ω with coefficients $a_\alpha \in C(\Omega)$, $a_0 \le 0$ in Ω. Then there exists a $C \in \mathbb{R}_+$, see [387], Theorem 5.1.9 and Exercise 5.1.10d, such that

$$\|u_0\|_{C(\Omega)} \le C\|f\|_{C(\Omega)} \forall f \in C(\Omega). \tag{2.91}$$

These are two reasons for studying and returning to (2.88) and (2.89).

For the following results we require A to be a uniformly elliptic differential operator in Ω. The B have to "fit" to A, a condition much more complicated for $m > 1$ than for

$m = 1$. The coefficients of A, B and the boundary have to be in C^∞, more precisely,

$$\begin{aligned} &A, B \text{ in (2.88) satisfy } a_\alpha \in C^\infty(\bar{\Omega}), b_{j\alpha} \in C^\infty(\partial\Omega)\ \forall |\alpha| \le 2m, \\ &j = 1 \ldots, m,\ \partial\Omega \in C^\infty,\ A \text{ is strongly (so uniformly) elliptic,} \\ &(B_j)_{j=1}^m \text{ satisfies the complementary condition, e.g. Dirichlet boundary} \\ &\text{conditions for } m \ge 1, \text{ see (2.80), or for } m = 1, \text{ let } b_0(x) \ne 0, \text{ see (2.83),} \end{aligned} \tag{2.92}$$

compare Theorem 2.50 below. These complementary conditions and the related concept of *normal boundary value problems*[9] are intensely studied, e.g. by Lions and Magenes [478], pp. 113 ff., and Oden and Reddy [518], pp. 152 ff. Equivalent (nonalgebraic) formulations via initial value problems of ODEs are described by Wloka [667], pp. 150 ff. and Taylor [620], pp. 379 ff. They are often handled more easily. As an example Wloka [667] shows that for a strongly elliptic operator A as in (2.76) with $m = 1$ the Dirichlet and natural boundary conditions are complementary and normal. The same is correct for the Laplacian $-\Delta$ combined with (2.83). For the biharmonic operator $(\Delta)^2$, hence $m = 2$, several cases are considered, see [667] and (2.88). So $m_1 = m_2 = 0$ is neither normal nor complementary. $m_1 = 0, m_2 = 1$ is normal and complementary if $b_{10} \times ((b_{1\alpha})_{|\alpha|=1}, \nu)_n(x) \ne 0 \forall x \in \partial\Omega$. This includes the Dirichlet boundary conditions $u|_{\partial\Omega} = g_1$, $(\partial u/\partial\nu)|_{\partial\Omega} = g_2$.

For the following regularity results we choose smoother Banach spaces

$$\begin{aligned} &\mathcal{X} := C^{2m+k,\gamma}(\bar{\Omega}),\ \mathcal{Y} := C^{k,\gamma}(\bar{\Omega}),\ \mathcal{Z} := C(\bar{\Omega}), \quad \text{with} \\ &\mathcal{X}_0 := \{u \in \mathcal{X} : Bu|_{\partial\Omega} = 0\} \text{ for } k \in \mathbb{N}_0, \gamma \in \mathbb{R}, 0 < \gamma < 1. \end{aligned} \tag{2.93}$$

So this $\mathcal{X}_0$ with homogeneous boundary conditions corresponds to our $\mathcal{V}$ above. Compared to (2.88) we change the spaces in $A : \mathcal{X} \to \mathcal{Y}$, $B_j : \mathcal{X} \to \mathcal{Y}_j := C^{2m+k-m_j,\gamma}(\bar{\Omega})$, but use the same notation A, B. The following results in Theorems 2.28–2.33 and the last statement in 2.35 are due to Zeidler [676], Chapter 6.3, and Problems 6.8, p. 259. They remain valid for the A, B in (2.98), (2.101).

Theorem 2.28. Fredholm alternative, regularity:

1. *Under the conditions (2.92), (2.93) the elliptic operator* $A \in \mathcal{L}(\mathcal{X}_0, \mathcal{Y})$ *is a Fredholm operator of* index $i = d - c := \dim\mathcal{N}(A) - \operatorname{codim}\mathcal{R}(A) < \infty$. *This* i, *determined by the pair* A, B, *is independent of* k. *This remains unchanged if in* A *and* B_j *terms of order* $< 2m$ *and* $< m_j$ *are changed, respectively. This remains correct for all the combinations in this section. For many*

[9] *Complementary condition:* Let $S_A(x, \vartheta)$ and $S_{B_j}(x, \vartheta)$, see (2.77), denote the principal parts of the symbols for A and B_j, respectively. These polynomials correspond to the terms with the highest derivatives. For all $x \in \partial\Omega$ let $\nu = \nu(x)$ denote the outer normal in $x \in \partial\Omega$, let $\vartheta \ne 0$ lie in the tangent plane of $\partial\Omega$ in x and let A be an elliptic operator with real coefficients. Then $S_A(x, \vartheta + z\nu(x))$ is a polynomial in $z \in \mathbb{C}$ of degree $2m$. It can be shown that it admits exactly m solutions $z_k \in \mathbb{C}$ with positive imaginary part $\forall x \in \partial\Omega$. Let $P(z) := \Pi_{k=1}^m (z - z_k)$. Then the system $A, B_j, j = 1, \ldots, m$ satisfies the complementary condition if the polynomials $S_{B_j}(x, \vartheta + z\nu(x))$ in z are linearly independent modulo $P(z)$, i.e. divide the $S_{B_j}(x, \vartheta + z\nu(x))$ by $P(z)$. Then the remainders $S_{B_j}^r(x, \vartheta + z\nu(x))$ of degree $< m$ of this division are linearly independent.

important cases i vanishes, e.g. compare the conditions in Theorem 2.50 below.

2. *If* $\operatorname{index}(A) = i = 0$ *and* $\dim\mathcal{N}(A) = 0$ *then* $A \in \mathcal{L}(\mathcal{X}_0, \mathcal{Y})$ *is boundedly invertible, hence* $\exists_1 u_0$ *for* $Au_0 = f$ *and* $\|u_0\|_{\mathcal{X}_0} \leq C\|f\|_{\mathcal{Y}}$.
3. *Finally, the Fredholm alternative in Theorem 2.21 for* $Au = f$ *has to be modified: There exist* $c = \operatorname{codim}\mathcal{R}(A)$ *and* $d = \dim\mathcal{N}(A)$ *linearly independent elements* $f_i^* \in \mathcal{Y}^* \perp \mathcal{R}(A), i = 1, \dots, c$ *and* $u_i \in \mathcal{N}(A) \subset \mathcal{X}_0, i = 1, \dots, d$ *such that:* Either $f_i^*(f) \neq 0$ *for at least one* $i \in \{1, \dots, c\}$. *Then* $Au = f$ *is not solvable. Otherwise* $Au = f$ *is solvable. If* u_s *is a special solution, the general solution* u_0 *is obtained as*

$$u_0 = u_s + \sum_{i=1}^{d} c_i u_i \text{ with } c_i \in \mathbb{R}.$$

Remark 2.29.

1. *Compared to Theorem 2.21 the* eigenvalue *formulation does not make sense here. For* $d > 0$ *we have* $\mathcal{N}(A) \neq 0 \;\forall\; \lambda$. *So the unique existence of an exact solution* u_0 *for any* $f \in \mathcal{Y}$, $g \equiv 0 \in \Pi_{j=1}^m \mathcal{Y}_j$ *in (2.89) is correct if and only if* $\mathcal{N}(A|_{\mathcal{X}_0}) = \mathcal{N}(A) = \{0\}$ *and index* $i = 0$.

2. *The above Remark 2.22 has to be updated for the new situation. The linear space* $\mathcal{V}$ *in Theorem 2.21 usually corresponds to our* $\mathcal{X}_0$ *with homogeneous boundary conditions, see above and Lions and Magenes [478], Oden and Reddy [518], Wloka [667], and Runst and Sickel [561], Section 3.5.3; cf. Section 8.6 in this book as well.*

3. *Some results for the* index i *can be found in these books as well.*

The next theorems summarize a priori estimates and regularity results for arbitrary functions $u \in \mathcal{X}_0$ and for solutions $u_0 \in C^{2m}(\bar{\Omega})$. In part this is related to the last line of Theorem 2.21 or to Theorem 2.45:

Theorem 2.30. Estimates in $\mathcal{X}_0$: *For (2.92), (2.93) any* $u \in \mathcal{X}_0$ *satisfies*

$$\|u\|_{\mathcal{X}} \leq C(\|Au\|_{\mathcal{Y}} + \|u\|_{\mathcal{Z}}) \text{ and } \|u\|_{\mathcal{X}} \leq C\|Au\|_{\mathcal{Y}} \text{ for } \mathcal{N}(A) = \{0\}. \tag{2.94}$$

For $\mathcal{N}(A) = \{0\}$ *the* $A : \mathcal{X}_0 = C^{2m+k,\gamma}(\bar{\Omega})) \to \mathcal{R}(A)$ *is boundedly invertible. If additionally* index $i = 0$ *then* $\mathcal{R}(A) = \mathcal{Y} = C^{k,\gamma}(\bar{\Omega})$ *with* $A^{-1} \in \mathcal{L}(\mathcal{Y}, \mathcal{X}_0)$ *and* $A^{-1} \in \mathcal{L}(\mathcal{Y}, \mathcal{Y})$ *is compact.*

Theorem 2.31. Regularity in $\mathcal{X}_0$: *For (2.92), (2.93) with* $g = 0$ *in (2.89) any solution* u_0 *for* $Au_0 = 0$ *satisfies* $u_0 \in C^\infty(\bar{\Omega})$. *For* $f \in \mathcal{Y}$ *in (2.93), let the classical solution* u_0 *of (2.89) be in* $C^{2m}(\bar{\Omega})$. *Then again* $u_0 \in \mathcal{X}_0$.

Now we turn to problems with inhomogeneous boundary conditions. Choose

$$\mathcal{X},\ \mathcal{Y},\ \mathcal{Z},\ \text{ as in (2.88), (2.93)},\ \ \mathcal{Y}_j := C^{k+2m-m_j,\gamma}(\partial\Omega),\ \text{and} \tag{2.95}$$

$$Au(x) = f(x)\forall x \in \Omega,\ Bu(x) = g(x) = (g_j(x))_{j=1}^m\ \forall\ x\ \in\ \partial\Omega. \tag{2.96}$$

Theorem 2.32. Estimates, regularity and Fredholm operator in $\mathcal{X}$:

1. *Let* $(A,B) \in \mathcal{L}\left(\mathcal{X}, \mathcal{Y} \times \Pi_{j=1}^m \mathcal{Y}_j\right)$ *be a* normal boundary value problem *and satisfy (2.92), (2.95), (2.96). Then it defines a Fredholm operator of* index i. *It allows us the a priori estimate*

$$\|u\|_{\mathcal{X}} \le C\left(\|Au\|_{\mathcal{Y}} + \|u\|_{\mathcal{Z}} + \sum_{j=1}^m \|B_j u\|_{\mathcal{Y}_j}\right)\ \ \forall u \in \mathcal{X}\ \textit{and}$$

$$\|u\|_{\mathcal{X}} \le C\left(\|Au\|_{\mathcal{Y}} + \sum_{j=1}^m \|B_j u\|_{\mathcal{Y}_j}\right)\ \textit{for}\ \mathcal{N}(A,B) = \{0\}. \tag{2.97}$$

For $\mathcal{N}(A,B) = \{0\}$ *and* $i = 0$, *the* $(A,B) : \mathcal{X} \to \mathcal{R}(A,B)$ *is boundedly invertible and we find* $\mathcal{R}(A,B) = \mathcal{Y} \times \Pi_{j=1}^m \mathcal{Y}_j$.
2. *If the classical solution* u_0 *of (2.96) is in* $C^{2m,\gamma}(\bar\Omega)$, *and* $f \in \mathcal{Y}$, $g_j \in \mathcal{Y}_j, j = 1, \ldots, m$ *with* $k > 0$ *in (2.95), then* $u_0 \in \mathcal{X}$ *with* k *in (2.93),(2.95).*

Now we replace the operators $A := A, B := (B_j)_{j=1}^m$ in (2.88), with the Hölder spaces in (2.93), (2.95), by the following Sobolev spaces:

$$A : W^{2m,p}(\Omega) \to L^p(\Omega), B_j : W^{2m,p}(\Omega) \to W^{2m-m_j-1/p,p}(\partial\Omega), 1 < p < \infty, \tag{2.98}$$

and the corresponding spaces for homogeneous boundary conditions

$$\mathcal{X} := W^{k+2m,p}(\Omega), \mathcal{X}_0 := \{u \in \mathcal{X} : Bu = 0\}, \mathcal{Y} := W^{k,p}(\Omega), \mathcal{Z} := L^1(\Omega)$$
$$\text{and for inhomogeneous boundary conditions:} \tag{2.99}$$

$$\mathcal{X}, \mathcal{Y} \text{ as in (2.99)},\ \mathcal{Z} := L^p(\Omega),\ \mathcal{Y}_j := W^{k+2m-m_j-1/p,p}(\partial\Omega). \tag{2.100}$$

We define solutions, for details see Subsections 2.4.4, 2.5.6 and Zeidler [676], Problems 6.8, p. 261, as

$$u_0 \in W^{2m,p}(\Omega),\ Au_0 = f \text{ in } \Omega,\ Bu_0 = 0 \text{ or } g = (g_j)_{j=1}^m \text{ on } \partial\Omega. \tag{2.101}$$

Theorem 2.33. Results for the Sobolev case: *In Theorems 2.28–2.32 and the operators in (2.88), replace the Hölder spaces (2.93), (2.95) by the Sobolev spaces (2.98), (2.99), (2.100). Correspondingly, replace the homogeneous and inhomogeneous boundary value problems (2.89), (2.96) by (2.101), require (2.92) and a normal boundary* *value problem. Finally, in Theorems 2.30–2.32 replace the conditions for* $C^{j,\gamma}(\bar{\Omega})$ *in (2.93), (2.95) by* $W^{j,p}(\Omega)$ *in (2.99), (2.100). Then Theorems 2.28–2.32 remain valid in this modification and the* index *remains unchanged.*

This astonishing similarity of the results is due to the fact that Hölder and Sobolev spaces are special cases of the so-called Besov–Triebel–Lizorkin spaces. For these spaces the above results are correct, see Runst and Sickel [561].

For our later applications to numerical methods we reformulate the original linear differential operator in (2.76) into the so-called divergence form.[10] The $a_\eta \in C^\infty(\bar{\Omega})$ and ∂^η in (2.88) and (2.73) define appropriate $a_{\alpha,\beta} \in C^\infty(\bar{\Omega})$ with

$$Au := \sum_{|\eta| \le 2m} a_\eta \partial^\eta u = \sum_{|\alpha|,|\beta| \le m} (-1)^{|\beta|} \partial^\beta (a_{\alpha,\beta} \partial^\alpha u), \tag{2.102}$$

e.g. with $a_\eta = (-1)^m a_{\alpha,\beta}$ for $\alpha + \beta = \eta$ and $|\alpha| = |\beta| = m$.

In Subsection 2.4.3 we discuss, for $2m = 2$, *general boundary operators*, see (2.82). However here, to avoid too many technicalities, we restrict the discussion to special *Dirichlet systems,* see Theorem 2.50 and (2.159) below, satisfying

$$\text{on } \partial\Omega \text{ for } j = 1, \ldots, m : B_j u := \sum_{|\alpha| \le j-1} b_{j,\alpha}(x) \partial^\alpha u = g_j \Longleftrightarrow \partial^{j-1} u / \partial \nu^{j-1} = \tilde{g}_j$$

$$\Longleftrightarrow (\partial^\alpha u)_{|\alpha| \le m-1} = (\tilde{g}_\alpha)_{|\alpha| \le m-1} \text{ for } \partial\Omega \in C^{m-1} \text{ and } \tilde{g}_j \in C^{m-j}. \tag{2.103}$$

For more general cases combine Oden and Reddy [518], pp. 184ff. and Zeidler [677], pp. 331 ff. By Lemma 2.26 we use the standard trick for modifying $f := \tilde{f}$ in (2.101) for obtaining homogeneous boundary conditions

$$Au(x) = f(x) \ \forall \ x \in \Omega, \ (\partial^{j-1} u / \partial \nu^{j-1})_{j=1}^m = 0 \ \forall \ x \in \partial\Omega, \tag{2.104}$$

or abbreviated into $u \in W_0^{m,p}(\Omega)$. So (2.88), (2.89), (2.92) has the simplified form for the a_η or $a_{\alpha,\beta}$,

$$\begin{aligned} &A \text{ in (2.102) or (2.102) satisfies } a_\eta \text{ or } a_{\alpha,\beta} \in C^\infty(\bar{\Omega}) \forall |\eta| \le 2m \text{ or } |\alpha| = |\beta| = m, \\ &A \text{ is strongly (so uniformly) elliptic and } \partial\Omega \in C^\infty. \end{aligned} \tag{2.105}$$

With (2.79), (2.102), uniform ellipticity means, noting that $\vartheta^\alpha \in \mathbb{R}, \vartheta \in \mathbb{R}^n, |\vartheta|_n \in \mathbb{R}$,

$$\gamma |\vartheta|_n^{2m} \le \sum_{|\alpha|,|\beta| = m} a_{\alpha,\beta}(x) \vartheta^\alpha \vartheta^\beta \le \Gamma |\vartheta|_n^{2m} \text{ for } 0 < \gamma, \Gamma < \infty \ \forall \ x \in \mathcal{D}. \tag{2.106}$$

The special form of f in (2.107) is proved by Adams [1], pp. 47 ff.

[10] As in the literature, we use different notations for the standard and the divergence form as indicated, e.g. in (2.102) systematically with consequences for the different ellipticity conditions; compare, e.g. $(-1)^m$ in (2.79) versus (2.106).

Proposition 2.34. Continuous forms, $f \in \mathcal{V}_b' = W_0^{-m,p'}(\Omega)$: *For* $1 < p < \infty,\ 1/p + 1/p' = 1$, *we obtain a* special unique representation:

$$\langle f, v\rangle_{\mathcal{V}_b' \times \mathcal{V}_b} := \int_\Omega \sum_{|\beta|\le m} f_\beta \partial^\beta v dx \ \forall\ v\ \in\ \mathcal{V}_b := W_0^{m,p}(\Omega) \text{ with} \tag{2.107}$$

$$f_\beta \in L^{p'}(\Omega) \text{ and } \|f\|_{\mathcal{V}'} = \left(\sum_{|\beta|\le m} \|f_\beta\|^{p'}_{L^{p'}(\Omega)}\right)^{1/p'}. \tag{2.108}$$

For $p = 1$ *or* $p = \infty$ *(2.108) has to be replaced by* $\|f\|_{\mathcal{V}'} := \min_{f_\beta} \left(\sum_{|\beta|\le m} \cdots\right)^{1/p'}$.
For $f \in \mathcal{V}' = W^{-m,p'}(\Omega)$ *we obtain more generally*

$$\langle f, v\rangle_{\mathcal{V}' \times \mathcal{V}} := \int_\Omega \sum_{|\beta|\le m} f_\beta \partial^\beta v dx + \int_{\partial\Omega} \sum_{|\beta|< m} \varphi_\beta \partial^\beta v ds \ \forall\ v\ \in\ \mathcal{V} := W^{m,p}(\Omega) \text{ with}$$

$$\varphi_\beta \in L^{p'}(\partial\Omega) \text{ and } \|f\|_{\mathcal{V}'} = \left(\sum_{|\beta|\le m} \|f_\beta\|^{p'}_{L^{p'}(\Omega)} + \sum_{|\beta|< m} \|\varphi_\beta\|^{p'}_{L^{p'}(\partial\Omega)}\right)^{1/p'}. \tag{2.109}$$

Note that for $f \in \mathcal{V}' = W^{-m,p'}(\Omega)$ *the* $f_0 := f|_{W_0^{-m,p'}(\Omega)}$ *is reduced to an* $f_0 \in W_0^{-m,p'}(\Omega)$ *as in (2.107). This is the reason that the following*

$$a(u_0, v) = \langle Au_0, v\rangle_{\mathcal{V}' \times \mathcal{V}} = \langle f, v\rangle_{\mathcal{V}' \times \mathcal{V}}\ \forall v \in \mathcal{V} = H_0^m(\Omega) \ni u_0$$

tested $\forall v \in H_0^m(\Omega)$ *makes sense for* $f \in W^{-m,p'}(\Omega)$ *in (2.109) as well.*

We combine the standard partial integration with Green's formula:

Theorem 2.35. Solutions for the divergence form:

1. *Assume the form (2.102), the conditions (2.103)–(2.106) and define the bilinear form* $a(u, v)$, *the principal part* $a_p(u, v)$, *and the linear* f *as in (2.107)*

$$a(u, v) := \langle Au, v\rangle_{\mathcal{V}' \times \mathcal{V}} := \int_\Omega \sum_{|\alpha|,|\beta|\le m} a_{\alpha,\beta} \partial^\alpha u \partial^\beta v dx, \tag{2.110}$$

$$a_p(u, v) := \langle A_p u, v\rangle_{\mathcal{V}' \times \mathcal{V}} := \int_\Omega \sum_{|\alpha|=|\beta|=m} a_{\alpha,\beta} \partial^\alpha u \partial^\beta v dx, \forall u, v \in \mathcal{V} = H_0^m(\Omega).$$

   These $a(u, v), a_p(u, v), \langle f, v\rangle_{\mathcal{V}' \times \mathcal{V}}$ *are continuous.*
2. *Any solution* $u_0 \in C^{k+2m,\gamma}(\Omega)$ *or* $u_0 \in W^{k+2m,p}(\Omega) \subset H^m(\Omega), k + m \ge n(1/p - 1/2)$ *of (2.104) necessarily satisfies the standard equation*

$$u_0 \in H_0^m(\Omega) : \quad a(u_0, v) = \langle f, v\rangle_{\mathcal{V}' \times \mathcal{V}} \forall v \in \mathcal{V} = H_0^m(\Omega). \tag{2.111}$$

Vice versa, under the conditions (2.92) in (2.93) and (2.99), (2.100), every such solution $u_0 \in H_0^m(\Omega)$ even satisfies $u_0 \in C^{k+2m,\gamma}(\Omega)$ and $u_0 \in W^{k+2m,p}(\Omega)$, respectively.

Theorem 2.36. Ellipticity, and Fredholm alternative for the divergence form: *In addition to the conditions in Theorem 2.35 let the coefficients $a_{\alpha,\beta}$ with $|\alpha| = |\beta| = m$ in (2.102), (2.106), be continuous in $\bar{\Omega}$. Then for a uniformly elliptic A in (2.105) the $a_p(u,v)$ is $H_0^m(\Omega)$-coercive. For an $H_0^m(\Omega)$-coercive* principal part $a_p(u,v)$, *the above $a(u,v)$ is an $H_0^m(\Omega)$-elliptic bilinear form. So the Fredholm alternative in Theorem 2.21 applies, with* index $(A,B) = 0$. *The unique solution of the* variational problem *(2.61) for $a_p(u,v)$ satisfies (2.111) for $a_p(u,v)$.*

Proof. $a(u,v) - a_p(u,v)$ are lower order terms. So a combination of (2.63) and Proposition 2.24 shows that since for $a_{\alpha,\beta} \in C(\bar{\Omega})$ with $|\alpha| = |\beta| = m$ the $a_p(u,v)$ by (2.105) is $H_0^m(\Omega)$-coercive, see Theorem 2.43 below, the $a(u,v)$ is $H_0^m(\Omega)$-elliptic and the claim is proved, compare Theorems 2.89, 2.104. □

2.4.3 Linear operators of order 2 under C^k conditions

In Subsection 2.4.2 we studied results under strong C^∞-conditions for coefficients and the boundary of the differential and boundary operators. For the most important special case of second order equations it is possible to relax this C^∞-condition and still get very similar results. We list the essential regularity results and the Fredholm alternative in Gilbarg and Trudinger [346]. It is appropriate to arrange, see (2.76), the a_α and ∂^α as $a_{i,j}, \partial^0 u = u$ and $\vartheta^0 := 1,\ \vartheta^i, \vartheta^j \in \mathbb{R},\ i,j = 0,\dots,n$. With ν, the outer normal unit vector for $\partial\Omega$, the differential and one of the three boundary operators (2.88) and (2.89), the boundary value problem has the form

$$Au := \sum_{i,j=0}^{n} a_{ij}\partial^i\partial^j u : C^2(\Omega) \to C(\Omega), \tag{2.112}$$

$$Bu = B_1 u = u, \text{ or } B_2 u = \partial u(x)/\partial\nu, \text{ or } B_3 u = \ b_0 u + \sum_{i=1}^{n} c_i\partial^i u \text{ on } \partial\Omega, \tag{2.113}$$

$$\text{determine } u_0 \in C^{2,\gamma}(\bar{\Omega}) \text{ such that} Au_0 = f\ \forall x \in \Omega, Bu_0 = \varphi\forall\ x \in \partial\Omega. \tag{2.114}$$

Then the $\lambda(x)$ and $\Lambda(x)$ in (2.79) are the minimal and the maximal (real) *eigenvalues of the (symmetric) matrix A_p* defining the different types of ellipticity:[11]

$$A_p(x) := (a_{i,j}(x))_{i,j=1}^n\ \forall\ x \in \Omega \text{ and (2.79) has the form} \tag{2.115}$$

$$\lambda(x)|\vartheta|_n^2 \le -\sum_{i,j=1}^{n} a_{ij}(x)\vartheta^i\vartheta^j \le \Lambda(x)|\vartheta|_n^2\ \forall\vartheta \in \mathbb{R}^n. \tag{2.116}$$

[11] Note the – sign in (2.106) compared to [346]; formula (6.2) yields $a_{00}(x) \ge 0$ in (2.118).

For the following results we require A to be a uniformly elliptic differential operator with smooth coefficients, boundary and inhomogeneities; more precisely, let

$$
\begin{aligned}
&A, B \text{ in (2.112), (2.113) satisfy } f, a_{ij} \in C^{\gamma}(\bar{\Omega}), b_0, c_j, \varphi \in C^{1,\gamma}(\bar{\Omega}),\ 0 < \gamma < 1,\\
&\forall i, j = 0, \dots, n,\ \partial\Omega \in C^{2,\gamma},\ \text{for } B_1 \text{ let } \varphi \in C^{2,\gamma}(\bar{\Omega}),\ \text{for } B_3 \text{and } c = (c_1, \dots, c_n) \text{ let}\\
&|(c(x), \nu(x))_n| > \kappa > 0\ \forall\, x \in \partial\Omega, A \text{ is strongly (so uniformly) elliptic,} \qquad (2.117)
\end{aligned}
$$

cf. (2.26), (2.33). The following results are proved by Gilbarg and Trudinger [346], see Theorems 6.14, 6.15, 6.17, 6.19, 6.30, 6.31 and the last lines in Section 6.7. We summarize the existence, uniqueness and regularity results in the interior Ω, its closure $\bar{\Omega}$ and the Fredholm alternative:

Theorem 2.37. Existence, (non-)uniqueness and estimates: *Let the operators A, B in (2.112), (2.113) satisfy (2.116), (2.117), and*

$$
\begin{aligned}
&a_{00}(x) \geq 0\ \forall\ x\ \in \Omega \text{ for } B_1 u \text{ and for } B_3 u \text{ with } b_0(x) > 0\ \forall\ x \in \partial\Omega,\\
&a_{00}(x) > 0\ \forall\ x\ \in \Omega \text{ for } B_2 u. \qquad (2.118)
\end{aligned}
$$

1. *Then $(A, B) : C^{2,\gamma}(\bar{\Omega}) \to C^{\gamma}(\bar{\Omega}) \times C^{1,\gamma}(\partial\Omega)$ for $B = B_2, B_3$ and $(A, B) : C^{2,\gamma}(\bar{\Omega}) \to C^{\gamma}(\bar{\Omega}) \times C^{2,\gamma}(\partial\Omega)$ for $B = B_1$.*
2. *If $\lambda = 0$ is not an eigenvalue, the $(A, B)u = (f, \varphi)$ is boundedly invertible, hence there exists a unique solution $u_0 \in C^{2,\gamma}(\bar{\Omega})$ for (2.114).*
3. *If $\lambda = 0$ is not an eigenvalue, e.g. for violated (2.118), (A, B) is not boundedly invertible, any solution $u_0 \in C^{2,\gamma}(\bar{\Omega})$ for (2.114) can be estimated as, compare (2.94),*

$$
\begin{aligned}
&\|u\|_{C^{2,\gamma}(\bar{\Omega})} \leq C(\|u\|_{C(\bar{\Omega})} + \|f\|_{C^{\gamma}(\bar{\Omega})} + \|\varphi\|_{C^{1,\gamma}(\Omega)}) \text{ for } B = B_2, B_3, \text{but} \quad (2.119)\\
&\|u\|_{C^{2,\gamma}(\bar{\Omega})} \leq C(\|f\|_{C^{\gamma}(\bar{\Omega})} + \|\varphi\|_{C^{1,\gamma}(\Omega)}) \text{ for a boundedly invertible } (A, B).
\end{aligned}
$$

For $B = B_1$ the $\|\varphi\|_{C^{1,\gamma}(\Omega)}$ in (2.119) has to be replaced by $\|\varphi\|_{C^{2,\gamma}(\Omega)}$.

Theorem 2.38. Regularity in Ω: *For the A, B_1 in (2.112), (2.113) modify (2.117) by replacing $\bar{\Omega}$ by Ω, C^{γ} by $C^{k,\gamma}$-conditions for the coefficients, no conditions for the boundary and φ, and assume that $u_0 \in C^2(\Omega)$ solves the Dirichlet boundary value problem (2.114). Then $u_0 \in C^{k+2,\gamma}(\Omega)$; note in Ω. Here $k = \infty$ is allowed.*

Corresponding results for weak solutions $u_0 \in H_0^1(\Omega)$ are formulated in Subsection 2.4.4. Regularity in $\bar{\Omega}$ requires additional conditions for the boundary.

Theorem 2.39. Regularity in $\bar{\Omega}$: *For the operator A, B_1 in (2.112), (2.113) modify (2.117) by replacing only the $C^{\gamma}, C^{2,\gamma}$-conditions by $C^{k,\gamma}, C^{k+2,\gamma}$, let $\partial\Omega$ be in $C^{k+2,\gamma}$ and assume that $u_0 \in C^0(\bar{\Omega}) \cap C^2(\Omega)$ solves the Dirichlet boundary value problem (2.114), see Theorem 2.37. Then $u_0 \in C^{k+2,\gamma}(\bar{\Omega})$.*

Note that (2.118) is not imposed in the last two theorems. Corresponding $u_0 \in W^{k+2,p=2}(\Omega) = H^{k+2}(\Omega)$ results are formulated in Subsection 2.4.4, cf. [346], Chapter 8.

Theorem 2.40. Fredholm alternative: *Let the operators A, B in (2.112), (2.113) satisfy (2.117). Then*

$$A \in \mathcal{L}(\mathcal{X}_0, \mathcal{Y}) \text{ with } \mathcal{X}_0 := \{u \in C^{2,\gamma}(\Omega) : Bu = 0\},\ \mathcal{Y} := C^{\gamma}(\Omega), \tag{2.120}$$

is a Fredholm operator of index $i = 0$, *satisfying the Fredholm alternative in Theorem 2.21.*

As in the last subsection we want to relate the original boundary value problem (2.112), (2.113) to the corresponding *strong divergence form.* In particular, we aim for existence and uniqueness results and the Fredholm alternative for different combinations of second order equations (2.112), and boundary conditions (2.113). With $\tilde{a}_{ij} := a_{ij}$ in (2.112), and appropriate new a_{ij}, let, see (2.16),

$$Au := A_s u := \sum_{i,j=0}^{n} \tilde{a}_{ij}\partial^i\partial^j u = \sum_{i,j=0}^{n} (-1)_{j>0}\partial^j(a_{ij}\partial^i u) : C^2(\Omega) \to C(\Omega),$$

$$a_{ij} \in C^{1,\gamma}(\bar{\Omega}), a_{i0} \in C^{\gamma}(\bar{\Omega})\ i,j = 0,\ldots,n, \tilde{a}_{ij} = -a_{ij}, i,j \geq 1. \tag{2.121}$$

In this Subsection we have studied $u \in C^{k+2,\gamma}(\bar{\Omega})$, so $u \in H^2(\Omega)$. We modify (2.117) by (2.121) and get instead of (2.116) for the uniform ellipticity for (2.121)

$$\lambda'|\vartheta|_n^2 \leq \sum_{i,j=1}^{n} a_{ij}(x)\vartheta^i\vartheta^j \leq \Lambda'|\vartheta|_n^2\ \forall \vartheta \in \mathbb{R}^n,\ x \in \Omega \text{ for } 0 < \lambda'. \tag{2.122}$$

As in (2.24) the *strong* and *weak bilinear forms* are related, with $A = A_s$ in (2.121):

$$a(u,v) = \int_\Omega \sum_{i,j=0}^{n} a_{ij}\partial^i u\partial^j v dx = \int_\Omega (A_s\ u)v dx + \int_{\partial\Omega} (B_a\ u)v ds \tag{2.123}$$

with $B_a\ u := \sum_{i=0,\ldots,n}^{j=1,\ldots,n} \nu_j a_{ij}\partial^i\ u$ and the principal parts

$$a_p(u,v) = \int_\Omega \sum_{i,j=1}^{n} a_{ij}\partial^i u\partial^j v dx = \int_\Omega (A_{s,p}\ u)v dx + \int_{\partial\Omega} (B_{a,p}\ u)v ds \tag{2.124}$$

with $B_{a,p}\ u := \sum_{i,j=1}^{n} \nu_j a_{ij}\partial^i\ u$ so $B_a - B_{a,p} = h_a := \sum_{j=1}^{n} \nu_j a_{0j}$,

where here B_a and $B_{a,p}$ are the *natural boundary operators* induced by A and A_p, respectively The special $Au = -\Delta u + cu$ induces $B_a\ u = B_{a,p}\ u = \partial u/\partial\nu$.

We discuss the mutual dependency of coercivity properties and boundary conditions for the *first (Dirichlet), the second or natural (generalized Neumann)* and the *third (generalized Robin) boundary value problem:* With a function $h \geq h_0$ on $\partial\Omega$ and $\mathcal{V}_i, f_{-i}, i = 1,2,3$, below let, cf. (2.123)–(2.124),

$$Bu = B_1 u = u, B_2 u = B_a\ u, B_3 u = B_{a,p}\ u + hu \text{ on } \partial\Omega \text{ and determine} \tag{2.125}$$

$$u_0 \in H^2(\Omega) \cap \mathcal{V}_i \text{ such that } Au_0 = f_{-i} \text{ a.e.} \forall x \in \Omega, B_i u_0 = \varphi \forall\ x \in \partial\Omega, i = 1 \text{ or } 2 \text{ or } 3.$$

These generalized Neumann and Robin boundary operators, B_2 and B_3, modify the standard forms $\partial u/\partial\nu$ and $au + b\partial u/\partial\nu$.

For different boundary conditions in (2.124), (2.125), compared to (2.113), we aim for compact perturbations of the original bilinear forms $a(u, v)$. They will turn out to be coercive for the above boundary operators B_i and the chosen function spaces $\mathcal{V}_i$, $i = 1, 2, 3$, cf. (2.126) ff. For $i = 1$, the *principal part* has this property; for $i = 2, 3$, it has to be modified. These $\mathcal{V}_i$-coercive $a_c^i(u, v)$ yield unique existence and are combined with the Fredholm alternative for the $\mathcal{V}_i$-elliptic $a(u, v)$, and the linear forms $\langle f_{-i}, v\rangle_{\mathcal{V}'\times\mathcal{V}}$, cf. Theorems 2.20 and 2.21. The $\mathcal{V}_i$-coercive $a_c^i(u, v)$ induce the linear operators A_c^i as compact perturbations of A induced by the original $a(u, v)$ in (2.121). We consider:

BVPs: $A_c^i u = f_{-i} \in \mathcal{V}'$ in Ω, $B_i u|_{\partial\Omega} = \varphi$, $i = 1, 2, 3$, and $\forall v \in \mathcal{V} := \mathcal{V}_i$. Let for

$$i = 1:\ a_c^1(u, v) := a_p(u, v) = \langle f_{-1}, v\rangle_{\mathcal{V}'\times\mathcal{V}}\ \mathcal{V} := \mathcal{V}_1 := H_0^1(\Omega), \tag{2.126}$$

$$\text{with } \langle f_{-1}, v\rangle_{\mathcal{V}'\times\mathcal{V}} := \int_\Omega \Big(f_0 v + \sum\nolimits_{j=1}^n f_j \partial^j v\Big)dx,\ B_1 u = u = 0 \text{ for the first BVP,}$$

$$i = 2:\ a_c^2(u, v) := a_p(u, v) + c(u, v)_{L^2} = \langle f_{-2}, v\rangle_{\mathcal{V}'\times\mathcal{V}}, B_2 u = B_a u, c > 0, \tag{2.127}$$

$$\langle f_{-2}, v\rangle_{\mathcal{V}'\times\mathcal{V}} := \langle f_{-1}, v\rangle_{\mathcal{V}'\times\mathcal{V}} + \int_{\partial\Omega} \varphi v ds, \mathcal{V} := \mathcal{V}_2 := \{v \in H^1(\Omega) : B_a v = 0\}$$

$$i = 3:\ a_c^3(u, v) := a_p(u, v) + \int_{\partial\Omega} huvds = \langle f_{-2}, v\rangle_{\mathcal{V}'\times\mathcal{V}}, \mathcal{V}_3 := H^1(\Omega), \tag{2.128}$$

$$B_3 u = B_{a,p}\ u + hu, f_{-3} := f_{-2}, \text{with } h(x)|_{\partial\Omega} \geq h_0 > 0 \text{ for the third BVP.}$$

The case $i = 2$ and the two alternatives have to be discussed. For $B_a\ 1 = 0$ and $\langle f_2, 1\rangle = 0$ with h_a in (2.124), $a_p(u, v) = \langle f_2, v\rangle \forall v \in H^1(\Omega), B_a\ u = 0$ and its dual problem admit the nontrivial solution $u_0 \equiv 1$, thus the Fredholm alternative applies. Enforcing unique existence for $B_a\ 1 = 0 = \langle f_2, 1\rangle$ is possible by defining $\mathcal{V}_2 := \{v \in H^1(\Omega) : \langle 1, v\rangle_{\mathcal{V}'\times\mathcal{V}} = 0\}$. This excludes the nontrivial solution $u_0 \equiv 1$. Then the equivalent Sobolev norms in Proposition 1.25 allow the following generalization of Zeidler [677], Sections 22.2 ff. for (2.126)–(2.128).

Theorem 2.41. Unique existence, and Fredholm alternative for (2.126)–(2.128):

1. *Assume the form (2.121), the conditions (2.122), (2.125), for B_a, c, h, f_{-i}, and define $a_c^i(u, v), \mathcal{V} = \mathcal{V}^i$, and $\langle f_{-i}, v\rangle_{\mathcal{V}'\times\mathcal{V}}$ as in (2.126)–(2.128). Then these bilinear and linear forms are continuous $\forall u, v \in \mathcal{V}$ and the $a_c^i(u, v)$ are $\mathcal{V}_i$-coercive for $i = 1, 2, 3$ with the conditions for B_a, h_a, f_{-i} listed in (2.126)–(2.128). Then unique solutions u_0^i exist for*

$$\exists_1 u_0^i = u_0 \in H^1(\Omega):\ a_c^i(u_0, v) = \langle f_{-i}, v\rangle_{\mathcal{V}'\times\mathcal{V}}, B_i u_0 = \varphi, \forall v \in \mathcal{V}_i = \mathcal{V}. \tag{2.129}$$

   The unique solution of the minimal problems *(2.61) for these $a_c^i(u, v)$ satisfies the variational problem (2.129).*
2. *Define the corresponding $a_i(u, v)$ by replacing the principal part of $a(u, v)$ by the above $a_c^i(u, v)$ and impose the previous conditions. Then the $a_i(u, v)$ are*

$\mathcal{V}_i$-elliptic bilinear forms. So the Fredholm alternative in Theorem 2.21 applies to (A, B_i), even with index $(A, B_i) = 0, i = 1, 2, 3$.

3. *Any solution $u_0 \in C^{k+2,\gamma}(\Omega) \subset H^1(\Omega)$ of (2.125) with the original $a_i(u, v)$, and the previous $\langle f_{-i}, v\rangle_{\mathcal{V}' \times \mathcal{V}}$ satisfies the standard equation*

$$u_0 \in H^1(\Omega): \quad a_i(u_0, v) = \langle f_{-i}, v\rangle_{\mathcal{V}' \times \mathcal{V}} \forall v \in \mathcal{V} = \mathcal{V}_i, \ B_i u_0 = \varphi, \ i = 1, 2, 3.$$

Vice versa, under the conditions (2.117), (2.120), every such solution $u_0 \in \mathcal{V}$ even satisfies $u_0 \in C^{k+2,\gamma}(\Omega)$.

Proof. As in Theorem 2.35 obviously the $a(u, v)$, $a_c^i(u, v)$ are bounded, except for $i = 3$, and the $\langle f_{-i} v\rangle$, except for $i = 2, 3$, in $H^1(\Omega)$. For the exceptional cases, Remark 2.1 and the Trace Theorem 1.38 for $v \in H^1(\Omega)$ imply $v|_{\partial\Omega} \in L^2(\partial\Omega)$

$$\|v\|_{L^2(\partial\Omega)}^2 = \int_{\partial\Omega} v^2 ds \le C\|v\|_{H^1(\Omega)}^2 \Longrightarrow |\int_{\partial\Omega} v\varphi ds| \le C\|\varphi\|_{L^2(\partial\Omega)}\|v\|_{H^1(\Omega)},$$

and hence the boundedness of all the $\langle f_{-i}, v\rangle_{\mathcal{V}' \times \mathcal{V}}, i = 1, 2, 3$. For $i = 3$ we estimate

$$\begin{aligned} |\int_{\partial\Omega} huv ds| &\le \|h\|_{C(\partial\Omega)} \int_{\partial\Omega} |uv| ds \\ &\le \|h\|_{C(\partial\Omega)}\|u\|_{L^2(\partial\Omega)}\|v\|_{L^2(\partial\Omega)} \le C_1\|h\|_{C(\partial\Omega)}\|u\|_{H^1(\Omega)}\|v\|_{H^1(\Omega)}. \end{aligned} \tag{2.130}$$

Finally, by Proposition 1.25, the $\left(a_c^i(u, u)\right)^{1/2}$ define norms equivalent to $\|u\|_{\mathcal{V}_i}$, $i = 1, 2, 3$.

The following proof, that the $a_c(u, v) := a_c^i(u, v)$ are $\mathcal{V}_i$-coercive, is a straightforward generalization of Zeidler [677], pp. 331 ff. Relations (2.125), (2.123), (2.124) imply

$$\begin{aligned} 0 &= \int_\Omega A_s\, u_0 v dx - \langle f_{-1}, v\rangle_{\mathcal{V}' \times \mathcal{V}} + \int_{\partial\Omega} (B_i\, u_0 - \varphi) v ds \\ &= a(u_0, v) - \langle f_{-1}, v\rangle_{\mathcal{V}' \times \mathcal{V}} + \int_{\partial\Omega} ((B_i - B_a)u_0 - \varphi) v ds \\ &= a_c(u_0, v) - \langle f_{-1}, v\rangle_{\mathcal{V}' \times \mathcal{V}} + \int_{\partial\Omega} ((B_i - B_a)u_0 - \varphi) v ds \\ &\quad + (a(u_0, v) - a_c(u_0, v)). \end{aligned} \tag{2.131}$$

Now we drop the lower order terms $(a(u_0, v) - a_c(u_0, v))$ and are left with $a_c(u_0, v)$. Then (2.121), (2.122) imply, with $h_a(x) \ge 0$,

$$\begin{aligned} a_c(u, u) &= \int_\Omega \sum_{i,j=1}^n a_{ij}\partial^i u \partial^j u + h_a u^2 dx \ \ge \gamma \int_\Omega \sum_{i=1}^n (\partial^i u)^2 dx + \min_{x \in \bar\Omega}\{h_a(x)\}\|u\|_{L^2(\Omega)}^2 \\ &= \gamma\|\partial u\|_{L^2(\Omega)}^2 + \min_{x \in \bar\Omega}\{h_a(x)\}\|u\|_{L^2(\Omega)}^2 \ge \gamma_i\|u\|_{H^1(\Omega)}^2, \ \forall u \in \mathcal{V}_i, i = 1, 3. \end{aligned} \tag{2.132}$$

The last inequality is correct for $i = 1$, since the norms $\|a_c(u, u)\|^{1/2}$ and $\|u\|_{H^1(\Omega)}$ are equivalent on $\mathcal{V}_1 = H_0^1(\Omega)$. For the different combinations of $\mathcal{V}_i, i = 2$, the $\mathcal{V}_i$-coercivity of $a_c(u,u)$ is a consequence of the discussion preceding Theorem 2.41, based upon the definition of the $\mathcal{V}_i$ and by, Proposition 1.25, the equivalence of the norms $\|u\|_{H^1(\Omega)}$ and $(a_c(u,u) + |\int_\Omega u dx|^2)^{1/2}$. For $i = 3$ it is correct with $\gamma_3 \geq \min\{\gamma, h_0\}$.

Again $a_i(u,v) - a_c^i(u,v)$ are only lower order terms. Since $a_c^i(u,v)$ is $\mathcal{V}_i$-coercive, a combination of (2.63) and Proposition 2.24 implies a $\mathcal{V}$-elliptic $a_i(u,v)$. That these solutions satisfy the differential equation and the boundary conditions is shown as indicated in Remarks 2.6 and 2.7. A combination of (2.63) and Proposition 2.24 implies a $\mathcal{V}$-elliptic $a_i(u,v)$, and thus the claim. □

We discuss in more detail the implications of the Fredholm alternative for one of the above examples. We consider for $i = 2$ the special case, cf. Section 8.6,

$$a(u,v) = \int_\Omega \sum_{i=1}^n \partial^i u \partial^i v dx = \int_\Omega \nabla u \nabla v dx \quad \forall v \in \mathcal{V}_2, \text{ inducing } B_a u = \partial u / \partial \nu,$$

with $a(1,v) = 0 \ \ \forall v \in \mathcal{V}_2, \ \ B_a 1 = 0$ implying $\mathcal{N}(A) = \mathcal{N}(A^d) = \mathbb{R}$.

Consequently, a (nontrivial) solution u_0 for

$$a(u_0, v) = \langle f_{-2}, v \rangle_{\mathcal{V}' \times \mathcal{V}} = \int_\Omega f_0 v dx + \int_{\partial\Omega} \varphi v ds \ \ \forall v \in \mathcal{V}_2,$$

exists, by the Fredholm alternative, iff $\langle f_{-2}, 1 \rangle_{\mathcal{V}' \times \mathcal{V}} = 0$ and every $u_0 + c, c \in \mathbb{R}$ solves this problem. Otherwise the trivial solution exists only for the homogeneous problem $a(u_0, v) = 0, f_{-2} = 0$.

As we have seen, enforcing a uniquely existing solution for $B_a \ 1 = 0 = \langle f_2, 1 \rangle$ is possible by defining, according to the Fredholm alternative, $\mathcal{V}_2 := \{v \in H^1(\Omega) : \int_\Omega v dx = 0\}$. This excludes the nontrivial solution $u_0 \equiv 1$.

2.4.4 Weak elliptic equation of order $2m$ in Hilbert spaces

In Subsection 2.4.2, we studied strong and weak forms of elliptic operators in general Sobolev spaces $W^{m,p}(\Omega), 1 < p < \infty$, under continuity conditions for the coefficients of the differential and boundary operators. The results in this subsection remain correct, in particular with respect to $H^m(\Omega)$-coercivity with appropriate modifications, for $W^{m,p}(\Omega), 2 \leq p \leq \infty$ as well, see below. This is most important in our context for the later linearization of operators and their stability, cf. Section 2.7 and Corollary 2.44. Regularity results in textbooks are usually formulated in the $H^{m+s}(\Omega)$ setting.

Beyond the previous $W^{m,p}(\Omega)$ results, we turn therefore in this subsection to weak forms in Hilbert spaces $H^m(\Omega) = W^{m,2}(\Omega)$. This modifies the approach in Section 2.2 for studying differential operators of order $2m$. We test the differential operator of order $2m$ with elements in $H^m(\Omega)$, often in $H_0^m(\Omega)$, and

$$A : \mathcal{V} \to \mathcal{V}', \ \mathcal{V} = H^m(\Omega) \subset \mathcal{W} = L^2(\Omega) = \mathcal{W}' \subset \mathcal{V}' = H^{-m}(\Omega), \tag{2.133}$$

with Ω satisfying (2.5) and the scalar product $(u,v)_{L^2(\Omega)}$. For *Sobolev spaces*, see Subsection 1.4.3, we use the *scalar product* and *norm* in (2.74). In this context it is appropriate to reformulate Au in divergence[12] form and the corresponding ellipticity condition.

Definition 2.42. Divergent linear differential operator: *For* $a_{\alpha\beta} \in W^{|\beta|,\infty}(\Omega)$ *and* $a_{\alpha\beta} \in L^\infty(\Omega)$ *the strong and weak form* $A = A_s$ *and* A *are, see Definition 2.25,*

$$A_s : H^{2m}(\Omega) \to L^2(\Omega),\ A_s u := \sum_{|\alpha|,|\beta|\le m} (-1)^{|\beta|}\partial^\beta(a_{\alpha\beta}\partial^\alpha u) \tag{2.134}$$

and $A : H^m(\Omega) \to H^{-m}(\Omega)$, *defined by*

$$a(u,v) := \langle Au, v\rangle_{H^{-m}(\Omega)\times H^m(\Omega)} := \int_\Omega \sum_{|\alpha|,|\beta|\le m} a_{\alpha\beta}\partial^\alpha u \partial^\beta v dx. \tag{2.135}$$

The symbol, S, principal part, A_p, *and the different types of* ellipticity *for* A, *are defined as obvious modifications of (2.77)–(2.79), possibly in* $\mathcal{D} \subset \Omega$, *as*

$$\lambda(x)|\vartheta|_n^{2m} \le \sum_{|\alpha|=|\beta|=m} a_{\alpha,\beta}(x)\vartheta^\alpha\vartheta^\beta \le \Lambda(x)|\vartheta|_n^{2m}\ \forall\vartheta \in \mathbb{R}^n,\quad a.e.\ x \in \mathcal{D}. \tag{2.136}$$

We repeat the three sets of m boundary conditions imposed in the trace sense, see Remark 2.1 and Theorem 1.37, $\forall x \in \partial\Omega$, or different conditions in complementary sets:

$$\textit{Dirichlet conditions: } \partial^i u/\partial\nu^i = 0\ \forall\ 0 \le j \le m-1, \tag{2.137}$$

$$\textit{Neumann conditions: } \partial^i u/\partial\nu^i = 0\ \forall\ m \le j \le 2m-1, \tag{2.138}$$

$$\textit{General conditions: } B_j u := \sum_{|\alpha|\le m_j} b_{j,\alpha}\partial^\alpha u, \text{ order } m_j < 2m. \tag{2.139}$$

$$\textit{Natural conditions: for } m = 1 : B_a\ u := \sum_{i=0\dots,n}^{j=1,\dots,n} \nu_j a_{ij}\partial^i\ u. \tag{2.140}$$

The latter are generalized to $m > 1$, in Lions and Magenes [478], p. 120, and Oden and Reddy [518], p. 290 ff., see below. For smooth $\partial\Omega$, the Dirichlet boundary conditions (2.137) are equivalent to, analogously for $W_0^{m,p}(\Omega)$,

$$u \in \mathcal{V} := H_0^m(\Omega) = \{u \in H^m(\Omega) : \partial^\alpha u(x) = 0\quad \forall|\alpha| \le m-1,\ x \in \partial\Omega\} \text{ for } \Omega \in C^{m-1}.$$

In (2.41) and (2.63) we had introduced the concepts of $\mathcal{V}$*-coercive* and $\mathcal{V}$*-elliptic* bilinear forms. We recall that, e.g. for $\mathcal{V} = H_0^m(\Omega)$ the *bounded* bilinear form $a(\cdot,\cdot) : \mathcal{V}\times\mathcal{V} \to \mathbb{R}$ is called $\mathcal{V}$*-coercive*, and $\mathcal{V}$*-elliptic*, respectively, if constants $\alpha \in \mathbb{R}_+, C_c \in \mathbb{R}$ exist such that, e.g. for $\mathcal{V} = H_0^m(\Omega), \mathcal{W} = L^2(\Omega)$

$$a(u,u) > \alpha\|u\|_{\mathcal{V}}^2 \text{ and } \ge \alpha\|u\|_{\mathcal{V}}^2 - C_c\|u\|_{\mathcal{W}}^2\ \forall u\ \mathcal{V}. \tag{2.141}$$

The second inequality is a so-called Gårding inequality.

[12] The $(-1)^{|\beta|}$ in (2.134) usually yields an opposite sign compared to (2.76) and (2.112). This $(-1)^{|\beta|}$ produces the standard sign for $a(\cdot,\cdot)$ in (2.135).

For the following coercivity results for the principal part A_p, we have to impose either the Dirichlet conditions on a subset $\partial\Omega_1$ of $\partial\Omega$ of *positive measure*, hence $\text{meas}_{n-1}(\partial\Omega_1) := \int_{\partial\Omega_1} 1ds > 0$, such that $\partial\Omega = \partial\Omega_1 \cup \partial\Omega_2$, otherwise for natural boundary conditions the coercivity results holds for $A_p + c,\ c > 0$. We impose conditions on the $a_{\alpha\beta}$ in (2.135) and for Ω as in (2.5) for guaranteeing the $H_0^m(\Omega)$-ellipticity of the $a(\cdot,\cdot)$ in (2.135). We summarize Hackbusch's [387] Theorems 7.2.2, 7.2.3, 7.2.11, 7.2.13, Lemma 7.2.12 and extend his Theorem 7.2.7. Some of the following results have already been formulated under different assumptions:

Theorem 2.43. Gårding: $\mathcal{V}$-coercivity and ellipticity, Fredholm alternative for Dirichlet and natural boundary conditions: *Let A in (2.134) have a strongly elliptic principal part, see (2.79), and satisfy $a_{\alpha\beta} \in L^\infty(\Omega)$. Then*

1. *For $m \geq 1$, and $a_{\alpha\beta} \in L^\infty(\Omega)$, the $a(\cdot,\cdot)$ in (2.135) are bounded.*
2. *For $m = 1$, the principal part $a_p(\cdot,\cdot)$ (or for $a_{00}(x) \geq 0$ the $a_p(u,v) + \int_\Omega a_{00}uvdx$) and $a(\cdot,\cdot)$ are $H_0^1(\Omega)$-coercive and $H_0^1(\Omega)$-elliptic, respectively*
3. *For $m > 1$ let, additionally, the $a_{\alpha\beta}$ be continuous in $\bar{\Omega}$ for $|\alpha| = |\beta| = m$. Then $a_p(\cdot,\cdot)$ and $a(\cdot,\cdot)$ are $H_0^m(\Omega)$-coercive and $H_0^m(\Omega)$-elliptic, respectively*
4. *For boundedly invertible $A : H_0^1(\Omega) \to H_0^{-1}(\Omega)$ or $H_0^1(\Omega)$-coercive $a(u,v)$ we obtain*

$$\begin{aligned} &\forall f \in H_0^{-1}(\Omega) : \exists_1 u_0 \in H_0^1(\Omega) : \ Au_0 = f \\ \iff &a(u_0, v) = \langle f, v\rangle_{H_0^{-1}(\Omega)\to H_0^1(\Omega)} \forall v \in H_0^1(\Omega). \end{aligned} \tag{2.142}$$

5. *If in $a(\cdot,\cdot) = a'(.,.) + a''(.,.)$ the $a'(.,.)$ is $H_0^m(\Omega)$-coercive or -elliptic and*

$$a''(u,v) = \sum_{|\alpha|,|\beta|\leq m}^{|\alpha|+|\beta|<2m} \int_\Omega a_{\alpha\beta}\partial^\alpha u \partial^\beta v dx,$$

contains only derivatives of orders $\leq 2m-1$, then $a(\cdot,\cdot)$ is $H_0^m(\Omega)$-elliptic.

6. *These results remain correct, if $\mathcal{V} = H_0^m(\Omega)$ is replaced by $H_{0,p}^m(\Omega) \subset H^m(\Omega)$*

$$\mathcal{V} := H_{0,p}^m(\Omega) = \{u : \partial^\alpha u(x) = 0 \ \ \forall |\alpha| \leq m-1,\ x \in \partial\Omega_1 \subset \partial\Omega\} \tag{2.143}$$

for, e.g. a connected $\partial\Omega_1 \in C^{m-1}$ with $\text{meas}_{n-1}(\partial\Omega_1) = \int_{\partial\Omega} ds > 0$. In $\partial\Omega_2 := \partial\Omega \setminus \partial\Omega_1$ natural or other boundary conditions and possibly a modified functional, f, cf. (2.38), might be imposed.

7. *For* natural boundary conditions, *explicitly formulated in Section 2.2 only for $m = 1$, the $a_p(\cdot,\cdot)$ and $\mathcal{V} = H_0^m(\Omega)$ are replaced by the $H^m(\Omega)$-coercive $a_p(u,v) + \int_\Omega c_0 uvdx$ with $c_0 \geq \epsilon > 0$ and $\mathcal{V}$ by $H^m(\Omega)$. Then the previous results remain correct.*

Vice versa, if $a_{\alpha\beta} \in L^\infty(\Omega)$ holds and $a_{\alpha\beta} \in C(\bar{\Omega})$ for $|\alpha| = |\beta| = m$, then the $H_0^m(\Omega)$-ellipticity of $a(u,v)$ implies a strongly elliptic principal part, see (2.79).

Proof. For $m = 1$, we indicate the nearly trivial proof. The strong ellipticity condition (2.136) yields, for $\partial^i u(x) = \vartheta^i \in \mathbb{R}, \vartheta = (\vartheta^1, \ldots, \vartheta^n) \in \mathbb{R}^n$,

$$\epsilon|\nabla u(x)|_n^2 = \epsilon|\vartheta|_n^2 \leq \sum_{i,j=1}^{n} a_{ij}(x)\vartheta^i\vartheta^j = \sum_{i,j=1}^{n} a_{ij}(x)\partial^i u(x)\partial^j u(x) \ \forall\ x \in \Omega. \quad (2.144)$$

Integration over Ω yields $a(u,u) \geq \int_\Omega \epsilon|\nabla u(x)|^2 dx \geq \epsilon\|u(x)\|^2_{H^1(\Omega)}$ by the equivalence of the semi- or energy-norm and $\|\cdot\|_{H^1(\Omega)}$ in $H_0^1(\Omega)$.

For $m > 1$, Fourier transforms, cf. Hackbusch's [387] proof of Theorem 7.2.7, are combined with the technique of frozen coefficients. In Theorem 2.104 we will give the proof for systems for the case of constant (frozen) $a_{\alpha\beta}$ for $|\alpha| = |\beta| = m$, and $a_{\alpha\beta} \equiv 0$ for $0 \leq |\alpha| + |\beta| \leq 2m - 1$. We do not give all the details, cf. Wloka [667], p. 282 ff. or Hörmander [403], but present only the main idea. By the continuity of the $a_{\alpha\beta}$ in $\bar{\Omega}$ we can choose, for an $\epsilon > 0$, a finite number of points $x_i \in \Omega, i = 1, \ldots, N$, and their neighborhoods $\mathcal{U}_{x_i}$ such that $\forall |\alpha| + |\beta| = 2m$

$$\bar{\Omega} \subset \cup_{i=1}^{N} \mathcal{U}_{x_i} \text{ with } a_{\alpha\beta}^{c,i} := a_{\alpha\beta}(x_i) : \left\| a_{\alpha\beta}^{c,i} - a_{\alpha\beta} \right\|_{C(\mathcal{U}_{x_i})} < \epsilon\ \forall i = 1, \ldots, N. \quad (2.145)$$

We choose a first generation of the i such that, after an appropriate renumbering, $\mathcal{U}_{x_i} \cap \partial\Omega \neq \emptyset, i = 1, \ldots, N_1 \leq N$. Then the standard technique allows us to prove, for the strongly elliptic principal part, see (2.79), and for $u \in H_0^m(\Omega)$, the coercivity of a_p on each of the $\mathcal{U}_{x_i}, i = 1, \ldots, N_1$. Now choose the next inner generation of $\mathcal{U}_{x_i}$ intersecting the union of the first generation, so

$$\mathcal{U}_{x_i} \cap \left(\cup_{i=1}^{N_1} \mathcal{U}_{x_i} \right) \neq \emptyset \ \ \forall i = N_1 + 1, \ldots, N_2.$$

The normal derivatives $\|\partial^k u/\partial\nu^k\|_{H^{m-k-1/2}(\partial\mathcal{U}_{x_i})}, i = 1, \ldots, N_1,\ k = 0, \ldots, m-1$ can be estimated by Remark 2.1 and the Trace Theorem 1.38, (1.67), by $\|u\|_{H^m(\mathcal{U}_{x_i})}$, $i = 1, \ldots, N_1$. Each of these $\partial^k u/\partial\nu^k|_{H^{m-k-1/2}(\partial\mathcal{U}_{x_i})}, k = 0, \ldots, m-1$, can be employed proving, for each $\mathcal{U}_{x_i}, i = N_1 + 1, \ldots, N_2$ in the next generation, the coercivity of a_p for the above nontrivial boundary values $\|\partial^k u/\partial\nu^k\|_{H^{m-k-1/2}(\partial\mathcal{U}_{x_i})}, k = 0, \ldots, m-1$, etc.

The other claims are proved in [387]. We only indicate the case of natural boundary conditions. Then the strong elliptic principal part and $\int_\Omega c_0 u^2 dx$ induce norms equivalent to the seminorm in $H^m(\Omega)$ and to $L^2(\Omega)$, their sum induces a norm equivalent to that in $H^m(\Omega)$, hence $a_p(u,u) + \int_\Omega c_0 u^2 dx$ is $W^{m,p}(\Omega)$-coercive. □

Corollary 2.44. $\mathcal{V}$-coercive and $\mathcal{V}$-elliptic $a(\cdot,\cdot)$ induce a boundedly invertible A and imply the Fredholm alternative, respectively, cf. Theorem 2.21:

1. *For, $\mathcal{V} = W^{m,p}(\Omega)$ or $\mathcal{V} = W_0^{m,p}(\Omega)$, and a $\mathcal{V}$-coercive $a(\cdot,\cdot)$, cf. 3., 4 below, Theorem 2.15 guarantees the unique existence of a solution u_0 for*

$$Au_0 = f \in \mathcal{V}', \ \textit{cf. (2.135),(2.107)}. \quad (2.146)$$

If $a(\cdot,\cdot)$ is even symmetric, it solves, by Theorem 2.16, the minimal problem *(2.61) as well.*

2. *For $\mathcal{V}$-elliptic $a(\cdot,\cdot)$, e.g. induced by an elliptic operator A with $\mathcal{V}$-coercive* principal part $a_p(\cdot,\cdot)$, *Theorem 2.21 implies the* Fredholm alternative *with* index $A = 0$. *So, for an* eigenvalue 0 *of A, infinitely many solutions for $Au_0 = f$ exist under additional conditions for f, otherwise exactly one solution exists and $\|A^{-1}\|_{\mathcal{V}\hookleftarrow\mathcal{V}'} < \infty$, hence*

$$\forall\ f \in \mathcal{V}'\ \exists_1 \text{ solution } u_0 \in \mathcal{V} \text{ s.t. } Au_0 = f \in \mathcal{V}' \text{ and } \|u_0\|_{\mathcal{V}} \le C\|f\|_{\mathcal{V}'}. \tag{2.147}$$

3. *All $H^m(\Omega)$-coercivity and ellipticity based existence, uniqueness and Fredholm results, including the numerical stability of all our discretization methods, remain correct with the following modifications: As a consequence of the continuous embedding of the $W^{m,p}(\Omega) \hookrightarrow H^m(\Omega), 2 \le p \le \infty$, cf. Theorem 1.26, elliptic operators defined on $W^{m,p}(\Omega)$ have a $H_0^m(\Omega)$-coercive principal part. For the $H^m(\Omega)$-coercivity we add $\mu \int_\Omega uvdx$ with sufficiently large μ.*
4. *These $W^{m,p}(\Omega)$, $2 \le p \le \infty$, results are most important for the later linearization of operators and their stability, cf. Section 2.7. Consequently, we will get for our discretization methods stability and convergence with respect to the discrete $H^m(\Omega)$ norms for $2 \le p \le \infty$. This applies to each of the different cases in Theorem 2.43.*

In fact, these last claims are a consequence of the continuous embeddings. So let the bilinear form $a(u, v)$ be $H^m(\Omega)$-coercive. Then the embeddings

$$W^{m,p}(\Omega) \hookrightarrow H^m(\Omega) \hookrightarrow W^{m,q}(\Omega),\ 2 \le p \le \infty, 1 \le q \le 2, \text{ imply} \tag{2.148}$$
$$\|u\|_{W^{m,q}(\Omega)} \le C\|u\|_{H^m(\Omega)} \le C'\|u\|_{W^{m,p}(\Omega)} \text{ for } C, C' \in \mathbb{R}_+.$$

So for $1 \le p \le 2$, the $W^{m,p}(\Omega)$-coercivity would be a consequence of the $H^m(\Omega)$-coercivity. Unfortunately the bilinear forms, cf. (2.135), usually are not defined for this case. For $2 \le p \le \infty$ we still obtain $H^m(\Omega)$-coercivity, and well-defined bilinear forms.

Again regularity results are valid for smooth or convex Ω with corners and for the interior of Ω and under conditions relaxed compared to Subsections 2.4.2, 2.4.3. For our Theorems 2.45–2.49 see [386], Theorems 9.1.16, 9.1.21, 9.1.22, 9.1.26. For $a(\cdot,\cdot)$ and its coefficients we impose the conditions:

$$\text{Let } a(\cdot,\cdot)\text{in (2.135) be } H_0^m(\Omega) - \text{elliptic, and } \forall \alpha, \beta \text{ with } |\alpha|, |\beta| \le m \tag{2.149}$$
$$\partial^\gamma a_{\alpha,\beta} \in L^\infty(\Omega)\ \forall\ \gamma \text{ with } |\gamma| \le \max\{0, t + |\beta| - m\} \text{ for } t \in \mathbb{N},$$
$$\text{or for } t \notin \mathbb{N} \text{ let } a_{\alpha,\beta} \in C^{t+|\beta|-m}(\bar{\Omega}) \text{ for } |\beta| > m - t, \text{ else } a_{\alpha,\beta} \in L^\infty(\Omega),$$
$$\text{finally, let } \Omega \in C^{t+m} \text{ in (2.5) for some } t \ge 0.$$

Theorem 2.45. Regular solutions and estimates:

1. *Let $a(\cdot,\cdot)$ in (2.135) and its coefficients satisfy (2.149). Let $s \ge 0$ and for $t \in \mathbb{N}$ let $s + 1/2 \notin \{1, 2, \ldots, m\}$ and $0 \le s \le t$ or for $t \notin \mathbb{N}$ let $0 \le s < t$. Every (not*

necessarily unique) weak solution $u_0 \in H_0^m(\Omega)$ *of*

$$a(u_0, v) = \langle f, v\rangle_{\mathcal{V}'\times\mathcal{V}} \ \forall\ v \in \mathcal{V} = H_0^m(\Omega), \ \text{for } f \in H^{-m+s}(\Omega), \tag{2.150}$$

see Corollary 2.44, is in $H^{m+s}(\Omega) \cap H_0^m(\Omega)$ *and satisfies the estimate*

$$\|u_0\|_{H^{m+s}(\Omega)} \le C_s\big(\|f\|_{H^{-m+s}(\Omega)} + \|u_0\|_{H^m(\Omega)}\big). \tag{2.151}$$

2. *If* $u_0 \in H^m(\Omega)$ *solves the inhomogeneous Dirichlet problem*

$$\partial^l u_0/\partial\nu^l = \varphi_l \in H^{m+s-l-1/2}(\partial\Omega), \ l = 0, \ldots, m-1, \tag{2.152}$$

then again $u_0 \in H^{m+s}(\Omega)$ *and we have to add to (2.151) the last term*

$$\|u_0\|_{H^{m+s}(\Omega)} \le C_s\left(\|f\|_{\ldots} + \|u_0\|_{\ldots} + \sum_{l=0}^{m-1}\|\varphi_l\|_{H^{m+s-l-1/2}(\partial\Omega)}\right). \tag{2.153}$$

3. *Specifically, for* $m = 1$ *and* $f \in H^{-1+s}(\Omega)$*, let* $u_0 \in \mathcal{V} = H^1(\Omega)$ *solve the inhomogeneous* natural boundary value problem, *cf. (2.150),*

$$a(u_0, v) = \langle f, v\rangle_{\mathcal{V}'\times\mathcal{V}} \forall v \in \mathcal{V}, f \in H^{-1+s}(\Omega), B_a u_0 = \varphi \in H^{s-1/2}(\partial\Omega). \tag{2.154}$$

Then $u_0 \in H^{1+s}(\Omega)$ *satisfies, cf. (2.151), (2.153)*

$$\|u_0\|_{H^{1+s}(\Omega)} \le C_s\big(\|f\|_{H^{-1+s}(\Omega)} + \|u_0\|_{H^1(\Omega)} + \|\varphi\|_{H^{s-1/2}(\partial\Omega)}\big).$$

Theorem 2.46. Regular solutions and estimates under weaker conditions, Nečas [509, 510]: *Let* Ω *in* $C^{0,1}$*, and* $a(\cdot,\cdot)$ *in (2.135) and its coefficients satisfy (2.149) for* $0 < s < t \le 1/2$ *and* $f \in H^{-m+s}(\Omega)$*. Then every (not necessarily unique) weak solution* $u_0 \in H_0^m(\Omega)$ *of (2.150) is in* $H^{m+s}(\Omega) \cap H_0^m(\Omega)$ *and satisfies the estimate (2.151).*

Theorem 2.47. Regular solutions for convex Ω: *Let* Ω *in (2.5) be* convex, *hence* $\Omega \in C^{0,1}$ *and* $a(\cdot,\cdot)$ *in (2.135) and its coefficients satisfy (2.149) for the case* $|\alpha| + |\beta| < 2m$, $1 = t \in \mathbb{N}$, *and*

$$a_{\alpha,\beta} \in C^{0,1}(\bar{\Omega}) \text{ for } |\alpha| = |\beta| = m. \tag{2.155}$$

Then the above result (2.151) is valid for $s = 1$.

The transition from the weak to the strong form, so from $Au_0 = f \in H^{-m}(\Omega)$ to $A_s u_0 = f_0 \in L^2(\Omega)$, is only possible for smooth enough coefficients and right-hand side. So for $m = 1$ and the boundary condition $B_a u_0 = \varphi_b$ this requires instead of $f \in H^{-1}(\Omega)$ the special form, cf. (2.109),

$$\langle f, v\rangle_{\mathcal{V}'\times\mathcal{V}} := \int_\Omega f_0 v dx + \int_{\partial\Omega} \varphi_b v ds \ \forall\ v \ \in \ \mathcal{V} := H^1(\Omega).$$

For our later studies of so-called nonconforming methods we have to be able to restrict the normal derivatives or more general boundary operators of functions to the boundary of or to interior curves in Ω. Then Remark 2.1 and Theorem 1.39 consider the continuous tracing $\gamma\partial^\alpha u$ in the sense

$$\Omega \in C^t, \text{ and } |\alpha| + 1/2 < s \le t \in \mathbb{N}, \text{ or } s < t \notin \mathbb{N} \Longrightarrow \qquad (2.156)$$
$$\gamma\partial^\alpha \in \mathcal{L}(H^s(\Omega), H^{s-|\alpha|-1/2}(\partial\Omega)).$$

Theorem 2.48. Traces of $u \in H^s(\Omega)$ to $H^{s-|\alpha|-1/2}(\partial\Omega)$: *Let* $\Omega_0 \in C^{0,1}$, $u \in H^{3/2+\epsilon}(\Omega)$, *and* $|\alpha| \le 1$. *Then* $\gamma\partial^\alpha u \in H^\epsilon(\partial\Omega)$.

Related results are found in, e.g. Taylor [618], Chapter 4, Proposition 4.5; Jonsson and Wallin [425]; Jerison and Kenig [419]; Schwab [575]; and Grisvard [373].

For the analysis of nonconforming methods we restrict the discussion to second order problems, so to $m = 1$. Then we usually assume either $H^2(\Omega)$ or $H^{3/2+\epsilon}(\Omega)$, cf. Remark 2.1.

In contrast to Theorem 2.45 the following result relates data and interior regularity, cf. [618], Chapter 5, Theorem 11.1.

Theorem 2.49. Interior regularity: *Let* $\Omega_0 \subset\subset \Omega_1 \subset\subset \Omega$ *in (2.5), let* $a(\cdot,\cdot)$ *in (2.135) be* $H_0^m(\Omega)$*-elliptic. Let the coefficients satisfy (2.149) with* Ω *replaced by* Ω_1, $s \ge 0$, *and* $t \ge s \in \mathbb{N}$ *or* $t > s$. *For* $f|_{\Omega_1} \in H^{-m+s}(\Omega_1)$ *let* u_0 *be a weak solution of (2.150). Then* $u_0|_{\Omega_0} \in H^{m+s}(\Omega_0)$ *satisfies a modified (2.151),*

$$\|u_0\|_{H^{m+s}(\Omega_0)} \le C(s, \Omega_0, \Omega_1, \Omega)\big(\|f\|_{H^{-m+s}(\Omega_1)} + \|u_0\|_{H^m(\Omega)}\big). \qquad (2.157)$$

The following result, answering the question of existence and uniqueness for a large class of boundary value problems, is a simple consequence of the equivalence of norms in Sobolev spaces. We only indicate its proof and assume

$$\text{A } \textit{special Dirichlet system } B_j u := \sum_{|\alpha| \le j-1} b_{j,\alpha}(x)\partial^\alpha u \text{ hence} \qquad (2.158)$$
$$m_1 = 0 < m_2 = 1 < \cdots < m_m = m - 1, \text{ and let } b_{j,\alpha} \in L^\infty(\Omega).$$

Theorem 2.50. Modified boundary conditions:

1. *Let* $a(\cdot,\cdot)$ *in (2.135) be* $H_0^m(\Omega)$*-elliptic, cf. Theorem 2.43. Assume (2.158) such that the* principal part of the boundary operator, *the* $B_{p,j}$, *defined by the highest derivatives* $\partial^\alpha, |\alpha| = j - 1$, *multiplied by the* $b_{j,\alpha}$, *in the boundary operator yields the rearranged*[13] *form (2.159). The normal and the* complementary tangential derivatives $\partial^{j-1}u/\partial\nu^{j-1}$ *and* $\partial^\beta u/\partial\, t^\beta$, *cf. (2.26), (2.33), are split as*

$$B_{p,j}u(x) := \sum_{|\alpha|=j-1} b_{j,\alpha}(x)\partial^\alpha u = b_j(x)\frac{\partial^{j-1}u}{\partial\nu^{j-1}} + \sum_{|\beta|=j-1} c_{j,\beta}(x)\frac{\partial^\beta u}{\partial\, t^\beta}$$
$$0 \ne b_j \in C(\partial\Omega),\ j = 1, \cdots, m, b_j \textit{ unique}. \qquad (2.159)$$

[13] The rearranged form is obtained by a transformation of the original coordinates into a system of ν and orthogonal elements $t_1, \ldots, t_{n-1}$ in the tangential plane of $\partial\Omega$. The $\partial^\alpha u : |\alpha| = j - 1$ are transformed into $\partial^{j-1}u/\partial\nu^{j-1}$ and the complementary $\partial^\beta u/\partial\, t^\beta$, and $\partial\, t^\beta = \partial\, t_1^{\beta_1} \cdots \partial\, t_{n-1}^{\beta_{n-1}} \partial\, \nu^{\beta_n}$ with $\beta_1 \cdots + \beta_{n-1} > 0$.

Then for a given $g = (B_j u = g_j)_{j=1}^m$ *unique* φ_j *exist, such that, in the trace sense,*

$$g = (B_j u = g_j)_{j=1}^m \iff (\partial^j u)/\partial\nu^j = \varphi_j, j = 0, \cdots, m-1.$$

2. *The* $a(\cdot,\cdot)$ *in (2.135) is simultaneously* $H_0^m(\Omega)$*- and* $H_B^m(\Omega)$*-elliptic with*

$$H_B^m(\Omega) := \{u \in H^m(\Omega) : B_j(u)(x) = 0 \ \forall x \in \partial\Omega, j = 1, \cdots, m\}.$$

3. *A combination with the index, regularity and embedding results in Theorems 2.28 and 2.45, 2.47 and 1.26, respectively, shows:* index $A = 0$ *for a* $H_0^m(\Omega)$*-elliptic* $a(\cdot,\cdot)$ *in (2.135) and for all the linear operators,* A*, induced by* $a(\cdot,\cdot)$*, studied in this section.*

Proof. By Theorem 2.43 it suffices to prove the claim for the principal part of the differential and boundary operators. Obviously the Dirichlet conditions in the trace sense, see Remark 2.1 and Theorem 1.37, $(\partial^j u/\partial\nu^j) = 0,\ j = 0, \ldots, m-1$, determine $\partial^\alpha u = 0,\ |\alpha| \le m-1$, and hence the conditions $B_j(u) = 0$. This remains correct for nonhomogeneous boundary conditions as well.

Vice versa we prove by induction that $B_j(u) = 0$ or $= g$ imply the homogeneous or inhomogeneous Dirichlet conditions $(\partial^j(u)/\partial\nu^j) = 0$ or $= \varphi_j, j = 0, \cdots, m-1$: in fact, $B_1(u) = 0$ or $= g_1$ is equivalent to $u = 0$ or $= (g_1/b_1)$. This information allows us to compute all the tangential partial derivatives $\partial^\alpha u/\partial\, t^\alpha, |\alpha| = 1$ along $\partial\Omega$, vanishing for homogeneous Dirichlet conditions. We combine these known $\partial^\alpha u/\partial\, t^\alpha, |\alpha| = 1$ with $B_2(u) = 0$ or $= g_2$ determining the $\partial u/\partial\nu$, again vanishing along $\partial\Omega$ for homogeneous Dirichlet conditions. Next, we combine the known $u, \partial u/\partial\nu,\ \partial^\alpha u/\partial\ t^\alpha, |\alpha| = 1$, for determining in a first step all the $\partial^\beta u/\partial\, t^\beta$ with $|\beta| = 2$. Then $B_3(u) = 0$ or $= g_3$ yields $\partial^2 u/\partial\nu^2$ along $\partial\Omega$, etc. Finally we use Lemma 2.26 to transform the inhomogeneous to homogeneous Dirichlet conditions. Then by Theorem 2.43, the *principal part* $a_p(\cdot,\cdot)$ is $H_0^m(\Omega)$-coercive. The next claims are an immediate consequence of this result. □

We finish by indicating the relation between A_s and A for the order $2m$ and general boundary conditions. Similarly to Section 2.2 we have extended $A = A_s$ in (2.134) to A in (2.135). For finding this relation we would have to determine a complementary system $S_j, j = 1, \ldots, m$, of boundary operators for the $B_j, j = 1, \ldots, m$. For these $B_j, S_j, j = 1, \ldots, m$, a second pair of $C_j, T_j, j = 1, \ldots, m$, can be determined such that a generalized Green's formula is valid, see e.g. Lions and Magenes [478] and Oden and Reddy [518], pp. 162 ff.

If we restrict the discussion to Dirichlet boundary conditions we can avoid all these technicalities, since these boundary operators are multiplied with the vanishing $\partial^\alpha u(x) = 0\ \forall|\alpha| \le m-1,\ x \in \partial\Omega$ anyway. Multiplying $A_s u$ in (2.134) with $v \in H_0^m(\Omega)$, integrating over Ω and iteratively applying the Green's formula we get, generalizing the special case $m = 1$ in (2.9), (2.16)–(2.18), the *weak form* A. Note that we only require boundary conditions for the v, but not for u in

$$\begin{aligned}
\int_\Omega vA_s u dx &= \int_\Omega v \sum_{|\alpha|,|\beta|\le m} (-1)^{|\beta|}\partial^\beta(a_{\alpha\beta}\partial^\alpha u)dx \ \forall\ u \in H^{2m}(\Omega),\ \ v\in L^2(\Omega),\\
&= (A_s u, v)_{L^2(\Omega)} \text{ with } a_{\alpha\beta}\in W^{|\beta|,\infty}(\Omega) \qquad (2.160)\\
&= a(u,v) := \langle Au, v\rangle_{\mathcal{V}'\times\mathcal{V}} = \int_\Omega \sum_{|\alpha|,|\beta|\le m} a_{\alpha\beta}\partial^\alpha u \partial^\beta v dx\\
&\qquad \forall\ u\in\mathcal{V}=H^m(\Omega),\ \ v\in\mathcal{V}_b = H_0^m(\Omega),\ \ a_{\alpha\beta}\in L^2(\Omega)\\
\text{with} \quad & |\langle Au, v\rangle_{\mathcal{V}'\times\mathcal{V}}| \le \max_{|\alpha|,|\beta|\le m} \|a_{\alpha\beta}\|_{L^\infty(\Omega)}\|u\|_{H^m(\Omega)}\|v\|_{H^m(\Omega)}.
\end{aligned}$$

By once more using the *Green's formula* we obtain for $v\in H_0^m(\Omega)\cap H^{2m}(\Omega)$

$$(A_s u, v)_{L^2(\Omega)} = a(u,v) = \langle Au, v\rangle_{\mathcal{V}'\times\mathcal{V}} = \langle u, A^d v\rangle_{\mathcal{V}\times\mathcal{V}'} = \left(u, A_s^d v\right)_{L^2(\Omega)} \qquad (2.161)$$

$$\forall u, v \in H^{2m}(\Omega)\ \cap\ H_0^m(\Omega),\ \text{with}\ A_s^d v = \sum_{|\beta|,|\alpha|\le m} ((-1)^{|\alpha|}\partial^\alpha(a_{\alpha\beta}\partial^\beta v),$$

for $a_{\alpha,\beta}\in W^{\max\{|\alpha|,|\beta|\},\infty}(\Omega)$. Summarizing these results we get, see (2.160),

$$A_s, A_s^d : \mathcal{V}_s := H^{2m}(\Omega)\cap H_0^m(\Omega) \to \mathcal{W} := L^2(\Omega) \text{ for } a_{\alpha,\beta}\in W^{\max\{|\alpha|,|\beta|\},\infty}(\Omega)$$

$$\begin{aligned}
\text{with}\quad \|A_s\|_{\mathcal{W}\leftarrow\mathcal{V}_s} &\le \sup_{\|u\|_{H^{2m}(\Omega)}=1} \left\{ \sum_{|\alpha|,|\beta|\le m} \int_\Omega |\partial^\beta(a_{\alpha\beta}(\partial^\alpha u))|dx \right\} \qquad (2.162)\\
&\le C \sum_{|\gamma|\le|\beta|,|\alpha|\le m} \|\partial^\gamma a_{\alpha\beta}\|_{L^\infty(\Omega)} \le C \sum_{|\alpha|,|\beta|\le m} \|a_{\alpha\beta}\|_{W^{\max\{|\alpha|,|\beta|\},\infty}(\Omega)}
\end{aligned}$$

$$\text{and similarly}\ \left\|A_s^d\right\|_{\mathcal{W}\leftarrow\mathcal{V}_s} \le \sum_{|\alpha|,|\beta|\le m} \|a_{\alpha\beta}\|_{W^{|\alpha|,\infty}(\Omega)},$$

$$A, A^d : \mathcal{V} := H_0^m(\Omega) \to \mathcal{V}' := H^{-m}(\Omega) \text{ for } a_{\alpha,\beta}\in L^\infty(\Omega) \text{ and}$$

$$\|A^d\|_{\mathcal{V}'\leftarrow\mathcal{V}}, \|A\|_{\mathcal{V}'\leftarrow\mathcal{V}} \le C \sum_{|\alpha|,|\beta|\le m} \|a_{\alpha\beta}\|_{L^\infty(\Omega)}. \qquad (2.163)$$

2.5 Nonlinear elliptic equations

2.5.1 Introduction

There is a vast literature on this extremely active topic, e.g. we mention that a search for regular solutions of nonlinear elliptic problems yields $\approx$ 1000 hits.[14]

[14] We list some of them: Alt and Luckhaus [23], Amann and Quittner [25], Bensoussan and Frehse [81], Boccuto and Mitidieri [99], Bochniak and Sändig [128], Chemin and Xu [169], Dachkovski [232], De Donno [271], Deng [283], Garello [337], Giaquinta, Necas, John and Stará [343], Girardi, Mastroeni

Already in Section 2.4 we have indicated the need for different concepts. We want to demonstrate this for the the case of nonlinear elliptic operators:

$$-\Delta u = f(u) \in \Omega, \ u|_{\partial\Omega} = 0 \text{ with } A = -\Delta. \tag{2.164}$$

This is a special case of the so-called *semilinear elliptic operators.* For $f(u) = e^u$ we obtain the famous *Bratu equation,* a standard example for bifurcation problems, e.g. Allgower and Jepson [22].

Either the f and Ω are smooth enough and $A = -\Delta$ is boundedly invertible, see Section 2.4. Then the preceding linear theory for A can be combined via $u = A^{-1}f(u)$ with fixed-point theorems to get results for specific semilinear equations. This is possible in Hölder spaces $A : C^{2,\gamma}(\Omega) \to C^\gamma(\Omega)$. No growth conditions for the $f(u)$ for $u \to \infty$ are necessary. This is well presented by Zeidler [676], e.g. Sections 6.5, 7.16, 8.12. We may relax the above smoothness conditions for A and impose growth conditions for the $f(u)$. In some cases this still allows us to apply the strong form $A : H^2(\Omega) \to L^2(\Omega)$.

Or the f and Ω are not smooth. Then we have to generalize the weak theory in the last sections to (2.164). The very powerful tool of weak solutions $u \in H^1(\Omega)$ for the corresponding weak linear operator A requires $Au - f(u) \in H^{-1}(\Omega)$. This is usually no longer correct for a nonlinear f. So we have to generalize $A : H^1(\Omega) \to H^{-1}(\Omega)$ to $A : W^{1,p}(\Omega) \to W^{-1,q}(\Omega), 1/p + 1/q = 1$, and have to impose appropriate growth conditions on the nonlinear function $f(u)$ for $u \to \infty$. This is necessary since the $W^{1,p}(\Omega)$ is not a Banach algebra, so, e.g. $u \in W^{1,p}(\Omega)$ does not imply $u^2 \in W^{1,p}(\Omega)$. A generalization of Sobolev spaces, e.g. towards $H^r(\Omega), W^{r,p}(\Omega), r \in \mathbb{R}$, or $A : W^{m,p}(\Omega) \to W^{-m,q}(\Omega)$, or to Besov or Besov–Triebel–Lizorkin spaces allows a deeper analysis, see e.g. Runst and Sickel [561].

A major topic in this direction are general semilinear, quasilinear and fully nonlinear elliptic equations, more generally the so-called monotone and coercive and noncoercive operators. A thorough presentation in the Sobolev and Hölder space setting, appropriate for our context, is given by Zeidler [676–678] and Taylor [618–620]. Second order equations are studied in the classical Gilbarg and Trudinger [346], and in Showalter [589] and Chen and Wu [170]. Here we essentially study homogeneous Dirichlet boundary conditions. In [589] extensions to more general quasilinear equations and special systems and to non-Dirichlet boundary conditions are discussed. Showalter [589] considers Neumann type conditions, and boundary and interior inequality constraints for the solutions. Here we combine, for special cases, the classical, strong or weak formulation for linear operators, e.g. $A : C^{2,\gamma}(\Omega) \to C^\gamma(\Omega)$, with $Au - f(u) = 0$ for

and Matzeu [347], Grunau and Sweers [377], Johnsen [424], Lakkis [468], Lassoued [470], Le Dung [308], Laurence and Stredulinsky [468], Luckhaus [481], Müller [506], Naniewicz and Panagiotopoulos [507], Necas, John and Stará [511], Nedev [512], Nguyen [633], Pacard [522], Shahgholian [582], Guo and Li [379], Uchida [639], Zhang [682] Yang [672]; compare Kozlov, Maz'ya and Rossmann [452]. Additional papers and books consider solution techniques, e.g. Altman [24]. A rather important feature of this theory of nonlinear evolution equations in Banach spaces is the following: Nonlinear problems can be solved via linearized evolution equations with many additional tricks. Dong [299], lists solvability and regularity of solutions for many applications. In his Chapter 9, fully nonlinear parabolic equations are studied.

interesting existence results. Estimates for the solutions, in particular with respect to higher norms, hence regularity, are highly complicated and often seem to be missing in textbooks. Sometimes a complete discussion in Hilbert spaces $H^1(\Omega)$ is still possible due to the specific structure of the nonlinearity. A famous example is the Navier–Stokes operator in Section 2.8.

We present in this section a short introduction to semilinear, quasilinear and fully nonlinear elliptic equations, strongly inspired by Zeidler [676–678]. Following the essential definitions in Subsection 2.5.2 we discuss special semilinear equations in Subsection 2.5.3, generalizing the above (2.164). Subsection 2.5.4 studies quasilinear equations of second order. We continue with the so-called Nemyckii operators in Subsection 2.5.5. They generalize the above $f(u)$ and allow the discussion of semilinear and quasilinear elliptic operators in Subsection 2.5.6 of order $2m$. We finish with fully nonlinear operators in Subsection 2.5.7. They are mainly studied here for specific cases, see e.g. Example 2.78 and Gilbarg and Trudinger [346].

Quasilinear problems have been studied very intensively. So their existence and uniqueness results, applied to semilinear problems, might be more general than those formulated for semilinear problems, see e.g. Theorem 2.61.

Recently nonlinear elliptic problems have been considered with more than one solution for specific nonlinearities, e.g. with continuation methods, cf. Allgower, Cruceanu and Tavener [18, 19].

For the many gaps in our knowledge of solutions of nonlinear elliptic problems, the results of this book open up new numerical techniques. Continuation and discretization methods with the "stability implies invertibility of linear operators" in Section 3.6 allow at least indications for solutions and their properties under appropriate conditions for these problems.

2.5.2 Definitions for nonlinear elliptic operators

For orders 2 and $2m$, we recall the notations, variables, and differential operators in (2.73) as, e.g.

$$\vartheta^0 = 1, \vartheta^\alpha, \Theta^0 \in \mathbb{R},\ \vartheta, \Theta = \Theta^1 \in \mathbb{R}^n, \Theta^k \in \mathbb{R}^{n_k}$$

$$\text{and } \partial^0 u = \nabla^0 u = u, \partial^\alpha u, \nabla^k = (\partial^\alpha)_{|\alpha|=k}. \tag{2.165}$$

In ∇^2 the u_{xy} and u_{yx} are different. So $\alpha = (1,1)$ appears twice, similarly in ∇^{2m}, thus defining the n^{2m} and N_{2m} below. We extend the previous linear to fully nonlinear equations and introduce their ellipticity. For nonlinear operators we consider real valued functions of the form

$$G^w : \mathcal{D}(G^w) \to \mathbb{R},\ w = (x, \Theta^0, \ldots, \Theta^{2m}) \in \mathcal{D}(G^w) \subset \Omega \times \mathbb{R}^{N_{2m}} \tag{2.166}$$

$$\text{where } \Omega \subset \mathbb{R}^n \text{ satisfies (2.5)}, \mathbb{R}^{N_{2n}} = \mathbb{R} \times \mathbb{R}^n \times \cdots \times \mathbb{R}^{n^{2n}}, \text{ and}$$

$$N_{2m} := (n^{2m+1} - 1)/(n-1) \text{ for } n > 1, \text{ and } N_{2m} := 2m+1 \text{ for } n = 1.$$

For the important case $2m = 2$ this reads as

$$G^w : \mathcal{D}(G^w) \to \mathbb{R}, w = (x, .., \Theta^2) \in \mathcal{D}(G^w) \subset \mathcal{U}_o := \Omega \times \mathbb{R} \times \mathbb{R}^n \times \mathbb{R}^{n\times n}. \quad (2.167)$$

Sometimes we replace $x, \Theta^0, \Theta^1, \Theta^2$ by x, z, p, r, and usually we need only the $n(n+1)/2$-dimensional subspace of symmetric real valued matrices in $\mathbb{R}^{n\times n}$.

In all the following nonlinear problems only functions, u, are considered with

$$u : \Omega \to \mathbb{R} \text{ and } w(x) := w_u(x) := (x, u(x), \nabla u(x), \ldots, \nabla^{2m} u(x)) \in \mathcal{D}(G^w). \quad (2.168)$$

Preparing the following definitions, we explicitly formulate different (strong) forms G of quasi- and semilinear operators of order 2. The generalization to order $2m$ is obvious.[15]

$G(\cdot) : C^2(\Omega) \to C(\Omega)$, for $u, \nabla u, G(u)$ we simplify the notation, writing

$$G(u) := \sum_{i,j=1}^{n} a_{ij}(\cdot, u, \nabla u)\partial^i\partial^j\, u + a(\cdot, u, \nabla u) \quad (2.169)$$

is the quasilinear operator, usually with $(a_{ij})_{i,j=1}^n$ symmetric,

$$G(u) := -\sum_{j=1}^{n} \partial^j (b_j(\cdot, u, \nabla u)) + b(\cdot, u, \nabla u) \quad (2.170)$$

is the quasilinear divergent operator,

$$G(u) := \sum_{i,j=1}^{n} a_{ij}(\cdot, u)\partial^i\partial^j\, u + b(\cdot, u, \nabla u), \text{ is the semilinear operator,} \quad (2.171)$$

$$G(u) := \sum_{i,j=0}^{n} a_{ij}(\cdot)\partial^i\partial^j\, u - f(u), \text{ is a special semilinear operator,} \quad (2.172)$$

$$G(u) := G^w(\cdot, u, \nabla u, \nabla^2 u), \text{ is fully nonlinear, if it is none of the above.} \quad (2.173)$$

For $G(\cdot) : C^2(\Omega) \to C(\Omega)$ we need, for the quasilinear case, real valued $a_{ij}, a, b \in C$ or $b_j \in C^1$. An extension $G(\cdot)$ to $G(\cdot) : W^{2,p}(\Omega) \to L^q(\Omega)$ requires appropriate growth conditions for the $a_{ij}(\cdot, u, \nabla u), b_j(\cdot, u, \nabla u), a(\cdot, u, \nabla u), b(\cdot, u, \nabla u), G^w$. We will come back to this point below.

Definition 2.51. Nonlinear differential operators and ellipticity:

1. *A nonlinear PDE of order $2m$ (analogously for order m) for a function u as in (2.168) and G^w as in (2.166) has the form and the solution u_0*

$$G(u) := G^w(w_u) = G^w(\cdot, u, \nabla u, \ldots, \nabla^{2m} u) \text{ and } G(u_0)(x) = 0\ \forall\ x \in \Omega \quad (2.174)$$

or a.e. for $x \in \Omega$. It is quasilinear *if G^w is an affine function of the variables Θ^{2m}, see (2.169)–(2.170). It is* semilinear *if Θ^{2m} and its coefficients in G^w, the* principal part, *define a linear operator in Θ^{2m}, hence, these coefficients only*

[15] Note the $-$ sign in (2.170) in contrast to the $+$ sign in [346], p. 87.

depend upon $x, u, \ldots, \nabla^{2m-1}u$, *see (2.171)–(2.172). In all other cases it is* fully nonlinear.

2. *Let* $x \in \mathcal{D} \subset \Omega$ *and* $w(x) = w_u(x) \in \mathcal{D}(G^w)$. *Then we require* $G \in C^1(\mathcal{D}(G^w))$ *(or e.g.* $\in W^{1,\infty}(\mathcal{D}(G^w))$*) and a* principal part *(or the symbol) of the linearized operator, compare (2.73), (2.79), with the corresponding property. So it satisfies, with the Euclidean norm,* $|\vartheta| = |\vartheta|_n$, *and (2.79)*

$$\lambda(w)|\vartheta|_n^{2m} \leq (-1)^m \sum_{|\alpha|=2m} a_\alpha(w)\vartheta^\alpha \leq \Lambda(w)|\vartheta|_n^{2m} \;\forall\; \vartheta \;\in\; \mathbb{R}^n, \tag{2.175}$$

$$\forall x \in \Omega \text{ or, a.e.} \forall x \in \Omega, \text{ with } w = w_u(x) \in \mathcal{D}(G^w), a_\alpha(w) := \left(G^w_{\vartheta^\alpha}\right)(w(x));$$

hence, $0 \leq \lambda(w)$ *for* elliptic, $0 < \epsilon \leq \lambda(w)$ *for* strongly elliptic *and* $0 < \epsilon \leq \lambda(w), \Lambda(w)/\lambda(w)$ *bounded for a specific* $w = w(x) \in \mathcal{D}(G^w)$, *or* $w \in \mathcal{D}$, *or* $\forall\; w \in \mathcal{D}(G^w)$ *for* uniformly elliptic *operators, G. The* $G^w_{\vartheta^\alpha}$ *denotes the partial derivative of the real valued function* G^w *with respect to* $\vartheta^\alpha, |\alpha| = 2m$. *For* $G^w_{\vartheta^\alpha}(w) \in L^\infty(\mathcal{D}(G))$ *this condition is only required a.e. in* Ω.

3. *For* $2m = 2$, $G^w \in C^2(\mathcal{D}(G^w))$ *we get the* symmetric *matrix*

$$-G^w_{\Theta^2}(w) := -\left(G^w_{\vartheta^\alpha}(w)\right)_{|\alpha|=2} := -\left(G^w_{\vartheta^i\vartheta^j}(w)\right)^n_{i,j=1}, \vartheta^\alpha = \vartheta^i\vartheta^j, \forall x \in \mathcal{D}. \tag{2.176}$$

Then $\lambda(w)$ *and* $\Lambda(w)$ *can be chosen as the minimal and the maximal (real)* eigenvalues *of* $G^w_{\Theta^2}(w(x))$.

2.5.3 Special semilinear and quasilinear operators

The topic of this subsection is special equations of the form (2.169)–(2.172) and their generalization to quasilinear elliptic equations of order $2m$. We study three types of results. The common structure of the first two types is their definition in Hölder spaces. This allows applying the Schauder fixed point theorem and the Schauder principle. In the special semilinear case, the nonlinearity only occurs in the lowest order term as $f(u)$ as in (2.172). For the quasilinear case, we restrict the presentation to $2m = 2$ with special coefficients a_{ij}, a, b_j, b in (2.169), (2.170), and (2.193), (2.194). The third type presents semilinear equations, where again the nonlinearity only occurs in the lowest order term as $f(u)$ as in (2.172), but only defined on nonlinear spaces.

We repeat: the existence and uniqueness results for quasilinear problems might be more general than those for semilinear problems, see e.g. Theorem 2.61.

Special semilinear operators in Sobolev and Hölder spaces

We study the orders $2m$ and 2 and impose $C^\infty(\bar{\Omega})$ and $C^{k,\gamma}(\bar{\Omega})$ conditions, respectively, with the norms $\|u\|_{C^{k,\gamma}(\Omega)}$, see (2.87). So the operators, considered here, use the linear operator A as in Section 2.4, homogeneous boundary and smoothness conditions as, see (2.88), (2.93), (2.117),

$$G(u) := Au - cf(u) = 0 \text{ in } \Omega,\ Bu = (B_j u)_{j=1}^m = 0 \text{ on } \partial\Omega \text{ with } k \in \mathbb{N}_0 \tag{2.177}$$

$$G : C^{2m+k,\gamma}(\bar\Omega) \to C^{k,\gamma}(\bar\Omega),\ B_j : C^{2m+k,\gamma}(\partial\Omega) \to C^{2m+k-m_j,\gamma}(\partial\Omega), m_j < 2m,$$

$$Au := \sum_{|\alpha|\le 2m} a_\alpha \partial^\alpha u,\ B_j u = \sum_{|\alpha|\le m_j} b_{j\alpha}\partial^\alpha u,\ j = 1,\dots,m, \text{ and}$$

B satisfies the *complementary condition,* either Dirichlet or as (2.183), with

$$C_{B,0}^{2m+k,\gamma}(\bar\Omega) := \{u \in C^{2m+k,\gamma}(\bar\Omega) : Bu|_{\partial\Omega} = 0\},\ k \in \mathbb{N}_0, \tag{2.178}$$

$$a_\alpha \in C^\infty(\bar\Omega), b_{j\alpha} \in C^\infty(\partial\Omega),\ \forall |\alpha| \le 2m, j = 1,\dots,m,\ \partial\Omega \in C^\infty, \tag{2.179}$$

$$A : C_{B,0}^{2m+k,\gamma}(\bar\Omega) \to C^{k,\gamma}(\bar\Omega) \text{ is strongly (so uniformly) elliptic and of } \textit{index } 0.$$

We use the notation $C_{B,0}^{2m+k,\gamma}(\bar\Omega)$ indicating the more general boundary conditions allowed in this context. The conditions (2.177)–(2.179) and the following (2.180)–(2.181) are formulated for $k \in \mathbb{N}_0$. In both cases we start with $k = 0$. The general $k \in \mathbb{N}_0$ will be needed for higher regularity results, see Theorem 2.56 below. Examples for (2.178), (2.179) are (2.80) and Theorem 2.50.

Similarly as above, it is possible to reduce the conditions for the coefficients and domain for the special case of differential equations of order 2, cf. Theorem 2.37 ff. and (2.83), (2.112). We assume for

$$2m = 2, m_1 = 1 \text{ let } A : C_{B,0}^{2,\gamma}(\bar\Omega) \to C^\gamma(\bar\Omega) \text{ with index } A = 0, \tag{2.180}$$

$$Au := \sum_{i,j=0}^n a_{ij}(x)\partial^i\partial^j u,\ Bu = b_0 u + \sum_{i=1}^n c_i \partial^i u,$$

$$a_{ij} \in C^{k,\gamma}(\bar\Omega), b_0, c_j \in C^{1+k,\gamma}(\bar\Omega),\ 0 < \gamma\ < 1\ \forall i,j = 0,\cdots,n, \tag{2.181}$$

$$\partial\Omega \in C^{2+k,\gamma},\ k \in \mathbb{N}_0,\ \text{for } c = (c_1,\cdots,c_n) \equiv 0 \text{ let } b_0(x) \ne 0, \text{ otherwise}$$

$$|(c(x),\nu(x))_n| > \kappa > 0\ \forall\ x \in \partial\Omega, A \text{ is strongly (so uniformly) elliptic.}$$

Leung [473] discusses different examples of interacting population reaction diffusion systems. These are PDEs of this type coupled with ODEs.

For solvability results for the semilinear equations $Au - f(u) = 0$ in (2.177) we have to extend some results for the linear equations $Au - g = 0$. We present the main idea for the proofs of the following theorems, but only indicate the necessary modifications of the proofs for the $\mathbb{R}^2$ results in [676], Proposition 6.4, to the $\mathbb{R}^n$ case: We start with a boundedly invertible $A : C_{B,0}^{2m,\gamma}(\bar\Omega) \to C^\gamma(\bar\Omega)$, see (2.178), and Proposition 2.54, with $k = 0$. Its inverse can be extended to a $K : C(\bar\Omega) \to C_{B,0}^{1,\beta}(\bar\Omega)$, hence $K|_{C^\gamma(\bar\Omega)} = (A)^{-1}$. This yields the two equivalent formulations

$$G(u)(x) = Au - cf(u) = 0 \ \Leftrightarrow\ u = Kcf(u), \tag{2.182}$$

for c, f, see Theorem 2.55. Then Schauder type results for the linear $K : C(\bar{\Omega}) \to C^{1,\beta}_{B,0}(\bar{\Omega})$ are combined with fixed-point theorems. This yields the existence and uniqueness of solutions for (2.182).

We combine the results in Subsections 2.4.2 and 2.4.3 for *linear equations of orders* $2m$ and 2 in $\mathbb{R}^n$, respectively, to obtain the continuous extension $K : C(\bar{\Omega}) \to C^{1,\beta}_{B,0}(\bar{\Omega})$, similarly to Zeidler [676]. We start summarizing conditions from Subsections 2.4.2 and 2.4.3 which guarantee a boundedly invertible *linear* $A : C^{2m,\gamma}_{B,0}(\bar{\Omega}) \to C^{\gamma}(\bar{\Omega})$. Theorem 2.28 shows that for order $2m$ we cannot directly guarantee a boundedly invertible A. For order 2 the situation is modified due to the maximum principle reflected in Theorem 2.37. For orders $2m$ and 2 and $k = 0$ we obtain Theorems 2.52 and 2.53 by combining Theorems 2.21, 2.28, 2.33, 2.39, 2.40, 2.43, Corollary 2.44 and Theorem 2.50.

In Theorem 2.50 we found a *special Dirichlet system* satisfying the complementary condition with index $A = 0$. We had transformed it into the form $(j = 1, \ldots, m)$,

$$B_j u = b_j \frac{\partial^{j-1} u}{\partial \nu^{j-1}} + \sum_{|\beta|=j-1} c_{j,\beta} \frac{\partial^{\beta} u}{\partial\, t^{\beta}} + \sum_{|\alpha|\le j-2} b_{j,\alpha} \partial^{\alpha} u, \text{ with } 0 \neq b_j(x) \forall x \in \partial\Omega. \quad (2.183)$$

Theorem 2.52. Results for the boundary conditions in (2.177)–(2.179):

1. *For an* $A : C^{2m,\gamma}_{B,0}(\bar{\Omega}) \to C^{\gamma}(\bar{\Omega})$ *of order* $2m$ *assume (2.177)–(2.179) and let* index $(A) = 0$. *If* $\lambda = 0$ *is not an* eigenvalue *of* A, *it is boundedly invertible. Otherwise* $(A + \mu)$ *with sufficiently small or large enough* $\mu \in \mathbb{R}$ *is boundedly invertible. The following estimates are valid for* $Au_0 = g \in C^{\gamma}(\bar{\Omega})$: *If* A *is boundedly invertible, compare Theorem 2.37, the* $\|u_0\|_{C^{\gamma}(\bar{\Omega})}$ *is dropped in*

$$\|u_0\|_{C^{2m,\gamma}(\bar{\Omega})} \le C\big(\|u_0\|_{C^{\gamma}(\bar{\Omega})} + \|g\|_{C^{\gamma}(\bar{\Omega})}\big). \quad (2.184)$$

2. *The preceding results are correct if the complementary or Dirichlet conditions are replaced by these modified conditions* B_j *in (2.183).*

For the particularly important case of second order equations, $2m = 2$, the boundary conditions in (2.183) can be generalized, by Theorem 2.37, as one of the

$$Bu = B_1 u = u, \text{ or } B_2 u = \partial u(x)/\partial\nu, \text{ or } B_3 u = bu + \sum_{i=1}^{n} c_i \partial^i u \text{ on } \partial\Omega \quad (2.185)$$

$$\text{with } C^{2,\gamma}_{B,0}(\bar{\Omega}) := \{u \in C^{2,\gamma}(\bar{\Omega}) : B_i u|_{\partial\Omega} = 0\}, \text{ for one of the } i = 1, 2, 3. \quad (2.186)$$

Theorem 2.53. Results for $2m = 2$ and B_i in (2.185), (2.186): *For* $A : C^{2,\gamma}_{B,0}(\bar{\Omega}) \to C^{\gamma}(\bar{\Omega})$ *of order* 2 *assume (2.180)–(2.181) with* $k = 0$, *(2.185), (2.186). Then* index $(A) = 0$ *and the results in Theorem 2.52 are valid for* $2m = 2$. *If additionally,* $\lambda = 0$ *is not an eigenvalue of* (A, B), *e.g. the* a_{ij} *satisfy* $a_{ij} \equiv 0$, *for* $ij = 0, i + j > 0$, *and*

$$a_{00}(x) \ge 0 \ \forall\ x\ \in \Omega \textit{ for } B_1 u \textit{ and for } B_3 u \textit{ with } b(x) > 0 \ \forall\ x \in \partial\Omega, \textit{ or} \quad (2.187)$$

$$a_{00}(x) > 0 \ \forall\ x\ \in \Omega \textit{ for } B_2 u \ \forall\ x \in \partial\Omega,$$

then $A : C^{2,\gamma}_{B,0}(\bar{\Omega}) \to C^{\gamma}(\bar{\Omega})$ *is boundedly invertible and (2.184) holds for* $2m = 2$.

We combine these results, in particular Theorems 2.32, 2.37, 2.52 and 2.53, with the results in Zeidler [676], pp. 233 ff., see his Proposition 6.4.

Proposition 2.54. *Let* $A : C^{2m,\gamma}_{B,0}(\bar{\Omega}) \to C^{\gamma}(\bar{\Omega})$ *in (2.177), (2.186), with* B *in (2.178), be boundedly invertible, and yield* $K = (A)^{-1} : C^{\gamma}(\bar{\Omega}) \to C^{2m,\gamma}_{B,0}(\bar{\Omega})$. *This is correct under the conditions in Theorems 2.52 and 2.53, hence for orders* $2m$ *and* 2, *respectively Then* K *can be uniquely extended to a continuous linear operator*

$$K : C(\bar{\Omega}) \to C^{1,\beta}_{B,0}(\bar{\Omega}) \text{ for a fixed } \beta,\ 0 < \beta < 1. \tag{2.188}$$

Proof. There are only minor differences compared to the proof in Zeidler [676], pp. 238, for $2m = 2$. We combine Theorems 2.30 and 2.33, in particular (2.94) and (2.98) to generalize Zeidler's condition (A) as

$$\|Kg\|_{W^{2m,p}(\Omega)} \leq C\|g\|_{L^p(\Omega)} \quad \forall g \in L^p(\Omega),\ 1\ <\ p. \tag{2.189}$$

By Theorem 1.26 the embeddings

$$C(\bar{\Omega}) \hookrightarrow L^p(\Omega) \text{ and } W^{2m,p}(\Omega) \hookrightarrow C^{1,\beta}(\Omega)$$

are continuous for $p > n/(2m - 1 - \beta)$. So (2.189) is modified into

$$\|Kg\|_{C^{1,\beta}(\bar{\Omega})} \leq C\|g\|_{C(\bar{\Omega})} \forall g \in C^{\gamma}(\bar{\Omega}).$$

By the Weierstrass approximation theorem, $C^{\gamma}(\bar{\Omega})$ is dense in $C(\bar{\Omega})$. So K can be extended as claimed. □

Now we are ready for applying this result to special semilinear equations, see Zeidler [676], pp. 233 ff., his Proposition 6.7, and Corollaries 6.8, 6.9:

Theorem 2.55. Existence and uniqueness of solutions for semilinear equations:

1. *Let* $A : C^{2m,\gamma}_{B,0}(\bar{\Omega}) \to C^{\gamma}(\bar{\Omega})$ *in (2.177), with* B *in (2.178), satisfy the conditions of Proposition 2.54 and assume:*

$$\text{for fixed } R > 0 \text{ let } f : [-R, R] \to \mathbb{R},\ f \in C^{\beta},\ 0 < \beta < 1. \tag{2.190}$$

Then for every sufficiently small $|c|, c \in \mathbb{R}$, *the semilinear boundary value problem*

$$(Au - cf(u))(x) = 0 \forall x \in \Omega, B_j u|_{\partial\Omega} = 0,\ j = 1, \ldots, m, \tag{2.191}$$

has a unique solution $u_0 \in C^{2m,\gamma}(\bar{\Omega}) \cap C^m_{B,0}(\bar{\Omega})$ *with* $0 < \gamma < 1$, *see (2.186).*

2. *For* $c \geq 0$ *and* $f(u) \geq 0$ *on* $[0, R]$, *the solution is nonnegative,* $u_0 \geq 0$ *on* $\bar{\Omega}$.
3. *For a Lipschitz-continuous* f *there exists exactly one solution with* $|u_0(x)| \leq R$ *on* $\bar{\Omega}$. *It can be determined by the convergent iteration process*

$$u^k \in C^{2m,\gamma}_{B,0}(\bar{\Omega}) := C^{2m,\gamma}(\bar{\Omega}) \cap C^m_{B,0}(\bar{\Omega}),\ u^0 := 0,\quad Au^{k+1} = cf(u^k),\ k \in \mathbb{N}_0.$$

Zeidler's proof [676], pp. 240,241, remains unchanged, if only his Proposition 6.4 is replaced by our Proposition 2.54. Modifications for f are discussed by von Wahl [658], and Leung [473], pp. 82 ff.

Higher regularity for this and later theorems is obtainable via different kinds of so-called *bootstrap arguments or inherited regularity,* cf. [676]. They are chosen according to the results available for the specific situation, usually based upon inductive arguments. Our approach here uses the regularity results for the corresponding linear elliptic equations, presented in Section 2.4, see Theorems 2.28, 2.30–2.33, 2.38, 2.39. We modify a technique of Hackbusch [387], Remark 9.1.2. We will refer to this result later on, hence, we formulate it as

Theorem 2.56. Regularity, inherited to compact and semilinear perturbations:

1. *Choose $k \in \mathbb{N}_0$ in (2.177)–(2.179) for A of order $2m$ or (2.180)–(2.181) for order 2. Then Theorems 2.52, 2.53 and the estimates (2.184) have the stronger form* $A : C^{2m+t,\gamma}_{B,0}(\bar\Omega) \to C^{t,\gamma}(\bar\Omega), 0 \le t \le k,$

$$Au_0 = g \text{ satisfies } \|u_0\|_{C^{2m+t,\gamma}(\bar\Omega)} \le C\big(\|u_0\|_{C^{2m,\gamma}(\bar\Omega)} + \|g\|_{C^{t,\gamma}(\bar\Omega)}\big). \tag{2.192}$$

   Note that for boundedly invertible A the term $\|u_0\|_{C^{2m,\gamma}(\bar\Omega)}$ is deleted.
2. *Let δA be a linear or nonlinear differential operator of order $\le 2m-1$, such that $\delta A : C^{r+2m,\gamma}(\bar\Omega) \cap C^{2m,\gamma}_{B,0}(\bar\Omega) \to C^{r+1,\gamma}(\bar\Omega)$ is bounded for all $r = 0,1,\ldots,k-1$. This implies $A + \delta A : C^{2m+t,\gamma}_{B,0}(\bar\Omega) \to C^{t,\gamma}(\bar\Omega)$ and (2.192) for the solution u_0 of $(A+\delta A)u_0 = g$.*
3. *Under the conditions of Theorem 2.55 let the f in (2.190) satisfy*

$$f^{(r)} : u \in C^{r+2m,\gamma}(\bar\Omega) \to f^{(r)}(u) := f(u(\cdot)) \in C^{r+1,\gamma}(\bar\Omega), \quad r = 0,1,\ldots,k-1,$$

   e.g. these $f^{(r)}$ are in $C^{r+1,\gamma}(\mathbb{R})$. Then the solutions u_0 of (2.191) are $u_0 \in C^{2m+t,\gamma}_{B,0}(\bar\Omega)$. The norms of u_0 can be estimated as in (2.192).

Proof. By induction: For $k = t = 0$ the result and estimates (2.192) are a consequence of Theorems 2.45 and 2.53. Now we inductively assume (2.192) for the solution $(A+\delta A)u_0 = g \in C^{t-1,\gamma}(\bar\Omega)$ and for $0 \le t-1 \le k-1$ for $k \in \mathbb{N}$. Consequently, $u_0 \in C^{t-1+2m,\gamma}(\bar\Omega)$ solves $Au_0 = \tilde g := g - \delta A u_0$ as well. By our inductive assumption for $A + \delta A$ and finally the condition for A we estimate

$$\begin{aligned}
\|u_0\|_{C^{t-1+2m,\gamma}(\bar\Omega)} &\le C'_{t-1}\big(\|u_0\|_{C^{2m,\gamma}(\bar\Omega)} + \|g\|_{C^{t-1,\gamma}(\bar\Omega)}\big), \text{ hence,}\\
\|\tilde g\|_{C^{t,\gamma}(\bar\Omega)} &\le \|\delta A\|_{C^{t,\gamma}(\bar\Omega)\leftarrow C^{t-1+2m,\gamma}}\|u_0\|_{C^{t-1+2m,\gamma}} + \|g\|_{C^{t,\gamma}(\bar\Omega)}\\
&\le C''_{t-1}\big(\|u_0\|_{C^{2m,\gamma}(\bar\Omega)} + \|g\|_{C^{t,\gamma}(\bar\Omega)}\big) \text{ and, by (2.192) for } A,\\
\|u_0\|_{C^{t+2m,\gamma}(\bar\Omega)} &\le C'_t\big(\|u_0\|_{C^{2m,\gamma}(\bar\Omega)} + \|g\|_{C^{t,\gamma}(\bar\Omega)}\big), \text{ hence, the claim.}
\end{aligned}$$

The final remark about the solutions u_0 of (2.192) is obvious. □

Special quasilinear operators of order 2 in Hölder spaces

The previous results were, modulo the differences in Theorems 2.52 and 2.53, valid for orders $2m$ and 2. The following theorem is restricted to quasilinear uniformly elliptic equations of order 2 in $\mathbb{R}^2$, a special form of (2.169). So we assume

$$n = 2: \quad G(u) := \sum_{i,j=1}^{2} a_{ij}(x,u,\nabla u)\partial^i\partial^j\, u, \quad (a_{ij})_{i,j=1}^2 \text{ symmetric.} \tag{2.193}$$

Similarly to (2.180), (2.181) we assume for the domain, the defining coefficients and the inhomogeneous boundary function, g,

$$\begin{aligned}&\text{for } 0<\gamma<1, \text{ let } \Omega \text{ in (2.5)}, \partial\Omega \in C^{2,\gamma} \text{ be a single curve with} \\ &\text{positive curvature } \forall\, x\in\partial\Omega, \text{ and in (2.193), let } a_{ij}\in C^{\gamma}(\bar{\Omega}\times\mathbb{R}^3), \\ &\forall i,j=1,2,\quad G \text{ be strongly (hence uniformly) elliptic, and } g\in C^{2,\gamma}(\bar{\Omega}).\end{aligned} \tag{2.194}$$

Theorem 2.57. Existence and uniqueness of solutions for (2.193), Zeidler [676], pp. 246 ff., Gilbarg and Trudinger [346], pp. 312: *Let the nonlinear G in (2.193) satisfy (2.194). Then the boundary value problem*

$$G(u)(x)=0\ \forall\ x\ \in\ \Omega,\ u(x)=g(x)\ \forall\ x\ \in\ \partial\Omega \tag{2.195}$$

has a solution $u_0\in C^{2,\gamma}(\Omega)\cap C^0(\bar{\Omega})$ with $\sup_{x\in\Omega}|\nabla u(x)|\le K$.

For regularity results we refer to the discussion at the end of this subsection and of 2.5.4 and 2.5.6. Many other related results are listed, e.g. in Zeidler [676], Gilbarg and Trudinger [346], and Runst and Sickel [561]; or the linearization techniques in Section 2.7 can be combined with the Fredholm alternative and the inherited regularity in Theorem 2.56 extending these results.

Special semilinear operators in nonlinear spaces

For this last type of results we return to semilinear equations. For $a_{\alpha\beta}\in W^{|\beta|,\infty}(\Omega)$ we assume the linear operator A in *strong divergence form,* cf. (2.134)

$$A=A_s:H^{2m}(\Omega)\to L^2(\Omega),\ Au=\sum_{|\alpha|,|\beta|\le m}(-1)^{|\beta|}\partial^{\beta}(a_{\alpha\beta}\partial^{\alpha}u). \tag{2.196}$$

We want to solve the *Dirichlet boundary value problem*

$$G(u)(x):=(Au-f(u))(x)=g(x)\ \forall\ x\in\ \Omega \tag{2.197}$$

$$\partial^{\alpha}u(x)=0\ \forall|\alpha|\le m-1,\ \forall\ x\in\ \partial\Omega. \tag{2.198}$$

As usual we multiply (2.197) with $v\in H_0^m(\Omega)$, integrate over Ω and iteratively use partial integration, see (2.9). We find the following *generalized form,* related to

$$\begin{aligned}&Au_0-f(u_0)=g, u_0\in\mathcal{V}=H_0^m(\Omega),\\ &(A_s u-f(u)-g,v)_{L^2(\Omega)}=a(u,v)-a_1(u,v)-(g,v)\forall u\in H^{2m}(\Omega), v\in\mathcal{V}\\ &a(u,v)=\langle Au,v\rangle_{\mathcal{V}'\times\mathcal{V}}=\int_{\Omega}\sum_{|\alpha|,|\beta|\le m}a_{\alpha\beta}\partial^{\alpha}u\partial^{\beta}v dx \text{ with } a_{\alpha\beta}\in L^{\infty}(\Omega),\end{aligned} \tag{2.199}$$

$$a_1(u,v)=\int_{\Omega}f(u(x))v(x)dx,\ (g,v)=\int_{\Omega}g(x)v(x)dx. \tag{2.200}$$

This shows that we certainly have to impose conditions upon f,g and choose appropriate u,v such that $a_1(u,v),(g,v)$ are well defined. So we assume for G and $a(\cdot,\cdot)$ in

(2.197) and (2.199),

$$a(\cdot,\cdot) \text{ is } \mathcal{V}-\text{coercive}, \ a_{\alpha\beta}\in L^\infty(\Omega)\ \forall\ |\alpha|,|\beta|\le m,\ \Omega \text{ satisfies } (2.5),$$

$$f:\mathbb{R}\to\mathbb{R} \text{ continuous with } a\in\mathbb{R} \text{ fixed}, \ (f(u)+a)u\le 0\ \forall u\in\mathbb{R}, \tag{2.201}$$

$$\mathcal{V}=H_0^m(\Omega), \mathcal{D}(G):=\{u\in\mathcal{V}: h:=(f(u)+a)u\in L^1(\Omega)\}\ne\mathcal{V} \text{ and} \tag{2.202}$$

$$\mathcal{W}=H_0^m(\Omega)\cap H^k(\Omega), k>n/2, k':=\max\{k,m\} \text{ and } \|v\|_{\mathcal{W}}=\|v\|_{H^{k'}(\Omega)}.$$

We still require (2.198), but instead of (2.197), we solve the generalized problem

$$\text{determine } u_0\in\mathcal{D}(G) \text{ s.t. } a(u_0,v)-a_1(u_0,v)-(g,v)=0\ \forall v\ \in\mathcal{W}. \tag{2.203}$$

The $a(u_0,v), a_1(u_0,v), (g,v)$ are well defined for $u_0\in\mathcal{D}(G), v\in\mathcal{W}$ but the domain $\mathcal{D}(G)$ is not a linear space. Then Zeidler's proof of his Proposition 27.21, [678], p. 605, for $Au=-\Delta u$ can be easily generalized to our situation:

Theorem 2.58. Existence and uniqueness for solutions of (2.197), (2.198): *Under the conditions (2.201), (2.202), e.g. for $f(u)=-e^u$, this boundary value problem has for all $g\in L^2(\Omega)$ a generalized solution in the sense of (2.203).*

Remark 2.59. *The condition $(f(u)+a)u\le 0\ \forall u\in\mathbb{R}$ in (2.201) is satisfied if $f:\mathbb{R}\to\mathbb{R}$ is continuous and monotonically decreasing, e.g. for $f(u)=-e^u, a=f(0)$. For $Au=-\Delta u$ this is a standard example in (numerical) bifurcation, the so-called* Bratu equation, *cf. e.g. Allgower and Jepson [22] and Govaerts [357], p. 21 ff.:*

$$-\Delta e+e^u=0 \ \textit{in}\ \Omega \ \textit{and}\ u=0 \ \textit{on}\ \partial\Omega. \tag{2.204}$$

We do not go into details, but only indicate some properties: This equation has mainly be discussed on the basis of very rough difference methods. So it is not quite clear whether the bifurcations scenarios in the many papers go back to spurious solutions, *which do not correspond to solutions of the exact equation, or reflect the exact (2.204) scenarios. In fact Peitgen, Saupe and Schmitt [525] have shown that difference methods applied to (2.204) always posess* spurious solutions, *moving to ∞ with the step size $h\to 0$. For $n=1,2$ one finds the situation indicated in Govaerts [357], p. 21 ff., with a turning and a bifurcation point. For $n=3$ the solutions oscillate around a specific value of λ infinitely often with decreasing amplitudes. For $n\ge 4$ there are no longer any solutions. For analytical results, cf. Joseph and Lundgren [428].*

Here the situation is very similar to Theorem 2.56. There we studied $A: C_0^{2m+t,\gamma}(\bar\Omega)\to C^{t,\gamma}(\bar\Omega)$; here we consider the weak form $A: H^{m+s}(\Omega)\cap H_0^m(\Omega)\to H^{-m+s}(\Omega)$. We nevertheless formulate, for the reader's convenience, the updated result once more. We omit the proof, since it is nearly identical in both cases, except that the $C_0^{2m+t,\gamma}(\bar\Omega)$ have to be replaced by $H^{m+s}(\Omega)$, etc.

Theorem 2.60. Regular solutions for linear and semilinear equations:

1. *Choose* $t \in \mathbb{N}_0$ *such that the conditions (2.149) and in Theorem 2.45 are satisfied. Then* $A : H^{m+s}(\Omega) \cap H_0^m(\Omega) \to H^{-m+s}(\Omega)$ *for* $0 \le s \le t$ *(or even for* $0 \le s < t \notin \mathbb{N}$*) and we obtain for the solution*

$$u_0 \text{ of } Au_0 = g \text{ that } \|u_0\|_{H^{m+s}(\Omega)} \le C\big(\|u_0\|_{H^m(\Omega)} + \|g\|_{H^{-m+s}(\Omega)}\big). \quad (2.205)$$

   Note that for boundedly invertible A *the term* $\|u_0\|_{H^m(\Omega)}$ *is deleted.*
2. *Let* δA *be a linear or nonlinear differential operator of order* $\le 2m-1$ *such that* $\delta A : H^{r+m}(\Omega) \cap H_0^m(\Omega) \to H^{r+1-m}(\Omega)$ *is bounded for all* $r = 0, 1, \ldots, s-1$. *This implies the same* $A + \delta A : H^{m+s}(\Omega) \cap H_0^m(\Omega) \to H^{s-m}(\Omega)$ *and (2.205) for the solution* u_0 *of* $(A + \delta A)u_0 = g$. *In particular, the condition for* δA *is satisfied, if the linear* δA *is induced by the lower order terms* $|\alpha| + |\beta| < 2m$ *of the bilinear form (2.199) and the* $a_{\alpha\beta}$ *are smooth enough,* $a_{\alpha\beta} \in C^{\max\{0, s-2m+|\beta|\}}(\Omega)$.
3. *Under the conditions of Theorem 2.58 let the* f *in (2.201) and its derivatives satisfy*

$$f^{(r)} : u \in H^{r+m}(\Omega) \to f^{(r)}(u(\cdot)) \in L^\infty(\Omega), \quad r = 0, 1, \ldots, s-1,$$

   e.g. these $f^{(r)}$ *are bounded. Then the solutions* u_0 *of (2.191) are* $u_0 \in C_0^{2m+t,\gamma}(\bar{\Omega})$. *The norms of* u_0 *can be estimated as in (2.205).*

2.5.4 Quasilinear elliptic equations of order 2

Throughout this subsection γ *always denotes a real* $0 < \gamma < 1$. *Furthermore, we will use* $W^{m,q}(\Omega)$ *instead of* $W^{m,p}(\Omega)$, *since we need the* p *in the triple* x, z, p *below.*

For quasilinear equations of second order, two competing approaches have been worked out in the monographs consulted here, essentially Ladyženskaja and Uralceva [464], Gilbarg and Trudinger [346], Zeidler [678], Taylor [620], Showalter [589], and Chen and Wu [170]. We present in this subsection results based upon the ellipticity of these equations in the sense of Definition 2.51. Then maximum principles, fixed point theorems, Hölder estimates for the gradient, and boundary gradient estimates are combined yielding global, interior and boundary estimates and corresponding existence results. These comparison results yield fewer uniqueness results than the second approach in Subsection 2.5.6. This uses Nemyckii and monotone operators. It replaces the ellipticity of the equations, strongly related to the coercivity in the sense of Definition 2.65, by a generalized coercivity in the sense of (2.286) and monotonicity in (2.289). Relation (2.289) implies the standard ellipticity condition only for $q = 2$, hence in $H^m(\Omega)$, cf. Proposition 2.121. The theory is in many respects independent of the order 2 or $2m$. So we will formulate it in Subsection 2.5.6 below, only for the order $2m$. It allows us existence results for a large class of operators and uniqueness under natural additional conditions. This difference influences our numerical methods. The approach via ellipticity allows the proofs for stability and convergence via linearization. For the monotonic situation this remains possible for all cases as well, but modifications for stabiliy have to be imposed, see Corollary 2.44 and Section 2.7 below.

We repeat for order 2 the different quasilinear equations, cf. (2.170), (2.169),

$$Gu = \sum_{i,j=1}^{n} a_{ij}(x,u,\nabla u)\partial^i\partial^j u + a_{00}(x,u,\nabla u) = 0,\ \ u \in C^{2,\gamma}(\Omega), \text{general} \tag{2.206}$$

$$Gu = \sum_{i=0}^{n}(-1)_{i>0}\partial^i(A_i(x,u,\nabla u)) = 0 \text{ on } \Omega, u \in W^{2,q}(\Omega), \text{divergent form,} \tag{2.207}$$

$$\langle Gu, v\rangle_{\mathcal{V}'\times\mathcal{V}} = \int_\Omega \sum_{i=0}^{n} A_i(x,u,\nabla u)\partial^i v dx = 0\ \forall\ v \in \mathcal{V} = W_0^{1,q}(\Omega),\ \text{weak} \tag{2.208}$$

$$\text{form, with } u \in \mathcal{V} = W_0^{1,q}(\Omega),\quad \text{sometimes } u \in W^{1,q}(\Omega), u|_{\partial\Omega} = \varphi|_{\partial\Omega}.$$

The following results are examples of existence and uniqueness results, serving as an illustration of the scope of the theory in Gilbarg and Trudinger [346]. They are obtained by appropriate combinations of tools formulated in the introductory remarks of this subsection. For special cases, [346], Theorems 10.1 and 10.7 immediately imply uniqueness. This is combined below with uniqueness results from Taylor [620]. Simplifying the notation we use, in this subsection, the standard $(z,p) \in \mathbb{R}^{n+1}$ instead of $(\Theta^0, \Theta) \in \mathbb{R}^{n+1}$ with $N_1 = n+1$, cf. (2.73). Denote the minimal and maximal *real eigenvalues* of $\{a_{ij}(x,z,p)\}_{i,j=1}^{n}$, cf. (2.206), an assumed positive matrix, as $\lambda(x,z,p)$ and $\Lambda(x,z,p)$ and $\vartheta \in \mathbb{R}^n$. We called (2.206) *elliptic*, etc. in (x,z,p), iff

$$0 < \lambda(x,z,p)|\vartheta|^2 < \sum_{i,j=1}^{n} a_{ij}(x,z,p)\vartheta^i\vartheta^j < \Lambda(x,z,p)|\vartheta|^2 \forall 0 \neq \vartheta \in \mathbb{R}^n, \tag{2.209}$$

with $a_{ij}(x,z,p), A_i(x,z,p) \in \mathbb{R}, (x,z,p) \in \mathcal{U}_o \subset \Omega \times \mathbb{R} \times \mathbb{R}^n, p = (p_1, \cdots, p_n)$.

Obviously (2.207) implies (2.206) for $A_i(\cdot,\cdot,\cdot) \in C^1$ with $a_{ij}(x,z,p) = -\partial(A_i(x,z,p))/(\partial p_j)$. Conversely, it is usually not possible to transform (2.206) into (2.207).

If in (2.209) the (x,z,p) is replaced by $(x,u(x),\nabla u(x))$ then this (uniform) ellipticity holds (locally) with respect to u. For the following growth conditions, we introduce the operators:

$$\text{For } u \in C^1(\Omega \times \mathbb{R} \times \mathbb{R}^n) \text{ let } \delta u := \left(\partial_z + |p|^{-2}\sum_{i=1}^{n} p_i\partial_{x_i}\right)u,\ \ \bar{\delta}u := \left(\sum_{i=1}^{n} p_i\partial_{p_i}\right)u. \tag{2.210}$$

Uniformly elliptic equations

We distinguish uniformly and nonuniformly elliptic equations in general and in divergence form. For the different forms we require growth and regularity conditions, assuming for (2.211)–(2.219), $|p| \to \infty$, uniformly for $x \in \Omega$, and bounded z, with

$(x, z, p) \in \mathcal{U}_o$, cf. (2.209), (2.210).

For (2.206): G unif. ell. on $\bar{\Omega}$, $a_{00}, a_{ij} \in C^1(\bar{\Omega} \times \mathbb{R} \times \mathbb{R}^n)$, $\partial\Omega \in C^{2,\gamma}$, (2.211)

$\varphi \in C^{2,\gamma}(\partial\Omega)$ and assume the estimates:

$$a_{ij}, \bar{\delta} a_{ij} = \mathcal{O}(\lambda),\ \delta a_{ij} = o(\lambda),\ a_{00}, \bar{\delta} a_{00} = \mathcal{O}(\lambda|p|^2),\ \delta a_{00} = o(\lambda|p|^2).$$

For (2.207): G unif. ell. on $\bar{\Omega}$, $A_i \in C^{1,\gamma}(\bar{\Omega} \times \mathbb{R} \times \mathbb{R}^n)$, $A_0 \in C^\gamma(\bar{\Omega} \ldots)$, (2.212)

$\partial\Omega \in C^{2,\gamma}, \varphi \in C^{2,\gamma}(\partial\Omega)$, and assume the estimates: $\exists \tau > -1:$

$$|p|^\tau = \mathcal{O}(\lambda),\ \sum_{i=1}^n \partial_{p_i} A_i = \mathcal{O}(|p|^\tau),\ |p|\partial_z A_i,\ \sum_{i=1}^n \partial_{x_i} A_i, A_0 = \mathcal{O}(|p|^{\tau+2})$$

Specific results are possible for the following special case of the general form (2.206), noting that $i, j \geq 1$:

$$Gu = Gu = \sum_{i,j=1}^n a_{ij}(x, u)\partial^i \partial^j u + a_{00}(x, u, \nabla u) = 0 \text{ on } \Omega, a_{ij} \in C^1(\Omega, \mathbb{R}). \tag{2.213}$$

Then a transformation into divergence form (2.207) is possible by choosing

$$A_i(x, z, p) := -\sum_{j=1}^n a_{ij}(x, z) p_j, \text{ for } i = 1, \ldots, n,$$

$$A_0(x, z, p) = -a_{00}(x, z, p) + \sum_{i,j=1}^n \partial_z a_{ij}(x, z) p_i p_j + \sum_{i,j=1}^n \partial_{x_i} a_{ij}(x, z) p_j.$$

For (2.213): G unif. ell. on $\bar{\Omega}$, $a_{ij} \in C^1(\bar{\Omega} \times \mathbb{R})$, $a_{00} \in C^\gamma(\bar{\Omega} \times \mathbb{R} \times \mathbb{R}^n)$ (2.214)

$\partial\Omega \in C^{2,\gamma}, \varphi \in C^{2,\gamma}(\partial\Omega)$, and the above p, x, z, assume $a_{00} = \mathcal{O}(|p|^2)$,

$$\exists \mu_1, \mu_2 > 0: \frac{a_{00}(x, z, p)\text{sign } z}{\mu_1|p| + \mu_2}|p|^2 \leq \mathcal{E}(x, z, p) := \sum_{i,j=1}^n a_{ij}(x, z, p) p_i p_j. \tag{2.215}$$

Note that for (2.213) these $a_{ij}(x, z, p), i, j = 1, \ldots, n$, are independent of p.

Nonuniformly elliptic equations

We no longer require G to be uniformly elliptic on $\bar{\Omega}$, and Λ/λ is no longer bounded. This situation is important, e.g. for minimal surface operators. We assume the original $a_{ij}(x, z, p)$ can be split, according to

$$a_{ij}(x, z, p) = a^*_{ij}(x, z, p) + (p_i c_j(x, z, p) + p_j c_i(x, z, p))/2, \quad i, j = 1, \ldots, n, \tag{2.216}$$

with $a_{ij}, a^*_{ij}, c_i \in C^1(\bar{\Omega} \times \mathbb{R} \times (\mathbb{R}^n \setminus \{0\})$, and a positive symmetric matrix $\left(a^*_{ij}\right)^n_{i,j=1}$. Furthermore, let λ^* indicate the lower bound in (2.209) defined by the matrix $\left(a^*_{ij}\right)^n_{i,j=1}$

instead of the original $(a_{ij})_{i,j=1}^n$ and use $\mathcal{E}$ in (2.215). We impose the structure and regularity conditions, modifying the above (2.211), (2.212)

For (2.206): The first two lines in (2.211) remain, except delete "unif.ell.on",

$$\text{change the last line into } |p|a_{ij}, a_{00}, \bar{\delta}a_{00} = \mathcal{O}(\mathcal{E}),\ \bar{\delta}a_{ij}^* = \mathcal{O}(\sqrt{\lambda^*\mathcal{E}}/|p|),$$

$$\delta a_{ij}^* = o(\sqrt{\lambda^*\mathcal{E}}/|p|),\ \delta a_{00} = o(\mathcal{E}),\ \mu_1, \mu_2 > 0 \text{ exist, s.t.} \tag{2.217}$$

$$\frac{a_{00}(x,z,p)\text{sign } z}{\mu_1|p| + \mu_2}|p|^2 \le \mathcal{E}(x,z,p) = \sum_{i,j=1}^n a_{ij}(x,z,p)p_ip_j\ \forall(x,z,p\neq 0) \in \mathcal{U}_o.$$

If we replace uniformly elliptic by elliptic, we impose

For (2.206): The first two lines in (2.211) remain, but require G elliptic on $\bar{\Omega}$,

$$\text{change the last line into } \Omega \text{ is uniformly convex} \tag{2.218}$$

$$a_{00} = o(|p|\Lambda),\ \delta a_{00} \le |\mathcal{O}\left(a_{00}^2/|p|^2\mathcal{T}^*\right), \mathcal{T}^* := \text{trace}\left[a_{ij}^*\right].$$

Finally we consider the divergence form (2.207). In the last two lines of (2.219) we use the general form (2.206) and introduce a family of equations, depending upon σ with $0 \le \sigma \le 1$. We assume:

For (2.207): The first two lines in (2.212) remain, but require G elliptic on $\bar{\Omega}$,

$$\text{change the last line into } \Omega \text{ is uniformly convex} \tag{2.219}$$

$$\exists\tau \in \mathbb{R} \text{ s.t. } |p|^\tau = \mathcal{O}(\lambda),\ |p|\partial_z A_i, \sum_{i=1}^n \partial_{x_i} A_i, A_0 = o(|p|^{\tau+1}),\ \text{for}$$

$$G_\sigma u := \sum_{i,j=1}^n a_{ij}(x,u,\nabla u)\partial^i\partial^j u + \sigma a_{00}(x,u,\nabla u) = 0 \text{ in } \Omega, u = \sigma\varphi \text{ on } \partial\Omega,$$

$$\text{let the set of solutions } u_0 = u_{0,\sigma} \in C^{2,\gamma}(\partial\Omega) \text{ be uniformly bounded in } \Omega. \tag{2.220}$$

Existence for these quasilinear elliptic equations

We summarize the existence results of the main cases in Gilbarg and Trudinger's [346] Theorems 15.10–15.15. We omit his results for problems with boundary curvature conditions.

Theorem 2.61. Existence for quasilinear operators: *For (2.206), (2.207) or (2.213) on a bounded Ω in (2.5), let one of the corresponding conditions be satisfied: For* uniformly elliptic equations *impose one of (2.211), (2.212) or (2.214), for* nonuniformly elliptic equations *impose one of (2.217), (2.218), (2.219). Then there exists a solution $u_0 \in C^{2,\gamma}(\bar{\Omega})$ of the Dirichlet problem $Gu_0 = 0$ in Ω, and $u_0|_{\partial\Omega} = \varphi$.*

For regularity results we refer to the end of this section and to subsection 2.5.6.

Uniqueness results

For the following results, we have to impose additional conditions:

$$\textbf{For (2.206)}: G \text{ unif. ell. with respect to } u, \text{consider solutions} u \in \mathcal{V} = C(\bar{\Omega}) \cap C^2(\Omega),$$
$$a_{ij} \text{ independent of } z, a_{00} \text{ nonincreasing in } z \text{ for each } (x,p) \in \Omega \times \mathbb{R}^n,$$
$$\text{and } a_{00}, a_{ij} \in C^1 \text{ with respect to the } p \text{ variable in } \mathcal{U}_o \subset \Omega \times \mathbb{R} \times \mathbb{R}^n. \qquad (2.221)$$

A combination of (2.211) with (2.221) shows that $\delta a_{ij} = o(\lambda), \delta a_{00} = o(\lambda|p|^2)$ in (2.211) would be implied by $\sum_{k=1}^n \partial_{x_k} a_{ij} = \mathcal{O}(\lambda)$, $\sum_{i=1}^n \partial_{x_i} a_{00} = \mathcal{O}(\lambda|p|^2)$.

$$\textbf{For (2.207)}: \quad G \text{ elliptic in } \Omega, \quad \text{consider solutions } u \in \mathcal{V} = C^1(\bar{\Omega}), \qquad (2.222)$$
$$A_0, A_i \in C^1 \text{ with respect to the } z, p \text{ variables in } \bar{\Omega} \times \mathbb{R} \times \mathbb{R}^n,$$
$$A_0 \text{ is nonincreasing in } z \text{ for each fixed } (x,p) \in \Omega \times \mathbb{R}^n$$

finally, one of the following conditions (1), (2), (3) is satisfied:

(1) $A_i, i = 1, \ldots, n,$ is independent of z; or
(2) A_0 is independent of p; or
(3) the following $(n+1) \times (n+1)$ matrix satisfies componentwise

$$\begin{pmatrix} [\partial_{p_j} A_i(x,z,p) & -\partial_{p_j} A_0(x,z,p)]_{i,j=1}^n \\ \partial_z A_i(x,z,p) & -\partial_z A_0(x,z,p) \end{pmatrix} \geq 0 \text{ in } \Omega \times \mathbb{R} \times \mathbb{R}^n.$$

By considering a special case of (2.206), Taylor [620] obtains uniqueness under, in some sense, less restrictive conditions, see (2.223)–(2.5.4). We define the

$$\text{Taylor case}: \quad Gu = \sum_{i,j=1}^n a_{ij}(\nabla u)\partial^i \partial^j u = 0 \text{ on } \Omega, u \in \mathcal{V} = C^\infty(\bar{\Omega}), \qquad (2.223)$$

cf. [620], Chapter 14, (10.1). It is combined with the following different conditions (2.224), (2.225), (2.226) for (2.223), or a generalization with (2.5.4). So we assume

$$\textbf{For (2.223)}: \text{ellipticity } 0 < \lambda(p)|\vartheta|^2 < \sum_{i,j=1}^n a_{ij}(p)\vartheta^i\vartheta^j < \Lambda(p)|\vartheta|^2 \forall \vartheta \neq 0,$$
$$a_{ij} \in C^\infty(\mathbb{R}^n), \ \partial\Omega \in C^\infty, \varphi \in C^\infty(\partial\Omega), \text{ and let } (2.224)$$
$$\text{the solutions } u_\sigma \text{ for } \sum_{i,j=1}^n a_{ij}(\nabla u)\partial^i\partial^j u = 0 \text{ on } \Omega, u|_{\partial\Omega} = \sigma\varphi$$
$$\text{satisfy } |\partial_{x_i} u_\sigma(x)| \leq K \ \ \forall 1 \leq i \leq n, \ x \in \partial\Omega, \forall 0 \leq \sigma \leq 1.$$

For (2.223): For strictly convex Ω, retain lines 1,2 in (2.224) (2.225)

and cancel lines 3,4.

For (2.223): For uniformly elliptic G, hence $0 < \Lambda(p)/\lambda(p) \leq K \; \forall p$ (2.226)

retain lines 1,2 in (2.224) and cancel lines 3,4.

Allowing an additional zero order term $a_{00}(u)$, cf. [620], Chapter 14, (10.49), we consider the

General Taylor case: $Gu = \sum_{i,j=1}^{n} a_{ij}(\nabla u)\partial^i\partial^j u + a_{00}(u) = 0$ on $\Omega, u \in \mathcal{V} = C^\infty(\bar{\Omega})$,

$$\text{unif. ellipt.: } 0 < \Lambda(p)/\lambda(p) \leq K \; \forall p \in \mathbb{R}^{n^2}, \text{ let } |a_{00}(z)| \leq A_1 + A_2|z|, A_1, A_2 > 0,$$

$$a'_{00}(z) \geq 0 \forall z \in \mathbb{R}, \; a_{ij} \in C^\infty(\mathbb{R}^n), a_{00} \in C^\infty(\mathbb{R}), \; \partial\Omega \in C^\infty, \varphi \in C^\infty(\partial\Omega). \quad (2.227)$$

Here we summarize Gilbarg and Trudinger's [346] uniqueness results from Theorem 10.7 and the remarks following Theorem 15.10, see (2.221), (2.222), and Taylor [620], cf. Chapter 14, Theorems 10.2, 10.6, 10.7, Proposition 10.9, see (2.224)–(2.5.4).

Theorem 2.62. Uniqueness*: For the quasilinear operators* (2.206) *or* (2.207) *or* (2.208) *or* (2.223) *on a bounded domain* $\Omega \subset \mathbb{R}^n$*, let one of the corresponding conditions (2.221), (2.222) or (2.224)–(2.5.4) be satisfied. Then the solution* $u_0 \in \mathcal{V}$ *of the Dirichlet problem* $Gu_0 = 0$ *in* Ω*, and* $u_0|_{\partial\Omega} = \varphi$ *exists, possibly except for (2.222), and is unique.*

Regularity results under Ladyženskaja–Uralceva conditions

Similarly to Subsection 2.6.2 below, see Theorems 2.106, 2.107, regularity results distinguish estimates for the interior or for the entire Ω. To give a flavor for the type of available results, for second order equations, we summarize Theorems 1.1, 2.1, 3.1., 5.1., 6.4., 6.5 in Chapter 4, from the great old ladies Ladyženskaja and Uralceva [464]. We formulate the necessary modifications of the later growth and coerciveness conditions (2.284), (2.285), (2.286) in Subsection 2.5.6, here for the special case $m = 1$ instead of a general m. For avoiding misunderstandings with $p \in \mathbb{R}^n$ we again formulate the results for $W^{1,q}(\Omega)$ with $1 < q < \infty$, cf. the beginning of this subsection. We consider the strong divergence form

$$Gu_0 = \sum_{i=0}^{n}(-1)_{i>0}\partial^i(A_i(x, u_0, \nabla u_0)) = 0 \text{ on } \Omega, \text{ and its weak form} \quad (2.228)$$

$$\langle Gu_0, v\rangle_{\mathcal{V}'\times\mathcal{V}} = \int_\Omega \sum_{i=0}^{n} A_i(x, u_0, \nabla u_0)\partial^i v dx = 0 \; \forall \; v \in \mathcal{V} = W_0^{1,q}(\Omega), \quad (2.229)$$

for $u_0 \in \mathcal{V} = W^{2,q}(\Omega)$ and $u_0 \in \mathcal{V} = W^{1,q}(\Omega)$, respectively The $A_i : (x, z, p) \to \mathbb{R}$ are sufficiently differentiable in $x, z, p \in \bar{\Omega} \times \mathbb{R}^{1+n}$. Boundary conditions $u|_{\partial\Omega} = \varphi|_{\partial\Omega}$ are imposed below for some of the results. We impose the following conditions for $x \in \bar{\Omega}, z \in \mathbb{R}$, and $p = (p_1, \ldots, p_n), p' = (p'_1, \ldots, p'_n) \in \mathbb{R}^n$: Let μ and ν be positive nondecreasing and nonincreasing functions.

$$1.\ \textit{Growth cond.} \sum_{i=1}^{n} |A_i(x,z,p)|(1+|p|) + |A_0(x,z,p)| \le \mu(|z|)(1+|p|)^q, \quad (2.230)$$

$$\textit{Coerciveness:} \sum_{i=1}^{n} A_i(x,z,p)p_i \ge \nu(|z|)|p|^q - \mu(|z|), \text{ see } (2.269), \quad (2.231)$$

$$\textit{Ellipticity:} \quad \nu(|z|) \le (1+|p'|)^{2-q}|p|^{-2} \sum_{i,j=1}^{n} \frac{\partial A_i(x,z,p')}{\partial p'_j} p_i p_j \ \le \mu(|z|), \quad (2.232)$$

$$2.\ \textit{Growth cond.} \left[\sum_{i=1}^{n} \left(|A_i| + \left|\frac{\partial A_i}{\partial z}\right| + \left|\frac{\partial A_0}{\partial p'_i}\right| \right) (1+|p|), \right.$$

$$\left. +|A_0| + \left|\frac{\partial A_0}{\partial z}\right| \right] (x,z,p) \le \mu(|z|)(1+|p|)^q, \quad (2.233)$$

$$3.\ \textit{Growth cond.} \left[\sum_{i=1}^{n} \left(|A_i| + \left|\frac{\partial A_i}{\partial z}\right| \right) (1+|p|) \right.$$

$$\left. + \sum_{i,j=1}^{n} \left|\frac{\partial A_i}{\partial p'_j}\right| + |A_0| \right] (x,z,p) \ \le \mu(|z|)(1+|p|)^q, \quad (2.234)$$

$$4.\ \textit{Growth cond.} \left[\left(\sum_{i,j=1}^{n} \left|\frac{\partial A_i}{\partial x_j}\right| + \sum_{i=1}^{n} \left|\frac{\partial A_0}{\partial p'_i}\right| \right) (1+|p|), \right.$$

$$\left. \left|\frac{\partial A_0}{\partial z}\right| + \sum_{i=1}^{n} \left|\frac{\partial A_0}{\partial x_i}\right| \right] (x,z,p) \ \le \mu(|z|)(1+|p|)^q, \quad (2.235)$$

$$\textit{Boundedness:} \ \max_{x\in\Omega, i,j=1}^{n} \left\{ \left|\frac{\partial A_i}{\partial p'_j}\right|, \left|\frac{\partial A_i}{\partial z}\right|, \left|\frac{\partial A_i}{\partial x_j}\right|, |A_0| \right\} (x,z,p) \le \mu_1 < \infty, \quad (2.236)$$

$$\textit{Condition A for } \ \partial\Omega: \ \exists\, 0 < a_1, a_2, \ \ a_2 < 1: \ \forall \mathcal{B}_\rho(x), \ x \in \partial\Omega, \ \rho \le a_1,$$

$$\forall \text{ components } \hat\Omega \subset \mathcal{B}_\rho(x) \cap \Omega: \ \ \text{measure } \hat\Omega \le (1-a_2) \text{ measure } \mathcal{B}_\rho(x), \quad (2.237)$$

$$\textit{Integr. cond.} \int_\Omega (1+|\nabla u|)^{q-2} \left(\sum_{i,j=1}^{n} \left| \frac{\partial^2 u}{\partial x_i \partial x_j} \right|^2 \right) dx, \int_\Omega (1+|\nabla u|)^{q+2} dx < \infty. \quad (2.238)$$

These (2.231)–(2.235) are not independent. For example, (2.232), (2.233) imply parts of (2.231), in particular, the standard coercivity, see Ladyženskaja and Uralceva [464], p. 255. This fact is generalized in Lemma 2.77 below. The above (2.232), and (2.289) below, are in some sense competing. Obviously, the above (2.232) indicates the ellipticity for the linearized differential equation. By different combinations of these conditions different results are achieved, see Theorems 1.1, 2.1, 5.2, 6.4, 6.5 in Chapter 4 [464]. The existence of solutions $u_0 \in W^{1,q}(\Omega)$ is discussed in Chapter 4, Section 8 [464], under highly technical additional conditions. There are some interesting differences between the conditions (2.230)–(2.232), imposed for $m = 1$, and for $m \geq 1$, see (2.284)–(2.289) below.

Theorem 2.63. Regular, bounded solutions and uniqueness on small balls: *We assume a generalized bounded solution $u_0 \in W^{1,q}(\Omega)$ in the sense of (2.229) with $M_0 := \|u_0\|_{L^\infty(\Omega)} < \infty, 0 \leq \alpha \leq 1$.*

1. Bounded generalized solutions: *Under the conditions (2.230) and (2.231), we get $u_0 \in C^{0,\alpha}(\Omega)$ with α depending upon $M_0, q, \mu(M_0), \nu(M_0)$. For an arbitrary $\Omega_1 \subset\subset \Omega$ the $\|u_0\|_{C^{0,\alpha}(\Omega_1)}$ can be estimated from above by a constant depending upon the previous constants and* $\operatorname{dist}(\Omega_1, \partial\Omega)$.
If additionally $\partial\Omega$ satisfies (2.237) and if $u|_{\partial\Omega} = \varphi|_{\partial\Omega} \in C^{0,\beta}(\partial\Omega)$ for $\alpha \leq \beta$, then $\|u_0\|_{C^{0,\alpha}(\Omega)}$ over the entire *Ω can be estimated from above by a constant depending upon the previous constants and the a_1, a_2, in (2.237), β and $\|\varphi\|_{C^{0,\beta}(\partial\Omega)}$ (Theorem 1.1).*
2. Uniqueness in the small: *Assume (2.232), (2.233) and let $u_1, u_2 \in W^{1,q}(\Omega)$ be two bounded solutions of (2.229), coinciding on the surface of a ball $\mathcal{B}_\rho \subset \Omega_1 \subset\subset \Omega$ with radius $\rho < \rho_1$, and* $\max\{\|u_1\|_{L^\infty(\partial\Omega)}, \|u_2\|_{L^\infty(\partial\Omega)}\} =: M_1 < \infty$. *Then they coincide in $\mathcal{B}_\rho$ as well. This ρ_1 is determined by the $M_1, q, \mu(M_1)$, $\nu(M_1)$ and* $\operatorname{dist}(\Omega_1, \partial\Omega)$ *(Theorem 2.1).*
3. A bound for $\|\nabla u_0\|_{L^\infty(\Omega_1)}$ over $\Omega_1 \subset\subset \Omega$ and $u_0 \in H^2(\Omega_1)$: *Assume differentiable $A_i(x, z, p), i = 0, \ldots, n$, and (2.232), (2.234), (2.235), and for $x \in \bar{\Omega}, |u| \leq M$, and arbitrary q the (2.238). Then for an arbitrary $\Omega_1 \subset\subset \Omega$ we get $u_0 \in H^2(\Omega_1)$. It satisfies (2.228) almost everywhere in Ω and $\|\nabla u_0\|_{L^\infty(\Omega_1)}$ can be estimated by constants depending only upon $M_0, q, \mu(M_0), \nu(M_0)$, $\partial\Omega$, $\|\varphi\|_{C^{2,0}(\Omega)}$, and* $\operatorname{dist}(\Omega_1, \partial\Omega)$ *(Theorem 5.2).*
4. Bounds for $\|u_0\|_{C^{l,\alpha}(\Omega)}, l \geq 2$. *Let $M_2 \geq \|\nabla u_0\|_{L^\infty(\Omega)}$ and $\mathcal{M} := \{(x, z, p) \in \bar{\Omega} \times \mathbb{R} \times \mathbb{R}^n : |z| \leq M_0, |p| \leq M_2\}$. If, in addition, $A_i(x, z, p) \in C^{l-1,\alpha}(\mathcal{M})$, $i = 1, \ldots, n$, $A_0(x, z, p) \in C^{l-2,\alpha}(\mathcal{M})$, and $\partial\Omega \in C^{l,\alpha}$, $\varphi \in C^{l,\alpha}(\bar{\Omega})$, then $u_0 \in C^{l,\alpha}(\bar{\Omega})$ and the $\|u_0\|_{C^{l,\alpha}(\bar{\Omega})}$ are bounded by constants depending upon $M_0, M_2, q, \mu(M_0), \nu(M_0)$, $\|\varphi\|_{C^{l,\alpha}(\bar{\Omega})}$, $\partial\Omega$ and $\|A_i\|_{C^{l-1,\alpha}(\mathcal{M})}$, $\|A_0\|_{C^{l-2,\alpha}(\mathcal{M})}$ (Theorem 6.5).*

Finally, we summarize regularity results from Chen and Wu [170] for quasilinear equations (2.206). Essentially based upon the Leray–Schauder theorem, Chen and Wu get solutions $u_0 \in C^{2,\gamma}(\bar{\Omega})$ under modified conditions. We present their Theorem 5.1 in Chapter 5, for divergent quasilinear second order equations, but only refer to Theorem 4.1 in Chapter 7 for nondivergent quasilinear second order equations. We use the remark on p. 117 in Chapter 7 for fully nonlinear second order equations formulating the higher regularity results $u_0 \in C^{2+k,\gamma}(\bar{\Omega}), k > 0$, in the last part of the following Theorem 2.64, compare [170], p. 78. Their Theorem 7.4. in Chapter 7 for fully nonlinear second order equations will be discussed as Theorem 2.82 below.

Additional results are available as special cases of Theorem 2.91 below. Chen and Wu study divergent equations of the form (2.207) and require *ellipticity* and other conditions in the form

$$A_i(x,z,p) \in \mathbb{R}, \forall\ (x,z,p) \in \mathcal{U}_o \subset \Omega \times \mathbb{R} \times \mathbb{R}^n,\ 0 \neq p' = (p'_1,\dots,p'_n) \in \mathbb{R}^n,$$

$$\exists \Lambda > \lambda > 0:\ 0 < \lambda |p'|^2 < \sum_{i,j=1}^{n} \frac{\partial A_i}{\partial p_j}(x,z,p) p'_i p'_j < \Lambda |p'|^2, \tag{2.239}$$

$$|A_i(x,z,0)| \leq g(x)\ \forall i = 1,\cdots,n, g \in L^q(\Omega), q > n, \text{ and finally}$$

$$\Bigg[(1+|p|^2)\left|\frac{\partial A_i}{\partial p_j}(\cdot)\right| + (1+|p|)|\left(\left|\frac{\partial A_i}{\partial z}(\cdot)\right| + |A_i(\cdot)|\right) + \left|\frac{\partial A_i}{\partial x_j}(\cdot)\right| \tag{2.240}$$

$$+ |A_0(\cdot)|\Bigg](x,z,p)\ \leq\ \mu(|z|)(1+|p|^2)\ \forall i,j = 1,\cdots,n, \forall (x,z,p) \in \mathcal{U}_o,$$

$$-A_0(x,z,p) \text{ sign } z \leq \Lambda[|p| + f(x)] \text{ for an } f \in L^{q^*}(\Omega), q^* := nq/(n+q).$$

The relation for the ellipticity conditions in (2.232) and (2.239) is discussed in Lemma 2.77 below for order $2m$.

Theorem 2.64. Existence and regularity of solutions:

1. *Let the conditions (2.239)–(2.240) be satisfied and assume* $\partial\Omega \in C^{2,\gamma}$, $A_i \in C^{1,\gamma}(\bar{\Omega} \times \mathbb{R} \times \mathbb{R}^n)$, $A_0 \in C^{0,\gamma}(\bar{\Omega} \times \mathbb{R} \times \mathbb{R}^n)$, $\varphi \in C^{2,\gamma}(\bar{\Omega})$ *with* $0 < \gamma < 1$. *Then there exists a solution* $u_0 \in C^{2,\gamma}(\bar{\Omega})$ *for the Dirichlet problem (2.206).*
2. *If additionally* $A_i \in C^{k+1,\alpha}(\mathcal{U}_o), A_0 \in C^{k,\alpha}(\mathcal{U}_o), k \geq 1,$ *and* $\partial\Omega \in C^{2+k,\gamma}$, $\varphi \in C^{2+k,\gamma}(\bar{\Omega})$, *then the solution* $u_0 \in C^{2,\gamma}(\bar{\Omega}) \cap C^{k+2,\alpha}(\Omega)$, *for* $A_i \in C^{k+1,\alpha}(\bar{\mathcal{U}}_o), A_0 \in C^{k,\alpha}(\bar{\mathcal{U}}_o)$, *even* $u_0 \in C^{k+2,\gamma}(\bar{\Omega})$.

Further regularity results are formulated in Subsection 2.5.7 and are studied by Ladyženskaja and Uralceva [463, 464, 466], Koshelev and Chelkak [449–451], Morrey [500], Nečas [510], and Giaquinta and Hildebrandt [341, 342].

These quasilinear equations are elliptic equations in the sense of Definition 2.51, since the linearized equation satisfies one of the ellipticity conditions (2.209), (2.214), (2.217), (2.224), (2.232), (2.239). In Lemma 2.77 below, we will discuss the relation of growth and coercivity conditions versus ellipticity for quasilinear equations of order $2m$.

2.5.5 General nonlinear and Nemyckii operators

In the last Subsections 2.5.3, 2.5.4 we studied nonlinear operators. We had to evaluate, e.g. $f(u)$ or $a_{ij}(x,u,\nabla u)$ usually for given continuous functions. Therefore the evaluation was obvious. The situation changes if $u \in W^{m,p}(\Omega), 1 \leq p \leq \infty$. This is necessary for studying monotone operators. They are, in some sense, evaluations of quasilinear elliptic operators and the main tool for analytical results for their solutions. So we define different monotone and Nemyckii operators and summarize their essential properties, see Skrypnik [590, 591], Zeidler [676–678], and Runst and Sickel [561].

For the next definition we need a function, see Zeidler [678], pp. 500 ff.

$$a:\mathbb{R}_+\to\mathbb{R}_+ \text{ used in (2.244), (2.253), continuous and strictly increasing with}$$
$$a(0)=0,\ \lim_{t\to\infty}a(t)=\infty,\ \text{e.g. } a(t)=c|t|^{p-1},\ 0<c,\ 1<p. \tag{2.241}$$

Definition 2.65. *Let $\mathcal{X}$ be a Banach space, $A:\mathcal{X}\to\mathcal{X}'$ an operator. The following conditions are imposed $\forall u,v\in\mathcal{X}$. Then we call A*[16]

$$\text{monotone} \Leftrightarrow \langle Au-Av,u-v\rangle \ \geq\ 0, \tag{2.242}$$
$$\text{strictly monotone} \Leftrightarrow \langle Au-Av,u-v\rangle \ >\ 0 \textit{ for } u\neq v, \tag{2.243}$$
$$\text{uniformly monotone} \Leftrightarrow \langle Au-Av,u-v\rangle \tag{2.244}$$
$$\geq a(\|u-v\|_{\mathcal{X}})\|u-v\|_{\mathcal{X}},\ \textit{e.g.,}\ \geq c\|u-v\|_{\mathcal{X}}^p,$$
$$\text{strongly monotone} \Leftrightarrow \langle Au-Av,u-v\rangle \ \geq\ c\|u-v\|_{\mathcal{X}}^2,\ c\in\mathbb{R}_+, \tag{2.245}$$
$$\text{(nonlinear) coercive} \Leftrightarrow \lim_{\|u\|_{\mathcal{X}}\to\infty}\frac{\langle Au,u\rangle}{\|u\|_{\mathcal{X}}}=\infty, \tag{2.246}$$
$$\text{weakly coercive} \Leftrightarrow \lim_{\|u\|_{\mathcal{X}}\to\infty}\|Au\|_{\mathcal{X}'}=\infty \textit{ for Hilbert spaces,} \tag{2.247}$$
$$\textit{where we identify } \mathcal{X}=\mathcal{X}',$$
$$\text{Lipschitz-continuous} \Leftrightarrow \exists L>0:\|Au-Av\|_{\mathcal{X}'}\leq L\|u-v\|_{\mathcal{X}}, \tag{2.248}$$
$$\text{continuous} \Leftrightarrow \forall\epsilon>0\exists\delta>0:\|Au-Av\|_{\mathcal{X}'}<\epsilon\ \ \forall\|u-v\|_{\mathcal{X}}<\delta, \tag{2.249}$$
$$\text{hemicontinuous} \Leftrightarrow t\to\langle A(u+tv),w\rangle \textit{ is continuous on } [0,1], \tag{2.250}$$
$$\text{strongly continuous} \Leftrightarrow u_n\rightharpoonup u\ \Rightarrow\ Au_n\to Au \textit{ for } n\to\infty, \tag{2.251}$$
$$\text{bounded} \Leftrightarrow A \textit{ maps bounded sets into bounded sets,} \tag{2.252}$$
$$\text{stable} \Leftrightarrow \|Au-Av\|_{\mathcal{X}'}\geq a(\|u-v\|_{\mathcal{X}}). \tag{2.253}$$

In (nonlinear) coercivity we often omit the "nonlinear".

Remark 2.66. *Obviously strong monotonicity implies uniform implies strict and implies monotonicity, so (2.245) $\Rightarrow$ (2.244) $\Rightarrow$ (2.243) $\Rightarrow$ (2.242). Lipschitz-continuity implies continuity and hemicontinuity, so (2.248) $\Rightarrow$ (2.249) $\Rightarrow$ (2.250). Obviously, a uniformly monotone A is coercive and stable.*

The preceding properties of A allow important consequences, see Zeidler [677], pp. 455 ff: Let $\mathcal{X}$ be a reflexive Banach space. We replace the bilinear forms $a(\cdot,\cdot)$ and the induced linear differential operators $A:\mathcal{X}\to\mathcal{X}'$ by nonlinear operators $A:\mathcal{X}\to\mathcal{X}'$, and the corresponding nonlinear forms. This is motivated by the results in

[16] Note that for (2.251), $u_n\rightharpoonup u\in\mathcal{X}$ is defined by $\langle u_n-u,v\rangle\to 0\forall v\in\mathcal{X}'$.

Subsection 2.5.3. For generalizing Proposition 2.10 we start, cf. [677], pp. 314 ff., with a

$$\text{nonlinear form } a(\cdot,\cdot):\mathcal{X}\times\mathcal{X}\to\mathbb{R},\ \text{s.t. } |a(u,v)|\le C(u)\|v\|_{\mathcal{X}}\ \ \forall u,v\in\mathcal{X}. \tag{2.254}$$

Proposition 2.67. *Under the condition (2.254) there exists a usually nonlinear but unique operator, such that, for all $u\in\mathcal{X}$,*

$$\langle Au,v\rangle_{\mathcal{X}'\times\mathcal{X}} := a(u,v)\ \forall\ v\in\mathcal{X},\ \textit{denoted as } A:\mathcal{X}\to\mathcal{X}',\ \ Au\in\mathcal{X}'. \tag{2.255}$$

Then for any $f\in\mathcal{X}'$ the following three problems are equivalent:

$$\begin{aligned} \textit{Determine } u_0\in\mathcal{X} \textit{ s.t. } Au_0 = f\in\mathcal{X}' &\Leftrightarrow a(u_0,v) = \langle f,v\rangle_{\mathcal{X}'\times\mathcal{X}}\forall v\in\mathcal{X} \\ &\Leftrightarrow \langle Au_0,v\rangle_{\mathcal{X}'\times\mathcal{X}} = \langle f,v\rangle_{\mathcal{X}'\times\mathcal{X}}\forall v\in\mathcal{X}. \end{aligned} \tag{2.256}$$

Obviously (2.254) and (2.255) and the following considerations can be generalized to $a(\cdot,\cdot):\mathcal{D}(a)\subset\mathcal{X}\times\mathcal{X}\to\mathbb{R}$, and $A:\mathcal{D}(A)\subset\mathcal{X}\to\mathcal{R}(A)\subset\mathcal{X}'$, respectively.

Theorem 2.68. Properties of nonlinear operators, Zeidler [678], based upon Zarantonello, Browder, and Minty [150, 491, 674]: *Let $\mathcal{X}$ be a real Banach space and $A:\mathcal{X}\to\mathcal{X}'$ a (nonlinear) operator. Then*

1. *If A is linear and monotone, then A is continuous, see [678], Proposition 26.4.*
2. *If A is monotone, coercive, and hemicontinuous on a reflexive Banach space, then A is surjective and, for each $f\in\mathcal{X}'$, the set of solutions of $Au=f$ is bounded, convex and closed, cf. [678], Theorem 26A(a) [150, 491].*
3. *If A is additionally strictly monotone, the above solution is unique. Then $A^{-1}:\mathcal{X}'\to\mathcal{X}$ exists and, cf. [678], Theorem 26A(c), (d) [150, 491],*
   a uniformly monotone A implies a continuous A^{-1};
   a strongly monotone A implies a Lipschitz-continuous A^{-1}.
4. *Let $\mathcal{X}$ be a real Hilbert space and A be strongly monotone and Lipschitz-continuous with constant L. Then for each f the $Au=f\in\mathcal{X}'$ has a unique solution, depending continuously upon f, more precisely, see [674], [678], Theorem 25B,*

$$\|u_1-u_2\|_{\mathcal{X}}\le L^{-1}\|f_1-f_2\|_{\mathcal{X}'}\ \textit{for } Au_1=f_1, Au_2=f_2, f_1\in\mathcal{X}'.$$

Definition 2.69. *For given functions f and appropriate $u\in\mathcal{D}(F)$, define $F(u)$ as*

$$f:\mathcal{D}(f)\subset\Omega\times\mathbb{R}^k\to\mathbb{R}, u:\Omega\to\mathbb{R}^k:\ \ F(u)(x):=f(x,u_1(x),\cdots,u_k(x)), \tag{2.257}$$

with $(x,u_1(x),\cdots,u_k(x))\in\mathcal{D}(f)$. This F is called a Nemyckii operator.

Now we formulate two conditions guaranteeing that $F(u)(x)$ makes sense. The *Carathéodory* and the *growth condition* enforce measurability and appropriate growth of the composite function F; more precisely

Carathéodory condition: For f in (2.257) let (2.258)

$\Omega \subset \mathbb{R}^n$ be *nonempty, measurable* and $x \mapsto f(x,u)$ *measurable* on $\Omega \ \forall u \in \mathbb{R}^k$

$u \mapsto f(x,u)$ is continuous on $\mathbb{R}^k$ almost everywhere (a.e.) for $x \in \Omega$.

$$\textit{Growth condition: } \forall (x,u) \in \Omega \times \mathbb{R}^k \text{ let } |f(x,u)| \le a(x) + b\sum_{i=1}^{k} |u_i|^{p_i/q}, \tag{2.259}$$

with fixed reals $0 < b, 1 \le p_i, q < \infty$ and fixed $a \in L^q(\Omega),\ 0 \ \le\ a(\cdot)$a.e.

Theorem 2.70. Estimates for nonlinear operators, Zeidler [678], Proposition 26.6: *If (2.258), (2.259) are satisfied, then*

$$F : \Pi_{i=1}^k L^{p_i}(\Omega) \to L^q(\Omega)$$

is continuous and bounded (this is not equivalent *for nonlinear operators!) such that*

$$\|F(u)\|_{L^q(\Omega)} \le C\left(\|a\|_{L^q(\Omega)} + \sum_{i=1}^{k} \|u_i\|_{L^{p_i}(\Omega)}^{p_i/q}\right) \quad \forall u \in \Pi_{i=1}^k L^{p_i}(\Omega). \tag{2.260}$$

Particularly important is the case $k = 1$. Then the Nemyckii operator has the form

$$\begin{aligned} &F(u)(x) := f(x,u(x)) \text{ and } f : \mathcal{D}(f) \subset \Omega \times \mathbb{R} \to \mathbb{R}, \\ &\text{mainly } F : \mathcal{X} = L^p(\Omega) \to \mathcal{X}' = L^q(\Omega), 1/p + 1/q = 1. \end{aligned} \tag{2.261}$$

We summarize appropriate conditions for f. The Carathéodory condition is only modified by replacing $\mathbb{R}^k$ by $\mathbb{R}$. We specialize the growth condition as

$$\begin{aligned} &\textit{Growth condition: } \text{For fixed reals } 0 < b, 1 < p, q < \infty, \\ &\quad 1/p + 1/q = 1 \Leftrightarrow p - 1 = p/q,\ a \in L^q(\Omega),\ 0 \ \le\ a(\cdot) \textit{ a.e.}, \\ &\quad \text{let } |f(x,u)| \le a(x) + b|u|^{p-1}\ \forall\ (x,u) \in \Omega \times \mathbb{R}. \end{aligned} \tag{2.262}$$

We assume $u, v \in \mathbb{R}$, and $\forall x \in \Omega$, and denote the function f in (2.261) to be

$$\textit{monotone} \Leftrightarrow f(x,u) \le f(x,v) \text{ for } u \le v, \tag{2.263}$$

$$\textit{strictly monot.} \Leftrightarrow f(x,u) < f(x,v) \text{ for } u < v, \tag{2.264}$$

$$\textit{coercive} \Leftrightarrow \exists d > 0, g \in L^1(\Omega) : f(x,u)u \ge d|u|^p + g(x), \tag{2.265}$$

$$\textit{positive} \Leftrightarrow f(x,u)u \ge 0, \tag{2.266}$$

$$\textit{asympt. positive} \Leftrightarrow \exists R > 0 : f(x,u)u \ge 0 \text{ for } |u| \ge R, \text{meas } \Omega < \infty. \tag{2.267}$$

Theorem 2.71. Properties of Nemyckii operators, Zeidler [678], Proposition 26.7: *For the function f in (2.261) we assume (2.258) with $\mathbb{R}^k$ replaced by $\mathbb{R}$, (2.262) and let $F : \mathcal{X} = L^p(\Omega) \to \mathcal{X}'$ be the Nemyckii operator defined by f, cf. (2.261).*

1. *Then F is continuous and bounded with*

$$\|F(u)\|_{L^q(\Omega)} \leq C\left(\|a\|_{L^q(\Omega)} + \|u\|^{p-1}_{L^p(\Omega)}\right) \quad \textit{and} \tag{2.268}$$

$$\langle F(u), u\rangle_{\mathcal{X}'\times\mathcal{X}} = \int_\Omega f(x, u(x))u(x)dx \forall\ u \in \mathcal{X}.$$

2. *A (strictly) monotone f, cf. (2.263), (2.264), implies a (strictly) monotone F.*
3. *A coercive f, see (2.265), implies a coercive F and*

$$\langle F(u), u\rangle_{\mathcal{X}'\times\mathcal{X}} \geq d\|u\|^p_{L^p(\Omega)} + \int_\Omega g(x)dx \forall\ u \in \mathcal{X}.$$

4. *A positive f, see (2.266), implies a positive F, hence*

$$\langle F(u), u\rangle_{\mathcal{X}'\times\mathcal{X}} \ \geq\ 0 \ \forall\ u \in \mathcal{X}.$$

5. *An asymptotic positive f, see (2.267), implies an estimate for F in the form*

$$\langle F(u), u\rangle_{\mathcal{X}'\times\mathcal{X}} \ \geq\ \ -c \ \forall\ u \in \mathcal{X} \textit{ for a constant } c\ >\ 0.$$

These properties of general nonlinear and Nemyckii operators are essential tools in the study of divergent quasilinear elliptic equations.

2.5.6 Divergent quasilinear elliptic equations of order $2m$

We start with the strong form of a quasilinear elliptic operator, G. The presentation is motivated by the properties of Nemyckii operators in the last Subsection 2.5.5. At the end of this subsection, we will show that, for the most important cases, under slightly stronger conditions, these quasilinear equations are elliptic equations in the sense of Definition 2.51. This allows us to prove convergence of our discrete solutions to the exact solutions by our general techniques with the correct rate of convergence for all the classes of space discretization methods, considered here, and the problems in $W^{m,p}(\Omega)$, with $2 \leq p \leq \infty$. A general convergence theory for monotone and quasilinear operators with its "pros and cons" is elaboreted in Section 4.5 for $1 < p < \infty$.

We only consider Dirichlet boundary conditions on bounded domains Ω, compare Theorem 2.50, and recall the standard notation in (2.73), (2.165), (2.166):

$$\vartheta \in \mathbb{R}^n,\ \Theta^0,\ \vartheta^i,\ \vartheta^\alpha \in \mathbb{R},\ \Theta^k = (t_\alpha \in \mathbb{R})_{|\alpha|=k} \in \mathbb{R}^{n_k},\ \text{often } = (\vartheta^\alpha)_{|\alpha|=k}, \tag{2.269}$$

$$A_\alpha \text{ defined as } A_\alpha : \Omega \times \mathbb{R}^{N_m} \to \mathbb{R}, w = (x, \Theta^0, \ldots, \Theta^m) =: (x, \Theta^{\leq m}) \in \Omega \times \mathbb{R}^{N_m},$$

$$\text{with } \mathbb{R}^{N_m} = \mathbb{R} \times \mathbb{R}^n \times \cdots \times \mathbb{R}^{n_m},\ \text{norms, e.g. } |\vartheta|, |\Theta^m|,\ \text{and}$$

$$N_m := (n^{m+1} - 1)/(n-1) \text{ for } \ n > 1,\ N_m := m+1 \text{ for } n = 1,\ \text{partials}$$

$$\partial^\alpha u := \frac{\partial^{|\alpha|}}{\partial x_1^{\alpha_1} \ldots \partial x_n{}^{\alpha_n}} u, \nabla^0 u := \partial^\alpha u := u \text{ for } |\alpha| = 0, \nabla^k u := (\partial^\alpha u)_{|\alpha|=k}.$$

We discuss the general form of quasilinear elliptic equations with Dirichlet boundary conditions, see Skrypnik [590, 591] and Zeidler [678]:

$$Gu - f = G_s u - f = \sum_{|\alpha|\le m} (-1)^{|\alpha|}\partial^\alpha (A_\alpha(x,u,\cdots,\nabla^m u) - f_\alpha) = 0 \text{ on } \Omega, \tag{2.270}$$

$$= \sum_{|\alpha|,|\beta|\le m} A_{\alpha,\beta}(x,u,\cdots,\nabla^m u)\partial^\alpha\partial^\beta u - (-1)^{|\alpha|}\partial^\alpha f_\alpha = 0 \text{ on } \Omega, \quad \text{with, e.g.}$$

$$A_{\alpha,\beta}(x,u,\cdots,\nabla^m u) = (-1)^{|\alpha|}\frac{\partial^m A_\alpha(x,u,\cdots,\nabla^m u)}{\partial\vartheta^\beta}\ \forall |\alpha| = |\beta| = m$$

$$\Omega \text{ a bounded nonempty measurable domain in } \mathbb{R}^n \text{ and} \tag{2.271}$$

$$\textit{Dirichlet boundary conditions } \partial^i u(x)/\partial\nu^i = 0 \ \ \forall\, 0 \le i \le m-1, \text{ on } \partial\Omega$$

$$\Leftrightarrow \partial^\alpha u|_{\partial\Omega} = 0\ \forall |\alpha| \le m-1 \text{ in the trace sense for smooth enough } \partial\Omega. \tag{2.272}$$

A possible condition $u \in W^{2m,p}(\Omega)$ is still too restrictive. Below, we will list conditions allowing us to multiply (2.270) with $v \in \mathcal{V}$, and to integrate over Ω, where

$$\mathcal{V} := W_0^{m,p}(\Omega) := \{v \in W^{m,p}(\Omega) : \partial^\alpha v = 0\ \forall |\alpha| \le m-1 \text{ on } \partial\Omega\}, 1 < p < \infty.$$

We iteratively use the *fundamental Green's formula* (2.9). For smooth enough w_1, w_2 and $|\alpha| \le m-1$ for w_1 we get

$$w_1 = \partial^\alpha v, w_2 = \partial^\alpha(A_\alpha(x,u,\cdots,\nabla^m u)) : \int_\Omega \partial(w_1 w_2)/\partial x_i dx = \int_{\partial\Omega} (w_1 w_2)\nu_i ds.$$

For $v, w \in H^1(\Omega)$ we obtain by partial integration and with the outer normal $\nu = (\nu_1, \ldots, \nu_n)^\top$ on a Lipschitz-continuous manifold $\partial\Omega$, the *fundamental Green's formula*, cf. Ciarlet [174], p. 14,

$$\int_\Omega w\partial v/\partial x_i dx + \int_\Omega v\partial w/\partial x_i dx = \int_\Omega \partial(vw)/\partial x_i dx = \int_{\partial\Omega} vw\nu_i ds, \tag{2.273}$$

$$\partial\Omega \in C_L, \forall w, v \in H^1(\Omega).$$

This yields the following *weak form for G* defined for $u, v \in \mathcal{V} = W_0^{m,p}(\Omega)$.

$$a_s(u,v) = \int_\Omega vG_s u dx = \int_\Omega \sum_{|\alpha|\le m} v(-1)^{|\alpha|}\partial^\alpha(A_\alpha(x,u,\cdots,\nabla^m u))dx, u \in W^{2m,p}$$

$$= \int_\Omega \sum_{|\alpha|\le m} A_\alpha(x,u,\cdots,\nabla^m u)\partial^\alpha v dx =: a(u,v) =: \langle Gu, v\rangle_{\mathcal{V}'\times\mathcal{V}}\ \forall\ u,v \in \mathcal{V} \tag{2.274}$$

$$\langle Gu_0, v\rangle_{\mathcal{V}'\times\mathcal{V}} = \langle f, v\rangle_{\mathcal{V}'\times\mathcal{V}} = \int_\Omega \sum_{|\alpha|\le m} f_\alpha \partial^\alpha v dx\ \forall\ u_0 \in \mathcal{V}\ \forall\ v \in \mathcal{V}. \tag{2.275}$$

The $A_\alpha(x, u, \ldots, \nabla^m u)$ are Nemyckii operators, studied in the last subsection.

These weak forms of quasilinear equations are often obtained as Euler equations for variational problems. For a smooth function $F(x, u, \dots, \nabla^m u)$, a functional

$$J(u) := \int_\Omega F(x, u, \dots, \nabla^m u)dx \tag{2.276}$$

has to be (locally) minimized. The first variation of this functional applied to v yields a form as in (2.274), for more details cf. Subsection 2.6.4. This is one of the reasons that for the class of quasilinear differential equations the most complete and final results are available.

Example 2.72.

Important model problems of second order:

1. For this example it is important that [678], p. 592 ff., $H_0^1(\Omega)$ is compactly embedded into $L^4(\Omega)$ for dimension $n = 1, 2, 3$:

$$Gu = -\Delta u + \alpha \sum_{i=1}^n sin(u)\partial^i u = f \in L^2(\Omega) \text{ on } \Omega, \tag{2.277}$$

$$n = 1, 2, 3, \text{ and the } \textit{Dirichlet condition } u(x) = 0 \text{ on } \partial\Omega. \tag{2.278}$$

   We define *linear, bilinear and nonlinear forms* related to $Gu = f$ as

$$\begin{aligned} b(v) &:= \int_\Omega f v dx, \quad a_1(u, v) := \int_\Omega \sum_{i=1}^n \partial^i u \partial^i v dx \quad \forall u, v, w \in H_0^1(\Omega) = \mathcal{V}, \\ c(u, w, v) &:= \int_\Omega \sum_{i=1}^n (sinu)\partial^i w v dx, \quad a_2(u, v) := c(u, u, v), \text{ and} \end{aligned} \tag{2.279}$$

$$\text{determine } u_0 \in H_0^1(\Omega) \text{ s.t. } a_1(u_0, v) + \alpha a_2(u_0, v) = b(v) \ \forall v \in H_0^1(\Omega).$$

2. We assume $2 \le p < \infty, 1/p + 1/q = 1, \quad \mathbb{R} \ni t \ge 0$, and Dirichlet boundary conditions, see [678], p. 567 ff., for

$$Gu = G_s u = -\sum_{i=1}^n \partial^i(|\partial^i u|^{p-2}\partial^i u) + tu = f \in L^q(\Omega) \text{ on } \Omega. \tag{2.280}$$

   With the above $b \in L^2(\Omega)$ we obtain the *weak form of G*

$$\begin{aligned} \int_\Omega v G_s u dx &= \int_\Omega v \left(-\sum_{i=1}^n \partial^i(|\partial^i u|^{p-2}\partial^i u) + tu \right) dx \\ &= a(u, v) := \langle Gu, v \rangle_{W^{-1,q}(\Omega) \times W^{1,p}(\Omega)} \qquad (2.281) \\ &:= \int_\Omega \left(\sum_{i=1}^n |\partial^i u|^{p-2}\partial^i u \partial^i v + tuv \right) dx \text{ and determine} \end{aligned}$$

$$u_0 \in W_0^{1,p}(\Omega) \text{ s.t. } a(u_0, v) = \langle f, v \rangle_{\mathcal{V}' \times \mathcal{V}} \ \forall v \in W_0^{1,p}(\Omega). \tag{2.282}$$

3. *Minimal surface equation,* see Gilbarg and Trudinger [346], p. 339: Let the minimal surface be represented by the function $u : \Omega \to \mathbb{R}$. Then, with prescribed boundary conditions $u = \varphi$ on $\partial\Omega$, it has the form

$$Gu = Gu = (1 + |\nabla u|^2)\Delta u + \sum_{i,j=1}^{n} \partial^i u \partial^j u \partial^i \partial^j u = 0 \text{ on } \Omega. \tag{2.283}$$

This is an example where the Schauder techniques with boundary gradient estimates are applied, see [346]. ■

Now we impose conditions which guarantee that $G(u)$ in (2.270), (2.277), (2.280), (2.283), and its weak form (2.274), (2.283) are well defined. The *Carathéo-dory* and *growth conditions* enforce measurability and appropriate growth of the composite function A_α, see [678], p. 571 ff. and (2.269) for the notation. More generally than the coercivity for bilinear forms we introduce a nonlinear coercivity, monotonicity and growth condition for these quasilinear operators, cf. (2.230)–(2.235), (2.242)–(2.247), (2.258), (2.259). This is again essential for the existence of solutions.

Carathéodory condition: Ω is bounded and measurable in $\mathbb{R}^n$,

$$\begin{aligned} &A_\alpha : (x, \Theta^{\leq m}) \in \ \Omega \times \mathbb{R}^{N_m} \to \mathbb{R}, \forall\ |\alpha| \leq m, \\ &\Theta^{\leq m} \mapsto A_\alpha(x, \Theta^{\leq m}) \text{ is continuous in } \mathbb{R}^{N_m} \text{ a.e. for } x \in \Omega, \\ &x \mapsto A_\alpha(x, \Theta^{\leq m}) \text{ is measurable in } \Omega \ \ \forall\ \Theta^{\leq m} \in \mathbb{R}^{N_m}, \text{e.g. for continuous } A_\alpha. \end{aligned} \tag{2.284}$$

For the following conditions, we assume $1 < p < \infty, 1/p + 1/q = 1$, the special form $\Theta^{\leq m} = (\vartheta^\gamma)_{|\gamma| \leq m}$, $\Theta_1^{\leq m} = (\vartheta_1^\gamma)_{|\gamma| \leq m} \in \mathbb{R}^{N_m}$ and $\exists\ c, d, C \in \mathbb{R}_+$, such that

$$\exists\ a \in L^q(\Omega), h \in L^1(\Omega), \text{ s.t. } \forall\ x \in \Omega, \Theta^{\leq m}, \Theta_1^{\leq m} \in \mathbb{R}^{N_m}, \vartheta^0, \vartheta^\alpha, \vartheta_1^\alpha \in \mathbb{R}.$$

Growth condition: $$|A_\alpha(x, \Theta^{\leq m})| \leq C(a(x) + \sum_{|\gamma| \leq m} |\vartheta^\gamma|^{p-1})\ \forall |\alpha| \leq m. \tag{2.285}$$

(Nonlinear) coercivity: $$\sum_{|\alpha| \leq m} A_\alpha(x, \Theta^{\leq m})\vartheta^\alpha \geq c \sum_{|\gamma| = m} |\vartheta^\gamma|^p - h(x). \tag{2.286}$$

Monotonicity: $$\sum_{|\alpha| \leq m} (A_\alpha(x, \Theta^{\leq m}) - A_\alpha(x, \Theta_1^{\leq m}))\,(\vartheta^\alpha - \vartheta_1^\alpha) \geq 0. \tag{2.287}$$

Strict monotonicity: $\sum_{|\gamma| \leq m} |\vartheta^\gamma - \vartheta_1^\gamma| > 0$ implies (2.288)

$$\sum_{|\alpha| \leq m} (A_\alpha(x, \Theta^{\leq m}) - A_\alpha(x, \Theta_1^{\leq m}))\,(\vartheta^\alpha - \vartheta_1^\alpha) > 0.$$

Uniform monotonicity: (2.289)

$$\sum_{|\alpha| \leq m} (A_\alpha(x, \Theta^{\leq m}) - A_\alpha(x, \Theta_1^{\leq m}))\,(\vartheta^\alpha - \vartheta_1^\alpha) \geq C \sum_{|\gamma| = m} |\vartheta^\gamma - \vartheta_1^\gamma|^p.$$

Other types of coercivity estimates towards (2.293) are discussed, e.g. by Showalter [589], p. 38 ff. By [678], Proposition 26.12, (2.284)–(2.287) imply the conditions of his Theorem 26A.

Theorem 2.73. Generalized solutions for (2.270), (2.275), [678], pp. 572 ff.: *Let (2.284)–(2.285) be satisfied and choose* $\mathcal{V} = W_0^{m,p}$. *Let*

$$F_\alpha : \mathcal{V} = W_0^{m,p}(\Omega) \to L^q(\Omega),\ (F_\alpha u)(x) := A_\alpha(x, u(x), \cdots, \nabla^m u(x)).$$

1. *Then for all* $|\alpha| \le m, \forall x \in \Omega$, *these Nemyckii operators* F_α *are well defined, continuous and bounded, see [678], Proposition 26.12, Proof.*
2. *With the above* $A_\alpha(x, u(x), \ldots, \nabla^m u(x))$ *define the* $a(u, v)$ *as in (2.274). Then there exists exactly one bounded operator* G*, cf. Proposition 2.67, such that, see [678], Proposition 26.12,*

$$G : \mathcal{V} \to \mathcal{V}' : a(u, v) = \langle Gu, v\rangle_{\mathcal{V}'\times\mathcal{V}}\ \forall u, v \in \mathcal{V} = W_0^{m,p}(\Omega) \tag{2.290}$$

$$\text{and } \|F_\alpha u\|_{L^q(\Omega)}, \|Gu\|_{W^{-m,q}(\Omega)} \le C\left(\|a\|_{L^q(\Omega)} + \left(\|u\|_{W_0^{m,p}(\Omega)}\right)^{p/q}\right).$$

Similarly to (2.275) and with $\langle f, v\rangle_{\mathcal{V}'\times\mathcal{V}}$, *the original problem (2.270) corresponds to the generalized problem: Determine for* $a(u, v)$ *in (2.290)*

$$u_0 \in \mathcal{V} = W_0^{m,p}(\Omega) \text{ s.t. } a(u_0, v) = \langle f, v\rangle_{\mathcal{V}'\times\mathcal{V}} \forall v \in \mathcal{V} \iff Gu_0 = f. \tag{2.291}$$

3. *Let, additionally, all the* A_α *be coercive and monotone, see (2.286) and (2.287). Then* G *is monotone, coercive, continuous and bounded, see [678], Proposition 26.12. So by Remark 2.66, parts 2.–4. of Theorem 2.68 are valid for for (2.291).*
4. *In particular, for any* $f \in W^{-m,q}(\Omega) = \mathcal{V}'$ *the (2.291) has a solution* $u_0 \in W_0^{m,p}(\Omega)$ *and the set of solutions is bounded, convex and closed. If all the* A_α *are additionally strictly monotone, see (2.288), Theorem 2.68, then* G *is strictly monotone as well and the above solution is unique. So the inverse* $G^{-1} : \mathcal{V}' \to \mathcal{V}$ *exists and is strictly monotone and bounded, see [678], Theorem 26A.*
5. *If all the* A_α *are, and hence* G *is uniformly monotone, see (2.289), or if* G *is strongly monotone, see (2.245), this* G^{-1} *is continuous or even Lipschitz-continuous.*

Remark 2.74. *For our above model problems (2.277), (2.280) and (2.283) with homogeneous Dirichlet boundary conditions we find:*

1. *There exists an* $\alpha_0 > 0$, *such that for all* $\alpha \in \mathbb{R}, |\alpha| \le \alpha_0$, *(2.279), corresponding to (2.277), has exactly one solution* $u_0 \in H_0^1(\Omega)$, *see Proposition 27.11 [678].*
2. *(2.281), corresponding to (2.280), has exactly one solution* $u_0 \in W_0^{1,p}(\Omega)$. *(2.281) is equivalent to* $Gu_0 = b$ *and* G *is uniformly monotone, coercive, continuous and bounded, see Proposition 26.10 [678].*
3. *More complicated conditions for (2.283) are discussed in [346], Chapters 12–15.*

We modify the preceding results, see Zeidler [678], Problem 27.6, by requiring, for a quasilinear operator G, some conditions only for the principal part:

Theorem 2.75. Modified coercivity and monotonicity only for the principal part: *For (2.270), let Ω be bounded, and choose $\mathcal{V} = W_0^{m,p}(\Omega)$. Assume the Carathéodory, the growth and the coercivity conditions, (2.284)–(2.286). Additionally impose the following modified coercivity (2.293) and the special case (2.294) of the strict monotonicity (2.288), both only for the principal part: With, cf. (2.269),*

$$\Theta^{\leq m} = (\Theta^0, \ldots, \Theta^{m-1}, \Theta^m) =: (d, \Theta^m), \Theta^m = (\vartheta^\alpha)_{|\alpha|=m}, \tag{2.292}$$

let a.e. $\forall x \in \Omega\ \forall \Theta^m\ \forall$ *bounded sets* $H \subset \mathbb{R}^{N_m - n_m}$:

$$\lim_{|\Theta^m|\to\infty} \sup_{d\in H} \sum_{|\alpha|=m} \frac{A_\alpha(x, d, \Theta^m)\vartheta^\alpha}{|\Theta^m| + |\Theta^m|^{p-1}} = \infty, \tag{2.293}$$

$$\sum_{|\alpha|=m} \left(A_\alpha(x, d, \Theta^m) - A_\alpha\left(x, d, \Theta_1^m\right)\right)\left(\vartheta^\alpha - \vartheta_1^\alpha\right) > 0, \forall \Theta^m \neq \Theta_1^m \in \mathbb{R}^{n_m}. \tag{2.294}$$

Then $\forall f \in L^q(\Omega)$ *the* $Gu = f$ *has a generalized solution* u_0 *in the sense of*

$$u_0 \in W_0^{m,p}(\Omega) : a(u_0, v) = b(v)\ \forall v \in W_0^{m,p}(\Omega). \tag{2.295}$$

Uniqueness results are available via linearization in Section 2.7.

The following result, again for *special semilinear operators,* is based upon the Carathéodory and growth conditions for the involved Nemyckii operator. We combine the linear operator A in strong divergence form, see (2.196),

$$Au = \sum_{|\alpha|,|\beta|\leq m} (-1)^{|\beta|}\partial^\beta(a_{\alpha\beta}\partial^\alpha u), u \in H^{2m}(\Omega) \tag{2.296}$$

with *Dirichlet boundary conditions.* We replace in (2.197) the $f(u)$ by $f(x,u)$ and determine u_0 such that

$$G(u_0)(x) := (A + \mu)u_0(x) + f(x, u_0(x)) = 0 \text{ a.e. } \forall x \in \Omega, u \in H_0^m \cap H^{2m}(\Omega). \tag{2.297}$$

As usual we find the following *generalized form,* for (2.297)

$$\text{with } a_\mu(u, v) = \int_\Omega \sum_{|\alpha|,|\beta|\leq m} a_{\alpha\beta}\partial^\alpha u\partial^\beta v + \mu uv dx \text{ determine} \tag{2.298}$$

$$u_0 \in \mathcal{V} := H_0^m(\Omega) : a_\mu(u_0, v) - \int_\Omega f(x, u_0(x))v(x)dx = 0\ \forall v \in \mathcal{V}. \tag{2.299}$$

For A, $a_\mu(\cdot,\cdot)$, f in (2.297), (2.298), and the domain Ω we impose the conditions

$$\begin{aligned} &\Omega \text{ bounded, } a_{\alpha\beta} : \Omega \to \mathbb{R} \text{ measurable and bounded } \forall |\alpha|, |\beta| \leq m, \\ &a(\cdot,\cdot) \text{ is } \mathcal{V}\text{-elliptic, } \iff a_\mu(\cdot,\cdot) \text{ is } \mathcal{V}\text{-coercive, for sufficiently large } \mu, \\ &f : \Omega \times \mathbb{R} \to \mathbb{R} \text{ satisfies the Carathéodory condition (2.258) for } n = 1, \\ &\text{(e.g. } f \text{ continuous) and the growth condition (2.262) for } p = 2, \text{ so fixed} \\ &b \in \mathbb{R}_+, 0 \leq a \in L^2(\Omega) \text{ exist, s.t. } |f(x, u)| \leq a(x) + b|u|\ \forall (x, u) \in \Omega \times \mathbb{R}. \end{aligned} \tag{2.300}$$

Theorem 2.76. Existence and uniqueness for semilinear problems, [678], Proposition 28.9: *Assume the conditions (2.300) for a $\mathcal{V}$-coercive $a_\mu(\cdot,\cdot)$, and for the boundary value problem (2.297) or its generalized form (2.299).*

1. Existence and uniqueness. *Let the function $u \to f(x,u)$ be monotone increasing on $\mathbb{R}$ for all $x \in \Omega$. Then (2.299) has exactly one solution.*
2. Existence. *Equation (2.299) has a solution, if the function $(x,u) \to f(x,u)$ satisfies*

$$\exists\ R > 0 \ \text{ s.t. } \forall u \text{ with } |u| \geq R: \ f(x,u)u \ \geq \ 0 \ \forall\ (x,u) \ \in \ \Omega \times \mathbb{R}.$$

We return to the question of whether quasilinear equations are elliptic equations in the sense of Definition 2.51, hence, the linearized equation is an elliptic equation. This has several important implications for our later discussion of the convergence of discretization methods and for the Fredholm alternative for quasilinear equations, see Section 2.7. For analytical results and our applications in discretizations, different types of conditions are imposed. We want to show that the above monotonicity implies ellipticity and we relate two important concepts.

We prove that an ellipticity condition for the linearization of the principal part essentially implies the nonlinear coercivity in the sense of (2.293) for the original quasilinear equation, cf. Theorem 2.75. Generalizing a remark by Ladyženskaja and Uralceva [464], p. 255, we formulate the following lemma, cf. (2.285), (2.289), (2.292)–(2.294).

Lemma 2.77. Quasilinear elliptic and nonlinear coercive equations: *Let the equation (2.270), (2.274) satisfy the following simplified growth and nonstandard ellipticity condition:*[17] *There exist positive continuous functions $\gamma, \Gamma : \mathbb{R}^{N_{m-1}} \to \mathbb{R}_+$ and $\mu : \mathbb{R}^{n_m} \to \mathbb{R}_+$ such that with*

$$|A_\alpha(x,d,\Theta^m = 0)| \leq \Gamma(d) \ \forall |\alpha| \leq m, \ \text{ and } \mu(\Theta^m) := (1 + |\Theta^m|)^{p-2}, \ p \geq 2, \quad (2.301)$$

$$\gamma(|d|)\mu(\Theta^m) \sum_{|\gamma|=m} |\vartheta^\gamma|^2 \leq \sum_{|\alpha|=|\beta|=m} \frac{\partial A_\alpha}{\partial \vartheta^\beta}(x,d,\Theta^m)\vartheta^\beta\vartheta^\alpha \leq \Gamma(|d|)\mu(\Theta^m) \sum_{|\gamma|=m} |\vartheta^\gamma|^2.$$

1. *If $0 \leq \gamma \leq \gamma(|d|)$ and $\Gamma(|d|)/\gamma(|d|)$ is bounded, we obtain uniform ellipticity. Then the* principal part *of the linearization evaluated in u_0, hence in $(x,d,\Theta^m) = (x, \partial^\alpha u_0, |\alpha| \leq m)$, is uniformly elliptic, cf. Definition 2.51.*
2. *For $p \geq 2$, the principal part is nonlinear coercive for $m \geq 1$, cf. (2.293) and, for $m = 1$ in the sense of (2.231) only for bounded $|\Theta^m|$. More precisely,*

$$\sum_{|\alpha|=m} A_\alpha(x,d,\Theta^m)\vartheta^\alpha \geq C\left(\gamma(|d|)|\Theta^m|^p - \Lambda(d)|\Theta^m|\right). \quad (2.302)$$

Proof. The uniform ellipticity of the principal part of the linearization is an immediate consequence of (2.301) combined with the continuity of the γ, Γ, μ evaluated in a bounded subdomain defined by $(x,d,\Theta^m) = (x, \partial^\alpha u_0, |\alpha| \leq m), x \in \Omega$.

[17] We have called (2.301) a nonstandard ellipticity condition, since we have chosen the nonstandard form of $\gamma(|d|)\mu(\Theta^m)$ instead of the usual ϵ.

For the coercivity, we modify the proof in [464], p. 255, and start with

$$(1+x)^\alpha \geq 1 + x^\alpha, \ \forall x \geq 0, 1 \leq \alpha \in \mathbb{R}. \tag{2.303}$$

We rescale, for $1 < x$, the $(1+x)^\alpha = x^\alpha(1+1/x)^\alpha$ and $1 + x^\alpha = x^\alpha(1+(1/x)^\alpha)$. So it suffices to apply the Taylor formula and find, with $0 < \theta < 1$, for

$$\forall 0 \leq x \leq 1, 1 \leq \alpha \in \mathbb{R}: \quad (1+x)^\alpha - 1 = \alpha(1+\theta x)^{\alpha-1}x \geq x \geq x^\alpha.$$

The last inequality $\alpha(1+\theta x)^{\alpha-1}x \geq x^\alpha$ certainly does not hold $\forall 0 \leq x, \alpha \leq 1$. In fact for a fixed $x > 0$ we get

$$\lim_{\alpha \to 0+} \alpha(1+\theta x)^{\alpha-1}x = 0 < \lim_{\alpha \to 0+} x^\alpha = 1.$$

Combining (2.303) with directional derivatives, integration and (2.301) we find with $\mu(\Theta^m) = (1+|\Theta^m|)^{p-2}$ and the mean value theorem 1.43 for operators, with $\Theta^m = (\vartheta^\beta | \beta| = m)$,

$$A_\alpha(x,d,\Theta^m) = A_\alpha(x,d,0) + \sum_{|\beta|=m} \vartheta^\beta \int_0^1 \frac{\partial A_\alpha}{\partial \vartheta^\beta}(x,d,\tau\Theta^m)d\tau, \text{hence,}$$

$$\sum_{|\alpha|=m} A_\alpha(x,d,\Theta^m)\vartheta^\alpha$$

$$\begin{aligned}
&= \sum_{|\alpha|=m} \Big(A_\alpha(x,d,0) + \sum_{|\beta|=m} \int_0^1 \frac{\partial A_\alpha}{\partial \vartheta^\beta}(x,d,\tau\Theta^m)d\tau\vartheta^\beta\Big)\vartheta^\alpha \\
&\geq -C\Gamma(|d|)|\Theta^m| + \gamma(|d|)\int_0^1 ((1+\tau|\Theta^m|)^{p-2})d\tau \sum_{|\gamma|=m} |\vartheta^\gamma|^2 \\
&\geq -C\Gamma(|d|)|\Theta^m| + \gamma(|d|)\frac{1}{|\Theta^m|(p-1)}(1+\tau|\Theta^m|)^{p-1}|_0^1 \sum_{|\gamma|=m} |\vartheta^\gamma|^2 \\
&\geq -C\Gamma(|d|)|\Theta^m| + \gamma(|d|)\frac{1}{|\Theta^m|(p-1)}((1+|\Theta^m|)^{p-1} - 1^{p-1}) \sum_{|\gamma|=m} |\vartheta^\gamma|^2 \\
&\overset{(2.303)}{\geq} -C\Gamma(|d|)|\Theta^m| + \gamma(|d|)\frac{|\Theta^m|^{p-1}}{|\Theta^m|(p-1)}|\Theta^m|^2, \text{ for } p \geq 2, \quad p \in \mathbb{R}, \\
&\geq -C\Gamma(|d|)|\Theta^m| + \gamma(|d|)|\Theta^m|^p.
\end{aligned}$$

γ, Γ are continuous with respect to d, so they are bounded for all bounded sets $H \subset \mathbb{R}^{N_m - n_m}$, see (2.73), implying (2.293) since with $p \geq 2$

$$\lim_{|\Theta^m| \to \infty} \sup_{d \in H} C\frac{\gamma(|d|)|\Theta^m|^p - \Gamma(|d|)|\Theta^m|}{|\Theta^m| + |\Theta^m|^{p-1}} = \infty \ \forall \text{ bounded } H \subset \mathbb{R}^{N_m - n_m}.$$

This proves the claim. □

General regularity results of an as "simple" form as in Section 2.4 do not seem to exist or did not find their way into the general monographs summarized here. In Subsection 2.5.4 we presented some examples for order 2, and in Subsection 2.6.6 and Theorem 2.116 we will present results for general systems of order $2m$.

For a lot of other results we refer to the literature. Morrey [500] and Giaquinta and Hildebrandt [341, 342] study quasilinear equations induced by variational problems. Giaquinta and Hildebrandt [341, 342] essentially consider second order problems. Then the functional in (2.276) is often assumed in the form $J(u) := \int_a^b F(u, \nabla u)dt$ where $u = u(t), a \leq t \leq b$, indicates a curve on a surface. Under appropriate conditions a minimizing curve is then $u_0 \in C^2([a, b])$, cf. [341, 342], Chapter 1, Section 3, Proposition 3 and Chapter 8, Section 4, Proposition 2. Relations of minima of convex functionals or duality and conjugate functionals and quasilinear equations of orders 2 and $2m$ are discussed as well by Zeidler [675], Sections 42.7 and 51.7. Koshelev and Chelkak [449, 450] study regularity for systems, and Ladyženskaja and Uralceva [463, 464, 466], for second order problems of elliptic and parabolic type. Ladyženskaja [462] discusses boundary value problems of mathematical physics. Morrey in his many papers uses multiple integrals and the calculus of variations, and studies differentiability of solutions for nonlinear elliptic equations and systems, e.g. [500–504].

2.5.7 Fully nonlinear elliptic equations of orders 2, m and $2m$

We turn to the fully nonlinear case and start with the most important order 2, see Gilbarg and Trudinger [346] and Chen and Wu [170]. We require $\Omega \subset \mathbb{R}^n$ satisfying (2.5) and a real valued function, G^w, cf. (2.165), (2.166), of the form

$$G^w : \mathcal{D}(G^w) \to \mathbb{R}, w = (x, z, p, r) \in \mathcal{D}(G^w) \subset \mathcal{U}_o = \Omega \times \mathbb{R} \times \mathbb{R}^n \times \mathbb{R}^{n^2}; \quad (2.304)$$

we restrict the elements in $\mathbb{R}^{n \times n}$ to the $n(n+1)/2$-dimensional subspace of symmetric real valued $n \times n$ matrices. Choose functions $u : \Omega \to \mathbb{R}$, such that

$$\mathcal{D}(G) := \{u \in H^2(\Omega) : w_u(x) := (x, u(x), \nabla u(x), \nabla^2 u(x)) \in \mathcal{D}(G^w) \forall x \in \Omega\}. \quad (2.305)$$

Remark. Fully nonlinear PDE of order 2 and 2m

1. *A fully nonlinear PDE of order 2 is none of the above special cases (2.169)–(2.172). It has, for a Dirichlet boundary value problem, the strong classical form, hence*

$$G : \mathcal{D}(G) \subset H^2(\Omega) \to L^2(\Omega), \text{ with } u, u_0 \in \mathcal{D}(G) \text{ s.t.} \quad (2.306)$$
$$L^2(\Omega) \ni G(u_0) := G^w(\cdot, u_0(\cdot), \nabla u_0(\cdot), \nabla^2 u_0(\cdot)) = 0 \text{ on } \Omega, u_0 = \varphi \text{ on } \partial\Omega.$$

   In particular, the spaces of ansatz functions, $u, u_0 \in \mathcal{D}(G) \subset H^2(\Omega)$, and of test functions, $v \in L^2(\Omega)$, are different.
2. *The same situation holds for the equations of order $2m$ and m at the end of this subsection and the systems in Subsection 2.6.8.*

In Definition 2.51 we have introduced *(uniformly) elliptic* G. For second order equations, $m = 1$, the -1 in (2.175), (2.176) is often omitted. Choose $w = (x, z, p, r) \in \mathcal{D}(G^w)$ as in (2.304), a u_1 with its $w_1(x) := (x, u_1(x), \nabla u_1(x), \nabla^2 u_1(x)) \in \mathcal{D}(G^w)$, and $w \in \mathcal{U}_{w_1}$, a neighborhood of w_1. We assume for $0 \neq \vartheta \in \mathbb{R}^n$

$$0 < \lambda|\vartheta|^2 \leq \sum_{i,j=1}^{n} \frac{\partial G^w}{\partial r_{ij}}(w)\vartheta^i\vartheta^j \leq \Lambda|\vartheta|^2 \forall w = (x, u, p, r) \in \mathcal{U}_{w_1} \subset \mathcal{D}(G^w), \quad (2.307)$$

for its principal part. For a *real symmetric* matrix $\partial G^w / \partial r_{ij}(w)$ with its real *eigenvalues* the minimal and maximal ***eigenvalues*** satisfy $\lambda \leq \lambda(w) \leq \Lambda(w) \leq \Lambda$.

Example 2.78. The two of the most important examples:

1. *Monge–Ampère equation*
   This is, for $u : \Omega \to \mathbb{R}$ and $n \geq 2$

   $$0 = G(u) := \det \nabla^2 u - f(x, u, \nabla u) = u_{xx}u_{yy} - u_{xy}^2 - f(x, u, \nabla u) \text{ for } n = 2. \quad (2.308)$$

   We claim that this G is uniformly elliptic in u_1 if $\nabla^2 u_1$ is strictly positive definite and symmetric in Ω: standard textbooks show that, for a strictly convex $u_1 \in C^2(\Omega)$ on a convex Ω, this is correct. Furthermore, a matrix and its inverse are simultaneously strictly positive definite and have a positive determinant. We define the cofactor of the element $r_{i,j} = \partial^i\partial^j u = \partial^2 u/\partial x_i \partial x_j$ of the matrix $\nabla^2 u = r = [r_{ij} = \partial^i\partial^j u]_{i,j=1}^n$, symmetric for $u \in C^2(\Omega)$: cancel row i and column j in $\nabla^2 u$ to get R_{ij} and define, independently of $r_{i,j} = \partial^i\partial^j u$, the cofactor of $\partial^i\partial^j u$ as $(-1)^{i+j} \det R_{ij}$. Then

   $$\det \nabla^2 u = \sum_{j=1}^{n} \partial^i\partial^j u \text{ cofactor } \partial^i\partial^j u = \sum_{j=1}^{n} \partial^i\partial^j u \ \frac{\partial \det \nabla^2 u}{\partial r_{ij}}(u)$$

   $$\Rightarrow \quad (\det \nabla^2)'(u_1)u = \sum_{j=1}^{n} \partial^i\partial^j u \ \frac{\partial \det \nabla^2 u}{\partial r_{ij}}(u_1). \quad (2.309)$$

   The inverse of a nonsingular matrix is obtained by the matrix of its cofactors as:

   $$\text{with } B := [\text{ cofactor } \partial^i\partial^j u_1]_{i,j=1}^n \text{ we get } (\nabla^2 u_1)^{-1} = B^T / \det \nabla^2 u_1.$$

   This (2.309) implies with (2.175), (2.307),

   $$\vartheta^T(\det \nabla^2)'(u_1)\vartheta = \sum_{i,j=1}^{n} \vartheta^i\vartheta^j \ \frac{\partial \det \nabla^2 u}{\partial r_{ij}}(u_1) = \det \nabla^2 u_1 \ \ \vartheta^T \ (\nabla^2 u)^{-1}(u_1) \ \vartheta,$$

   hence our claim by (2.307).
   So the $G(u)$ in (2.308) is uniformly elliptic in u_1 for a symmetric strictly positive matrix $r = \nabla^2 u_1$ in Ω, hence for a function $u_1 \in C^2(\Omega)$, strictly convex $\forall x \in \Omega$ on a convex Ω. With $\det(\nabla^2 u) > 0$, uniformly convex solutions are only possible for positive f, [345], pp. 441 ff. For partially nonelliptic generalizations, see [345], p. 467ff. and Courant and Hilbert [214], p. 324. According to Haltiner and Kasahara the leading term in the balance equation in dynamical meteorology

has the form (2.308) [389,437]. In Westcott [664] it is related to geometric optics. Among the many applications in geometry we mention the

2. *Equation for a surface with prescribed Gauss curvature* $K = K(x)$ at x
This is closely related to the Monge–Ampère Equation. Equation (2.308) has to be modified as

$$G(u) := \det \nabla^2 u - K(1 + |\nabla u|^2)^{(n+2)/2} = 0, \ K = K(x) > 0. \tag{2.310}$$

Again $G(u)$ is elliptic only for uniformly convex functions $u \in C^2(\Omega)$ in a convex Ω. Our FEMs for this class of problems are interesting for aspects of global geometry. The existence of surfaces with prescribed Gauss curvature is known for appropriate K. Specific properties would become more flexibly available via our FEMs than by the available finite difference methods, see e.g. Crandall and Lions [227]. Our FEM allows us FEs in $C^1 \cap \mathcal{W}$ on quasiuniform grids. It yields higher orders of convergence, cf. Theorem 5.18, than the difference methods of order 2, e.g. in Chapter 8 or in [227]. Both methods are only defined on equidistant grids. ■

For existence and uniqueness results we have to impose *growth conditions.* We distinguish the cases $n = 2$ and $n \geq 2$ and evaluate the derivatives G^w_x, G^w_z, G^w_p, for G^w_r, compare (2.307). We assume nonnegative $\lambda = \lambda(\cdot)$ and $\Lambda = \Lambda(\cdot), \mu = \mu(\cdot)$ in (2.307), (2.311)–(2.312) depending upon $|z|$, nonincreasing and nondecreasing in $z \in [0, \infty)$, respectively, and require for $w = (x, z, p, r), x \in \partial\Omega$,

$$\begin{aligned} \text{for } n = 2: \ & \left|G^w_z(w)\right|, \left|G^w_p(w)\right| \leq \lambda\mu \\ & \left|G^w_x(w)\right| \leq \mu\lambda(1 + |p| + |r|) \text{ and} \end{aligned} \tag{2.311}$$

$$\begin{aligned} \text{for } n \geq 2: \ & \left|G^w_z(w)\right|, \left|G^w_p(w)\right|, \left|G^w_{rx}(w)\right|, \left|G^w_{px}(w)\right|, \left|G^w_{zx}(w)\right| \leq \lambda\mu \\ & \left|G^w_x(w)\right|, \left|G^w_{xx}(w)\right| \leq \mu\lambda(1 + |p| + |r|). \end{aligned} \tag{2.312}$$

Theorem 2.79. Existence and uniqueness for fully nonlinear problems in $\mathbb{R}^2$, [346], Theorem 17.12: *Assume* $n = 2$, $\partial\Omega \in C^3$, $\varphi \in C^3(\bar{\Omega})$. *Let* G *satisfy the conditions (2.307), (2.311), and* $G^w_z \leq 0$ *in* $\mathcal{U}_o$, *see (2.304). Then the classical Dirichlet problem (2.306) has a unique solution* $u_0 \in C^{2,\alpha}(\bar{\Omega})$ *for all* $0 < \alpha < 1$.

Ω is called satisfying an *exterior sphere condition* for every $\xi \in \partial\Omega$, if there exist a ball $B = B_R(y) \subset \mathbb{R}^n \setminus \Omega$ such that $\bar{B} \cap \bar{\Omega} = \{\xi\}$.

Theorem 2.80. Regular solutions for fully nonlinear problems in $\mathbb{R}^n, n \geq 2$, [346], Theorem 17.17: *Assume* $n \geq 2$, $\varphi \in C(\partial\Omega)$. *Let* Ω *satisfy an exterior sphere condition for every* $\xi \in \partial\Omega$, $G^w \in C^2(\mathcal{U}_o)$ *be concave (or convex) with respect to* (z, p, r), *nonincreasing with respect to* z *and let* G *satisfy the conditions (2.307), (2.312). Then the classical Dirichlet problem (2.306) has a unique solution* $u_0 \in C^2(\Omega) \cap C(\bar{\Omega})$.

More detailed results for the above special cases in Example 2.78 are presented in [346].

Skrypnik [591], Theorem 3.2.3, considers the Monge–Ampère equation (2.308) on the two-dimensional ball $\Omega = B_R(0)$ with center 0 and imposes the boundary condition $(u_0 - g)|_{\partial B_R(0)} = 0$. He assumes the existence of functions

$$\phi : B_R(0) \to \mathbb{R}_+, \; f_0 : \mathbb{R}^2 \to \mathbb{R}_+, \text{ s.t., with } m := \max_{x \in B_R(0)} g(x)$$

$$f(x,u,p) \leq \phi(\|x\|^2) f_0(\|p\|^2) \text{ for } x,p,s \in \mathbb{R}^2, \; \mathbb{R} \ni u \leq m \text{ and let} \tag{2.313}$$

$$\int_{B_R(0)} \phi(\|x\|^2)dx \leq \int_{\mathbb{R}^2} \inf_{\|s-p\|^2 < M_H} 1/f_0(\|s\|^2)dp,$$

where M_H is the lower winding number of the curve defined by $(u_0 - g)|_{\partial B_R(0)} = 0$, see, e.g. Bakelman [55]. Finally, let $\mathcal{G} := \{(x,u,p) : x \in B_R(0), u \in \mathbb{R}, p \in \mathbb{R}^2\}$ and

$$H^4_+(\Omega) := \{u \in H^4(\Omega) : (\partial^{20}u\partial^{02}u - (\partial^{11}u)^2)(x) > 0, \partial^{20}u(x) > 0 \forall x \in \overline{B_R(0)}\}.$$

Theorem 2.81. Regular solutions for the Monge–Ampère equation, Skrypnik [591], Theorem 3.2.3: *Let (2.313) be satisfied and $f \in C^{2,\gamma}(\mathcal{G}), g \in C^{4,\gamma}(\partial\Omega)$. Then (2.308) has at least one solution in $H^4_+(\Omega)$.*

We continue with some existence and regularity results, elaborated by Chen and Wu [170], Chapter 7, Theorem 7.4 and Remark on p. 117, for general fully nonlinear equations (2.306) of order 2. They impose six conditions $(F_1) - (F_6)$ for (2.306):

$$\text{Let} \quad G(x,z,p,r) \in \mathbb{R}, (x,z,p,r) \in \mathcal{D}(G) \subset \mathcal{U}_o = \Omega \times \mathbb{R} \times \mathbb{R}^n \times \mathbb{R}^{n^2}, \text{ and} \tag{2.314}$$

assume a constant $\bar{\mu}$ and nonnegative, nondecreasing $\mu_i : [0,\infty) \to \mathbb{R}_+, i = 1,2,3,$ s.t.

$$(F_1) : \exists 0 < \lambda = \lambda(x,z,p), \Lambda = \Lambda(x,z,p) \text{ s.t., with the identity matrix } I$$

$$\lambda \; I \leq \left(\frac{\partial G}{\partial r_{ij}}\right)^n_{i,j=1} \leq \Lambda \; I, \text{ and } \Lambda/\lambda \leq \mu_1(|z|) \text{ on } \mathcal{U}_o;$$

$$(F_2) : \quad |G(x,z,p,0)| \leq \lambda\mu_2(|z|)(1+|p|^2);$$

$$(F_3) : \quad \big[(1+|p|)^{-1}|G_x| + |G_z| + (1+|p|)|G_p|\big](x,z,p,r)$$

$$\leq \lambda\mu_3(|z|)(1+|p|^2+|r|)\forall(x,z,p,r) \in \mathcal{U}_o;$$

$$(F_4) : \quad \text{on } \mathcal{U}_o : G(x,z,p,r) \text{ is concave with respect to } r;$$

$$(F_5) : \quad G(x,z,p,0) \text{ sign } z \leq \lambda\bar{\mu}(1+|p|);$$

$$(F_6) : \quad G_z(x,z,p,r) \leq 0 \; \forall(x,z,p,r) \in \mathcal{U}_o. \tag{2.315}$$

Theorem 2.82. Regular solution for fully nonlinear problems:

1. *Let the conditions (2.314)–(2.315) be satisfied, Λ be bounded on $\Omega \times \mathbb{R} \times \mathbb{R}^n$. Then there exists $0 < \alpha < 1$ with the following property: For any $0 < \gamma < \alpha < 1$ with $\partial\Omega \in C^{2,\gamma}$ $\varphi \in C^{2,\gamma}(\bar{\Omega})$, for the Dirichlet problem (2.306) there exists a unique solution $u_0 \in C^{2,\gamma}(\bar{\Omega})$. This α depends only upon $n, \mu_1, \mu_2, \mu_3, \|\varphi\|_{C^{2,\gamma}(\bar{\Omega})}$ and Ω.*

2. *If* $G \in C^{k,\alpha}(\mathcal{U}_o), k \geq 1$, *and* $\partial\Omega \in C^{2+k,\gamma}$, $\varphi \in C^{2+k,\gamma}(\bar{\Omega})$, *then the solution* $u_0 \in C^{2,\gamma}(\bar{\Omega}) \cap C^{k+2,\alpha}(\Omega)$. *For* $G \in C^{k,\alpha}(\bar{\mathcal{U}}_o)$, *it is* $u_0 \in C^{k+2,\gamma}(\bar{\Omega})$.

We turn to some existence and regularity results for fully nonlinear equations of order $2m$. Skrypnik [591] seems to be the author of the only textbook studying these problems. He combines monotone operators and degree theory to get his results for quasilinear and fully nonlinear elliptic equations. He discusses some of these problems on narrow strips and in perforated domains. To give a flavor of his existence and regularity results we cite one of his theorems, a "conditional" result. For the original problem, he formulates a family of parameter dependent problems ($t \in [0,1]$) with $t = 0$ and $t = 1$ for the original problem and a problem with appropriate degree properties. He determines a solution u_0 for the nonlinear Dirichlet problem. Let a function G be given such that, with N_{2m} as in (2.285), cf. Remark concerning (2.306)

$$G^w \in C(\mathcal{D}(G^w)),\ \mathcal{D}(G^w) := [0,1] \times \mathcal{D}(G_0^w) = [0,1] \times \Gamma \subset [0,1] \times \bar{\Omega} \times \mathbb{R}^{N_{2m}},$$

$$\text{with appropriate } N_{2m},\ \text{let } G^w(0,x,-\Theta^{N_{2m}}) = -G^w(0,x,\Theta^{N_{2m}}),\ \text{and}$$

$$\forall t \in [0,1]:\ G^w(t,\ldots) \in C^{l+1}(\mathcal{D}(G_0^w)),\ l \geq [n/2]+1, \quad \text{where} \tag{2.316}$$

$$u, u_1 \in \mathcal{D}(G) \text{ if, e.g. } w_1(x) := (x, \partial^\alpha u_1(x), |\alpha| \leq 2m)\ \in \mathcal{D}(G_0^w)\, \forall x \in \Omega,$$

$$G(u) :=\ G(t=0,u) := G^w(t=0,\cdot,u(\cdot),\ldots,\nabla^{2m}u(\cdot))$$

$$G: \mathcal{D}(G) \subset \mathcal{U} := H_0^m(\Omega) \cap H^{2m}(\Omega) \to \mathcal{V} := L^2(\Omega),\ \text{determine } u_0 \text{ from} \tag{2.317}$$

$$G(u_0(\cdot)) = G^w(t=0,\cdot,u_0,\ldots,\nabla^{2m}u_0)(\cdot) = 0, u_0 \in \mathcal{X}_0 := H_0^m(\Omega) \cap H^{2m+l}(\Omega).$$

He assumes the existence of constants $0 < k, 0 < \gamma < 1, 0 < \Lambda < \infty$ for (2.316), and modifies the *ellipticity condition,* compare (2.232), such that

$$\exists k \in \mathbb{R}_+, 0 < \gamma < 1:\ \forall t \in [0,1], v \in \mathcal{X}_0:\ G(t,v) = 0 \,\text{implies}\, \|v\|_{C^{2m,\gamma}(\bar{\Omega})} \leq k,\ \text{and}$$

$$k^{-1}|\vartheta|^{2m} \leq \sum_{|\alpha|=2m} \left[\partial^\alpha\left(G^w(t,\cdot,v,\cdots,\nabla^{2m}v)\right)(x)\right]\vartheta^\alpha \leq \Lambda|\vartheta|^{2m}\ \forall x \in \Omega. \tag{2.318}$$

Theorem 2.83. Regular solutions, Skrypnik [591], Theorem 3.2.2: *Let G satisfy (2.316), (2.318) and let $\partial\Omega \in C^\infty$ for a bounded domain in $\mathbb{R}^n$. Then (2.317) has, for $t = 0$, at least one solution $u_0 \in \mathcal{U}$.*

We finish with a *nonlinear PDE of order m,* see Definition 2.85, in the form

$$G(u) := G(x,u,\nabla u,\ldots,\nabla^m u) \text{ and } G(u)(x) = g(x)\ \forall\ x \in \Omega. \tag{2.319}$$

Theorem 2.84. Local solvability and regularity result – Taylor, [620], Chapter 14, Theorem 3.1: *Let $g \in C^\infty(\Omega)$, and let $u^1 \in C^\infty(\Omega)$ satisfy (2.319) at $x = x_0$,*

$$G(u^1)(x) = g(x) \textit{ for } x = x_0$$

and let G be m-elliptic at u^1. Then for any ℓ there exists a $u \in C^\ell(\Omega)$ such that

$$G(u)(x) = g(x) \textit{ in a neighborhood of } x_0.$$

2.6 Linear and nonlinear elliptic systems

2.6.1 Introduction

In this section we discuss systems of q elliptic differential equations essentially in the Hilbert Sobolev space setting for two reasons: Practically all available results in the textbooks consulted are formulated for the Hilbert case. It is straightforward as in Subsection 2.4.4 to extend the following results, based upon the H^m-coercivity, to the $W^{m,p}$ case for $2 \leq p < \infty$. The $1 \leq p < 2$ usually have to be excluded. As in Corollary 2.44, 3., 4., we modify them for $2 < p \leq \infty$: these $W^{m,p}(\Omega) \hookrightarrow H^m(\Omega)$ are densely and continuously embedded. Consequently, all H^m-coercivity based existence, uniqueness and Fredholm results, including the numerical stability and convergence of all our discretization methods, remain correct for the $W^{m,p}$ case for $2 < p \leq \infty$ with respect to (discrete) H^m norms. We will explicitly formulate this fact for the particularly ineresting Theorems 2.89 and 2.104.

Comparing equations with systems shows an astonishing difference: the definition of ellipticity for linear and nonlinear equations coincides for nearly all aspects. For order $2m$ they are direct generalizations of those for second order. This is no longer correct for systems. In Subsection 2.6.2, Definition 2.85, we introduce the concept of systems of elliptic differential equations, a concept of ellipticity, pretty different from those previously introduced. It is motivated by important examples, including the Navier–Stokes and von Kármán equations. Starting with Subsection 2.6.3, and excluding Navier–Stokes equations, we turn back to the definitions of ellipticity, similar to or directly generalizing those for the case $m = q = 1$. However, depending upon the chosen goal, e.g. existence and uniqueness or regularity, different types of ellipticity become relevant. Subsection 2.6.3 is devoted to linear second order systems, cf. (2.343), extended to quasilinear second order systems in Subsection 2.6.4, cf. (2.370), (2.371) for existence and (2.384) for regularity. The last three Subsections 2.6.5, 2.6.6 and 2.6.8 generalize some of these results to order $2m$ of linear, cf. (2.394), (2.401), quasilinear, cf. (2.437), and fully nonlinear elliptic systems, cf. (2.446). Not much seems to be known about unified conditions for the different aspects and orders.

Systems of elliptic differential equations are intensively studied in different monographs, e.g. Hörmander [402], Garabedian [336], Lions and Magenes [478–480], Chen and Wu [170], Taylor [618–620], and Koshelev and Chelkak [449–451]. For systems the ellipticity is not uniformly defined. In particular, existence and regularity are discussed by Ladyženskaja and Uralceva [463–466], Koshelev and Chelkak [450, 451], Skrypnik [590, 591], Morrey [500], Agmon, Douglis and Nirenberg [3], Taylor [620], Zeidler [675–678], Giaquinta and Hildebrandt [341], Nečas [510], and Kozlov Maz'ya and Rossmann [452]. Here we include some papers as well: Wahl and Grunau [378,659], Luckhaus and Alt [23, 481], Nedev [512], and Boccuto and Mitidieri [99]. These books and papers show, e.g. [23,99,481,659], that usually generalizations of regularity results from the linear to the nonlinear case require new techniques and a careful analysis. Regularity for nonlinear elliptic systems with interesting applications to stochastic games, ergodic control, semiconductors, plasticity, and the stationary Navier–Stokes equations are studied by Bensoussan and Frehse [81]. The regularity results for the stationary Navier–Stokes equations would be interesting for Section 2.8. However, in [81] the main goals are specific "maximum-like solutions" for higher dimensions $n > 3$, which might give insight into the nonstationary Navier–Stokes equations. We summarize some essential results of Koshelev and Chelkak [449–451] and Skrypnik [591], in Subsections 2.6.4 and 2.6.6, 2.6.8.

2.6.2 General systems of elliptic differential equations

In this subsection we give information for general systems of elliptic equations in the sense of Agmon, Douglis and Nirenberg [3]. The stationary Navier–Stokes equation is one of the most important of these systems of elliptic differential equations. It certainly motivated the new definition of ellipticity in this and the next subsections. Although we discuss the Navier–Stokes equation in a separate section, we memorize it here. We want to motivate the general systems, discussed here in contrast to the following subsections. As before we usually omit the index s for the strong form, possible for a regular enough situation. The context indicates the appropriate form. In this subsection essentially all operators are in strong form.

Already in Chapter 1 we introduced the Navier–Stokes equation. For the stationary form, the solution, $\vec{u}^0, p_0$ satisfies

$$G(\vec{u}^0, p_0) := \begin{pmatrix} -\nu\Delta\vec{u}^0 + \sum\limits_{i=1}^{n} u_i^0 \partial^i \vec{u}^0 + \nabla\, p_0 \\ -\operatorname{div}\, \vec{u}^0 \end{pmatrix} = \begin{pmatrix} \vec{f} \\ 0 \end{pmatrix} \quad \text{in } \Omega, \tag{2.320}$$

$$\vec{u}^0 = 0 \text{ on } \partial\Omega, \quad \int_\Omega p_0 dx = 0, \quad \Omega \text{ satisfies (2.5) and}$$

$$\vec{u} = (u_1, \ldots, u_n)^\mathsf{T}, \ \vec{v} = (v_1, \ldots, v_n)^\mathsf{T}, \vec{f} = (f_1, \ldots, f_n)^\mathsf{T} : \Omega \subset \mathbb{R}^n \to \mathbb{R}^n,$$
$$\operatorname{div}\, \vec{u} := \partial u_1/\partial x_1 + \cdots + \partial u_n/\partial x_n, \ p : \Omega \to \mathbb{R}, \ \vec{u}, p, \ldots \text{ sufficiently smooth,}$$

where $\vec{u}, p$ and $\vec{f}$ denote the velocity, pressure and forcing term of an incompressible medium and $\vec{v}$ a test function. The condition $\int_\Omega p_0 dx = 0$ enforces a unique p_0. Note that (2.320) is a system of $n+1$ differential equations for $n+1$ unknown functions submitted to the boundary conditions and an integral constraint. For physical applications only $n \le 3$ is interesting, for higher n, cf. [81].

The derivative G' of G, evaluated at $(\vec{u}, p), \vec{u} \ne 0$, and applied to an increment $(\vec{w}, r)$, is the *linearized Navier–Stokes operator,* again an operator with $n+1$ components,

$$G'(\vec{u}, p)(\vec{w}, r) = \begin{pmatrix} -\nu\Delta\vec{w} + \sum_{i=1}^{n}(w_i\partial^i\vec{u} + u_i\partial^i\vec{w}) + \nabla r \\ -\operatorname{div}\, \vec{w} \end{pmatrix}. \tag{2.321}$$

The special case $G'(\vec{u} \equiv 0, p_0)$ and $\nu = 1$ applied to an increment $(\vec{u}, p)$ yields the *Stokes operator,* S, and the corresponding *Stokes equation,*

$$G_{\vec{u}}(0, p_0)(\vec{u}, p) = S(\vec{u}, p) = \begin{pmatrix} -\Delta\vec{u} + \nabla p \\ -\operatorname{div}\, \vec{u} \end{pmatrix} = \begin{pmatrix} \vec{f} \\ 0 \end{pmatrix} \quad \text{in } \Omega, \tag{2.322}$$
$$\vec{u} \quad = 0 \text{ on } \partial\Omega, \quad \textstyle\int_\Omega p dx = 0.$$

The nonlinear system (2.320) and its linearizations (2.321) and (2.322) do not allow a direct generalization of the previous framework (in contrast to the situation in the next subsections). Again, the following notation and even some results for linear systems of q equations for q functions $\vec{u} = (u_1, \dots, u_q)$ are often similar to the concepts of linear algebra, cf. (2.73), (2.269), for $\vartheta \in \mathbb{R}^n$ and $\vartheta^\alpha, \partial^\alpha u_j$.

Definition 2.85. Elliptic systems – Agmon *et al.* [2,3]:

1. *A linear system of differential equations has the form,*

$$L\vec{u} := \left(\sum_{j=1}^{q} L^{ij} u_j \right)_{i=1}^{q} = \vec{f} := (f^i)_{i=1}^q, \vec{u} = (u_1, \dots, u_q)^\mathsf{T} \text{ defined in } \Omega \subset \mathbb{R}^n,$$

$$L^{ij} := \sum_{|\alpha| \le k_{ij}} a_\alpha^{ij}(x) \partial^\alpha, \ i, j = 1, \dots, q, \ \ \partial^\alpha u_j = u_j \text{ for } |\alpha| = 0, \tag{2.323}$$

usually with $a_\alpha^{ij} \in C(\Omega)$.[18] *We choose obviously not unique*

$$m_1, \dots, m_q, \ m_1', \dots, m_q' \text{ s.t. } k_{ij} \le m_i + m_j', \ i, j = 1, \dots, q, \tag{2.324}$$

with the $2m$ *in (2.328).* The principal part *of* L^{ij} *or its symbol are defined by the components of its "highest order" tems*

$$L_p^{ij} := \sum_{|\alpha| = m_i + m_j'} a_\alpha^{ij} \partial^\alpha \text{ or } L_p^{ij}(x, \vartheta) := \sum_{|\alpha| = m_i + m_j'} a_\alpha^{ij}(x) \vartheta^\alpha, \tag{2.325}$$

for $i, j = 1, \dots, q$. *If* $k_{ij} < m_i + m_j'$, *these* $a_\alpha^{ij} = 0$, *for* $|\alpha| = m_i + m_j'$, *hence* $L_p^{ij} = 0$.

2. *We define the* symbol, L_p, *and* characteristic polynomial, $\det L_p$, *as*

$$L_p(x, \vartheta) := \left(L_p^{ij}(x, \vartheta)\right)_{i,j=1}^q \forall x \in \Omega, \ \det L_p(x, \vartheta) = \det \left(L_p^{ij}(x, \vartheta)\right)_{i,j=1}^q. \tag{2.326}$$

3. *A linear system (2.323) is called* elliptic *in* $x \in \Omega$, *if* $L_p(x, \vartheta)$ *is invertible,*

$$\text{the } q \times q \text{ matrix } L_p(x, \vartheta) \text{is invertible} \Leftrightarrow \det L_p(x, \vartheta) \ne 0 \ \forall \ 0 \ne \vartheta \in \mathbb{R}^n. \tag{2.327}$$

Often, stronger than (2.327), $\epsilon > 0$ *exists, such that (only a.e. for* $a_\alpha \in L^\infty(\Omega)$*)*

$$|\det L_p(x, \vartheta)| \ge \epsilon |\vartheta|^{2m} \ \forall \times \epsilon \Omega, \ 0 \ne \vartheta \in \mathbb{R}^n \text{ with } 2m := \sum_{i=1}^{q} (m_i + m_i'). \tag{2.328}$$

4. *We call* $2m : = \max_{i,j=1,\dots,q} (m_i + m_j')$ *the* order *of an elliptic system.*
5. *As in Definition 2.51 we call a* nonlinear system elliptic *if its linearization is elliptic. Obviously, we obtain the original ellipticity for* $1 = q \le m = m_1 = m_1'$.

[18] Several types of notation are used in the literature, each with pros and cons. For linear systems (2.323) the notation $L\vec{u}$ as a matrix of operators $(L^{ij})_{i,j=1}^q$ or coefficients $(a_\alpha^{ij})_{i,j=1}^q$ applied to a vector $\vec{u} = (u_1, \dots, u_q)^\mathsf{T}$ of functions is advantageous. This does not work too well for quasilinear systems. In the Einstein convention the same index and exponent i indicates the sum over i. In (2.323) and (2.339) this could be used directly for the j and k, l, so $\sum_{j=1}^q$ and $\sum_{k,l=1}^n$, respectively, would be omitted. This Einstein convention is not too common for numerical methods and can be applied only in parts of this section. So we decided to use the componentwise or vectorial notation.

6. *A modification is discussed by Taylor [618], pp. 245 ff. and pp. 379 ff.: Replace in (2.323) the* $|\alpha| \le k_{ij}$ *and in (2.325) the* $|\alpha| = m_i + m'_j$ *by* $|\alpha| \le m$ *and* $|\alpha| = m$. *We denote this modified linear system (2.323) as* m-elliptic, *if* $L_p(x, \vartheta)$ *is invertible* $\forall x \in \Omega, \vartheta \in \mathbb{R}^n$.

In the literature "strong ellipticity" is used in different ways. We use "strong ellipticity" for (2.343) below, but not for (2.328).

Example 2.86. We list several examples of elliptic systems, in particular two systems introduced in Chapter 1. ■

1. *The Stokes operator* $L = S$ *and the linearized Navier–Stokes operator:* In (2.322) we choose

$$q := n+1, \vec{u} = (u_1, \dots, u_n, p)^\mathsf{T}, \ \vec{f} = (f_1, \dots, f_n, 0)^\mathsf{T}, \tag{2.329}$$
$$L^{ii} := -\Delta, \ L^{iq} := -L^{qi} = \partial^i = \partial/\partial x_i \text{ for } i = 1, \dots, n, \text{ otherwise } L^{ij} := 0.$$

Then $k_{ii} = 2, k_{iq} = 1 = k_{qi}$ for $i = 1, \dots, n$ and $k_{ij} = 0$ otherwise. Hence the following m_i, m'_i satisfy (2.324):

$$m_i = m'_i = 1 \text{ for } i = 1, \dots, n = q-1, m_q = m'_q = 0.$$

Furthermore L_p^{ij} are independent of x and we get

$$L_p^{ii}(\vartheta) = -|\vartheta|^2, L_p^{iq}(\vartheta) = -L_p^{qi}(\vartheta) = \vartheta^i, \text{ for } i = 1, \dots, n, \text{ else } L_p^{ij}(\vartheta) = 0.$$

This yields finally, with the $2m$ in (2.328)

$$|\det L_p| = \left| \det \begin{pmatrix} -|\vartheta|^2 & & \mathbf{0} & \vartheta^1 \\ & \ddots & & \vdots \\ \mathbf{0} & & -|\vartheta|^2 & \vartheta^n \\ -\vartheta^1 & \dots & -\vartheta^n & 0 \end{pmatrix} \right| = |\vartheta|^{2m} \text{ with } 2m = 2n. \tag{2.330}$$

Note that this $2m = 2n$ and the order 2 for the Stokes operator introduced in Definition 2.85, 4. and Definition 2.102 are different.
If we linearize the Navier–Stokes operator in an arbitrary $(\vec{u}, p)$, the additional terms are of lower order, and the corresponding $L_p^{ij}(\vartheta)$ are $\nu\times$ those of the Stokes operator. Hence, (2.328) is satisfied with $\epsilon = 1$ and $\epsilon = \nu^n$ for the Stokes and the linearized Navier–Stokes operator. So both are elliptic, with $2m = 2n$.
This $\epsilon = \nu^n$ indicates one of the most fascinating properties of the Navier–Stokes equations: for $\nu \to 0$ the elliptic aspects are dominated by convection phenomena, caused by $\sum_{i=1}^n u_i \partial^i \vec{u}$ terms in (2.320) and their linear counterparts in (2.321). For the time-dependent form of the equations this corresponds to a transition from a parabolic to a hyperbolic system. We will come back to (2.320) in Section 2.8.

2. *The von Kármán equations*, see Example 1.4: Berger and Five [82–84] have elaborated the appropriate spaces for these equations and proved the existence of solutions. Ciarlet [175] has identified these equations as the first term in an

asymptotic expansion, describing the mechanical bending of a plate. He analyzes these equations and the corresponding FEMs.
An external load λ compresses a plate located at Ω. The Airy stress function $w(x,y)$ at the point (x,y) and the deviation or deflection $u(x,y)$ of the plate from its flat state $u(x,y) \equiv 0$, $u, w : \Omega \to \mathbb{R}$, are described by

$$G(u,w) := \begin{pmatrix} \Delta^2 u & -[u,w] & +\lambda u_{xx} \\ & & \\ & \Delta^2 w & +\frac{1}{2}[u,u] \end{pmatrix}, \quad G(u_0,w_0) = \begin{pmatrix} f \\ \\ 0 \end{pmatrix} \tag{2.331}$$

in Ω. Here $\Delta^2 = \Delta\Delta$ is the biharmonic operator in $\mathbb{R}^2$, in some papers the λu_{xx} is missing. The bracket operator $[\,\cdot,\cdot\,]$ is defined by

$$[u,v] = [v,u] = u_{xx}v_{yy} - 2u_{xy}v_{xy} + u_{yy}v_{xx}.$$

The derivative $G'(u_1,w_1)$ applied to (v,z) has the form

$$G'(u_1,w_1)(v,z) := \begin{pmatrix} \Delta^2 v & -([u_1,z]+[v,w_1]) & +\lambda v_{xx} \\ & & \\ & \Delta^2 z & +[u_1,v] \end{pmatrix}.$$

Here we choose

$$\begin{aligned} &q := 2, L^{ii} := \Delta^2 \text{ for } i = 1,2, \ L^{ij} := 0 \text{ for } i \neq j \text{ so} \\ &k_{ii} := 4 \text{ for } i = 1,2, k_{ij} := 0 \text{ for } i \neq j, \ m_i = m_i' = 2, \text{ and} \\ &L_p^{ii}(\vartheta) := |\vartheta|^4 \text{ for } i = 1,2, \ L_p^{ij}(\vartheta) := 0 \text{ for } i \neq j, \ q = 2. \end{aligned} \tag{2.332}$$

This yields finally

$$|\det L_p(\vartheta)| = |\det L_p| = \left|\det \begin{pmatrix} |\vartheta|^4 & 0 \\ 0 & |\vartheta|^4 \end{pmatrix}\right| = |\vartheta|^8. \tag{2.333}$$

Hence, the von Kármán equations are strongly ($\epsilon = 1$) elliptic and $2m = 8$. We will come back to this in Subsection 2.6.6.

3. *The system of Lamé:* This describes the deflection function $\vec{u}$ of an elastic plate by a linear system: with $\mu, \lambda > 0$ we determine the solution $\vec{u}^0$ from

$$\vec{u}^0 : \Omega \subset \mathbb{R}^3 \to \mathbb{R}^3, \ \mu\Delta\vec{u}^0 + (\lambda+\mu)\nabla \operatorname{div}\vec{u}^0 = f \text{ in } \Omega, \quad = \varphi \text{ on } \partial\Omega. \tag{2.334}$$

Similarly as above one shows that $|\det L_p| = \mu^2(2\mu+\lambda)|\vartheta|^6$, so $2m = 6$. ■

Agmon *et al.* [2, 3] strongly stimulated research on partial differential equations and systems. It is absolutely impossible to appropriately summarize these results in this chapter. Furthermore, some of these general systems require mixed discretization methods, e.g. the Navier–Stokes equations in Section 2.8, requiring special treatment. So, in the following subsections we formulate for special elliptic systems the results for existence and regularity, similar to those for elliptic equations. We give a flavor for general systems, by summarizing a few examples on local existence and regularity from Taylor [618] for linear systems in Subsections 2.6.3 and 2.6.5.

2.6.3 Linear elliptic systems of order 2

In this and the following subsections we will study special elliptic systems of order 2 and $2m$. The linear systems are obtained, very similarly to elliptic equations, as compact perturbations of the usually H_0^m-coercive principal parts of these operators. As in Corollary 2.44, 3., 4., this implies only the H_0^m-coercivity for $2 \le p \le \infty$. Then existence, uniqueness and Fredholm alternatives are easily obtainable by Section 2.3, see e.g. Theorem 2.21.

We elaborate the corresponding results for linear and divergent quasilinear elliptic systems. Our presentation generalizes Chen and Wu [170] with respect to existence and regularity. Their regularity results are only valid for equations defined by the principal part, but will be extended to the general case by the inherited regularity technique in Theorem 2.56. We refer to Subsections 2.6.4 and 2.6.6 for more general types of regularity results, even for quasilinear equations, see Taylor [618–620], Koshelev and Chelkak [449–451] and Skrypnik, [591]. For these systems we can replace the $|\alpha| \le k_{ij}$ in (2.323) by $|\alpha| \le m$. So we do not need the m_i, m_j' in (2.324) any more. This allows a simpler matrix formulation for the equations. We start with the discussion of second order linear problems and extend it to quasilinear systems of second and then to higher order.

Before we introduce the following setting for homogeneous Dirichlet boundary conditions, we want to recall the above generalizations. For a nonhomogeneous Dirichlet boundary function, extended to a function on Ω, Lemma 2.26, and the standard trick, reduces the nonhomogeneous to homogeneous Dirichlet boundary conditions. Then similarly to the cases following (2.124), generalized second (modified Neumann), third or natural boundary value problems, compare (2.125), can be introduced and treated for second order problems. Similarly to (2.38), different boundary conditions on different parts of $\partial\Omega$ are possible, but will not be formulated here.

In this subsection, considering systems of order 2, we have to apply the partial derivatives $\partial^l, l = 1, \dots, n$, to q components $u_j, v_i, i, j = 1, \dots, q$. We modify the original notation in (2.73), replacing reals $\partial^l u(x), \partial^\alpha u(x)$, $\vartheta^l, \vartheta^\alpha \in \mathbb{R}$, and n-vectors $\partial u(x)$, $\vartheta = (\vartheta^1, \dots, \vartheta^n) \in \mathbb{R}^n$ by vectors $\partial^l \vec{u}(x), \partial^\alpha \vec{u}(x)$ $\vec{\vartheta}^l, \vec{\vartheta}^\alpha \in \mathbb{R}^q$, and the corresponding $n \times q$ matrices $\partial\vec{u}(x), \vec{\vartheta} \in \mathbb{R}^{n\times q}$, using just one vector sign to keep consistency with, but indicate the difference to (2.73). Evaluated in x (ev. at x) we get

$$\begin{aligned}
&\vec{u} := (u_1, \cdots, u_q), \partial^l \vec{u} := (\partial^l u_1, \cdots, \partial^l u_q), \partial^\alpha \vec{u} := (\partial^\alpha u_1, \cdots, \partial^\alpha u_q) \in \mathbb{R}^q, \text{ ev. at } x,\\
&\nabla \vec{u} = (\nabla u_1, \cdots, \nabla u_q) = (\partial u_1, \cdots, \partial u_q) = \partial \vec{u} = (\partial^1 \vec{u}, \cdots, \partial^n \vec{u}) \in \mathbb{R}^{n\times q}, \text{ ev. at } x,\\
&\nabla^k \vec{u} := (\partial^\alpha \vec{u})_{|\alpha|=k} \text{ev. in } x :\in \mathbb{R}^{n_k \times q}, \nabla^{\le k}\vec{u} := (\partial^\alpha \vec{u})_{|\alpha|\le k} \in \mathbb{R}^{N_k\times q}, \text{ ev. at } x,\\
&\qquad \text{with } \nabla^k \vec{u} = \partial^l \vec{u} = \partial^\alpha \vec{u} = \vec{u} \text{ for } k = l = |\alpha| = 0 \text{ and} \qquad (2.335)\\
&\vec{\vartheta}_j = \left(\vartheta_j^1, \dots, \vartheta_j^n\right) \in \mathbb{R}^n, j = 1, \dots, q, \quad \vec{\vartheta}^l = \left(\vartheta_1^l, \dots, \vartheta_q^l\right) \in \mathbb{R}^q, l = 1, \dots, n, \text{ with}\\
&\vec{\vartheta} := (\vec{\vartheta}^1, \cdots, \vec{\vartheta}^n) := (\vec{\vartheta}_1, \cdots, \vec{\vartheta}_q) \in \mathbb{R}^{n\times q}, \quad \text{and } |\vec{\vartheta}| = |\vec{\vartheta}|_{nq} \in \mathbb{R},\\
&\vec{\vartheta}^\alpha := \left(\vartheta_1^\alpha, \cdots, \vartheta_q^\alpha\right) \in \mathbb{R}^q, \quad \vec{\Theta}^k := (\vec{\vartheta}^\alpha)_{|\alpha|=k} \in \mathbb{R}^{n_k\times q}, \ \vec{\Theta}^{\le k} := (\vec{\vartheta}^\alpha)_{|\alpha|\le k} \in \mathbb{R}^{N_k \times q}.
\end{aligned}$$

We recall and extend the usual Sobolev spaces and norms for Ω in (2.5). *Sobolev (Hilbert) spaces* and their *inner products, norms, and seminorms* are defined as

$$W^{m,p}(\Omega,\mathbb{R}^q) := (W^{m,p}(\Omega))^q, W_0^{m,p}(\Omega,\mathbb{R}^q), H^m(\Omega,\mathbb{R}^q) := W^{m,2}(\Omega,\mathbb{R}^q), \quad (2.336)$$

$$(\vec{u},\vec{v})_m := (\vec{u},\vec{v})_{H^m(\Omega,\mathbb{R}^q)} := \sum_{i=1}^{q}\sum_{|\alpha|\le m}(\partial^\alpha u_i,\partial^\alpha v_i)_{L^2(\Omega)} \text{ and} \quad (2.337)$$

$$\|\vec{u}\|_{W^{m,p}(\Omega)} := \|\vec{u}\|_{W^{m,p}(\Omega,\mathbb{R}^q)} := \left(\sum_{i=1}^{q}\sum_{|\alpha|\le m}\|\partial^\alpha u_i\|^p_{L^p(\Omega)}\right)^{1/p} \text{ in particular}$$

$$W^{m,p}(\Omega,\mathbb{R}^q) \hookrightarrow H^m(\Omega,\mathbb{R}^q) \text{ densely and continuously embedded for } 2\le p.$$

We return to the second order equation with a functional, f, in general form:

$$\vec{f} := (f_0,\cdots,f_n)\in L^{p'}(\Omega,\mathbb{R}^{q(n+1)}), f_k(x) = \left(f_k^1,\cdots,f_k^q\right)(x), \partial^k\vec{v}(x)\in\mathbb{R}^q, \quad (2.338)$$

$$\langle\vec{f},\vec{v}\rangle_{\mathcal{V}'\times\mathcal{V}} := \int_\Omega\sum_{k=0}^{n}(f_k,\partial^k\vec{v})_q dx,\ \forall\vec{f}\in\mathcal{V}' := W^{-1,p'}(\Omega,\mathbb{R}^q), 1/p+1/p'=1,$$

cf. (2.107), (2.339), (2.340), noting that the $f_k(x),\ldots,$ are only defined a.e.

Remark 2.87. *As a consequence of (2.337), (2.338), the following integrals are defined for $\vec{u},\vec{v}\in W_0^{1,p}(\Omega,\mathbb{R}^q)$, $\vec{f}\in W_0^{-1,p'}(\Omega,\mathbb{R}^q)$, $p\ge 2$. Hence the following arguments remain valid for this case as well, possibly with a $H_0^1(\Omega,\mathbb{R}^q)$-coercivity.*

The strong form L is defined for $\vec{u},\vec{u}^0\in H^2(\Omega,\mathbb{R}^q)$ and with $\vec{f}$ instead of f_0:[19]

$$f_0 = \left[f_0^i\right]_{i=1}^q = L_s\vec{u}^0 = L\vec{u}^0 := \left[\sum_{j=1}^{q}\left(\sum_{k,l=0}^{n}(-1)_{k>0}\partial^k\left(a_{kl}^{ij}\partial^l u_j^0\right)\right)\right]_{i=1}^q \quad (2.339)$$

$$= \sum_{k,l=0}^{n}(-1)_{k>0}\partial^k(A_{kl}\partial^l\vec{u}^0) = \left[\sum_{j=1}^{q}L_s^{ij}u_j^0\right]_{i=1}^q, \text{ with } A_{kl}(x) = \left(a_{kl}^{ij}\right)_{i,j=1}^q,$$

$$\text{for } A_{kl}\in W^{1-\delta_{k,0},\infty}(\Omega,\mathbb{R}^{q\times q}) \text{ and } |\partial^k\left(a_{kl}^{ij}\right)(x)|\le\Lambda_0.$$

For Dirichlet boundary conditions we take the standard scalar product of (2.339) with $\vec{v}=(v_1,\ldots,v_q)\in H_0^1(\Omega,\mathbb{R}^q)$ and integrate by parts. Then the weak solution $\vec{u}^0=(u_1^0,\ldots,u_q^0)\in H_0^1(\Omega,\mathbb{R}^q)$ for (2.339) is determined, cf. (2.274), by equating the

[19] Trying to keep the notation as for quasilinear equations, we have changed the order of differentiation in the strong form (2.339) compared to (2.134).

following linear and bilinear forms, $\langle \vec{f}, \vec{v} \rangle$, and $a(\vec{u}^0, \vec{v})$:

$$\vec{u}^0 \in \mathcal{V} := H_0^1(\Omega, \mathbb{R}^q) \text{ s.t. } \langle \vec{f}, \vec{v} \rangle = \left\langle \left[\sum_{k=0}^n f_k^i \partial^k \right]_{i=1}^q, \vec{v} \right\rangle = \int_\Omega \sum_{k=0}^n (f_k, \partial^k \vec{v})_q dx \quad (2.340)$$

$$= a(\vec{u}^0, \vec{v}) := (L\vec{u}^0, \vec{v})_{\mathcal{V}} := \int_\Omega \sum_{k,l=0}^n (A_{kl} \partial^l \vec{u}^0, \partial^k \vec{v})_q dx \forall\, \vec{v} \in \mathcal{V}, \text{ with}$$

$$A_{kl} \in L^\infty(\Omega, \mathbb{R}^{q\times q}), |a_{kl}^{ij}(x)| \le \Lambda_0 \quad \text{(a.e.) } \forall x \in \Omega, f_k^i \in L^2(\Omega)\ \forall i,j,k,l, \text{ or} \quad (2.341)$$

$$\Longrightarrow \vec{f} \in H^{-1}(\Omega, \mathbb{R}^{q(n+1)}), \text{ with } \|\vec{f}\|_{H^{-1}(\Omega, \mathbb{R}^{q(n+1)})} \le \max_{k=0,\dots,n}^{i=1,\dots,q} \{\|f_k^i\|_{L^2(\Omega)}\}.$$

Note the change in notation for $\vec{f} \in H_0^{-1}(\Omega, \mathbb{R}^{q(n+1)})$ here to $f_k^i \in L^2(\Omega)$ above and in Chen and Wu. The *principal part and principal symbol* L_p of L are defined as

$$\textit{principal part: } (L_p\vec{u}, \vec{v})_{\mathcal{V}} := \int_\Omega \sum_{k,l=1}^n (A_{kl} \partial^l \vec{u}, \partial^k \vec{v})_q dx, \quad \text{and } \textit{symbol:} \quad (2.342)$$

$$L_p(x, \vec{\vartheta}) := \left(L_{kl}^{ij}(x, \vec{\vartheta}) \right)_{i,j=1}^q = \left(\sum_{k,l=1}^n a_{kl}^{ij}(x) \vartheta_j^l \vartheta_i^k \right)_{i,j=1}^q = \sum_{k,l=1}^n (A_{kl}(x) \vec{\vartheta}^l, \vec{\vartheta}^k)_q.$$

Accordingly, for the Stokes operator the matrix in (2.330) would have to be modified by replacing the last row and column by zeros. For our special linear system of divergent differential equations we further update the above ellipticity in Definition 2.85, see (2.328), towards the Legendre, sometimes called the strong Legendre condition, and the Legendre–Hadamard condition.

Definition 2.88. *L_s or L in (2.339) or (2.340) are called* strongly elliptic *and* uniformly elliptic, *respectively, if they satisfy the* strong *and* uniform Legendre condition. *That is, there exist $0 < \lambda$ and $0 < \lambda < \Lambda < \infty$, respectively, independent of x (sometimes only $0 < \lambda < \lambda(x) < \Lambda(x)$, $\Lambda(x)/\lambda(x) < C < \infty$), such that a.e. $\forall x \in \Omega$, cf. (2.342),*

$$\lambda|\vec{\vartheta}|^2 = \lambda|\vec{\vartheta}|_{nq}^2 \le \sum_{i,j=1}^q L_p^{ij}(x, \vec{\vartheta}) = \sum_{k,l=1}^n (A_{kl}(x)\vec{\vartheta}^l, \vec{\vartheta}^k)_q \le \Lambda|\vec{\vartheta}|^2. \quad (2.343)$$

So the Stokes operator violates (2.343). For our $a_{kl}^{ij} \in L^\infty(\Omega)$, the inequality in (2.343) indeed only makes sense a.e. $\forall\ x \in \Omega$. We will discuss the relation of the different types of ellipticity in Remark 2.90.

For proving existence and uniqueness results for (2.340) we reinterpret the bilinear form $a(\vec{u}, \vec{v})$ as the principal part, see (2.342), and additional lower order terms

$$a(\vec{u}, \vec{v}) = a_p(\vec{u}, \vec{v}) + b(\vec{u}, \vec{v}), \text{ with } b(\vec{u}, \vec{v}) := \int_\Omega \sum_{k,l=0,k\times l=0}^n (A_{kl} \partial^l \vec{u}, \partial^k \vec{v})_q dx. \quad (2.344)$$

We repeat the concepts of $\mathcal{V}$-*coercive* and $\mathcal{V}$-*elliptic bounded* bilinear forms, see (2.41) and (2.63). For $\mathcal{V} = H_0^m(\Omega, \mathbb{R}^q), m \geq 1$, $a(\cdot,\cdot)$ is called $\mathcal{V} = H_0^m(\Omega, \mathbb{R}^q)$-*coercive*, and $\mathcal{V}$-*elliptic*, respectively, if constants $\alpha \in \mathbb{R}_+, C_c \in \mathbb{R}$ exist such that

$$a(\vec{u},\vec{u}) \geq \alpha \|\vec{u}\|_{\mathcal{V}}^2 \text{ and } \geq \alpha\|\vec{u}\|_{\mathcal{V}}^2 - C_c\|\vec{u}\|_{\mathcal{W}}^2 \; \forall \vec{u} \in \mathcal{V} = H_0^m(\Omega, \mathbb{R}^q), \mathcal{W} = L^2(\Omega). \quad (2.345)$$

Applying Theorems 2.21 and 2.16 we find

Theorem 2.89. $\mathcal{V}$-coercivity and ellipticity:

1. *Let (2.340)–(2.341) be a uniformly elliptic system, hence let (2.343) with* $0 < \lambda \leq \Lambda < \infty$ *be satisfied. Then* $\langle \vec{f}, \vec{v}\rangle$, $a_p(\vec{u},\vec{v}), a(\vec{u},\vec{v})$ *in (2.340), (2.344) are bounded with respect to* $\mathcal{V}$, *and* $a_p(\vec{u},\vec{v})$ *and* $a(\vec{u},\vec{v})$ *are* $\mathcal{V}$-*coercive and* $\mathcal{V}$-*elliptic, respectively.*
2. *All these results remain correct, if the above strong Legendre condition (2.343) is replaced by the weaker* Legendre–Hadamard condition *(2.348).*
3. *For the principal part,* L_p, *and its* $a_p(\vec{u},\vec{v})$, *the unique solution* $\vec{u}^0$ *of (2.346) solves the variational problem (2.61).*
4. *Finally, all* H^1-*coercivity based existence, uniqueness and Fredholm results remain correct for the* $W^{1,p}$ *case for* $2 \leq p \leq \infty$, *however with respect to the* H^1 *norm.*

Under these conditions Corollary 2.44 applies. It implies for a $\mathcal{V}$-coercive $a(\cdot,\cdot)$ the *existence of a unique generalized or weak solution* $\vec{u}^0$ and a boundedly invertible L, and for a $\mathcal{V}$-elliptic $a(\cdot,\cdot)$ the Fredholm alternative with its consequences. Hence, for a boundedly invertible L, with the induced $a(\vec{u},\vec{v})$,

$$\forall \vec{f} \in \mathcal{V}' = H_0^{-1}(\Omega, \mathbb{R}^q) \; \exists_1 \vec{u}^0 \in \mathcal{V} : a(\vec{u}^0, \vec{v}) = \langle \vec{f}, \vec{v}\rangle \forall \vec{v} \in \mathcal{V}. \quad (2.346)$$

Proof. For completeness we include this nearly obvious proof: by (2.341) the $\langle \vec{f}, \vec{v}\rangle, a_p(\vec{u},\vec{v}), a(\vec{u},\vec{v})$ are bounded with respect to $\mathcal{V}$. Furthermore (2.343) implies

$$a_p(\vec{u},\vec{u}) \geq \lambda \int_\Omega |\partial\vec{u}|^2 dx \text{ with } |\partial \vec{u}|^2 = \sum_{i=1}^q \sum_{k=1}^n |\partial^k u_i|^2 \; \forall \vec{u} \in \mathcal{V}. \quad (2.347)$$

The $\|\vec{u}\|_{\mathcal{V}}$ and $\int_\Omega |\partial\vec{u}|^2 dx$ are equivalent norms in $\mathcal{V}$, so $a_p(\vec{u},\vec{v})$ is $\mathcal{V}$-coercive. Since $a(\vec{u},\vec{v}) - a_p(\vec{u},\vec{v})$ are lower order terms, a combination of Theorem 1.26 and Proposition 2.24 yields the $\mathcal{V}$-ellipticity of $a(\vec{u},\vec{v})$. Theorems 2.15 and 2.21 thus verify the above claims.

The final remark concerning the strong Legendre–Hadamard condition is a special case of Theorem 2.104 below. The $W^{1,p}$-coercivity is a consequence of the initial discussion of Remark 2.87. □

Remark 2.90. *We list some properties, relating the different concepts of elliptic systems* $(q \geq 1)$ *to elliptic equations* $(q = 1)$ *in (2.327)–(2.343).*

1. *We start by showing that (2.343) for* $2m = 2$ *implies the Taylor modification of (2.327): In fact let (2.327) be violated for our special case. Then a nontrivial*

solution $y = (y_1, \cdots, y_q)$ *for the linear homogeneous system exists:*

$$\sum_{j=1}^{q} L_p^{ij}(x, \vartheta) y_j = 0, y = (y_1, \cdots, y_q) \neq 0, \text{ hence,}$$

$$\sum_{i,j=1}^{q} \sum_{k,l=1}^{n} \alpha_i a_{kl}^{ij}(x) \vartheta_j^l \vartheta_i^k y_j = 0 \; \forall \; \alpha_i \in \mathbb{R}, \quad a.e. \; x \in \Omega.$$

By an appropriate rescaling of the $\alpha_i \vartheta_j^l \vartheta_i^k y_j$ *into new* $\vartheta_j^l := \vartheta_j^l y_j, \vartheta_i^k := \alpha_i \vartheta_i^k$ *we obtain the desired contradiction to (2.343).*

2. *The linearized von Kármán and Navier–Stokes operators satisfy and violate (2.343), respectively. We will see in Section 2.8 that the Stokes operator (and all of its compact perturbations) is not coercive. So the results of this and the next subsection are not applicable. Hence the Navier–Stokes operator and the related saddle point problems require a separate treatment.*
3. *For* $q = 1$ *the coercivity in (2.347) for the principal part and the two preceding ellipticity conditions (2.328) and (2.343) are essentially equivalent. This is no longer correct for* $q > 1$. *Chen and Wu [170], see Chapter 8, Theorems 1.1, and 1.2, show (compare the Taylor form of 2m-ellipticity in Definition 2.85, 6.): under the condition (2.341) for the* $A_{kl} = \left(a_{kl}^{ij}\right)_{i,j=1}^{q}$ the strong Legendre condition *implies the coercivity in (2.345) for the principal part.*
4. *Conversely, the coercivity of the principal part and (2.341) imply the so-called* strong Legendre–Hadamard condition: *with* $L_p(x, \vartheta)$ *in (2.326)*

$$\exists \lambda > 0 : \; \forall x \in \Omega, \;\; \vartheta \in \mathbb{R}^n, \eta \in \mathbb{R}^q,$$

$$\eta^T L_p(x, \vartheta) \eta = \sum_{i,j=1}^{q} \sum_{k,l=1}^{n} a_{kl}^{ij}(x) \vartheta^l \vartheta^k \eta_j \eta_i \geq \lambda |\vartheta|^2 |\eta|^2. \tag{2.348}$$

So instead of the above equivalence for $q = 1$ *we find for* $q > 1$ *the following implication*

For $q > 1$ *assume (2.341) for* L_p. *Then: (2.343)* $\Longrightarrow$ *(2.347)* $\Longrightarrow$ *(2.348)*(2.349)

or verbatim, for $A_{kl} = (a_{kl}^{ij} \in L^\infty(\Omega))_{i,j=1}^q$, *the strong Legendre condition* $\Longrightarrow$

coercivity for $a_p(\cdot, \cdot) \Longrightarrow$ *strong Legendre–Hadamard condition for* A_{kl}.

By the proof of Theorem 2.104 we get for $a_{kl}^{ij} \in C(\bar{\Omega}), k, l = 1, \ldots, n,$ *even the equivalence (2.347)* $\Longleftrightarrow$ *(2.348).*

For our numerical applications again regularity results $\vec{u}^0 \in H^s(\Omega, \mathbb{R}^q), s \geq 1$, for the general equation (2.340) are important. Similarly to the preceding sections, different regularity results are known. In the remainder of this subsection we combine Chen and Wu [170], and the proof of Theorems 2.56 and 2.60. For linear L, Chen and Wu prove their results only for the solution $\vec{u}^0$ of the equation (2.350) defined by the *principal part* L_p and with the general $\vec{f}$ in (2.340)

$$a_p(\vec{u}^0, \vec{v}) = \langle \vec{f}, \vec{v} \rangle \forall \vec{v} \in H_0^1(\Omega, \mathbb{R}^q). \tag{2.350}$$

Then the proof of Theorem 2.56 allows us the extension to the general problem (2.340).

We summarize the results, obtained by this combination with Chen and Wu [170]: Theorem 2.91 1. and 2. correspond to [170], Chapter 8, Theorems 2.3, 2.6, 2.7, and to Corollary 2.4. Theorem 2.91 1. yields $\vec{u}^0 \in H^{s+2}(\Omega, \mathbb{R}^q), s \geq -1$, Theorem 2.91 3. generalizes [170], Chapter 9, Theorems 2.7, 2.8. The resulting $\vec{u}^0 \in C^{s,\delta}(\Omega, \mathbb{R}^q), s \geq 1$ is needed in Subsection 2.6.4.

Theorem 2.91. Regular solution $\vec{u}^0 \in H^{s+2}(\Omega, \mathbb{R}^{n\times q})$ and $\partial\vec{u}^0 \in C^{s,\delta}(\bar{\Omega}, \mathbb{R}^{n\times q})$:

1. *Let $a(\cdot,\cdot)$ be strongly elliptic, see (2.343), and $\vec{f}$ of the form (2.340). For $s \in \mathbb{Z}, s \geq -1$, let*

$$\partial\Omega \in C^{s+2},\ a_{kl}^{ij} \in W^{s+1,\infty}(\Omega),\ f_k^i \in H^{s+1}(\Omega)\forall i,j,k,l, \tag{2.351}$$

equivalent to $\vec{f} \in H^{s+1}(\Omega, \mathbb{R}^n)$. Then any weak solution $\vec{u}^0 \in H_0^1(\Omega, \mathbb{R}^q)$ of (2.340) satisfies $\vec{u}^0 \in H^{s+2}(\Omega, \mathbb{R}^q)$ and

$$\|\vec{u}^0\|_{H^{s+2}(\Omega,\mathbb{R}^q)} \leq C\big(\|\vec{u}^0\|_{H^1(\Omega,\mathbb{R}^q)} + \|\vec{f}\|_{H^{s+1}(\Omega,\mathbb{R}^q)}\big). \tag{2.352}$$

2. *We even get $\vec{u}^0 \in C_0^\infty(\Omega, \mathbb{R}^q)$ if (2.351) is replaced by*

$$a_{kl}^{ij}, f_k^i \in C^\infty(\Omega)\ \forall i,j,k,l,\ \partial\Omega \in C^\infty. \tag{2.353}$$

3. *For a strongly elliptic $a(\cdot,\cdot)$ and the Dirichlet problem let*

$$a_{kl}^{ij}, f_k^i \in C^{s,\delta}(\bar{\Omega}), \partial\Omega \in C^{s+1,\delta},\ 0 < \delta < 1, 0 \leq s,\ \forall i,j,k,l. \tag{2.354}$$

Then for any weak solution $\vec{u}^0 \in H_0^1(\Omega, \mathbb{R}^q)$ of (2.340) the $\partial\vec{u}^0 \in C^{s,\delta}(\bar{\Omega}, \mathbb{R}^{n\times q})$. Finally, the operator L, induced by $a(.,.)$, with $L : H^{2+k}(\Omega, \mathbb{R}^q) \cap H_0^1(\Omega, \mathbb{R}^q) \to H^k(\Omega, \mathbb{R}^q)$ in (2.340) is Fredholm.

Proof. We only prove Theorem 2.91, 1.; for 3. the proof is analogous. For $s = 0$ and $a_p(\cdot,\cdot)$ the (2.352) is Chen and Wu's Theorem 2.6, however with $\|\vec{u}^0\|_{H^1(\Omega,\mathbb{R}^q)}$ replaced by $\|\vec{u}^0\|_{L^2(\Omega',\mathbb{R}^q)}$. For $s > 0$ and $\vec{u}^0 \in H^{s+2}(\Omega, \mathbb{R}^q)$ it is their Theorem 2.7 with (2.352) missing, for $s = \infty$ their Corollary 2.4. Applying the arguments in the modification of the proof of Theorem 2.56 now for Theorem 2.60 verifies these results for the general $a(\cdot,\cdot)$ in (2.340). In fact, (2.339) shows that the a_{kl}^{ij} are derived at most once, hence any $\vec{u} \in H^{s+2}(\Omega, \mathbb{R}^q)$ yields $L\vec{u} \in H^s(\Omega, \mathbb{R}^q)$ and thus (2.352). □

Since Hackbusch's [387] proof for Remark 9.1.1 remains valid, we can improve Theorem 2.91 1. as

Theorem 2.92. *Let (2.340) with $\vec{f} \in \mathcal{V}'$ have a unique solution $\vec{u}^0 \in \mathcal{V}$, let $a(\cdot,\cdot)$ induce $L \in \mathcal{L}(\mathcal{V}, \mathcal{V}')$, and let the conditions of Theorem 2.91 be satisfied. Then $L^{-1} \in \mathcal{L}(H^s(\Omega, \mathbb{R}^q), H^{s+2}(\Omega, \mathbb{R}^q) \cap \mathcal{V})$ for $s \in \mathbb{Z}, s \geq -1$, and the following estimates are equivalent for $\vec{f} \in V'_{-s} := V' \cap H^s(\Omega, \mathbb{R}^q)$:*

$$\|L^{-1}\|_{H^{s+2}(\Omega,\mathbb{R}^q)\cap\mathcal{V}\hookleftarrow H^s(\Omega,\mathbb{R}^q)} \leq C \Longleftrightarrow \|\vec{u}^0\|_{H^{s+2}(\Omega,\mathbb{R}^n)} \leq C\|\vec{f}\|_{H^s(\Omega,\mathbb{R}^n)}. \tag{2.355}$$

We finish with some Schauder type results. Again we generalize Chen and Wu's results for the principal part to the full operator. They distinguish regularity in the interior and the closure of Ω. Theorem 2.93 1. combines the results [170], Chapter 9, Theorems 2.5, 2.6, Theorem 2.93 2. with [170], Chapter 10, Theorems 2.1, 2.2.

Theorem 2.93. Interior regularity for $\partial\vec{u}^0 \in C^{0,\delta}(\bar{\Omega}_1, \mathbb{R}^{n\times q})$ and $\in L^p(\Omega_1, \mathbb{R}^{n\times q})$:

1. *Assume $a_p(\cdot,\cdot)$ is strongly elliptic, see (2.343), and*

$$\partial\Omega \in C^{1,\delta}, a_{kl}^{ij}, f_i^k \in C^{0,\delta}(\Omega), 0 < \delta < 1, \left|a_{kl}^{ij}(x)\right| \leq \Lambda \ \forall x \in \Omega, \ \forall i,j,k,l. \quad (2.356)$$

Then for any weak solution $\vec{u}^0 \in H_0^1(\Omega, \mathbb{R}^q)$ of the original equation (2.340) its derivative is $\partial\vec{u}^0 \in C^{0,\delta}(\Omega, \mathbb{R}^{n\times q})$. For any $\Omega_1 \subset\subset \Omega$ it is estimated as

$$\|\partial\vec{u}^0\|_{C^{0,\delta}(\bar{\Omega}_1,\mathbb{R}^{n\times q})} \leq C\big(\|\partial\vec{u}^0\|_{L^2(\Omega',\mathbb{R}^{n\times q})} + \|\vec{f}\|_{C^{0,\delta}(\bar{\Omega}',\mathbb{R}^{n\times q})}\big), \quad (2.357)$$

where $\Omega' := \{x \in \Omega : \operatorname{dist}(x, \partial\Omega) > \operatorname{dist}(\Omega_1, \partial\Omega)/2\}$ and C depends on n, q, λ, Λ, $\delta, \operatorname{dist}(\Omega_1, \partial\Omega)$, $\left\|a_{kl}^{ij}\right\|_{C^{0,\delta}(\Omega')}$, and $\operatorname{diam}\ \Omega$.

2. *Assume $a_p(\cdot,\cdot)$ is strongly elliptic and replace in (2.351) $a_{kl}^{ij}, f_k^i \in C^0(\Omega)$ by $a_{kl}^{ij} \in C^0(\bar{\Omega}), \implies \left|a_{kl}^{ij}(x)\right| \leq \Lambda \ \forall x \in \Omega, \ f_k^i \in L^{p'}(\Omega), p \geq 2$. Then for any weak solution $\vec{u}^0 \in H_0^1(\Omega, \mathbb{R}^q)$ of (2.340) its derivative is $\partial\vec{u}^0 \in L^p(\Omega, \mathbb{R}^{n\times q})$. For any $\Omega_1 \subset\subset \Omega$ it can be estimated as*

$$\|\partial\vec{u}^0\|_{L^p(\Omega_1,\mathbb{R}^{n\times q})} \leq C\big(\|\vec{u}^0\|_{H^1(\Omega,\mathbb{R}^q)} + \|\vec{f}\|_{L^p(\Omega,\mathbb{R}^{n\times q})}\big),$$

with $C = C\big[n, q, p, \lambda, \Lambda, \operatorname{dist}(\Omega_1, \partial\Omega)$, modulus of continuity of the $a_{kl}^{ij}\big]$.

Similarly to Theorem 2.47 we obtain regularity results for convex domains, however only for systems of order 2 and only in $\mathbb{R}^n$ with $n = 2, 3$. For $a(\cdot,\cdot)$ and its coefficients we impose the conditions, cf. Rossmann [558], and Kozlov, Maz'ya and Rossmann [452], Sections 8.6 and 11.4:

$$\begin{aligned}&\text{Let } a(\cdot,\cdot) \text{ and its coefficient matrices } A_{kl} \in \mathbb{R}^{q\times q} \text{in (2.344) satisfy} \quad (2.358)\\ &A_{kl} = A_{lk}^*, A_{kl} \in W^{1,\infty}(\Omega), A_{kl} \in C(\overline{U_v}) \ \ \forall 1 \leq k,l \leq n, \text{ and chosen}\\ &\text{neighborhoods } U_v \subset \mathbb{R}^n \ \ \forall \text{ vertices } v \in \partial\Omega, \text{ finally } \vec{f} \in W^{1,\infty}(\Omega, \mathbb{C}^q),\\ &\text{for } \vec{f^{(j)}}(x) \in \mathbb{C}^q, j \leq n, \text{ let } \sum_{k,l=1}^n (A_{kl}\vec{f^{(l)}}, \vec{f^{(k)}})_q \geq c_0 \sum_{j=1}^n |\vec{f^{(j)}}|_q^2.\end{aligned}$$

This last inequality obviously implies the coercivity of the principal part.

Theorem 2.94. Regular solutions for convex Ω: *Let $\Omega \subset \mathbb{R}^n$, in (2.5) be convex, $n = 2$ or $n = 3$, and $\Omega \in C^2$, except in the vertices $v \in \partial\Omega$, and $a(\cdot,\cdot)$ in (2.344) and its coefficients satisfy (2.358). Then the exact solution is $\vec{u}^0 \in H^2(\Omega, \mathbb{R}^q)$.*

Proof. This result is not an immediate consequence of results in Subsections 8.6 and 11.4 of [452]. So we *indicate* the missing arguments, by locally modifying Ω into $\tilde{\Omega}$, in the neighborhoods, U_v, of the vertices, v. Similarly to [452], the system and coefficients are modified such that Theorem 2.91 1. yields for $s = 1$ the regularity $\vec{u}^0 \in H^2(\tilde{\Omega})$. For the local analysis, the condition $A_{kl} \in C(\overline{U_e})$ allows us the technique of frozen coefficients near the vertex, v. Then the local $H^2(\Omega)$-regularity is correct, if the frozen system does not admit a singularity function $r^\lambda U(\varphi)$ with $0 < \text{Re}\lambda \leq 1$ and $-1/2 < \text{Re}\lambda \leq 1/2$ for two and three dimensional problems. In [452], Sections 8.6 and 11.4, it is shown that $1 < \text{Re}\lambda$ and $1/2 < \text{Re}\lambda$ if the angle between edges is $< \pi$. That is exactly the situation in vertices of convex domains under the conditions of our theorem. □

Regularity estimates of a different type, not used in our numerical context, are due to Morrey [500], Chen and Wu [170], pp. 137 ff., and Koshelev [449], pp. 108 ff. For example, for $1 \leq p, 0 \leq \mu$ and $\Omega(x,\rho) := \Omega \cap B(x,\rho)$ we define a norm as

$$\|\vec{u}\|_{L^{p,\mu}(\Omega)} := \left[\sup_{x\in\Omega, 0<\rho<\text{diam }\Omega} \rho^{-\mu} \int_{\Omega(x,\rho)} |\vec{u}(z)|^p dz \right]^{1/p}. \tag{2.359}$$

Then the space

$$L^{p,\mu}(\Omega) := \{\vec{u} \in L^p(\Omega) : \|\vec{u}\|_{L^{p,\mu}(\Omega)} < \infty\}$$

is a normed linear space, the *Morrey space,* with respect to the norm $\|\vec{u}\|_{L^{p,\mu}(\Omega)}$.

Theorem 2.95. Regularity in Morrey spaces: *Assume the a_{kl}^{ij} in (2.342) satisfy (2.343). If*

$$a_{kl}^{ij}(x) \in C^0(\bar{\Omega}), f_k^i \in L^{2,\mu}(\Omega) \ \forall i,j,k,l, \tag{2.360}$$

(hence (2.341) is valid!), then for any weak solution $\vec{u}^0 \in H_0^1(\Omega, \mathbb{R}^q)$ of (2.350) its derivative is in $\partial\vec{u}^0 \in L^{2,\mu}(\Omega, \mathbb{R}^{n\times q})$. For $\Omega_1 \subset\subset \Omega$ it can be estimated as

$$\|\partial\vec{u}^0\|_{L^{2,\mu}(\Omega_1,\mathbb{R}^{n\times q})} \leq C\big(\|\partial\vec{u}^0\|_{L^2(\Omega,\mathbb{R}^{n\times q})} + \|\vec{f}\|_{L^{2,\mu}(\Omega,\mathbb{R}^{n\times q})}\big),$$

with $C = C\big(n,q,\lambda,\mu,\Lambda,\text{dist}(\Omega_1,\partial\Omega)$, and the modulus of continuity of the $a_{kl}^{ij}\big)$.

2.6.4 Quasilinear elliptic systems of order 2 and variational methods

We generalize some of the results from the last Subsection 2.6.3 to quasilinear systems and use the notation in (2.335). Our presentation again follows Chen and Wu [170], Chapter 11, and finally indicates some results from Leung [473] and Koshelev [449]. We consider the following divergent quasilinear system in strong form:

$$G_s\vec{u}^0(x) := G\vec{u}^0(x) = \left[\sum_{k=0}^{n}(-1)_{k>0}\partial^k\left(a_k^i(x,\vec{u}^0(x),\partial\vec{u}^0(x))\right)\right]_{i=1}^{q} \tag{2.361}$$

$$= \sum_{k=0}^{n}(-1)_{k>0}\partial^k(\vec{a}_k(x,\vec{u}^0(x),\partial\vec{u}^0(x))) = 0,$$

$$\text{with } \vec{u}^0, \vec{u} \in H^2(\Omega,\mathbb{R}^q),\ \vec{a}_k = \left[a_k^i\right]_{i=1}^q,$$

and $a_k^i : \Omega \times \mathbb{R}^q \times \mathbb{R}^{n\times q} \ni w = (x,z,p) \to a_k^i(x,z,p) \in \mathbb{R}$. Often this divergent quasilinear system characterizes a critical point of a variational problem as in (2.366). The corresponding strong form of the nondivergent quasilinear system, (2.361), originates, if (2.366) can be differentiated, see (2.369). In this sense, Taylor [620], pp. 198 ff., presents his similar results on surfaces, and for smooth enough coefficients.

The standard multiplication of (2.361) with $\vec{v} \in \mathcal{V} := H_0^1(\Omega, \mathbb{R}^q)$, the scalar product $(\cdot,\cdot)_q$ in $\mathbb{R}^q$, and integration by parts yields the corresponding weak system,

$$\vec{u}^0 \in \mathcal{V} : \; \langle G\vec{u}^0, \vec{v}\rangle_{\mathcal{V}'\times\mathcal{V}} := \int_\Omega \sum_{i=1}^{q} \left(\sum_{k=0}^{n} a_k^i(x, \vec{u}^0(x), \partial\vec{u}^0(x)), \partial^k v_i(x) \right) dx \tag{2.362}$$

$$= \int_\Omega \sum_{k=0}^{n} (\vec{a}_k(x, \vec{u}^0(x), \partial\vec{u}^0(x)), \partial^k \vec{v}(x))_q dx =: a(\vec{u}^0, \vec{v}) = 0 \;\forall \vec{v} \in \mathcal{V} = H_0^1(\Omega, \mathbb{R}^q).$$

The most important examples for (2.361) or (2.362) are the so-called *Euler systems* of a corresponding variational problem: We start with a function F, a functional $J(\vec{u})$, an admissible set $\mathcal{A}$ and formulate the following conditions essentially with respect to F (note that the following F_p etc. indicates the partial with respect to p etc.)

$$F : \Omega \times \mathbb{R}^q \times \mathbb{R}^{n\times q} \ni w = (x, \vec{z}, p) \to F(x, \vec{z}, p) \in \mathbb{R} \tag{2.363}$$

$$J(\vec{u}) := \int_\Omega F(x, \vec{u}(x), \partial\vec{u}(x))dx, \quad J : H^1(\Omega, \mathbb{R}^q) \to \mathbb{R} \tag{2.364}$$

with boundary conditions in the form $\vec{g} \in H^1(\Omega, \mathbb{R}^q)$ s.t. $J(\vec{g}) < \infty$

e.g. choose $\mathcal{A} := \{\vec{u} \in H^1(\Omega, \mathbb{R}^q) : \vec{u} - \vec{g} \in \mathcal{V} = H_0^1(\Omega, \mathbb{R}^q)\}$ and (2.365)

determine $\vec{u}^0 \in \mathcal{A}$ s.t. $J(\vec{u}^0) \le J(\vec{v}) \quad \forall \vec{v} \in \mathcal{A}$. (2.366)

If F is convex in p, hence $F_p(x, \vec{z}, \hat{p})(p - \hat{p}) \le F(x, \vec{z}, p) - F(x, \vec{z}, \hat{p})$, then (2.366) and J in (2.364) are called a *regular variational problem* and a *regular integral functional.* Obviously the *admissible set* $\mathcal{A}$ in (2.365) is a closed convex subset of $H^1(\Omega, \mathbb{R}^q)$. The uniqueness of the solution, $\vec{u}^0$, is an immediate consequence of the arguments on top of page 166 in [170].

Theorem 2.96. Variational problems, [170], Chapter 11, Theorem 2.1*:*

1. *Let F, F_p be continuous in $\Omega \times \mathbb{R}^q \times \mathbb{R}^{n\times q}$, F convex in p, and $F(x, \vec{z}, p) \ge \lambda |p|^2$ with $\lambda > 0$. Then the* regular variational problem *(2.366) has a solution $\vec{u}^0 \in H^1(\Omega, \mathbb{R}^q)$.*
2. *If F is strictly convex in p, hence $F_p(x, \vec{z}, \hat{p})(p - \hat{p}) < F(x, \vec{z}, p) - F(x, \vec{z}, \hat{p})$ for $p \ne \hat{p}$, then $\vec{u}^0$ is unique.*

So Chen and Wu [170] have shown the existence and uniqueness of a solution $\vec{u}^0 \in H^1(\Omega, \mathbb{R}^q)$ for the regular variational problem (2.366). In their Chapter 11, 2.2

they elaborate the relation between this variational problem and quasilinear equations.

For the following considerations we need the derivatives of $J(\vec{u})$. We start with formal derivatives and present conditions in Theorem 2.97, justifying these derivatives. We introduce, similarly to (2.336), the notation for the components of $\vec{z} \in \mathbb{R}^q, p \in \mathbb{R}^{n\times q}$, without completely indicating their vector structure, as

$$\begin{aligned}
&\vec{u}(x) = (u_1, \ldots, u_q)(x), \vec{z} = (z_1, \ldots, z_q) \in \mathbb{R}^q, \text{ and corresponding} \\
&p = (p_1, \ldots, p_q) = (p^1, \ldots, p^n),\ \partial\vec{u}(x) = (\partial u_1, \ldots, \partial u_q)(x) \in \mathbb{R}^{n\times q}, \text{ where} \\
&p_i = \left(p_i^1, \ldots, p_i^n\right),\ \partial u_i(x) = \left(\partial^1 u_i, \ldots, \partial^n u_i\right)(x), x \in \mathbb{R}^n, p_i^k \in \mathbb{R}, \\
&p^k = \left(p_1^k, \ldots, p_q^k\right),\ \partial^k \vec{u}(x) = \left(\partial^k u_1, \ldots, \partial^k u_q\right)(x) \in \mathbb{R}^q, \text{ and for} \\
&w := (x, \vec{z}, p) \in \Omega \times \mathbb{R}^q \times \mathbb{R}^{n\times q},\quad i = 1, \ldots, q,\quad k = (0), 1, \ldots, n, \\
&F_{\vec{z}}(w)\vec{v} = \sum_{i=1}^{q} F_{z_i}(w)v_i,\ F_p(w)\partial\vec{v} = \sum_{i=1}^{q}\sum_{k=1}^{n} F_{p_i^k}(w)\partial^k v_i, \qquad (2.367) \\
&F'(w)\vec{v} = F_{\vec{z}}(w)\vec{v} + F_p(w)\partial\vec{v},\quad \text{note that for } w(x) = (x, \vec{u}(x), \partial\vec{u}(x)), \\
&[\partial F_p(w(x))]\vec{v}(x) = [\partial F_p(x, \vec{u}(x), \partial\vec{u}(x))]\vec{v}(x) = \sum_{i=1}^{q}\sum_{k=1}^{n}[\partial^k F_{p_i^k}(w(x))]v_i(x) \\
&\qquad = \sum_{i,j=1}^{q}\sum_{k,l=1}^{n}\Big[F_{p_i^k p_j^l}(w(x))\partial^k\partial^l u_j(x) \\
&\qquad\qquad + F_{p_i^k z_j}(w(x))\partial^k u_j(x) + F_{p_i^k x_j}(w(x))\Big] v_i(x).
\end{aligned}$$

Chen and Wu [170] derive two necessary conditions by the standard and a modified trick in the classical calculus of variations:

Firstly, under appropriate growth condition, a solution $\vec{u}^0 \in \mathcal{A} \cap C^1(\Omega, \mathbb{R}^q)$ of (2.366) satisfies the *weak and strong form of the Euler system for J,* see (2.368), (2.369) and compare Theorem 2.73. We formulate them componentwise and in a short form. With (2.367), (2.361) or (2.362) and $\vec{a}_k : \Omega \times \mathbb{R}^q \times \mathbb{R}^{n\times q} \to \mathbb{R}^q$, we get

$$\vec{a}_k(w) = \left(a_k^i(w) := F_{p_i^k}(w)\right)_{i=1}^{q}, \text{ for } k > 0, \text{ and } \vec{a}_0(w) = \left(a_0^i(w) := F_{z_i}(w)\right)_{i=1}^{q}.$$

With $w^0(x) = (x, \vec{u}^0(x), \partial\vec{u}^0(x))$, we obtain the weak form

$$\begin{aligned}
a(\vec{u}^0, \ \vec{v}) : &= (J'(\vec{u}^0), \vec{v})_{H_0^1(\Omega,\mathbb{R}^q)} = \int_\Omega \big(F_p(w^0(x))\partial\vec{v} + F_{\vec{z}}(w^0(x))\vec{v}(x)\big)dx \qquad (2.368) \\
&= \int_\Omega (F'(w^0)\vec{v})(x)dx
\end{aligned}$$

$$= \int_\Omega \sum_{k=0}^{n} (\vec{a}_k(w^0(x)), \partial^k \vec{v}(x))_q dx = 0 \ \forall \vec{v} \in \mathcal{V} = H_0^1(\Omega, \mathbb{R}^q) \ni \vec{u}^0.$$

By standard partial integration (2.368) is transformed into $\int_\Omega (\cdots)\vec{v}(x)dx = 0 \ \forall \vec{v} \in H_0^1(\Omega, \mathbb{R}^q)$. This implies the system in its strong form as

$$(\cdots) = -\partial F_p(w^0(x)) + F_{\vec{z}}(w^0(x)) = \sum_{k=0}^{n} (-1)_{k>0} (\partial^k (\vec{a}_k(w^0(x)))$$

$$= \Bigg[F_{z_i}(w^0(x)) - \sum_{j=1}^{q} \sum_{k,l=1}^{n} \Big(F_{p_i^k p_j^l}(w^0(x)) \partial^k \partial^l u_j$$

$$+ F_{p_i^k z_j}(w^0(x)) \partial^k u_j + F_{p_i^k x_j}(w^0(x)) \Big) \Bigg]_{i=1}^{q} = 0. \tag{2.369}$$

So the quasilinear equations (2.361), (2.369) and (2.362), (2.368) represent the strong and weak Euler system for J for the solution $\vec{u}^0$. These $F_{p_i^k}(w^0(x))$, and their gradients $\partial^k(F_{p_i^k})(w^0(x))), k > 0, \ldots$, are Nemyckii operators of a type similar to those studied in Subsections 2.5.5 and 2.5.6. We will impose appropriate conditions to have them well defined.

Secondly, for $F \in C^2$, a solution $\vec{u}^0$ of (2.366) necessarily satisfies, cf. [170], p. 169, the condition for local minima, the so-called *Legendre–Hadamard condition* for the corresponding principal part, see (2.348).

$$\sum_{i,j=1}^{q} \sum_{k,l=1}^{n} F_{p_i^k p_j^l}(x, \vec{u}^0(x), \partial \vec{u}^0(x)) \vartheta^k \vartheta^l \zeta^i \zeta^j \geq 0 \ \forall \ \vartheta \in \mathbb{R}^n, \zeta \in \mathbb{R}^q. \tag{2.370}$$

For our later linearization we assume, stronger than (2.370), cf. (2.349), that (2.368), (2.369) satisfies for $\vec{u}^0$ the so-called *strong Legendre condition,* compare (2.342), (2.343), so F is strictly convex in $w(x)$,

$$\sum_{i,j=1}^{q} \sum_{k,l=1}^{n} F_{p_i^k p_j^l}(w^0(x)) \vartheta_i^k \vartheta_j^l \geq \lambda |\vec{\vartheta}|_{nq}^2 \ \forall \ \vec{\vartheta} \in \mathbb{R}^{n \times q}. \tag{2.371}$$

This implies, for $F_{p_i^k p_j^l}(w^0(x)) \in L^\infty(\Omega, \mathbb{R}^{q \times q})$ by Theorem 2.89, the coercivity of the linearized operator. Similarly to quasilinear equations, Chen and Wu [170], pp. 163 ff., guarantee solutions by imposing *Carathéodory type* and *controllable and natural growth conditions* for F, measurable in x, cf. [170], pp. 171 ff.

$$\begin{aligned} &F : \Omega \times \mathbb{R}^q \times \mathbb{R}^{n \times q} \to \mathbb{R} \ \text{ measurable in } x, \\ &\text{let } F \in C^1 \text{with respect to} (\vec{z}, p) \text{ a.e.} x \in \Omega. \end{aligned} \tag{2.372}$$

Controllable growth conditions: With constants $\lambda, \Lambda, C \in \mathbb{R}_+$ and functions[20]

$$\begin{aligned}
&g_i \geq 0, g_i \in L^1(\Omega),\ \ f_k^i, f_i \geq 0, f_k^i \in L^2(\Omega), f_i \in L^{r/(r-1)}(\Omega), \\
&\text{define } V := (1 + |\vec{z}|^2 + |p|^2). \text{ We assume for } F, F_{x_k}, F_{p_i^k}: \qquad (2.373)\\
&\lambda V - g_1(x) \leq F(x, \vec{z}, p) \leq \Lambda V + g_2(x),\ \text{ mind our } V \text{ is } V^2 \text{ in [170]}, \\
&\left|F_{x_k}(x, \vec{z}, p)\right| \leq C(|p|^{2(1-1/r)} + |\vec{z}|^{r-1} + f_k(x)), k = 1, \dots, n, \\
&\left|F_{p_i^k}(x, \vec{z}, p)\right| \leq C\left(|p| + |\vec{z}|^{r/2} + f_k^i(x)\right), i = 1, \dots, q. \qquad (2.374)
\end{aligned}$$

Natural growth conditions: For $|\vec{z}| \leq M$, and constants $\lambda, \mu, \Lambda = \Lambda(M) \in \mathbb{R}_+$ let

$$\lambda |p|^2 - \mu \leq F(x, \vec{z}, p) \leq \Lambda(1 + |p|^2) \qquad (2.375)$$

$$|F_{x_k}(x, \vec{z}, p)| \leq \Lambda(1 + |p|^2), \qquad \left|F_{p_i^k}(x, \vec{z}, p)\right| \leq \Lambda(1 + |p|). \qquad (2.376)$$

Theorem 2.97. $J'(\vec{u})v$ is a weak Euler system, Chen and Wu, [170], Chapter 11, Theorems 2.3, 2.4.: *Let (2.372)–(2.374) or (2.372), (2.375)–(2.376) be satisfied for* $\vec{u} \in H^1(\Omega, \mathbb{R}^q) \cap L^\infty(\Omega, \mathbb{R}^q)$. *Then* $J(\vec{u})$ *is differentiable in* $\vec{u}$ *and* $J'(\vec{u})v$, *a quasilinear operator, has the form of the* weak Euler system *(2.368).*

We call $\vec{u}^0 \in \mathcal{A} \subset L^\infty(\Omega, \mathbb{R}^q)$, cf. (2.365), a *generalized solution* of (2.362) if for controllable growth conditions, (2.368) is satisfied for $\vec{u}^0$ and for all $\vec{v} \in \mathcal{A}$. For natural growth conditions, we have to test (2.368) with all $\vec{v} \in L^\infty(\Omega, \mathbb{R}^q)$. It is well known that F is strictly convex in p iff the strong Legendre condition (2.371) is satisfied.

Theorem 2.98. Generalized solution of (2.362), [170], Chapter 11, Summary*: Under the conditions of Theorems 2.96 and 2.97 and for* F *convex in* p*, there exists, and for* F *strictly convex there uniquely exists, a generalized solution* $\vec{u}^0$ *of (2.362), characterized as a weak solution of an elliptic quasilinear equation.*

Now, we formulate regularity results, see Chen and Wu [170], Chapter 12, pp. 173 ff., directly for the quasilinear systems in (2.361) or (2.362). Accordingly we introduce modifications of the above *strong Legendre and growth conditions,* obviously stronger than (2.373), (2.374), and (2.375), (2.376), as

$$A := \left(a_k^i,\ \text{e.g. } a_k^i = F_{p_i^k}\right)_{i=1,k=1}^{q\quad n}, \vec{b} := \left(a_0^i,\ \text{e.g. } a_0^i = F_{z_i}\right)_{i=1}^q, A, \vec{b} \in C^1, \qquad (2.377)$$

$$\exists \lambda, \Lambda > 0: \sum_{i,j=1}^q \sum_{k,l=1}^n \left(a_k^i\right)_{p_j^l}(w)\vartheta_i^k \vartheta_j^l \geq \lambda |\vec{\vartheta}|^2\ \forall \vec{\vartheta} \in \mathbb{R}^{n\times q}, \text{with } w = (x, \vec{z}, p), (2.378)$$

$$|A(w)|, |A_x(w)|, |\vec{b}(w)|, |\vec{b}_x(w)| \leq \Lambda(1 + |\vec{z}|^2 + |p|^2)^{1/2}, \qquad (2.379)$$

$|A_{\vec{z}}(w)|, |A_p(w)|, |\vec{b}_{\vec{z}}(w)|, |\vec{b}_p(w)| \leq \Lambda$ for the *controllable growth conditions, and*

[20] r is the so-called Sobolev conjugate of 2, $r = 2n/(n-2)$ for $n > 2$, else $r \in [2, \infty)$.

$$|A(w)|, |A_x(w)|, |A_{\vec{z}}(w)|, |\vec{b}_p(w)| \leq \Lambda(1+|p|), |A_p(w)| \leq \Lambda, \tag{2.380}$$
$$|\vec{b}(w)|, |\vec{b}_x(w)|, |\vec{b}_{\vec{z}}(w)| \leq \Lambda(1+|p|^2)\forall|\vec{z}| \leq M, \text{for the } \textit{natural growth conditions.}$$

The constants λ, Λ may depend upon M for natural growth conditions. We have to slightly modify the concept of generalized solutions: For the two cases of controllable and natural growth conditions (2.379) and (2.380), we test a possible generalized solution for (2.362), $\vec{u}^0 \in H^1(\Omega, \mathbb{R}^q)$ and $\vec{u}^0 \in H^1(\Omega, \mathbb{R}^q) \cap C^0(\Omega, \mathbb{R}^q)$, by $\forall \vec{v} \in \mathcal{V} := \mathcal{V}_1 := H_0^1(\Omega, \mathbb{R}^q)$ and $\vec{v} \in \mathcal{V} := \mathcal{V}_1 \cap L^\infty(\Omega, \mathbb{R}^q)$, respectively. Independently of boundary conditions we get:

Theorem 2.99. Regularity of solutions, [170], pp 173 ff., Theorems 1.1, 1.2:

1. *Let the quasilinear equation (2.362) have smooth coefficients, $A, \vec{b} \in C^1$, and satisfy the strong (linearized) Legendre condition for $\vec{u}^0$, see (2.377), (2.378), and the controllable or natural growth condition (2.379) or (2.380). Then a generalized solution $\vec{u}^0 \in H^1(\Omega, \mathbb{R}^q)$ is in $H^2(\Omega, \mathbb{R}^q)$.*
2. *The derivatives $\partial^\ell \vec{u}^0, \ell = 1, \ldots, n$ satisfy some kind of linearized form of (2.362):*

$$\int_\Omega \sum_{i,j=1}^{q} \sum_{k,l=1}^{n} \Big[\left(a_k^i\right)_{p_j^l} w^0(x) \partial^j \left(\partial^\ell u_l^0\right) + \left(a_k^i\right)_{z_l} w^0(x) \left(\partial^\ell u_l^0\right) + \left(a_k^i\right)_{x_\ell} w^0(x)$$

$$-\delta_{i,\ell} a_k^0 w^0(x) \Big] \partial^i v_k dx = 0 \forall \vec{v} \in \mathcal{V} \textit{ with } w^0(x) := (x, \vec{u}^0(x), \partial \vec{u}^0(x)). \tag{2.381}$$

For the two cases of controllable and natural growth conditions the chosen $\mathcal{V}$ are dense in $H_0^1(\Omega, \mathbb{R}^q)$. Hence, the system (2.381) is, by (2.378), a linear elliptic system for each of the $\partial^\ell \vec{u}^0, \ell = 1, \ldots, n$. So we can achieve higher regularity by induction, cf. Theorems 2.38, 2.39, 2.45. We only discuss Theorem 2.91 1. Applications of 2. are only indicated.

Theorem 2.100. Regular solution of (2.362) with $\partial^\ell \vec{u}^0 \in H^{s+2}(\Omega, \mathbb{R}^q)$:

1. *Under the conditions of Theorem 2.99 the relation (2.381) is strongly elliptic. We assume for a $w^1(x) := (x, \vec{u}^1(x), \partial \vec{u}^1(x))$ with $\partial \vec{u}^1(x)) \in H^{s+1}(\Omega, \mathbb{R}^{n \times q})$ that, cf. (2.351),*

$$\partial\Omega \in C^{s+2}, \ \left(a_k^i\right)_{p_j^l} (w^1(x)), \left(a_k^i\right)_{z_j} (w^1(x)) \in W^{s+1,\infty}(\Omega), \tag{2.382}$$

$$\vec{f} := \left(\sum_{k,l=1}^{n} \left(a_k^i\right)_{x_\ell} (w^1(x)) - \delta_{k,\ell} a_i^0(w^1(x)) \in H^{s+1}(\Omega) \right)_{i=1,k=1}^{q,\ n}.$$

2. *Then any weak solution $\vec{u}^0 \in H^1(\Omega, \mathbb{R}^q)$ and $\vec{u}^0 \in H^1(\Omega, \mathbb{R}^q) \cap C^0(\Omega, \mathbb{R}^q)$ of (2.362) satisfies $\partial^\ell \vec{u}^0 \in H^{s+2}(\Omega, \mathbb{R}^{n \times q}), \ell = 1, \ldots, n$, and*

$$\|\partial^\ell \vec{u}^0\|_{H^{s+2}(\Omega, \mathbb{R}^n)} \leq C(\|\partial^\ell \vec{u}^0\|_{L^2(\Omega', \mathbb{R}^n)} + \|\vec{f}\|_{H^{s+1}(\Omega, \mathbb{R}^n)}). \tag{2.383}$$

3. *Theorem 2.91 2., 3. implies* $\vec{u}^0 \in C_0^\infty(\Omega, \mathbb{R}^{n\times q})$ *and* $\partial^\ell \vec{u}^0 \in C^{s,\delta}(\bar{\Omega}, \mathbb{R}^{n\times q})$, $\ell = 1, \ldots, n$, *if (2.382) is replaced by the C-conditions in (2.353) and (2.354), respectively*

Specific systems allow $\vec{u}^0 \in C^{0,\delta}(\Omega_1, \mathbb{R}^q)$ for $\Omega_1 = \Omega$, or $\Omega_1 \subset\subset \Omega$, [170], pp. 178, 181 ff. is extended to the general form by our Theorem 2.91. Indirect methods, reverse estimates for $\partial\vec{u}^0$ and singular sets, $\Omega \setminus \Omega_1$, are studied.

Leung [473], Theorem 3.2-1, shows $\vec{u}^0 \in C^2(\Omega, \mathbb{R}^q) \cap C^1(\bar{\Omega}, \mathbb{R}^q)$ for the solution of a special case of our quasilinear systems, the semilinear systems of the form

$$\Delta\vec{u} + f(x, \vec{u}) = 0 \ \forall x \in \Omega, \partial\vec{u}/\partial\nu + g(x, \vec{u}) = b(x) \ \forall x \in \partial\Omega.$$

Koshelev and Chelkak [449,450] present many different regularity results for elliptic and parabolic equations. For example, for special cases of the quasilinear equation (2.362), Hölder continuity for the derivatives of solutions in weighted spaces, see Theorem 3.2.1, is proved. A combination of Liouville's theorem, generalized to elliptic problems, with the method of elastic solutions describes appearing cracks. For the Navier–Stokes problem [450] estimates smoother norms for time dependent solutions. We give one example of [450] results. Again, we consider the above quasilinear boundary value problem (2.361), (2.365), see (2.16). In [449] he assumes a modified *Carathéodory condition* (2.284) for all terms, so all $a_i^k(x, \vec{z}, p), i = 1, \ldots, q, k = 0, \ldots, n$, are measurable with respect to x for all finite $(\vec{z}, p) \in \mathbb{R}^{n+nq}$ and continuously differentiable with respect to $(\vec{z}, p)$ a.e. for $x \in \Omega$. As growth conditions he requires a sharpened *strong Legendre condition* for the linearized operator, compare (2.378). For all $\vec{\vartheta} \in \mathbb{R}^{n\times q}$, and with $w = (x, \vec{z}, p)$, let

$$\begin{aligned} &\lambda(1 + |(\vec{z}, p)|^2)^{(s-1)/2}|\vec{\vartheta}|^2 \\ &\leq \sum_{i,j=1}^{q} \sum_{k,l=0}^{n} \left(\partial a_k^i/\partial p_l^j\right)(w)\vartheta_i^k\vartheta_j^l \leq \Lambda(1 + |(\vec{z}, p)|^2)^{(s-1)/2}|\vec{\vartheta}|^2, \end{aligned} \tag{2.384}$$

for $1 < s \leq 2, \lambda, \Lambda =$ const $. > 0$ and

$$\left|\left(\partial a_k^i/\partial p_l^j\right)(w)\right| \leq \Lambda(1 + |(\vec{z}, p)|^2)^{(s-1)/2} \ \forall\ i, j, k, l, \text{ and for some} \tag{2.385}$$

$1 < p' \ \ \forall \vec{u} \in W^{1,p'}(\Omega')$ and $\Omega' \subset \Omega$ let $a_k^i(x, \vec{u}(x), \partial\vec{u}(x)) \in L^{p'/(s-1)}(\Omega')$.

Choose $g \in H_\alpha^1(\bar{\Omega})$ and for $0 < \gamma < 1$ let $\alpha := 2 - n - 2\gamma$ and K satisfy

$$1 - \frac{\alpha(\alpha + n - 2)}{2(m-1)} > 0, K^2\left[1 - \frac{\alpha(n-2)}{m-1}\right]\left[1 - \frac{\alpha(\alpha + n - 2)}{2(m-1)}\right]^{-1} < 1, \tag{2.386}$$

$$\begin{aligned} H_\alpha^1(\bar{\Omega}) := \Bigg\{ &\vec{u} \in H^1(\Omega) : \\ &\|\vec{u}\|_{H_\alpha^1(\Omega)} := \sup_{x_0 \in \bar{\Omega}} \int_\Omega \left(|\vec{u}(x)|^2 + \sum_{|\alpha|=1} |\partial^\alpha \vec{u}(x)|^2 |x - x_0|^\alpha\right) dx < \infty \Bigg\}. \end{aligned} \tag{2.387}$$

Theorem 2.101. Solution of (2.362) in $H^1_\alpha(\bar{\Omega})$: *Under the conditions (2.384)–(2.386), the nonhomogeneous Dirichlet problem (2.362), (2.365) has a unique solution $\vec{u}^0 \in H^1_\alpha(\bar{\Omega})$ and $\vec{u}^0 \in C^{(2-m-\alpha)/2}(\Omega)$. Stronger results are possible for weighted norms in Sobolev spaces, extending the $H^1_\alpha(\bar{\Omega})$ setting.*

2.6.5 Linear elliptic systems of order $2m$, $m \geq 1$

We essentially study weak forms of systems of differential operators of order $2m$. We start with coercivity and the related existence and Fredholm results. Again the following results are valid for $W^{m,p}(\Omega, \mathbb{R}^n), p \geq 2$, in the sense of Corolary 2.44. But we essentially restrict the discussion here to the Hilbert space setting and extend it to Banach spaces in the next subsection. For the necessary higher order Fourier transforms we need Hilbert spaces anyway. For the case of second order systems, Theorem 2.89 shows that the coercivity is a consequence of the strong Legendre condition (2.343). This implies the strong Legendre–Hadamard condition, see (2.349). This simple situation does not seem to remain correct for higher order $2m$. In fact, the strong Legendre–Hadamard condition is needed to prove the coercivity of the principal part. We combine

$$\mathcal{V} = H_0^m(\Omega, \mathbb{R}^q) \subset \mathcal{W} = L^2(\Omega, \mathbb{R}^q) = \mathcal{W}' \subset \mathcal{V}' = H_0^{-m}(\Omega, \mathbb{R}^q), \tag{2.388}$$

with the *scalar products* and *norms* in (2.337) and $L\vec{u}$ in divergence form.

We recall and update the notation, cf. (2.73), (2.79), (2.102), (2.336), (2.335). Instead of $\vec{\vartheta}$ in (2.335) the η, ϑ are here more appropriate:

$$\vec{u}(x) = (u_1, \ldots, u_q)(x), \eta = (\eta_1, \ldots, \eta_q) \in \mathbb{R}^q, \vartheta = (\vartheta^1, \ldots, \vartheta^n) \in \mathbb{R}^n, \tag{2.389}$$

$$\vartheta^\alpha = (\vartheta^1)^{\alpha_1} \ldots (\vartheta^n)^{\alpha_n} \in \mathbb{R}, \vartheta^\alpha = 1 \text{ for } |\alpha| = 0,\ \partial^\alpha u_i = \frac{\partial^{|\alpha|}}{\partial x_1^{\alpha_1} \ldots \partial x_n{}^{\alpha_n}} u_i,$$

$$\partial^\alpha \vec{u} = (\partial^\alpha u_1, \ldots, \partial^\alpha u_q), \text{ with } \partial^\alpha u_i(x) \in \mathbb{R}, \partial^\alpha \vec{u}(x) \in \mathbb{R}^q,\ \partial^\alpha \vec{u} = \vec{u} \text{ for } |\alpha| = 0.$$

Similarly to our Lemma 2.26 above, see (2.82), and the beginning of Subsection 2.6.3, generalizations of Dirichlet boundary conditions to *Dirichlet system* are possible. Several cases are studied by Taylor [620], Chapter 5, Section 11. For avoiding the many technicalities we have restricted the discussion to *homogeneous Dirichlet boundary conditions* and choose $\mathcal{V} = H_0^m(\Omega, \mathbb{R}^q)$ instead of a possible $H_0^m(\Omega_1, \mathbb{R}^q), \Omega_1 \subset \Omega$. For smooth enough Ω we get the equivalence

$$\frac{\partial^i \vec{u}(x)}{\partial \nu^i} = 0\ \forall 0 \leq j \leq m-1 \Leftrightarrow \partial^\alpha \vec{u}(x) = 0\ \forall |\alpha| \leq m-1\ \forall x \in \partial\Omega, \tag{2.390}$$

in the trace sense; the usual *weak form* below and the strong form are well defined

$$L : H^{2m} \cap \mathcal{V} \to \mathcal{W} = L^2(\Omega, \mathbb{R}^q), L\vec{u} := \sum_{|\alpha|,|\beta| \leq m} (-1)^{|\alpha|} \partial^\alpha (A_{\alpha\beta} \partial^\beta \vec{u}) = \vec{f} \tag{2.391}$$

for smooth enough $A_{\alpha\beta}$. For the weak form we take the standard scalar product with $\vec{v} = (v_1, \ldots, v_q) \in \mathcal{V}$ and integrate by parts. A weak solution $\vec{u}^0 = (u_1^0, \ldots, u_q^0) \in \mathcal{V}$ for

(2.391) is defined by equating the corresponding linear and bilinear forms, $\langle \vec{f}, \vec{v}\rangle$ and $a(\vec{u}^0, \vec{v})$, compare (2.146), (2.274), (2.275),

$$\vec{u}^0 \in \mathcal{V} = H_0^m(\Omega, \mathbb{R}^q): \ \langle \vec{f}, \vec{v}\rangle = \left\langle \left(\sum_{|\alpha|\le m} f_\alpha^i \partial^\alpha \right)_{i=1}^q, \vec{v} \right\rangle = \int_\Omega \sum_{|\alpha|\le m} (\vec{f}_\alpha, \partial^\alpha \vec{v})_q dx$$

$$= a(\vec{u}^0, \vec{v}) := \langle L\vec{u}^0, \vec{v}\rangle_{\mathcal{V}'\times\mathcal{V}} := \int_\Omega \sum_{|\alpha|,|\beta|\le m} (A_{\alpha\beta}\partial^\beta \vec{u}^0, \partial^\alpha \vec{v})_q dx \tag{2.392}$$

$$= \sum_{i,j=1}^q \sum_{|\alpha|,|\beta|\le m} \int_\Omega \left(a_{\alpha\beta}^{ij} \partial^\beta u_j^0 \partial^\alpha v_i \right) dx, \ \forall\, \vec{v} \in \mathcal{V}, A_{\alpha\beta}(x) \in \mathbb{R}^{q\times q}, a_{\alpha\beta}^{ij} \in L^\infty(\Omega),$$

$$\vec{f} = (f_\alpha^i)_{i=1,|\alpha|\le m}^q = (\vec{f}_\alpha)_{|\alpha|\le m}, f_\alpha^i \in L^2(\Omega) \forall i, j, \alpha, \beta, \text{ or } \vec{f} \in \mathcal{V}'. \tag{2.393}$$

We generalize the above *strong Legendre–Hadamard condition* for $m = 1$, cf. (2.348), to order $2m$ and will combine it with variable coefficients $a_{\alpha\beta}^{ij}$, see Theorem 2.43.

Definition 2.102. *A system of linear differential operators of* order *2m is called* elliptic, *if its* principal part, L_p, *satisfies the* strong Legendre–Hadamard condition

$$a_p(\vec{u}, \vec{v}) := \int_\Omega L_p \vec{u}\vec{v} dx := \int_\Omega \sum_{|\alpha|=|\beta|=m} (A_{\alpha\beta}\partial^\beta \vec{u}, \partial^\alpha \vec{v})_q dx, \ \textit{and for} \tag{2.394}$$

$$L_p(x, \vartheta) := \left(\sum_{|\alpha|=|\beta|=m} a_{\alpha\beta}^{ij}(x) \vartheta^\beta \vartheta^\alpha \right)_{i,j=1}^q = \left(\sum_{|\alpha|=|\beta|=m} A_{\alpha\beta}(x) \vartheta^\beta \vartheta^\alpha \right) \ \textit{let}\ \exists \lambda > 0:$$

$$\bar{\eta}^T L_p(x, \vartheta)\eta = \sum_{i,j=1}^q \sum_{|\alpha|=|\beta|=m} a_{\alpha\beta}^{ij}(x) \vartheta^\beta \eta_j \vartheta^\alpha \bar{\eta}_i \ge \lambda |\vartheta|^{2m} |\eta|^2 \ \forall \vartheta \in \mathbb{R}^n, \eta \in \mathbb{C}^q, x \in \Omega.$$

For $1 = m < q$ *the condition (2.343) was required, cf Remark 2.90.*

For proving existence and uniqueness results for (2.392) we reinterpret the bilinear form as $a(\vec{u}, \vec{v}) =: a_p(\vec{u}, \vec{v}) + b(\vec{u}, \vec{v})$. As above we prove that $a(\cdot,\cdot), a_p(\cdot,\cdot) : \mathcal{V}\times\mathcal{V} \to \mathbb{R}$ and $\vec{f} \in W^{-m,p'}(\Omega, \mathbb{R}^q) = \mathcal{V}' \to \mathbb{R}$ are *bounded* bilinear and linear forms and $a_p(\cdot,\cdot)$ is $\mathcal{V}$-coercive. We summarize

Theorem 2.103. Strong and weak solutions of (2.392): *For a system of the form (2.392) define the bilinear and linear forms* $a(\vec{u}, \vec{v})$ *and* $\langle \vec{f}, \vec{v}\rangle_{\mathcal{V}'\times\mathcal{V}}$ *, and the principal part* $a_p(\vec{u}, \vec{v})$ *as in (2.392), (2.394). Under the conditions (2.393), these* $a(\vec{u}, \vec{v}), a_p(\vec{u}, \vec{v}), \langle \vec{f}, \vec{v}\rangle_{\mathcal{V}'\times\mathcal{V}}$ *are continuous in* $\mathcal{V} = H_0^m(\Omega, \mathbb{R}^q)$. *For* $\left(a_{\alpha\beta}^{jk}\right)_{i,j=1}^q$ $= A_{\alpha\beta} \in W^{|\beta|,\infty}(\Omega, \mathbb{R}^q)$, *any solution* $\vec{u}^0 \in H_0^{2m}(\Omega, \mathbb{R}^q)$ *of (2.391) necessarily satisfies the weak equation, cf. (2.392),*

$$\vec{u}^0 \in \mathcal{V}: \ a(\vec{u}^0, \vec{v}) = \langle \vec{f}, \vec{v}\rangle_{\mathcal{V}'\times\mathcal{V}} \forall \vec{v} \in \mathcal{V}. \tag{2.395}$$

Vice versa, every smooth solution $\vec{u}^0 \in H^{2m}(\Omega, \mathbb{R}^q) \cap H_0^m(\Omega, \mathbb{R}^q)$ of (2.392) satisfies (2.391).

Our next goal are analogies to Theorems 2.43, 2.89 and Corollary 2.44. The key result towards this goal is the $\mathcal{V}$-coercivity of the principal part.

Theorem 2.104. Coercivity, unique solutions, Fredholm alternative: *In (2.392) let the coefficients $a_{\alpha\beta}^{jk} \in L^\infty(\Omega)$, but for $|\alpha| = |\beta| = m > 1$ be continuous in $\bar{\Omega}$, see Theorem 2.43. Furthermore, let (2.394), the strong Legendre–Hadamard condition, be satisfied. Then the* principal part $a_p(\vec{u}, \vec{v})$ *is $\mathcal{V} = H_0^m(\Omega, \mathbb{R}^q)$-coercive.*

Under these conditions Corollary 2.44 implies for a $\mathcal{V}$-coercive $a(\cdot, \cdot)$ the *existence of a unique generalized or weak solution* $\vec{u}^0$ and a boundedly invertible L. Hence, for a $\mathcal{V}$-elliptic $a(\cdot, \cdot)$ the Fredholm alternative with its consequences is valid. For a $\mathcal{V}$-coercive, symmetric $a(\cdot, \cdot)$, the unique $\vec{u}^0$ solves the variational problem, the updated (2.61). Finally, all H^m-coercivity based existence, uniqueness and Fredholm results, including numerical stability and index $L = 0$, remain correct for the $W^{m,p}$ case and for $p \geq 2$, with the modifications in Corollary 2.44 3., 4..

Proof. We cannot repeat the argument in the proof of Theorem 2.89, see (2.347). The reason is the inequality $(\partial u_j/\partial x_i)^{\alpha_i} \neq \partial^{\alpha_i} u_j/\partial x_i^{\alpha_i}$ for $\alpha_i > 1$. Instead we use the standard Fourier transforms, see, e.g. Hackbusch [388], Theorem 7.2.7. We start the proof by assuming constant highest coefficients. But the same ideas as in the proof of Theorem 2.43 can be used here as well to obtain the coercivity for continuous highest coefficients. So for constant highest coefficients we get

$$a_p(\vec{u}, \vec{u}) = \int_\Omega \left(\sum_{j,k=1}^{q} \sum_{|\alpha|=|\beta|=m} a_{\alpha\beta}^{jk} \partial^\beta u_k \partial^\alpha u_j \right) dx \ \forall\ \vec{u}, \vec{v} \in \mathcal{V}. \tag{2.396}$$

We extend the $\vec{u}, \vec{v} \in \mathcal{V}$ by $\vec{u}(x) = 0\ \forall\ x \in \mathbb{R}^n \setminus \Omega$. This $\vec{u}$ has compact support $\bar{\Omega}$. Thus the *Fourier transform*, see, e.g. [388], Subsection 6.2.3, is well defined[21] for this

[21] It is defined, with the imaginary unit ι with $\iota^2 = -1$, via the components $u = u_j \in C_0^\infty(\mathbb{R}^n)$:

$$F : C_0^\infty(\mathbb{R}^n) \to L^2(\mathbb{R}^n), \quad \text{with } (\xi, x)_n = \sum_{k=1}^{n} \xi_k x_k \ \forall\ \xi, x \in \mathbb{R}^n, \text{ as}$$

$$\hat{u}(\xi) := (Fu)(\xi) := (2\pi)^{-n/2} \int_{\mathbb{R}^n} e^{-\iota(\xi,x)_n} u(x) dx$$

$$:= \lim_{R\to\infty} (2\pi)^{-n/2} \int_{|\xi|_\infty \leq R} e^{-\iota(\xi,x)_n} u(x) dx. \tag{2.397}$$

Since $C_0^\infty(\mathbb{R}^n)$ is dense in $L^2(\mathbb{R}^n)$, this F can be extended to $L^2(\mathbb{R}^n) \to L^2(\mathbb{R}^n)$ and inverted on $C_0^\infty(\mathbb{R}^n)$, dense in $L^2(\mathbb{R}^n)$, and again extended to $L^2(\mathbb{R}^n) \to L^2(\mathbb{R}^n)$ as

$$(F^{-1}\hat{u})(x) := (2\pi)^{-n/2} \int_{\mathbb{R}^n} e^{\iota(\xi,x)_n} \hat{u}(\xi) d\xi \ \forall\ u \in C_0^\infty(\mathbb{R}^n), \text{ extended to } L^2(\mathbb{R}^n).$$

Its characteristics are summarized in

$\vec{u}$. For $a_p(\vec{u}, \vec{u})$ we obtain for constant coefficients

$$a_p(\vec{u}, \vec{u}) = \sum_{j,k=1}^{q} \sum_{|\alpha|=|\beta|=m} a_{\alpha\beta}^{jk} \int_{\mathbb{R}^n} (\partial^\beta u_k \cdot \partial^\alpha u_j) dx \; \forall \; \vec{u} \in \mathcal{V}.$$

We apply the Fourier transform to each term $\partial^\beta u_k$ and $\partial^\alpha u_j$. By the invariance of the scalar products in (2.398), the strong Legendre–Hadamard condition (2.394), and with $\eta_k = \hat{u}_k(\xi) = F(u)(\xi)$ we get

$$\begin{aligned} a_p(\vec{u}, \vec{u}) &= \sum_{j,k=1}^{q} \sum_{|\alpha|=|\beta|=m} a_{\alpha\beta}^{jk} \int_{\mathbb{R}^n} \left([\iota^{|\beta|} \xi^\beta \hat{u}_k(\xi)] \overline{[\iota^{|\alpha|} \xi^\alpha \hat{u}_j(\xi)]}\right) d\xi \\ &= \int_{\mathbb{R}^n} \sum_{j,k=1}^{q} \sum_{|\alpha|=|\beta|=m} \left(a_{\alpha\beta}^{jk} \xi^\beta \iota^{|\beta|} \hat{u}_k(\xi) \xi^\alpha \overline{[\iota^{|\alpha|} \hat{u}_j(\xi)]}\right) d\xi \qquad (2.400) \\ &\geq \int_{\mathbb{R}^n} \lambda \left(\sum_{\ell=1}^{n} (\xi_\ell)^2\right)^m \sum_{i=1}^{q} |\hat{u}_i(\xi)|^2 d\xi \end{aligned}$$

For a fixed $m > 1$ there exists an $\epsilon > 0$, such that $\left(\sum_{\ell=1}^n (\xi_\ell)^2\right)^m > \epsilon \sum_{|\alpha|=m} (\xi^\alpha)^2$, see (2.75). So we find with (2.399)

$$a_p(\vec{u}, \vec{u}) > \lambda\epsilon \sum_{i=1}^{q} \int_{\mathbb{R}^n} \left(\sum_{|\alpha|=m} (\xi^\alpha)^2\right) |\hat{u}_i(\xi)|^2 d\xi = \lambda\epsilon |\vec{u}|^2_{H^m(\Omega, \mathbb{R}^q)}.$$

Since $|\vec{u}|_{H^m(\Omega,\mathbb{R}^q)}$ and $\|\vec{u}\|_{H^m(\Omega,\mathbb{R}^q)}$ are equivalent norms on $\mathcal{V} = H_0^m(\Omega, \mathbb{R}^q)$, the $\mathcal{V}$-coercivity of $a_p(\vec{u}, \vec{u})$ is proved. The extension to "continuous coefficients" is obtained as in Theorem 2.43. □

Theorem 2.105. *The Fourier transform and its inverse are linear isometries:*

$$\begin{aligned} &F, F^{-1} : L^2(\mathbb{R}^n) \to L^2(\mathbb{R}^n), \text{ with} \\ &\quad \|F\|_{L^2(\mathbb{R}^n) \hookrightarrow L^2(\mathbb{R}^n)} = \|F^{-1}\|_{L^2(\mathbb{R}^n) \hookrightarrow L^2(\mathbb{R}^n)} = 1, \\ &\quad (u, v)_{L^2(\mathbb{R}^n)} = (\hat{u}, \hat{v})_{L^2(\mathbb{R}^n)} \; \forall \; u, v \in L^2(\mathbb{R}^n), \qquad (2.398) \\ &\quad \forall \; u \in H^k(\mathbb{R}^n), |\alpha| \leq k \;\; F(\partial^\alpha u)(\xi) = \iota^{|\alpha|} \xi^\alpha \hat{u}(\xi), \text{ and} \\ &\quad \|u\|_{H^k(\mathbb{R}^n)} = \left\| \sqrt{\sum_{|\alpha| \leq k} |\xi^\alpha|^2} \hat{u}(\xi) \right\|_{L^2(\mathbb{R}^n)} \quad \text{and} \\ &\quad \text{the seminorm } |u|_{H^k(\mathbb{R}^n)} = \left\| \sqrt{\sum_{|\alpha|=k} |\xi^\alpha|^2} \hat{u}(\xi) \right\|_{L^2(\mathbb{R}^n)}. \qquad (2.399) \end{aligned}$$

By applying F componentwise, this F can be extended to $F : L^2(\mathbb{R}^n, \mathbb{R}^q) \to L^2(\mathbb{R}^n, \mathbb{R}^q)$, noting that then $(u_i, v_j)_{L^2(\mathbb{R}^n)} = (\hat{u}_i, \hat{v}_j)_{L^2(\mathbb{R}^n)} \; \forall \; i, j = 1, \ldots, q$.

For an elliptic linear system in the sense of Definition 2.85 6., we give a flavor of some results for local existence and regularity, by summarizing a few examples from Taylor [618]. For regularity results we combine several Propositions in Taylor [618], Chapter 5, Section 11 with our previous results. We formulate an interior estimate, a near boundary "collar" estimate, and a local solvability and regularity result. We simplify Taylor [618], Chapter 5, Theorem 11.1 ff., originally for elliptic systems on manifolds, to our $\Omega \subset \mathbb{R}^n$.

Taylor [618] discusses regular boundary value problems, cf. Chapter 5, Propositions 11.9–11.10. For example, a system (2.392) with Dirichlet boundary conditions is regular, if it satisfies a *modified strong Legendre condition,* cf. Chapter 5, (11.79),

$$\exists\lambda > 0 : \forall x \in \Omega, \forall\vartheta \in \mathbb{R}^n : \left(\sum_{|\alpha|=|\beta|=m} a^{ij}_{\alpha\beta}\vartheta^\beta\vartheta^\alpha \right)^q_{i,j=1} \geq \lambda|\vartheta|^{2m} I,\ I \in \mathbb{R}^{q\times q}, \quad (2.401)$$

with the idendity matrix, I. This implies the strong Legendre–Hadamard condition (2.394) and a Gårding inequality, cf. (2.141), [618], Chapter 5, Exercise 1. Other types of boundary conditions are discussed as well Taylor [618], Chapter 5, Proposition 11.10 and Theorem 11.11 apply to our situation:

Theorem 2.106. Interior estimate: *Let L, a system of linear differential operators of order m, be m-elliptic, cf. Definition 2.85, 6., and $\vec{u} \in H^m(\Omega, \mathbb{R}^q)$ (a distribution in [618], Theorem 11.1) and $L\vec{u} = f \in H^s(\Omega, \mathbb{R}^q), 0 \leq s \in \mathbb{N}_0$. Then $\vec{u} \in H^{m+s}(\Omega, \mathbb{R}^q)$. Furthermore, for any $\Omega_1 \subset\subset \Omega_2 \subset\subset \Omega$ and $\sigma < m + s$ the following estimate holds*

$$\|\vec{u}\|_{H^{m+s}(\Omega_1,\mathbb{R}^q)} \leq C(\|L\vec{u}\|_{H^s(\Omega_2,\mathbb{R}^q)} + \|\vec{u}\|_{H^\sigma(\Omega_2,\mathbb{R}^q)}).$$

For the next theorem we modify some properties related to the atlas of a domain Ω, see Proposition 1.32, (1.63). In a first step we choose the $\mathcal{V}^i, i = 1, \dots, N$, with $\partial\Omega \subset \cup_{i=1}^N \mathcal{V}^i$. Let $\hat{\mathcal{V}}^i := \bar{\Omega} \cap \mathcal{V}^i$, $\hat{\mathcal{V}} := \cup_{i=1}^N \hat{\mathcal{V}}^i$. For $\Omega \in C^{0,1}$ Proposition 1.32, implies that there exist $\Omega_3 \subset\subset \Omega_4 \subset\subset \hat{\mathcal{V}}$ with $\Omega_3 \cap \partial\Omega = \Omega_4 \cap \partial\Omega = \partial\Omega$, such that $b_3 : \partial\Omega \times [0, 2] \to \Omega_3$, and $b_4 : \partial\Omega \times [0, 1] \to \Omega_4$ are bijections. Now boundary conditions and "collar" norms are relevant, see [618], Chapter 5, Propositions 11.14, 15, 16. With $\|\vec{u}(\cdot, y)\|_{H^s(\partial\Omega)}, 0 \leq y \leq 2$, and ∇^j_y, the jth partial derivative into the y-direction, let the norms

$$\|\vec{u}\|^2_{H_{0,s}(\Omega_4,\mathbb{R}^q)} := \int_0^1 \|\vec{u}(\cdot, y)\|^2_{H^s(\partial\Omega)} dy,\ s, k \in \mathbb{N}_0,$$

$$\|\vec{u}\|^2_{H_{k,s}(\Omega_4,\mathbb{R}^q)} := \sum_{j=0}^k \int_0^1 \left\|\nabla^j_y \vec{u}(\cdot, y)\right\|^2_{H^{k-j+s}(\partial\Omega)} dy, \quad (2.402)$$

define the corresponding Banach spaces $H_{k,s}(\Omega_4, \mathbb{R}^q)$, and similarly $H_{k,s}(\Omega_3, \mathbb{R}^q)$ with $\int_0^2$ replacing $\int_0^1$. Additionally, let L, a system of linear differential operators now of order $2m$, be *strongly $2m$-elliptic,* cf. Definition 2.85, 6. By Taylor [618], Chapter 5, Proposition 11.10. a *strongly $2m$-elliptic* system (2.392) or (2.323) with Dirichlet boundary conditions is *regular.* For other types of boundary conditions see [618].

Theorem 2.107. "Collar" estimate, Taylor, [618], Chapter 5, Propositions 11.14: *Under these conditions let* $\vec{u_0} \in H_{2m,\sigma}(\Omega_3)$ *solve the (nonhomogeneous) "collar" Dirichlet boundary problem*[22]

$$L\vec{u_0} = \vec{f} \in H_{k,s}(\Omega_3, \mathbb{R}^q), \partial^j \vec{u_0}/\partial\nu^j = \varphi_j \in H^{2m+k-j+1/2+s}(\partial\Omega, \mathbb{R}^q), j = 0, \ldots, m-1,$$

for some $0 \leq \sigma < s \in \mathbb{R}$. *Then we even obtain* $\vec{u_0} \in H_{2m+k,s}(\Omega_4, \mathbb{R}^q)$ *and it satisfies the estimate*

$$\|\vec{u_0}\|_{H_{2m+k,s}(\Omega_4,\mathbb{R}^q)} \leq C \left(\|L\vec{u_0}\|_{H_{k,s}(\Omega_3,\mathbb{R}^q)} + \sum_{j=0}^{m-1} \|\partial^j \vec{u_0}/\partial\nu^j\|_{H^{2m+k-j+1/2+s}(\partial\Omega,\mathbb{R}^q)} \right).$$

Proposition 11.16 of [618] applies to our situation and states, compare Zeidler [676], Section 4.17:

Theorem 2.108. Regularity and Fredholm alternative for order 2m*: Let the coefficients* $a^{ij}_{\alpha\beta}$, $\partial\Omega$ *in (2.392) be in* C^∞, *and let Definition 2.85 6. be satisfied. Then the operator* $L : H^{2m+k}(\Omega, \mathbb{R}^q) \cap H_0^m(\Omega, \mathbb{R}^q) \to H^k(\Omega, \mathbb{R}^q)$, $q \geq 1$, *in (2.392) is Fredholm, hence, for* $f \in H^k(\Omega, \mathbb{R}^q), k \geq 0$, *any solution for (2.392), if it exists, is* $\vec{u^0} \in H^{2m+k}(\Omega)$.

2.6.6 Divergent quasilinear elliptic systems of order $2m$

In this subsection we study strong and weak forms of systems of quasilinear differential operators of order $2m$. The results for this case are not available in a similarly general form as before. We summarize parts of Koshelev and Chelkak [450] but start with one of the most important examples:

von Kármán equations

Different types of existence and regularity results for this system are discussed by John and Necas [423], John and Naumann [422], Ciarlet [175], Giaquinta *et al.* [343], Kondrat'ev and Olejnik [447, 448], Horn [404], and Bock [100]. In John and Naumann [422], the strong form of the von Kármán equations is studied:

$$G_s(u, w) := \begin{pmatrix} \Delta^2 u & -[u, w] \\ \Delta^2 w & +\frac{1}{2}[u, u] \end{pmatrix} = \begin{pmatrix} f \\ 0 \end{pmatrix} \text{ in } \Omega \subset \mathbb{R}^2, \tag{2.403}$$

with $\Delta^2 = \Delta\Delta$ and the bracket operator $[u, v] := u_{xx}v_{yy} - 2u_{xy}v_{xy} + u_{yy}v_{xx} = [v, u]$.

Specific existence and uniqueness results for the von Kármán equations are discussed in the papers of Hlavácek and Naumann [399, 508], John *et al.* [422, 423], even with different boundary conditions on different parts of $\partial\Omega$, and Giaquinta *et al.*, [343]. For a full formulation, many technical details would have to be discussed. So we just refer

22 Or let L with the boundary operators B be regular elliptic, see [620], pp. 379 ff.; essentially L, B satisfy a generalized complementary condition.

to [343, 399, 422, 423, 508], and only combine John and Naumann [422] in (2.403) with some of our earlier results.

Existence and uniqueness can also be elaborated numerically. The above (2.331) and (2.403) represent compact perturbations of its strong principal part

$$G_p(u,w) := \left(\Delta^2 u, \Delta^2 w\right)^T = (f,0)^T \text{in} \quad \Omega. \tag{2.404}$$

The weak form of (2.404) is obviously $H_0^2(\Omega, \mathbb{R}^2)$-coercive, so the linearized (2.403) is $H_0^2(\Omega, \mathbb{R}^2)$-elliptic. Furthermore, (2.404) satisfies the conditions in Subsection 2.6.5. Hence, existence and uniqueness for (2.331) and (2.403) can be guaranteed by combining the Fredholm alternative for the linearized form of (2.403) with discretizations and numerical continuation techniques, starting with (2.404) and ending with (2.331) or (2.403).

Equation (2.403) is slightly simplified compared to (2.331). In the first equation the $+\lambda u_{xx}$ is missing. As a consequence of the special form of (2.331), and since the $+\lambda u_{xx}$ term represents a compact perturbation, the arguments for Theorem 2.56 3. remain valid for this nonlinear case as well. Hence, we formulate John and Naumann [422] directly for the more general (2.331) case. They discuss two sets of non-Dirichlet boundary conditions. Using the unit outer normal vector $\nu = (\nu_x, \nu_y)$, the boundary tangent vector $\sigma = (-\nu_y, \nu_x)$ and the Poisson ratio of the plate material, μ, they introduce the operators

$$\begin{aligned} Mu &:= \mu\Delta u + (1-\mu)\left(u_{xx}\nu_x^2 + 2u_{xy}\nu_x\nu_y + u_{yy}\nu_y^2\right) \text{ with constant } \mu \in (0, 1/2), \text{ or} \\ Tu &:= -\mu\frac{\partial}{\partial\nu}\Delta u + (1-\mu)\frac{\partial}{\partial\sigma}\left(u_{xx}\nu_x\nu_y - u_{xy}\left(\nu_x^2 - \nu_y^2\right) - u_{yy}\nu_x\nu_y\right). \end{aligned} \tag{2.405}$$

This allows us to formulate the two sets of boundary conditions, considered in [422],

$$u = Mu = 0, w = \frac{\partial}{\partial\nu}w = 0 \text{ on } \partial\Omega \text{ and} \tag{2.406}$$

$$Mu = Tu = 0, w = \frac{\partial}{\partial\nu}w = 0 \text{ on } \partial\Omega. \tag{2.407}$$

We denote (2.403), (2.406) as Problem I, and (2.403), (2.407) as Problem II.

For weak solutions of (2.403), John and Naumann [422] modify the standard approach. With

$$\begin{aligned} a(u,v) &:= \int_\Omega [u_{xx}v_{xx} + 2(1-\mu)u_{xy}v_{xy} + u_{yy}v_{yy} + \mu(u_{xx}v_{yy} + u_{yy}v_{xx})]dxdy \\ &\text{and } \exists c_1, c_2 : c_1\|u\|^2_{H^2(\Omega)} \le a(u,u) \le c_2\|u\|^2_{H^2(\Omega)}, \forall u, v \in H^2(\Omega), \end{aligned} \tag{2.408}$$

they get a modified Green's formula, based upon M, T

$$a(u,v) = \int_\Omega u\Delta^2 v dxdy + \int_{\partial\Omega} uTvds + \int_{\partial\Omega} \frac{\partial u}{\partial\nu}Mvds \;\; \forall u, v \in H^4(\Omega), \tag{2.409}$$

or combined as

$$\int_\Omega (u\Delta^2 v - \Delta^2 uv)dxdy = + \int_{\partial\Omega} (vTu - uTv)ds + \int_{\partial\Omega} \left(\frac{\partial v}{\partial \nu} Mu - \frac{\partial u}{\partial \nu} Mv\right) ds.$$

Correspondingly we have to determine the solution

$$(u_0, w_0) \in \big(W := \{u \in H^2(\Omega) : u|_{\partial\Omega} = 0\}\big) \times H_0^2(\Omega)\big), \tag{2.410}$$

by testing (2.403) in the first and second equation as

$$a(u_0, \varphi) = \int_\Omega [u_0, w_0]\varphi dxdy + \int_\Omega f\varphi dxdy \forall \varphi \in W \tag{2.411}$$

$$(w_0, \psi)_{H_0^2(\Omega)} := \sum_{|\alpha|=2} \int_\Omega \partial^\alpha u_0 \partial^\alpha v dxdy = -\int_\Omega [u_0, u_0]\psi dxdy \quad \forall \psi \in H_0^2(\Omega). \tag{2.412}$$

Equation (2.331) has to be tested by slightly changing (2.411) into (2.413) and leaving (2.412) unchanged:

$$a(u_0, \varphi) = \int_\Omega [u_0, w_0]\varphi dxdy - \int_\Omega \lambda u_{0,xx}\varphi dxdy + \int_\Omega f\varphi dxdy \quad \forall \varphi \in W. \tag{2.413}$$

Obviously with a solution (u_0, w_0) of Problem II, (2.403), (2.407), the $(u_0 + p_1, w_0), \forall p_1 \in \Pi_1$, the polynomials of degree 1, is again a solution. We eliminate the nonuniqueness for u_0 in (2.403), (2.407) by defining K as $H^2(\Omega) =: K \oplus \Pi_1$. So the quotient space $H^2(\Omega)/\Pi_1$ is the space of all classes $\tilde{u}$, such that $u, v \in \tilde{u}$ iff $u - v \in \Pi_1$. Each $u \in \tilde{u}$ has a unique decomposition $u = u_K + p_1$ with $u_k \in K$, $p_1 \in \Pi_1$ and u_K uniquely determines $\tilde{u}$.

A $(u_0, w_0) \in W \times H_0^2(\Omega)$, cf. (2.410), is a weak solution of Problem I, (2.403), (2.406), if it satisfies (2.411) $\forall \varphi \in W$, and (2.412) $\forall \psi \in H_0^2(\Omega)$, and (2.406). Similarly, $(u_0, w_0) \in H^2(\Omega) \times H_0^2(\Omega)$ weakly solves Problem II, if it satisfies (2.411) $\forall v \in H^2(\Omega)$, and (2.412) $\forall \psi \in H_0^2(\Omega)$, and (2.406). If we consider (2.331), we have to replace (2.411) by (2.413).

Next, we present regularity results for the solutions of (2.331), (2.403). We start with the result in [422], for (2.403).

Theorem 2.109. Regular Kármán solutions:

1. *For* $f \in L^p(\Omega), 1 < p < \infty$ *let* (mind $[f] \neq |f|$)

$$[f] := \|f\|_{L^p(\Omega)} \left(1 + \|f\|^2_{L^p(\Omega)}\right).$$

Let (u_0, w_0) be a solution of Problem I or II. Then

$$u_0 \in W^{4,p}(\Omega), Mu_0 = 0 \text{ pointwise on } \partial\Omega, w_0 \in W^{6,p}(\Omega) \cap H_0^2(\Omega). \tag{2.414}$$

2. If (u_0, w_0) solves Problem I, then additionally $u_0 \in W^{4,p}(\Omega) \cap W$. For Problem II we get $Tu_0 = 0$ a.e. on $\partial\Omega$ in the trace sense. Moreover, for

$$1 < p < 2: \quad \|u_0\|_{W^{4,p}(\Omega)} \leq C[f], \|w_0\|_{W^{6,p}(\Omega)} \leq C[f]^2$$

and for

$$2 \leq p < \infty: \|u_0\|_{W^{4,p}(\Omega)} \leq C(\|f\|_{L^p(\Omega)} + [f]^3), \|w_0\|_{W^{6,p}(\Omega)} \leq C(\|f\|_{L^p(\Omega)} + [f]^3)^2.$$

Finally, we combine Theorem 2.91 with Corollaries 3.1 and 3.2 in [422] to get

Corollary 2.110. *Enforce the conditions of Theorem 2.109 by requiring $f \in W^{k,p}(\Omega)$. Then the solutions are $u_0 \in W^{4+k,p}(\Omega), w_0 \in W^{6+k,p}(\Omega)$. If even $f \in C^\infty(\bar{\Omega})$ then we even get $u_0, w_0 \in C^\infty(\bar{\Omega})$. In all these cases, $Tu_0 = 0$ pointwise on $\partial\Omega$ for Problem II. These results remain correct for (2.331).*

2.6.7 Nemyckii operators and quasilinear divergent systems of order $2m$

Generalized solutions of quasilinear divergent systems of order $2m$

Zeidler [678] does not study quasilinear divergent *systems*. But his results for quasilinear equations, cf. Zeidler [678], pp. 500 ff., 554 ff., Theorem 25B, Proposition 26.6 and pp. 572 ff., are easily generalized to the present case. So we update the necessary definitions and results from Subsections 2.5.5 and 2.5.6.

For the Banach space $\mathcal{X}$ we use here the product space $\mathcal{X} = W^{m,p}(\Omega, \mathbb{R}^q) = (W^{m,p}(\Omega))^q$. For $1 < p < \infty$ this $\mathcal{X}$ is reflexive and for

$$\mathcal{X} = W^{m,p}(\Omega, \mathbb{R}^q) \text{ with } 1 < p < \infty \text{ let } \langle A\vec{u} - A\vec{v}, \vec{u} - \vec{v}\rangle = \langle A\vec{u} - A\vec{v}, \vec{u} - \vec{v}\rangle_{\mathcal{X}' \times \mathcal{X}}.$$

Then equations (2.242)–(2.253) remain unchanged, and we have called

$$A \text{ monotone} \Leftrightarrow \langle A\vec{u} - A\vec{v}, \vec{u} - \vec{v}\rangle \geq 0, \tag{2.415}$$

$$A \text{ strongly monotone} \Leftrightarrow \langle A\vec{u} - A\vec{v}, \vec{u} - \vec{v}\rangle \geq c\|\vec{u} - \vec{v}\|^2_{\mathcal{X}}, \ c \in \mathbb{R}_+, \tag{2.416}$$

$$A \text{ stable} \Leftrightarrow \|A\vec{u} - A\vec{v}\|_{\mathcal{X}'} \geq a(\|\vec{u} - \vec{v}\|_{\mathcal{X}}), \text{ with} \tag{2.417}$$

$$a : \mathbb{R}_{+,0} \to \mathbb{R}_{+,0} \quad \text{continuous and strictly increasing, cf. (2.418), with}$$

$$a(0) = 0, \quad \lim_{t\to\infty} a(t) = \infty, \text{ e.g. } a(t) = c|t|^{p-1}, \ 0 < c, \ 1 < p. \tag{2.418}$$

Then Proposition 2.67–Theorem 2.70 remain valid. We only reformulate the most important parts of Theorem 2.68:

Theorem 2.111. Properties of nonlinear operators: *Let $\mathcal{X}$ be a real Banach space and $A : \mathcal{X} \to \mathcal{X}'$ a (nonlinear) operator. Then*

1. *If A is monotone, coercive, and hemicontinuous on a reflexive Banach space, then A is surjective and, for each $f \in \mathcal{X}'$, the set of solutions of $Au = f$ is bounded, convex and closed, cf. [678], Theorems 25B and 26A(a).*
2. *If A is additionally strictly monotone, the above solution is unique. Then $A^{-1} : \mathcal{X}' \to \mathcal{X}$ exists and, cf. [678], Theorem 26A(c), (d),*
   a uniformly monotone A implies a continuous A^{-1},
   a strongly monotone A implies a Lipschitz–continuous $A^{-1} \in C_L$.

We have to update the definition of Nemyckii operators, cf. Definition 2.69:

Definition 2.112. *For given functions $\vec{f}$ and appropriate $\vec{u} \in \mathcal{D}(\vec{F})$, define $\vec{F}(\vec{u})$ as*

$$\vec{f} = (f_i)_{i=1}^q : \mathcal{D}(\vec{f}) \subset \Omega \times \mathbb{R}^{N_m \times q} \to \mathbb{R}^q, \vec{u} : \Omega \to \mathbb{R}^q : \quad \vec{F}(\vec{u})(x) := \vec{f}(x, \partial^{\leq m}\vec{u}(x)), \tag{2.419}$$

cf. (2.423), with $(x, \partial^{\leq m}\vec{u}(x)) \in \mathcal{D}(\vec{f})$. This $\vec{F}$ is called a Nemyckii operator.

Now we formulate two conditions guaranteeing that $\vec{F}(\vec{u})(x)$ makes sense. The *Carathéodory* and the *growth condition* enforce measurability and appropriate growth of the composite function, $\vec{F}$. More precisely, we assume

Carathéodory condition: Ω is bounded and measurable in $\mathbb{R}^n$,

$$\vec{f} : (x, \vec{\vartheta}^{\leq m}) \in \mathcal{D}(\vec{f} \subset \ \Omega \times \mathbb{R}^{N_m \times q} \to \mathbb{R}^q, \forall\ |\alpha| \leq m, \text{ cf. (2.423)},$$
$$\vec{\vartheta}^{\leq m} \mapsto \vec{f}(x, \vec{\vartheta}^{\leq m}) \text{ is continuous in } \mathbb{R}^{N_m \times q} \text{ a.e. for } x \in \Omega, \tag{2.420}$$
$$x \mapsto \vec{f}(x, \vec{\vartheta}^{\leq m}) \text{ is measurable in } \Omega \ \ \forall\ \vec{\vartheta}^{\leq m} \in \mathbb{R}^{N_m \times q}, \text{e.g. for continuous } \vec{f},$$

and with $1 < p < \infty, 1/p + 1/p' = 1, p - 1 = p/p'$ and $\exists\ a \in L^{p'}(\Omega), 0 \leq a(\cdot)$ a.e.

$\exists\ C \in \mathbb{R}_+,$ s.t. $\forall\ x \in \Omega, \vec{\vartheta}^{\leq m} \in \mathbb{R}^{N_m \times q}, \vec{\vartheta}^{\gamma} \in \mathbb{R}^q$:

Growth condition: $$|\vec{f}(x, \vec{\vartheta}^{\leq m})| \leq C(a(x) + \sum_{|\gamma| \leq m} |\vec{\vartheta}^{\gamma}|^{p-1}) \ \forall |\alpha| \leq m. \tag{2.421}$$

Theorem 2.113. Estimates for nonlinear operators: *If (2.420), (2.421) are satisfied, then $\vec{F} : W^{m,p}(\Omega, \mathbb{R}^q) \to L^{p'}(\Omega, \mathbb{R}^q)$ in (2.419) is continuous and bounded (this is* not equivalent *for nonlinear operators!) such that*

$$\|\vec{F}(\vec{u})\|_{(L^{p'}(\Omega))^q} \leq C \left(\|a\|_{L^{p'}(\Omega)} + \sum_{i=1}^{k} \|\partial^{\gamma}\vec{u}\|^{p-1}_{(L^p(\Omega))^q} \right) \ \forall \vec{u} \in W^{m,p}(\Omega, \mathbb{R}^q). \tag{2.422}$$

The following general quasilinear systems are studied by Koshelev and Chelkak [450]. They consider Sobolev spaces $W^{m,p}(\Omega, \mathbb{R}^q) \subset L^2(\Omega, \mathbb{R}^q) \subset W^{-m,p'}(\Omega, \mathbb{R}^q)$,

$1/p + 1/p' = 1$. We extend their existence results to the case $1 < p < \infty$, by generalizing our results in Subsections 2.5.5 and 2.5.6, cf. Zeidler [678].

We recapitulate the notation for partials and vectors so, cf. (2.367), (2.389), (2.73), (2.79), (2.102), (2.336), (2.335):

$$\vec{u} \in W^{m,s}(\Omega, \mathbb{R}^q),\ \vec{u}(x), \qquad \partial^\beta \vec{u}(x), \vec{\vartheta}^\beta = \left(\vartheta_1^{\beta_1}, \cdots, \vartheta_q^{\beta_q}\right), \ \in \mathbb{R}^q \tag{2.423}$$
$$\partial^{\leq m}\vec{u}(x) = (\partial^\beta \vec{u}(x))_{0\leq|\beta|\leq m}, \qquad \vec{\vartheta}^{\leq m} = (\vec{\vartheta}^\beta)_{0\leq|\beta|\leq m}, \in \mathbb{R}^{N_m\times q}.$$

We discuss the general form of quasilinear elliptic equations on bounded domains Ω, compare Theorem 2.50, with Dirichlet boundary conditions, see Skrypnik [590,591] and Zeidler [678], and recall the standard notation in (2.165), (2.166), (2.269), (2.16), (2.361). The strong quasilinear system is

$$\begin{aligned} G\vec{u}^0 = G_s\vec{u}^0 &:= \sum_{|\alpha|\leq m} (-1)^{|\alpha|}\partial^\alpha \vec{A}_\alpha(\cdot, \partial^{\leq m}\vec{u}^0) = \left[\sum_{|\alpha|\leq m} (-1)^{|\alpha|}\partial^\alpha A_\alpha^i(\cdot, \partial^{\leq m}\vec{u}^0)\right]_{i=1}^q \\ &= 0,\ \vec{u}^0 \in W^{2m,p}(\Omega, \mathbb{R}^q) \cap W_0^{m,p}(\Omega, \mathbb{R}^q),\ \ \vec{A}_\alpha = \left[A_\alpha^i\right]_{i=1}^q : \Omega \times \mathbb{R}^{N_m\times q} \to \mathbb{R}^q. \end{aligned} \tag{2.424}$$

Here, we consider the weak solution $\vec{u}^0 \in W_0^{m,p}(\Omega, \mathbb{R}^q)$,

$$a(\vec{u^0}, \vec{v}) := \langle G\vec{u}^0, \vec{v}\rangle = \int_\Omega \sum_{|\alpha|\leq m} (\vec{A}_\alpha(\cdot, \partial^{\leq m}\vec{u}^0), \partial^\alpha \vec{v})_q dx = 0\ \forall \vec{v} \in W_0^{m,p}(\Omega, \mathbb{R}^q). \tag{2.425}$$

These integrals are well defined under the following sharper *Carathéodory and growth conditions*, compare (2.420), (2.384), (2.385).

The $\vec{A}_\alpha(x, \partial^{\leq m}\vec{u})$ are Nemyckii operators, generalizing Subsection 2.5.5. Now we impose conditions which guarantee that the $G(u)$ in (2.424) and its weak form (2.425) are well defined. The *Carathéodory* and *growth conditions* enforce measurability and appropriate growth of the composite function $\vec{A}_\alpha$, cf. Theorems 2.73, 2.113, and [678], pp. 571 ff. (2.426). The following nonlinear coercivity, monotonicity and growth conditions for these quasilinear operators generalize (2.230)–(2.235), (2.242)–(2.247). They are, with the equivalence of all norms, $|\vec{\vartheta}^\gamma|$ in $\mathbb{R}^q$, essential for the existence of solutions.

Carathéodory condition: Ω is bounded and measurable in $\mathbb{R}^n$,

$$\begin{aligned} &\forall\ |\alpha| \leq m: \quad \vec{A}_\alpha : (x, \vec{\vartheta}^{\leq m}) \in\ \Omega \times \mathbb{R}^{N_m\times q} \to \mathbb{R}^q, \\ &\vec{\vartheta}^{\leq m} \mapsto \vec{A}_\alpha(x, \vec{\vartheta}^{\leq m}) \text{ is continuous in } \mathbb{R}^{N_m\times q} \text{ a.e. for } x \in \Omega, \\ &x \mapsto \vec{A}_\alpha(x, \vec{\vartheta}^{\leq m}) \text{ is measurable in } \Omega\ \ \forall\ \vec{\vartheta}^{\leq m} \in \mathbb{R}^{N_m\times q}, \text{e.g. for continuous } \vec{A}_\alpha. \end{aligned} \tag{2.426}$$

For the following conditions, we assume $1 < p < \infty, 1/p + 1/p' = 1$, the special form $\vec{\vartheta}^{\leq m} = (\vec{\vartheta}^{\gamma})_{|\gamma|\leq m}, \vec{\vartheta}_1^{\leq m} = (\vec{\vartheta}_1^{\gamma})_{|\gamma|\leq m} \in \mathbb{R}^{N_m \times q}$ and $\exists\, c, d, C \in \mathbb{R}_+$, such that

$$\exists\, a \in L^{p'}(\Omega), h \in L^1(\Omega), \text{ s.t. a.e. } \forall\, x \in \Omega, \vec{\vartheta}^{\leq m}, \vec{\vartheta}_1^{\leq m} \in \mathbb{R}^{N_m\times q}, \vec{\vartheta}^0, \vec{\vartheta}^{\alpha}, \vec{\vartheta}_1^{\alpha} \in \mathbb{R}^q :$$

Growth condition: $$|\vec{A}_\alpha(x, \vec{\vartheta}^{\leq m})| \leq C(a(x) + \sum_{|\gamma|\leq m} |\vec{\vartheta}^{\gamma}|^{p-1}) \ \forall |\alpha| \leq m. \tag{2.427}$$

(Nonlinear) coercivity: $$\sum_{|\alpha|\leq m} (\vec{A}_\alpha(x, \vec{\vartheta}^{\leq m}), \vec{\vartheta}^{\alpha})_q \geq c \sum_{|\gamma|=m} |\vec{\vartheta}^{\gamma}|^p - h(x). \tag{2.428}$$

Monotonicity: $$\sum_{|\alpha|\leq m} \left(\vec{A}_\alpha(x, \vec{\vartheta}^{\leq m}) - \vec{A}_\alpha(x, \vec{\vartheta}_1^{\leq m}), \vec{\vartheta}^{\alpha} - \vec{\vartheta}_1^{\alpha}\right)_q \geq 0. \tag{2.429}$$

Strict monotonicity: $$\sum_{|\gamma|\leq m} \left|\vec{\vartheta}^{\gamma} - \vec{\vartheta}_1^{\gamma}\right| > 0 \text{ implies} \tag{2.430}$$

$$\sum_{|\alpha|\leq m} \left(\vec{A}_\alpha(x, \vec{\vartheta}^{\leq m}) - \vec{A}_\alpha(x, \vec{\vartheta}_1^{\leq m}), \vec{\vartheta}^{\alpha} - \vec{\vartheta}_1^{\alpha}\right)_q > 0.$$

Uniform monotonicity: (2.431)

$$\sum_{|\alpha|\leq m} \left(\vec{A}_\alpha(x, \vec{\vartheta}^{\leq m}) - \vec{A}_\alpha(x, \vec{\vartheta}_1^{\leq m}), \vec{\vartheta}^{\alpha} - \vec{\vartheta}_1^{\alpha}\right)_q \geq C \sum_{|\gamma|=m} \left|\vec{\vartheta}^{\gamma} - \vec{\vartheta}_1^{\gamma}\right|^p.$$

For regularity results, $p = 2$ and additional conditions have to be imposed, cf. Theorem 2.116.

By [678], Proposition 26.12, the conditions (2.426)–(2.429) imply the conditions of his Theorem 26A. So we refer in the next theorem to the corresponding results for equations in [678].

Theorem 2.114. Generalized solutions for (2.425): *Let (2.426)–(2.427) be satisfied and choose* $\mathcal{V} = W_0^{m,p}(\Omega, \mathbb{R}^q)$. *Let*

$$\vec{F}_\alpha : \mathcal{V} = W_0^{m,p}(\Omega, \mathbb{R}^q) \to L^{p'}(\Omega, \mathbb{R}^q), \ (\vec{F}_\alpha \vec{u})(x) := \vec{A}_\alpha(x, \partial^{\leq m}\vec{u}(x)).$$

1. *Then for all* $|\alpha| \leq m$, *a.e.* $\forall x \in \Omega$ *these Nemyckii operators* $\vec{F}_\alpha$ *are well defined, continuous and bounded, see Theorem 2.113 and [678], Proposition 26.12, proof.*
2. *With the above* $\vec{A}_\alpha(x, \partial^{\leq m}\vec{u}(x))$ *define the* $a(\vec{u}, \vec{v})$ *as in (2.425). Then there exists exactly one bounded operator* G, *cf. Proposition 2.67, [678], Proposition 26.12,*

$$G : \mathcal{V} \to \mathcal{V}' : a(\vec{u}, \vec{v}) = \langle G\vec{u}, \vec{v}\rangle_{\mathcal{V}'\times\mathcal{V}} \ \forall \vec{u}, \vec{v} \in \mathcal{V} = W_0^{m,p}(\Omega, \mathbb{R}^q) \text{ and} \tag{2.432}$$

$$\|F_\alpha \vec{u}\|_{L^{p'}(\Omega,\mathbb{R}^q)}, \|G\vec{u}\|_{W^{-m,p'}(\Omega,\mathbb{R}^q)} \leq C\left(\|a\|_{L^{p'}(\Omega,\mathbb{R}^q)} + \left(\|\vec{u}\|_{W_0^{m,p}(\Omega,\mathbb{R}^q)}\right)^{p/p'}\right).$$

Similarly to (2.425) and with $\langle\vec{f},\vec{v}\rangle_{\mathcal{V}'\times\mathcal{V}}$ *replacing* 0*, the original problem (2.424) corresponds to the generalized problem: Determine for* $a(\vec{u},\vec{v})$ *in (2.432)*

$$\vec{u}_0 \in \mathcal{V} = W_0^{m,p}(\Omega,\mathbb{R}^q) \text{ s.t. } a(\vec{u}_0,\vec{v}) = \langle\vec{f},\vec{v}\rangle_{\mathcal{V}'\times\mathcal{V}} \forall\vec{v}\in\mathcal{V} \iff G\vec{u}_0 = \vec{f}. \tag{2.433}$$

3. *Additionally let all the* $\vec{A}_\alpha$ *be coercive and monotone, see (2.428) and (2.429). Then* G *is monotone, coercive, continuous and bounded, see [678], Proposition 26.12. So by Remark 2.66, Theorem 2.68 2.–4. are valid for (2.425).*
4. *In particular, for any* $\vec{f}\in W^{-m,p'}(\Omega,\mathbb{R}^q) = \mathcal{V}'$*, (2.433) has a solution* $\vec{u}_0 \in W_0^{m,p}(\Omega,\mathbb{R}^q)$ *and the set of solutions is bounded, convex and closed. If all the* $\vec{A}_\alpha$ *are additionally strictly monotone, see (2.430), Theorem 2.68, then* G *is strictly monotone as well and the above solution is unique. So the inverse* $G^{-1}:\mathcal{V}'\to\mathcal{V}$ *exists and is strictly monotone and bounded, see [678], Theorem 26A. If all the* $\vec{A}_\alpha$ *are, and hence* G *is uniformly monotone, see (2.431), this* G^{-1} *is continuous.*
5. *If* G *is strongly monotone, cf (2.416), Theorem 2.111, then* G^{-1} *is Lipschitz–continuous.*

Koshelev–Chelkak existence and regularity results for quasilinear systems

As mentioned above, these problems are discussed by Koshelev and Chelkak [450]. Here the Sobolev spaces $W^{m,s}(\Omega,\mathbb{R}^q)\subset L^2(\Omega,\mathbb{R}^q)\subset W^{-m,s'}(\Omega,\mathbb{R}^q)$, $1/s+1/s'=1$, in (2.388) have to be combined with the *scalar products* and *norms* in (2.337) and $G_s\vec{u}$ in divergence form. We introduce, similarly to the partials and vectors $\vec{\vartheta}^\beta, \vec{u}(x), \partial^\beta\vec{u}(x)\in\mathbb{R}^q$ in (2.423), (2.367), (2.389), the notation for the components $\vec{p}^\beta$:

$$\vec{p} := (p_1,\cdots,p_q),\quad \vec{p}^\beta = \left(p_1^{\beta_1},\cdots,p_q^{\beta_q}\right)\in\mathbb{R}^q,\quad \vec{p}^{\,\le m} := (p^\beta)_{0\le|\beta|\le m}\in\mathbb{R}^{N_m\times q}. \tag{2.434}$$

We consider and determine the solution, $\vec{u}^0\in W^{2m,s}(\Omega,\mathbb{R}^q)\cap W_0^{m,s}(\Omega,\mathbb{R}^q)$, of the strong quasilinear system (2.424), see (2.16), (2.361). Koshelev and Chelkak [450] restrict (2.424) essentially to the case $1<s\le 2$. In their introduction they intensively discuss other possibilities and aspects of the problem (2.424) and give references.

The weak solution $\vec{u}^0\in W_0^{m,s}(\Omega,\mathbb{R}^q)$ is determined from (2.425). The integrals in (2.425) are well defined under the following modified *Carathéodory and growth conditions*, compare (2.426)–(2.431), and (2.284), (2.384), (2.385). We impose a *strong Legendre condition* for the linearized operator, compare (2.384). In fact, compared to (2.378), $\sum_{|\alpha|,|\beta|=m}$ is replaced by $\sum_{|\alpha|,|\beta|\le m}$ in (2.437). This appears again in the growth condition (2.437): For $w=x,\vec{p}^{\,\le m}$ and with $|\vec{p}^{\,\le m}|^2=\sum_{|\beta|\le m}\sum_{i=1}^q|p_i^{\beta_i}|^2$ and $\lambda,\Lambda=$ const. >0, let, for almost all $x\in\Omega$,

$$A_\alpha^i(x,\vec{p}^{\,\le m}) \text{ are measurable and continuouslydifferentiable with respect to } \vec{p}^{\,\le m}$$
$$\forall|\alpha|\le m \text{ in any domain of boundedvariation of } \vec{p}^{\,\le m}=\partial^{\le m}\vec{u} \text{ of } \vec{u}, \tag{2.435}$$

$$\lambda(1+|\vec{p}^{\,\le m}|^2)^{(s-2)/2}|\vec{\vartheta}^{\le m}|^2 \tag{2.436}$$
$$\le \sum_{i,j=1}^{q}\sum_{|\alpha|,|\beta|\le m}\left(\partial A^i_\alpha/\partial p^j_\beta\right)(w)\vartheta^\alpha_i\vartheta^\beta_j \le \Lambda(1+|\vec{p}^{\,\le m}|^2)^{(s-2)/2}|\vec{\vartheta}^{\le m}|^2,$$
$$\text{for } 1<s\le 2,$$

$$|\left(\partial A^i_\alpha/\partial p^{\beta_j}_j\right)(w)| \le \Lambda(1+|\vec{p}^{\,\le m}|^2)^{(s-1)/2}\ \forall\ i,j,\alpha,\beta, \text{ and for any } 1<q', \tag{2.437}$$

$$\text{domain } \Omega'\subset\Omega \text{ and } \vec{u}\in W^{1,q'}(\Omega',\mathbb{R}^q): \vec{A}_\alpha(\cdot,\partial^{\le m}\vec{u}(x))\in L^{q'/(s-1)}(\Omega'). \tag{2.438}$$

Koshelev and Chelkak call them *systems with bounded nonlinearities.* Similarly to our Lemma 2.26, see (2.82), generalizations of Dirichlet boundary conditions to *Dirichlet systems* are possible. To avoid the many technicalities we do not present the iteration process, which plays an important role in [450], for determining the solution. We restrict the discussion to *Dirichlet boundary conditions*

$$\frac{\partial^j\vec{u}(x)}{\partial\nu^j} = g_j\ \forall 0\le j\le m-1 \Leftrightarrow \partial^\alpha\vec{u}(x) = g^\alpha\ \forall|\alpha|\le m-1\ \forall x\in\partial\Omega. \tag{2.439}$$

Theorem 2.115. Existence – Theorems 2.1.1 and 2.3.1 in [450]:

1. *Assume $s=2$, and the above modified* Carathéodory and growth conditions *(2.435)–(2.438), implying $|(\partial A^i_\alpha/\partial p^{\beta_j}_j)(x,\vec{p}^{\,\le m})|\le C$. Then the quasilinear system (2.425) with the Dirichlet boundary conditions (2.439) admits a solution $\vec{u}^0\in H^m(\Omega,\mathbb{R}^q)$.*
2. *If, more generally, for $1<s\le 2$, a weak solution $\vec{u}^0$ for (2.425), (2.439) exists and satisfies the estimate*

$$\int_\Omega\sum_{|\alpha|\le m}|A_\alpha(x,\partial^{\le m}\vec{u}^0(x))|^{s/(s-1)}dx \le C_1\left\|\vec{u}^0\right\|^s_{W^{m,s}(\Omega,\mathbb{R}^q)} + C_2, \tag{2.440}$$

then this $\vec{u}^0$ is the unique solution of the problem.

To get regularity results, again $s=2$ and additional conditions have to be imposed. Whenever conditions for partials, e.g. $\partial\vec{A}_\alpha/\partial x_k$ are required, the existence of these partials is silently assumed. For $\Omega'\subset\subset\Omega$ we introduce

$$\|\vec{u}\|^2_{H^m_\delta(\Omega',\mathbb{R}^q)} := \sup_{x_0\in\Omega'}\int_{\Omega'}\left(\sum_{|\alpha|<m}|\partial^\alpha\vec{u}(x)|^2+\sum_{|\alpha|=m}|\partial^\alpha\vec{u}(x)|^2|x-x_0|^\delta\right)dx \tag{2.441}$$

$$H^m_\delta(\Omega',\mathbb{R}^q) := \{\vec{u}\in H^m(\Omega',\mathbb{R}^q): \|\vec{u}\|_{H^m_\delta(\Omega',\mathbb{R}^q)}<\infty\}, L^2_\delta(\Omega',\mathbb{R}^q) := H^0_\delta(\Omega',\mathbb{R}^q).$$

We impose three sets of conditions for the three cases: (2.443) for $\vec{u}\in H^m(\Omega',\mathbb{R}^q)$, (2.444) for $\vec{u}\in H^m_\delta(\Omega',\mathbb{R}^q)$, and (2.445) for $\vec{u}\in W^{m+1,q'}(\Omega',\mathbb{R}^q)$:

$$\text{for all three cases let } \Omega'\subset\subset\Omega: \exists\ C(R) \text{ independent of } \vec{u}, \text{ such that} \tag{2.442}$$

$$\forall \vec{u} \in H^m(\Omega', \mathbb{R}^q) \text{ with } \|\vec{u}\|_{H^m(\Omega',\mathbb{R}^q)} \le R, \tag{2.443}$$
$$\forall\, 0 \le |\alpha| \le m: \ \|\vec{A}_\alpha(\cdot, \partial^{\le m}\vec{u}(\cdot))\|_{L^2(\Omega',\mathbb{R}^q)} \le C(R), \text{ and}$$
$$\forall |\alpha| = m, k = 1, \ldots, n, \text{let } \left\| \frac{\partial \vec{A}_\alpha}{\partial x_k}(\cdot, \partial^{\le m}\vec{u}(\cdot)) \right\|_{L^2(\Omega',\mathbb{R}^q)} \le C(R) \text{ or}$$

$$\forall\, \vec{u} \in H^m_\delta(\Omega', \mathbb{R}^q) \text{ with } \|\vec{u}\|_{H^m_\delta(\Omega',\mathbb{R}^q)} \le R, \ \delta \in [2 - m - 2\gamma, 0], \tag{2.444}$$
$$0 < \gamma < 1, \ \forall\, 0 \le |\alpha| \le m \text{ let } \|\vec{A}_\alpha(\cdot, \partial^{\le m}\vec{u}(\cdot))\|_{L^2_\delta(\Omega',\mathbb{R}^q)} \le C(R) \text{ and}$$
$$\forall |\alpha| = m, k = 1, \ldots, n, \text{ let } \left\| \frac{\partial \vec{A}_\alpha}{\partial x_k}(\cdot, \partial^{\le m}\vec{u}(\cdot)) \right\|_{L^2_\delta(\Omega',\mathbb{R}^q)} \le C(R), \text{ finally}$$

$$\forall \vec{u} \in W^{m+1,q'}(\Omega', \mathbb{R}^q), m < q', \frac{\partial \vec{A}_\alpha}{\partial x_k}(\cdot, \nabla^{\le m}\vec{u}(\cdot)) \in L^{q'}(\Omega', \mathbb{R}^q). \tag{2.445}$$

Theorem 2.116. Regularity, Theorem 4.3.1 in [450]: *Assume the conditions (2.442)–(2.445), those in Theorem 2.115 (note $s = 2$), except (2.440), and some highly technical conditions, see (1.3.6), (1.3.7), (1.3.18), (1.3.20), (1.3.46), (4.1.11), (4.3.3) in [450]. Let the boundary values in (2.439) be defined by one function g as*

$$g \in H^m(\Omega, \mathbb{R}^q): \quad g_j = \frac{\partial^j g(x)}{\partial \nu^j} \quad \forall 0 \le j \le m - 1.$$

Then the weak solution $\vec{u}^0$, for (2.425), (2.439) belongs to $\vec{u}^0 \in C^{m,\gamma}(\Omega', \mathbb{R}^q)$.

The techniques for proving higher regularity as in Theorem 2.91 3. do not apply.

2.6.8 Fully nonlinear elliptic systems of orders 2 and $2m$

In contrast to the earlier subsections, only few results for fully nonlinear systems have found their way into the monographs consulted here. Formally, they are obtained by the obvious generalizations of (2.174), (2.306), (2.319) as c.f. Remark around (2.306),

$$G(\cdot) : \mathcal{D}(G) \subset \mathcal{U} := H^1_0(\Omega, \mathbb{R}^q) \cap H^2(\Omega, \mathbb{R}^q) \to \mathcal{V} := L^2(\Omega, \mathbb{R}^q), \text{ with}$$
$$G(\vec{u}(\cdot)) := G^w(\cdot, \vec{u}(\cdot), \partial\vec{u}(\cdot), \nabla^2\vec{u}(\cdot)), \ G(\vec{u}^0) = 0 \tag{2.446}$$

for systems of order 2. The corresponding equations are updated by replacing the original $\mathcal{U}, \mathcal{V}, \mathcal{W}$. For systems of order $2m$ we obtain

$$G(\cdot) : \mathcal{D}(G) \subset \mathcal{U} := H^m_0(\Omega, \mathbb{R}^q) \cap H^{2m}(\Omega, \mathbb{R}^q) \to \mathcal{V} := L^2(\Omega, \mathbb{R}^q), \text{ with}$$
$$G(\vec{u}) := G^w(\cdot, \vec{u}(\cdot), \ldots, \nabla^{2m}\vec{u}(\cdot)), \ G(\vec{u}^0) = G^w(\cdot, \partial^\gamma\vec{u}^0(\cdot), |\gamma| \le 2m) = 0. \tag{2.447}$$

Often $H^{2m}(\Omega, \mathbb{R}^q)$ is replaced by $C^{2m,\gamma}(\Omega, \mathbb{R}^q)$ or $C^\infty(\Omega, \mathbb{R}^q)$, disregarding boundary conditions. In Taylor [620], pp. 108 ff., some "local existence results" are proved: if at a point $x_0 \in \Omega$ the equation $G^w(\cdot, \partial^\gamma\vec{u}^1(\cdot), |\gamma| \le 2m)(x_0) = 0$ is satisfied for a $\vec{u}_1$, then there exists a solution $\vec{u}$ in a neighborhood of x_0. For second order some results can be found, e.g. in [79, 80, 311, 475, 671].

One of the few exceptional results is proved by Taylor [620], Chapter 14, Theorem 3.1, for *nonlinear systems of PDEs of order m*, see Definitions 2.51 and 2.85. It extends verbatim (2.319) and Theorem 2.84 to systems, by simply replacing u by $\vec{u}$. *Nonlinear systems of PDEs of order m* have the form

$$G(\vec{u}) := G(x, \vec{u}, \nabla\vec{u}, \ldots, \nabla^m\vec{u}) \text{ and } G(\vec{u})(x) = g(x)\ \forall\ x \in \Omega. \tag{2.448}$$

Theorem 2.117. Local solvability result for nonlinear elliptic systems: *Let $g \in C^\infty(\Omega)$, and let $\vec{u}^1 \in C^\infty(\Omega)$ satisfy (2.448) at $x = x_0$, hence*

$$G(\vec{u}^1)(x) = g(x) \text{ for } x = x_0$$

and let G be m-elliptic at $\vec{u}^1$. Then for any ℓ there exists a $\vec{u} \in C^\ell(\Omega)$ such that

$$G(\vec{u})(x) = g(x) \text{ in a neighborhood of } x_0.$$

In this situation with only few known results, a combination of the linearized operators in Subsection 2.7.3, our FE or other discretization methods below and continuation techniques, allows us to study existence of solutions of these problems. So we give some details in Subsections 2.7.3 and 2.7.4 below.

2.7 Linearization of nonlinear operators

2.7.1 Introduction

Linearization of operators including systems plays an essential role in our approach. We are mainly interested in numerical methods for these PDEs and their bifurcation scenarios. The stability of numerical methods can be proved via the stability of the method for the linearized operator. For the numerical approximation of their bifurcation scenarios we do need the Fredholm alternative, again formulated for the linearized operator. For this reason we have not included the many Fredholm type results, directly available for nonlinear differential equations, see, e.g. Zeidler [676–678].

In many contexts, nonlinear parameter-dependent operators $G(u,\lambda)$ are studied. It is important to start with a parameter value λ_0 such that the existence and (local) uniqueness of a solution u_0 of $G(u_0,\lambda_0) = 0$ is guaranteed. The results in Sections 2.4–2.6 yield this information for solutions u_0 for a fixed parameter λ_0. Based upon this knowledge, the Implicit Function Theorem 1.46 defines a uniquely determined solution curve of the nonlinear problem $G(u,\lambda) = 0$ through (u_0,λ_0). Note that, e.g. for an operator $G : \mathcal{V} \times \mathbb{R}^p \to \mathcal{V}'$ the derivative $G'(u,\lambda) = (G_u(u,\lambda), G_\lambda(u,\lambda))$ always maps $\mathcal{V} \times \mathbb{R}^p \to \mathcal{V}'$, whenever it exists as a bounded operator. The corresponding continuation process again is based upon the Fréchet derivative $G'(u,\lambda) = (G_u(u,\lambda), G_\lambda(u,\lambda))$ at a solution point (u,λ). The continuation breaks down and the solution curve usually bifurcates, whenever $G'(u,\lambda)$ is not surjective in a solution point (u,λ), cf. Böhmer [120]. Then usually $G_u(u,\lambda)$ has a nontrivial kernel. So results from the Fredholm alternative for the linearized operator $G_u(u,\lambda)$ apply in combination with the Liapunov–Schmidt methods studied in [120]. For both aspects we have to verify that for the different types of nonlinear operators $G(u)$, studied in the last Sections 2.5, 2.6, the linearized operators $G'(u_0)$ or the $G_u(u_0, \lambda_1)$ are well defined and that Fredholm alternatives are available for them. By appropriate combinations of the corresponding Banach spaces, linearizations for a wide class of nonlinear elliptic equations and systems make sense. In this monograph we use discretization concepts different from the Brezzi–Rappaz–Raviart [147–149] ideas. So, most of Santos's [566] negative arguments concerning numerical methods as in [147–149], related to linearization, do not apply.

Prima vista, linearization does not seem to be appropriate for some of the quasilinear elliptic equations and systems. In fact the results for quasilinear equations from Subsection 2.5.6 strongly depend upon monotony and nonlinear coercivity arguments. But we show in Subsection 2.7.3 that the linearization approach works for most quasilinear equations. *Noncoercive equations,* e.g. of the form

$$Au_0 + Nu_0 = b, \text{ with } A \in \mathcal{L}(\mathcal{U},\mathcal{U}') \text{ linear, } N \text{ nonlinear } \lim_{\|u\|\to\infty} \|Nu\|/\|u\| = 0,$$

require specific handling. However, some type of Fredholm alternative still holds. This equation has a solution $u_0 \forall b \in \mathcal{U}'$ if $Au = 0$ has the unique solution $u = 0$, otherwise solutions u_0 exist only for special $b \in \mathcal{U}'$, see Zeidler [678], Chapter 29.

The situation for systems is different, since for orders 2 and $2m$ the ellipticity of (the linearized) operator plays a dominant role compared to analogous monotony arguments. The situation changes if viscosity solutions are admitted instead, as studied in an impressive series of papers by Crandall *et al.* [215, 217–228], which we did not consider here.

Above we have documented mainly two sets of results for nonlinear equations, either based upon (linear) coercivity arguments and the corresponding $H^m(\Omega)$ or $W^{m,p}(\Omega)$ setting, combined with Schauder results for second order equations. Or they are based upon monotony arguments in the $W^{m,p}(\Omega)$ setting. In both cases the linearization usually yields operators as discussed in Section 2.4 and Subsections 2.6.3, 2.6.5, and satisfies the Fredholm alternative. An essential fact here is the strong ellipticity or the Legendre condition of the linearizations and, for systems of order $2m, m > 1$, the strong Legendre–Hadamard condition.

In this section we will show that the linearization of all nonlinear problems studied before in the $H^m(\Omega)$ or $W^{m,p}(\Omega)$ setting, $2 < p \leq \infty$, yields bounded linear operators, with the corresponding bounded bilinear forms $a(u, v)$. They are compact perturbations of their principal parts. For $2 < p \leq \infty$, we use the compact embedding of the $W^{m,p}(\Omega)$ into $H^m(\Omega)$. Then the principal parts are $H^m(\Omega)$-coercive, and the convergence for numerical methods is measured with respect to a discrete $H^m(\Omega)$ norm, cf. Corollary 2.44. For $1 \leq p < 2$ corresponding bilinear forms $a(u, v)$ are not even defined. So we have to exclude this case. Finally, for the fully nonlinear systems in Subsection 2.6.8 no existence results are available. However, linearizations may be combined with the Fredholm alternative and discretization and continuation techniques to obtain existence results.

In Section 4.5 we will apply the general discretization theory from Chapter 3 to all our general quasilinear problems. This is formulated for FEMs, but is valid for all other methods as well. It is based upon monotony and related properties and allows the previously excluded cases $1 < p < \infty$.

For all the following cases we *do not repeat,* but only refer to, the conditions for the defining coefficients and the domain imposed for the original nonlinear problem. We formulate the stronger conditions where they are necessary for the linearization. To simplify the notation we always omit parameters λ and use the abbreviations

$$\frac{\partial G}{\partial u}(u) := \frac{\partial G}{\partial u}(u,\lambda),\ \frac{\partial f}{\partial u}(u) := \frac{\partial f}{\partial u}(u,\lambda),\ \text{etc., often for } (u,\lambda) = (u_0,\lambda_0). \tag{2.449}$$

We summarize the main result of this section as

Theorem 2.118. *For the different types of nonlinear equations and systems $G(u)$, studied in Sections 2.5, 2.6, the corresponding linearized operators are well defined and bounded under the conditions in the following theorems, throughout assuming existing derivatives. Then Corollary 2.44 implies for $2 \le p \le \infty$ and the H^m-coercive essentially principal parts, the existence and uniqueness of solutions and a boundedly invertible A, for a H_0^m-elliptic $a(\cdot,\cdot)$, the Fredholm alternative and, in Chapter 3, the numerical stability, for the discertization methods studied in our books, including bifurcation and the related dynamics with respect to discrete H^m norms. For natural limitations see the Introduction in Subsection 2.7.3.*

2.7.2 Special semilinear and quasilinear equations

In contrast to the general quasilinear differential equations in Subsection 2.5.6 the linearization of the equations of the special cases of semilinear and quasilinear differential equations in Subsection 2.5.3 is straightforward.

A: Special semilinear operators of order 2m

For the equations of order $2m$ in (2.177), (2.191) with boundary conditions (2.178) and the conditions in (2.179) and (2.190) for f we get the strong form (the linear boundary operators are unchanged and satisfy the same complementary condition),

$$\begin{aligned} &Gu = A - f(u),\ \frac{\partial G}{\partial u}(u_0)v := Av - \frac{\partial f}{\partial u}(u_0)v : C^{2m,\gamma}(\Omega) \to C^\gamma(\Omega),\\ &Av = \sum_{|\alpha|\le 2m} a_\alpha \partial^\alpha u,\ \text{unif. elliptic of index } 0,\ f_u = \frac{\partial f}{\partial u} \in C^\infty(\mathbb{R}^2). \end{aligned} \tag{2.450}$$

For[23] order 2 we only have to replace $2m$ by 2 in (2.450), reduce the smoothness of f_u, and choose Au, Bu and the conditions by the special form with

$$Av := \sum_{i,j=0}^{n} a_{ij}(x)\partial^i\, \partial^j\, u,\ \text{with } Bu \text{ as in (2.185), and } f_u \in C^\gamma(\mathbb{R}^2). \tag{2.451}$$

[23] According to (2.449), $f_u \in C^\infty(\mathbb{R}^2)$ often means, here and below, $f_u(u,\lambda) \in C^\infty(\mathbb{R}^3)$.

These are exactly the linear operators in Theorems 2.37 ff. with (2.83), (2.88), (2.93), (2.117), (2.181). They are studied in Subsections 2.4.2 and 2.4.3. In Theorems 2.32, 2.41, 2.52, and Proposition 2.54, the corresponding results and Fredholm alternatives are formulated.

B: Special quasilinear operators of order 2

For the linearization of the *special quasilinear operators in (2.193) and (2.452)* we combine (2.194) and (2.452a), (2.453) of order $2m = 2$ and $n = 2$. We recall the notation in (2.269), $\partial v/\partial\vartheta^k, k \geq 0$ with $= v$ for $k = 0$, and $\partial a_{ij}/\partial\vartheta^0 = \partial a_{ij}/\partial u$. The original form and its linearization are

$$n = 2: \; G(u) = \sum_{i,j=1}^{2} a_{ij}(x, u, \nabla u)\partial^i\partial^j \; u, \frac{\partial G}{\partial u}(u_0) : C^{2,\gamma}(\bar{\Omega}) \to C^{\gamma}(\bar{\Omega}), \tag{2.452}$$

$$\frac{\partial G}{\partial u}(u_0)v = \sum_{i,j=1}^{2} \Big(a_{ij}(x, u_0, \nabla u_0)\partial^i\partial^j \; v + \sum_{k=0}^{2} \frac{\partial a_{ij}}{\partial\vartheta^k}(x, u_0, \nabla u_0)\partial^i\partial^j \; u_0\partial^k v\Big), \tag{2.452a}$$

see Theorem 2.57, but with stronger condition for coefficients, than (2.194),

$$a_{ij}, \quad \frac{\partial a_{ij}}{\partial\vartheta^k} \quad \in C^{\gamma}(\bar{\Omega} \times \mathbb{R}^3), \forall i, j = 1, 2, k = 0, 1, 2. \tag{2.453}$$

C: Special semilinear operators in nonlinear spaces

For the last case in Subsection 2.5.3, (2.197), (2.198), we require (2.196), (2.199)–(2.202) and (2.455). Then we consider

$$G(u) = Au - f(u) - g = \sum_{|\alpha|,|\beta| \leq m} (-1)^{|\beta|}\partial^{\beta}(a_{\alpha\beta}\partial^{\alpha}u) - f(u) - g \tag{2.454}$$

in (2.196) and (2.199) in nonlinear spaces. The linearized strong form is now defined, with appropriate $f_u(u_0)$, as $(\partial G/\partial u)(u_0) : H^{2m}(\Omega) \to L^2(\Omega)$. Applied to v we get

$$\frac{\partial G}{\partial u}(u_0)v = \left(A - \frac{\partial f}{\partial u}(u_0)\right) v, \;\text{ with } f_u(u_0) \in L^{\infty}(\Omega). \tag{2.455}$$

The conditions for the weak form are listed in (2.201), (2.202) and in Theorem 2.58.

D: Semilinear nonautonomous operators

For this (2.296), (2.297) we replace the previous $f(u) + g(x)$ by $f(x, u)$ and obtain, with A as in (2.454), the linearized strong form

$$G, A, \frac{\partial G}{\partial u}(u_0) : H^{2m}(\Omega) \to L^2(\Omega), \text{ with } f_u(\cdot, u_0(\cdot)) \in L^{\infty}(\Omega) \tag{2.456}$$

$$G(u) = Au - f(x, u), \quad \frac{\partial G}{\partial u}(u_0)v = \left(A - \frac{\partial f}{\partial u}(\cdot, u_0(\cdot))\right) v, f_u \in L^{\infty}(\mathbb{R}^2),$$

with *Dirichlet boundary conditions.* The conditions for the weak form are listed in (2.201), (2.202), (2.298)–(2.300) and Theorem 2.76. $f(x, u)$ has to be monotone in

$u\ \forall x \in \Omega$ for existence and uniqueness (or, only for existence, $f(x,u)u \geq 0\ \forall x \in \Omega, u \in \mathbb{R}$ for large enough $|u|$) to get the linearization in (2.456).

We *summarize the conditions for linearization* of the *special cases of semilinear and quasilinear differential equations* in Subsection 2.5.3 and the last example (2.297) in Subsection 2.5.6.

Theorem 2.119. Linearization for different problems: *We consider the special semi- and quasilinear equations in the above cases* $\boldsymbol{A}, \boldsymbol{B}, \boldsymbol{C}, \boldsymbol{D}$ *and the necessary conditions for their elliptic and bounded linearizations listed above. Then the original problems and their linearizations have the form*

$$G(u) = Au - f(u) = g, \ \frac{\partial G}{\partial u}(u_0)v = Av - \frac{\partial f}{\partial u}(u_0)v, \ \text{for } \boldsymbol{A}, \boldsymbol{C}, \text{ and} \tag{2.457}$$

$$G(u) = Au - f(\cdot, u) = g, \ \frac{\partial G}{\partial u}(u_0)v = Av - \frac{\partial f}{\partial u}(\cdot, u_0)v \ \text{for } \boldsymbol{D}, \ \text{and (2.452) for } \boldsymbol{B}.$$

Under the listed conditions all the linearizations in (2.457) are bounded linear operators, induced by compact perturbations of H_0^m*-coercive bilinear forms, hence Corollary 2.44 holds, e.g. implying the* Fredholm alternative.

2.7.3 Divergent quasilinear and fully nonlinear equations

We begin with the general divergent quasilinear equations studied in Subsections 2.5.4 and 2.5.6, see, e.g. (2.206)–(2.208), (2.228)–(2.229), (2.270)–(2.275). Here the linearization for these equations makes sense if we exclude some cases for the following reasons: Since many of the existence, uniqueness and regularity results are only known for the case $H^m(\Omega)$, generalizations to the $W^{m,p}(\Omega, \mathbb{R}^q)$ situation would be worthwhile. But for the appropriate $W^{m,p}(\Omega)$ setting and for $1 \leq p < 2$, the bilinear forms, $a(u,v)$, for the linearized operator, $G'(u_0)$, usually are not even defined, cf. (2.459). So we exclude $1 \leq p < 2$. For $2 \leq p \leq \infty$, and a satisfied ellipticity condition (2.461), the bilinear form, $a_p(u,v)$, for the *principal part* of the linearized operators turns out to be only H_0^m-coercive. The compact embedding of $W^{m,p}(\Omega)$ into $H^m(\Omega)$ allows us to formulate numerical results for a discrete $H^m(\Omega)$ norm, cf. Corollary 2.44. But, the above (linear) H_0^m-coercivity arguments and the corresponding $C^{2,\gamma}(\Omega)$ for a Schauder type approach do not always apply to $m > 1$. The original conditions for these quasilinear elliptic equations are related, but not identical with those garanteeing the H_0^m-coercivity of their linearized principal part, see Lemma 2.77 and the following discussion. So, for some cases, based upon monotony arguments, we impose modified conditions, cf. Proposition 2.121 and Theorem 2.122. We discuss the relations between these conditions.

We had assumed Ω as a bounded nonempty measurable domain in $\mathbb{R}^n$. The weak form, see (2.274), for general m is defined for $u, v \in \mathcal{V} = W_0^{m,p}(\Omega)$, as

$$a(u,v) = \langle Gu, v \rangle_{\mathcal{V}' \times \mathcal{V}} = \int_\Omega \sum_{|\alpha| \leq m} A_\alpha(x, u, \ldots, \nabla^m u) \partial^\alpha v dx = \langle f, v \rangle_{\mathcal{V}' \times \mathcal{V}} \tag{2.458}$$

with the exact solution u_0. We linearize analogously to (2.455) and get

$$\frac{\partial G}{\partial u}(u_0) : \mathcal{V} := W_0^{m,p}(\Omega) \to \mathcal{V}' = W_0^{-m,q}(\Omega), 1 < p < \infty, 1/p + 1/q = 1, \text{ with } \quad (2.459)$$

$$\left\langle \frac{\partial G(u_0)}{\partial u} w, v \right\rangle_{\mathcal{V}' \times \mathcal{V}} = \int_\Omega \sum_{|\alpha|\le m} \left(\sum_{|\beta|\le m} \frac{\partial A_\alpha}{\partial \vartheta^\beta}(x, u_0, \ldots, \nabla^m u_0) \partial^\beta w \right) \partial^\alpha v dx, \ u_0, v,$$

$$w \in \mathcal{V}.$$

Although the condition $G, \partial G/\partial u(u_0) : \ \mathcal{V} := W_0^{m,p}(\Omega) \to \mathcal{V}' = W_0^{-m,q}(\Omega)$, makes sense, the previous $\langle (\partial G(u_0)/\partial u)w, v\rangle_{\mathcal{V}'\times\mathcal{V}}$ needs $\partial A_\alpha / \partial \vartheta^\beta (x, u_0, \ldots, \nabla^m u_0) \, \partial^\beta w \partial^\alpha v$ $\in L^1(\Omega) \ \forall u_0, v, w \in W_0^{m,p}(\Omega)$, requiring $p \ge 2$.

The usual ellipticity condition, and sometimes the not quite usual ellipticity condition (2.301) is imposed, with $d \in \mathbb{R}^{N_{m-1}}$, $\Theta^m \in \mathbb{R}^{n_m} \lambda(d), \Lambda(d), \mu(\Theta^m)$, defined there. This implies the usual ellipticity. Recalling (2.75), (2.79) with

$$|\vartheta|_{n,\alpha}^{2m} := \sum_{|\alpha|=m} \vartheta^{2\alpha} \text{ and } |\vartheta|_n^{2m}/C \le |\vartheta|_{n,\alpha}^{2m} \le |\vartheta|_n^{2m} \le C\,|\vartheta|_{n,\alpha}^{2m} \text{ we have } \quad (2.460)$$

$$\lambda(d)\mu(\Theta^m)\,|\vartheta|_n^{2m}/C \le \lambda(d)\mu(\Theta^m)\,|\vartheta|_{n,\alpha}^{2m}$$

$$\le \sum_{|\alpha|=|\beta|=m} \frac{\partial A_\alpha}{\partial \vartheta^\beta}(x, d, \Theta^m)\vartheta^\beta \vartheta^\alpha \le \Lambda(d)\mu(\Theta^m)\,|\vartheta|_{n,\alpha}^{2m}.$$

We have to linearize in the exact solution u_0. The preceding regularity and boundedness results show that the special values for the $d = (\vartheta^\beta = \partial^\beta u_0, |\beta| < m)$, $\Theta^m = (\vartheta^\beta = \partial^\beta u_0, |\beta| = m)$, and hence the $\lambda(d), \mu(\Theta^m), \Lambda(d)$ are bounded. Then (2.460) has the form of a standard ellipticity condition for the linearized operator, evaluated in u_0, formulated either with $|\vartheta|_n$ or $|\vartheta|_{n,\alpha}$, cf. (2.75) and the following Definition 2.25:

$$\lambda\,|\vartheta|_n^{2m}/C \le \lambda\,|\vartheta|_{n,\alpha}^{2m} \le \sum_{|\alpha|=|\beta|=m} \frac{\partial A_\alpha}{\partial \vartheta^\beta}(x, u_0(x), \ldots, \nabla^m u_0(x))\vartheta^\beta \vartheta^\alpha \le \Lambda|\vartheta|_{n,\alpha}^{2m}. \quad (2.461)$$

For the above Example 2.72 the derivatives do exist. Similarly to the Monge–Ampère equation, the ellipticity of $G'(u_0)$ depends upon u_0.

Example 2.120.

1. For $n = 1, 2, 3$, and $u \in H^2(\Omega) \cap H_0^1(\Omega)$ the linearization of

$$Gu = -\Delta u + \alpha \sum_{i=1}^{n} \sin(u)\partial^i u = f \in L^2(\Omega) \text{ on } \Omega,$$

yields

$$G'(u_0)v = -\Delta v + \alpha \sum_{i=1}^{n} \left(\sum_{j=1}^{n} cos(u_0)\partial^i u_0 \partial^j v \; + \; sin(u_0)\partial^i v \right).$$

2. For $2 \leq p < \infty, 1/p + 1/q = 1, \; s \in \mathbb{R}, s \geq 0$, we obtain

$$Gu = -\sum_{i=1}^{n} \partial^i (|\partial^i u|^{p-2} \partial^i u) + su = f \in L^q(\Omega) \text{ on } \Omega,$$

$$G'(u_0)v = -\sum_{i=1}^{n} \partial^i \Big(|\partial^i u_0|^{p-2} (1 + (p-2)(\partial^i u_0)^{-1}) \partial^i v \Big)$$

$$+\text{lower order derivatives for } v, \text{ with a singularity in } \partial^i u_0(x) = 0.$$

3. The linearization of the *minimal surface equation,*

$$Gu = (1 + |\nabla u|^2)\Delta u + \sum_{i,j=1}^{n} \partial^i u \partial^j u \partial^i \partial^j u = 0 \text{ on } \Omega$$

yields

$$G'(u_0)v = (1 + |\nabla u_0|^2)\Delta v + \sum_{i,j=1}^{n} \partial^i u_0 \partial^j u_0 \partial^i \partial^j v$$

$$+\text{lower order derivatives for } v.$$

■

Some of the results for the general divergent quasilinear equations of order 2 in Subsection 2.5.4 are based upon the ellipticity conditions in the form of (2.460) with $m = 1$, see (2.209), (2.239), for the linearized operators. In Theorems 2.61 and 2.62 existence and uniqueness results are summarized under appropriate growth conditions listed there. Theorems 2.63 and 2.64 yield existence and regularity results in the $C^{m,\gamma}(\Omega)$ setting, combining appropriate growth and H_0^m-coercivity conditions.

General divergent quasilinear equations of order $2m$ are studied in Subsection 2.5.6, see, e.g. (2.270)–(2.275). Conditions (2.284)–(2.289) are chosen in different combinations. Comparing condition (2.461), see Lemma 2.77, its evaluation in (2.461) is related to a modified H_0^m monotony (2.294) in Theorem 2.75. This is relevant for the existence of a solution u_0 for (2.295), but not for the finer results of Theorem 2.73 based upon the nonlinear H_0^m-coercivity in (2.286). Hence, if we impose the conditions of Theorem 2.75 and Lemma 2.77, in particular (2.301), we obtain an $H_0^m(\Omega)$-coercive linearization of the principal part. In fact we find

Proposition 2.121. *The monotonicity conditions (2.287), (2.288) and (2.289) for quasilinear equations imply, for* $p = 2$, $\Theta^{\leq m} = (d, \Theta^{\leq m})$, *and* $A_\alpha(x, d, \Theta^m) \in C^1(\mathbb{R}^{n_m})$

with respect to Θ^m, *in a neighborhood of* Θ^m, *classical and strong and uniform* $H_0^m(\Omega)$-*ellipticity. This implies, for uniform ellipticity, an* $H_0^m(\Omega)$-*coercive linearization of the principal part.*

Proof. We only consider (2.289); the other cases are similar. In (2.289) we choose the above $\Theta^{\leq m}$ and $\Theta_1^{\leq m} = \left(d, \Theta_1^{\leq m}\right)$, and obtain

$$\sum_{|\alpha|\leq m} = \sum_{|\alpha|=m} \left(A_\alpha\left(x, d, \Theta_1^m\right) - A_\alpha(x, d, \Theta^m)\right)\left(\vartheta_1^\alpha - \vartheta^\alpha\right) \geq C \sum_{|\gamma|=m} |\vartheta_1^\gamma - \vartheta^\gamma|^p .$$

This implies, by Theorem 1.43 and for sufficiently small $(\vartheta^\gamma)_{|\gamma|=m} := \Theta_1^m - \Theta^m$,

$$\sum_{|\alpha|=|\beta|=m} (\partial A_\alpha(x, d, \Theta^m)/\partial\beta)\vartheta^\alpha\vartheta^\beta \geq C/2 \sum_{|\gamma|=m} |\vartheta^\gamma|^p \geq C_{2,p}C/2 \left(\sum_{|\gamma|=m} |\vartheta^\gamma|^2\right)^{p/2},$$

since $\sum_{|\gamma|=m} |\vartheta^\gamma|^p \geq C_{2,p}\big(\sum_{|\gamma|=m} |\vartheta^\gamma|^2\big)^{p/2}$ for the equivalent norms in $\mathbb{R}^{n_m}$. Skipping the middle term in this equation and dividing it by $\big(\sum_{|\gamma|=m} |\vartheta^\gamma|^2\big)$ yields

$$\sum_{|\alpha|=|\beta|=m} (\partial A_\alpha(x, d, \Theta^m)/\partial\beta)\vartheta^\alpha\vartheta^\beta / \left(\sum_{|\gamma|=m} |\vartheta^\gamma|^2\right) \geq C_{2,p}C/2 \left(\sum_{|\gamma|=m} |\vartheta^\gamma|^2\right)^{p/2-1}.$$

For $p \neq 2$ the right-hand side in this inequality can become arbitrarily small or large. So (2.289) implies uniform ellipticity only for $p = 2$. □

We have excluded the case $W_0^{m,p}(\Omega), 1 \leq p < 2$. The continuous embedding of $W_0^{m,p}(\Omega), 2 \leq p \leq \infty$, into $H_0^m(\Omega)$, allows, under slightly enforced conditions, a $H_0^m(\Omega)$-coercive linearization and full convergence results with respect to discrete $H^m(\Omega)$ norms later on for the different discretization methods.

For quasilinear equations of order $2m, m \geq 1$, in Subsection 2.5.6 the results refer to the nonlinear H_0^m-coercivity in (2.286) and the monotonicity conditions (2.287)–(2.289), but not to ellipticity. In particular, linearization via the standard (2.461) is certainly not equivalent to (2.289). If the derivatives of the A_α are bounded and $\Theta_1^{N_m} - \Theta^{N_m}$ is sufficiently small, then (2.461) implies with the Taylor formula, a modified (2.289) with $\sum_{|\gamma|\leq m} |\vartheta^\gamma - \vartheta_1^\gamma|^p$ on the right-hand side, instead of the usual $\sum_{|\gamma|=m} |\vartheta^\gamma - \vartheta_1^\gamma|^2$, cf. the proof of Proposition 2.121. In this sense, (2.461) implies uniform monotonicity (2.294), and H_0^m-coercivity for the principal part of $G'(u_0)$. With the standard techniques and Corollary 2.14 this shows that (2.59) is satisfied, hence G' is boundedly invertible.

We have seen that for quasilinear problems of order 2, ellipticity conditions as in (2.461) are required anyway. For orders $2m, m \geq 1$, they are pretty close to the standard conditions. We now prove that for $2 \leq p \leq \infty$, the derivatives $G'(u_0)$ of the nonlinear $G(u)$ in $W_0^{m,p}(\Omega)$ are bounded and the principal parts are $H_0^m(\Omega)$-coercive.

Theorem 2.122. Bounded linearization in $W_0^{m,p}(\Omega)$, $2 \leq p \leq \infty$, $H_0^m(\Omega)$-coercive principal part, Fredholm alternative: *In addition to the conditions (2.284), (2.285),* *for the original quasilinear operator in (2.290), we impose for its linearization (2.459) the Legendre condition (2.461). Either for the solution* $u_0 \in W^{m,p}(\Omega)$ *of (2.295) or close to it,*[24] *cf. (1.49),*

$$\text{let for } 2 \leq p \leq \infty: \; h := \frac{\partial A_\alpha}{\partial \vartheta^\beta}(\cdot, u_0, \ldots, \nabla^m u_0) \in L^\infty(\Omega), \; \text{or } h \in L^{p/(p-2)}(\Omega),$$

$$\text{or let } \left\| \frac{\partial A_\alpha}{\partial \vartheta^\beta}(\cdot, \vartheta^{\beta'}, |\beta'| \leq m) \right\|_{L^\infty(\Omega)} \leq C \; \forall \vartheta^{\beta'} \in \mathbb{R}, |\vartheta\beta' - \partial^{\beta'} u_0| \leq \rho > 0. \quad (2.462)$$

Then the linearization in (2.459) is bounded.

We use the compact embedding of $W^{m,p}(\Omega), 2 \leq p \leq \infty$, *into* $H^m(\Omega)$, *and obtain* $G'(u_0)$ *and its principal part as* $H^m(\Omega)$*-elliptic and* $H_0^m(\Omega)$*-coercive, respectively.*

Remark 2.123.

1. *Any types of estimates based upon linearization for* $u \in W^{k,p}(\Omega), 2 \leq p \leq \infty$, *are therefore still possible, however with respect to the* $H^k(\Omega)$ *norm. This is particularly important for discretization methods. Then stability, consistency, and convergence are valid with respect to the discrete, e.g.* $H^k(\Omega^h)$ *norm.*
2. *The results of Corollary 2.44 imply, e.g. existence and uniqueness of solutions and a boundedly invertible* A, *for the* H_0^m*-coercive and the* Fredholm alternative *for the elliptic case. Regularity results as indicated in Subsections 2.5.4, 2.5.7, and 2.5.6, are valid under the corresponding conditions.*

Proof. With $\partial^\beta w, \; \partial^\alpha v \in L^p(\Omega), \; \partial A_\alpha / \partial \vartheta^\beta(\cdot, u_0, \ldots, \nabla^m u_0) \in L^\infty(\Omega)$ or, for $p \geq 2$, in $L^{p/(p-2)}(\Omega)$ and the continuous embedding of $L^p(\Omega)$ into $L^2(\Omega)$, the following $\int_\Omega$ are defined. By (1.49),

$$h = \frac{\partial A_\alpha}{\partial \vartheta^\beta}(x, u_0, .., \nabla^m u_0) \in L^{p/(p-2)}(\Omega), \partial^\beta w, \; \partial^\alpha v \in L^p(\Omega) \Rightarrow h \partial^\beta w \partial^\alpha v \in L^1(\Omega)$$

$$\text{and } \left| \int_\Omega \frac{\partial A_\alpha}{\partial \vartheta^\beta}(x, u_0, .., \nabla^m u_0) \partial^\beta w \partial^\alpha v dx \right|$$
$$\leq C \left\| \frac{\partial A_\alpha}{\partial \vartheta^\beta}(x, u_0, .., \nabla^m u_0) \right\|_{L^{p/(p-2)}(\Omega)} \|\partial^\beta w\|_{L^p(\Omega)} \|\partial^\alpha v\|_{L^p(\Omega)} \text{ hence,}$$

$$\sum_{|\alpha|,|\beta| \leq m} \left| \int_\Omega \frac{\partial A_\alpha}{\partial \vartheta^\beta}(x, u_0, .., \nabla^m u_0) \partial^\beta w \partial^\alpha v dx \right|$$
$$\leq C \sum_{|\alpha|,|\beta| \leq m} \|\partial^\beta w\|_{L^p(\Omega)} \|\partial^\alpha v\|_{L^p(\Omega)} \leq C \|w\|_{W^{m,p}(\Omega)} \|v\|_{W^{m,p}(\Omega)}, \quad (2.463)$$

hence $(\partial G / \partial u)(u_0) = G'(u_0)$ is bounded for $2 \leq p \leq \infty$, but not for $1 \leq p < 2$.

[24] Since the $W^{m,p}(\Omega)$ are continuously embedded into $W^{j,s}(\Omega)$ for $1/p - 1/s \leq (m-j)/n$ the conditions for $\partial A_\alpha / \partial \vartheta^\beta$ can be relaxed for $|\alpha|, |\beta| \leq m - 1$.

Theorem 2.43 with (2.461) for $m \geq 1$ imply the $H_0^m(\Omega)$-coercivity of the principal part, $G'_p(u_0)$, and hence the $H_0^m(\Omega)$-ellipticity of the operator. The continuous embedding of $W^{m,p}(\Omega) \hookrightarrow H^m(\Omega), p \geq 2$, for bounded Ω allows considering $u \in W^{m,p}(\Omega)$ as $u \in H^m(\Omega)$, with the $H^m(\Omega)$ norm. These H_0^m-coercivity results imply all the other claims. □

In other cases the Fredholm alternative can be directly verified for the quasilinear problem as formulated by Zeidler [677, 678].

For *fully nonlinear equations* in (2.306), the $G(u)$ and $G'(u_0)u$ have the form

$$G, G'(u_0) : \mathcal{U} = W^{2,p}(\Omega) \cap W_0^{1,p}(\Omega) \to \mathcal{V}' = L^q(\Omega), u, u_0 \in \mathcal{D}(G) \subset \mathcal{U} \text{ s.t.} \tag{2.464}$$
$$w = (x, z, p, r),\ w_0(x) := (x, u_0(x), \nabla u_0(x), \nabla^2 u_0(x)) \in \mathcal{D}(G^w)\ \forall x \in \Omega$$
$$G(u_0(\cdot)) = G^w(\cdot, u_0(\cdot), \nabla u_0(\cdot), \nabla^2 u_0(\cdot)) = 0, \text{ hence,}$$
$$G'(u_0)u = \frac{\partial G^w}{\partial z}(w_0)u + \sum_{i=1}^n \frac{\partial G^w}{\partial p_i}(w_0)\partial^i u + \sum_{i,j=1}^n \frac{\partial G^w}{\partial r_{ij}}(w_0)\partial^i \partial^j u,$$

$$\text{for } \frac{\partial G^w}{\partial z}(w_0(\cdot)), \frac{\partial G^w}{\partial p_i}(w_0(\cdot)), \frac{\partial G^w}{\partial r_{ij}}(w_0(\cdot)) \in L^{p/(p-2)}(\Omega) \subset L^\infty(\Omega). \tag{2.465}$$

The term $\sum_{i,j=1}^n$ in (2.464), its principal part, is used in (2.307) defining the uniform ellipticity of G: For $w \in \mathcal{U}_{w_0}$ a neighborhood of w_0, we required for $0 \neq \vartheta \in \mathbb{R}^n$

$$0 < \lambda|\vartheta|^2 \leq \sum_{i,j=1}^n \frac{\partial G^w}{\partial r_{ij}}(w)\vartheta_i\vartheta_j \leq \Lambda|\vartheta|^2 \forall w = (x, u, p, r)\ \in \mathcal{U}_{w_0} \subset\ \mathcal{D}(G^w). \tag{2.466}$$

We do not explicitly formulate the case of *nondivergent quasilinear equations.* This is a special case of (2.306) or (2.464) and, compared to (2.459), $G'(u_0)$ requires only slight changes.

For our later studies in discretization methods it is important that, in contrast to the original nonlinear problem, it is possible to transform the linearization in (2.464) and in (2.467) below into the standard weak form.

For the Skrypnik [591] setting, the derivative $G'(u_0)u$ for the fully nonlinear operator $G(u) = G(0, u)$ of order $2m$ is obtained. In a *Hilbert space,* evaluated at u_0 applied to u, see (2.316) and the notation in (2.269), it has the form

$$u, u_0 \in \mathcal{D}(G),\quad u, u_0 \in \mathcal{U} \text{ with } w_0(x) := (0, x, \partial^\alpha u_0(x), |\alpha| \leq 2m) \tag{2.467}$$

hence,

$$G'(u_0) : \mathcal{U} = H^{2m}(\Omega) \cap H_0^m(\Omega) \to \mathcal{V}' = L^2(\Omega),\ G'(u_0)u = \sum_{|\alpha|\leq 2m} \partial^\alpha\big(G^w(w_0)\big)\partial^\alpha u$$

and we need, for the weak form,

$$\partial^\alpha\big(G^w(w_0(\cdot)\big) \in L^\infty(\Omega)\ \forall \alpha : |\alpha| \leq 2m. \tag{2.468}$$

We recapitulate the *modified ellipticity condition* (2.318): There exist constants $0 < k, 0 < \mu < 1, 0 < \Lambda < \infty$ such that

$$\forall t \in [0,1], v \in \mathcal{U}: \ G(t,v) = 0 \text{ implies } \|v\|_{C^{2m,\gamma}(\bar{\Omega})} \leq k, \text{ and in } \Omega \text{ the} \tag{2.469}$$
$$\textit{ellipticity condition } k^{-1}|\vartheta|^{2m} \leq \sum_{|\alpha|=2m} \partial^\alpha \big(G^w(t,\cdot,v,\ldots,\nabla^{2m}v)(x)\big)\vartheta^\alpha \leq C|\vartheta|^{2m}.$$

A combination of the last proof with these results and Theorem 2.33 yields

Theorem 2.124. Elliptic bilinear form, Fredholm alternative:

1. *For the linearization (2.464) of the fully nonlinear differential operator of order $2m = 2$ in (2.306) we require: In addition to the conditions for the data of the original problem, listed in Theorems 2.79 and 2.80, e.g. (2.469), assume $2 \leq p \leq \infty$, and (2.465), for any $u_0 \in \mathcal{U}$ in (2.464) with $w_0(x) \in \mathcal{D}(G^w)\forall x \in \Omega$. Then the linearization in (2.464) is bounded in $W_0^{m,p}(\Omega,\mathbb{R}^q)$ with an $H_0^m(\Omega,\mathbb{R}^q)$-coercive principal part for $2 \leq p \leq \infty$.*
2. *This remains correct for order $2m, m > 1, p \geq 2$ in (2.467) with the corresponding (2.468), (2.469) for $\partial^\alpha\big(G^w\big)(w^0(\cdot)) \in C(\bar{\Omega})$ for $|\alpha| = 2m$. In both cases, Corollary 2.44 is valid.*

2.7.4 Quasilinear elliptic systems of orders 2 and $2m$

Here we are in a situation similar to Subsection 2.7.3, now starting in the $H_0^m(\Omega,\mathbb{R}^q)$ setting. Most of the existence, uniqueness and regularity results are known only for this case. Generalizations to the $W_0^{m,p}(\Omega,\mathbb{R}^q)$ situation are possible and worthwhile. In fact results as in Theorem 2.122 form the basis for numerical methods for the $W_0^{m,p}(\Omega,\mathbb{R}^q)$ setting, opening the possibility for research into new problems beyond the previous scope. The same extensions are possible for fully nonlinear equations and systems.

We have studied quasilinear elliptic systems as Euler equations of a variational problem and the general case. We start with the Euler equations and formally linearize this system (2.362) in a *generalized solution* $\vec{u}^0$ using $\vec{w}^0(x) := (x, \vec{u}^0(x), \nabla\vec{u}^0(x))$ or $\vec{w}(x)$ to get, compare (2.377) ff.,

$$\langle G(\vec{u}),\vec{v}\rangle = \int_\Omega \big[F_p(\vec{w}(x))\nabla\vec{v} + F_z(\vec{w}(x))\vec{v}\big]dx \ \forall \vec{v} \in H_0^1(\Omega,\mathbb{R}^q)$$
$$\begin{aligned}\langle G'(\vec{u}^0)\vec{u},\vec{v}\rangle &= \int_\Omega \big[F_{pz}(\vec{w}^0(x))\vec{u} + F_{pp}(\vec{w}^0(x))\nabla\vec{u}\big]\nabla\vec{v} \\ &\quad + \big[F_{zz}(\vec{w}^0(x))\vec{u} + F_{zp}(\vec{w}^0(x))\nabla\vec{u}\big]\vec{v}dx \qquad (2.470) \\ &= \int_\Omega \sum_{i,j=1}^q \sum_{k=1}^n \left[\left(a_k^i\right)_{z_j}(\vec{w}^0(x))u_j + \sum_{l=1}^n \left(a_k^i\right)_{p_j^l}(\vec{w}^0(x))\partial^l u_j\right]\partial^k v_i \\ &\quad + \Big[\left(a_0^i\right)_{z_j}(\vec{w}^0(x))u_j + \sum_{l=1}^n \left(a_0^i\right)_{p_j^l}(\vec{w}^0(x))\partial^l u_j\Big]v_i dx = 0.\end{aligned}$$

We have to impose, similarly to (2.462), slightly stronger conditions than the original (2.372)–(2.374) or (2.372), (2.375)–(2.376), and (2.377)–(2.379) or (2.377), (2.378), (2.380) for the differentiability of J and the regularity of solutions of (2.362). For general $\vec{u}^1$ or our solution $\vec{u}^0$ and the above $\vec{w}^1(x), \vec{w}^0(x)$ we require for $p = q = 2$ and the usual $\partial^0 u = u, \vartheta^\beta, \vartheta_j^l \in \mathbb{R}$, $\vec{\vartheta} \in \mathbb{R}^{nq}$: Assume there exists, similarly to (2.462) and (2.371), a neighborhood $\mathcal{U}_{w^0}$ of $\vec{w}^0$, for the solution, such that

$$\vec{u}^0 \in H^1(\Omega, \mathbb{R}^q): \quad F_{pp}, F_{pz}, F_{zz}(\vec{w}^0(x)) \in L^\infty(\Omega), \text{ and for } \vec{w}^1 \in \mathcal{U}_{w^0} \exists \lambda > 0 \qquad (2.471)$$

$$\text{s.t.} \sum_{i,j=1}^{q} \sum_{k,l=1}^{n} F_{p_i^k p_j^l}(\vec{w}(x)) \vartheta_i^k \vartheta_j^l \geq \lambda |\vec{\vartheta}|_{nq}^2 \ \forall\, \vec{\vartheta} \in \mathbb{R}^{nq} \text{ (ellipticity condition).} \qquad (2.472)$$

Theorem 2.125. Quasilinear operators of order 2 with H_0^1-coercive linearization: *For the quasilinear operator $G(\vec{u})$ and its linearization, $G'(\vec{u}^0)$, cf. (2.470), assume (2.471). Then $G'(\vec{u}^0) : H_0^1 \to H_0^{-1}$ is a bounded linear operator. Its* principal part *is uniformly $H_0^1(\Omega, \mathbb{R}^q)$-coercive if (2.472) is satisfied, hence $G'(\vec{u}^0)$ is uniformly elliptic. As in Theorem 2.122 this remains correct for $W_0^{2,p}(\Omega, \mathbb{R}^q), 2 \leq p \leq \infty$. The full machinery of Corollary 2.44 applies, e.g. with respect to existence and uniqueness.*

Similarly as before we now differentiate the weak form (2.425) of a general quasilinear elliptic system. This yields

$$\langle G'(\vec{u}^0)\vec{u}, \vec{v}\rangle = \int_\Omega \sum_{|\alpha|,|\beta| \leq m} \left(\frac{\partial^{|\beta|} \vec{a}_\alpha}{\partial \vartheta^\beta}(\cdot, \vec{\partial}^{\leq m} \vec{u}^0(x)) \partial^\beta \vec{u}(x), \partial^\alpha \vec{v}(x) \right)_q dx. \qquad (2.473)$$

This is the linear form (2.392) of the system of order $2m$, now restricted to $\vec{u}, \vec{u}^0 \in H^m(\Omega)$ instead of the above $\vec{u} \in W^{m,p}(\Omega)$ for $m = 1$. The differentiation is correct if for the solution

$$\vec{u}^0 \in H^m(\Omega) : A_{\alpha,\beta}(\cdot) := \frac{\partial \vec{a}_\alpha}{\partial \vartheta^\beta}(\cdot; \vec{\partial}^{\leq m} \vec{u}^0(\cdot)) \in L^\infty(\Omega), \text{ e.g. } |A_{\alpha,\beta}(\cdot)| < \Lambda.$$

Theorem 2.126. Quasilinear operators of order $2m$ with H_0^m-coercive linearization: *Under these conditions, the operator G of the quasilinear system (2.425) of order $2m$ is differentiable in $\vec{u}^0 \in H^m(\Omega)$. The weak linearized operator $G'(\vec{u}^0) : H_0^m(\Omega) \to H_0^{-m}(\Omega)$ has the form (2.473). Let the strong Legendre–Hadamard condition (2.394) be satisfied for the above $(\partial \vec{a}_\alpha / \partial \vartheta^\beta)(\cdot, \vec{\partial}^{\leq m} \vec{u}^0(x)) \in C(\bar{\Omega})$ for $|\alpha| = |\beta| = m$. Then the* principal part *of $G'(\vec{u}_0)$ is $H_0^m(\Omega)$-coercive and $G'(\vec{u}_0)$ is elliptic. As in Theorem 2.125 this remains correct for $W_0^{m,p}(\Omega, \mathbb{R}^q), 2 \leq p \leq \infty$, and allows us the full machinery of Corollary 2.44.*

2.7.5 Linearizing general divergent quasilinear and fully nonlinear systems

In Subsection 2.6.8 we have already indicated that for systems of q fully nonlinear equations no results seem to be listed in the consulted monographs. We had only hinted at Yang [671], Fanciullo [311], Belopolskaya [79, 80], Chen and Wu [170] and Li and Liu [475].

In this situation, linearization cannot be used for theoretical results. However, it might be the only possibility, for obtaining existence and bifurcation results by combining the FEMs later on for these problems with numerical continuation techniques. In fact, the linearized operators do fit the linear elliptic systems studied in Subsection 2.6.5

For convenience, we repeat the necessary generalizations, compare (2.446), (2.447). For $\vec{v}, \vec{u} = (u_1, \dots, u_q) \in \mathcal{W} = H_0^m(\Omega, \mathbb{R}^q)$ we have to apply the $\partial^l, l = 0, \dots, n, \partial^\alpha, |\alpha| \le 2m$, to the $\vec{u}, u_j, v_i, i, j = 1, \dots, q$ and use the notation in (2.389). Accordingly, (2.306) and (2.317) have to be reinterpreted as q equations in strong form for the q components of $\vec{u}$, defined in

$$\mathcal{U} := H_0^m(\Omega, \mathbb{R}^q) \cap H^{2m}(\Omega, \mathbb{R}^q), \mathcal{W} := H_0^m(\Omega, \mathbb{R}^q), \mathcal{V} := L^2(\Omega, \mathbb{R}^q) \tag{2.474}$$

with the obvious norms. For systems of order 2 with $m = 1$ in (2.474), we obtain

$$G : \mathcal{D}(G) \subset \mathcal{U} \to \mathcal{V}, G(\vec{u}(\cdot)) := G^w(\cdot, \vec{u}(\cdot), \nabla\vec{u}, \nabla^2\vec{u}(\cdot)),\ G(\vec{u}_0) = 0. \tag{2.475}$$

The linearization $G'(\vec{u}^0) : \mathcal{U} \to \mathcal{V}'$ in $\vec{u}^0$ applied to $\vec{u}$ has the form, with $\vec{w}^0(x) := (x, \vec{u}^0(x), \nabla\vec{u}^0(x), \nabla^2\vec{u}^0(x)) \in \mathcal{D}(G^w)\ \forall x \in \Omega$, $p = (p_k \in \mathbb{R}^q)_{k=1}^n$, $r = (r_{kl} \in \mathbb{R}^q)_{k,l=1}^n$,

$$G'(\vec{u}^0)\vec{u} = \frac{\partial G^w}{\partial z}(\vec{w}^0)\vec{u} + \sum_{k=1}^n \frac{\partial G^w}{\partial p_k}(\vec{w}^0)\partial^k\vec{u} + \sum_{k,l=1}^n \frac{\partial G^w}{\partial r_{kl}}(\vec{w}^0)\partial^k\partial^l\vec{u}, \tag{2.476}$$

and we require

$$\frac{\partial G^w}{\partial z}(\vec{w}^0(\cdot)),\ \frac{\partial G^w}{\partial p_k}(\vec{w}^0(\cdot)), \frac{\partial G^w}{\partial r_{kl}}(\vec{w}^0(\cdot)) \in L^\infty(\Omega, \dots).$$

Again, in contrast to the original nonlinear problem, it is possible to transform the linearization in (2.476), with the usual partial integration, into the standard weak form. For the strong operator $G'(\vec{u}^0) : \mathcal{U} \to \mathcal{V}'$ we introduce new coefficients A_{kl}. This yields the *strong bilinear form* $a_s(\cdot, \cdot)$, and conditions for the partials of G^w:

$G'(\vec{u}^0) : \mathcal{U} \to \mathcal{V} = \mathcal{V}'$, $a_s(\cdot, \cdot) : \mathcal{U} \times \mathcal{V} \to \mathbb{R}$, and with $\delta_{0k}, (-1)_{k>0}$ we find

$$\begin{aligned}
G'(\vec{u}^0)\vec{u} &:= \sum_{k,l=0}^n (-1)_{k>0}\partial^k(A_{kl}\partial^l\ \vec{u}) \\
\text{with} A_{kl} &= A_{kl}(\vec{w}^0(\cdot)) \in W^{1-\delta_{k,0},\infty}(\Omega, \mathbb{R}^{q\times q}), \\
A_{lk} &= A_{kl} = -\frac{\partial G^w}{\partial r_{kl}}(\vec{w}^0) \text{ for } k, l = 1, \dots, n, \\
A_{0l} &= \frac{\partial G^w}{\partial p_l}(\vec{w}^0) + \sum_{k=1}^n \partial^k \frac{\partial G^w}{\partial r_{kl}}(\vec{w}^0), \\
A_{l0} &= 0 \text{ for } l = 1, \dots, n,\ A_{00} = \frac{\partial G^w}{\partial z}(\vec{w}^0),\ \text{then the}
\end{aligned} \tag{2.477}$$

$G'(\vec{u}^0) : \mathcal{U} \to \mathcal{V} = \mathcal{V}'$ and $a_s(\vec{u}, \vec{v}) := (G'(\vec{u}^0)\vec{u}, \vec{v})_\mathcal{V}\ \forall\ \vec{u} \in \mathcal{U}, \vec{v} \in \mathcal{V}$ are bounded.

With $(\cdot,\cdot) := (\cdot,\cdot)_\mathcal{V}$ and $\langle\cdot,\cdot\rangle := \langle\cdot,\cdot\rangle_{\mathcal{W}\times\mathcal{W}'}$ the strong and weak solutions $\vec{u}^0 \in \mathcal{U}$ and $\vec{u}^0 \in \mathcal{W}$, respectively, of the second order linearized systems are determined by, cf. (2.476),

$$(G'(\vec{u}^0)\vec{u},\vec{v}) = \Big(\sum_{k,l=0}^{n}(-1)_{k>0}\partial^k(A_{kl}\partial^l\vec{u}),\vec{v}\Big)_q = (\vec{f},\vec{v}) \quad \forall\vec{v}\in\mathcal{V}, \tag{2.478}$$

$$a(\vec{u},\vec{v}) := \langle G'(\vec{u}^0)\vec{u},\vec{v}\rangle := \int_\Omega \sum_{k,l=0}^{n}(A_{kl}\partial^l\vec{u},\partial^k\vec{v})_q dx = \langle\vec{f},\vec{v}\rangle, \forall\vec{v}\in\mathcal{W}$$

$$\text{and with } A_{kl} = A_{kl}(w^0)\in L^\infty(\Omega,\mathbb{R}^{q\times q}) \text{ for the weak form.} \tag{2.479}$$

The *uniform Legendre condition* defines a uniformly elliptic system:

$$\lambda|\vec{\vartheta}|^2 \le \sum_{k,l=1}^{n}(A_{kl}(x)\vartheta^l)\vartheta^k \le \Lambda|\vec{\vartheta}|^2 \ \forall\vec{\vartheta}\in\mathbb{R}^{nq}, |\vec{\vartheta}|^2 = \sum_{i=1}^{q}\sum_{k=1}^{n}\left(\vartheta_i^k\right)^2. \tag{2.480}$$

This implies the $H_0^1(\Omega,\mathbb{R}^{q\times q})$-coercivity of the *principal part* $a_p(\vec{u},\vec{v})$.

For order $2m, m>1$, we again reinterpret, now (2.317), as a nonlinear system of q equations in $\mathcal{U} := H_0^m(\Omega,\mathbb{R}^q)\cap H^{2m}(\Omega,\mathbb{R}^q)$. We update the above notations in (2.335) as:

$$\vec{w} := (x,\vartheta^\gamma,|\gamma|\le 2m)\in\mathcal{D}(G^w),\ \ \vec{u},\vec{u}^0\in\mathcal{D}(G), \vec{w}^0(x) := (x,\partial^\gamma\vec{u}^0(x),|\gamma|\le 2m).$$

Then a systems of order $2m$ has the form $G:\mathcal{U}\to\mathcal{V}$

$$G(\vec{u}(\cdot)) := G^w(\cdot,\vec{u}(\cdot),\cdots,\nabla^{2m}\vec{u}(\cdot)) = G^w(\cdot,\partial^\gamma\vec{u}(\cdot),|\gamma|\le 2m),\ G(\vec{u}_0)=0. \tag{2.481}$$

The linearization $G'(\vec{u}^0):\mathcal{U}\to\mathcal{V}=\mathcal{V}'$ in $\vec{u}^0$ applied to $\vec{u}$ has the form,

$$G'(\vec{u}^0)\vec{u} = \sum_{|\gamma|\le 2m}\frac{\partial G^w}{\partial\vartheta^\gamma}(\vec{w}^0)\partial^\gamma\vec{u}, \ \text{ with } \frac{\partial G^w}{\partial\vartheta^\gamma}(\vec{w}^0(\cdot))\in L^\infty(\Omega,\mathbb{R}^q)\ \forall|\gamma|\le 2m. \tag{2.482}$$

Introducing new coefficients $A_{\alpha\beta}$ yields the *strong and weak bounded linear operators and bilinear forms* $G'_s(\vec{u}^0), G'(\vec{u}^0)$, and $a_s(\cdot,\cdot), a(\cdot,\cdot)$, and conditions for the partials of G^w:

$$a_s(\cdot,\cdot):\mathcal{U}\times\mathcal{V}\to\mathbb{R}, a(\cdot,\cdot):\mathcal{W}\times\mathcal{W}\to\mathbb{R}, G'(\vec{u}^0):\mathcal{W}\to\mathcal{W}',$$

$$G'_s(\vec{u}^0)\vec{u} := \sum_{|\alpha|,|\beta|\le m}(-1)^{|\alpha|}\partial^\alpha(A_{\alpha\beta}\partial^\beta\,\vec{u}) := \sum_{|\gamma|\le 2m}\frac{\partial G^w}{\partial\vartheta^\gamma}(\vec{w}^0)\partial^\gamma\vec{u}, \text{ with}$$

$$A_{\alpha\beta} = A_{\alpha\beta}(\vec{w}^0(\cdot))\in W^{|\alpha|,\infty}(\Omega,\mathbb{R}^{q\times q}), \forall|\alpha|,|\beta|\le m, \text{ e.g.,} \tag{2.483}$$

$$A_{\alpha\beta} = (-1)^m\frac{\partial G^w}{\partial\vartheta^{\alpha\beta}}(\vec{w}^0) = A_{\beta\alpha} \text{ for } |\alpha|=|\beta|=m, \text{ and with}$$

$(\cdot,\cdot) := (\cdot,\cdot)_\mathcal{V},\ \langle\cdot,\cdot\rangle := \langle\cdot,\cdot\rangle_{\mathcal{W}\times\mathcal{W}'}$ we get the strong solution $\vec{u}^1\in\mathcal{U}$ from

$(\vec f, \vec v) = a_s(\vec u^1, \vec v) := (G'_s(\vec u^0)\vec u^1, \vec v)\ \forall\ \vec v \in \mathcal{V}$ or the weak solution $\vec u^1 \in \mathcal{W}$, from

$$a(\vec u^1, \vec v) := \langle G'(\vec u^0)\vec u^1, \vec v\rangle = \int_\Omega \sum_{|\alpha|,|\beta|\le m} (A_{\alpha\beta}\partial^\beta \vec u^1, \partial^\alpha \vec v)dx = \langle \vec f, \vec v\rangle,$$

$$\forall \vec v \in H_0^m(\Omega, \mathbb{R}^q) \text{ and with } A_{\alpha\beta} \in L^\infty(\Omega, \mathbb{R}^{q\times q}) \text{ for this weak form.} \qquad (2.484)$$

Here we have to impose the *uniform Legendre–Hadamard condition:* $\exists \Lambda > \lambda > 0:$

$$\lambda|\eta|^2|\vartheta|^{2m} \le \sum_{|\alpha|,|\beta|=m}^{n} (\bar\eta A_{\alpha\beta}(\vec w^0(x))\eta)\vartheta^\beta\vartheta^\alpha \le \Lambda|\eta|^2|\vartheta|^{2m}\ \forall \vartheta \in \mathbb{R}^n, \eta \in \mathbb{C}^q. \quad (2.485)$$

To enforce the H_0^m-coercivity of the principal part, see Theorem 2.104, we need, in addition to the above $A_{\alpha\beta} \in W^{|\alpha|,\infty}(\Omega, \mathbb{R}^{q\times q})$, $\forall |\alpha|, |\beta| \le m$,

$$A_{\alpha\beta}(\vec w^0(\cdot)) \in C(\bar\Omega, \mathbb{R}^{q\times q})\ \forall |\alpha| = |\beta| = m > 1. \qquad (2.486)$$

A combination of Theorems 2.104 and 2.33 yields the following

Theorem 2.127. Bounded linearization, Fredholm alternative for fully nonlinear systems in $W^{2m,p}(\Omega, \mathbb{R}^{q\times q})$, $2 \le p \le \infty$:

1. *For the linearizations in (2.478), and (2.483), (2.484) of the fully nonlinear systems in (2.475), and (2.481), respectively, we require, for the solution $\vec u^0$ in (2.478) and (2.481), (2.483), the conditions in (2.477),(2.483). Furthermore, let*

$$A_{ij} \in L^\infty(\Omega, \mathbb{R}^{q\times q}) \textit{ for } m = 1, \textit{ and } A_{\alpha\beta} \in L^\infty(\Omega, \mathbb{R}^{q\times q}) \textit{ for } |\alpha| = |\beta| \le m, \quad (2.487)$$

   and assume the conditions for the data of the original problem in Subsection 2.6.8. Then the above linearizations are bounded.
2. *Additionally, we assume the strong Legendre condition (2.480) for $m \ge 1 = q$ and $q \ge 1 = m$ with (2.486) for $m > 1$ and the strong Legendre–Hadamard condition, (2.485), and (2.486) for $m, q > 1$. Then the above linearizations are even elliptic and the principal parts are $H_0^m(\Omega, \mathbb{R}^q)$-coercive. Thus Corollary 2.44 is valid for $p \ge 2$. The strong and weak bilinear forms coincide in $u \in H^{2m}(\Omega, \mathbb{R}^q) \cap H_0^m(\Omega, \mathbb{R}^q), m \ge 1$, and $\vec v \in H_0^m(\Omega, \mathbb{R}^q)$ if furthermore $A_{kl} \in W^{1-\delta_{k0},\infty}(\Omega, \mathbb{R}^{q\times q})$ for $m = 1$ and $A_{\alpha\beta} \in W^{|\alpha|,\infty}(\Omega, \mathbb{R}^{q\times q})$ for $m > 1$.*

We discuss the problem of how nonexactly satisfied differential and boundary equations perturb the exact solution. This is an important problem for discretization methods. In particular, the mesh-free methods in [120] usually aim for a discrete solution of an approximate original problem. Similarly to Section 5.2, we introduce an operator $F := (G, B) : \mathcal{D}(G) \cap \mathcal{D}(B) \subset \mathcal{U} \to \mathcal{V} \times \mathcal{Z}$ with differential and boundary operators, G amd B as components. We restrict the discussion here to, cf. (2.158), linear *special Dirichlet systems.* In Theorems 1.29, 2.50 we have extended functions defined on the boundary to the whole domain, Ω. In Lemmas 2.26, 2.27, this was used for reducing the inhomogeneous to homogeneous boundary conditions. Similar strategies are possible for some more general, in particular nonlinear boundary conditions as well. We leave that to the interested reader. Note that for fully nonlinear problems (in strong form) we have $\mathcal{U} = H^{2m}(\Omega, \mathbb{R}^q)$ and $\mathcal{V} = L^2(\Omega, \mathbb{R}^q)$, $\mathcal{Z} = \Pi_{j=1}^m H^{m-j+1-1/2}(\partial\Omega, \mathbb{R}^q)$. The solutions are denoted for all cases $q \geq 1$, as u_0. For all the other problems (in weak form) we have $\mathcal{U} = W^{m,p}(\Omega, \mathbb{R}^q)$ and $\mathcal{V} = W^{-m,p'}(\Omega, \mathbb{R}^q), 1/p + 1/p' = 1$, $\mathcal{Z} = \Pi_{j=1}^m W^{m-j+1-1/p,p}(\partial\Omega, \mathbb{R}^q)$. We combine Theorems 1.48 and 1.43 with our results for boundedly invertible linearizations.

Therefore we present, in the next theorem, estimates for the norm of the difference $\|u_0 - \tilde{u}_0\|_{\mathcal{U}}$, with the exact and the perturbed solutions, hence for

$$F(u_0) = (G(u_0), B(u_0)) = (0,0) \text{ and } F(\tilde{u}_0) = (G(\tilde{u}_0), B(\tilde{u}_0)) = (g, \varphi). \quad (2.488)$$

For many applications, e.g. for discretization methods, results for boundedly invertible $G'(u_0)$, and bounds for its norm, are more directly available than for $F'(u_0)$. Therefore we present in the next theorem estimates for $\|u_0 - \tilde{u}_0\|_{\mathcal{U}}$, in terms of $\|(F'(u_0))^{-1}\|_{\mathcal{U} \leftarrow \mathcal{V} \times \mathcal{Z}}$ and of $\|(G'(u_0))^{-1}\|_{\mathcal{U} \leftarrow \mathcal{V}}$.

Theorem 2.128. Error estimates for perturbed elliptic problems: *With the differential and boundary operators, G and B, let the operator $F := (G, B) : \mathcal{D}(G) \cap \mathcal{D}(B) \subset \mathcal{U} \to \mathcal{V} \times \mathcal{Z}$ have a locally unique solution u_0 for $F(u_0) = 0$, and let $F \in C^1$ in a neighborhood of u_0, hence $F'(u_0) = (G'(u_0), B'(u_0)) : \mathcal{U} \to \mathcal{V} \times \mathcal{Z}$ exists and is boundedly invertible. Let u_0 and $\tilde{u}_0$ be the solutions of the exact and the perturbed problems in (2.488) with small enough $(g, \varphi) \in \mathcal{V} \times \mathcal{Z}$. Then*

$$\|u_0 - \tilde{u}_0\|_{\mathcal{U}} \leq 2\|(F'(u_0))^{-1}\|_{\mathcal{U} \leftarrow \mathcal{V} \times \mathcal{Z}} \|(g, \varphi)\|_{\mathcal{V} \times \mathcal{Z}}. \quad (2.489)$$

$G(u) = 0, B(u) = 0$ has the same locally unique solution u_0, and $G'(u_0) : \mathcal{U} \cap \mathcal{N}(B) \to \mathcal{V}$ exists in a neighborhood of u_0, and is also boundedly invertible. Then

$$\|\tilde{u}_0 - u_0\|_{\mathcal{U}} \leq 2\|(G'(u_0))^{-1}\|_{\mathcal{U} \leftarrow \mathcal{V}} (\|g\|_{\mathcal{V}} + C|\varphi\|_{\mathcal{Z}}(1 + 2\|G'(u_0)\|_{\mathcal{V} \leftarrow \mathcal{U}}\|)). \quad (2.490)$$

Proof. F' is continuous in a neighborhood of u_0, hence equi continuous in a smaller closed neighborhood of u_0. Therefore, (2.489) is an immediate consequence of Theorems 1.48 and 1.43.

We employ the lemmas and theorems mentioned above. In the first step, we extend $\varphi \in \mathcal{Z}$, defined along $\partial\Omega$, to a $\tilde{u}$, defined on Ω. By lemma 2.26 this yields

$$B\tilde{u} = \varphi \text{ on } \partial\Omega \Longrightarrow \|\tilde{u}\|_{\mathcal{U}} \leq C\|\varphi\|_{\mathcal{Z}}. \quad (2.491)$$

Next we estimate for the linear B, and with (2.491) and $\mathcal{O}(???)$ as defined by (2.493),

$$F(\tilde{u}_0 - \tilde{u}) - F(u_0) = \big(G(\tilde{u}_0 - \tilde{u}) - G(u_0),\ B(\tilde{u}_0 - \tilde{u}) - B(u_0)\big) \quad (2.492)$$

$$= (G(\tilde{u}_0 - \tilde{u}) - 0, 0) = (G'(u_0)(\tilde{u}_0 - \tilde{u} - u_0) + \mathcal{O}(???), 0) \text{ with}$$

$$\|G(\tilde{u}_0 - \tilde{u})\|_{\mathcal{V}} = \|G(\tilde{u}_0) - G'(u_0)\tilde{u} + \mathcal{O}(???)\|_{\mathcal{V}} \leq (\|g\|_{\mathcal{V}} + 2\|G'(u_0)\|_{\mathcal{V} \leftarrow \mathcal{U}}\|C\|\varphi\|_{\mathcal{Z}}).$$

We have modified (1.78), e.g. with $x_0 = u_0$ near $\overline{u_0\tilde{u}_0}$, for the last inequality as

$$\|G(\tilde{u}_0) - G(u_0) - G'(x_0)(\tilde{u}_0 - u_0)\| \leq \|\tilde{u}_0 - u_0\| \sup_{x \in \overline{u_0\tilde{u}_0}} \|G'(x) - G'(x_0)\| =: \|\mathcal{O}(???)\|_{\mathcal{V}}. \quad (2.493)$$

The right-hand term (2.493) is indicated in (2.492) as $\mathcal{O}(???)$. It gets very small and implies the previous factors 2 in the last two lines in (2.492) and in (2.494). The combination of the last two lines in (2.492) yields

$$\|\tilde{u}_0 - \tilde{u} - u_0\|_{\mathcal{U}} \leq 2\|(G'(u_0))^{-1}\|_{\mathcal{U}\hookleftarrow\mathcal{V}}\big(\|g\|_{\mathcal{V}} + 2\|G'(u_0)\|_{\mathcal{V}\hookleftarrow\mathcal{U}}\|C|\varphi\|_{\mathcal{Z}}\big). \quad (2.494)$$

The triangle inequality

$$\|\tilde{u}_0 - u_0\|_{\mathcal{U}} \leq \|\tilde{u}_0 - \tilde{u} - u_0\|_{\mathcal{U}} + \|\tilde{u}\|_{\mathcal{U}}$$

finally yields the claim. □

2.8 The Navier–Stokes equation

2.8.1 Introduction

This is one of the most stimulating equations for research in mathematics, physics, and engineering. It models fluid mechanics, e.g. the flow of water around a vessel, and the air around a wing of an airplane. In this context (2.495) with small values of $0 < \nu$ are interesting. Then boundary layers and turbulence develop, modeling the up-wind properties carrying an airplane. Mathematically, it was one of the first extremely important systems of nonlinear differential equations. It required an update of the definition of elliptic or parabolic equations, and for small $0 < \nu$ the borders between parabolic and hyperbolic problems are no longer well defined. Furthermore, by varying $0 < \nu$ and other parameters, extremely interesting regimes of bifurcation and the corresponding dynamics and chaotic phenomena originate. Still, for a lot of these problems satisfactory answers are missing. Some answers can be found in [120].

We only consider incompressible Navier-Stokes equations. From the huge number of publications we have to choose a very narrow selection of results.

Further results can be found in, e.g. Kreiss and Lorenz [454], von Wahl and Grunau [378, 658, 659], Kötter [453], Bensoussan and Frehse [81], and in Ladyženskaja and Uralceva [463–466]. Further regularity results for Navier-Stokes equations are due, e.g. to Cannone and Karch [157], Ladyzhenskaya [465], and Zubelevich [683].

Models of Navier-Stokes equations as Newtonian compressible fluids in the steady and unsteady regime are studied, e.g. by Feistauer *et al.* [313, 315], and Novotný and Straškraba [515]. The steady regime is discussed in Subsection 7.15.2 as well.

2.8.2 The Stokes operator and saddle point problems

We start with the *Stokes operator* and its properties, see Hackbusch [386, 387]. It is closely related to the stationary Navier-Stokes equation, cf. (2.320),

$$G(\vec{u}, p) := \begin{pmatrix} -\nu\Delta\vec{u} + \sum\limits_{i=1}^{n} u_i\partial^i\vec{u} + \nabla\ p \\ -\ \mathrm{div}\ \vec{u} \end{pmatrix} = \begin{pmatrix} \vec{f} \\ 0 \end{pmatrix} \text{ in } \Omega, \quad (2.495)$$

$$\vec{u} = 0 \text{ on } \Gamma = \partial\Omega, \qquad \int_\Omega p dx = 0, \quad \Omega \text{ satisfies (2.5)}$$

and $\vec{u}, p$ and $\vec{f}$ denote the velocity, pressure and forcing terms of an incompressible medium, $\vec{v}$ a test function with $\text{div } \vec{u} = \sum_{i=1}^{n} \partial u_i/\partial x_i$, and $\nu = 1/R$ with the *Reynolds number*, R. The Stokes operator is obtained from (2.495) for $\nu = 1$ and by deleting the nonlinear term as

$$S(\vec{u}, p) := \begin{pmatrix} -\Delta\vec{u} + \nabla\, p \\ -\text{ div }\ \vec{u} \end{pmatrix} = \begin{pmatrix} \vec{f} \\ 0 \end{pmatrix} \text{ in } \Omega, \quad \vec{u}|_{\partial\Omega} = 0, \int_{\Omega} p dx = 0. \quad (2.496)$$

We introduce $\mathcal{V}_b, \mathcal{W}$, see Subsection 2.6.2, as

$$\mathcal{V}_b := H_0^1(\Omega, \mathbb{R}^n), \mathcal{V} := H^1, \mathcal{W} := L_*^2(\Omega) := \left\{ p \in L^2(\Omega) : \int_{\Omega} p(x)dx = 0 \right\}. \quad (2.497)$$

Now we take the scalar product of the Stokes operator with $\vec{v} \in \mathcal{V}_b$ and apply the Green's formula for smooth enough $\vec{u}, \vec{v}, p$. With the Euclidean norm $|\cdot|_n$ and inner product $(\cdot,\cdot)_n$ in $\mathbb{R}^n$ and the product norm $\|\vec{v}\|_{\mathcal{V}} := |\left(\|v_i\|_{H^1(\Omega)}\right)_{i=1}^{n}|_n$ the first line in (2.496) yields

$$(-\Delta\vec{u} + \nabla p, \vec{v})_{L^2(\Omega,\mathbb{R}^n)} := \int_{\Omega} (-\Delta\vec{u} + \nabla p, \vec{v})_n dx = (\nabla\vec{u}, \nabla\vec{v})_{L^2} - (p, \text{ div } \vec{v})_{L^2(\Omega)}.$$

We define the bilinear and linear forms $a(\cdot,\cdot), b(\cdot,\cdot)$ and $(\vec{f}, \cdot)$ as

$$\begin{aligned} a(\vec{u}, \vec{v}) &:= \int_{\Omega} (\nabla\vec{u}(x), \nabla\vec{v}(x))_n dx \text{ for } \vec{u}, \vec{v} \in \mathcal{V}_b, \\ b(p, \vec{v}) &:= -\int_{\Omega} p(x) \text{ div } \vec{v}(x) dx \text{ for } p \in L_*^2(\Omega), \qquad (2.498) \\ (\vec{f}, \vec{v}) &:= \int_{\Omega} (f, \vec{v})_n dx \text{ for } \vec{f} \in L^2(\Omega, \mathbb{R}^n) \text{ or } \in \mathcal{V}' = H^{-1}(\Omega, \mathbb{R}^n), \text{ cf. (2.110).} \end{aligned}$$

This yields for the two equations in (2.496) and for $u \in H^2(\Omega, \mathbb{R}^n) \cap \mathcal{V}_b$

$$\begin{aligned} &(-\Delta\vec{u} + \nabla p, \vec{v})_{L^2(\Omega,\mathbb{R}^n)} = a(\vec{u}, \vec{v}) + b(p, \vec{v}) = (\vec{f}, \vec{v}), \ \forall \ \vec{u}, \vec{v} \in \mathcal{V}_b, p \in \mathcal{W} \text{ and} \\ &-\int_{\Omega} q(x) \text{ div } \vec{u}(x) dx = b(q, \vec{u}) = 0 \ \forall \ q \in \mathcal{W}, \vec{u} \in \mathcal{V}_b. \end{aligned}$$

This requires testing the first equation by $\vec{v}$ and $\text{div } \vec{u} = 0$ by $q \in \mathcal{W}$. For our bounded Ω in (2.5), the $a(\cdot,\cdot), b(\cdot,\cdot)$, and $(\vec{f}, \cdot)$ are continuous, cf. Theorem 2.43 and Proposition 2.34:

$$\begin{aligned} &|a(\vec{u}, \vec{v})| \le C_a \|\vec{u}\|_{\mathcal{V}} \cdot \|\vec{v}\|_{\mathcal{V}}, |b(p, \vec{v})| \le C_b \|\vec{v}\|_{\mathcal{V}} \cdot \|p\|_{\mathcal{W}}, \qquad (2.499) \\ &|(\vec{f}, \vec{v})| \le \|\vec{f}\|_{L^2(\Omega,\mathbb{R}^n)} \cdot \|\vec{v}\|_{\mathcal{V}}, \text{ with } \|\vec{f}\|_{L^2(\Omega,\mathbb{R}^n)} \text{ or } \|\vec{f}\|_{\mathcal{V}'}. \end{aligned}$$

In the weak formulation of (2.496), we replace $\vec{f} \in L^2(\Omega, \mathbb{R}^n)$ and 0 by $f^1 \in \mathcal{V}'$ and $f^2 \in \mathcal{W}'$, respectively, and obtain

$$\begin{aligned} &\text{for given } \vec{f} = (f^1, f^2) \in \mathcal{V}_b' \times \mathcal{W}' \text{ determine } \vec{u}_0 \in \mathcal{V}_b, p_0 \in \mathcal{W} \text{ such that} \\ &a(\vec{u}_0, \vec{v}) + b(p_0, \vec{v}) = (f^1, \vec{v}) \quad \forall \ \vec{v} \in \mathcal{V}_b, \\ &\qquad b(q, \vec{u}_0) = (f^2, q) \ \forall \ \ q \in \mathcal{W}. \end{aligned} \tag{2.500}$$

Saddle point problems

To obtain a situation similar to Section 2.3, we reinterpret (2.500): We replace the $\mathcal{V}_b, \mathcal{W}, a(\cdot,\cdot), b(\cdot,\cdot), (\vec{f},\cdot), 0$ in (2.497)–(2.500) by general Hilbert spaces $\mathcal{V}_b, \mathcal{W}$, continuous bilinear forms a, b and linear forms $f^1 \in \mathcal{V}_b', f^2 \in \mathcal{W}'$, respectively. We call the Stokes problem as in (2.498), (2.500) and this generalized version a *saddle point problem*. Furthermore, we introduce

$$\begin{aligned} &\mathcal{X} := \mathcal{V}_b \times \mathcal{W} \text{ and } \vec{x}^\mathsf{T} := (\vec{u}, p), \vec{y}^\mathsf{T} := (\vec{v}, q) \in \mathcal{X} \text{ to obtain} \\ &c(\vec{x}, \vec{y}) := c\left(\begin{pmatrix} \vec{u} \\ p \end{pmatrix}, \begin{pmatrix} \vec{v} \\ q \end{pmatrix} \right) := a(\vec{u}, \vec{v}) + b(p, \vec{v}) + b(q, \vec{u}), \\ &(\vec{f}, \vec{y}) := (f^1, \vec{v}) + (f^2, q), \quad \vec{f} = (f_1, f_2) \in \mathcal{X}' = \mathcal{V}_b' \times \mathcal{W}'. \end{aligned} \tag{2.501}$$

Then $c(\cdot,\cdot)$ and $(\vec{f},\cdot)$ are continuous bilinear and linear forms on $\mathcal{X} \times \mathcal{X}$ and $\mathcal{X}$. Let $A \in \mathcal{L}(\mathcal{V}_b, \mathcal{V}_b'), B \in \mathcal{L}(\mathcal{W}, \mathcal{V}_b')$ and $C \in \mathcal{L}(\mathcal{X}, \mathcal{X}')$ be the linear operators induced by $a(\cdot,\cdot), b(\cdot,\cdot)$ and $c(\cdot,\cdot)$, respectively. Then

$$\begin{aligned} C &:= \begin{pmatrix} A & B \\ B^d & 0 \end{pmatrix} \in \mathcal{L}(\mathcal{X}, \mathcal{X}'), \ C\vec{x} = \begin{pmatrix} A & B \\ B^d & 0 \end{pmatrix} \begin{pmatrix} \vec{u} \\ p \end{pmatrix} = \begin{pmatrix} A\vec{u} + Bp \\ B^d\vec{u} \end{pmatrix}, \\ c(\vec{x}, \vec{y}) &= \langle C\vec{x}, \vec{y} \rangle_{\mathcal{X}' \times \mathcal{X}} = \left\langle \begin{pmatrix} A\vec{u} + Bp \\ B^d\vec{u} \end{pmatrix}, \begin{pmatrix} \vec{v} \\ q \end{pmatrix} \right\rangle_{\mathcal{X}' \times \mathcal{X}} \\ &= \langle A\vec{u}, \vec{v} \rangle_{\mathcal{V}_b' \times \mathcal{V}_b} + \langle Bp, \vec{v} \rangle_{\mathcal{V}_b' \times \mathcal{V}_b} + \langle B^d\vec{u}, q \rangle_{\mathcal{W}' \times \mathcal{W}} = a(\vec{u}, \vec{v}) + b(p, \vec{v}) + b(q, \vec{u}). \end{aligned} \tag{2.502}$$

Obviously, $c(\cdot,\cdot)$ is bounded, but it is not coercive, since $c(\vec{x}, \vec{x}) = 0$ for all $\vec{x} = (0, p)$.
We formulate this *saddle point problem* in three equivalent forms as

$$\begin{aligned} &\text{for given} \vec{f}, = (f^1, f^2) \in \mathcal{X}' \text{ determine } \vec{x}_0 = (\vec{u}_0, p_0) \in \mathcal{X} \text{ such that} \\ &c(\vec{x}_0, \vec{y}) = \langle \vec{f}, \vec{y} \rangle_{\mathcal{X}' \times \mathcal{X}} \text{ for all } \vec{y} \in \mathcal{X} \Leftrightarrow C\vec{x}_0 = \vec{f} \in \mathcal{X}' \\ &\quad \Leftrightarrow a(\vec{u}_0, \vec{v}) + b(p_0, \vec{v}) = (f^1, \vec{v}) \ \forall \ \vec{v} \in \mathcal{V}_b, b(q, \vec{u}_0) = (f^2, q) \ \forall \ \ q \in \mathcal{W}, \end{aligned} \tag{2.503}$$

compare (2.55). The following theorem explains why we call this a saddle point problem. Let

$$\begin{aligned} J(\vec{x}) &:= J(\vec{u}, p) = a(\vec{u}, \vec{u}) + 2b(p, \vec{u}) - 2(f^1, \vec{u}) - 2(f^2, p) \\ &= c(\vec{x}, \vec{x}) - 2(\vec{f}, \vec{x}) \text{ for } \vec{x} = (\vec{u}, p) \in \mathcal{V}_b \times \mathcal{W}. \end{aligned} \tag{2.504}$$

This $J(\vec{u},p)$ is neither bounded from above nor from below. So a solution $\vec{x}_0 = (\vec{u}_0, p_0) \in \mathcal{X}$ of (2.503) cannot be characterized by a minimum as in Theorem 2.16, but only as a saddle point, Hackbusch [387], Theorem 12.2.4.

Theorem 2.129. Saddle point problem: *Let $a(\cdot,\cdot)$ be symmetric and $\mathcal{V}_b$-coercive. Then $\vec{x}_0 = (\vec{u}_0, p_0) \in \mathcal{V}_b \times \mathcal{W}$ solves one of the equivalent problems in (2.503), if and only if*

$$J(\vec{u}_0, p) \le J(\vec{u}_0, p_0) \le J(\vec{u}, p_0) \text{ for } \vec{x} = (\vec{u}, p) \in \mathcal{X} = \mathcal{V}_b \times \mathcal{W}$$
$$\Leftrightarrow J(\vec{u}_0, p_0) = \min_{u \in \mathcal{V}_b} J(\vec{u}, p_0) = \max_{p \in \mathcal{W}} \min_{u \in \mathcal{V}_b} J(\vec{u}, p). \tag{2.505}$$

Existence and uniqueness of solutions $\vec{x}_0 = (\vec{u}_0, p_0) \in \mathcal{X}$ for (2.503) cannot be obtained with our previous methods. So we introduce a closed $\mathcal{V}_0$ and its orthogonal complement $\mathcal{V}^\perp$ such that

$$\mathcal{V}_0 = \ker B^d = \{\vec{v} \in \mathcal{V}_b : B^d \vec{v} = 0\} = \{\vec{v} \in \mathcal{V}_b : b(q, \vec{v}) = 0 \forall q \in \mathcal{W}\}, \mathcal{V}_b = \mathcal{V}_0 \oplus \mathcal{V}^\perp.$$

Conditions for the unique solvability of (2.503) require a slight reinterpretation. Let

$$A := \begin{pmatrix} A_{00} & A_{0\perp} \\ A_{\perp 0} & A_{\perp\perp} \end{pmatrix} \text{ with} \tag{2.506}$$
$$A_{00} \in \mathcal{L}(\mathcal{V}_0, \mathcal{V}_0'), \ A_{0\perp} \in \mathcal{L}(\mathcal{V}_\perp, \mathcal{V}_0'), \ A_{\perp 0} \in \mathcal{L}(\mathcal{V}_0, \mathcal{V}_\perp'), \ A_{\perp\perp} \in \mathcal{L}(\mathcal{V}_\perp, \mathcal{V}_\perp'),$$

and an analogous $B^* = (B_0^*, B_\perp^*) = (0, B^*)$, since $B_0^* = 0$.

Theorem 2.130. Unique solution for a saddle point problem: *Let $a(\cdot,\cdot) : \mathcal{V} \times \mathcal{V} \to \mathbb{R}$, $b(\cdot,\cdot) : \mathcal{W} \times \mathcal{V} \to \mathbb{R}$,, and $f^1 \in \mathcal{V}'$, $f^2 \in \mathcal{W}'$ be bounded and $\mathcal{V}_0, \mathcal{V}^\perp$ be defined as above. Then (2.503) or equivalently (2.505) is uniquely solvable if and only if the inverses of A_{00} and B exist, hence*

$$\mathcal{V}_0 = \ker B^d = \{\vec{v} \in \mathcal{V}_b : B^d \vec{v} = 0\} = \{\vec{v} \in \mathcal{V}_b : b(q, \vec{v}) = 0 \forall q \in \mathcal{W}\},$$
$$\mathcal{V}_b = \mathcal{V}_0 \oplus \mathcal{V}^\perp, \ \mathcal{V}_0, \mathcal{V}^\perp \subset \mathcal{V}_b. \tag{2.507}$$

Now we are able to modify Theorem 2.12 to this new situation. With the above A, B, C the C is boundedly invertible, if and only if the following *modified Brezzi–Babuska condition* or *Ladyženskaja–Brezzi–Babuska condition* is fulfilled:

$$C \in \mathcal{L}(\mathcal{X}, \mathcal{X}') \text{ is boundedly invertible } \Leftrightarrow$$
$$\exists\ \alpha, \beta > 0 \text{ s.t. } \quad \sup_{0 \ne \vec{v}_0 \in \mathcal{V}_0} |a(\vec{v}_0, \vec{v}_1)| / \|\vec{v}_0\|_{\mathcal{V}_b} > 0 \ \ \forall 0 \ne \vec{v}_1 \in \mathcal{V}_0, \tag{2.508}$$
$$\sup_{0 \ne \vec{v}_1 \in \mathcal{V}_0} |a(\vec{v}_0, \vec{v}_1)| / \|\vec{v}_1\|_{\mathcal{V}_b} \ge \alpha \|\vec{v}_0\|_{\mathcal{V}_b} \ \ \forall \vec{v}_0 \in \mathcal{V}_0, \tag{2.509}$$
$$\sup_{0 \ne \vec{v} \in \mathcal{V}_b} |b(p, \vec{v})| / \|\vec{v}\|_{\mathcal{V}_b} \ge \beta \|p\|_{\mathcal{W}} \ \ \forall p \in \mathcal{W}. \tag{2.510}$$

We do not always need all these conditions. If $a(\cdot,\cdot)$ is symmetric or Proposition 2.19 holds, then (2.508) can be skipped. The second line (2.509) implies the $\mathcal{V}_0$-coercivity of $a(\cdot,\cdot)$. For other combinations, see Hackbusch [387], Lemma 12.2.9 ff.

Solvability and regularity of the Stokes problem

Hackbusch's Lemmas 12.2.12 and 13 [387], yield (2.508), (2.509), (2.510). He refers to Nečas [509] for $n \geq 2$, Ladyženskaja [461], and Kellog and Osborn [443], see [387], Theorems 12.2.14, 18 and 19, for the following

Theorem 2.131. Unique and regular solution for the Stokes problem:

1. *Let $\Omega \in C^{0,1}$ be bounded, see (2.5), C be defined by (2.502) for the Stokes case with A, B as induced by (2.498). Then C is boundedly invertible, $C^{-1} \in \mathcal{L}(\mathcal{X}', \mathcal{X})$. So the Stokes problem*

$$-\Delta\vec{u} + \nabla p = f^1 \in H^{-1}(\Omega), -\ div\ \vec{u} = f^2 \in L^2_*(\Omega) \ on\ \Omega, \vec{u} = 0 \ on\ \partial\Omega$$

has a unique weak solution $(\vec{u}_0, p_0) \in \mathcal{V}_b \times \mathcal{W} = H^1_0(\Omega, \mathbb{R}^3) \times L^2_*(\Omega)$ for (2.500) satisfying

$$\|\vec{u}_0\|_{H^1(\Omega)} + \|p_0\|_{L^2(\Omega)} \leq C_\Omega(\|f^1\|_{H^{-1}(\Omega)} + \|f^2\|_{L^2(\Omega)}). \tag{2.511}$$

2. *Smoother data imply smoother solutions: If, in the above Stokes problem, Ω is smooth enough and $f^1 \in H^k(\Omega), f^2 \in H^{k+1}(\Omega) \cap L^2_*(\Omega), k \in \mathbb{N}_0$, we get for the solution $\vec{u}_0 \in H^{k+2}(\Omega) \cap H^1_0(\Omega), p_0 \in H^{k+1}(\Omega) \cap L^2_*(\Omega)$. There exists a constant $C > 0$, only depending upon Ω, such that*

$$\|\vec{u}_0\|_{H^{k+2}(\Omega)} + \|p_0\|_{H^{k+1}(\Omega)} \leq C_\Omega(\|f^1\|_{H^k(\Omega)} + \|f^2\|_{H^{k+1}(\Omega)}).$$

For $n = 2, k = 0$ a bounded and convex $\Omega \subset \mathbb{R}^2$ suffices.

2.8.3 The Navier–Stokes operator and its linearization

We want to study the linearization of the stationary Navier–Stokes operator in (2.495). Its derivative, the *linearized Navier-Stokes operator,* is

$$G'(\vec{u}, p)(\vec{w}, r) = \begin{pmatrix} -\nu\Delta\vec{w} + \sum_{i=1}^{n}(w_i\partial^i\vec{u} + u_i\partial^i\vec{w}) + \nabla r \\ -\ \text{div}\ \ \vec{w} \end{pmatrix}. \tag{2.512}$$

The special cases $\vec{u} \equiv 0, \nu = 1$, yields the above *Stokes operator, S,* see (2.496), and

$$\vec{u} \equiv 0, \nu \neq 1,\ S_\nu := G'(0, p) \text{ we call the } \nu\text{-Stokes operator.} \tag{2.513}$$

The results of the last subsection remain correct for this ν-Stokes operator, if throughout, the $a(\vec{u}, \vec{v})$ are replaced by $\nu a(\vec{u}, \vec{v})$ and ν is not too small.

We present the stationary Navier-Stokes operator following Temam *et al.* [349, 623, 624]. They discuss the local existence of solutions for the nonstationary Navier-Stokes equations and many open problems, e.g. related to nonlocal existence. As in the transition from (2.497) to (2.498), we multiply the first equation in (2.495) with the test function $\vec{v} = (v_1, \ldots, v_n)$ and the second with q and integrate. We use Green's formula to obtain, with $a(\cdot, \cdot), b(\cdot, \cdot)$ in (2.498), for given $(f^1, f^2) \in \mathcal{V}_b' \times \mathcal{W}'$ determine $\vec{u}_0 \in \mathcal{V}_b, p_0 \in \mathcal{W}$ such that

$$\int_\Omega \left(-\nu\Delta\vec{u}_0 + \sum_{i=1}^{n} (\vec{u}_0)_i \partial^i \vec{u}_0 + \nabla\ p_0, \vec{v} \right)_n dx \tag{2.514}$$

$$= \nu a(\vec{u}_0, \vec{v}) + d(\vec{u}_0, \vec{u}_0, \vec{v}) + b(p_0, \vec{v}) = (f^1, \vec{v}) \quad \forall \quad \vec{v} \in \mathcal{V}_b \text{ for the first, and}$$

$$b(q, \vec{u}_0) = (f^2, q) \quad \forall \quad q \in \mathcal{W} \text{ for the second equation;}$$

here we use the previous $a(\vec{u}, \vec{v}), b(q, \vec{v}), (\vec{f}, \vec{v})$ in (2.498) and introduce the trilinear

$$d(\vec{u}, \vec{w}, \vec{v}) := \sum_{i=1}^{n} \int_\Omega (u_i(\partial^i \vec{w}), \vec{v})_n dx = \sum_{i,j=1}^{n} \int_\Omega u_i(\partial^i w_j) v_j dx. \tag{2.515}$$

For bounded Ω, and $n \le 4$, see Temam [624] Lemma 1.2, Chapter II, Section 1.

$$d(\vec{u}, \vec{v}, \vec{w}) \text{ is a bounded trilinear form on } \mathcal{V}_b \times \mathcal{V}_b \times \mathcal{V}_b \tag{2.516}$$
$$\text{with } \mathcal{V}_b \hookrightarrow \mathcal{W}^* := L^2(\Omega, \mathbb{R}^3) \text{ compactly embedded.}$$

To linearize (2.514) with respect to $\vec{u}$, we consider, for fixed $\vec{u}, \vec{v}$ and small $\vec{w}$,

$$d(\vec{u} + \vec{w}, \vec{u} + \vec{w}, \vec{v}) - d(\vec{u}, \vec{u}, \vec{v}) = d(\vec{u}, \vec{w}, \vec{v}) + d(\vec{w}, \vec{u}, \vec{v}) + \mathcal{O}\left(\|\vec{w}\|_{\mathcal{V}}^2 \|\vec{v}\|_{\mathcal{V}} \right) \Longrightarrow$$

$$\left(((G'(\vec{u}, p)(\vec{w}, r)), (\vec{v}, q))_{n+1} \right)_{L^2(\Omega)} \tag{2.517}$$

$$= \begin{pmatrix} \nu a(\vec{w}, \vec{v}) + d(\vec{u}, \vec{w}, \vec{v}) + d(\vec{w}, \vec{u}, \vec{v}) + b(r, \vec{v}) \\ b(q, \vec{w}) \end{pmatrix} \forall (\vec{v}, q) \in \mathcal{V}_b \times \mathcal{W} = \mathcal{X}.$$

Now we consider the $d(\vec{u}, \vec{w}, \vec{v}), d(\vec{w}, \vec{u}, \vec{v})$ in (2.515). For fixed $\vec{u}$, both terms $d(\vec{u}, \vec{w}, \vec{v})$, $d(\vec{w}, \vec{u}, \vec{v})$ are continuous bilinear forms in $\vec{w}$ and $\vec{v}$, corresponding to the operators $\int_\Omega \sum_{i,j=1}^{n} \left(u_i \partial^i \vec{w} + \sum_{i,j=1}^{n} w_i \partial^i \vec{u} \right) dx$ in (2.515). Thus, the $d(\vec{u}, \vec{w}, \vec{v})$ and $d(\vec{w}, \vec{u}, \vec{v}) \in \mathbb{R}$ define elements in $\mathcal{V}_b'$:

$$d(\vec{u}, \vec{w}, \cdot), d(\vec{w}, \vec{u}, \cdot) \in \mathcal{V}_b' \text{ for variable } \vec{w} \in \mathcal{V}_b \text{ and for fixed } \vec{u} \in \mathcal{V}_b.$$

Hence, by Section 2.3 they induce linear continuous operators, see (2.514),

$$D_1, D_2 \in \mathcal{L}(\mathcal{V}_b, \mathcal{V}_b') \text{ as } D_1\vec{w} := d(\vec{u}, \vec{w}, \cdot), D_2\vec{w} := d(\vec{w}, \vec{u}, \cdot), \ \vec{u} \text{ fixed} \tag{2.518}$$

$$\text{with } \langle D_1\vec{w}, \vec{v} \rangle_{\mathcal{V}_b' \times \mathcal{V}_b} = d(\vec{u}, \vec{w}, \vec{v}), \ \langle D_2\vec{w}, \vec{v} \rangle_{\mathcal{V}_b' \times \mathcal{V}_b} = d(\vec{w}, \vec{u}, \vec{v}) \forall \vec{v} \in \mathcal{V}_b. \tag{2.519}$$

The embedding $I : H_0^1(\Omega, \mathbb{R}^n) \to L^2(\Omega, \mathbb{R}^n)$ is continuous and compact, see Theorem 1.26. So (2.519) shows that

$$D_1\vec{w} = D_1 I\vec{w}, D_2\vec{w} = D_2 I\vec{w} \ \forall \ \ \vec{w} \in H_0^1(\Omega, \mathbb{R}^n).$$

Hence, as a product of a compact and a continuous operator, $D_1 = D_1 I$ and $D_2 = D_2 I$ are compact operators.

Similarly to the transformation from (2.500) to (2.503) we use here, with the factor ν and the same A, B as in (2.502),

$$C_\nu := \begin{pmatrix} \nu A & B \\ B^d & 0 \end{pmatrix}, \quad D := \begin{pmatrix} D_1 + D_2 & 0 \\ 0 & 0 \end{pmatrix}. \tag{2.520}$$

Consequently, $D_1 + D_2$ is a compact perturbation of νA and D of C_ν. As indicated above, this observation is only appropriate if ν is not too small, we call it *moderate* ν. Then $(C_\nu)^{-1}(D_1 + D_2)$ is compact and allows us to apply Theorem 2.20. This shows that for the linearized Navier-Stokes operator Corollary 2.44, implying the Fredholm alternative, is valid. For $0 < \nu \approx 0$ the convection terms $D_1 + D_2$ dominate the behavior of the equation.

We apply $C_\nu + D$ to the above $\vec{x} = (\vec{u}, p)^\mathsf{T}$, multiply by $\vec{y} = (\vec{v}, q)^\mathsf{T}$ and integrate:

$$\text{so} \quad (C_\nu + D)\vec{x} = \begin{pmatrix} \nu A\vec{u} + Bp + D_1\vec{u} + D_2\vec{u} \\ B^d\vec{u} \end{pmatrix} \text{ yields}$$

$$\begin{aligned} \langle (C_\nu + D)\vec{x}, \vec{y}\rangle_{\mathcal{X}'\times\mathcal{X}} &= \nu\langle A\vec{u}, \vec{v}\rangle_{\mathcal{V}_b'\times\mathcal{V}_b} + \langle Bp, \vec{v}\rangle_{\mathcal{V}_b'\times\mathcal{V}_b} \\ &\quad + \langle D_1\vec{u} + D_2\vec{u}, \vec{v}\rangle_{\mathcal{V}_b'\times\mathcal{V}_b} + \langle B^d\vec{u}, q\rangle_{\mathcal{W}'\times\mathcal{W}}. \end{aligned} \tag{2.521}$$

The exact solution $\vec{x}_0 = (\vec{u}_0, p_0)$ has to be determined from

$$\begin{aligned} &\text{for given } \vec{f} \in \mathcal{X}' \text{ determine } \vec{x}_0 = (\vec{u}_0, p_0) \text{ such that} \\ &\langle (C_\nu + D)\vec{x}_0, \vec{y}\rangle_{\mathcal{X}'\times\mathcal{X}} = \langle \vec{f}, \vec{y}\rangle_{\mathcal{X}'\times\mathcal{X}} \quad \forall \ \vec{y} \in \mathcal{X} \\ &\Leftrightarrow \ (C_\nu + D)\vec{x}_0 = \vec{f} \in \mathcal{X}' \\ &\Leftrightarrow \ C_\nu(I + (C_\nu)^{-1}D)\vec{x}_0 = \vec{f} \in \mathcal{X}' \\ &\Leftrightarrow \ (I + (C_\nu)^{-1}D)\vec{x}_0 = (C_\nu)^{-1}f \in \mathcal{X}. \end{aligned} \tag{2.522}$$

Applying Theorems 2.20 and 2.131, we can guarantee the Fredholm alternative for $C_\nu + D$.

Theorem 2.132. Unique existence, Fredholm alternative:

1. *For $\nu > 0$ not too close to 0 the linearized Navier-Stokes operator $C_\nu + D$ in (2.512) is a compact perturbation of the ν-Stokes operator C_ν in (2.513). This implies the spectrum of results of Corollary 2.44.*
2. *These statements remain correct if the Navier-Stokes operator is replaced by a generalized nonlinear saddle point problem, according to (2.521), where the nonlinearity satisfies (2.516) with the special $\mathcal{V}_b = H_0^1(\Omega, \mathbb{R}^n), \mathcal{W}^* := L^2(\Omega)$ replaced by general compactly embedded $\mathcal{V}_b \hookrightarrow \mathcal{W}^*$.*

So whenever -1 is not an *eigenvalue* of the compact operator $C_\nu^{-1}D$, or equivalently $\mathcal{N}(C_\nu + D) = \{0\}$, the linearized Navier-Stokes operator $C_\nu + D$ is boundedly invertible, and thus excludes bifurcation phenomena. On the other hand, this allows us to use continuation techniques for determining solutions of the nonlinear Navier-Stokes equations (2.495) and localizing bifurcation.

Finally we inherit the regularity from the Stokes to the Navier-Stokes operator by combining Theorem 2.56 with the second part of Theorem 2.131 to yield:

Theorem 2.133. Smooth data imply smooth solutions: *If, in the above Navier-Stokes problem, Ω is smooth enough and $f^1 \in H^k(\Omega), f^2 \in H^{k+1}(\Omega) \cap L^2_*(\Omega), k \in \mathbb{N}_0$, we get for the solution $\vec{u}_0 \in H^{k+2}(\Omega) \cap H^1_0(\Omega), p_0 \in H^{k+1}(\Omega) \cap L^2_*(\Omega)$. There exists a constant $C > 0$, only depending upon Ω, such that*

$$\|\vec{u}_0\|_{H^{k+2}(\Omega)} + \|p_0\|_{H^{k+1}(\Omega)} \leq C_\Omega(\|f^1\|_{H^k(\Omega)} + \|f^2\|_{H^{k+1}(\Omega)}).$$

Part II

Numerical Methods

3

A general discretization theory

3.1 Introduction

In this chapter we combine general Petrov–Galerkin methods, the standard approach for many elliptic discretizations, with a modification of one of the classical general discretization methods, cf. Stetter [596]. This is a new way of proving stability for the general class of elliptic equations discussed in Chapter 2, cf. Böhmer and Sassmannshausen [128] and Böhmer [116]. It probably applies to all up-to-date discretization methods. As examples we demonstrate this for six different types of discretization methods: we choose the still very alive grandfather, the difference methods, the most popular, the FEM including variational crimes, and their adaptive version, the DCGMs, then two methods still strongly in the process of development, the wavelets and the mesh–free methods, and finally, the spectral methods. The two last methods are particularly useful for bifurcation problems with their underlying symmetries. So they are described by Böhmer [120]. The proof of the convergence of mesh–free methods applied to general nonlinear problems was an open problem for two decades and is solved for the first time in [120]. Proofs of different versions of linear problems are given by Schaback and his group, e.g. [327, 328, 401, 568, 569]. Specific properties, e.g. the adaptivity for wavelets, have to be studied additionally.

We combine compact perturbations of coercive linear, hence boundedly invertible operators with additional approximation properties, see Böhmer [116] and Böhmer Sassmannshausen [128]. This technique applies to Navier–Stokes problems and to the bordered systems, required for numerical bifurcation. Or we directly apply this theory to nonlinear via monotone operators.

This mix of different concepts is strongly influenced by many earlier papers and books. Discrete convergence has been studied by Stummel [607–610], Reinhardt [547] and Vainikko [644]. Stetter [596] studies admissible discretization methods; the (inner and outer) admissible approximation schemes due to Petryshyn [528–530, 532] are nicely collected, presented and extended by Zeidler [677]. Keller [441] has combined, for special problems, linearization with stability and consistency for proving convergence. Anselone and Ansorge [28–31] strongly elaborate the functional analytic side of general discretizations, based in the concept of compact perturbations. In Ansorge's *iterated discretization* [32] additional iteration steps in improved projection methods for more general discretization procedures, not only projection methods, and more general classes of problems, sometimes even yield superconvergence.

Brezzi, Rappaz and Raviart have applied a related version to finite dimensional approximations of nonlinear problems, including limit and simple bifurcation points, however excluding many operator equations and, necessarily, missing the new discretization methods and higher singularities [147–149]. Rappaz and Raugel extended this approach to special cases of finite dimensional approximation of bifurcation problems at a multiple eigenvalue [542–544], and to symmetric problems. Crouzeix and Rappaz gave a short survey book on numerical approximation in bifurcation theory [229]. Griewank and Reddien [366–370] prove convergence for general bifurcation scenarios under the conditions of [147–149]. Jepson and Spence [418] study bifurcation for perturbed operators, not covering discretization methods. Calos and Rappaz presented numerical approximations in nonlinear and bifurcation problems in [156]. Here, in contrast to the earlier [147–149, 542, 543], two projectors for the analysis of FEMs for nonlinear problems are appropriately introduced, but are not employed for the bifurcation analysis. Finally, Cliffe, Spence and Tavener [180] and Calos and Rappaz [156] give a well–written introduction to the numerical analysis of bifurcation problems with application to fluid mechanics.

All these papers, starting with [147], do not prove convergence for higher bifurcations and sometimes of more complicated problems such as porous media nor the new methods. Often the Navier–Stokes equation is studied. Extensions to higher bifurcation scenarios and the previously omitted other methods are discussed by Böhmer *et al.* [122, 124, 128]. Approaches to these more complicated cases use the concepts of consistent differentiability and modified or bordered stability, introduced in [111], cf. [8, 41, 42, 113, 114, 114, 115, 127–129].

We have discussed many different elliptic equations and systems. We consider linear and nonlinear, often parameter–dependent, problems of the form

$$G : \mathcal{D}(G) \subset \mathcal{U} \to \mathcal{V}',\ G(u_0) = 0,\ \mathcal{U}, \mathcal{V}\ \text{ Banach spaces, with duals } \mathcal{U}', \mathcal{V}'. \quad (3.1)$$

For some problems solutions with special features can be obtained by analytical methods. For general equations or interesting parameters in the problem this is usually impossible. So we introduce discretization methods. Let

$$G^h : \mathcal{D}(G^h) \subset \mathcal{U}^h \to \mathcal{V}'^h,\ G^h(u^h) = 0,\ \mathcal{U}^h, \mathcal{V}^h \text{ discrete Banach spaces,} \quad (3.2)$$

represent a corresponding discrete problem. The classical convergence results for (nonlinear) problems (3.1), (3.2) are obtained by combining the well-known concepts of stability and consistency. Consistency requires small errors in

$$0 = G(u_0) \approx G^h\left(u_0^h\right) + \text{errors, with } u_0^h \text{ an "approximation" for } u_0.$$

The following Petrov–Galerkin and difference methods have to be included in the general discretization methods. ***Stability and consistency,*** cf. Subsections 3.3 and 3.4, are the basic concepts. They guarantee, with some technical conditions, the existence and uniqueness of the discrete solutions for uniquely existing original solutions and convergence for the discrete compared to the exact solution. Variational consistency for linear operators is more familiar in the FE and Petrov–Galerkin community, but classical consistency is more appropriate in combination with stability and nonlinear problems. It is unavoidable for fully nonlinear problems.

For the many different discretization methods in the following chapters we choose the neutral notation of $\mathcal{U}, \mathcal{V}$ and $\mathcal{U}^h, \mathcal{V}^h$ for the Banach spaces and their discrete approximations. $\mathcal{U}_b, \mathcal{V}_b$ and $\mathcal{U}_b^h, \mathcal{V}_b^h$ is used for contrasting the full space to the subspace of functions satisfying the boundary conditions, e.g. $\mathcal{U} = H^1(\Omega), \mathcal{U}_b = H_0^1(\Omega)$. Often we work with the weak form. The strong form is important for all kinds of *variational crimes*, e.g. in FEMs or DCGMs or difference or mesh–free methods, and for fully nonlinear elliptic problems. Boundary conditions are imposed for unique existence. In contrast to many other books, we maintain this $\mathcal{U}$ versus $\mathcal{U}_b$ form, since nonconforming FE, DCG, and mesh–free methods violate boundary conditions and, e.g. continuity, hence $\mathcal{U}_b^h \not\subset \mathcal{U}_b$. The linear and nonlinear operators and their discretizations are denoted as A, G and A^h, G^h. In the previous chapter we indicated, for most classes of problems, appropriate conditions for higher regularity of the solutions, which will allow better convergence.

We will formulate the following Examples 3.1–3.4 as FEMs in Petrov–Galerkin form or as difference methods in classical form. We are aiming for a definition of a general discretization theory, applicable to all intended analytical problems. It will include these examples and all the discretization methods, considered in both books, and very probably most of the future methods as well. In addition, we will be able to avoid the dichotomy between interior and exterior approximation methods.

3.2 Petrov–Galerkin and general discretization methods

Our (stationary) nonlinear problems are usually formulated in weak, and sometimes in strong form. For unique existence, we have to impose boundary conditions, indicated by $u \in \mathcal{U}_b \subset \mathcal{U}$. So we determine $u_0 \in \mathcal{U}_b$ by testing with $v \in \mathcal{V}_b$:

$$G : \mathcal{U}_b \to \mathcal{V}_b', \quad \text{determine } u_0 \in \mathcal{U}_b \text{ s.t. } G(u_0) = 0 \in \mathcal{V}_b' \iff G(u_0) \perp \mathcal{V}_b$$
$$e.g.\ \mathcal{V}_b = H_0^1(\Omega) \iff < G(u_0), v >_{\mathcal{V}' \times \mathcal{V}} = 0\ \forall\ v \in \mathcal{V}_b. \qquad (3.3)$$

In Remark 2.3, we have summarized the fact that for the problems in weak form in Chapter 2, $\mathcal{V}' = \mathcal{V}_b'$. However, here we have to use both duals for the discrete $\mathcal{V}_b'^h \neq \mathcal{V}'^h$, and sometimes use $\mathcal{V}', \mathcal{V}_b'$ as well, thus maintaining $\mathcal{U}$ versus $\mathcal{U}_b$.

Reducing the technical difficulties we start with a *bounded linear* operator, e.g. the derivative $A := G_u(u_1)$, in either weak or strong form. Nonlinear operators will be discussed below. For a second order linear A and f,

$$A \in \mathcal{L}(\mathcal{U}, \mathcal{V}'), \quad \text{determine } u_0 \in \mathcal{U}_b \text{ s.t. } Au_0 - f = 0 \in \mathcal{V}_b' \qquad (3.4)$$
$$\Leftrightarrow Au_0 - f \perp \mathcal{V}_b \iff a(u_0, v) = \langle f, v \rangle_{\mathcal{V}' \times \mathcal{V}} = \int_\Omega \sum_{j=0}^n f^j \partial^j v dx\ \forall\ \ v \in \mathcal{V}_b,$$

with the standard $a(u, v) = \langle Au, v \rangle_{\mathcal{V}' \times \mathcal{V}}\ \forall\ v \in \mathcal{V},\ \forall u \in \mathcal{U}$. The boundary conditions $\mathcal{U}_b$ and $\mathcal{V}_b$, necessary for *unique* solvability of (3.3) and (3.4) are realized by continuous trace operators, e.g. $B(u) = 0$.

$$u \in \mathcal{U}_b := \{u \in \mathcal{U} : B(u) = 0\} \subset \mathcal{U}, \quad v \in \mathcal{V}_b := \{v \in \mathcal{V} : B_2(v) = 0\} \subset \mathcal{V},$$

with closed subspaces $\mathcal{U}_b, \mathcal{V}_b$, and usually $\mathcal{V}_b' = \mathcal{V}'$, *often* $B(u) = B_2(u)$.

So, $A \in L(\mathcal{U}, \mathcal{V}')$ is a bounded linear operator, defined, but usually only invertible as $A \in L(\mathcal{U}_b, \mathcal{V}_b')$.

Before we give some general definitions and properties for the intended discretization methods, we formulate as motivation the simplest cases of the *finite element and finite difference method for the Laplacian* with Dirichlet boundary conditions. We choose, for the weak bilinear form, $\mathcal{U}_b, \mathcal{V}_b = H_0^1(\Omega)$, $\mathcal{U} = H^1(\Omega)$, and for the strong form, $\mathcal{U}_b = H^2(\Omega) \cap H_0^1(\Omega), \mathcal{V} = L^2(\Omega)$, similarly for natural boundary conditions.

Example 3.1. *Finite element method for the Laplacian*
Its weak and strong form, A and A_s, respectively, with the same boundary conditions for $\mathcal{U}_b = \mathcal{V}_b$, are

$$u_0 \in \mathcal{U}_b = \mathcal{V}_b = H_0^1(\Omega) : \quad \langle Au_0, v\rangle_{\mathcal{V}' \times \mathcal{V}} = a(u_0, v) = \int_\Omega (\nabla u_0, \nabla v)_n dx \tag{3.5}$$

$$= \langle f, v\rangle_{\mathcal{V}' \times \mathcal{V}} \forall v \in \mathcal{V}_b, \text{ and for } f \in L^2(\Omega), u_0 \in H^2(\Omega) \cap \mathcal{U}_b : A_s u_0 = -\Delta u_0 = f.$$

Our goal is the reduction of the original problems, defined and tested on infinite dimensional spaces $\mathcal{U}_b, \mathcal{V}_b$, to related discrete problems, defined and tested on spaces $\mathcal{U}_b^h, \mathcal{V}_b^h$ of the same finite dimension. We choose two grids Ω^h and Ω_0^h in $\Omega = [0,1]^2$:

$$\Omega_0^h := \{P_{i,j} := (x_i, y_j),\ x_i := i/N,\ y_j := j/N, 0 \le i,j \le N\} \quad \text{and} \tag{3.6}$$

$$\Omega^h := \{P_{i,j},\ 1 \le i,j \le N-1\}, \partial\Omega^h := \Omega_0^h \setminus \Omega^h, \ \textit{cf. Figure 3.1.}$$

The following approximating spaces, $\mathcal{U}_b^h$, are defined as the set of piecewise linear hat functions, $\phi_{i,j}$, with 1 and 0 in the grid points, hence

$$\mathcal{U}_b^h := \Big\{u^h := \sum_{i,j=1}^{N-1} C_{i,j}\phi_{i,j} : \quad C_{i,j} \in \mathbb{R}, \phi_{i,j}(P_{k,l}) = 1 \quad \text{for } (i,j) = (k,l), \tag{3.7}$$

$$\text{or } P_{i,j} \in \Omega^h, \quad \text{and } = 0 \text{ else, so } \forall 0 \le k,l \le N, (i,j) \ne (k,l)\Big\} \subset H_0^1(\Omega).$$

Since $\mathcal{U}_b^h \subset H_0^1(\Omega)$, we call the u^h *conforming FEs*. Then the weak form in (3.5) is reduced to determining the $C_{i,j}$ for $u_0^h \in \mathcal{U}_b^h$ in (3.7), such that

$$\langle Au_0^h, v^h\rangle_{\mathcal{V}' \times \mathcal{V}} = a\left(u_0^h, v^h\right) = \int_\Omega \left(\nabla u_0^h, \nabla v^h\right)_n dx = \langle f, v^h\rangle_{\mathcal{V}' \times \mathcal{V}} \forall v^h \in \mathcal{V}_b^h,$$

$$\iff a\left(u_0^h, \phi_{k,l}\right) = \int_\Omega \left(\nabla u_0^h, \nabla \phi_{k,l}\right)_n dx = \langle f, \phi_{k,l}\rangle_{\mathcal{V}' \times \mathcal{V}} \ \forall 1 \le k,l \le N-1. \tag{3.8}$$

We expect this u_0^h to be a good approximation for u_0. ■

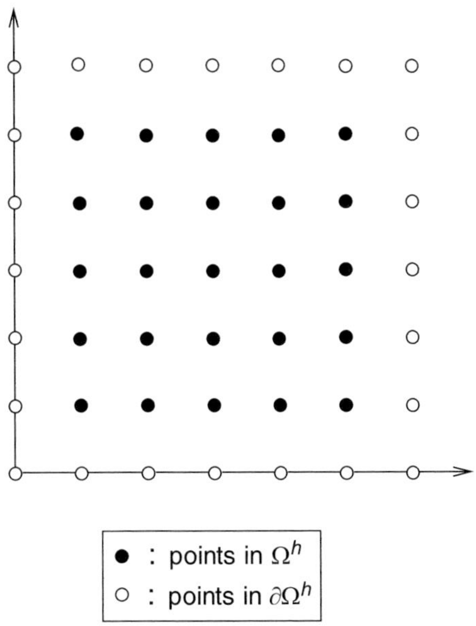

Figure 3.1 Two-dimensional grid for Dirichlet boundary conditions.

Example 3.2.

Nonconforming finite element methods are characterized by $\mathcal{U}_b^h \not\subset H_0^1(\Omega)$. In modified forms this ansatz is used in DCGMs as well. The basic case are the *discontinuous Crouzeix–Raviart elements,* cf. Figure 4.10. They are defined as follows: For all $P_{i,j} \in \Omega_0^h$ let

$$\Omega_m^h := \{Q_{k,l} \text{ are the midpoints between all neighboring } P_{i,j} \in \Omega_0^h\} \tag{3.9}$$

$$\mathcal{U}_b^h := \text{span}\{\psi_{i,j} : \ \psi_{i,j}(Q_{k,l}) = 1 \text{ for } (i,j) = (k,l), \text{ and } \ Q_{i,j} \in \Omega, \\ \text{and } \ = 0 \text{ else and, in particular, } \forall Q_{k,l} \in \partial\Omega\} \not\subset H_0^1(\Omega).$$

Then obviously the $a\left(u_0^h, \phi_{k,l}\right)$ in (3.8) are no longer defined. So we define a triangulation, $\mathcal{T}^h$, of Ω: through all points $P_{i,j} \in \Omega_0^h$ the parallels to the x- and y-axes and to the diagonal $x = y$ are drawn. Then let

$$\mathcal{T}^h := \{\text{all these constructed open rectangular triangles } T\}.$$

We introduce broken Sobolev spaces of functions and functionals on $\mathcal{T}^h$ by

$$\mathcal{U}_{\mathcal{T}^h} : = \mathcal{V}_{\mathcal{T}^h} := H^1(\mathcal{T}^h) := \{u : \Omega \to \mathbb{R} : \forall T \in \mathcal{T}^h : u|_T \in H^1(T)\} \supset \mathcal{U}, \tag{3.10}$$

$$\mathcal{U}'_{\mathcal{T}^h} : = \mathcal{V}'_{\mathcal{T}^h} := H^{-1}(\mathcal{T}^h) := \{f_h \in \mathcal{L}(\mathcal{V}_{\mathcal{T}^h}, \mathbb{R}) : \forall T \in \mathcal{T}^h : f_h|_T \in H^{-1}(T)\} \subset \mathcal{U}'.$$

Then we generalize $A, a(u^h, v^h)$, and $\int_\Omega f v^h dx$ in (3.8) by defining

$$A_h : \mathcal{U}_{\mathcal{T}^h} \to \mathcal{U}'_{\mathcal{T}^h}, \quad a_h(\cdot,\cdot) : \mathcal{U}_{\mathcal{T}^h} \times \mathcal{U}_{\mathcal{T}^h} \to \mathbb{R}, f_h(\cdot) : \mathcal{U}_{\mathcal{T}^h} \to \mathbb{R}, \tag{3.11}$$

$$A = A_h|_{\mathcal{U}}, \quad a(\cdot,\cdot) = a_h(\cdot,\cdot)|_{\mathcal{U}\times\mathcal{U}}, \quad a^h(\cdot,\cdot) := a_h(\cdot,\cdot)|_{\mathcal{U}^h\times\mathcal{U}^h},$$

$$\langle A_h u^h, v^h \rangle_{\mathcal{U}'_{\mathcal{T}^h}\times\mathcal{U}_{\mathcal{T}^h}} := a_h(u^h, v^h) := \sum_{T\in\mathcal{T}^h} \int_T (\nabla u^h, \nabla v^h)_n dx,$$

$$f_h \in \mathcal{U}'_{\mathcal{T}^h} : \langle f_h, v^h \rangle := \langle f_h, v^h \rangle_{\mathcal{U}'_{\mathcal{T}^h}\times\mathcal{U}_{\mathcal{T}^h}} := \sum_{T\in\mathcal{T}^h} \int_T \sum_{j=0}^n f_h^j \partial^j v dx$$

$$\text{with } f(\cdot) = f_h(\cdot)|_{\mathcal{U}}, \quad f^h(\cdot) := f_h(\cdot)|_{\mathcal{U}^h}.$$

Finally, determine $u_0^h \in \mathcal{U}_b^h = \mathcal{V}_b^h = \text{ span } \{\psi_{k,l}, Q_{k,l} \in \Omega\} \subset \mathcal{U}_{\mathcal{T}^h}$, cf. (3.9), from

$$a^h\left(u_0^h, \psi_{k,l}\right) = \sum_{T\in\mathcal{T}^h} \int_T \left(\nabla u_0^h, \nabla \psi_{k,l}\right)_n dx = \langle f^h, \psi_{k,l} \rangle_{\mathcal{U}'_{\mathcal{T}^h}\times\mathcal{U}_{\mathcal{T}^h}} \; \forall \psi_{k,l} \in \mathcal{U}_b^h. \tag{3.12}$$

Note that for conforming FEMs all these new forms coincide, hence, for

$$\textit{conforming FEMs } A = A_h, \; a(\cdot,\cdot) = a_h(\cdot,\cdot) = a^h(\cdot,\cdot), \; f(\cdot) = f_h(\cdot) = f^h(\cdot). \tag{3.13}$$

Again this u_0^h in (3.12) should be a good approximation for u_0. ■

Example 3.3. *Finite difference method for the Laplacian in its strong form,* $A_s u = Au = \Delta u$, on $\Omega = [0,1]^2$

We consider the strong form as $A : \mathcal{U}_b = C^2(\Omega) \cap H_0^1 \to \mathcal{V}' = C(\Omega)$. This unusual notation $\mathcal{V}' = C(\Omega)$ is a consequence of our aim, keeping the general discretization approach as close as possible to the usual Petrov–Galerkin methods, cf. below.[25] We evaluate Au in (3.5) in $P_{i,j} \in \Omega^h$, cf. (3.6), and approximate $-(u_{xx} + u_{yy})(P_{i,j})$ by the usual five-point star. This yields the difference approximation A^h, defining an approximate solution $u_0^h \in \mathcal{U}_b^h$ with $u_{i,j}^h := u_0^h(P_{i,j})$ as indicated in Table 3.1. Let

$$\mathcal{U}^h := \{u^h : \Omega^h \to \mathbb{R}\} \supset \mathcal{U}_b^h := \{u^h \in \mathcal{U}^h : u^h|_{\partial\Omega^h} = 0\}, \mathcal{V}^h = \mathcal{V}'^h := \{u^h : \Omega_0^h \to \mathbb{R}\}$$

$$\textit{and } \|u^h\|_{\mathcal{U}_b^h} := \max_{P_{i,j}\in\Omega_0^h} \{|u^h(P_{i,j})|\}, \quad \|f^h\|_{\mathcal{V}_b'^h} := \max_{P_{i,j}\in\Omega_0^h} \{|(f^h)(P_{i,j})|\}, \tag{3.14}$$

$$u_0^h \in \mathcal{U}_b^h : \; \left(A^h u_0^h\right)(P_{i,j}) := \frac{1}{N^2}\left(4u_{i,j}^h - u_{i-1,j}^h - u_{i+1,j}^h - u_{i,j-1}^h - u_{i,j+1}^h\right)$$

$$= f(P_{i,j}), 1 \le i,j \le N-1, h := 1/N, \quad u^h(P_{i,j}) = 0 \text{ for } i = 0, N \text{ or } j = 0, N.$$

[25] Then $\mathcal{V}$ would be $\{f : \Omega \to \mathbb{R}, f$ of bounded variation$\}$. At the same time, this example provides the possibility of indicating Stetter's [596] approach in the original form: He avoids testing by $v \in \mathcal{V}_b$ and studies directly $A : \mathcal{U}_b \to \mathcal{V} = C(\Omega)$, with the norms in (3.14).

Table 3.1: Stencil for the five-point star around $P_{i,j} \in \Omega^h \subset \mathbb{R}^2$ for $-h^2\Delta_h$.

0	$-u^h_{i,j+1}$	0		0	-1	0
$-u^h_{i-1,j}$	$4u^h_{i,j}$	$-u^h_{i+1,j}$	*abbreviated as*	-1	4	-1
0	$-u^h_{i,j-1}$	0		0	-1	0

Note that (3.14) directly defines an $A^h : \mathcal{U}_b^h \to \mathcal{V}'^h$ without any detour as for nonconforming FEMs. It will turn out to be obtainable as the strong form of a weak analogue by a less familiar testing process in Chapter 8. To this end, we will discuss discrete Sobolev spaces of grid functions. This will yield the more or less standard approach.

A similar construction as for FEMs is possible: In the first step, we reduce Ω to $\Omega_r := [h, 1-h]^2$. Next, we approximate the previous A for functions $u \in \mathcal{U}_b = C^2(\Omega) \cap H_0^1$ by applying the five-point star in (3.14), thus defining

$$(\tilde{A}_h u)(P) := \big(\textit{five point star for } u\big)(P)\forall P \in \Omega_r \Rightarrow A^h u(P_{i,j}) = (\tilde{A}_h u)(P_{i,j}), \quad (3.15)$$

again well defined for $u^h \in \mathcal{U}_b^h$. This will be modified in Example 3.8 and in Chapter 8. ■

Example 3.4. *Summary of these discretizations*

For all three cases the original $A : \mathcal{U}_b \to \mathcal{V}_b'$, tested by elements of the space $\mathcal{V}_b$, is transformed into $A^h : \mathcal{U}_b^h \to \mathcal{V}_b'^h$. All equations are tested with elements of $\mathcal{V}_b^h$ and spaces $\mathcal{U}_b^h, \mathcal{V}_b^h$ with the same finite dimension. For the above difference method this is realized by evaluation in $P_{i,j}$; a more natural setting is presented in Chapter 8. There are still essential differences between Examples 3.1–3.3.

As in (2.20), (2.22), the $a(u,v)$ in Example 3.1, (3.5) is $\mathcal{U}_b$-coercive, and, by $\mathcal{U}_b^h \subset \mathcal{U}_b$, the $a(u,v)$ in (3.8) is $\mathcal{U}_b^h$-coercive as well. This will imply the stability of the discrete A^h with respect to the $\|\cdot\|_{H^1(\Omega)}$. For conforming FEMs the *variational consistency error* will be shown to vanish, and the *classical consistency error* tends to 0. This will imply the convergence in the sense of $\left\|u_0 - u_0^h\right\|_{H^1(\Omega)} \to 0$ for $h \to 0$.

For the nonconforming FEMs in Example 3.2, and with $H_0^1(\mathcal{T}^h) := \{u \in H^1(\mathcal{T}^h) : u(Q_{i,j}) = 0 \forall Q_{i,j} \in \partial\Omega\}$ we show similarly as above that the $a(u,v)$ in (3.11) is now $\mathcal{U}_b^h$-coercive. However the proof for *variational or classical consistency* is much more complicated, cf. Subsections 5.5.2, 5.5.6, and 5.5.7. Still, we get the same convergence, now with respect to the discrete $\|\cdot\|_{H^1(\mathcal{T}^h)}$.

For the difference methods in Example 3.3, (3.14), the matrix for A^h is a *weakly diagonal dominant matrix*, a special case of an *M-matrix.* This implies the stability, cf. Definition 3.19, of A^h with respect to the discrete $\|\cdot\|_{L^\infty(\Omega^h)}$. The *classical consistency* is an immediate consequence of the local errors of the difference approximations, cf. Lemma 8.2. This allows convergence results with respect to the max norm, so

$\max_{P_{i,j}\in\Omega^h}\{|u_0(P_{i,j}) - u_0^h(P_{i,j})|\} \to 0$ for $h \to 0$, cf. Definition 3.18. In contrast to Examples 3.1 and 3.2, with the optimal discrete $\|\cdot\|_{H^1(\mathcal{T}^h)}$ norm, we use here, as in [596], the discrete $\|\cdot\|_{L^\infty(\mathcal{T}^h)}$ norm. However, we will prove in Chapter 8 convergence for difference methods even in a discrete H^1 Sobolev norm, $\|\cdot\|_{H^1(\Omega^h)}$, obtaining a much better $\|u_0 - u_0^h\|_{H_h^1(\Omega^h)} \to 0$. This is possible by transforming the strong form (3.14) into a corresponding weak form, similar to (3.8), combined with general discretization methods. ■

Examples 3.1–3.4 motivate the definition of a general discretization theory, applicable to all intended analytical problems and their discretization methods. The previous forms of FEMs are special cases of the so-called Petrov–Galerkin methods.

Definition 3.5. Approximating spaces: *Let $\mathcal{U}$ be a Banach space, and $\mathcal{U}_b \subset \mathcal{U}$, its closed subspace. Let the sequence $\{\mathcal{U}^h\}_{h\in H}$ assign to h, $0 < h \in H \subset \mathbb{R}, \inf_{h\in H} h = 0$ the*[26] *spaces $\mathcal{U}^h$. Often we use the short notation of $\mathcal{U}^h$. Then we call $\mathcal{U}^h$* (a sequence of) approximating spaces for *$\mathcal{U}$ if for well defined $||u||_{\mathcal{U}^h}$, e.g. for conforming and nonconforming FEMs, and for DCGMs,*

$$dist(u,\mathcal{U}^h) := \inf_{u^h\in\mathcal{U}^h} ||u - u^h||_{\mathcal{U}^h} \to 0 \quad \forall u \in \mathcal{U} \text{ for } h \to 0. \tag{3.16}$$

This is sometimes only required for $u \in \mathcal{U}_b$ and hence $u^h \in \mathcal{U}_b^h$. The (sequence of) approximating spaces $\mathcal{U}_b^h$ for $\mathcal{U}_b$, with $_b$ indicating the boundary conditions, are called conforming spaces, *if $\mathcal{U}_b^h \subset \mathcal{U}_b$ for all $h \in H$. Otherwise they are called* nonconforming spaces. *In the latter case, either the $\mathcal{U}_b^h$ violate the boundary conditions, or $\mathcal{U}^h \not\subset \mathcal{U}$.*

The modifications for difference methods are discussed in Proposition 8.15 and Remark 8.16.

For difference methods $||u||_{\mathcal{U}^h}$ is not defined, so $\inf_{u^h\in\mathcal{U}^h} ||u - u^h||_{\mathcal{U}^h} \to 0$ does not make sense. Nevertheless, the grid functions there do approximate Sobolev spaces in an appropriate sense as well, cf. Subsection 8.5.2. This problem motivates introducing the concept of *external in contrast to internal approximation schemes,* cf. Petryshyn [528–530, 532], Temam [621] and Zeidler [677]. These concepts only play a minor role for the approach chosen here.

Definition 3.6. Admissible, conforming, nonconforming Petrov–Galerkin approximating spaces and general Petrov–Galerkin methods*:*

1. A pair of admissible approximating spaces $\mathcal{U}_b^h \subset \mathcal{U}^h$ and $\mathcal{V}_b^h \subset \mathcal{V}^h$ *for $\mathcal{U}_b$* and *$\mathcal{V}_b$, respectively, has to have the same finite dimensions*
$$dim\, \mathcal{U}_b^h = dim\, \mathcal{V}_b^h < \infty \;\forall\; h. \tag{3.17}$$
2. *For admissible approximating spaces, the terms* conforming *and* Galerkin *and* Petrov–Galerkin approximating spaces *are used if $\mathcal{U}_b^h \subset \mathcal{U}_b$, $\mathcal{V}_b^h \subset \mathcal{V}_b$ and $\mathcal{U}_b^h = \mathcal{V}_b^h$ and $\mathcal{U}_b^h \neq \mathcal{V}_b^h$, respectively. We denote $\mathcal{U}_b^h \not\subset \mathcal{U}_b$, $\mathcal{V}_b^h \not\subset \mathcal{V}_b$ or $\mathcal{U}_b^h \not\subset \mathcal{U}$, $\mathcal{V}_b^h \not\subset \mathcal{V}$ as* nonconforming (Petrov–Galerkin) approximating spaces. *Sometimes, all these*

[26] We do not want to over-formalize the notation and have chosen $h \in H$, instead of the possible $h^n = (h_1, \cdots, h_n) \in H^n$.

cases are denoted as generalized Petrov–Galerkin approximating spaces. *For a discretization method, see Definition 3.12, we use the same name as for its pair of approximating spaces.*

3. *As in Examples 3.1, 3.2, let the weak form of a linear, or more general quasilinear problem be given as in (3.5). Discretize it by replacing the* $\mathcal{U}_b, \mathcal{V}_b$ *in (3.5) by its approximating spaces* $\mathcal{U}_b^h, \mathcal{V}_b^h$ *as in (3.8) or (3.12). Then we call this transformation a* Petrov–Galerkin (discretization) method.
4. *If, in addition to (3.16) for* $\mathcal{U}_b$*, the* $\mathcal{V}^h$ *satisfy (3.16)* $\forall\ \ v \in \mathcal{V}_b$ and $\forall\ \ v'' \in \mathcal{V}''$, *the bi–dual space of* $\mathcal{V}$*, then we choose* $\mathcal{V}^h = \mathcal{V}''^h$ *and call* $\mathcal{U}^h,\ \mathcal{V}^h$ a bi–dual pair of conforming or nonconforming (Petrov–Galerkin) approximating spaces.

Remark 3.7. *We will need the bi–duality only in Section 3.6. There we prove that for a bi–dual pair of approximating spaces, defining a stable* A^h*, this stability and technical conditions imply the invertibility of* A*. This bi–duality condition is essentially satisfied in separable reflexive spaces.*[27] *The most important reflexive examples are Sobolev spaces,* $W_p^m(\Omega)$*, with* $1 < p < \infty$*. In these separable spaces the bi–duality is correct for all discretization methods studied here.*

Example 3.8. *Structure of generalized Petrov–Galerkin methods*

1. The form of Petrov–Galerkin methods in Examples 3.1, 3.2 are not directly applicable to finite difference methods, e.g. in Example 3.3. However, we observe that the arguments and images of the original A and their discrete counterparts A^h are related by two types of projectors: $P^h : \mathcal{U}_b \to \mathcal{U}_b^h$ map functions onto their FE approximations or restrict them to their grid values on Ω^h. Similarly, we reinterpret the right-hand sides of the conforming and nonconforming FE equations (3.8) and (3.12) as an orthogonal projection of Au or f onto, e.g.

$$Q'^h : \mathcal{V}_b' \to \mathcal{V}_b'^h : \quad \langle Q'^h f - f, v^h\rangle_{\mathcal{V}'^h \times \mathcal{V}^h} = 0 \forall v^h \in \mathcal{V}_b'^h, \ \textit{generally}, \qquad (3.18)$$

$$Q'^h : \mathcal{V}_b' \cup \mathcal{V}_{\mathcal{T}^h}' \to \mathcal{V}_b'^h : \quad \langle Q'^h f - f, v^h\rangle_{\mathcal{V}_{\mathcal{T}^h}' \times \mathcal{V}_{\mathcal{T}^h}} = 0 \forall v^h \in \mathcal{V}_b'^h.$$

2. For the difference methods in (3.14), we use restriction operators, cf. Definition 8.13,

$$P^h : \mathcal{U} \to \mathcal{U}^h : \quad P^h u := \{u(P_{i,j}) : 0 \le i,j \le N, h := 1/N\} \in \mathcal{U}^h, \quad (3.19)$$

$$Q'^h : \mathcal{V}' \to \mathcal{V}_b'^h : \quad Q'^h f := \{f(P_{i,j}) : 1 \le i,j \le N-1\} \in \mathcal{V}_b'^h.$$

3. Then the discrete operators A^h for Examples 3.1–3.3, can be formulated as

$$A^h = Q'^h A_h|_{\mathcal{U}_b^h} \ \textit{cf. (3.8),(3.12), and } A^h = Q'^h \tilde{A}_h|_{\mathcal{U}_b^h} \ \ \textit{for (3.15).} \qquad (3.20)$$

This $\tilde{A}_h$ could, as well, be obtained by quadrature formulas for A.

[27] Certainly, separability is not enough. This is shown by the well-known example of spaces of infinite sequences: $c^0 \subset l^1 \subset l^\infty = (c^0)''$, where a countable basis for c^0 is not a basis for l^∞. However, a statement for $W_p^m(\Omega)$, analogous to the $c^0 \subset l^1 \subset l^\infty = (c^0)''$ situation does not seem to be known.

4. In a similar way, we define the nonlinear discrete operators, G^h, replacing A by G, as

$$G^h = Q'^h G_h|_{\mathcal{U}_b^h} \ \textit{cf. (3.8),(3.12), and } G^h = Q'^h \tilde{G}_h|_{\mathcal{U}_b^h} \ \textit{ for (3.15).} \qquad (3.21)$$

Here $A_h : \mathcal{U}^h \cup \mathcal{U}_{\mathcal{T}^h} \to \mathcal{V}'_{\mathcal{T}^h}$ is either an extension, or $\tilde{A}_h : \mathcal{U}^h \cup \mathcal{U} \to \mathcal{V}^h \cup \mathcal{V}'$ is an approximation for $A : \mathcal{U} \to \mathcal{V}'$. We will denote $\Phi^h A := A^h$ as the discrete operators generated by Petrov–Galerkin methods as general discretization methods. Similarly, $G_h : \mathcal{U}^h \cup \mathcal{U}_{\mathcal{T}^h} \to \mathcal{V}'_{\mathcal{T}^h}$ is an extension, or $\tilde{G}_h : \mathcal{U}^h \cup \mathcal{U} \to \mathcal{V}^h \cup \mathcal{V}'$ is an approximation for $G : \mathcal{U} \to \mathcal{V}'$, or both. ■

The above $Q'^h : \mathcal{V}'_b \to \mathcal{V}'^h_b$ or $\mathcal{V}' \to \mathcal{V}'^h_b$ will be shown to satisfy the necessary conditions for a general discretization theory, cf. Definition 3.12, (3.24), even for nonconforming FEMs. However, for discussing their variational consistency and with $A_h : \mathcal{U}_{\mathcal{T}^h} \to \mathcal{V}'_{\mathcal{T}^h}$ we have extended these Q'^h into $Q'^h : \mathcal{V}'_b \cup \mathcal{V}'_{\mathcal{T}^h} \to \mathcal{V}'^h_b$.

For Petrov–Galerkin methods we sometimes obtain useful properties for the projectors and approximating spaces.

Lemma 3.9. *Let $\mathcal{V} = \mathcal{V}''$ and $Q^h_{\inf} : \mathcal{V} \to \mathcal{V}^h$ be uniquely defined by*

$$\left\| v - Q^h_{\inf} v \right\|_{\mathcal{V}} = \inf_{v^h \in \mathcal{V}^h} \| v - v^h \|_{\mathcal{V}} = \min_{v^h \in \mathcal{V}^h} \| v - v^h \|_{\mathcal{V}}. \qquad (3.22)$$

Then $Q'^h = Q^h_{\inf}$.

Proof. For $\mathcal{V} = \mathcal{V}''$ we have $\|v\| = \sup_{v' \in \mathcal{V}'} \langle v, v' \rangle_{\mathcal{V}^h \times \mathcal{V}'}$. Now, we choose in addition to $\mathcal{V}'^h$, annihilating $\mathcal{V}^h$, an additional nontrivial $v'_v \in \mathcal{V}'$ annihilating $\mathrm{span}\{v, \mathcal{V}^h\}$ and let $\mathcal{V}'_v := \mathrm{span}\{\mathcal{V}'^h, v'_v\}$. Then $\mathcal{V}'_v$ annihilates $\mathcal{V}^h_v := \mathrm{span}\{\mathcal{V}^h, v\}$. This implies, with $\mathcal{V}' = \mathcal{V}'_v \oplus \mathcal{V}'_{\mathrm{comp}}$ where $\langle v - v', v' \rangle = 0 \ \ \forall v' \in \mathcal{V}'_{\mathrm{comp}}$,

$$\begin{aligned} \left\| v - Q^h_{\inf} v \right\|_{\mathcal{V}} &= \min_{v^h \in \mathcal{V}^h} \sup_{v' \in \mathcal{V}'} |\langle v - v^h, v' \rangle| / \|v'\|_{\mathcal{V}'} \\ &= \min_{v^h \in \mathcal{V}^h} \sup_{v' \in \mathcal{V}'_v} |\langle v - v^h, v' \rangle| / \|v'\|_{\mathcal{V}'} \\ &= \min_{v^h \in \mathcal{V}^h} \sup_{v' \in \mathcal{V}'^h} |\langle v - v^h, v' \rangle| / \|v'\|_{\mathcal{V}'} = \|v - Q'^h v\|_{\mathcal{V}}, \end{aligned}$$

hence, the claim. □

Lemma 3.10. *Let $\mathcal{U}$ be a* separable Banach space, *e.g. our Sobolev spaces. Let $\mathcal{U}^h$ be approximating spaces for a dense subspace $\mathcal{U}_d$ of $\mathcal{U}$ with respect to the norm $\| \cdot \|_{\mathcal{U}}$. Then $\mathcal{U}^h$ are approximating spaces for $\mathcal{U}$ as well.*

Proof. Let $u \in \mathcal{U}$ be given. Then $\forall \epsilon > 0$ we can find $u_d \in \mathcal{U}_d$ with $\|u - u_d\|_{\mathcal{U}} < \epsilon/2$. For this u_d there exists a $u^h \in \mathcal{U}^h$ such that $\|u^h - u_d\|_{\mathcal{U}} < \epsilon/2$. □

We can extend this to the dual spaces and define the corresponding projectors: usually, the dual space $\mathcal{V}'$ of $\mathcal{V}$ consists of functions which admit approximating subspaces, $\mathcal{V}'^h$. This is the background for

Lemma 3.11. *Let $\mathcal{V}$ be a* separable *Banach space and $\mathcal{V}'^h$ be approximating spaces for $\mathcal{V}'$. Finally, define $\hat{P}^h : \mathcal{V} \to \mathcal{V}^h := (\mathcal{V}'^h)'$ by*

$$\text{for } f \in \mathcal{V} \text{ let } \langle \hat{P}^h f - f, v'^h \rangle_{\mathcal{V}\times\mathcal{V}'} = 0 \quad \forall\, v'^h \in \mathcal{V}'^h. \tag{3.23}$$

Then $\mathcal{V}^h$ are approximating spaces for $\mathcal{V}$, $\hat{P}^h$ is a projector and

$$\lim_{h\to 0} \|\hat{P}^h f - f\|_{\mathcal{V}} = 0 \quad \forall f \in \mathcal{V}.$$

Proof. With $\mathcal{V}$ the $\mathcal{V}'$ is separable as well. As a consequence of the separable $\mathcal{V}'$, we can and do choose a basis $\varphi_i' \in \mathcal{V}'$ such that

$$\mathcal{V}'^h = \operatorname{span}\{\varphi_i', \ldots, \varphi_{i(h)}'\}.$$

Furthermore, define a dual basis φ_i for $\mathcal{V}$, hence

$$\varphi_i' \in \mathcal{V}', \quad \varphi_j \in \mathcal{V} \text{ and } \langle \varphi_i, \varphi_j' \rangle_{\mathcal{V}\times\mathcal{V}'} = \delta_{i,j}.$$

Obviously, $\mathcal{V}^h := \operatorname{span}\{\varphi_i, \ldots, \varphi_{i(h)}\}$. Then

$$\begin{aligned}
&f \in \mathcal{V} \Rightarrow f = \sum_i \langle f, \varphi_i' \rangle \varphi_i, \text{ i.e.} \\
&\forall\ \epsilon > 0\ \exists\ j = j(\epsilon) \text{ and a } h(\epsilon) : \epsilon > \|f - \sum_{i\le j(\epsilon)} \langle f, \varphi_i' \rangle \varphi_i\|_{\mathcal{V}} \\
&\quad = \|\sum_{i>j(\epsilon)} \langle f, \varphi_i' \rangle \varphi_i\|_{\mathcal{V}} \ge \|\sum_{i>j(\epsilon)} \langle f, \varphi_i' > \varphi_i : \varphi_i \notin \mathcal{V}^{h(\epsilon)} \rangle\|_{\mathcal{V}}.
\end{aligned}$$

This $\mathcal{V}^{h(\epsilon)}$ includes at least the $\varphi_i, i \le j(\epsilon)$, but usually some more. The above (3.23) implies

$$\begin{aligned}
&0 = \langle \hat{P}^h f - f, \varphi_i' \rangle \ \ \forall\ \ \varphi_i' \in \mathcal{V}'^h \Leftrightarrow \varphi_i \in \mathcal{V}^h, \text{hence we obtain} \\
&\|\hat{P}^h f - f\|_{\mathcal{V}} \le \|\sum \langle f, \varphi_i' \rangle \varphi_i : \varphi_i \notin \mathcal{V}^h\|_{\mathcal{V}} \to 0 \text{ for } h \to 0.
\end{aligned}$$

□

Summarizing our examples, we have defined an operator Φ^h such that the original A is discretized into $A^h = \Phi^h A$. This process has to be generalized to linear and nonlinear operators A and G, respectively. Formalizing, we introduce general discretization methods. We modify Stetter's [596] approach to fit the testing as in Petrov–Galerkin methods and to the finite difference methods.

Definition 3.12. A discretization method $\mathcal{M}$, *applicable to a problem $G : \mathcal{D}(G) \subset \mathcal{U}_b \to \mathcal{V}_b'$ is defined by a sequence of quintuples, $\mathcal{M} := \left\{\mathcal{U}_b^h, \mathcal{V}_b^h, P^h, Q'^h, \Phi^h\right\}_{h\in H}$:*

1. *Let $\mathcal{U}, \mathcal{V}$ be Banach spaces, $\mathcal{U}_b \subset \mathcal{U}$, $\mathcal{V}_b \subset \mathcal{V}$ closed subspaces. To h, $0 < h \in H \subset \mathbb{R}, \inf_{h\in H} h = 0$ assign sequences $\{\mathcal{U}^h, \mathcal{V}^h\}_{h\in H}$, of finite dimensional*[28] *spaces. Often we use the short notation of $\mathcal{U}^h$. The relation between the spaces is realized by the bounded approximation and projection operators, P^h and Q'^h, usually*

[28] We do not want to over-formalize the notation and have chosen $h \in H$.

$$
\begin{array}{llcccc}
\mathcal{U} \supset & \mathcal{U}_b & \xrightarrow{A,G} & \mathcal{V}_b' & \xleftrightarrow{\text{tested by}} & \mathcal{V}_b \subset \mathcal{V} \\
& P^h \downarrow & \Phi^h \downarrow & Q'^h \downarrow & & \\
& \mathcal{U}_b^h & \xrightarrow{A^h, G^h} & \mathcal{V}_b'^h & \xleftrightarrow{\text{(approximately) tested by}} & \mathcal{V}_b^h
\end{array}
\qquad (3.27)
$$

Figure 3.2 General discretization methods.

linear and generalized approximation and projection operators, respectively, such that, cf. Proposition 4.50, We use the notation $\mathcal{V}'^h = \mathcal{V}^{h'}$, $Q'^h = Q^{h'}$ *equivalently throughout the book.*

$$P^h : \mathcal{U}_b \to \mathcal{U}_b^h, \text{ and } Q'^h : \mathcal{V}_b' \to \mathcal{V}_b'^h, \text{ s.t. } P^h 0 = 0, Q'^h 0 = 0, \text{ and} \qquad (3.24)$$
$$\lim_{h\to 0} \|P^h u\|_{\mathcal{U}^h} = \|u\|_{\mathcal{U}} \forall u \in \mathcal{U}, \ \lim_{h\to 0} \|Q'^h f\|^h_{\mathcal{V}_b'^h} = \|f\|_{\mathcal{V}_b'} \forall f \in \mathcal{V}_b',$$

2. *We need the mappings* Φ^h *defined for nonlinear mappings denoted as* $\mathcal{NL}(\mathcal{U}_b, \mathcal{V}_b') = \left(\mathcal{U}_b \to \mathcal{V}_b'\right)$. *They might be defined on subsets of* $\mathcal{U}_b$.

$$\Phi^h : \ (\mathcal{U}_b \ \to \ \mathcal{V}_b') \to \left(\mathcal{U}_b^h \ \to \ \mathcal{V}_b'^h\right). \qquad (3.25)$$

If $G \in \mathcal{D}(\Phi^h) \ \forall\, h \in H, h < h_0$, *then* Φ^h *or* $\mathcal{M}$ *is called a* discretization applicable to G, *and* $G^h := \Phi^h G$ *the* discretization or discrete operator for G.
3. *Let the restriction of* Φ^h *to linear operators* $\mathcal{L}$ *be linear and satisfy*

$$\Phi^h : \mathcal{NL}(\mathcal{U}_b, \mathcal{V}_b') \to \mathcal{NL}\left(\mathcal{U}_b^h, \mathcal{V}_b'^h\right), \Phi^h|_{\mathcal{L}} : \mathcal{L}(\mathcal{U}_b, \mathcal{V}_b') \to \mathcal{L}\left(\mathcal{U}_b^h, \mathcal{V}_b'^h\right) \text{ with} \qquad (3.26)$$
$$\Phi^h(cA) = c\Phi^h A, \Phi^h(A + A_1) = \Phi^h A + \Phi^h A_1,$$
$$\Phi^h(A + f) = \Phi^h A + \Phi^h f \text{ and } \Phi^h(G + G_1) = \Phi^h G + \Phi^h G_1$$
$$\forall c \in \mathbb{R}, A, A_1 \in \mathcal{L}(\mathcal{U}_b, \mathcal{V}_b'), f \in \mathcal{V}', G, G_1 \in \mathcal{NL}(\mathcal{U}_b, \mathcal{V}_b'), \text{ all in } \mathcal{D}(\Phi^h)$$

Then we call Φ^h a linear discretization method.
4. *Often we use the short notation (for the infinite sequence of)* Φ^h *for indicating the discretization method* $\mathcal{M}$.
5. *Whenever in the next chapters a problem is transformed and solved only in its weak form we will choose* $\mathcal{U} = \mathcal{V}$, $\mathcal{U}_b = \mathcal{V}_b$. *This is no longer correct, e.g. for fully nonlinear problem, cf. Subsections 2.5.7, 2.6.8 and 5.2.3 ff.*

Summarizing, we obtain the *diagram* in Figure 3.2 with uniformly bounded P^h, Q'^h, cf. e.g. Proposition 4.50.

Remark 3.13.

1. *It has to be pointed out that the choice of* P^h, Q'^h *for the given* $\mathcal{U}_b^h, \mathcal{V}_b'^h$ *and problem (3.3) or (3.4) is certainly not unique. However the combination* $\mathcal{U}_b^h, \mathcal{V}_b'^h, P^h, Q'^h, \Phi^h$ *should be chosen appropriately to yield consistency, see*

below, with the highest possible order p. Otherwise the results are no longer optimal, see, e.g. (3.32).

2. *In Example 3.8, we have formulated the discrete operators A^h for Examples 3.1–3.3 with appropriate $P^h, Q'^h, \Phi^h, A_h, \tilde{A}_h$, as*

$$\Phi^h A = A^h = Q'^h A_h|_{\mathcal{U}_b^h} \text{ or } \Phi^h A = A^h = Q'^h \tilde{A}_h|_{\mathcal{U}_b^h}. \tag{3.28}$$

3. *We have to show that the P^h, Q'^h satisfy (3.24) for the case of* generalized Petrov–Galerkin methods. *Then these $\Phi^h A = A^h, \Phi^h G = G^h$ define general discretization methods, in the sense of Definition 3.12.*

Obviously, the generalized discretization methods and some of the following results in this chapter include nonlinear methods. The later applications to bifurcation numerics, however, require linear methods. All the *generalized discretization methods,* studied here, applicable to linear and nonlinear elliptic problems, are *linear discretization methods.* It is important that Φ^h acts on the different u dependent terms in A or G by applying linear operators. This holds even for more complicated cases, e.g. the Fourier collocation derivative in spectral methods, cf. [114, 115, 120, 123, 124, 158, 631] for details. The above linearity condition (3.26) is violated for the (nonlinear) *Runge–Kutta methods.* They are general discretization methods in the above sense, but are not relevant for our space discretizations in bifurcation and the small dimensional center manifolds. Therefor we avoid all the difficulties with parabolic problems. Our definition allows all forms of finite difference, finite element, discontinuous Galerkin, spectral, wavelet, and mesh–free methods, and approximating the discrete operators by quadrature formulas or difference approximations. These methods are necessary for studying effects in space discretization for numerical bifurcation and center manifolds in partial differential equations.

Definition 3.12 is a modification of [596] and is related to [607–609, 677]. Differently from these, we study operators $G : \mathcal{U}_b \rightarrow \mathcal{V}_b'$ and require, in some form, approximating spaces $\mathcal{U}^h$, $\mathcal{V}^h$ for $\mathcal{U}$, $\mathcal{V}$. This emphasizes the testing with the $\mathcal{V}_b$, $\mathcal{V}_b^h$ and avoids, appropriately for all our methods, a direct consideration of the dual space $\mathcal{V}'$. So we do not have to distinguish the *inner and outer discretization schemes,* important in [528, 531, 532, 677].

Next, we have to study the relations between operators and their discretizations.

3.3 Variational and classical consistency

We want to know how well the discrete solution, u_0^h, approximates the exact solution, u_0. One tool is an estimate of how well u_0^h satisfies the original equation. We motivate by our previous examples, two different, however, closely related concepts, the so-called variational and the classical consistency (errors) for linear and nonlinear operators.

For conforming FEMs we have $\mathcal{U}_b^h \subset \mathcal{U}_b$, $\mathcal{V}_b^h \subset \mathcal{V}_b$; for nonconforming FEMs and for difference methods the $\mathcal{U}_b^h$, $\mathcal{V}_b^h$ have to be modified as in Example 3.8. Accordingly, we have defined the exact and discrete solutions u_0 and u_0^h, respectively,

$$u_0 \in \mathcal{U}_b : \langle Au_0, v\rangle_{\mathcal{V}'\times\mathcal{V}} = a(u_0, v) = \langle f, v\rangle_{\mathcal{V}'\times\mathcal{V}} \ \forall\ v \in \mathcal{V}_b \text{ and} \tag{3.29}$$
$$u_0^h \in \mathcal{U}_b^h : \left\langle Au_0^h, v^h\right\rangle_{\mathcal{V}'\times\mathcal{V}} = a\left(u_0^h, v^h\right) = \left\langle f, v^h\right\rangle_{\mathcal{V}\times\mathcal{V}} \forall v^h \in \mathcal{V}_b^h, \text{ conform. FE}$$
$$u_0^h \in \mathcal{U}_b^h : \left\langle A_h u_0^h, v^h\right\rangle_{\mathcal{V}'^h\times\mathcal{V}^h} = a^h\left(u_0^h, v^h\right) = \left\langle f^h, v^h\right\rangle_{\dots} \forall v^h \in \mathcal{V}_b^h, \text{ nonc. FE, so}$$
$$u_0^h \in \mathcal{U}_b^h : Q'^h\left(Au_0^h - f\right) = 0 \textit{ for conf. FEs or } Q'^h\left(A_h u_0^h - f\right) = 0 \textit{ for nonconf. FEs}$$
$$u_0^h \in \mathcal{U}_b^h : Q'^h\left(\tilde{A}_h u_0^h - f\right) = 0, \text{ for diff. meth., with the } P^h, Q'^h \text{ in (3.18), (3.19).}$$

Hence, the second equation represents a subset of the conditions in the first equation for conforming FEs, the third and fourth equations for nonconforming Petrov–Galerkin and difference methods don't.

For conforming FEs with $a(u,v) = a_h(u,v) \forall u \in \mathcal{U}, v \in \mathcal{V}$, we obtain

$$a\left(u_0 - u_0^h, v^h\right) = 0\ \forall\ \ v^h \in \mathcal{V}_b^h,\ Q'^h\left(Au_0 - Au_0^h\right) = 0, \tag{3.30}$$

hence, these *variational consistency errors* vanish, see below.

For nonconforming FEs a direct comparison of Au_0 and $A^h u_0^h$ is impossible. So we have to employ the extended $a_h\left(u_0 - u_0^h, v^h\right)$ or $A_h u_0 - A_h u_0^h$. This *variational consistency error* is nontrivial for nonconforming methods, and with $A^h = Q'^h A_h|_{\mathcal{U}_b^h}$

$$\sup_{0\neq v^h \in \mathcal{V}_b^h} \left\{\left|a_h\left(u_0 - u_0^h, v^h\right)\right| / \|v^h\|_{\mathcal{V}^h}\right\} = \left\|Q'^h A_h u_0 - A^h u_0^h\right\|_{\mathcal{V}_b'^h}. \tag{3.31}$$

This *variational consistency error* is the usual concept in the FE community. For the linear case the results in Lemma 5.52 will allow estimates of the form

$$\left\|u_0^h - u_0\right\|_{\mathcal{U}^h} \le C\left(\text{dist }\left(u_0, \mathcal{U}_b^h\right) + \sup_{0\neq v^h\in\mathcal{V}_b^h} \frac{\left|a_h\left(u_0 - u_0^h, v^h\right)\right|}{\|v^h\|_{\mathcal{V}^h}}\right). \tag{3.32}$$

However, this variational consistency error differs from the *classical consistency error*, $Q'^h Au_0 - A^h P^h u_0$, cf. Definition 3.14, (3.33), (3.35), and [596]. The latter is more appropriate for using the whole nonlinear machinery. We aim for the relation of the variational consistency to the classical consistency error. Again, we use the general notation of A_h for the different cases, unless the difference has to be emphasized.

To compare both "consistencies", we modify (3.31) in operator form as $Q'^h A_h u_0 - A^h u_0^h$. This can be generalized to arbitrary (smooth enough) u, if we introduce $f_u := Au$ and the discrete solution u^h of $A^h u^h = Q'^h f_u := Q'^h Au$:

$$\begin{aligned}\underbrace{Q'^h A_h u - A^h u^h}_{\text{var.cons.error}} &= A^h(P^h u - u^h) + Q'^h A_h u - Q'^h A_h P^h u \\ &= A^h P^h u - Q'^h f_u + Q'^h A_h u - Q'^h A_h P^h u \\ &= \underbrace{A^h P^h u - Q'^h Au}_{\text{class.cons.error}} + \underbrace{Q'^h A_h u - Q'^h A_h P^h u}_{\text{related to interpol.error}}.\end{aligned} \tag{3.33}$$

As indicated, the term $Q'^h A_h u - A^h u^h$ left in the first line in (3.33) and the first term in the last line, $A^h P^h u - Q'^h Au$, are called the *variational* and *classical consistency error* for the linear operator A. For linear Q'^h we find $Q'^h A_h u - Q'^h A_h P^h u = Q'^h A_h(u - P^h u)$, hence $\|Q'^h A_h(u - P^h u)\| \le C\|u - P^h u\|$ is bounded by the interpolation error. For difference methods we simply replace the A_h in (3.33) by $\tilde{A}_h$, however, the *variational consistency error*, $Q'^h A_h u - A^h u^h$, is only defined for smooth enough u. There is no problem with the classical consistency error, $A^h P^h u - Q'^h Au = Q'^h \tilde{A}_h P^h u - Q'^h Au$.

This (3.33) has to be slightly modified for nonlinear operators. Again, we introduce $f_u := Gu$ and assume u^h, a discrete solution of $G^h u^h = Q'^h f_u = Q'^h Gu$. From the possibly many solutions u^h for the nonlinear G we choose that with small $\|u^h - u\|$ for an appropriate norm. In general notation we find with $G^h = Q'^h G_h|_{\mathcal{U}_b^h}$ in (3.21),

$$\underbrace{Q'^h G_h u - G^h u^h}_{\text{var.cons.error}} = \underbrace{(G^h P^h u - Q'^h Gu)}_{\text{class.cons.error}} + \underbrace{(Q'^h G_h u - Q'^h G_h P^h u)}_{\text{related to interpol.error}} \tag{3.34}$$

Again, for difference methods the $Q'^h \tilde{G}_h u$ only makes sense for smooth enough u. Similarly, for the general discretization methods in Definition 3.12, with $G^h = \Phi^h G$, the classical discretization error, $G^h P^h u - Q'^h Gu = (\Phi^h G)P^h u - Q'^h Gu$, is always defined, but the other terms are usually not defined. So we introduce in the following definition the classical discretization error for general discretization methods. Next, we discuss, for generalized Petrov–Galerkin methods, the variational and classical discretization errors in Example 3.15.

Definition 3.14. Classical consistency, local discretization error:

1. *Under the conditions of Definition 3.12 for a general discretization method*

$$(\Phi^h A)P^h u - Q'^h Au = A^h P^h u - Q'^h Au \text{ and } (\Phi^h G)P^h u - Q'^h Gu \tag{3.35}$$

$$\text{or its norm } \|A^h P^h u - Q'^h Au\|_{\mathcal{V}_b'^h} \text{ and } \|G^h P^h u - Q'^h Gu\|_{\mathcal{V}_b'^h},$$

for A and G, respectively, are called the (classical) consistency error or local discretization error *in $u \in \mathcal{D}(G) \cap \mathcal{U}_b$. Note that for the exact solution u_0 of $G(u_0) = 0$ the classical consistency error reduces to $\|G^h P^h u_0\|_{\mathcal{V}_b'^h}$.*

2. *A method is called* classically consistent *for A or G in u, if*

$$\|A^h P^h u - Q'^h Au\|_{\mathcal{V}_b'^h} \to 0 \text{ or } \|G^h P^h u - Q'^h Gu\|_{\mathcal{V}'} \to 0 \text{ for } h \to 0.$$

3. *If even*

$$\|A^h P^h u - Q'^h Au\|_{\mathcal{V}_b'^h} \text{ or } \|G^h P^h u - Q'^h G_h u\|_{\mathcal{V}'} = \mathcal{O}(h^p),$$

we call it consistent of order p in u. *These $\mathcal{O}(h^p)$ are usually estimated by $Ch^p\|u\|_{W^{k,q}(\Omega)}$ with appropriate p, k, q.*

Example 3.15. *Consistency errors*

1. As in Example 3.8 we choose $a_h(\cdot,\cdot)$ and the discrete operators $A^h = Q'^h A_h|_{\mathcal{U}_b^h}$ and $G^h = Q'^h G_h|_{\mathcal{U}_b^h}$ by applying a generalized Petrov–Galerkin method to A and

G. Let u and u^h be related as

$$A^h u^h = Q'^h f_u := Q'^h Au \text{ and } G^h u^h = Q'^h f_u := Q'^h Gu$$

for the linear and nonlinear operator A and G, respectively. Then

$$\sup_{0 \neq v^h \in \mathcal{V}_b^h} \{|a_h(u-u^h, v^h)|/\|v^h\|\} = \|Q'^h A_h u - A^h u^h\|_{\mathcal{V}_b'^h} \text{ and } \quad (3.36)$$

$$\|Q'^h G_h u - G^h u^h\|_{\mathcal{V}_b'^h}, \text{ for } A \text{ and } G, \text{respectively,}$$

is called the *variational consistency error* in u.[29] A method is called *variationally consistent* for A in u, if

$$\sup_{0 \neq v^h \in \mathcal{V}_b^h} \{|a^h(u-u^h, v^h)|/\|v^h\|\} = \|Q'^h A_h u - A^h u^h\|_{\mathcal{V}_b'^h} \to 0, \quad (3.37)$$

similarly for the nonlinear G, e.g. $\|G^h P^h u - Q'^h Gu\|_{\mathcal{V}'^h} \to 0$ for $h \to 0$.

2. Note that for all these cases the variational errors vanish for conforming FEMs.

■

The variational consistency error usually is only considered and estimated for the exact solution, u_0, or if necessary for an approximant $P^h u_0$, see Section 5.5 for estimates. A generalization to arbitrary u is formulated in (3.33). The classical consistency errors are often studied for general u. There is the simple relation for the difference of variational and classical consistency errors in (3.33) and (3.34). In fact, see Proposition 5.47 and Theorem 3.17, this difference is, under appropriate conditions, bounded as $\mathcal{O}(\|u - P^h u\|_{\mathcal{U}^h}) \to 0\ \forall u \in \mathcal{U}_b\ \forall h \to 0$. Hence, they even have *the same size* if both dominate or have the same size as the interpolation error $\|u_0 - P^h u_0\|_{\mathcal{U}}$. Therefore it suffices to estimate the more convenient of the two consistency errors.

The general case needs the continuity or even Lipschitz-continuity of the G_h, a condition satisfied for all the examples studied here, simultaneously with G. We assume equicontinuous and equibounded Q'^h, cf. Theorem 4.54.

Theorem 3.16. Conforming classical consistency, and interpolation errors:

1. *Let $\Phi^h G = G^h = Q'^h G|_{\mathcal{U}_b^h} : \mathcal{D}(G^h) = \mathcal{U}_b^h \cap \mathcal{D}(G)$ be a conforming FEM and let $u^h \in \mathcal{D}(G^h)$ be given. Assume G is Lipschitz-continuous in $B_r(u)$ and Q'^h equicontinuous and equibounded. Then the classical consistency error satisfies*

$$\|G^h P^h u - Q'^h Gu\|_{\mathcal{V}_b'^h} = \|Q'^h G P^h u - Q'^h Gu\|_{\mathcal{V}_b'^h} \leq L\|(P^h u - u)\|_{\mathcal{U}^h}. \quad (3.38)$$

2. *If $P^h : \mathcal{U}_b \to \mathcal{U}_b^h$ is chosen as the interpolation or approximation operator, then the classical consistency error vanishes as, e.g. the interpolation error.*

[29] We sometimes use the notation $\|Q'^h A_h u - A^h u^h\|_{\mathcal{V}_b'^h}$ instead of $\|Q'^h A_h u - A^h u^h\|_{\mathcal{V}_b'^h}$ to emphasize that this $Q'^h A_h u - A^h u^h$ is tested by the $v^h \in \mathcal{V}_b^h$. This notation in (3.36) is used similarly for the isoparametric variants.

Theorem 3.17. Nonconforming variational, classical consistency, interpolation errors:

1. *For the case of* nonconforming FEs *we extend the original linear* $A:\mathcal{U}\to\mathcal{V}'$ *and the nonlinear* $G:\mathcal{D}(G)\subset\mathcal{U}\to\mathcal{V}'$ *as* $A_h:\mathcal{U}_{\mathcal{T}^h}\to\mathcal{V}'_{\mathcal{T}^h}$ *and* $G_h:\mathcal{D}(G_h)\subset\mathcal{U}_{\mathcal{T}^h}\to\mathcal{V}'_{\mathcal{T}^h}$. *Assume* A_h *and* Q'^h *to be equibounded and equicontinuous and* ($G\in C_L(\mathcal{D}(G))$ *usually implying*) $G_h\in C_L(\mathcal{D}(G_h))$, *with Lipschitz constant* L. *Furthermore, let* $A^h=Q'^hA_h|_{\mathcal{U}_{\mathcal{T}^h}}$, $G^h=Q'^hG_h|_{\mathcal{U}_{\mathcal{T}^h}}$ *and* $u\in\mathcal{D}(G)$, $u^h,P^hu\in\mathcal{D}(G_h)$, *and* u^h *be the discrete approximation for* u *from* $G^hu^h=Q'^hG_hu^h=Q'^hf_u=Q'^hGu$ *or its analogue for the linear* A,A_h.
2. *Then the* difference between the variational and classical discretization errors
$$(Q'^hG_hu-G^hu^h)-(G^hP^hu-Q'^hGu)=(Q'^hG_hu-Q'^hG_hP^hu)\quad(3.39)$$
$$\forall\ u\in\mathcal{D}(G),u^h,P^hu\ \in\ \mathcal{D}(G_h)$$
can be estimated $\forall\ u\ \in\ \mathcal{U}$ *as*
$$\|Q'^hG_hu-Q'^hG_hP^hu\|_{\mathcal{V}'^h_b},\|Q'^hA_hu-Q'^hA_hP^hu\|_{\mathcal{V}'^h_b}\le\mathcal{O}(\|P^hu-u\|_{\mathcal{U}_{\mathcal{T}^h}}).$$
3. *If only* $G\in C(\mathcal{D}(G))$, $G_h\in C(\mathcal{D}(G_h))$, *then*
$$\|Q'^hG_hu-Q'^hG_hP^hu\|_{\mathcal{V}'^h_b}\le\omega_{G_h}(\|P^hu-u\|_{\mathcal{U}_{\mathcal{T}^h}}),$$
with the modulus of continuity of G_h,
$$\omega_{G_h}(t):=\sup_{\|x-y\|_{\mathcal{U}_{\mathcal{T}^h}}\le t,\ \ x,y\in\mathcal{D}(G_h)}\|G_hx-G_hy\|_{\mathcal{V}'^h_b}.$$
4. *For nonconforming FEs and DCGMs the estimates for variational and classical discretization errors are rather technically involved, cf. Section 5.5 and Chapter 7.*

Proof. With the equibounded and equicontinuous Q'^h and $G_h\in C_L\left(\mathcal{U}_{b,c}+\mathcal{U}_b^h\right)$, $u\in\mathcal{D}(G)$, $u^h,P^hu\in\mathcal{D}(G_h)$, the first estimate is obvious. The $\omega_{G_h}(\|P^hu-u\|_{\mathcal{U}_c})$ modification is straightforward. □

3.4 Stability and consistency yield convergence

We want to unfold this well-known fact for the general discretization methods introduced in Definition 3.12. We combine consistency introduced in Section 3.3 with the missing convergence and stability to be defined now. An appropriate interplay of the different concepts of consistency, convergence, stability, the operators P^h,Q'^h, the norms, the different approximations by spaces and operators is the basic condition for the success of this approach. These choices are by no means unique. Stetter [596] gives some examples for failing combinations. In Theorem 3.21 we prove the existence of a discrete solution and its convergence towards the exact solution essentially for a stable and consistent discretization as in Definition 3.12. This includes our general Petrov–Galerkin methods, cf. Remark 3.13. Except in Section 3.7, we skip, in the remainder of this chapter, the condition of a "linear" Φ^h, see (3.25), and study the general discretization methods.

Convergence is expected to describe a discrete solution, u_0^h, well approximating the exact solution, u_0. In some methods we can directly compare u_0^h and u_0 since for both $\|\cdot\|_{\mathcal{U}^h}$, and hence $\|u_0 - u_0^h\|_{\mathcal{U}^h}$ is well defined with hopefully $\|u_0 - u_0^h\|_{\mathcal{U}^h} \to 0$ for $h \to 0$. Theoretically, two other measures are possible as well. We might consider $\|P^h u_0 - u_0^h\|_{\mathcal{U}^h}$ with our known $P^h : \mathcal{U} \to \mathcal{U}^h$, e.g. for difference methods. Or we might study $\|u_0 - P_h u_0^h\|_{\mathcal{U}}$ and need an additional $P_h : \mathcal{U}^h \to \mathcal{U}$. We apply this P_h, essentially defined by an *anticrime transformation,* cf. Lemma 5.77, to nonconforming FEs for proving stability of linearized elliptic operators.

Definition 3.18. Global discretization error, convergence:

1. *Let* $G : \mathcal{D}(G) \subset \mathcal{U}_b \to \mathcal{V}_b'$ *and its discretization* $G^h = \Phi^h G : \mathcal{D}(G^h) \subset \mathcal{U}_b^h \to \mathcal{V}_b^{'h}$ *have the (locally) unique exact and approximate solutions* u_0 *and* u_0^h*, respectively. Then (the sequence)*
$$u_0^h - u_0 \text{ or } u_0^h - P^h u_0 \tag{3.40}$$
is called the (global) discretization error *of* Φ^h *for* G.
2. *The* u_0^h *are called* convergent *to* u_0 *and* convergent of order p *(sometimes,* $G^h = \Phi^h G$ *is called* convergent *to* G *in* u_0*) if* for $h \to 0, h \in H$, respectively
$$\|u_0 - u_0^h\|_{\mathcal{U}^h} \to 0 \text{ or } \|P^h u_0 - u_0^h\|_{\mathcal{U}^h} \to 0 \quad \text{and} \tag{3.41}$$
$$\|u_0 - u_0^h\|_{\mathcal{U}^h} = \mathcal{O}(h^p) \text{ or } \|P^h u_0 - u_0^h\|_{\mathcal{U}^h} = \mathcal{O}(h^p).$$
3. *As in Definition 3.5, modifications are necessary for difference methods.*

We want to prove convergence. This will not be possible directly. So we try to derive it from the local discretization error. Consistency alone will not do it, cf. [596] for counterexamples. Let us motivate what is missing: $G(u_0) = 0$ and $G^h(u_0^h) = 0$ have to be solved and $P^h u_0$ and u_0^h have to be compared. The classical consistency, see (3.35), in u_0 with $G u_0 = 0$ implies

$$G^h P^h u_0 - Q'^h G u_0 = G^h P^h u_0 - Q'^h 0 = G^h P^h u_0 \ \to \ 0 \ \forall\, h \ \to \ 0, \ h \in H. \tag{3.42}$$

If $(G^h)^{-1}$ is defined in and close to $G^h P^h u_0$, then with $G^h u_0^h = 0$,

$$P^h u_0 - u_0^h = (G^h)^{-1} G^h P^h u_0 - (G^h)^{-1} 0.$$

For a vanishing consistency error in u_0 and a continuous or even Lipschitz-continuous $(G^h)^{-1}$ this implies

$$\|P^h u_0 - u_0^h\|_{\mathcal{U}^h} \to 0 \text{ or } \le L \|G^h(P^h u_0)\|_{\mathcal{V}_b'^h} \ \to \ 0 \ \forall\, h \ \to \ 0, \ h \in H. \tag{3.43}$$

This kind of perturbation insensitivity is called stability.

Definition 3.19. Stability: *Under the conditions of Definition 3.12 we assume a non-linear and linear operator* $G^h : \mathcal{D}(G^h) \subseteq \mathcal{U}_b^h \to \mathcal{V}_b^{'h}$*, and* $B^h : \mathcal{U}_b^h \to \mathcal{V}_b^h$*, respectively,* *to be assigned to every* $h \in H$ *(usually defined by a discretization method* $\mathcal{M}$*). For* u^h *and* $h_0 > 0$ *let*

$$B_r(u^h) = \{v^h \in \mathcal{U}^h : \|v^h - u^h\|_{\mathcal{U}^h} < r\} \subset \mathcal{D}(G^h)\ \forall h < h_0, h \in H.$$

Furthermore, let $h_0, r, S \in \mathbb{R}_+$ *be fixed constants, such that,* uniformly in $h \in H, h < h_0$:

$$u_i^h \in B_r(u^h), i = 1, 2, \Longrightarrow \left\|u_1^h - u_2^h\right\|_{\mathcal{U}^h} \leq S \left\|G^h\left(u_1^h\right) - G^h\left(u_2^h\right)\right\|_{\mathcal{V}_b'^h}. \tag{3.44}$$

For B^h *or* $B^h \cdot + f^h$ *this reduces to (for any* r *and* u^h*, but for* $h < h_0$*)*

$$(B^h)^{-1} \in \mathcal{L}(\mathcal{V}_b'^h, \mathcal{U}_b^h) \text{ exists and } \|(B^h)^{-1}\|_{\mathcal{U}_b^h \leftarrow \mathcal{V}_b'^h} \leq S.$$

Then G^h *and* B^h *are called* stable in u^h *and* stable, *respectively. The* S *and* r *are called* stability bound and stability threshold, *respectively.*

The proof of stability is simple for some cases; for general problems it is usually rather complicated. For example, for the conforming FEs in Example 3.1 it is an immediate consequence of the coercivity and the Brezzi–Babuska condition in Theorem 2.12. For more general cases we will discuss in Theorems 3.23–3.29, conditions which guarantee stability. For example, it allows reducing stability for the nonlinear problem to that of the linearized problem, and from a linear *coercive operator* to its compact perturbations. In Theorem 2.122 we discuss the implication of $H_0^m(\Omega) \Rightarrow W_0^{m,p}(\Omega)$-coercivity for $p < 2$ and $\not\Rightarrow W_0^{m,p}(\Omega)$-coercivity for $2 < p$. In modified form these arguments are applicable to the (linearized) Navier–Stokes operator as well.

In [596] the first inequality in (3.44), $u_i^h \in B_r(u^h)$, is replaced by the weaker

$$\|G^h\left(u_i^h\right) - G^h(u^h)\|_{\mathcal{V}_b'^h} < R, \ i = 1, 2. \tag{3.45}$$

Since for nonlinear G (and hence G^h) this (3.45) may be satisfied for large $\left\|u^h - u_i^h\right\|_{\mathcal{U}^h}$ we have chosen the stronger assumption in (3.44), cf. [596], Corollary 1.2.2.

Proposition 3.20. *In the discretization* $G^h : \mathcal{D}(G^h) \subset \mathcal{U}_b^h \to \mathcal{V}_b'^h$ *let the* G^h *be continuous in* $B_r(u^h) \subset \mathcal{D}(G^h)$ *and satisfy (3.44). Then these* u_i^h *satisfy (3.45) with* $R = r/S$*, and thus are stable in this weaker sense.*

Assuming continuity, consistency and stability and some technicalities, we will now prove the existence of a discrete solution u_0^h for $G^h\left(u_0^h\right) = 0$, defined by the general discretization methods in Definition 3.12, cf. [596].

Theorem 3.21. *Let the original problem (3.3) have the exact solution* u_0*. Let* $G^h : \mathcal{D}(G^h) \subset \mathcal{U}_b^h \to \mathcal{V}_b'^h$*,* $h \in H$*, be its discretization, see Definition 3.12 (here nonlinear* Φ^h *are allowed) and satisfy the following conditions:*

1. $G^h = \Phi^h G : \mathcal{D}(G^h) \subset \mathcal{U}_b^h \to \mathcal{V}_b'^h$ *is defined and continuous in* $B_r(P^h u_0) \subset \mathcal{D}(G^h)$ *with* $r > 0$ *independent of* h*;*
2. G^h *is (classically) consistent with* G *in* $P^h u_0$*;*
3. G^h *is stable for* $P^h u_0$.

Then the discrete problem $G^h(u^h) = 0$ possesses the unique solution $u_0^h \in \mathcal{D}(G^h) \subset \mathcal{U}_b^h$ near u_0 for all sufficiently small h and u_0^h converges to u_0. If G^h is consistent and consistent of order p, then u_0^h converges and converges of order p, hence

$$\|u_0^h - u_0\|_{\mathcal{U}^h} \leq S\|G^h(P^h u_0)\|_{\mathcal{V}'^h} \text{ and } \leq \mathcal{O}(h^p) \text{ for } h \to 0, h \in H, \text{ respectively} \quad (3.46)$$

As essential tool for the proof is the following.

Proposition 3.22. *For a fixed $r > 0$ and a sequence $\bar{u}^h$, let the discrete*

$$G^h : B_r := B_r(\bar{u}^h) \to \mathcal{V}_b'^h, \ G^h \in C(B_r(\bar{u}^h)) \text{ be continuous.}$$

Furthermore for $S \in \mathbb{R}_+$ let

$$u_i^h \in B_r, i = 1, 2, \text{ imply } \|u_1^h - u_2^h\|_{\mathcal{U}^h} \leq S \ \|G^h u_1^h - G^h u_2^h\|_{\mathcal{V}_b'^h}. \quad (3.47)$$

Then G^h is invertible on

$$B'_R = \left\{v'^h \in \mathcal{V}_b'^h : \|v'^h - G^h\bar{u}^h\|_{\mathcal{V}_b'^h} < R\right\}, \text{ with } R := r/S. \quad (3.48)$$

This $(G^h)^{-1} : B'_R \to \mathcal{U}_b^h$ is Lipschitz-continuous with Lipschitz-bound S.

Proof. By (3.47) the G^h are invertible on every set $M \subset G^h(B_r)$. In fact, if $u_1^h, u_2^h \in B_r$ have the same images $G^h\left(u_i^h\right) = v'^h \in M, i = 1, 2$, then (3.47) implies

$$\|u_2^h - u_1^h\|_{\mathcal{U}^h} \leq S\left\|G^h u_2^h - G^h u_1^h\right\|_{\mathcal{V}_b'^h} = S \cdot 0, \text{ such that } u_2^h = u_1^h. \quad (3.49)$$

Hence, it suffices to prove

$$B'_R \subset G^h(B_r). \quad (3.50)$$

If this were incorrect, a $v_1'^h \in B'_R \backslash G^h(B_r)$ would exist. We define

$$v'^h(\lambda) := (1 - \lambda)G^h\bar{u}^h + \lambda v_1'^h, \lambda \geq 0 \quad \text{for } v_1'^h \in B'_R \backslash G^h(B_r). \quad (3.51)$$

This implies $v'^h(0) = G^h\bar{u}^h \in G^h(B_r), v'^h(1) = v_1'^h \notin G^h(B_r)$. With

$$\Lambda := \{\lambda' > 0 : v'^h(\lambda) \in G^h(B_r) \quad \text{for } \lambda \in [0, \lambda')\}$$

we define

$$\bar{\lambda} := \begin{cases} \sup_{\lambda \in \Lambda} \lambda & \text{for } \Lambda \neq \emptyset, \\ 0 & \text{for } \Lambda = \emptyset. \end{cases} \quad (3.52)$$

The inclusion (3.50) would be proved if we could show that $\bar{\lambda} > 1$. We construct a contradiction to $\bar{\lambda} \leq 1$ in two steps: first we show $\bar{v}^h := v'^h(\bar{\lambda}) \in G^h(B_r)$. This allows constructing the contradiction in step 2.

For $\bar{\lambda} = 0$ we automatically have $\bar{v}^h = v'^h(0) = G^h\bar{u}^h \in G^h(B_r)$. For the complementary case $0 < \bar{\lambda} \leq 1$ we have by definition $v'^h(\bar{\lambda} - \epsilon) \in G^h(B_r)$, see (3.52) for any $0 < \epsilon < \bar{\lambda}$ and, see (3.51),

$$v'^h(\bar{\lambda} - \epsilon) \in B'_R, \text{ since } \|v'^h(\bar{\lambda} - \epsilon) - G^h\bar{u}^h\|_{\mathcal{V}_b'^h} < \left\|v_1'^h - G^h(\bar{u}^h)\right\|_{\mathcal{V}_b'^h} < R.$$

Therefore an $u_\epsilon^h \in B_r$ exists such that $G^h u_\epsilon^h = v'^h(\bar\lambda - \epsilon)$, see (3.52). A combination with (3.47) and (3.51) shows the property of a Cauchy sequence

$$\left\|u_\epsilon^h - u_{\bar\epsilon}^h\right\|_{\mathcal{U}^h} \le S\left\|G^h u_\epsilon^h - G^h u_{\bar\epsilon}^h\right\|_{\mathcal{V}_b'^h} = S\|v'^h(\bar\lambda-\epsilon) - v'^h(\bar\lambda-\bar\epsilon)\|_{\mathcal{V}_b'^h} \to 0$$

$$\text{for } \epsilon, \bar\epsilon > 0 \text{ and } \epsilon \to \bar\epsilon.$$

Hence the $(G^h)^{-1}(v'^h(\bar\lambda - \epsilon))$ exist and we have convergence for the u_ϵ^h, $v'^h(\bar\lambda-\epsilon)$, and, by (3.47), for the $(G^h)^{-1}(v'^h(\bar\lambda - \epsilon))$ as well for $\epsilon \to +0$. This yields

$$u_0^h := \lim_{\epsilon\to+0} u_\epsilon^h = \lim_{\epsilon\to+0} (G^h)^{-1}(v'^h(\bar\lambda-\epsilon)).$$

This u_0^h is $\in B_r$: in fact,

$$\begin{aligned}\|(G^h)^{-1}(v'^h(\bar\lambda-\epsilon)) - \bar u^h\|_{\mathcal{U}^h} &\le S\|v'^h(\bar\lambda-\epsilon) - G^h\bar u^h\|_{\mathcal{V}_b'^h} \\ &= S\|(\bar\lambda-\epsilon)\left(v_1'^h - G^h\bar u^h\right)\|_{\mathcal{V}_b'^h}.\end{aligned}$$

and, see (3.51), $\|v_1'^h - G^h\bar u^h\|_{\mathcal{V}_b'^h} < R$, independent of ϵ. With $0 < \bar\lambda - \epsilon < 1$, we obtain the following inequality

$$\|(G^h)^{-1}(v'^h(\bar\lambda-\epsilon)) - \bar u^h\|_{\mathcal{U}^h} < S(\bar\lambda-\epsilon)R \le (\bar\lambda-\epsilon)r \le r, \text{ hence } \left\|u_0^h - \bar u^h\right\|_{\mathcal{U}^h} < r.$$

Finally G^h is continuous in B_r and we have reached our first goal

$$\bar v^h := v'^h(\bar\lambda) = G^h u_0^h \in G^h(B_r).$$

This allows the contradiction: since $u_0^h \in B_r$ and G^h is continuous in B_r, there exists a closed ball $\bar B = \overline{B_\delta\left(u_0^h\right)} \subset B_r$ with $G^h(\bar B) \subset B'_R$. For this $\bar B$ the $G^h : \bar B \to G^h(\bar B) \subset B'_R$ is injective, as we have seen already in (3.49).

Since $\dim \mathcal{U}^h < \infty$ then $\bar B$ is compact. Now $G^h|_{\bar B}$ is an injective continuous mapping with compact domain $\bar B$. Then $(G^h)^{-1} : G^h(\bar B) \to \bar B$ is defined and continuous as well, see e.g. [560], [74], p. 56. In particular, with compact $\bar S := \left\{u^h \in \mathcal{U}_b^h : \left\|u^h - u_0^h\right\|_{\mathcal{U}^h} = \delta\right\}$ the injectivity implies

$$\rho := \min_{u^h\in\bar S} \left\|G^h u_0^h - G^h u^h\right\| > 0,$$

and hence

$$G^h u_0^h = \bar v^h \in B_\rho^0 = B_\rho^0(\bar v^h) \subset G^h(\bar B) \subset G^h(B_r) \text{ with open } B_\rho^0.$$

Then the above $\bar\lambda$ cannot be the supremum defined in (3.52). This is our contradiction. So either $\bar\lambda > 1$ or $v_1'^h \in G^h(B_r)$.

The Lipschitz-continuity for $(G^h)^{-1}$ and the Lipschitz constant S are immediate consequences from (3.47). □

Proof of Theorem 3.21. The conditions 1, 3 imply by Proposition 3.22 the existence of a unique discrete solution $u_0^h \in B_r(P^h u_0)$: in fact we use $\bar u^h = P^h u_0$ and $G^h\bar u^h = G^h P^h u_0$ in (3.48), the local discretization error $G^h P^h u_0$, see Definition 3.14.

Furthermore, R is independent of h since r, S are independent of h. The consistency of the method implies

$$\|G^h P^h u_0\|_{\mathcal{V}_b'^h} \to 0 \quad \text{for} \quad h \to 0.$$

Therefore $0 \in B_R^0(G^h P^h u_0)$ for $h < h_1$ sufficiently small and $u_0^h = (G^h)^{-1}0$ is uniquely determined for $h < h_1$.

In the next step we show convergence and convergence of order p for this discrete solution. For $h < h_1$ let u_0^h be the unique solution of $G^h u_0^h = 0$. Now we use the global discretization error $\epsilon_h = u_0^h - P^h u_0$, see Definition 3.18, obtaining

$$\underbrace{G^h u_0^h}_{=0} - G^h P^h u_0 = \underbrace{G^h(P^h u_0 + \epsilon_h)}_{=0} - G^h P^h u_0 = -G^h P^h u_0.$$

G^h is stable for $P^h u_0$ hence

$$\|\epsilon_h\|_{\mathcal{U}^h} = \left\|u_0^h - P^h u_0\right\|_{\mathcal{U}^h} \le S\|G^h P^h u_0\|_{\mathcal{V}_b'^h} \quad \text{for} \quad \|G^h P^h u_0\|_{\mathcal{V}_b'^h} < R.$$

Since G^h is consistent with G in $P^h u_0$, then $\|G^h P^h u_0\|_{\mathcal{V}_b'^h} \to 0$. This implies, for $h < h_2$, that $\|G^h P^h u_0\|_{\mathcal{V}_b'^h} < R$ and hence

$$\|\epsilon_h\|_{\mathcal{U}^h} \le S\|G^h P^h u_0\|_{\mathcal{V}_b'^h} \to 0 \quad \text{for} \quad h \to 0.$$

This shows the convergence and convergence of order p for this method. □

3.5 Techniques for proving stability

The following theorems formulate known and well applicable criteria for stability. We start discussing the relation between stability of a nonlinear operator and its linearized form and with perturbed nonlinear problems. Most importantly, we combine a coercive principal part with its compact perturbations, cf. Theorem 3.29. This allows, for all problems and methods treated in both books, the proof of stability for (nonlinear) operators. Their linearizations have to be boundedly invertible in the exact solution, implying locally unique solutions. Natural extensions to the bifurcation situation are valid as well. Here we prove Theorem 3.29 only for the case of conforming FEMs. Generalizations to nonconforming methods require specific additional information, available in the corresponding chapters.

Theorem 3.23. *Under the conditions of Definition 3.12 let a discretization Φ^h, $G^h = \Phi^h G : \mathcal{D}(G^h) \subset \mathcal{U}_b^h \to \mathcal{V}_b'^h$ and $u^h \in \mathcal{D}(G^h) \subset \mathcal{U}^h$ be given. Assume $G^h \in C^1(B_r(u^h))$, $B_r(u^h) \subset \mathcal{D}(G^h)$, with fixed $r > 0$. Furthermore, for all $v^h \in B_r(u^h)$ let the inverse of the derivative, $((G^h)'(v^h))^{-1}$, exist for $h < h_0$ and*

$$\|((G^h)'(v^h))^{-1}\|_{\mathcal{U}_b^h \leftarrow \mathcal{V}_b'^h} \le S \ \forall v^h \in B_r(u^h) \text{ uniformly for } h_0 > h \in H. \qquad (3.53)$$

1. *Then the discretization G^h is stable at u^h with stability bound S and stability threshold $r_0 = r/S$.*

2. *The above* $((G^h)'(v^h))^{-1}$ *in* $B_r(u^h)$ *can be replaced by* $((G^h)'(u^h))^{-1}$ *in* $B_s(u^h)$ *with* $s < r$ *and, for bounded* Φ^h, *the* $G^h \in C^1(B_r(u^h))$ *by* $G \in C^1(B_r(u)), B_r(u) \subset \mathcal{D}(G)$ *and* S *by* $S/2$ *in (3.53).*

Proof. The existence of an inverse operator $(G^h)^{-1} : B'_R \to \mathcal{U}_b^h$ in the balls $B'_R = B'_R(G^h u^h), R = r/S$, is shown as for Proposition 3.22. $B'_R \subset \mathcal{V}_b^{'h}$ is an open neighborhood of $G^h u^h$ with respect to $\mathcal{V}_b^{'h}$. Then Theorem 1.48 guarantees the differentiability of this inverse operator $(G^h)^{-1}$. Moreover it yields, uniformly for $h_0 > h \in H$,

$$\|((G^h)^{-1})'(G^h v^h)\|_{\mathcal{U}_b^h \leftarrow \mathcal{V}_b^{'h}} = \|((G^h)'(v^h))^{-1}\|_{\mathcal{U}_b^h \leftarrow \mathcal{V}_b^{'h}} \le S \tag{3.54}$$

implying

$$\|((G^h)')^{-1}(\delta)\|_{\mathcal{U}_b^h \leftarrow \mathcal{V}_b^{'h}} \le S \ \ \forall \ \ \delta \in B'_R(G^h u^h), \ \ h_0 > h \in H. \tag{3.55}$$

Now $B'_R(G^h u^h)$ is convex, hence, $G^h u_2^h + \lambda \left(G^h u_1^h - G^h u_2^h\right) \in B'_R(G^h u^h)$ for $0 \le \lambda < 1 \ \forall \ G^h u_1^h, G^h u_2^h \in B'_R(G^h u^h)$. For these u_i^h and with the existence of $((G^h)^{-1})'$ Theorem 1.43 yields

$$\begin{aligned}
\|u_1^h - u_2^h\|_{\mathcal{U}_b^h} &= \left\|(G^h)^{-1}\left(G^h u_1^h\right) - (G^h)^{-1}\left(G^h u_2^h\right)\right\|_{\mathcal{U}_b^h} \\
&\le \max_{\delta \in B'_R(G^h u^h)} \|((G^h)^{-1})'(\delta)\|_{\mathcal{U}_b^h \leftarrow \mathcal{V}_b^{'h}} \left\|(G^h u_1^h) - G^h u_2^h\right\|_{\mathcal{V}_b^{'h}} \\
&\le S \left\|G^h u_1^h - G^h u_2^h\right\|_{\mathcal{V}_b^{'h}} \ \text{ by (3.55)},
\end{aligned}$$

hence the claim. Theorem 1.46 and bounded Φ^h imply the modifications. □

Theorem 3.24. Stability of a perturbed nonlinear operator: *Under the conditions of Definition 3.12 let* $G, F, G + F \in \mathcal{D}(\Phi^h)$, *the discretization* $\bar{G}^h := G^h + F^h : \mathcal{D}(G^h) \cap \mathcal{D}(F^h) \subset \mathcal{U}_b^h \to \mathcal{V}_b^{'h}$ *and a sequence* $u^h \in \mathcal{U}_b^h$ *be given. Assume* $S, r, L > 0$ *exist, independent of* h, *such that* $B_r(u^h) \subset \mathcal{D}(G^h) \cap \mathcal{D}(F^h)$ *and*

1. G^h *are continuous and stable in* $B_r(u^h)$ *with the stability bound* S*;*
2. F^h *are Lipschitz-continuous in* $B_r(u^h)$, *with*

$$\left\|F^h\left(u_1^h\right) - F^h\left(u_2^h\right)\right\|_{\mathcal{V}_b^{'h}} \le L\left\|u_1^h - u_2^h\right\|_{\mathcal{U}^h} \ \ \textit{and} \ L \cdot S < 1.$$

Then $\bar{G}^h$ *is stable in* u^h, *with bound* $S/(1 - LS)$ *and threshold* $(1 - LS)r/S$.

Proof. For all $h \in H$ and $u_i^h \in B_r(u^h)$ we have

$$\begin{aligned}
\left\|\bar{G}^h u_1^h - \bar{G}^h u_2^h\right\|_{\mathcal{V}_b^{'h}} &= \left\|\left(G^h u_1^h - G^h u_2^h\right)\right\|_{\mathcal{V}_b^{'h}} - \left\|F^h u_1^h - F^h u_2^h\right\|_{\mathcal{V}_b^{'h}} \\
&\ge (1/S - L)\left\|u_1^h - u_2^h\right\|_{\mathcal{U}^h},
\end{aligned}$$

hence the claim. □

Stetter [596] proves the following modification:

Theorem 3.25. *Under the conditions in Theorem 3.24 replace 1., 2. there by*

1. $G^h : B_r(u^h) \to G^h(B_r(u^h))$ *are differentiable, one-to-one and* $G^h(B_r(u^h))$ *are convex with*

$$\|((G^h)'(v^h))^{-1}\|_{\mathcal{U}_b^h \leftarrow \mathcal{V}_b'^h} \le S \ \forall\ v^h \in B_r(u^h). \tag{3.56}$$

2. *For all* $v^h, u_1^h, u_2^h \in B_r(u^h)$ *and with* $L'S < 1$ *let*

$$\|((G^h)'(v^h))^{-1}\left(F^h\left(u_1^h\right) - F^h\left(u_2^h\right)\right)\|_{\mathcal{U}^h} \le L'\left\|u_1^h - u_2^h\right\|_{\mathcal{U}^h}. \tag{3.57}$$

Then again $\bar{G}^h$ *is stable in* u^h*, with bound* $S/(1-L')$ *and threshold* $(1-L')r/S$.

In the previous sections two related conditions compete with each other for nonlinear operators, G. We required well defined continuous G for consistency results and G^h for convergence. The continuity of $G : \mathcal{D}(G) \subset \mathcal{U} \to \mathcal{V}'$ can often be avoided, since consistency requires smoother functions $u \in \mathcal{D}(G) \cap \mathcal{U}_s$, where G is in fact well defined and continuous, cf. Theorems 4.53 and 5.4 We will discuss some possibilities in Example 3.26 below. The continuity of $G^h : \mathcal{D}(G^h) \subset \mathcal{U}^h \to \mathcal{V}'^h$ is much less restrictive than that of G, since the $u^h \in \mathcal{U}^h$ have a very special structure, e.g. piecewise polynomials for FE, DCG and wavelet methods on appropriate triangulations, or algebraic or trigonometric polynomials of a finite degree on the whole or domain decomposed Ω. For these concepts we refer to the corresponding chapters below.

The condition of continuous $G^h : \mathcal{D}(G^h) \subset \mathcal{U}^h \to \mathcal{V}'^h$ is unavoidable in the context of piecewise polynomial methods for fully nonlinear elliptic operators G on Ω. But it is applicable as well for other nonlinear operators G. E.g. a combination with the stability results in Section 3.6, implying boundedly invertible derivatives $G'(u_0)$, allows extended search for discrete and related exact solutions of quasilinear problems. These results might relax the conditions imposed, e.g. in Chapters 4 and 5.

Example 3.26. The G, G^h dilemma for the Monge-Ampère operator. G, G^h *have to be well defined and continuous*, cf. Example 2.78: We recall

$$G(u) := \det \nabla^2 u - f(x,u,\nabla u) \ \textit{and}\ = u_{xx}u_{yy} - u_{xy}^2 - f(x,u,\nabla u) \text{ for } n=2. \tag{3.58}$$

Let $\mathcal{D}_f := \{u : f(x,u(x),\nabla(x))$ *is defined a.e. in* $\Omega\}$. Obviously, this $G(u)$ is well defined for $u \in \mathcal{D}(G) := \mathcal{D}_f \cap C^2(\Omega)$, but certainly is $\notin L^2(\Omega)$ for $u \in \mathcal{D}_f \cap H^2(\Omega)$. So for (3.59), we choose the domain, $\mathcal{D}(G) \subset \mathcal{D}_f$, as dense subset, here as $C^2(\Omega)$ or $H^4_+(\Omega)$ of $H^2(\Omega)$ and the range, $\mathcal{R}(G) \subset L^2(\Omega)$, cf. Theorems 2.79 ff. Another possibility is $\mathcal{D}(G) = \mathcal{D}_f \cap \{u \in H^2(\Omega) \cap W^{2,\infty}(\Omega)\}$. Then the exact solution, u_0, satisfies the differential equation, and the boundary condition in the trace sense, cf. Subsections 5.2.4, 5.2.5. We assume $\mathcal{D}(G)$ to be defined s.t.

$$\begin{aligned}
&G : \mathcal{D}(G) \subset \mathcal{U} := H^2(\Omega) \to \mathcal{R}(G) \subset \mathcal{V} := L^2(\Omega) = \mathcal{V}' : G(u) := G^w(\cdot, u, \nabla u, \nabla^2 u),\\
&s.t.\ \forall u \in \mathcal{D}(G) : G(u) \in L^2(\Omega), G(u_0)(x) = 0 \forall\ a.e.\ x \in \Omega, \Leftrightarrow (G(u_0), v)_{L^2(\Omega)} = 0 \forall v \in \mathcal{V},\\
&\quad \text{and } u_0|_{\partial\Omega} = 0 \iff (u_0|_{\partial\Omega}, v_b)_{L^2(\partial\Omega)} = 0 \quad \forall v_b \in L^2(\partial\Omega),
\end{aligned} \tag{3.59}$$

with $G^w : \overline{\Omega} \times \mathbb{R} \times \mathbb{R}^n \times \mathbb{R}^{n^2} \to \mathbb{R}$, (see (2.304)). For this second order fully nonlinear problem (3.59) on a polyhedral Ω, we will choose $\mathcal{U}^h \subset C^1(\Omega)$, with piecewise poly nomials, $\mathcal{P}_d^n$ of degree d in n variables, and a triangulation, $\mathcal{T}^h$ e.g. with simplices T, cf. Definition 4.2. So

$$\textit{for a polyhedral } \Omega \textit{ and a triangulation for } \overline{\Omega} = \cup_{T \in \mathcal{T}^h} \overline{T} \subset \mathbb{R}^n \textit{ let}$$

$$\mathcal{U}^h = \{u^h : u^h|_T \in \mathcal{P}_d^n\} \cap C^1(\Omega), \quad \mathcal{U}_b^h = \{u^h \in \mathcal{U}^h : u^h|_{\partial\Omega} = 0\}. \tag{3.60}$$

For this Ω, and $u_0^h \in \mathcal{U}_b^h \subset H^2(\Omega) \cap H_0^1(\Omega)$, the FE solution, u_0^h, is determined by testing only the differential equation with respect to $v^h \in \mathcal{V}^h$, cf. Section 5.2.

$$u_0^h \in (\mathcal{U}_b^h = \mathcal{V}^h) \cap \mathcal{D}(G) \text{ s.t. } \left(G\left(u_0^h\right), v^h\right)_{L^2(\Omega)} = 0 \quad \forall v^h \in \mathcal{V}^h. \tag{3.61}$$

■

Proposition 3.27. *For the three cases of $u \in \mathcal{D}(G) \subset \mathcal{D}_f$ in the previous Example the $G(u) \in L^2(\Omega)$ and for $u^h \in \mathcal{U}^h \cap \mathcal{D}_f$ in (3.60) the $G(u^h) \in L^2(\Omega)$ are well defined a.e. for the Monge-Ampère equation in (3.58) and both G and G^h are continuous w.r.t. the $G : H^2(\Omega) \to L^2(\Omega)$ norms. Generalized to curved Ω, and to the order $2m$ with $\mathcal{U}^h \subset C^{2m-1}(\Omega)$, this remains correct and is proved exactly the same way.*

Proof. We formulate the proof only for $G(u^h)$ and $n = 2$. It is well known that for $u \in H^2(\Omega)$ the $u_{xx}u_{yy} \notin L^2(\Omega)$. But for the piecewise polynomials u^h in (3.60) we obtain $u^h, u_x^h, u_y^h, u_{xx}^h, u_{yy}^h, u_{xy}^h, u_{xx}^h u_{yy}^h, (u_{xy}^h)^2, \ldots \in L^2(\Omega) \cap L^\infty(\Omega)$. For $u_1^h, u_2^h \in \mathcal{U}^h$, we apply the Cauchy-Schwarz inequality (1.48) to any $T \in \mathcal{T}^h$:

$$\Big(\int_T (u_1^h|_T u_2^h|_T)dx\Big)^2 \le \int_T (u_1^h|_T)^2 dx \int_T (u_2^h|_T)^2 dx \textit{ implying} \tag{3.62}$$

$$\|u_1^h u_2^h\|_{L^2(T)}^4 = \Big(\int_T (u_1^h u_2^h|_T)^2 dx\Big)^2 = \Big(\int_T |u_1^h|_T^2 |u_2^h|_T^2 dx\Big)^2 \le \int_T (u_1^h|_T)^4 dx \int_T (u_2^h|_T)^4 dx.$$

These integrals are defined as limits of Riemann sums with appropriately distributed N knotes $x_i \in T$. The maximal distance between them vanishes and the covering with open volume elements $M_i \ni x_i$ and measure $\mu(M_i)$ satisfies $\overline{T} = U_{i=1}^N \overline{M}_i$.

$$\int_T (u^h|_T)^4 dx = \lim_{N\to\propto} \sum_{i=1}^N (u^h(x_i))^4 \mu(M_i) \tag{3.63}$$

$$\le \lim_{N\to\propto} \Big(\sum_{i=1}^N (u^h(x_i))^2 \mu(M_i)\Big) \max_{i=1}^N (u^h(x_i))^2 = \|u^h\|_{L^2(T)}^2 \|u^h\|_{L\propto(T)}^2,$$

and similarly for all other up to second order derivatives of the u^h, hence continuity. □

We prove stability for linear coercive, hence boundedly invertible operators and their compact perturbations. These linear coercive operators are usually the principal parts of elliptic operators, sometimes, e.g. for natural boundary conditions a term such as $c(u,v)_{L^2(\Omega)}, c>0$ has to be added. It is important to realize that for all elliptic problems discussed in this book, the weak principal parts always have the form

$$a_p(\cdot,\cdot):\mathcal{V}_b\times\mathcal{V}_b\to\mathbb{R}\ or\ \mathcal{V}\times\mathcal{V}\to\mathbb{R}.$$

In a few exceptional cases, e.g. for mesh–free methods, in one of the $\mathcal{V}\times\mathcal{V}$ a dense subspace has to be chosen. For our generalized Petrov–Galerkin methods, always the $\mathcal{U}_b^h,\mathcal{V}_b^h$, even for $\mathcal{U}_b^h\neq\mathcal{V}_b^h$, are approximating subspaces for the $\mathcal{U}_b=\mathcal{V}_b$, with the same norms $\|v^h\|_{\mathcal{V}^h}=\|v^h\|_{\mathcal{U}^h}$. So we additionally require

$$\text{for generalized Petrov–Galerkin methods: } \forall\epsilon>0\ \exists\delta<\epsilon\in\mathbb{R}_+: \tag{3.64}$$
$$\forall v^h\in\mathcal{V}_b^h\exists u^h\in\mathcal{U}_b^h:\|u^h-v^h\|_{\mathcal{U}^h}\le\delta\|v^h\|_{\mathcal{U}^h},\ \text{always correct for}\ \mathcal{U}_b^h=\mathcal{V}_b^h.$$

Fully nonlinear elliptic problems cannot be solved by generalized Petrov–Galerkin methods. In that case, (3.64) only plays the role for the weak linearized form of these problems. However, it has to be modified there for handling the violated boundary conditions.

For nonconforming FE, the DCG, the difference, and the mesh–free methods, the following proof has to be generalized by employing specific techniques. So we will formulate the corresponding stability results in those chapters.

So we restrict the discussion to conforming methods, cf. Examples 3.4, 3.8. These are, in our two books, the conforming FE and the spectral methods. Then the orthogonal projection Q'^h in (3.18) and the A^hu^h and f^h in (3.20) are strongly simplified. For

$$\mathcal{U}_b^h\subset\mathcal{U}_b,\mathcal{V}_b^h\subset\mathcal{V}_b:\ Q'^h:\mathcal{V}_b'\to\mathcal{V}_b'^h:\ \langle Q'^hf-f,v^h\rangle_{\mathcal{V}_b'\times\mathcal{V}_b}=0\forall v^h\in\mathcal{V}_b^h, \tag{3.65}$$
$$A^h=Q'^hA|_{\mathcal{U}_b^h},\quad with\ \left(Q'^hA|_{\mathcal{U}_b^h}\right)^{-1}:\mathcal{V}_b'^h\to\mathcal{U}_b^h,\ f^h=Q'^hf|_{\mathcal{V}_b^h}.$$

For $\mathcal{U}_b$-*coercive* $A\in\mathcal{L}(\mathcal{U}_b,\mathcal{U}_b')$, the stability of conforming generalized Petrov–Galerkin methods can be verified directly. Often this $A\in\mathcal{L}(\mathcal{U}_b,\mathcal{U}_b')=A|_{\mathcal{U}_b}$ of an $A\in\mathcal{L}(\mathcal{U},\mathcal{U}')$.

Theorem 3.28. $\mathcal{U}_b$-coercive operators induce stable Petrov–Galerkin methods:

1. *Let* $A\in\mathcal{L}(\mathcal{U}_b,\mathcal{U}_b')$ *be a linear* $\mathcal{U}_b$-coercive operator, *i.e.*
$$\exists M>0: a(u,u)=\langle Au,u\rangle_{\mathcal{U}'\times\mathcal{U}}\ge M\|u\|_{\mathcal{U}}^2,\quad \text{for all } u\in\mathcal{U}_b. \tag{3.66}$$
Then $A^{-1}:\mathcal{U}_b'\to\mathcal{U}_b$ *exists and* $\|A^{-1}\|_{\mathcal{U}_b\leftarrow\mathcal{U}_b'}\le 1/M$.
2. *For any conforming generalized Petrov–Galerkin method, satisfying (3.64) and with* Q'^h *in (3.65), the* $A^h=Q'^hA|_{\mathcal{U}_b^h}$ *is stable and*
$$\left\|\left(Q'^hA|_{\mathcal{U}_b^h}\right)^{-1}\right\|_{\mathcal{U}_b^h\leftarrow\mathcal{V}_b^h}\le 2/M\quad \text{for sufficiently small } h. \tag{3.67}$$

Proof. Relations (3.66) and (3.67) follow from Theorem 2.12 and (3.64). □

In general the existence of a bounded inverse of A is not sufficient to ensure that $A^h=Q'^hA|_{\mathcal{U}_b^h}$ is stable. However, the next result allows all types of elliptic equation, bordered systems, hence numerical methods for bifurcation, and Navier-Stokes equations, see the following chapters.

Theorem 3.29. For conforming generalized Petrov–Galerkin methods invertible A imply stable A^h: *For the approximating spaces* $(\mathcal{U}_b^h, \ \mathcal{V}_b^h)_{h \in H}$ *for* $\mathcal{U}_b, \mathcal{V}_b$, *satisfying (3.64), let* $B \in \mathcal{L}(\mathcal{U}_b, \mathcal{V}_b')$ *be boundedly invertible and the* $Q'^h B|_{\mathcal{U}_b^h}$ *be stable. Let* $A := B + C$, *with* $C \in \mathcal{C}(\mathcal{U}_b, \mathcal{V}_b')$, *the set of compact operators from* $\mathcal{U}_b \to \mathcal{V}_b'$. *Then:*

$$A^{-1} \in \mathcal{L}(\mathcal{V}_b', \mathcal{U}_b) \Rightarrow Q'^h A|_{\mathcal{U}_b^h} \ \text{ is stable.}$$

Remark 3.30. *The existence of* $A^{-1} \in \mathcal{L}(\mathcal{V}_b', \mathcal{U}_b)$ *has to be discussed in a little bit more detail. Since* $B^{-1} \in \mathcal{L}(\mathcal{V}_b', \mathcal{U}_b)$ *exists, this implies* $B^{-1}A = I + B^{-1}C \in \mathcal{L}(\mathcal{U}_b, \mathcal{U}_b)$ *and, as a product of a continuous and a compact operator,* $B^{-1}C \in \mathcal{C}(\mathcal{U}_b, \mathcal{U}_b)$ *is a compact perturbation of the identity. Therefore the Riesz–Schauder theory applies, cf. Theorem 2.20. Hence for* $\lambda \in C \setminus \{0\}$ *either* $(B^{-1}C - \lambda I)^{-1} \in \mathcal{L}(\mathcal{U}_b, \mathcal{U}_b)$ *or* λ *is one of the at most countably many eigenvalues of* $B^{-1}C$. *This implies that* A^{-1} *exists if and only if* $\lambda = -1$ *is no eigenvalue of* $B^{-1}C$.

Proof. We have to show that $A^{-1} \in \mathcal{L}(\mathcal{V}_b', \mathcal{U}_b)$ implies the stability of $Q'^h A|_{\mathcal{U}_b^h}$. We determine for an arbitrary $u \in \mathcal{U}_b$ and $v := Cu$ the, by assumption, unique exact and discrete solutions, $\hat{u}$ and $\hat{u}^h$, of the equations $B\hat{u} = v$ and $Q'^h B|_{\mathcal{U}_b^h} \hat{u}^h = Q'^h B \hat{u}^h = Q'^h v \in \mathcal{V}_b'^h$, cf. (3.4). With the notation $B^h := Q'^h B|_{\mathcal{U}_b^h}$, $T := B^{-1} \in \mathcal{L}(\mathcal{V}_b', \mathcal{U}_b)$ and $T^h := B^{h^{-1}} Q'^h \in \mathcal{L}(\mathcal{V}_b', \mathcal{U}_b^h)$, Theorems 3.16 and 3.21 imply that

$$\lim_{h \to 0} ||\hat{u} - \hat{u}^h||_{\mathcal{U}_b} = ||(T - T^h)Cu||_{\mathcal{U}_b} = 0 \text{ and } u \in \mathcal{U}_b \ \textit{or} \tag{3.68}$$

$$\textit{for any } v \in \mathcal{V}_b' \text{ and } B\tilde{u} = v, B^h \tilde{u}^h = Q'^h v : \lim_{h \to 0} ||\tilde{u} - \tilde{u}^h||_{\mathcal{U}_b} = ||(T - T^h)v||_{\mathcal{U}_b} = 0.$$

By Lemma 3.31 this second limit implies

$$||(T - T^h)C||_{\mathcal{U}_b \leftarrow \mathcal{U}_b} \to 0 \quad \text{ for } h \to 0. \tag{3.69}$$

Now let $u^h \in \mathcal{U}_b^h \subset \mathcal{U}_b$. Because A is boundedly invertible, we can estimate

$$\begin{aligned} ||u^h||_{\mathcal{U}_b} &\le ||A^{-1}||_{\mathcal{U}_b \leftarrow \mathcal{V}_b'} ||Au^h||_{\mathcal{V}_b'} = ||A^{-1}||_{\mathcal{U}_b \leftarrow \mathcal{V}_b'} ||B(I + TC)u^h||_{\mathcal{V}_b'} \\ &\le ||A^{-1}||_{\mathcal{U}_b \leftarrow \mathcal{V}_b'} ||B||_{\mathcal{V}_b' \leftarrow \mathcal{U}_b} ||(I + TC)u^h||_{\mathcal{U}_b}, \text{ hence,} \end{aligned}$$

$$||(I + TC)u^h||_{\mathcal{U}_b} \ge \left(||A^{-1}||_{\mathcal{U}_b \leftarrow \mathcal{V}_b'} ||B||_{\mathcal{V}_b' \leftarrow \mathcal{U}_b}\right)^{-1} ||u^h||_{\mathcal{U}_b}. \tag{3.70}$$

For the following we recall that $Q'^h Bu$ is defined for every $u \in \mathcal{U}_b$. We apply the stability of B^h to $w^h := B^h u^h$ to find $\|u^h\|_{\mathcal{U}_b} \le \|(B^h)^{-1}\|_{\mathcal{U}_b^h \leftarrow \mathcal{V}_b'^h} \cdot \|w^h\|_{\mathcal{V}_b'}$. Furthermore

$$\begin{aligned} \|(I + T^h C)u^h\|_{\mathcal{U}_b} &= \|(B^h)^{-1} B^h (I + T^h C)u^h\|_{\mathcal{U}_b} \\ &\le \|(B^h)^{-1}\|_{\mathcal{U}_b^h \leftarrow \mathcal{V}_b'^h} \|B^h (I + T^h C)u^h\|_{\mathcal{V}_b'} \end{aligned}$$

implies

$$\|B^h(I+T^hC)u^h\|_{\mathcal{V}'_b} \geq \frac{1}{\|(B^h)^{-1}\|_{\mathcal{U}^h_b\leftarrow\mathcal{V}'^h_b}} \cdot \|(I+T^hC)u^h\|_{\mathcal{U}_b}. \tag{3.71}$$

We combine (3.70), (3.71) for estimating

$$\begin{aligned}
\left\|Q'^h Au^h\right\|_{\mathcal{V}'_b} &= \|B^h(I+TC)u^h\|_{\mathcal{V}'_b} \geq \|B^h(I+T^hC)u^h\|_{\mathcal{V}'_b} - \|Q'^h B(T-T^h)Cu^h\|_{\mathcal{V}'_b} \\
&\overset{(3.71)}{\geq} \|(I+T^hC)u^h\|_{\mathcal{U}_b}/\|(B^h)^{-1}\|_{\mathcal{U}^h_b\leftarrow\mathcal{V}'^h_b} - \|Q'^h B(T-T^h)Cu^h\|_{\mathcal{V}'_b} \\
&\geq 1/\|(B^h)^{-1}\|_{\mathcal{U}^h_b\leftarrow\mathcal{V}'^h_b} \left(\|(I+TC)u^h\|_{\mathcal{U}_b} - \|(T-T^h)Cu^h\|_{\mathcal{U}_b}\right) \\
&\quad -\|Q'^h B(T-T^h)C\|_{\mathcal{V}'_b\leftarrow\mathcal{U}_b}\|u^h\|_{\mathcal{U}_b} \\
&\overset{(3.70)}{\geq} \frac{1}{\|B^{h^{-1}}\|_{\mathcal{U}^h_b\leftarrow\mathcal{V}'^h_b}} \left[\frac{1}{\left(\|A^{-1}\|_{\mathcal{U}_b\leftarrow\mathcal{V}'_b}\|B\|_{\mathcal{V}'_b\leftarrow\mathcal{U}_b}\right)} \right. \\
&\quad \left. -\mathcal{O}\left(\|(T-T^h)C\|_{\mathcal{V}'_b\leftarrow\mathcal{U}_b}\right)\right] \|u^h\|_{\mathcal{U}_b}.
\end{aligned}$$

Because of (3.69) and the stability of $(B^h)_{h\in H}$ there exists a positive constant K, independent of h, such that for all $h \leq h_0$ the following holds:

$$\|Q'^h Au^h\|_{\mathcal{V}'_b} \geq K\|u^h\|_{\mathcal{U}_b} \quad \text{for all } u^h \in \mathcal{U}^h_b.$$

This and $\dim \mathcal{U}^h_b = \dim \mathcal{V}'^h_b < \infty$ shows that $Q'^h A|_{\mathcal{U}^h_b}$ is invertible for $h \leq h_0$. Moreover, we obtain $\left\|\left(Q'^h A|_{\mathcal{U}^h_b}\right)^{-1}\right\|_{\mathcal{U}^h_b\leftarrow\mathcal{V}'^h_b} \leq 1/K$, i.e. the stability. □

Lemma 3.31. Raasch [539]: *With the notation and conditions introduced in the previous proof, the second line of (3.68) implies (3.69).*

Proof. We have to prove the implication that for any

$$\forall v \in \mathcal{V}'_b \text{ and } B\tilde{u} = v, B^h\tilde{u}^h = Q'^h v$$

$$\forall v \in \mathcal{V}'_b : \lim_{h\to 0} \|\tilde{u} - \tilde{u}^h\|_{\mathcal{U}_b} = \|(T-T^h)v\|_{\mathcal{U}_b} = 0 \tag{3.72}$$

$$\Longrightarrow \lim_{h\to 0} \|(T-T^h)C\|_{\mathcal{U}_b\leftarrow\mathcal{U}_b} = 0. \tag{3.73}$$

This limit in (3.72) is a consequence of the stability and consistency of B^h with B, hence the convergence of the method.

For the compact operator $C : \mathcal{U}_b \to \mathcal{V}'_b$, and with the previous $v := Cu$, we claim that

$$\mathcal{M} := \{u \in \mathcal{U}_b : \|u\|_{\mathcal{U}_b} = 1\} \ni u \text{ implies } \|(T-T^h)Cu\|_{\mathcal{U}_b} \leq C'\epsilon = C'\epsilon\|u\|_{\mathcal{U}_b} \tag{3.74}$$

implying (3.73). With C, the closed image $\overline{C\mathcal{M}}$ is compact. Therefore by Definition 1.8 ff. the following *finite open coverings* exist, such that

$$\forall \epsilon > 0 \exists N = N(\epsilon) : v_1, \ldots, v_N \in \mathcal{V}_b' \ s.t. \tag{3.75}$$

$$\overline{C\mathcal{M}} \subset \bigcup_{i=1}^{N} \left\{ v \in \mathcal{V}_b' : \|v - v_i\|_{\mathcal{V}_b'} \leq \epsilon \right\} \ \textit{and by (3.72)}$$

$$\Longrightarrow \exists h_0 : \ ||(T - T^h)v_i||_{\mathcal{U}_b} \leq \epsilon \ \ \forall h < h_0, i = 1, \ldots, N.$$

As a consequence of (3.73) and the equiboundedness of T, T^h, hence $||(T - T^h)||_{\mathcal{U}_b \hookleftarrow \mathcal{V}_b'} \leq C^*$, we find

$$\forall u \in \mathcal{M} \exists i = i_u \in \{1, \ldots, N\} \ \textit{with} \ \|Cu - v_i\|_{\mathcal{V}_b'} \leq \epsilon \ \textit{hence} \tag{3.76}$$

$$\begin{aligned} ||(T - T^h)Cu||_{\mathcal{U}_b} &\leq ||(T - T^h)(Cu - v_i)||_{\mathcal{U}_b} + ||(T - T^h)v_i||_{\mathcal{U}_b} \\ &\leq C^*\epsilon + \epsilon \leq (1 + C^*)\epsilon \leq C'\epsilon\|u\|_{\mathcal{U}_b}, \end{aligned}$$

hence the claim by (3.73). □

An important example for operators, which are compact perturbations of coercive operators, are $A \in \mathcal{L}(\mathcal{U}_b, \mathcal{U}_b')$ satisfying a Gårding inequality, cf. (2.141).

Theorem 3.32. Gårding inequality and compact perturbation: *In the Gelfand triple of Banach spaces $\mathcal{U}_b \subset \mathcal{W} = \mathcal{W}' \subset \mathcal{U}'$, cf. Definition 2.17, let $\mathcal{U}_b$ be (continuously, densely) and compactly embedded into the Hilbert space $\mathcal{W}$. Assume that $A \in \mathcal{L}(\mathcal{U}_b, \mathcal{U}_b')$ fulfills a Gårding inequality, i.e. $\exists M > 0$ and $m \in \mathbb{R}$, such that*

$$< Au, u >_{\mathcal{U}_b' \times \mathcal{U}_b} \geq M||u||_{\mathcal{U}_b}^2 - m||u||_{\mathcal{W}}^2 \ \ \forall \ \ u \in \mathcal{U}_b. \tag{3.77}$$

Then A is a compact perturbation of a coercive operator.

Proof. With the identified $\mathcal{W}$ with $\mathcal{W}'$, the above assumptions yield $\mathcal{U}_b \subset \mathcal{W} \subset \mathcal{U}_b'$. Additionally the embedding $I_{\mathcal{U}_b \to \mathcal{U}_b'}$ is compact, see Zeidler [677]. We obtain the splitting

$$A = \left(A + mI_{\mathcal{U}_b \to \mathcal{U}_b'}\right) - mI_{\mathcal{U}_b \to \mathcal{U}_b'}, \text{ with coercive } \left(A + mI_{\mathcal{U}_b \to \mathcal{U}_b'}\right),$$

hence the claim. □

Finally, we discuss variational crimes of the easiest type. We allow inexact evaluations of the operators $Q'^h A|_{\mathcal{U}_b^h}, Q'^h G|_{\mathcal{U}_b^h}$, and $Q'^h A_h|_{\mathcal{U}_b^h}, Q'^h G_h|_{\mathcal{U}_b^h}$, and of $Q'^h f$, cf. Example 3.8. This is necessary, since the exact values often are not available or are too expansive. So they are approximated, e.g. by numerical integration. The influence of this type of perturbation is analyzed in the following theorem. We only present the basic result for conforming Petrov–Galerkin methods. More detailed results including nonconforming FE and spectral methods will again be discussed in Chapters 4 and in [120].

Theorem 3.33. *Let $\mathcal{U}_b^h, \mathcal{V}_b^h$ be approximating spaces for $\mathcal{U}_b, \mathcal{V}_b$ in a conforming Petrov–Galerkin method, and $f \in \mathcal{V}_b'$. Assume $\tilde{A}^h \in \mathcal{L}(\mathcal{U}_b^h, \mathcal{V}_b'^h)$ and $\tilde{f}^h \in \mathcal{V}_b'^h$ are* *"proper" perturbations of* $A^h = Q'^h A|_{\mathcal{U}_b^h}$ *and* $Q'^h f$*, respectively, i.e.*

$$||\tilde{A}^h - Q'^h A|_{\mathcal{U}_b^h}||_{\mathcal{V}_b'^h \leftarrow \mathcal{U}_b^h} \to 0 \text{ and } ||\tilde{f}^h - Q'^h f||_{\mathcal{V}_b^{h\prime h}} \to 0 \text{ for } h \to 0. \tag{3.78}$$

1. *Then, for sufficiently small* h*,*

$$\tilde{A}^h \text{ is stable } \Leftrightarrow \left(Q'^h A|_{\mathcal{U}_b^h}\right) \text{ is stable.} \tag{3.79}$$

2. *Furthermore, if* $A^h = Q'^h A|_{\mathcal{U}_b^h}$ *is stable, the uniquely determined solution* $\tilde{u}_0^h$ *of the perturbed discrete equation*

$$\tilde{A}^h \tilde{u}_0^h = \tilde{f}^h, \quad \tilde{u}_0^h \in \mathcal{U}_b^h$$

converges to the uniquely determined solution u_0^h *of* $A^h u_0^h = f^h$*, more precisely, for sufficiently small* h*, we obtain finally, with* $\|(A^h)^{-1}\|_{\mathcal{U}_b^h \leftarrow \mathcal{V}_b'^h} \le S$*,*

$$\|\tilde{u}^h - u^h\|_{\mathcal{U}_b^h} \le 2S^2 \|\tilde{A}^h - A^h\|_{\mathcal{V}_b'^h \leftarrow \mathcal{U}_b^h} \|f\|_{\mathcal{V}_b'^h} + S\|\tilde{f} - f\|_{\mathcal{V}_b'^h}. \tag{3.80}$$

Proof. The condition (3.78) immediately shows that for small enough h the relation (3.79) is correct. For proving the estimate we consider

$$\begin{aligned}
A^h u^h &= f, \quad u^h = (A^h)^{-1} f \\
\tilde{A}^h \tilde{u}^h &= (A^h + (\tilde{A}^h - A^h))\tilde{u}^h = A^h (I + (A^h)^{-1}(\tilde{A}^h - A^h))\tilde{u}^h = \tilde{f} \\
\tilde{u}^h &= (I + (A^h)^{-1}(\tilde{A}^h - A^h))^{-1}(A^h)^{-1}\tilde{f} \\
\tilde{u}^h - u^h &= (I + (A^h)^{-1}(\tilde{A}^h - A^h))^{-1}(A^h)^{-1}\tilde{f} - (A^h)^{-1}\tilde{f} + (A^h)^{-1}(\tilde{f} - f).
\end{aligned}$$

We assume, e.g. for sufficiently small h, that $\|(A^h)^{-1}(\tilde{A}^h - A^h)\|_{\mathcal{U}_b^h \leftarrow \mathcal{U}_b^h} < 1/n$. With the Neumann series we obtain

$$(I + (A^h)^{-1}(\tilde{A}^h - A^h))^{-1} = \sum_{j=0}^{\infty} (-(A^h)^{-1}(\tilde{A}^h - A^h))^j \Longrightarrow$$

$$(I + (A^h)^{-1}(\tilde{A}^h - A^h))^{-1} = (I + (A^h)^{-1}(\tilde{A}^h - A^h)) + \mathcal{O}(\|(A^h)^{-1}(\tilde{A}^h - A^h)\|^2_{\mathcal{U}_b^h \leftarrow \mathcal{U}_b^h}).$$

This implies

$$\begin{aligned}
\tilde{u}^h - u^h = &\ \Big((I + (A^h)^{-1}(\tilde{A}^h - A^h)) + \mathcal{O}(\|(A^h)^{-1}(\tilde{A}^h - A^h)\|^2_{\mathcal{U}_b^h \leftarrow \mathcal{U}_b^h})\Big)(A^h)^{-1}\tilde{f} \\
- &\ (A^h)^{-1}\tilde{f} + (A^h)^{-1}(\tilde{f} - f) \\
= &\ \Big((A^h)^{-1}(\tilde{A}^h - A^h) + \mathcal{O}(\|(A^h)^{-1}(\tilde{A}^h - A^h)\|^2_{\mathcal{U}_b^h \leftarrow \mathcal{U}_b^h})\Big) \\
&\ (A^h)^{-1}\tilde{f} + (A^h)^{-1}(\tilde{f} - f).
\end{aligned}$$

So we obtain finally with $\|(A^h)^{-1}\|_{\mathcal{U}_b^h \leftarrow \mathcal{V}_b'^h} \le S$

$$\|\tilde{u}^h - u^h\|_{\mathcal{U}_b^h} \le 2S^2 \|\tilde{A}^h - A^h\|_{\mathcal{V}_b'^h \leftarrow \mathcal{U}_b^h} \|\tilde{f}\|_{\mathcal{V}_b'^h} + S\|\tilde{f} - f\|_{\mathcal{V}_b'^h},$$

hence the claim. □

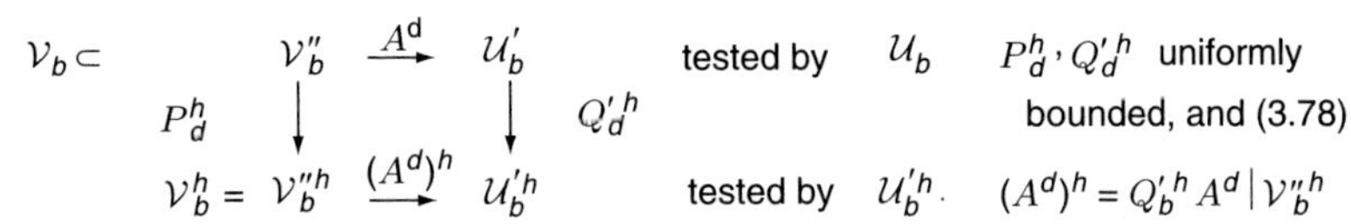

Figure 3.3 General discretization methods for dual problems.

3.6 Stability implies invertibility

In this section we study the opposite relation: under which conditions is the invertibility of an operator A a consequence of the stability of its discretization A^h. In fact we show that stability is, for some cases, the stronger condition, but both properties are equivalent under rather general conditions. Again, we restrict the discussion to *conforming Petrov–Galerkin methods,* with $A^h = Q'^h A$. Extensions to the other methods would be possible as well.

We have to simultaneously study $Au_0 = f$, cf. (3.4), and its dual problem. So the original equation and its dual form

$$A : \mathcal{U}_b \to \mathcal{V}_b', \ Au = f \text{ and } A^d : \mathcal{V}_b'' \to \mathcal{U}_b', \ A^d v_0'' = f' \tag{3.75}$$

induce a pair of corresponding discrete equations of the form

$$\begin{aligned} A^h : \mathcal{U}_b^h \subset \mathcal{U}_b \to \mathcal{V}_b'^h, \quad &\text{and} \quad (A^d)^h : \mathcal{V}_b^h = \mathcal{V}_b''^h \subset \mathcal{V}_b'' \to \mathcal{U}_b'^h, \\ A^h u_0^h = Q'^h A u_0^h = Q'^h f \quad &\text{and} \quad (A^d)^h v_0'^h = Q_d'^h A^d v_0'^h = Q_d'^h f'. \end{aligned} \tag{3.76}$$

This $\mathcal{V}_b^h = \mathcal{V}_b''^h \subset \mathcal{V}_b$ is a consequence of the assumed conforming Petrov–Galerkin method with the bi-dual pair of approximating spaces. For P_d^h we choose an interpolation or approximation operator, similar to the previous P^h for $\mathcal{U}_b^h$. The $Q_d'^h$ are defined as special case of (3.18), as

$$Q_d'^h : \mathcal{U}_b' \to \mathcal{U}_b'^h : \ \left\langle Q_d'^h g' - g', u^h \right\rangle_{\mathcal{U}_b'^h \times \mathcal{U}_b''^h} = 0 \forall u^h \in \mathcal{U}_b''^h. \tag{3.77}$$

Therefore the situation in Figure 3.2 implies a corresponding diagram for $A^d \in \mathcal{L}(\mathcal{V}_b'', \mathcal{U}_b')$ in Figure 3.3.

For the proof of Theorem 3.37 we need the following proposition for the bounded invertibility of a linear operator between Banach spaces. It is a consequence of Theorem 2.12, however it can be proved directly via Theorem 1.19.

Proposition 3.34. *Let $\mathcal{U}$, $\mathcal{V}$ be Banach spaces and $A \in \mathcal{L}(\mathcal{U}, \mathcal{V}')$. Then the following conditions are equivalent:*

(1) $A^{-1} \in \mathcal{L}(\mathcal{V}', \mathcal{U})$;
(2) $(A^d)^{-1} \in \mathcal{L}(\mathcal{U}', \mathcal{V}'')$;
(3) *there exist positive constants C_1, C_2 such that*

   (a) $||Au||_{\mathcal{V}'} \geq C_1 ||u||_{\mathcal{U}}$ *for all $u \in \mathcal{U}$ and*
   (b) $||A^d v''||_{\mathcal{U}'} \geq C_2 ||v''||_{\mathcal{V}''}$ *for all $v'' \in \mathcal{V}''$.*

If one of the above conditions is satisfied, we obtain

$$||A^{-1}||_{\mathcal{U}\leftarrow\mathcal{V}'} = ||(A^d)^{-1}||_{\mathcal{V}''\leftarrow\mathcal{U}'} \leq \min(1/C_1, 1/C_2).$$

Remark 3.35. *The three conditions (1), (2), (3) are not symmetric: with $\mathcal{D}(A), \mathcal{N}(A)$, $\mathcal{R}(A)$ denoting the domain, null-space and range of A, respectively, note that $A \in \mathcal{L}(\mathcal{U}, \mathcal{V}')$ implies $\mathcal{D}(A) = \mathcal{U}$. Therefore $A^d \in L(\mathcal{V}'', \mathcal{U}')$ is defined and $A^{-1} \in L(\mathcal{V}', \mathcal{U})$ exists if and only if $(A^d)^{-1} \in L(\mathcal{U}', \mathcal{V}'')$ exists. Hence, we only need one of the equivalent conditions (1) or (2). However, in (3) we indeed need booth conditions, since (3)(a) still admits $A\mathcal{U} \neq \mathcal{V}'$.*

Proof. The equivalence "(1) $\Leftrightarrow$ (2)" is well known. To show "(1) $\Rightarrow$ (3)" let $A^{-1} \in \mathcal{L}(\mathcal{V}', \mathcal{U})$ (and therefore $(A^d)^{-1} \in \mathcal{L}(\mathcal{U}', \mathcal{V}'')$). This yields immediately condition (3) with $C_1 = 1/||A^{-1}||$ (and $C_2 = 1/||((A^d)^{-1}|| = C_1$).

"(1) $\Leftarrow$ (3)" 3(a) and 3(b) imply that A and A^d are injective. 3(a) yields, for a convergent sequence $\{Au_k\}$, $u_k \in \mathcal{U}$, that $\{u_k\}$ is a Cauchy sequence and consequently converges to an $u \in \mathcal{U}$. A is continuous by assumption, so we obtain $Au_k \to Au$ and can conclude that $\mathcal{R}(A)$ is closed in $\mathcal{V}'$. The closed range theorem and the injectivity of A^d yield that A is surjective. Finally, the inverse mapping theorem implies the first condition. □

Remark 3.36. *Applying Proposition 3.34 to the bi-dual Petrov–Galerkin approximation case $A^h : \mathcal{U}_b^h \to \mathcal{V}_b'^h$, and $(A^d)^h : \mathcal{V}_b^h = \mathcal{V}_b''^h \to \mathcal{U}_b'^h$, shows that A^h is stable if and only if $(A^d)^h$ is stable, and for both sequences the same stability constants can be chosen. For applying Theorem 3.21 we need a solution, u_0, for $Au_0 = f$, and the properties* 1., 2., 3.

Theorem 3.37. Stability implies the existence of A^{-1}:

1. *For $A \in \mathcal{L}(\mathcal{U}_b, \mathcal{V}_b')$ let the pair of bi-dual approximating spaces $(\mathcal{U}_b^h,\ \mathcal{V}_b^h = \mathcal{V}_b''^h)$, define a generalized Petrov–Galerkin method. Then the stability of $A^h = Q'^h A|_{\mathcal{U}_b^h}$ implies the existence of A^{-1}.*
2. *If $A^h = Q'^h A|_{\mathcal{U}_b^h}$ is replaced, according to Theorem 3.33, by an approximation $\hat{A}^h$ the corresponding discrete operator $(\hat{A}^d)^h$ for the dual of A has to satisfy (3.78) as well. In this sense the stability of $\hat{A}^h$ again implies the existence of $A^{-1} \in \mathcal{L}(\mathcal{V}_b', \mathcal{U}_b)$.*

Proof. We have to show that the existence of a bounded inverse of A follows from the stability of $Q'^h A|_{\mathcal{U}_b^h}$. Let $u_0 \in \mathcal{U}_b$ and $v_0'' \in \mathcal{V}_b''$ be given. Obviously, they solve $f := Au_0 \in \mathcal{V}_b'$, $g' := A^d v_0'' \in \mathcal{U}_b'$, respectively. The corresponding discrete solutions u_0^h, $v_0''^h$, see (3.75), (3.76), uniquely exist, since Remark 3.36 and the conforming bi-dual Petrov–Galerkin method imply $\|Q'^h\| = \|Q_d'^h\| = 1$, stability and consistency for A^h and $(A^h)^d = (A^d)^h$, hence $\|u_0^h - u_0\|_{\mathcal{U}_b} \to 0$, $\|v_0''^h - v_0''\|_{\mathcal{V}_b'} \to 0$ for $h \to 0$. One obtains the following estimates:

$$\|u_0^h\|_{\mathcal{U}_b} = \|(Q'^h A|_{\mathcal{U}_b^h})^{-1} Q'^h f\|_{\mathcal{U}_b} = \|(Q'^h A|_{\mathcal{U}_b^h})^{-1} Q'^h A u_0\|_{\mathcal{U}_b}$$

$$\leq \|(Q'^h A|_{\mathcal{U}_b^h})^{-1}\|_{\mathcal{U}_b^h \leftarrow \mathcal{V}_b'^h} \|Q'^h\|_{\mathcal{V}_b'^h \leftarrow \mathcal{V}_b'} \cdot \|A u_0\|_{\mathcal{V}_b'} \leq C \|A u_0\|_{\mathcal{V}_b'} \text{ and}$$

$$\|v_0''^h\|_{\mathcal{V}_b''} = \|(Q_d'^h A^d|_{\mathcal{V}_b''^h})^{-1} Q_d'^h g'\|_{\mathcal{V}_b''} = \|(Q_d'^h A^d|_{\mathcal{V}_b''^h})^{-1} Q_d'^h A^d v_0''\|_{\mathcal{V}_b''}$$

$$\leq \|(Q'^h A|_{\mathcal{U}_b^h})^{-1}\|_{\mathcal{U}_b^h \leftarrow \mathcal{V}_b'^h} \|A^d v_0''\|_{\mathcal{U}_b'} \leq C \|A^d v_0''\|_{\mathcal{U}_b'}.$$

Theorem 3.21 applied to the bi-dual Petrov–Galerkin method yields the convergence of u_0^h and $v_0''^h$ to the chosen solutions u_0 and v_0'', respectively. Consequently, we obtain with the above estimates and the continuity of the norms

$$\|A u_0\|_{\mathcal{V}_b''} \geq K \|u_0\|_{\mathcal{U}_b} \quad \text{and} \quad \|A^d v_0''\|_{\mathcal{U}_b'} \geq K \|v_0''\|_{\mathcal{V}_b'}.$$

Therefore Proposition 3.34 implies that $A^{-1} \in \mathcal{L}(\mathcal{V}_b'', \mathcal{U}_b)$ does exist. If, instead of $A^h = Q'^h A|_{\mathcal{U}_b^h}$ a sequence $\hat{A}^h$ is chosen, the corresponding $(\hat{A}^d)^h$ has to satisfy (3.78) as well. □

3.7 Solving nonlinear systems: Continuation and Newton's method based upon the mesh independence principle (MIP)

3.7.1 Continuation methods

For the many discretization methods discussed in our two books much more specific solution techniques, in particular for linear problems, have been studied. The joint structure of these methods is less strong than for general discretization methods. So we do not aim at a synopsis of these methods. Rather we indicate two principles, allowing for all presented cases a simultaneous solution technique. We determine the discrete solutions, u_0, by combing the well-known continuation methods with the *mesh independence principle* (MIP). Proving convergence results for Newton's method for many discrete problems avoiding the MIP often require highly technical efforts.

In continuation methods we embed the discrete problem $G^h(u_0^h) = 0$ into a one-parameter problem. We identify u_0^h and $x_1 \in \mathbb{R}^m$ and define

$$F^h : \mathbb{R}^{m+1} \to \mathbb{R}^m \text{ with } F^h(x,\lambda) := (1-\lambda)G_0^h(x) + \lambda G_1^h(x) \text{ where} \tag{3.79}$$

$$G_1^h := G^h \text{ and a } G_0^h : \mathbb{R}^m \to \mathbb{R}^m \textit{ with a known approximate solution } x_0 \in \mathbb{R}^m.$$

We start we the continuation process with $\lambda = 0$ and the known (approximate) solution x_0 aiming for $\lambda = 1$: we only consider the easiest case and refer for other strategies to, e.g. Allgower and Georg [20, 21], Decker and Keller [278, 439, 440] Rheinboldt and Burkardt [550, 551], and Seydel [580, 581].

This allows formulating the following algorithm:

Algorithm 3.1.
Continuation method by parametrization

INPUT: $(x_0, \lambda_0 = 0) \in \mathbb{R}^m \times \mathbb{R}, G_0^h, G_1^h, G_0^h(x_0) = G^h(x_0, 0) \approx 0)$; starting point with $h > 0$ (Initial step size);

OUTPUT: (x_i, λ_i), $i = 1, 2, \ldots$, (Stop at $\lambda_{i_e} = 1$ with $\left(x_{i_e} \approx u_0^h, \lambda_{i_e} = 1\right)\big)$;

ITERATION: FOR $(i = 1$; Stopping criterion)

(i) $\lambda_i = \min\{\lambda_{i-1} + h, 1\}$; $\quad x_i = x_{i-1}$;
(ii) With Newton-like methods (MIP) improve this x_i into $x_i \in \mathbb{R}^m$, such that $F^h(x_i, \lambda_i) \approx 0$;
(iii) Stop for $\lambda_i = 1$, else goto (iv);
(iv) Adapt the stepsize $h > 0$, goto (i).

Figure 3.4 shows the predictor and the trace of a few corrector steps.

3.7.2 MIP for nonlinear systems

As a consequence of the general discretization approach, the number of iterations in the Newton methods for the original G in (3.81) and its discrete counterpart G^h in (3.84) are strongly related. The number of iterations for a given tolerance, $\|u_i^h - u_0^h\|_{\mathcal{U}^h}^2 \le \epsilon$, for G and G^h, is nearly the same, independent of h. This is achieved by the MIP, cf. Theorem 3.40, Allgower and Böhmer [9], and Allgower *et al.* [17].

We only present this main form of the MIP. There are many other papers on this subject, dealing with different aspects of discrete Newton methods: Deuflhard and Potra [285] discuss the “affine” Newton method, and Böhmer [112] combines it with the defect correction method. In particular, Argyros [36, 37] discusses Newton methods, where the derivative of the operator is Hölder instead of the original

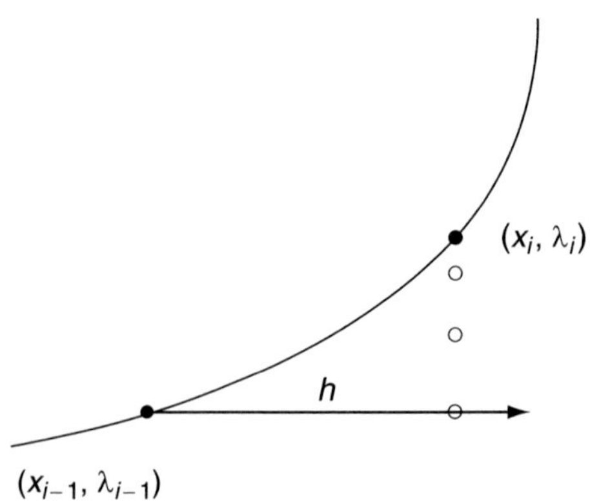

Figure 3.4 Natural parametrization.

Lipschitz-continuous, or the discretized secant iterations. In both cases the quadratic convergence is reduced. Instead of discretizations he allows perturbed or inexact Newton methods or directly works in Banach spaces, sometimes based upon majorant principles.

Efficient strategies combining the MIP with sequences of increasingly finer meshes have been studied by Allgower and Böhmer [9, 10]. They can be combined with the nested sequences of, e.g. FE spaces $\mathcal{S}_d^1\left(\mathcal{T}_c^{h,1}\right) \sqsubset \mathcal{S}_d^1\left(\mathcal{T}_c^{h,2}\right) \sqsubset \cdots$ yielding multiscaling or multiresolution techniques.

For formulating the MIP we recall the main condition for the existence of a unique solution u_0 for the original $G(u) = 0$ and the convergence for a general discretization: it is the bounded invertibility of the linearized $G'(u_0)$. For a $G \in C_L^1$, this is simultaneously the main condition for the (quadratic) convergence of the Newton–Kantorovich method applied to G. We do not want to formulate the sharpest possible conditions for the following theorems, but rather give their flavor. We compare Newton's method applied to G and to G^h. We additionally assume a Lipschitz-continuous derivative, so

$$\text{Let } u_0 \text{ be an isolated solution of } G(u_0) = 0, \text{ s.t. } G'(u_0) : \mathcal{U} \to \mathcal{V}' \tag{3.80}$$

is boundedly invertible and

$$\|G'(u) - G'(v)\|_{\mathcal{V}' \leftarrow \mathcal{U}} \leq L\|u - v\|_{\mathcal{U}} \quad \forall u, v \in \overline{B_R(u_0)}.$$

Theorem 3.38. Newton–Kantorovich method for G, Ortega and Rheinboldt [520]: *Under the condition (3.80) and for small enough $\|u_0 - u_1\|_{\mathcal{U}}$ the method in (3.81) is well defined and converges quadratically:*

$$u_{i+1} := u_i - G'(u_i)^{-1}G(u_i), \ i = 1, \ldots, \ \text{and } \|u_0 - u_{i+1}\|_{\mathcal{U}} \leq C\|u_i - u_0\|_{\mathcal{U}}^2, \tag{3.81}$$

or $\leq C'\|u_i - u_{i-1}\|_{\mathcal{U}}^2$. *These C, C' depend upon $\|u_0 - u_1\|_{\mathcal{U}}, L, \|(G'(u_0))^{-1}\|$.*

For maintaining the consistency of G^h of order $p > 0$ for $u \in \mathcal{U}_s$ a smooth subspace of $\mathcal{U}$, during the Newton process we have to require

$$G : \mathcal{D}(G) \cap \mathcal{U}_s \to \mathcal{V}_s', \ \text{and } G'(u_0) : \mathcal{U}_s \to \mathcal{V}_s' \text{ is boundedly invertible.} \tag{3.82}$$

Furthermore, we have to extend the concept of consistency in Definition 3.14 to the first derivative.

Definition 3.39. Differentiable consistency: *Under the conditions of Definition 3.14 a general discretization method is called* differentiably consistent and of order p, *if it is classically consistent and if for a Frechet differentiable nonlinear operator G,*

$$\|(G^h)'(P^h u)P^h v - Q'^h(G'(u)v)\|_{\mathcal{V}_b'^h} \to 0 \text{ and} \tag{3.83}$$

$$\|(G^h)'(P^h u)P^h v - Q'^h(G'(u)v)\|_{\mathcal{V}_b'^h} \leq Ch^p(1 + \|u\|_{\mathcal{U}_s})\|v\|_{\mathcal{U}_s} \text{ for } h \to 0.$$

These $\|v\|_{\mathcal{U}_s}$ are usually $\|v\|_{W^{k,q}(\Omega)}$ and p, k, q appropriately chosen.

Theorem 3.40. Newton method for the discrete G^h, cf. [9, 10, 17]: *Under the conditions (3.80), (3.82), and for a linear discretization method $\mathcal{M}$ assume a stable G^h, consistent and differentiably consistent of order $p > 0$ for $u_0 \in \mathcal{U}_s \subset \mathcal{U}$, a smooth* *subspace, still with Lipschitz-continuous* $(G^h)'$. *Start the discrete Newton process for* $i = 1, \ldots,$ *with* $u_1^h := P^h u_1, u_1 \in \mathcal{U}_s$, *as*

$$u_{i+1}^h := u_i^h - \left(G^h\left(u_i^h\right)'\right)^{-1} G^h\left(u_i^h\right). \tag{3.84}$$

Then, for small enough $\|u_0 - u_1\|_{\mathcal{U}}$ *and* h, *the* u_{i+1}^h *in (3.84) uniquely exist and converge quadratically to* u_0^h, *such that*

$$\begin{aligned} \left\|u_{i+1}^h - u_i^h\right\|_{\mathcal{U}^h} &\le C\left\|u_i^h - u_{i-1}^h\right\|_{\mathcal{U}^h}^2 \text{ and} \\ \left\|u_i^h - P^h u_i\right\|_{\mathcal{U}^h} &\le Ch^p\|u_i\|_{\mathcal{U}_s},\ i = 1, \ldots, \\ \|G^h\left(u_i^h\right) - Q'^h G(u_i)\|_{\mathcal{V}'^h} &\le Ch^p\|u_i\|_{\mathcal{U}_s},\ i = 1, \ldots, \\ \|\left(u_i^h - u_0^h\right) - \left(P^h(u_i - u_0)\right)\|_{\mathcal{U}^h} &\le Ch^p\|u_i\|_{\mathcal{U}_s},\ i = 1, \ldots. \end{aligned} \tag{3.85}$$

Corollary 3.41. MIP, cf. [9, 17]: *Under the conditions of the previous theorem there exists* $\overline{h}$ *such that for all* $h \in (0, \overline{h}]$

$$\left|\min\{i \ge 1 : \left\|u_i^h - P^h u_0\right\|_{\mathcal{U}^h} < \epsilon\} - \min\{i \ge 1 : \|u_i - u_0\|_{\mathcal{U}} < \epsilon\}\right| \le 1. \tag{3.86}$$

This result is the reason for the MIP notation and our remark concerning the essential h independence of the convergence.

Now we are able to formulate the discrete Newton algorithm, based upon the MIP. We start with a moderately good approximation for u_0^h obtained by Algorithm 3.2

Algorithm 3.2.
Discrete Newton method and MIP

INPUT: Starting approximation $u_1^h \in \mathbb{R}^m$ with $G^h\left(u_1^h\right) \approx 0$ and

tolerance $\epsilon > 0$; $h > 0$;

OUTPUT: Approximation u_n^h for u_0^h s.t. $\left\|u_n^h - u_0^h\right\|_{\mathcal{U}^h} < \epsilon$.

ITERATION: FOR ($i = 1$; Stopping criterion)

(i) Calculate $u_{i+1}^h := u_i^h - \left(G^h\left(u_i^h\right)'\right)^{-1} G^h\left(u_i^h\right)$, see (3.84);

(ii) Update $i := i + 1$;

(iii) Estimate $\left\|u_i^h - u_0^h\right\|_{\mathcal{U}^h}$ (with standard Newton techniques);

(iv) Stop if $\left\|u_i^h - u_0^h\right\|_{\mathcal{U}^h} < \epsilon$, else goto (i).

4

Conforming finite element methods (FEMs)

4.1 Introduction

Finite element methods (FEMs) are nowadays the most advanced methods, applicable to nearly all types of PDEs. As throughout this book, we restrict the discussion to elliptic problems, from linear, to semilinear, to quasilinear to fully nonlinear equations and systems of orders 2 and $2m$. The goal in this chapter is the proof of stability and convergence for different types of *conforming FEMs* and to generalize many of the known the results. In Chapter 3, we developed appropriate tools for proving results for these operator equations and a relatively general class of discretization methods, based upon consistency and stability. Conforming FEMs are consistent for continuous problems. We proceed in two steps to stability for nonlinear a consequence of stability for the linearized operator. Let a boundedly invertible linear operator yield a stable discretization, e.g. for an operator with coercive bilinear form. Under some technical conditions, this stability is inherited to its compactly perturbed linear operators, if these are still boundedly invertible and consistent. These results are applicable essentially to all the elliptic problems in Chapter 2, if, for nonlinear problems, the linearization is bounded.

As mentioned, a stable discretization of a boundedly invertible operator is a consequence of a coercive principal part or a Gårding inequality, cf. Theorem 3.32. This is correct, except for (nonlinear) elliptic systems in the sense of Agmon *et al.* [3], with different highest orders of the derivatives in different equations, cf. Definition 2.85. The most famous example is the Navier–Stokes operator. The stability and bounded invertibility of the Stokes operator has been proved by other techniques. Then the above arguments show the stability of the (linearized) Navier–Stokes operator for moderate Reynolds numbers. This is applicable to other examples as well. Convergence for monotone operator is available without linearization.

Excellent books are available on FEMs. We list only some of them, best suited to support specific points in the text. We need discretizations of general problems and their convergence properties. Hackbusch's [386, 387] studies of elliptic, including higher order PDEs, with finite difference and FEMs, and Ciarlet's [174] Brenner and Scott's [141] books on FEMs have very strongly influenced this book and this and the next chapters. Different numerical methods and the corresponding analysis are studied by Grossmann *et al.* [374–376], for second order PDEs. FEMs, again combined with many interesting applications, are discussed by Babuška and Strouboulis [51] and Oden and Reddy [518]. Other important applications with more or less theory for linear and nonlinear equations and systems are studied, e.g. by Braess [135], Brenner and Scott [141], Debnath *et al.* [277, 638], Leung [473], DuChateau and Zachmann [305–307], Larsson and Thomee [469], Quarteroni and Vali [537], and Thomas [626]. In all these books, nonlinear problems essentially are discussed via some practically important examples. Zenísek [681] approaches nonlinear problems with monotony methods. Haslinger *et al.* [393] study unilateral linear boundary value, mainly mechanical contact, problems via inequalities. The nonlinear problems in these books are so-called semilinear problems. The analytic theory, special Galerkin methods and a general discretization theory for nonlinear, here semilinear and quasilinear, problems are presented by Zeidler [675–678]. FEMs for fully nonlinear problems have not been available before. So they are not included in previous books, c.f. Section 5.2.

In this chapter we only consider and approximately solve elliptic and monotone problems formulated as weak forms. Therefore we assume throughout this chapter that our operator and approximating subspaces satisfy

$$A : \mathcal{U}_b \to \mathcal{U}_b^{'}, G : \mathcal{D}(G) \subset \mathcal{U}_b \to \mathcal{U}_b^{'}, \quad \mathcal{U}_b^h = \mathcal{V}_b^h \ \forall \ h$$

usually A and G are defined on $\mathcal{U}$ as well.

The basic idea of FEMs is simple. We demonstrate it for a linear operator

$$A : \mathcal{U} = \mathcal{V} := H^1(\Omega) \to \mathcal{V}' := H^{-1}(\Omega), \text{ s.t. } A : \mathcal{U}_b = \mathcal{V}_b := H_0^1(\Omega) \to \mathcal{V}_b' \text{ bijective.} \quad (4.1)$$

The index and exponent, $_b$ and ^{h}, e.g. in $\mathcal{U}_b$ and $\mathcal{U}_b^h$, indicate boundary conditions and finite dimensional subspaces. The exact solution, u_0, is determined by testing

$$u_0 \in \mathcal{U}_b \subset \mathcal{U} : \ a(u_0, v) = \langle Au_0, v\rangle_{\mathcal{V}'\times\mathcal{V}} \ = \ \langle f, v\rangle_{\mathcal{V}'\times\mathcal{V}} \quad (4.2)$$
$$\forall \ \ v \in \mathcal{V}_b \subset \mathcal{V} \text{ often with } \mathcal{U}_b = \mathcal{V}_b \subset \mathcal{U} = \mathcal{V}.$$

In FEMs the $\mathcal{U}_b, \mathcal{V}_b$ in (4.2) are replaced by finite dimensional approximating subspaces $\mathcal{U}_b^h, \mathcal{V}_b^h$ of piecewise, e.g. linear polynomials, defined on a triangulation of Ω, cf. Definition 4.2, with $\mathcal{U}_b^h = \mathcal{V}_b^h$ for conforming methods. We determine the discrete or approximate solution, u_0^h, such that $\mathcal{U}_b^h \subset \mathcal{U}_b, \mathcal{V}_b^h \subset \mathcal{V}_b$,

$$u_0^h \in \mathcal{U}_b^h \subset \mathcal{U}_b \subset \mathcal{U} : a\left(u_0^h, v^h\right) = \left\langle Au_0^h, v\right\rangle_{\mathcal{V}'\times\mathcal{V}} = \langle f, v\rangle_{\mathcal{V}'\times\mathcal{V}} \forall v^h \in \mathcal{V}_b^h \subset \mathcal{V}_b. \quad (4.3)$$

The $\mathcal{U}_b, \mathcal{U}_b^h$ and $\mathcal{V}_b, \mathcal{V}_b^h$, represent ansatz and test functions, respectively. If the functions $u^h \in \mathcal{U}_b, \mathcal{U}_b^h$ are continuous and exactly satisfy the boundary conditions, we call (4.3) a *conforming FEM*. For a bounded and coercive $a(u, u)$, so $\exists \alpha > 0 \ \forall u \in \mathcal{U}_b : \ a(u, u) \geq \alpha \|u\|_{\mathcal{U}}^2$, the convergence of the u_0^h to u_0 via the Cea Lemma 4.48 is surprisingly simply proved and yields for piecewise polynomial FEs of degree $d - 1$

$$\left\|u_0 - u_0^h\right\|_{\mathcal{U}} \leq (C/\alpha)\text{dist}(u_0, \mathcal{U}^h) = C^* h^{d-1} \|u_0\|_{H^d(\Omega)}. \quad (4.4)$$

We do not study super convergence results, c.f. e.g. Bank and Xu, Griewank and Reddien, Jiang *et al.* [66, 368, 417]

The conditions of continuous FEs, exactly satisfying the boundary conditions, is too restrictive for many applications. If we relax them we obtain the so-called *nonconforming FEMs*, cf. Chapter 5.

Our approach to FEMs differs from the usual versions in several respects. We generalize known results and modify proofs in many directions. The main tool is the systematic use of the general theory of discretization methods as presented in Chapter 3. This contrasts with the standard coercivity and ellipticity arguments for mainly linear problems with their bilinear forms and trilinear modifications as, e.g. for the Navier–Stokes equations, cf. Section 4.6, and to the monotony arguments for higher nonlinearities. In fact, *conforming* and *nonconforming FEMs,* as usual with one reference element and here with fixed polynomial degree, are special cases of *generalized Petrov–Galerkin methods.* We have to ensure stability and convergence for FEMs for elliptic problems. Stability is a consequence of the fact that for a linear A the (modified) principal part A_p induces a stable A_p^h. For conforming FEMs the (linear) discrete A^h are (classically) consistent with the original A.

The goal is an extension to nonlinear problems and to higher order conforming and nonconforming FEMs for the large class of linear, semilinear, quasilinear and fully nonlinear elliptic equations mentioned above. This is achieved, including quasilinear problems, for FEMs based upon the classical weak form of the problems. For fully nonlinear problems we have to use nonconforming FEMs for their strong form.

Applying FEMs in the context of path-following of parameter-dependent solutions of nonlinear problems and avoiding spurious solutions for turning and bifurcation points requires high order methods. They can conveniently and with a wide range of applications be defined as conforming and nonconforming FEMs.

We organize this chapter on conforming FEMs as follows. In Section 4.2 we start with the standard approximation theory for FEs. We only prove results not included in standard textbooks, e.g. [141], and in easily accessible papers, e.g. [362]. As preparation for the later nonconforming FEMs, we present FEs on curved domains Ω and smooth FEs $\in C^r(\Omega), r \geq 1$ in Subsections 4.2.7, and with Davydov, in 4.2.6. In Section 4.3 a simple example, the Helmholz equation, motivates the definition of conforming FEMs. They are extended to general linear and in Section 4.4 to quasilinear elliptic differential equations and systems of orders 2 and $2m$.

Another type of generalization represents the very active area of *mixed FEMs,* presented here only for the *Navier–Stokes equation* in Section 4.6. For the case of moderate Reynolds numbers, it is a compact perturbation of the well studied boundedly invertible Stokes operator with stable discretization. Consequently a boundedly invertible linearized Navier–Stokes operator implies stability for the linearized and the nonlinear operator. This stability result for the Navier–Stokes operator seems to be the state of the art in discretizations via FEM, spectral, difference and wavelet methods anyway. Extensions to nonconforming and adaptive forms are not considered here. In mixed FEMs the different components of the solution, here the velocity, $\vec{u}$, and the pressure, p, live in different spaces and satisfy side conditions, $\operatorname{div} \vec{u} = 0$. Correspondingly, for $\vec{u}^h, p^h$, different and appropriate spaces, $\mathcal{U}^h, \mathcal{W}^h$, mixed FEMs have to be chosen.

In particular, for the later studies of bifurcation, FEMs for eigenvalue problems are important. We discuss them in Section 4.7. The extension of the results to nonlinear scenarios is not interesting here, since the eigenvalues that decide bifurcation are always defined by the linearized operators.

Adaptive FEMs are very well advanced these days. So the main tools and results are presented in Chapter 6 by W. Doerfler.

We have already mentioned that we delay the study of FEMs with crimes, or nonconforming FEMs to the next chapter. Another possibility for violated continuity and boundary conditions, the *discontinuous Galerkin methods (DCGMs)* follows.

We *exclude* the following important areas. The *hp*-methods and their discontinuous versions are found, e.g. in Babuska *et al.* [52, 53, 517]. Even some of the recent results towards superconvergence in Babuska and Strouboulis [51], and exponential accuracy as in Schwab and Melenk [485–487], would probably fit our approach. Multigrid methods, multigrid solutions of bifurcation problems, hierarchical bases and all kinds of a posteriori and reliability estimates are studied by, e.g. Hackbusch [385], Babuska and Strouboulis [51], Brenner and Scott [142], Mittelmann and Weber [492–494, 660, 661], and Babuska *et al.* [51, 59–61, 63–66]. Special high order FEMs are studied, e.g. by Šolín, Segeth and Doležel [593].

Throughout the chapter, all convergence results require a sufficiently small discretization parameter, h. In fact, we characterize, strongly simplifying, an FE discretization method by a parameter h. In the following we indicate the essential *constants depending upon essential parameters,* by indices listing these constants, e.g. $C_{h,\mu}$ depends upon h, μ. Sometimes, we omit parameters to show the *independence of constants*, e.g. we would write C instead of $C_{h,\mu}$ to indicate the independence of C from h, μ.

4.2 Approximation theory for finite elements

The following introduction to the approximation theoretic properties of finite elements (FEs) and many other topics is strongly influenced by the classical Ciarlet [174], and by Hackbusch, Braess, Brenner and Scott [136, 141, 386, 387]. The basis for the success of FEMs are the good approximation properties and the high flexibility of FEs. We present the basic results. After introducing subdivisions and general FEs, we consider the most interesting example of piecewise polynomials. We formulate general FE spaces and their approximation properties for functions. For vectors of functions, e.g. solutions of systems of elliptic equations, they remain unchanged if each component shares the same structural properties and has the same smoothness. We summarize interpolation errors, inverse estimates and extensions to curved boundaries. The results in this section are strictly local, except Subsection 4.2.6 with O. Davydov. Only here do we present *and prove* the essentially new results for smooth FEs on curved domains with the necessary splitting properties.

We *assume throughout the whole chapter* that the (open) domain Ω satisfies

$$\Omega \subset \mathbb{R}^n \text{ bounded domain, piecewise smooth Lipschitz-continuous } \partial\Omega. \qquad (4.5)$$

4.2.1 Subdivisions and finite elements

The power of FEs is based upon approximating functions, $u^h : \Omega \subset \mathbb{R}$, by splitting the domain into subsets $T_i \subset \Omega$. For simplified computations we assume each of these T_i

to be affinely (or isoparametrically, see Section 4.2.7), mapped onto the same reference domain, K. In the *simplest case* an approximating function u^h, restricted to a triangle $T = T_i$, is $u^h \in \mathcal{P}_1$, hence a linear polynomial, defined by its function values in the three vertices of T. This is generalized as, see [174]:

Definition 4.1. *We assume that*

1. *the* element *or* reference domain $K \subseteq \mathbb{R}^n$ *is an open domain with piecewise smooth boundary;*
2. *the* space of shape functions $\mathcal{P}$ *is a finite-dimensional space of functions defined on* K*; and*
3. *the set,* $\mathcal{N} = \{N_1, N_2, \ldots, N_{d'}\}$, *of* nodal variables *are linear, linearly independent, usually interpolation functionals defined on* $\mathcal{P}$*. They represent a basis for* $\mathcal{P}'$*, hence,* $\mathcal{P}' = \operatorname{span} \mathcal{N}$.

We need the same dimension $\dim \mathcal{P} = \dim \mathcal{P}' = d'$*. Then* $(K, \mathcal{P}, \mathcal{N})$ *is called a* reference element *or a* finite element. *There are many possible combinations of* $\mathcal{P}, \mathcal{N}$*. If* $\mathcal{P} = \operatorname{span}\{\phi_1, \phi_2, \ldots, \phi_{d'}\}$ *is a basis for* $\mathcal{P}$ *dual to* $\mathcal{N}$*, hence* $N_i(\phi_j) = \delta_{ij}$*, then* $\phi_1, \ldots, \phi_{d'}$ *is called the* nodal basis *for* $\mathcal{P}$*. A set of nodal variables,* $\mathcal{N}$*, is called* unisolvent *for* $\mathcal{P}$*, or* $K, \mathcal{P}, \mathcal{N}$ *a unisolvent FE, compare* 3.*, if*

$$\forall (\beta_i)_{i=1}^{d'} \in \mathbb{R}^{d'} \ \ \exists_1 \ \ \phi \in \mathcal{P} : N_i(\phi) = \beta_i, \ \ i = 1, \ldots, d'.$$

This choice of a reference element $(K, \mathcal{P}, \mathcal{N})$ has the important consequence that much of the following analysis and computations can be performed in the reference domain K.

In the previous *simplest case,* motivating Definition 4.1, K is a (unit) triangle with vertices P_1, P_2, P_3, the $\mathcal{P} = \mathcal{P}_1$ and $\mathcal{N}$ are the point evaluations in the P_1, P_2, P_3. It is implicitly assumed that the nodal variables, N_i, lie in the dual space of some larger function space, $\mathcal{U}$, allowing $N_i(\phi) \forall \phi \in \mathcal{U}$. The N_i usually denote evaluation of functions or derivatives in given interior and boundary points of K. Obviously $\mathcal{N}$ is unisolvent for $\mathcal{P}$ if and only if $\dim \mathcal{P} = \dim \operatorname{span} \mathcal{N} = d'$ and $\phi \in \mathcal{P}, N_i(\phi) = 0, i = 1, \ldots, d'$ implies $\phi \equiv 0$.

For a given $u \in \mathcal{U}$, a nodal basis of shape functions $\mathcal{P} = \operatorname{span}\{\phi_1, \ldots, \phi_{d'}\}$ and a unisolvent $\mathcal{N} = \{N_1, N_2, \ldots, N_{d'}\}$, applicable to functions in a space $\mathcal{U} : \overline{K} \to \mathbb{R}$, a *local interpolation operator* is uniquely defined, as

$$I_K := I : \mathcal{U} \to \mathcal{P}, Iu := \sum_{i=1}^{d'} N_i(u)\phi_i, N_i(u) = N_i(Iu), i = 1, \ldots, d'. \tag{4.6}$$

We use the standard notation of the *Dirac delta function* $\delta(P_i)$ and $\delta(P_i) \in \mathcal{N}$ if $N_i(u) = u(P_i) = \delta(P_i)u$. We mainly consider these or more general interpolations.

Usually the K is chosen as a triangle, tetrahedron, an n-dimensional simplex, rectangle, parallelogram or an n-dimensional affine cuboid. Usually the following T_i are affinely (or isoparametrically), see below, mapped onto this same K. Then properties 2.–4. in Definition 4.2 make sense. We introduce subdivisions for general finite elements and will combine it below with the above local interpolation operator on K.

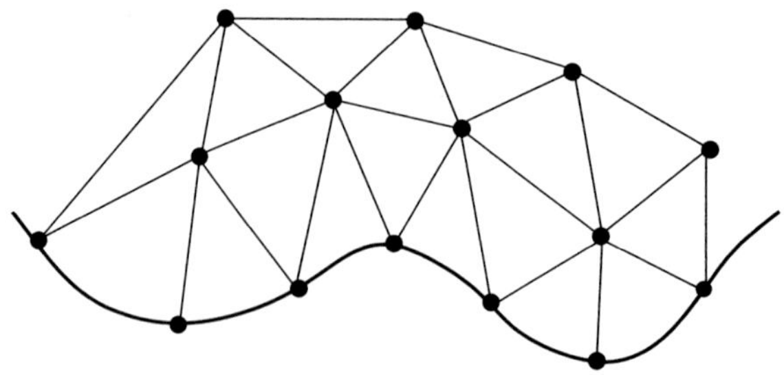

Figure 4.1 Admissible triangulation for Ω with curved boundary.

Definition 4.2. Subdivision: *A set* $\mathcal{T}^h = \{T_1, \ldots, T_M\}$ *of open subsets* $T_i \subset \Omega \subset \mathbb{R}^n$ *is called a* subdivision *and an* admissible subdivision *for* Ω *if it satisfies* 1. *and* 1.–4., *respectively.*

1. $T_i \cap T_j = \emptyset$ *for* $i \neq j$ *and* $\overline{\Omega} = \bigcup_{i=1}^{M} \overline{T}_i$;
2. *each* $T \in \mathcal{T}^h$ *has at least* $n+1$ *and at most* $\overline{n} \geq n+1$ vertices, *e.g. triangles and quadrangles with straight or curved edges;*
3. $\overline{T}_i \cap \overline{T}_j$ *for* $i \neq j$ *is either empty or has one vertex or a common edge or one, in general* $(n-1)$*-dimensional surface spanned by joint vertices of* T_i *and* T_j, *see Figure 4.1;*
4. *for* paralellotopes *or* cuboids, *hence* $Q_i = \Pi_{j=1}^n [a_j, b_j]$, *generally* T_i, *3. can be relaxed, e.g. for rectangles,* $\bar{T}_i \cap \bar{T}_j$ *may be one half of the larger edge of the larger* T_j, *see Figure 4.15.*

$\mathcal{T}^h$ *is called a* triangulation *when the* $T \in \mathcal{T}^h$ *are triangles, tetrahedrons or n-dimensional simplices, possibly with curved edges, cf. Figure 4.1.*

4.2.2 Polynomial finite elements, triangular and rectangular K

Chosing piecewise polynomials as $\mathcal{P}$ are certainly the most important FEs. We start choosing K as triangles and quadrangles for $n = 2$ and their generalizations to $n > 2$. To avoid difficulties near the boundary, we start with

$$\text{a polyhedral domain } \Omega, \text{ for } n = 2 \text{ called a polygon.} \tag{4.7}$$

Curved boundaries Ω are treated by interpolation, approximation or isoparametric FEs in Subsections 4.2.7 and 4.2.6. We choose the reference domain K as a unit rectangular n-triangle or a unit n-cube, hence diam $K = 1$ or diam $K = \sqrt{n}$. We give some examples of FE triples $(K, \mathcal{P}, \mathcal{N})$. The most important choice for $\mathcal{P}$ are *polynomials with degree d*:

$$\mathcal{P} = \mathcal{P}_d^n := \left\{ \sum_{|i| \leq d} \alpha_i x^i : x = (x_1, \ldots, x_n) \in \mathbb{R}^n \right\}, \mathcal{P}_d := \mathcal{P}_d^2 \,, \mathcal{P}^d := \mathcal{P}_d^1, \tag{4.8}$$

with $i = (i_1, \ldots, i_n) \geq 0,\ |i| = i_1 + \cdots + i_n,\ x^i = x_1^{i_1} \cdots x_n^{i_n},\ \alpha_i \in \mathbb{R}$.

Sometimes, we allow $\mathcal{P} \subset \mathcal{P}_d^n$. The dimensions of these $\mathcal{P}_d^n$ are

$$\dim \mathcal{P}_d^1 = \dim \mathcal{P}^d = (d+1), \dim \mathcal{P}_d^2 = \dim \mathcal{P}_d = \frac{(d+1)(d+2)}{2}, \tag{4.9}$$

and in general

$$\dim \mathcal{P}_d^n = \binom{n+d}{n} = \binom{n+d}{d}.$$

A convenient criterion for checking the unisolvence property for n-triangles combines the edges $L \in \{L_1, L_2, \ldots, L_{n+1}\}$ of the triangle K with Proposition 4.3, see [141]. Let a *nondegenerate hyperplane* L be defined as

$$L := \left\{x \in \mathbb{R}^n : L(x) = \sum_{|i|=1} \alpha_i x^i - \beta = 0\right\} \text{ with } \alpha_i \in \mathbb{R}, \sum |\alpha_i| > 0. \tag{4.10}$$

Proposition 4.3. *Let a polynomial $P \in \mathcal{P}_d^n$ vanish on L. Then there exists a $Q \in \mathcal{P}_{d-1}^n$, such that $P(x) = L(x)Q(x)$. Otherwise apply computer algebraic tools.*

Triangular finite elements

In Figures 4.2–4.11 we show combinations of K with $\mathcal{P}_d$ for $d = 1, \ldots, 5$, and appropriate $\mathcal{N}$, unisolvent for $\mathcal{P}_d$. The different cases are indicated by points z_i in K or on the boundary of K according to

z_i is marked by $\bullet$ if $N_i(\phi) = \phi(z_i)$, $N_i \in \mathcal{N}$, for $\phi \in \mathcal{P}_d$,

and by $\odot$ if $N_i(\phi) = \phi(z_i)$ and $N_i^g(\phi) = \text{grad } \phi(z_i)$, $N_i, N_i^g \in \mathcal{N}$.

If only function values, $\bullet$, are required, we have a *Langrange finite element.* If function values and first ($\odot$) or, even additionally, second derivatives, indicated by two circles are required, e.g. the Argyris element, see below, we have *Hermite finite elements.* Proposition 4.3 is visualized in Figures 4.2–4.5.

In some cases, e.g. for the Argyris $\mathcal{P}_5$-element, see Figure 4.9, in addition to the values of the function and first and second derivative, normal derivatives are

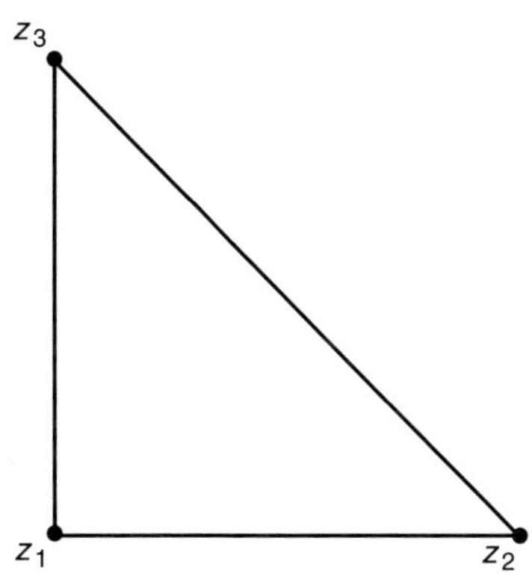

Figure 4.2 $\mathcal{P}_1$: Linear Lagrange FEs.

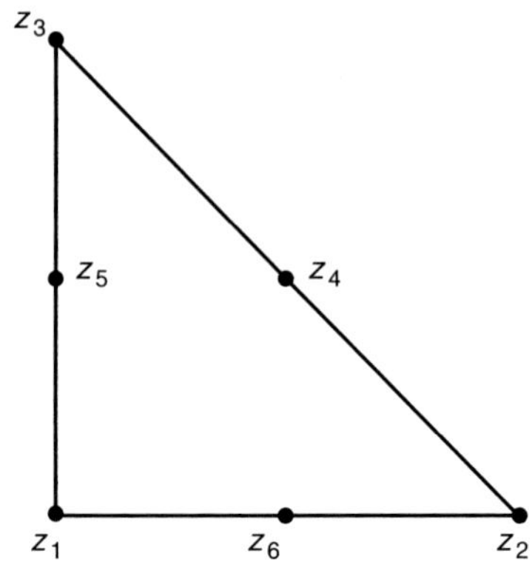

Figure 4.3 $\mathcal{P}_2$: Quadratic Lagrange FEs.

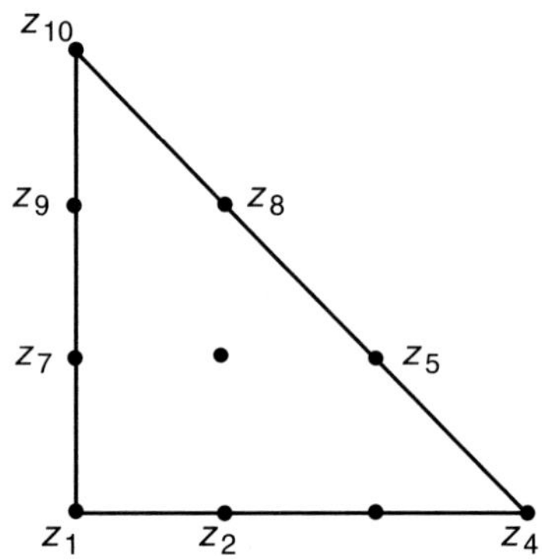

Figure 4.4 $\mathcal{P}_3$: Cubic Lagrange FEs.

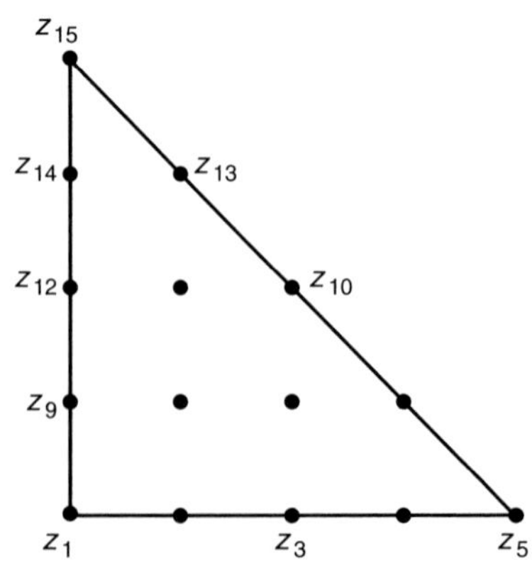

Figure 4.5 $\mathcal{P}_4$: Quartic Lagrange FEs.

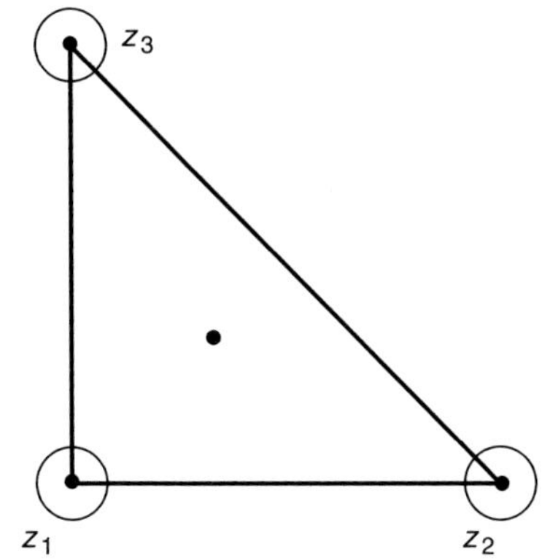

Figure 4.6 $\mathcal{P}_3$: Cubic Hermite FEs.

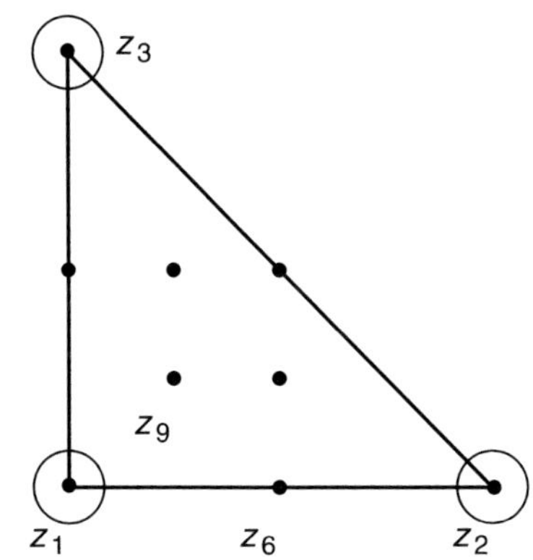

Figure 4.7 $\mathcal{P}_4$: Quartic Hermite FEs.

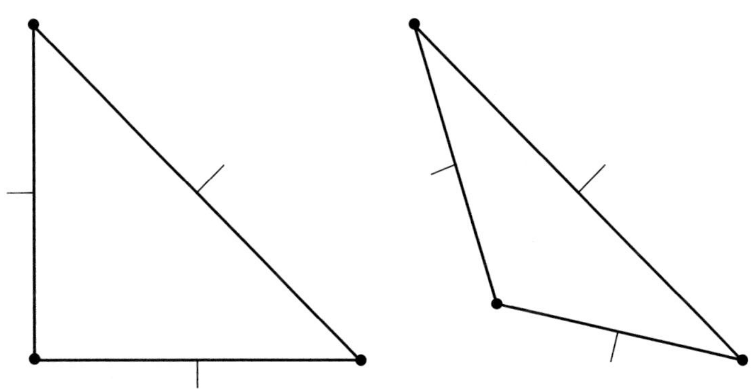

Figure 4.8 Inequivalent cubic Hermite FEs: Incompatible normal derivatives.

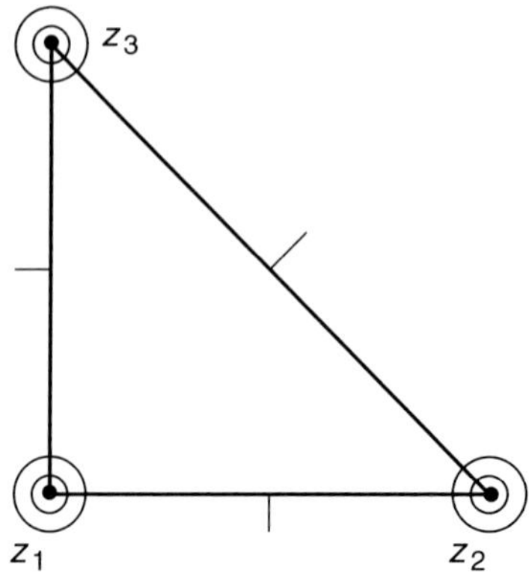

Figure 4.9 Quintic Argyris FEs.

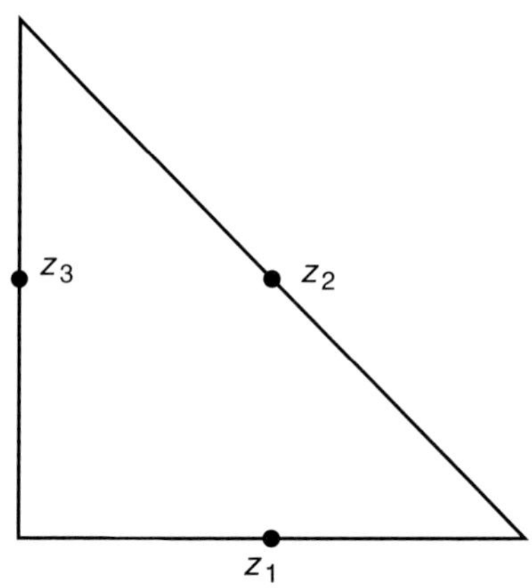

Figure 4.10 $\mathcal{P}_1$: Nonconforming linear Lagrange FEs (Crouzeix–Raviart).

prescribed. So in the sense of Definition 4.6 below, the *Argyris elements* are not affine equivalent, but only nearly affine equivalent [135, 174]. On the other hand, see Subsection 4.2.6, there are C^1 FEs as modified Argyris elements, important for the later FEMs. The *Bell element* corresponds to an 18-dimensional subset of $\mathcal{P}_5$.

For unisolvence we need the same number of values of the function (and/or its derivative) as $\dim \mathcal{P}$, e.g. $\dim \mathcal{P}_d = (d+1)(d+2)/2$. All figures show unisolvent, hence $\dim \mathcal{P}_d = (n+1)(n+2)/2 = \dim \mathcal{N}$, and *conforming finite elements.* Figures 4.10 and 4.11 show *nonconforming elements,* see Section 5.5. The case $\mathcal{P}_1$ for the linear conforming and nonconforming Lagrange elements shows that the $\mathcal{N}$ is certainly not uniquely determined by $\mathcal{P}$.

Remark 4.4. *All the following figures show a high symmetry with respect to the boundary (and interior) points. For general nondegenerate subdivisions, see Definition 4.12 below, we have to impose $N_i \in \mathcal{N}$ for boundary points symmetric and of the same kind on all edges, corresponding to those in neighboring subtriangles. For highly*

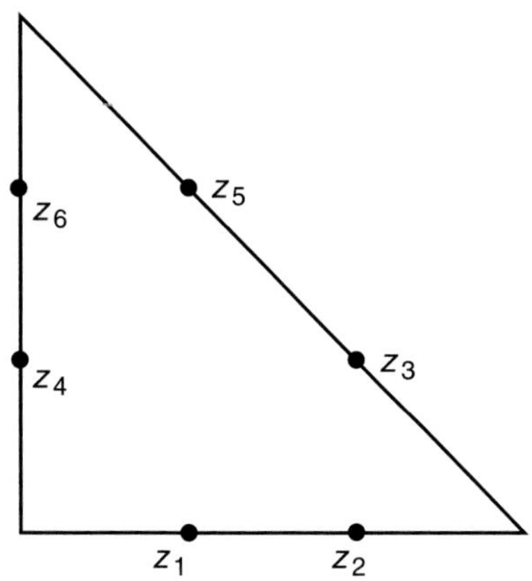

Figure 4.11 $\mathcal{P}_2$: Nonconforming quadratic Lagrange FEs.

symmetric subdivisions, e.g. rectangular triangles in a rectangle, relaxations of this symmetry of the boundary points are possible.

Remark 4.5. Simple inductive computation of the FE-interpolant: *Let the knots be chosen in a triangle K on $d+1$ parallel lines with altogether $s = 1 + 2 + \cdots (d+1)$ points $z_1, z_2, \ldots, z_s$, see [135]. We determine*

$$\phi \in \mathcal{P}_d \ s.t.\ \phi(z_i) = f_i, i = 1, \ldots, s = \frac{(d+1)(d+2)}{2} = dim\ \mathcal{P}_d. \tag{4.11}$$

Using the affine equivalence, see Definition 4.6, we can choose the parallel lines as $y =$ constant. The ϕ for the $d+1 = 1$ line is obvious. So we apply the following (4.12) inductively. We assume the first d lines for $\phi \in \mathcal{P}^2_{d-1}$ in (4.11). Let the last line, with the first $d+1$ knots, $z_1, \cdots, z_{d+1}$, be located at $y = 0$.

$$\textit{Determine } p_0 \in \mathcal{P}^1_d : p_0(z_i) = f_i,\ i = 1, \ldots, d+1, \textit{ and, by induction,} \tag{4.12}$$

$$\textit{determine the unique } q \in \mathcal{P}^2_{d-1} : q(z_i) = \frac{1}{y_i}(f_i - p_0(z_i)),\ i = d+2, \ldots, s.$$

$$\textit{Then } p \in \mathcal{P}^2_d = \mathcal{P}_d, \textit{ with } p(x,y) := p_0(x) + y \quad q(x,y) \textit{ solves (4.11).}$$

A similar approach is possible for higher dimensions and some rectangular elements.

Bilinear, biquadratic, . . . finite elements

For rectangular K we often use instead of $\mathcal{P}_d$ the *tensor products* of *bilinear, biquadratic, . . . polynomials*, see Figures 4.12–4.15 and [141, 572]

$$Q^2_d = \mathcal{P}^1_d \otimes \mathcal{P}^1_d = \left\{ u = \left(\sum_{0 \le i \le d} c_i\ x^i \right) \left(\sum_{0 \le j \le d} d_j\ y^j \right),\ {}_i, d_j, x, y \in \mathbb{R} \right\} \subset \mathcal{P}^2_{2d}$$

$$\mathcal{P}^n_d \subset Q^n_d = \prod_{i=1}^{n} \mathcal{P}^1_d \subset \mathcal{P}^n_{nd},\ Q_d = Q^2_d,\ \ \dim Q^n_d = \left(\dim \mathcal{P}^1_d\right)^n = (d+1)^n. \tag{4.13}$$

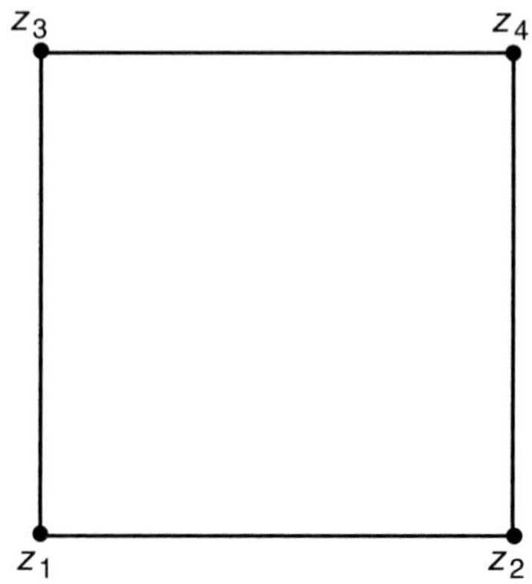

Figure 4.12 $\mathcal{Q}_1$: Bilinear Lagrange FEs.

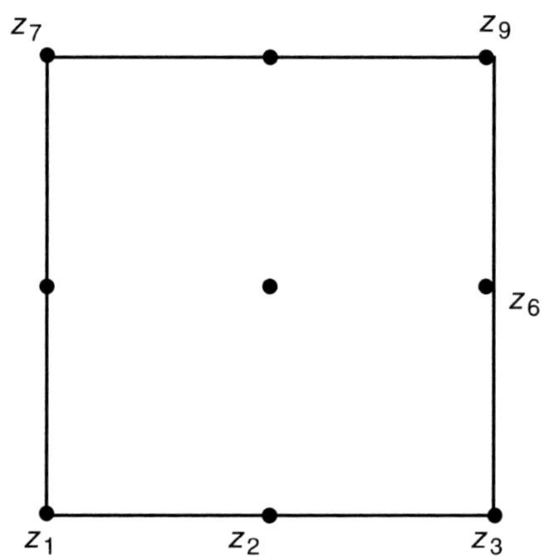

Figure 4.13 $\mathcal{Q}_2$: Biquadratic Lagrange FEs.

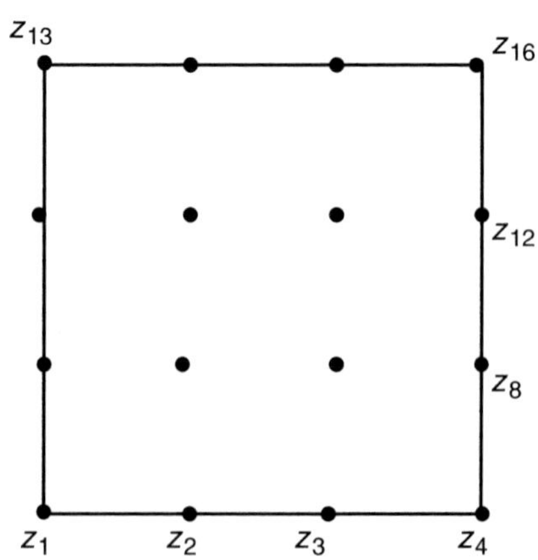

Figure 4.14 $\mathcal{Q}_3$: Bicubic Lagrange FEs.

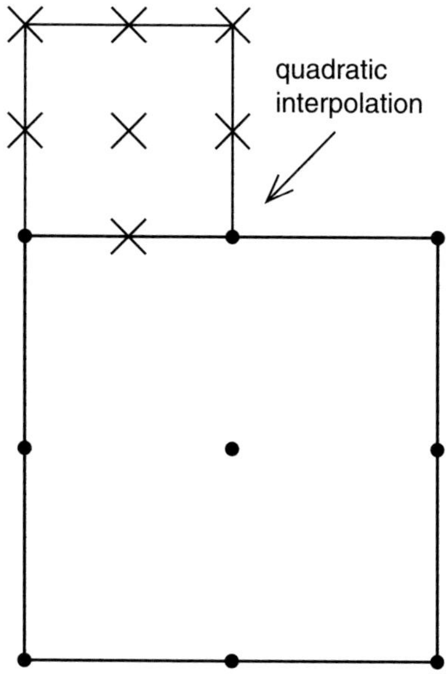

Figure 4.15 Quadratic subdivision (mesh-refinement), violating Definition 4.2, 3., see 4..

4.2.3 Interpolation in finite element spaces, an example

We want to interpolate on general subdivisions of Ω. This requires the extension from the $(K,\mathcal{P},\mathcal{N})$ situation to the collection of the $\hat{K} = T \in \mathcal{T}^h$, denoted as $(\hat{K},\hat{\mathcal{P}},\hat{\mathcal{N}})$. For efficient computations they have to be closely related to the original $(K,\mathcal{P},\mathcal{N})$. Therefore we introduce

Definition 4.6. Affine equivalent FEs: *Let $(K,\mathcal{P},\mathcal{N})$ and $(\hat{K},\hat{\mathcal{P}},\hat{\mathcal{N}})$ be two finite elements, let $F : K \to \hat{K}, F(x) = Ax + b, A \in \mathbb{R}^{n\times n}$ nonsingular, $b \in \mathbb{R}^n$, be an affine (or isoparametric) map such that, see Figure 4.16,*

1. $F(K) = \hat{K}$,
2. $\hat{\mathcal{P}} \circ F = \mathcal{P}$ *or* $\hat{\mathcal{P}} = \mathcal{P} \circ F^{-1}$ *and*
3. $\hat{\mathcal{N}}(\hat{f}) = \mathcal{N}(\hat{f} \circ F) = \mathcal{N}(f)$ *for* $\hat{f} : \hat{K} \to \mathbb{R}$ *and* $f := \hat{f} \circ F$.

Then $(K,\mathcal{P},\mathcal{N}) \underset{F}{\cong} (\hat{K},\hat{\mathcal{P}},\hat{\mathcal{N}})$ *and* $(\hat{K},\hat{\mathcal{P}},\hat{\mathcal{N}})$ *is called* (affine) equivalent *to* $(K,\mathcal{P},\mathcal{N})$.

Obviously affine equivalence is an equivalence relation. If $\mathcal{P} = \text{span}\{\phi_1, \phi_2, \ldots, \phi_{d'}\}$, cf. Definition 4.1, is a dual basis for $\mathcal{N}$ with $N_i(\phi_j) = \delta_{ij}$, then $\hat{\mathcal{P}} = \text{span}\{\hat{\phi}_1 = \phi_1 \circ F^{-1}, \ldots, \hat{\phi}_{d'} = \phi_{d'} \circ F^{-1}\}$ is a dual basis for $\hat{\mathcal{N}}$ and $\hat{N}_i\hat{v} = N_i(\hat{v} \circ F)$.

The elements in Figures 4.8 and 4.17 are inequivalent.

Now we can handle the local interpolation in $(K,\mathcal{P},\mathcal{N})$ and its affine equivalent elements, e.g. the $(T,\mathcal{P}_T,\mathcal{N}_T)$. If (4.7) is satisfied, we assume for simplicity the same $(K,\mathcal{P},\mathcal{N})$ and appropriate F_T and $\mathcal{P}_T \circ F_T = \mathcal{P}$. A generalization to different K is straightforward, but we need additional technical tools, e.g. combining triangular

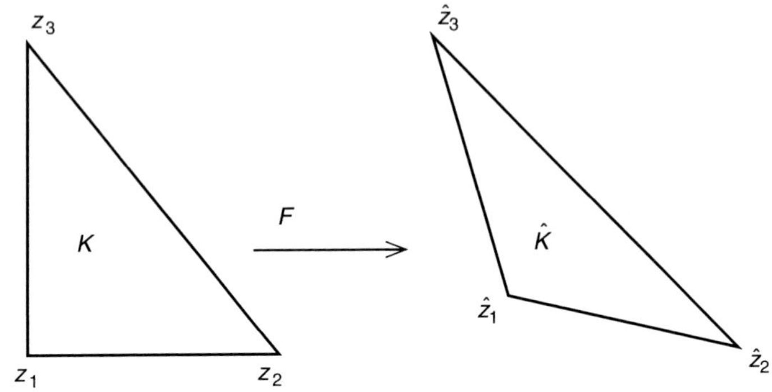

Figure 4.16 Affine transformation.

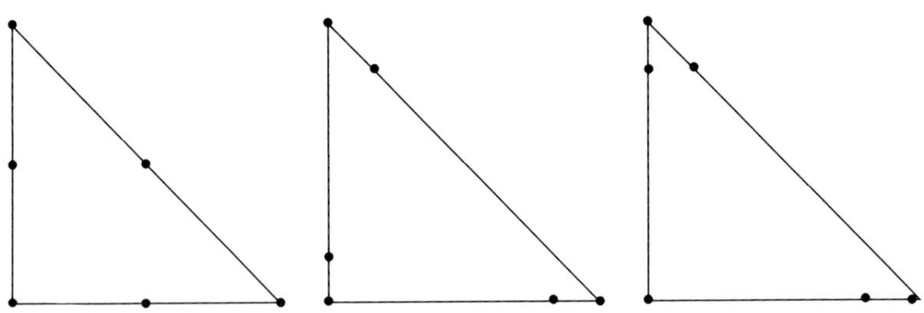

Figure 4.17 Inequivalent quadratic Lagrange FEs: Affine mappings impossible.

and rectangular elements. For introducing the space of FEs, we need $K, \mathcal{P}, \mathcal{N}$, the subdivision $\mathcal{T}^h$ and the affine (or isoparametric) mappings

$$F = F_T : K \to T, Fx = Ax + b, A = A_T \in \mathbb{R}^{n\times n}, b = b_T \in \mathbb{R}^n \forall\, T \in \mathcal{T}^h. \quad (4.14)$$

Definition 4.7. *Let $\mathcal{T}^h$ be a subdivision for Ω and let $F_T : K \to T$ be chosen as in (4.14). Then*

$$\mathcal{U}^h := \{u^h \in L^\infty(\Omega) : u^h|_T \in \mathcal{P}_T := \mathcal{P} \circ (F_T)^{-1}\} \quad (4.15)$$

is denoted as (approximating) space of FEs, u^h *as* FEs. *Since the $T \in \mathcal{T}^h$ are open, (4.15) does not imply transition properties for $u^h|_T$ and $u^h|_{T_1}$ if $\bar{T} \cap \bar{T}_1 \neq \emptyset$. For nonconforming FEs, u^h, see below, and an edge $e \subset \bar{T} \cap \bar{T}_1$, usually $u^h|_{\bar{T}|_e} \neq u^h|_{\bar{T}_1|_e}$. Therefore, in many cases additional properties are imposed such as*

$$\mathcal{U}^h \cap L^2(\Omega),\ \mathcal{U}^h \cap C^0(\Omega),\ \mathcal{U}^h \cap H_0^1(\Omega).$$

For the following *local and global interpolation operators* we combine K, $\mathcal{P} = \{\phi_1, \ldots, \phi_{d'}\}$, and $\mathcal{N} = \{N_1, \ldots, N_{d'}\}$, with the bijection F_T, see Definitions 4.2 ff. Choose $\mathcal{U} \ni v : \Omega \to \mathbb{R}$ such that $\forall T \in \mathcal{T}^h$ the $N_i^T(v) := N_i(v \circ F_T)$ are defined.

Definition 4.8. *Choose the FE space $\mathcal{U}^h : \Omega \to \mathbb{R}$ with the underlying $(K, \mathcal{P}, \mathcal{N})$, $\mathcal{P}$ a nodal basis for $\mathcal{N}$, the subdivision $\mathcal{T}^h$ and a function $v : \Omega \to \mathbb{R}$:*

1. *Let $(T, \mathcal{P}_T, \mathcal{N}_T)$ and the fixed $(K, \mathcal{P}, \mathcal{N})$ be affine equivalent $\forall T \in \mathcal{T}^h$ such that $F_T x = A_T x + b_T$ satisfies $F_T(K) = T$. cf. Figure 4.18.*
2. *Define the interpolation operator $I^h = I^h_{\mathcal{T}^h} : \mathcal{U} \to \mathcal{U}^h$ with $N_i^T(v) := N_i(v \circ F_T)\ \big(= N_i^T(I^h v \circ F_T)\big)$ and $\phi_i^T := \phi_i \circ F_T^{-1}$, as, see (4.6),*
$$I^h v|_T = I^h_{\mathcal{T}^h} v|_T = I_T v = \sum_{i=1}^{d'} N_i^T(v)\phi_i^T : T \to \mathbb{R}\ \forall\ T \in \mathcal{T}^h. \tag{4.16}$$
3. *The $\phi_i^T = \phi_i \circ F_T^{-1}$, $i = 1, \dots d'$, $\forall\ \ T \in \mathcal{T}^h$ are called an* interpolation basis *for $\mathcal{P}_T$ for a fixed $T \in \mathcal{T}^h$ (for $\forall T \in \mathcal{T}^h$ not a basis for $\mathcal{U}^h$, see 4.*
4. *Assume $P \in \bar{T} \cap \bar{T}_1$ to be a joint vertex or a point on an edge e or an $(n-1)$-dimensional boundary surface of $\bar{T}$ with, e.g. $\delta(P) \in \mathcal{N}_T$. Then $\forall\ \ T_j \in \mathcal{T}^h$ with $P \in \bar{T}_j$ the $\delta(P) \in \mathcal{N}_{T_j}$ and identical function values in this point P have to be used for all these $\bar{T}_j \ni P$ defining, e.g. an $N(v \circ F_{T_j}) = \delta(P)v$. (This implies continuity of the interpolation function in P from T to T_j, but not on e and in Ω and $\delta(P)$ yields exactly one basis element for $\mathcal{U}^h$.) Analogous restrictions are to be imposed for derivatives. Exceptions are the DCGMs in Chapter 7.*
5. *Let l be the highest derivative required in $\mathcal{N}$. A reference element $(K, \mathcal{P}, \mathcal{N})$, $K \subset \Omega$ is said to be a C^r* element *if r is the largest nonnegative integer such that*
$$\mathcal{U}^h = I^h(C^l(\bar{\Omega})) \subseteq C^r(\Omega) \cap W_\infty^{r+1}(\Omega).$$
6. *Let $\mathcal{P}^n_{d-1} = \mathcal{P} \subset W^d_\infty(K)$. Then K or the FEs are called* polynomial FEs of (total) degree $d-1$ or order d *and denoted as $\mathcal{P}^n_{d-1}(\mathcal{T}^h)$. For $\mathcal{P}^n_{d-1} \subseteq \mathcal{P} \subseteq \mathcal{P}^n_{d+\tau}$, $\mathcal{P} \subset W^d_\infty(K)$, with the smallest possible $\tau \geq -1$, e.g. the bilinear FEs, we call them* FEs of at least degree $d-1$.

Proposition 4.9. *Under the conditions of Definition 4.8 we get, cf. (4.15),*
$$I_K v = v\ \ \forall\ \ v \in \mathcal{P} \text{ and } I^h v^h = v^h\ \ \forall\ \ v^h \in \mathcal{U}^h \text{ implying} \tag{4.17}$$
$$(I_K)^2 = I_K \text{ on } \mathcal{P} \text{ and } (I^h)^2 = I^h \text{ on } \mathcal{U}^h, \text{ hence, } I_K, I^h \text{ are projectors.}$$

Remark 4.10. *There are different reasons to use the reference elements $(K, \mathcal{P}, \mathcal{N})$. Most theoretical results can be proved by combining the specific situation $(K, \mathcal{P}, \mathcal{N})$ with the affine equivalence.*

For the numerical praxis another point is even more important: The $N_i(v)$ and ϕ_i, $i = 1, \dots, d'$ can be computed once. Then the $N_i^T(v) := N_i(v \circ F_T)$ and $\phi_i^T = \phi_i \circ F_T^{-1}$ are obtained by simple transformation formulas via the F_T. This applies to the terms needed in the FEMs below, e.g. elements of the mass and stiffness matrices as well, cf. (4.121).

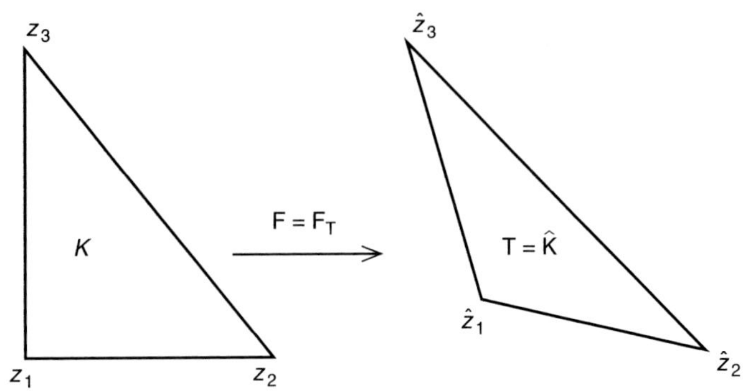

Figure 4.18 Affine transformations here applied to linear FEs.

Example 4.11.
We want to demonstrate this procedure for a triangulation applied to two cases of linear and cubic two-dimensional polynomials defined by the values indicated in Figures 4.19 and 4.20, see [576], p. 68 ff. We start with the transformation from the reference triangle K with the vertices $z_1 := (0,0),\ z_2 := (1,0),\ z_3 := (0,1),$ and the coordinates $(\xi\ ,\eta) \in K$ onto the triangle T. We denote edges of T as $\hat{z}_i := (x_i, y_i),\ i = 1,2,3$ and its coordinates as $(x\ ,y) \in T$, thus defining $F_T = ((F_T)_1, (F_T)_2)^T$ as

$$\begin{aligned} x = (F_T)_1\ (\xi\ ,\eta) &:= x_1 + (x_2 - x_1)\ \xi + (x_3 - x_1)\ \eta \\ y = (F_T)_2\ (\xi\ ,\eta) &:= y_1 + (y_2 - y_1)\ \xi + (y_3 - y_1)\ \eta. \end{aligned} \tag{4.18}$$

Since F_T is linear its derivative is constant and

$$F_T' = \begin{pmatrix} (x_2 - x_1) & (x_3 - x_1) \\ (y_2 - y_1) & (y_3 - y_1) \end{pmatrix}.$$

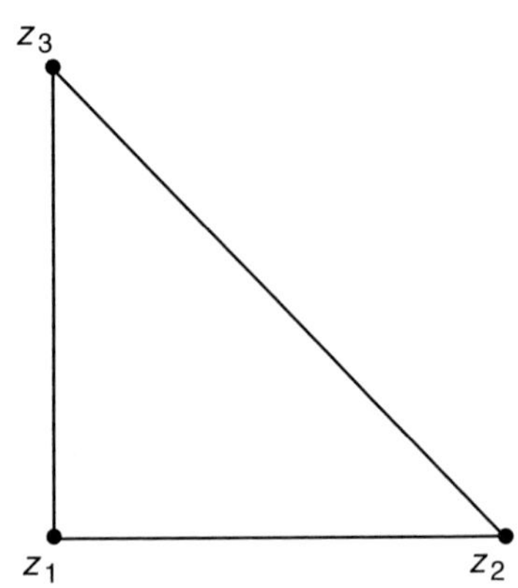

Figure 4.19 $\mathcal{P}_1$: Again linear Lagrange FEs.

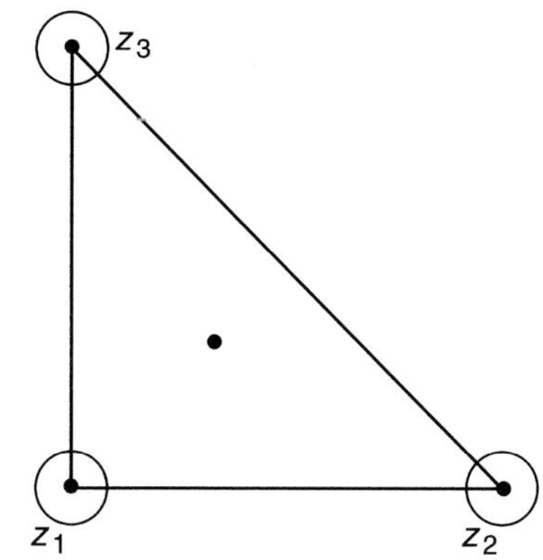

Figure 4.20 $\mathcal{P}_3$: Again cubic Hermite FEs.

This F_T is bijective if and only if the Jacobian, $\mathbf{J}$, of F_T' is

$$\mathbf{J} = \det(F_T') = \det\begin{vmatrix} (x_2 - x_1) & (x_3 - x_1) \\ (y_2 - y_1) & (y_3 - y_1) \end{vmatrix}$$

$$= (x_2 - x_1)(y_3 - y_1) - (y_2 - y_1)(x_3 - x_1) \neq 0. \quad (4.19)$$

Then F_T is invertible. $(F_T)^{-1}$ has the form

$$\xi = (F_T)_1^{-1}(x, y) := a_1 + a_2 x + a_3 y$$
$$\eta = (F_T)_2^{-1}(x, y) := b_1 + b_2 x + b_3 y. \quad (4.20)$$

By equating the coefficients in $F_T \circ (F_T)^{-1} = I$ we determine the $a_1, \ldots, b_3$ as

$$a_1 = -(x_1(y_3 - y_1) - y_1(x_3 - x_1))/\mathbf{J}, \; a_2 = (y_3 - y_1)/\mathbf{J},$$
$$a_3 = -(x_3 - x_1)/\mathbf{J}, \; b_1 = (x_1(y_2 - y_1) - y_1(x_2 - x_1))/\mathbf{J},$$
$$b_2 = -(y_2 - y_1)/\mathbf{J}, \; b_3 = (x_2 - x_1)/\mathbf{J}. \quad (4.21)$$

We use the notation

$$u : (\xi, \eta) \in K \to \mathbb{R} \text{ and } v : (x, y) \in T \to \mathbb{R}.$$

For the later computations of the FE terms in mass and stiffness matrices, see Example 4.51 below, we need the corresponding transformations of the partials via the chain rule:

$$\text{For } v := u \circ F_T^{-1} \text{ we find } v_x = u_\xi \xi_x + u_\eta \eta_x$$
$$v_y = u_\xi \xi_y + u_\eta \eta_y.$$

For computing the partials of the ξ, η we differentiate (4.18) with respect to x and find

$$\begin{aligned} 1 &= (x_2 - x_1)\,\xi_x + (x_3 - x_1)\,\eta_x \\ 0 &= (y_2 - y_1)\,\xi_x + (y_3 - y_1)\,\eta_x. \end{aligned}$$

Analogously, we differentiate with respect to y. Both systems yield the solutions

$$\begin{aligned} \xi_x &= (y_3 - y_1)/\mathbf{J}, \quad \xi_y = -(x_3 - x_1)/\mathbf{J} \\ \eta_x &= -(y_2 - y_1)/\mathbf{J}, \quad \eta_y = (x_2 - x_1)/\mathbf{J}. \end{aligned} \tag{4.22}$$

To give a feeling for the interplay of the transformations in $N_i(v \circ F_T) \cdot \left(\phi_i \circ F_T^{-1}\right)$ we will study below a special case of (4.18), the scaling

$$\begin{aligned} F(\xi, \eta) := F_T(\xi, \eta) &:= (x = h\xi,\ y = h\eta) \text{ with} \\ \xi_x &= 1/h,\ \xi_y = 0,\ \eta_x = 0,\ \eta_y = 1/h \end{aligned} \tag{4.23}$$

implying

$$v_x = u_\xi/h,\ v_y = u_\eta/h. \tag{4.24}$$

We apply these results to two special cases:

Linear Lagrange FEs: We choose $\mathcal{P} = \mathcal{P}_T = \mathcal{P}_1$ and v and u have the form

$$\begin{aligned} v = v(x, y) &= c_1 + c_2\,x + c_3\,y, \\ u = u(\xi, \eta) &= \alpha_1 + \alpha_2\,\xi + \alpha_3\,\eta. \end{aligned}$$

For the $z_1 = (0,0),\ z_2 = (1,0),\ z_3 = (0,1),$ and $\mathcal{N} = \{\delta(z_1), \delta(z_2), \delta(z_3)\}$ we want to determine a nodal basis $\mathcal{P} =$ span $\{\phi_1, \phi_2, \phi_3\}$. More generally, to determine a ϕ interpolating in the z_i or the ϕ_i in (4.6), we compute in a first step the α_j for given $u_i := u(z_i)$ in Figure 4.19. We introduce the vectors $\alpha := (\alpha_1, \alpha_2, \alpha_3)^T$ and $u := (u_1, u_2, u_3)^T$ and obtain the system

$$\begin{array}{lllcllll} \alpha_1 & & = u_1 & & u_1 & & = \alpha_1 \\ \alpha_1 & +\alpha_2 & = u_2 & \text{implying} & -u_1 & +u_2 & = \alpha_2 \\ \alpha_1 & \quad +\alpha_3 & = u_3 & & -u_1 & \quad +u_3 & = \alpha_3. \end{array} \tag{4.25}$$

This (4.25) has the form

$$u = B\alpha \text{ with } B := \begin{pmatrix} 1 & 0 & 0 \\ 1 & 1 & 0 \\ 1 & 0 & 1 \end{pmatrix}, \quad \text{or } \alpha = Au \text{ with } A := \begin{pmatrix} 1 & 0 & 0 \\ -1 & 1 & 0 \\ -1 & 0 & 1 \end{pmatrix}, \tag{4.26}$$

where $AB = I$.

We employ the unit vectors $e^i, i = 1, 2, 3$, e.g. $e^1 := (1, 0, 0)^T$, for determining the nodal basis ϕ_i. We obtain the coefficients $\alpha^i = \left(\alpha_1^i, \alpha_2^i, \alpha_3^i\right)^T$ for the ϕ_i as

$$\alpha^i = Ae^i, \quad \text{e.g. } \alpha^1 = (1, -1, -1)^T,\ \phi_1(\xi, \eta) = 1 - \xi - \eta.$$

For determining the $\phi_i(F_T^{-1})$ we insert in $\phi_i(\xi,\eta)$ for the ξ,η the terms in (4.20) with the $a_1,\dots,b_3$ determined in (4.21). Finally $I_K u = \sum_{j=1}^3 \phi_j u(P_j)$ and $I_T v = \sum_{j=1}^3 \phi_j \left(F_T^{-1}\right) v(\hat{P}_j)$.

In cases of simple $T,\mathcal{P},\mathcal{N}$ we can often determine the ϕ_j and $\phi_j\left(F_T^{-1}\right)$ and the $I^h v|_T$ directly, e.g. for our linear Lagrange FEs and the scaling in (4.23). With $T = hK$, we get obviously $\phi_1 = 1-\xi-\eta, \phi_2 = \xi, \phi_3 = \eta$ from (4.26) and with F_T^{-1} in (4.23) we find, e.g. $\phi_1^T = \phi_1 \circ F_T^{-1} = (h-x-y)/h$. The $z_i = P_i$ and $\hat{z}_i = \hat{P}_i$ show $N_i^T v = N_i(v \circ F_T) = v(F_T(P_i)) = v(\hat{P}_i)$. Then (4.16) yields

$$I^h v|_T = [v(\hat{P}_1)(h-x-y) + v(\hat{P}_2)x + v(\hat{P}_i)y]/h$$
$$(I^h v|_T)' = ((v(\hat{P}_2) - v(\hat{P}_1)),\ (v(\hat{P}_3) - v(\hat{P}_1)))/h$$

with

$$\|I^h v|_T\|_\infty \le \max\{|v(\hat{P}_i)| : i = 1,2,3\} \le \|v|_{\bar{T}}\|_\infty,$$
$$\|(I^h v|_T)'\|_\infty \le \max\{|(v(\hat{P}_2) - v(\hat{P}_1))|,\ |(v(\hat{P}_3) - v(\hat{P}_1))|\}/h \le \|v'|_T\|_\infty.$$

Cubic Hermite FEs: The ansatz for v and u is now

$$\begin{aligned} v = v(x,y) &= c_1 + c_2\,x + c_3\,y + c_4\,x^2 + c_5\,xy + c_6\,y^2 \\ &\quad + c_7\,x^3 + c_8\,x^2y + c_9\,xy^2 + c_{10}\,y^3, \\ u = u(\xi,\eta) &= \alpha_1 + \alpha_2\,\xi + \alpha_3\,\eta + \alpha_4\,\xi^2 + \alpha_5\,\xi\,\eta + \alpha_6\,\eta^2 \\ &\quad + \alpha_7\,\xi^3 + \alpha_8\,\xi^2\,\eta + \alpha_9\,\xi\,\eta^2 + \alpha_{10}\,\eta^3. \end{aligned} \tag{4.27}$$

In addition we need the partials of u

$$\begin{aligned} u_\xi &= \alpha_2 + 2\alpha_4\xi + \alpha_5\eta + 3\alpha_7\xi^2 + 2\alpha_8\xi\eta + \alpha_9\eta^2 \\ u_\eta &= \alpha_3 + \alpha_5\xi + 2\alpha_6\eta + \alpha_8\xi^2 + 2\alpha_9\xi\eta + 3\alpha_{10}\eta^2. \end{aligned}$$

With $z_i = P_i, \hat{z}_i = \hat{P}_i, z_4 := (z_1+z_2+z_3)/3$, we determine the FEs in (4.6), see Figure 4.20, by the following function values and derivatives:

$$u_i := u(z_i),\ i = 1,\dots,4,\ p_i := u_\xi(z_i),\ q_i := u_\eta(z_i),\ i = 1,2,3. \tag{4.28}$$

We determine any ϕ in (4.6), by evaluating in the first step, the u and its derivatives in (4.27), then compute, in the next step, the $\alpha := (\alpha_1,\dots,\alpha_{10})^T$ from the system

$$u := (u_1,p_1,q_1,u_2,p_2,q_2,u_3,p_3,q_3,u_4)^T = B\alpha,$$

with a matrix B obtained as follows. Every row of B, e.g. numbers 1 and 5, yield the values of the corresponding components of u. For numbers 1 and 5 these are the u_1

and p_2. Then (4.28) tells us $u_1 = u(z_1 = (0,0)), p_2 = u_\xi(z_2 = (1,0))$. The evaluation according to (4.27) and the following u_ξ then yields rows 1 and 5 of B, and so on.

$$B := \left(\begin{array}{ccc|ccc|cccc}
1 & 0 & 0 & 0 & 0 & 0 & 0 & 0 & 0 & 0 \\
0 & 1 & 0 & 0 & 0 & 0 & 0 & 0 & 0 & 0 \\
0 & 0 & 1 & 0 & 0 & 0 & 0 & 0 & 0 & 0 \\
- & - & - & - & - & - & - & - & - & - \\
1 & 1 & 0 & 1 & 0 & 0 & 1 & 0 & 0 & 0 \\
0 & 1 & 0 & 2 & 0 & 0 & 3 & 0 & 0 & 0 \\
0 & 0 & 1 & 0 & 1 & 0 & 0 & 1 & 0 & 0 \\
- & - & - & - & - & - & - & - & - & - \\
1 & 0 & 1 & 0 & 0 & 1 & 0 & 0 & 0 & 1 \\
0 & 1 & 0 & 0 & 1 & 0 & 0 & 0 & 1 & 0 \\
0 & 0 & 1 & 0 & 0 & 2 & 0 & 0 & 0 & 3 \\
- & - & - & - & - & - & - & - & - & - \\
1 & 1/3 & 1/3 & 1/9 & 1/9 & 1/9 & 1/27 & 1/27 & 1/27 & 1/27
\end{array}\right).$$

For computing the ϕ_i we recall $N_i(\phi_j) = \delta_{i,j}$, hence, we determine α^i from $B\alpha^i = e^i$, with the unit vectors $e^i, i = 1, \ldots, 10$, or we invert B as $A := B^{-1}$. Then the coefficient vectors α for a FE and the α^i for the ϕ_i are obtained from the systems $\alpha = Au$ and $\alpha^i = Ae^i$, e.g. for $i = 8$ as column number 8 of A, as

$$\alpha = Au \text{ with } \alpha^i = Ae^i, \quad \text{e.g., } \phi_8 = -\xi\eta + \xi^2\eta + 2\xi\eta^2 \text{ and}$$

$$A := \left(\begin{array}{ccc|ccc|cccc}
1 & 0 & 0 & 0 & 0 & 0 & 0 & 0 & 0 & 0 \\
0 & 1 & 0 & 0 & 0 & 0 & 0 & 0 & 0 & 0 \\
0 & 0 & 1 & 0 & 0 & 0 & 0 & 0 & 0 & 0 \\
- & - & - & - & - & - & - & - & - & - \\
-3 & -2 & 0 & 3 & -1 & 0 & 0 & 0 & 0 & 0 \\
-13 & -3 & -3 & -7 & 2 & -1 & -7 & -1 & 2 & 27 \\
-3 & 0 & -2 & 0 & 0 & 0 & 3 & 0 & -1 & 0 \\
- & - & - & - & - & - & - & - & - & - \\
2 & 1 & 0 & -2 & 1 & 0 & 0 & 0 & 0 & 0 \\
13 & 3 & 2 & 7 & -2 & 2 & 7 & 1 & -2 & -27 \\
13 & 2 & 3 & 7 & -2 & 1 & 7 & 2 & -2 & -27 \\
2 & 0 & 1 & 0 & 0 & 0 & -2 & 0 & 1 & 0
\end{array}\right).$$

Then we proceed as for the Lagrange case, however we do not directly compute the ϕ_i^T.

In addition we do need the transformation of the partials in the $\hat{z}_i$ to the z_i. We explicitly formulate this for the above special case of scaling in (4.23). With (4.24) and, e.g. $N_8(v \circ F_T) \cdot \left(\phi_8 \circ F_T^{-1}\right)$, we obtain

$$\begin{aligned}
\left(\phi_8 \circ F_T^{-1}\right)(x,y) &= -xy/h^2 + (x^2y + 2xy^2)/h^3 \\
N_8(v \circ F_T) &= (v \circ F_T)_\xi(0,1) = v'(F_T(0,1)) \cdot (F_T)_\xi(0,1) \\
&= ((v_x, v_y)(0,h))(h,0)^T = hv_x(0,h)
\end{aligned}$$

hence

$$N_8(v \circ F_T) \cdot \left(\phi_8 \circ F_T^{-1}\right)(x,y) = v_x(0,h)(-xy/h + (x^2y + 2xy^2)/h^2)$$
$$\text{for } 0 \le x, y, x+y \le h.$$

This shows, in particular, that the contribution of $N_8(v \circ F_T) \cdot \left(\phi_8 \circ F_T^{-1}\right)(x,y)$ in T is $\mathcal{O}(h)$.[30] Similarly the other terms $N_i(v \circ F_T) \cdot \left(\phi_i \circ F_T^{-1}\right)$ can be computed. ■

We do not intend formulating the sharpest results under the weakest possible conditions for the different cases. Rather we formulate Condition 4.16 below, which allows the following implications: The interpolation operator I^h is a bounded operator, a good approximation and the FEs allow inverse estimates. We introduce

Definition 4.12. *Let* $\{\mathcal{T}^h\}, 0 < h \le 1$, *be a family of* subdivisions *for* Ω *such that*

$$\max\{\text{diam } T : T \in \mathcal{T}^h\} \le h \text{ diam } \Omega.$$

Then h *is denoted as* the maximal step size. *The family is said to be* quasiuniform *if there exists* $\chi > 0$ *such that*

$$\min\{\text{diam } B_T : B_T \subset T \in \mathcal{T}^h\} \ge \chi\, h \tag{4.29}$$

for all $h \in (0,1]$; *here* B_T *is the largest ball such that* T *is* star-shaped *with respect to* B_T, *cf. Definition 4.14. The family is said to be* nondegenerate *if there exists* $\chi > 0$ *such that*

$$\text{diam } B_T \ge \chi \text{ diam } T \quad \forall\, T \in \mathcal{T}^h,\ h_T := \text{diam} T,\ h \in (0,1]. \tag{4.30}$$

Remark 4.13.

1. *The sequence of* $\{\mathcal{T}^h\}$ *is* nondegenerate *if and only if the chunkiness parameter* $\gamma = \gamma_T$, *see (4.32), is uniformly bounded for all* $T \in \mathcal{T}^h$ *and for all* $h \in (0,1]$, *hence* $\gamma_T = 1/\chi$. *Let all* $T \in \mathcal{T}^h$ *be triangles. Then* $\mathcal{T}^h$ *is nondegenerate if and only if for every* $T \in \mathcal{T}^h$ *the interior angles* $\alpha \ge \alpha_0 > 0$. *This is used as the definition for quasiuniform* $\mathcal{T}^h$ *e.g. in [135, 387]. A quasiuniform family is nondegenerate, but not conversely.*
2. *If we start with an arbitrary triangulation in two dimensions and repeatedly subdivide by connecting midpoints of edges, we obtain a* nondegenerate *family of subdivisions. For a well-defined alternative, see [58].*
3. *For quasiuniform* $\mathcal{T}^h$ *each* T *is* uniformly star-shaped.

4.2.4 Interpolation errors and inverse estimates

We have already formulated the general interpolating $u^h = I^h u$ on Ω in (4.16). It reproduces function values and/or derivatives of the original u according to $\mathcal{N}$. We want to study the question of how well these $I^h u$ approximate the original u. We collect

[30] $f(h) = \mathcal{O}(h^p)$ and $f(h) = o(h^p)$ iff $|f(h)|/h^p \le C$ and $|f(h)|/h^p \to 0$ for $h \to 0$.

the appropriate assumptions in Condition 4.16. Mainly for the study of variational crimes we need estimates of higher Sobolev norms of FEs by lower Sobolev norms, the so-called inverse estimates.

We need the standard and broken *Sobolev norms*, cf. (4.36), and some lemmas. We use the notation for *multi-indices*, α, and *partials*, ∂^α, see (2.73),

$$\|v\|_{W^{d,p}(\Omega)} := \Big(\sum_{j=0}^{d}(|v|_{W^{j,p}(\Omega)})^p\Big)^{1/p}, \quad |v|_{W^{j,p}(\Omega)} := \Big(\sum_{|\alpha|=j}\|\partial^\alpha v\|^p_{L^p(\Omega)}\Big)^{1/p}. \tag{4.31}$$

For $w \in W^{m,p}(\Omega)$ the classical derivatives and hence their point evaluation and Taylor polynomials $T_y^m w(x) = \sum_{|\alpha|<m} \partial^\alpha w(y)(x-y)^\alpha/\alpha!$ are not defined. So we use the $\partial^\alpha\, w \in L^2(\Omega)$, $|\alpha| < m$, directly and introduce the *averaged Taylor polynomials* $Q^m w$ for $w \in W^{m,p}(\Omega)$, see [141], in

Definition 4.14. *Let $G \subset \mathbb{R}^n$, e.g. $G = \Omega$, have finite* diam G *and let $B := B_\rho(x_0)$ be a ball of radius ρ. We assume $\bar{B} \subset G$ such that for all $x \in G$ the closed convex hull of $\{x\} \cup \bar{B} \subset G$. Then G is called* star-shaped with respect to B. *With*

$$\rho_{\max} := \sup\{\rho : G \textit{ star-shaped with respect to } B\}, \textit{ and } \gamma = \text{diam } G/\rho_{\max}, \tag{4.32}$$

this γ is called the chunkiness parameter *of G. Let G be* star-shaped *with respect to B and let ϕ be a* cut-off function *for B, that is (i) supp $\phi = \bar{B}$ and (ii) $\int_{\mathbb{R}^n} \phi(x)dx = 1$. Then the following* averaged Taylor polynomials *$Q^d w$ of degree $d-1$, and the $T_y^d w(x)$ are well defined, $\forall x \in B$ and $\forall y \in B \setminus M$, M a set of Lebesgue measure 0, as*

$$Q^d w(x) := \int_B T_y^d w(x)\phi(y)dy \quad \textit{with } T_y^d w(x) = \sum_{|\alpha|<d} \partial^\alpha w(y)(x-y)^\alpha/\alpha!. \tag{4.33}$$

Obviously, triangles are star-shaped. Note that only $\partial^\alpha u$ with $|\alpha| < d$ are employed in (4.33). A standard example for a cut-off function on $B = B_\rho(x_0)$ is

$$\phi(x) = c \exp\left(-(1-(|x-x_0|/\rho)^2)^{-1}\right) \text{ for } x \in B \text{ and } \phi(x) \equiv 0 \text{ outside } B.$$

We only consider *star-shaped* G (or $T \in \mathcal{T}^h$, see below) with bounded *chunkiness parameter*, γ, see (4.32). We obtain, see [141],

Lemma 4.15. Bramble–Hilbert lemma: *Let G be star-shaped with respect to B of radius ρ such that $\rho > (1/2)\rho_{\max}$, G has the chunkiness parameter γ, and $Q^d w$ be the averaged Taylor polynomial. Then for all $w \in W^{d,p}(G), \ell = 0, 1, \ldots, d$, $1 \le p \le \infty$, there exist constants $C_{d,n,\gamma}$ such that*

$$|w - Q^d w|_{W^{\ell,p}(G)} \le \|w - Q^d w\|_{W^{\ell,p}(G)} \le C_{d,n,\gamma}(\textit{diam } G)^{d-\ell}|w|_{W^{d,p}(G)}. \tag{4.34}$$

Since the $u^h \in \mathcal{U}^h$ might not be continuous across common edges $e \subset \overline{T} \cap \overline{T}_1$, and hence $u^h \notin H^1(\Omega)$ [174], pp. 207–208, the usual Sobolev norms $\|\cdot\|_{W^{d,p}(\Omega)}$ might not be defined. However, for all practically important cases the $\|u^h|_T\|_{W^{d,p}(T)}$ and

$\|u|_{\bar{T}}\|_{C^d(\bar{T})}$. So the *broken Sobolev spaces and norms* $\|u^h\|_{W^{d,p}(\mathcal{T}^h)}$ make sense:

$$\mathcal{U}_{\mathcal{T}^h} := W^{k,p}(\mathcal{T}^h) := \{u^h : \Omega \to \mathbb{R} : \forall\, T \in \mathcal{T} u^h|_T \in W^{k,p}(T)\}, \text{ with} \tag{4.35}$$

$$\|u^h\|^h_{k,q} := \|u^h\|_{W^{k,q}(\mathcal{T}^h)} = \Big(\sum_{T\in\mathcal{T}^h} \sum_{|\alpha|\le k} \|\partial^\alpha u^h|_T\|^q_{L^q(T)} \Big)^{1/q},$$

$$\text{for } 1 \le q < \infty, \tag{4.36}$$

$$\|u^h\|_{W^k_\infty(\mathcal{T}^h)} := \operatorname{ess\,sup}\{|\partial^\alpha u^h(x)| \ : \forall \alpha\ |\alpha| \le k \forall x\ \in\ T\ \forall\ T\ \in\ \mathcal{T}^h\}. \tag{4.37}$$

These norms satisfy *Hölder inequalities* as the standard norms:

$$\|u^h v^h\|_{W^{k,1}(\mathcal{T}^h)} \le \|u^h\|_{W^{k,p}(\mathcal{T}^h)} \|v^h\|_{W^{k,q}(\mathcal{T}^h)} \text{ for } 1 \le p, q \le \infty, 1/p + 1/q = 1 \tag{4.38}$$

The corresponding *seminorms* with $\sum_{|\alpha|=k}$ in (4.36) are

$$|u^h|_{W^{k,q}(\mathcal{T}^h)} = \Big(\sum_{T\in\mathcal{T}^h} \sum_{|\alpha|=k} \|\partial^\alpha u^h|_T\|^q_{L^q(T)} \Big)^{1/q} = \Big(\sum_{T\in\mathcal{T}^h} |u^h|^q_{W^{k,q}(T)} \Big)^{1/q}. \tag{4.39}$$

These norms coincide with the original norm $\|u^h\|_{W^{k,q}(\mathcal{T}^h)}$ for $u^h \in W^{d,p}(\Omega)$. Analogous definitions and equalities may have to be used for scalar products and bilinear forms as well. The decision about which of the following choices of d, p in the spaces

$$\mathcal{U}^h_{W^{d,p}(\Omega)} := \{u^h \in \mathcal{U}^h : \|u^h\|_{W^{d,p}(\mathcal{T}^h)} < \infty\} \text{ for } h \to 0 \tag{4.40}$$

is appropriate, depends upon the interplay of ansatz- and test-functions and the (non)linear elliptic problem. They are related by the corresponding bilinear or other forms. Certainly, for each fixed h we want the $\|u^h\|_{W^{d,p}(\mathcal{T}^h)} < \infty$. We start with a polyhedral domain $\Omega \subset \mathbb{R}^n$, for curved $\partial\Omega$ see Subsections 4.2.7, 4.2.6.

Condition 4.16. *For the reference FE* $(K, \mathcal{P}, \mathcal{N})$ *and the* subdivision $\mathcal{T}^h$ *assume:*

1. K *is star-shaped with respect to some ball.*
2. *Choose the smallest* $\tau \ge -1$ *such that* $\mathcal{P}^n_{d-1} \subseteq \mathcal{P} \subseteq \mathcal{P}^n_{d+\tau}$, $\mathcal{P} \subset W^d_\infty(K)$, *usually* $\tau = -1$. *For* $\tau \ge 0$, *let* $\mathcal{P} \subsetneq \mathcal{P}^n_{d+\tau}$, *hence, there exist* $v \in \mathcal{P} \cap \mathcal{P}^n_{d+\tau}$.
3. $\mathcal{N} \subset (C^l(\bar{K}))'$, *hence the elements of* $\mathcal{N}$ *are evaluations of* u *and derivatives of* u *up to the order* $\le l$ *in (different) points* P *for* $P \in \bar{K}$ *and additionally let* $1 \le p \le \infty$ *and*
$$\begin{cases} d - l - n/p > 0 & \text{for } p > 1 \text{ or} \\ d - l - n \ge 0 & \text{for } p = 1. \end{cases}$$
4. *Alternatively to 3. we assume* $\mathcal{N} \subset (W^{d,p}(K))'$.
5. $\{\mathcal{T}^h\}, 0 < h \le 1$, *is a* nondegenerate *family of admissible subdivisions of a bounded polyhedral domain* $\Omega \subset \mathbb{R}^n$ *with the parameter* χ, *see (4.30).*
6. *For all* $T \in \mathcal{T}^h, 0 < h \le 1$, *the* $(T, \mathcal{P}_T, \mathcal{N}_T)$ *is affine equivalent to* $(K, \mathcal{P}, \mathcal{N})$.
7. *The combination of 1. and 5. shows that* T *is uniformly star-shaped* $\forall\, T \in \mathcal{T}^h$ *and that the chunkiness parameter* γ *of all* $T \in \mathcal{T}^h$ *is bounded, see e.g. [141], Theorem 4.4.20.*

The unusual condition $\tau \geq -1$ such that $\mathcal{P}^n_{d-1} \subseteq \mathcal{P} \subsetneq \mathcal{P}^n_{d+\tau}$ allows us to study, e.g the families of tensor product FEs as in (4.13), the serendipity FEs, see [141], pp. 88 ff. and pp. 112 ff. For all these families and for Davydov's C^1 splines, see Subsection 4.2.6 and [263], specific error estimates are possible. For linear and quadratic tensor product FEs we get $\tau = 1$ and $\tau = 3$ for $n = 2$ or $\tau = 4$ and $\tau = 17$ for $n = 3$. The following results are proved in [141], Theorems 4.4.4, 4.4.20, 4.6.11 and 4.6.14, compare [135, 387] as well. Note that the conditions, relating d, l, n, p in [141], Theorems 4.4.4 and 4.4.20 are imposed to guarantee Condition 4.16, 3. and to apply the Sobolev embedding lemma. They are not necessary for 4. Furthermore, these results extend to a combination of $v \in W^{d,p}(\Omega)$ and FEs of higher order.

Theorem 4.17. Interpolation errors: *Let the $\{\mathcal{T}^h\}, 0 < h \leq 1$, and the reference element, $(K, \mathcal{P}, \mathcal{N})$, satisfy Condition 4.16 for some $l, d, p,\ 1 \leq p \leq \infty$, cf. (4.36), (4.37). Let I^h be the (local) interpolation operator in (4.16) defined by $(K, \mathcal{P}, \mathcal{N})$.*

1. *Then there exists a constant $C > 0$, such that $\forall\ 0 \leq s \leq d$, the following* local and global interpolation error estimates *hold $\forall v \in W^{d,p}(\Omega)$:*

$$\|v - I^h v\|_{W^{s,p}(T)} \leq C\ (\text{diam } T)^{d-s} |v|_{W^{d,p}(T)},$$
$$\|I^h v\|_{W^{s,p}(T)} \leq \|v\|_{W^{s,p}(T)} + C (\text{diam } T)^{d-s} |v|_{W^{d,p}(T)}. \tag{4.41}$$
$$\|v - I^h v\|_{W^{s,p}(\mathcal{T}^h)} \leq C\ h^{d-s} |v|_{W^{d,p}(\Omega)},$$
$$\|I^h v\|_{W^{s,p}(\mathcal{T}^h)} \leq \|v\|_{W^{s,p}(\Omega)} + C\ h^{d-s} |v|_{W^{d,p}(\Omega)} \leq C \|v\|_{W^{d,p}(\Omega)}. \tag{4.42}$$

   C depends on $(K, \mathcal{P}, \mathcal{N}), n, d, p$ and the number χ in (4.30).

2. *For $0 \leq s \leq l < d$, see 3. in Condition 4.16, we have*

$$\|v - I^h v\|_{W^{s,\infty}(\mathcal{T}^h)} \leq C\ h^{d-s-n/p} |v|_{W^{d,p}(\Omega)} \quad \forall\ v \in W^{d,p}(\Omega). \tag{4.43}$$

3. *For the case of* tensor product FEs *the estimates for the interpolation errors in (4.41)-(4.43) can be modified by replacing the seminorms $|v|_{W^{d,p}(T)}, |v|_{W^{d,p}(\Omega)}$ by $|v|^t_{W^{d,p}(T)}, |v|^t_{W^{d,p}(\Omega)} := \left(\sum_{i=1}^n \|\partial^d v / \partial x_i^d\|^p_{L^p(\Omega)}\right)^{1/p}$.*

Proposition 4.18. *Our FE spaces $\mathcal{U}^h$ are approximating spaces for $\mathcal{U} = W^{s,p}(\Omega)$ and $d > s \geq 0$, in the sense of Definition 3.5, hence,*

$$\lim_{h \to 0} \text{dist}\ (u, \mathcal{U}^h) = \inf_{u^h \in \mathcal{U}^h} ||u - u^h||_{\mathcal{U}^h} = 0\ \forall\ u \in \mathcal{U}\ \Rightarrow \lim_{h \to 0} ||I^h u||_{\mathcal{U}^h} = ||u||_{\mathcal{U}}. \tag{4.44}$$

Proof. For fixed $u \in W^{s,p}(\Omega)$ and arbitrary $\epsilon > 0$ we choose a $v \in W^{d,p}(\Omega)$ such that $\|v - u\|_{W^{s,p}(\mathcal{T}^h)} < \epsilon/2$. Now by $\|v - I^h v\|_{W^{s,p}(\mathcal{T}^h)} \leq C\ h^{d-s} |v|_{W^{d,p}(\Omega)}$ we can choose h_0 such that $\|v - I^h v\|_{W^{s,p}(\mathcal{T}^h)} < \epsilon/2$ for $h < h_0$, so finally $\|u - I^h v\|_{W^{s,p}(\mathcal{T}^h)} < \epsilon$ for $h < h_0$, Hence, our FE spaces for $d > s$ are approximating spaces. For the last claim we combine (4.42) with $\|u - I^h u\| < \|u - v\| + \|v - I^h v\| + \|I^h v - I^h u\|$. □

Inverse estimates, presented in the sequel, relate various norms $\|v^h\|_{W^{s,p}(\mathcal{T}^h)}$ for finite element spaces for $s \geq d$. This is necessary to handle variational crimes. We again present local and global estimates. Note that for the next two results, see [141], pp. 110 ff., we need $\mathcal{T}^h$ and the affine (isoparametric) maps $F_T : K \to T$. However, no reference to $\mathcal{N}, \mathcal{N}_T$ nor the unisolvence is made. Since $\mathcal{P}$ always satisfies $\mathcal{P} \subseteq W^{j,p}(T) \cap W_q^l(T)$, $1 \le q \le p, 0 \le l \le j$, we get

Theorem 4.19. Inverse estimates: *Let $\{\mathcal{T}^h\}, 0 < h \le 1$, $(K, \mathcal{P}, \mathcal{N}), \mathcal{P}_T = \mathcal{P}$ satisfy Condition 4.16, let $j \le d + \tau$, $1 \le q \le p, 0 \le l \le j$, and*

$$\mathcal{U}^h = \{v^h : \Omega \to \mathbb{R}, v^h|_T \in \mathcal{P}_T \ \ \forall \ \ T \in \mathcal{T}^h\}. \tag{4.45}$$

Then there exists $C = C(j, p, q, \chi)$, with χ in (4.30), such that

$$\|v^h\|_{W^{j,p}(T)} \le C \ (\text{diam } T)^{l-j+n/p-n/q} \|v^h\|_{W_q^l(T)}, \ \forall v^h \in \mathcal{U}^h \tag{4.46}$$

and

$$\|v^h\|_{W^{j,p}(\mathcal{T}^h)} \le C \ h^{l-j+\min(0,n/p-n/q)} \|v^h\|_{W_q^l(\mathcal{T}^h)}, \tag{4.47}$$

where $\{\mathcal{T}^h\}, 0 < h \le 1$, is quasiuniform *for (4.47), and for all $v^h \in \mathcal{U}^h$. These inequalities only make sense for $j \le d + \tau$, otherwise $\|v^h\|_{W^{j,p}(\mathcal{T}^h)} = \|v^h\|_{W^{d+\tau,p}(\mathcal{T}^h)}$.*

4.2.5 Inverse estimates on nonquasiuniform triangulations

This condition of quasiuniformity for (4.47) can be relaxed by applying the results for boundary and finite element methods in Graham, Hackbusch and Sauter [360–364], and with Dahmen, Faermann and Grasedyck in [244, 359]. We present some of their wide range of inverse-type inequalities, here applications to finite element functions on general classes of meshes, including degenerate meshes obtained by an isotropic refinement. These are obtained for Sobolev norms of positive, zero and negative order. In contrast to classical inverse estimates, negative powers of the minimum mesh diameter are avoided.

In the whole subsection the functions $u = u^h$ are always FEs.

For $n = 2$ or 3, let $\Omega \subset \mathbb{R}^n$ denote a bounded domain. Suppose that Ω is decomposed into a mesh, $\mathcal{T} = \mathcal{T}^h$, of *tetrahedra/bricks* $(n = 3)$ or *curvilinear triangles/ quadrilaterals* $(n = 2)$. Then classical inverse estimates give, cf. (4.47) for the left inequality, for $\sim$, $\lesssim$ compare the index,

$$\|u\|_{H^s(\mathcal{T}^h)} \lesssim h_{\min}^{-s} \|u\|_{L^2(\mathcal{T}^h)} \lesssim h_{\min}^{-2s} \|u\|_{H^{-s}(\mathcal{T}^h)}, \tag{4.48}$$

for a suitable range of positive s and for all functions $u \in H^s(\mathcal{T}^h)$ which are piecewise polynomials of degree $\le d - 1$ with $d - 1 \ge 0$ with respect to this mesh. The quantity $h_{\min}$ is the minimum diameter of all the elements of the mesh and (4.48) holds under the assumption of *shape regularity*, i.e. $\rho_\tau \gtrsim h_\tau$ for each $\tau \in \mathcal{T}^h$ where h_τ is the diameter of τ and $\rho_\tau = \rho$ is the diameter of the largest inscribed sphere, cf. Definition 4.14. We will use such estimates regularly in the finite element analysis, particularly for nonconforming FEMs, cf. Chapter 5. When the mesh is quasiuniform, cf. Definition 4.12, so $h \lesssim h_{\min}$, with h the maximum diameter of all the elements, they can be used to obtain convergence rates in powers of h for various quantities in various norms.

However, practical meshes are often nonquasiuniform and then the negative powers of $h_{\min}$ in (4.48) may give rise to overly pessimistic convergence rates. In [361], less pessimistic replacements for (4.48) have been derived, a particular case being

$$\|u\|_{H^s(\mathcal{T}^h)} \lesssim \|h^{-s}u\|_{L^2(\mathcal{T}^h)} \lesssim \|h^{-2s}u\|_{H^{-s}(\mathcal{T}^h)}, \tag{4.49}$$

where $h : \Omega \longrightarrow \mathbb{R}$ is now a continuous piecewise linear *mesh function* whose value on each element τ reflects the diameter of that element, i.e. $h_\tau \lesssim h|_\tau \lesssim h_\tau$.

Estimates (4.49) have several applications, e.g. to the analysis of the *Mortar element method* [361]. In fact [361] contains more general versions of (4.49), e.g. in the Sobolev space $W^{s,p}(\mathcal{T}^h)$ and in related Besov spaces. While the left-hand inequality in (4.49) is well-known, at least in the Sobolev space case, the right-hand inequality requires rather delicate analysis.

In this subsection we obtain more general versions of (4.49) which do not require the *mesh sequence to be shape-regular.* A typical estimate is

$$\|u\|_{H^s(\mathcal{T}^h)} \lesssim \|\rho^{-s}u\|_{L^2(\mathcal{T}^h)} \lesssim \|\rho^{-2s}u\|_{H^{-s}(\mathcal{T}^h)}, \tag{4.50}$$

where the mesh function $\rho : \Omega \longrightarrow \mathbb{R}$ is now a continuous piecewise linear function whose value on each element τ reflects the diameter of the largest inscribed sphere, introduced in Definition 4.12. Estimates (4.50) hold under the rather weak assumptions that (i) the quantities h_τ and ρ_τ are *locally quasiuniform* (i.e. $h_\tau/h_{\tau'} \lesssim 1$ and $\rho_\tau/\rho_{\tau'} \lesssim 1$ for all neighboring elements τ, τ') and (ii) the number of elements which touch any element remains bounded as the mesh is refined (see Assumption 4.26). These assumptions admit degenerate meshes, containing *long thin "stretched" elements,* which are typically used for *approximating edge singularities or boundary layers* in solutions of PDEs. Our estimates (4.50) hold true when all the elements τ of a mesh are obtained by suitable maps from a single unit element, as for the above finite element spaces. For the purpose of a readable introduction we have here written our estimates (4.50) in a very compact form. In fact the range of s for which the right-hand inequality in (4.50) holds may be greater then that for which the left-hand inequality holds and we shall give precise ranges in the last subsubsection.

Meshes and finite elements

Throughout, Ω will denote a bounded n-dimensional subset of $\mathbb{R}^n$, for $n = 2$ or 3. We define the Sobolev spaces $H^s(\Omega)$, and $H^s(\mathcal{T}^h), s \geq 0$, as in Subsection 1.4.3. Throughout, we let $[-k, k]$ denote the range of Sobolev indices for which we are going to prove the inverse estimates (where k is a positive integer), and we assume that $H^s(\Omega)$ is defined for all $s \in [-k, k]$, and that $H^{-s}(\Omega)$ is the dual of $H^s(\Omega)$, for $s > 0$. We assume that Ω is decomposed as above into a mesh $\mathcal{T} = \mathcal{T}^h$ of relatively open pairwise-disjoint finite elements $\tau \subset \mathcal{T}^h$ with the property $\overline{\mathcal{T}^h} = \cup\{\bar{\tau} : \tau \in \mathcal{T}\}$.

Definition 4.20. Mesh parameters: *For each $\tau \in \mathcal{T}, |\tau|$ denotes its n-dimensional measure, h_τ denotes its diameter and ρ_τ is the diameter of the largest sphere centered at a point in τ whose intersection with $\overline{\mathcal{T}^h}$ lies entirely inside $\bar{\tau}$. For any other simplex or cube $t \in \mathbb{R}^n$ (not necessarily an element of $\mathcal{T}$) we define h_t and ρ_t in the same way.*

In order to impose a simple geometric character on the mesh τ, we assume that each $\tau \in \mathcal{T}$ is diffeomorphic to a simple unit element. More precisely, let $\hat{\sigma}^3$ in $\mathbb{R}^3$ denote the unit simplex with vertices $(0,0,0), (1,0,0), (0,1,0), (0,0,1)$, and let $\hat{\kappa}^3$ denote the unit cube with vertices $\{(x,y,z) : x,y,z \in \{0,1\}\}$. In $\mathbb{R}^2$, define $\hat{\sigma}^2$ to be the unit simplex with vertices $(0,0), (1,0)$ and $(0,1)$ and define $\hat{\kappa}^2$ to be the unit square with vertices $\{(x,y) : x,y \in \{0,1\}\}$.

Then we assume, as above, that for each $\tau \in \mathcal{T}$, there exists a unit element $\hat{\tau} = \hat{\sigma}^n$ or $\hat{\kappa}^n$ and a bijective map $\chi_\tau : \hat{\tau} \to \tau$, with χ_τ and χ_τ^{-1} both smooth, here $\in C^\infty$. Since χ_τ is smooth, each element $\tau \in \mathcal{T}$ is either a curvilinear tetrahedron/brick ($n = 3$) or a curvilinear triangle/rectangle ($n = 2$). The mesh $\mathcal{T}$ is allowed to contain both types of elements. Each element has vertices and edges. In the $\mathbb{R}^3$ case the element also has faces, comprising the images of the faces of the unit element. For a suitable index set $\mathcal{Q}$, we let $\{\mathbf{x}_p : p \in \mathcal{Q}\}$ denote the *set of all vertices* $\mathcal{T}$. We assume the mesh is *conforming*, cf. Definition 4.2, i.e. for each $\tau, \tau' \in \mathcal{T}$ with $\tau \neq \tau', \bar{\tau} \cap \overline{\tau'}$ is allowed to be either empty, a node, an edge or (when $n = 3$) a face of both τ and τ'. The requirement that χ_τ is smooth ensures that edges of $\mathcal{T}^h$ ($n = 2$) and edges $\partial\mathcal{T}^h$ ($n = 3$) are confined to edges of elements $\tau \in \mathcal{T}$. Let J_τ denote the $n \times n$ *Jacobian* of χ_τ. Then

$$g_\tau := \left\{\det J_\tau^T J_\tau\right\}^{1/2}$$

is the *Gram determinant* of the map χ_τ, which appears in the change of variable formula: $\int_\tau f(\mathbf{x})dx = \int_{\hat{\tau}} f(\chi_\tau(\hat{\mathbf{x}}))g_\tau(\hat{\mathbf{x}})d\hat{x}$. To ensure that the map χ_τ is sufficiently regular we shall make the following assumptions on J_τ:

Assumption 4.21. Mapping properties:

$$\begin{aligned} D^{-1}|\tau|^2 \leq \det\{J_\tau(\hat{\mathbf{x}})^T J_\tau(\mathbf{x})\} \ &\leq\ D|\tau|^2, \\ E\rho_\tau^2 \leq \lambda_{\min}\{J_\tau(\hat{\mathbf{x}})^T J_\tau(\mathbf{x})\} \end{aligned} \tag{4.51}$$

uniformly in $\hat{\mathbf{x}} \in \hat{\tau}$, *for all* $\tau \in \mathcal{T}$, *with positive constants* D, E *independent of* τ.

(Throughout this section, for a symmetric matrix $A, \lambda_{\min}(A)$ and $\lambda_{\max}(A)$ denote respectively the minimum and maximum eigenvalues of A.) Assumption 4.21 is satisfied in a number of standard cases.

Example 4.22.

Suppose $\tau \subset \Omega$, where τ is for $n = 2$ and $n = 3$ a polygon and a polyeder, respectively. Then χ_τ can be chosen as an affine map. Then the Jacobian J_τ is identical to an $n \times n$ constant matrix and it is well known that $\det J_\tau = |\tau|/|\hat{\tau}|$ and that $\|J_\tau^{-1}\|_2 \leq h_{\hat{\tau}}\rho_\tau^{-1}$, from which the estimates (4.51) follow. ■

Example 4.23.

In many applications *quadrilateral or hexahedral elements* are important. Consider, for example, quadrilateral elements τ obtained by mapping from the unit element $\hat{\kappa}^2 = (0,1)^2$. If the map is affine, then the estimates for (4.51) obtained in Example 4.22 carry over verbatim. However only parallelograms can be obtained by applying affine maps to $\hat{\kappa}^2$. More general quadrilaterals can be obtained using bilinear maps and it turns out that, under quite moderate assumptions, (4.51) still hold. ■

This is a consequence of the following considerations for quadrilateral elements τ. A parallelogram π (considered as an open subset of $\mathbb{R}^2$) will be called an *inscribed parallelogram* for τ if $\pi \subset \tau$ and if two adjacent edges of π are identical with two adjacent edges of τ. It is easy to show that any convex quadrilateral τ has at least one inscribed parallelogram.

Proposition 4.24. cf. [364]: *Let τ be a convex planar quadrilateral obtained by applying a bilinear mapping to $\hat{\kappa}^2$. Let π be any inscribed parallelogram for τ. Then the estimates (4.51) hold uniformly in $\hat{\mathbf{x}} \in \hat{\tau}$ with D, E depending continuously on the ratio $r := |\pi|/|\tau| \in (0, 1]$.*

Remark 4.25. *If follows that if the ratio $|\pi|/|\tau|$ is bounded below by some constant $\gamma > 0$ say, for all elements τ as the mesh is refined, then (4.51) hold (with D and E dependent on γ). There are obvious degenerate elements which satisfy this, for example any parallelograms (no matter how small the smallest interior angles are) satisfy it. Similarly "moderately" distorted parallelograms also satisfy it.*

Assumption 4.21 describes the quality of the maps which take the unit element $\hat{\tau}$ to each τ. We also need assumptions on how the size and shape of neighboring elements in our mesh may vary. Here we impose *only very weak local conditions which require the meshes to be neither quasiuniform nor shape-regular.* Throughout the rest of this subsubsection we make the following assumption.

Assumption 4.26. Mesh properties: *For some $K, L \in \mathbb{R}_+$ and $M \in \mathbb{N}$, we assume that, for all $\tau, \tau' \in \mathcal{T}$ with $\bar{\tau} \cap \overline{\tau'} \neq \emptyset$,*

$$h_\tau \leq K h_{\tau'}, \qquad \rho_\tau \leq L \rho_{\tau'}, \tag{4.52}$$

$$\max_{p \in \mathcal{Q}} \#\{\tau \in \mathcal{T} : \mathbf{x}_p \in \bar{\tau}\} \leq M. \tag{4.53}$$

Note that condition (4.52) requires that h_τ and π_τ do not vary too rapidly between neighboring elements. This allows elements with large aspect ratio, provided their immediate neighbors have a comparable aspect ratio.

Example 4.27.

Shape-regular meshes are easily shown to satisfy Assumption 4.26 with moderate K, L, M. Also, *meshes which are anisotropically graded towards an edge* typically lie in this class. Near an edge, but away from the corners, the solution typically is badly behaved only in the direction orthogonal to the edge and efficient approximations require meshes which are anisotropically graded.

For example, for the square screen $[0, 1] \times [0, 1]$, a typical tensor product anisotropic mesh would be: $\mathbf{x}_{i,j} = (t_i, t_j)$, where $t_i = (i/n)^g/2$ and $t_{2n-i} = 1 - (i/n)^g/2$ for $i = 0, \ldots, n$, for some grading exponent $g \geq 1$. In this case the elements become very long and thin near smooth parts of edges. In the *hp* version of the finite element method similar meshes but with more extreme grading may be used and these also satisfy Assumption 4.26. ■

We denote the class of meshes which satisfy Assumptions 4.21 and 4.26 as $\mathcal{M}_{\mathcal{D},\mathcal{E},\mathcal{K},\mathcal{L},\mathcal{M}}$. From now on, if $A(\mathcal{T})$ and $B(\mathcal{T})$ are two mesh-dependent quantities, then the inequality $A(\mathcal{T}) \lesssim B(\mathcal{T})$ means that $A(\mathcal{T}) \le CB(\mathcal{T})$ with the constant C independent of $\mathcal{T} \in \mathcal{M}_{D,E,K,L,M}$, but not necessarily independent of D, E, K, L, M, similarly for $A(\mathcal{T}) \sim B(\mathcal{T})$ and $B(\mathcal{T}) \lesssim A(\mathcal{T})$.

Now we generalize the above definition of finite element spaces on the mesh $\mathcal{T}$.

Definition 4.28. Finite element spaces: *For $d-1 \ge 0$ and $\hat{\tau} \in \{\hat{\sigma}^n, \hat{\kappa}^n\}$, we define*

$$\mathbb{P}^{d-1}(\hat{\tau}) = \begin{cases} \textit{polynomials of total degree } \le d-1 \text{ on } \hat{\tau} & \text{if } \hat{\tau} = \hat{\sigma}^n, \\ \textit{polynomials of coordinate degree } \le d-1 \text{ on } \hat{\tau} & \text{if } \hat{\tau} = \hat{\kappa}^n, \end{cases}$$

with the $\hat{\sigma}^n$, $\hat{\kappa}^n$ on top of p. 234. Then we define

$$\mathcal{S}_0^{d-1}(\mathcal{T}) = \{u \in L^\infty(\mathcal{T}^h) : u \circ \chi_\tau \in \mathbb{P}^{d-1}(\hat{\tau}), \tau \in \mathcal{T} \quad \text{for } d-1 \ge 0.$$
$$\mathcal{S}_1^{d-1}(\mathcal{T}) = \{u \in C^0(\mathcal{T}^h) : u \circ \chi_\tau \in \mathbb{P}^{d-1}(\hat{\tau}), \tau \in \mathcal{T} \quad \text{for } d-1 \ge 1.$$

Inverse estimates

In this subsubsection we summarize the inverse estimates from [364], which were motivated above (see (4.50)). To define the scaling function ρ, recall the parameters ρ_τ introduced in Definition 4.20. From these we construct a continuous mesh function $\rho \in \mathcal{S}_1^1$ on $\mathcal{T}^h$ as follows.

Definition 4.29. Mesh function: *For each $p \in \mathcal{Q}$, set $\rho_p = \max\{\rho_\tau : \mathbf{x}_p \in \bar{\tau}\}$. The mesh function ρ is the unique function in $\mathcal{S}_1^1(\mathcal{T})$ such that $\rho(\mathbf{x}_p) = \rho_p$, for each $p \in \mathcal{Q}$.*

Clearly ρ is a positive, continuous function on Ω and, by Assumption 4.26, it follows that $\rho(\mathbf{x}) \sim \rho_\tau$ for $\mathbf{x} \in \tau$, and all $\tau \in \mathcal{T}$. The main results of this section are Theorems 4.30, 4.32 and 4.33. The first two of these provide inverse estimates in positive Sobolev norms for functions $u \in \mathcal{S}_i^{d-1}(\mathcal{T})$ with continuity index $i = 1, 0$, respectively. The third theorem provides inverse estimates in negative norms.

Theorem 4.30. Inverse estimates: *Let $0 \le s \le 1$ and $-\infty < \underline{\alpha} < \bar{\alpha} < \infty$. Then*

$$\|\rho^\alpha u\|_{H^s(\mathcal{T}^h)} \lesssim \|\rho^{\alpha - s} u\|_{L^2(\mathcal{T}^h)},$$

uniformly in $\alpha \in [\underline{\alpha}, \bar{\alpha}]$, $u \in \mathcal{S}_1^{d-1}(\mathcal{T})$.

Remark 4.31. *Since $\mathcal{S}_1^{d-1}(\mathcal{T}) \subset H^s(\mathcal{T}^h)$ for all $s < 3/2$, it may be expected that the range of Sobolev indices for which Theorem 4.30 holds may be extended. Such an extension has been obtained in [361] for shape-regular meshes at the expense of working in Besov norms. We have avoided such extensions here in order to simplify the present subsubsection.*

Theorem 4.32. Inverse estimates: *Let $0 \le s < 1/2$ and $-\infty < \underline{\alpha} < \bar{\alpha} < \infty$. Then*

$$\|\rho^\alpha u\|_{H^s(\mathcal{T}^h)} \lesssim \|\rho^{\alpha - s} u\|_{L^2(\mathcal{T}^h)},$$

uniformly in $\alpha \in [\underline{\alpha}, \bar{\alpha}], u \in \mathcal{S}_0^{d-1}(\mathcal{T})$.

Theorem 4.33. Inverse estimates, negative norms: *Let* $i \in \{0,1\}, d-1 \geq i, 0 \leq s \leq 1$ *and* $-\infty < \underline{\alpha} < \bar{\alpha} < \infty$. *Then the inequality*

$$\|\rho^{s+\alpha} u\|_{L^2(\mathcal{T}^h)} \lesssim \|\rho^\alpha u\|_{H^{-s}(\mathcal{T}^h)}, \tag{4.54}$$

holds uniformly in $u \in \mathcal{S}_i^{d-1}(\mathcal{T})$ *and* $\alpha \in [\underline{\alpha}, \bar{\alpha}]$. *If* χ_τ *is affine for all* τ *then (4.54) holds for all* $0 \leq s \leq k$, *where* k *is as described in the first paragraph of the last subsubsection.*

Remark 4.34. *Note that the test function* w *constructed in the proof of Theorem 4.33 in Graham et al. [364] vanishes at the boundaries of elements. Hence* w *belongs to the closure of the space* $C_0^\infty(\Omega)$ *with respect to the* $H^s(\Omega)$ *norm (this space is usually denoted* $H_0^s(\Omega)$*). Thus the result of Theorem 4.33 also holds if* $H^{-s}(\Omega)$ *was defined as the dual of* $H_0^s(\Omega)$, *although we have not so defined it here.*

4.2.6 Smooth FEs on polyhedral domains, with O. Davydov

This subsection is an adapted version of Davydov [264]. In the context of this book, we need smooth FEs for two kinds of problems. Conforming FEMs for weak $2m$-th order equations require FEs (piecewise polynomial splines) in $C^{m-1}(\Omega)$. Clearly such piecewise polynomial functions belong to $H^m(\Omega)$. The new FEM for fully nonlinear strong $2m$-th order elliptic problems can only be used on (curved) domains with C^{2m} boundary, and require FEs in $C^{2m-1}(\Omega)$ on such domains in the case $m > 1$. For $m = 1$ we still need $C^1(\Omega)$ splines, but can admit convex domains, which allows convex polyhedral domains in particular. Moreover, applications to fully nonlinear equations in Section 5.2.5 require certain stable splitting of the spline space. Consequently, we consider smooth FEs in three different settings:

(1) Approximation theory of $C^r(\Omega)$ splines of local degree d, for $r \geq 1$ on polyhedral domains in $\mathbb{R}^n$ with $d \geq r2^n + 1$. For this setting we prove inverse estimates (Theorem 4.35) and error bounds for the quasi-interpolation operators (Theorem 4.36).
(2) $C^1(\Omega)$ splines on polygonal domains in $\mathbb{R}^2$. A modification of the Argyris FE presented below is shown to admit a stable splitting, see Theorem 4.39.
(3) $C^{2m-1}(\Omega)$ splines, for $m \geq 1$, on curved C^{2m} domains in $\mathbb{R}^n$ as needed in FEMs for fully nonlinear problems. Although the question of constructing such splines remains widely open, we formulate Conjecture 4.41 as a basis for these methods.

In this and the next chapter we, in particular, prove the full convergence theory for elliptic equations and systems of order $2m$ with FEs in $H^m(\Omega)$. Smooth FEs in two variables have been studied, e.g. in the classical book by Ciarlet [174]. For $n > 2$, see Zenísek [679] and Le Méhauté [484]. For most FEMs, smooth FEs or splines are successfully avoided nowadays by using the standard FEMs in $C(\Omega)$ and in $H^1(\Omega)$. Then equations or systems of order $2m$ are transformed into a system of m second order equations, analogously for systems, sometimes losing information. The corresponding variational crimes, see e.g. Sections 5.5, allow further generalizations. This is not yet possible for fully nonlinear elliptic problems.

In the literature, several different approaches for smooth FEs are discussed. Either FEs on a polyhedral or on approximate curved domains approximate the functions on the original polyhedral or curved domain. $C^1(\Omega)$ FEs on polyhedral or C^2 domains, and $C^{m-1}(\Omega)$ FEs for $m \geq 1$, are studied in many papers.

C^1 FEs on curved domains are studied and applied to the Kármán equations by Bernadou, e.g. [85,86] for the "curved" Argyris and Bell FEs, hence with degree $d = 5$. The results in [85, 86, 680, 681] and the compilations only apply to $\mathbb{R}^2$ and provide partial answers in the above setting (3). A totally different approach is based on the weighted extended B-splines, see e.g. Höllig, Reif and Wipper [400] They allow the incorporation of curved boundaries into splines on uniform grids with a new technique. The results about stable splitting mentioned in (2) above are new (see Davydov [264]).

On polyhedral domains many results are available for *bivariate polynomial splines*, see e.g. the recent book [467]. However, certain important questions have not been addressed, in particular the error bounds for multivariate smooth piecewise polynomials on general triangulations in $\mathbb{R}^n$.

The error bounds below are formulated for a triangulation $\mathcal{T}^h$ of a polyhedral domain Ω. Thus, we deal with a family of triangulations $\{\mathcal{T}^h\}$ and spaces $\{\mathcal{S}^h\}$ parametrized by h. It is assumed that the other parameters, e.g. of a basis for $\mathcal{S}^h$ characterizing its *stability* and *locality*, or $n, d, p, \omega_{\mathcal{T}^h}$, remain independent of h. In this sense, one obtains respective asymptotic estimates and rates of convergence.

Splines on polyhedral domains in $\mathbb{R}^n$

The triangulation $\mathcal{T}^h$ for polyhedral domains has been introduced in Definition 4.2. $\mathcal{T}^h$ has to satisfy the standard conditions, see Definition 4.2. In Definition 4.12 we introduced the maximal and local step sizes h and $h_T = \operatorname{diam} T$. For a *quasiuniform* and *nondegenerate* $\mathcal{T}^h$, see (4.30), we require

$$\exists \chi > 0 : \quad \frac{\min_{T\in\mathcal{T}^h} \operatorname{diam} B_T}{h} \geq \chi \quad \text{and} \quad \frac{1}{\omega_{\mathcal{T}^h}} := \min_{T\in\mathcal{T}^h} \frac{\operatorname{diam} B_T}{h_T} \geq \chi, \tag{4.55}$$

respectively, and $\omega_{\mathcal{T}^h}$ is called the *shape regularity constant* of $\mathcal{T}^h$.

For any $d \geq 0$, we consider the spaces of all (including discontinuous) piecewise polynomials $S_d(\mathcal{T}^h)$ and the space $\mathcal{S}_d^r(\mathcal{T}^h)$ of splines of smoothness $r \geq 0$, defined by

$$\begin{aligned} s \in S_d(\mathcal{T}^h) \Leftrightarrow s|_T \in \mathcal{P}_d^n \ \forall T \in \mathcal{T}^h \qquad & \text{and} \\ \mathcal{S}_d^r(\mathcal{T}^h) = S_d(\mathcal{T}^h) \cap C^r(\Omega^h), \ 0 \leq r < d, & \end{aligned} \tag{4.56}$$

where $\mathcal{P}_d^n$ denotes the space of n-variate polynomials of total degree $\leq d$. In particular, $\mathcal{S}_d^0(\mathcal{T}^h)$ is the classical space of continuous FEs.

The spaces $\mathcal{S}_d^r(\mathcal{T}^h)$ are subspaces of the $W^{r+1,p}(\mathcal{T}^h)$ with norms and subnorms in (4.35)–(4.39) and corresponding scalar products for $p = 2$.

As a consequence of the differentiability of the $\mathcal{S}_d^r(\mathcal{T}^h)$ we have to change the assumptions here and in Section 5.2 compared with the earlier sections. The differentiability imposes a semilocal condition, cf. (4.62), not directly comparable with the preceding $\mathcal{P}_{d-1} \subset \mathcal{P} \subset \mathcal{P}_{d+\tau}$, cf. Definition 4.8, 6. Condition (4.62) allows us for example to include reduced FEs, e.g. the Bell element into consideration.

Note that for nested sequences of triangulations $\mathcal{T}_1^h \sqsubset \mathcal{T}_2^h \sqsubset \cdots$, nested spline spaces $\mathcal{S}_d^\tau(\mathcal{T}_1^h) \subset \mathcal{S}_d^\tau(\mathcal{T}_2^h) \subset \cdots$ are available, allowing multiresolution techniques, see Davydov and Petrushev [265] and Davydov and Stevenson [268].

Inverse estimates

We start by establishing the inverse estimates needed in the proof of the interpolation error bound and of independent interest in the FEM.

The *multivariate Markov inequality* for a simplex $T \subset \mathbb{R}^n$, see Coatmélec [181],

$$\|D^\alpha s\|_{L_\infty(T)} \le c \frac{nd^2}{\operatorname{diam} B_T} \|s\|_{L_\infty(T)}, \quad s \in \mathcal{P}_d^n, \quad |\alpha| = 1,$$

implies the following *inverse estimates.*

Theorem 4.35. Inverse estimates: *For any $1 \le p \le \infty$, $0 \le k < \mu \le d$, we have*

$$|s|_{W^{\mu,p}(T)} \le \frac{K}{h_T^{\mu-k}} |s|_{W^{k,p}(T)}, \quad s \in S_d(\mathcal{T}^h) \quad \text{for simplices } T \in \mathcal{T}^h, \tag{4.57}$$

where the constant K depends only on $n, d, p, \omega_{\mathcal{T}^h}$, see (4.55). For a quasiuniform $\mathcal{T}^h$, $h_T = \operatorname{diam} T$ can be replaced by h.

Proof. Let $|\alpha| = \mu$ and $g = s|_T \in \mathcal{P}_d^n$. Since $D^\alpha g = D^\beta D^\gamma p$ for some β, γ with $|\beta| = \mu - k, |\gamma| = k$, and since $D^\gamma g \in \mathcal{P}_{d-k}^n$, a repeated application of the Markov inequality implies

$$\|D^\alpha g\|_{L_\infty(T)} \le \frac{a_1}{\operatorname{diam} B_T^{\mu-k}} \|D^\gamma g\|_{L_\infty(T)} \le \frac{a_2}{h_T^{\mu-k}} \|D^\gamma g\|_{L_\infty(T)},$$

where a_1 depends only on n, d, and $a_2 = a_1 \omega_{\mathcal{T}^h}^{\mu-k}$. This already proves (4.57) in the case $p = \infty$. For $1 \le p < \infty$, a simple scaling argument shows that

$$a_3^{-1} \operatorname{vol}^{1/p}(T) \|q\|_{L_\infty(T)} \le \|q\|_{L_p(T)} \le a_4 \operatorname{vol}^{1/p}(T) \|q\|_{L_\infty(T)}, \quad q \in \mathcal{P}_d^n, \tag{4.58}$$

where $\operatorname{vol}(T)$ is the n-dimensional volume of the simplex T, and the constants a_3, a_4 depend only on n, d, p. Therefore, we get

$$\|D^\alpha g\|_{L_p(T)} \le \frac{a_2 a_3 a_4}{h_T^{\mu-k}} \|D^\gamma g\|_{L_p(T)},$$

and (4.57) follows. The statement for the quasiuniform case is obvious. □

Error bounds

For smooth splines on a polyhedral domain Ω in $\mathbb{R}^n$ we now prove error bounds of quasi-interpolation. For FE subspaces these bounds can be found e.g. in [141], and for bivariate spline spaces in [467], Theorem 5.18. Anisotropic triangulations have been considered in [265].

We introduce the *star of a vertex* v of $\mathcal{T}^h$, denoted by $\operatorname{star}(v)$, as the closure of the union of all n-simplices $T \in \mathcal{T}^h$ with a vertex at v. We set $\operatorname{star}^1(v) := \operatorname{star}(v)$, and define $\text{star}^\gamma(v)$, $\gamma \geq 2$, recursively as the union of the stars of all vertices of $\mathcal{T}^h$ contained in $\text{star}^{\gamma-1}(v)$.

Assuming that $\mathcal{S}^h$ is a subspace of $\mathcal{S}_d(\mathcal{T}^h)$, we choose a basis for $\mathcal{S}^h$ and its dual $(\mathcal{S}^h)'$, such that

$$\mathcal{S}^h = \text{span}\,[s^1, \dots, s^N],\ (\mathcal{S}^h)' = \text{span}\,[\lambda^1, \dots, \lambda^N],\ \lambda^i s^j = \delta_{ij},\ i,j = 1, \dots, N. \quad (4.59)$$

The basis $[s^1, \dots, s^N]$ is said to be *stable and local* if for each $k = 1, \dots, N$, a set $E_k \subset \Omega$ exists, such that

$$\text{supp}\, s^k \subset E_k \subset \text{star}^\gamma(v_k) \quad \text{for an appropriate vertex } v_k \text{ of } \mathcal{T}^h, \quad (4.60)$$

$$\|s^k\|_{L_\infty(\Omega)} \leq C_1 \quad \text{and} \quad |\lambda^k s| \leq C_2 \|s\|_{L_\infty(E_k)},\ \forall s \in \mathcal{S}^h, \quad (4.61)$$

for some C_1, C_2 and γ. Indeed, (4.60) implies that the basis functions s^k have local support, and it follows from (4.60), (4.61) that they are *stable* in L_∞ in the sense that for any real $\alpha_1, \dots, \alpha_N$,

$$K_1 \max_{1 \leq k \leq N} |\alpha_k| \leq \Big\| \sum_{k=1}^N \alpha_k s^k \Big\|_{L_\infty(\Omega)} \leq K_2 \max_{1 \leq k \leq N} |\alpha_k|,$$

with K_1, K_2 depending only on $n, d, \omega_{\mathcal{T}^h}, \gamma, C_1, C_2$.

It can also be shown [263], Lemma 6.2, that, after renorming, the new basis $\tilde{s}_1, \dots, \tilde{s}_N$ with $\tilde{s}^k := \text{vol}^{-1/p}(E_k)\, s^k$, is stable in L_p, $1 \leq p < \infty$, that is

$$K_1 \left(\sum_{k=1}^N |\alpha_k|^p \right)^{1/p} \leq \Big\| \sum_{k=1}^N \alpha_k \tilde{s}^k \Big\|_{L_p(\Omega)} \leq K_2 \left(\sum_{k=1}^N |\alpha_k|^p \right)^{1/p}.$$

In addition, we assume that $\mathcal{S}^h$ contains the space of polynomials of certain degree, such that

$$\mathcal{P}^n_{\ell-1} \subset \mathcal{S}^h \subset S_d(\mathcal{T}^h) \quad \text{for some } 0 \leq \ell - 1 \leq d. \quad (4.62)$$

The following theorem provides error bounds for certain *quasi-interpolation* operator $I^h : L_1(\Omega) \to \mathcal{S}^h$. For $\mathcal{S}^r_{d-1}(\mathcal{T}^h)$ in (4.56), we have $\mathcal{P}^n_{d-1} \subset \mathcal{S}^r_{d-1}(\mathcal{T}^h)$. Therefore the role of the earlier d in the context of interpolation errors is taken over by $\ell - 1$ here. Note that in applications to FEMs usually only the *existence* of an operator I^h with desired approximation properties is important. Hence it is acceptable to use the Hahn–Banach Theorem 1.14 as a tool in the definition of I^h.

Theorem 4.36. Quasi-interpolation error on polyhedral domains Ω in $\mathbb{R}^n$:

1. *Under conditions (4.59)–(4.62) there exists a linear operator $I^h : L_1(\Omega) \to \mathcal{S}^h$, such that for any $T \in \mathcal{T}^h$, $1 \leq p \leq \infty, 0 \leq |\alpha| \leq \ell - 1$, $u \in L_1(\Omega)$ with $|u|_{W^{\ell,p}(\Omega^\gamma_T)} < \infty$,*

$$\|D^\alpha(u - I^h u)\|_{L^p(T)} \leq K h_T^{\ell - |\alpha|} |u|_{W^{\ell,p}(\Omega^\gamma_T)}, \quad (4.63)$$

where $\Omega_T^\gamma := \bigcup_{v\in T} \operatorname{star}^{2\gamma-1}(v)$, $K = K(n, d, \omega_{\mathcal{T}^h}, \gamma, C_1C_2, L_{\partial\Omega})$, *and* $L_{\partial\Omega}$ *is the Lipschitz constant of the boundary* $\partial\Omega$ *of* Ω. *Moreover, if* $u \in W^{\ell,p}(\Omega)$, *then for all* $1 \le p \le \infty$, $0 \le |\alpha| \le \ell - 1$,

$$\|D^\alpha(u - I^h u)\|_{L^p(\mathcal{T}^h)} \le K h^{\ell-|\alpha|} |u|_{W^{\ell,p}(\Omega)}. \tag{4.64}$$

2. *Rewriting (4.64) for Sobolev norms, we obtain for all* $0 \le k \le \ell - 1$,

$$\|u - I^h u\|_{W^{k,p}(\mathcal{T}^h)} \le K h^{\ell-k} \|u\|_{W^{\ell,p}(\Omega)} \text{ and} \tag{4.65}$$

$$\lim_{h\to 0} \|I^h u\|_{W^{k,p}(\mathcal{T}^h)} = \|u\|_{W^{k,p}(\Omega)}.$$

3. *Moreover, it follows that for all* $0 \le k \le \ell - 1$,

$$\|u - I^h u\|_{W^{k,p}(\partial\Omega)} \le K h^{\ell-k-1/p} \|u\|_{W^{\ell,p}(\Omega)}, \text{ and} \tag{4.66}$$

$$\lim_{h\to 0} \|I^h u\|_{W^{k,p}(\partial\Omega)} = \|u\|_{W^{k,p}(\partial\Omega)},$$

where $p > 1$ *if* $k = \ell - 1$.

Proof.

(1) Let us define the operator I^h. In view of (4.61), each functional λ^k, $k = 1, \dots, N$, is well defined on $\mathcal{S}^h|_{E_k}$. By the Hahn–Banach Theorem 1.14, we extend λ^k from $\mathcal{S}^h|_{E_k}$ to $S_d(\mathcal{T}^h)|_{E_k}$, such that

$$|\lambda^k s| \le C_2 \|s\|_{L_\infty(E_k)}, \quad \text{for all } s \in S_d(\mathcal{T}^h)|_{E_k},$$

and define $\hat{I}^h : S_d(\mathcal{T}^h) \to \mathcal{S}^h$ by

$$\hat{I}^h s = \sum_{k=1}^N \lambda^k(s|_{E_k}) s^k, \quad s \in S_d(\mathcal{T}^h). \tag{4.67}$$

Let T be a simplex in $\mathcal{T}^h$, and let $T \subset E_k$. In view of (4.60), $\operatorname{diam}(E_k) \le c_1 h_T$ and $\operatorname{vol}(E_k) \le c_2 \operatorname{vol}(T)$, with some constants c_1, c_2 depending only on n, γ and $\omega_{\mathcal{T}^h}$.[31] Hence, for any $s \in S_d(\mathcal{T}^h)$ we have in view of (4.58),

$$c_3^{-1} \operatorname{vol}^{1/p}(T) \|s\|_{L_\infty(E_k)} \le \|s\|_{L_p(E_k)} \le c_4 \operatorname{vol}^{1/p}(T) \|s\|_{L_\infty(E_k)},$$

where c_3, c_4 depend only on $n, d, p, \gamma, \omega_{\mathcal{T}^h}$. Since the basis splines s^k are also piecewise polynomials, we have by (4.61),

$$\|s^k\|_{L_p(T)} \le c_5 \operatorname{vol}^{1/p}(T) \|s^k\|_{L_\infty(T)} \le c_5 \operatorname{vol}^{1/p}(T) C_1,$$

[31] This is easy to show considering that the number of simplices in E_k is bounded by a constant depending only on n, γ and $\omega_{\mathcal{T}^h}$, and that any two simplices T', T'' with a common facet satisfy $h_{T''} \le \tilde{c}_1 h_{T'}$ and $\operatorname{vol}(T'') \le \tilde{c}_2 \operatorname{vol}(T')$, for some $\tilde{c}_1, \tilde{c}_2$ depending only on n and $\omega_{\mathcal{T}^h}$.

where c_5 depends only on n, d and p. Hence, assuming $s \in S_d(\mathcal{T}^h)$ and applying (4.60) and (4.61), we obtain for any $T \in \mathcal{T}^h$,

$$\begin{aligned}\|\hat{I}^h s\|_{L_p(T)} &= \left\| \sum_{\substack{k=1\\ T\subset E_k}}^{N} \lambda^k(s|_{E_k}) s^k \right\|_{L_p(T)} \leq \sum_{\substack{k=1\\ T\subset E_k}}^{N} |\lambda^k(s|_{E_k})| \|s^k\|_{L_p(T)} \\ &\leq c_5 C_1 C_2 \sum_{\substack{k=1\\ T\subset E_k}}^{N} \|s\|_{L_\infty(E_k)} \mathrm{vol}^{1/p}(T) \\ &\leq c_3 c_5 C_1 C_2 \sum_{\substack{k=1\\ T\subset E_k}}^{N} \|s\|_{L_p(E_k)} \leq c_6 c_3 c_5 C_1 C_2 \|s\|_{L_p(\Omega_T^\gamma)},\end{aligned}$$

where the last inequality follows from

$$\bigcup_{\substack{k=1\\ T\subset E_k}}^{N} E_k \subset \Omega_T^\gamma, \quad T \in \mathcal{T}^h,$$

and

$$\begin{aligned}\#\{k : T \subset E_k\} &\leq \dim S_d(\mathcal{T}^h)|_{\Omega_T^\gamma} = \binom{n+d}{n} \# \left\{T' \in \mathcal{T}^h : T' \subset \Omega_T^\gamma\right\} \\ &\leq \binom{n+d}{n} c_6,\end{aligned}$$

where c_6 depends only on γ and $\omega_{\mathcal{T}^h}$. Thus, we have shown that

$$\|\hat{I}^h s\|_{L_p(T)} \leq c\|s\|_{L_p(\Omega_T^\gamma)}, \quad s \in S_d(\mathcal{T}^h), \quad T \in \mathcal{T}^h, \tag{4.68}$$

with c depending only on n, p, d, $\omega_{\mathcal{T}^h}$, C, and γ.

(2) In order to extend $I^h = \hat{I}^h$ from $S_d(\mathcal{T}^h)$ to $L_1(\Omega)$, we consider, for any $u \in L_1(\Omega)$ and any $T \in \mathcal{T}^h$, the *averaged Taylor polynomial* $p_T(u)$ of degree $\ell - 1$ with respect to the inscribed ball of T (see Definition 4.14). By Lemma 4.15,

$$|u - p_T(u)|_{W^{k,p}(T)} \leq B h_T^{\ell-k} |u|_{W^{\ell,p}(T)}, \quad 0 \leq k \leq \ell - 1, \tag{4.69}$$

where B depends only on $\ell, n, \omega_{\mathcal{T}^h}$. We define $\hat{s}(u) \in S_d(\mathcal{T}^h)$ by

$$\hat{s}(u)|_T = p_T(u), \quad T \in \mathcal{T}^h,$$

and set

$$I^h u := \hat{I}^h \hat{s}(u), \quad u \in L_1(\Omega).$$

Clearly, I^h is a projector onto $\mathcal{S}^h$, and, in particular, in view of (4.62),

$$I^h p = p, \qquad \text{for any } p \in \mathcal{P}_{\ell-1}^n. \tag{4.70}$$

(3) Let us prove (4.63). Suppose that $T \in \mathcal{T}^h$, $0 \leq |\alpha| \leq \ell - 1$, and $|u|_{W^{\ell,p}(\Omega_T^\gamma)} < \infty$. By using the Stein extension theorem (see e.g. [141], Theorem 1.4.5), we

extend $u|_{\Omega_T^\gamma}$ to a function $\tilde{u}$ defined on the convex hull U of Ω_T^γ such that

$$|\tilde{u}|_{W^{\ell,p}(U)} \leq c_7 |u|_{W^{\ell,p}(\Omega_T^\gamma)},$$

where c_7 depends only on $n, p, \ell, \omega_{\mathcal{T}_c^h}$ and, possibly, on the Lipschitz constant $L_{\partial\Omega}$ if Ω is nonconvex, and the boundary of Ω_T^γ contains a part of $\partial\Omega$, compare [467], Theorem 1.9. Now, let $p_T \in \mathcal{P}_{\ell-1}^n$ be the average Taylor polynomial for $\tilde{u}$ with respect to a ball in U of the greatest diameter. Again by the Bramble–Hilbert Lemma 4.15, we have[32]

$$|u - p_T(u)|_{W^{k,p}(\Omega_T^\gamma)} \leq c h_T^{\ell-k} |u|_{W^{\ell,p}(\Omega_T^\gamma)}, \quad 0 \leq k \leq \ell - 1, \tag{4.71}$$

which implies, in particular,

$$\|D^\beta(u - p_T(u))\|_{L_p(\Omega_T^\gamma)} \leq c_8 h_T^{\ell-|\beta|} |u|_{W^{\ell,p}(\Omega_T^\gamma)}, \quad 0 \leq |\beta| \leq \ell - 1, \tag{4.72}$$

with c_8 depending only on $n, p, \ell, \omega_{\mathcal{T}^h}, L_{\partial\Omega}$. Therefore,

$$\begin{aligned}
\|D^\alpha(u - I^h u)\|_{L_p(T)} &\leq \|D^\alpha(u - p_T(u))\|_{L_p(T)} + \|D^\alpha(p_T(u) - I^h u)\|_{L_p(T)} \\
&\leq c_8 h_T^{\ell-|\alpha|} |u|_{W^{\ell,p}(\Omega_T^\gamma)} + \|D^\alpha(p_T(u) - I^h u)\|_{L_p(T)}.
\end{aligned}$$

By (4.70), (4.57), (4.68), and (4.72),

$$\begin{aligned}
\|D^\alpha(p_T(u) - I^h u)\|_{L_p(T)} &= \|D^\alpha \hat{I}^h(p_T(u) - \hat{s}(u))\|_{L_p(T)} \\
&\leq \frac{A}{h_T^{|\alpha|}} \|\hat{I}^h(p_T(u) - \hat{s}(u))\|_{L_p(T)} \\
&\leq \frac{c}{h_T^{|\alpha|}} \|p_T(u) - \hat{s}(u)\|_{L_p(\Omega_T^\gamma)} \\
&\leq \frac{c}{h_T^{|\alpha|}} \|u - p_T(u)\|_{L_p(\Omega_T^\gamma)} + \frac{c}{h_T^{|\alpha|}} \|u - \hat{s}(u)\|_{L_p(\Omega_T^\gamma)} \\
&\leq c h_T^{\ell-|\alpha|} |u|_{W^{\ell,p}(\Omega_T^\gamma)} + \frac{c}{h_T^{|\alpha|}} \|u - \hat{s}(u)\|_{L_p(\Omega_T^\gamma)}.
\end{aligned}$$

Now, since $h_{T'} \leq c_9 h_T$, for all $T' \in \mathcal{T}^h$ such that $T' \subset \Omega_T^\gamma$, where c_9 depends only on n, γ and $\omega_{\mathcal{T}^h}$, we have by (4.69)

$$\begin{aligned}
\|u - \hat{s}(u)\|_{L_\infty(\Omega_T^\gamma)} &= \max_{\substack{T' \in \mathcal{T}^h \\ T' \subset \Omega_T^\gamma}} \|u - p_{T'}(u)\|_{L_\infty(T')} \\
&\leq B \max_{\substack{T' \in \mathcal{T}^h \\ T' \subset \Omega_T^\gamma}} h_{T'}^\ell |u|_{W^{\ell,\infty}(T')} \leq B c_9 h_T^\ell |u|_{W^{\ell,\infty}(\Omega_T^\gamma)}
\end{aligned}$$

[32] For a convex Ω it is possible to show (4.71) by arguments similar to the proof of Lemma 2.5 in [436], without any resort to the Stein extension theorem.

in the case $p = \infty$, and

$$\|u - \hat{s}(u)\|^p_{L_p(\Omega_T^\gamma)} = \sum_{\substack{T' \in \mathcal{T}^h \\ T' \subset \Omega_T^\gamma}} \|u - p_{T'}(u)\|^p_{L_p(T')}$$

$$\leq B^p \sum_{\substack{T' \in \mathcal{T}^h \\ T' \subset \Omega_T^\gamma}} h_{T'}^{\ell p} |u|^p_{W^{\ell,p}(T')} \leq B^p c_9^p h_T^{\ell p} |u|^p_{W^{\ell,p}(\Omega_T^\gamma)}$$

in the case $1 \leq p < \infty$, which completes the proof of (4.63).

(4) To show (4.64), we first consider the case $p = \infty$. For some $T^* \in \mathcal{T}^h$, we have by (4.63),

$$\|D^\alpha(u - I^h u)\|^{\mathcal{T}^h}_{L_\infty(\Omega)} = \|D^\alpha(u - I^h u)\|_{L_\infty(T^*)} \leq K h_{T^*}^{\ell-|\alpha|} |u|_{W^{\ell,\infty}(\Omega_{T^*}^\gamma)}$$

$$\leq K |\mathcal{T}^h|^{\ell-|\alpha|} |u|_{W^{\ell,\infty}(\Omega)}.$$

Assume now that $1 \leq p < \infty$. Then by (4.63),

$$\left(\|D^\alpha(u - I^h u)\|^{\mathcal{T}^h}_{L_p(\Omega)}\right)^p = \sum_{T \in \mathcal{T}^h} \|D^\alpha(u - I^h u)\|^p_{L_p(T)}$$

$$\leq K^p \sum_{T \in \mathcal{T}^h} h_T^{(\ell-|\alpha|)p} |u|^p_{W^{\ell,p}(\Omega_T^\gamma)}$$

$$\leq K^p |\mathcal{T}^h|^{(\ell-|\alpha|)p} \sum_{T \in \mathcal{T}^h} \sum_{\substack{T' \in \mathcal{T}^h \\ T' \subset \Omega_T^\gamma}} |u|^p_{W^{\ell,p}(T')}.$$

Now

$$\sum_{T \in \mathcal{T}^h} \sum_{\substack{T' \in \mathcal{T}^h \\ T' \subset \Omega_T^\gamma}} |u|^p_{W^{\ell,p}(T')} = \sum_{T \in \mathcal{T}^h} \#\left\{T' \in \mathcal{T}^h : T \subset \Omega_{T'}^\gamma\right\} |u|^p_{W^{\ell,p}(T)},$$

and since $T \subset \Omega_{T'}^\gamma \Leftrightarrow T' \subset \Omega_T^\gamma$, we have

$$\#\left\{T' \in \mathcal{T}^h : T \subset \Omega_{T'}^\gamma\right\} = \#\left\{T' \in \mathcal{T}^h : T' \subset \Omega_T^\gamma\right\} \leq c_6,$$

and, hence,

$$\left(\|D^\alpha(u - I^h u)\|^{\mathcal{T}^h}_{L_p(\Omega)}\right)^p \leq c_6 K^p |\mathcal{T}^h|^{(\ell-|\alpha|)p} \sum_{T \in \mathcal{T}^h} |u|^p_{W^{\ell,p}(T)}$$

$$= c_6 K^p |\mathcal{T}^h|^{(\ell-|\alpha|)p} |u|^p_{W^{\ell,p}(\Omega)},$$

which completes the proof of (4.64).

(5) The last statement of the theorem is a direct consequence of (4.64) and Theorem 1.38. Indeed, by (1.68),

$$\|u - I^h u\|_{W^{k,p}(\partial\Omega)} \le \left(\|u - I^h u\|_{W^{k,p}(\Omega)}\right)^{1-1/p} \left(\|u - I^h u\|_{W^{k+1,p}(\mathcal{T}^h)}\right)^{1/p}$$
$$\le K h^{\ell-k-1/p} \|u\|_{W^{\ell,p}(\Omega)},$$

and (4.66) follows. □

From the different types of extension operators mentioned above we refer to Treves [632]. He proves Theorem 26.A.3 for functions in $u \in C^k(\bar{\Omega})$ and extends them continuously to $u \in W^{k,p}(\Omega), 1 \le p < \infty$, in Corollary 26.A.1, yielding (4.73), (4.74). This technique can obviously be extended the following situation to show (4.75). For the FEMs for fully nonlinear elliptic problems and $\Omega \in C^2$ we will need approximate polyhedral Ω^h or curved Ω_c^h with dist $(\Omega, \Omega_c^h) \le Ch$.

Theorem 4.37. Modified Sobolev–Stein extension operator: *Let $\Omega \in C^k$, Ω_c^h, $\Omega_0 \supset (\Omega \cup \Omega_c^h)$, and $k \in \mathbb{N}$ be given. Then there exist a linear bounded extension $E_c : H^k(\Omega) \to H^k(\Omega \cup \Omega_c^h) \to H_0^k(\Omega_0)$ and a constant, $C = C(k, \Omega, \Omega_c^h, \Omega_0)$, such that*

$$E_c u|_\Omega = u \text{ on } \Omega \text{ and } \|E_c u\|_{H^k(\Omega\cup\Omega_c^h)} \le C\|u\|_{H^k(\Omega)} \quad \forall u \in H^k(\Omega). \tag{4.73}$$

The traces of the partials of u and $E_c u \in H^k(\Omega \cup \Omega_c^h)$ to the boundary $\partial\Omega$ coincide and for dist $(\Omega, \Omega_c^h) \le Ch$ *we obtain*

$$H^{k-|\beta|-1/2}(\partial\Omega) \ni \partial^\beta(u - E_c u)|_{\partial\Omega} = 0 \ \forall|\beta| \le k, \forall u \in H^k(\Omega) \tag{4.74}$$

$$\Longrightarrow \|\partial^\beta(E_c u(x)) - E_c(\partial^\beta u(x))\|_{L^2(\Omega\cup\Omega_c^h)} \le Ch\|u\|_{H^{|\beta|}(\Omega)} = 0 \forall u \in H_0^k(\Omega). \tag{4.75}$$

Based upon this theorem we immediately obtain

Theorem 4.38. Interpolation errors: *Under the conditions* (4.59)–(4.62), *those of Theorems* 4.36 and 4.37, *and for Ω piecewise in $C^p, 2 \le p \le 5$, a positive constant $K = K(n, d, \omega_{\mathcal{T}_c^h}, \gamma, C := C_1C_2, L_{\partial\Omega^h})$ exists, such that for $u \in H^\ell(\Omega)$, and $0 \le s \le \ell - 1$, cf.* (4.62), *we obtain for the operator*

$$P^h = I^h E_c : \quad \|E_c u - P^h u\|_{H^s(\mathcal{T}_c^h)} \le C\, h^{\ell-s}|u|_{H^\ell(\Omega)}, \ \|P^h u\|_{H^s(\mathcal{T}_c^h)} \le C\|u\|_{H^\ell(\Omega)}$$

$$\|E_c u - P^h u\|_{H^s(\partial\Omega_c^h)} \le C\, h^{\ell-s-1/2}|u|_{H^\ell(\Omega)}, \ \|P^h u\|_{H^s(\partial\Omega_c^h)} \le C\|u\|_{H^\ell(\Omega)}. \tag{4.76}$$

The $\mathcal{U}^h := \{P^h u : u \in H^\ell(\Omega)\}$ approximate the $H^s(\Omega), \ell \ge s$, and $H^s(\partial\Omega)$, $\ell \ge s + 1/2$, e.g. in the sense that

$$dist(E_c u, \mathcal{U}^h) := \inf_{u^h \in \mathcal{U}^h} \|E_c u - u^h\|_{H^s(\mathcal{T}_c^h)} \to 0 \quad \forall u \in H^s(\Omega) \quad \text{for } h \to 0. \tag{4.77}$$

Spaces of smooth piecewise polynomials with stable local bases

Bases with properties (4.60), (4.61), with $\gamma = 1$ and C_1, C_2 depending only on d, are available for $\mathcal{S}_d^r(\mathcal{T}^h)$ and certain subspaces of it on arbitrary $\mathcal{T}^h$ if $d \ge r2^n + 1$, see [263] and references therein. In the case of two variables, [267] provides a construction of bases for $\mathcal{S}_d^r(\mathcal{T}^h)$ satisfying (4.60)–(4.62), with $\gamma = 3$ and C_1, C_2 depending only on d, as soon as $d \geq 3r + 2$. There are many more results on stable local spline bases in the recent literature, especially in the contexts of high order macro-element and Lagrange interpolation methods, see e.g. [467, 516] and references therein.

Some classical spaces of smooth finite elements, for example those based on the Argyris element, see Figure 4.9, and [141] Example 3.2.10, belong to the class of so-called *superspline subspaces* of $\mathcal{S}_d^r(\mathcal{T}^h)$ [573], and their associated bases satisfy the requirements of Theorem 4.36.

In particular, for the Argyris FE we have in the notation of Theorem 4.36,

$$\mathcal{S}^h = \left\{s \in \mathcal{S}_5^1(\mathcal{T}^h) : s \text{ is } C^2 \text{ smooth at any vertex } v \text{ of } \mathcal{T}^h\right\}. \tag{4.78}$$

Clearly, (4.62) is satisfied with $n = 2$ and $\ell = 6$. The functionals $\lambda^k : \mathcal{S}^h \to \mathbb{R}$ are function evaluations, weighted first and second derivatives at the vertices,

$$s(v),\ h_T \frac{\partial s}{\partial x_1}(v), \quad h_T \frac{\partial s}{\partial x_2}(v), \quad h_T^2 \frac{\partial^2 s}{\partial x_1^2}(v), \quad h_T^2 \frac{\partial^2 s}{\partial x_2^2}(v), \quad h_T^2 \frac{\partial^2 s}{\partial x_1 \partial x_2}(v), \tag{4.79}$$

and weighted first order normal derivatives at the midpoints of the edges of $\mathcal{T}^h$,

$$h_T \frac{\partial s}{\partial n}\big((v_1 + v_2)/2\big), \tag{4.80}$$

where h_T is the diameter of a triangle in $\mathcal{T}^h$ containing the corresponding evaluation point v or $(v_1 + v_2)/2$ from (4.79), (4.80). The sets $E_k = \operatorname{supp} s^k$ are either star(v) for the functionals of type (4.79), or the unions of two triangles sharing the edge $[v_1, v_2]$ in case (4.80). Hence, (4.60) is satisfied with $\gamma = 1$. Furthermore, by estimating the norms of local Hermite interpolation operators, it can be shown that $\|s^k\|_{L_\infty(\Omega)} \leq C_1$, where C_1 is an absolute constant, see [263, Lemma 3.3]. Similarly, in view of the inverse estimates (4.57), $|\lambda^k s| \leq C_2 \|s\|_{L_\infty(E_k)}$, for any $s \in \mathcal{S}^h$, where C_2 is again an absolute constant, see [263, p. 292]. Thus, (4.61) is correct with $C = C_1 C_2$ being an absolute constant.

Stable splitting $\mathcal{S}^h = \mathcal{S}_0^h \oplus \mathcal{S}_b^h$

In the finite element method an important role is played by the spaces of finite elements vanishing on (parts of) the boundary. We set

$$\mathcal{S}_0^h = \{s \in \mathcal{S}^h : s|_{\partial\Omega} = 0\}.$$

For the sake of applications to the numerical solution of fully nonlinear problems, we construct a splitting of $\mathcal{S}^h$ into a direct sum

$$\mathcal{S}^h = \mathcal{S}_0^h \oplus \mathcal{S}_b^h$$

such that there exists a basis $\{s^1, \ldots, s^N\}$ for $\mathcal{S}^h$ satisfying the hypotheses (4.60), (4.61) of Theorem 4.36 (i.e. a stable local basis), where $\{s^1, \ldots, s^{N_0}\}$ and $\{s^{N_0+1}, \ldots, s^N\}$ are bases for $\mathcal{S}_0^h$ and $\mathcal{S}_b^h$, respectively.

We provide a construction for $n = 2$ and $\mathcal{S}^h \subset \mathcal{S}_5^1(\Omega)$ based on a modified Argyris finite element. We replace (4.78) by

$$\mathcal{S}^h = \left\{s \in \mathcal{S}_5^1(\mathcal{T}^h) : s \text{ is } C^2 \text{ smooth at any } \textit{interior} \text{ vertex } v \text{ of } \mathcal{T}^h\right\}. \tag{4.81}$$

Thus, in contrast to the Argyris element, the functions in $\mathcal{S}^h$ are not necessarily C^2 at the boundary vertices.

We now describe a set of functionals $\{\lambda^1, \dots, \lambda^N\} \subset (\mathcal{S}^h)'$, such that the desired basis $\{s^1, \dots, s^N\}$ for $\mathcal{S}^h$ will be uniquely defined by duality

$$\lambda^i s^j = \delta_{i,j}.$$

The set $\{\lambda^1, \dots, \lambda^N\}$ includes

(a) the functionals (4.79) for all *interior* vertices v of $\mathcal{T}^h$;
(b) the functionals (4.80) for all edges of $\mathcal{T}^h$; and
(c) the following functionals for each boundary vertex v of $\mathcal{T}^h$:

$$s(v),\ h_T \frac{\partial s}{\partial e_0}(v), \quad h_T \frac{\partial s}{\partial e_0^\perp}(v), \quad h_T^2 \frac{\partial^2 s}{\partial e_0^2}(v), \dots, \quad h_T^2 \frac{\partial^2 s}{\partial e_n^2}(v), \quad h_T^2 \frac{\partial^2 s}{\partial e_0 \partial e_1}(v),$$

where $e_0, \dots, e_n$ are all edges of $\mathcal{T}^h$ emanating from v, in counterclockwise order, with e_0 and e_n being the boundary edges.

Here the symbol $\partial/\partial e$ denotes the usual directional derivative in the direction of edge e, and $\partial/\partial e^\perp$ in the orthogonal direction. The above second order edge derivatives

$$\frac{\partial^2 s}{\partial e_0^2}(v), \dots, \quad \frac{\partial^2 s}{\partial e_n^2}(v), \quad \frac{\partial^2 s}{\partial e_0 \partial e_1}(v)$$

are well defined despite s being only C^1 at the boundary vertices, see [266]. Note that these second order derivatives are independent of each other even if some edges are collinear. This choice of the degrees of freedom at boundary vertices in (c) is motivated by the construction of the Morgan–Scott basis [497] and is shown to be stable in [266].

Following the argument in [266], one can see that the basis $\{s^1, \dots, s^N\}$ for $\mathcal{S}^h$, defined by duality, satisfies (4.60), (4.61) with $\gamma = 1$ and bounded C_1, C_2. Hence it is a stable local basis. Moreover, (4.62) is obviously true for $\mathcal{S}^h$ with $n = 2$ and $\ell = 6$. Therefore, Theorem 4.36 applies to this basis.

To determine the subsets of $\{s^1, \dots, s^N\}$ which generate $\mathcal{S}_0^h$ and $\mathcal{S}_b^h$, respectively, we now split the functionals in (c) into two groups (c1) and (c2) as follows.

(c1) The first group includes

$$h_T^2 \frac{\partial^2 s}{\partial e_1^2}(v), \dots, \quad h_T^2 \frac{\partial^2 s}{\partial e_{n-1}^2}(v), \quad h_T^2 \frac{\partial^2 s}{\partial e_0 \partial e_1}(v),$$

for all boundary vertices, and, in addition, $h_T \frac{\partial s}{\partial e_0^\perp}(v)$ for those boundary vertices, where e_0 and e_n are collinear.

(c2) The second group includes

$$s(v),\ h_T \frac{\partial s}{\partial e_0}(v), \quad h_T^2 \frac{\partial^2 s}{\partial e_0^2}(v), \quad h_T^2 \frac{\partial^2 s}{\partial e_n^2}(v), \tag{4.82}$$

for all boundary vertices, and, in addition, $h_T \dfrac{\partial s}{\partial e_0^\perp}(v)$ for those boundary vertices, where e_0 and e_n are *not* collinear.

Now let $\{\lambda^1, \ldots, \lambda^{N_0}\}$ list all functionals λ^i in (a), (b), and (c1), and let $\{\lambda^{N_0+1}, \ldots, \lambda^N\}$ be those in (c2). It is easy to see that

$$\mathcal{S}_0^h = \{s \in \mathcal{S}^h : \lambda^{N_0+1}s = \cdots = \lambda^N s = 0\}. \tag{4.83}$$

Therefore $\mathcal{S}_0^h = \text{span}\,\{s^1, \ldots, s^{N_0}\}$, and $\mathcal{S}_b^h := \text{span}\,\{s^{N_0+1}, \ldots, s^N\}$ is its complement in $\mathcal{S}^h$ as required.

Clearly, both $\{s^1, \ldots, s^{N_0}\}$ and $\{s^{N_0+1}, \ldots, s^N\}$ are *stable local bases* as subsets of the stable local basis $\{s^1, \ldots, s^N\}$.

Note that $h_T \dfrac{\partial s}{\partial e_0^\perp}(v)$ belongs to (c1) or (c2) depending on whether e_0 and e_n are *exactly* collinear or not. In particular it is in (c2) if e_0 and e_n are near-collinear, but not collinear, a situation which may appear quite often when a polygonal domain is an approximation of a smooth domain. As e_0 and e_n become exactly collinear, $h_T \dfrac{\partial s}{\partial e_0^\perp}(v)$ is moved into (c1). Thus, the dimensions of $\mathcal{S}_0^h$ and $\mathcal{S}_b^h$ jump if a vertex shared by collinear boundary edges is slightly perturbed. This 'dimension instability' is not a new phenomenon in the theory of multivariate splines. Indeed, it is related to the well-known fact that the dimension formulas for the spline spaces may depend on some geometric information about the placement of the vertices, see [467]. This behavior is compatible with the availability of stable bases, see for example the discussion in [267], Remark 13.1.

Since λ^i are function evaluations or derivatives of at most second order, they may be applied to any sufficiently smooth functions, thus leading to the *interpolation operator* $I^h : C^2(\Omega) \to \mathcal{S}^h$, defined by

$$I^h(v) = \sum_{i=1}^{N} \lambda^i(u)s^i, \quad u \in C^2(\Omega). \tag{4.84}$$

The following is an obvious property of this operator:

$$v|_{\partial\Omega} = 0 \quad \Longrightarrow \quad I^h(v) \in \mathcal{S}_0^h. \tag{4.85}$$

It is easy to see that the operator I^h can be used in the proof of Theorem 4.36 in place of the quasi-interpolation operator assuming that u is sufficiently smooth.

Finally, consider the operator $I_b^h : C^2(\Omega) \to \mathcal{S}_b^h$, defined by

$$I_b^h(u) = \sum_{i=N_0+1}^{N} \lambda^i(u)s^i. \tag{4.86}$$

Clearly,

$$I_b^h(u)|_{\partial\Omega} = I^h(u)|_{\partial\Omega}. \tag{4.87}$$

Hence, by Trace Theorem 1.38 we obtain for any $1 \le p \le \infty$,

$$\left\|u - I_b^h(u)\right\|_{L_p(\partial\Omega)} \le C\|u - I^h(u)\|_{L_p(\Omega)}^{1-1/p}\|u - I^h(u)\|_{W^{1,p}(\Omega)}^{1/p},$$

where C is a constant depending only on p and the Lipschitz constant $L_{\partial\Omega}$ of $\partial\Omega$. This allows us to find a bound for $\left\|u - I_b^h(u)\right\|_{L_p(\partial\Omega)}$, see (4.90) below, using the estimates available for $\|u - I^h(u)\|_{L_p(\Omega)}$ and $\|D^\alpha(v - I^h(u))\|_{L_p(\Omega)}$, $|\alpha| = 1$.

We summarize the results for the stable splitting in the following theorem.

Theorem 4.39. Stable splitting: *For the space $\mathcal{S}^h$ defined in (4.81), there exists a basis $\{s^1, \ldots, s^N\}$ satisfying (4.60)–(4.62), such that $\{s^1, \ldots, s^{N_0}\}$ is a basis for $\mathcal{S}_0^h = \{s \in \mathcal{S}^h : s|_{\partial\Omega} = 0\}$. The interpolation operators $I^h : C^2(\Omega) \to \mathcal{S}^h$ and $I_b^h : C^2(\Omega) \to \mathcal{S}_b^h$ defined by (4.84) and (4.86), respectively, have the following properties, cf. (4.63), (4.64).*

1. *If $u|_{\partial\Omega} = 0$, then $I^h(u) \in \mathcal{S}_0^h$.*
2. *Let $1 \le p \le \infty$, $\ell \le 6$, be such that $W^{\ell,p}(\Omega) \subset C^2(\Omega)$. Then for any $0 \le |\alpha| \le \ell - 1$,*

$$\|D^\alpha(u - I^h(u))\|_{L_p(T)} \le K h_T^{\ell-|\alpha|}|u|_{W^{\ell,p}(\Omega_T^1)}, \quad \text{for any } T \in \mathcal{T}^h, \tag{4.88}$$

and, as a consequence

$$\|D^\alpha(u - I^h(u))\|_{L_p(\mathcal{T}^h)} \le K' h^{\ell-|\alpha|}|u|_{W^{\ell,p}(\Omega)}. \tag{4.89}$$

3. *Moreover,*

$$\left\|u - I_b^h(u)\right\|_{L_p(\partial\Omega)} \le K'' h^{\ell-1/p}|u|_{W^{\ell,p}(\Omega)}. \tag{4.90}$$

The constants K, K', K'' depend only on p, $\omega_{\mathcal{T}^h}$, and the Lipschitz constant $L_{\partial\Omega}$.

Conjecture 4.40. *The results in Theorems 4.35 and 4.36 are proved in $\mathbb{R}^n$ and for $\mathcal{S}^h = \mathcal{S}_d^r(\mathcal{T}^h)$ and n, r if $d \ge r2^n + 1$. We* conjecture *that the techniques of [263]* can *be used to generalize Theorem 4.39 as well to $\mathcal{S}_d^r(\mathcal{T}^h)$ and appropriate subspaces of it.*

Conjecture 4.41. *We extend the previous Conjecture 4.40 to suitable nonpolyhedral (curved) boundaries $\partial\Omega_c^h \approx \partial\Omega$ with $\operatorname{dist}\left(\partial\Omega_c^h, \partial\Omega\right) = \mathcal{O}(h^q)$, $q > 2$, and the corresponding triangulations $\mathcal{T}_c^h$. We expect that Theorems 4.35–4.39 can be generalized in this case to appropriate subspaces of $\mathcal{S}^h = \mathcal{S}_d^r\left(\mathcal{T}_c^h\right)$ for any r if $d \ge r2^n + 1$, at least for $n = 2$.*

4.2.7 Curved boundaries

In our earlier discussion we have excluded curved boundaries. We study them in this and the next subsection. We will formulate corresponding FEMs in Subsections 5.5.6, 5.5.10. The results here are valid partially for $\mathbb{R}^n, n \ge 2$, or $\mathbb{R}^n, n \le 3$. So we consider

$$\Omega \subset \mathbb{R}^n,\ n \ge 2, \text{ with piecewise smooth } \partial\Omega \in C_p^t,\ t \ge 1,\ \ \mathcal{P} = \mathcal{P}_{d-1}^n, d \ge 1. \tag{4.91}$$

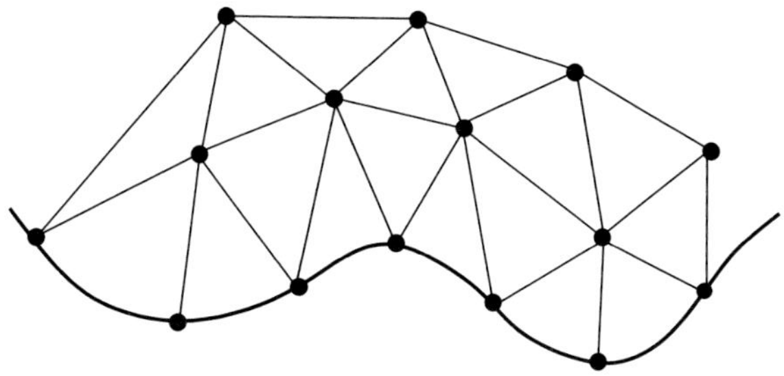

Figure 4.21 Admissible triangulation for a curved boundary.

To simplify the presentation, we *restrict the construction to two-variate FEs* in this subsection. Extensions to $\mathbb{R}^n, n \geq 3$, are theoretically pretty obvious, in contrast to the practical realization. In a first step we define an approximating Ω^h and the corresponding $\mathcal{T}^h$:

> Choose points $P_j \in \partial\Omega$ with distance $\leq h$ for neighboring points. Include all "non-smooth" points on $\partial\Omega \in C_p^t$ into these P_j. Replace $\partial\Omega$ and Ω by $\partial\Omega^h$ and Ω^h : $\partial\Omega^h$ is obtained by connecting the neighboring $P_j \in \partial\Omega$ by straight lines, thus defining the new edges $e \subset \partial\Omega^h$. The *polygonal* Ω^h is then the interior of $\partial\Omega^h$. Choose a *nondegenerate triangulation* $\mathcal{T}^h$ for Ω^h. (4.92)

Then automatically the $T \in \mathcal{T}^h$ are star-shaped. Figure 4.21 shows the original $\partial\Omega$ with an interior $\mathcal{T}^h$. We have to be prepared that the necessarily modified FEs will satisfy the boundary conditions only approximately. The discussion here is restricted to triangulations, possibly curved at the boundary, and to

$$K \text{ triangle}, \forall\, \bar{e} \subset \bar{K} \text{ only function evaluations and no derivatives } \in \mathcal{N}. \quad (4.93)$$

Note that vertices $P_j \in \bar{e}$, see (4.98), are included.

Here, we present two different possibilities to handle Ω^h and $\mathcal{T}^h$. For the "boundary FEs" $(T, \mathcal{P}_T, \mathcal{N}_T)$ with $|\bar{T} \cap \partial\Omega| \geq 2$ we change $\mathcal{N}_T$ or $\mathcal{P}_T$, thus modifying Definition 4.6. In the first case, see Figure 4.22, we replace, along the boundary,

$$\mathcal{N}_T \text{ by interpolation along the curved } \partial\Omega \text{ instead of the straight } \partial\Omega^h. \quad (4.94)$$

Secondly, we use the original F_T for T with respect to Ω^h, see (4.92), Figures 4.22, 4.23, but,

$$\text{replace } F_T : K \to T \text{ affine, by } F^h|_T \circ F_T : K \to T_c, F^h|_T \in \mathcal{P}_{d-1}^n, \quad (4.95)$$

for the $T \in \mathcal{T}^h$ at the boundary and with the original polynomials $\mathcal{P} = \mathcal{P}_{d-1}^n$. This yields the so-called *isoparametric FEs*. We describe in Definition 4.42 and the following pages, how these F_c^h, T_c are constructed. Obviously the basis functions $\phi_i \circ (F^h|_T \circ F_T)^{-1} \notin \mathcal{P}$ are more complicated nonpolynomial functions. For a third choice see Subsection 4.2.6.

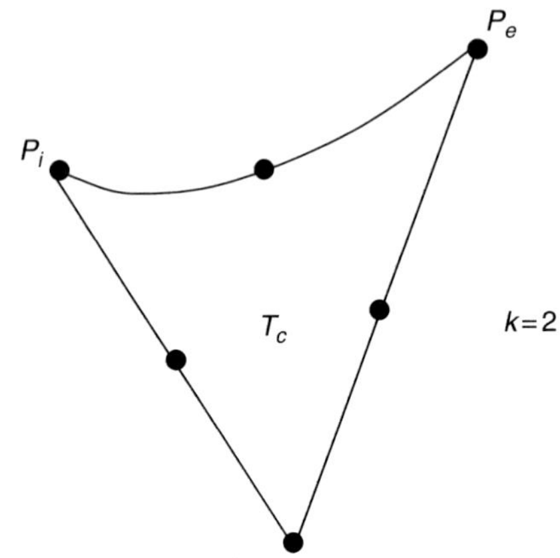

Figure 4.22 Polynomial interpolation on $\partial\Omega$.

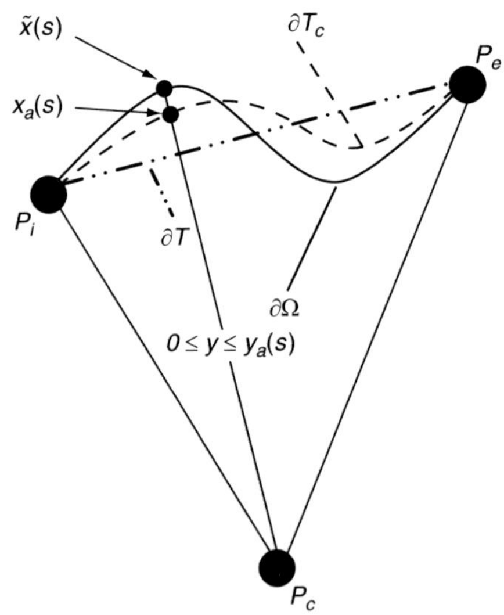

Figure 4.23 Isoparametric interpolation on $\partial\Omega$.

Polynomial approximation and interpolation in points of $\partial\Omega$

For the *new triangulation*, $\mathcal{T}^h$, we modify the T and $\mathcal{N}_T$ at the boundary with $|\bar{T} \cap \partial\Omega| \geq 2$, see (4.94): We denote the vertex points of $\bar{T}$ as $\bar{T}_v$ and define

$$\bar{T}_v \cap \partial\Omega^h = \{P_i, P_e\}, \quad P_i \neq P_e, \; \bar{T}_v := \{\text{vertices of } T\}. \tag{4.96}$$

According to (4.94), we modify these triangles T into T_c by replacing the straight line $\overline{P_i, P_e}$ by that part of the boundary $\partial\Omega$ between P_i and P_e, see Figure 4.22.

The modified triangulation we denote as $\mathcal{T}_c^h$. So we still obtain, in contrast to the isoparametric FEs below,

$$\bar{\Omega} = \bigcup_{T_c \in \mathcal{T}_c^h} \overline{T_c}. \tag{4.97}$$

For modifying $\mathcal{N}_T$ we proceed in two steps. Choose a smooth parametrization $x = x(s)$ of $\partial\Omega$ between P_i and P_e with respect to the arc length $s, 0 \leq s \leq h_e$, such that $s \in C^{2d-1}[0, h_e]$. Now we choose d Gauss–Lobatto points, cf. (5.307)–(5.311), with $\rho = 2$ and the $y_j^\rho = y_j^2$ there. Let $\xi_0, \ldots, \xi_{d-1}, \xi_j := h_e\left(1 + y_j^2\right)/2,\ \xi_0 = 0, \xi_{d-1} = h_e$ in $[0, h_e]$ (or Gauss points in Subsubsections 5.5.5 and 5.5.6, 5.5.7). Determine, on the curved edge $e = \partial T$, cf. Figure 4.22,

$$\partial\Omega \ni P_j^e := P_j := x(\xi_j) \approx P_j^s := P_i + \xi_j(P_e - P_i)/\|(P_e - P_i)\|_{\mathbb{R}^2},$$
$$j = 0, \ldots, d-1,\ \ P_i = P_0,\ \ P_e = P_{d-1}. \tag{4.98}$$

In principle, we could use any d interpolation points on e. The choice of the Gauss-type points is motivated by the applications to FEMs in Section 5.5.

Now, we replace the d function evaluations in the P_j^s along the straight edge $\overline{P_i P_e}$ for the original T, by the d function evaluations in the $P_j^e = P_j,\ \ j = 0, \ldots, d-1$. Thus we replace $\mathcal{N}_T$ by a modified $\mathcal{N}_T^c$ corresponding to one curved edge, see Figure 4.22. For Gauss and Gauss–Radau points we obtain the corresponding $0 < \xi_0 \leq \xi_{d-1} < h_e$ and $\xi_0 = 0, \xi_{d-1} < h_e$, respectively. Then discontiuous $\mathcal{U}^h$ are generated. We discuss these cases in Subsections 5.5.6, 5.5.7. For small enough h, the edge $\overline{P_i P_e}$ will be at most $\mathcal{O}(h^2)$ away from the curved part of $\partial\Omega$ between P_i and P_e, see [141], 8 ex.3. Hence $\mathcal{P}_T, \mathcal{N}_T^c$ will be unisolvent simultaneously with $\mathcal{P}_T, \mathcal{N}_T$ for small enough h. These perturbation arguments show that the results in Subsection 4.2.4 remain essentially valid.

For $\mathbb{R}^3$ the above Gauss-type points could be replaced by the cubature points of the best types of cubature formulas in tetrahedra in $\mathbb{R}^3$. Their order of convergence is smaller than for the $\mathbb{R}^2$ case. This is inherited by the FEMs. We do not present those modifications here, since we consider discontinuous Galerkin methods in Chapter 7, avoiding these difficulties. For details we refer to Cools [208–210].

Isoparametric nonpolynomial approximation

The Gauss type points decrease, as far as possible, interpolation and, for applications in FEMs, the quadrature errors along the relevant boundary part of $\partial\Omega$. There is another method, well established in engineering applications, allowing much more freedom, but with the same order of convergence as before. In addition this method is applicable in FEMs to $\mathbb{R}^n, n \geq 2$. Particularly efficient is the *isoparametric polynomial approach* in (4.95). As in (4.96) we assume one curved and two straight edges P_i, P_e and $P_0\,P_i,\ \ P_0\,P_e$ for T_c. Well known results of Ciarlet and Raviart [176], Ciarlet [174] and Lenoir [471], construct a bijective, piecewise polynomial mapping $F^h : \Omega^h \to \Omega_c^h \approx \Omega \subset \mathbb{R}^2$, see [471] for $\mathbb{R}^n, n \geq 2$, such that $F^h = id$ away from the boundary. We present a very rough idea of this construction with the goal of showing that the isoparametric

FEMs fit into our general discretization framework. Figure 4.23 motivates the next definition. In 3. below we need $\mathrm{dist}(F^h(\partial\Omega^h), \partial\Omega)$.[33]

Definition 4.42. *We define the* triangulation $\mathcal{T}_c^h$, *the bijective* isoparametric mapping $F^h : \Omega^h \to \Omega_c^h \approx \Omega$, *and the* boundary $\partial\Omega_c^h := F^h(\partial\Omega^h)$: *Choose* Ω^h, $\partial\Omega^h$ *and a* nondegenerate triangulation $\mathcal{T}^h$ *for* Ω^h *as in (4.92), (4.96). Then, correlating* $\mathcal{T}^h$ *with* $\mathcal{T}_c^h$, *we define* $F^h : \Omega^h \to \Omega_c^h := F^h(\Omega^h) \subset \mathbb{R}^n, n \geq 2$, $\mathcal{T}_c^h$ *and* $T_c \in \mathcal{T}_c^h$ *as:*

1. *Let* $(F^h)|_T = id|_T \ \forall T \in \mathcal{T}^h$ *with* $|\bar{T}_v \cap \partial\Omega| \leq 1$, *see (4.96), and let* $F^h|_{\bar{e}} = id|_{\bar{e}}$, $\forall \bar{e} \subset \bar{T}$ *with* $|\bar{e} \cap \partial\Omega| \leq 1$.
2. *For the components of* F^h, *the* $(F^h)_i, i = 1, 2$, *let* $(F^h)_i|_T \in \mathcal{P}_{d-1}^n \ \forall T \in \mathcal{T}^h$ *with* $|\bar{T}_v \cap \partial\Omega| \geq 2$, *see Figure 4.23, and* $\mathcal{T}_c^h \ni T_c := F^h T \ \forall T \in \mathcal{T}^h$.
3. dist $(F^h(\partial\Omega^h), \partial\Omega) = \mathcal{O}(h^d)$, dist $(F^h(\Omega^h), \Omega) = \mathcal{O}(h^d)$.
4. $\|(F^h)'\|_{W^{d,\infty}(T_c) \hookrightarrow W^{d,\infty}(T)}$, $\|((F^h)')^{-1}\|_{\dots} \leq C$, *h independent,* $\forall T_c \in \mathcal{T}_c^h$.
5. *Define the* T_c *as* $T_c = F^h(T) \ \forall \ T \in \mathcal{T}^h$ *with* $T = T_c \ \forall \ T \in \mathcal{T}^h$ *for* $|\bar{T} \cap \partial\Omega| \leq 1$ *and* $T_c \neq T$ *for* $|\bar{T} \cap \partial\Omega| \geq 2$, *unless* $\partial\Omega$ *between* P_i, P_e *is a straight line.*
6. *Define* $\mathcal{T}_c^h := \{T_c^h\}$ *for the* T_c^h *in (v), and* $\bar{\Omega}_c^h := \bigcup_{T_c \in \mathcal{T}_c^h} \overline{T_c}$.

Note that the conditions 1.–6. are strictly local, including 3.

We only describe this construction, see above, for a triangle $T \in \mathcal{T}_c^h$ with $2 = |\bar{T}_v \cap \partial\Omega| = \{P_i, P_e\}$, see Figure 4.23. For all other cases, $(F^h)|_T = id|_T$ by 1. Our chosen $T \in \mathcal{T}^h$ is defined by the vertices P_i, P_e, P_c. The solid line $\partial\Omega$ marks the original part of the boundary between P_i and P_e. Then $F^h|_T : T \to T_c$, by 2. The isoparametric triangle T_c has the straight edges $\overline{P_c, P_i}$, $\overline{P_c, P_e}$ and the curved edge ∂T_c connecting P_i, P_e. By 2. the components of F^h are $(F^h)_i|_T \in \mathcal{P}_{d-1}^n$ with $F^h|_{\bar{e}} = id|_{\bar{e}}$, for the two straight edges $\overline{P_c, P_i}$ $\overline{P_c, P_e}$. By 3. we require dist $(F^h(\partial\Omega^h), \partial\Omega) = \mathcal{O}(h^d)$. For the two points marked as $x_a(s), \tilde{x}(s)$ on the ray from P_c the distance in Figure 4.23 is $\mathcal{O}(h^d)$. We did not mark any intersection point of $\partial\Omega$ and ∂T_c, since only the distance $\mathcal{O}(h^d)$ and 4. are required. Finally 5., 6. yield $T_c \in \mathcal{T}_c^h$.

Definition 4.42 introduces an *isoparametric subdivision*, $\mathcal{T}_c^h$. We proceed by introducing *isoparametric FEs*, $\mathcal{U}^h$, and, later on, an *isoparametric interpolation operator*, I^h. We combine the affine mapping $F = F_T : K \to T \in \mathcal{T}^h$ with the polynomials $F^h|_T : T \to T_c$, see Figure 4.24. The construction of this F^h is a main point in this subsection. So we get nonconforming FEs violating the boundary conditions. This will be treated in Subsection 5.5.10, see Theorem 5.85.

We start with the original $\mathcal{T}^h$ defined for Ω^h in (4.92), the F^h, $\mathcal{T}_c^h$ and Ω_c^h in Definition 4.42, and the points P_i^e on curved T_c with edges e, approximating $\partial\Omega$, for d' cf. Definition 4.1:

[33] For compact A, B, the *distance* between A and B is defined as $\mathrm{dist}(A, B) = \max\{\max_{x \in A} : \{\min\{|x - y| : y \in B\}\}, \max_{y \in B} : \{\min\{|x - y| : x \in A\}\}\}$. It represents the maximal distance from any point of one set to the nearest point of the other set.

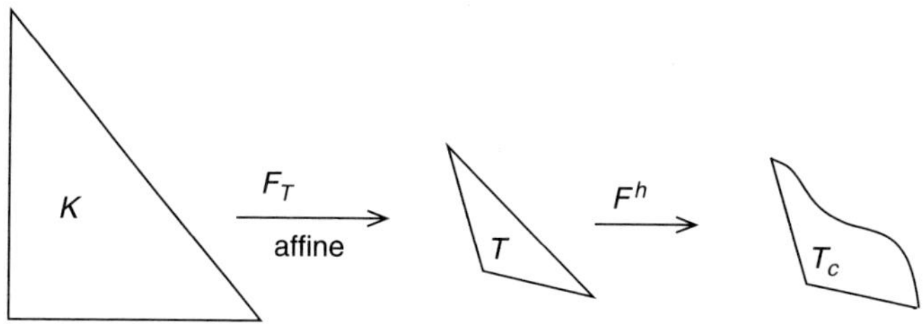

Figure 4.24 Isoparametric mapping $F_T : K \to T, F^h \circ F_T : K \to T_c$.

$$\mathcal{U}^h := \{u^h : \Omega_c^h = F^h(\Omega^h) \to \mathbb{R} \; : \; u^h|_{T_c} = \sum_{i=1}^{d'} \alpha_i \cdot \phi_i \circ (F^h|_T \circ F_T)^{-1},$$

$$\text{with (non)polynomials } \phi_i \circ (F^h|_T \circ F_T)^{-1}, \; \alpha_i \in \mathbb{R} \;\; \forall \;\; T_c = F^h(T) \in \mathcal{T}_c^h\},$$

$$\|u^h\|_{W^{k,q}(\mathcal{T}_c^h)} \text{ as in (4.36) with } \Omega, \; \mathcal{T}^h \text{ replaced by} F^h(\Omega^h), \; \mathcal{T}_c^h \tag{4.99}$$

$$\mathcal{U}_b^h := \{u^h \in \mathcal{U}^h : \; \forall e \subset T_c \; u^h(P_i^e) = 0, i = 0, \ldots, d'-1, \text{interpolating in } e \approx \partial\Omega_c^h\},$$

where these $P_0^e = P_i$, $P_{d'-1}^e = P_e$, but $P_i^e, i = 1, \ldots, d'-2$ are $d'-2$ points on e, unequal to those in (4.98).

Let $e \subset \bar{K}$ denote the edge mapped by the original $F_T : K \to T$ onto $\overline{P_i P_e}$. Then $\partial\Omega_T^h := F^h(F_T(e))$, marked as ∂T_c in Figure 4.23, only approximates that part of $\partial\Omega$ between P_i and P_e up to $\mathcal{O}(h^d)$. For the following interpolation process this might have the consequence that some of the interpolation points $P_j \in \partial\Omega_T^h$, $j = 1, \ldots, d-2$, are $P_j \notin \Omega \cup \partial\Omega$ and so $u(P_j)$ might not be defined. In this case, Theorem 4.37 yields an appropriate extension u^e of u, allowing the evaluation of $u^e(P_j)$, see [141]. More appropriate is a kind of piecewise interpolation, F_c, of the above F^h.

The complicated construction of F_c and the following properties of E_c are presented for $\mathbb{R}^n, n \geq 2$, in [471]. We only indicate the $\mathbb{R}^2$ results and summarize the essential properties in (4.100) and (4.101). They allow formulating error and inverse estimates and the proofs for the FEMs below in $\mathbb{R}^n, n \geq 2$, cf. Theorem 4.43 and 5.85 ff. This auxiliary F_c exactly maps $\Omega^h \to \Omega$.

$$F_c = ((F_c)_i)_{i=1}^2 : \Omega^h \to \Omega \subset \mathbb{R}^n \text{ s.t. } \forall T \in \mathcal{T}^h : F_c|_T : T \to \tilde{T}_c \approx T_c, \text{ and} \tag{4.100}$$

$$(F_c|_T)^{-1} : \tilde{T}_c \to T \text{ are } C^d \text{ diffeomorphism s.t with the operator norm } |||\cdot|||_\infty,$$

$$|||D^s(F_c|_T - id_T)|||_\infty, |||D^s((F_c|_T)^{-1} - id_{\tilde{T}_c})|||_\infty \leq Ch^{d-s} \forall s \leq d$$

$$|J(D(F_c|_T) - 1|_\infty, |J(D(F_c|_T)^{-1}) - 1|_\infty \leq Ch^d$$

$$\text{and } \partial F_c(\Omega^h) = F_c(\partial\Omega^h) = \partial\Omega \text{ and } F^h(\partial\Omega^h) = \partial\Omega + \mathcal{O}(h^d),$$

and, replacing F^h by F_c, this F_c satisfies Definition 4.42 1.; here $|||D^s(F_c|_T - id_T)|||_\infty$ indicates the sup-norm of the sth derivative and $J(D(F_c|_T)$ the absolute value of the

Jacobian. Thus F_c has the property that in Figure 4.23 the part of $\partial\Omega$ from P_i to P_e is the image $F_c(\partial T)$ where ∂T indicates the straight segment $\overline{P_i, P_e}$.

We introduce E_c, transforming functions defined on the original Ω to those defined on the approximate $\Omega_c^h = F^h(\Omega^h) \approx \Omega$ and its inverse $(E_c)^{-1}$:

1. $E_c : \Omega \to \Omega_c^h = F^h(\Omega^h)$ as $E_c := F^h \circ F_c^{-1} : t \in \Omega \to x \in F^h(\Omega^h) \approx \Omega$,
2. $(E_c)^{-1} : F^h(\Omega^h) \to \Omega$ with $t - E_c(t) = \mathcal{O}(h^d); x - (E_c)^{-1}(x) = \mathcal{O}(h^d)$ and
3. $(E_c)' - Id_\Omega = \mathcal{O}(h^{d-1}), ((E_c)^{-1})' - Id_{F^h(\Omega^h)} = \mathcal{O}(h^{d-1})$, as in (4.100),
4. $E_c|_T = Id_T$ and $E_c|_e = Id_e \ \forall\ T, e$ in Definition 4.42 1. (4.101)

With this E_c, the u^h,$\mathcal{U}^h$ and $\Omega_c^h = F^h(\Omega^h)$ in (4.99) we start with

$$\begin{aligned} &u^h : F^h(\Omega^h) \to \mathbb{R},\ u^h \in \mathcal{U}^h, \quad \text{and we define the new } \hat{u}^h \text{ as} \\ &\hat{u}^h : \Omega \to \mathbb{R},\ \hat{u}^h(t) := (u^h \circ E_c)(t), \text{ and } \widehat{\mathcal{U}_b^h} = \left\{\hat{u}^h : u^h \in \mathcal{U}_b^h\right\}. \end{aligned} \tag{4.102}$$

Note that $\widehat{\mathcal{U}_b^h} \subset \mathcal{U}$, but trivial *Dirichlet boundary conditions are satisfied only approximately on the exact* $\partial\Omega$ for $\hat{u} \in \widehat{\mathcal{U}_b^h}$. Vice versa

$$\begin{aligned} &\text{for } f : \Omega \to \mathbb{R} \text{ the } \check{f} : F^h(\Omega^h) \to \mathbb{R} \text{ is defined as} \\ &\qquad \check{f}(x) := f((E_c)^{-1}(x)) = (f \circ (E_c)^{-1})(x), \end{aligned} \tag{4.103}$$

With F^h in Definition 4.42 the *isoparametric interpolation operator,* $\widehat{I^h}$, is

$$\begin{aligned} \widehat{I^h}u &:= (I^h u) \circ E_c \in \hat{\mathcal{U}}^h, \text{ with } I^h : \mathcal{U} := \{u : \Omega \to \mathbb{R}\} \to \mathcal{U}^h := \left\{u^h :\ \Omega_c^h \to \mathbb{R}\right\}, \\ I^h u|_T &= I^h u|_T \text{ as in (4.16) } \forall \text{ inner } T = T_c \in \mathcal{T}_c^h \text{ with } |\bar{T}_c \cap \partial\Omega^h| \le 1, \quad \text{else} \qquad (4.104) \\ I^h u|_{T_c} &= \sum_{i=1}^{d'} N_i(u \circ F^h|_{T_c} \circ F_{T_c}) \cdot (\phi_i \circ (F^h|_{T_c} \circ F_{T_c})^{-1}) = \sum_{i=1}^{d'} N_i^{T_c}(u)\phi_i^{T_c}. \end{aligned}$$

We have replaced the affine F_T by isoparametric generalizations with components $(F^h|_{T_c} \circ F_{T_c})_i \in \mathcal{P}_{d-1}^n, i = 1, 2,$ for T_c with two boundary vertices. Obviously, I^h is a bounded linear operator as the original I^h, if the norms are now defined with respect to the $\mathcal{T}_c^h$ above. Furthermore, the original $\phi_i \circ (F_T)^{-1}$ and $N_i(u \circ F_T)$ have to be replaced by the more complicated (non)polynomial $\phi_i \circ (F^h|_{T_c} \circ F_{T_c})^{-1}$ but still linear $N_i(u \circ F^h|_{T_c} \circ F_{T_c})$. Since, in mapping a polyhedral to a smooth domain, a C^1 mapping is not appropriate, we only choose C^0 reference elements, cf. [141], p.118.

Theorem 4.43. Interpolation errors: *For Ω as in (4.91) define Ω^h as in (4.92). Let, for $0 < h \le 1$, the $\mathcal{T}^h$ for Ω^h satisfy Condition 4.16 for some l, d, p. We combine F^h and $\mathcal{T}_e^h$ from Definition 4.42, with F^h, piecewise of degree $d-1$. For a C^0 reference element $(K, \mathcal{P}, \mathcal{N})$, define $\mathcal{U}^h$,$\mathcal{U}_b^h$, I^h, $\widehat{I^h}u$ as in (4.99), (4.102), (4.104). Then there* *exists a positive constant* $C = C((K,\mathcal{P},\mathcal{N}),n,d,p,\chi)$, *for* χ *cf. (4.30), such that for* $0 \le s \le d$, *and* $\forall u \in W^{d,p}(\Omega)$,

$$\|u - \widehat{I^h u}\|_{W^{s,p}(T)} \le C\ (\text{diam } T)^{d-s}|u|_{W^{d,p}(T)}, \tag{4.105}$$

$$\|\widehat{I^h u}\|_{W^{s,p}(T)} \le (1 + \mathcal{O}((\text{diam } T)^{d-s}))\|u\|_{W^{d,p}(T)} \ \textit{with}$$

$$\max_{T\in\mathcal{T}^h} \|u - \widehat{I^h u}\|_{W^{s,\infty}(T)} \le C\ h^{d-s-n/p}|u|_{W^{d,\infty}(\Omega)}\ \forall\ u \in W^{d,\infty}(\Omega),\ \textit{and}$$

$$\|u - \widehat{I^h u}\|_{W^{s,p}(\mathcal{T}^h)} \le C\ h^{d-s}|u|_{W^{d,p}(\Omega)},$$

$$\|\widehat{I^h u}\|_{W^{s,p}(\mathcal{T}^h)} \le (1 + C\ h^{d-s})\|u\|_{W^{d,p}(\Omega)}.$$

Since $\mathcal{P}$ always satisfies $\mathcal{P} \subseteq W^{j,p}(T) \cap W^l_q(T)$, $1 \le q \le p, 0 \le l \le j$, we get

Theorem 4.44. Inverse estimates: *Under the conditions of Theorem 4.43 and for* $T \in \mathcal{T}_c^h$ *of a quasiuniform triangulation* $\mathcal{T}^h$ *let*

$$\mathcal{U}^h = \left\{u^h : u^h \textit{ is measurable and } u^h \in \mathcal{U}^h \textit{ according to (4.99) } \forall T_c \in \mathcal{T}_c^h\right\}.$$

Then there exists a constant $C = C(l,p,q,\chi)$ *such that*

$$\|u^h\|_{W^{j,p}(\mathcal{T}_c^h)} \le Ch^{l-j-(n/q-n/p)}\|u^h\|_{W^l_q(\mathcal{T}_c^h)}\ \forall u^h \in \mathcal{U}^h,\ \forall j \ge l. \tag{4.106}$$

4.3 FEMs for linear problems

The usual approach in finite element methods considers the weak operator, A, G and its corresponding weak bilinear or nonlinear form, $a(\cdot,\cdot)$ as in this subsection and in Section 4.4. We will modify this approach for fully nonlinear problems in Section 5.2. In contrast to the following Chapters 5 and 7, we consider here *conforming FEMs,* i.e. the exact and approximate solutions and test functions are elements of the same spaces, so $u_0, u_0^h \in \mathcal{U}_b$, $v, v^h \in \mathcal{V}_b$, here with $\mathcal{U}_b = \mathcal{V}_b = H_0^1(\Omega)$ and $\mathcal{U}_b = \mathcal{V}_b = H_0^m(\Omega)$ for elliptic differential equations of order 2 in this and order $2m$ in Subsection 4.3.2. A straightforward modification allows us to include these equations and systems of order $2m$. Extensions to quasilinear problems are studied in Section 4.4.

For conforming FEMs and our problems here, we have $\mathcal{U}_b = \mathcal{V}_b$ and $\mathcal{U}_b^h = \mathcal{V}_b^h$, so from here on we use the notation $\mathcal{V}_b$ and $\mathcal{V}_b^h$, the standard notation.

More involved discussions would be necessary to study the different combinations of natural and induced boundary conditions, see e.g. [387] and Chapter 2.

We start with the main idea for FEMs, presented for a simple example in Subsection 4.3.1. In particular, we show how the definition of FEs via the triple $K, \mathcal{P}, \mathcal{N}$ can be used efficiently to determine the elements of the *mass* and *stiffness matrices* for a FEM. Next, general elliptic operators of order 2 and $2m$ and their corresponding coercive and elliptic bilinear forms and the induced boundary operators are recapitulated as a basis for FEMs in Subsection 4.3.2. Finally, the convergence for conforming FEMs for these linear elliptic boundary value problems is formulated in Theorem 4.57, with the Aubin–Nitsche improvement in Theorem 4.60.

4.3.1 Finite element methods: a simple example, essential tools

The main idea for FEs is very simple and is demonstrated for a simple example. We choose the Helmholz operator and Dirichlet boundary conditions:

Example 4.45. *Homogeneous Dirichlet boundary value problem*
For $u \in H_0^1(\Omega) := \{u \in H^1(\Omega) : u|_{\partial\Omega} = 0\}$, $f \in H_0^{-1}(\Omega)$, cf. Proposition 2.34, we have defined the bilinear form $a(\cdot,\cdot)$ and the corresponding operator A as

$$a(\cdot,\cdot) : H_0^1(\Omega) \times H_0^1(\Omega) \to \mathbb{R} \text{ and } A : H_0^1(\Omega) \to H_0^{-1}(\Omega) \text{ are defined by}$$

$$a(u,v) := \int_\Omega (\nabla u, \nabla v)_n + cuvdx \;\; =: \langle Au, v\rangle_{H^{-1}(\Omega)\times H^1(\Omega)} \;\forall v \in H_0^1(\Omega),$$

with the Euclidean product $(\nabla u(x), \nabla v(x))_n = (\nabla u(x))^T \nabla v(x)$ in $\mathbb{R}^n$. For $f \in L^2(\Omega)$ or $f \in H_0^{-1}(\Omega)$, the exact (weak) solution u_0 is then defined by

$$u_0 \in \mathcal{U}_b = \mathcal{V}_b := H_0^1(\Omega) : a(u_0, v) = \int_\Omega (\nabla u_0, \nabla v)_n + c\, u_0\, v\, dx \tag{4.107}$$
$$= \langle Au_0, v\rangle_{H^{-1}(\Omega)\times H^1(\Omega)} = (f,v) \text{ or } \langle f, v\rangle \forall\, v \in \mathcal{V}_b. \qquad \blacksquare$$

Basic idea for conforming finite elements methods
For conforming FEs we replace $\mathcal{V}_b$ in (4.107) by a sequence of finite dimensional spaces $\{\mathcal{V}_b^h\}_{h\in H}$. For nonconforming FEMs, we sometimes even choose two sequences $\{\mathcal{U}_b^h\}_{h\in H}$ and $\{\mathcal{V}_b^h\}_{h\in H}$. The indices h satisfy $0 < h \in H$, with $\inf_{h\in H} h = 0$.

Definition 4.46. Conforming FEs *are characterized, for second order problems, by $\mathcal{V}_b^h$ as subsets of solution and test spaces, exactly satisfying the boundary conditions, here the trivial Dirichlet conditions, and*

$$u^h, v^h \in \mathcal{V}_b^h \subset \mathcal{V}_b = H_0^1(\Omega),\; \forall T \in \mathcal{T}^h : u^h|_T, v^h|_T \in \mathcal{P}, \mathcal{P}_{d-1} \subset \mathcal{P} \subset \mathcal{P}_{d+\tau}, d > 1,$$
$$\tau \geq -1, \textit{ thus } a(u^h, v^h) : \mathcal{V}_b^h \times \mathcal{V}_b^h \to \mathbb{R}, \textit{ and } \langle f, v^h\rangle : \mathcal{V}_b^h \to \mathbb{R} \textit{ for } f \in \mathcal{V}_b' = H_0^{-1}(\Omega)$$

are well defined and $\mathcal{V}_b^h \subset H_0^1(\Omega) \cap C(\Omega)$. Conforming FEMs *are defined with conforming FEs.*

For problems of order $2m$, we require $\mathcal{V}_b^h \subset H_0^m(\Omega)$.

We determine the approximate and the exact solutions $u_0^h \in \mathcal{V}_b^h$ and u_0 from

$$u_0^h \in \mathcal{V}_b^h : a\left(u_0^h, v^h\right) = \int_\Omega \left(\nabla u_0^h, \nabla v^h\right)_n + c\, u_0^h\, v^h\, dx = \left\langle Au_0^h, v^h\right\rangle_{\mathcal{V}_b'\times\mathcal{V}_b} \tag{4.108}$$
$$= \langle f, v^h\rangle_{\mathcal{V}_b'\times\mathcal{V}_b}\; \forall\, v^h \in \mathcal{V}_b^h \subset \mathcal{V}_b = H_0^1(\Omega) \text{ with } A^h \in \mathcal{L}(\mathcal{V}_b^h, \mathcal{V}_b'^h)$$
$$\text{and } u_0 \in \mathcal{V}_b \text{ by omitting all }^h. \tag{4.109}$$

Remark 4.47. *We want to discuss an* often used notation. *For practically all spaces, $\mathcal{V}$, discussed here, we find $\mathcal{V} \neq \mathcal{V}_b$, and $\mathcal{V}' \neq \mathcal{V}_b'$, cf. Proposition 2.34. The same holds for the discrete spaces. Nevertheless, for test functions $v \in \mathcal{V}_b \neq \mathcal{V}$ and functionals $f \in \mathcal{V}' \neq \mathcal{V}_b'$ we find $f|_{\mathcal{V}_b} \in \mathcal{V}_b'$. We summarize*

$$\mathcal{V} \neq \mathcal{V}_b \Rightarrow \mathcal{V}' \neq \mathcal{V}_b', \text{ e.g., } H^{-1}(\Omega) = (H^1(\Omega))' \neq \left(H_0^1(\Omega)\right)' = H_0^{-1}(\Omega) \tag{4.110}$$

$$f \in \mathcal{V}' : \quad \langle f, v\rangle_{\mathcal{V}'\times\mathcal{V}} = \langle f, v\rangle_{\mathcal{V}'\times\mathcal{V}_b} \ \forall\, v \in \mathcal{V}_b \Rightarrow f|_{H_0^1(\Omega)} = f|_{\mathcal{V}_b} \in \mathcal{V}_b' = H_0^{-1}(\Omega)$$

$$\text{and similarly } \mathcal{V}^{h'} \neq \mathcal{V}_b^{h'}, \mathcal{V}^{h'} \neq \mathcal{V}_b^{h'}, f^h \in \mathcal{V}^{h'} \neq \mathcal{V}_b^{h'} \Rightarrow f^h|_{\mathcal{V}_b^h} \in \mathcal{V}_b^{h'}.$$

Usually we will use the notation $\langle f^h, v^h\rangle_{\mathcal{V}_b^{h'}\times\mathcal{V}_b^h}$, similarly $\langle f, v\rangle_{\mathcal{V}_b'\times\mathcal{V}_b}$, but now and then, $\langle f^h, v^h\rangle_{\mathcal{V}^{h'}\times\mathcal{V}^h}$ $\forall v^h \in \mathcal{V}_b^h$ as well. By (4.110) there is not much difference for conforming FEs, $\mathcal{V}_b^h \subset \mathcal{V}_b$. For nonconforming FEs, $\mathcal{V}_b^h \not\subset \mathcal{V}_b$, cf. Chapter 5, this will change.

For general cases we will prove the convergence via the results of Chapter 3. Nevertheless, honoring Cea, we include the surprisingly simple proof for convergence $\|u_0 - u_0^h\|_{H^1(\Omega)} \to 0$ for $h \to 0$ for his lemma. It applies not only to (4.109), but to all "conforming" FEMs for "coercive" bilinear forms.

Lemma 4.48. Cea lemma: *Let $\mathcal{V}_b^h \subset \mathcal{V}_b \subset \mathcal{V}$, with* dim *$\mathcal{V}_b^h < \infty$ and assume a (bounded) $\mathcal{V}_b$-coercive bilinear form, hence, cf. Proposition 1.25,*

$$|a(u,v)| \leq C\|u\|_{\mathcal{V}}\|v\|_{\mathcal{V}}, \quad |a(u,u)| \geq \alpha\|u\|_{\mathcal{V}}^2 \geq \alpha'\|u\|_{\mathcal{V}_b}^2, \forall u, v \in \mathcal{V}_b \tag{4.111}$$

with constants $C, \alpha, \alpha' > 0$. Then, with the projector Q'^h in (4.117) the discrete $A^h := (Q'^h A|_{\mathcal{V}_b^h})^{-1} \in \mathcal{L}(\mathcal{V}_b^{h'}, \mathcal{V}_b^h)$ is bounded and boundedly invertible with $\|A^h\|_{\mathcal{V}_b^{h'}\leftarrow\mathcal{V}_b^h} < C$, $\|(A^h)^{-1}\|_{\mathcal{V}_b^h\leftarrow\mathcal{V}_b^{h'}} < 1/\alpha$. Then unique exact solutions $u_0 \in \mathcal{V}_b$ and discrete solutions $u_0^h \in \mathcal{V}_b^h$ of (4.107) and (4.109) uniquely exist, and satisfy the error estimate:

$$\left\|u_0 - u_0^h\right\|_{\mathcal{V}} \leq \frac{C}{\alpha} \min_{u^h\in\mathcal{V}_b^h} \|u^h - u_0\|_{\mathcal{V}}. \tag{4.112}$$

Proof. By (4.111) the $a(.,.) : \mathcal{V}_b \times \mathcal{V}_b \to \mathbb{R}$, and $a(.,.) : \mathcal{V}_b^h \times \mathcal{V}_b^h \to \mathbb{R}$, satisfy the inf–sup conditions in Theorem 2.12. Thus $\|A\|_{\mathcal{V}_b\leftarrow\mathcal{V}_b'} < C$, $\|A^{-1}\|_{\mathcal{V}_b\leftarrow\mathcal{V}_b'} < 1/\alpha$ and $\|A^h\|_{\mathcal{V}_b^h\leftarrow\mathcal{V}_b^{h'}} < C$, $\|(A^h)^{-1}\|_{\mathcal{V}_b^{h'}\leftarrow\mathcal{V}_b^h} < 1/\alpha$ are bounded and boundedly invertible. So the solutions $u_0 \in \mathcal{V}_b$ and $u_0^h \in \mathcal{V}_b^h$ of (4.107) and (4.109) uniquely exist and imply

$$a\left(u_0 - u_0^h, v^h\right) = 0 \ \forall\ v^h \in \mathcal{V}_b^h,$$

and furthermore

$$\begin{aligned}
\alpha\left\|u_0 - u_0^h\right\|_{\mathcal{V}}^2 &\leq a\left(u_0 - u_0^h, u_0 - u_0^h\right) && \text{by coercivity}\\
&= a\left(u_0 - u_0^h, u_0 - u^h\right) + a\left(u_0 - u_0^h, u^h - u_0^h\right)\\
&= a\left(u_0 - u_0^h, u_0 - u^h\right) && \text{with } u^h - u_0^h \in \mathcal{V}^h\\
&\leq C\left\|u_0 - u_0^h\right\|_{\mathcal{V}}\|u_0 - u^h\|_{\mathcal{V}} && \text{by continuity.}
\end{aligned}$$

This implies

$$\|u_0 - u_0^h\|_{\mathcal{V}} \le \frac{C}{\alpha} \|u_0 - u^h\|_{\mathcal{V}} \ \forall\, u^h \in \mathcal{V}_b^h.$$

Since $\dim \mathcal{V}_b^h < \infty$ we find, with inf = min,

$$\|u_0 - u_0^h\|_{\mathcal{V}} \le \frac{C}{\alpha} \min_{u^h \in \mathcal{V}_b^h} \|u^h - u_0\|.$$

This estimate is exactly the same as that obtained below via Chapter 3. □

We re-interpret (4.109) by defining and applying a projector Q'^h. For later more general applications we formulate the following Proposition for broken Sobolev spaces $\mathcal{V}_b = W_0^{1,p}(\mathcal{T}^h), \mathcal{V}_b' = W_0^{-1,q}(\mathcal{T}^h), 1/q + 1/p = 1, \mathcal{V}_b^h \subset W_0^{1,p}(\mathcal{T}^h), \mathcal{V}_b^{h'} := (\mathcal{V}_b^h)'$, and a generalized bounded $f_h \in \mathcal{V}_b'$ in the form, so without boundary terms, cf. Proposition 2.34 and Remark 4.47,

$$\langle f, v^h\rangle := \langle f_h, v^h\rangle_{\mathcal{V}_b'\times\mathcal{V}_b} := \sum_{T\in\mathcal{T}^h}\int_T\left(\sum_{i=0}^n f_i\partial^i v^h\right)dx \ \forall v^h \in \mathcal{V}_b \cup \mathcal{V}_b^h. \tag{4.113}$$

Proposition 4.49. *For this $f = f_h \in \mathcal{V}_b' = W_0^{-\ell,q}(\mathcal{T}^h)$ and the restriction $f^h = f_h|_{\mathcal{V}_b^h}$ the norms are asymptotically equal $\forall\, 0 \le \ell,\ 1 \le q \le \infty, 1/p + 1/q = 1$,*

$$\begin{aligned}\|f^h\|_{\mathcal{V}_b^{h'}} &= \|\langle f^h, \cdot\rangle_{\mathcal{V}_b^{h'}\times\mathcal{V}_b^h}\| \le \|f\|_{\mathcal{V}_b'} = \|f_h\|_{\mathcal{V}_b^{h'}}(1 + o(1)), \ \textit{and} \\ \|f^h\|_{\mathcal{V}^{h'}} &= \|\langle f^h, \cdot\rangle_{\mathcal{V}^{h'}\times\mathcal{V}^h}\| \le \|f\|_{\mathcal{V}'} = \|f_h\|_{\mathcal{V}^{h'}}(1 + o(1)).\end{aligned} \tag{4.114}$$

For smoother f, e.g. $f \in W_{0,k}^{-\ell,q}(\mathcal{T}^h), k > 0$, cf. (4.137), satisfying homogeneous boundary conditions, we can estimate the $\|f\|_{L^q(T)}, \|f\|_{L^q(\mathcal{T}^h)}$ in (4.116) by $\|f\|_{W_{0,k}^{-\ell,q}(T)}, \|f\|_{W_{0,k}^{-\ell,q}(\mathcal{T}^h)}$ and still obtain the estimate (4.114).

Particularly interesting are the $\langle f, v\rangle_{W_0^{-1,q}(\Omega)\times W_0^{1,p}(\Omega)} = \int_\Omega (f_{-1}, \nabla v)_n + f_0 v dx$. For nonconforming FEMs in Sections 5.2, 5.5 we need the piecewise form as in (4.116). In the quadrature context, cf. Section 5.4, (4.114) applies to "smoother" $f \in W_k^{-\ell,q}(\mathcal{T}^h), k > 0$, cf. Theorem 5.28 and (5.186).

Proof. We start with the $\mathcal{V} = L^p(\mathcal{T}^h)$ result: It is well known that

$$|(f, v)_{L^2(\mathcal{T}^h)}| \le \|f\|_{L^q(\mathcal{T}^h)}\|v\|_{L^p(\mathcal{T}^h)}, \ \|f\|_{L^q(\mathcal{T}^h)} = \left(\int_{\mathcal{T}^h}(|f|^q)dx\right)^{1/q}, \tag{4.115}$$

valid for $\mathcal{T}^h$ and T. Combined with $v^h \in \mathcal{V}^h$, dense in $\mathcal{V}$ we get $\|f\|_{\mathcal{V}^{h'}} \le \|f\|_{L^q(\mathcal{T}^h)}$. This is combined with Proposition 4.18. For any $\epsilon > 0$ there is a $v_1 \ne 0$ such that

$$|\|f\|_{L^q(\mathcal{T}^h)} - (f, v_1)_{L^2(\mathcal{T}^h)}/\|v_1\|_{L^p(\mathcal{T}^h)}| \le \epsilon/2.$$

Proposition 4.18 implies the existence of a v^h such that for small enough h

$$\|v_1 - v^h\|_{L^p(\mathcal{T}^h)} < \epsilon/(2\|f\|_{L^q(\mathcal{T}^h)}) \Rightarrow$$
$$\|f\|_{L^q(\mathcal{T}^h)} < |(f, v^h)_{L^2(\mathcal{T}^h)}|/\|v^h\|_{L^p(\mathcal{T}^h)} + \epsilon \le \|f\|_{\mathcal{V}^{h\prime}} + \epsilon.$$

Now (4.115) shows the asymptotic equality, hence (4.114) for the $L^q(\mathcal{T}^h)$ case.

For the $W_0^{-\ell,q}(\mathcal{T}^h)$ case we formulate the first step, $W_0^{-1,q}(\mathcal{T}^h)$. We apply the discrete Hölder inequality (1.44) to the $f \in W_0^{-1,q}(\mathcal{T}^h)$, cf. Proposition 2.34,

$$\begin{aligned}
&\langle f, v^h\rangle_{W_0^{-1,q}(\mathcal{T}^h)\times W_0^{1,p}(\mathcal{T}^h)} \qquad (4.116)\\
&= \sum_{T\in\mathcal{T}^h} \int_T (f_{-1},\ \nabla\ v^h)_n + f_0 v dx\\
&\le \sum_{T\in\mathcal{T}^h} \sum_{i=1}^n \|f_{-1}^i\|_{L^q(T)} \|\partial^i v^h\|_{L^p(T)} + \|f_0\|_{L^q(T)} \|v^h\|_{L^p(T)}\\
&\le \left(\sum_{T\in\mathcal{T}^h} \|f_{-1}\|^q_{L^q(T)} + \|f_0\|^q_{L^q(T)}\right)^{1/q} \left(\sum_{T\in\mathcal{T}^h} \|\nabla\ v^h\|^p_{L^p(T)} + \|v^h\|^p_{L^p(T)}\right)^{1/p}\\
&= \|f\|_{W_0^{-1,q}(\mathcal{T}^h)} \|v^h\|_{W_0^{1,p}(\mathcal{T}^h)}\ \forall\ v^h\ \in\ W_0^{1,p}(\Omega) \cup W_0^{1,p}(\mathcal{T}^h).
\end{aligned}$$

yielding (4.114). □

For our special case $\mathcal{V} = H^1(\Omega), \mathcal{V}' = H^{-1}(\Omega), \mathcal{V}_b = H_0^1(\Omega),\ \mathcal{V}_b^h \subset H_0^1(\Omega),\ \mathcal{V}_b^{h'}$, we define for the $f \in \mathcal{V}_b'$ in (4.113), but valid for $f \in \mathcal{V}_b' = W_0^{-m,q}(\Omega)$ as well,

$$Q'^h \in \mathcal{L}\left(\mathcal{V}_b', \mathcal{V}_b^{h'}\right) \quad \text{as} \quad \langle Q'^h f - f, v^h\rangle_{\mathcal{V}_b'\times\mathcal{V}_b} = 0\ \forall\ v^h \in \mathcal{V}_b^h \iff Q'^h f - f \perp \mathcal{V}_b^h,$$

$$\|Q'^h f\|_{\mathcal{V}_b^{h'}} = \sup_{v^h\in\mathcal{V}_b^h, \|v^h\|_{\mathcal{V}}=1} \langle Q'^h f, v^h\rangle_{\mathcal{V}_b'\times\mathcal{V}_b} = \sup_{v^h\in\mathcal{V}_b^h, \|v^h\|_{\mathcal{V}}=1} \langle f, v^h\rangle_{\mathcal{V}_b'\times\mathcal{V}_b}$$

$$= \|f\|_{\mathcal{V}_b^{h'}} \le \sup_{v\in\mathcal{V}_b, \|v\|_{\mathcal{V}}=1} \langle f, v\rangle_{\mathcal{V}_b'\times\mathcal{V}_b} = \|f\|_{\mathcal{V}_b'} \Longrightarrow \|Q'^h\|_{\mathcal{L}(\mathcal{V}_b', \mathcal{V}_b^{h'})} \le 1, \qquad (4.117)$$

and, with dense $\mathcal{V}_b^h \subset \mathcal{V}_b$, $\quad \lim_{h\to 0} \|Q'^h f\|_{\mathcal{V}_b^{h'}} = \|f\|_{\mathcal{V}_b'}$.

Proposition 4.50. *These projectors Q'^h in (4.117) satisfy the conditions in Definition 3.12 for convergence of general discretization methods.*

Consequently (4.109) can be re-interpreted as

$$u_0^h \in \mathcal{V}_b^h:\ \langle (Au_0^h - f), v^h\rangle_{\mathcal{V}_b'\times\mathcal{V}_b} = 0 \forall\ v^h \in \mathcal{V}_b^h \iff Q'^h A u_0^h = Q'^h f. \qquad (4.118)$$

Since $\mathcal{V}_b^h \subset \mathcal{V}_b$, the (4.107), (4.109) imply

$$a\left(u_0 - u_0^h, v^h\right) = (f, v^h) - (f, v^h) = 0\ \ \forall\ v^h \in \mathcal{V}_b^h. \qquad (4.119)$$

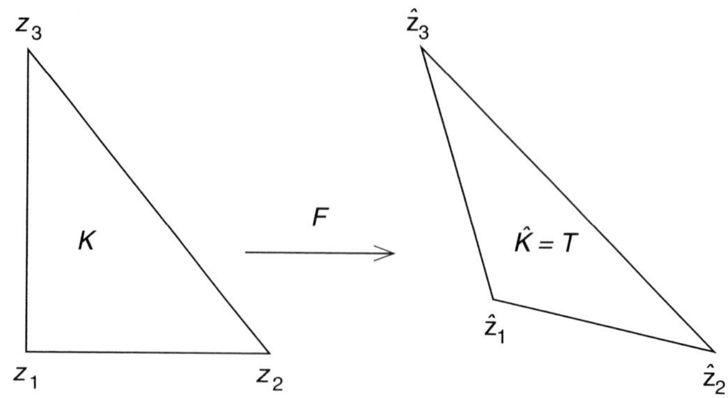

Figure 4.25 Affine transformations applied to a triangle.

In Subsection 4.3.3 and Section 5.5 we will call $a(u_0 - u_0^h, v^h)$ the *(variational) consistency error* for u_0. So it vanishes for conforming FEMs.

Reducing the computational costs

By choosing a basis $\{\phi_i\}_{i=1}^d$ for $\mathcal{P}$ and using the affine equivalence in Definition 4.6, it is possible to strongly reduce the cost for computing the

$$\int_T (\nabla v, \nabla w)_n dxdy, \int_T vwdxdy, \int_T vdxdy, \text{ mainly for } v = v^h, w = w^h \in \mathcal{V}^h \quad (4.120)$$

by computing the integrals over the reference domain K only once, cf. Proposition 5.41 and Figures 4.25 and 4.26. An analogous procedure applies to the general case and, modified, to nonconforming FEs as well. The values for $i, j = 1, \ldots d'$,

$$\int_T \left(\nabla u_i^h, \nabla u_j^h\right)_n dxdy \text{ and } \int_T u_i^h, u_j^h dxdy, \text{ with } d' = \dim \mathcal{V}_b^h, \quad (4.121)$$

are denoted as elements of the *stiffness and mass matrices*, respectively

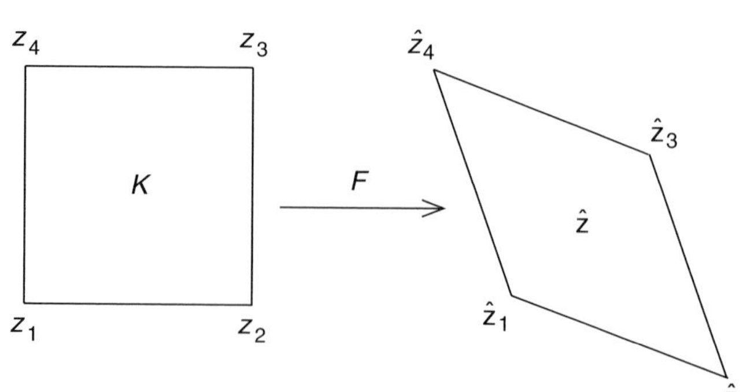

Figure 4.26 Affine transformations applied to a rhombus.

Example 4.51. We demonstrate this procedure for Example 4.11 and a subdivision into triangles or rhombuses. Obviously, it can be extended to many other cases as well. K is either the unit triangle or the square. We memorize the transformation F_T, its Jacobian and its partials from (4.14), (4.19), (4.22).

$$\begin{aligned} x &= (F_T)_1\,(\xi\,,\eta) := x_1 + (x_2 - x_1)\,\xi + (x_3 - x_1)\,\eta \\ y &= (F_T)_2\,(\xi\,,\eta) := y_1 + (y_2 - y_1)\,\xi + (y_3 - y_1)\,\eta. \end{aligned} \tag{4.122}$$

For the substitution in the integrals we need the Jacobian

$$\mathbf{J} = \det(F'_T) = (x_2 - x_1)(y_3 - y_1) - (y_2 - y_1)(x_3 - x_1) \neq 0 \tag{4.123}$$

and the partials (note that $\mathbf{J}, \xi_x, \xi_y, \eta_x, \eta_y$ are constant for each T)

$$\begin{aligned} v_x &= u_\xi \xi_x + u_\eta \eta_x,\ \xi_x = (y_3 - y_1)/\mathbf{J},\ \ \xi_y = -(x_3 - x_1)/\mathbf{J} \\ v_y &= u_\xi \xi_y + u_\eta \eta_y,\ \eta_x = -(y_2 - y_1)/\mathbf{J},\ \ \eta_y = (x_2 - x_1)/\mathbf{J}. \end{aligned} \tag{4.124}$$

With the notation

$$v = u \circ F_T^{-1},\ u = v \circ F_T,\ w = z \circ F_T^{-1},\ z = w \circ F_T,$$

we obtain for the integrals in (4.120) the following relations cf. Figure 4.25

$$\begin{aligned} \int_T (\nabla v)^2 dxdy &= \int_T (v_x^2 + v_y^2)\, dxdy \\ &= \int_K [(u_\xi \xi_x + u_\eta \eta_x)^2 + (u_\xi \xi_y + u_\eta \eta_y)^2]\mathbf{J} d\xi d\eta \\ &= a\int_K u_\xi^2 d\xi d\eta + 2b\int_K u_\xi u_\eta d\xi d\eta + c\int_K u_\eta^2 d\xi d\eta \end{aligned} \tag{4.125}$$

with

$$\begin{aligned} a &:= [(x_3 - x_1)^2 + (y_3 - y_1)^2]/\mathbf{J}, \\ b &:= -[(x_3 - x_1)(x_2 - x_1) + (y_3 - y_1)(y_2 - y_1)]/\mathbf{J}, \\ c &:= [(x_2 - x_1)^2 + (y_2 - y_1)^2]/\mathbf{J}. \end{aligned}$$

Similarly, we obtain

$$\begin{aligned} &\int_T (\nabla v, \nabla w)_n dxdy = \int_T (v_x w_x + v_y w_y) dxdy \\ &= \int_K [(u_\xi \xi_x + u_\eta \eta_x)(z_\xi \xi_x + z_\eta \eta_x) \\ &+ (u_\xi \xi_y + u_\eta \eta_y)(z_\xi \xi_y + z_\eta \eta_y)]\mathbf{J} d\xi d\eta \\ &= a\int_K u_\xi z_\xi d\xi d\eta + b\int_K (u_\xi z_\eta + z_\xi u_\eta) d\xi d\eta + c\int_K u_\eta z_\eta d\xi d\eta \end{aligned} \tag{4.126}$$

and

$$\int_T vw dxdy = \mathbf{J}\int_K uzd\xi d\eta, \int_T vdxdy, \quad \mathbf{J}\int_K ud\xi d\eta. \tag{4.127}$$

Whenever we use polynomial FEs, we are thus left with the evaluation of the integrals $\int_K \xi^p\eta^q d\xi d\eta$.

$$\int_K \xi^p\eta^q d\xi d\eta =: I_{pq}^{\text{triang}} = p!q!/(p+q+2)! \text{ for the unit triangle } K \tag{4.128}$$

$$\int_K \xi^p\eta^q d\xi d\eta =: I_{pq}^{\text{square}} = 1/(p+1)(q+1) \text{ for the unit square } K.$$

Finally, a basis for the polynomials in $\mathcal{P}$ has to be inserted into (4.125), (4.126), (4.127) to determine the system (4.109). This allows us in an efficient way to compute the elements of the *stiffness and mass matrices* for our system. ■

4.3.2 Finite element methods for general linear equations and systems of orders 2 and $2m$

These methods represent a straightforward generalization of the preceding (4.107), (4.117), (4.118), (4.119). So we do not repeat the original forms of the equations and systems, cf. Sections 2.3, 2.4, 2.6, but rather formulate the FE equations for these cases. For the convenience of the reader we recapitulate the notation: For the *multi-indices* α, we defined, see (2.73),

$$\partial^i u = \frac{\partial u}{\partial x_i}, \nabla u = (\partial^1 u, \ldots, \partial^n u) \text{ and } \partial^\alpha u = \partial_1^{\alpha_1}\ldots\partial_n^{\alpha_n} u = \frac{\partial^{|\alpha|}}{\partial x_1^{\alpha_1}\ldots\partial x_n{}^{\alpha_n}} u,$$

$$\nabla^0 u = u, \nabla^k u = (\partial^\alpha u)_{|\alpha|=k}, \nabla u = \nabla^1 u = (\partial^\alpha u)_{|\alpha|=1}, \text{ and vectors and reals}$$
$$\vartheta = (\vartheta^1,\ldots,\vartheta^n) \in \mathbb{R}^n, \vartheta^\alpha = (\vartheta^1)^{\alpha_1}\cdots(\vartheta^n)^{\alpha_n}, \vartheta^i, \Theta^0 \in \mathbb{R}, i \geq 0, \text{ choose the}$$
$$n_k, N_k, \text{ s.t. } \Theta^k = (\vartheta^\alpha)_{|\alpha|=k} \in \mathbb{R}^{n_k}, (\vartheta^\alpha)_{|\alpha|\leq k} \in \mathbb{R}^{N_k}, \Theta = \Theta^1, \tag{4.129}$$
$$\text{with } \partial^i u(x), \partial^\alpha u(x) \in \mathbb{R}, \ \nabla u(x) \in \mathbb{R}^n, \ \nabla^k u(x) \in \mathbb{R}^{n_k}.$$

The *FEM for a linear second order elliptic equation*, see (2.160), determines u_0^h such that

$$u_0^h \in \mathcal{U}_b^h: \ a\left(u_0^h, v^h\right) = \left\langle Au_0^h, v^h\right\rangle_{\mathcal{V}_b'\times\mathcal{V}_b} = \int_\Omega \sum_{i,j=0}^n a_{ij}\partial^i u_0^h \partial^j v^h dx = \langle f, v^h\rangle_{\mathcal{V}_b'\times\mathcal{V}_b}$$
$$\forall\, v^h \in \mathcal{V}_b^h, \text{ with } a_{ij} \in L^\infty(\Omega), \ \mathcal{V}_b^h \subset \mathcal{V}_b = H_0^1(\Omega). \tag{4.130}$$

The linear form f and our elliptic bilinear form are bounded, so

$$|\langle f, v\rangle_{\mathcal{V}_b'\times\mathcal{V}_b}| \leq C'\|v\|_{\mathcal{V}}, \quad |a(u,v)| \leq C\|u\|_{\mathcal{U}}\|v\|_{\mathcal{V}} \ \forall u, v \in \mathcal{V}_b.$$

The principal part, $a_p(u,v)$, satisfies for all $\vartheta \in \mathbb{R}^n$

$$\lambda|\vartheta|_n^2 \le \lambda(x)|\vartheta|_n^2 \le (-1)^m \sum_{i,j=1}^{n} a_{ij}(x)\vartheta^i\vartheta^j \le \Lambda|\vartheta|_n^2 \ \forall\, x \in \Omega \text{ with } \lambda > 0. \quad (4.131)$$

By (2.22) the principal part, $a_p(u,v)$, is $\mathcal{V}_b = H_0^1(\Omega)$-coercive, see (2.79), hence,

$$a_p(u,u) = \int_\Omega \sum_{i,j=1}^{n} a_{ij}\partial^i u \partial^j u dx \ge \lambda^* \|u\|_{\mathcal{U}}^2, \lambda^* > 0, \forall u \in \mathcal{U}_b \supset \mathcal{U}_b^h. \quad (4.132)$$

Conforming FEMs for elliptic equations of order $2m$ require $\mathcal{V}_b^h \subset \mathcal{V}_b = H_0^m(\Omega)$, hence smooth FEs, cf. (2.160) the references in Subsections 4.2.6, and 4.3.2. For linear A, determine $u_0^h \in \mathcal{U}_b^h \subset H_0^m(\Omega)$

$$a\left(u_0^h, v^h\right) = \left\langle Au_0^h, v^h\right\rangle_{\mathcal{V}_b'\times\mathcal{V}_b} = \int_\Omega \sum_{|\alpha|,|\beta|\le m} a_{\alpha\beta}\partial^\alpha u_0^h \partial^\beta v^h dx = \langle f, v^h\rangle_{\mathcal{V}_b'\times\mathcal{V}_b} \quad (4.133)$$

$$\forall v^h \in \mathcal{V}_b^h \subset \mathcal{V}_b \text{ with } a_{\alpha\beta} \in L^\infty(\Omega), \text{ and for } 1 < m = |\alpha| = |\beta| : \ a_{\alpha\beta} \in C(\bar{\Omega}).$$

This conformity condition $\mathcal{U}_b^h \subset H_0^m(\Omega)$ is somehow cumbersome, since C^{m-1} FEs are not too common. In fact, we have only given some references and the corresponding interpolation error and inverse estimates in Subsection 4.2.6.

We had assumed an elliptic equation, hence $a(u,v)$ is bounded and the principal part, $a_p(u,v)$, satisfies with $\lambda > 0$ and $\forall\vartheta \in \mathbb{R}^n$, see (2.79),

$$\lambda|\vartheta|_n^{2m} \le \lambda(x)|\vartheta|_n^{2m} \le (-1)^m \sum_{|\alpha|=|\beta|=m} a_{\alpha\beta}(x)\vartheta^\alpha\vartheta^\beta \le \Lambda|\vartheta|_n^{2m} \ \forall\, x \in \Omega. \quad (4.134)$$

By (2.160), and Theorem 2.43, the principal part, $a_p(u,v)$, is $\mathcal{V}_b$-coercive, hence

$$a_p(u,u) = \int_\Omega \sum_{|\alpha|=|\beta|=m} a_{\alpha\beta}\partial^\alpha u_0 \partial^\beta u dx \ge \lambda^* \|u\|_{\mathcal{V}}^2, \lambda^* > 0, \forall u \in \mathcal{U}_b \supset \mathcal{U}_b^h. \quad (4.135)$$

For systems we have to extend and modify the above notation: We have to apply the partials $\partial^l, l = 1,\ldots,n$, to q components $u_j, v_i, i,j = 1,\ldots,q$ and modify (4.129) from reals $\vartheta^l, \partial^l u(x) \in \mathbb{R}$, to vectors $\vec{\vartheta}^l, \vec{\vartheta}^\alpha, \partial^l\vec{u}(x), \partial^\alpha\vec{u}(x) \in \mathbb{R}^q$, and from n-vectors $\vartheta = (\vartheta^1,\ldots,\vartheta^n), \nabla u(x) \in \mathbb{R}^n$ to $n \times q$-matrices $\vec{\vartheta}, \nabla\vec{u}(x) \in \mathbb{R}^{n\times q}$, see (2.335), (2.389). In addition to the notation in (4.129) we need:[34]

$$\begin{aligned}
&\vec{u} = (u_1,\ldots,u_q), \partial^\alpha\vec{u} = (\partial^\alpha u_1,\ldots,\partial^\alpha u_q), \vec{\vartheta}^l = \left(\vartheta_1^l,\ldots,\vartheta_q^l\right), \vec{\vartheta}^\alpha = \left(\vartheta_1^\alpha,\ldots,\vartheta_q^\alpha\right),\\
&\vec{\vartheta} = (\vec{\vartheta}^1,\ldots,\vec{\vartheta}^n), \text{ with } |\vec{\vartheta}| = |\vec{\vartheta}|_{nq} \in \mathbb{R}, \vec{u}(x), \ \partial^\alpha\vec{u}(x), \vec{\vartheta}^l, \vec{\vartheta}^\alpha \in \mathbb{R}^q, \quad (4.136)\\
&\nabla\vec{u}(x) = (\partial^1\vec{u},\ldots,\partial^n\vec{u})(x), \vec{\Theta} = (\vec{\vartheta}^1,\ldots,\vec{\vartheta}^n) \in \mathbb{R}^{n\times q}, \ \partial^\alpha\vec{u} = \vec{u} \text{ for } |\alpha| = 0.
\end{aligned}$$

[34] The usual notation for θ and $\vec{u}$ motivates the certainly not optimal notation $\vec{\vartheta}^l$, $\vec{\Theta}$, which turns out to be appropriate below.

The *FEM for a linear second order elliptic system* requires ansatz and test-functions $\vec{u}^h = (u_1^h, \cdots, u_q^h)\ \vec{v}^h \in \mathcal{V}_b^h(\mathbb{R}^q) \subset \mathcal{V}_b = H_0^1(\Omega, \mathbb{R}^q)$, with $u_j^h, v_j^h \in \mathcal{V}_b^h$. With the Euclidean product $(f_k, \partial^k \vec{v}^h)_q$ in $\mathbb{R}^q$, see (2.340), (2.341), we obtain the FE solution $\vec{u}_0^h \in \mathcal{V}_b^h(\mathbb{R}^q)$, and the exact solution $\vec{u}_0 \in \mathcal{V}_b(\mathbb{R}^q)$ by omitting all the ^{h}, as

$$\langle \vec{f}, \vec{v}^h \rangle = \int_\Omega \sum_{k=0}^n (f_k, \partial^k \vec{v}^h)_q dx = a\left(\vec{u}_0^h, \vec{v}^h\right) = \int_\Omega \sum_{k,l=0}^n \left(A_{kl}\partial^l \vec{u}_0^h, \partial^k \vec{v}^h\right)_q dx$$

$$\forall\ \vec{v}^h \in \mathcal{V}_b^h \subset H_0^1(\Omega, \mathbb{R}^q),\ \text{with } A_{kl} \in L^\infty(\Omega, \mathbb{R}^{q\times q}), f_k \in L^2(\Omega, \mathbb{R}^{q\times q}), \qquad (4.137)$$

$$\text{and } \|\vec{f}\|_{H_{0,j}^{-1}(\Omega)} := \|\vec{f}\|_{H_{0,j}^{-1}(\Omega,\mathbb{R}^{q\times q})} := \max_{k=0,\ldots,n} \{\|f_k\|_{H^j(\Omega,\mathbb{R}^{q\times q})}\},$$

with bounded $\langle \vec{f}, \vec{v}^h \rangle$, $a(\vec{u}^h, \vec{v}^h)$. The $\mathcal{V}_b$-coercivity of the principal part is guaranteed for *uniformly elliptic systems*, satisfying the *uniform Legendre condition*, see (2.343), (2.345), hence

$$\exists\ \lambda, \Lambda,\ \text{s.t. } 0 < \lambda < \Lambda < \infty:\ \ \lambda|\vec{\vartheta}|_{nq}^2 \leq \sum_{k,l=1}^n (A_{kl}(x)\vec{\vartheta}^l)\vec{\vartheta}^k \leq \Lambda|\vec{\vartheta}|^2 \forall x \in \Omega \Rightarrow \qquad (4.138)$$

$$a_p(\vec{u}, \vec{u}) > \lambda^* \|\vec{u}\|_{\mathcal{V}}^2 \text{ and } a(\vec{u}, \vec{u}) \geq \lambda^{**}\|\vec{u}\|_{\mathcal{V}}^2 - C_c\|\vec{u}\|_{\mathcal{W}}^2\ \forall \vec{u} \in \mathcal{V}_b, \mathcal{W} = L^2(\Omega, \mathbb{R}^q). \qquad (4.139)$$

For the *FEM for a linear elliptic system of order* $2m$ we choose $\vec{u}^h, \vec{v}^h = (v_1, \ldots, v_q) \in \mathcal{V}_b^h \subset \mathcal{V}_b = H_0^m(\Omega, \mathbb{R}^q)$, and determine $\vec{u}_0^h \in \mathcal{U}_b^h$ such that, see (2.392),

$$\vec{u}_0^h \in \mathcal{V}_b^h \subset \mathcal{V}_b = H_0^m(\Omega, \mathbb{R}^q) : \langle \vec{f}, \vec{v}^h \rangle = \int_\Omega \sum_{|\alpha|\leq m} (f_\alpha, \partial^\alpha \vec{v}^h)_q dx = a\left(\vec{u}_0^h, \vec{v}^h\right) = \qquad (4.140)$$

$$\int_\Omega \sum_{|\alpha|,|\beta|\leq m} \left(A_{\alpha\beta}\partial^\beta \vec{u}_0^h, \partial^\alpha \vec{v}^h\right)_q dx \forall\ \vec{v}^h \in \mathcal{V}_b^h, A_{\alpha\beta}(x) \in L^\infty(\Omega, \mathbb{R}^{q\times q}), f_\alpha \in L^2(\Omega, \mathbb{R}^q).$$

We assume the *strong Legendre–Hadamard condition*

$$\exists \lambda > 0 \forall \vartheta \in \mathbb{R}^n, \eta \in \mathbb{C}^q, x \in \Omega:\ \sum_{|\alpha|=|\beta|=m} (A_{\alpha\beta}(x)\vartheta^\beta\eta, \vartheta^\alpha\bar{\eta})_q \geq \lambda|\vartheta|^{2m}|\eta|^2. \qquad (4.141)$$

By Theorem 2.104 the *principal part* is $\mathcal{U}_b$- and $\mathcal{U}_b^h$-coercive

$$a_p(\vec{u}^h, \vec{u}^h) = \int_\Omega \sum_{|\alpha|=|\beta|=m} (A_{\alpha\beta}\partial^\beta \vec{u}^h, \partial^\alpha \vec{u}^h)_q dx \geq \lambda^* \|\vec{u}^h\|_{\mathcal{V}}^2 \qquad (4.142)$$

if additionally $A_{\alpha\beta} \in C(\bar{\Omega})\ \forall |\alpha| = |\beta| = m$.

4.3.3 General convergence theory for conforming FEMs

For all the previous cases (4.109), (4.130), (4.133), (4.137), (4.140), the *conforming FEMs* have a very similar structure: we have replaced $\mathcal{V}_b$ by a sequence of finite dimensional spaces $\{\mathcal{V}_b^h\}_{h\in H}$. These $\mathcal{V}_b^h \subset \mathcal{V}_b$ do satisfy the boundary conditions and the appropriate regularity conditions, e.g. $\mathcal{V}_b^h \subset \mathcal{V}_b = H_0^m(\Omega) \subset \mathcal{V} = H^m(\Omega)$, e.g. $\mathcal{V}_b^h \subset C^{m-1}(\Omega)$. Thus $a(u^h, v^h) : \mathcal{V}_b^h \times \mathcal{V}_b^h \to \mathbb{R}$ is well defined. We consider the exact and the FE solutions $u_0 \in \mathcal{V}_b = H_0^m(\Omega)$, and $u_0^h \in \mathcal{V}_b^h \subset H_0^m(\Omega)$ of equations and $\vec{u}_0 \in \mathcal{V}_b \subset H_0^m(\Omega, \mathbb{R}^q)$, $\vec{u}_0^h \in \mathcal{V}_b^h \subset H_0^m(\Omega, \mathbb{R}^q)$ of systems. We denote the approximate solution $u_0^h \in \mathcal{V}_b^h$, using the

$$\text{same symbol for } u_0 \in \mathcal{V}_b = H_0^m(\Omega,\mathbb{R}^q), u_0^h \in \mathcal{V}_b^h \subset \mathcal{V}_b = H_0^m(\Omega,\mathbb{R}^q), q \geq 1,$$
$$\text{so } u_0^h \in \mathcal{V}_b^h \subset \mathcal{V}_b: \ a\left(u_0^h, v^h\right) = \left\langle A^h u_0^h, v^h \right\rangle_{\mathcal{V}_b' \times \mathcal{V}_b} = \langle f, v^h \rangle_{\mathcal{V}_b' \times \mathcal{V}_b} \ \forall\, v^h \in \mathcal{V}_b^h. \quad (4.143)$$

Here $A^h u^h$ is induced by the bilinear form $a(u^h, v^h)$, analogous for quasilinear problems.

All these methods are special cases of the general discretization methods in Chapter 3. For the proof of convergence we have to verify the conditions in Subections 3.3–3.4 for the different spaces and operators constituting these methods. We summarize the conditions and definitions in Summary 4.52. The wavelet and spectral methods in Chapter 9 and in [120] are conforming variational methods as well. So the following summary is applicable to these cases and to non conforming FEMs as well. Their relations are sketched in the diagram in Figure 4.27.

Summary 4.52. *We formulate it in a more general form to include FEMs with variational crimes, by using the necessary spaces $\mathcal{U}, \mathcal{U}_b, \mathcal{V}, \mathcal{V}_b, \mathcal{U}_b^h, \mathcal{V}_b^h$ and extensions A_h, f_h, G_h, cf. Chapter 5. For conforming methods, we have $A_h = A, f = f_h, G_h = G$.*

1. *The method has to be applicable to the problem $Au_0 = f$ or $Gu_0 = 0$, see Definition 3.12. This requires, throughout with $\forall h \in H, h < h_0$, and fixed h_0,* a sequence *of approximating spaces $\mathcal{U}_b^h, \mathcal{V}_b^h$ for the $\mathcal{U}_b, \mathcal{V}_b$, projectors P^h, Q'^h, and the discrete problems A^h or G^h, with $\inf\{0 < h \in H\} = 0$, such that*

$$\{\mathcal{U}^h, \mathcal{V}^h, P^h, Q'^h, G^h\}_{h\in H}, \quad \dim \mathcal{U}_b^h = \dim \mathcal{V}_b^h < \infty, \quad (4.144)$$

thus $\lim_{h\to 0} = \lim_{h\in H, h\to 0}$ *makes sense. Definition 3.12 does not require that, but for our methods the $\mathcal{U}_b^h, \mathcal{V}_b^h$ are approximating spaces, such that*

$$\forall u \in \mathcal{U}: \ \lim_{h\to 0} \text{dist}\,(u, \mathcal{U}^h) = 0, \ \textit{e.g.} \ \lim_{h\to 0} ||u - P^h u||_{\mathcal{U}^h} = 0, \quad (4.145)$$

similarly for $\mathcal{V}$, with $P^h = I^h$ for conforming methods, cf. Proposition 4.18.

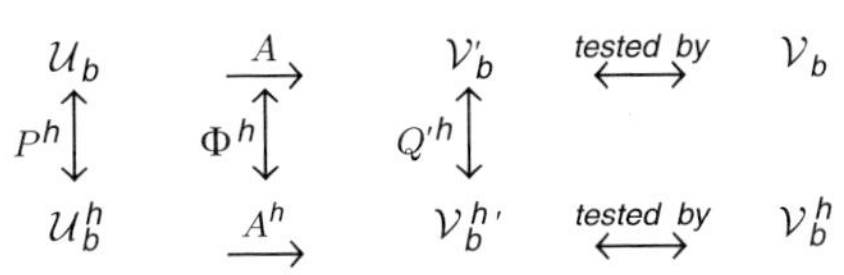

Figure 4.27 Our methods: spaces, operators, often $\mathcal{U}_b = \mathcal{V}_b, \mathcal{U}_b^h = \mathcal{V}_b^h$.

2. *The* $P^h \in \mathcal{L}(\mathcal{U}_b, \mathcal{U}_b^h)$, $Q'^h \in \mathcal{L}(\mathcal{V}_b', \mathcal{V}_b^{h'})$, *are linear approximation and projection operators, cf. (4.117), Proposition 4.50 and Definitions 3.6, 3.12, such that*

$$\lim_{h\to 0} \|P^h u\|_{\mathcal{U}^h} = \|u\|_{\mathcal{U}} \text{ by (4.145) and } \lim_{h\to 0} \|Q'^h f\|_{\mathcal{V}_b^{h'}} = \|f\|_{\mathcal{V}_b'}. \tag{4.146}$$

3. *Our methods define a mapping* $\Phi^h : \mathcal{D}(\Phi^h) \subset (\mathcal{U} \to \mathcal{V}') \to (\mathcal{U}^h \to \mathcal{V}^{h'})$, *transforming* $Au_0 = f$ *or* $Gu_0 = 0$ *into* $A^h u_0^h = f^h$ *or* $G^h u_0^h = 0$. *For nonconforming FEs, we have to extend the* $A, G : \mathcal{U} \to \mathcal{V}'$ *onto the corresponding spaces,* $\mathcal{U}_{\mathcal{T}^h}, \mathcal{V}'_{\mathcal{T}^h}$, *piecewise defined on the subdivision,* $\mathcal{T}^h$, *cf. (4.35), (4.36), and (5.251) ff., (7.57) ff.*

$$A, G : \mathcal{U} \to \mathcal{V}' \text{ into } A_h, G_h : \mathcal{U} \cup \mathcal{U}_{\mathcal{T}^h} \to \mathcal{V}' \cup \mathcal{V}'_{\mathcal{T}^h} \text{ with} \tag{4.147}$$
$$A_h, G_h|_{\mathcal{U}} = A, G, \text{ and } A_h = A, G_h = G, \text{ for conforming methods.}$$

In fact, the A, G, A^h, G^h *do not, but the unique solvability does need the boundary conditions. For our methods,* Φ^h *is defined as*

$$A^h := \Phi^h A := Q^{h'} A_h|_{\mathcal{U}_b^h}, \text{ and } f^h := \Phi^h f := Q^{h'} f_h|_{\mathcal{U}_b^h}, \text{ hence} \tag{4.148}$$
$$\Phi^h(A\cdot - f) = Q^{h'}(A_h \cdot - f_h)|_{\mathcal{U}_b^h} \text{ or } G^h := \Phi^h G := Q^{h'} G_h|_{\mathcal{U}_b^h}.$$

4. *According to Definition 3.14, we choose* u, u^h, *related as* $A^h u^h = Q'^h A_h u^h = Q'^h A u$ *and* $G^h u^h = Q'^h G_h u^h = Q'^h G u$. *Then*

$$Q'^h A_h u - A^h u^h = Q'^h (A_h u - A_h u^h),\ Q'^h G_h u - G^h u^h = Q'^h (G_h u - G_h u^h) \tag{4.149}$$

are the so-called (variational) consistency errors.

5. *The* (classical) consistency error or local discretization error *in* u *is*

$$A^h P^h u - Q'^h A u = A^h P^h u - Q'^h A_h u \text{ or } G^h P^h u - Q'^h G u. \tag{4.150}$$

As a consequence of the above extension A, G *to* A_h, G_h *we always have* $Au = A_h u, Gu = G_h u$, *cf. Theorem 4.54, (4.157).*

6. *The* G^h *are called* stable in u^h, *and* A^h stable, *if for fixed* $h_0, r, S \in \mathbb{R}_+$, *and* $\forall h \in H$

$$u_i^h \in B_r(u^h), i = 1, 2, \Rightarrow \left\|u_1^h - u_2^h\right\|_{\mathcal{U}^h} \le S \left\|G^h\left(u_1^h\right) - G^h\left(u_2^h\right)\right\|_{\mathcal{V}^{h'}}. \tag{4.151}$$

For A^h *or* $A^h - f^h$ *this reduces to (independent of* u^h, *but for* $\forall r > 0, h < h_0$*)*

$$(A^h)^{-1} \in \mathcal{L}\left(\mathcal{V}_b^{h'}, \mathcal{U}_b^h\right) \text{ exists and } \|(A^h)^{-1}\|_{\mathcal{U}_b^h \leftarrow \mathcal{V}_b^{h'}} \le S.$$

The stability of G^h *is essentially implied by the stability of its derivative* $(G'(u_0))^h$, *cf. Theorem 3.23.*

7. *If the exact and discrete solutions,* u_0 *and* u_0^h *exist, the* u_0^h *are called* convergent and convergent of order p, *if*

$$\lim_{h\to 0} \left\|u_0^h - u_0\right\|_{\mathcal{U}} = 0 \text{ and } \left\|u_0^h - u_0\right\|_{\mathcal{U}} \le C h^p. \tag{4.152}$$

For nonconforming FEMs $\|u_0^h - u_0\|_{\mathcal{U}}$ *have to be replaced by* $\|u_0^h - u_0\|_{\mathcal{U}^h}$.

8. *For quadrature approximations the* Q'^h, A^h, G^h *are replaced by* $\tilde{Q}'^h, \tilde{A}^h, \tilde{G}^h$.
9. *For the nonconforming FEMs in Chapters 5 and 7, (4.150) remains correct.*
10. *For isoparametric FEMs* Q'^h *are replaced by* $Q_c'^h$, *applied to functionals defined on a curved approximation* $\Omega_c^h \approx \Omega$.
11. *For fully nonlinear problems in strong form,* $F := (G, B_D)$, *cf. (5.41),*

$$\{\mathcal{U}^h, \mathcal{V}_\Pi^h = \mathcal{V}^h \times \mathcal{V}_b^h, P^h, Q_\Pi^h, Q_{c,\Pi}^h, F^h\}_{h\in H}, \text{ with } 0 < h \in H, \quad (4.153)$$

replaces (4.144) and modifies all the following concepts, cf. Section 5.2. Here G *and* B_D *indicate the differential and the Dirichlet boundary operators, respectively, approximated by* G^h *and* B_D^h *and* Q_Π^h *projects both components* (G, B_D) *to* $F^h = (G^h, B_D^h)$.

12. *All our methods are so-called* linear methods *in the sense*

$$\Phi^h(B + C) = \Phi^h B + \Phi^h C, \text{ and } \Phi^h(G + f) = \Phi^h G + \Phi^h f. \quad (4.154)$$

We reformulate Theorem 3.21 in Chapter 3 for FEMs. In fact, we only have to update the previous $G_h, \mathcal{U}_b^h, \mathcal{V}_b^{h'}, P^h, Q'^h, G^h$, to the following definitions. The obvious updates for the fully nonlinear case, according to (4.153), are not explicitly formulated in the next theorem.

Theorem 4.53. Unique existence and convergence of discrete solutions: *Let the original problems* $Gu = 0$ *have the exact solution* $u_0 \in \mathcal{U}$. *Let the FE discretization method be applicable to* G *and yield* $G^h = Q'^h G_h|_{\mathcal{U}^h} : \mathcal{U}^h \to \mathcal{V}^{h'}$ *in (5.62), satisfying*

1. $G^h : \mathcal{U}^h \to \mathcal{V}^{h'}$ *is defined and continuous in* $B_r(P^h u_0)$, $r > 0, h$*-independent;*
2. G^h *is consistent with* G *in* $P^h u_0$;
3. G^h *is stable for* $P^h u_0$.

Then the discrete problems $G^h u^h = 0$ *possess unique solutions* $u_0^h \in \mathcal{U}^h$ *near* u_0 *for all sufficiently small* $h \in H$ *and* u_0^h *converge to* u_0. *If* G^h *is consistent and consistent of order* p, *then* u_0^h *converge and converge of order* p, *respectively:*

$$\|u_0^h - u_0\|_{\mathcal{U}} \leq S\|Q'^h G_h(P^h u_0)\|_{\mathcal{V}_b^{h'}} \to 0 \text{ and } \|u_0^h - u_0\|_{\mathcal{U}} \leq \mathcal{O}(h^p). \quad (4.155)$$

Hence, good convergence, essentially determined by the consistency, is usually only satisfied if u_0 is smoother than $u_0 \in \mathcal{U}$.

The exact formulation for the different nonconforming FE and other methods is delayed to Chapters 5, 7 and 8. In particular, Theorem 4.53 will be specified for the other cases.

We have restricted the discussion in this chapter to conforming FEMs: the *variational consistency errors* vanish by (4.148) for the exact solution u_0. This is a simple consequence of $\mathcal{V}_b^h \subset \mathcal{V}_b$, e.g. (4.143) implies

$$\langle Au_0 - A^h u_0^h, v^h \rangle_{\mathcal{V}^{h'} \times \mathcal{V}} = a\left(u_0 - u_0^h, v^h\right) = (f, v^h) - (f, v^h) = 0 \;\forall v^h \in \mathcal{V}_b^h. \quad (4.156)$$

Theorem 4.54. Variational and classical consistency errors for conforming FEMs:

1. *For* conforming FEMs *applied to linear and nonlinear problems,* $Au - f = 0$ *and* $Gu = 0$*, the variational consistency errors in* u *vanish, cf. (4.143), (4.149).*
2. *For* $A \in \mathcal{L}(\mathcal{U}, \mathcal{V}')$*, and a nonlinear* $G : \mathcal{D}(G) \subset \mathcal{U} \to \mathcal{V}'$ *with Lipschitz-continuous* $G \in C_L(\mathcal{D}(G))$ *the*classical consistency or local discretization error *is*

$$\|(\Phi^h A)(I^h u) - Q'^h Au\|_{\mathcal{V}_b'} = \|Q'^h A(I^h u - u)\|_{\mathcal{V}_b'} \leq C\|A\|_{\mathcal{V}_b' \leftarrow \mathcal{U}_b}\|I^h u - u\|_{\mathcal{U}}.$$
$$\|(\Phi^h G)(I^h u) - Q'^h Gu\|_{\mathcal{V}_b'} = \|Q'^h (GI^h u - Gu)\|_{\mathcal{V}_b'} \leq CL_{\mathcal{V}_b' \leftarrow \mathcal{U}_b}\|I^h u - u\|_{\mathcal{U}}. \tag{4.157}$$

3. *By Proposition 4.18, the classical consistency error tends to* 0*, according, e.g. to* $\|Q'^h A(I^h u - u)\|_{\mathcal{V}_b'} \leq C\|A\|_{\mathcal{V}_b' \leftarrow \mathcal{U}_b}\|I^h u - u\|_{\mathcal{U}} \to 0$ *for* $u \in \mathcal{U}$*.*

Proof. For the classical consistency error we use $\mathcal{V}_b^h \subset \mathcal{V}_b$ and $(\Phi^h A)u^h = A^h u^h = Q'^h Au^h$. We find $\|A^h I^h u - Q'^h Au\|_{\mathcal{V}_b^{h'}} \leq C\|A\|_{\mathcal{V}_b' \leftarrow \mathcal{U}_b}\|I^h u - u\|_{\mathcal{U}}$. □

We formulate for later reference the changes for nonconforming FEMs:

Remark 4.55. Nonconforming variational and classical consistency errors:

1. Nonconforming FEMs *applied to linear and nonlinear problems,* $Au - f = 0$ *and* $Gu = 0$*, yield nontrivial variational consistency errors, since* $A \neq A_h,\ \ G \neq G_h$*, cf. (4.149).*
2. *The* nonconforming classical consistency error *is estimated as the sum of these nontrivial variational consistency errors and with* $\mathcal{U}, \mathcal{V}, \mathcal{U}_b^h, \mathcal{V}_b^h$*, cf. (4.35), (4.36),*

$$\|Q'^h A_h(I^h u - u)\|_{\mathcal{V}_b'} \leq C\|A_h\|_{\mathcal{V}' \cup \mathcal{V}'_{\mathcal{T}^h} \leftarrow \mathcal{U} \cup \mathcal{U}_{\mathcal{T}^h}}\|I^h u - u\|_{\mathcal{U}_{\mathcal{T}^h}} \tag{4.158}$$
$$\text{or } \|Q'^h G_h I^h u - G_h u)\|_{\mathcal{V}_b'} \leq CL\|I^h u - u\|_{\mathcal{U}_{\mathcal{T}^h}},$$

with a Lipschitz constant L *for* $G_h : \mathcal{D}(G^h) \subset \mathcal{U} \cup \mathcal{U}_{\mathcal{T}^h} \to \mathcal{V}' \cup \mathcal{V}'_{\mathcal{T}^h}$
3. *The classical consistency error tends to* 0*, with the variational consistency and the interpolation error.*

To admit nonconforming methods we again consider $\mathcal{U}, \mathcal{V}, \mathcal{U}_b^h, \mathcal{V}_b^h$, the operators $P^h = I^h$ and A, A^h are already defined in (4.109), (4.117), (4.130), (4.133), (4.137), (4.140).

So Q'^h in (4.117) and Proposition 4.50 have to be generalized. The arguments (4.113)–(4.113) are still valid for $f \in W_0^{-m,p'}(\Omega)$, $1/p + 1/p' = 1$, so for

$$\langle f, v^h \rangle = \langle f, v^h \rangle_{\mathcal{V}_b' \times \mathcal{V}_b^h} := \int_\Omega \sum_{|\alpha| \leq m} (f_\alpha, \partial^\alpha v^h) dx \ \forall v^h \in \mathcal{V}_b^h \subset \mathcal{V}_b. \tag{4.159}$$

The intended $Q'^h f \in \mathcal{V}_b^{h'}$ implies that $\langle Q'^h f, v^h \rangle_{\mathcal{V}_b' \times \mathcal{V}_b}$ is only defined for $v^h \in \mathcal{V}_b^h$:

$$Q'^h \in \mathcal{L}(\mathcal{V}', \mathcal{V}^{h'}) \text{ with } \langle Q'^h f - f, v^h \rangle_{\mathcal{V}_b' \times \mathcal{V}_b^h} = 0 \forall v^h \in \mathcal{V}_b^h \Leftrightarrow Q'^h f - f \perp \mathcal{V}_b^h. \tag{4.160}$$

Lemma 4.56. *For $f \in \mathcal{V}_b'$ in (4.159), and the projector Q'^h in (4.160), induced by the approximating spaces $\mathcal{V}_b^h$ for $\mathcal{V}_b$, we get, $\forall\ 0 \le m,\ 1 \le p' \le \infty, 1/p + 1/p' = 1,$*

$$\|f\|_{\mathcal{V}_b^{h'}} \le \|f\|_{\mathcal{V}_b'} = \|f\|_{\mathcal{V}_b^{h'}}(1 + o(1)),\ \mathcal{V}_b' = W_0^{-m,p'}(\mathcal{T}^h) \text{ and} \tag{4.161}$$

$$Q'^h f = f|_{\mathcal{V}_b^h},\ \ \|f\|_{\mathcal{V}_b^{h'}} = \|Q'^h f\|_{\mathcal{V}_b^{h'}} \le \|f\|_{\mathcal{V}_b'}, \lim_{h \to 0} \|Q'^h f\|_{\mathcal{V}_b^{h'}} = \|f\|_{\mathcal{V}_b'},$$

$$\|Q'^h\|_{\mathcal{V}_b^{h'} \hookrightarrow \mathcal{V}_b'} = 1 + o(1),\ \ \lim_{h \to 0} \|Q'^h\|_{\mathcal{V}^{h'} \hookrightarrow \mathcal{V}_b'} = 1,\ \text{hence (4.146).} \tag{4.162}$$

These results remain correct for smooth $f \in W_{0,k}^{-m,p'}(\Omega), k > 0$, see (4.137) and essentially for non-conforming FEMs as well, cf. Proposition 5.55.

After these preparations, we re-interpret (4.143) as

$$\left\langle \left(Au_0^h - f\right), v^h \right\rangle_{\mathcal{V}' \times \mathcal{V}} = 0 \forall\ v^h \in \mathcal{V}_b^h \Longleftrightarrow A^h u_0^h = Q'^h f \,\text{with}\, A^h := Q'^h A|_{\mathcal{U}_b^h}. \tag{4.163}$$

As a consequence of Theorem 3.21, the convergence for conforming FEMs for these linear elliptic boundary value problems is proved by the following summarizing theorem. The Aubin–Nitsche improvements are formulated in Theorem 4.60.

We study the FEMs in (4.130), (4.133), (4.137), (4.140), with piecewise polynomials of degree $\ge d - 1$, and the spaces $\mathcal{V}, \mathcal{V}_b, \mathcal{V}_b^h$. We assume the conditions (4.131), (4.133), (4.134) for equations, and (4.137), (4.138), (4.141) for systems of orders $2, m = 1$ and $2m$. Note that conforming FEMs for *order* $2m$ require the smooth FEs in Subsection 4.2.6. The special (4.109) always satisfies (4.131).

Theorem 4.57. Stability and convergence for linear and quasilinear problems:

1. *Under these conditions the principal part $a_p(\cdot,\cdot)$ is $\mathcal{U}_b^h$-coercive, and the original form $a(\cdot,\cdot)$ is $\mathcal{U}_b^h$-elliptic. For a boundedly invertible $A : \mathcal{V}_b \to \mathcal{V}'$, and under Condition 4.16, $A^h = Q'^h A|_{\mathcal{U}_b}$ is stable by Theorem 3.29.*
2. *Assume an exact solution $u_0 \in H^s(\Omega, \mathbb{R}^q), q \ge 1, m \le \min\{s, d\}$, and $\mathcal{V} = H^m(\Omega, \mathbb{R}^q)$. Then the FEM (4.163) has unique solutions $u_0^h \in \mathcal{U}_b^h$, converging as*

$$\left\|u_0 - u_0^h\right\|_{\mathcal{V}} \le C \text{ dist } \left(u_0, \mathcal{U}_b^h\right) \le Ch^{\min\{s,d\}-m}\|u_0\|_{H^s(\Omega,\mathbb{R}^q)}. \tag{4.164}$$

3. *This remains correct for quasilinear problems $Gu_0 = 0$ and their FEMs for a boundedly invertible $G'(u_0)$, satifying the above conditions for A.*

Remark 4.58. *These convergence estimates, (4.164) in Theorems 4.57, and 5.71, 5.66, are formulated for functions, e.g. $u \in H^s(\Omega), s \ge d$ smooth enough to yield the best order of accuracy. If this* smoothness is reduced, *then the interpolation errors are modified according to*

$$u \in H^{d-p}(\Omega) \Longrightarrow \|I^h u - u\|_{H^m(\Omega)} \le Ch^{d-m-p}|u|_{H^{d-p}(\Omega)} \tag{4.165}$$

$$\text{or, in (5.348) for } m = 1 : \|A^h I^h u - Q'^h Au\|_{H^m(\Omega)} \le Ch^{d-1-p}|u|_{H^{d-p}(\Omega)}.$$

This is a consequence of the fact that these results, including the interpolation errors, are proved with the Bramble–Hilbert Lemma 4.15. Analogous reductions are valid for all the other cases as well. We do not always refer to this remark.

Proof. Now we have available all the necessary results to apply the general converegence theory summarized in Theorems 4.53, 4.54. In fact, the conditions (4.144)–(4.151) are verified, ecxcept (4.151): The $\mathcal{V}_b^h$ are *admissible Galerkin approximating spaces* for $\mathcal{V}_b$. Their Φ^h is *applicable to all the above problems* yielding the FEMs listed preceding Theorem 4.57. By (4.150), (4.156), the consistency errors vanish. The above $u_0 \in H^s(\Omega), s > m$, yields a consistency error as in (4.164). The discrete operator A^h defined in (4.163) is continuous in $\mathcal{U}_b^h$. So the only missing condition, (4.151), for Theorem 4.53 is the stability of A^h. We have seen above that in all cases the principal parts $a_p(.,.)$ are $\mathcal{V}_b$- and $\mathcal{V}_b^h$-coercive by the inf–sup condition and Theorem 2.12. So these principal parts generate an invertible A_p with stable A_p^h. Theorem 3.29 implies then, for a consistent A^h in u_0, and an invertible $A : \mathcal{U}_b \to \mathcal{V}'$, in fact a compact perturbation of A_p, a stable A^h. So these FE equations have a unique solution, u_0^h, converging as in (4.164). □

Under conditions, which usually are satisfied, it is possible to increase the order of convergence with respect to a weaker norm. In addition to the original problem $a(u_0, v) = (f, v)\forall v \in \mathcal{V}_b$ we need the solution, ϕ_g, of its *dual problem,* see (5.352),

$$\phi_g \in \mathcal{V}_b \text{ s.t. } a(w, \phi_g) = (g, w)\forall w \in \mathcal{U}_b.$$

The proof of this lemma and the following theorem is presented for the more general case of FEMs violating continuity in Lemma 5.73 and Theorem 5.74.

Lemma 4.59. Aubin–Nitsche lemma: *Let $a(\cdot,\cdot) : \mathcal{U}_b \times \mathcal{V}_b \to \mathbb{R}$ be a continuous bilinear form, and let the conditions of Theorem 4.57 be satisfied. Then the original and its* dual problem *have unique exact and FE solutions u_0, u_0^h and ϕ_g, ϕ_g^h. For the Banach space $\mathcal{W} \supset \mathcal{V}_b$ with inner product and norm (g, w) and $\|g\|_{\mathcal{W}}$, respectively, let the embedding $\mathcal{V}_b \hookrightarrow \mathcal{W}$ be continuous. Then the error for the FE solution can be estimated in the $\mathcal{W}$ norm as*

$$\left\|u_0 - u_0^h\right\|_{\mathcal{W}^h} \le C \left\|u_0 - u_0^h\right\|_{\mathcal{U}^h} \sup_{0 \ne g \in \mathcal{W}} \left\|\phi_g - \phi_g^h\right\|_{\mathcal{V}^h} / \|g\|_{\mathcal{W}}. \tag{4.166}$$

Theorem 4.60. Aubin–Nitsche convergence for $\|u_0 - u_0^h\|_{L^2(\Omega)}$: *Under the conditions of Theorem 4.57 and Lemma 4.59 and for an exact solution $u_0 \in H^d(\Omega)$, the solution of the FEM equation satisfies*

$$\left\|u_0 - u_0^h\right\|_{L^2(\Omega)} \le Ch \left\|u_0 - u_0^h\right\|_{\mathcal{V}} \le Ch^d \|u_0\|_{H^d(\Omega)}. \tag{4.167}$$

Instead of the Cea lemma the discrete analogue of the inf–sup Theorem 2.12 can be employed, an important tool for proving the stability of the discrete A^h.

Theorem 4.61. Discrete Brezzi–Babuska or discrete inf–sup condition: *Let $\mathcal{V}_b^h$ be finite dimensional Banach spaces, $A^h \in \mathcal{L}(\mathcal{U}_b^h, \mathcal{V}_b'^h)$, and $a^h(\cdot,\cdot) : \mathcal{U}_b^h \times \mathcal{V}_b^h \to \mathbb{R}$ the associated operator and continuous bilinear form, defined*

$$\textit{for fixed } u^h \in \mathcal{U}_b^h \textit{ by } \langle A^h u^h, v^h \rangle_{\mathcal{V}_b'^h \times \mathcal{V}_b^h} = a^h(u^h, v^h) \quad \forall v^h \in \mathcal{V}_b^h. \tag{4.168}$$

Then the four statements (4.169), (4.170), (4.171) and (4.172) are equivalent.

$$(A^h)^{-1} \in \mathcal{L}\left(\mathcal{V}_b'^h, \mathcal{U}_b^h\right) \quad \text{and } C \text{ exists, s.t.} \|(A^h)^{-1}\|_{\mathcal{V}_b'^h \leftarrow \mathcal{U}_b^h} \leq C, \tag{4.169}$$

$$\exists \;\; \epsilon, \epsilon' > 0 \text{ s.t } \sup_{0 \neq v^h \in \mathcal{V}_b^h} |a^h(u^h, v^h)|/\|v^h\|_{\mathcal{V}^h} \geq \epsilon \|u^h\|_{\mathcal{U}} \forall u^h \in \mathcal{U}_b^h \tag{4.170}$$

$$\text{and} \quad \sup_{0 \neq u^h \in \mathcal{U}_b^h} |a^h(u^h, v^h)|/\|u^h\|_{\mathcal{U}^h} \geq \epsilon' \|v^h\|_{\mathcal{V}^h} \;\; \forall v^h \in \mathcal{V}_b^h,$$

$$\exists \;\; \epsilon > 0 \text{ s.t } \sup_{0 \neq v^h \in \mathcal{V}_b^h} |a^h(u^h, v^h)|/\|v^h\|_{\mathcal{V}^h} \geq \epsilon \|u^h\|_{\mathcal{U}} \;\; \forall u^h \in \mathcal{U}_b^h \tag{4.171}$$

$$\text{and} \quad \sup_{0 \neq u^h \in \mathcal{U}_b^h} |a^h(u^h, v^h)|/\|u^h\|_{\mathcal{U}^h} > 0 \;\; \forall v^h \in \mathcal{V}_b^h,$$

$$\exists \;\; \epsilon > 0 \text{ s.t. } \sup_{0 \neq v^h \in \mathcal{V}_b^h} |a^h(u^h, v^h)|/\|v^h\|_{\mathcal{V}^h} \geq \epsilon \|u^h\|_{\mathcal{U}^h} \;\forall u^h \in \mathcal{U}_b^h \tag{4.172}$$

$$\text{and} \quad \forall \; v^h \in \mathcal{V}_b^h \; \exists \; u^h \in \mathcal{U}_b^h : a^h(u^h, v^h) \neq 0.$$

For stability arguments the following constants C, ϵ, ϵ' have to be valid only for sufficiently small $h < h_0$, $h_0 > 0$ *and have to be* independent of h, but uniformly with respect to h.

Remark 4.62. *We want to point out that the* inf−sup *conditions for $A, a(\cdot,\cdot)$, $\mathcal{U}, \mathcal{V}, \ldots$ in Theorem 2.12 in Chapter 2 do not imply the corresponding* inf−sup *conditions for the corresponding $A^h, a^h(\cdot,\cdot)$. However the equivalence of the four statements is still correct. The original $a(\cdot,\cdot) : \mathcal{U}^h \to \mathcal{V}^h$ or later generalizations $a^h(\cdot,\cdot) : \mathcal{U}^h \to \mathcal{V}^h$ have to satisfy these equations* uniformly *with respect to h. Hence, $\epsilon > 0, \epsilon' > 0$, have to be independent of h. For stability arguments the constants C, ϵ, ϵ' have to be valid only for* sufficiently small $h < h_0$, $h_0 > 0$ *and have to be* independent of h *in the discrete counterpart of Theorem 2.12. Then the previous references (4.169), (4.170), (4.171) and (4.172) in Theorem 4.61 correspond to (2.51), (2.52), (2.53) and (2.54) in Theorem 2.12.*

4.4 Finite element methods for divergent quasilinear elliptic equations and systems

For divergent semilinear and quasilinear elliptic equations, FEMs are known, see, e.g. Zeidler [678]. We extend them to systems. Many of the existence, uniqueness and regularity results are only known in $H_0^m(\Omega)$. Generalizations to the $W_0^{m,p}(\Omega, \mathbb{R}^q)$ situation are possible and worthwhile. In fact linearization results as in Theorem 2.122 form the basis for numerical methods for the $W_0^{m,p}(\Omega, \mathbb{R}^q)$ setting for all types of nonlinear and even fully nonlinear problems. This opens the possibility for research into new problems outside of the previous scope.

We use the standard notation (4.129), and again the *same symbol*

$$u, u^h \text{ for functions } \vec{u} \in W_0^{m,p}(\Omega, \mathbb{R}^q), \vec{u}^h \in \mathcal{V}_b^h \subset W_0^{m,p}(\Omega, \mathbb{R}^q), m, q \geq 1, p \geq 2,$$

of equations and systems. On the basis of Section 2.7 we restrict the discussion to $2 \le p$. But in Section 4.5 we will study these quasilinear problems as monotone operators. Then we prove existence and convergence for $W_0^{m,p}(\Omega, \mathbb{R}^q)$ for $1 < p < \infty$. To emphasize the similarity between orders 2 and $2m$, we we use the notation (4.173) with $|\alpha| \le 1$ for second order equations. The necessary existence, regularity and linearization results are quoted before we formulate Theorem 4.63. This includes the conditions for the $W_0^{m,p}(\Omega, \mathbb{R}^q)$-coercivity of the principal part. The divergent quasilinear elliptic equation or system with Dirichlet boundary conditions in the trace sense and Ω in (4.5), see Skrypnik [591], Zeidler [678] and Böhmer [117], is defined and solved by $u_0 \in \mathcal{U}_b$ such that with $\langle \cdot, \cdot \rangle := \langle \cdot, \cdot \rangle_{\mathcal{V}_b' \times \mathcal{V}_b}$ and $\mathcal{V} := W^{m,p}(\Omega, \mathbb{R}^q)$

$$
\begin{aligned}
m, q \ge 1: \quad & \textit{Dirichlet condition } (\partial^j u/\partial \nu^j)|_{\partial\Omega} = 0, j = 0, \cdots, m-1, \text{ hence} \qquad (4.173)\\
& G : \mathcal{D}(G) \subset \mathcal{V}_b := W_0^{m,p}(\Omega, \mathbb{R}^q) \to \mathcal{V}' = W_0^{-m,p'}(\Omega, \mathbb{R}^q), 1/p + 1/p' = 1,\\
& \forall v \in \mathcal{V}_b : \langle G u_0, v \rangle = a(u_0, v) := \int_\Omega \sum_{|\alpha| \le m} (A_\alpha(\cdot, u_0, \ldots, \nabla^m u_0), \partial^\alpha v)_q dx = \langle f, v \rangle.
\end{aligned}
$$

The $A_\alpha(\cdot, u, \ldots, \nabla^m u) : \mathcal{V}_b \to \mathcal{V}'$ are so-called Nemyckii operators. $a(u, v)$ is well defined under the conditions discussed in Theorems 2.63, 2.73, 2.75 and 2.76.

In Section 2.7, Theorems 2.122, 2.124–2.126, we have formulated conditions for bounded $G'(u_0) : \mathcal{V}_b \to \mathcal{V}'$. In particular, $G'(u_0) : H_0^m(\Omega, \mathbb{R}^q) \to H^{-m}(\Omega, \mathbb{R}^q)$ yields a $H_0^m(\Omega, \mathbb{R}^q)$-coercive principal part. This implies, for $2 \le p < \infty$, only a $H_0^m(\Omega, \mathbb{R}^q)$-coercive principal part, cf. Theorem 2.122. For $1 \le p < 2$, in $W_0^{m,p}(\Omega, \mathbb{R}^q)$ the bilinear $a(u^h, v^h)$ are not even defined, hence $1 \le p < 2$ has to be excluded. This difference will be reflected in the convergence results in Theorem 4.63. However, we will develop in Section 4.5 a full convergence theory for quasilinear problems in $W_0^{m,p}(\Omega, \mathbb{R}^q)$ for all cases $1 < p < \infty$.

However, there is an essential difference between equations and systems. For $q = 1$ analytic results for the Banach space setting, $G : \mathcal{D}(G) \subset W_0^{m,p}(\Omega, \mathbb{R}^q) \to W^{-m,p'}(\Omega, \mathbb{R}^q)$, are available, see Subsections 2.5.3, 2.5.4, 2.5.6. One of the conditions is *uniform monotonicity* for the quasilinear operator or its principal part, see (2.289) or (2.294). This is very loosely related to the $W_0^{m,p}(\Omega, \mathbb{R}^q)$-coercivity of the principal part of $G'(u_0)$, the basic result for the stability of the FEM for the quasilinear systems. We have discussed this relation in Subsections 2.7.3 and 2.7.4.

For $q > 1 = m$ analytic results are available only for the Hilbert space setting, $G : \mathcal{D}(G) \subset H_0^m(\Omega, \mathbb{R}^q) \to H^{-m}(\Omega, \mathbb{R}^q)$; for $q, m > 1$ only little seems to be known, see Subsections 2.6.4, 2.6.6.

The proof of numerical stability follows our standard lines. Combining the different types of coercivity for the principal part with the compact perturbation argument in Theorem 3.29 implies the corresponding stability of the FEM for a boundedly invertible $G'(u_0)$.

For this weak form (4.173) it is straightforward to formulate a conforming FEM. As test and approximation spaces $\mathcal{V}_b^h$ we use the standard continuous FEs vanishing along $\partial\Omega$. Since $\mathcal{V}_b^h \subset W_0^{1,p}(\Omega,\mathbb{R}^q), q \geq 1$, we simply replace in (4.173) the $u, v \in \mathcal{V}_b$ by $u^h, v^h \in \mathcal{V}_b^h$. In this conforming FEM we determine, for $m, q \geq 1$

$$u_0^h \in \mathcal{V}_b^h \text{ s.t. } a\left(u_0^h, v^h\right) = \left\langle Gu_0^h, v^h\right\rangle_{\mathcal{V}_b'\times\mathcal{V}_b}$$

$$:= \int_\Omega \sum_{|\alpha|\leq m} \left(A_\alpha\left(x, u_0^h, \nabla u_0^h\right), \partial^\alpha v^h\right)_q dx$$

$$= \langle f, v^h\rangle_{\mathcal{V}_b'\times\mathcal{V}_b} = \sum_{|\alpha|\leq m}\int_\Omega f_\alpha \partial^\alpha v^h dx \forall\ v^h \in \mathcal{V}_b^h \subset \mathcal{V} = W_0^{m,p}(\Omega,\mathbb{R}^q), \quad (4.174)$$

cf. Theorems 2.63, 2.73, and 2.76. Again, the variational consistency error vanishes by Theorem 4.54.

The generalization of $Q'^h \in \mathcal{L}(\mathcal{V}', \mathcal{V}^{h'})$ in (4.160) and of Lemma 4.56 to the present situation is obvious. This includes the necessary limits of the norms and Summary 4.52 applies. So we mainly need the results guaranteeing the unique existence and regularity of solutions, and the bounded linearization for the original problem, for the many different cases included in (4.174). These results are available in the unique existence in Theorems 2.55, 2.57, 2.58, 2.61, 2.62, 2.73, 2.75, 2.76, 2.96, 2.115 or for the regularity of solutions, cf. Theorem 4.57, in Theorems 2.56, 2.60, 2.63, 2.64,2.99, 2.91, 2.101, 2.109, 2.116, and finally the bounded linearization in Theorems 2.119, 2.122, 2.99, 2.125, 2.126. Similarly we have summarized in Theorems 2.122, 2.124–2.126 the conditions for $W_0^{m,p}(\Omega,\mathbb{R}^q)$, the $H_0^m(\Omega,\mathbb{R}^q)$-coercive principal part, and have to modify it as above. The FE solutions, with the same symbol $u_0, u_0^h \in H^s(\Omega,\mathbb{R}^q)$, converge according Theorems 4.17, 4.54; compare Theorem 4.57 and (4.164). However, we have to choose the appropriate norms.

Theorem 4.63. Unique u_0^h converge for quasilinear problems and $2 \leq p < \infty$: *We assume for the different cases the conditions in the previous lines, $2 \leq p < \infty$, and a locally unique exact solution $u_0 \in W^{s,p}(\Omega), s \geq m$, for the original problems (4.173), with a bounded and boundedly invertible linearization $G'(u_0) \in \mathcal{L}(\mathcal{V}_b, \mathcal{V}_b')$. Again we have to exclude $1 \leq p < 2$.*

1. *For the* conforming FEMs *in (4.174), with $d \geq m$ the variational consistency errors for u_0 vanish, $Q'^h G u_0 - G^h u_0^h \equiv 0$.*
2. *The principal part $G_p'(u_0)$ of $G'(u_0) \in \mathcal{L}(\mathcal{V}_b, \mathcal{V}_b')$ in $\mathcal{V}_b = W_0^{m,p}(\Omega)$ is $H_0^m(\Omega)$-coercive, by Theorem 2.122. So the discrete $\left(G'(u_0)\right)^h$ is stable and consistency and convergence have to be and are formulated with respect to this $\mathcal{V} = H^m(\Omega)$ norm.*
3. *For $G \in C_L(\mathcal{D}(G))$, with the Lipschitz constant L for G, and $u, I^h u$, $u_0, I^h u_0 \in \mathcal{D}(G)$ the* classical consistency or local discretization error *is estimated by*

$$\|(\Phi^h G)(I^h u) - Q'^h G u\|_{\mathcal{V}'} = \|(\Phi^h G)(I^h u_0)\|_{\mathcal{V}'} \leq CL\|I^h u - u\|_{\mathcal{V}}.$$

4. *By Theorem 4.54 the FE equations (4.174) are stable and have unique solutions* $u_0^h \in \mathcal{V}_b^h$. *These* u_0^h *converge as*

$$\left\|u_0 - u_0^h\right\|_{H^m(\Omega)} \leq C \text{ dist } \left(u_0, \mathcal{V}_b^h\right) \leq Ch^{\min\{s,d\}-m}\|u_0\|_{H^s(\Omega)}. \quad (4.175)$$

Proof. G, hence G^h, is continuous near $I^h u_0$. For the classical consistency error we find, with $\mathcal{V}_b^h \subset \mathcal{V}_b$, by the mean value Theorem 1.43,

$$\begin{aligned}(\Phi^h G)(I^h u) - Q'^h(Gu) &= Q'^h(G(I^h u) - G(u)) \quad (4.176)\\ &= Q'^h \int_0^1 G'(u + t(I^h u - u))dt(I^h u - u)\\ &= \mathcal{O}(\|I^h u - u\|_{\mathcal{V}}).\end{aligned}$$

The stability, and consequently consistency and convergence, is for $2 \leq p < \infty$, guaranteed by the above conditions for the linearization $G'(u_0)$, see Theorems 3.16 and 3.23, hence the convergence by Theorem 3.21. □

Pousin and Rappaz [534] study from another point of view consistency, stability, convergence and a priori and a posteriori errors for Petrov-Galerkin methods applied to other nonlinear problems.

The conforming FEMs are *linear*, cf. (4.154). Hence all the quasilinear equations in (4.174) can be solved with the methods in Section 3.7: Continuation methods yield good approximations u_1^h for u_0^h. We reformulate the Newton method for the discrete problem, $G^h\left(u_0^h\right) = 0$, based upon the mesh independence principle in Theorem 3.40 and Corollary 3.41. We start with a moderately good approximation for u_0^h obtained, e.g. by continuation. Let the Newton method for the exact problem $u_{i+1} := u_i - (G(u_i)')^{-1}G(u_i)$, start with u_1. The Newton method for the discrete problem, starting with $u_1^h := P^h u_1$, yields for $i = 1, \ldots,$

$$u_{i+1}^h := u_i^h - \left(G^h\left(u_i^h\right)'\right)^{-1} G^h\left(u_i^h\right), \text{ with } \left\|u_0^h - u_{i+1}^h\right\|_{\mathcal{V}} \leq C^h\left\|u_0^h - u_i^h\right\|_{\mathcal{V}}^2. \quad (4.177)$$

We require the following Lipschitz-continuity for G',

$$\begin{aligned}&G'(\cdot) \in C_L(B_r(u_0) \cup B_r(I^h u_0)), \quad (4.178)\\ &G|_{\mathcal{V}_s} : \mathcal{D}(G) \cap \left(\mathcal{V}_s := H^s(\Omega)\right) \to H^{s-m}(\Omega).\end{aligned}$$

Then Theorem 3.40 shows, for the iterates u_i^h of the Newton method, a locally quadratic convergence, essentially independent of h for small enough h. For $u_0 \in H^s(\Omega, \mathbb{R}^q), s > m \geq 1$, we obtain convergence of the discrete to the exact iterates of order $\min\{s, d\} - m > 0$ by (4.175).

Theorem 4.64. Newton's method for the discrete problem and mesh independence principle: *In addition to the conditions in Theorem 4.63 we assume (4.178). Start the Newton process for* G *and* G^h *in (4.177) with* $u_1 \in B_r(u_0) \cap \mathcal{D}(G) \cap \mathcal{V}_s$ *and* $u_1^h := I^h u_1$. *Then, for small enough* $\|u_0 - u_1\|_{\mathcal{V}}$ *and* h, *both methods converge quadratically and the sequences* u_i, u_i^h *satisfy for* $i = 1, \ldots,$

$$\left\|u_{i+1}^h - u_i^h\right\|_{\mathcal{V}} \le C\left(\left\|u_i^h - u_{i-1}^h\right\|_{\mathcal{V}}\right)^2 \text{ and } \left\|u_i^h - I^h u_i\right\|_{\mathcal{V}} \le C h^{\min\{s,d\}-m} \|u_i\|_{\mathcal{V}_s}. \tag{4.179}$$

4.5 General convergence theory for monotone and quasilinear operators

The linearization technique for the previous FEMs and all the following methods does not allow convergence results for monotone operators and quasilinear operators in $W^{m,p}(\Omega)$ for $1 < p < 2$, cf. Subsection 4.3.3 and Lemma 2.77, Section 2.7, and Chapters 4 and 5. However it yields, for $2 \le p < \infty$, convergence of the expected order, e.g. $h^{\min\{s,d\}-m}$, with respect to the discrete $H^m(\Omega)$ norms for all types of elliptic problems. The conditions for the analytical approaches strongly vary, e.g. the condition of a strongly monotone or an elliptic operator, cf. Chapter 2.

On the other hand, the monotone operator techniques yield convergence results for $1 < p < \infty$ with respect to discrete $W^{m,p}(\Omega)$ norms, with different orders of convergence, cf. Theorem 4.67. Strongly monotone operators or quasilinear operators in $H^2(\Omega)$ allow specific results.

So we do not want to miss the chance for embedding the monotone operator approach into our general discretization theory, cf. Chapter 3. We do not aim to give a full proof for this area. Rather we combine the excellent presentation in Zeidler [678], mainly his Theorems 25.B and 26.A, cf. Theorem 2.68, with Chapter 3. Additionally, we formulate the results such that the nonconforming methods, indicated in Summary 4.52, are included.

We only formulate results yielding unique existence of discrete solutions, u_0^h, their convergence, and converging iteration methods for computation.

The basis for the following results is the obvious

Proposition 4.65. *All approximation spaces $\mathcal{U}^h, \mathcal{V}^h$, in particular $\mathcal{U}^h, \mathcal{V}^h \subset W^{m,p}(\Omega)$ or $\subset W^{m,p}(\mathcal{T}^h), 1 \le p < \infty$, in this and the next book [120], are reflexive Banach spaces.*

We need the operators $P^h : \mathcal{U}_b \to \mathcal{U}_b^h$ and the Q'^h such that for $f \in \mathcal{V}_b'$ and $f \in \mathcal{V}_b' \cup \mathcal{V}_{\mathcal{T}^h}'$

$$Q'^h : \mathcal{V}_b' \to \mathcal{V}_b'^h : \left\langle Q'^h f - f, v^h \right\rangle_{\mathcal{V}'^h \times \mathcal{V}^h} = 0 \;\forall v^h \in \mathcal{V}_b^h, \text{ for conforming, and}$$

$$Q'^h : \mathcal{V}_b' \cup \mathcal{V}_{\mathcal{T}^h}' \to \mathcal{V}_b'^h : \left\langle Q'^h f - f, v^h \right\rangle_{\mathcal{V}_{\mathcal{T}^h}' \times \mathcal{V}_{\mathcal{T}^h}} = 0 \;\forall v^h \in \mathcal{V}_b^h \text{ nonconf. meth.} \tag{4.180}$$

In Summary 4.52 we had introduced FEMs by applying a mapping $\Phi^h : \mathcal{D}(\Phi^h) \subset (\mathcal{U} \to \mathcal{V}') \to (\mathcal{U}^h \to \mathcal{V}^{h'})$. It transforms $Au_0 = f$ or $Gu_0 = 0$ into $A^h u_0^h = f^h$ or $G^h u_0^h = 0$. For nonconforming FEs, we have to extend the $A, G : \mathcal{U} \to \mathcal{V}'$ to the corresponding spaces, $\mathcal{U}_{\mathcal{T}^h}, \mathcal{V}_{\mathcal{T}^h}'$, piecewise defined on the subdivision, $\mathcal{T}^h$, cf. (4.35), (4.36), and (5.251) ff., (7.57) ff.

$$\text{extend } A, G : \mathcal{U} \to \mathcal{V}' \text{ into } A_h, G_h : \mathcal{U} \cup \mathcal{U}_{\mathcal{T}^h} \to \mathcal{V}' \cup \mathcal{V}_{\mathcal{T}^h}' \text{ piecewise with} \tag{4.181}$$
$$A_h, G_h|_{\mathcal{U}} = A, G, \text{ and } A_h, G_h = A, G, \text{ for conforming FEMs.}$$

These A, G, A_h, G_h, A^h, G^h do not, but the unique solvability does need the boundary conditions. With the previous $Q^{h'}, A_h, G_h, \mathcal{U}_b^h$, the Φ^h is defined as

$$A^h := \Phi^h A := Q^{h'} A_h|_{\mathcal{U}_b^h}, \text{ and } f^h := \Phi^h f := Q^{h'} f_h|_{\mathcal{U}_b^h}, \text{ hence} \tag{4.182}$$
$$\Phi^h(A \cdot -f) = Q^{h'}(A_h \cdot -f_h)|_{\mathcal{U}_b^h} \text{or } G^h := \Phi^h G := Q^{h'} G_h|_{\mathcal{U}_b^h}.$$

Now we restrict the discussion to the case

$$\mathcal{U}_b^h = \mathcal{V}_b^h. \tag{4.183}$$

We modify the results in Subsection 2.5.5 to show that Theorem 2.68 in Chapter 2 applies to G^h. We formulate the properties for $G^h : \mathcal{D}(G^h) \subset \mathcal{X}^h = \mathcal{U}_b^h \to \mathcal{U}_b^{'h}$, with $\mathcal{X}^h$ to emphasize the similarity to (2.242)–(2.253) The definition of $Q^{'h}$ in (4.180) is combined with the essential properties of Definition 2.65, cf (2.241), and the function

$$a : \mathbb{R}_+ \to \mathbb{R}_+, \ a(0) = 0, \ \lim_{t\to\infty} a(t) = \infty, \text{ e.g. } a(t) = c|t|^{p-1}, \ 0 < c, \ 1 < p. \tag{4.184}$$

So $G^h : \mathcal{D}(G^h) \subset \mathcal{U}_b^h \to \mathcal{U}_b^{'h}$ is an operator, defined on a Banach space, $\mathcal{X}^h = \mathcal{U}_b^h$. The following conditions are imposed $\forall u^h, v^h \in \mathcal{X}^h = \mathcal{U}_b^h$. Then G^h is, cf. (2.242)–(2.253),

$$\textit{monotone} \Leftrightarrow \langle G^h u^h - G^h v^h, u^h - v^h\rangle \ \geq \ 0, \ \ \forall u^h, v^h \in \mathcal{X}^h = \mathcal{U}_b^h, \ldots, \tag{4.185}$$
$$\textit{strictly mon.} \Leftrightarrow \langle G^h u^h - G^h v^h, u^h - v^h\rangle \ > \ 0 \ \ \forall u^h \neq v^h, \tag{4.186}$$
$$\textit{uniform. mon.} \Leftrightarrow \langle G^h u^h - G^h v^h, u^h - v^h\rangle \ \geq a(\|u^h - v^h\|_{\mathcal{X}^h})\|u^h - v^h\|_{\mathcal{X}^h}, \tag{4.187}$$
$$\text{e.g. } \geq c\|u^h - v^h\|_{\mathcal{X}^h}^p,$$
$$\textit{strongly mon.} \Leftrightarrow \langle G^h u^h - G^h v^h, u^h - v^h\rangle \ \geq \ c\|u^h - v^h\|_{\mathcal{X}^h}^2, \ c \in \mathbb{R}_+, \tag{4.188}$$
$$\textit{coercive} \ \Leftrightarrow \ \lim_{\|u^h\|_{\mathcal{X}^h}\to\infty} \frac{\langle G^h u^h, u^h\rangle}{\|u^h\|_{\mathcal{X}^h}} = \infty, \tag{4.189}$$
$$\textit{continuous} \Leftrightarrow \forall \epsilon > 0 \exists \delta > 0 : \|G^h u^h - G^h v^h\|_{\mathcal{X}'^h} < \epsilon \forall \|u^h - v^h\|_{\mathcal{X}^h} < \delta, \tag{4.190}$$
$$\textit{hemi cont.} \Leftrightarrow \ t \to \langle G^h(u^h + tv^h), w\rangle \text{ is continuous on } [0,1], \tag{4.191}$$
$$\textit{stable} \Leftrightarrow \ \|G^h u^h - G^h v^h\|_{\mathcal{X}'^h} \geq a(\|u^h - v^h\|_{\mathcal{X}^h}). \tag{4.192}$$

We claim that the properties (4.185)–(4.192) for the discrete G^h are inherited from (2.242)–(2.253) for the original G. This is a consequence of (3.18), (4.147). Combined with (2.242), we obtain for *conforming* methods, hence $G = G_h$, $u^h, v^h \in \mathcal{U}_b^h \in \mathcal{U}_b$

$$\begin{aligned}\forall u^h, v^h \in \mathcal{U}_b^h : \ \ \langle G^h u^h - G^h v^h, u^h - v^h\rangle &= \langle Q^{'h} G u^h - Q^{'h} G v^h, u^h - v^h\rangle \\ &= \langle G u^h - G v^h, u^h - v^h\rangle \ \geq \ 0,\end{aligned} \tag{4.193}$$

with the same relation for all the other (4.185)–(4.192).

For extending this argument to nonconforming methods, cf. Definition 3.5 and Summary 4.52, we distinguish the cases of $\mathcal{U}_b^h \not\subset \mathcal{U}_b$ violating the boundary conditions, or $\mathcal{U}^h \not\subset \mathcal{U}$. The first case can be handled by combining Theorem 2.128 with the technique applied in Section 5.2.6. We do not give further details.

For $\mathcal{U}^h \not\subset \mathcal{U}$, we refer to the conditions for the coefficient functions of G, simultaneously of G^h, e.g. in (2.262)–(2.267) and (2.284)–(2.289). Then Theorems 2.71 and 2.73 can be immediately applied to the broken Sobolev spaces, $W^{m,p}(\mathcal{T}^h) \supset \mathcal{U}^h$, cf. (4.35) ff. They yield the desired properties for G^h by Theorem 2.68, e.g. (4.185)–(4.192). The previous ellipticity conditions for the different equations and systems of orders 2 and $2m$ imply strong monotonicity. We summarize this discussion in

Theorem 4.66. Existence of $(G^h)^{-1}$ and converging discrete solutions:

1. *The properties of monotony-stability of G are inherited by G^h for conforming and nonconforming methods, cf. (4.185)–(4.192).*
2. *The previous properties of monotony-stability for G and G^h, (4.185)–(4.192), are valid for equations and systems, where the u^h, v^h have to be replaced by $\vec{u}^h, \vec{v}^h$.*
3. *If G^h is strictly monotone, coercive and hemicontinuous, cf. (4.186), (4.189), (4.191), Theorem 2.68 3., implies the existence of the inverse operator $(G^h)^{-1}$: $\mathcal{U}'^h \to \mathcal{U}^h$ and $\forall f^h \in \mathcal{U}'^h$ of unique solutions u_0^h for $G^h\left(u_0^h\right) = f^h$.*
4. *If G^h is additionally uniformly and strongly monotone, respectively, cf. (4.187) and (4.188), then the inverse $(G^h)^{-1}$ is continuous and Lipschitz-continuous.*
5. *A strongly monotone operator is coercive. Hence, if G^h is strongly monotone and Lipschitz-continuous, then the inverse $(G^h)^{-1}$ exists and is Lipschitz-continuous.*

The classical consistency in u_0 with $Gu_0 = 0$ implies, see (3.35), (3.42)–(3.43),

$$G^h P^h u_0 - Q'^h G u_0 = G^h P^h u_0 - Q'^h 0 = G^h P^h u_0 \;\to\; 0 \;\forall\, h \;\to\; 0,\; h \in\; H. \quad (4.194)$$

By Theorem 4.66 $(G^h)^{-1}$ is defined in and close to $0 = G^h u_0^h$. So we obtain

$$P^h u_0 - u_0^h = (G^h)^{-1} G^h P^h u_0 - (G^h)^{-1} 0.$$

For a vanishing consistency error in u_0 and a continuous or even Lipschitz-continuous $(G^h)^{-1}$ this implies the following estimate and theorem:

$$\left\|P^h u_0 - u_0^h\right\|_{\mathcal{U}^h} \;\to\; 0 \text{ or } \le L \|G^h(P^h u_0)\|_{\mathcal{U}_b'^h} \;\to\; 0 \;\forall\, h \;\to\; 0,\; h \in\; H. \quad (4.195)$$

Theorem 4.67. Unique solution u_0^h and convergence for all monotone operators and quasilinear equations and systems of order $2m, m \ge 1$, in $W_0^{m,p}(\mathcal{T}^h), 1 < p < \infty$, and all discretization methods in this and the next book [120]: *Let (2.242)–(2.253) or (2.284)–(2.288) be satisfied for $\mathcal{X}^h, G : \mathcal{X}^h \!\to \mathcal{X}'^h$, consequently for $\mathcal{U}_b^h, G^h : \mathcal{U}_b^h \to \mathcal{U}_b'^h$ as well. Here $\mathcal{U}^h = \mathcal{X}^h$ is a sequence of approximating subspaces for monotone operators and $\mathcal{U}_b^h \subset W_0^{m,p}(\Omega)$ for quasilinear equations and systems or $\subset W_0^{m,p}(\mathcal{T}^h)$ for the nonconforming cases. Choose any of the conforming or nonconforming discretization methods in this or the next book [120], cf. (4.181), (4.182), yielding G^h for G.*

1. *Then both $G(u) = 0$ and $G^h(u^h) = 0$ have unique solutions u_0 and u_0^h, respectively.*
2. *These solutions converge according to $\lim_{h\to 0, h\in H} \left\|P^h u_0 - u_0^h\right\|_{\mathcal{U}_b^h} = 0$ for conforming methods. For nonconforming methods with $m = 1$, they converge as*

$\|G^h P^h u_0\|_{\mathcal{U}_b'^h}$. *These classical consistency errors* $\|G^h P^h u_0\|_{\mathcal{U}_b'^h}$ *are estimated in Chapters 5 and 7.*

3. *If the function a in (4.184) has the form* $a(t) = c|t|^{p-1},\ 0 < c,\ 1 < p,$ *then*

$$\|P^h u_0 - u_0^h\|_{\mathcal{U}_b^h} \leq \left(L\|G^h(P^h u_0)\|_{\mathcal{U}_b'^h}/c\right)^{1/(p-1)}. \tag{4.196}$$

4. *For FEMs of local degree* $\geq d-1$*, a Lipschitz-continuous G, hence* G^h*, and* $u_0 \in W_0^{d,p}(\Omega), d > m$*, cf. (4.175), (4.41), we estimate for conforming methods*

$$\|P^h u_0 - u_0^h\|_{\mathcal{U}_b^h} \leq \left(h^{d-m}L\|u_0\|_{W^d(\Omega)}\right)^{1/(p-1)}. \tag{4.197}$$

For the nonconforming methods in Chapters 5 and 7 this holds for $m = 1$.

5. *If, additionally, G is strongly monotone,* $\mathcal{U}^h \subset \mathcal{U} = H_0^m(\Omega)$*, and* $u_0 \in H_0^d(\Omega)$*,* $d > m$*, we obtain*

$$\|P^h u_0 - u_0^h\|_{\mathcal{U}^h} \leq \left(h^{d-m}L\|u_0\|_{H^d(\Omega)}\right). \tag{4.198}$$

6. *The case* $\mathcal{U}_b^h \not\subset \mathcal{U}_b$*, violating the boundary conditions, can be handled by combining Theorem 2.128 with the technique applied in Section 5.2.6.*

Remark 4.68. *Quasilinear PDEs cannot yet be evaluated for wavelets, hence wavelet methods are impossible. For all other methods, these results remain valid with slight modifications as well. For difference methods this is formulated in Remark 8.27.*

We finish with formulating the following *generalized gradient iteration method* for computing the u_0^h, cf. Zeidler [678], Theorem 26.B. For the above

$$\begin{aligned} &G^h : \mathcal{D}(G^h) \subset \mathcal{U}_b^h \to \mathcal{U}_b'^h, \quad \mathcal{U}_b^h \text{ a Hilbert space, and} \\ &G^h\left(u_0^h\right) = 0, \quad u_0^h \in \mathcal{U}_b^h \text{ define} \\ &u_{k+1}^h := u_k^h - tLG\left(u_k^h\right), \quad k = 1, 2, \ldots. \end{aligned} \tag{4.199}$$

The following assumption $G^h(0) \neq 0$ makes sense, since otherwise 0 would be the desired solution.

Theorem 4.69. Convergent gradient iteration: *In (4.199) let* $G^h(0) \neq 0$*, and* G^h *be uniformly monotone, cf. (4.187), locally Lipschitz-continuous, hence for* $u^h, v^h \in \mathcal{D}(G^h)$*,*

$$\forall r > 0\ \exists M(r) : \forall \|u^h\|_{\mathcal{U}^h}, \|v^h\|_{\mathcal{U}^h} \leq r : \ \|G^h u^h - G^h v^h\|_{\mathcal{U}_b'^h} \leq M(r)\|u^h - v^h\|_{\mathcal{U}^h}.$$

Furthermore, let L be linear, bounded, self-adjoint, and strongly monotone with $(Lu^h, u^h) \geq b\|u^h\|_{\mathcal{U}^h}^2, \forall u^h \in \mathcal{U}_b^h, b > 0$*, and define the numbers with a in (4.187),*

$$c := \|L\|_{\mathcal{U}_b^h \to \mathcal{U}_b'^h}, \quad r_0 := \left(1 + \sqrt{c/b}\right)\|G(0)\|/a, \quad t_0 := 2a/\left(cM(r_0)^2\right).$$

Then for every $t \in (0, t_0)$*, the sequence* u_k^h *in (4.199) converges to* u_0^h *according to*

$$\lim_{k\to\infty} \|u_k^h - u_0^h\|_{\mathcal{U}_b^h} = 0 \text{ and } \|u_k^h - u_0^h\|_{\mathcal{U}_b^h} \leq \|G^h u_k^h\|_{\mathcal{U}_b'^h}/a. \tag{4.200}$$

4.6 Mixed FEMs for Navier–Stokes and saddle point equations

4.6.1 Navier–Stokes and saddle point equations

In Section 2.8 we presented the necessary theoretical results for the Navier–Stokes and saddle point equations, cf. (2.320)–(2.522) and (2.501). So we repeat here only those aspects necessary for the mixed FEMs in Subsection 4.6.2. This area has collected a huge amount of very substantial work. For examples, the numerical methods for the Navier–Stokes equations are listed 3,731 times in the Zentralblatt in late 2009. So the author has decided to present one of the classical approaches. This shows many of the essential ideas for these methods very clearly.

The *stationary Navier–Stokes equation* has the form, cf. (2.320),

$$G(\vec{u},p) := \begin{pmatrix} -\nu\Delta\vec{u} + \sum\limits_{i=1}^{n} u_i\partial_i\vec{u} + \text{grad } p \\ \text{div } \vec{u} \end{pmatrix} = \begin{pmatrix} \vec{f} \\ 0 \end{pmatrix} \quad \text{in } \Omega$$

$$\vec{u} = 0 \text{ on } \quad \partial\Omega, \int_\Omega p dx = 0, \ \Omega \in C_L^1 \text{ bounded} \quad \text{and } p : \Omega \to \mathbb{R}, \quad (4.201)$$

$$\vec{u} = (u_1, \dots, u_n), \vec{v} = (v_1, \dots, v_n), \vec{f} = (f_1, \dots, f_n) : \Omega \subset \mathbb{R}^n \to \mathbb{R}^n, n \le 3.$$

The $\vec{u}, p$ and $\vec{f}$ denote the velocity, pressure and forcing terms of an incompressible medium, $\vec{v}$ is a test function with $\text{div } \vec{u} = \sum_{i=1}^n \partial u_i/\partial x_i$, and $\nu = 1/R$ with the *Reynolds number*, R. Its linearization in $(\vec{u}_0 \equiv 0, p_0)$, applied to an increment $(\vec{u}, p), \vec{u} \neq 0$, and for $\nu = 1$, is the Stokes operator, S, cf. (2.496),

$$S(\vec{u},p) := \begin{pmatrix} -\Delta\vec{u} + \nabla p \\ -\text{div } \vec{u} \end{pmatrix} = \begin{pmatrix} \vec{f} \\ 0 \end{pmatrix} \quad \text{in } \Omega, \vec{u} = 0 \text{ on } \partial\Omega, \quad \int_\Omega p dx = 0. \qquad (4.202)$$

For proving convergence for our conforming FEMs, we recall that the variational consistency errors vanish, hence the classical consistency error can be estimated by the interpolation errors. The stability results are obtained by applying the results in Chapter 3, cf. Summary 4.52 to S and its generalization, a saddle point problem, see (4.208). In Theorems 2.129, 4.70, this is shown to be stable, essentially if the well-known Brezzi–Babuska conditions are satisfied, see (4.214). We show that, for moderate ν, the linearized Navier–Stokes operator represents a compact perturbation of S, hence satisfies the conditions of Theorem 3.29. So we obtain convergence for the stationary Navier–Stokes equations as well. This seems to be the state of the art anyway.

We have introduced the spaces and continuous bilinear forms, cf (2.497),(2.498),

$$\mathcal{V}_b = \left(H_0^1(\Omega)\right)^n, \ \mathcal{W} = L_*^2(\Omega) = \left\{p \in L^2(\Omega) : \int\limits_\Omega p(x)dx = 0\right\}, \ \mathcal{X} = \mathcal{V}_b \times \mathcal{W} \quad (4.203)$$

$$a(\vec{u},\vec{v}) = \int_\Omega (\nabla\vec{u}(x), \nabla\vec{v}(x))_n dx \text{ with } (\nabla\vec{u}(x), \nabla\vec{v}(x))_n = \sum_{i,j=1}^{n} \frac{\partial u_i}{\partial x_j}\frac{\partial v_i}{\partial x_j} \qquad (4.204)$$

$$\text{and } b(p,\vec{v}) = -\int_\Omega p(x) div\ \vec{v}(x)dx \text{ for } p \in L_*^2(\Omega), \ \vec{u},\vec{v} \in \mathcal{V}_b,$$

$$\text{with bounded } |a(\vec{u},\vec{v})| \le C_a\|\vec{u}\|_\mathcal{V} \cdot \|\vec{v}\|_\mathcal{V}, |b(p,\vec{v})| \le C_b\|\vec{v}\|_\mathcal{V} \cdot \|p\|_\mathcal{W}. \quad (4.205)$$

We generalized this weak Stokes form (4.203)–(4.205), cf. (2.500), to a *saddle point problem* by admitting continuous linear and bilinear forms, $\vec{f}^1(\vec{v}), f^2(\vec{v}), a(\vec{u},\vec{v}), b(p,\vec{v})$, and Hilbert spaces $\mathcal{V}_b, \mathcal{W}$: For

$$\begin{aligned}
&\text{continuous } a(\cdot,\cdot):\mathcal{V}_b\times\mathcal{V}_b\to\mathbb{R}, b(\cdot,\cdot):\mathcal{W}\times\mathcal{V}_b\to\mathbb{R}, \vec{f}=(\vec{f}^1,f^2):\mathcal{V}_b\times\mathcal{W}\to\mathbb{R}^2,\\
&\text{determine } \vec{u}_0\in\mathcal{V}_b, p_0\in\mathcal{W} \text{ s.t. } a(\vec{u}_0,\vec{v})+b(p_0,\vec{v})=\vec{f}^1(\vec{v}) \quad \forall\ \vec{v}\in\mathcal{V}_b,\\
&\qquad\qquad b(q,\vec{u}_0)=f^2(q)\ \forall\ q\in\mathcal{W}. \qquad (4.206)
\end{aligned}$$

To obtain the situation in Chapter 3 and Summary 4.52, we introduce $\mathcal{X}:=\mathcal{V}_b\times\mathcal{W}$, $\vec{x}:=(\vec{u},p), \vec{y}:=(\vec{v},q)\in\mathcal{X}$ and the continuous bilinear form $c(\vec{x},\vec{y})=a(\vec{u},\vec{v})+b(p,\vec{v})+b(q,\vec{u})$. Let $A\in\mathcal{L}(\mathcal{V}_b,\mathcal{V}_b'), B\in\mathcal{L}(\mathcal{W},\mathcal{V}_b')$ and $C\in\mathcal{L}(\mathcal{X},\mathcal{X}')$ be the linear operators induced by $a(\cdot,\cdot), b(\cdot,\cdot)$ and $c(\cdot,\cdot)$, respectively. Then, cf (2.502),

$$C:=\begin{pmatrix}A & B\\ B^d & 0\end{pmatrix}\in\mathcal{L}(\mathcal{X},\mathcal{X}'),\ C\vec{x}=\begin{pmatrix}A & B\\ B^d & 0\end{pmatrix}\begin{pmatrix}\vec{u}\\ p\end{pmatrix}=\begin{pmatrix}A\vec{u}+Bp\\ B^d\vec{u}\end{pmatrix}$$

$$\begin{aligned}
c(\vec{x},\vec{y}) &= \langle C\vec{x},\vec{y}\rangle_{\mathcal{X}'\times\mathcal{X}} = \left\langle\begin{pmatrix}A\vec{u}+Bp\\ B^d\vec{u}\end{pmatrix},\begin{pmatrix}\vec{v}\\ q\end{pmatrix}\right\rangle_{\mathcal{X}'\times\mathcal{X}} \qquad (4.207)\\
&= \langle A\vec{u},\vec{v}\rangle_{\mathcal{V}_b'\times\mathcal{V}_b}+\langle Bp,\vec{v}\rangle_{\mathcal{V}_b'\times\mathcal{V}_b}+\langle B^d\vec{u},q\rangle_{\mathcal{W}'\times\mathcal{W}}\\
&= a(\vec{u},\vec{v})+b(p,\vec{v})+b(q,\vec{u})=c(\vec{x},\vec{y}).
\end{aligned}$$

For solving the *saddle point and the Stokes problem,* cf. (2.503) and Chapter 9,

$$\begin{aligned}
&\text{determine } \vec{x}_0=(\vec{u}_0,p_0)\in\mathcal{X} \text{ for given } \vec{f}\in\mathcal{X}' \text{ such that} \qquad (4.208)\\
&c(\vec{x}_0,\vec{y})=\langle\vec{f},\vec{y}\rangle_{\mathcal{X}'\times\mathcal{X}} \text{ for all } \vec{y}\in\mathcal{X} \iff C\vec{x}_0=\vec{f}\in\mathcal{X}'.
\end{aligned}$$

In Theorem 2.131 we formulated results for the bounded invertibility of C and the existence, uniqueness and regularity of solutions $\vec{x}_0$.

4.6.2 Mixed FEMs for Stokes and saddle point equations

We formulate the corresponding stability, consistency and convergence properties in this subsection, based upon the results in Chapter 3. In the next subsection we will combine these results with Summary 4.52. The $C^{-1}\in\mathcal{L}(\mathcal{X}',\mathcal{X})$, and the stability of C^h in (4.211) yield the stability for the (linearized) Navier–Stokes operator, cf. Theorems 2.131 and 3.22.

By (4.201) the orders of the highest derivatives of $\vec{u}$ and p are 2 and 1, respectively. This is one of the reasons for choosing the FE spaces approximating $\vec{u}$ and p differently. The corresponding FEMs are called mixed FEMs. Obviously, approximating spaces $\left(\mathcal{V}_b^h\right)_{h\in H}, (\mathcal{W}^h)_{h\in H}$ for $\mathcal{V}_b, \mathcal{W}$ yield $\left(\mathcal{V}_b^h\times\mathcal{W}^h\right)_{h\in H}$, again approximating $\mathcal{V}_b\times\mathcal{W}$. They are Petrov–Galerkin approximations for $\mathcal{X}$ simultaneously with $\mathcal{V}_b^h$ and $\mathcal{W}^h$ for $\mathcal{V}_b$ and $\mathcal{W}$. One can show that $\dim\mathcal{V}_b^h\geq\dim\mathcal{W}^h$ is a necessary criterion for the stability of the discrete operator, cf. [387]. According to [374, 387], we formulate a *conforming FEM* for (4.206) and (4.208): we choose approximating subspaces $\mathcal{V}_b^h, \mathcal{W}^h$

and a right-hand side $\vec{f}$:

$$\left(\mathcal{V}_b^h\right)_{h\in H}, (\mathcal{W}^h)_{h\in H}\forall h\in H : \dim \mathcal{V}_b^h \geq \dim \mathcal{W}^h \text{ and } \mathcal{V}_b^h \subset \mathcal{V}_b, \mathcal{W}^h \subset \mathcal{W} \quad (4.209)$$

$$\text{for } \vec{f} = (\vec{f}^1, f^2) \in \mathcal{V}_b' \times \mathcal{W}' = \mathcal{X}' \text{ determine } \vec{x}_0^h = \left(\vec{u}_0^h, p_0^h\right) \in \mathcal{V}_b^h \times \mathcal{W}^h = \mathcal{X}^h$$

$$\text{by } a\left(\vec{u}_0^h, \vec{v}^h\right) + b\left(p_0^h, \vec{v}^h\right) = \vec{f}^1(\vec{v}^h) \ \ \forall \ \ \vec{v}^h \in \mathcal{V}_b^h, \quad (4.210)$$

$$b\left(q^h, \vec{u}_0^h\right) = f^2(q^h) \ \ \forall \ \ q^h \in \mathcal{W}^h \qquad \Longleftrightarrow$$

$$c\left(\vec{x}_0^h, \vec{y}^h\right) = \vec{f}(\vec{y}^h) \ \forall \ \vec{y}^h = (\vec{v}^h, q^h) \in \mathcal{X}^h \quad \Longleftrightarrow \ C^h \vec{x}_0^h = \vec{f}^h. \quad (4.211)$$

In Section 2.8 we introduced in (2.507) the closed subspace $\mathcal{V}_0 \subset \mathcal{V}_b$. This $\mathcal{V}_0$ and the corresponding discrete spaces $\mathcal{V}_0^h$ are

$$\mathcal{V}_0 = \{\vec{v} \in \mathcal{V}_b : b(q, \vec{v}) = 0 \forall q \in \mathcal{W}\}, \mathcal{V}_0^h = \left\{\vec{v^h} \in \mathcal{V}_b^h : b\left(q^h, \vec{v^h}\right) = 0 \forall q^h \in \mathcal{W}^h\right\} \not\subset \mathcal{V}_0.$$

For $f^2 \equiv 0$ every solution u_0^h of (4.209)–(4.211)

$$\vec{u}_0^h \in \mathcal{V}_0^h \text{ solves } a\left(\vec{u}_0^h, \vec{v}^h\right) = \vec{f}^1(\vec{v}^h) \ \ \forall \ \ \vec{v}^h \in \mathcal{V}_0^h. \quad (4.212)$$

Now $\vec{v^h} \in \mathcal{V}_0^h \not\subset \mathcal{V}_0$ usually violates div $\vec{v^h} = 0$. So this form (4.212) is, cf. [386], a *nonconforming FEM* for

$$\vec{u}_0 \in \mathcal{V}_0 \text{ solves } a(\vec{u}_0, \vec{v}) = \vec{f}^1(\vec{v}) \ \ \forall \ \ \vec{v} \in \mathcal{V}_0. \quad (4.213)$$

This observation is the starting point for intensive research combining mixed FEMs with nonconformity.

Here we use (4.209)–(4.211) directly for the conforming FEMs for our saddle point problems. According to Theorem 4.53 we need $C^{-1} \in \mathcal{L}(\mathcal{X}', \mathcal{X})$, see Theorem 3.29, for the unique existence of an exact solution, u_0, the consistency, a consequence of Theorem 4.54 and the approximating subspaces $\mathcal{V}_b^h, \mathcal{W}^h, \mathcal{X}^h$ for $\mathcal{V}_b, \mathcal{W}, \mathcal{X}$, and the stability of C^h in (4.211). The latter is related to the *Ladyženskaja–Brezzi–Babuska condition.* It has the form, [387], Theorem 12.3.6,

$$\inf\left\{\sup\left\{|a(\vec{u}^h, \vec{v}^h)| : \vec{v}^h \in \mathcal{V}_0^h, \|\vec{v}^h\|_\mathcal{V} = 1\right\} : \vec{u}^h \in \mathcal{V}_0^h, \|\vec{u}^h\|_\mathcal{V} = 1\right\} = \alpha_h > 0,$$

$$\inf\left\{\sup\left\{|b(q^h, \vec{v}^h)| : \vec{v}^h \in \mathcal{V}_b^h, \|\vec{v}^h\|_\mathcal{V} = 1\right\} : q^h \in \mathcal{W}^h, \|q^h\|_\mathcal{W} = 1\right\} = \beta_h > 0. \quad (4.214)$$

For these saddle point problems, Hackbusch's Theorem 12.3.6 [386] yields:

Theorem 4.70. Unique discrete solutions and their convergence for saddle point problems:

1. *Let $\Omega \in C^{0,1}$, let the $a(\cdot,\cdot), b(\cdot,\cdot)$ and $\vec{f}_1, f_2$, in (4.206) be continuous bilinear and linear forms on $\mathcal{V}_b, \mathcal{W}$, choose the approximating spaces $(\mathcal{X}^h)_{h\in H}$ and the mixed FEM as in (4.209)–(4.211), and let (4.214) be satisfied. Then the discrete problem (4.210) is uniquely solvable.*

2. *If in (4.214) additionally* $\alpha_h \geq \alpha > 0$, $\beta_h \geq \beta < 0 \ \forall\ h \in H$, *then* C^h *in (4.211) is stable and, with* $C_h = C_h(\alpha, \beta, C_a, C_b)$ *and* C_a, C_b *in (4.205),*

$$\left(\|\vec{u}_0^h\|_{\mathcal{V}}^2 + \|p_0^h\|_{\mathcal{W}}^2\right)^{1/2} \leq C_h \left(\|\vec{f}_1\|_{\mathcal{V}_b'}^2 + \|f_2\|_{\mathcal{W}'}^2\right)^{1/2}, \quad and \tag{4.215}$$

$$\|\vec{u}_0 - \vec{u}_0^h\|_{\mathcal{V}} + \|p_0 - p_0^h\|_{\mathcal{W}} \leq C\left(\inf_{\vec{v}^h \in \mathcal{V}_b^h} \|\vec{u}_0 - \vec{v}^h\|_{\mathcal{V}} + \inf_{q^h \in \mathcal{W}^h} \|p_0 - q^h\|_{\mathcal{W}}\right)$$

$$e.g. \ \leq Ch(\|\vec{u}_0\|_{H^2(\Omega)} + \|p_0\|_{H^1(\Omega)})$$

by (4.220) for the approximations in Examples 4.71 and 4.72

For the specific case of the Stokes operator we only present three examples for $\mathcal{X}^h = \mathcal{V}_b^h \times \mathcal{W}^h$, see [386], satisfying the above conditions. For other examples cf. e.g. Brenner and Scott [141], Brezzi and Fortin [144], Grossmann and Roos [374], Girault and Raviart [348, 349] and Temam [622, 623].

Let $\Omega \subset \mathbb{R}^2$ be a polygon, hence $\mathcal{V}_b = H_0^1(\Omega) \times H_0^1(\Omega)$ and $\mathcal{T}^h$ a quasiuniform triangulation for Ω, such that, there exists a fixed $\sigma_0 > 0$ with the quotient $h/\dim T \leq \sigma_0$ for every triangle $T \in \mathcal{T}^h$.

For quadrature approximations or curved $\partial\Omega$ the modifications from Sections 5.2–5.5 can be used.

Example 4.71. *Piecewise linear elements and bubble functions*:
We define

$$\mathcal{W}^h := \left\{q^h : \ \forall\ T \in \mathcal{T}^h : q^h|_T \in \mathcal{P}_1, \ \int_\Omega q^h dx = 0\right\} \subset L_*^2(\Omega). \tag{4.216}$$

To introduce the bubble functions $\vec{u}$ on $\mathcal{T}^h$ let $\tilde{T}$ represent the reference triangle $\tilde{T} = \{(\xi\eta) : \xi, \eta > 0, \xi + \eta < 1\}$ and Φ the bijective affine mapping $\Phi : \tilde{T} \to T$. We define, for every $T \in \mathcal{T}^h$ and the components u of $\vec{u}$

$$\tilde{u}(\xi,\eta) := \xi \cdot \eta(1 - \xi - \eta) \text{ for } (\xi,\eta) \in \tilde{T}, \quad \text{and } := 0 \text{ otherwise,} \tag{4.217}$$

$$u_T(x,y) := \tilde{u}(\Phi^{-1}(x,y)) \text{ for } (x,y) \in T \text{ and } := 0 \text{ otherwise, } \vec{u}_T \in \left(H_0^1(\Omega)\right)^2 \text{ and}$$

$$\mathcal{V}_{b1}^h := \{v^h \in H_0^1(\Omega) : \ \forall\ T \in \mathcal{T}^h : v|_T^h \text{ linear combinations of } \mathcal{P}_1$$

$$\text{and bubble functions on } \mathcal{T}^h\}, \quad \mathcal{V}_b^h := \mathcal{V}_{b1}^h \times \mathcal{V}_{b1}^h, \ \mathcal{X}^h = \mathcal{V}_b^h \times \mathcal{W}^h. \qquad \blacksquare$$

Example 4.72.

1. *Piecewise linear elements on* $\mathcal{T}_{h/2}^h$ *and* $\mathcal{T}^h$: Let $\mathcal{T}_{h/2}^h$ be defined by replacing each $T \in \mathcal{T}^h$ by four congruent subtriangles. Then

$$\mathcal{V}_{b1}^h := \left\{v \in H_0^1(\Omega) : \text{ linear elements on } \mathcal{T}_{h/2}^h\right\}, \quad \mathcal{V}_b^h := \mathcal{V}_{b1}^h \times \mathcal{V}_{b1}^h, \tag{4.218}$$

$$\mathcal{W}^h := \{q \in L_*^2(\Omega) : \text{ linear elements on } \mathcal{T}^h\}, \quad \mathcal{X}^h = \mathcal{V}_b^h \times \mathcal{W}^h.$$

2. *Piecewise quadratic and linear elements on* $\mathcal{T}^h$: Let

$$\mathcal{V}_{b,2}^h := \left\{ v \in H_0^1(\Omega) : \text{quadratic elements on } \mathcal{T}^h \right\}, \mathcal{V}_b^h := \mathcal{V}_{b,2}^h \times \mathcal{V}_{b,2}^h, \qquad (4.219)$$
$$\mathcal{W}^h := \left\{ q \in L_*^2(\Omega :) \text{ linear elements on } \mathcal{T}^h \right\}, \quad \mathcal{X}^h = \mathcal{V}_{b,2}^h \times \mathcal{W}^h. \qquad \blacksquare$$

These approximations (4.217)–(4.219) satisfy

$$\inf \left\{ \|v - v^h\|_{\mathcal{V}} : v^h \in \mathcal{V}_b^h \right\} \le h\|v\|_{H^2(\Omega)} \forall v \in \mathcal{V}_b \qquad (4.220)$$
$$\inf\{\|p - p^h\|_{\mathcal{W}} : p^h \in \mathcal{W}^h\} \le h\|p\|_{H^1(\Omega)} \forall p \in \mathcal{W}.$$

The following theorem, cf. Hackbusch [387], Theorem 12.3.11, needs an Ω, such that the Poisson problem is $H^2(\Omega)$-regular. That means, the unique solution, u_0, satisfies

$$u_0 \in H_0^1(\Omega) : a(u_0, v) = \int_\Omega (\nabla u_0, \nabla v)_n dx = \int_\Omega f v dx \forall v \in H_0^1(\Omega) \qquad (4.221)$$
$$\text{and } f \in L^2(\Omega) \Longrightarrow u_0 \in H^2(\Omega). \qquad (4.222)$$

By Theorems 2.47, 2.131, this is correct if Ω is convex. We summarize [387], Theorems 12.3.11–12.3.15 and Corollary 12.3.16:

Theorem 4.73. Unique discrete solutions and their convergence for the Stokes problem:

1. *Let* $\Omega \subset \mathbb{R}^2$ *be a bounded polygon, hence* $\Omega \in C^{0,1}$, *let* $\mathcal{V}_b = \left(H_0^1(\Omega)\right)^2$ *and* $\mathcal{T}^h$ *be a quasiuniform triangulation for* Ω *and let (4.221) be satisfied, e.g. for a convex* Ω. *Then Examples 4.71, 4.72 satisfy (4.214) with* $\alpha^h > \alpha > 0, \beta^h > \beta > 0$.
2. *Then we obtain the following convergence estimates for the discrete solutions* $\vec{u}_0^h, p_0^h$ *of (4.210), (4.211), toward the exact solution of (4.202). For* $p_0^h - p_0$ *we use the dual norm* $\left\|p_0^h - p_0\right\|_{(H^1(\Omega)\cap\mathcal{W})^*}$ *with respect to* $\mathcal{W} \cap H^1(\Omega)$. *The convergence is of order* 1 *or* 2, *depending upon the smoothness of the solutions:*

$$\left\|\vec{u}_0^h - \vec{u}_0\right\|_{(L^2(\Omega))^2} + \left\|p_0^h - p_0\right\|_{(H^1(\Omega)\cap\mathcal{W})^*} \le C^* h\big(\|\vec{u}_0\|_{(H^1(\Omega))^2} + \|p_0\|_{L^2(\Omega)}\big)$$
$$\le C^{**} h^2 \big(\|\vec{u}_0\|_{(H^2(\Omega))^2} + \|p_0\|_{H^1(\Omega)}\big) \qquad (4.223)$$
$$\left\|\vec{u}_0^h - \vec{u}_0\right\|_{(L^2(\Omega))^2} + h\left\|p_0^h - p_0\right\|_{L^2(\Omega)} \le C^{***} h^2 \big(\|\vec{u}_0\|_{(H^2(\Omega))^2} + \|p_0\|_{H^1(\Omega)}\big).$$

Proof. We only show that the additional condition in [387], Theorem 12.3.15, is a consequence of (4.221). In fact, for $f^2 \in \mathcal{W} \cap H^1(\Omega)$ we can determine a $\vec{u}_1 \in \mathcal{W} \cap H^2(\Omega)$, such that div $\vec{u}_1 = f^2$. Then $\vec{u} - \vec{u}_1$ satisfies $\vec{f}_3 := \Delta(\vec{u} - \vec{u}_1) \in (L^2(\Omega))^2$. So we have to discuss for every $\vec{f} - \vec{f}_3 \in (L^2(\Omega))^2$ and $f^2 = 0 \in \mathcal{W} \cap H^1(\Omega)$ the Stokes problem $\Delta\vec{u} + \nabla p = \vec{f}$, div $\vec{u} = 0.0$ By (4.221) this has a solution $\vec{u}_0 \in (H^2(\Omega))^2 \cap \left(H_0^1(\Omega)\right)^2$, $p_0 \in H^1(\Omega)^2 \cap \mathcal{W}$, cf. Theorem 2.131, with

$$\|\vec{u}_0\|_{(H^2(\Omega))^2} + \|p_0\|_{H^1(\Omega)} \le C\big(\|\vec{f} - \vec{f}_3\|_{(L^2(\Omega))^2}\big),$$

and we are done. $\square$

4.6.3 Mixed FEMs for the Navier–Stokes operator

Now we discuss the linearized Navier–Stokes operator again for $\Omega \subset \mathbb{R}^n$ and follow [349, 623, 623]. Again we multiply by the test function $\vec{v} = (v_1, \ldots, v_n)$ and use the Green formula to obtain, with $a(\cdot,\cdot), b(\cdot,\cdot), \mathcal{V}_b, \mathcal{W}$ in (4.204)

$$\text{For given } (\vec{f}^1, 0) \in \mathcal{V}_b' \times \mathcal{W}' \text{ determine } \vec{u}_0 \in \mathcal{V}_b, p_0 \in \mathcal{W} \text{ such that}$$

$$\int_\Omega \left(-\nu\Delta\vec{u}_0 + \sum_{i=1}^n (\vec{u}_0)_i \partial_i \vec{u}_0 + \text{grad } p_0, \vec{v} \right)_n dx \tag{4.224}$$

$$= \nu a(\vec{u}_0, \vec{v}) + d(\vec{u}_0, \vec{u}_0, \vec{v}) + b(p_0, \vec{v}) = \vec{f}^1(\vec{v}) \quad \forall \ \vec{v} \in \mathcal{V}_b \text{ and}$$

$$b(p_0, \vec{v}) = 0 \quad \forall \ \vec{v} \in \mathcal{V}_b; \text{ here}$$

$$d(\vec{u}, \vec{v}, \vec{w}) := \sum_{i,j=1}^n \int_\Omega \vec{u}_i (\partial_i v_j) w_j dx.$$

For bounded Ω, and $n \le 4$, see Temam [623] Lemma 1.2, Chapter II, Section 1,

$$d(\vec{u}, \vec{v}, \vec{w}) \text{ is a bounded trilinear form on } \mathcal{V}_b \times \mathcal{V}_b \times \mathcal{V}_b.$$

To linearize we consider for fixed $\vec{u}, \vec{v}$ and small $\vec{w}$

$$d(\vec{u} + \vec{w}, \vec{u} + \vec{w}, \vec{v}) - d(\vec{u}, \vec{u}, \vec{v}) = d(\vec{u}, \vec{w}, \vec{v}) + d(\vec{w}, \vec{u}, \vec{v}) + o(\vec{w}).$$

An analogous results holds for the nonlinear (4.201) with (4.224): We obtain

$$G'(\vec{u}, p)(\vec{w}, r) = \begin{pmatrix} -\nu\Delta\vec{w} + \sum_{i=1}^n (u_i \partial_i \vec{w} + w_i \partial_i \vec{u}) + grad\ r \\ \text{div } \vec{w} \end{pmatrix} \tag{4.225}$$

and, with the $a(\cdot,\cdot), b(\cdot,\cdot), d(\cdot,\cdot)$ in (4.204) and (4.224).

$$\big((G'(\vec{u}, p)(\vec{w}, r)), (\vec{v}, q)\big)_{(L^2(\Omega))^2} := \begin{pmatrix} \nu a(\vec{w}, \vec{v}) + d(\vec{u}, \vec{w}, \vec{v}) + d(\vec{w}, \vec{u}, \vec{v}) + b(r, \vec{v}) \\ b(q, \vec{w}) \end{pmatrix}$$

$$\forall (\vec{v}, q) \in \mathcal{V}_b \times \mathcal{W} = \mathcal{X}. \tag{4.226}$$

Now we consider the $d(\vec{u}, \vec{w}, \vec{v}), d(\vec{w}, \vec{u}, \vec{v})$

$$d(\vec{u}, \vec{w}, \vec{v}) = \sum_{i,j=1}^n \int_\Omega u_i (\partial_i w_j) v_j dx, \quad \text{for } \vec{u}, \vec{v}, \vec{w} \in \mathcal{V}_b \tag{4.227}$$

$$d(\vec{w}, \vec{u}, \vec{v}) = \sum_{i,j=1}^n \int_\Omega w_i (\partial_i u_j) v_j dx = -\sum_{i,j=1}^n \int_\Omega w_i u_j \partial_i v_j dx.$$

For fixed $\vec{u}$, the $d(\vec{u}, \vec{w}, \vec{v}) + d(\vec{w}, \vec{u}, \vec{v})$ are continuous bilinear forms, corresponding to the (sum of) operator(s) $\sum_{i,j=1}^n (u_i \partial_i \vec{w} + w_i \partial_i \vec{u})$ in (4.225). This is, for fixed $\vec{u}$, linear and bounded in $\vec{v}, \vec{w} \in \mathcal{V}_b$. Then the continuous bilinear forms $d(\vec{u}, \vec{w}, \vec{v})$ and

$d(\vec{w}, \vec{u}, \vec{v}) \in \mathbb{R}$ define elements

$$d(\vec{u}, \vec{w}, \cdot), d(\vec{w}, \vec{u}, \cdot) \in \mathcal{V}_b',$$

hence define linear continuous operators

$$D_1, D_2 \in \mathcal{L}\left(\mathcal{V}_b, \mathcal{V}_b'\right) \text{ as } D_1\vec{w} := d(\vec{u}, \vec{w}, \cdot), \quad D_2\vec{w} := d(\vec{w}, \vec{u}, \cdot), \text{ for } \vec{u} \text{ fixed.} \tag{4.228}$$

By (4.224), note that $\vec{u}$ is fixed, see (4.227)

$$\langle D_1\vec{w}, \vec{v}\rangle_{\mathcal{V}_b' \times \mathcal{V}_b} = d(\vec{u}, \vec{w}, \vec{v}), \quad \langle D_2\vec{w}, \vec{v}\rangle_{\mathcal{V}_b' \times \mathcal{V}_b} = d(\vec{w}, \vec{u}, \vec{v}). \tag{4.229}$$

Now the embedding $I : H_0^1(\Omega) \to L^2(\Omega)$ is continuous and compact, cf. Theorem 1.26, and (4.229) shows that

$$D_1\vec{v} = D_1 I\vec{v} \;\forall\; \vec{v} \in H_0^1(\Omega).$$

Hence, as a product of a compact and a continuous operator, $D_1 = D_1 I$ is a compact operator. The same is correct for D_2 as well.

Similarly to the transformation from (4.206) to (4.208) we use here, with the factor ν in (4.201),

$$C := \begin{pmatrix} \nu A & B \\ B^d & 0 \end{pmatrix}, D := \begin{pmatrix} D_1 + D_2 & 0 \\ 0 & 0 \end{pmatrix}, \tag{4.230}$$

and, with slightly different $\vec{x} = (\vec{w}, r), \vec{y} = (\vec{v}, q)$,

$$(C + D)\vec{x} = \begin{pmatrix} \nu A\vec{w} + Br + D_1\vec{w} + D_2\vec{w} \\ B^d\vec{w} \end{pmatrix},$$

$$\begin{aligned} \langle (C + D)\vec{x}, \vec{y}\rangle_{\mathcal{X}' \times \mathcal{X}} &= \nu\langle A\vec{w}, \vec{v}\rangle_{\mathcal{V}_b' \times \mathcal{V}_b} + \langle Br, \vec{v}\rangle_{\mathcal{V}_b' \times \mathcal{V}_b} \\ &\quad + \langle D_1\vec{w} + D_2\vec{w}, \vec{v}\rangle_{\mathcal{V}_b' \times \mathcal{V}_b} + \langle B^d\vec{w}, q\rangle_{\mathcal{W}' \times \mathcal{W}}. \end{aligned}$$

We formulate

$$\begin{aligned} &\text{for given } \vec{f} \in \mathcal{X}' \text{ determine } \vec{x}_0 = (\vec{w}_0, r_0) \text{ such that} \\ &\langle (C + D)\vec{x}_0, \vec{y}\rangle_{\mathcal{X}' \times \mathcal{X}} = \langle \vec{f}, \vec{y}\rangle_{\mathcal{X}' \times \mathcal{X}} \quad \forall\; \vec{y} = (\vec{v}, q) \in \mathcal{X} \\ &\qquad \text{or equivalently } (C + D)\vec{x}_0 = \vec{f} \in \mathcal{X}'. \end{aligned} \tag{4.231}$$

Again Remark 3.30 applies. So we can guarantee the existence of the inverse of $C + D$ if -1 is not an eigenvalue of the compact operator $C^{-1}D$. The corresponding mixed FEMs are then defined by straightforward modification of the above (4.209)–(4.211) for (4.230)–(4.231).

The mixed FEM for the nonlinear Navier–Stokes equation requires: Find a pair $\left(u_0^h, p_0^h\right) \in \mathcal{V}_b^n \times \mathcal{W}^h$ such that

$$\begin{aligned} \nu a\left(u_0^h, v^h\right) + d\left(u_0^h, u_0^h, v^h\right) + b\left(v^h, p_0^h\right) &= \langle f, v^h\rangle \quad \text{for all } v^h \in \mathcal{V}_b^h, \\ b\left(u_0^h, q^h\right) &= \langle g, q^h\rangle \quad \text{for all } q^h \in \mathcal{W}^h. \end{aligned} \tag{4.232}$$

Combining this with Summary 4.52 we obtain the following.

Theorem 4.74. Convergence of mixed FEMs for the (nonlinearized) Navier–Stokes equations, and solution of the nonlinear systems, cf. Theorems 4.70 and 4.73:

1. *For $\Omega \in C^{0,1}$, the approximating spaces $\left(\mathcal{X}^h = \left(\mathcal{V}_b^h, \mathcal{W}^h\right)\right)_{h\in H}$ and the mixed FEM for (4.230)–(4.231) analogous to (4.209)–(4.211), let (4.214) be satisfied, and let $G'(\vec{u}, p) : \mathcal{X} \to \mathcal{X}'$ be boundedly invertible. Then the mixed FE equations (4.230)–(4.231) are uniquely solvable.*
2. *If additionally $\alpha^h \geq \alpha > 0, \beta^h \geq \beta > 0$, then (4.230)–(4.231) are stable, consistent, and the discrete solutions converge as in Theorem 4.70.*
3. *Let $\Omega \subset \mathbb{R}^2$ be a bounded polygon, hence $\Omega \in C^{0,1}$, let $\mathcal{V}_b = \left(H_0^1(\Omega)\right)^2$ and $\mathcal{T}^h$ be a quasiuniform triangulation for Ω and let (4.221) be satisfied, e.g. for a convex Ω. Then Examples 4.71, 4.72 satisfy (4.214) with $\alpha^h > \alpha > 0, \beta^h > \beta > 0$.*
4. *Then we obtain the following convergence estimates for the discrete solutions $\vec{u}_0^h, p_0^h$ of (4.210), (4.211), and of (4.232) toward the exact solution of the Stokes, the linearized Navier–Stokes and the nonlinear Navier–Stokes equations (4.202), (4.231), and (4.224). For $p_0^h - p_0$ we use the dual norm $\|p_0^h - p_0\|_{(H^1(\Omega)\cap\mathcal{W})^*}$ with respect to $\mathcal{W} \cap H^1(\Omega)$. The convergence is of order 1 or 2, depending upon the smoothness of the solutions:*

$$\begin{aligned}\left\|\vec{u}_0^h - \vec{u}_0\right\|_{(L^2(\Omega))^2} + \left\|p_0^h - p_0\right\|_{(H^1(\Omega)\cap\mathcal{W})^*} &\leq C^* h\left(\|\vec{u}_0\|_{(H^1(\Omega))^2} + \|p_0\|_{L^2(\Omega)}\right)\\ &\leq C^{**} h^2\left(\|\vec{u}_0\|_{(H^2(\Omega))^2} + \|p_0\|_{H^1(\Omega)}\right) \qquad (4.233)\\ \left\|\vec{u}_0^h - \vec{u}_0\right\|_{(L^2(\Omega))^2} + h\left\|p_0^h - p_0\right\|_{L^2(\Omega)} &\leq C^{***} h^2\left(\|\vec{u}_0\|_{(H^2(\Omega))^2} + \|p_0\|_{H^1(\Omega)}\right).\end{aligned}$$

5. *The numerical solution of the high dimensional nonlinear systems in this theorem is efficiently computed by combining continuation and discrete Newton's methods, cf. Theorem 3.40.*

Proof. We again apply Summary 4.52. The consistency is proved in Theorem 4.54. The stability follows from Theorem 4.57. Hence, Theorem 4.53 implies the unique existence of discrete solutions, their convergence and the quadratic convergence of the discrete Newton's method, cf. Theorem 3.40. □

Other types of FEMs based upon nonlinear stability, bubbles, and a posteriori estimates are studied by Brezzi, Hughes, Marini, Houston, Rannacher, Russo, and Süli. [145, 406, 612].

4.7 Variational methods for eigenvalue problems

4.7.1 Introduction

Eigenvalue problems play an important role in many scientific and technical applications. Examples are stability, buckling and breaking, bifurcation of, e.g. mechanical structures, and originating of periodic solutions and vibration, e.g. in reaction–diffusion models, and biological, geophysical, fluid dynamical, phenomena, cf. Böhmer [120]. A detailed discussion of important applications, emphasizing linear problems, is presented by e.g., Dautray and Lions [256–261], and nonlinear problems by Zeidler [675–678].

We study only *eigenvalue problems for linear operators* for two reasons. The numerical methods for nonlinear eigenvalue problems are far less developed than for the linear case. Our main motivation for including eigenvalue problems is the computation of a bifurcation as a starting point for some types of bifurcation and/or evolving dynamical scenarios. If G depends upon a parameter, these scenarios are related to the eigenvalues of its Fréchet derivative $G'(u_0)$, evaluated at u_0.

In this section we formulate the results for eigenvalue problems in this chapter on conforming FEMs, based upon the variational characterization of the eigenvalues. However this remains correct for the different variational methods, presented in this book and in [120]. So all the following results remain valid for conforming and nonconforming FEMs, the DCGMs, difference, wavelet, mesh-free and spectral methods. This is due to the unified treatment of these methods, based upon their similar structure.

In fact Hackbusch's [387], Chapter 11, proofs for the eigenvalue results remain valid with extremely minor changes. So we only indicate these modifications and omit the proofs. So we generalize his results from elliptic equations of order 2 and conforming FEMs to equations and systems of order 2 and $2m$ and all the above methods. For nonconforming FEMs and DCGMs the results are restricted to $m = 1$.

For *finite difference methods,* Hackbusch [387], Chapter 11, employed several technical definitions and lemmas for proving convergence for one second order equation. Here we only refer to these results in [387], but do not formulate or even generalize them. Partly these results are included or modified in our variational methods for eigenvalues and difference methods.

There are many interesting papers and books on problems with higher eigenvalues as well. Without trying to present a complete list, we give some examples: Ahues [5], Alimov and Pulatov [7], Arias [38], Aslanov [43], Banerjee and Suri [57], Bramble and Osborn [140], Osborn [521], Parlett *et al.* [514, 523, 524], Chatelin [167, 168], Babuska and Osborn [46–48], Ciarlet and Lions [177], Davies [262], Delic *et al.* [279], Dencker *et al.* [282], Graham *et al.* [365], Jörgens [427], Levendorskij [474], Mertins [488, 489], Saad [562, 563], Schechter [570], Sun [614], Vanmaele and Zenísek [651], Vanmaele and Van Keer [650] Beattie [76], and Toyonaga *et al.* [630].

We only consider eigenvalue problems with simple eigenvalues. The study of multiple eigenvalues and their numerical methods is much less advanced than for simple eigenvalues. The main reason for our restriction here is the fact that the numerical study here yields the tools for determining bifurcation. In this context, the invariant subspaces, a generalization of eigenspaces of higher eigenvalues, are important. So this class of problems is delayed to Böhmer [120]. Again we quote some papers and books: Beyn [92], Beyn *et al.* [93, 94], Bošek [133],Cliffe *et al.* [179], Dieci and Friedman [287], Govaerts [357], Joshi [429]. Sebastian [579] and Schwarzer [577].

4.7.2 Theory for eigenvalue problems

The classical EVP for elliptic PDEs is here defined by a linear elliptic operator A of order $2m, m \geq 1$, and its boundary operators B_j as

$$\text{determine } \lambda \in \mathbb{C} \text{ s.t.} \exists e \neq 0 : Ae = \lambda e \text{ in } \Omega . B_j e = 0, j = 0, \ldots, m-1, \text{on } \partial\Omega. \quad (4.234)$$

So for a linear operator, A, we determine those nontrivial elements $0 \neq e \in \mathcal{V}$ which are mapped into a multiple of themselves.

This A represents either one equation or a system of q equations. For a simplified notation we usually employ the same symbols $e, u, v, \ldots$ for the functions in equations and systems and avoid the $\vec{e}, \vec{u}, \vec{v}, \ldots$ form for systems. Similarly, $W^{m,p}(\Omega)$ indicates $W^{m,p}(\Omega)$ and $W^{m,p}(\Omega, \mathbb{R}^q)$ as well. We include the boundary conditions in $\mathcal{V}$, hence $\forall v \in \mathcal{V}_b \iff B_j v = 0, j = 0, \ldots, m-1$. Again we consider the bounded bilinear form $a(\cdot,\cdot)$, induced by A as in Section 2.3, and the Gelfand triple $\mathcal{V}_b \subset \mathcal{W} \subset \mathcal{V}_b'$, cf. (2.62). Then we

$$\text{determine } \lambda \in \mathbb{C} \text{ s.t. } \exists 0 \neq e : a(e,v) = \lambda(e,v)_{\mathcal{W}} \ \forall\, v \in \mathcal{V}_b \subset \mathcal{W} \subset \mathcal{V}_b', \quad (4.235)$$

usually with $(u,v)_{\mathcal{W}} = \int_\Omega u\bar{v}dx$. For discussing eigenvalue problems appropriately, we have to *complexify* $\mathcal{V}, \mathcal{W}, A$, cf. Definition 1.21, Theorem 1.22. The adjoint eigenvalue problem requires

$$\text{determine } \lambda \in \mathbb{C} \text{ s.t. } \exists 0 \neq e : a(v,e) = \bar{\lambda}(v,e)_{\mathcal{W}} \ \forall\, v \in \mathcal{V}_b \subset \mathcal{W} \subset \mathcal{V}_b'. \quad (4.236)$$

The different A and $a(\cdot,\cdot)$ are introduced for equations of order 2, in (2.16), (2.18), of order $2m$ in (2.76), (2.88), (2.102), (2.110), for systems of order 2 in (2.339), (2.340), and order $2m$ in (2.391), (2.392).

In Theorems 2.20 and 2.21 we have already introduced the concepts of *eigenvalues, eigenfunctions,* and *eigenspaces* $\mathcal{E}(\lambda)$ for A and for $a(\cdot,\cdot)$, and $\mathcal{E}^d(\bar{\lambda})$ for A^d. Note that we use the notation $\mathcal{E}^d(\bar{\lambda})$ in contrast to $\mathcal{E}^d(\lambda)$ in Hackbusch [387].

Example 4.75. *Chladny sound figures*

This nonlinear problem satisfies, with a parameter λ, cf. Example 1.3, (1.16),

$$G(u) = \Delta u + \lambda \sin u = 0 \text{ in } \Omega = [0,L] \times [0,L] \text{ and } \frac{\partial u}{\partial n} = 0 \text{ on } \partial\Omega. \quad (4.237)$$

We linearize this equation with respect to u at $u = 0$ and apply $G'(u)$ to v, cf. (1.18), obtaining

$$G'(0)v = \Delta v + \lambda v = 0 \iff \Delta v = -\lambda v \text{ in } \Omega \text{ with } \frac{\partial u}{\partial n} = 0 \text{ on } \partial\Omega. \quad (4.238)$$

The $-\lambda$, allowing nontrivial solutions of (4.238), are the *eigenvalues,* $\lambda_0 = -\lambda$, of the Laplacian with the corresponding *eigenfunctions,* $v_{m,n}, v_{n,m}$,

the eigenvalues $\lambda_0 = -(m^2+n^2)\pi^2/L^2 \ \ \forall\, m,n \in \mathbb{N}_0$ with the eigenfunctions,

$$v(x,y) = v_{m,n}(x,y) := \cos\frac{m}{L}\pi x\, \cos\frac{n}{L}\pi y, \text{ and } v(x,y) = v_{n,m}(x,y). \quad (4.239)$$

Therefore, we obtain, e.g. for $(m,n) = (1,1)$, $(m,n) = (2,3)$ and $(m,n) = (5,5)$, $(m,n) = (1,7)$ one, two and three linearly independent solutions $v_{m,n}, v_{n,m}$ for the simple, double and triple eigenvalues $\lambda_0 L^2/\pi^2 = -2, -13$, and -50, respectively. ■

The relevant theoretical results for eigenvalue problems in elliptic PDEs are summarized in the Riesz–Schauder theory in Theorem 2.20 and the *Fredholm alternative* in Theorem 2.21, the latter formulated for the operator A. Here, we essentially use the bilinear form $a(\cdot,\cdot)$, and recapitulate the most important results.

Theorem 4.76. Alternatives for eigenvalue problems: *We assume a Gelfand triple, see (2.62), $\mathcal{V}_b \subset \mathcal{W} = \mathcal{W}' \subset \mathcal{V}_b'$ with a continuous dense and additionally compact embedding $\mathcal{V}_b \hookrightarrow \mathcal{W}$, a reflexive Banach space $\mathcal{V}_b$ and a Hilbert space $\mathcal{W}$. For the bounded elliptic bilinear form $a(\cdot,\cdot) : \mathcal{V}_b \times \mathcal{V}_b \to \mathbb{R}$ we get:*

1. *For every $\lambda \in \mathbb{C}$ one of the following alternatives is valid: either the induced*
   (i) $(A - \lambda I)^{-1} \in \mathcal{L}(\mathcal{V}_b', \mathcal{V}_b)$*;* or
   (ii) *λ is an eigenvalue of A or of $a(\cdot,\cdot)$.*
2. *λ is an eigenvalue for (4.235) $\Longleftrightarrow$ $\bar{\lambda}$ is an eigenvalue for (4.236). Then the finite dimensional* eigenspaces *$\mathcal{E}(\lambda)$ and $\mathcal{E}^d(\bar{\lambda})$ satisfy* $\dim \mathcal{E}(\lambda) = \dim \mathcal{E}^d(\bar{\lambda}) < \infty$.
3. *The* spectrum *of $a(\cdot,\cdot)$, or A, the $\sigma(A) := \{\lambda$ eigenvalue for $A\}$, is at most countable and admits no limit points in $\mathbb{C}$.*
4. *If $a(\cdot,\cdot)$ is symmetric, hence $a(u,v) = a(v,u) \forall u,v \in \mathcal{V}_b$, then the eigenvalues $\lambda = \bar{\lambda}$ in (4.235) and (4.236) are real and equal and $\mathcal{E}(\lambda) = \mathcal{E}^d(\bar{\lambda}) = \mathcal{E}^d(\lambda)$.*

From here on, we proceed in line with the presentation in Hackbusch [386, 387]. As indicated above, the reason is the following. His proofs for the results for the eigenvalues of A and their numerical approximation are based upon the structural results for $\mathcal{V}_b$-coercive and $\mathcal{V}_b$-elliptic bilinear forms and regularity results for the solutions of linear elliptic problems. So they remain valid for our more general situation as well. Note that Hackbusch's definition of $\mathcal{V}_b$-coercive and $\mathcal{V}_b$-elliptic bilinear forms is chosen according to the originally more analytic version. We have exchanged their role, according to the usual version in FEMs. So we formulate his results for our more general case of $\mathcal{V}_b$-coercive and $\mathcal{V}_b$-elliptic bilinear forms with $\mathcal{V}_b = W^{m,p}(\Omega, \mathbb{R}^q), m \geq 1, q \geq 1, 2 \leq p < \infty$, and again exclude $1 \leq p < 2$, cf. Sections 4.4 and 4.5. Instead of repeating Hackbusch's proofs, we indicate which of his tools (exercises, lemmas, theorems) have to be replaced by our results. We only formulate those results, that are important for the regularity of the exact solutions, and the computation and convergence of the numerical approximations of eigenvalues and eigenfunctions. For Hackbusch's other results, necessary for his proofs, we only indicate the corresponding replacement of his tools by ours. Note that we only formulate these results for Dirichlet, but not for natural boundary conditions. Correspondingly, our results are only valid for these cases, cf. [386, 387], Theorem 11.1.5:

Theorem 4.77. Regular eigenfunctions: *Assume $\mathcal{V}_b = W_0^{m,p}(\Omega, \mathbb{R}^q), m, q \geq 1, 2 \leq p < \infty$, and the conditions for regular solutions for different types of problems in Theorems 2.28, 2.31, 2.32, 2.33, and 2.38,2.39, and 2.45,2.47,2.49 for equations, and 2.91, 2.92, and 2.108 for systems of order* 2m *and* 2 *and* 2m*, respectively Then for appropriate $s > 0$ we find $\mathcal{E}(\lambda) \subset W^{m+s,p}(\Omega, \mathbb{R}^q), m, q \geq 1$, usually omitting $,\mathbb{R}^q$.*

Proof. In Hackbusch's proof we only have to replace his Corollary 9.1.19 by our generalized versions. Hackbusch's (obvious) proof of his Corollary 9.1.19 uses his

Theorem 9.1.16, which is our Theorem 2.45. Similarly the other regularity results can be employed for proving Theorem 4.77. □

4.7.3 Different variational methods for eigenvalue problems

The usual approach in variational methods, applied to (4.235) and (4.236), yields

$$\begin{aligned}&\text{determine } \lambda^h \in \mathbb{C} \text{ s.t. } \exists 0 \neq e^h \in \mathcal{V}_b^h : a^h(e^h, v^h) = \lambda^h (e^h, v^h)_{\mathcal{W}} \ \forall\, v^h \in \mathcal{V}_b^h, \quad (4.240)\\ &\text{determine } \bar{\lambda}^h \in \mathbb{C} \text{ s.t. } \exists 0 \neq e^{h^*} \in \mathcal{V}_b^h : a^h(v^h, e^{h^*}) = \bar{\lambda}^h (v^h, e^{h^*})_{\mathcal{W}} \ \forall\, v^h \in \mathcal{V}_b^h,\\ &\text{with } \mathcal{V}_b^h \subset \mathcal{V}_b \text{ or } \mathcal{V}_b^h \not\subset \mathcal{V}_b \text{ and } (e^h, v^h)_{\mathcal{W}}, (v^h, e^{h^*})_{\mathcal{W}} \text{ defined } \forall\, e^h, e^{h^*}, v^h \in \mathcal{V}_b^h,\end{aligned}$$

with real $\lambda^h = \bar{\lambda}^h$ for a symmetric $a^h(u^h, v^h)$. There is one obvious difficulty: The original eigenvalue problems for elliptic PDEs, (4.235) and (4.236), have infinitely many eigenvalues. Their discrete counterparts, (4.240), represent (generalized) eigenvalue problems of finite dimension, and thus only have finitely many eigenvalues. Furthermore, we cannot expect $\dim \mathcal{E}(\lambda) = \dim \mathcal{E}(\lambda^h)$ for an approximation $\lambda^h \approx \lambda$. In fact the above Example 4.75 yields for a square and a symmetric difference method approximate eigenvalues λ^h with multiplicities different from the exact eigenvalues λ. So for simple or double eigenvalues λ we get $\dim \mathcal{E}(\lambda) = \dim \mathcal{E}(\lambda^h) = k = 1, 2$, with λ^h, near λ. For a multiple eigenvalue λ with $\dim \mathcal{E}(\lambda) = k > 2$ we find $\lambda_j^h \approx \lambda, j = 1, \ldots, k$ with $\dim \mathcal{E}(\lambda) = \sum_{j=1}^k \dim \mathcal{E}\left(\lambda_j^h\right)$. Then only the mean value $\hat{\lambda}^h := \sum_{j=1}^k \lambda_j^h / k$ converges comparably well to λ as for simple eigenvalues, cf. Babuška and Aziz [45], p. 338.

We start with the conditions in Theorem 4.76, cf. Hackbusch [387], (11.2.2a) ff. We have seen in Sections 4.4, 4.5 that a numerical convergence theory for $\mathcal{V}_b = W_0^{m,p}(\Omega), 2 \leq p < \infty$, has to employ the embedding $\mathcal{V}_b \hookrightarrow H^m(\Omega)$, and yields convergence with respect to discrete $H^m(\Omega)$ norms. We assume,

$$\begin{aligned}&\mathcal{V}_b = W_0^{m,p}(\Omega) \hookrightarrow H^m(\Omega), 2 \leq p < \infty,\ a(\cdot,\cdot) : \mathcal{V}_b \times \mathcal{V}_b \to \mathbb{R} \text{ is } H_0^m(\Omega)\text{-elliptic, so}\\ &\qquad \mathcal{V}_b \subset L^2(\Omega) \text{ is continuously, densely and compactly embedded.} \qquad (4.241)\end{aligned}$$

Then all our FE, DCG, difference, wavelet, mesh-free and spectral approximations introduce sequences of approximating spaces with the corresponding conforming or nonconforming norms

$$\begin{aligned}&\mathcal{V}_b \text{ approximated by } \mathcal{V}_b^h \subset H^m(\mathcal{T}^h) \text{ with the norm } \|v - v^h\|_{\mathcal{V}_b^h} := \|v - v^h\|_{H^m(\mathcal{T}^h)}\\ &\qquad \text{and } \forall v \in \mathcal{V}_b : \liminf_{h \to 0} \left\{ \|v - v^h\|_{\mathcal{V}_b^h} : v^h \in \mathcal{V}_b^h \right\} = 0. \qquad (4.242)\end{aligned}$$

We still use the notation $\mathcal{V}_b^h$, standard in this chapter, although for nonconforming cases we might have $\mathcal{V}_b^h \not\subset H_0^1(\Omega)$. Except for conforming FE, difference and wavelet methods, we have to introduce extensions, $a_h(\cdot,\cdot)$, of the original, $a(\cdot,\cdot)$,

$$\begin{aligned}&a_h(\cdot,\cdot) : \left(\mathcal{V}_b \cup \mathcal{V}_b^h\right) \times \left(\mathcal{V}_b \cup \mathcal{V}_b^h\right) \to \mathbb{R} \text{ s.t.} \qquad (4.243)\\ &a(\cdot,\cdot) = a_h(\cdot,\cdot)|_{\mathcal{V}_b \times \mathcal{V}_b} \text{ and } a^h(\cdot,\cdot) = a_h(\cdot,\cdot)|_{\mathcal{V}_b^h \times \mathcal{V}_b^h}.\end{aligned}$$

For conforming FEMs the original $a(\cdot,\cdot)$ suffices, and for difference methods we directly work with the discrete $a^h(\cdot,\cdot)$. So for conforming FEMs we simply omit the index h, for difference methods we only discuss the nonobvious modifications.

But for all cases the $\mathcal{V}_b$-ellipticity of $a(\cdot,\cdot)$ and the $\mathcal{V}_b$-coercivity of its principal part implies the $\mathcal{V}_b^h$-ellipticity of $a^h(\cdot,\cdot)$ and the $\mathcal{V}_b^h$-coercivity of its principal part with respect to the discrete norms $\|v^h\|_{\mathcal{V}_b^h}$ and $\|g^h\|_{\mathcal{W}^h}$. For nonconforming FEMs we have to add $c(u^h,v^h)_{L^2(\omega)}$, for DCGMs the appropriate penalty terms. These are necessary as well for defining the $\|v^h\|_{\mathcal{V}_b^h}$.

Now we are ready to introduce the $a_\lambda(\cdot,\cdot), a_{\lambda,h}(\cdot,\cdot), a_\lambda^h(\cdot,\cdot)$ as

$$a_\lambda(\cdot,\cdot):\mathcal{V}_b\times\mathcal{V}_b\to\mathbb{C} \text{ by } a_\lambda(u,v):=a(u,v)+\lambda(u,v)_{\mathcal{W}^h} \tag{4.244}$$

$$\omega(\lambda):=\inf_{u\in\mathcal{V}_b,\ \|u\|_{\mathcal{V}_b}=1}\ \sup_{v\in\mathcal{V}_b,\ \|v\|_{\mathcal{V}_b}=1}\big|a_\lambda(u,v)\big| \text{ and}$$

$$a_\lambda^h(\cdot,\cdot):\mathcal{V}_b^h\times\mathcal{V}_b^h\to\mathbb{C} \text{ by } a_\lambda^h(u^h,v^h):=a^h(u^h,v^h)+\lambda(u^h,v^h)_{\mathcal{W}},$$

$$\omega^h(\lambda):=\inf_{u^h\in\mathcal{V}_b^h,\ \|u^h\|_{\mathcal{V}_b^h}=1}\ \sup_{v^h\in\mathcal{V}_b^h,\ \|v^h\|_{\mathcal{V}_b^h}=1}\big|a_\lambda^h(u^h,v^h)\big|.$$

Note $(u^h,v^h)_{\mathcal{W}^h}$ is well defined, and $=(u^h,v^h)_{\mathcal{W}}$, except for difference methods. Analogously to $a_\lambda^h(\cdot,\cdot)$ we define $a_{\lambda,h}(\cdot,\cdot)$.

Then [387], Exercise 11.2.4. remains correct, since his Lemmas 6.5.3 and 6.5.17 correspond to our Theorem 2.12 and Corollary 2.14 with Theorem 2.20 yielding his Lemma 6.5.17. His Theorem 6.5.15, our Theorem 2.20, allows relating the $\omega(\lambda)$ and $\omega^h(\lambda)$ as in Remark 11.2.5 and Lemmmas 11.2.6 and 11.2.7. Here Remark 11.2.5 and Lemmma 11.2.6 are rather obvious. For the technical proof of his important Lemmma 11.2.7 we have to keep in mind that his definition of z^h in his (11.2.6b) requires the well-defined $a_\mu^h(z^h,v^h)=(\lambda-\mu)(u,v^h)_{\mathcal{W}}\forall v^h\in\mathcal{V}_b^h$. The reference to his Theorem 8.2.2 following (11.2.6f) has to replaced by our convergence results for the variational methods, mentioned above. His estimate $\sup|a_\lambda(u-u^h,v)|\le C_S\|u-u^h\|_{\mathcal{V}_b}$ has to be replaced by $\sup|a_{\lambda,h}(u-u^h,v)|\le C_S\|u-u^h\|_{\mathcal{V}_b^h}$. This last inequality remains correct with $a_{\lambda,h}(u-u^h,v)$ indicating the bounded extension of $a_\lambda(\cdot,\cdot)$ in (4.243), (4.244), with still coercive $a_{\mu,h}(u^h,v^h)$, and the consistency of the $a_{\lambda,h}(u^h,v^h)$ with the $a_\lambda(u,v)$. The bounds in (11.2.6a)–(11.2.6f) in [387] have to be slightly modified. For difference methods the coercive $a_\mu^h(u^h,v^h)$, and the consistency of the $a_\lambda^h(u^h,v^h)$ with the $a_\lambda(u,v)$ allow maintaining these [387] arguments.

Similarly his Theorem 11.2.8, Lemma 11.2.9, Theorems 11.2.10–11.2.11 and Exercise 11.2.12 remain correct. Here the proofs of Theorem 11.2.8, Lemma 11.2.9, and Theorem 11.2.10 are simple generalizations. The proof of Theorem 11.2.11 remains verbatim correct for conforming variational methods. Only the reference to [387], Theorem 8.2.2, after (11.2.6f) has to replaced by our convergence results for the variational methods. For the nonconforming FEMs and the DCGMs the [387] approximate eigenfunctions e^h are no longer in $\mathcal{V}_b$. Then convergence results for eigenfunctions have to be formulated with respect to $\|\cdot\|_{\mathcal{V}_b^h}$. Additionally, the *anticrime transformation*, $E^h:\mathcal{U}^h\to\mathcal{U}$, cf. Lemma 5.77 for nonconforming FEMs and for the DCGMs, have to be employed.

Then the [387] e^h are approximated by $E^h e^h \in \mathcal{V}_b$ and again allow the proof in [387], and consequently of Exercise 11.2.12 as well.

We summarize the qualitative convergence results from the [387], Theorems 11.2.8, 11.2.10, 11.2.11, Exercise 11.2.12 in.

Theorem 4.78. Convergence of discrete eigenvalues and eigenfunctions: *For all cases we assume (4.241)–(4.242), or the modifications for conforming FE and difference methods.*

1. *Choose, for the sequence of* $\mathcal{V}_b^h$ *in (4.242), the approximate eigenvalues* λ^h *in (4.240), and assume* $\lim_{h\to 0} \lambda^h = \lambda_0$. *Then* λ_0 *is an eigenvalue of (4.235).*
2. *For* λ_0*, an eigenvalue of (4.235), there exist discrete eigenvalues* λ_0^h *of (4.240) with* $\lim_{h\to 0} \lambda^h = \lambda_0$.
3. *For* $e^h \in \mathcal{E}(\lambda^h)$ *in (4.240) with* $\|e^h\|_{\mathcal{V}_b^h} = 1$ *and* $\lim_{h\to 0} \lambda^h = \lambda_0$*, a subsequence* h_i *exists such that for an eigenfunction* $e \in \mathcal{E}(\lambda)$ *in (4.235), the* e^{h_i} *converge, so* $\lim_{h_i\to 0} \|e^{h_i} - e\|_{\mathcal{V}_b^{h_i}} = 0$.
4. *If* $\dim \mathcal{E}(\lambda_0) = 1$*, we get* $\lim_{h\to 0} \dim \mathcal{E}(\lambda^h) = \dim \mathcal{E}(\lambda_0)$*, for these* $\lambda^h, e^h, \lambda_0, e$.
5. *For* $\dim \mathcal{E}(\lambda_0) = 1$*, choose the* e^h *as in (3). and* $\hat{e}^h := e^h/(e^h, e)_{\mathcal{V}}$ *if* $(e^h, e)_{\mathcal{V}} \geq 1/2$ *and otherwise* $\hat{e}^h := e^h$. *Then* $\lim_{h\to 0} \|e^h - e\|_{\mathcal{V}_b^h} = 0$.

We turn to the quantitative convergence results. In [387] the purely analytic results in Lemmas 11.2.13, 11.2.14 and Exercise 11.2.15 obviously remain valid. Lemmas 11.2.16, 11.2.18 and Exercise 11.2.17 remain correct for the conforming method and have to be modified for the other methods here and below: $\|v\|_{\mathcal{V}_b}$ for $v \in \mathcal{V}_b^h$ and $a_\lambda(\cdot,\cdot)$ have then to be replaced by $\|v\|_{\mathcal{V}_b^h}$ and $a_\lambda^h(\cdot,\cdot)$ or $a_{\lambda,h}(\cdot,\cdot)$. In particular for Lemma 11.2.16, the proof of part (1) with its $L^2(\Omega)$ arguments remain correct, since $\mathcal{V}_b^h \subset L^2(\Omega)$. For all other cases the previous replacements yield the proofs. In particular, for difference methods the $a_\lambda^h(\cdot,\cdot)$ and the consistency have to be combined. For the next three theorems see [387], Theorems 11.2.19, 11.1.20.

Theorem 4.79. Quantitative convergence results for discrete eigenvalues: *We assume (4.241)–(4.242), an eigenvalue* λ_0 *for (4.235) with* $\dim \mathcal{E}(\lambda_0) = 1$ *and* geometric equal algebraic multiplicity *of* λ_0*, cf. [387], Lemma 11.2.13, hence*

$$\dim \mathcal{N}((A - \lambda_0 I)) = \dim \mathcal{N}((A - \lambda_0 I)^2). \tag{4.245}$$

Then span $e = \mathcal{E}(\lambda_0)$ *and span* $e^* = \mathcal{E}^d(\bar{\lambda}_0)$ *can be normed such that* $(e, e^*)_{\mathcal{W}} = 1$. *Then there exist discrete eigenvalues* λ^h *of (4.240), such that*

$$|\lambda^h - \lambda_0| \leq C \operatorname{dist}\left(e, \mathcal{V}_b^h\right) \operatorname{dist}\left(e^*, \mathcal{V}_b^h\right) \text{ with } \operatorname{dist}\left(e, \mathcal{V}_b^h\right) = \inf_{w\in\mathcal{V}_b^h} \|w - e\|_{\mathcal{V}_b^h}. \tag{4.246}$$

Theorem 4.80. Convergence estimates for discrete eigenfunctions: *Under the conditions of Theorem 4.79 and for the above* $e \in \mathcal{E}(\lambda_0)$ *and* $e^* \in \mathcal{E}^d(\bar{\lambda}_0)$ *there exist discrete approximations* $e^h \in \mathcal{E}^h\left(\lambda_0^h\right)$ *and* $e^{h,*} \in \mathcal{E}^{h,d}\left(\bar{\lambda}_0^h\right)$ *such that*

$$\|e - e^h\|_{\mathcal{V}_b^h} \leq C \operatorname{dist}\left(e, \mathcal{V}_b^h\right) \text{ and } \|e^* - e^{h,*}\|_{\mathcal{V}_b^h} \leq C \operatorname{dist}\left(e^*, \mathcal{V}_b^h\right). \tag{4.247}$$

More detailed convergence results are possible for smooth eigenfunctions. In Theorem 4.77 we have formulated conditions guaranteeing

$$\mathcal{E}(\lambda_0), \mathcal{E}^d(\lambda_0) \subset W^{m+s,p}(\Omega, \mathbb{R}^q), m, q \geq 1 \text{ for } \mathcal{V}_b = W^{m,p}(\Omega, \mathbb{R}^q). \tag{4.248}$$

Then the error estimates for the different methods, e.g. Theorems 4.17, 4.43, 4.36, 7.7, and Lemma 9.7, with appropriate modifications for mesh-free and spectral methods in Böhmer [120], imply

$$\operatorname{dist}\left(u, \mathcal{V}_b^h\right) \leq Ch^s \|u\|_{W^{m+s,p}(\Omega,\mathbb{R}^q)} \forall u \in \mathcal{E}(\lambda_0) \cup \mathcal{E}^d(\bar{\lambda}_0). \tag{4.249}$$

Theorem 4.81. Order of convergence for discrete eigenvalues and eigenfunctions: *Under the conditions of Theorems 4.79, 4.80 and (4.248), (4.249) there exist* $\lambda^h, e^h \in \mathcal{E}^h(\lambda^h)$ *and* $e^{h,*} \in \mathcal{E}^{h,d}(\bar{\lambda}^h)$ *such that*

$$|\lambda^h - \lambda_0| \leq Ch^{2s}, \|e - e^h\|_{\mathcal{V}_b^h} \leq Ch^s, \|e^* - e^{h,*}\|_{\mathcal{V}_b^h} \leq Ch^s. \tag{4.250}$$

For difference methods this holds for $s = 1$ *and* $s = 2$ *for unsymmetric and symmetric formulas.*

Sometimes the eigenfunctions have better regularity properties than in Theorem 4.77. This allows for $m = 1$, cf. [387], Theorem 11.2.22, better estimates of the form $\|e - e^h\|_{\mathcal{V}_b^h} \leq C'h^2, \|e^* - e^{h,*}\|_{\mathcal{V}_b^h} \leq C'h^2$.

Remark 4.82. *These results remain valid with slight modifications for the other methods as well. In particular, we even might consider cases, where* $\mathcal{U}_b^h \neq \mathcal{V}_b^h$.

5

Nonconforming finite element methods

5.1 Introduction

In Chapter 4 we discussed conforming FEMs and their strong requirements with respect to ansatz and test functions, exactly evaluated test integrals, and restrictions to weak forms. We avoid this drawback by turning to nonconforming FEMs in this chapter. But then the proofs for consistency and stability become much more complicated.

Proceding along the lines of Chapter 3 requires, for nonconforming FEMs, replacing the *variational consistency errors* by *classical consistency errors.* The variational consistency error is the usual concept in the finite element community. It is, for a linear operator A, based upon the bilinear form $a(.,.)$, and for a quasilinear operator based upon its weak form and vanishes for conforming FEMs. This contrasts with the classical consistency error, directly related to operators and very appropriate for any nonlinear problems as well. We have to discuss both concepts for the following variational crimes.

We start with the most convincing example for this general approach. For two decades FEMs for general fully nonlinear elliptic problems were an open problem. They were, for the first time and on the basis of nonconforming FEMs, published by Böhmer [118,119]. Before results for special equations of order 2 in $\mathbb{R}^2$ in Fulton [330], Oliker and Prussner [519], Dean and Glowinski [272–276], have been known. The latter formulate for Monge–Ampère and Pucci equations in $\mathbb{R}^2$ equivalent variational forms or use regularization techniques, but see below. Linear or quasilinear elliptic problems in divergence form are studied in the weak form and solved numerically this way. However, fully nonlinear and nondivergent quasilinear elliptic problems do not allow a weak form. So FEMs for these equations and systems, $G(u_0) = 0$, are based upon the strong form and are studied in Section 5.2.

As a consequence we do need in this chapter the full machinery of discretization methods with

$$G : \mathcal{D}(G) \subset \mathcal{U} \to \mathcal{V}', G^h : \mathcal{D}(G^h) \subset \mathcal{U}^h \to \mathcal{V}'^h \text{ and possibly } \mathcal{U} \neq \mathcal{V}, \mathcal{U}^h \neq \mathcal{V}^h. \quad (5.0)$$

Quite recently the author discovered a totally different approach in a paper which seems to be forgotten: Barles and Souganidis [73] prove the convergence for different discretization methods, based upon the concept of *viscosity solutions* going back to Crandall and Lions [217–222, 227] and Crandall *et al.* [215, 216, 228]. [73]

gives completely new results concerning the convergence of numerical schemes for stochastic differential games. This approach is up to now not extendable to higher order equations and systems.

For general fully nonlinear elliptic problems, conforming FEMs are not (yet?) possible. In fact, we need new regularity results for the FEM solutions of these $G^h(u_0^h) = 0$, requiring convex polyhedral domains or $\Omega \in C^1$ for $m = 1$. Thus $u_0^h \in C^1(\Omega)$ will violate the boundary conditions. So we have to turn to nonconforming FEMs. The standard techniques for handling violated boundary conditions or continuity of the FEMs, cf. Section 5.5, do not apply in this case. Combining very unusual proofs for stability, consistency and Davydov's FEs in $C^1(\Omega)$, cf. Subsection 4.2.6, we are able to show that for a continuous G with boundedly invertible linearization $G'(u_0)$, the discrete solutions u_0^h for the nonlinear discretization $G^h\left(u_0^h\right) = 0$ converge to u_0 in a discrete $H^2(\Omega)$ norm. We generalize this result to elliptic equations and systems of orders 2 and $2m$.

Fortunately, the numerical solution methods can be reduced in this case to the well established methods for weak forms. However, the standard FEs in $C(\Omega)$ there have to be replaced by FEs in $C^1(\Omega)$ here.

In many of the preceding and following cases the integrals in (4.3) and their appropriate generalizations cannot be computed exactly. So quadrature and cubature formulas have to be employed. These approximate FEMs, the next kind of variational crimes, are studied in Section 5.4.

The following Section 5.5 is devoted to other types of nonconformity. The exact boundary conditions and/or the continuity required in conforming FEs, are violated here, cf. Feistauer and Zenísek [317]. Since the fully nonlinear problems are discussed above and FEs $\notin C(\Omega)$ are excluded for them, only linear and quasilinear equations are left. The convergence proofs for violated boundary conditions and/or the continuity, use, generalizing Brenner and Scott [141], the detour from the weak form to its strong counterpart. For FEMs violating these conditions, there are two options: either we use the approach via quadrature approximations along the boundaries of the subtriangles or polytopes for polynomial FEs. This allows piecewise polynomial FEs of degree $d-1$ in $\Omega \subset \mathbb{R}^n$ with $n = 2$, and converging with high order. For violated boundary conditions the second alternative is isoparametric FEs in $\mathbb{R}^n, n \geq 2$. For all these cases we extend the previously known results from linear to quasilinear elliptic equations and systems of order 2. As a consequence of the above violated continuity a direct comparison of $a(\cdot,\cdot)$ and $a^h(\cdot,\cdot)$ is not possible. So, in contrast to [129, 567], we have to prove stability via coercivity of the principal part *and* estimate the consistency error and finally combine both to get the desired convergence.

Another possibility for violated continuity and boundary conditions, the *discontinuous Galerkin methods (DCGMs)*, combine duality arguments with penalty terms, see e.g. Rachford and Wheeler [540], Douglas and Dupont [303], Wheeler [665], Arnold [39], Riviere *et al.* [554, 555, 555], Arnold *et al.* [40]. We present them in Chapter 7 with V. Dolejší, including *hp*- methods. DCGMs eliminate some of the above limitations, but impose others.

5.2 Finite element methods for fully nonlinear elliptic problems

5.2.1 Introduction

This section is certainly one of the highlights of this book. We present a FEM for the general case of fully nonlinear elliptic differential equations of second order in $\mathbb{R}^n, n \geq 2$ and its extension to order $2m$ and to systems of orders 2 and $2m$. To the knowledge of many specialists, consulted by the author, his papers [118, 119] seem to be the first papers solving these general cases, see below.

The basis is the coercivity of the principal part of the linearized operator. This is directly available via Legendre and Legendre–Hadamard conditions. So we require these conditions. This does not work for the most general cases of nonlinear systems in the sense of Agman *et al.* [3]. Here other methods, e.g. as used for the Navier-Stokes equations, cf. Section 4.6, will have to be developed. We apply our general discretization methods in Chapter 3. The necessary quadrature approximations are studied in Subsubsection 5.4.4 and 5.4.5.

For these "nonstandard" semiconforming FEMs (based upon conforming FEs and strong forms of the operator) and (fully) nonconforming FEMs, none of the standard proofs for converegence of FEMs work, see below. So we start giving a short summary of the main ideas behind the formulation and the proofs for our FEMs. For simplifying the presentation we *start with one equation of order* 2:

- Our FEM for general fully nonlinear elliptic differential equations requires a bounded, convex, polyhedron with conforming FEs, or a C^2 domain in $\mathbb{R}^n$ with nonconforming C^1 FEs. Both cases are presented.
- Only FEMs for strong, not for weak, forms make sense. This requires FEs in C^1, satisfying Dirichlet boundary conditions on polyhedra and violating them on C^2 domains.
- The classical theory of discretization methods, cf. Chapter 3, has to be applied to the differential operator and, on C^2 domains, to the differential and the boundary operator.
- The most difficult claim, the stability of the nonlinear operator, is proved via the stability of the linearized operator, with coinciding strong and weak bilinear FE forms. Compact perturbation and a new regularity result for FE solutions of weak linear FE equations lift the "weak stability" back to the "strong stability", cf. Theorem 3.29 and Summary 4.52.
- For computations the weak linear forms, coinciding with the strong forms, are used. Thus the many known powerful numerical methods apply for our FEM as well.
- The mesh independence principle (MIP) guarantees a quadratically convergent solution process via the Newton method for highly nonlinear systems of equations.

Our FEM is a natural translation of the nonstandard situation for fully nonlinear elliptic differential equations, including quasilinear equations not in divergence form; we abbreviate this as *nondivergent quasilinear equations.* As for fully nonlinear equations, weak forms are not possible. The solution has to be determined from the original strong form with derivatives up to second order. This situation is directly translated into our strong form of a FEM; see Subsection 5.2.2 for more details. This FEM, based upon classical discretization theory, strongly differs from the earlier approaches.

Difference methods are studied by Crandall and Lions [217] and Crandall *et al.* [216], by modifying their concept of *exact viscosity solutions.* Kuo and Trudinger [456] employ discrete maximum principles.

For FEMs, there are only a few papers with results for special equations, e.g. the Monge–Ampère equations in $\mathbb{R}^2$ or modifications. Oliker and Prussner [519] prove, by convexity arguments, that their sequences converge monotonically to the solution. Fulton [330] studies multigrid solutions as models of nonlinear balance equations in meteorology. Glowinski with Caffarelli, Dean, Guidoboni, Juárez, and Pan [154, 272–276, 352] reformulate it in $\mathbb{R}^2$ as a problem of calculus of variations involving the biharmonic (or a related) operator. A reinterpretation as a saddle-point problem for a well-chosen augmented Lagrangian functional leads to iterative methods such as that of Uzawa, Douglas, and Rachford with viscosity solutions. This method can be applied to a related problem, the Pucci equation. Other techniques are based upon operator-splitting methods. Similarly, Feng and Neilan [318] solve the Monge-Ampère equations and prove convergence of Galerkin methods via a regularization with $\epsilon\Delta^2$. This yields a fourth order again quasilinear equation with a solution u_0^ϵ and an error estimate, including a term $c(\epsilon)\,\|u_0^\epsilon\|_{H^l}\, h^l/\sqrt{\epsilon}$. Finally, [155, 230, 671] are loosely related to our problem: Cai *et al.* [155] present a FEM for fully nonlinear, three-dimensional water surface waves, described by standard potential theory. Their method is based on a transformation of the dynamic water volume onto a fixed domain. The transformation brings about an elliptic boundary value problem. This equation is solved by a preconditioned conjugate gradient method. Numerical experiments indicate convergence of the iterative solver as well as convergence of the entire solution towards a reference solution computed by an another scheme. Extensions to parabolic equations are possible and studied, e.g. by Yang [671] and Cui [230].

The other FE approaches are based upon either a combinations of the special properties of a specific problem with the proposed numerical method or upon maximum and/or monotonicity arguments. But none of them seems to be applicable nor has been applied to the general fully nonlinear elliptic differential equations and systems as treated in this paper. This seems to indicate the need for new ideas.

5.2.2 Main ideas and results for the new FEM: An extended summary

We propose an approach totally different from those in the above papers. As mentioned, a major part of the analysis for fully nonlinear differential equations is based upon the strong form of these equations; see, e.g. Gilbarg and Trudinger [346], Showalter [589] and Taylor [618–620]. This motivates a FEM, based upon the strong form of the differential equation. For proving existence, uniqueness, and related results for the exact solution, e.g. monotonicity, maximum principles, and the Schauder fixed point theorem are needed. We avoid this, but employ, as for many other numerical approaches, these results as the basis for our method.

We simplify the presentation by starting with semiconforming and the nonconforming FEMs for second order equations; cf. Subsections 5.2.4, 5.2.5 and 5.2.8 for the general case. We prove convergence in the framework of general discretization theory, based upon consistency and stability, cf. Chapter 3, in a first step fitted to the present problem. We need the strong form for the FEM and (see below) a convex, bounded (and we assume additionally) polyhedral Ω or an $\Omega \in C^2$; both cases are discussed here. The strong form requires C^1 FEs for an L^2 evaluation of the differential equation. For a polyhedral Ω, (trivial) boundary conditions are satisfied exactly by the FEs, for curved domains they violate the boundary conditions. For the first case and the FEs in C^1 we need only Theorem 4.36, for the second case Theorems 4.35–4.39. Hence, the first case allows the full order of convergence in Theorem 4.36. For $\Omega \in C^2$, the proved results in Subsection 4.2.6 yield second order in $\mathbb{R}^2$. Conjecture 4.40, probably proved soon, allows an extension to $\mathbb{R}^n$, those in Conjecture 4.41 higher order as well. So we anticipate Conjecture 4.40 and correspondingly formulate the following results, sometimes hinting at Conjecture 4.41.

The second step is the proof of consistency. For polyhedral Ω and their conforming FEs and a continuous differential operator this is a straightforward consequence of a trivial variational error $= 0$ and a vanishing interpolation error.

For smooth $\Omega \in C^2$, requiring nonconforming FEs, consistency is more complicated. We formulate the results on the basis of Conjecture 4.40 and indicate modifications by Conjecture 4.41. The earlier approaches for curved boundaries, cf. e.g. Lenoir [471] and Brenner and Scott [142] for linear problems, do not seem to be applicable to fully nonlinear problems. So other nonstandard, not surprisingly similarly complicated considerations are necessary. The clue is an operator with two components, the differential and the boundary operator. We will use this idea for other problems as well, e.g. for nonlinear boundary operators in Section 5.3. This requires a more complicated version of general discretization theory. Specifically, we have to carefully introduce the necessary projectors and spaces for including the violated differential operator *and* the boundary conditions into the classical discretization theory. Then the "classical consistency error" for the differential and the boundary operator is estimated in (5.65) and Theorem 5.4 below. Essentially we split this error into the FE error for a semiconforming FEM on an $\Omega_c^h \approx \Omega$ and its complementary *geometric error,* due to $\partial\Omega \neq \partial\Omega_c^h$, according to a new idea of Tiihonen [629].

The third step is the most difficult, the proof of stability. We did not see a direct possibility for the strong form. So we combine linearization with regularity of FE solutions. This is another difference of our FEM to the standard methods and those in the previous papers. Stability is first proved for the linearized weak form. Unfortunately, we lose the powerful machinery of weak bilinear forms and their FEMs, *unless* we find a kind of *connection* between the two forms. This is achieved by Proposition 5.7 and Lemma 5.10. The first shows coinciding weak and strong FE bilinear forms in our case. The latter states, similarly to the exact solution of a "smooth" linear elliptic problem, a new regularity and stability result for our FE solutions. This and the available existence, uniqueness, and regularity results require a convex, bounded, polyhedral Ω implying satisfied or an $\Omega \in C^2$, implying violated boundary conditions for our FEM. We generalize our earlier compact perturbation techniques, cf. Chapter 3, Theorem 3.29, and Böhmer and Sassmannshausen [128] and Böhmer [114, 115], to the above satisfied or violated boundary conditions. This allows lifting the stability from coercive to general elliptic bilinear forms. A combination of these results lifts the weak back to the strong stability and allows the convergence proof.

We do need convex, bounded (polyhedral) or smooth $\Omega \in C^2$ for existence, convergence, and the regularity lemma, Lemma 5.10. The smooth $\Omega \in C^2$ requires and allows the extension operator in Theorem 4.37. We require FEs in C^1, since for FEs in C^0 we did not see a possibility to handle the interior variational crimes. The standard interplay between the weak and strong form is impossible for fully nonlinear problems and *excludes, e.g. discontinuous Galerkin methods.* Thus we choose the unusual C^1 FEs, conforming for polyhedral and nonconforming for C^2 domains.

Davydov [264], cf. Subsection 4.2.6, has proved complete approximation results on bounded, polyhedral Ω in $\mathbb{R}^n$. This allows convergence results similarly to the the standard conforming FEMs, cf. Theorem 5.2.

For $\Omega \in C^2$ the situation changes. We need the new "stable splitting" of C^1-FEs into interior and boundary elements. Presently, these results in Subsection 4.2.6 are available for modified Argyris piecewise polynomials of degree 5 on polygonal $\Omega^h \approx \Omega \subset \mathbb{R}^2$. This restricts, for $\Omega \in C^2$, the order of convergence of our FEM to 2. Davydov is extending his results to $\mathbb{R}^n$ and to higher polynomial degrees. Our proof is formulated for including the extensions in Conjecture 4.40, and indicating those in Conjecture 4.41.

The FE equations have to be set up. They are solved numerically by a quadratically converging discrete Newton method. Its local quadratic convergence is guaranteed by the MIP; see Section 3.7, Subsection 5.2.10 and Allgower and Böhmer [9, 10] and Allgower *et al.* [17]. It starts with a good approximation, e.g. obtained by a continuation method. Again linear problems are essential to solve the highly nonlinear systems of equations. As a welcome consequence of our coinciding weak and strong FE bilinear forms, the standard solution techniques for weak linear problems are applicable to the fully nonlinear equations as well. However, the standard C^0 FEs have to be replaced by C^1 FEs.

With the notation introduced below, let G be a uniformly elliptic operator on Ω, see (5.18), (5.20). Homogeneous Dirichlet conditions are obtained by the standard techniques. Violated boundary conditions for the nonconforming FEM below will play a dominant role. The exact solution, u_0, is determined by the differential equation and the boundary condition in the trace sense, cf. Subsections 5.2.4, 5.2.5. Mind that $G(u_0) \in L^2(\Omega)$ is satisfied for a smooth solution $u_0 \in \mathcal{U}_s$, *e.g.* $\mathcal{U}_s = H^4_+(\Omega)$, cf. Example 3.26 and Proposition 3.27. Smoothness will be required for good convergence anyway.

$$\begin{aligned} & u_0 \in \mathcal{D}(G) \cap \mathcal{U}_s \subset H^2(\Omega) : G(u_0)(x) := G^w(x, u_0(x), \nabla u_0(x), \nabla^2 u_0(x)) = 0 \qquad (5.1) \\ & \forall \text{ } a.e. \text{ } x \in \Omega, \text{ } G(u_0) \in L^2(\Omega) \iff (G(u_0), v)_{L^2(\Omega)} = 0 \quad \forall v \in L^2(\Omega) = (L^2(\Omega))', \\ & \qquad \text{and } u_0|_{\partial\Omega} = 0 \iff (u_0|_{\partial\Omega}, v_b)_{L^2(\partial\Omega)} = 0 \quad \forall v_b \in L^2(\partial\Omega), \end{aligned}$$

with $G^w : \mathcal{D}(G^w) \subset \Omega' \times \mathbb{R} \times \mathbb{R}^n \times \mathbb{R}^{n^2} \to \mathbb{R}$, $\overline{\Omega}$, $\overline{\Omega^h_c} \subset \Omega'$ (see (5.16)). We assume $G(u^h)(x) := G^w(x, u^h(x), \nabla u^h(x), \nabla^2 u^h(x))$ to be defined as well, cf. Proposition 3.27. This condition has to be proved for any specific G, e.g. the Monge-Ampere equations.

a.e. for $x, u^4(x), \nabla^2 u^4(x) \in D(G^w)$. Since $H^{3/2}(\partial\Omega)$ is dense in $L^2(\partial\Omega)$ only testing for all $v_b \in L^2(\partial\Omega)$ is correct. For a systematic study of the differential and the boundary operator, we abbreviate (5.1) into an equivalent form, determining

$$u_0 \in \mathcal{U}_s \cap \mathcal{D}(G) : 0 = F(u_0) := (G(u_0)|_\Omega = 0, u_0|_{\partial\Omega} = 0) \tag{5.2}$$

$$\iff (G(u_0), v)_{L^2(\Omega)} + (u_0|_{\partial\Omega}, v_b)_{L^2(\partial\Omega)} = 0 \quad \forall v \in L^2(\Omega), \quad v_b \in L^2(\partial\Omega).$$

The difference between quasilinear and fully nonlinear problems for a second order equation is discussed in Subsection 5.2.3. This motivated our new FEM. The standard ellipticity condition for nonlinear systems via linear systems of order 2 has to be modified for equations and systems of higher order. They have to imply the coercivity of the linearized principal part, see Subsections 5.2.3, 2.4.4, 2.6.3, 2.6.5. There we formulated existence, uniqueness and regularity results as a basis for the corresponding FEMs for fully nonlinear equations and systems of orders 2 and $2m$ in Subsection 5.2.8.

We introduce the C^1 FE spaces of local degree $d \geq n^2 + 1$ in $\mathbb{R}^n$, the $\mathcal{S}_d^1\left(\mathcal{T}_c^h\right)$, for polyhedral, and for C^2 domains. For a unified notation we assume that they are defined on a triangulation $\mathcal{T}_c^h$ of the original polyhedral $\Omega = \Omega_c^h$, or a curved $\Omega_c^h \approx \Omega \in C^2$. In *this subsection we use the same notation* $\mathcal{T}_c^h, \Omega_c^h$ *for all cases;* cf. Subsections 4.2.6 and 5.2.5 ff. for $S^h \subset \mathcal{S}_d^1\left(\mathcal{T}_c^h\right)$:

$$\mathcal{U}^h := S^h \subset \mathcal{S}_d^1\left(\mathcal{T}_c^h\right) \text{ and } \mathcal{U}_b^h := \mathcal{V}^h := \left\{u^h \in S^h : u^h|_{\partial\Omega_c^h} = 0\right\} \subset \mathcal{U}^h \neq \mathcal{V}^h. \tag{5.3}$$

For convex polyhedral Ω, these FEs $u^h \in \mathcal{U}_0 \subset \mathcal{S}^h \subset H^2(\Omega)$ satisfy the trivial boundary conditions exactly. So they define a semiconforming FEM. For $\Omega \in C^2$ they violate the boundary conditions, so they yield a nonconforming FEM.

For convex polyhedral Ω, the FE solution, $u_0^h \in \mathcal{U}_b^h$, is determined by testing only the differential equation with respect to $v^h \in \mathcal{V}^h$, again in the L^2 sense:

$$\mathcal{G}:\mathcal{U}_b^h \cap \mathcal{D}(G) \to L^2(\Omega) \text{ determine } u_0^h \in \mathcal{U}_b^h = \mathcal{V}^h \text{ s.t. } \left(G\left(u_0^h\right), v^h\right)_{L^2(\Omega)} = 0 \; \forall v^h \in \mathcal{V}^h. \tag{5.4}$$

For this semiconforming FEM, $u^h \in \mathcal{U}_b^h = \mathcal{V}^h \subset H^2(\Omega) \cap H_0^1(\Omega)$.

For smooth $\Omega \in C^2$ we have to test both operators. For $\Omega \in C^2$, comparing u_0^h and u_0 requires extension operators, e.g. $E_c : H^2(\Omega) \to H^2\left(\Omega \cup \Omega_c^h\right)$, cf. Subsection 4.2.6, Theorem 4.37. A simplified notation is possibel by (realistically) assuming

$$G \text{ in (5.1) defined on } \mathcal{D}(G) \subset \mathcal{U}_s \cap \mathcal{U}_c \to \mathcal{V}_c \text{ with } \mathcal{U}_c := H^2\left(\Omega \cup \Omega_c^h\right), \mathcal{V}_c := L^2\left(\Omega \cup \Omega_c^h\right). \tag{5.5}$$

So the FEM, cf. (5.2), defines F^h, and determines, under the Condition **H**, see below, the FE solution $u_0^h \in \mathcal{U}^h$, cf. Theorems 4.39, 5.6, from

$$u_0^h \in \mathcal{U}^h : 0 = F^h\left(u_0^h\right) := \left(G^h\left(u_0^h\right), u_0^h|_{\partial\Omega_c^h}\right) = 0 \iff$$

$$\left(G\left(u_0^h\right), v^h\right)_{L^2(\mathcal{T}_c^h)} + \left(u_0^h|_{\partial\Omega_c^h}, v_b^h\right)_{L^2(\partial\Omega_c^h)} = 0$$

$$\forall v^h \in \mathcal{V}^h, v_b^h \in \mathcal{V}_b^h := \left\{u^h|_{\partial\Omega_c^h} : u^h \in \mathcal{U}^h\right\}. \tag{5.6}$$

Our FEM transforms $F(u_0) = 0$ in (5.2) into $F^h\left(u_0^h\right) = 0$ in (5.6). It is a classical discretization method in the sense of Chapter 3 and Summary 4.52, cf. e.g. Stetter [596], Stummel [607–609], Vainikko [644], Keller [441], Reinhardt [547], Zeidler [676–678] and Böhmer [113–115]. In particular, Theorems 3.21, 3.25 provide the tools for showing convergence, if consistency and stability can be proved. Thus Subsections 5.2.4–5.2.7 are the core of the section.

The test integrals for nonlinear operators, see (5.6), can be exactly computed only for special cases. So usually they have to be approximated. We extend our results to the necessary quadrature and cubature approximations, see Subsections 5.4.4 and 5.4.5: We prove the corresponding convergence results for quadrature approximated FEMs.

In the general case, we will update the $H^1(\Omega), H^2, H_0^1$ and C^1 FEs into $H^m(\Omega, \mathbb{R}^q)$, H^{2m}, H_0^m and C^{2m-1} FEs, where $2m$ indicates the order and q the number of equations of the operators. Accordingly, (5.2) and (5.6) have to be generalized. Finally, the boundary condition $u|_{\partial\Omega} = 0$ has to be replaced by $(\partial^k u)/(\partial\nu)^k|_{\partial\Omega} = 0,\ k = 0, \ldots, m-1$.

For our unusual FEM, we summarize the main ideas. Here, we formulate the results for the nonconforming FEM for $\Omega \in C^2$. The corresponding reduction to the semiconforming FEM for polyhedral Ω is obvious, cf. Subsection 5.2.4.

1. *The FEMs, transforming* (5.1) *into* (5.4), *and* (5.2) *into* (5.6), *are general discretization methods:*
   - The existence, uniqueness, and regularity results for the original problems in Subsection 5.2.3 are the basic results for a general discretization method.
   - The fully nonlinear elliptic problem (5.1), (5.2) is approximated by the *strong* form (5.4), (5.6), requiring FEs in C^1. For (5.4) identical ansatz and test functions, $\mathcal{U}_b^h = \mathcal{V}^h$, are possible. For (5.6) we choose as ansatz functions $\mathcal{U}^h = S^h$ and test the differential operator with $v^h \in \mathcal{V}^h = S^h \cap H_0^1\left(\Omega_c^h\right)$ and the boundary operator with the restrictions to the boundary, $v_b^h \in \mathcal{V}_b^h = S^h|_{\partial\Omega_c^h}$; see (5.3) and Subsection 4.2.6 for the relevant extension results in Theorem 4.37.
   - The FEM is a general linear discretization method; cf. Chapter 3 and Subsections 5.2.4 and 5.2.5 ff.
   - In Subsections 5.2.5–5.2.7 we reformulate the FEM, introduce the necessary projectors, and verify the required conditions. Thus the FEMs are convergent if they are stable and consistent; see Theorem 5.2.
   - For a "smooth" problem, G, the semiconforming FEMs for a polyhedron and the nonconforming FEMs for $\Omega \in C^2$ are consistent; see Theorems 4.54 and 5.4.
2. *Stability for the nonlinear* strong *equation* (5.4), (5.6), *is the most difficult problem; cf. Subsection* 5.2.7:
   - The nonlinear stability is a consequence of the stability for the *linearized strong* problem, again under the conditions (5.4), (5.5), Condition **H**, see (5.7) and Theorem 5.6. For given u_0^h and f, ϕ, determine the FE solution u_1^h from

$$u_1^h \in \mathcal{U}^h \text{ s.t. } (F^h)'\left(u_0^h\right) u_1^h = \left(G'\left(u_0^h\right) u_1^h, u_1^h|_{\partial\Omega_c^h}\right) = (f, \phi). \tag{5.7}$$

For this linearized problem, two FEMs complement each other; see below:

- The first component of the linearized *strong* and *weak* bilinear forms coincide:

$$\left(G'\left(u_0^h\right)u^h,\ v^h\right)_{L^2(\Omega_c^h)} = \left\langle G'\left(u_0^h\right)u^h,\ v^h\right\rangle \quad \forall\, u^h \in \mathcal{U}^h,\quad v^h \in \mathcal{V}^h, \quad (5.8)$$

  with $\langle \cdot,\cdot\rangle = \langle \cdot,\cdot\rangle_{H_0^{-1}(\Omega_c^h)\times H_0^1(\Omega_c^h)}$; see Proposition 5.7.

- For (extended) $(f,\phi) \in H^{-1}\left(\Omega_c^h\right) \times H^{1/2}\left(\partial\Omega_c^h\right)$, the linearized *weak* problem

$$(F^h)'\left(u_0^h\right)u_1^h = (f,\phi) \Leftrightarrow \left\langle G'\left(u_0^h\right)u_1^h - f, v^h\right\rangle + \left(\left(u_1^h - \phi\right)|_{\partial\Omega_c^h}, v_b^h\right)_{L^2(\partial\Omega_c^h)} = 0$$

  for all $v^h \in \mathcal{V}^h, v_b^h \in \mathcal{V}_b^h$ is stable, essentially if u_0 is an isolated solution of $F(u_0) = 0$ with boundedly invertible $F'(u_0)$. This is a consequence of the discussion at the end of Subsection 5.2.3 and the compactness arguments in Theorem 5.9.

- By the new Lemma 5.10 the regularity results for the exact solution of a linear elliptic problem are valid for the corresponding FE solutions. In the *weak linearized* problem, again under the conditions (5.4), (5.5), $(F^h)'\left(u_0^h\right)u_1^h = (f,\phi) \in H^{-1}\left(\Omega_c^h\right) \times H^{1/2}\left(\partial\Omega_c^h\right)$, we determine $u_1^h \in \mathcal{U}^h$ for the smooth $(f,\phi) \in L^2\left(\Omega_c^h\right) \times H^{3/2}\left(\partial\Omega_c^h\right)$; see (5.4), (5.6), (5.8). The stability of the *weak* problem implies the stability of the *strong* problem

$$\left(G'\left(u_0^h\right)u_1^h - f, v^h\right)_{L^2(\Omega_c^h)} + \left(\left(u_1^h - \phi\right)|_{\partial\Omega_c^h}, v_b^h\right)_{L^2(\partial\Omega_c^h)} = 0$$
$$\forall v^h \in \mathcal{V}^h, \quad v_b^h \in \mathcal{V}_b^h.$$

  This allows estimates for discrete solutions with respect to a piecewise H^2 norm instead of the usual H^1 norm for a right-hand side in $L^2\left(\mathcal{T}_c^h\right) \times H^{3/2}\left(\partial\Omega_c^h\right)$ instead of the usual $H^{-1}\left(\mathcal{T}_c^h\right) \times H^{1/2}\left(\partial\Omega_c^h\right)$. This requires smooth coefficients, necessary for good convergence, and a convex, bounded, polyhedral Ω, or $\Omega \in C^2$.

3. *Summary: Essentially simultaneously with a unique* u_0, *the FE solutions* u_0^h *of* (5.4) *and* (5.6) *uniquely exist, for small pnenough h, and converge, under the conditions (5.4), (5.5), and if* $G : (\mathcal{D}(G) \cap \mathcal{U}^h) \subset H^2(\Omega \cap \Omega_c^h) \to L^2(\Omega \cap \Omega_c^h)$ *is continuous; see* (5.9), *Condition* ***H****, and Theorems* 5.2 *and* 5.13.

4. *Setting up the FE equations* (5.4), (5.6), *and solving them; see Subsection* 5.2.10*:*

- Setting up the nonlinear system (5.4), (5.6), is formulated in (5.153).
- As a consequence of **1.–2.** the MIP is valid. So the Newton method for the discrete equation (5.4), (5.6), with a good enough initial guess $u_1^h = P^h u_1$, e.g. obtained by a continuation method, converges quadratically to u_0^h equally well as the Newton–Kantorovich method, started in u_1, for the original problem (5.1); see Theorem 5.21 and (5.10).
- The linearization of G is elliptic. Thus the efficient solvers for linear elliptic problems in weak form (see (5.7)) can be applied in the continuation and Newton processes, replacing C^0 FEs by C^1 FEs. Results of Davydov [263] and Davydov and Stevenson [268] for Ω^h allow extensions to nested sequences of triangulations $\mathcal{T}_1^h \sqsubset \mathcal{T}_2^h \sqsubset \cdots$, hence even multiresolution techniques.

5. *Quadrature approximations for (5.4), (5.6), (5.7), see Subsections 5.4.4, 5.4.5 :*
  - The exact equations (5.4), (5.6), (5.7) usually have to be approximated by good enough quadrature to maintain convergence, see Theorems 5.33 and 5.40.

We formulate the convergence of the FE and the discrete Newton method: let (5.1), for smooth enough G and Ω, have a smooth isolated solution $u_0 \in H^\ell(\Omega)$ with bounded $(F'(u_0))^{-1}$ and choose FEs of local degree d in C^1 with $0 \leq \ell - 1 \leq d$, $d \geq n^2 + 1$; cf. (4.62). Then the discrete solutions u_0^h of (5.6) converge as (see Theorem 5.13)

$$\left\| E_c u_0 - u_0^h \right\|_{H^2(\Omega_c^h)}^h \leq C h^{\min\{\ell-2,p\}} \|u_0\|_{H^\ell(\Omega)} \quad \text{for large enough } \ell,\ d. \tag{5.9}$$

For convex, bounded, polyhedral domains and semiconforming FEMs we only need the results from Theorem 4.36 in $\mathbb{R}^n$, cf. Subsection 4.2.6, and we get $p = \ell - 2$. For $\Omega \in C^2$ and the nonconforming FEMs we obtain $p = 2$ for polyhedral Ω_c^h in $\mathbb{R}^2$, by Theorem 4.39, in $\mathbb{R}^n$, by Conjecture 4.40 and with $p > 2$ for future Ω_c^h by Conjecture 4.41 with appropriate curved boundary approximations. Davydov has fully proved the case $\mathbb{R}^2, d = 5, p = 2$, cf. Subsection 4.2.6, and he is working on higher $\mathbb{R}^n, p > 2$. The proofs in this paper are based upon these generalizations in Conjecture 4.40.

The Newton method for the discrete problems (5.4), (5.6) has, for (5.6), the form (5.10) and converges locally quadratically; see Theorem 5.21:

$$\begin{aligned} & u_{\nu+1}^h \in \mathcal{U}^h : \ (F^h)' \left(u_\nu^h\right) \left(u_{\nu+1}^h - u_\nu^h\right) = -F^h \left(u_\nu^h\right) \\ & \Longrightarrow \left\| u_{\nu+1}^h - u_\nu^h \right\|_{H^2(\Omega_c^h)}^h \leq C \left(\left\| u_\nu^h - u_{\nu-1}^h \right\|_{H^2(\Omega_c^h)}^h \right)^2 . \end{aligned} \tag{5.10}$$

Finally, we indicate, but do not explicitly formulate, two further extensions. In Subsection 5.2.7 the linearized weak problem turns out to be a compact perturbation of a coercive bilinear form. Hence, the usual Fredholm alternative results apply. The above FEM can easily be extended to parameter dependent problems and for studying bifurcation scenarios. Some general results prove the convergence of the bifurcation characteristics of the discrete FEM to those of the original problem, starting with Brezzi *et al.* [147–149], and, e.g. Böhmer [113–115] and Cliffe *et al.* [180]. Extensions of these results to convergence of the characteristics of center manifolds for the discrete FEM to those for the original problem are, for the first time, studied in Böhmer [115, 118], cf. [120].

Remark 5.1. *Except for DCGMs and wavelet methods these results remain valid with slight modifications for the other methods as well.*

5.2.3 Fully nonlinear and general quasilinear elliptic equations

We recapitulate the standard notation for derivatives, with $(-1)_{j>0} := 1,\ j = 0$, and $:= -1$ for $j \geq 1$,

$$\partial^i u = \frac{\partial u}{\partial x_i}, \ \nabla u = (\partial^i u)_{i=1}^n, \ \nabla^2 u = (\partial^i \partial^j u)_{i,j=1}^n, \ \partial^0 u := u. \tag{5.11}$$

We formulate the conditions for the domain and introduce four Hilbert spaces with $\mathcal{W}_0$ for trivial Dirichlet boundary conditions:

$$\Omega \text{ is an (open) bounded domain in } \mathbb{R}^n \text{ with Lipschitz-continuous } \partial\Omega \text{ and} \tag{5.12}$$
$$\mathcal{U} := H^2(\Omega) \subset \mathcal{W} := H^1(\Omega) \subset \mathcal{V} := L^2(\Omega) = \mathcal{V}', \mathcal{V}_D := H^{3/2}(\partial\Omega), \mathcal{W}_D := H^{1/2}(\partial\Omega),$$
$$\mathcal{W}_0 := H_0^1(\Omega) := \{v \in H^1(\Omega) : v = 0 \text{ on } \partial\Omega\},\ \mathcal{U}_0 := \mathcal{W}_0 \cap \mathcal{U}, \mathcal{W}' := H^{-1}(\Omega),$$

with the standard Sobolev norms, seminorms, inner products for $\mathcal{U}, \mathcal{W}, \mathcal{V}$, and Sobolev–Sloboditskii norms for $\mathcal{V}_D, \mathcal{W}_D$; cf. (1.60), Theorem 1.30 [387, 518].

For a quasilinear elliptic equation in divergence form a weak semiconforming FEM determines the FE solution $u_0^h \in \mathcal{V}^h = \mathcal{U}_b^h$ as

$$\langle Gu_0^h, v^h\rangle_{\mathcal{W}'\times\mathcal{W}} := \int_\Omega \sum_{j=0}^{n} a_j\left(\cdot, u_0^h, \nabla u_0^h\right)\partial^j v^h dx = \langle f, v^h\rangle_{\mathcal{W}'\times\mathcal{W}} \quad \forall v^h \in \mathcal{V}^h. \tag{5.13}$$

For a general quasilinear or a fully nonlinear G a weak form is impossible:

$$u_0 \in \mathcal{U}_0 :\ Gu_0 - f = \sum_{i,j=0}^{n} a_{i,j}(\cdot, u_0, \nabla u_0)\partial^i\partial^j u_0 - f(\cdot) = 0 \in \mathcal{V}, \text{ or} \tag{5.14}$$

$$u_0 \in \mathcal{U}_0 :\ G(u_0) = G(\cdot, u_0(\cdot), \nabla u_0(\cdot), \nabla^2 u_0(\cdot)) = 0 \in \mathcal{V}. \tag{5.15}$$

Consequently, a weak form of a FEM is impossible, thus motivating our FEM.

The general theory of discretization methods used in this section, cf. Chapter 3, can be applied if the original problem has two properties: It admits a locally unique solution u_0, strongly related to the stability by the condition of a boundedly invertible $G'(u_0)$, and u_0 is smooth enough, implying the consistency and good enough convergence of our method, cf. Condition **H**. So we refer to the existence and regularity results for fully nonlinear elliptic equations in Subsection 2.5.7. We assume (5.12) for $\Omega \subset \mathbb{R}^n$, a slightly larger open $\Omega' \subset \mathbb{R}^n$ with $\bar{\Omega} \subset \Omega'$, $\overline{\Omega_c^h} \subset \Omega'$, where Ω_c^h indicates later approximations for Ω for the $u^h \in \mathcal{U}^h$. A real valued function G^w is defined such that

$$G^w : w = (x, z, p, r) \in \mathcal{D}(G^w) = \Omega' \times \Gamma \to \mathbb{R}, \Gamma \subset \mathbb{R} \times \mathbb{R}^n \times \mathbb{R}^{n^2}, \Omega', \Gamma \text{ open}, \tag{5.16}$$

with r restricted to symmetric real valued $n \times n$ matrices. We consider the strong elliptic BVP, cf. (5.1), (5.15), (5.16), often with $\phi = 0$ and

$$w(x) := w_u(x) := (x, u(x), \nabla u(x), \nabla^2 u(x)), \quad u \to G(u) := G^w(w_u) \text{ for} \tag{5.17}$$
$$u \in \mathcal{D}(G) := \{u \in \mathcal{U} : w_u(x) \in \mathcal{D}(G^w) \quad \forall x \in \Omega \text{ a.e., and } G(u) \in \mathcal{V}\},$$

$$\text{the solution satisfies } u_0 \in \mathcal{D}(G) :\ G(u_0) = 0 \text{ a.e. on } \Omega, \quad u_0 = \phi \text{ on } \partial\Omega. \tag{5.18}$$

Ellipticity is defined by the linear $G'(u_0)$ in $w_0 := w_{u_0}$ applied to u:

$$G'(u_0)u = \frac{\partial G^w}{\partial z}(w_0)u + \sum_{i=1}^{n} \frac{\partial G^w}{\partial p_i}(w_0)\partial^i u + \sum_{i,j=1}^{n} \frac{\partial G^w}{\partial r_{ij}}(w_0)\partial^i\partial^j u \forall u, u_0 \in \mathcal{D}(G). \tag{5.19}$$

This will be important below for local uniqueness and stability. The operator G is called *uniformly elliptic* in $u_0 \in \mathcal{D}(G)$ and all of $\mathcal{D}(G)$ if its linearization $G'(u)$ is uniformly elliptic for $u = u_0$ and for all $u \in \mathcal{D}(G)$, respectively. In the first case its principal part, $a_p(u, v)$ (see (5.21)), satisfies, for $0 < \lambda < \Lambda$, and $w_0 = w_{u_0}$ in (5.17),

$$0 < \lambda|\vartheta|^2 \leq -\sum_{i,j=1}^{n} \frac{\partial G^w}{\partial r_{ij}}(w_0)\vartheta^i\vartheta^j \leq \Lambda|\vartheta|^2 \quad \forall 0 \neq \vartheta = (\vartheta^1, \ldots, \vartheta^n) \in \mathbb{R}^n, \tag{5.20}$$

sometimes only for a.e. $x \in \Omega$ in $w_0(x)$. For fully nonlinear second order elliptic equations for the smooth $u, u_0 \in \mathcal{D}(G)$ this implies the $\mathcal{W}_0$-coercivity of the bilinear form induced by the principal part, $a_p(u, v)$ of $G'(u_0)$. By the standard $\vartheta^i = \partial^i u(x)$, integrating over Ω, and using partial integration we obtain, with the equivalence in $\mathcal{W}_0$ of the energy norm $(\sum_{i,j=1}^n \|\partial^i u\|^2_{\mathcal{V}})^{1/2}$ and $\|\cdot\|_{H^1(\Omega)}$ and again with $0 < \lambda_1 < \Lambda_1 \in \mathbb{R}$,

$$\lambda_1\|u\|^2_{H^1(\Omega)} \leq a_p(u, u) := \int_\Omega \sum_{i,j=1}^{n} \frac{\partial G^w}{\partial r_{ij}}(w_0)\partial^i u\partial^j u dx \leq \Lambda_1\|u\|^2_{H^1(\Omega)} \quad \forall u \in \mathcal{W}_0. \tag{5.21}$$

Therefore we choose the Hilbert space setting, $\mathcal{W} = H^1(\Omega)$. A Banach space setting, $W^{1,p}(\Omega)$, might be appropriate if in addition to (5.20) some monotonicity conditions were considered. Furthermore, since $W^{2,p}(\Omega)$ is continuously embedded into $H^2(\Omega)$, for $p \geq 2$, our new FEM even works for fully nonlinear equations in $W^{2,p}(\Omega)$, e.g. $G(u) = (\Delta u)^p + f(u)$; however it yields only discrete $H^2(\Omega)$ estimates.

In Example 2.78, some of the most important examples are listed, the Monge–Ampère equation and the *equation for a surface with prescribed Gauss curvature,* see (2.308), (2.310). The existence of surfaces with prescribed Gauss curvature is known for appropriate given curvature. Specific properties would become more flexibly available via our FEMs than by our finite difference methods in Chapter 8, and e.g. Crandall and Lions [227] and Barles and Souganidis [73]: our FEM allows C^1 FEs on quasiuniform grids and yields, for the forthcoming better boundary approximations, higher orders of convergence than the difference methods in [227], which are only defined on equidistant grids.

We have previously observed that we need the following conditions **H**:

Condition H: Essential conditions for well-defined and differentiable G: *These very important properties (5.1), (5.4), (5.5) have to be discussed in more detail. We have realized that for the Monge-Ampere equation, cf. Example 3.26, the $G(u)$ is not defined for $u \in \mathcal{U}$, but only for $u \in \mathcal{D}(G)$, a subset of smooth functions $u \in \mathcal{U}_s$ fitting to the growth properties of G: Smooth solutions u_0 are necessary anyways for good convergence. For the Monge-Ampere equation Proposition 3.27 shows that $G(u^h) \in L^2(\Omega)$, easily extended to $L^2(\Omega \cup \Omega_c^h)$. More generally, we require, cf. (5.5)*

($\mathbf{H}_1$) *Assume G in (5.1) is well defined on*

$$G : \mathcal{D}(G) \subset \mathcal{U}_s \subset \mathcal{U}_c = H^2(\Omega \cup \Omega_c^h) \to \mathcal{V}_c = L^2(\Omega \cup \Omega_c^h),$$

so $G(u^h)$ is well defined.

($\mathbf{H}_2$) *Below, $G'(u_0)$ is important for stability. So we will assume*

$$G : B_r(u_0) \subset \mathcal{D}(G) \subset \mathcal{U}_s \to L^2(\Omega \cup \Omega_c^h), G \in C^1(B_r(u_0))$$

w.r.t. the $\|\cdot\|_{\mathcal{U}_s}$, s.t. $\forall u \in B_r(u_0) : G'(u) \in \mathcal{L}(H^2(\Omega), L^2(\Omega))$, and can be extended as $\in \mathcal{L}(\mathcal{U}_c, \mathcal{V}_c)$ and $\in \mathcal{L}(\mathcal{W}_c, \mathcal{W}_c')$, e.g. by imposing (5.19)–(5.21), or in more detail (5.73)–(5.75).

These properties have to be verified for each combination of $G, \mathcal{U}, \mathcal{V}, \mathcal{U}^h, \mathcal{V}^h$.

Existence, uniqueness and regularity results are summarized in Theorems 2.79, 2.80, 2.81, and 2.83. They are obtained by methods, usually independent of linearization. However, preparing the next subsections, we assume additionally smooth $\mathcal{D}(G)$, compare Condition **H**

$$\text{for } F : \mathcal{D}(G) \subset \mathcal{U} \to \mathcal{V} \times \mathcal{V}_D, \text{in (5.2), with a locally unique solution } u_0 \text{ of}$$
$$F(u_0) = (0, \phi), \text{let } F'(u_0) \in \mathcal{L}(\mathcal{U}, \mathcal{V} \times \mathcal{V}_D) \text{ be boundedly invertible.} \qquad (5.22)$$

For systems of q fully nonlinear equations of order $2m$ of the form (5.111), (5.112) below, nearly nothing seems to be known, for the order 2 some results exist, see Subsection 2.6.8. Now, a combination of our FEMs below and continuation techniques allows studying this class of problems.

5.2.4 Existence and convergence for semiconforming FEMs

In Subsection 5.2.2 we have seen that the standard FEM approaches, based on weak forms, do not seem to be possible for the general case of fully nonlinear elliptic equations. We formulated the new FEM for fully nonlinear equations of second order with trivial Dirichlet boundary conditions, see (5.4), (5.6) and introduced the necessary approximating FE spaces in Subsection 4.2.6.

We apply the general theory of discretization methods, cf. Chapter 3 and Summary 4.52, with its standard "consistency and stability implies convergence" result to our FEMs, here the *semiconforming and the nonconforming FEMs* in the next subsection. Semiconforming FEMs are based upon strong forms and conforming FEs. We use the notation in (5.11), (5.16), and (5.17). Let (5.18) have a locally unique solution (see Subsection 5.2.3) with boundedly invertible derivative $G'(u_0)$.

We recall that our program for proving convergence requires several steps, summarized in Subsection 5.2.2 **1.**–**4.**: Until now we have worked through the first two bullets of **1.**, and we turn, in this and the next subsection, to the last three bullets of **1.**

The *conforming FEs,* defined on a triangulation with straight edges, $\mathcal{T}^h$, require

$$\Omega \text{ convex, bounded and polyhedral with } \bar{\Omega} = \cup_{T \in \mathcal{T}^h} \bar{T}. \qquad (5.23)$$

In Subsection 5.2.5 we will return to *nonconforming FEs,* defined on $\Omega \in C^2$. This yields the more complicated case of violated differential and boundary equations.

The original problem and its semiconforming discretization

We need the following spaces, note no Gelfand triple:

$$\mathcal{U} = H^2 \subset \mathcal{V} = \mathcal{V}' = L^2, \quad \mathcal{U}_0 = \mathcal{U} \cap H_0^1 \text{ all defined on } \Omega. \qquad (5.24)$$

G and its exact solution $u_0 \in \mathcal{D}(G) \cap \mathcal{U}_0$ satisfy, cf. Condition **H**

$$G : \mathcal{D}(G) \subset \mathcal{U} \to \mathcal{V} \text{ and } G(u_0) = 0 \in \mathcal{V} \Longleftrightarrow (G(u_0), v)_{\mathcal{V}} = 0 \quad \forall v \in \mathcal{V}. \quad (5.25)$$

We choose the discrete G^h for G as in (5.4). The *sequences of approximating subspaces* $\mathcal{U}^h = \mathcal{V}^h = S_0^h \subset H^2(\Omega) = \mathcal{U}$ in (5.3), defined on the original Ω in (5.23), satisfy the boundary conditions on the exact boundary. So we call this FEM a *semiconforming FEM.*

More precisely we need

$$\{\mathcal{U}^h, \mathcal{V}^h, P^h, Q^{h'}, G^h\}_{h\in H}, \ \inf\{0 < h \in H\} = 0, \ \mathcal{S}_d^1(\mathcal{T}^h) \supset \mathcal{U}^h = \mathcal{V}^h = S_0^h. \quad (5.26)$$

For these spaces with identical elements we require (asymptotically) *nonequivalent norms and scalar products,* defined as those of the original spaces, cf. (5.24), e.g.

$$\|u^h\|_{\mathcal{U}^h} := \|u^h\|_{\mathcal{U}}, \quad \|v^h\|_{\mathcal{V}^h} := \|v^h\|_{\mathcal{V}}. \quad (5.27)$$

So we might call this method a Petrov–Galerkin method as well. By Theorem 4.36 these sequences *approximate the Banach spaces* $\mathcal{U}, \mathcal{V}$.

For the final FE version of (5.4) we need $G^w, w_{u^h}(x), \mathcal{D}(G^w)$ in (5.16), (5.18), (5.25):

$$\mathcal{D}(G^h) := \{u^h \in \mathcal{U}^h \cap \mathcal{D}(G) : w_{u^h}(x) \in \mathcal{D}(G^w) \quad \forall x \text{ a.e. in } \Omega, G(u^h) \in \mathcal{V}\}. \quad (5.28)$$

So we define the FE solution $u_0^h \in \mathcal{U}^h \cap \mathcal{D}(G^h) \subset \mathcal{U}$ such that

$$G^h : \mathcal{D}(G^h) \to \mathcal{V}^h, \text{ and } G^h\left(u_0^h\right) = 0 \Longleftrightarrow \left(G\left(u_0^h\right), v^h\right)_{\mathcal{V}} = 0 \quad \forall v^h \in \mathcal{V}^h \subset \mathcal{V}. \quad (5.29)$$

The necessary projectors

Hence, we introduce *sequences* of projectors, $\{P^h, Q^{h'}\}_{h\in H}$, as appropriate tools for our FEM. The P^h have to satisfy, cf. Theorem 4.36,

$$P^h \in \mathcal{L}(\mathcal{U},\mathcal{U}^h) : \ \forall u \in \mathcal{U} \ \lim_{h\to 0} \|P^h u\|_{\mathcal{U}^h} = \|u\|_{\mathcal{U}}, \text{ we choose } P^h u := I^h u. \quad (5.30)$$

Here we have chosen I^h in (4.63) as in Theorem 4.36, cf. Davydov [264]. Hence, interpolating values of functions or derivatives or other kinds of quasi- or averaging-interpolation or best approximation operators could be used, sometimes called Clément interpolation; cf. e.g. Clément [178], de Boor [270], Scott and Zhang [578].

More delicate than P^h are the projectors to the image spaces, the $Q^{h'} \in \mathcal{L}(\mathcal{V}, \mathcal{V}^{h'})$, motivated by G^h. We test $G(u_0) = 0$ as in (5.25) and $G^h\left(u_0^h\right) = 0$ in (5.29). We keep the concepts for the original and the discrete problem as close as possible and use the function spaces and norms in (5.24), (5.26), (5.27). In (5.25) and (5.29) we have satisfied the differential equation and its discrete counterpart by testing with $\mathcal{V}$ and $\mathcal{V}^h$. We define,[35] cf. (4.73)–(4.75) and Theorem 4.38, for all $f \in \mathcal{V} = \mathcal{V}'$,

$$Q^{h'} \in \mathcal{L}(\mathcal{V}, \mathcal{V}^{h'}) : \ \langle Q^{h'} f - f, v^h\rangle_{(\mathcal{V}^{h'}\cup\mathcal{V}')\times\mathcal{V}^h} = 0 \ \forall v^h \in \mathcal{V}^h; \ \lim_{h\to 0} \|Q^{h'} f\|_{\mathcal{V}^{h'}} = \|f\|_{\mathcal{V}}. \quad (5.31)$$

[35] Systematically we use the following identification: For $\mathcal{U} \subset \mathcal{V} = \mathcal{V}'$, for all $u \in \mathcal{U}$, for all $v \in \mathcal{V} \subset \mathcal{U}'$, and the corresponding discrete terms, we obtain $(u, v)_{\mathcal{V}} = \langle u, v\rangle_{\mathcal{U}\times\mathcal{U}'} = \langle u, v\rangle_{\mathcal{U}'\times\mathcal{U}}$, similarly for $\mathcal{V}_D \subset \mathcal{V}_b = \mathcal{V}_b' \subset \mathcal{V}_D', \ldots$, and $\mathcal{W} \subset \mathcal{V} = \mathcal{V}' \subset \mathcal{W}'$ below. This is applied for the following projectors $Q^{h'}, \ldots$ as well.

$$\begin{array}{ccccccc}
\mathcal{U} & \xrightarrow{G} & \mathcal{V} & \xleftrightarrow{\text{tested by}} & \mathcal{V}=\mathcal{V}' \\
P^h\downarrow & \Phi^h\downarrow & Q^{h'}\downarrow & & \\
\mathcal{U}^h & \xrightarrow{\mathcal{U}^h} & \mathcal{V}^{h'} & \xleftrightarrow{\text{tested by}} & \mathcal{V}^h \neq \mathcal{V}^{h'}
\end{array}$$

Figure 5.1 Semiconforming FEMs for fully nonlinear problems.

The last limit is obtained as a consequence of (4.64). By (4.82), (4.83) with $\mathcal{V}^h \neq \mathcal{V}^{h'}$ it makes sense that for a function, e.g. $f \in \mathcal{V}$ or $G(u^h) \notin \mathcal{V}^h$, the $Q^{h'} f \in \mathcal{V}^{h'}$ is not a function. We did not identify $\mathcal{V}^h \neq \mathcal{V}^{h'}$ for emphasizing this situation more clearly.

The choice of $P^h, Q^{h'}$, for the given $\mathcal{U}^h, \mathcal{V}^h$, and problem (5.25) certainly is not unique. The combination $\mathcal{U}, \mathcal{V}, \mathcal{U}^h, \mathcal{V}^h, P^h, Q^{h'}, G^h$ should be chosen appropriately to yield classical consistency (see below) with the highest possible order; see (4.150).

Projectors and discretization

For the nonlinear operator, $G : \mathcal{D}(G) \subset \mathcal{U} \to \mathcal{V}$, in (5.25), the previously introduced projectors $P^h, Q^{h'}$ allow a shorthand notation of the corresponding discrete operator, $G^h : \mathcal{D}(G^h) \subset \mathcal{U}^h \to \mathcal{V}^{h'}$; see (5.29). So we determine the approximate solution $u_0^h \in \mathcal{U}^h$ such that

$$\left\langle G\left(u_0^h\right), v^h \right\rangle_{\mathcal{V}\times\mathcal{V}^h} = 0 \quad \forall v^h \in \mathcal{V}^h \Longleftrightarrow G^h\left(u_0^h\right) = 0 \text{ with } G^h := Q^{h'} G|_{\mathcal{U}^h}. \quad (5.32)$$

This brings about a mapping

$$\mathcal{D}(\Phi^h) := \{G \text{ in } (5.25)\} \text{ and } \Phi^h \text{ defined by } \Phi^h(G) := G^h := Q^{h'} G|_{\mathcal{U}^h}. \quad (5.33)$$

This Φ^h is even linear in the sense

$$\Phi^h(c_1 G^1 + c_2 G^2) = c_1 \Phi^h(G^1) + c_2 \Phi^h(G^2) \quad \forall c_1, c_2 \in \mathbb{R},\ \forall G^1, G^2 \in \mathcal{D}(\Phi^h). \quad (5.34)$$

Linearity is essential for the MIP in Subsection 5.2.10, allowing the proof of a quadratically convergent discrete Newton method.

Our general discretization theory in Chapter 3, in contrast to Stetter [596] and Vainikko [644], emphasizes the testing with the $\mathcal{V}$, $\mathcal{V}^h$: summarizing yields Figure 5.1 with uniformly bounded $P^h, Q^{h'}$.

In Summary 4.52 we have recalled discretization methods for FEMs for $G(u) = 0$. Essential are *consistency*, *consistency of order* k and *stability* in u_0. Our semiconforming FEM is consistent for a continuous or Lipschitz-continuous G, by Theorem 4.54. For $G \in C_L$, the consistency order is the same as the order of the interpolation error, $\|I^h u_0 - u_0\|_{H^2(\Omega)}$, cf. Theorem 4.36, (4.66). It is nearly obvious how to simplify the highly technical proof for the stability of the nonconforming $F^h\left(u_0^h\right)$ in Subsection 5.2.7. So we refer to this modification yielding stability for the semiconforming $G^h\left(u_0^h\right)$ in Theorem 5.12. With Theorem 3.21, we get the following Theorem:

Theorem 5.2. Existence and convergence for the semiconforming FEM for convex polyhedral domains Ω in $\mathbb{R}^n$: *Let the original problem* (5.25) *be defined on* Ω *in (5.23), satisfy Condition* **H** *and* (5.22) *and have the exact isolated solution* $u_0 \in H^\ell(\Omega), \ell - 1 \le d$ *with a boundedly invertible* $G'(u_0)$. *Let* G *be Lipschitz-continuous* *in* $B_r(u_0) \cap \mathcal{D}(G)$ *as* $\|G(u) - G(v)\| \le L\|u - v\|_{\mathcal{U}}$, *recall* $\mathcal{U} = H^2, \mathcal{V} = L^2(\Omega)$. *Then its discretization* $G^h = Q^{h'} G|_{\mathcal{U}^h} : \mathcal{U}^h \to \mathcal{V}^{h'}$ *in* (5.32) *with* $\mathcal{S}_d^1(\mathcal{T}^h) \supset \mathcal{U}^h = \mathcal{V}^h = S_0^h$ *in (5.26)* has the following properties:

1. *The projectors* P^h *and* $Q^{h'}$ *are (equi-)bounded and satisfy* (5.30) *and* (5.31).
2. $G^h : \mathcal{U}^h \to \mathcal{V}^{h'}$ *is defined and continuous in* $B_r(P^h u_0)$, $r > 0, h$*-independent.*
3. G^h *is consistent with* G *in* $P^h u_0$, *so, by (4.66),*

$$\|G^h P^h u_0 - Q^{h'} G u_0\|_{\mathcal{V}^{h'}} \le L\|u - I^h u\|_{H^2(\Omega)} \le KLh^{\ell-2}|u|_{H^\ell(\Omega)}. \tag{5.35}$$

4. G^h *is stable for* $P^h u_0$.

Therefore the discrete problem $G^h(u^h) = 0$ *possesses a unique solution* $u_0^h \in \mathcal{U}^h$ *near* u_0 *for all sufficiently small* $h \in H$, *and* u_0^h *converges to* u_0, *according to*

$$\left\|u_0^h - P^h u_0\right\|_{\mathcal{U}^h} \le SKLh^{\ell-2}|u|_{H^\ell(\Omega)}. \tag{5.36}$$

We will come back to these conditions **1.–4.** in Theorem 5.2 in Remark 5.3 below.

5.2.5 Definition of nonconforming FEMs

Now we return to *nonconforming FEs.* We choose a triangulation, $\mathcal{T}_c^h$, possibly with *curved edges near* $\partial\Omega$ and require and define, for Ω' cf. (5.16),

$$\Omega \in C^2, \text{ a triangulation, } \mathcal{T}_c^h, \text{ with curved edges, s.t. } \overline{\Omega_c^h} := \cup_{T \in \mathcal{T}_c^h} \overline{T} \tag{5.37}$$

$$\textit{and a sequence of open } \Omega_0^h \textit{ s.t. } \Omega' \supset \Omega_0^h \supset \Omega_c^h \cap \Omega \textit{ and } \operatorname{dist}\left(\partial\Omega_0^h, \partial\Omega\right) \le Ch^p,$$

with $p = 2$ for the situation in Conjecture 4.40, and possibly $p \ge 2$ for Conjecture 4.41. As a consequence of Theorems 4.37 and 4.38 for the $\mathcal{U}, \mathcal{W}, \mathcal{V} = \mathcal{V}', G, B, G', \ldots$ *extensions from* Ω *to* Ω_0^h are possible. The following estimates then have to be multiplied by the constant C in Theorems 4.37 and 4.38, C independent of h. We maintain the notation $\mathcal{U}, \mathcal{W}, \mathcal{V} = \mathcal{V}', G, B, G', \ldots$ defined on Ω, and only refer to Ω_0^h if necessary.

Again, the first two paragraphs in Subsection 5.2.4 remain valid. The $\mathcal{U}, \mathcal{W}, \mathcal{V} = \mathcal{V}'$, defined on the original Ω, are now approximated by *sequences of nonconforming subspaces* $\mathcal{U}^h, \mathcal{W}^h, \mathcal{V}^h$, defined on the approximate Ω_c^h. This implies *violated boundary conditions* on the exact boundary; cf. (5.38). The FEM changes into a so-called *nonconforming Petrov–Galerkin method,* requiring a more complicated treatment than in Subsection 5.2.4.

The original problem and its nonconforming discretization

We need, cf. (5.24),

$$\mathcal{U} = H^2 \subset \mathcal{W} = H^1 \subset \mathcal{V} = \mathcal{V}' = L^2 \subset \mathcal{W}' = H^{-1}, \ \mathcal{W}_0 = H_0^1 \text{ all defined on } \Omega \tag{5.38}$$

$$\text{and } \mathcal{U}_c = H_c^2 \subset \mathcal{W}_c = H_c^1 \subset \mathcal{V}_c = L_c^2 \subset \mathcal{W}_c' = H_c^{-1}, \ \mathcal{W}_{c,0} = H_{c,0}^1 \text{ all defined on } \Omega_c^h.$$

We have indicated above that the realization of the differential equation *and* the boundary condition and their violation is a deciding step of this method. Hence, the discretization method has to care about both aspects. This is achieved by the new operator $F = (G, B_D)$, with the differential operator and the boundary operator as its two components; see (5.1), (5.2). The boundary operator, B_D, maps, by the Trace Theorem 1.38,

$$B_D : \mathcal{U} \to \mathcal{V}_D := H^{3/2}(\partial\Omega) = B_D\mathcal{U}, \quad B_D u := u|_{\partial\Omega}, \; \mathcal{V}_D \subset \mathcal{V}_b' = \mathcal{V}_b := L^2(\partial\Omega). \quad (5.39)$$

We consider here mainly trivial Dirichlet conditions. Thus we have to characterize

$$u|_{\partial\Omega} = 0 \Leftrightarrow \langle v', u|_{\partial\Omega}\rangle_{\mathcal{V}_D' \times \mathcal{V}_D} = 0 \; \forall v' \in \mathcal{V}_D' \Leftrightarrow (u|_{\partial\Omega}, v)_{L^2(\partial\Omega)} = 0 \; \forall v \in \mathcal{V}_b \qquad (5.40)$$

for $u \in \mathcal{U}$, with $\mathcal{V}_D = H^{3/2}(\partial\Omega) \hookrightarrow \mathcal{V}_b = \mathcal{V}_b' \hookrightarrow \mathcal{V}_D'$, a dense (compact) embedding.

The transformation from nontrivial to trivial Dirichlet conditions will be combined in Theorems 5.4, 5.9, and 5.13 obtaining stability and convergence for the inhomogeneous case as well.

$F = (G, B_D)$ is called *elliptic* simultaneously with G, cf. (5.20). F and its Fréchet derivative $F'(u)$ map $\mathcal{D}(G)$ and $\mathcal{U}$ into and onto, respectively, the Cartesian product $\mathcal{V} \times \mathcal{V}_D$:

$$F := (F_1, F_2) := (G, B_D) : \mathcal{D}(G) \subset \mathcal{U} \to \mathcal{V}_\Pi := \mathcal{V} \times \mathcal{V}_D \subset \hat{\mathcal{V}}_\Pi := \mathcal{V} \times \mathcal{V}_b = \hat{\mathcal{V}}_\Pi'; \quad (5.41)$$

see (5.17). Since $\mathcal{V}_D$ is dense in $\mathcal{V}_b$, the smooth exact solution u_0 is hence characterized by

$$\begin{aligned} F(u_0) = 0 \in \mathcal{V}_\Pi &\Longleftrightarrow (F(u_0), v_\Pi)_{\hat{\mathcal{V}}_\Pi} = 0 \quad \forall v_\Pi := (v, v_b) \in \hat{\mathcal{V}}_\Pi \qquad (5.42)\\ &\Longleftrightarrow (G(u_0), v)_{\mathcal{V}} = 0 \; \forall v \in \mathcal{V}, \text{ and } (B_D u_0, v_b)_{\mathcal{V}_b} = 0 \quad \forall v_b \in \mathcal{V}_b,\\ \text{abbreviated } &\Longleftrightarrow (G(u_0), v)_{\mathcal{V}} + (B_D u_0, v_b)_{\mathcal{V}_b} = 0 \quad \forall v \in \mathcal{V}, \quad v_b \in \mathcal{V}_b. \end{aligned}$$

For the standard weak form for $F'(u)$ in Subsection 5.2.7, we need, cf. (5.39), (5.40)

$$\begin{aligned} B_D : \mathcal{W} \to \mathcal{W}_D := H^{1/2}(\partial\Omega) = B_D\mathcal{W}, \quad u \to B_D u = u|_{\partial\Omega}, \text{ with} \qquad (5.43)\\ u \in \mathcal{W} \text{ and } B_D u = 0 \Longleftrightarrow (u|_{\partial\Omega}, v)_{L^2(\partial\Omega)} = 0 \quad \forall v \in \mathcal{V}_b \text{ dense in } \mathcal{W}_D'. \end{aligned}$$

$B_D : \mathcal{U} \to \mathcal{V}_D$ and $B_D : \mathcal{W} \to \mathcal{W}_D$ are continuous with respect to the corresponding Sobolev–Sloboditskii norms, cf. (1.60), Theorem 1.30. Similarly to (5.38), we introduce

$$\begin{aligned} \mathcal{V}_{c,D} := H^{3/2}\left(\partial\Omega_c^h\right) \subset \mathcal{W}_{c,D} := H^{1/2}\left(\partial\Omega_c^h\right) \subset \mathcal{V}_{c,b} := L^2\left(\partial\Omega_c^h\right) \text{ and} \qquad (5.44)\\ B_{c,D} : \mathcal{U}_c \to \mathcal{V}_{c,D}, u_c \to B_{c,D} u_c := u_c|_{\partial\Omega_c^h}; \text{ similarly, } B_{c,D} : \mathcal{W}_c \to \mathcal{W}_{c,D}. \end{aligned}$$

To apply the general discretization theory in Chapter 3, we introduce the discrete F^h for F, and the corresponding projectors $P^h \in \mathcal{L}(\mathcal{U}, \mathcal{U}^h)$, and $Q_\Pi^{h'} \in \mathcal{L}(\mathcal{V}_\Pi, \mathcal{V}_\Pi^{h'})$, $Q_{c,\Pi}^{h'} \in \mathcal{L}(\mathcal{V}_{c,\Pi}, \mathcal{V}_\Pi^{h'})$; see (5.62). This yields the standard "consistency and stability implies convergence" result, (5.37); note $S_0^h = S^h \cap H_0^1(\mathcal{T}_c^h)$ and the stable splitting in Theorem 4.36:

$$S_d^1\left(\mathcal{T}_c^h\right) \supset \mathcal{U}^h = \mathcal{W}^h = S^h = S_0^h \oplus S_b^h, \quad \mathcal{V}^h = S_0^h, \quad \mathcal{V}_D^h = \mathcal{W}_D^h = \mathcal{V}_b^h := S_b^h|_{\partial\Omega_c^h}. \qquad (5.45)$$

Again these spaces require *nonequivalent norms and scalar products.* With the norms of the original spaces, now defined on Ω_c^h and $\partial\Omega_c^h$, cf. (5.38), (5.44), let

$$\|u^h\|_{\mathcal{U}^h} := \|u^h\|_{\mathcal{U}_c}, \|w^h\|_{\mathcal{W}^h} := \|w^h\|_{\mathcal{W}_c}, \|v^h\|_{\mathcal{V}^h} := \|v^h\|_{\mathcal{V}_c}, \left\|u^h|_{\partial\Omega_c^h}\right\|_{\mathcal{V}_D^h} := \left\|u^h|_{\partial\Omega_c^h}\right\|_{\mathcal{V}_{c,D}},$$

$$\left\|w^h|_{\partial\Omega_c^h}\right\|_{\mathcal{W}_D^h} := \left\|w^h|_{\partial\Omega_c^h}\right\|_{\mathcal{W}_{c,D}}, \left\|v^h|_{\partial\Omega_c^h}\right\|_{\mathcal{V}_b^h} := \left\|v^h|_{\partial\Omega_c^h}\right\|_{\mathcal{V}_{c,b}}. \tag{5.46}$$

By Theorems 4.36–4.38 these sequences *approximate the Banach spaces and subsets,* cf. (4.73), $E_c\mathcal{U}$, $E_c\mathcal{W}$, $E_c\mathcal{V}$, $E_c\mathcal{V}_D$, $E_c\mathcal{W}_D$, $E_c\mathcal{V}_b$, $E_c\mathcal{D}(G)$ defined on Ω_c^h and $\partial\Omega_c^h$ with respect to the corresponding norms. The $u^h \in S^h$, and $v^h \in \mathcal{V}_b^h$, violate the original boundary conditions. Since E_c maintains smoothness properties, we can choose this E_c s.t. (5.5) is maintained to these $E_c\mathcal{U}, E_c\mathcal{W}, \ldots$.

The product operator $F : \mathcal{D}(G) \subset \mathcal{U} \to \mathcal{V} \times \mathcal{V}_D$ and the corresponding $F^h : \mathcal{D}(G) \cap \mathcal{U}^h \to \mathcal{V}^{h'} \times \mathcal{V}_D^h$ cause some unpleasant technicalities in the definition of $Q_\Pi^{h'}$, $Q_{c,\Pi}^{h'}$, compared to Subsection 5.2.4. Subsequently, the consistency theory becomes relatively simple again; see Figure 5.2 and Subsection 5.2.6. These technicalities are not surprising compared to some other approaches for violated boundary conditions, e.g. in Lenoir [471] and Bernadou [85–88]. The unavoidable complications are piled up in Subsection 5.2.7.

The terms u^h, $G(u^h)$, and $u^h|_{\partial\Omega_c^h}$ are, by (5.17), defined on Ω_c^h and $\partial\Omega_c^h$ instead of the original Ω and $\partial\Omega$. Thus we start updating the preliminary FE version (5.6). With $G^w, \mathcal{D}(G^w), w_{u^h}(x)$ in (5.16), (5.17), (5.18), and $\mathcal{U}^h$ in (5.45), let

$$\mathcal{D}(G^h) := \left\{u^h \in \mathcal{U}^h : w_{u^h}(x) \in \mathcal{D}(G^w) \quad \forall x \in \Omega \cup \Omega_c^h, G(u^h) := G^w(w_{u^h}) \in \mathcal{V}_c\right\}. \tag{5.47}$$

This condition has to be verified similarly to Proposition 3.27, cf. Condition **H**. The extended definition of G^w was the reason for introducing the Ω' in (5.16).

We replace the differential equation $G(u_0) = 0 \in \mathcal{V}, u_0 \in \mathcal{D}(G)$, in (5.42) by the FE equations in equivalent formulations; see (5.6), (5.41), (5.46): define

$$\text{the sequence } F_1^h\left(u_0^h\right) = G^h\left(u_0^h\right) = 0 \Longleftrightarrow \left(G\left(u_0^h\right), v^h\right)_{\mathcal{V}_c} = 0 \quad \forall v^h \in \mathcal{V}^h \subset \mathcal{V}_c. \tag{5.48}$$

Similarly, the boundary conditions (5.40) are tested (see the equivalence (5.6)) as

$$u^h \in \mathcal{U}^h : F_2^h(u^h) = B_D^h u^h = u^h|_{\partial\Omega_c^h} = 0 \in \mathcal{V}_D^h \Leftrightarrow \left(v_b^h, u^h\right)_{\mathcal{V}_b^h} = 0 \quad \forall v_b^h \in S_b^h, \tag{5.49}$$

with $\cup_{h\in H} S_b^h$ dense in $\mathcal{V}'_{c,D}$. Consequently we introduce, cf. F in (5.41),

$$F^h = \left(F_1^h, F_2^h\right) = \left(G^h, B_D^h\right) \text{ with } \mathcal{D}(F^h) := \mathcal{D}(G^h). \tag{5.50}$$

The nonconforming projectors

Again, projectors represent an appropriate tool for $F^h = \left(F_1^h, F_2^h\right)$. They have to handle the discrepancy between u^h, $G(u^h)$, and $u^h|_{\partial\Omega_c^h}$, by (5.17), defined on Ω_c^h and $\partial\Omega_c^h$ instead of the original Ω and $\partial\Omega$. We combine the extension operator E_c introduced in Theorem 4.37 with inverse estimates and the estimates for interpolation errors on Ω in Theorems 4.36 and 4.38. The *sequence* of P^h has to satisfy

$$P^h \in \mathcal{L}(\mathcal{U},\mathcal{U}^h) : \forall u \in \mathcal{U} \lim_{h\to 0} \|P^h u\|_{\mathcal{U}^h} = \|u\|_{\mathcal{U}}, \text{ we choose } P^h u := I^h E_c u. \quad (5.51)$$

E_c is unnecessary for the Argyris FEs but possibly needed for other schemes.

Again more delicate than P^h are the projectors on the product space, $Q_\Pi^{h'} \in \mathcal{L}(\mathcal{V}_\Pi, \mathcal{V}_\Pi^{h'})$. They live on the original and discrete image spaces, listed in (5.52)–(5.54), with their duals, scalar products, and pairings; cf. (5.41), (5.42). For the reader's convenience, we summarize and introduce the spaces defined on Ω and Ω_c^h; cf. (4.82), (4.83), Theorem 4.36, (5.38), (5.40), (5.45), (5.48), (5.49). We mark functions and functionals, according to their definition, by obvious indices. For functions defined on Ω_c^h we use $_c$, for $\partial\Omega$ we use $_b$, for $\partial\Omega_c^h$ we use $_{c,b}$, for the images of B_D on $\partial\Omega_c^h$ we use $_{c,D}$, and on the product of the domain and boundary on Ω and Ω_c^h we use $_\Pi$ or $_{c,\Pi}$:

$$\begin{aligned} \mathcal{V}_\Pi &:= \mathcal{V} \times \mathcal{V}_D \subset \hat{\mathcal{V}}_\Pi := \mathcal{V} \times \mathcal{V}_b, \text{ with } \mathcal{U} = H^2(\Omega), \mathcal{V} = L^2(\Omega) = \mathcal{V}', \quad (5.52)\\ \mathcal{V}_b &= L^2(\partial\Omega) = \mathcal{V}_b' \supset \mathcal{V}_D = B_B\mathcal{U} = H^{3/2}(\partial\Omega), \text{ defined on } \Omega, \text{ with} \end{aligned}$$

$$\big((u,u_b),(v,v_b)\big)_{\mathcal{V}_\Pi} := (u,v)_\mathcal{V} + (u_b,v_b)_{\mathcal{V}_D}, \ \big((u,u_b),(v,v_b)\big)_{\hat{\mathcal{V}}_\Pi} := (u,v)_\mathcal{V} + (u_b,v_b)_{\mathcal{V}_b}.$$

Similarly, we define the spaces and scalar products on the curved approximations

$$\mathcal{V}_c = L^2\left(\Omega_c^h\right), \quad \mathcal{V}_{c,D} = B_B\mathcal{U}_c = H^{3/2}\left(\partial\Omega_c^h\right), \quad \mathcal{V}_{c,\Pi}, \mathcal{V}_{c,b}, \ldots, \text{ defined on } \Omega_c. \quad (5.53)$$

Correspondingly, the discrete analog spaces and their scalar products are

$$\begin{aligned} &\mathcal{V}_\Pi^h := \mathcal{V}^h \times \mathcal{V}_D^h := S_0^h \times S_b^h|_{\partial\Omega_c^h}, \ \mathcal{V}^{h'} \neq \mathcal{V}^h, \ \mathcal{V}_\Pi^{h'} = \mathcal{V}^{h'} \times \mathcal{V}_D^h := S_0^{h'} \times S_b^h|_{\partial\Omega_c^h} \neq \mathcal{V}_\Pi^h,\\ &\hat{\mathcal{V}}_\Pi^h := \mathcal{V}^h \times \mathcal{V}_b^h, \left\langle \left(u^{h'}, u_b^h\right), \left(v^h, v_b^h\right)\right\rangle_{\mathcal{V}_\Pi^{h'} \times \mathcal{V}_\Pi^h} := \langle u^{h'}, v^h\rangle_{\mathcal{V}^{h'} \times \mathcal{V}^h} + \left(u_b^h, v_b^h\right)_{\mathcal{V}_D^h}. \end{aligned} \quad (5.54)$$

In (5.42), (5.48) and (5.40), (5.49) we have satisfied the differential equation and boundary condition or their discrete counterparts by testing with $\mathcal{V}$ and $\mathcal{V}_b$ or (see the equivalence (5.6)) with $\mathcal{V}^h$ and $\mathcal{V}_b^h$. The two components (5.42), (5.40) and (5.48), (5.49) of F and F^h, defined on Ω and Ω_c^h, respectively, are now combined with the extension operator E_c in (4.73) for motivating the appropriate definitions of the projection operators. For the discrete spaces $\mathcal{V}^{h'} \neq \mathcal{V}^h$ and $\mathcal{V}_b^{h'} = \mathcal{V}_b^h$ the FE equations are always tested with $v^h \in \mathcal{V}^h$ and $v_b^h \in \mathcal{V}_b^h$, but we use for the boundary condition the equivalence (5.49). Thus corresponding to the product spaces $\mathcal{V}_\Pi$ and $\mathcal{V}_{c,\Pi}$ we have to introduce two projectors $Q_\Pi^{h'} := (Q^{h'}, Q_b^{h'}) \in \mathcal{L}(\mathcal{V}_\Pi, \mathcal{V}_\Pi^{h'})$ and $Q_{c,\Pi}^{h'} := (Q_c^{h'}, Q_{c,b}^{h'}) \in \mathcal{L}(\mathcal{V}_{c,\Pi}, \mathcal{V}_\Pi^{h'})$. We start with the component of the differential operator, (5.48) (note $\mathcal{V} = \mathcal{V}', \mathcal{V}_b = \mathcal{V}_b'$), and define (see (4.73)–(4.75) and Theorem 4.38)

$$Q^{h'} \in \mathcal{L}(\mathcal{V}, \mathcal{V}^{h'}) \text{ for } f \in \mathcal{V} \text{ by } \langle Q^{h'} f, v^h\rangle_{\mathcal{V}^{h'} \times \mathcal{V}^h} - (E_c f, v^h)_{\mathcal{V}_c} = 0 \quad \forall v^h \in \mathcal{V}^h. \quad (5.55)$$

The approximate domains require

$$Q_c^{h'} \in \mathcal{L}(\mathcal{V}_c, \mathcal{V}^{h'}) \text{ for } f_c \in \mathcal{V}_c : \left\langle Q_c^{h'} f_c, v^h \right\rangle_{\mathcal{V}^{h'} \times \mathcal{V}^h} - (f_c, v^h)_{\mathcal{V}_c} = 0 \quad \forall v^h \in \mathcal{V}^h.$$

Both projectors yield the norm limits, the last limit as a consequence of (4.64):

$$\lim_{h\to 0} \|Q^{h'} f\|_{\mathcal{V}^{h'}} = \|f\|_{\mathcal{V}}, \text{ and } \lim_{h\to 0} \left\|Q_c^{h'} f_c\right\|_{\mathcal{V}^{h'}} = \lim_{h\to 0} \|f_c\|_{\mathcal{V}_c}. \tag{5.56}$$

Since the S_0^h and $S_0^{h'}$ form a pair of dual spaces (see Theorem 4.36), we alternatively formulate corresponding projectors (see Theorem 4.38) and use them, e.g. in (5.103):

$$Q_d^{h'} \in \mathcal{L}(\mathcal{V}, \mathcal{V}^h), \text{ by } f \in \mathcal{V}: \ Q_d^{h'} f := \sum_{j=1}^{m_0} \langle Q^{h'} f, s^j \rangle_{\mathcal{V}^{h'} \times \mathcal{V}^h} s^j \in \mathcal{V}^h, \text{ again with}$$

$$\lim_{h\to 0} \left\|Q_d^{h'} f\right\|_{\mathcal{V}^h} = \|f\|_{\mathcal{V}}; \text{ similarly, e.g. } \lim_{h\to 0} \left\|Q_{c,d}^{h'} f_c\right\|_{\mathcal{V}^h} = \lim_{h\to 0} \|f_c\|_{\mathcal{V}_c}. \tag{5.57}$$

We turn to the projectors for the traces on the boundary. We always have to recall that the $u_b \in \mathcal{V}_D$ and $u_b^h \in \mathcal{V}_b^h \subset \mathcal{V}_b$ have the forms $u_b = u|_{\partial\Omega}, u \in \mathcal{U}$ and $u_b^h = u^h|_{\partial\Omega_c^h}, u^h \in \mathcal{V}^h$, respectively. It suffices to test vanishing boundary conditions with respect to the L^2 scalar product $(\cdot,\cdot)_{\mathcal{V}_{c,b}}$. This and the dense embeddings $\mathcal{V}_D \subset \mathcal{V}_b$ and $\mathcal{V}_{c,D} \subset \mathcal{V}_{c,b}$ motivate the following projectors:

$$\begin{aligned}
&Q_b^{h'} \in \mathcal{L}\left(\mathcal{V}_D, \mathcal{V}_D^h\right) \quad \text{for} \quad u_b := u|_{\partial\Omega} \in \mathcal{V}_D \subset \mathcal{V}_b : \left(Q_b^{h'} u_b - (E_c u)|_{\partial\Omega_c^h}, v_b^h\right)_{\mathcal{V}_{c,b}} = 0,\\
&Q_{c,b}^{h'} \in \mathcal{L}\left(\mathcal{V}_{c,D}, \mathcal{V}_D^h\right), \ u_{c,b} := u|_{\partial\Omega_c^h} \in \mathcal{V}_{c,D} : \left(Q_{c,b}^{h'} u_{c,b} - u_{c,b}, v_b^h\right)_{\mathcal{V}_{c,b}} = 0 \quad \forall v_b^h \in \mathcal{V}_b^h,
\end{aligned} \tag{5.58}$$

and by (5.45) we obtain for all $v_b^h \in \mathcal{V}_b^h = \mathcal{V}_D^h$ that $v_b^h = v^h|_{\partial\Omega_c^h}$ for an appropriate $v^h \in \mathcal{V}^h$ and $v_b^h = Q_{c,b}^{h'} v^h|_{\partial\Omega_c^h}$. The convergence results along $\partial\Omega_c^h$ in (4.66), (5.6), and Theorems 4.36 and 4.38 imply the limits of the boundary norms

$$\lim_{h\to 0} \left\|Q_{c,b}^{h'} u_{c,b}\right\|_{\mathcal{V}_D^h} = \lim_{h\to 0} \|u_{c,b}\|_{\mathcal{V}_{c,D}}, \quad \lim_{h\to 0} \left\|Q_b^{h'} u_b\right\|_{\mathcal{V}_D^h} = \lim_{h\to 0} \|u_b\|_{\mathcal{V}_D}. \tag{5.59}$$

We summarize the product projectors and norms in (5.55)–(5.59) obtaining

$$Q_\Pi^{h'} := \left(Q^{h'}, Q_b^{h'}\right) \in \mathcal{L}(\mathcal{V} \times \mathcal{V}_D, \mathcal{V}^{h'} \times \mathcal{V}_D^h) = \mathcal{L}(\mathcal{V}_\Pi, \mathcal{V}_\Pi^{h'}), \tag{5.60}$$

$$Q_{d,\Pi}^{h'} := \left(Q_d^{h'}, Q_b^{h'}\right) \in \mathcal{L}(\mathcal{V}_\Pi, \mathcal{V}_\Pi^h), \text{with } \|(f, u|_{\partial\Omega})\|_{\mathcal{V}_\Pi} = \|f\|_{\mathcal{V}} + \|u|_{\partial\Omega}\|_{\mathcal{V}_D}, \text{ and}$$

$$\lim_{h\to 0} \left\|Q_\Pi^{h'}(f, u|_{\partial\Omega})\right\|_{\mathcal{V}_\Pi^{h'}} = \lim_{h\to 0} \left\|Q_{d,\Pi}^{h'}(f, u|_{\partial\Omega})\right\|_{\mathcal{V}_\Pi^h} = \|(f, u|_{\partial\Omega})\|_{\mathcal{V}_\Pi}.$$

The approximate domains yield

$$Q_{c,\Pi}^{h'} := \left(Q_c^{h'}, Q_{c,b}^{h'}\right) \in \mathcal{L}(\mathcal{V}_c \times \mathcal{V}_{c,D}, \mathcal{V}_\Pi^{h'}) = \mathcal{L}(\mathcal{V}_{c,\Pi}, \mathcal{V}_\Pi^{h'}), \tag{5.61}$$

$$\lim_{h\to 0} \left\| Q_{c,\Pi}^{h'} \left(f_c, E_c u|_{\partial\Omega_c^h}\right) \right\|_{\mathcal{V}_\Pi^{h'}} = \lim_{h\to 0} \left\| \left(f_c, E_c u\,|_{\partial\Omega_c^h}\right) \right\|_{\mathcal{V}_{c,\Pi}}.$$

Again, the choice of $P^h, Q_\Pi^{h'}, Q_{c,\Pi}^{h'}, \ldots$ is not unique.

Nonconforming projectors and discretization

We need only a few changes compared to the corresponding semiconforming subsubsection. The operator, spaces, and projectors there, $G : \mathcal{D}(G) \subset \mathcal{U} \to \mathcal{V}, Q^{h'}$, have to be replaced by the product terms, $F = (F_1, F_2) = (G, B_D) : \mathcal{D}(F) \subset \mathcal{U} \to \mathcal{V}_\Pi = \mathcal{V} \times \mathcal{V}_D$, $Q_\Pi^{h'}, Q_{c,\Pi}^{h'}$. Again we obtain a short definition of the corresponding discrete operator, $F^h : \mathcal{U}^h \to \mathcal{V}^{h'} \times \mathcal{V}_b^h$, and a linear mapping, Φ^h; see (5.47), (5.48) and (5.40), (5.49), (5.52)–(5.54), (5.60), (5.33). We finally get

$$F^h(u^h) = (G^h(u^h), u^h|_{\partial\Omega_c^h}) = \left(Q_c^{h'} G(u^h), Q_{c,b}^{h'} u^h|_{\partial\Omega_c^h}\right). \tag{5.62}$$

We determine the approximate solution u_0^h from

$$u_0^h \in \mathcal{U}^h \text{ s.t. } F^h\left(u_0^h\right) = 0 \iff \left(F\left(u_0^h\right), v_\Pi^h\right)_{\mathcal{V}_{c,\Pi}} = 0 \quad \forall v_\Pi^h = \left(v^h, v_b^h\right) \in \mathcal{V}_\Pi^h \tag{5.63}$$

$$\iff \left(G\left(u_0^h\right), v^h\right)_{\mathcal{V}^h} + \left(u_0^h, v_b^h\right)_{\mathcal{V}_b^h} = 0 \quad \forall v^h \in S_0^h = \mathcal{V}^h \subset \mathcal{V}_c \quad \forall v_b^h \in S_b^h = \mathcal{V}_b^h.$$

By (5.62), (5.63) the new FEM is defined for the $F = (G, B_D)$ in (5.41), cf. (5.44), as

$$\mathcal{D}(\Phi^h) := \{F = (G, B_D)\} : \Phi^h(F) := F^h = Q_{c,\Pi}^{h'} F|_{\mathcal{U}^h} = \left(Q_c^{h'} G|_{\mathcal{U}^h}, Q_{c,b}^{h'} B_{c,D}|_{\mathcal{U}^h}\right). \tag{5.64}$$

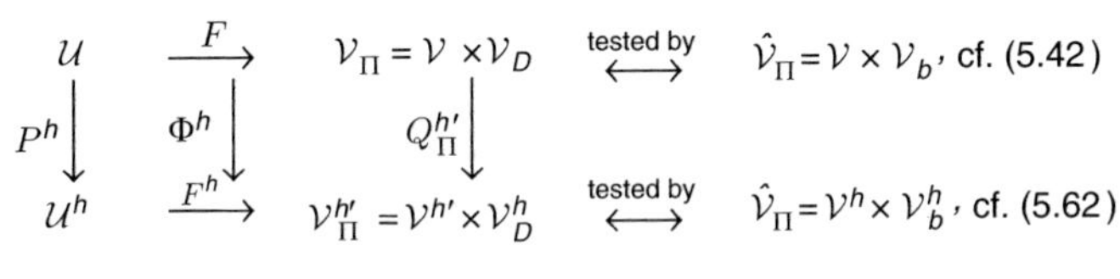

Figure 5.2 Nonconforming FEMs for fully nonlinear problems.

We have thus defined a *nonconforming Petrov–Galerkin method applicable* to every F in (5.41). With uniformly bounded $P^h, Q_\Pi^{h'}, Q_{c,\Pi}^{h'}$ we modify the above Figure 5.1 into Figure 5.2. We reformulate only the not quite trivial consistency:

$$\left\| F^h P^h u - Q_\Pi^{h'} F u \right\|_{\mathcal{V}_\Pi^{h'}} = \left\| \left(Q_c^{h'} G(P^h u) - Q^{h'} G(u), P^h u\,|_{\partial\Omega_c^h} - E_c u|_{\partial\Omega_c^h} \right) \right\|_{\mathcal{V}_\Pi^{h'}}. \tag{5.65}$$

Remark 5.3.

1. *The above changes will modify Theorem* 5.2. *We still obtain* convergence, and consistency of reduced order, stability, on modifies domains.

2. *We need for our approach only equicontinuous operators and the limits of the norms as listed in Definition 3.12 and (3.24). They are satisfied according to (5.51) and (5.55) ff. Stetter [596] requires for his theory linear bounded operators* P^h *and* $Q^{h'}_{c,\Pi} := \left(Q^{h'}_c, Q^{h'}_{c,b}\right)$ *in (5.51) and (5.60). The convergence properties for the spaces in (5.45) are not required here, since usually the consistency, hence Condition 3. in Theorem* 5.2*, is impossible without them.*
3. *Condition 2. is correct if* F *or* G *are continuous in* $B_r(P^h u_0)$ *or* $B_r(u_0)$.
4. *The consistency condition 3. will be verified in Theorem 5.4 with reduced order. Due to the careful preparations this is a relatively simple proof.*
5. *Then the stability condition 4. is the only missing and hard condition in Theorems 5.2 and 5.13. In our approach the stability of the linearized problem and its regularity are the essential tools for the proof of stability; see Theorem 5.6. This makes sense only if the original* F *is continuously differentiable. Then* F *is Lipschitz-continuous as well.*

5.2.6 Consistency for nonconforming FEMs

Verifying the missing conditions analogous to Theorem 5.2 we start with the easier consistency. Similarly to the Ciarlet, Raviart, Lenoir, Bernadou and Boisserie approaches (see [85–89, 174, 176, 471]) we again use the extension operator, E_c in Theorem 4.37. It allows estimates of the interpolation errors on Ω, $\partial\Omega$, and Ω^h_c and $\partial\Omega^h_c$, cf. (4.76), (4.77), and thus is essential for the following proof.

Theorem 5.4. Consistency for nonconforming FEM: *We require the conditions for Theorems* 4.36*, with* Ω *in* $C^q, q = 2$ *and* $2 < q \le 5$ *for future curved approximations* $\partial\Omega^h_c$ *for* $\partial\Omega$ *as in Conjecture 4.41, an exact solution* $u_0 \in H^\ell(\Omega)$, $\ell > 2$, *with Condition* **H** *satisfied, a Lipschitz-continuous* G^w *in* (5.16) *with a global constant* L,

$$G^w(\cdot) \in C_L\big(\overline{(\Omega^h_c \cup \Omega) \times \Gamma'}\ \big) \text{ with } \overline{(\Omega^h_c \cup \Omega) \times \Gamma'} := \{(x, z, p, r) : \quad (5.66)$$
$$x \in \overline{\Omega^h_c \cup \Omega}, (z, p, r) = \big((E_c u)(x), \nabla(E_c u)(x), \nabla^2(E_c u)(x)\big) \text{ with } u \in \overline{B_r(u_0) \cap \mathcal{D}(G)}\},$$

and $\mathcal{U}^h \subset S^1_d\left(\mathcal{T}^h_c\right)$ *with* $d \ge n^2 + 1$. *Then* F^h *is consistent with* F *in* $P^h u_0$ *and*

$$\left\|F^h P^h u_0 - Q^{h'}_\Pi F u_0\right\|_{\mathcal{V}^{h'}_\Pi} = \left\|\left(Q^{h'}_c G(P^h u_0), -E_c u_0|_{\partial\Omega^h_c}\right)\right\|_{\mathcal{V}^{h'}_\Pi} \quad (5.67)$$
$$\le C L h^{\min\{\ell-2, q\}} \|u_0\|_{H^\ell(\Omega)}.$$

Remark 5.5. $d \ge n^2 + 1$ *(see (5.3)), and* $\ell - 1 \le d$ *are related by (4.62). For inhomogeneous boundary conditions as in (5.39), a modified estimate (5.67) remains valid.*

Proof. We estimate the first component in (5.67) with (5.55), L in (5.66), $G(u_0) \equiv 0$, and (4.76):

$$\left\|Q_c^{h'} G(P^h u_0) - Q^{h'} G(u_0)\right\|_{\mathcal{V}^{h'}} \tag{5.68}$$
$$= \left\|Q_c^{h'} G(P^h u_0) - 0\right\|_{\mathcal{V}^{h'}}$$
$$\leq \left\|Q_c^{h'} G(P^h u_0) - Q_c^{h'} G(E_c u_0)\right\|_{\mathcal{V}^{h'}} + \left\|Q_c^{h'} G(E_c u_0)\right\|_{\mathcal{V}^{h'}}$$
$$\leq CL\|P^h u_0 - E_c u_0\|_{\mathcal{U}^h} + \left\|Q_c^{h'} G(E_c u_0)\right\|_{\mathcal{V}^{h'}} \text{ by (4.65)}$$
$$\leq CLh^{\ell-2}\|u_0\|_{H^\ell(\Omega)} + \left\|Q_c^{h'} G(E_c u_0)\right\|_{\mathcal{V}^{h'}}.$$

We test $Q_c^{h'} G(E_c u_0) - Q^{h'} G(u_0)$ with $v^h \in \mathcal{V}^h$, combine it with (5.16), (5.55), and get

$$\left|\left\langle Q_c^{h'} G(E_c u_0), v^h \right\rangle_{\mathcal{V}^{h'} \times \mathcal{V}^h}\right| \tag{5.69}$$
$$= \left|\int_{\Omega_c^h} [G(E_c u_0)(x)] v^h(x) dx\right| \quad \forall v^h \in \mathcal{V}^h$$
$$\leq \left|\int_{\Omega_c^h} [G(E_c u_0)(x)] v^h(x) dx - \int_\Omega [0 \equiv G(u_0)(x)] E_c v^h(x) dx\right|$$
$$\leq \int_{\Omega_c^h \setminus \Omega} |G(E_c u_0)(x) v^h(x)| dx, \text{ since } \left(\int_{\Omega_c^h} - \int_\Omega\right)\Big|_{\Omega_c^h \cap \Omega} = 0,$$

where $G(E_c u_0), v^h \in L^2(\Omega_c^h)$ are well defined; cf. (5.5), and Condition **H**, (5.47). By $[0 \equiv G(u_0)(x)] E_c v^h(x)$, we do not need to explicitly define the extension $E_c v^h$. Now (5.66) and $u_0 \in H^\ell(\Omega)$, $\ell > 2$, imply $|G(E_c u_0)(x)| \leq CL\|u_0\|_{H^\ell(\Omega)}$. This is combined with $\operatorname{dist}(\Omega_c^h, \Omega) \leq C' h^q$, cf. Conjecture 4.41, and the bounded $S_{\partial\Omega} :=$ measure of the surface $(\partial\Omega) < \infty$ yielding

$$G(E_c u_0)(x) \text{ satisfies } G(E_c u_0) \equiv 0 \text{ in } \Omega \Longrightarrow$$
$$\|G(E_c u_0)\|_{L^2(\Omega_c^h \setminus \Omega)} \tag{5.70}$$
$$\leq CL\|u_0\|_{H^\ell(\Omega)} \|1\|_{L^2(\Omega_c^h \setminus \Omega)}$$
$$\leq CL\|u_0\|_{H^\ell(\Omega)} S_{\partial\Omega} C' h^q \leq C^* h^q \|u_0\|_{H^\ell(\Omega)}.$$

We combine (5.68), (5.69), (5.70) with the Cauchy–Schwarz inequality

$$\|Q_c^{h'} G(P^h u_0)\|_{\mathcal{V}^h} \leq CLh^{\ell-2}\|u_0\|_{H^\ell(\Omega)} + \|Q_c^{h'} G(E_c u_0)\|_{\mathcal{V}^h} \tag{5.71}$$
$$\leq CLh^{\ell-2}\|u_0\|_{H^\ell(\Omega)} + C'' h^q \|u_0\|_{H^\ell(\Omega)}$$
$$\leq \hat{C} h^{\min\{\ell-2,q\}} \|u_0\|_{H^\ell(\Omega)};$$

thus we have estimated the first component in (5.67).

To simplify estimating the second component in (5.67), we assume $n < 2(\ell - 1)$. Then the Sobolev embedding Theorem 1.26, implies the compact embedding of $H^\ell(\Omega)$

into $C^1(\bar{\Omega})$. Then $u_0 \in H^\ell(\Omega) \cap H_0^1(\Omega)$ implies $C(\partial\Omega) \ni u_0|_{\partial\Omega} \equiv 0$. Now choose any $P \in \partial\Omega_c^h$, a ray through $P \perp \partial\Omega_c^h$, the intersection point, P_b of this ray with $\partial\Omega$, and the directional derivative $\partial_P E_c u_0$ of $E_c u_0$ from P_b to P. Then

$$|E_c u_0(P)| = \left|0 + \int_{P_b}^{P} \partial_P E_c u_0\right| \le \|u_0\|_{H^\ell(\Omega)} C' h^q.$$

So we obtain for the second component in (5.67)

$$\|E_c u_0\|_{L^2(\partial\Omega_c^h)} \le C'' h^q \|u_0\|_{H^\ell(\Omega)}.$$

For inhomogeneous boundary conditions, cf. (5.39), we employ the extension of ϕ, necessarily here $\phi \in H^{\ell-1/2}(\partial\Omega)$, to $\phi \in H^\ell(\Omega)$; we still have $G(u_0) = 0$, and so (5.68), (5.69) remain correct. Boundary conditions for the exact and the discrete problem are $(u_0 - \phi)|_{\partial\Omega} \equiv 0$ and $\left(u_0^h - P^h\phi\right)|_{\partial\Omega_c^h} \equiv 0$ with $P^h\phi$ extended to Ω_c^h. We replace the exact boundary conditions by $(u_0 - P^h\phi)|_{\partial\Omega} \equiv 0$. Then the error estimates in Theorem 4.38 and the stability results for linear elliptic operators with respect to the right-hand side $(f, \phi) \in \mathcal{V} \times \mathcal{V}_D$, cf. Theorem 2.45, Hackbusch [387] Theorem 9.1.16, again yield (5.67). □

5.2.7 Stability for the linearized operator and convergence

Fortunately, there is a standard result strongly simplifying the proof of stability to that for the derivative of an operator; see Theorem 3.23, and e.g. Stetter [596] Theorem 1.2.5, Stummel [607–609] and Keller [441]: we impose the condition of a continuously differentiable F^h in $B_r(P^h u), u \in \mathcal{D}(G)$. This slightly enforced condition is usually satisfied by the equations considered here. It is even simplified in our case, since $F \in C^1(B_r(u))$ implies for smooth u, $F^h \in C^1(B_{r-\delta}(P^h u))$ with a small $\delta > 0$ for $h_0 > h \in H$. We formulate it for our F^h.

Theorem 5.6. From linear to nonlinear stability: *Assume Condition* **H**, *so the nonlinear discrete F^h, is continuously differentiable in $B_r(P^h u)$ for $u \in \mathcal{D}(G)$, and let the sequence of linearized $(F^h)' : \mathcal{U}^h \to \mathcal{V}_{\Pi}^{h'}$ be stable at $P^h u$, satisfying*

$$\left\|\left((F^h)'(P^h u)\right)^{-1}\right\|_{\mathcal{U}^h \hookleftarrow \mathcal{V}_{\Pi}^{h'}} \le S \text{ uniformly } \quad \forall h \in H, \quad h_0 > h. \qquad (5.72)$$

Then the nonlinear F^h is stable at $P^h u$ as well with stability bound and threshold, $2S$ and $r_0 = r/(2S)$, respectively.

According to Subsection 5.2.2 **3.**, Theorems 5.2, 5.4, and 5.6 and Remark 5.3 show that the only missing claim for convergence is the stability for the linearized operator for $u^h = P^h u$, $u \in \mathcal{D}(G)$, here $u = u_0$. So we have to elaborate the program in Subsection 5.2.2 in **3**. We indicate only the semiconforming FEM on convex polyhedral domains in (5.23) and mainly discuss the nonconforming FEM on C^2 domains in (4.5). For the conforming FEs it suffices to discuss the differential operator G. Then the boundary operators, B_D, the extension, E_c, and all the product machinery can be avoided. This would considerably simplify the proofs, presented here only for the nonconforming FEMs. The interested reader can easily verify this claim. For our $F(u)$

with $F'(u_1)u = (G'(u_1)u, u|_{\partial\Omega})$ the linearization of G is essential. Thus a major part of the following discussion is concerned with $G'(u_1)$. We obtain the claimed stability, $\|\big((F^h)'(u_1^h)|_{\mathcal{U}^h \to \mathcal{V}_\Pi^{h'}}\big)^{-1}\|_{\mathcal{V}_\Pi^{h'} \leftarrow \mathcal{U}^h} \le C$, by the end of this subsection.

The linearized operator

We start with $G : \mathcal{D}(G) \subset \mathcal{U} \to \mathcal{V} = L^2(\Omega)$, assuming a continuously differentiable operator G or its defining function G^w; see (5.17), (5.18). Here we sometimes integrate the boundary conditions into the spaces $\mathcal{U}_0, \mathcal{W}_0$ and $\mathcal{U}_b^h, \mathcal{W}_b^h$. In the literature the same notation $G'(u_0) : \mathcal{U} \to \mathcal{V}$ and $G'(u_0) : \mathcal{W} \to \mathcal{W}'$ is used for the strong and weak forms of the linearized operators, $G_s'(u_0) = G'(u_0)$ and $G'(u_0)$, respectively. For a smooth enough situation, the strong form is just the restriction of the weak form, and so $G'(u_0)|_{\mathcal{U}} = G_s'(u_0) : \mathcal{U} \to \mathcal{V}$ under the condition (5.73). The proofs in this subsection fundamentally depend upon a systematic interplay between these strong and weak forms of the linearized operators. To avoid misunderstandings, it is necessary to distinguish at least the strong and weak bilinear forms as $a_s(\cdot,\cdot)$ and $a(\cdot,\cdot)$. Later on, we even use the notation $A = G'(u_0), A_s = G_s'(u_0)$ and A^h, A_s^h for the weak and strong linear operators. Correspondingly, we formulate two different FEMs for the linearized problem. To its strong form, $a_s(\cdot,\cdot)$, we apply the FEM (5.48), (5.78); to its weak form, $a(\cdot,\cdot)$, the standard FEM; see (5.79) below. We will see that for the linear case both bilinear forms coincide, and so $a_s(u^h, v^h) = a(u^h, v^h)$ for all $u^h \in \mathcal{U}^h = \mathcal{W}^h \supset \mathcal{U}_b^h = \mathcal{V}^h \ni v^h$; see (5.77) and Proposition 5.7.

We called the original nonlinear problem G *uniformly elliptic* in u_0 if its derivative $G'(u_0)u$ in (5.19) satisfies (5.20). In contrast to G, it is possible to transform the linearization in (5.19) into the standard weak form. We start with the strong operator $G'(u_0) : \mathcal{U} \to \mathcal{V}$ (see (5.12)) and introduce new coefficients a_{ij}. Condition This yields the *strong bilinear form* $a_s(\cdot,\cdot)$ and conditions for the partials of G^w; see (5.12), (5.16), (5.17). We already have included the (5.73), (5.74) into Condition **H**.

$$G'(u_0) : \mathcal{U} \to \mathcal{V} \text{ and } a_s(u,v) := (G'(u_0)u, v)_{\mathcal{V}} \quad \forall\, u \in \mathcal{U}, v \in \mathcal{V} \text{ with} \tag{5.73}$$

$$\begin{aligned}
G'(u_0)u &= \sum_{i,j=0}^{n} (-1)_{j>0} \partial^j (a_{ij} \partial^i\, u) \text{ and } w_0(\cdot) = (\cdot, u_0(\cdot), \nabla u_0(\cdot), \nabla^2 u_0(\cdot)) :\\
a_{ij} &= a_{ij}(u_0) = \frac{\partial G^w}{\partial r_{ij}}(w_0) \text{ for } i,j \ge 1, a_{00} = a_{00}(u_0) = \frac{\partial G^w}{\partial z}(w_0),\\
a_{i0} &= a_{i0}(u_0) = \frac{\partial G^w}{\partial p_i}(w_0) + \sum_{j=1}^{n} \partial^j \frac{\partial G^w}{\partial r_{ij}}(w_0), \quad a_{0i} = 0 \text{ for } i \ge 1;\\
\text{and assume} \quad a_{ij} &= a_{ij}(w_0(\cdot)) \in W^{1-\delta_{0j},\infty}(\Omega), \text{ so this strong form } G'(u_0) \text{ is}\\
&\quad \text{bounded.}
\end{aligned}$$

Partial integration or Green's formula now combined with $v \in \mathcal{W} \cap H_0^1(\Omega)$ yields the (standard) weak problem or the *coinciding bounded weak and strong bilinear form*, $a(\cdot,\cdot) : \mathcal{W} \times \mathcal{W} \to \mathbb{R}$, and the induced weak, $G'(u_0) : \mathcal{W} \to \mathcal{W}'$ (see (5.12), (5.13)), as

$$a(u,v) := \int_\Omega \sum_{i,j=0}^n a_{ij}\partial^i u \partial^j v dx \quad =: \langle G'(u_0)u, v\rangle_{\mathcal{W}'\times\mathcal{W}} \ \forall v \in \mathcal{W}_0 = \mathcal{W}\cap H_0^1(\Omega)$$

$$\forall u \in \mathcal{W},\ \ a(\cdot,\cdot), G'(u_0) \text{ bounded for } a_{ij} = a_{ij}(u_0) = a_{ij}(w_0(\cdot)) \in L^\infty(\Omega)$$
$$\text{for } i,j = 0,\ldots,n, \text{ and with } a_s(u,v) = a(u,v) \quad \forall u \in \mathcal{U},\ \forall v \in \mathcal{W}_0. \tag{5.74}$$

The principal part $a_p(u,v)$ of $a(u,v)$ is $\mathcal{W}_0$ *coercive*, by (5.20), and, since $\|u\|_{H_0^1(\Omega)}$ and $\|u\|_{\mathcal{W}}$ are equivalent norms on $\mathcal{W}_0$,

$$a_p(u,u) \geq \lambda \int_\Omega \sum_{i=1}^n (\partial^i u)^2 dx = \lambda \|u\|^2_{H_0^1(\Omega)} \geq \alpha\|u\|^2_{\mathcal{W}} \quad \forall u \in \mathcal{W}_0. \tag{5.75}$$

The original $a(u,v)$, inducing $A = G'(u_0)$, is the sum of $a_p(u,v)$, inducing B, and its complement $c(u,v)$, inducing C, a compact perturbation of $A = B + C$, with $A, B, C \in \mathcal{L}(\mathcal{W},\mathcal{W}')$:

$$a(u,v) = a_p(u,v) + c(u,v) = \langle G'(u_0)u, v\rangle_{\mathcal{W}'\times\mathcal{W}} = \langle (B+C)u, v\rangle_{\mathcal{W}'\times\mathcal{W}}. \tag{5.76}$$

B is boundedly invertible on $\mathcal{W}_0$ by (5.75). Thus $G'(u_0)$ satisfies the *Fredholm alternative* on $\mathcal{W}_0$. This compact perturbation $C \in \mathcal{L}(\mathcal{W},\mathcal{W}')$ will play a dominant role for the following stability results.

In addition to the above approximating spaces (see (5.45), (5.46), (5.54)), we need for $\mathcal{W}_0$

$$\mathcal{U}^h = \mathcal{W}^h \supset \mathcal{V}^h \text{ and } \mathcal{U}_b^h = \mathcal{W}_b^h := \mathcal{V}^h = \mathcal{U}^h \cap H_0^1\left(\Omega_c^h\right) \qquad \text{for } \mathcal{U}_0 \subset \mathcal{W}_0, \tag{5.77}$$

again with the corresponding nonequivalent norms $\|\cdot\|_{\mathcal{U}^h}, \|\cdot\|_{\mathcal{W}^h}, \|\cdot\|_{\mathcal{V}^h}$.

We want to contrast the two different FEMs for the linear problem. We start with (5.48) and then turn to the standard FEM based upon the *weak bilinear form* $a(\cdot,\cdot)$. In both cases we have to determine, cf. (5.73), $u_1^h \in \mathcal{U}_b^h$ and $u_1^h \in \mathcal{W}_b^h$ s.t.

$$a_s\left(u_1^h, v^h\right) = \left(G'\left(P^h u_0\right) u_1^h, v^h\right)_{\mathcal{V}_c} := \left(\sum_{i,j=0}^n (-1)_{j>0}\partial^j\left(a_{ij}(P^h u_0)\partial^i\ u_1^h\right), v^h\right)_{\mathcal{V}_c}$$
$$= (E_c f, v^h)_{\mathcal{V}_c} \quad \forall v^h \in \mathcal{V}^h, \quad \text{with } f \in \mathcal{V}, a_{ij} \in W^{1-\delta_{0j},\infty}(\Omega). \tag{5.78}$$

Similarly as G^w we assume the $a_{ij}(P^h u_0)$ to be well defined on the extended $\Omega \cup \Omega_c^h$. The standard FEM for the *weak bilinear form* $a(\cdot,\cdot)$ with its solution $u_1^h \in \mathcal{W}_b^h$ is

$$u_1^h \in \mathcal{W}_b^h : a\left(u_1^h, v^h\right) = \int_{\Omega_c^h} \sum_{i,j=0}^n a_{ij}(P^h u_0)\partial^i u_1^h \partial^j v^h dx = \langle G'(P^h u_0)u_1^h, v^h\rangle_{\mathcal{W}_c'\times\mathcal{W}_c}$$
$$= \langle E_c f, v^h\rangle_{\mathcal{W}_c'\times\mathcal{W}_c} \text{ or}$$
$$= \langle E_c f, v^h\rangle_{\mathcal{V}_c} \text{ for } f \in \mathcal{V}\forall v^h \in \mathcal{W}_b^h, \text{ with } a_{ij} \in L^\infty(\Omega). \tag{5.79}$$

As in (5.48) for G, we use here the same notation for the discrete $a_s(u^h, v^h), a(u^h, v^h)$ and the original $a_s(u,v), a(u,v)$. These bilinear forms are, for small enough h, simultaneously bounded and elliptic. Equations (5.78), (5.79) can and will be similarly reinterpreted by using the above projectors and defining new ones.

Weak and strong linear operators coincide on $\mathcal{U}^h \times \mathcal{V}^h$

The techniques in this and the next subsubsections are strongly related to the standard approach in spectral methods for collocation (see Böhmer [120] and Canuto *et al.* [158]) and to regularity for difference equations; see Chapter 8 and, e.g. Hackbusch [387], Theorem 9.2.26.

Proposition 5.7. *Choose the above $\mathcal{U}^h, \dots, \subset \mathcal{S}_d^1\left(\mathcal{T}_c^h\right)$ in (5.77) for $d \geq 2^n+1$ as approximating spaces for $\mathcal{U}, \dots,$ and let $a_{ij} \in W^{1-\delta_{0j},\infty}(\Omega)$. Then for all $u^h \in \mathcal{U}^h, v^h \in \mathcal{V}^h = \mathcal{U}_b^h$,*

$$a_s(u^h, v^h) = (G'(P^h u_0)u^h, v^h)_{\mathcal{V}_c} = a(u^h, v^h) = \langle G'(P^h u_0)u^h, v^h\rangle_{\mathcal{W}_c' \times \mathcal{W}_c}. \quad (5.80)$$

Proof. We apply partial integration for every $T \in \mathcal{T}_c^h$ with $\overline{\Omega_c^h} = \cup_{T\in\mathcal{T}_c^h} \bar{T}$. We use the notation $\nu = \nu_T$ for the outer unit normal vector of T and B_a and u for the boundary operator, induced by $A = G'(u_0)$; see, e.g. [387]:

$$B_a\; u = \sum_{i,j=1}^{n} \nu_j a_{ij} \partial^i\; u + \sum_{j=1}^{n} \nu_j a_{0j}\; u, \text{ e.g. } B_a\; u = \partial u/\partial \nu \quad \text{for } A_s u = -\Delta u. \quad (5.81)$$

All ∂T have a *positive orientation* with respect to T. Hence in (5.83) every interior edge $e \in \bar{T}$ will be obtained twice in opposite directions. Edges $e \subset \partial\Omega_c^h$ will appear once. Let $\nu = \nu_T$ be one of the normal vectors for an interior edge $e \subset \overline{T_r}$. It is oppositely oriented for neighboring $T_r\,, T_l \in \mathcal{T}_c^h$ with $e \subset \overline{T_r} \cap \overline{T_l}$ and ν the outer normal[36] for $e \subset \partial\Omega_c^h$. To consider the transition from a triangle T_l to its neighboring T_r, we introduce the restriction of v to $\overline{T_l}, \overline{T_r}, \partial\Omega_c^h$ and the standard notation; see [665]:

$$v_l = v|_{\overline{T_l}}, \quad v_r = v|_{\overline{T_r}}, \quad [v] := v_l|_e - v_r|_e, \quad \text{and } \{v\} := (v_l|_e + v_r|_e)/2, \quad (5.82)$$

for the corresponding jumps and arithmetic means of v across an interior e, and

$$[v] := \{v\} := v|_{\partial\Omega_c^h} \text{ along } e \in \partial\Omega_c^h, \quad \text{and} \quad v \text{ arbitrary in } \mathbb{R}^n \backslash \Omega_c^h.$$

The next transformation via the n-dimensional Green's formula is the deciding reason for the of C^1 FEs. Only for this choice do the following jump contributions, $\left[B_a u^h\right] = 0$, vanish (the $[v^h]$ already vanish for C FEs v^h):

[36] We will use the notation $e \in \partial\Omega_c^h$, although mostly $e \subset \partial\Omega_c^h$; similarly, $e \in \mathcal{T}_c^h \setminus \partial\Omega_c^h$.

$$a(u^h, v^h) = \langle G'(E_c u_0)u^h, v^h\rangle_{\mathcal{W}_c' \times \mathcal{W}_c}$$
$$= \sum_{T\in\mathcal{T}_c^h} \left(\int_T \sum_{i,j=0}^{n} a_{ij}\partial^i\ u^h \partial^j\ v^h dx\right)$$
$$= \sum_{T\in\mathcal{T}_c^h} \left(\int_T \sum_{i,j=0}^{n} (-1)_{j>0}\partial^j (a_{ij}\partial^i\ u^h) v^h dx\right) + \int_{\partial\Omega_c^h} v_l^h B_a u^h\, ds$$
$$+ \sum_{e\in\mathcal{T}_c^h\backslash\partial\Omega_c^h} \int_e (\{v^h\}\,[B_a u^h] + [v^h]\{B_a u^h\}\,)ds = \int_{\Omega_c^h} (G'(E_c u_0)_s\ u^h) v^h dx$$
$$= a_s(u^h, v^h) \text{ since } u^h \in \mathcal{U}^h \subset C^1\left(\Omega_c^h\right), v^h \in \mathcal{V}^h \subset H_0^1\left(\Omega_c^h\right);\ \text{hence,}$$
$$a(u^h, v^h) = a_s(u^h, v^h) \quad \forall u^h \in \mathcal{U}^h, v^h \in \mathcal{V}^h. \tag{5.83}$$

Discontinuous Galerkin methods, cf. Chapter 7, would certainly be possible for $a(u^h, v^h)$. But they would violate the necessary condition $a(u^h, v^h) = a_s(u^h, v^h)$ by their penalty terms. □

Stability of the weak linear operator

The previous subsubsections play the role of preparing this stability proof. We had essentially confined the discussion there to the differential equation and indicated the boundary conditions either by the index $_0$ or using $v^h \in \mathcal{V}^h$. Instead of the previous strong and weak forms, $A_s : \mathcal{U} \to \mathcal{V}$ and $A : \mathcal{W} \to \mathcal{W}'$ (see (5.38), (5.41)–(5.43)), we need their two component versions:

$$A_\mathcal{U} := (A_s, B_D) : \mathcal{U} \to \mathcal{V} \times \mathcal{V}_D \subset \hat{\mathcal{V}}_\Pi = \mathcal{V} \times \mathcal{V}_b,\ A_\mathcal{U} \in \mathcal{L}(\mathcal{U}, \mathcal{V} \times \mathcal{V}_D) \text{ and} \tag{5.84}$$
$$A_\mathcal{W} := (A, B_D) : \mathcal{W} \to \mathcal{W}' \times \mathcal{W}_D \subset \mathcal{W}_\Pi' = \mathcal{W}' \times \mathcal{V}_b,\ A_\mathcal{W} \in \mathcal{L}(\mathcal{W}, \mathcal{W}' \times \mathcal{W}_D).$$

With the dense (compact) embedding $\mathcal{W}_D \hookrightarrow \mathcal{V}_b = L^2(\partial\Omega)$ we test the $\mathcal{W}' \times \mathcal{W}_D \subset \mathcal{W}_\Pi'$ by the following $\mathcal{W}_\Pi$. We emphasize that the stable splitting $S^h = S_0^h + S_b^h$ in Theorem 4.36 plays a dominant role. We choose, cf. (4.65), (5.6), (5.52)–(5.54),

$$\mathcal{W} \times \mathcal{W}_D \subset \mathcal{W}_\Pi := \mathcal{W} \times \mathcal{V}_b \neq \mathcal{W}_\Pi' = \mathcal{W}' \times \mathcal{V}_b, \text{ with } \mathcal{V}_b \supset \mathcal{W}_D \text{ and} \tag{5.85}$$
$$\langle (u', u_b), (v, v_b)\rangle_{\mathcal{W}_\Pi' \times \mathcal{W}_\Pi} := \langle u', v\rangle_{\mathcal{W}' \times \mathcal{W}} + (u_b, v_b)_{\mathcal{V}_b}, \text{ and on}$$
$$\Omega_c^h \text{ the } \mathcal{W}_{c,\Pi} = \mathcal{W}_c \times \mathcal{V}_{c,b} \neq \mathcal{W}_{c,\Pi}' \text{ with } \langle (u', u_b), (v, v_b)\rangle_{\mathcal{W}_{c,\Pi}' \times \mathcal{W}_{c,\Pi}}.$$

The discrete counterparts are

$$\mathcal{W}_\Pi^h = \mathcal{W}^h \times \mathcal{V}_b^h := S^h \times S_b^h|_{\partial\Omega_c^h} \neq \mathcal{V}_\Pi^h,\ \ \mathcal{W}_\Pi^{h'} := \mathcal{W}^{h'} \times \mathcal{V}_b^h := S^{h'} \times S_b^h|_{\partial\Omega_c^h} \neq \mathcal{V}_\Pi^{h'},$$
$$\text{and } \left\langle \left(u^{h'}, u_b^h\right), \left(v^h, v_b^h\right)\right\rangle_{\mathcal{W}_\Pi^{h'} \times \mathcal{W}_\Pi^h} := \langle u^{h'}, v^h\rangle_{\mathcal{W}^{h'} \times \mathcal{W}^h} + \left(u_b^h, v_b^h\right)_{\mathcal{V}_b^h}. \tag{5.86}$$

The exact and discrete linear boundary conditions remain nearly unchanged, except replacing $\mathcal{V}_D$ by $\mathcal{W}_D$, and so they can be treated as in (5.40), (5.49); see (5.58), (5.59).

$$
\begin{array}{ccccc}
\mathcal{W} & \xrightarrow{A_{\mathcal{W}}} & \mathcal{W}'_\Pi = \mathcal{W}' \times \mathcal{W}_D & \overset{\text{tested by}}{\longleftrightarrow} & \mathcal{W}_\Pi = \mathcal{W} \times \mathcal{V}_b \\
P^h \big\downarrow & \Phi^h_{\mathcal{W}} \big\downarrow & Q^{h'}_\Pi \big\downarrow & & \\
\mathcal{W}^h & \xrightarrow{A^h_{\mathcal{W}}} & \mathcal{W}^{h'}_\Pi = \mathcal{W}^{h'} \times \mathcal{W}^h_D & \overset{\text{tested by}}{\longleftrightarrow} & \mathcal{W}^h_\Pi = \mathcal{W}^h \times \mathcal{V}^h_b
\end{array}
$$

Figure 5.3 Nonconforming FEMs for linear problems.

For the weak differential operator, we introduce a modified Ritz operator. The above extension operator E_c in Theorem 4.37 is continuously extended to $f \in H^k(\Omega), k < 0$. We define the projectors, $Q^{h'}, Q_c^{h'}$, and so on, indicating the weak form by the exponent $^{h'}$, in contrast to the strong projectors $Q^{h'}$ and so on with the exponent ^{h} in (5.60); cf. (4.64), (4.66), (4.65), (5.6):

$$Q^{h'} \in \mathcal{L}(\mathcal{W}', \mathcal{W}^{h'}) \text{ for } f \in \mathcal{W}' \text{ by } \langle Q^{h'} f, v^h \rangle_{\mathcal{W}^{h'} \times \mathcal{W}^h} - \langle (E_c f), v^h \rangle_{\mathcal{W}'_c \times \mathcal{W}_c} = 0, \quad (5.87)$$

$$Q_c^{h'} \in \mathcal{L}\left(\mathcal{W}'_c, \mathcal{W}^{h'}\right) : f_c \in \mathcal{W}'_c : \left\langle Q_c^{h'} f_c, v^h \right\rangle_{\mathcal{W}^{h'} \times \mathcal{W}^h} - \langle f_c, v^h \rangle_{\mathcal{W}'_c \times \mathcal{W}_c} = 0 \ \forall v^h \in \mathcal{W}^h.$$

This allows formulating the product projectors, and we obtain

$$Q_\Pi^{h'} := \left(Q^{h'}, Q_b^{h'}\right) \in \mathcal{L}(\mathcal{W}'_\Pi, \mathcal{W}_\Pi^{h'}), \ Q_{c,\Pi}^{h'} := \left(Q_c^{h'}, Q_{c,b}^{h'}\right) \in \mathcal{L}(\mathcal{W}'_{c,\Pi}, \mathcal{W}_\Pi^{h'}). \quad (5.88)$$

For both types of projectors we obtain the necessary norm limits:

$$\begin{aligned}
&\lim_{h \to 0} \|Q^{h'} f\|_{\mathcal{W}^{h'}} = \|f\|_{\mathcal{W}'}, \text{and } \lim_{h \to 0} \left\|Q_c^{h'} f_c\right\|_{\mathcal{W}^{h'}} = \lim_{h \to 0} \|f_c\|_{\mathcal{W}'_c} = \lim_{h \to 0} \left\|E_c f_c\right\|_{\mathcal{W}'}, \\
&\lim_{h \to 0} \|Q_\Pi^{h'}(f, u_1|_{\partial\Omega})\|_{\mathcal{W}_\Pi^{h'}} = \|(f, u_1|_{\partial\Omega})\|_{\mathcal{W}'_\Pi}, \text{ and} \\
&\lim_{h \to 0} \left\|Q_{c,\Pi}^{h'}\left(f_c, (E_c u)|_{\partial\Omega_c^h}\right)\right\|_{\mathcal{W}_\Pi^{h'}} = \lim_{h \to 0} \left\|\left(f_c, (E_c u)|_{\partial\Omega_c^h}\right)\right\|_{\mathcal{W}'_{c,\Pi}} \\
&= \lim_{h \to 0} \|(E_c f_c, (E_c u)|_{\partial\Omega})\|_{\mathcal{W}'_\Pi}. \qquad (5.89)
\end{aligned}$$

Similarly to the above $F^h = Q_{c,\Pi}^{h'} F|_{\mathcal{U}^h}$ we use the projectors for reformulating (5.78) and (5.79), cf. (5.84), for $a_{ij} \in C_L\left(\Omega \cup \Omega_c^h\right)$. For the strong $A_{\mathcal{U}}$ and $A_{\mathcal{U}}^h$ we get

$$A_{\mathcal{U}}^h = Q_{c,\Pi}^{h'} A_{\mathcal{U}}|_{\mathcal{U}^h} : u_0^h \in \mathcal{U}^h : A_{\mathcal{U}}^h u_0^h = Q_{c,\Pi}^{h'}\left(A_s u_0^h, u_0^h|_{\partial\Omega_c^h}\right) = Q_\Pi^{h'}(f, \phi). \quad (5.90)$$

Similarly, the weak linear operators $A_{\mathcal{W}}$ and $A_{\mathcal{W}}^h$ yield, for $(f, \phi) \in \mathcal{W}'_\Pi$,

$$A_{\mathcal{W}}^h = Q_{c,\Pi}^{h'} A_{\mathcal{W}}|_{\mathcal{W}^h} : u_0^h \in \mathcal{W}^h : A_{\mathcal{W}}^h u_0^h = Q_{c,\Pi}^{h'}\left(A u_0^h, u_0^h|_{\partial\Omega_c^h}\right) = Q_\Pi^{h'}(f, \phi). \quad (5.91)$$

We summarize the new situation with uniformly bounded $A^h, P^h, Q_{c,\Pi}^{h'}$ in Figure 5.3. For proving the stability for $A_{\mathcal{W}}^h$ in Theorem 5.9 assume the principal part induces the ($\mathcal{W}_0$-coercive) operator B; cf. (5.76). We use the notation and assume

$$\begin{aligned}
&A_{\mathcal{W}} = (A, B_D), \ B_{\mathcal{W}} = (B, B_D), \ \text{the principal part B is } \mathcal{W}_0\text{-coercive}, \qquad (5.92) \\
&C_{\mathcal{W}} = (C, B_D), C \text{ compact, and } A_{\mathcal{W}}^h, B_{\mathcal{W}}^h, C_{\mathcal{W}}^h \text{ are consistent with } A_{\mathcal{W}}, B_{\mathcal{W}}, C_{\mathcal{W}}.
\end{aligned}$$

Theorem 5.8. Stability and convergence for the principal part: *Assume* (5.74), *the ellipticity* (5.75), *the approximate spaces in* (5.77) *with the norm limits in* (5.89) *and* (5.60), *and* $B_{\mathcal{W}}$ *and* $B^h_{\mathcal{W}}$ *as in* (5.84), (5.92). *Then* $B_{\mathcal{W}}$ *is boundedly invertible, and* $B^h_{\mathcal{W}}$ *is stable. The unique solutions* u_1 *of* $B_{\mathcal{W}} u_1 = (f, \phi) \in \mathcal{W}'_\Pi$ *and* u^h_1 *of* $B^h_{\mathcal{W}} u^h_1 = Q^{h'}_\Pi (f, \phi) \in \mathcal{W}^{h'}_\Pi$ *exist and* u^h_1 *converge such that* $\|u^h_1 - E_c u_1\|_{\mathcal{W}^h} \to 0$ *for* $h \to 0$.

Proof. Equation (5.74) shows that $a^h_p(u^h, v^h)$ is bounded on $\mathcal{W}^h$. The ellipticity condition (5.75) implies with the usual arguments, cf. (5.75), that

$$a^h_p(u^h, u^h) \geq \lambda \int_{\Omega^h_c} \sum_{i=1}^{n} (\partial^i u^h)^2 dx = \lambda \|u^h\|^2_{H^1_0(\Omega^h_c)} \geq \alpha \|u^h\|^2_{\mathcal{W}^h} \quad \forall u^h \in \mathcal{W}^h_b, \tag{5.93}$$

hence the stability of $B^h_{\mathcal{W}}$ in $\mathcal{W}^h_b$. We reduce the in homogeneous problem

$$B^h_{\mathcal{W}} u^h_1 = Q^{h'}_\Pi (f, \phi) = \left(Q^{h'} f \in \mathcal{W}^{h'}, Q^{h'}_b \phi \in \mathcal{V}^{h'}_b \right) \in \mathcal{W}^{h'}_\Pi$$

to $\mathcal{W}^h_b$: As usual, we determine $u^h_2 \in \mathcal{W}^h$ with $\left(u^h_2 - Q^{h'}_b \phi \right) |_{\partial \Omega^h_c} = 0 \in \mathcal{V}^h_b$. Then

$$B^h_{\mathcal{W}} \left(u^h_1 - u^h_2 \right) = \left(Q^{h'} f - B^h u^h_2 \in \mathcal{W}^{h'}, 0|_{\partial \Omega^h_c} \in \mathcal{V}^h_b \right) \in \mathcal{W}^{h'}_\Pi.$$

This guarantees, by (5.93) and Theorems 5.2 and 5.4, the stability, consistency, and hence existence of the unique discrete solution $u^h_1 - u^h_2$ and its convergence to the corresponding exact solution $u_1 - u_2$, and hence of u^h_1 as well. □

Theorem 5.9. Stability of $A^h_{\mathcal{W}} = (A = G'(u_0), B_D)^h \in \mathcal{L}\left(\mathcal{W}^h, \mathcal{W}^{h'} \times \mathcal{V}^h_b\right)$: *Under the conditions of Theorem* 5.8, *assume a boundedly invertible* $A_{\mathcal{W}}$ *in* (5.84). *Then* $A^h_{\mathcal{W}}$ *in* (5.91) *is stable.*

Proof. This is a generalization to nonconforming FEMs of the Böhmer and Sassmannshausen results (see [114, 115, 567]) for conforming FEMs. For an arbitrary $u \in \mathcal{W}$ choose $v_w := C_{\mathcal{W}} u \in \mathcal{W}'_\Pi = \mathcal{W}' \times \mathcal{W}_D$. By assumption, unique exact and discrete solutions, $\hat{u}$ and $\hat{u}^h$, exist for the equations

$$B_{\mathcal{W}} \hat{u} = v_w \text{ and } B^h_{\mathcal{W}} \hat{u}^h = Q^{h'}_{c,\Pi} B_{\mathcal{W}}|_{\mathcal{W}^h} \hat{u}^h = Q^{h'}_\Pi B_{\mathcal{W}} \hat{u} = Q^{h'}_\Pi v_w \in \mathcal{W}^{h'}_\Pi. \tag{5.94}$$

With the notation

$$T_w := B^{-1}_{\mathcal{W}} \in \mathcal{L}(\mathcal{W}'_\Pi, \mathcal{W}) \text{ and } T^h_w := \left(B^h_{\mathcal{W}}\right)^{-1} Q^{h'}_\Pi \in \mathcal{L}\left(\mathcal{W}'_\Pi, \mathcal{W}^h\right).$$

Theorem 5.8 implies that

$$\|E_c \hat{u} - \hat{u}^h\|_{\mathcal{W}^h} = \left\| \left(E_c T_w - T^h_w\right) C_{\mathcal{W}} u \right\|_{\mathcal{W}^h} \to 0 \quad \text{for } h \to 0 \text{ and } \forall u \in \mathcal{W}.$$

Since, by assumption, $C_{\mathcal{W}}$ is compact and $\left(E_c T_w - T^h_w\right)$ are equibounded, this implies

$$\left\| \left(E_c T_w - T^h_w\right) C_{\mathcal{W}} \right\|_{\mathcal{W} \hookleftarrow \mathcal{W}} \to 0 \quad \text{for } h \to 0. \tag{5.95}$$

Now let $u^h \in \mathcal{W}^h \subset \mathcal{W}_c$. Because $A_{\mathcal{W}}$ has been extended to yield

$$A_{\mathcal{W}} : \mathcal{W}_c \to \mathcal{W}'_{c,\Pi} = \mathcal{W}'_c \times \mathcal{W}_{c,D} \text{ still boundedly invertible with } \mathcal{W}^h \subset \mathcal{W}_c, \tag{5.96}$$

we can estimate

$$\|u^h\|_{\mathcal{W}^h} \le \|A_{\mathcal{W}}^{-1}\|_{\mathcal{W}^h\leftarrow\mathcal{W}'_{c,\Pi}} \|A_{\mathcal{W}}u^h\|_{\mathcal{W}'_{c,\Pi}} = \|A_{\mathcal{W}}^{-1}\|_{\mathcal{W}^h\leftarrow\mathcal{W}'_{c,\Pi}} \|B_{\mathcal{W}}(I+T_wC_{\mathcal{W}})u^h\|_{\mathcal{W}'_{c,\Pi}}$$

$$\le \|A_{\mathcal{W}}^{-1}\|_{\mathcal{W}^h\leftarrow\mathcal{W}'_{c,\Pi}} \|B_{\mathcal{W}}\|_{\mathcal{W}'_{c,\Pi}\leftarrow\mathcal{W}^h}\|(I+T_wC_{\mathcal{W}})u^h\|_{\mathcal{W}^h}; \text{ hence,} \tag{5.97}$$

$$\|(I+T_wC_{\mathcal{W}})u^h\|_{\mathcal{W}^h} \ge \left(\|A_{\mathcal{W}}^{-1}\|_{\mathcal{W}^h\leftarrow\mathcal{W}'_{c,\Pi}} \|B_{\mathcal{W}}\|_{\mathcal{W}'_{c,\Pi}\leftarrow\mathcal{W}^h}\right)^{-1} \|u^h\|_{\mathcal{W}^h}.$$

The $Q_\Pi^{h'} B_{\mathcal{W}} u$ is defined for every $u \in \mathcal{W}$. We apply the stability of $B_{\mathcal{W}}^h$ to $w^h := B_{\mathcal{W}}^h u^h$:

$$\|u^h\|_{\mathcal{W}^h} \le \|(B_{\mathcal{W}}^h)^{-1}\|_{\mathcal{W}^h\leftarrow\mathcal{W}_\Pi^{h'}} \cdot \|w^h\|_{\mathcal{W}_\Pi^{h'}}.$$

Furthermore

$$\left\|\left(I+T_w^hC_{\mathcal{W}}\right)u^h\right\|_{\mathcal{W}^h} = \left\|\left(B_{\mathcal{W}}^h\right)^{-1}B_{\mathcal{W}}^h(I+T_w^hC_{\mathcal{W}})u^h\right\|_{\mathcal{W}^h}$$

$$\le \left\|\left(B_{\mathcal{W}}^h\right)^{-1}\right\|_{\mathcal{W}^h\leftarrow\mathcal{W}_\Pi^{h'}} \left\|B_{\mathcal{W}}^h\left(I+T_w^hC_{\mathcal{W}}\right)u^h\right\|_{\mathcal{W}_\Pi^{h'}}$$

implies

$$\left\|B_{\mathcal{W}}^h\left(I+T_w^hC_{\mathcal{W}}\right)u^h\right\|_{\mathcal{W}_\Pi^{h'}} \ge \frac{1}{\left\|\left(B_{\mathcal{W}}^h\right)^{-1}\right\|_{\mathcal{W}^h\leftarrow\mathcal{W}_\Pi^{h'}}} \cdot \left\|\left(I+T_w^hC_{\mathcal{W}}\right)u^h\right\|_{\mathcal{W}^h}. \tag{5.98}$$

We combine (5.94)–(5.98) and use the fact that $A_{\mathcal{W}}, B_{\mathcal{W}}, C_{\mathcal{W}}$ are extended to $\mathcal{W}^h$, such that $Q_{c,\Pi}^{h'}A_{\mathcal{W}}u^h$ is well defined for estimating

$$\begin{aligned}
&\left\|A_{\mathcal{W}}^h u^h\right\|_{\mathcal{W}_\Pi^{h'}} \\
&= \left\|Q_{c,\Pi}^{h'}A_{\mathcal{W}}|_{\mathcal{W}^h}u^h\right\|_{\mathcal{W}_\Pi^{h'}} = \left\|Q_{c,\Pi}^{h'}A_{\mathcal{W}}u^h\right\|_{\mathcal{W}_\Pi^{h'}} \qquad \forall u^h \in \mathcal{W}^h \\
&= \left\|Q_{c,\Pi}^{h'}B_{\mathcal{W}}(I+T_wC_{\mathcal{W}})u^h\right\|_{\mathcal{W}_\Pi^{h'}} \overset{(5.94)}{=} \left\|B_{\mathcal{W}}^h(I+T_wC_{\mathcal{W}})u^h\right\|_{\mathcal{W}_\Pi^{h'}} \\
&\ge \left\|B_{\mathcal{W}}^h\left(I+T_w^hC_{\mathcal{W}}\right)u^h\right\|_{\mathcal{W}^{h'}} - \left\|B_{\mathcal{W}}^h\left(E_cT_w-T_w^h\right)C_{\mathcal{W}}u^h\right\|_{\mathcal{W}_\Pi^{h'}} \\
&\overset{(5.98)}{\ge} \left\|\left(I+T_w^hC_{\mathcal{W}}\right)u^h\right\|_{\mathcal{W}^h} / \left\|\left(B_{\mathcal{W}}^h\right)^{-1}\right\|_{\mathcal{W}^h\leftarrow\mathcal{W}_\Pi^{h'}} - \left\|B_{\mathcal{W}}^h\left(E_cT_w-T_w^h\right)C_{\mathcal{W}}u^h\right\|_{\mathcal{W}_\Pi^{h'}} \\
&\ge \left(\left\|\left(I+T_wC_{\mathcal{W}}\right)u^h\right\|_{\mathcal{W}^h} - \left\|\left(E_cT_w-T_w^h\right)C_{\mathcal{W}}u^h\right\|_{\mathcal{W}^h}\right) / \left\|\left(B_{\mathcal{W}}^h\right)^{-1}\right\|_{\mathcal{W}^h\leftarrow\mathcal{W}_\Pi^{h'}} \\
&\quad - \left\|B_{\mathcal{W}}^h\left(E_cT_w-T_w^h\right)C_{\mathcal{W}}\right\|_{\mathcal{W}_\Pi^{h'}\leftarrow\mathcal{W}^h}\|u^h\|_{\mathcal{W}^h} \\
&\overset{(5.97)}{\ge} \left(\frac{1}{\left(\|A_{\mathcal{W}}^{-1}\|_{\mathcal{W}^h\leftarrow\mathcal{W}'_{c,\Pi}} \|B_{\mathcal{W}}^h\|_{\mathcal{W}_\Pi^{h'}\leftarrow\mathcal{W}^h} \|(B_{\mathcal{W}}^h)^{-1}\|_{\mathcal{W}^h\leftarrow\mathcal{W}_\Pi^{h'}}\right)}\right. \\
&\quad \left. -\mathcal{O}\left(\left\|(E_cT_w-T_w^h)C_{\mathcal{W}}\right\|_{\mathcal{W}_\Pi^{h'}\leftarrow\mathcal{W}^h}\right)\right)\|u^h\|_{\mathcal{W}^h}.
\end{aligned}$$

The last estimate shows the following: the $\|A_{\mathcal{W}}^{-1}\|_{\mathcal{W}^h\hookleftarrow\mathcal{W}'_{c,\Pi}}$ is the deciding factor if $B_{\mathcal{W}}^h$ has a moderate condition number. Because of (5.95) and the stability of $(B_{\mathcal{W}}^h)_{h\in H}$ there exists a positive constant K, independent of h, such that for all $h \le h_0$,

$$\|A_{\mathcal{W}}^h u^h\|_{\mathcal{W}^h} = \left\|Q_{c,\Pi}^{h'} A_{\mathcal{W}} u^h\right\|_{\mathcal{W}^h} \ge K\|u^h\|_{\mathcal{W}^h} \quad \forall u^h \in \mathcal{W}^h.$$

This and $\dim \mathcal{W}^h < \infty$ show that $Q_{c,\Pi}^{h'} A_{\mathcal{W}}|_{\mathcal{W}^h}$ is invertible for $h \le h_0$. Moreover, we obtain $\|(Q_{c,\Pi}^{h'} A_{\mathcal{W}}|_{\mathcal{W}^h})^{-1}\|_{\mathcal{W}^h\hookleftarrow\mathcal{W}^{h'}} \le 1/K$; i.e. $A_{\mathcal{W}}^h|_{\mathcal{W}^h}$ is stable. □

Regularity for FE solutions

Our next task is lifting the preceding stability result with respect to the weak forms back to the strong form. This is achieved by the regularity for FE solutions in this subsubsection. Until now, *regularity results for the solutions of FEMs* seem to be known only as oscillation results in the sense of De Giorgi, Nash, and Moser; see Aguilera and Caffarelli [4]. They are not applicable to our problem of stability. Here we need, for the solution of a FEM applied to a differential equation of order 2, results analogous to those for the exact solution. These are estimates of the $H^2(\Omega)$ instead of the usual $H^1(\Omega)$ norm for a right-hand side in $L^2(\Omega) \times H^{3/2}(\Omega)$ instead of the usual $H^{-1}(\Omega) \times H^{1/2}(\Omega)$. To the author's knowledge, the following type of lemma was not known for FEMs. It is a transformation to our FEM of Hackbusch's regularity result [387], Theorem 9.2.26, for difference methods. We employ inverse and interpolation error estimates (4.57), (4.64), (4.76), valid under the conditions of Theorems 4.43, 4.36, and 4.38. The lemma is based upon a regularity result for the exact solution u_1 of the corresponding linear problem $A_{\mathcal{W}} u_1 = (f, \phi) \in \mathcal{V} \times \mathcal{V}_D \subset \hat{\mathcal{V}}_\Pi, u_1 \in \mathcal{W}_0$. Indeed, known regularity results, e.g. in Hackbusch [387], Theorems 9.1.22, 9.1.16, show that, for a convex bounded or smooth domain and smooth coefficients, the restriction $A_{\mathcal{U}} = A_{\mathcal{W}}|_{\mathcal{U}} \in \mathcal{L}(\mathcal{U}, \mathcal{V} \times \mathcal{V}_D)$ is boundedly invertible for a boundedly invertible $A_{\mathcal{W}} \in \mathcal{L}(\mathcal{W}, \mathcal{W}' \times \mathcal{W}_D)$; hence we assume

$$\Omega \text{ bounded, and either convex, polyhedral or } \Omega \in C^2, \tag{5.99}$$

$$(A_{\mathcal{W}})^{-1} \in \mathcal{L}(\mathcal{W}' \times \mathcal{W}_D, \mathcal{W}), \quad a_{i,j} \in W^{1-\delta_{j0},\infty}(\Omega) \quad \forall 0 \le i, j \le n. \tag{5.100}$$

Lemma 5.10. Regularity for FE solutions:

1. *We assume* (5.99), *the bounded elliptic* $A_{\mathcal{W}} \in \mathcal{L}(\mathcal{W}, \mathcal{W}' \times \mathcal{W}_D)$ *and* $A_{\mathcal{U}} \in \mathcal{L}(\mathcal{U}, \mathcal{V} \times \mathcal{V}_D)$ *to be boundedly invertible, e.g. by* (5.100). *Choose the above FEs,* (5.90) *and* (5.91), *of local degree* $d \ge 2$, *on a quasiuniform triangulation; see* (5.3), (5.77). *Then* $A_{\mathcal{W}}^h$ *is regular with respect to stability and convergence; hence, there exist unique solutions* u_1 *for* $A_{\mathcal{W}} u_1 = (f, \phi)$, u_1^h, *and a* $C' > 0$, *independent of* h, *such that (see* (4.64), (4.76)*)*

$$A_{\mathcal{W}}^h u_1^h = Q_{c,\Pi}^{h'}(f, \phi), \; (f, \phi) \in \mathcal{V} \times \mathcal{V}_D \Rightarrow \left\|u_1^h\right\|_{\mathcal{U}^h} \le C'(\|f\|_{\mathcal{V}} + \|\phi\|_{\mathcal{V}_D}) \tag{5.101}$$

$$\text{with } \left\|E_c u_1 - u_1^h\right\|_{\mathcal{U}^h} \le C'(\|E_c u_1 - P^h u_1\|_{\mathcal{U}^h} + \|E_c \phi - P^h u_1\|_{\mathcal{V}_D^h}). \tag{5.102}$$

2. $A_{\mathcal{W}} \in \mathcal{L}(\mathcal{W}, \mathcal{W}'_\Pi)$ *and* $A_{\mathcal{U}} \in \mathcal{L}(\mathcal{U}, \hat{\mathcal{V}}_\Pi)$ *simultaneously satisfy the Fredholm alternative.*

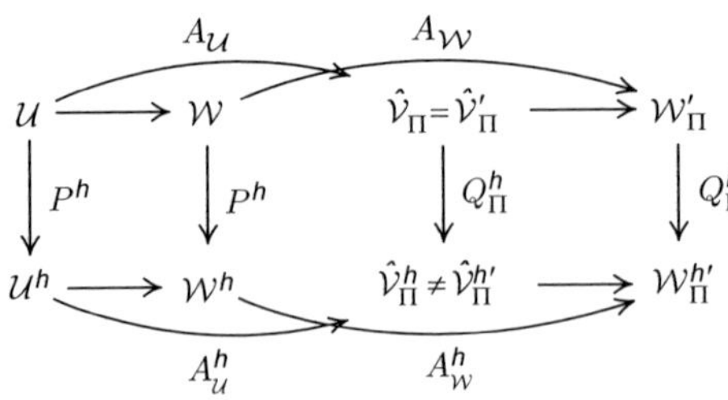

Figure 5.4 Strong and weak FEMs for linear problems.

Remark 5.11. *Hence, the FE error in the strong norm $\|E_c u_1 - u_1^h\|_{\mathcal{U}^h}$ is estimated by the interpolation errors of u_1 in $\mathcal{U}^h$ and $\mathcal{V}_D^h$. The FE spaces considered here need a quasiuniform triangulation guaranteeing inverse estimates in the global form, $\|v^h\|^h_{H^j(\Omega_c^h)} \le C\, h^{l-j}\|v^h\|^h_{H^l(\Omega_c^h)}$; see (4.57). In Subsection 4.2.5 we have summarized inverse estimates for $\mathcal{S}_d^1(\mathcal{T}_c^h)$ on specific degenerate triangulations, based upon the results of Graham, Hackbusch and Sauter [360–364]. The previous lemma remains valid for this case. However, since the smooth FEs in Subsection 4.2.6 still need nondegenerate triangulations,* we require nondegenerate triangulations in this Section. *Under the conditions of the lemma, (5.101) holds even for noninvertible $A_{\mathcal{W}}$ with $\|(f,\phi)\|_{\hat{\mathcal{V}}_\Pi}$ replaced by $\|(f,\phi)\|_{\hat{\mathcal{V}}_\Pi} + \|u_1\|_{\mathcal{W}}$; see [387].*

For Figure 5.4 and the following proof we contrast the strong and weak operators $A_{\mathcal{U}} \in \mathcal{L}(\mathcal{U} \cup \mathcal{U}_c, \hat{\mathcal{V}}_\Pi)$ and $A_{\mathcal{W}} \in \mathcal{L}(\mathcal{W} \cup \mathcal{W}_c, \mathcal{W}'_\Pi)$, extended as in (5.96); see (5.52)–(5.54), (5.85), (5.86) with the FE versions in (5.60), (5.88). The horizontal arrows in the first line of Figure 5.4 indicate compact embeddings and in the second line the decreasingly weaker norms in the spaces $\mathcal{U}^h = \mathcal{W}^h$ and $\hat{\mathcal{V}}^h_\Pi \subset \mathcal{W}^{h'}_\Pi$. The $\longrightarrow$ between $\mathcal{W}, \hat{\mathcal{V}}_\Pi$ is missing, since only $\mathcal{W} \longrightarrow \mathcal{V}$ would be appropriate.

Proof. For a boundedly invertible $A_{\mathcal{W}} \in \mathcal{L}(\mathcal{W}, \mathcal{W}'_\Pi)$, Theorem 5.9 implies the stability and thus the bounded invertibility of $A^h_{\mathcal{W}} = Q^{h'}_{c,\Pi} A_{\mathcal{W}}|_{\mathcal{W}^h} \in \mathcal{L}(\mathcal{W}^h, \mathcal{W}^{h'}_\Pi)$. This yields, with the consistency in Theorem 5.4, the existence of a discrete solution u_1^h for $A^h_{\mathcal{W}} u_1^h = Q^{h'}_\Pi(f,\phi) \in \mathcal{W}^{h'}_\Pi$ by Theorem 5.2. The solution $u_1^h \in \mathcal{W}^h$ for $(f,\phi) \in \mathcal{W}'_\Pi$ satisfies the estimate $\|u_1^h\|_{\mathcal{W}^h} \le C\|(f,\phi)\|_{\mathcal{W}'_\Pi}$. Now we use the same technique as in the proof of Theorem 5.8 for reducing the inhomogeneous to trivial Dirichlet boundary conditions. We have assumed $\phi \in \mathcal{V}_D = H^{3/2}(\partial\Omega)$ as a restriction of $\phi \in \mathcal{V} = H^2(\Omega)$. This is possible by combining Theorems 4.37, 1.37, 1.38, cf. [678], p. 1030. The last lines of the proof of Theorem 5.4 can then be updated to obtain the (5.101), (5.102) results for $\phi \not\equiv 0$ if they are correct for $\phi \equiv 0$. So we restrict the discussion for the remaining proof to trivial Dirichlet boundary conditions.

We find for $P^h = I^h E_c$, $A_{\mathcal{U}}$, and the $v^h_\Pi = (v^h, 0) \in \hat{\mathcal{V}}^h_\Pi$, cf. (5.96), (5.52)–(5.54),

$$P^h A_{\mathcal{U}}^{-1} : \mathcal{V}^h_\Pi \subset \mathcal{V}_{c,\Pi} \to \mathcal{U}^h \text{ or } P^h A_{\mathcal{U}}^{-1} v^h_\Pi \text{ is well defined } \forall v^h_\Pi = (v^h, 0) \in \hat{\mathcal{V}}^h_\Pi.$$

$\mathcal{V}^h \subset C^1(\overline{\Omega^h_c}) \subset C^\gamma(\overline{\Omega^h_c})$ for a $0 < \gamma < 1$ implies that $A_{\mathcal{U}}^{-1} v^h_\Pi \in C^{2,\gamma}(\overline{\Omega^h_c})$ by standard regularity results, cf. Theorems 2.38, 2.39, and, e.g. [677], Chapter 6.3, Problem 6.8,

p. 259. So the $P^h A_{\mathcal{U}}^{-1} v_\Pi^h$ is well defined for all $v_\Pi^h = (v^h, 0) \in \hat{\mathcal{V}}_\Pi^h$. Alternatively, we choose interpolation via the Bramble–Hilbert lemma, Theorem 4.15. Thus $P^h A_{\mathcal{U}}^{-1} v_\Pi^h$ is well defined for derivatives up to second order for P^h in (4.83), e.g. for the Argyris FEs. Higher derivatives defining P^h require quasi-interpolants as, e.g. in Theorem 4.36, Scott and Zhang [578], and Davydov [264].

For the estimate in (5.107) we need the embedding $I_{WU} : \mathcal{W}^h \to \mathcal{U}^h$, the bounded transformation, cf. (5.57), and [264]:

$$I_{\mathcal{V}^0}^h : \mathcal{V}_\Pi^{h',0} := \mathcal{V}^{h'} \times \{0\} \subset \mathcal{V}_\Pi^{h'} \to \mathcal{V}_{c,\Pi}^0 := \mathcal{V}_c \times \{0\} \subset \mathcal{V}_{c,\Pi}, \ (v^{h'}, 0) \to \left(Q_d^{h'} v^{h'}, 0\right). \tag{5.103}$$

We assumed a smooth $(f \in \mathcal{V}, \phi \equiv 0) \in \hat{\mathcal{V}}_\Pi$. By (5.58) and $\phi \equiv 0$ we restrict the discussion to $u^h \in \mathcal{W}^h \cap S_0^h = \mathcal{U}_b^h = \mathcal{W}_b^h = \mathcal{V}^h$. For the equalities between weak and strong problems we use special spaces and (5.52)–(5.55), (5.80), (5.85)–(5.89):

$$\forall u^h, v^h \in \mathcal{U}_b^h = \mathcal{W}_b^h = \mathcal{V}^h, \ \forall v_\Pi^h := (v^h, 0) \in \hat{\mathcal{V}}_\Pi^{h,0} := \mathcal{V}^h \times \{0\}, \ \hat{\mathcal{V}}_\Pi^{h',0}, \ \hat{\mathcal{W}}_\Pi^{h,0}, \ \mathcal{W}_\Pi^{h',0}.$$

We find for $f \in \mathcal{V}$ and the definition of A, A_s in $\Omega \cup \Omega_c^h$ that

$$\langle A_{\mathcal{W}}^h u^h, v_\Pi^h \rangle_{\mathcal{W}_\Pi^{h',0} \times \mathcal{W}_\Pi^{h,0}} = \langle A u^h, v^h \rangle_{\mathcal{W}_c' \times \mathcal{W}_c} = (A_s u^h, v^h)_{\mathcal{V}_c} = \langle A_{\mathcal{U}}^h u^h, v_\Pi^h \rangle_{\mathcal{V}_\Pi^{h',0} \times \mathcal{V}_\Pi^{h,0}},$$
$$(E_c f, v^h)_{\mathcal{V}_c} = \left\langle Q_\Pi^{h',0}(E_c f, 0), v_\Pi^h \right\rangle_{\mathcal{W}_\Pi^{h',0} \times \mathcal{W}_\Pi^{h,0}} = \left\langle Q_\Pi^{h,0}(E_c f, 0), v_\Pi^h \right\rangle_{\mathcal{V}_\Pi^{h',0} \times \mathcal{V}_\Pi^{h,0}}. \tag{5.104}$$

This implies coinciding weak and strong bilinear forms, since for all $v^h \in \mathcal{V}^h$,

$$\langle A_{\mathcal{W}}^h u_1^h, v_\Pi^h \rangle_{\mathcal{W}_\Pi^{h',0} \times \mathcal{W}_\Pi^{h,0}} = \left(A_s u_1^h, v^h\right)_{\mathcal{V}_c} = \langle A_{\mathcal{U}}^h u_1^h, v_\Pi^h \rangle_{\mathcal{V}_\Pi^{h',0} \times \mathcal{V}_\Pi^{h,0}} = (E_c f, v^h)_{\mathcal{V}_c}. \tag{5.105}$$

Hence, a weak solution u_1^h solves the strong problem as well:

$$A_{\mathcal{W}}^h u_1^h = Q_\Pi^{h',0}(f, 0), \quad (f, 0) \in \mathcal{V}_\Pi^0 := \mathcal{V} \times \{0\} \Longrightarrow A_{\mathcal{U}}^h u_1^h = Q_\Pi^{h,0}(f, 0). \tag{5.106}$$

For proving stability and consistency for the strong form, we start with the identity

$$\left(A_{\mathcal{U}}^h\right)^{-1}_{\mathcal{U}_b^h \leftarrow \mathcal{V}_\Pi^{h',0}} = \left[P^h_{\mathcal{U}_b^h \leftarrow \mathcal{U}_{c,0}} \left(A_{\mathcal{U}}^{-1}\right)_{\mathcal{U}_{c,0} \leftarrow \mathcal{V}_{c,\Pi}^0} - (I_{WU})_{\mathcal{U}_b^h \leftarrow \mathcal{W}_b^h}\right. \tag{5.107}$$
$$\left. \times \left(A_{\mathcal{U}}^h\right)^{-1}_{\mathcal{W}_b^h \leftarrow \mathcal{W}_\Pi^{h',0}} \left[\left(A_{\mathcal{U}}^h P^h - Q_{c,\Pi}^h A_{\mathcal{U}}\right)\right]_{\mathcal{W}_\Pi^{h',0} \leftarrow \mathcal{U}_{c,0}} \left(A_{\mathcal{U}}^{-1}\right)_{\mathcal{U}_{c,0} \leftarrow \mathcal{V}_{c,\Pi}^0}\right] I_{\mathcal{V}_{c,\Pi}^0, \mathcal{V}_\Pi^{h',0}}.$$

The inverse and the error estimates (4.57) and (4.64), (4.76), the approximation property (4.76), for the local degree $d \geq 2$, show the three inequalities

$$\|I_{WU}\|_{\mathcal{U}_b^h \leftarrow \mathcal{W}_b^h} \leq C_1 h^{-1}, \ \|P^h\|_{\mathcal{U}_b^h \leftarrow \mathcal{U}_{c,0}} \leq C_2, \ \|P^h u_1 - E_c u_1\|_{\mathcal{W}^h} \leq C_3 h \|u_1\|_{\mathcal{U}}. \tag{5.108}$$

With (4.76), the solution of $A_{\mathcal{U}} u_1 = (f, 0) \in \mathcal{V}_\Pi$, equal weak and strong bilinear forms, $a_{i,j}$, cf. (5.73), defined in $\Omega \cup \Omega_c^h$ and the special $v_\Pi^h = (v^h, 0) \in \mathcal{V}_\Pi^{h,0}$, we get by (4.75)

$$\begin{aligned}
&\left| \langle A_{\mathcal{U}}^h P^h u_1 - Q_\Pi^h A_{\mathcal{U}} u_1, v_\Pi^h \rangle_{\mathcal{W}_\Pi^{h',0} \times \mathcal{W}_\Pi^{h,0}} \right| \\
&\quad = |(A_s P^h u_1 - Q^{h'} f, v^h)_{\mathcal{V}_c}| \\
&\quad = |(A_s P^h u_1 - E_c f, v^h)_{\mathcal{V}_c}| = |(A_s P^h u_1 - E_c A_s u_1, v^h)_{\mathcal{V}_c}| \\
&\quad = \left| \left(\sum_{i,j=0}^{n} a_{ij} (\partial^i \, P^h u_1 - \; E_c \partial^i u_1), \partial^j v^h \right)_{\mathcal{V}_c} \right| + Ch \|u_1\|_{\mathcal{U}} \|v^h\|_{\mathcal{W}^h} \\
&\quad \le (C_3 + C) h \|u_1\|_{\mathcal{U}} \|v^h\|_{\mathcal{W}^h} \quad \forall v_\Pi^h = (v^h, 0) \in \mathcal{V}_\Pi^{h,0}.
\end{aligned}$$

With (5.106), this consistency, and the stability of $A_{\mathcal{W}}^h$, the $\|(A_{\mathcal{W}}^h)^{-1}_{\mathcal{W}_b^h \leftarrow \mathcal{W}_\Pi^{h',0}}\| \le C_4$, we estimate the first term in (5.107):

$$\begin{aligned}
&\left\| (A_{\mathcal{U}}^h)^{-1}_{\mathcal{W}_b^h \leftarrow \mathcal{W}_\Pi^{h',0}} \left[(A_{\mathcal{U}}^h P^h - Q_{c,\Pi}^h A_{\mathcal{U}}) \right]_{\mathcal{W}_\Pi^{h',0} \leftarrow \mathcal{U}_{c,0}} \right\|_{\mathcal{W}_b^h \leftarrow \mathcal{U}_{c,0}} \\
&\quad = \left\| (A_{\mathcal{W}}^h)^{-1}_{\mathcal{W}_b^h \leftarrow \mathcal{W}_\Pi^{h',0}} \left[(A_{\mathcal{U}}^h P^h - Q_{c,\Pi}^h A_{\mathcal{U}}) \right]_{\mathcal{W}_\Pi^{h',0} \leftarrow \mathcal{U}_{c,0}} \right\|_{\mathcal{W}_b^h \leftarrow \mathcal{U}_{c,0}} \\
&\quad \le C_4 \left\| (A_{\mathcal{U}}^h P^h - Q_\Pi^h A_{\mathcal{U}}) \right\|_{\mathcal{W}_\Pi^{h',0} \leftarrow \mathcal{U}_{c,0}} \\
&\quad \le C_4 \sup_{0 \ne u \in \mathcal{U}_{c,0}} \sup_{0 \ne v^h \in \mathcal{W}_b^h} \frac{\left| \langle A_{\mathcal{U}}^h P^h u - A_{\mathcal{U}} u, v_\Pi^h \rangle_{\mathcal{W}_\Pi^{h',0} \times \mathcal{W}_\Pi^{h,0}} \right|}{\|v^h\|_{\mathcal{W}^h} \|u\|_{\mathcal{U}}} \le C_3 C_4 h.
\end{aligned}$$

With (5.108), (5.107) and a boundedly invertible $\|A_{\mathcal{U}}^{-1}\|_{\mathcal{U}_{c,0} \leftarrow \mathcal{V}_{c,\Pi}^0} \le C_5$ and a bound for $\|I_{\mathcal{V}_{c,\Pi}^0, \mathcal{V}_\Pi^{h',0}}\|_{\mathcal{V}_{c,\Pi}^0 \leftarrow \mathcal{V}_\Pi^{h',0}} \le C_6$, this yields the stability of $A_{\mathcal{U}}^h : \mathcal{U}_b^h \to \mathcal{V}_\Pi^{h',0}$, since

$$\begin{aligned}
&\left\| \left(A_{\mathcal{U}}^h\right)^{-1} \right\|_{\mathcal{U}_b^h \leftarrow \mathcal{V}_\Pi^{h',0}} \le \Big[\|P^h\|_{\mathcal{U}_b^h \leftarrow \mathcal{U}_{c,0}} \|A_{\mathcal{U}}^{-1}\|_{\mathcal{U}_{c,0} \leftarrow \mathcal{V}_{c,\Pi}^0} + \|(I_{WU})\|_{\mathcal{U}_b^h \leftarrow \mathcal{W}_b^h} \\
&\quad \left\| (A_{\mathcal{U}}^h)^{-1}_{\mathcal{W}_b^h \leftarrow \mathcal{W}_\Pi^{h',0}} \left[A_{\mathcal{U}}^h P^h - Q_{c,\Pi}^h A_{\mathcal{U}} \right]_{\mathcal{W}_\Pi^{h',0} \to \mathcal{U}_{c,0}} \right\|_{\mathcal{W}_b^h \leftarrow \mathcal{U}_{c,0}} \|A_{\mathcal{U}}^{-1}\|_{\mathcal{U}_{c,0} \leftarrow \mathcal{V}_{c,\Pi}^0} \times \\
&\quad \|I_{\mathcal{V}_{c,\Pi}^0, \mathcal{V}_\Pi^{h',0}}\|_{\mathcal{V}_{c,\Pi}^0 \leftarrow \mathcal{V}_\Pi^{h',0}} \le \left[C_2 C_5 + C_1 h^{-1} C_3 C_4 h C_5 \right] C_6 \le C.
\end{aligned}$$

The combination of these inequalities yields the claim, since

$$\|u_1^h\|_{\mathcal{U}^h} = \left\| \left(A_{\mathcal{U}}^h\right)^{-1} Q_\Pi^h (f, 0) \right\|_{\mathcal{U}^h} \le \left\| \left(A_{\mathcal{U}}^h\right)^{-1} \right\|_{\mathcal{U}_b^h \leftarrow \mathcal{V}_\Pi^{h',0}} \|Q_\Pi^h\|_{\mathcal{V}_\Pi^{h',0} \leftarrow \mathcal{V}_\Pi^0} \|f\|_{\mathcal{V}} \le C \|f\|_{\mathcal{V}}.$$

The convergence in (5.102) is, by Theorem 5.2, an immediate consequence of the stability in (5.101) and the consistency in Theorem 5.4. $\square$

Stability for the strong linear operator and convergence

Combining Theorem 5.9, Proposition 5.7, and Lemma 5.10 we obtain:

Theorem 5.12. Stability for semiconforming and nonconforming FEMs: *For the linearized operator* $A_s = G'(u_0) \in \mathcal{L}(\mathcal{U}, \mathcal{V})$ *in* (5.73), *assume* (5.20), *a quasiuniform* $\mathcal{S}_d^1(\mathcal{T}^h), d \geq 2^n + 1$, *cf.* (5.3), *and let the corresponding* $A_s \in \mathcal{L}(\mathcal{U}_0, \mathcal{V})$, *cf., for the domain (5.100), and, hence,* $A_{\mathcal{U}} := (A_s, B_D) \in \mathcal{L}(\mathcal{U}, \mathcal{V} \times \mathcal{V}_D)$ *be boundedly invertible. Then for a convex polyhedral* Ω *in* $\mathbb{R}^n$ *and linear* $A_s^h u_1^h = Q^{h'} A_s|_{\mathcal{U}_0} u_1^h = Q^{h'} f, f \in \mathcal{V}$, *the semiconforming FEM* (5.78) *is stable. For an* $\Omega \in C^2$, *the nonconforming FEM* (5.90), $A_{\mathcal{U}}^h u_1^h = Q_{c,\Pi}^h A_{\mathcal{U}} u_1^h = Q_\Pi^h(f, \phi)$, $(f, \phi) \in \mathcal{V} \times \mathcal{V}_D$, *is stable as well.*

We discussed the relation between the bounded invertibility of $A_s = G'(u_0) \in \mathcal{L}(\mathcal{U}_0, \mathcal{V})$ and $A_{\mathcal{U}} = (A_s, B_D) \in \mathcal{L}(\mathcal{U}, \mathcal{V} \times \mathcal{V}_D)$ and the existence of a locally unique solution of $G(u_0) = 0$ and $F(u_0) = 0$ at the end of Subsection 5.2.3.

Proof. The boundedness of the strong $A_{\mathcal{U}} := (A_s, B_D) \in \mathcal{L}(\mathcal{U}, \mathcal{V} \times \mathcal{V}_D)$ and its weak counterpart $A_{\mathcal{W}} \in \mathcal{L}(\mathcal{W}, \mathcal{W}' \times \mathcal{W}_D)$ under the conditions in (5.100), (5.73), (5.74) is straightforward. We want to show that the bounded invertibility of $A_{\mathcal{U}}$ implies that of $A_{\mathcal{W}} \in \mathcal{L}(\mathcal{W}, \mathcal{W}' \times \mathcal{W}_D)$ as well. Indeed, by (5.73)–(5.74), the corresponding $G'(u_0) \in \mathcal{L}(\mathcal{U}, \mathcal{V}')$ induces a strong bounded bilinear form, by Proposition 5.7 simultaneously a $\mathcal{W}$-elliptic bilinear form, and thus finally $A_{\mathcal{W}} \in \mathcal{L}(\mathcal{W}, \mathcal{W}' \times \mathcal{W}_D)$. Thus the Fredholm alternative applies: if $A_{\mathcal{W}} \in \mathcal{L}(\mathcal{W}, \mathcal{W}' \times \mathcal{W}_D)$ were not boundedly invertible, each of the infinitely many solutions of $A_{\mathcal{W}} u_1 = (f, \phi) \in \mathcal{V} \times \mathcal{V}_D$ for appropriate (f, ϕ) would be $u_1 \in \mathcal{U}$ by (5.99) and the well-known regularity results; see, e.g. Hackbusch [387], Chapter 9. So $A_{\mathcal{U}}$ could not be boundedly invertible either.

Now we apply Theorem 5.9 to guarantee the stability for the strong form: the bounded invertibility of $Q_\Pi^{h'} A_{\mathcal{W}}|_{\mathcal{W}^h} \in \mathcal{L}(\mathcal{W}^h, \mathcal{W}_\Pi^{h'})$ and Lemma 5.10 guarantee the stability of $Q_\Pi^{h'} A_{\mathcal{U}}|_{\mathcal{U}^h} \in \mathcal{L}(\mathcal{U}^h, \mathcal{V}_\Pi^h)$ and thus the claim. □

Theorems 5.2, 5.4 and 5.6 and the last subsection combined with the last proof show that we have proved stability and convergence for the FEM in (5.62) if G is continuously differentiable near a locally unique smooth solution u_0, indicated by $A = G'(\cdot) \in C(B_{r,\mathcal{W}}(u_0))$ with boundedly invertible $F'(u_0)$ cf. Condition **H**. This is guaranteed, again in the interplay between the weak and strong form of the derivative, e.g. with $B_{r,\mathcal{W}}(u_0)$ the ball with respect to $\|u - u_0\|_{\mathcal{W}} < r$, by

$$a_{ij}(w(\cdot)) \in C^{1-\delta_{0j},\infty}(B_{r,\mathcal{W}}(u_0)) \text{ and } a_{ij}(w(\cdot)) \in C^{1-\delta_{0j},\infty}(B_{r,\mathcal{U}}(u_0)), \tag{5.109}$$

$$\text{yield } C(B_{r,\mathcal{W}}(u_0)) \ni A = G'(\cdot) : \mathcal{W} \to \mathcal{W}' \text{ and } C(B_{r,\mathcal{U}}(u_0)) \ni A_s = G'(\cdot) : \mathcal{U} \to \mathcal{V}',$$

where $i, j = 0, \ldots, n$, $a_{ij}(w(\cdot))$ are defined in (5.17). Alternatively, we might require $a_{ij}(\cdot) \in C^{1-\delta_{0j}}(\mathcal{D}(G^w))$, with $w, w_0 \in \mathcal{D}(G^w)$ as in (5.16).

Theorem 5.13. Existence and convergence for the nonconforming FEM: *Let the nonlinear elliptic problem in* (5.42) *satisfy Condition* **H**. *In particular, let* $G(u_0) = 0$, *for* $u_0 \in \mathcal{D}(G) \cap \mathcal{U}_0$, *or* $F(u_0) = (G, B_D)(u_0) = 0$, *for* $u_0 \in B_r(u_0) \cap \mathcal{D}(G) \subset \mathcal{U}$, *have an isolated solution, with boundedly invertible* $G'(u_0) : \mathcal{U}_0 \to \mathcal{V}$, *or* $F'(u_0) : \mathcal{U} \to \hat{\mathcal{V}}_\Pi$, *let* G *be continuously differentiable in* $B_r(u_0) \cap \mathcal{D}(G)$, *e.g. by* (5.109), *and let the* *domain satisfy* (5.99) *with* $\Omega \in C^2$. *Choose* $\mathcal{U}^h = \mathcal{S}_d^1\left(\mathcal{T}_c^h\right), d \geq 2^n + 1$ *(see* (5.3)*), on a quasiuniform triangulation.*

1. *Then the nonconforming FEM* (5.62), *so*

$$F^h = Q_{c,\Pi}^{h'} F|_{\mathcal{U}^h} : \mathcal{D}(F^h) \subset \mathcal{U}^h \cap \mathcal{D}(G) \to \mathcal{V}^{h'} \times \mathcal{V}_b^h, \quad u_0^h \in \mathcal{D}(F^h) \text{ s.t. } F^h u_0^h = 0,$$

is stable and consistent in $P^h u_0$, *and convergent. It has, for small enough* h, *a unique solution* u_0^h *near* u_0. *We obtain, cf.* (5.67), *for* $u_0 \in H^\ell(\Omega)$, $\ell > 2$, *for the nonconforming FEM, and* $C^p, 2 \leq p \leq 5$, *domains,*

$$\left\|P^h u_0 - u_0^h\right\|_{\mathcal{U}^h} \leq C h^{\min\{\ell-2,p\}} \|u_0\|_{H^\ell(\Omega)}, \tag{5.110}$$

respectively, with $p = 2$ *for polyhedral* Ω^h *and* $p > 2$ *for better curved* Ω_c^h *approximations for* $\Omega \in C^2$, *cf. Conjectures 4.40 and 4.41.*

2. $F'(u_0) \in \mathcal{L}(\mathcal{W}, \mathcal{W}' \times \mathcal{W}_D)$ *and* $F'(u_0) \in \mathcal{L}(\mathcal{U}, \mathcal{V} \times \mathcal{V}_D)$ *simultaneously satisfy the Fredholm alternative.*
3. *For quadrature approximations, exactly reproducing polynomials of degree* $< k$ *with* $k > 2(d-1) > d > 2$, *the solution* $\tilde{u}_0^h$ *of an approximate* $\tilde{F}^h\left(\tilde{u}_0^h\right) = 0$ *satisfies the same estimate as in (5.110); see Subsection 5.4.4 and 5.4.5 for* $m > 1$ *or for systems.*

Remark 5.14. *For the above* F *(see (5.18), (5.20), (5.73)),* ℓ *and* d *are related by (4.62). With the ellipticity assumption (5.20), the previous (5.73), (5.74), (5.100) are implied by (5.109).* Ω *convex polyhedral or in* C^2 *are necessary for the regularity and the extension operators.*

5.2.8 Discretization of equations and systems of order $2m$

The essential ideas have been presented above for equations of second order. So we only have to generalize our FEM formulated in (5.6) to equations and systems of orders 2 and $2m$. The basic facts for success are the existence of C^{2m-1} FEs with local support, combined with an unusual splitting. In Theorems 4.36–4.39 we have formulated and completely proved that for $m = 1$ in $\mathbb{R}^2$ on polyhedral domains. An extension to systems $q > 1$ is obvious. Extensions to $m > 1, n \geq 2$, are the goal for future research in Davydov's group. For the time being, we have formulated these results for smooth splines, $\mathcal{S}_d^r\left(\mathcal{T}_c^h\right)$, needed for $m > 1$, for the case $n \geq 2$, in Conjecture 4.40. Any smooth $\Omega \in C^{2m}$ have to be approximated now by polyhedral Ω^h in $\mathbb{R}^n$. For future results this is extended in Conjecture 4.41 to approximating curved boundaries, $\Omega_c^h \approx \Omega \in C^{2m}$ in $\mathbb{R}^n$.

The obvious difference in changing 1 to $m > 1$, is the change from the $H^1(\Omega)$, $H^2(\Omega)$ setting to the $H^m(\Omega, \mathbb{R}^q)$, $H^{2m}(\Omega, \mathbb{R}^q)$ setting and the updated product spaces. We again split F into the differential and a more complicated boundary operator.

The less obvious difference is the fact that for elliptic systems and equations of higher order, the simple equivalence of ellipticity (2.136) for $m = q = 1$ and $\mathcal{W}_0$-coercivity of the principal part, see Theorem 2.43 2., is no longer correct for the general case

$m, q \geq 1$. This has been discussed in Subsections 2.4.4, 2.6.3, 2.6.5, Theorems 2.43, 2.89, 2.104. We will repeat the essential conditions for ellipticity for $m, q \geq 1$ below, when we introduce the derivative of F in (5.128). Secondly, the conditions for the regularity of the exact solutions of linear systems considerably differ for the different cases, see the summary in (5.140) ff., and Subsections 2.4.4, 2.6.3, 2.6.5, Theorems 2.45, 2.47, 2.91–2.93, 2.95, 2.100, 2.101, 2.108.

Finally we generalize the semiconforming FEMs for convex polyhedral domains from Subsection 5.2.4 to $m = 1, q \geq 1$. In fact, for these domains only for $m = 1, q \geq 1$ is the regularity of the FE solutions correct. Compared to (5.32) only the functions u_0^h have to be replaced, for $q \geq 1$, by $\vec{u}_0^h$. So we reformulate the semiconforming FEM for $m = 1 < q$ in (5.149). For the other cases we have to impose specific conditions, in particular $\Omega \in C^{2m}$, cf. (5.140) ff.

These three aspects have to be considered in the following discussion.

We present the cases $m, q \geq 1$ all in parallel. The differences in this and the beginning of the next subsection are minimal. A direct comparison of the following differences in, e.g. (5.132) and (5.140) ff. is more informative this way, than it would be for separate formulations.

So we only define the updated differential and boundary operators, the original and approximating spaces, projectors, FEMs in the different equivalent forms and their solutions. The proofs of consistency and stability, now based upon the updated conditions, require only the minor modifications reported below. So we strongly restrict the presentation here.

Although we have presented the results for second order equations, we include them here once more as the case $m = q = 1$. This allows directly comparing all different cases.

Fully nonlinear elliptic systems and their discretization

In the following we use the notation $\vec{u}, \vec{v}, \ldots,$ for defining $\mathbb{R}^q$ valued functions. We need the following operators, extensions, cf. (5.5), and spaces defined on Ω and Ω_c^h :

$$G(\vec{u}(\cdot)) := G^w(\cdot, \vec{u}, \nabla\vec{u}, \ldots, \nabla^{2m}\vec{u})(\cdot), G : \mathcal{D}(G) \subseteq \mathcal{U} \cup \mathcal{U}_c \to \mathcal{V} \cup \mathcal{V}_c, \tag{5.111}$$

$$B_D : \mathcal{U} \to \mathcal{V}_D \text{ or } B_D : \mathcal{W} \to \mathcal{W}_D, B_D\vec{u}(x) := \left(\frac{\partial^j \vec{u}(x)}{\partial \nu^j}|_{\partial\Omega} \; \forall 0 \leq j \leq m-1\right) \tag{5.112}$$

$$\text{with } \mathcal{U} = H^{2m}(\Omega, \mathbb{R}^q) \subset \mathcal{W} = H^m \subset \mathcal{V} = L^2, \; \mathcal{W}_0 := \{\vec{u} \in \mathcal{W} : B_D\vec{u} = 0\},$$

$$\mathcal{U}_0 := \mathcal{W}_0 \cap \mathcal{U}, \; \mathcal{W}' = H^{-m}, \mathcal{V}_D = B_D\mathcal{U} \subset \mathcal{W}_D := B_D\mathcal{W} \subset \mathcal{V}_b = (L^2(\partial\Omega, \mathbb{R}^q))^m$$

$$\text{and similarly, defined on } \Omega_c^h : \; \mathcal{U}_c = H^{2m}\left(\mathcal{T}_c^h, \mathbb{R}^q\right) \subset \mathcal{W}_c \subset \mathcal{V}_c, \; \mathcal{W}_{c,0}, \; \mathcal{U}_{c,0},$$

$$\mathcal{W}_c', \; \mathcal{V}_{c,D} \subset \mathcal{W}_{c,D} \subset \mathcal{V}_{c,b} = \left(L^2\left(\partial\Omega_c^h, \mathbb{R}^q\right)\right)^m. \tag{5.113}$$

As a consequence of the consistent notation for $m = q = 1$ and $m, q \geq 1$, the product operator $F = (G, B_D)$, with the differential and the boundary operators, requires only obvious changes: There are essentially two differences: Functions $u, \ldots,$ have to be replaced by $\vec{u}(x) \in \mathbb{R}^q, \ldots,$ and $B_D\vec{u}(x) \in \mathbb{R}^{m \times q}$. This remains correct for all other operators as well, e.g. the projectors in (5.55), ... F is called elliptic simultaneously with G, and maps $\mathcal{D}(G) \subset \mathcal{U}$ onto the Cartesian product $\mathcal{V}_\Pi$, cf. (5.111), (5.112),

$$F := (G, B_D) : \mathcal{D}(G) \subset \mathcal{U} \to \mathcal{V}_\Pi := \mathcal{V} \times \mathcal{V}_D \subset \mathcal{V} \times \mathcal{V}_b; F(\vec{u}_0) := \big(G(\vec{u}_0), B_D\vec{u}_0\big) = \vec{0}, \tag{5.114}$$

with a locally unique solution, $\vec{u}_0$, and a boundedly invertible $F'(\vec{u}_0) \in \mathcal{L}(\mathcal{U}, \mathcal{V} \times \mathcal{V}_D)$.

We mainly consider trivial right-hand sides, $\vec{0} = (\vec{0}, \vec{0})$, characterizing vanishing $G(\vec{u}_0)$ and $B_D\vec{u}_0$ by testing with elements $\vec{v} \in \mathcal{V}$ and $\vec{v}_b \in \mathcal{V}_b$, respectively.

For applying general discretization methods, see Chapter 3, we choose the approximating spaces, $\mathcal{U}^h, \mathcal{V}_\Pi^h, \ldots,$ combine F^h with the corresponding projectors $P^h \in \mathcal{L}(\mathcal{U}, \mathcal{U}^h)$, and $Q_\Pi^{h'} \in \mathcal{L}(\mathcal{V}_\Pi, \mathcal{V}_\Pi^h)$, $Q_{c,\Pi}^{h'} \in \mathcal{L}(\mathcal{V}_{c,\Pi}, \mathcal{V}_\Pi^{h'})$ to get the *sequences*

$$\left\{\mathcal{U}^h, \mathcal{V}_\Pi^h = \mathcal{V}^h \times \mathcal{V}_b^h, P^h, Q_\Pi^{h'}, Q_{c,\Pi}^{h'}, F^h\right\}_{h\in H}, \text{with } \inf\{0 < h \in H\} = 0 \text{ for } \lim_{h\to 0}.$$

Theorems 4.36 and 4.38 show that the sequences, defined on Ω_c^h, see (4.77),

$$\begin{aligned} \mathcal{U}^h &= \mathcal{W}^h = \mathcal{S}_d^{2m-1}\left(\mathcal{T}_c^h, \mathbb{R}^q\right) = \mathcal{S}_d^{2m-1}\left(\mathcal{T}_c^h\right) = S_0^h \oplus S_b^h; \\ \mathcal{V}^h &= S_0^h = \mathcal{S}_d^{2m-1}\left(\mathcal{T}_c^h\right) \cap \mathcal{W}_{c,0}, \mathcal{V}_b^h := \left(S_b^h|_{\partial\Omega_c^h}\right)^m, \text{ or} \\ \mathcal{V}_D^h &:= B_D^h S_b^h,\ d \geq (2m-1)2^n + 1, h \in H, \end{aligned} \tag{5.115}$$

with the corresponding norms, cf. (5.27), *approximate the Banach spaces* $\mathcal{U}, \mathcal{W}, \mathcal{V}, \mathcal{V}_b$ defined on Ω and $\partial\Omega$. The $\vec{v}^h \in \mathcal{V}_b^h$ violate the original boundary conditions.

Again, the terms $\vec{u}^h$, $G(\vec{u}^h)$, cf. (5.5), and $B_D\vec{u}^h$ are defined on Ω_c^h and $\partial\Omega_c^h$ instead of the original Ω and $\partial\Omega$, cf. (5.5), motivating the choice our projectors.

$$P^h \in \mathcal{L}(\mathcal{U}, \mathcal{U}^h) : \ \forall\, \vec{u} \in \mathcal{U}\ \lim_{h\to 0} \|P^h\vec{u}\|_{\mathcal{U}^h} = \|\vec{u}\|_{\mathcal{U}}, \text{ we choose } P^h\vec{u} := I^h E_c\vec{u}. \tag{5.116}$$

For the discretization we define analogously new G^h, B_D^h in (5.117), (5.118). This will be combined in Theorems 5.15, 5.18 to obtain stability and convergence for the inhomogeneous case as well.

More delicate than P^h are the projectors to the image spaces, e.g. the $Q_\Pi^{h'} \in \mathcal{L}\left(\mathcal{V}_\Pi, \mathcal{V}_\Pi^h\right)$, motivated by F^h. Thus we start updating the previous FE version by the FE equations in equivalent formulations, see (5.114): define for $\vec{u}_0^h \in \mathcal{U}^h$ the

$$\begin{aligned} &\text{sequence } G^h\left(\vec{u}_0^h\right) = \vec{0} \Longleftrightarrow \left(G\left(\vec{u}_0^h\right), \vec{v}^h\right)_{\mathcal{V}_c} = 0\ \forall \vec{v}^h \in \mathcal{V}^h \subset \mathcal{V}_c \\ &\Longleftrightarrow \mathcal{V}^h \perp_{\mathcal{V}_c} G^h\left(\vec{u}_0^h\right) \neq G\left(\vec{u}_0^h\right) \not\ni \mathcal{V}^h \text{ with } \vec{u}_0^h \in \mathcal{D}(G^h) \subset \mathcal{U}_c. \end{aligned} \tag{5.117}$$

Similarly, the boundary conditions (5.114) are tested, see the equivalence (5.6), as

$$B_D^h\vec{u}_0^h := \left(\frac{\partial^j \vec{u}_0^h}{\partial \nu^j}|_{\partial\Omega_c^h}\ \forall 0 \leq j \leq m-1\right) = \vec{0} \Leftrightarrow \left(\vec{v}_b^h, B_D^h\vec{u}_0^h\right)_{\mathcal{V}_{c,b}} = 0 \forall \vec{v}_b^h \in \mathcal{V}_b^h. \tag{5.118}$$

Projectors and discretization

So we update the spaces and norms on Ω and Ω_c^h, omitting $\mathbb{R}^q$, cf. (5.38), (5.113):

$$For\ \mathcal{U},\mathcal{W},\mathcal{V},\mathcal{U}_0,\mathcal{V}_D,\mathcal{U}^h,\mathcal{W}^h,\mathcal{V}^h,\mathcal{U}_b^h,\mathcal{V}_D^h\ cf.\ (5.113)\text{–}(5.115),\ (5.46)\ for\ \|\cdot\|_{\mathcal{U}^h},$$
$$\mathcal{V}_\Pi := \mathcal{V}\times\mathcal{V}_D \subset \mathcal{V}\times\mathcal{V}_b, \mathcal{V}_D = B_B\mathcal{U}, \mathcal{V}_b := (L^2(\partial\Omega))^m \text{ with } \mathcal{V}=\mathcal{V}', \mathcal{V}_b = \mathcal{V}'\supset\mathcal{V}_D$$
$$\mathcal{V}_{c,\Pi} := \mathcal{V}_c\times\mathcal{V}_{c,D}\subset\mathcal{V}_c\times\mathcal{V}_{c,b}, \mathcal{V}_{c,D} = B_D^h\mathcal{U}_c, \mathcal{V}_{c,b} := \left(L^2\left(\partial\Omega_c^h\right)\right)^m, \mathcal{V}_c = \mathcal{V}_c', \ldots,$$
$$\text{with } \left((\vec{u},\vec{u}_b),(\vec{v}^h,\vec{v}_b)\right)_{\mathcal{V}_\Pi} := (\vec{u},\vec{v})_{\mathcal{V}} + (\vec{u}_b,\vec{v}_b)_{\mathcal{V}_b} \text{ and } \left((\vec{u},\vec{u}_b),(\vec{v}^h,\vec{v}_b)\right)_{\mathcal{V}_{c,\Pi}} \text{ versus}$$
$$\mathcal{V}_\Pi^h = \mathcal{V}^h\times\mathcal{V}_b^h := S_0^h\times\left(S_b^h|_{\partial\Omega_c^h}\right)^m,\ \mathcal{V}_\Pi^{h'} = \mathcal{V}^{h'}\times\mathcal{V}_b^h := S_0^{h'}\times\mathcal{V}_b^h \neq \mathcal{V}_\Pi^h \text{ with}$$
$$\mathcal{V}^{h'}\neq\mathcal{V}^h \text{ and } \left\langle\left(\vec{u}^{h'},\vec{u}_b^h\right),\left(\vec{v}^h,\vec{v}_b^h\right)\right\rangle_{\mathcal{V}_\Pi^{h'}\times\mathcal{V}_\Pi^h} := \langle\vec{u}^{h'},\vec{v}^h\rangle_{\mathcal{V}^{h'}\times\mathcal{V}^h} + \left(\vec{u}_b^h,\vec{v}_b^h\right)_{\mathcal{V}_b^h}. \tag{5.119}$$

The two components (5.117), (5.118) of F^h motivate the appropriate definitions of the projectors. Compared to the original form (5.55), …, we only replace functions $u,\ldots,$ by $\vec{u}\in\mathbb{R}^q,\ldots$ So we do not repeat these formulas, except $Q_{c,b}^{h'}$, similarly $Q_b^{h'}$, for demonstrating $B_D\vec{u}\in\mathbb{R}^{m\times q}$, e.g. by

$$Q_{c,b}^{h'}\in\mathcal{L}\left(\mathcal{V}_{c,b},\mathcal{V}_b^h\right), \vec{u}_{c,b} := B_D^hE_c\vec{u} \text{ or } := B_D^h\vec{u}^h\in\mathcal{V}_{c,D} \text{ as } \left(Q_{c,b}^{h'}\vec{u}_{c,b},\vec{v}_b^h\right)_{\mathcal{V}_{c,b}}$$
$$-\left(\vec{u}_{c,b},\vec{v}_b^h\right)_{\mathcal{V}_{c,b}} = 0\ \forall\vec{v}_b^h\in\mathcal{V}_b^h, \text{ now with } Q_{c,b}^{h'}B_D^h\vec{u}^h = B_D^h\vec{u}^h\ \forall\vec{u}^h\in\mathcal{U}^h \tag{5.120}$$
$$\text{and } \lim_{h\to0}\left\|Q_{c,b}^{h'}\vec{u}_{c,b}\right\|_{\mathcal{V}_b^h} = \lim_{h\to0}\|\vec{u}_{c,b}\|_{\mathcal{V}_{c,b}},\quad \lim_{h\to0}\left\|Q_b^{h'}\vec{u}_b\right\|_{\mathcal{V}_b^h} = \lim_{h\to0}\|\vec{u}_b\|_{\mathcal{V}_b}. \tag{5.121}$$

The above changes yield the updated definition of product projectors, see (5.52), e.g.

$$Q_\Pi^{h'} := \left(Q^{h'},Q_b^{h'}\right)\in\mathcal{L}\left(\mathcal{V}\times\mathcal{V}_D,\mathcal{V}^{h'}\times\mathcal{V}_b^h\right) = \mathcal{L}\left(\mathcal{V}\times\mathcal{V}_D,\mathcal{V}_\Pi^{h'}\right), \text{ similarly } Q_{c,\Pi}^{h'}, \tag{5.122}$$

with the corresponding limits of norms. The corresponding discrete operators, $F^h : \mathcal{U}^h\to\mathcal{V}^{h'}\times\mathcal{V}_b^h$, see (5.48)–(5.50), (5.52), (5.60) define the FE solutions $\vec{u}_0^h\in\mathcal{U}^h$ as

$$\vec{u}_0^h\in\mathcal{D}(G)\cap\mathcal{U}^h:\ F^h\left(\vec{u}_0^h\right) = \vec{0}\Leftrightarrow\langle F^h\left(\vec{u}_0^h\right),\vec{v}_\Pi^h\rangle_{\mathcal{V}_{c,\Pi}^{h'}\times\mathcal{V}_\Pi^h}$$
$$= \left(F_h\left(\vec{u}_0^h\right),\vec{v}_\Pi^h\right)_{\mathcal{V}_{c,\Pi}^h} = \vec{0}\ \forall\vec{v}_\Pi^h\in\mathcal{V}_\Pi^h$$
$$\Longleftrightarrow \left(G\left(\vec{u}_0^h\right),\vec{v}^h\right)_{\mathcal{V}_c} + \left(B_D^h\vec{u}_0^h,\vec{v}_b^h\right)_{\mathcal{V}_b^h} = 0\ \forall\vec{v}^h\in\mathcal{V}^h\subset\mathcal{V}_c\ \forall\vec{v}_b^h\in\mathcal{V}_b^h, \text{ where}$$
$$F^h:\mathcal{D}(F^h) = \mathcal{D}(G^h)\cap\mathcal{U}^h\to\mathcal{V}^{h'}\times\mathcal{V}_b^h \text{ with} \tag{5.123}$$
$$F^h(\vec{u}^h) = \left(G^h(\vec{u}^h),B_D^h\vec{u}^h\right) = Q_{c,\Pi}^{h'}F_h(\vec{u}^h) = Q_{c,\Pi}^{h'}\left(G(\vec{u}^h),B_D^h\vec{u}^h\right).$$

With these listed modifications the above Theorem 5.2 and Remark 5.3 remain correct, based upon the following

Condition Hm: Generalized Condition **H**: *In Condition **H** replace* $H^2(\Omega\cup\Omega_c^h)$ by $H^{2m}(\Omega\cup\Omega_\epsilon^h,\mathbb{R}^{2q})$ *and* (5.19)–(5.21), (5.73)–(5.75) by (5.128)–(5.132), (5.140)–(5.148), everything else remains unchanged.

5.2.9 Consistency, stability and convergence for $m,q \geq 1$

By Theorem 3.21, we have to prove consistency and stability, by Theorem 5.6, only for the linearized operator. The approach and the proofs are nearly identical with $m = q = 1$ up to the above minor modifications. So we reformulate the theorems and indicate the changes in the proofs. They are mainly caused by the modified conditions, guaranteeing the coercivity of and regularity results for the linearized operator. Wherever appropriate we refer to the conforming FEM for $m = 1, q \geq 1$, cf. Theorems 5.2 and 2.94, (2.358).

Theorem 5.15. Consistency for semiconforming and nonconforming FEs: *We require the conditions in Theorem 4.36, and Condition* **Hm**, *in particular an exact smooth solution* $\vec{u}_0 \in H^\ell(\Omega, \mathbb{R}^q)$, $\ell > 2m$, *a Lipschitz-continuous* G^w *in (5.16) with a global* L,

$$G^w(\cdot) \in C_L(\mathcal{D}(G^w) \cap (\overline{\Omega'} \times \overline{B_{r,\mathcal{U}}(\vec{u}_0)})) \text{ and } d \geq (2m-1)n^2 + 1. \tag{5.124}$$

1. *For* $m = 1, q \geq 1$ *and convex, polyhedral* Ω, *the* G^h *is consistent with* G *in* $P^h \vec{u}_0$ *and*

$$\|G^h P^h \vec{u}_0 - Q^{h'} G \vec{u}_0\|_{\mathcal{V}^{h'}} \leq K L h^{\ell-2} \|\vec{u}_0\|_{H^\ell(\Omega,\mathbb{R}^q)}. \tag{5.125}$$

2. *For* $m, q \geq 1$ *and* $\Omega \in C^{2m}$, *the* F^h *is consistent with* F *in* $P^h \vec{u}_0$ *and, with* $p = 2$ *for the polyhedral* Ω^h *and* $p > 2$ *for curved* Ω_c^h, *cf. Conjectures 4.40, 4.41,*

$$\left\| F^h P^h \vec{u}_0 - Q_\Pi^{h'} F \vec{u}_0 \right\|_{\mathcal{V}_\Pi^{h'}} \leq C L h^{\min\{\ell-2m,p\}} \|\vec{u}_0\|_{H^\ell(\Omega,\mathbb{R}^q)}. \tag{5.126}$$

Proof. Replace $h^{\ell-2}, \ldots,$ by $h^{\ell-2m}, \ldots,$ in the proof of Theorem 5.4. □

The linearized operator

We assume a continuously differentiable G^w. Again we sometimes integrate the boundary conditions into the spaces $\mathcal{U}_0, \mathcal{W}_0$ and $\mathcal{U}_b^h, \mathcal{W}_b^h$, see (5.113). We distinguish the strong and weak coinciding bilinear forms $a_s(\cdot,\cdot)$ and $a(\cdot,\cdot)$, see Proposition 5.16. We formulate two different FEMs for the linearized problem applying the new FEM (5.127), to its strong form and the standard FEM (5.128) to the weak form.

Equations and systems of order $2m$ *with* $1 \leq m, q$: We determine the solutions $\vec{u}_0$ (smooth) and $\vec{u}_0^h$ by

$$\begin{aligned} F(\vec{u}_0) &= (G(\vec{u}_0)(\cdot) = G^w(\cdot, \vec{u}_0(\cdot), \ldots, \nabla^{2m}\vec{u}_0(\cdot)), B_D\vec{u}_0) = 0 \in \mathcal{V} \times \mathcal{V}_D, \quad \text{and} \\ & F^h\left(\vec{u}_0^h\right) := \left(G^h\left(\vec{u}_0^h\right), B_D^h \vec{u}_0^h\right) = 0 \qquad (5.127) \\ &\iff \left(G\left(\vec{u}_0^h\right), \vec{v}^h\right)_{L^2(\mathcal{T}_c^h,\mathbb{R}^q)} + \left(B_D^h \vec{u}_0^h, \vec{v}_b^h\right)_{\mathcal{V}_b^h} = 0 \; \forall \vec{v}_\Pi^h = \left(\vec{v}^h, \vec{v}_b^h\right) \in \mathcal{V}_\Pi^h. \end{aligned}$$

The stability uses the linearization $(F^h)'(\vec{u}_0^h)\vec{u}^h = (G'(E_c\vec{u}_0)\vec{u}^h, B_D^h\vec{u}^h)$, starting with $G'(E_c\vec{u}_0)$ and $\vec{w}_0^h := \vec{w}_{\vec{u}_0^h}^h$ and $A_{\alpha\beta}^h = A_{\alpha\beta}^h(\vec{w}_0^h)$, cf. (5.17),

$$G'(E_c\vec{u}_0)\vec{u}^h = \sum_{|\gamma|\le 2m} \left(\partial_{\vartheta^\gamma} G^w(\cdot, \vec{\vartheta}^\beta, |\beta| \le 2m)\left(\vec{w}_0^h\right)\right)\partial^\gamma \vec{u}^h \tag{5.128}$$

$$=: \sum_{|\alpha|,|\beta|\le m} (-1)^{|\alpha|}\partial^\alpha \left(A_{\alpha\beta}^h \partial^\beta \vec{u}^h\right), a_s^h(\vec{u}^h, \vec{v}^h) := (G'(E_c\vec{u}_0)\vec{u}^h, \vec{v}^h)_{\mathcal{V}_c} \text{ and}$$

$$a^h(\vec{u}^h, \vec{v}^h) := \int_{\Omega_c^h} \sum_{|\alpha|,|\beta|\le m} \left(A_{\alpha\beta}^h \partial^\beta \vec{u}^h, \partial^\alpha \vec{v}\right)_q dx = \langle G'(E_c\vec{u}_0)\vec{u}^h, \vec{v}^h\rangle_{\mathcal{W}_c'\times\mathcal{W}_c}$$

$$\forall \vec{u}^h \in \mathcal{U}^h, \vec{v}^h \in \mathcal{V}^h \text{ with } A_{\alpha\beta} \in W^{|\alpha|,\infty} \text{ and } A_{\alpha\beta} \in L^\infty\left(\mathcal{T}_c^h, \mathbb{R}^{q\times q}\right).$$

The imposed conditions $A_{\alpha\beta} \in W^{|\alpha|,\infty}$ and $A_{\alpha\beta} \in L^\infty(\mathcal{T}_c^h, \mathbb{R}^{q\times q})$ imply well-defined continuous bilinear forms $a_s(\cdot,\cdot)$ and $a^h(\cdot,\cdot)$, respectively Again we split $a(u,v)$, induced by $A = G'(u_0)$, into the sum of its principal part, $a_p(u,v)$, inducing B, and its complement $c(u,v)$, inducing C, a compact perturbation as $A, B, C \in \mathcal{L}(\mathcal{W}, \mathcal{W}')$,

$$a(u,v) = a_p(u,v) + c(u,v) = \langle G'(u_0)u, v\rangle_{\mathcal{W}'\times\mathcal{W}} = \langle (B+C)u, v\rangle_{\mathcal{W}'\times\mathcal{W}}. \tag{5.129}$$

Here we come back to an essential difference to the case $m = q = 1$, with equivalent ellipticity and coercivity of the principal part. For $m, q \ge 1$, we impose

$$\forall x \in \Omega, \vec{\Theta} \in \mathbb{R}^{n\times q}: \sum_{|\alpha|=|\beta|=m} (A_{\alpha\beta}(x)\vec{\vartheta}^\beta)\vec{\vartheta}^\alpha \ge \lambda|\vec{\Theta}|^{2m}, \; or \tag{5.130}$$

$$\forall \vartheta \in \mathbb{R}^n, \eta \in \mathbb{C}^q : \bar{\eta}^T A_p(x,\vartheta)\eta = \bar{\eta}^T \sum_{|\alpha|=|\beta|=m} A_{\alpha\beta}(x)\vartheta^\beta\vartheta^\alpha\eta \ge \lambda|\vartheta|^{2m}|\eta|^2. \tag{5.131}$$

Notice that (5.130), (5.131) reduce to (2.136), for $m = q = 1$. Now we require

$$A_{\alpha\beta} \in L^\infty(\Omega, \mathbb{R}^{q\times q}) \text{ for } |\alpha|, |\beta| \le m \text{ with } A_{\alpha\beta} = A_{ij} \text{ for } m = 1,$$
$$\text{and } A_{\alpha\beta} = a_{\alpha\beta} \text{ for } q = 1, \text{with } A_{\alpha\beta} \in C(\bar{\Omega}) \text{ for } |\alpha| = |\beta| = m > 1, \tag{5.132}$$

the strong Legendre condition, (5.130), for $m \ge 1, q = 1$ and $m = 1, q \ge 1$, and the strong Legendre–Hadamard condition, (5.131), for $m > 1, q > 1$.

Then the principal part $a_p(\vec{u}, \vec{v})$ is $\mathcal{W}_0$-coercive and B is boundedly invertible on $\mathcal{W}_0$. This is a consequence of (2.22), Theorems 2.43, 2.89, 2.104. Thus $G'(u_0)$ satisfies the *Fredholm alternative* on $\mathcal{W}_0$.

In addition to (5.115), we need for $\mathcal{W}_0$

$$\mathcal{U}^h = \mathcal{W}^h \supset \mathcal{V}^h \text{ and } \mathcal{U}_b^h = \mathcal{W}_b^h := \mathcal{S}_0^h = \mathcal{V}^h = \mathcal{U}^h \cap H_0^m\left(\mathcal{T}_c^h, \mathbb{R}^q\right). \tag{5.133}$$

We contrast the two different FEMs for the *strong* $a_s(\cdot,\cdot)$ and the *weak bilinear form* $a(\cdot,\cdot)$ and its standard FEM. We determine $\vec{u}_1^h \in \mathcal{U}_b^h$ and $\vec{u}_1^h \in \mathcal{W}_b^h$ such that

$$\vec{u}_1^h \in \mathcal{U}_b^h : a_s^h\left(\vec{u}_1^h, \vec{v}^h\right) = \left(G'(E_c\vec{u}_0)\vec{u}_1^h, \vec{v}^h\right)_{\mathcal{V}_c} = (E_c\vec{f}, \vec{v}^h)_{\mathcal{V}_c} \; \forall \vec{v}^h \in \mathcal{V}^h, \vec{f} \in \mathcal{V}, \tag{5.134}$$

$$\vec{u}_1^h \in \mathcal{W}_b^h : a^h\left(\vec{u}_1^h, \vec{v}^h\right) = \left\langle G'(E_c\vec{u}_0)\vec{u}_1^h, \vec{v}^h\right\rangle = \langle E_c\vec{f}, \vec{v}^h\rangle \; \forall \vec{v}^h \in \mathcal{W}_b^h, \vec{f} \in \mathcal{W}', \tag{5.135}$$

with $\langle\cdot,\cdot\rangle = \langle\cdot,\cdot\rangle_{\mathcal{W}_c'\times\mathcal{W}_c}$. Equations (5.78), (5.79) can and will be similarly reinterpreted by replacing the above projectors, $Q_{\Pi}^{h'},\ldots$, by the previous new ones $Q_{\Pi}^{'h},\ldots$, cf. (5.120).

Coinciding strong and weak bilinear form and stability of the weak form

Proposition 5.16. *For $\mathcal{U}^h,\mathcal{V}^h$ in (5.119) and $A_{\alpha,\beta}\in W^{|\alpha|,\infty}(\mathcal{T}_c^h,\mathbb{R}^{q\times q})$ we obtain $\forall\vec{u}^h\in\mathcal{U}^h,\vec{v}^h\in\mathcal{V}^h=S_0^h=\mathcal{U}_b^h$*

$$a_s^h(\vec{u}^h,\vec{v}^h)=(G'(E_c\vec{u}_0)\vec{u}^h,\vec{v}^h)_{\mathcal{V}_c}=a^h(\vec{u}^h,\vec{v}^h)=\langle G'(E_c\vec{u}_0)\vec{u}^h,\vec{v}^h\rangle_{\mathcal{W}_c'\times\mathcal{W}_c}. \quad (5.136)$$

Proof. This is generalized by induction of the proof for Proposition 5.7. □

Remark 5.17. *The technique of the proof for Proposition 5.7, can be applied to quasilinear equations and systems as well. This yields instead of (5.106), and $\forall\,\vec{u}^h\in\mathcal{U}^h,\vec{v}^h\in\mathcal{V}^h\subset\mathcal{W}_0$,*

$$\begin{aligned}(G(\vec{u}^h),\vec{v}^h)_{\mathcal{V}_c} &= \int_{\Omega_c^h}\sum_{|\alpha|\le m}\left((-1)^{|\alpha|}\partial^\alpha(A_\alpha(\cdot,\vec{u}^h,\ldots,\nabla^m\vec{u}^h))\right)\vec{v}^h\,dx\\ &= \langle G(\vec{u}^h),\vec{v}^h\rangle_{\mathcal{W}_c'\times\mathcal{W}_c}\\ &:= \int_{\Omega_c^h}\sum_{|\alpha|\le 1}A_\alpha(\cdot,\vec{u}^h,\ldots,\nabla^m\vec{u}^h)\partial^\alpha\vec{v}^h\,dx. \end{aligned}\quad (5.137)$$

From the differential operators, we return to the two-component version of our operators

$$\begin{aligned}A_{\mathcal{U}} &:= (A_s,B_D):\mathcal{U}\to\mathcal{V}_{\Pi}=\mathcal{V}\times\mathcal{V}_D\subset\mathcal{V}\times\mathcal{V}_b,\ A_{\mathcal{U}}\in\mathcal{L}(\mathcal{U},\mathcal{V}\times\mathcal{V}_D)\text{ and}\\ A_{\mathcal{W}} &:= (A,B_D):\mathcal{W}\to\mathcal{W}_{\Pi}'=\mathcal{W}'\times\mathcal{W}_D\subset\mathcal{W}'\times\mathcal{V}_b,\ A_{\mathcal{W}}\in\mathcal{L}(\mathcal{W},\mathcal{W}'\times\mathcal{W}_D),\end{aligned}\quad (5.138)$$

where the "weak spaces" $\mathcal{W},\ldots$ and the following projectors represent obvious generalization of (5.85), (5.87), (5.88) to our case. Similarly to the above $F^h=Q_{c,\Pi}^{h'}F|_{\mathcal{U}^h}$ we reformulate (5.134) and (5.135) for the strong and weak linear $A_{\mathcal{U}}$ and $A_{\mathcal{W}}$ and determine, for $(\vec{f},\vec{\phi})\in\mathcal{W}_{\Pi}'$, the solution $\vec{u}_1^h$ from

$$\begin{aligned}A_{\mathcal{U}}^h &= Q_{c,\Pi}^{h'}A_{\mathcal{U}}|_{\mathcal{U}^h}:\vec{u}_1^h\in\mathcal{U}^h: A_{\mathcal{U}}^h\vec{u}_1^h=Q_{c,\Pi}^{h'}\left(A_s^h\vec{u}_1^h,B_D^h\vec{u}_1^h\right)=Q_{\Pi}^{h'}(\vec{f},\vec{\phi}),\\ A_{\mathcal{W}}^h &= Q_{c,\Pi}^{'h}A_{\mathcal{W}}|_{\mathcal{W}^h}:\vec{u}_1^h\in\mathcal{W}^h: A_{\mathcal{W}}^h\vec{u}_1^h=Q_{c,\Pi}^{'h}\left(A\vec{u}_1^h,B_D^h\vec{u}_1^h\right)=Q_{\Pi}^{'h}(\vec{f},\vec{\phi})\iff\\ &\left\langle Q_{c,\Pi}^{'h}A_{\mathcal{W}}\vec{u}_1^h-Q_{\Pi}^{'h}(\vec{f},\vec{\phi}),\vec{v}_{\Pi}^h\right\rangle_{\mathcal{W}_{\Pi}^{h'}\times\mathcal{W}_{\Pi}^h}=0\,\forall\vec{v}_{\Pi}^h\in\mathcal{W}_{\Pi}^h, A_{\alpha,\beta}\in C_L\left(\mathcal{T}_c^h\cup\Omega_c^h\right),\end{aligned}$$

see (5.138). Figure 5.2 remains unchanged.

For the bounded and equibounded $A_{\mathcal{W}}=(A,B_D),B_{\mathcal{W}}=(B,B_D),C_{\mathcal{W}}=(C,B_D)$, and $A_{\mathcal{W}}^h,B_{\mathcal{W}}^h,C_{\mathcal{W}}^h\in\mathcal{L}(\mathcal{W}^h,\mathcal{W}^{h'}\times\mathcal{V}_b^h)$ we use the notation in (5.79). Theorems 5.8 and 5.9, stating stability of the weak form and convergence for the principal part and "weak" stability for an invertible $A_{\mathcal{W}}$, remain correct if the ellipticity is replaced by

(5.132) and the approximation spaces in (5.77) by (5.133). Then both proofs remain unchanged except replacing functions, e.g. u by $\vec{u}$.

Regularity for FE solutions

For lifting the preceding stability result with respect to the weak forms back to the strong form, we need modified regularity results for the exact solutions of linear equations. These are estimates of the $H^{2m}(\mathcal{T}_c^h, \mathbb{R}^q)$ instead of the usual H^m norm for the right-hand side in $L^2 \times H^{2m-1/2} \times \cdots \times H^{m+1/2}(\mathcal{T}_c^h, \mathbb{R}^q)$ instead of the usual $H^{-m} \times H^{m-1/2} \times \cdots \times H^{1/2}(\mathcal{T}_c^h, \mathbb{R}^q)$. The basic assumptions for all cases are the bounded invertibility of $A_{\mathcal{W}}$ and the coercivity of its principal part, by the conditions in (5.132). So we require for A:

$$A \text{ strongly elliptic, } \mathcal{W}_0\text{-coercive principal part, } (A_{\mathcal{W}})^{-1} \in \mathcal{L}(\mathcal{W}' \times \mathcal{W}_D, \mathcal{W}_0). \quad (5.139)$$

For the different cases $m \geq 1, q \geq 1$ we impose, cf. Theorems 2.45, 2.46, 2.47, 2.91–2.93, 2.94, 2.108

$$\text{For } m = q = 1 \text{ let} \Omega \in C^2, a_{i,j} \in W^{1-\delta_{j0},\infty}\left(\mathcal{T}_c^h\right), 0 \leq i,j \leq n, (f,\phi) \in \mathcal{V} \times \mathcal{V}_D \quad (5.140)$$

$$\begin{aligned} m > q = 1 \text{ let } \Omega \in C^{2m}, \forall \alpha,\beta \text{ with } |\alpha|,|\beta| \leq m, \partial^\gamma a_{\alpha,\beta} \in L^\infty\left(\mathcal{T}_c^h\right) \forall \gamma \text{ with} \\ |\gamma| \leq |\beta|, \text{ else } a_{\alpha,\beta} \in L^\infty\left(\mathcal{T}_c^h\right), \ (f,\phi) \in \mathcal{V} \times \mathcal{V}_D, \quad (5.141) \\ m = 1 < q \text{ let } \Omega \in C^2, \ A_{kl} \in W^{1-\delta_{0k},\infty}\left(\mathcal{T}_c^h\right), \vec{f} \in \mathcal{V} \ \forall k,l, \ \textit{or} \quad (5.142) \\ m = 1 < q \text{ let } n = 2,3, \ \Omega \text{ be convex, polyhedral, and satisfy Theorem 2.94,} \\ \text{and } A_{kl} \text{ satisfy (2.358).} \quad (5.143) \end{aligned}$$

For systems of order $2m$, we need the *modified strong Legendre condition,* cf.(2.401),

$$\exists \lambda > 0 : \forall x \in \Omega, \forall \vartheta \in \mathbb{R}^n : \sum_{|\alpha|=|\beta|=m} A_{\alpha\beta} \vartheta^\beta \vartheta^\alpha \geq \lambda |\vartheta|^{2m} I, \quad (5.144)$$

with the identity matrix I.

$$\text{For } m,q > 1, \text{ assume (5.144), } \Omega \in C^\infty, \ A_{\alpha,\beta} \in C^\infty\left(\mathcal{T}_c^h\right) \ \forall \alpha,\beta, \vec{f} \in L^2\left(\mathcal{T}_c^h\right). \quad (5.145)$$

These analytic regularity results show that (5.139)–(5.145) imply

$$A_{\mathcal{W}} \vec{u}_1 = (\vec{f}, \vec{\phi}) \in \mathcal{V} \times \mathcal{V}_D \Rightarrow \|\vec{u}_1\|_{\mathcal{U}} \leq C\big(\|\vec{f}\|_{\mathcal{V}} + \|\vec{\phi}\|_{\mathcal{V}_D}\big) = C\|(\vec{f}, \vec{\phi})\|_{\mathcal{V}_\Pi}. \quad (5.146)$$

With these modifications Lemma 5.10 in particular (5.101), (5.102) hold. In the proof for our general case, we only have to replace h, h^{-1} in (5.108) by h^m, h^{-m}.

Stability for the strong linear operator and convergence

We are nearly done. Collecting the results from the last two subsubsections, we get the "strong" stability of the linear operator (only formulated analogously) and then finally the existence and convergence of discrete solutions for fully nonlinear problems.

After all these preparations we are ready to prove the stability of the linearized strong operator and thus the convergence of the method. We had to impose many different conditions for the different areas of ellipticity and $\mathcal{W}_0$-coercivity, approximation, coinciding weak and strong bilinear forms and regularity. We recall the notation $A_{\alpha\beta} = a_{\alpha\beta}$ for $m > 1, q = 1$, $A_{\alpha\beta} = A_{ij}$ for $m = 1 < q$, and $A_{\alpha\beta} = a_{ij}, i, j = 0, \ldots, n$ for $m = 1 = q$. For the operator and the domain, we imposed (5.130), the strong Legendre condition for $m \geq 1, q = 1$ and $m = 1, q \geq 1$, and (5.131) and (5.144), the strong Legendre–Hadamard and the modified Legendre condition for regularity, for $m > 1, q > 1$, summarized as

$$\forall m,\ q \geq 1 : \mathcal{U},\ \mathcal{U}^h = \mathcal{S}_d^{2m-1}\left(\mathcal{T}_c^h\right) .., \text{ in (5.113), (5.115), } d \geq (2m-1)2^n + 1 \quad (5.147)$$

$$\text{and (5.130)}, \Omega \in C^{2m}, A_{\alpha\beta} \in W^{|\alpha|,\infty}\left(\mathcal{T}_c^h, \mathbb{R}^{q\times q}\right)\ \forall |\alpha|, |\beta| \leq m, \text{ except } m, q > 1,$$

$$\text{equations } : m = q = 1 :\ a_{ij} \in W^{1-\delta_{i,0},\infty}(\Omega),\ \forall 0 \leq i, j \leq n,$$

$$\text{systems } : m = 1 < q :\ A_{ij} \in W^{1-\delta_{i,0},\infty}(\Omega, \mathbb{R}^{q\times q}), \forall 0 \leq i, j \leq n,$$

$$\text{equations } : m > q = 1 :, \Omega \in C^{2m}, a_{\alpha\beta} \in C(\bar{\Omega})\forall |\alpha|, |\beta| = m,$$

$$\text{systems } : m, q > 1 : \Omega \in C^{\infty}, (5.131), (5.144), A_{\alpha\beta} \in C^{\infty}(\bar{\Omega}, \mathbb{R}^{q\times q})\forall |\alpha| = |\beta| \leq m,$$

$$\text{finally let } m = 1 < q,\ n = 2, 3,\ \Omega \text{ be convex, polyhedral,}$$

$$\text{and satisfy Theorem 2.94, and let } A_{kl} \text{ satisfy (2.358).}$$

The consistency in Theorem 5.4 needed an exact solution $\vec{u}_0 \in H^{\ell}(\Omega, \mathbb{R}^q), \ell > 2m$, and a Lipschitz-continuous G^w in (5.16) with a global constant L, see (5.66).

With the notation in (5.73), (5.147), we require continuity with respect to the variable $\vec{u}$, via $\vec{w}_{\vec{u}}$, see (5.17), (5.111), for Theorem 5.13 below, such that

$$A_{\alpha\beta}(\vec{w}_{\vec{u}}) \in C^{|\alpha|,\infty}(\mathcal{W} \cap B_r(\vec{u}_0)) \text{ and } A_{\alpha\beta}(\vec{w}_{\vec{u}}) \in C^{|\alpha|,\infty}(\mathcal{U} \cap B_r(\vec{u}_0)), \quad (5.148)$$

$$\forall |\alpha|, |\beta| \leq m, \text{ yield } G'(u) : \mathcal{W} \to \mathcal{W}', \text{ and } G'(u) : \mathcal{U} \to \mathcal{V}' \in C(B_r(\vec{u}_0)).$$

Theorem 5.18. Discrete solutions uniquely exist and converge for $m, q \geq 1$: *Let the nonlinear elliptic problem in (5.114), $F(\vec{u}_0) = 0$ with $F = (G, B_D) : \mathcal{U} \to \mathcal{V}_\Pi = \mathcal{V} \times \mathcal{V}_D$, equivalently $G(\vec{u}_0) = 0$ with $G : \mathcal{U}_0 \to \mathcal{V}$, have an isolated solution $\vec{u}_0 \in \mathcal{D}(G) \cap \mathcal{U}_0$. Let G' and F' satisfy (5.132), (5.139)–(5.145) (5.147)–(5.148), let $G'(\vec{u}_0) : \mathcal{U}_0 \to \mathcal{V}$, hence $F'(\vec{u}_0) : \mathcal{U} \to \mathcal{V}_\Pi$ be boundedly invertible and G, hence F as well, be continuously differentiable in $B_r(\vec{u}_0) \cap \mathcal{D}(G)$, satisfying Condition* **Hm**. *In particular, define the FEM for $m = 1, q \geq 1$ and convex, polyhedral $\Omega \in \mathbb{R}^n$, with $n = 2, 3$ for $q > 1$, as, cf. (5.123),*

$$\vec{u}_0^h \in \mathcal{D}(G^h) \subset \mathcal{U}_b^h \text{ s.t. } \ G^h\left(\vec{u}_0^h\right) = 0, \quad G^h = Q^{h'} G|_{\mathcal{U}_b^h} : \mathcal{U}_b^h \to \mathcal{V}^{h'}, \quad (5.149)$$

and for the other cases in (5.147) as

$$\vec{u}_0^h \in \mathcal{D}(F^h) \subset \mathcal{U}^h \text{ s.t. } \ F^h\left(\vec{u}_0^h\right) = 0, \ F^h = Q_{c,\Pi}^{h'} F|_{\mathcal{U}^h} : \mathcal{U}^h \to \mathcal{V}^{h'} \times \mathcal{V}_b^h. \quad (5.150)$$

1. *Then these FEMs are stable in* $P^h\vec{u}_0$, *consistent and convergent. They have, for small enough* h, *a unique solution* $\vec{u}_0^h$ *near* $\vec{u}_0$.
2. *We get, see (5.67), for* $\vec{u}_0 \in H^\ell(\Omega, \mathbb{R}^q)$, $\mathcal{P}^n_{\ell-1} \subset \mathcal{S}^h, \ell > 2m$, *cf. (4.62),*

$$\left\|P^h\vec{u}_0 - \vec{u}_0^h\right\|_{\mathcal{U}^h} \leq Ch^{\min\{\ell-2m,p\}}\|\vec{u}_0\|_{H^\ell(\Omega,\mathbb{R}^q)}, \tag{5.151}$$

with $p = 2$ *for polygonal* Ω^h *and* $p > 2$ *for curved* Ω^h_c.
3. *For* $m = 1, q \geq 1$ *and convex, polyhedral* Ω *we obtain instead*

$$\left\|P^h\vec{u}_0 - \vec{u}_0^h\right\|_{\mathcal{U}^h} \leq KLh^{\ell-2}\|\vec{u}_0\|_{H^\ell(\Omega,\mathbb{R}^q)}. \tag{5.152}$$

4. *Simultaneously* $F'(\vec{u}_0) \in \mathcal{L}(\mathcal{W}, \mathcal{W}' \times \mathcal{W}_D)$ *and* $F'(\vec{u}_0) \in \mathcal{L}(\mathcal{U}, \mathcal{V} \times \mathcal{V}_D)$ *satisfy the Fredholm alternative.*

5.2.10 Numerical solution of the FE equations with Newton's method

We have indicated this problem for the (highly) nonlinear systems, cf. (5.29), (5.63), (5.149), (5.150), already in Subsection 5.2.2 **4**. We studied it for general operators and discretization methods in Section 3.7. So we discuss here only the specifics for the fully nonlinear case. We have to determine the FE solution $u_0^h \in \mathcal{U}^h \subset C^1(\Omega_c^h)$ such that

$$\left\langle F^h\left(u_0^h\right), v_\Pi^h\right\rangle = 0 \Longleftrightarrow \left(G\left(u_0^h\right), v^h\right)_{\mathcal{V}_c} = 0 \ \forall v^h \in \mathcal{V}^h, \left(u_0^h, v_b^h\right)_{\mathcal{V}_{c,b}} = 0 \ \forall v_b^h \in \mathcal{V}_b^h. \tag{5.153}$$

We restrict the presentation here to second order equations. Let Condition **H** be satisfied The general notation, introduced in (5.113), allow a relatively easy generalization to the other cases.

In this context again *local and stable bases* and *stable splittings* for S^h are important; see Theorems 4.36 and 4.36:

$$S^h = S_0^h + S_b^h = \left\{u^h - \sum_{\mu=1}^{m_0} c_\mu s^\mu + \sum_{\mu=m_0+1}^{m_1} c_\mu s^\mu\right\}. \tag{5.154}$$

We insert $u_0^h \in S^h$ into (5.153) and test $G(u_0^h)$ with $v^h = s^j, j = 1, \ldots, m_0$, and the boundary term with $v_b^h = s_b^i = s^{i+m_0}, i = 1, \ldots, m_1 - m_0$. So we obtain a nonlinear system of m equations for the m unknowns $c_\mu, \mu = 1, \ldots, m_1$.

This allows formulating the following algorithm:

Algorithm 5.1.
Nonstandard FEM for elliptic equations

INPUT: $G, \mathcal{T}_c^h, \mathcal{S}_d^1\left(\mathcal{T}_c^h\right) = S_0^h + S_b^h$ (Differential operator, subdivision and a stable splitting for the FE space of dimension m_1 on Ω_c^h);

COMPUTATION: FOR $j = 1, \ldots, m_0, i = 1, \ldots, m_1 - m_0$
/* Determine the m_1 equations in (5.153) */
compute the $\left\langle F_h\left(u_0^h\right), v_\Pi^h = \left(s^j, s_b^i\right)\right\rangle_{\mathcal{V}_\Pi^{h\prime} \times \mathcal{V}_\Pi^h} = \left(G\left(u_0^h\right), s^j\right)_{\mathcal{V}_c} + \left(u_0^h, s_b^i\right)_{\mathcal{V}_{c,b}}$;

OUTPUT: $\left\langle F_h\left(u_0^h\right), v_\Pi^h\right\rangle_{\mathcal{V}_\Pi^{h\prime} \times \mathcal{V}_\Pi^h}$, $j = 1, \ldots, m_0$, $i = 1, \ldots, m_1 - m_0$
(Nonlinear system of m_1 equations in (5.63)).

For computing the terms in (5.153) the multilevel results of Davydov [263] and Davydov and Stevenson [268], with multiresolution techniques can be used.

Sparse nonlinear systems

After calculating this system (5.153) has to be solved. The most important tools are continuation and discrete Newton methods. They require the solution of the corresponding *linear sparse FE equations* with the defects for the nonlinear FE equations, $F^h\left(u_\nu^h\right), u_\nu^h \approx u_0^h$ as a special input, see below. For these systems, we can employ the standard efficient solvers for weak formulations by Proposition 5.7. But the usual FEs in $C^0(\Omega^h)$ have to be replaced by FEs in $C^1\left(\Omega_c^h\right)$.

Nevertheless the question is interesting, whether (5.153) represents a *sparse nonlinear system,* with only a few nontrivial terms for the unknowns c_μ in each equation. Special conditions guarantee this, see Proposition 5.19.

K has the *special k-linear algebraic property* if

$$K(u) := g(a_1(u),\dots,a_s(u)) \text{ with } k_\ell\text{-linear } a_\ell(u_1,\dots,u_{k_\ell}) : \Omega_c^h \to \mathbb{R},\ k_\ell \le k', \tag{5.155}$$

$$u_j : \Omega_c^h \to \mathbb{R}, \ell = 1,\dots,s, a_\ell(u_1,\dots,u_{k_\ell}) = 0 \text{ if supp } u_1 \cap \cdots \cap \text{ supp } u_{k_\ell} = \emptyset,$$

with $k := \max_{\ell=1}^s\{k_\ell\} \le k'$, $a_\ell(u) := a_\ell(u,\dots,u)$ and a *special algebraic function*

$$g : \mathbb{R}^s \to \mathbb{R}, \text{ defined by sums of constants, the } a_\ell(u) \text{ and roots.} \tag{5.156}$$

For Proposition 5.19 and $\mathcal{S}_d^1\left(\mathcal{T}_c^h\right)$ we introduce

$$C_\mathcal{S}^1 := \max_{\mu=1,\dots,m_1}\{\text{ number of triangles } T \in \mathcal{T}_c^h \text{ s.t.supp } s^\mu \cap T \ne \emptyset\}$$

$$C_\mathcal{S}^k := \max_{\nu_i=1,\dots,m_1}\{\text{ number of } k-\text{ tuples } (\nu_1\dots,\nu_k) \text{ s.t.} \tag{5.157}$$

$$\text{supp } s^{\nu_1} \cap\cdots\cap \text{supp } s^{\nu_k} \ne \emptyset \text{ for } \nu_i \ne \nu_j \text{ for } i \ne j, \text{ and } i,j = 1,\cdots,k\}$$

Proposition 5.19. *For $\mathcal{S}_d^1\left(\mathcal{T}_c^h\right)$ we get the upper bound $C_\mathcal{S}^k \le \left(C_\mathcal{S}^1\right)^k$. If K has the special k-linear algebraic property (5.155), (5.156) and if $\left(C_\mathcal{S}^1\right)^{k+1} \ll m_1$, then (5.153) is a sparse nonlinear system. For $k = 1$, this yields sparse linear systems.*

Remark 5.20. *In some sense, it is disappointing that this proposition only discusses one special case. This is compensated by the* sparse linear FE equations *in continuation and discrete Newton methods, obtained by linearizing (5.6) into equations (5.19). Thus it does not matter too much that (5.155), (5.156) is a strong condition nor did we try to get the sharpest possible results. Products are excluded, since a_1a_2 would be (k_1+k_2)-linear. But Proposition 5.19 admits at least important examples: (2.308), for appropriate f and (2.310) are included. Generalizations are certainly possible.*

Proof. Obviously $C_\mathcal{S}^k \le \left(C_\mathcal{S}^1\right)^k$ and by testing a linear $K(u^h)$ with a fixed s^j, we find at most $C_\mathcal{S}^1 \ll m_1$ nontrivial coefficients, thus we obtain a sparse linear system. We restrict the proof to $k = 2; k > 2$ is then clear. A bilinear a_1 evaluated in x is:

$$a_1(u^h)(x) = a_1(u^h = \sum_{\mu=1}^{m_1} c_\mu s^\mu, u^h = \sum_{\nu=1}^{m_1} c_\nu s^\nu)(x) = \sum_{\mu,\nu=1}^{m_1}{}^{*} c_\mu c_\nu a_1(s^\mu, s^\nu)(x), \tag{5.158}$$

where * indicates that, for a fixed $x \in \Omega$, we only need those terms with $x \in \text{supp } s^\mu \cap \text{supp } s^\nu \neq \emptyset$, so at most $C_\mathcal{S}^2 = \left(C_\mathcal{S}^1\right)^2$ terms. For all other x, the $a_1(s^\mu, s^\nu)(x) = 0$ by (5.155), (5.156). Now we test with a fixed s^j and obtain

$$\int_{\Omega_c^h} a_1(u^h)(x)s^j(x)dx = \int_{\Omega_c^h} \sum_{\mu,\nu=1}^{m_1}{}^{*} c_\mu c_\nu a_1(s^\mu, s^\nu)(x)s^j(x)dx. \qquad (5.159)$$

Each of the $C_\mathcal{S}^1$ triangles T, with $\text{supp } s^j \cap T \neq \emptyset$, contributes nontrivially at most for $T \in \text{supp } s^\mu \cap \text{supp } s^\nu \neq \emptyset$. The at most $C_\mathcal{S}^2$ triangles in $\cup_{\mu,\nu=1}^{m_1} \text{supp } s^\mu \cap \text{supp } s^\nu \neq \emptyset$, yield at most $C_\mathcal{S}^3 = \left(C_\mathcal{S}^1\right)^3$ nontrivial terms in (5.159). □

Solving the nonlinear systems with Newton's method, MIP

In Section 3.7 we have presented continuation methods and the mesh independence priniple for Newton's method. So we give only a very short summary here.

Equation (5.153) is solved by the well-known continuation and Newton methods. We formulate the discrete Newton method and its quadratic convergence as special case of the preceding mesh independence principle (MIP), cf. Theorem 3.40 and Corollary 3.41. Both methods require the solution of the corresponding *linearized sparse FE equations,* with the defect or residuum, $F^h\left(u_\nu^h\right), u_\nu^h \approx u_0^h$. The $u^h|_{\partial\Omega_c^h} = 0$ and Proposition 5.7 allow standard efficient solvers with FEs in $C^0(\Omega^h)$ replaced by FEs $C^1\left(\Omega_c^h\right)$. For using these methods for differential equations and systems of order $2m$, they can be transformed into larger equations and systems of order 2. Otherwise the FEs in $C^0(\Omega^h)$ have to be replaced by $C^{2m-1}\left(\Omega_c^h\right)$ and appropriate solvers have to be used.

Efficient strategies combining continuation with the MIP on different discretization levels have been studied by Allgower and Böhmer [9, 10]. Currently, multiresolution techniques are available for polyhedral Ω in $\mathbb{R}^n$ and $d \geq 2^n + 1$; see Davydov's papers, e.g., [263, 267, 268]. For $\Omega \in C^2$ and the necessary Ω_c^h with curved boundaries for general $\mathcal{S}_d^1\left(\mathcal{T}_c^h\right)$ this is still an open problem. For the specific cases in Proposition 5.19, (5.153) represents a *sparse nonlinear system.*

The MIP, the Newton–Kantorowich method, and the convergence of our FEM require a unique solution u_0, such that $F'(u_0)$ is boundedly invertible. So we assume

$$F(u_0) = 0 \text{ with } F'(\cdot) \in C_L(B_r(u_0)), \quad (F'(u_0))^{-1} \in \mathcal{L}(\mathcal{V}, \mathcal{U}). \qquad (5.160)$$

Starting near $u_0^h \approx u_0$, we apply Newton's method to F^h. By Theorem 5.12, the $\left(F^h\left(u_i^h\right)\right)' : \mathcal{U}^h \to \mathcal{V}^h$ are equiboundedly invertible; hence

$$u_{i+1}^h := u_i^h - \left(F^h\left(u_i^h\right)'\right)^{-1} F^h\left(u_i^h\right), \quad i = 1, \ldots, \text{ is well defined.} \qquad (5.161)$$

In addition to (5.160), the MIP, cf. Subsection 3.7.2, and Allgower and Böhmer [9, 10] and Allgower *et al.* [17], requires a linear FEM as in (5.34), satisfied for our FEMs. Then the Newton method converges quadratically; cf. Theorem 3.40.

Equations (5.110) and (5.151) show that for $u_0 \in \mathcal{U}_s = H^\ell(\Omega), \ell > 2$, the FEM converges with orders $p' = \ell - 2m > 0$ and $p' := \min\{\ell - 2m, p\} > 0$. Then, with the above projectors $P^h = I^h E_c : \mathcal{U} \to \mathcal{U}^h$, $Q_\Pi^{h'} : \hat{\mathcal{V}}_\Pi \to \mathcal{V}_\Pi^{h'}$ (see (5.51), (5.60)), we impose the condition

$$F|_{\mathcal{U}_s} : \mathcal{D}(F) \cap \big(\mathcal{U}_s := H^\ell(\Omega)\big) \to H^{\ell-2m}(\Omega). \tag{5.162}$$

Theorem 5.21. MIP; see Theorem 3.40: *In addition to the conditions in Theorems* 5.13 *and* 5.18, *in particular Condion* **H**, *with convergence in* (5.110) *of order* $p' = \ell - 2m > 0$ *or* $p' = \min\{\ell - 2m, p\} > 0$ *we assume* (5.160), (5.162). *Start the Newton processes for* F, *by omitting all upper indices* ^{h} *in* (5.161), *and for* F^h *with the sufficiently good starting values* $u_1 \in B_r(u_0) \cap \mathcal{D}(F) \cap \mathcal{U}_s$ *and* $u_1^h := P^h u_1$. *Then, for* $h \le h_0 = (\min\{r, (SCL)^{-1}\}/(2Sc_0))^{1/p'}$, *cf. the following proof, both Newton methods converge quadratically to* u_0 *and* u_0^h, *respectively, and*

$$\begin{aligned}
&\left\|u_{\nu+1}^h - u_\nu^h\right\|_{\mathcal{U}^h} \le C\left(\left\|u_\nu^h - u_{\nu-1}^h\right\|_{\mathcal{U}^h}\right)^2 \quad \textit{and}\\
&\left\|u_i^h - P^h u_i\right\|_{\mathcal{U}^h} \le C h^{p'} \|u_i\|_{\mathcal{U}_s}, \quad i = 1, \ldots,\\
&\left\|F^h\left(u_i^h\right) - Q_\Pi^{h'} F(u_i)\right\|_{\mathcal{V}} \le C h^{p'} \|u_i\|_{\mathcal{U}_s}, \quad i = 1, \ldots,\\
&\left\|\left(u_i^h - u_0^h\right) - \left(P^h(u_i - u_0)\right)\right\|_{\mathcal{U}^h} \le C h^{p'} \|u_i\|_{\mathcal{U}_s}, \quad i = 1, \ldots.
\end{aligned} \tag{5.163}$$

Under these conditions there exists $\bar{h} = \bar{h}(\epsilon, u_0)$ *such that for all* $h \in (0, \bar{h}]$

$$\left|\min\left\{i \ge 1 : \left\|u_i^h - P^h u_0\right\|_{\mathcal{U}^h} < \epsilon\right\} - \min\left\{i \ge 1 : \|u_i - u_0\|_{\mathcal{U}} < \epsilon\right\}\right| \le 1. \tag{5.164}$$

This last result is the reason for the notation as MIP and our remark concerning the essential h independence of the convergence.

Proof. We have to show that the conditions in Theorem 3.40 are satisfied. By Subsection 5.2.5, $P^h = I^h E_c \in \mathcal{L}(\mathcal{U}, \mathcal{U}^h)$, $Q_\Pi^{h'} \in \mathcal{L}(\hat{\mathcal{V}}_\Pi = \hat{\mathcal{V}}_\Pi', \mathcal{V}_\Pi^{h'} \ne \mathcal{V}_\Pi^h)$ are bounded and linear, and $F^h = Q_{c,\Pi}^{h'} F|_{\mathcal{U}^h} : \mathcal{D}(F^h) = \mathcal{D}(F) \cap \mathcal{U}^h \to \mathcal{V}_\Pi^{h'}$, similar for $m > 1$. As a consequence of (5.162) and $\mathcal{U}^h$, we obtain the uniform Lipschitz-continuity of our $(F^h)'$:

$$\|(F^h)'(u^h) - (F^h)'(v^h)\|_{\mathcal{V}_\Pi^h \leftarrow \mathcal{U}^h} = \left\|Q_{c,\Pi}^{h'}(F'(u^h) - F'(v^h))\right\|_{\mathcal{V}_\Pi^h \leftarrow \mathcal{U}^h} \le CL\|u^h - v^h\|_{\mathcal{U}^h}$$

for all $u^h, v^h \in B_r(P^h u_0)$. As a consequence of $F'(\cdot) \in C_L(B_r(u_0))$ and the isolated solution u_0 of $F(u) = 0$ with boundedly invertible $F'(u_0)$ we obtain stability of $(F^h(u^h))'$, and hence $\|((F^h)'(u^h))^{-1}\|_{\mathcal{U}^h \leftarrow \mathcal{V}_\Pi^h} \le S$ for all $u^h \in B_{r/2}(P^h u_0)$ with a slightly reduced $r/2$; see Theorem 5.6. Finally we need the *consistent differentiability*

$$\forall u, v \in B_r(u_0) \cap \mathcal{U}_s : \quad \left\|Q_\Pi^{h'} F(u) - Q_{c,\Pi}^{h'} F(P^h u)\right\|_{\mathcal{V}_\Pi^h} \le C h^{p'} \|u\|_{\mathcal{U}_s} \le c_0 h^{p'}, \tag{5.165}$$

$$\text{and} \quad \left\|Q_\Pi^{h'} F'(u) v - Q_{c,\Pi}^{h'} F'(P^h u) P^h v\right\|_{\mathcal{V}_\Pi^h} \le C h^{p'} (1 + \|u\|_{\mathcal{U}_s}) \|v\|_{\mathcal{U}_s} \le c_1 h^{p'} \|v\|_{\mathcal{U}_s},$$

an immediate consequence of $F'(\cdot) \in C_L(B_r(u_0))$ and of (5.162).

For (5.164), the only missing conditions (see (3.25) in [17]) require the existence of a constant $\delta > 0$ such that

$$\liminf_{h>0} \|P^h u\|_{\mathcal{U}^h} \geq \delta \|u\|_{\mathcal{U}} \quad \forall u \in \mathcal{U}_s.$$

This is an immediate consequence of (4.64). □

5.3 FE and other methods for nonlinear boundary conditions

The techniques in the preceding section open up the possibility for formulating FEMs and prove their convergence for nonlinear boundary conditions in elliptic problems. Compared to the previous situation only few analytical existence and even less regularity results are available, however, cf. Zeidler [678], p. 537 ff.

A convergence analysis for numerical methods seems to be totally missing. We demonstrate this analysis only for FEMs, although all the other methods, considered in this and the next book, allow the same results. We assume the nonlinear boundary conditions, studied here, in a form such that the linearization yields either Dirichlet boundary conditions or the more general form of a special Dirichlet system, cf. (2.80) ff., (2.137) ff., (2.158), and Theorem 2.50. We formulate a nonlinear version of a Dirichlet boundary condition; extensions to special Dirichlet systems are obvious

$$b_j(\partial^{j-1}u/\partial\nu^{j-1}) = 0 \textit{ with appropriate nonlinear } b_j, j = 0, \dots m-1, \quad (5.166)$$

with obvious extensions to special Dirichlet systems.

For $m = 1$, we admit natural boundary condition, cf. (2.123). For example, the boundary operators, induced by quasilinear operators as in (5.286), might be imposed as nonlinear boundary operators

$$B_G\, u^h := B_G\,(x, u^h, \nabla u^h) := \sum_{i=1}^{n} \nu_i A_i(x, u^h, \nabla u^h). \quad (5.167)$$

Generalizations to other boundary conditions are possible, but are not studied here. We formulate the convergence result for problems of order $2m$.

Theorem 5.22. FE and other methods for nonlinear boundary conditions: *Let the nonlinear elliptic problem in* (5.42)*,* $F(u_0) = (G, B)(u_0) = 0, u_0 \in \mathcal{U}$*, have an isolated solution, with boundedly invertible* $F'(u_0) : \mathcal{U} \to \mathcal{V}_\Pi$*. We admit nonlinear* G *and/or* B *and assume* G *and* B *to be continuously differentiable in* $B_r(u_0)$*, e.g. by* (5.109)*, and an analogous condition for* B*. Let the domain satisfy* (4.5)*. Let* G^h *and* B^h *be consistent of the same order, usually* $\min\{s, d\} - m$*, for the following combinations.*

1. *If* G *is linear or quasilinear, choose* $\mathcal{U}^h, \mathcal{V}^h$ *as one of the above families of conforming or nonconforming FEs of local degree* $d-1$*, such that* $\mathrm{dist}(\partial\Omega, \partial\Omega^h) \leq Ch^d$*. Assume an exact solution* $u_0 \in H^s(\Omega), m < \min\{s, d\}$*. Then the FEM for* G *is stable, consistent in* $P^h u_0$*, and convergent. Then the same is valid for the FEM for* F *with the same order of convergence. So this* $F^h(u_0^h) = 0$ *has, for small*

enough h, a unique solution u_0^h near u_0, usually converging as

$$\left\| u_0 - u_0^h \right\|_{\mathcal{V}} \le C h^{\min\{s,d\}-m} \| u_0 \|_{H^s(\Omega)}. \tag{5.168}$$

2. *For a fully nonlinear G and a convex polyhedral or a C^2 domain Ω in (5.100), choose $\mathcal{U}^h = \mathcal{S}_d^1\left(\mathcal{T}_c^h\right)$, $d \ge 2^n + 1$ (see (5.3)), on a quasiuniform triangulation. Then the nonconforming FEM in (5.47) ff., now for nonlinear B,*

$$F^h = Q_{c,\Pi}^{h'} F|_{\mathcal{U}^h} : \mathcal{U}^h \to \mathcal{V}^{h'} \times \mathcal{V}_b^h, \quad u_0^h \in \mathcal{D}(F^h) \subset \mathcal{U}^h \text{ s.t. } F^h u_0^h = 0,$$

is stable and consistent in $P^h u_0$, and convergent. It has, for small enough h, a unique solution u_0^h near u_0. We obtain, cf. (5.67), for $u_0 \in H^\ell(\Omega)$, $\ell > 2$, for $m = 1$, the nonconforming FEM, and convex polyhedral or $C^p, 2 \le p \le 5$, domains,

$$\left\| P^h u_0 - u_0^h \right\|_{\mathcal{U}^h} \le C h^{\min\{\ell-2,p\}} \| u_0 \|_{H^\ell(\Omega)}, \tag{5.169}$$

respectively, with $p = 2$ for polyhedral Ω and $p \ge 2$ for polyhedral or curved $\Omega_c^h \approx \Omega \in C^p$.

3. *$F'(u_0) \in \mathcal{L}(\mathcal{W}, \mathcal{W}' \times \mathcal{W}_D)$ and $F'(u_0) \in \mathcal{L}(\mathcal{U}, \mathcal{V} \times \mathcal{V}_D)$ simultaneously satisfy the Fredholm alternative.*

Remark 5.23.

1. *For the above F (see (5.18), (5.20), (5.73)), ℓ and d are related by (4.62). With the ellipticity assumption (5.20), the previous (5.73), (5.74), (5.100) are implied by (5.109), except Ω convex polyhedral or in C^2. The latter are necessary for the regularity and the extension operators. The conforming FEM in (5.26) ff. does not apply, since nonlinear boundary conditions will be violated.*
2. *For quadrature approximations, exactly reproducing polynomials of degree $< k$ with $k > 2(d-1) > d > 2$, the solution $\tilde{u}_0^h$ of an approximate $\tilde{F}^h(\tilde{u}_0^h) = 0$ satisfies the same estimate as in (5.110); see Subsection 5.4.4 and 5.4.5 for $m > 1$ or for system.*
3. *Quadrature approximations for these problems follow the lines of Section 5.4.*

Proof. The proof follows essentially the lines of the last section with some straightforward modifications. The consistency in Theorem 5.4 has to be modified for quasilinear G if nonconforming FEMs are used. In any case these results have to be applied to B as well. For the stability we apply the technique to the product operator $F = (G, B)$, but need the regularity result in Lemma 5.10 only for fully nonlinear G. The remaining part of the proof remains essentially unchanged. □

5.4 Quadrature approximate FEMs

5.4.1 Introduction

In many cases it is impossible to evaluate exactly the linear, bilinear and higher nonlinear forms defining the FE equations. So we consider in this section, *quadrature and cubature or other approximations*, e.g. difference approximations, for the different terms in these FE equations. They are applied to elliptic problems of order $2m, m \geq 1$ in $\mathbb{R}^n$. These quadrature approximations will be chosen such that, the order of a given FEM is maintained in its quadrature approximate form. So the quadrature approximate forms have to be stable and consistent with the exact forms. These two conditions will sometimes have different implications, see e.g. Theorem 5.31.

Simplifying the notation, we assume throughout that the functionals, coefficients, nonlinear functions, $f, a_{ij}, A_{\alpha\beta}; G^w$ are already extended to $\Omega \cup \Omega_c^h$, so $(f, a_{ij}, A_{\alpha\beta}, G^w) = (E_c f, E_c a_{ij}, E_c A_{\alpha\beta}, E_c G^w)$. e.g. by the extension operator E_c, cf. Theorem 4.37. Thus, e.g. $\tilde{a}^h(u^h, v^h)$ in (5.181) is defined, in contrast to $\tilde{a}^h(u, v^h)$ for $u \in \mathcal{U}$, where the necessary function values of u might not be available. In particular, for the fully nonlinear problems as in Section 5.2 and Subsection 5.4.4 we sometimes omit the extension operator E_c. This is not always possible for functions, u, and the solution, u_0, e.g. Theorem 5.33.

All the following quadrature formulas are defined piecewise on every $T_c \in \mathcal{T}_c^h$, admitting broken Sobolev spaces for the different terms. Extensions to several neighboring $T_c \in \mathcal{T}_c^h$, similarly to Subsections 4.2.5, and 4.2.6 would be possible, but are not considered here. The FEs studied here are again piecewise polynomials with $\mathcal{P}_{d-1}^n \subset \mathcal{P} \subset \mathcal{P}_{d-\tau}^n, \tau \geq -1$. For including the violated boundary conditions, studied already in Section 5.2 and later on in Section 5.5, we use the notation $T_c \in \mathcal{T}_c^h$ indicating that we admit either a polyhedral or a curved approximation $\Omega_c^h \approx \Omega$ triangulated by $\mathcal{T}_c^h$ with straight or curved edges along the boundary $\partial\Omega$. For the isoparametric FEMs in Subsection 5.5.10 we will need other strategies anyway.

For the second type of variational crimes, studied in Subsection 5.5.7 and caused by violated continuity, the $a(u^h, v^h)$ is not, but the piecewise $a_h(\cdot,\cdot)\ \forall T_c \in \mathcal{T}_c^h$ in (5.179) is defined, cf. (5.78) above, due to $u^h \notin \mathcal{U}, v^h \notin \mathcal{V}$. We will in Subsection 5.5.9 study the total error introduced by the combination of quadrature and these nonconformity errors.

We denote the original linear, bilinear forms, linear, nonlinear operators as $f(\cdot), a(\cdot,\cdot),\ A, G$, their extensions and modifications due to variational crimes to broken Sobolev spaces defined on $\Omega_c^h \cup \Omega$ as $f_h(\cdot), a_h(\cdot,\cdot), A_h, G_h$, and their restriction to the approximating spaces $\mathcal{U}^h, \mathcal{V}^h, \ldots,$ as $f^h(\cdot), a^h(\cdot,\cdot),\ A^h, G^h$. Their quadrature approximations are denoted as $\tilde{f}^h(\cdot), \tilde{a}^h(\cdot,\cdot), \tilde{A}^h, \tilde{G}^h$, respectively Other, e.g. difference approximations of order k_a, and combinations with quadrature of order k, we indicate as $f^{\tilde{h}}(\cdot), a^{\tilde{h}}(\cdot,\cdot), A^{\tilde{h}},\ G^{\tilde{h}}$ and $\tilde{f}^{\tilde{h}}(\cdot), \tilde{a}^{\tilde{h}}(\cdot,\cdot), \tilde{A}^{\tilde{h}}, \tilde{G}^{\tilde{h}}$, respectively Explicitly, we will study and formulate the quadrature approximations in this Section. For the other appropriate approximations, $\tilde{a}^{\tilde{h}}(\cdot,\cdot), \tilde{A}^{\tilde{h}}, \tilde{G}^{\tilde{h}}$ with the new order $k := \min\{k, k_a\}$, the quadrature results essentially remain correct.

According to Theorem 3.21, we need two different results for proving convergence. For a linear problem with stable $\tilde{A}^h$, the inducing $\tilde{a}^h(\cdot,\cdot)$ has to satisfy a discrete inf–sup condition. Usually this is a consequence of the corresponding property for $a^h(\cdot,\cdot)$, see e.g. Theorem 5.30. For nonlinear problems its derivative, $G'(u_0) = 0$, has to satisfy this condition. Further we do need consistency results for the discretized equations.

It causes some problems that the $\tilde{a}^h(\cdot,\cdot), \langle f, \cdot\rangle^{\tilde{h}}, \tilde{A}^h, \tilde{G}^h, \tilde{F}^h$ are usually not defined for the original $u \in \mathcal{U}, v \in \mathcal{V}$, since quadrature formulas require $u(P_j)$ or even $\partial^i u(P_j)$, see (5.170), (5.311). This is sometimes circumvented by comparing the exact solution u_0 with u_0^h, solving the exact FE equation or the interpolant $I^h u_0$. Then the stability and estimates for $\tilde{a}^h(u^h, v^h) - a^h(u^h, v^h)$ or nonlinear analogues, see e.g. (5.188), (5.204), allow the final error estimates for $\|u_0 - \tilde{u}_0^h\|_{\mathcal{U}^h}$. The standard approach based upon the Strang lemmas and a neighboring $u_0 \approx z^h \in \mathcal{U}^h$ only works for linear problems. Alternatively, extensions as discussed in Theorem 4.36 could be used.

For the proofs of the following Theorems 5.30, 5.33, 5.36 5.37 and 5.40, we apply the results of the general discretization theory in Chapter 3, here as Summary 4.52, in Subsection 5.2.5. We will prove in this section that the $\tilde{a}^h(\cdot,\cdot),\ldots$ are stable and consistent in $P^h u$, simultaneously with the original or the modified $a^h(\cdot,\cdot),\ldots$, if the quadrature formulas, introduced in Subsection 5.4.2, are good enough and h is small enough. By Theorem 3.21, we thus get convergence, via stability and consistency of the quadrature approximate FEMs, see Definition 4.46.

The standard argumentation for the following convergence results is: the quadrature approximations represent discretizations satisfying all the conditions for the chosen spaces, their approximations, the projectors, the operators and their discretizations, $\tilde{Q}'^h$, $\tilde{Q}^h$, $\tilde{A}^h$, $\tilde{G}^h$, or $\tilde{F}^h$, guaranteeing a unique solution, $\tilde{u}_0^h$, and its convergence to the unique exact solution, u_0. In Subsection 5.4.3 we start with the spaces in (5.177), combined with the exact projectors and FEMs in (4.117), (4.132), or (4.135) for linear differential operators of order 2, and extend it to fully nonlinear equations of order 2, see (5.52), (5.60), (5.62) in Subsection 5.4.4. The next generalization to linear and nonlinear equations and systems of order $2m$ as in (4.133)–(4.140), (4.163), and (5.111)–(5.114), (5.123), is presented in Subsection 5.4.5. As a consequence of the perturbation or consistency results, e.g. in Theorem 5.28, the approximate quadrature projectors, $\tilde{Q}'^h$, $\tilde{Q}^h$, the linear and nonlinear $\tilde{A}^h$, $\tilde{G}^h$, $\tilde{F}^h$ simultaneously satisfy these conditions with the exact Q^h, A^h, G^h, F^h. In particular, $\tilde{A}^h$ and $\tilde{G}^h$, $\tilde{F}^h$ are continuous near $P^h u_0$ and consistent and stable in $P^h u_0$.

Remark 5.24. *These results remain valid with slight modifications for the other methods as well.*

5.4.2 Quadrature and cubature formulas

The main tools in this section are the following quadrature formulas in higher dimensions. Usually these quadrature formulas reproduce the integral of polynomials of a certain degree. For tensor product FEs, e.g. bilinear FEs, see Subsection 4.2.2, or the serendipity class, see [141], modifications are possible.

Proposition 5.25. *Assume the following quadrature formula*

$$\int_{T_c} f(x)dx \approx q_{T_c}^h(f) := \sum_{P_{j,T_c}\in\overline{T_c}} w_{j,T_c} f(P_{j,T_c}) \tag{5.170}$$

to be exact for polynomials of degree $k-1$, and let diam $T_c \leq h$. *Then* $\forall k \in \mathbb{N}, k > n/p'$, *there exists* $C = C_k \ \forall\, 0 < h < 1, 1 \leq p' \leq \infty, f \in W^{k,p'}(T_c) \cap C(\overline{T_c})$, *such that*

$$\left| \int_{T_c} f(x)dx - \sum_{P_{j,T_c} \in \overline{T_c}} w_{j,T_c} f(P_{j,T_c}) \right| \leq C_k\ h^{k+(1-1/p')} |f|_{W^{k,p'}(T_c)}. \qquad (5.171)$$

Proof. By $k > n/p'$ and the Sobolev embedding Theorem 1.26, $f \in C(\overline{\Omega})$, so all the $f(P_{j,T_c})$ are well defined. Now we proceed as in Proposition 5.62, adding and subtracting the averaged Taylor polynomial $Q^k f \in \mathcal{P}_{k-1}^n$ for f in (5.171). Then (5.170) implies with $f \in W^{k,p'}(\Omega)$, the Hölder inequality (1.45) and the Bramble–Hilbert lemma 4.15

$$\begin{aligned} &\left| \int_{T_c} f(x)dx - \sum_{P_{j,T_c} \in \overline{T_c}} w_{j,T_c} f(P_{j,T_c}) \right| \\ &\leq \left| \int_{T_c} f(x)dx - Q^k f + Q^k f dx - \sum_{P_{j,T_c} \in \overline{T_c}} w_{j,T_c} f(P_{j,T_c}) \right| \\ &\leq \left| \int_{T} 1 \cdot (f(x) - Q^k f) dx \right| \quad \text{with Hölder and Bramble–Hilbert} \\ &\leq C_k\ h^{1-1/p'} |f - Q^k f|_{L^{p'}(T_c)} \leq C_k\ h^{k+1-1/p'} |f|_{W^{k,p'}(T_c)}. \end{aligned}$$

□

Remark 5.26.

1. *In these quadrature approximations (5.170) the quadrature points P_{j,T_c} are usually chosen independently of the interpolation points for the FEs.*
2. *The $P_{j,T_c} \in \overline{T_c}, T_c \in \mathcal{T}_c^h$ in (5.170) are often determined as $P_{j,T_c} = F_{T_c}(P_j), P_j \in \overline{K}$. Then the computations are performed along the lines of Examples 4.11 and 4.51. So let all the $T_c \in \mathcal{T}_c^h$ be affinely equivalent to the reference element K, see Definition 4.6, hence, $F_{T_c}(K) = T_c, F_{T_c} = A_{T_c}x + b_{T_c}, A_{T_c}$ nonsingular. Then $q_{T_c}^h(f) = \sum_{P_{j,T_c} \in \overline{T_c}} w_{j,T_c} f(P_{j,T_c})$ so we obtain the $w_{j,T_c} := \det(A_{T_c}) w_j, P_{j,T_c} := F_{T_c}(P_j)$, from the quadrature formula on K, $q^h(f) = \sum_{P_j \in \overline{K}} w_j f(P_j)$. For T_c at the boundary slight modifications are necessary.*

We apply the above $q_{T_c}^h$ to the $T_c \in \mathcal{T}_c^h$, to yield a quadrature formula $q^h(f)$ for Ω or Ω_c^h. Note that for conforming and nonconforming FEs, satisfying the boundary conditions, we have $\Omega = \Omega_c^h$ and, hence, $\int_{\Omega_c^h} f(x)dx = \int_\Omega f(x)dx$, otherwise we use the extension $f_h : \Omega \cup \Omega_c^h \to \mathbb{R}$, cf. (5.178), (5.5), cf. (5.170) for $q_{T_c}^h(f_h)$, in

$$q^h(f_h) := \sum_{T_c \in \mathcal{T}_c^h} q_{T_c}^h(f_h) \approx \int_{\Omega_c^h} f_h(x)dx \approx \int_\Omega f(x)dx. \qquad (5.172)$$

Often we additionally find, sometimes with $C = 1$, if all weights are positive,

$$|q^h_{T_c}(f)| \le C \text{ meas } (T_c)\|f\|_{L^\infty(T_c)} \quad \forall T_c \in \mathcal{T}_c^h, \tag{5.173}$$

admitting modified relations. We estimate the total quadrature error in (5.172) with Proposition 5.25 and for $p' = 1$ as[37]

$$|q^h(f_h) - \int_{\Omega_c^h} f(x)dx| \le C_k h^k \sum_{T_c \in \mathcal{T}_c^h} |f_h|_{W^{k,1}(T_c)} = C_k h^k |f_h|_{W^{k,1}(\mathcal{T}_c^h)}. \tag{5.174}$$

Some of the following estimates are applied to $f_h = u^h v^h, u^h \in W^{k,p}, v^h \in W^{k,p'}$.

Proposition 5.27. *For $u^h \in W^{k,p}\left(\mathcal{T}_c^h\right), v^h \in W^{k,p'}\left(\mathcal{T}_c^h\right), 1/p + 1/p' = 1$ we get*

$$|q^h(u^h v^h) - \int_{\Omega_c^h} u^h v^h(x)dx| \le C_k h^k \|u^h\|_{W^{k,p}(\mathcal{T}_c^h)} \|v^h\|_{W^{k,p'}(\mathcal{T}_c^h)}. \tag{5.175}$$

Proof. The standard and discrete Hölder inequalities (1.45), (1.44) yield with $p' = 1$ in (5.171) and again with C_k indicating different constants, depending upon k,

$$\begin{aligned}
&|q^h(u^h v^h) - \int_{\Omega \text{ or } \Omega_c^h} u^h v^h(x)dx| \\
&\le C_k h^k \sum_{T_c \in \mathcal{T}_c^h} |u^h v^h|_{W^{k,1}(T_c)} \\
&\le C_k h^k \sum_{T_c \in \mathcal{T}_c^h} \sum_{\alpha,\beta \ge 0} \sum_{|\beta + \alpha| = k} \|\partial^\beta u^h \partial^{\alpha - \beta} v^h\|_{L^1(T_c)} \\
&\le C_k h^k \sum_{T_c \in \mathcal{T}_c^h} \sum_{\alpha,\beta \ge 0} \sum_{|\beta + \alpha| = k} \|\partial^\beta u^h\|_{L^p(T_c)} \|\partial^{\alpha - \beta} v^h\|_{L^{p'}(T_c)} \\
&\le C_k h^k \left(\sum_{T_c \in \mathcal{T}_c^h} \sum_{|\beta| = k} \|\partial^\beta u^h\|^p_{L^p(T_c)} \right)^{1/p} \left(\sum_{T_c \in \mathcal{T}_c^h} \sum_{|\alpha| = k} \|\partial^\alpha v^h\|^{p'}_{L^{p'}(T_c)} \right)^{1/p'} \\
&\le C h^k \|u^h\|_{W^{k,p}(\mathcal{T}_c^h)} \|v^h\|_{W^{k,p'}(\mathcal{T}_c^h)}.
\end{aligned}$$

□

5.4.3 Quadrature for second order linear problems

Here we consider weak forms of second order problems, hence we choose $\mathcal{U} = \mathcal{V}$. The basic condition, compare the *Sobolev Embedding Theorem* 1.26 is a

$$\text{bounded } \Omega \subset \mathbb{R}^n \text{ with Lipschitz boundary }, \ n \ge 1 \text{ and } f, a_{ij} \in C(\Omega). \tag{5.176}$$

$$u, u_0, v \in C^1(\Omega), \text{ e.g., } W^{k,p}(\Omega) \subset C^j(\Omega), 1 \le p < \infty, \ k > j + n/p \text{ for } \Omega,$$

see the modifications below for less smooth $u, u_0, v \notin C^1(\Omega)$.

[37] For some cases $p' = 1$ is the optimal choice: It gives, in combination with (1.44), $h^p \|v^h\|_{H^1(\mathcal{T}_c^h)}$ with largest p, e.g. $p' = \infty$ would yield at most $h^{p-1}\|v^h\|_{H^1(\mathcal{T}_c^h)}$.

The piecewise polynomials u^h, v^h are continuously differentiably extendable to each $\overline{T_c}$. For a $P_{j,T_{c1}} \in \overline{T_{c1}} \cap \overline{T_{c2}}$ the values $u^h(P_{j,T_{c1}})$ and $u^h(P_{j,T_{c2}})$ might differ. But it is no problem to exactly evaluate the $u^h, v^h, \partial^i u^h, \partial^j v^h$ in the points P_j and to exactly integrate their products and the $a^h(\cdot,\cdot)$. Usually the $\langle f, v^h\rangle$ have to be approximated by an $\langle f, v^h\tilde{\rangle}^h$. So there are good reasons to study both cases. Sometimes other forms of $\tilde{a}^h(\cdot,\cdot) : \mathcal{U}^h \times \mathcal{V}^h \to \mathbb{R}$ are defined by combining the exact $u^h, v^h, \partial^i u^h, \partial^j v^h$ with appropriate means of the a_{ij}, f. We do not discuss those methods here, cf. Grossmann, Roos and Stynes [374–376].

The spaces here and their approximating spaces are, cf. (4.35), (4.36)

$$\mathcal{U} = \mathcal{V} \subset H^1(\Omega), \mathcal{U}_b^h, \mathcal{V}_b^h \subset \left\{L^\infty\left(\mathcal{T}_c^h\right) : \forall T_c \in \mathcal{T}_c^h : u^h|_{T_c} \in \mathcal{P}\right\}, \mathcal{P}_{d-1} \subset \mathcal{P} \subset \mathcal{P}_{d+\tau},$$
$$\dim \mathcal{U}_b^h = \dim \mathcal{V}_b^h, \text{usually } \tau = -1, \text{ and norms } \|\cdot\|_{\mathcal{U}^h} := \|\cdot\|_{\mathcal{V}^h} := \|\cdot\|_{H^1(\mathcal{T}_c^h)}. \tag{5.177}$$

These $\mathcal{U}_b^h, \mathcal{V}_b^h$ indicate the FE spaces, satisfying the boundary conditions either exactly or approximately, corresponding to the chosen FEM. If we use $\mathcal{U}^h, \mathcal{V}^h$ we omit boundary conditions, often for the bilinear forms or operators. Sometimes $\mathcal{U} \subset W^{1,p}(\Omega), \mathcal{V} \subset W^{1,p'}(\Omega)$ or $\mathcal{U} \subset W^{m,p}(\Omega), \mathcal{V} \subset W^{m,p'}(\Omega), m > 1$, are appropriate. This can be handled similarly to below via the Sobolev Embedding Theorem 1.26.

In addition to the quadrature versions $\tilde{f}^h$, $\tilde{a}^h(\cdot,\cdot)$, $\tilde{A}^h$ we need the approximate projectors $\tilde{Q}'^h$, cf. Propositions 4.49, 4.50. All versions are obtained by applying the quadrature formulas (5.172) to different combinations, f, fv^h, $a_{ij}u^h v^h$, ... As in Section 5.2, we denote, e.g. the original and the sometimes necessary extensions, note (5.5) omitting E_c,

$$\text{extensions of } f, a(\cdot,\cdot), A, G, \text{ to } \Omega \cup \Omega_c^h \text{ as } f_h, a_h(\cdot,\cdot), A_h, G_h, \text{ a.s.o., so} \tag{5.178}$$

$$a_h(u^h, v^h) := \langle A_h u^h,\ v^h\rangle := \sum_{T_c \in \mathcal{T}_c^h} \int_{T_c} \left(\sum_{i,j=0}^n a_{ij}\partial^i\ u^h \partial^j\ v^h\right) dx\ \forall u^h \in \mathcal{U} \cup \mathcal{U}^h,$$

$$\tilde{a}_h(u^h, v^h) := \langle \tilde{A}_h u^h,\ v^h\tilde{\rangle}^h := \sum_{T_c \in \mathcal{T}_c^h} q_{T_c}^h \left(\sum_{i,j=0}^n a_{ij}\partial^i\ u^h \partial^j\ v^h\right), \text{ and} \tag{5.179}$$

$$\langle f_h, v^h\rangle := \sum_{T_c \in \mathcal{T}_c^h} \int_{T_c} \left(\sum_{i=0}^n f_i \partial^i v^h\right) dx\ \forall v^h \in \mathcal{V} \cup \mathcal{V}^h, f_j \in H^k\left(\mathcal{T}_c^h\right) \text{or} f \in H_k^{-1}\left(\mathcal{T}_c^h\right),$$

$$\langle f_h, v^h\tilde{\rangle}^h := \sum_{T_c \in \mathcal{T}_c^h} q_{T_c}^h \langle f_h\ v^h\rangle = \sum_{T_c \in \mathcal{T}_c^h} \sum_{P_{j,T_c} \in \overline{T_c}} w_{j,T_c} \left(\sum_{i=0}^n f_i \partial^i v^h\right)(P_{j,T_c})$$

$$\tilde{Q}'^h : \mathcal{V} \to \mathcal{V}^{h'},\ \langle \tilde{Q}'^h f - f, v^h\tilde{\rangle}^h = ((\tilde{Q}'^h f - f_h)v^h)^{\tilde{h}} = 0\ \ \forall v^h \in \mathcal{V}_b^h. \tag{5.180}$$

Note that for applying the quadrature formulas, we have to impose the additional conditions f or $f = (f_j)_{j=0}^n, a_{ij} \in H^k(\Omega)$, or $f \in H_k^{-1}(\Omega)$, k large enough for the

$$\begin{array}{ccccccc}
\mathcal{U}_b & \overset{G}{\longrightarrow} & & \mathcal{V}'_b & \overset{\text{tested by}}{\longleftrightarrow} & \mathcal{V} \\
P^h\big\downarrow & & \tilde{\Phi}^h\big\downarrow & \tilde{Q}'^h\big\downarrow & & \\
\mathcal{U}^h_b & \underset{\longrightarrow}{\tilde{G}^h} & & \mathcal{V}^{h\prime}_b & \underset{\longleftrightarrow}{\text{appropx. tested by}} & \mathcal{V}^h_b
\end{array}$$

Figure 5.5 Quadrature approximate FEMs.

required function evaluations, cf. (5.184), (5.185). Similarly to (5.5) we get

$$a^h(u^h, v^h) := \langle A^h u^h,\ v^h\rangle := \langle A_h u^h,\ v^h\rangle\Big|_{\mathcal{U}^h_b\times\mathcal{V}^h_b} = a_h(u^h, v^h)\Big|_{\mathcal{U}^h_b\times\mathcal{V}^h_b},$$

$$a(u,v) = a_h(u,v)|_{\mathcal{U}_b\times\mathcal{V}_b} = \langle Au,\ v\rangle, \text{ and} \tag{5.181}$$

$$\tilde{a}^h(u^h, v^h) := \langle \tilde{A}^h u^h,\ v^h\tilde{\rangle}^h := \tilde{a}_h(u^h, v^h)\Big|_{\mathcal{U}^h_b\times\mathcal{V}^h_b}, \langle\tilde{f}^h, v^h\tilde{\rangle}^h := \langle f_h, v^h\tilde{\rangle}^h\Big|_{\mathcal{V}^h},$$

$$\text{for } u, v \in H^k(\Omega) : \tilde{a}^h(u,v) := \langle \tilde{A}^h u,\ v\tilde{\rangle}^h := \tilde{a}_h(u,v)|_{\mathcal{U}\times\mathcal{V}} := \langle \tilde{A}_h u,\ v\tilde{\rangle}^h|_{\mathcal{U}\times\mathcal{V}}.$$

The approximate solutions $\tilde{u}^h_0$ are defined for these approximations $u^h_0 \in \mathcal{U}^h_b$

$$\left\langle A^h\tilde{u}^h_0, v^h\right\rangle = a^h\left(\tilde{u}^h_0, v^h\right) = \langle f^h, v^h\tilde{\rangle}^h \text{ or } = \langle\tilde{A}^h\tilde{u}^h_0, v^h\tilde{\rangle}^h = \tilde{a}^h\left(\tilde{u}^h_0, v^h\right) \forall v^h \in \mathcal{V}^h_b. \tag{5.182}$$

Reformulating with the exact and quadrature projectors yields

$$A^h = Q'^h A_h|_{\mathcal{U}^h_b},\ A^h\tilde{u}^h_0 = \tilde{Q}'^h f, \text{ and } \tilde{A}^h = \tilde{Q}'^h A_h|_{\mathcal{U}^h_b}, \tilde{A}^h\tilde{u}^h_0 = \tilde{Q}'^h f. \tag{5.183}$$

Again a slight modification of the general discretization theory in Chapter 3, see Figure 5.5, cf. Figures 4.27, 5.2, yields *convergence* for a consistent and stable method in $P^h u_0$ with uniformly bounded $\tilde{Q}^h$, since $P^h = I^h$ or $I^h E_c$ is again uniformly bounded, note Summary 4.52.

We will show in Theorems 5.28, 5.30 that $\tilde{a}^h(\cdot,\cdot), \langle f_h, \cdot\tilde{\rangle}^h$ in (5.182) are stable and consistent in $P^h u$, simultaneously with $a^h(\cdot,\cdot), \langle f_h, \cdot\rangle^h$ and the above P^h and the new $\tilde{Q}'^h$, $\tilde{Q}^h_\Pi$ are uniformly bounded for smooth enough data, hence if

$$k > \max\left\{d + \tau - 1, \frac{n}{2}\right\} \text{ for } \langle f_h, \cdot\tilde{\rangle}^h, f = \left(f_j \in H_k\left(\Omega\cup\mathcal{T}^h_c\right)\right)^n_{j=0} \in H^{-1}_k(\ldots) \tag{5.184}$$

$$k > 2(d + \tau - 1) \text{ for } \tilde{a}^h(\cdot,\cdot), a_{ij} \in W^{k,\infty}\left(\Omega\cup\Omega^h_c\right), \tau \geq -1, \text{ usually } \tau = -1. \tag{5.185}$$

Then, by the Sobolev Embedding Theorem 1.26, the quadrature formulas in (5.179)–(5.180) are well defined. We assume (5.184), (5.185), but do not show (5.185) to be necessary for a stable $\tilde{A}^h$; analogously to (4.137) we define $\|f\|_{H^{-1}_k(\mathcal{T}^h_c)}$.

Hence k is often pretty large. Solutions u_0 and general functions u will usually $\notin H^{k+1}(\Omega)$. So we combine in (5.190) the $E_c u$ with $I^h u$. If u is smooth enough, the quadrature formula can be applied to u directly. This is not a necessary condition, but is satisfied, e.g. if $u \in H^s(\Omega)$ with $s > n/2$, and the quadrature formula requires only function values, cf. Theorem 1.26. FEMs for fully nonlinear problems, their linearization and isoparametic FEs do need the extension operator E_c.

Quadrature approximate FEMs for nonconforming FEs are delayed to Section 5.5 Subsection 5.5.9 we omit the E_c here, except for fully nonlinear equations.

Theorem 5.28. Convergence of the $\tilde{f}^h$ and $\tilde{Q}'^h$, consistent $\tilde{a}^h(u^h, v^h)$:

1. *Assume (5.176), and for the quasiuniform subdivision $\mathcal{T}^h$ the Condition 4.16. Choose a quadrature formula as in (5.172), (5.174), satisfying (5.184) and (5.185) for $\tilde{\langle f_h, \cdot\rangle}^h$ and $\tilde{a}^h(\cdot,\cdot)$, with $f_j \in H^k(\Omega)$ or $f \in H_k^{-1}(\Omega)$, $a_{ij} \in W^{k,\infty}(\Omega)$, and let $h > 0$ be small enough.*
2. *Then the following consistency results hold, with $u \in H^{d-p}(\Omega), p, \geq 0$, in (5.190), notice the different norms in (5.188), (5.189),*

$$|\langle f_h, v^h\rangle - \tilde{\langle \tilde{f}^h, v^h\rangle}^h| \leq Ch^{k-(d+\tau-1)}\|f\|_{H_k^{-1}(\mathcal{T}_c^h)} \cdot \|v^h\|_{\mathcal{V}^h}, \tag{5.186}$$

$$\|Q'^h f - \tilde{Q}'^h f\|_{\mathcal{V}^{h'}} \leq Ch^{k-(d+\tau-1)}\|f\|_{H_k^{-1}(\mathcal{T}_c^h)}, \lim_{h\to 0}\|\tilde{Q}'^h f\|_{\mathcal{V}^{h'}} = \|f\|_{\mathcal{V}'}, \tag{5.187}$$

$$\frac{|a^h(u^h,v^h) - \tilde{a}^h(u^h,v^h)|}{\|v^h\|_{\mathcal{V}^h}} \leq Ch^{k-2(d+\tau-1)}M_{a_{ij}}\|u^h\|_{\mathcal{V}^h}, \textit{for } v^h \neq 0 \tag{5.188}$$

$$\leq Ch^{k-(d+\tau-1)} \cdot M_{a_{ij}}\|u^h\|_{H^{d+\tau}(\mathcal{T}^h)}; M_{a_{ij}} := \max_{i,j=0}^{n}\|a_{ij}\|_{W^{k,\infty}(\Omega_c^h)}. \tag{5.189}$$

$$\frac{|a_h(u,v^h) - \tilde{a}^h(I^h u, v^h)|}{\|v^h\|_{\mathcal{V}^h}} \leq C\big(h^{d-1-p} + h^{k-(d+\tau-1)}M_{a_{ij}}\big)\|u\|_{H^{d-p}(\Omega)}. \tag{5.190}$$

Remark 5.29. *The condition of a quasiuniform subdivision, $\mathcal{T}^h$, in this and the following theorems, can be slightly relaxed with the results in Subsection 4.2.5. This requires the restriction to the most important cases for FEMs, $n = 2, 3$, and requiring Assumptions 4.21 and 4.26. Then in the sum in (5.191) the terms have to be estimated for each $T \in \mathcal{T}^h$ separately. The results in Theorems 4.30–4.33 allow appropriate estimates for continuous and L^2 FEs. We leave the details to the interested reader.*

Proof. We start using (5.172),(5.174) estimating for $f \in H^k(\Omega)$

$$|\langle f_h, v^h\rangle - \tilde{\langle \tilde{f}^h, v^h\rangle}^h| = \left|\sum_{T_c\in\mathcal{T}_c^h}\sum_{j=0}^{n}\int_{T_c} f_j\partial^j v^h dx - q_{T_c}^h(f_j\partial^j v^h)\right|$$

$$\leq C_k h^k \sum_{T_c\in\mathcal{T}_c^h}\sum_{j=0}^{n}|f_j\partial^j v^h|_{W^{k,1}(T_c)}$$

$$= C_k h^k \sum_{j=0}^{n}|f_j\partial^j v^h|_{W^{k,1}(\mathcal{T}_c^h)}. \tag{5.191}$$

We estimate these derivatives and use (5.184)

$$
\begin{aligned}
&|f_j\partial^j v^h|_{W^{k,1}(T_c)} \\
&\quad = \sum_{|\alpha|=k} \|(f_j\partial^j v^h)^{(\alpha)}\|_{L^1(T_c)} \\
&\quad \le \sum_{|\alpha|=k}\sum_{0\le \ell\le\alpha} C_{\ell\alpha}\|f_j^{(\ell)}(\partial^j v^h)^{(\alpha-\ell)}\|_{L^1(T_c)} \text{ by (5.312)} \\
&\quad \le C\sum_{|\alpha|=k}\sum_{0\le \ell\le\alpha} \|f_j^{(\ell)}\|_{L^2(T_c)}\|(\partial^j v^h)^{(\alpha-\ell)}\|_{L^2(T_c)} \text{ by (1.44)} \\
&\quad \le C\left(\sum_{|\alpha|=k}\sum_{0\le \ell\le\alpha} \|f_j^{(\ell)}\|^2_{L^2(T_c)}\right)^{1/2}\left(\sum_{|\alpha|=k}\sum_{0\le \ell\le\alpha} \|(\partial^j v^h)^{(\alpha-\ell)}\|^2_{L^2(T_c)}\right)^{1/2} \\
&\quad = C\|f_j\|_{H^k(T_c)}\|\partial^j v^h\|_{H^{d+\tau}(T_c)} = C\|f_j\|_{H^k(T_c)}\|v^h\|_{H^{d+\tau-(1-\delta_{j0})}(T_c)}.
\end{aligned} \tag{5.192}
$$

Combining (5.191), (1.44) with the usual δ_{ij} yields, cf. (4.137), and [141], 8.x.15,

$$
|\langle f_h, v^h\rangle - \langle f_h, v^h\rangle^{\tilde{h}}| = |\langle f_h, v^h\rangle - \langle \tilde{f}^h, v^h\rangle^{\tilde{h}}| \overset{(1.44)}{\le} C_k h^k \tag{5.193}
$$

$$
\sum_{T_c\in\mathcal{T}_c^h}\sum_{j=0}^{n} \|f_j\|_{H^k(T_c)}\|v^h\|_{H^{d+\tau-(1-\delta_{j0})}(T_c)} \le C_k h^{k-(d+\tau-1)}\|f\|_{H_k^{-1}(\mathcal{T}_c^h)}\|v^h\|_{H^1(\mathcal{T}_c^h)}.
$$

thus (5.186) by the inverse estimate (4.47) for the quasiuniform $\mathcal{T}_c^h$. Relation (5.187) is an immediate consequence of (5.186).

Similarly we have to estimate the quadrature errors for the weak bilinear form, replacing (5.184) by (5.185). For modifying (5.192), (5.193), we use (5.185) and the Cauchy–Schwarz inequality, (1.48). Then

$$
\begin{aligned}
&|(a_{ij}\partial^i u^h\partial^j v^h)|_{W^{k,1}(T_c)} \\
&\qquad \le C\|a_{ij}\|_{W^{k,\infty}(\mathcal{T}_c^h)}\|\partial^i u^h\|_{H^k(T_c)}\|\partial^j v^h\|_{H^k(T_c)} \\
&\qquad \le C\|a_{ij}\|_{W^{k,\infty}(\mathcal{T}_c^h)}\|\partial^i u^h\|_{H^{d+\tau-(1-\delta_{i0})}(T_c)}\|\partial^j v^h\|_{H^{d+\tau-(1-\delta_{j0})}(T_c)}.
\end{aligned} \tag{5.194}
$$

u^h, v^h are always $\in H^k(\mathcal{T}_c^h)$. This implies with the discrete Hölder inequality

$$
\left|\sum_{T_c\in\mathcal{T}_c^h}\int_{T_c} a_{ij}\partial^i u^h\partial^j v^h dx - q^h_{T_c}(a_{ij}\partial^i u^h\partial^j v^h)\right| \tag{5.195}
$$

$$
\begin{aligned}
&\le C_k h^k\|a_{ij}\|_{W^{k,\infty}(\Omega_c^h)}\sum_{T_c\in\mathcal{T}_c^h}\|\partial^i u^h\|_{H^{d+\tau-(1-\delta_{i0})}(T_c)}\|\partial^j v^h\|_{H^{d+\tau-(1-\delta_{j0})}(T_c)} \\
&\le C_k h^k\|a_{ij}\|_{W^{k,\infty}(T_c)}\|\partial^i u^h\|_{H^{d+\tau-(1-\delta_{i0})}(\mathcal{T}_c^h)}\|\partial^j v^h\|_{H^{d+\tau-(1-\delta_{j0})}(\mathcal{T}_c^h)}
\end{aligned}
$$

with $\|u\|_{H^{k+1}(\Omega)}$ for $\|\partial^i u^h\|_{H^{d+\tau-(1-\delta_{i0})}(\mathcal{T}_c^h)}$ if u^h is replaced by u.

The analogous summation as in (5.193) and the inverse estimate (4.47) again yield (5.188), (5.190), since by (5.177), (5.180). (5.185), (5.195), we find

$$
\begin{aligned}
&|a^h(u^h,v^h) - \tilde{a}^h(u^h,v^h)| \\
&= \left| \sum_{T_c \in \mathcal{T}_c^h} \sum_{i,j=0}^{n} \left(\int_{T_c} a_{ij} \partial^i u^h \partial^j v^h dx - q_{T_c}^h \cdots \right) \right| \\
&\le Ch^k \max_{i,j=0}^{n} \|a_{ij}\|_{W^{k,\infty}(\Omega_c^h)} \left(\sum_{T_c \in \mathcal{T}_c^h} \sum_{i,j=0}^{n} \|\partial^i u^h\|_{H^{d+\tau-(1-\delta_{i0})}(T_c)} \|\partial^j v^h\|_{H^{d+\tau-(1-\delta_{j0})}(T_c)} \right) \\
&\le Ch^k \max_{i,j=0}^{n} \|a_{ij}\|_{W^{k,\infty}(\Omega_c^h)} \cdot \Big(\|\nabla u^h\|_{H^{d+\tau-1}(\mathcal{T}_c^h)} \|\nabla v^h\|_{H^{d+\tau-1}(\mathcal{T}_c^h)} \\
&\quad + \|\nabla u^h\|_{H^{d+\tau-1}(\mathcal{T}_c^h)} \|v^h\|_{H^{d+\tau}(\mathcal{T}_c^h)} + \|u^h\|_{H^{d+\tau}(\mathcal{T}_c^h)} \|\nabla v^h\|_{H^{d+\tau-1}(\mathcal{T}_c^h)} \\
&\quad + \|u^h\|_{H^{d+\tau}(\mathcal{T}_c^h)} \|v^h\|_{H^{d+\tau}(\mathcal{T}_c^h)} \Big);
\end{aligned}
\tag{5.196}
$$

note that ∇v^h indicates $\partial^j v^h \in \mathcal{P}^{d+\tau-1}, j = 1, \ldots, n$. We employ (4.47) for u^h, v^h, e.g. $\|\nabla v^h\|_{L^2(\mathcal{T}_c^h)} \le \|v^h\|_{H^1(\mathcal{T}_c^h)}$ and verify (5.188) by (5.185).

If we replace u^h by a smooth enough u, such that the quadrature formula is defined for u, we end up with (5.190). If only $u \in H^{d-p}(\Omega), p, \ge 0$, we use $I^h u$. Then $a_h(u - I^h u, v^h)$ causes the $h^{d+\tau-p}$ term by the interpolation error, the quadrature error, $a_h(I^h u, v^h) - \tilde{a}_h(I^h u, v^h)$ remains unchanged by (5.189). □

Theorems 5.30 and 5.31 imply stability and a convergent quadrature approximate FEM, see Definition 4.46.

Theorem 5.30. Stability of $\tilde{A}^h$: *Under the conditions of Theorem 5.28 the* $a^h(\cdot,\cdot)$ *and* $\tilde{a}^h(\cdot,\cdot)$ *are simultaneously bounded on* $\mathcal{U}^h \times \mathcal{V}^h$. *Let* $a^h(\cdot,\cdot)$ *be* $\mathcal{U}_b^h$*-coercive or, e.g. for* $\mathcal{U}_b^h \ne \mathcal{V}_b^h$, *satisfy a discrete inf–sup condition and (4.145) for* $\mathcal{U}^h$ *and* $\mathcal{V}^h$, *hence* A^h *is stable and let* $k > 2(d+\tau-1)$. *Then, for small enough* h, $\tilde{A}^h$ *and* $\tilde{a}^h(\cdot,\cdot)$ *have the same property.*

Proof. We only consider the case of a discrete inf–sup condition for $a^h(\cdot,\cdot)$. This implies with Theorem 5.28

$$
\begin{aligned}
\epsilon' \|u^h\|_{\mathcal{U}^h} &\le \inf_{0 \ne v^h \in \mathcal{V}_b^h} |a^h(u^h,v^h)|/\|v^h\|_{\mathcal{U}^h} \\
&\le \inf_{0 \ne v^h \in \mathcal{V}_b^h} |\tilde{a}^h(u^h,v^h) + (a^h(u^h,v^h) - \tilde{a}^h(u^h,v^h))|/\|v^h\|_{\mathcal{U}^h} \\
&\le \inf_{0 \ne v^h \in \mathcal{V}_b^h} |\tilde{a}^h(u^h,v^h)|/\|v^h\|_{\mathcal{U}^h} \\
&\quad + \sup_{0 \ne v^h \in \mathcal{V}_b^h} |(a^h(u^h,v^h) - \tilde{a}^h(u^h,v^h))|/\|v^h\|_{\mathcal{U}^h} \\
&\le \inf_{0 \ne v^h \in \mathcal{V}_b^h} |\tilde{a}^h(u^h,v^h)|/\|v^h\|_{\mathcal{U}^h} + Ch\|u^h\|_{\mathcal{U}^h},
\end{aligned}
\tag{5.197}
$$

so finally

$$(\epsilon' - Ch)\|u^h\|_{\mathcal{U}^h} \leq \inf_{0\neq v^h\in\mathcal{V}_b^h} |\tilde{a}^h(u^h,v^h)|/\|v^h\|_{\mathcal{U}^h}. \tag{5.198}$$

Analogously the inequalities for the $\inf_{0\neq u^h\in\mathcal{U}_b^h}$ are handled. This shows the discrete coercivity or discrete inf–sup condition for $\tilde{a}^h(\cdot,\cdot)$ for sufficiently small $h>0$. □

Theorem 5.31. Stability and convergence for different quadrature approximations:

1. *Let $a^h(\cdot,\cdot)$ induce a stable A^h and assume the conditions of Theorem 5.28. Then, for small enough h, simultaneously $\tilde{a}^h(\cdot,\cdot)$ induces a stable $\tilde{A}^h$. If furthermore (5.184) is correct, then $\tilde{a}^h(\tilde{u}_0^h,v^h) = \langle f,v^h\rangle^{\tilde{h}}\ \forall v^h\in\mathcal{V}_b^h$, defines a stable and convergent discretization for $Au_0=f$. For the other approximations, mentioned above, this result remains correct if the $\tilde{a}^h,\tilde{A}^h$ are replaced by $a^{\tilde{h}},\tilde{a}^{\tilde{h}}$, $A^{\tilde{h}},\tilde{A}^{\tilde{h}}$.*
2. *For* conforming FEMs *and $u_0\in H^s(\Omega)$, $M_{a_{ij}} = \max_{i,j=0}^{n}\|a_{ij}\|_{W^{k,\infty}(\Omega)}$, we obtain*

$$\left\|u_0-\tilde{u}_0^h\right\|_{\mathcal{U}^h} \leq Ch^{\min\{s-1,k-(d+\tau-1)\}} \tag{5.199}$$
$$\left((M_{a_{ij}}+1)\|u_0\|_{H^{\min\{s,k+1\}}(\Omega)} + \|f\|_{H_k^{-1}(\Omega)}\right).$$

3. *If only approximations for $\langle f,v^h\rangle^{\tilde{h}}$ and none for $\tilde{a}^h(u^h,v^h)$ are used in (5.182), then the term $(1+M_{a_{ij}})\|u_0\|_{H^{\min\{s,k+1\}}(\Omega)}$ in (5.199) is dropped.*

Proof. This theorem is an immediate consequence of (5.187), (5.190). The original (conforming) FEM is based upon an appropriate combination of spaces, projectors Q'^h, linear and bilinear forms. Then (5.187) shows that $\tilde{Q}'^h\in\mathcal{L}(H_k^{-1}(\mathcal{T}_c^h)\subset\mathcal{V}',\mathcal{V}^{h'})$ and with (5.190) we obtain for conforming FEMs and $u_0\in H^{k+1}(\Omega)$

$$\tilde{a}^h\left(\tilde{u}_0^h,v^h\right) - a(u_0,v^h) = \langle\tilde{f}^h,v^h\rangle^{\tilde{h}} - \langle f,v^h\rangle$$
$$\Longrightarrow \tilde{a}^h\left(\tilde{u}_0^h-u_0,v^h\right) = a(u_0,v^h) - \tilde{a}^h(u_0,v^h) + \langle\tilde{f}^h,v^h\rangle^{\tilde{h}} - \langle f,v^h\rangle$$
$$\Longrightarrow \left\|\tilde{u}_0^h-u_0\right\|_{\mathcal{U}^h} \leq Ch^k\|u_0\|_{H^{k+1}(\Omega)}$$
$$+\,Ch^{k-(d+\tau-1)}(M_{a_{ij}}\cdot\|u_0\|_{H^{k+1}(\Omega)} + \|f\|_{H_k^{-1}(\Omega)}).$$

For $u_0\in H^s(\Omega), s<k+1$, we combine the interpolating FE I^hu_0 with bounded $\|I^hu_0\|_{H^s(\mathcal{T}_c^h)} < C\|u_0\|_{H^s(\Omega)} < \infty$ and $\|u_0-I^hu_0\|_{H^1(\mathcal{T}_c^h)} < Ch^{s-1}\|u_0\|_{H^s(\Omega)}$, cf. (4.42), or (4.105), with Theorem 4.17 and 4.61, the discrete Brezzi–Babuska condition (4.169),

$$\begin{aligned}
&\epsilon \left\|I^h u_0 - \tilde{u}_0^h\right\|_{\mathcal{U}} \|v^h\|_{cV^h} \le \left|\tilde{a}^h\left(\tilde{u}_0^h - I^h u_0, v^h\right)\right| \\
&= |\left(\tilde{a}^h\left(\tilde{u}_0^h, v^h\right) - \tilde{a}^h(I^h u_0, v^h)\right) + \left(a^h(I^h u_0, v^h)\right. \\
&\quad - a^h(I^h u_0, v^h)) + (a(u_0, v^h) - a(u_0, v^h))| \\
&\le \left|\langle \tilde{f}^h, v^h\rangle^{\sim h} - \langle f, v^h\rangle\right| + \left|\tilde{a}^h(I^h u_0, v^h) - a^h(I^h u_0, v^h)\right| + \left|a^h(I^h u_0, v^h) - a(u_0, v^h)\right| \\
&\le C\big(h^{k-(d+\tau-1)}(\|f\|_{H_k^{-1}(\mathcal{T}_c^h)} + M_{a_{ij}} \cdot \|I^h u_0\|_{H^{d+\tau}(\mathcal{T}^h)}) + \|u_0 - I^h u_0\|_{\mathcal{U}}\big) \cdot \|v^h\|_{\mathcal{V}^h} \\
&\le C\big(h^{k-(d+\tau-1)}(\|f\|_{H_k^{-1}(\mathcal{T}_c^h)} + M_{a_{ij}} \|u_0\|_{H^s(\Omega)}) + h^{s-1}\|u_0\|_{H^s(\Omega)}\big) \cdot \|v^h\|_{\mathcal{V}^h}
\end{aligned}$$

With $\|u_0 - \tilde{u}_0^h\|_{\mathcal{U}} \le \|u_0 - I^h u_0\|_{\mathcal{U}} + \|I^h u_0 - \tilde{u}_0^h\|_{\mathcal{U}}$ we have proved the claim. □

5.4.4 Quadrature for second order fully nonlinear equations

We do not formulate the results for second order quasilinear problems, since the general case in Subsection 5.4.5 is nearly the same. Second order fully nonlinear equations require $\mathcal{U} \neq \mathcal{V}$. Since the boundary conditions remain unchanged, we start with the quadrature differential operator. The approximate solution has to satisfy cf. (5.3)

$$\tilde{u}_0^h \in \mathcal{U}_b^h : \left(G^w\left(\cdot, \tilde{u}_0^h, D\tilde{u}_0^h, D^2\tilde{u}_0^h\right), v^h\right)^{\sim h} = \sum_{T_c \in \mathcal{T}^h} q_{T_c}^h \left(G\left(\tilde{u}_0^h\right), v^h\right) = 0 \forall v^h \in \mathcal{V}_b^h \tag{5.200}$$

thus defining $\tilde{G}^h = \tilde{Q}_c^h G|_{\mathcal{U}^h}$. We need the spaces, approximate projectors $\tilde{Q}^h$, the nonlinear operators and the generalizations, see (5.77)–(5.79), and their linearization, cf. (5.73) with the a_{ij}, and the strong bilinear forms, $\tilde{a}_s^h(u^h, v^h)$,

$$\mathcal{U} \subset H^2(\Omega), \mathcal{V} \subset L^2(\Omega), \mathcal{U}_b^h = \mathcal{S}^h \subset \mathcal{S}_d^1\left(\mathcal{T}_c^h\right), \mathcal{V}_b^h = \mathcal{S}^h \cap C_0^1\left(\Omega_c^h\right), \tag{5.201}$$

$\mathcal{U}_c \subset H^2\left(\Omega_c^h\right), \ldots, \|\cdot\|_{\mathcal{U}^h} := \|\cdot\|_{H^2(\mathcal{T}_c^h)}, \|\cdot\|_{\mathcal{V}^h} := \|\cdot\|_{L^2(\mathcal{T}_c^h)}$, the linearization

$$(A_{s,h}u^h, v^h) = a_{s,h}(u^h, v^h) = \sum_{T_c \in \mathcal{T}_c^h} \int_{T_c} \left(\sum_{i,j=0}^{n} (-1)_{j>0} \partial^j (a_{ij} \partial^i u^h)\, v^h\right) dx \text{ with}$$

$$\left(\tilde{A}_s^h u^h, v^h\right) := \tilde{a}_s^h(u^h, v^h) = \sum_{T_c \in \mathcal{T}_c^h} q_{T_c}^h \left(\sum_{i,j=0}^{n} (-1)_{j>0} \partial^j (a_{ij} \partial^i u^h)\, v^h\right).$$

In Theorems 5.32 and 5.33 we will show the consistency and stability of our approximate FEM (5.200). The stability is equivalent to showing under which conditions $\tilde{a}_s^h(\cdot, \cdot)$ inherits the boundedness and discrete inf–sup condition from $a_s(\cdot, \cdot)$ on $\mathcal{U}_b^h, \mathcal{V}_b^h$. The proofs of these results are very similar to those in Subsection 5.4.3.

Note the consequences of the different norms in (5.202) and testing in the L^2 sense in (5.200) in contrast to the weak form in (5.182), $\|\cdot\|_{\mathcal{V}^h} := \|\cdot\|_{L^2(\mathcal{T}_c^h)}$ compared to $\|\cdot\|_{\mathcal{V}^h} := \|\cdot\|_{H^1(\mathcal{T}_c^h)}$ in the last subsection. So the exponents in the consistency estimates, e.g. $k-d$ in (5.202) compared to $k-(d+\tau-1)$ in (5.186), differ by 1. Furthermore, the $\ell-1$ in $\mathcal{P}_{\ell-1}^n \subset S^h \subset \mathcal{S}_d^1$, is relevant:

Theorem 5.32. Consistency for different quadrature approximations of fully nonlinear problems: *Assume (5.176), and for the quasiuniform subdivision $\mathcal{T}^h$ the Condition 4.16. Choose a quadrature formula as in (5.172), (5.174), satisfying (5.184) and (5.185). Let the fully nonlinear extension $G : H^{k+2}\left(\Omega \cup \Omega_c^h\right) \to H^k\left(\Omega \cup \Omega_c^h\right)$, be differentiable, $G_h'(u_0)$ be bounded with $a_{ij} \in W^{k+1,\infty}\left(\Omega \cup \Omega_c^h\right)$, see (5.73), $f, f_j \in H^k\left(\Omega \cup \Omega_c^h\right)$, and let a $u \in H^\ell(\Omega)$, yield $E_c u \in H^\ell\left(\Omega \cup \Omega_c^h\right)$, e.g. for $u = u_0$. Choose the approximating spaces $\mathcal{U}_b^h, \mathcal{V}_b^h \subset S^h$, $\mathcal{P}_{\ell-1}^n \subset S^h \subset \mathcal{S}_d^1$, $d \geq 2^n+1$, as in (5.201), with $k > 2(d-1)$, hence $k > d$, and let $h > 0$ be small enough. Finally, let u^h be extended into $u^h \in H^{k+2}\left(\Omega \cup \Omega_c^h\right)$.*

Then the following consistency results hold with $M_{a_{ij}} = \max_{i,j=0}^{n} \|a_{ij}\|_{W^{k+1,\infty}(\Omega)}$

$$|\langle f, v^h\rangle - \langle \tilde{f}^h, v^h\tilde{\rangle}^h| \leq C_k h^{k-d} \|f\|_{H^k(\mathcal{T}_c^h)} \|v^h\|_{\mathcal{V}^h}, \; k > d, \tag{5.202}$$

$$\|Q^h f - \tilde{Q}^h f\|_{\mathcal{V}^{h\prime}} \leq C_k h^{k-d} \|f\|_{H_k^{-1}(\Omega)}, \lim_{h\to 0} \|\tilde{Q}^h f\|_{\mathcal{V}^{h\prime}} = \|f\|_{H_k^{-1}(\Omega)}, \tag{5.203}$$

$$|(G(u^h), v^h) - (G(u^h), v^h)^{\tilde{h}}| \leq C h^{k-d} \|G(u^h)\|_{H^k(\mathcal{T}_c^h)} \|v^h\|_{\mathcal{V}^h}, \text{ and}$$

$$\left\|Q_c^h G(u^h) - \tilde{Q}_c^h G(u^h)\right\|_{\mathcal{V}^{h\prime}} \leq C h^{k-d} \|G(u^h)\|_{H^k(\mathcal{T}_c^h)}, \text{ or for } E_c u, I^h u : \tag{5.204}$$

$$\left\|Q_c^h G(E_c u) - \tilde{Q}_c^h G(I^h u)\right\|_{\mathcal{V}^{h\prime}} \leq C h^{k-d} \|G(u^h)\|_{H^k(\mathcal{T}_c^h)},$$

$$\textit{and for } G_h' : \quad \left|a_s^h(u^h, v^h) - \tilde{a}_s^h(u^h, v^h)\right| \leq C h^{k-2(d-1)} M_{a_{ij}} \|u^h\|_{\mathcal{U}^h} \|v^h\|_{\mathcal{V}^h} \tag{5.205}$$

$$|a^h(E_c u, v^h) - \tilde{a}^h(I^h u, v^h)| \leq C h^{\min\{k-d, \ell-2\}} M_{a_{ij}} \|u\|_{H^{\min\{k+2,\ell\}}(\Omega)} \|v^h\|_{\mathcal{V}^h}.$$

These inequalities only estimate the quadrature errors for the differential and not for boundary operators. For the complete consistency errors the terms in Theorem 5.4 and $L\times$ the interpolation errors have to be added, cf. Theorem 5.33. L denotes the Lipschitz constant for G.

Proof. The above conditions $d \geq 2^n + 1 > 2, n \geq 2$, for $\mathcal{S}_d^1\left(\mathcal{T}_c^h\right) \subset C^1(\Omega)$ and $k > 2(d-1)$, imply $k > 2(d-1) \geq 2^{n+1} > n/2$, $n \geq 2$, $k > d$. Hence, by the Sobolev embedding theorem the $W^{k,\infty}(\Omega), H^k(\Omega)$ are compactly embedded into $C(\bar{\Omega})$. Thus all the quadrature formulas for a_s and $(G(u^h), v^h))$, see e.g. (5.200), are well defined. If u or u_0 should not be smooth enough, we use the interpolating $I^h u_0$. Since the $u^h, v^h \in \mathcal{S}_d^1\left(\mathcal{T}_c^h\right) \subset H^k\left(\mathcal{T}_c^h\right) \cap C\left(\mathcal{T}_c^h\right)$, the estimates in (5.174) can be applied to (5.201)–(5.200) under the above conditions to different choices of f or f_h. Without

$$\begin{array}{ccccc}
\mathcal{U} & \xrightarrow{F} & \mathcal{V}_\Pi=\mathcal{V}\times\mathcal{V}_D & \overset{\text{tested by}}{\longleftrightarrow} & \mathcal{V}_\Pi=\mathcal{V}\times\mathcal{V}_b \\
P^h\big\downarrow & \tilde{\Phi}^h\big\downarrow & \tilde{Q}^h_\Pi \big\downarrow & & \\
\mathcal{U}_b & \xrightarrow{F^h} & \mathcal{V}^{h\prime}_\Pi=\mathcal{V}^{h\prime}\times\mathcal{V}^h_b & \overset{\text{approx. tested by}}{\longleftrightarrow} & \mathcal{V}^h_\Pi=\mathcal{V}^h_b\times\mathcal{V}^h_b
\end{array}$$

Figure 5.6 Nonconforming quadrature FEMs for fully nonlinear problems.

repeating all the details we get

$$\begin{aligned}
|\langle f_h, v^h\rangle - \langle \tilde{f}^h, v^h\tilde{\rangle}^h| &= \left|\sum_{T_c\in\mathcal{T}_c^h}\sum_{j=0}^{n}\int_{T_c} f_j\partial^j v^h dx - q^h_{T_c}(f_j\partial^j v^h)\right| \\
&\le C_k h^k \sum_{T_c\in\mathcal{T}_c^h}\sum_{j=0}^{n} |f_j\partial^j v^h|_{W^{k,1}(T_c)} \qquad (5.206)\\
&\le C_k h^{k-d}\|f\|_{H_k^{-1}(\mathcal{T}_c^h)}\|v^h\|_{\mathcal{V}^h}.
\end{aligned}$$

As in the proof of Theorem 5.28, the last inequality follows by the inverse estimates (4.57) for the piecewise polynomials v^h of degree $\le d$ with $d > 2$, the discrete Hölder inequality and $\|f_h v^h\|_{L^1(\mathcal{T}_c^h)} \le \|v^h\|_{\mathcal{V}^h}\ \|f_h\|_{L^2(\Omega_c^h)}$. Again (5.203) is an obvious consequence of (5.202). We get for $k > d$

$$\begin{aligned}
|(G(u^h), v^h) - (G(u^h), v^h)^{\tilde{h}}| &\le Ch^k\|G(u^h)\|_{H^k(\mathcal{T}_c^h)}\|v^h\|_{H^d(\mathcal{T}_c^h)} \qquad (5.207)\\
&\le C_k h^{k-d}\|G(u^h)\|_{H^k(\mathcal{T}_c^h)}\|v^h\|_{\mathcal{V}^h}.
\end{aligned}$$

The other estimates follow by similar modifications compared to above. □

Now we consider the fully nonlinear problem, including the approximate boundary conditions. In particular, we update the original product projectors and the FEM, see (5.62),(5.63), into the quadrature version, based upon $\tilde{Q}_c^h$ and the unchanged boundary projector Q_b^h and $F_h = \left(G, B_D^h\right)$, cf. Figure 5.6.

$$\tilde{Q}^h_{c,\Pi} := \left(\tilde{Q}^h_c, Q^h_b\right) : \mathcal{V}_c\times\mathcal{V}_{c,D}\to\mathcal{V}_\Pi^{h'}, \tilde{u}_0^h\in\mathcal{U}_b^h : \tilde{F}^h\left(\tilde{u}_0^h\right) := \tilde{Q}^h_{c,\Pi}F_h|_{\mathcal{U}^h}\left(\tilde{u}_0^h\right) = 0.$$

Theorem 5.33. Convergence for different quadrature approximations of fully nonlinear problems: *Under the conditions of Theorem 5.32 ($k > 2(d-1), d \ge 2^n + 1, \mathcal{P}^n_{\ell-1}\subset \mathcal{S}^h, \ell > 2$) with $G \in C_L(B_r(P^h u_0))$ and boundedly invertible $F'(u_0)$, this quadrature approximate FEM $\tilde{F}^h : \mathcal{U}_b^h \to \mathcal{V}_\Pi^{h'}$ is stable and consistent in $P^h u_0$, and has a unique solution $\tilde{u}_0^h$ for $\tilde{F}^h(\tilde{u}_0^h) = 0$. It converges, for $u_0 \in H^\ell(\Omega)$ with $E_c u_0 \in H^\ell(\Omega\cup\Omega_c^h)$, $p = 2$ and > 2 for polyhedral Ω^h and curved Ω_c^h, and $\|u^h\|_{\mathcal{U}^h} = \|u^h\|_{H^2(\mathcal{T}_c^h)}$, cf. (5.67)(5.110), (5.205), Conjectures 4.40, 4.41,*

$$\left\|E_c u_0 - \tilde{u}_0^h\right\|_{\mathcal{U}^h} \le Ch^{\min\{\ell-2,p,k-d\}}\|u_0\|_{H^{\min\{\ell,p+2,k-d+2\}}(\Omega)}, \qquad (5.208)$$

for the nonconforming, and with $p, p+2$ deleted in the exponents in (5.208) for the conforming FEM.

Remark 5.34. *The combination of the inequalities for k and d requires large values for k, e.g. $k > 8$ for $n = 2$ and $k > 16$ for $n = 3$. If the stability of $\tilde{G}^h$ is proved via inverse estimates, as in (5.205), this is unavoidable. Otherwise, the quadrature formulas could be applied to subtriangles of the $T_c \in \mathcal{T}_c^h$ and k could be reduced.*

Proof. We show that the $a_s(\cdot,\cdot)$ and $\tilde{a}_s^h(\cdot,\cdot)$ are simultaneously bounded and, for small enough h, satisfy the discrete inf–sup condition on $\mathcal{U}_b^h \times \mathcal{V}_b^h$ with equiboundedly invertible $\tilde{A}^h : \mathcal{U}_b^h \to \mathcal{V}^{h'}$. This implies the stability of the corresponding $(\tilde{G}^h(u_0))' : \mathcal{U}_b^h \to \mathcal{V}^{h'}$, and hence of $\tilde{G}^h : \mathcal{U}_b^h \to \mathcal{V}^{h'}$ in $P^h u_0$. By Lemma 5.10 we can assume the discrete inf–sup condition for $a_s(\cdot,\cdot)$ and obtain

$$
\begin{aligned}
\epsilon' \|u^h\|_{\mathcal{U}^h} &\le \inf_{0 \ne v^h \in \mathcal{V}_b^h} |a_s(u^h, v^h)| / \|v^h\|_{\mathcal{V}^h} \ \forall u^h \in \mathcal{U}_b^h \\
&= \inf_{0 \ne v^h \in \mathcal{V}_b^h} \left|\tilde{a}_s^h(u^h, v^h) + (a_s(u^h, v^h) - \tilde{a}_s^h(u^h, v^h))\right| / \|v^h\|_{\mathcal{V}^h} \\
&\le \inf_{0 \ne v^h \in \mathcal{V}_b^h} \left|\tilde{a}_s^h(u^h, v^h)\right| / \|v^h\|_{\mathcal{V}^h} \qquad (5.209) \\
&\quad + \sup_{0 \ne v^h \in \mathcal{V}_b^h} \left|(a_s(u^h, v^h) - \tilde{a}_s^h(u^h, v^h))\right| / \|v^h\|_{\mathcal{V}^h} \\
&\le \inf_{0 \ne v^h \in \mathcal{V}_b^h} \left|\tilde{a}_s^h(u^h, v^h)\right| / \|v^h\|_{\mathcal{V}^h} + Ch \|u^h\|_{\mathcal{U}^h},
\end{aligned}
$$

with (5.205) and $k > 2(d-1)$, so finally

$$
(\epsilon' - Ch) \|u^h\|_{\mathcal{U}^h} \le \inf_{0 \ne v^h \in \mathcal{V}_b^h} \left|\tilde{a}_s^h(u^h, v^h)\right| / \|v^h\|_{\mathcal{V}^h}. \qquad (5.210)
$$

The inequalities for the $\inf_{0 \ne u^h \in \mathcal{U}_b^h}$ are handled analogously. This shows the discrete inf–sup condition for $\tilde{a}_s^h(\cdot,\cdot)$ for sufficiently small $h > 0$.

The classical consistency error, see (5.205), is combined with the preceding (5.204) yielding with $k > d$, $u_0 \in H^{k+2}(\Omega)$ and Theorem 5.13

$$
\begin{aligned}
&\left\|\tilde{F}^h P^h u_0 - \tilde{Q}_\Pi^h F u_0\right\|_{\mathcal{V}_\Pi^{h'}} \\
&\quad = \left\|(\tilde{Q}_c^h G(P^h u_0), -Q_b^h B_D u_0)\right\|_{\mathcal{V}_\Pi^{h'}} \\
&\quad \le \left\|((\tilde{Q}_c^h G(P^h u_0) - Q_c^h G P^h u_0) + Q_c^h G P^h u_0, -Q_b^h B_D u_0)\right\|_{\mathcal{V}_\Pi^{h'}} \\
&\quad \le C h^{k-d} \|u_0\|_{H^{k+2}(\Omega)} + C L h^{\min\{\ell, k+2, p+2\} - 2} \|u_0\|_{H^{k+2}(\Omega)} \\
&\quad \le C' h^{\min\{\ell, p+2, k-d+2\} - 2} \|u_0\|_{H^{\min\{\ell, k+2\}}(\Omega)}.
\end{aligned}
$$

and hence, the claimed convergence.

We finally discuss the implications of the two inequalities for k and d.

$$d \geq 2^n + 1 \Longrightarrow \mathcal{S}_d^1\left(\mathcal{T}_c^h\right) \subset C^1(\Omega) \text{ and } k \geq 2(d-1)+1 \Longrightarrow$$

$$\text{inf–sup condition for } \tilde{a}_s^h(u^h, v^h), \text{ both } \Longrightarrow k \geq 2^{n+1}+1 > 0.$$

This indeed requires the large values for k mentioned above. □

5.4.5 Quadrature FEMs for equations and systems of order $2m$

Starting with linear and quasilinear cases we generalize the above set of spaces and approximations as in (5.177), with the notation in (4.129), (4.136), cf. Figure 5.5,

$$\begin{aligned}
&\mathcal{U} = \mathcal{V} \subset W^{m,p}(\Omega,\mathbb{R}^q), \mathcal{U}_b^h, \mathcal{V}_b^h \subset \left\{L^\infty\left(\mathcal{T}_c^h,\mathbb{R}^q\right) : \forall T_c \in \mathcal{T}_c^h : u^h|_{T_c} \in \mathcal{P}, u^h\right.\\
&\qquad \left.\text{vanishes exactly or approximately along } \partial\Omega_c^h\right\}, \mathcal{P}_{d-1} \subset \mathcal{P} \subset \mathcal{P}_{d+\tau}, \text{and}\\
&\|\cdot\|_{\mathcal{U}^h} := \|\cdot\|_{\mathcal{V}^h} := \|\cdot\|_{H^m(\mathcal{T}_c^h,\mathbb{R}^q)}, \text{ with } \dim \mathcal{U}_b^h = \dim \mathcal{V}_b^h \text{ and} \qquad (5.211)\\
&p = 2 \textit{ and } p \geq 2 \text{ for linear and quasilinear cases, respectively.}
\end{aligned}$$

cf. Theorem 5.38 3. for $1 < p < \infty$. The generalized weak linear FEM form and the projectors $\tilde{Q}'^h$ are

$$\text{For } A_{\alpha\beta}(x) \in H^{k,\infty}(\Omega,\mathbb{R}^{q\times q}), f_\alpha \in H^k(\Omega,\mathbb{R}^{q\times q}) \text{ define} \qquad (5.212)$$

$$\sum_{T_c\in\mathcal{T}_c^h} q_{T_c}^h\left(\sum_{|\alpha|,|\beta|\leq m}\left(A_{\alpha\beta}\partial^\beta\tilde{\vec{u}}_0^h, \partial^\alpha\vec{v}^h\right)_q\right) =: \langle A_h\tilde{\vec{u}}_0^h,\ \vec{v}^h\tilde{\rangle}^h \text{ and determine } \tilde{\vec{u}}_0^h \in \mathcal{U}_b^h \text{ s.t.}$$

$$\langle \tilde{A}^h\tilde{\vec{u}}_0^h, \vec{v}^h\tilde{\rangle}^h = \tilde{a}^h\left(\tilde{\vec{u}}_0^h, \vec{v}^h\right) = \langle \vec{f}_h, \vec{v}^h\tilde{\rangle}^h = \sum_{T_c\in\mathcal{T}_c^h} q_{T_c}^h\left(\sum_{|\alpha|\leq m}(\vec{f}_\alpha, \partial^\alpha\vec{v}^h)_q\right) \forall\, \vec{v}^h \in \mathcal{V}_b^h,$$

$$\text{and } \tilde{Q}'^h : \mathcal{V} \to \mathcal{V}^{h\prime},\ \langle \tilde{Q}'^h\vec{f} - \vec{f}, \vec{v}^h\tilde{\rangle}^h = ((\tilde{Q}'^h\vec{f} - \vec{f})\vec{v}^h)^{\tilde{h}} = 0\ \forall \vec{v}^h \in \mathcal{V}_b^h.$$

The previous Theorem 5.28 is generalized to these new linear problems. As a consequence of the modified spaces in (5.211) the exponent $d+\tau-1$ in Theorem 5.28 has to be replaced by $d+\tau-m$ here. So we update (5.184) and (5.185) for order $2m$ and systems in $\mathbb{R}^q$, usually with $\tau = -1$, as

$$\begin{aligned}
&k > \max\left\{d+\tau-m, \frac{n}{2}\right\} \text{ for } \langle \vec{f}_h, \cdot\tilde{\rangle}^h, \vec{f} = (\vec{f}_\alpha \in H^k(\ldots))_{|\alpha|\leq m} \in H_k^{-m}\left(\Omega\cup\mathcal{T}_c^h,\mathbb{R}^q\right)\\
&k > 2(d+\tau-m) \text{ for } \tilde{a}^h(\cdot,\cdot), A_{\alpha\beta} \in W^{k,\infty}\left(\Omega\cup\Omega_c^h,\mathbb{R}^{q\times q}\right), |\alpha|,|\beta| \leq m. \qquad (5.213)
\end{aligned}$$

On this basis Theorems 5.30 and 5.31 remain verbatim and again guarantee convergence of the quadrature approximation.

Theorem 5.35. Consistency of different quadrature approximations of linear equations and systems of order $2m$:

1. *Assume (5.176), (5.211), and for the quasiuniform subdivision $\mathcal{T}^h$ the Condition 4.16. Choose a quadrature formula as in (5.172), (5.174), satisfying (5.213) for*

vector valued functions, $\langle\vec f,\cdot\rangle^{\tilde{h}}$ *and* $\tilde a^h(\cdot,\cdot)$, *with (extended)* $\vec f\in H^k\left(\Omega\cup\mathcal{T}_c^h,\mathbb{R}^q\right)$, $A_{\alpha,\beta}\in W^{k,\infty}\left(\Omega\cup\mathcal{T}_c^h,\mathbb{R}^{q\times q}\right)$, $\vec u\in H^s(\Omega,\mathbb{R}^q), s\geq d$, *in (5.217).*

2. *Then, for small enough* $h>0$, *the following consistency results hold*

$$|(\vec f_h,\vec v^h)-(\tilde{\vec f}_h,\vec v^h)^{\tilde h}|\leq Ch^{k-(d+\tau-m)}\|\vec f\|_{H_k^{-m}(\Omega,\mathbb{R}^q)}\cdot\|\vec v^h\|_{\mathcal{V}^h},\quad(5.214)$$

$$\|Q'^h\vec f-\tilde Q'^h\vec f\|_{\mathcal{V}^{h\prime}}\leq Ch^{k-(d+\tau-m)}\|\vec f\|_{H_k^{-m}(\Omega,\mathbb{R}^q)},\quad\lim_{h\to0}\|\tilde Q'^h\vec f\|_{\mathcal{V}^{h\prime}}=\|\vec f\|_{\mathcal{V}'},\quad(5.215)$$

$$\frac{|a^h(\vec u^h,\vec v^h)-\tilde a^h(\vec u^h,\vec v^h)|}{\|\vec v^h\|_{\mathcal{V}^h}}\leq Ch^{k-2(d+\tau-m)}M_{A_{\alpha,\beta}}\|\vec u^h\|_{\mathcal{V}^h},\ \textit{for } v^h\neq0 \textit{ and}\quad(5.216)$$

$$\leq Ch^{k-(d+\tau-m)}M_{A_{\alpha,\beta}}\|\vec u^h\|_{H^s(\mathcal{T}_c^h,\mathbb{R}^q)}$$

$$\text{with } M_{A_{\alpha,\beta}}:=\max_{|\alpha|,|\beta|\leq m}\|A_{\alpha,\beta}\|_{W^{k,\infty}(\Omega,\mathbb{R}^{q\times q})},$$

and

$$\frac{|a_h(\vec u,\vec v^h)-\tilde a^h(I^h\vec u,\vec v^h)|}{\|\vec v^h\|_{\mathcal{V}^h}}\leq C\big(h^{d-1}+h^{k-(d+\tau-1)}M_{A_{\alpha,\beta}}\big)\|\vec u\|_{H^s(\Omega,\mathbb{R}^q)}.\quad(5.217)$$

Theorem 5.36. Convergence for different quadrature approximations of linear equations and systems of order $2m$:

1. *Let* $a^h(\cdot,\cdot)$ *induce a stable* A^h *and assume (5.212) and the conditions in Theorem 5.35. Then, simultaneously* $\tilde a^h(\cdot,\cdot)$ *induces a stable* $\tilde A^h$. *If furthermore (5.213) is correct, then* $\tilde a^h(\tilde{\vec u}_0^h,\vec v^h)=\langle\vec f,\vec v^h\rangle^{\tilde h}\,\forall\vec v^h\in\mathcal{V}_b^h$, *defines a stable and convergent discretization for* $A\vec u_0=\vec f$. *For the other approximations, mentioned above, this result remains correct if the* $\tilde a^h,\tilde A^h$ *are replaced by good enough approximations* $a^{\tilde h},\tilde a^{\tilde h},A^{\tilde h},\tilde A^{\tilde h}$.
2. *For* conforming FEMs, *the above* $M_{A_{\alpha,\beta}}$, *and* $C^*:=C(1+M_{A_{\alpha,\beta}})$, $\vec u_0\in H^s(\Omega)$, *we obtain*

$$\left\|\vec u_0-\tilde{\vec u}_0^h\right\|_{\mathcal{U}^h}\leq C^*h^{\min\{s-m,k-(d+\tau-m)\}}\big(\|\vec u_0\|_{H^{\min\{s,k+m\}}(\Omega,\mathbb{R}^q)}+\|\vec f\|_{H^k(\Omega,\mathbb{R}^q)}\big).$$

3. *If only approximations for* $\langle\vec f,\vec v^h\rangle^{\tilde h}$ *and none for* $\tilde a^h(\vec u^h,\vec v^h)$ *are used, then*

$$\left\|\vec u_0-\tilde{\vec u}_0^h\right\|_{\mathcal{U}^h}\leq Ch^{k-(d+\tau-m)}\|\vec f\|_{H^k(\Omega,\mathbb{R}^q)}\ \textit{for } k>d+\tau-m.$$

We turn to quasilinear systems with a “quasibilinear” $\tilde a^h(\vec u^h,\vec v^h)$, where

$$A_\alpha(\cdot,\vec u^h,\ldots,\nabla^m\vec u^h):W^{m,p}(\Omega,\mathbb{R}^q)\to L^{p'}(\Omega,\mathbb{R}^q),1/p+1/p'=1,\quad(5.218)$$

$$\tilde a^h\left(\vec u_0^h,\vec v^h\right)=\sum_{T_c\in\mathcal{T}_c^h}\left(\sum_{|\alpha|\leq m}q_{T_c}^h\left\langle A_\alpha\left(\cdot,\vec u_0^h,\ldots,\nabla^m\vec u_0^h\right),\partial^\alpha\vec v^h\right\rangle_q\right)=:\langle G_h\vec u_0^h,\vec v^h\rangle^{\tilde h}.$$

The previous estimate (5.214) is not needed, but (5.215) is still needed here, and (5.216), (5.217) have to be reformulated for the quasilinear operator and its linearized form, providing consistency and stability, hence convergence.

Theorem 5.37. Consistency of different quadrature approximations of quasilinear systems of order $2m$ in $W^{s,p}(\Omega), p \geq 2$:

1. *Assume (5.176), the updated (5.211), and for the quasiuniform subdivision $\mathcal{T}^h$ the Condition 4.16. Choose a quadrature formula as in (5.174), satisfying (5.213), (5.212), (5.218), a differentiable G_h, $k > m + n/2, s \geq d, m$, and*

$$\frac{1}{p} + \frac{1}{p'} = 1, \forall \vec{u} \in W^{s,p}(\Omega) \quad h := \frac{\partial A_\alpha}{\partial \vartheta^\beta}(\cdot, \vec{u}_0, .., \nabla^m \vec{u}_0) \in W^{k,p/(p-2)}(\Omega), \quad (5.219)$$

with $\|\vec{u} - \vec{u}_0\|_{W^{m,p}(\Omega)} < \delta$ and $h < h_0$.

2. *Then the following consistency results hold*

$$|a_h(\vec{u}, \vec{v}^h) - \tilde{a}^h(I^h\vec{u}, \vec{v}^h)| = |\langle G_h(\vec{u}), \vec{v}^h\rangle - \langle G_h(I^h\vec{u}), \vec{v}^h\rangle^{\tilde{h}}| \leq C\|\vec{v}^h\|_{\mathcal{V}^h} \quad (5.220)$$

$$\left(h^{d-m}\|\vec{u}\|_{W^{s,p}(\Omega,\mathbb{R}^q)} + h^{k-(d+\tau-m)} \cdot \max_{|\alpha|\leq m} \|A_\alpha(\cdot, \vec{u}, \ldots, D^m\vec{u})\|_{W^{k,p'}(\Omega,\mathbb{R}^q)}\right),$$

and for $\|\vec{u}_1^h - I^h\vec{u}_0\|_{\mathcal{V}^h} < \delta$:

$$\left|\langle (G_h)'\left(\vec{u}_1^h\right)\vec{u}^h, \vec{v}^h\rangle - \langle (G_h)'\left(\vec{u}_1^h\right)\vec{u}^h, \vec{v}^h\rangle^{\tilde{h}}\right| \leq Ch^{k-2(d+\tau-m)} \cdot \quad (5.221)$$

$$\max_{|\alpha|,|\beta|\leq m}\left\{\left\|\frac{\partial A_\alpha}{\partial\vartheta^\beta}\left(\cdot, \vec{u}_1^h, .., \nabla^m\vec{u}_1^h\right)\right\|_{W^{k,p/(p-2)}(\mathcal{T}_c^h)}\right\} \cdot \|\vec{u}^h\|_{\mathcal{V}^h}\|\vec{v}^h\|_{\mathcal{V}^h}.$$

Theorem 5.38. Convergence of different quadrature approximations for quasilinear systems of order $2m$ in $W^{s,p}(\Omega), p \geq 2$:

1. *Under the conditions of Theorem 5.37, and for a locally unique exact solution, $\vec{u}_0$, of $G(\vec{u}) = 0$, with boundedly invertible $G'(\vec{u}_0)$, the FE equation $\tilde{G}^h(\vec{u}^h) = 0, \tilde{G}^h = \tilde{Q}'^h G_h|_{\mathcal{U}^h}$ in (5.218) has a unique solution $\vec{u}_0^h$ converging to $\vec{u}_0$.*
2. *With $M_{A_\alpha} := \max_{|\alpha|\leq m} \|A_\alpha(\cdot, \ldots, D^m\vec{u}_0)\|_{W^{k,p'}(\Omega,\mathbb{R}^q)}$ and for conforming FEMs we get with $\vec{u}_0 \in W^{s,p}(\Omega, \mathbb{R}^q)$*

$$\left\|\vec{u}_0 - \tilde{\vec{u}}_0^h\right\|_{\mathcal{U}^h} \leq Ch^{\min\{s-m,k-(d+\tau-m)\}}\|\vec{u}_0\|_{W^{\min\{k,s\},p}(\Omega,\mathbb{R}^q)}(1 + M_{A_\alpha}). \quad (5.222)$$

3. *With Theorem 4.67 and (4.194), (4.195) analogous results are possible for the monotone operator approach for quasilinear problems even for $1 < p < \infty$ and discrete $W^{m.p}(\mathcal{T}^h)$ norms.*

Proof. Relation (5.221) implies the simultaneous stability of G^h and $\tilde{G}^h$. For the consistency of $\tilde{G}^h$ with G in $P^h\vec{u}_0$ we obtain by (5.174)

$$\int_{T_c} \left\langle A_\alpha \left(\cdot, \vec{u}_0^h, \ldots, D^m \vec{u}_0^h\right), \partial^\alpha \vec{v}^h \right\rangle_q dx - q_{T_c}^h\left(\left\langle A_\alpha \left(\cdot, \vec{u}_0^h, \ldots, D^m \vec{u}_0^h\right), \partial^\alpha \vec{v}^h \right\rangle_q \right)$$
$$\leq C_k h^k \left\| A_\alpha \left(\cdot, \ldots, D^m \vec{u}_0^h\right) \right\|_{W^{k,p'}(T_c)} \|v^h\|_{W^{k,p}(T_c)}, \qquad (5.223)$$

hence, (5.220). For the stability we have to linearize our quasilinear operator, see (2.459), and get for $\vec{u}_1^h, \vec{u}^h, \vec{v}^h \in \mathcal{V}_b^h \subset W^{m,p}\left(\mathcal{T}_c^h, \mathbb{R}^q\right)$, and with $\langle \cdot, \cdot \rangle := \langle \cdot, \cdot \rangle_{\mathcal{V}^{h'} \times \mathcal{V}^h}$,

$$\left\langle \frac{\partial G_h\left(\vec{u}_1^h\right)}{\partial \vec{u}^h} \vec{u}^h, \vec{v}^h \right\rangle = \sum_{T_c \in \mathcal{T}_c^h} \int_{T_c} \sum_{|\alpha|, |\beta| \leq m} \left\langle \frac{\partial A_\alpha}{\partial \vartheta^\beta} \left(x, .., \nabla^m \vec{u}_1^h\right) \partial^\beta \vec{u}^h, \partial^\alpha \vec{v}^h \right\rangle_q dx. \qquad (5.224)$$

With the above condition (5.219), we get

$$\left| \left\langle \frac{\partial A_\alpha}{\partial \vartheta^\beta} \left(\cdot, .., \nabla^m \vec{u}_1^h\right) \partial^\beta \vec{u}^h, \partial^\alpha \vec{v}^h \right\rangle_q \right|_{W^{k,1}(T_c, \mathbb{R})}$$
$$\leq C \left\| \frac{\partial A_\alpha}{\partial \vartheta^\beta} \right\|_{W^{k,p/(p-2)}(\mathcal{T}_c^h, \mathbb{R}^{q \times q})} \|\partial^\beta \vec{u}^h\|_{W^{d+\tau-|\beta|,p}(T_c, \mathbb{R}^q)} \|\partial^\alpha \vec{v}^h\|_{W^{d+\tau-|\alpha|,p}(T_c, \mathbb{R}^q)}$$

We modify (5.223) and estimate

$$\left| \int_{\mathcal{T}_c^h} \left\langle \frac{\partial A_\alpha}{\partial \vartheta^\beta} \partial^\beta \vec{u}^h, \partial^\alpha \vec{v}^h \right\rangle_q dx - q^h \left\langle \frac{\partial A_\alpha}{\partial \vartheta^\beta} \partial^\beta \vec{u}^h \partial^\alpha \vec{v}^h \right\rangle_q \right| \qquad (5.225)$$
$$\leq C h^k \left\| \frac{\partial A_\alpha}{\partial \vartheta^\beta} \right\|_{W^{k,p/(p-2)}(\mathcal{T}_c^h)} \sum_{T_c \in \mathcal{T}_c^h} \|\partial^\beta \vec{u}^h\|_{W^{d+\tau-|\beta|,p}(T_c)} \|\partial^\alpha \vec{v}^h\|_{W^{d+\tau-|\alpha|,p}(T_c)}$$
$$\leq C_k h^k \left\| \frac{\partial A_\alpha}{\partial \vartheta^\beta} \left(\cdot, .., \nabla^m \vec{u}_1^h\right) \right\|_{W^{k,p/(p-2)}(\mathcal{T}_c^h)} \|\partial^\beta \vec{u}^h\|_{W^{d+\tau-|\beta|,p}(\mathcal{T}_c^h)} \|\partial^\alpha \vec{v}^h\|_{W^{d+\tau-|\alpha|,p}(\mathcal{T}_c^h)}.$$

The inverse estimates show, with (5.212), the stability, hence the claim (5.222). □

Considering the many cases of semilinear equations as special cases of the preceding quasilinear problems with appropriately simplified conditions, the only missing class of elliptic equations are the nondivergent quasilinear and the fully nonlinear equations and systems of orders 2 and $2m$. So we recall the spaces, the extension operators, E_c, the approximate projectors $\tilde{Q}^h$, the strong bilinear forms, nonlinear operators and the generalizations, see (5.77), (4.73), (5.78), (5.47), (5.128), cf. Figure 5.6. We replace in (5.200), (5.201), the $2m = 2$ and $q = 1$ by general m and q and obtain, note $\mathcal{U} \neq \mathcal{V}$, from here on

$$\mathcal{U} \subset H^{2m}(\Omega, \mathbb{R}^q), \mathcal{V} \subset L^2(\Omega, \mathbb{R}^q), \mathcal{U}_b^h = \mathcal{S}^h = S_0^h \oplus S_b^h \subset \mathcal{S}_d^{2m-1}\left(\mathcal{T}_c^h, \mathbb{R}^q\right), \qquad (5.226)$$
$$\mathcal{V}_b^h = S_0^h = \mathcal{S}^h \cap C_0^{m-1}\left(\mathcal{T}_c^h, \mathbb{R}^q\right), \mathcal{V}_b^h = S_b^h, \|\cdot\|_{\mathcal{U}^h} := \|\cdot\|_{H^{2m}(\mathcal{T}_c^h, \mathbb{R}^q)}, \|\cdot\|_{\mathcal{V}^h} := \|\cdot\|_{L^2}.$$

The FE solution $\tilde{\vec{u}}_0^h$ has to satisfy the quadrature approximate differential operator

$$\tilde{\vec{u}}_0^h \in \mathcal{U}_b^h \text{ s.t.} \langle G^w\left(\cdot, \dots, D^{2m}\tilde{\vec{u}}_0^h\right), \vec{v}^h \tilde{\rangle}_q^h = \sum_{T_c \in \mathcal{T}_c^h} q_{T_c}^h \left(G\left(\tilde{\vec{u}}_0^h\right) \vec{v}^h\right) = 0 \forall \vec{v}^h \in \mathcal{V}_b^h. \tag{5.227}$$

For the stability we need the linearized form

$$\tilde{a}_s^h(\vec{u}^h, \vec{v}^h) = \sum_{T_c \in \mathcal{T}_c^h} q_{T_c}^h \sum_{|\gamma| \le 2m} \left\langle \left[\partial_{\vartheta^\gamma} G^w(\cdot, \vec{\vartheta}^\beta, |\beta| \le 2m)\right] \left(\vec{w}_0^h\right) \partial^\gamma \vec{u}^h, \vec{v}^h \right\rangle_q. \tag{5.228}$$

So we replace the original approximation $G^h = Q_c^h G|_{\mathcal{U}^h}$, by the quadrature approximation $\tilde{G}^h = \tilde{Q}_c^h G|_{\mathcal{U}^h}$ in (5.227). We start as in Subsection 5.4.4 with $\tilde{G}^h, \tilde{Q}^h, \tilde{Q}_c^h$:

Theorem 5.39. Consistency of different quadrature approximations for fully nonlinear systems of order $2m$:

1. *For $m, q \ge 1, k > 2(d-m)$, hence, $k - d > 0$, let*

$$\forall |\gamma| \le 2m: \ \partial_{\vartheta^\gamma}\left(G^w(\cdot, \vec{\vartheta}^\beta, |\beta| \le 2m)\right)\left(\vec{w}_0^h\right) \in W^{k+|\gamma|,\infty}(\Omega \times \mathbb{R}^{N_{2m}}),$$

   cf (5.73), (5.17) (5.5), $G : \mathcal{D}(G) \cap H^{k+2m}(\Omega) \to H^k(\Omega)$ and $G : \mathcal{D}(G) \cap H^{k+2m}\left(\Omega \cup \Omega_c^h\right) \to H^k\left(\Omega \cup \Omega_c^h\right)$. Choose the approximating $\mathcal{U}_b^h = S^h \supset \mathcal{V}_b^h = S_0^h$, $d \ge (2m-1)2^n + 1$ as in (5.226). Let the quadrature formula satisfy (5.172), (5.174), and let $h > 0$ be small enough. Again let, $p = 2$ for the conforming FEM, and for the nonconforming FEM $2 \le p \le 5$, with $p = 2$ for polyhedral Ω^h and $p > 2$ for better curved Ω_c^h approximations for $\Omega \in C^{2m}$, cf. Conjectures 4.40 and 4.41.
2. *Then (5.202) holds and*

$$\|Q^h f - \tilde{Q}^h f\|_{\mathcal{V}^{h'}} \le C_k h^{k-d} \|f\|_{H_k^{-m}(\Omega)}, \quad \lim_{h\to 0} \|Q^h f\|_{\mathcal{V}^{h'}} = \|f\|_{H_k^{-m}(\Omega)}, \tag{5.229}$$

$$|\langle G(\vec{u}^h), \vec{v}^h\rangle_{\mathcal{V}_c} - \langle G(\vec{u}^h), \vec{v}^h \tilde{\rangle}^h| \le C h^{k-d} \|G(\vec{u}^h)\|_{H_c^k(\mathcal{T}_c^h)} \|\vec{v}^h\|_{\mathcal{V}^h}, \ or$$

$$\left\|Q_c^h G(\vec{u}^h) - \tilde{Q}_c^h G(\vec{u}^h)\right\|_{\mathcal{V}} \le C h^{k-d} \|G(\vec{u}^h)\|_{H_c^k(\mathcal{T}_c^h)}, \ and \tag{5.230}$$

$$\left|a_s^h(\vec{u}^h, \vec{v}^h) - \tilde{a}_s^h(\vec{u}^h, \vec{v}^h)\right| \le C h^{k-2(d-m)} \|\vec{u}^h\|_{\mathcal{U}^h} \|\vec{v}^h\|_{\mathcal{V}^h}$$

$$\times \max_{|\gamma| \le 2m} \|\partial_{\vartheta^\gamma}\left(G^w(\cdot, \vec{\vartheta}^\beta, |\beta| \le 2m)\right)\left(\vec{w}_0^h\right)\|_{W^{k+|\gamma|,\infty}(\mathcal{T}_c^h)}. \tag{5.231}$$

3. *Again, these inequalities only estimate the quadrature errors for the differential and not for boundary operators. For the complete consistency errors the terms in Theorem 5.15 and $L\times$ the interpolation errors have to be added, cf. Theorem 5.40. L denotes the Lipschitz constant for G.*

Proof. Since $\vec{u}^h, \vec{v}^h \in \mathcal{S}_d^{2m-1}\left(\mathcal{T}_c^h\right) \subset H_c^k\left(\mathcal{T}_c^h\right)$, the estimates in (5.174) again can be applied to different choices of f_c. All the above estimates follow as in the proof of Theorem 5.32, by replacing $d-1$ by $d-m$ and updating the coefficients.

The above conditions $d \geq (2m-1)2^n + 1 > 2, n \geq 2$, for $\mathcal{S}_d^{2m-1}(\mathcal{T}_c^h) \subset C^{2m-1}(\Omega)$ and $k > 2(d-m)$, imply $k > 2(d-m) \geq 2^{n+1} > n/2,\ n \geq 2,\ k > d$. Hence, by the Sobolev embedding theorem the $W^{k,\infty}(\Omega), H^k(\Omega)$ are compactly embedded into $C(\bar{\Omega})$. So all the quadrature formulas, even for a $\vec{u} \in W^{k,\infty}(\Omega), H^k(\Omega)$, are well defined. □

Now we add the (unchanged) boundary conditions, cf. (5.118), the

$$B_D^h \vec{u}_0^h := \left(\left. \frac{\partial^j \vec{u}_0^h}{\partial \nu^j} \right|_{\partial \Omega_c^h} \right)_{j=0}^{m-1} = \vec{0} \Leftrightarrow \left(\vec{v}_b^h, B_D^h \vec{u}_0^h \right)_{\mathcal{V}_{c,b}} = 0\ \forall \vec{v}_b^h \in \mathcal{V}_b^h \Leftrightarrow Q_{c,b}^h B_D^h \vec{u}_0^h = \vec{0}.$$

Replacing the previously introduced projectors $Q_{c,\Pi}^h = (Q_c^h, Q_{c,b}^h)$, by $\tilde{Q}_{c,\Pi}^h = (\tilde{Q}_c^h, Q_{c,b}^h)$, we define the discrete quadrature operator and the FE solution $\vec{\tilde{u}}_0^h \in \mathcal{U}_b^h$ as

$$\tilde{Q}_{c,\Pi}^h := \left(\tilde{Q}_c^h, Q_b^h \right) : \mathcal{V}_c \times \mathcal{V}_{c,D} \to \mathcal{V}_\Pi^{h'}, \tilde{F}^h \left(\vec{\tilde{u}}_0^h \right) := \tilde{Q}_{c,\Pi}^h F_h|_{\mathcal{U}^h} \left(\vec{\tilde{u}}_0^h \right) = 0 \in \mathcal{V}_\Pi^{h'}. \tag{5.232}$$

Theorem 5.40. Convergence of different quadrature approximations for fully nonlinear systems of order $2m$:

1. *Assume the conditions in Theorem 5.39, e.g.* $k > 2(d-m), \mathcal{P}_{\ell-1}^n \subset \mathcal{S}^h, d \geq (2m-1)2^n + 1$, *a locally unique solution* $\vec{u}_0 \in H^\ell(\Omega), \ell > 2m$, *and boundedly invertible* $F'(\vec{u}_0)$. *Choose for the conforming FEMs for* $m = 1$ *and a convex polyhedral domain, cf. (5.67), and for the nonconforming FEM for domains in* $C^p, 2 \leq p \leq 5$, *the* $p = 2$ *for polyhedral approximation* Ω^h *and* $p > 2$ *for better curved* Ω_c^h *approximations for* $\Omega \in C^2$, *cf. Conjectures 4.40 and 4.41.*
2. *Then the quadrature approximate FEM* $\tilde{F}^h : \mathcal{U}_b^h \to \mathcal{V}_\Pi^{h'}$, $\tilde{F}^h(\vec{\tilde{u}}_0^h) = 0$ *in (5.232) has a unique solution* $\vec{\tilde{u}}_0^h$, *is stable and consistent in* $P^h \vec{u}_0$, *hence convergent, such that, see (5.67),(5.110),*

$$\left\| E_c \vec{u}_0 - \vec{\tilde{u}}_0^h \right\|_{\mathcal{U}^h} \leq C h^{\min\{\ell, p+2m, k-d+2m\}-2m} \|\vec{u}_0\|_{\in H^{\min\{\ell,k+2m\}}(\Omega)}. \tag{5.233}$$

Proof. The proof of Theorem 5.32 remains essentially unchanged. The u^h are turned into $\vec{u}^h$, etc., and the last inequality, with $k > 2(d-m), d > (2m-1)2^n$, has to be modified into

$$\left\| \tilde{F}^h P^h \vec{u}_0 - \tilde{Q}_\Pi^h F \vec{u}_0 \right\|_{\mathcal{V}_\Pi^{h'}} \leq C' h^{\min\{\ell, p+2m, k-d+2m\}-2m} \|\vec{u}_0\|_{H^{\min\{\ell,k+2m\}}(\Omega)}.$$

We finally discuss the implications of the two inequalities for k and d. We consider $m \geq 1$ and obtain

$$d \geq (2m-1)2^n + 1 \text{ to guarantee } \mathcal{S}_d^{2m-1}\left(\mathcal{T}_c^h\right) \subset C^{(2m-1)}(\Omega) \text{ and} \tag{5.234}$$

$$k \geq 2(d-m) + 1, m \geq 1 \Longrightarrow \text{inf–sup condition for } \tilde{a}_s^h(\vec{u}^h, v^h), \tag{5.235}$$

$$\text{consequently } k - d \overset{(5.235)}{\geq} d - 2m + 1 \overset{(5.234)}{>} 0. \tag{5.236}$$

This indeed requires the large values for k mentioned above. □

5.4.6 Two useful propositions

At the end of this section we want to present two results which are important for integrations and for applications in the final stability proof for nonconforming FEMs. Sometimes we have to rescale the integral over an interval $[-1,1]$ to $[0,h]$ or a more general reference domain K to T. Examples are (5.171), (5.172) and (5.173). We prove the following result on *scaled Sobolev norms* and apply it to the *interpolation basis*. Let $\mathcal{P} = \text{span}\{\phi_1, \phi_2, \dots, \phi_d\}$ be a basis dual to $\mathcal{N}$, hence $N_i(\phi_j) = \delta_{ij}$ and $F_T : K \to T$ the affine bijection. Then, see Definition 4.8,

$$\begin{aligned}
&\phi_i^T = \phi_i \circ F_T^{-1} = \phi_i \circ \varphi : T \to \mathbb{R},\ i = 1, \dots, d, \forall\ T \in \mathcal{T}^h \text{ with the affine} \\
&\varphi := F_T^{-1} : T \ni t \to x := Bt/h + b \in K,\ B \in \mathbb{R}^{n \times n}, b \in \mathbb{R}^n. \qquad (5.237)
\end{aligned}$$

is called an interpolation basis for $\mathcal{U}_b^h$ or $\mathcal{V}_b^h$. It is not necessarily a basis for $\mathcal{U}_b^h$ or $\mathcal{V}_b^h$, but for a fixed $T \in \mathcal{T}^h$ the ϕ_i^T, $i = 1, \dots, d$ are a basis for $\mathcal{P}_T$.

Proposition 5.41. Transformation theorem: *Let $K \subset \mathbb{R}^n$ be a reference domain, see Definition 4.1, $0 < h \leq 1$ and φ in (5.237) an affine scaling mapping. Choose $u \in W^{s,p}(K)$ and $\alpha \in \mathbb{N}^n$ a multi-index with $|\alpha| \leq s$. Then*

$$\left(\int_T (u \circ \varphi)\right) = h^n |det(B)|^{-1} \left(\int_K u\right), \quad \textit{and}$$

$$\left(\int_T |\nabla^\alpha (u \circ \varphi)|^p\right)^{\frac{1}{p}} = h^{n/p - |\alpha|} \left(\int_K |\nabla^\alpha u|^p\right)^{\frac{1}{p}}, \quad \textit{for } B = I, \textit{ else}$$

$$\left(\int_T |\nabla^\alpha (u \circ \varphi)|^p\right)^{\frac{1}{p}} \leq h^{n/p - |\alpha|} \|B\|^{|\alpha|} |det(B)|^{-1/p} \left(\int_K |\nabla^\alpha u|^p\right)^{\frac{1}{p}}, \qquad (5.238)$$

$$\text{specifically } |u \circ \varphi|_{W^{s,p}(T)} \leq C\, h^{n/p-s} (\max\{1, \|B\|\})^s |det(B)|^{-1/p} |u|_{W^{s,p}(K)},$$

$$\|u \circ \varphi\|_{W^{s,p}(T)} \leq C\, h^{n/p-s} (\max\{1, \|B\|\})^s |\det(B)|^{-1/p} \|u\|_{W^{s,p}(K)} \quad \textit{and}$$

$$\|\phi_i^T\|_{W^{s,p}(T)} \leq C\, h^{n/p-s} (\max\{1, \|B\|\})^s |\det(B)|^{-1/p} \|\phi_i\|_{W^{s,p}(K)},$$

$$\textit{with } \|\phi_i^T\|_{H^1(\Omega^h)} \leq C\, h^{n/2-1} \max_{j=1}^{d} \{\|\phi_j\|_{H^2(K)}\},$$

so the $\|\phi_i^T\|_{H^1(\mathcal{T}^h)}$ are bounded for $n \geq 2$. Defining a quadrature formula as in (5.172) for T, we start with a quadrature formula for K with the corresponding w_j, P_j. Then the $P_{j,T} := F_T(P_j)$ and the $w_{j,T} := h^n |det(B)|^{-1} w_j$.

We omit the straightforward proof, but prove that for $f^h \in \mathcal{V}^{h'}$ there exists an appropriate $f_0^h \in \mathcal{V}'$ such that $f^h = Q'^h f_0^h$. By Theorem 2.35, (2.110f), (2.108) we know that f^h has the form (5.239) with bounded $\|f_j\|_{L^{p'}(\mathcal{T}_c^h) = L^{p'}(\Omega)}$, so $f \in W^{-1,p'}(\Omega) = \mathcal{V}'$. We need this property for the later proof of stability of nonconforming FEMs for

$f^h = A^h u^h$, for Q'^h cf. (5.294).

$$\langle f^h, v^h \rangle := \sum_{T_c \in \mathcal{T}_c^h} \int_{T_c} \left(\sum_{i=0}^{n} f_i \partial^i v^h \right) dx \; \forall v^h \in \mathcal{V} \cup \mathcal{V}_b^h, f^h \in W^{-1,p'}\left(\mathcal{T}_c^h\right), \quad (5.239)$$

$$f_j \in L^{p'}\left(\mathcal{T}_c^h\right), \text{with } Q'^h : \mathcal{V}' \to \mathcal{V}^{h'}, \; \langle Q'^h f - f, v^h \rangle = 0 \;\; \forall v^h \in \mathcal{V}_b^h \in W^{1,p}(\Omega).$$

Proposition 5.42. *Assume the local and nonlocal interpolation operators* I^h *in Definition 4.8 (4.67) are well defined and allow the corresponding interpolation bases, the* ϕ_i^T *in (5.237) and let* $\mathcal{V} = W^{1,p}(\Omega)$. *Then*

$$\forall f^h \text{ in (5.239)}, \;\; \exists\, f_e^h \in W^{-1,p'}(\Omega) \subset \mathcal{V}', \;\; 1 \le p' \le \infty, \; s.t. \; f^h = Q'^h f_e^h, \; so$$

$$\langle f_e^h, v^h \rangle = \langle f^h, v^h \rangle \forall v^h \in \mathcal{V}_b^h, \lim_{h\to 0} \left(\|f^h\|_{\mathcal{V}^{h'}} = \|Q'^h f_e^h\|_{\mathcal{V}^{h'}} \right) = \|f_e^h\|_{\mathcal{V}'}. \quad (5.240)$$

Proof. We aim for $f_e^h \in W^{-1,p'}(\Omega) \subset \mathcal{V}', 1 \le p, p' \le \infty, 1/p + 1/p' = 1$ for f^h in (5.239) with $\mathcal{V}^h \subset W^{1,p}\left(\mathcal{T}_c^h\right)$. With the restrictions $v^h|_{T_c}$ and $f_e^h|_{T_c}$, we have

$$v^h \in \mathcal{V}^h \subset W^{1,p}\left(\mathcal{T}_c^h\right), f^h \in W^{-1,p'}\left(\mathcal{T}_c^h\right) \Rightarrow \partial^i v^h|_{T_c} \in L^p(T_c), \; f_i|_{T_c} \in L^{p'}(T_c),$$

$\forall T \in \mathcal{T}_c^h, i = 0, \ldots, n$. For this type of $f^h \in W^{-1,p'}\left(\mathcal{T}_c^h\right) \subset \mathcal{V}^{h'}$ the $f_i|_{T_c} \in L^{p'}(T_c)$ imediately extend to an $f_{i,e} \in L^{p'}(\Omega)$ s.t. $f_{i,e}|_{T_c} \in L^{p'}(T_c)$. Then obviously the extended

$$\langle f_e^h, v \rangle := \int_{\Omega} \left(\sum_{i=0}^{n} f_{i,e} \partial^i v \right) dx \in \mathcal{V}'$$

satifies (5.240). □

5.5 Consistency, stability and convergence for FEMs with variational crimes

5.5.1 Introduction

Similarly to the conforming FEMs in Section 4.3 we restrict the discussion here to $\mathcal{U} = \mathcal{V}, \mathcal{U}_b = \mathcal{V}_b, \mathcal{U}_b^h = \mathcal{V}_b^h$. These we had exactly evaluated the weak bilinear and higher nonlinear forms, e.g. $a(\cdot,\cdot) : \mathcal{V}^h \times \mathcal{V}^h \to \mathbb{R}$. So they are well defined and exactly available. However, $\mathcal{V} = H^1(\Omega)$ for $m = 1$ requires $\mathcal{V}^h \subset C(\Omega)$ (and $\mathcal{V} = H^2(\Omega)$ for $m = 2$ even $\mathcal{V}^h \subset C^1(\Omega)$), so often too restrictive conditions. We have to allow violations of these conditions. The $\mathcal{V}_b^h \not\subset \mathcal{V}_b$, admit violated boundary conditions and $\mathcal{V}_b^h \not\subset \mathcal{V}$ violated continuity (and differentiability) across the boundary edges, e of $T \in \mathcal{T}^h$. These are the so-called *nonconforming FEs*, see cases (1), (2) in Subsection 5.5.2. Except partially in Subsection 5.5.7, we consider $\mathcal{T}^h$ with curved triangles at the boundary. So we again use the notation $\mathcal{T}_c^h$. As for quadrature approximations, the goal here is fitting the violated boundary and continuity conditions, such that the original convergence rate is not invalidated. This will motivate the appropriate choice of parameters for these methods. We will come back to this point in (5.325) ff. below. Before we discussed $\mathcal{P}_{d-1}^2 \subseteq \mathcal{P} \subseteq \mathcal{P}_{d+\tau}^2$ with $\tau \geq -1$. This allowed including, as the most important case, the tensor product FEs Q_d^n, cf. (4.13) with $\tau = d-2$. The following convergence results would remain correct for $\tau > -1$, or even for $\tau = d-2$ only if unrealistic conditions were imposed. So we restrict the FEs, studied here, to

$$\mathcal{V}^h \subset \left\{v : \Omega \subset \mathbb{R} \to \mathbb{R}^2 : v|_T \in \mathcal{P} = \mathcal{P}_{d-1}^2 \ \forall\, T \ \in \mathcal{T}_c^h\right\},\ \bar{\Omega} = \cup_{T \in \mathcal{T}_c^h} \overline{T} \qquad (5.241)$$

the piecewise polynomials of degree $d-1$. For reasons discussed below, we restrict the discussion in this section to $\mathbb{R}^2$ to, except for isoparametric FEs.

We unify the presentation again by using the same symbols $u, u^h, \ldots \in H^1(\Omega), \ldots$ instead of $\in H^1(\Omega, \mathbb{R}^q)$ and, as coefficients, a_{ij} instead of A_{ij}, for real and vector valued functions as in (5.285).

Already in the last Sections 5.2–5.4 we studied different types of nonconforming FEMs. For fully nonlinear problems the available C^1 FEs are not able to satisfy the boundary conditions exactly. In fact, we had to introduce an $\Omega_c^h \approx \Omega$ for the new FEs. We will similarly discuss violated boundary conditions in Subsection 5.5.6 ff., and isoparametric FEMs in Subsection 5.5.10. In the Subsections before 5.5.10 we will, however, only study polynomial FEs, defined on a subsdivision of the original domain Ω or a curved approximation Ω_c. For more complicated data the integrals in the FE equations have to be computed by quadrature or more general approximations as in the previous Section 5.4.

The interplay between the weak and *strong* operators yields an essential tool for handling nonconforming FEMs. Here we start with linear operators, estimating the violation of $\mathcal{V}_b^h \not\subset \mathcal{V}_b$ or $\mathcal{V}_b^h \not\subset \mathcal{V}$. Later on we will extend this approach to nonlinear problems as well. We systematically indicate by appropriate indexing, the weak and *strong* operators as A, A^h and A_s, A_s^h, and the corresponding bilinear forms as $a(\cdot,\cdot), a_h(\cdot,\cdot), a^h(\cdot,\cdot)$ and $a_s(\cdot,\cdot), a_{s,h}(\cdot,\cdot), a_s^h(\cdot,\cdot)$.

The following technique in Subsections 5.5.6 and 5.5.7 is a generalization of Brenner and Scott [141]. It extends their results for the Laplacian in $\mathbb{R}^2$ to general second order linear and quasilinear equations and systems. We systematically develop high accuracy employing the high order Gauss-type quadrature formulas along the one-dimensional edges. They exist no longer for the two-dimensional edges, e, of $T \in \mathcal{T}^h \subset \mathbb{R}^3$. So even for the few examples of two-dimensional cubature rules of higher order, the corresponding FEMs lose much of their higher order convergence. So we omit them here. The $\mathbb{R}^2$ results are closely related to the conditions for patch tests as discussed by Stummel, Shi *et al.* [583–585, 587, 611].

High accuracy FEMs with variational crimes are important for path-following of parameter-dependent solutions of nonlinear problems and the study of turning and bifurcation points, in particular for avoiding spurious solutions. Furthermore, we will use a similar approach for the spectral methods in [120].

Other possibilities are the DCGMs and the direct tensor product versions for rectangular subdivisions and collocation due to Bialecki and Fairweather [95], not discussed here.

We start with the Helmholz operator, Example 5.44 in Subsection 5.5.2, see Example 4.45. This simple case motivates deriving the necessary extensions for bilinear forms and operators and are discuss the consequences of the nonconformity for the bilinear form. We elaborate the impact of variational crimes via the difference of the weak and strong bilinear forms and prove the basic Strang Lemma 5.52. In Subsection 5.5.3 we generalize these results to second order linear and quasilinear problems as appropriate general discretization concepts. These are again generalized Petrov–Galerkin methods.

Subsection 5.5.4 shows the coercivity of the principal part as a first step towards stability. It formulates, as a guideline for the further subsections, the variational consistency error for linear problems. The variational consistency errors are estimated via the quadrature errors of quadrature formulas applied to the edges of neighboring triangles. So we introduce in Subsection 5.5.5 different types of high order Gauss quadrature formulas. Now we are ready to estimate, in Subsections 5.5.6 and 5.5.7, the classical and variational consistency for violated bondary conditions and continuity in linear and quasilinear problems. In Subsection 5.5.7 and for linear problems, the Aubin–Nitsche trick for increasing the order of convregence by one with respect to the $L^2(\Omega)$ norm is applied. By combining the results from the last three subsections, we prove stability for boundedly invertible linear operators with consistent discretization in Subsection 5.5.8. We use here an anticrime transformation for a modification of the original proof in Section 3.5. The next subsection, 5.5.9, proves convergence for the exact FEMs and their quadrature approximations for linear and quasilinear problems. For efficiently solving the nonlinear equations, we again apply the mesh independence principle for our discrete problems. The last Subsection 5.5.10 considers isoparameteric FEMs and strongly extends them to quasilinear problems.

In this section we assume that

$$\Omega \subset \mathbb{R}^2 \text{ is a bounded domain, and the solution is } u_0 \in H^2(\Omega). \tag{5.242}$$

For nonsmooth $u_0 \notin H^2(\Omega)$ or nonconvex Ω, singular adapted FEs are added $\mathcal{V}^h$. Then the strong regularity conditions can be reduced and still the usual high accuracy can be guaranteed, for details cf., e.g. Blum and Rannacher [98].

5.5.2 Variational crimes for our standard example

Often, these nonconforming FEMs are only studied for cases, particularly interesting for applications. This holds, e.g. for the results of Graham, Hackbusch and Sauter, [361–364] Their results, relevant for our approach, are summarized in Subsection 4.2.5. On this basis it is possible, generalizing the following results for nonconforming FEMs on quasiuniform to more general triangulations.

Remark 5.43. *Again, the condition of a quasiuniform subdivision, $\mathcal{T}^h$, in this and the following theorems, can be relaxed, to more general triangulations cf. Subsection 4.2.5. Here we study only FEMs in $\mathbb{R}^2$, and require Assumptions 4.21 and 4.26. Again in the messy sums below, $\sum_{T\in\mathcal{T}^h}$, the terms have to be estimated for each $T \in \mathcal{T}^h$ separately.*

The results in Theorems 4.30–4.33 allow appropriate estimates for continuous and L^2 FEs. We leave the details to the interested reader.

Here we intend a discussion of general nonconforming FEMs, where we impose Condition 4.16.

For presenting the essental ideas more clearly, we start studying, for our simple Example 4.45, the impact of violated boundary conditions and continuity. Later on, we have to generalize linear and bilinear forms, operators and projectors, similarly to Sections 5.2 and 5.4.

Necessary extensions towards variational crimes

Example 5.44. For *Dirichlet boundary conditions,* $u \in H_0^1(\Omega),\ f \in H^{-1}(\Omega)$, and the Helmholtz operator we memorize, see (2.2) ff, (4.107), (4.109), (5.242), the bounded operators and the corresponding bilinear forms and obtain by partial integration

$A : \mathcal{V} := H^1(\Omega) \to H^{-1}(\Omega),\ A_s : H^2(\Omega) \to L^2(\Omega)$ bounded, and for

$$u \in H^2(\Omega) : a(u,v) := \int_\Omega \langle \nabla u,\, \nabla v\rangle + cuvdx =: \langle Au, v\rangle_{H^{-1}(\Omega)\times H^1(\Omega)} \quad (5.243)$$

$$= a_s(u,v) := (A_s u, v)_{L^2(\Omega)} = \int_\Omega (-\Delta u + cu)vdx\ \forall v \in \mathcal{V}_b := H_0^1(\Omega).$$

The weak and FE solutions $u_0 \in \mathcal{V}_b = H_0^1(\Omega)$ and $u_0^h \in \mathcal{V}_b^h \subset \mathcal{V}_b$ are defined by

$$a(u_0, v) = \langle f, v\rangle_{H^{-1}(\Omega)\times H_0^1(\Omega)} \forall v \in \mathcal{V}_b \quad (5.244)$$

$$a\left(u_0^h, v^h\right) = \int_\Omega \left(\nabla u_0^h,\, \nabla v^h\right)_2 + c\, u_0^h\, v^h\, dx = \langle f, v^h\rangle_{\dots}\ \forall\ v^h \in \mathcal{V}_b^h \subset \mathcal{V}_b, \quad (5.245)$$

with $u_0 \in H^2(\Omega)$ for $f \in L^2(\Omega), \Omega \in C^2$, and the inner product $(\cdot,\cdot)_2$ in $\mathbb{R}^2$. Indeed, by Theorems 2.43 and 2.45 u_0, u_0^h uniquely exist for $c \geq 0$. Obviously the integral is well defined for conforming FEs $\mathcal{V}_b^h \subset H_0^1(\Omega)$. But here we want to study nonconforming finite elements. ■

Nonconforming finite elements

1. *Finite elements violating the boundary conditions:*
   In Section 5.2 we have discussed this problem of nonconforming FEs already for fully nonlinear problems. It is always possible to apply this technique to linear and quasilinear operators and their nonlinear forms. However, here we aim for the higher accuracy obtainable with polynomial FEs on curved domains in Subsection 4.2.7. For these FEs we recall (4.97), $\overline{\Omega} = \cup_{T\in\mathcal{T}_c^h}\overline{T}$ and restrict the discussion to homogeneous Dirichlet boundary conditions. So the FEs $u^h : \Omega \to \mathbb{R}$ satisfy $u^h(P_j) = 0$ only at certain points $P_j \in \partial\Omega$ or even only close to $\partial\Omega$ for curved boundaries by (4.98). For homogeneous Dirichlet boundary conditions $Bv = v|_{\partial\Omega} = 0$, this implies

$$\mathcal{V}_b^h = \{u^h \in \mathcal{V}^h \subset H^1(\Omega) = \mathcal{V} : B(u^h) \approx 0\}, H_0^1(\Omega) = \mathcal{V}_b \not\supset \mathcal{V}_b^h, \quad (5.246)$$

This implies that (5.243) is violated if u, v are replaced by u^h, v^h, yielding an estimate for the error due to violated boundary conditions in the corresponding FEM. For natural boundary conditions, only specific test functions $v^h \in \mathcal{V}_b \not\supset \mathcal{V}_b^h \subset \mathcal{V}$ will cause problems. But the original bilinear $a(\cdot,\cdot)$ and linear forms $\langle f, \cdot\rangle$ are defined in $\mathcal{V}^h \times \mathcal{V}^h$ and $\mathcal{V}^h$. We postpone isoparametric FEs to Subsection 5.5.10.

2. *Finite elements violating the continuity conditions:*
Let $\mathcal{V}_b^h \not\subset H^1(\Omega)$, e.g. we have discontinuous FEs, or more generally, cf. (5.300)

$$\mathcal{V}_b^h \not\subset \mathcal{V} \text{ e.g. discontinuous across the edges } e. \tag{5.247}$$

This happens for nonconforming finite elements, e.g. the Crouzeix–Raviart FE, see Figure 4.10 [135,141]. So, the u^h are continuous only in all the, say k, affinely equivalent points $P_j^e \in e \subset T \in \mathcal{T}_c^h$ on every e. Here k is so small that $u_l^h(P_j) = u_r^h(P_j), j = 1, \ldots, k$, allows $(u^h|_{\overline{T_l}})|_e \not\equiv (u^h|_{\overline{T_r}})|_e$. Usually, even for curved $\partial\Omega$, the number of points $P_j \in e \subset T$ on the edges interior to Ω and along $\partial\Omega$ is the same. Then violated continuity implies violated boundary conditions as well.

In these so-called nonconforming FEMs, (1), (2), we study polynomial FEs,

$$\begin{aligned} &\text{FE spaces with } \mathcal{V}_b^h \not\subset \mathcal{V}_b \text{ or } \mathcal{V}_b^h \not\subset \mathcal{V}, \qquad (5.248)\\ &u^h|_T,\ v^h|_T \in \mathcal{P}\ \forall u^h, v^h \in \mathcal{V}^h,\ T \in \mathcal{T}_c^h \text{ with } \mathcal{P}_{d-1} = \mathcal{P}, \text{ of fixed degree,} \end{aligned}$$

with a sequence $\{\mathcal{V}_b^h\}_{h\in H}$ indicated by $h,\ 0 < h \in H, \inf_{h\in H} h = 0$.

We combine these crimes with *approximations for projectors, operators* and *linear, bilinear and higher nonlinear forms,* mainly by quadrature, cf Section 5.4. In this subsubsection we mainly study the impact of nonconforming FEs onto necessary modifications of the linear and bilinear forms.

For the following results we memorize, see Subsection 4.2.4: the $v^h \in \mathcal{V}^h$ usually are discontinuous across common edges $e \subset \overline{T} \cap \overline{T_1}$ of neighboring $T, T_1 \in \mathcal{T}_c^h$, and hence $v^h \notin H^1(\Omega)$, [174], pp. 207–208, or for short

$$v^h \notin C(\Omega) \Rightarrow v^h \notin H^1(\Omega) \text{ (but } C(\Omega) \not\subset H^1(\Omega)). \tag{5.249}$$

The original linear, bilinear and higher nonlinear forms, e.g. $\langle f, \cdot\rangle$ and $a(\cdot,\cdot)$, $a_s(\cdot,\cdot)$, have to be modified for *the nonconforming FEs.* This implies that the modified (5.243) is violated for u^h, v^h. We will estimate the errors due to nonconforming FEMs, caused by $\mathcal{V}_b^h \not\subset H_0^1(\Omega)$ or $\not\subset H^1(\Omega)$. As in Section 5.4 the $\langle f, \cdot\rangle$, $a(\cdot,\cdot), a_s(\cdot,\cdot), A, A_s, \ldots,$ have to be defined piecewise according to (5.251). We start the discussion, assuming $\overline{\Omega} = \overline{\Omega^h} = \cup_{T\in\mathcal{T}_c^h}\overline{T}$. We summarize these cases, cf. (5.178), under the notation $v^h \in \mathcal{V}^h \subset W^{1,p}\left(\mathcal{T}_c^h\right)$, $f_h \in W^{-1,p'}\left(\mathcal{T}_c^h\right), \ldots,$ with $L^p(\Omega) = L^p\left(\mathcal{T}_c^h\right)$. For the important cases $f \in \mathcal{V}'$, $f_h \in W^{-1,p'}\left(\mathcal{T}_c^h\right)$, including $A_h u^h$, we obtain the general form

for $f \in \mathcal{V}'$, and

$$\overline{\Omega} = \overline{\Omega^h} = \cup_{T \in \mathcal{T}_c^h} \overline{T},\ 1/p + 1/p' = 1,\ \mathcal{V} = W^{1,p}(\Omega), f \in \mathcal{V}' \subset W^{-1,p'}(\mathcal{T}_c^h):$$

$$\langle f, v \rangle_{W^{-1,q}(\Omega) \times W^{1,p}(\Omega)} = \int_\Omega (f_{-1}, \nabla\ v)_2 + f_0 v\ dx, \text{ cf. Propos. 2.34, extended}$$

$$\langle f_h, v^h \rangle_{W^{-1,q}(\mathcal{T}_c^h) \times W^{1,p}(\mathcal{T}_c^h)} := \sum_{T \in \mathcal{T}_c^h} \int_T (f_{-1}, \nabla\, v^h)_2 + f_0 v^h dx \tag{5.250}$$

for $v^h \in \mathcal{V} + \mathcal{V}^h$, $f_h \in \mathcal{V}' + \mathcal{V}^{h'}$, with $f = f_h|_{\mathcal{V}}$, $f^h := f_h|_{\mathcal{V}^h}$.

Correspondingly to f, we define the following *bounded extensions of the weak and strong bilinear forms and the operators* $a_h(u^h, v^h)$, $A_h u^h$, and $a_{s,h}(u^h, v^h)$, $A_{s,h}$. For a simplified presentation we use the Hilbert–Sobolev version with $p = 2$ for linear problems:

$$a_h(u^h, v^h) := \sum_{T \in \mathcal{T}_c^h} \int_T (\nabla\, u^h, \nabla\, v^h)_2 + c\, u^h\, v^h\, dx =: \langle A_h u^h, v^h \rangle_{H^{-1}(\mathcal{T}_c^h) \times H^1(\mathcal{T}_c^h)}$$

$$\forall u^h, v^h \in \mathcal{V} + \mathcal{V}^h, \text{with } a(\cdot,\cdot) = a_h(\cdot,\cdot)|_{\mathcal{V}\times\mathcal{V}}, a^h(\cdot,\cdot) := a_h(\cdot,\cdot)|_{\mathcal{V}^h \times \mathcal{V}^h} \tag{5.251}$$

and $$a_{s,h}(u^h, v^h) := \sum_{T \in \mathcal{T}_c^h} \int_T (-\Delta\, u^h + c\, u^h)v^h\, dx =: (A_{s,h} u^h, v^h)_{L^2(\mathcal{T}_c^h)} \tag{5.252}$$

$\forall u^h \in H^2(\Omega) \cap \mathcal{V}^h, \forall v^h \in L^2(\Omega) + \mathcal{V}^h$

with $a_s(\cdot,\cdot) = a_{s,h}(\cdot,\cdot)|_{H^2(\Omega)\times L^2(\Omega)},\quad a_s^h(\cdot,\cdot) = a_{s,h}(\cdot,\cdot)|_{\mathcal{V}^h \times \mathcal{V}^h}$.

Remark 5.45. We prevent a well-motivated misunderstanding*: obviously the* $a_h(u, v^h), a_{s,h}(u, v^h)$ *are well defined for* $u \in \mathcal{V}_b$. *Now,* $a_h(u, v^h) \neq \langle Au, v^h \rangle_{H^{-1}(\mathcal{T}_c^h) \times H^1(\mathcal{T}_c^h)}$ $\forall\ v^h \in \mathcal{V}_b^h$ *for the variational crimes with* $\mathcal{V}_b^h \not\subset \mathcal{V}_b$. *This inequality will be specified in Proposition 5.47. Note that* $\langle Au, v^h \rangle_{H^{-1}(\mathcal{T}_c^h) \times H^1(\mathcal{T}_c^h)}$ *always makes sense for* $u \in H^1(\Omega)$*: the* $f := Au \in \mathcal{V}'$ *according to (5.243), is defined by testing* $\forall\ v \in \mathcal{V}_b$. *This* f *induces, by (5.250), an* f_h*, now tested* $\forall\ v^h \in \mathcal{V}_b^h \cup \mathcal{V}_b$ *and a* f^h*, now tested* $\forall\ v^h \in \mathcal{V}_b^h$. *So for* $u \in H^1(\Omega)$ *the* $A_h u$ *is certainly well defined by*

$$a_h(u, v^h) = \langle A_h u, v^h \rangle_{H^{-1}(\mathcal{T}_c^h) \times H^1(\mathcal{T}_c^h)}, \text{but for variational crimes} \tag{5.253}$$
$$\neq \langle Au, v^h \rangle_{H^{-1}(\mathcal{T}_c^h) \times H^1(\mathcal{T}_c^h)}\ \forall\ v^h \in \mathcal{V}_b^h \text{ or } Q'^h Au \neq Q'^h A_h u,$$

for the projectors Q'^h *introduced in (5.294) below. This situation is essential for the case of the later* classical consistency errors.

Impact of variational crimes for the standard example

The relation between *strong and weak bilinear* forms is the essential tool for studying the impact of variational crimes in FEMs. They are reflected by the discontinuity of the FEs crossing the interior edges $e \subset \overline{T} \in \mathcal{T}_c^h$ and violated boundary conditions along the boundary edges $e \subset \partial\mathcal{T}_c^h$.

Generalizing (5.245), we determine the *(weak) discrete solution* as

$$u_0^h \in \mathcal{V}_b^h : a^h\left(u_0^h, v^h\right) = \left\langle A_h u_0^h, v^h \right\rangle_{H^{-1}(\mathcal{T}_c^h)\times H^1(\mathcal{T}_c^h)} = \langle f^h, v^h\rangle_{\dots}\ \forall\, v^h \in \mathcal{V}_b^h,$$

$$\text{with } \mathcal{V}_b^h \not\subset H_0^1(\Omega) \text{ or } \not\subset H^1(\Omega). \tag{5.254}$$

cf. (5.248). For natural boundary conditions only $\mathcal{V}_b^h, \not\subset H^1(\Omega)$ is problematic.

Similarly as in Section 5.2 we now systematically study the relation between weak and strong operators and forms, here for our standard example. We apply the first Green's formula (2.23) to every T or $T = \Omega$ finding

$$\int_T -v\ \Delta\, u\, dx = \int_T (\nabla u, \nabla\, v)_2 dx - \int_{\partial T} v \frac{\partial u}{\partial \nu} ds. \tag{5.255}$$

This implies for the exact bilinear forms, see (5.243) for $c \neq 0$ as well,

$$a_s(u,v) = a(u,v)\ \forall\ u \in H^2(T), \text{ and either } \partial u^h/\partial\nu|_{\partial T} = 0 \text{ or } v \in H_0^1(T). \tag{5.256}$$

Hence, (5.244), and a corresponding strong problem $a_s(u_0, v) = (f, v) \forall\ v \in \mathcal{V}_b$ would have identical solutions $u_0 \in H^2(\Omega)$, e.g. for $f \in L^2(\Omega), \Omega \in C^2$. This is satisfied for either natural or Dirichlet boundary conditions.

For the extended bilinear forms $a_h(u,v)$ and $a_{s,h}(u,v)$ in (5.251) and (5.252), now for $c = 0$, we apply the Green's formula (5.255) for every $T \in \mathcal{T}_c^h$. With the surface integrals $\int_e \cdots ds, e \in \bar{T}$ and ν_e indicating the outer normal vector for T, perpendicular to e, we get $\forall u \in (\mathcal{V} + \mathcal{V}^h) \cap H^2(\mathcal{T}_c^h), v \in \mathcal{V} + \mathcal{V}^h$,

$$\begin{aligned} a_{s,h}(u,v) &= \sum_{T\in\mathcal{T}_c^h} \int_T -v\Delta u dx = \sum_{T\in\mathcal{T}_c^h} \Big(\int_T (\nabla v, \nabla u)_2 dx - \sum_{e\in\bar{T}} \int_e v(\nabla u, \nu_e)_2 ds \Big) \\ &= a_h(u,v) - \sum_{T\in\mathcal{T}_c^h} \sum_{e\in\bar{T}} \int_e v(\nabla u, \nu_e)_2 ds \\ \text{with } B_\Delta u :&= (\nabla u, \nu_e)_2 = \partial u/\partial \nu_e. \end{aligned} \tag{5.257}$$

Obviously (5.257) reduces to (5.243) for $u \in (\mathcal{V} \cap H^2(\Omega)),\ v \in \mathcal{V}_b$:

In (5.257) the surface integral $\int_e$ for every interior edge $e \in \overline{T}$ will be obtained twice with opposite normal directions for neighboring $T_r\ , T_l \in \mathcal{T}_c^h$ and $e \subset \overline{T_r} \cap \overline{T_l}$. The edges[38] $e \in \partial\Omega$ will be summed up once.

We consider the transition from a triangle T_l to its neighboring T_r. Let $\nu = \nu_l = \nu_e$ be the outward normal for e as the face of T_l. Then $\nu_r = -\nu_l$ is the outward normal for e as the face of T_r. We denote the restriction of v, analogously for $\nabla u, (\nabla u, \nu_e)_2$, to T_l, T_r, and their extension to $\overline{T_l}, \overline{T_r}$ as v_l, v_r, ∇u_l, $(\nabla u, \nu_e)_2$. We introduce the standard notation, see [665],

$$v_l = v|_{\overline{T_l}},\ v_r = v|_{\overline{T_r}},\ [v] := v_l|_e - v_r|_e \text{ and } \{v\} := (v_l|_e + v_r|_e)/2, \tag{5.258}$$

[38] We use the notation $e \in \partial\Omega$, although either $e \subset \partial\Omega$ or even less for curved bouindaries, similarly $e \in \mathcal{T}_c^h \setminus \partial\Omega$.

for the corresponding jumps and arithmetic means of v across an interior e, and

$$[v] := \{v\} := v|_{\partial\Omega} \text{ along } \partial\Omega, \quad \text{and} \quad v \text{ arbitrary in } \quad \mathbb{R}^2\backslash\Omega.$$

For (5.257) we need, with $e \in \mathcal{T}_c^h \setminus \partial\Omega$ or $e \in \partial\Omega$, cf. [665], p. 153, [141], p. 206,

$$\sum_{T\in\mathcal{T}_c^h}\sum_{e\in\overline{T}}\int_e v\partial u/\partial\nu_e ds = \sum_{T\in\mathcal{T}_c^h}\sum_{e\in\overline{T}}\int_e v(\nabla u,\nu_e)_2 ds \tag{5.259}$$

$$= \sum_{e\in\mathcal{T}_c^h\backslash\partial\Omega}\int_e (v_l(\nabla u_l,\nu_l)_2 + v_r(\nabla u_r,\nu_r)_2)\, ds + \sum_{e\in\partial\Omega}\int_e v(\nabla u,\nu_e)_2 ds$$

$$= \sum_{e\in\mathcal{T}_c^h\backslash\partial\Omega}\int_e (v_l(\nabla u_l,\nu_e)_2 - v_r(\nabla u_r,\nu_e)_2)\, ds + \sum_{e\in\partial\Omega}\int_e v(\nabla u,\nu_e)_2 ds$$

$$= \sum_{e\in\mathcal{T}_c^h\backslash\partial\Omega}\int_e ([v](\{\nabla u\},\nu_e)_2 + \{v\}([\nabla u],\nu_e)_2)\, ds + \sum_{e\in\partial\Omega}\int_e v(\nabla u,\nu_e)_2 ds,$$

$$\text{since } ([v]\{\nabla u\} + \{v\}[\nabla u]) \tag{5.260}$$

$$= \frac{1}{2}(v_l - v_r)(\nabla u_l + \nabla u_r) + \frac{1}{2}(\nabla u_l - \nabla u_r)(v_l + v_r) = (v_l\nabla u_l - v_r\nabla u_r).$$

This yields the following relation between the weak and strong bilinear forms

$$a_{s,h}(u,v) = a_h(u,v) - \sum_{e\in\mathcal{T}\backslash\partial\Omega}\int_e ([v](\{\nabla u\},\nu_e)_2 + \{v\}([\nabla u],\nu_e)_2)\, ds$$

$$- \sum_{e\in\partial\Omega}\int_e v(\nabla u,\nu_e)_2 ds \;\forall u \in (\mathcal{V}+\mathcal{V}^h)\cap H^2(\Omega), v\in\mathcal{V}+\mathcal{V}^h. \tag{5.261}$$

We will mainly estimate the absolute value of the last messy sums for $u, v \in \mathcal{V}_b^h$, see below and Subsubsections 5.5.2 and 5.5.4. So we can choose any one of the two different $\nu_T = \nu_{T_l}$ or $\nu_T = \nu_{T_r}$. The following relations for the $a_s(u,v), a(u,v)$ and $a_s^h(u,v), a^h(u,v)$ are basic to study the impact of variational crimes. This (5.261) implies several cases: for $u \in \mathcal{V}_b \cap H^2(\Omega), v \in \mathcal{V}_b$ we find again $a_s(u,v) = a(u,v)$. More interesting are the $u^h, v^h \in \mathcal{V}_b^h$ and we obtain from (5.261)

Proposition 5.46.

1. *For our standard Example 5.44, we obtain for the bounded $a_h(u^h,v^h)$ and $a_{s,h}(u^h,v^h)$ in (5.251) and (5.252), respectively, now with $\nu = \nu_e$,*

$$a_{s,h}(u^h,v^h) = a_h(u^h,v^h) - \sum_{e\in\mathcal{T}\backslash\partial\Omega}\int_e ([v^h](\{\nabla u^h\},\nu)_2 + \{v^h\}([\nabla u^h],\nu)_2)\, ds$$

$$- \sum_{e\in\partial\Omega}\int_e v^h(\nabla u^h,\nu)_2 ds \;\forall u^h \in (\mathcal{V}+\mathcal{V}^h)\cap H^2(\mathcal{T}_c^h), v^h\in\mathcal{V}+\mathcal{V}^h. \tag{5.262}$$

2. *More generally, if the $\mathcal{V}_b^h$ satisfy exact Neumann and Dirichlet boundary conditions, respectively, then $v^h(\nabla u^h, \nu)_2 = 0 \;\forall\; e \in \partial\Omega,\; u^h, v^h \in \mathcal{V}_b^h$ and the last term $\sum_{e\in\partial\Omega}\int_e v^h(\nabla u^h, \nu)_2 ds$ in (5.262) vanishes.*

Three special cases are important. Relation (5.262) simplifies by (1.67) as

$$a^h(u^h, v^h) = a_s^h(u^h, v^h) - \sum_{e\in\partial\Omega}\int_e v^h(\nabla u^h, \nu)_2 ds \tag{5.263}$$

$$\text{for } v^h \in \mathcal{V}_b^h \subset H^1(\Omega) \text{ and } u^h \in \mathcal{V}_b^h \cap H^2(\Omega),$$

$$a^h(u^h, v^h) = a_s^h(u^h, v^h) - \sum_{e\in\mathcal{T}_c^h\setminus\partial\Omega}\int_e v^h([\nabla u^h], \nu)_2 ds \tag{5.264}$$

$$- \sum_{e\in\partial\Omega}\int_e v^h(\nabla u^h, \nu)_2 ds \text{ for } v^h \in \mathcal{V}^h \subset H^1(\Omega),$$

$$a^h(u^h, v^h) = a_s^h(u^h, v^h) \text{ for } v^h \in \mathcal{V}_b^h \subset H_0^1(\Omega), u^h \in \mathcal{V}_b^h \subset H^2(\mathcal{J}_c^h). \tag{5.265}$$

For the different types of variational crimes we combine (5.262)–(5.265), with the nonconformity $\mathcal{V}_b^h \not\subset \mathcal{V}_b$ and $\mathcal{V}_b^h \not\subset \mathcal{V}$.

Our final goal is an estimate for $\left\|u_0 - u_0^h\right\|_{\mathcal{V}^h}$ for the exact and discrete solutions u_0 and u_0^h, respectively. For general discretization methods, the classical and variational consistency errors are relevant. If $a^h(\cdot,\cdot)$ satisfies an inf–sup condition, this is achieved by estimates for $a_h\left(u_0 - u_0^h, v^h\right)$ or $a_h(u - u^h, v^h)$, cf. Subsection 5.5.4, based upon the following proposition.

Proposition 5.47. *For FEs with violated boundary conditions and/or continuity choose an arbitrary $u \in H^2(\Omega)$, $f_u := Au \in L^2(\Omega)$ and the corresponding discrete solution $A^h u^h = Q'^h Au$, e.g. exact and the discrete solutions $u_0 \in H^2(\Omega)$ and u_0^h for (5.244) and (5.254). Then we obtain, in particular for $(u, u^h) = \left(u_0, u_0^h\right)$,*

$$\begin{aligned}
&\langle A_h u, v^h\rangle_{H^{-1}(\mathcal{T}_c^h)\times H^1(\mathcal{T}_c^h)} - \langle Au, v^h\rangle_{H^{-1}(\mathcal{T}_c^h)\times H^1(\mathcal{T}_c^h)}\\
&= \langle A_h u - Au, v^h\rangle_{\dots}\\
&= a_h(u - u^h, v^h) = \sum_{e\in\mathcal{T}_c^h\setminus\partial\Omega}\int_e [v^h]\nabla u\cdot\nu ds + \sum_{e\in\partial\Omega}\int_e v^h(\nabla u, \nu)_2 ds.
\end{aligned} \tag{5.266}$$

Remark 5.48. *For variational crimes (5.263)–(5.265)and (5.266) imply $a_s^h(u^h, v^h) \neq a^h(u^h, v^h)$ and $a_h(u_0 - u_0^h, v^h) \neq 0$. So for small errors, hence a close relation between the $a^h(\cdot,\cdot)$ and $a_s^h(\cdot,\cdot) : \mathcal{V}_b^h \times \mathcal{V}_b^h \to \mathbb{R}$, the FEs have to be chosen such that along the common edges e the $[v^h]$ and on $\partial\Omega$ the v^h disappear sufficiently often. Choosing these zeros among appropriate quadrature points, the corresponding quadrature rules applied to the edges e disappear and consequently the integrals, e.g. $\int_e v^h([\nabla u^h], \nu)_2 ds$, are very small. Extensions to $\mathbb{R}^3$ are possible using results as in Cools [208–210]. Since these formulas have only lower order of converge, and since we consider DCGMs in Chapter 7, we do not discuss these modifications here.*

Remark 5.49. *In Section 5.5.4 we will estimate the consistency errors introduced in Definition 3.14. The standard form, for appropriate p, k, q, will be of the form*

$$\sup_{0 \neq v^h \in \mathcal{V}_b^h} \{|a_h(P^h u - u, v^h)|/\|v^h\|_{\mathcal{V}^h}\} = C^\mu h^p \|u\|_{W_q^k(\Omega)}. \tag{5.267}$$

Now this $u \in W_q^k(\Omega)$ is much smoother than the usual $u \in \mathcal{V} = H^1(\Omega) \supset W_q^k(\Omega)$. Nevertheless we get vanishing consistency errors for $u \in H^1(\Omega)$ as well. For a given $u \in \mathcal{V} = H^1(\Omega)$ and any $\epsilon > 0$ we have to show that

$$|a_h(P^h u - u, v^h)|/\|v^h\|_{\mathcal{V}^h} < \epsilon \ \forall h < h_\epsilon.$$

We choose $u_s \in W_q^k(\Omega)$ such that $\|u - u_s\|_{\mathcal{V}} < C'\delta$ and determine δ such that the previous $< \epsilon$ is satisfied. We combine (5.267), the boundedness of a_h and (4.42) to get

$$\begin{aligned}
&|a_h(P^h u - u, v^h)| \\
&\quad \leq |a_h(P^h u_s - u_s, v^h)| + |a_h(P^h u - P^h u_s, v^h)| + |a_h(u_s - u, v^h)| \\
&\quad \leq |a_h(P^h u_s - u_s, v^h)| + C''\|P^h u - P^h u_s\|_{\mathcal{V}^h}\|v^h\|_{\mathcal{V}^h} + C''\|u_s - u\|_{\mathcal{V}^h}\|v^h\|_{\mathcal{V}^h} \\
&\quad \leq \big(C^\mu h^p \|u_s\|_{W_q^k(\Omega)} + C''(CC' + 1)\delta\big)\|v^h\|_{\mathcal{V}^h} \\
&\quad \leq \epsilon \|v^h\|_{\mathcal{V}^h} \text{ for sufficiently small } h, \delta.
\end{aligned}$$

Due to the testing of the exact and discrete solutions u_0, u_0^h by $v \in \mathcal{V}_b, v^h \in \mathcal{V}_b^h$ with $\mathcal{V}_b^h \not\subset \mathcal{V}, \mathcal{V}_b$, this argument is not sufficient for variational crimes.

Therefore, violated continuity and boundary conditions and approximations imply more or less complicated errors. The corresponding estimates for these nonconforming FEMs are highly technical. We will study the variational consistency errors in Subsections 5.5.4 ff. A first step in this direction is to update Proposition 5.47 for the generalized elliptic operators considered in (2.135). We only have to replace the $\nabla u_0 \cdot \nu$ in (5.269) by $B_a u_0$. This yields

Proposition 5.50. *For FEs with violated boundary conditions and/or continuity the exact solution $u_0 \in H^2(\Omega)$ and the discrete (weak) solutions $u_0^h \in \mathcal{V}_b^h$ are related by*

$$a_h\left(u_0 - u_0^h, v^h\right) = \sum_{e \in \mathcal{T}_c^h \setminus \partial\Omega} \int_e [v^h] B_a u_0 ds + \sum_{e \in \partial\Omega} \int_e v^h B_a u_0 ds \ \forall v^h \in \mathcal{V}_b^h. \tag{5.268}$$

Remark 5.51. *For evaluating these $\partial u/\partial\nu|_{\partial\Omega}$ or $B_a u|_{\partial\Omega}$, and formulating the nonconforming FEMs, the situation is more complicated. An arbitrary $u \in H^1(\Omega)$ does not allow a meaningful restriction to $\partial\Omega$ or to e to obtain $\partial u/\partial\nu \in L^2(\partial\Omega)$ and $\partial u/\partial\nu \in L^2(e)$. We omit the reference to $B_a u|_{\partial\Omega}$. There are two ways out of this dilemma. Either Theorems 1.37, 1.39 are used for replacing the above $u \in H^1(\Omega)$ by $u \in H^{3/2+\epsilon}(\Omega)$ with any $\epsilon > 0$. Then $\partial u/\partial\nu|_{\partial\Omega} \in H^\epsilon(\partial\Omega) \subset L^2(\partial\Omega)$ or $\partial u/\partial\nu|_e \in L^2(e)$ is valid. Hence, $\int_{\partial\Omega} v\nabla u dS$, and $\int_e v\nabla u dS$, are well defined, cf. (7.42). Or $u \in H^1(\Omega)$ is a function in the domain, e.g. of the Laplacian; then again $\int_e v\nabla u dS$ are well defined. This problem of boundary traces has attracted some attention recently, cf. Taylor [618], Chapter 4, Proposition 4.5, and Jonsson and Wallin [425], Jerison,* *and Kenig [419], Schwab [575], and Grisvard [373]. Readers not so familiar with the $H^{3/2+\epsilon}(\Omega)$ results can in this chapter nearly everywhere replace $H^{3/2+\epsilon}(\Omega)$ by $H^2(\Omega)$, with $\partial u/\partial\nu|_{\partial\Omega} \in L^2(\partial\Omega)$ and $\partial u/\partial\nu|_e \in L^2(e)$, thus requiring a bit more than necessary.*

Proof. For $u \in H^2(\Omega)$ we modify (5.262) and use $\int_e \{v^h\}([\nabla u], \nu)_2 ds = 0$ along edges $e \in \mathcal{T}_c^h$ finding

$$
\begin{aligned}
\langle A_h u - Au, v^h\rangle_{\dots} &= \langle A_h u, v^h\rangle_{\dots} - \langle Au, v^h\rangle_{\dots} \text{ with } A^h u^h = Q'^h Au \\
&\overset{(5.251)}{=} a_h(u, v^h) \quad - \; a^h(u^h, v^h) \\
&\qquad u \text{ incorrectly} - u^h \text{ correctly tested by } v^h \in \mathcal{V}_b^h \\
&= a_h(u - u^h, v^h) \qquad (5.269) \\
&\overset{(5.254)}{=} \sum_{T \in \mathcal{T}_c^h} \int_T (\nabla u \,\nabla v^h + c\, u\, v^h) dx - a^h(u^h, v^h) \\
&\overset{(5.262)}{=} \sum_{T \in \mathcal{T}_c^h} \int_T (-\Delta u + c\, u) v^h\, dx \\
&\quad + \sum_{e \in \mathcal{T}_c^h \setminus \partial\Omega} \int_e (\{v^h\}([\nabla u], \nu)_2 + [v^h](\{\nabla u\}, \nu)_2) ds \\
&\quad + \sum_{e \in \partial\Omega} \int_e v^h (\nabla u, \nu)_2 ds - a^h(u^h, v^h) \\
&\overset{u \in H^2(\Omega)}{=} (f, v^h)_{L^2(\Omega)} - (f, v^h)_{L^2(\Omega)} + \\
&\quad \sum_{e \in \mathcal{T}_c^h \setminus \partial\Omega} \int_e [v^h](\{\nabla u\}, \nu)_2 ds + \sum_{e \in \partial\Omega} \int_e v^h (\nabla u, \nu)_2 ds,
\end{aligned}
$$

hence the claim. □

In Subsection 5.5.4 we will estimate the terms in the last sum for general problems and thus estimate $a_h\left(u_0 - u_0^h, v^h\right)$. That these estimates are crucial is shown in the following *Strang lemma,* generalizing the Cea Lemma 4.48, see [135, 141]. For general cases we will proof convergence via the results of the general discretization theory in Chapter 3. Nevertheless, we include Strang's proof for the convergence $\lim_{h\to 0} \left\|u_0 - u_0^h\right\|_{H^1(\Omega)} \to 0$. It essentially applies to "coercive" bilinear forms.

Lemma 5.52. Strang lemma: Violated boundary and/or continuity conditions: *Let $\mathcal{V}_b^h \not\subset \mathcal{V}_b$, and/or $\mathcal{V}_b^h \not\subset \mathcal{V}$, hence the $v^h, u^h \in \mathcal{V}_b^h$ violate the Dirichlet boundary conditions $v^h|_{\partial\Omega} = 0$ and the Neumann boundary conditions $\nabla u^h|_{\partial\Omega} = 0$, and/or are discontinuous. Let the bounded $a(\cdot,\cdot)$ and for more general cases let $a(\cdot,\cdot)$ satisfy an inf-sup condition and $a_h(\cdot,\cdot)$ a uniform inf-sup condition. Let $u_0 \in \mathcal{V}_b$ and $u_0^h \in \mathcal{V}_b^h$*

be the exact and the unique approximate solution, defined by

$$a(u_0, v) = \langle f, v\rangle_{\mathcal{V}'\times\mathcal{V}}\ \forall\ v \in \mathcal{V}_b,\ \text{ and } a^h\left(u_0^h, v^h\right) = \langle f, v^h\rangle_{\mathcal{V}^{h\prime}\times\mathcal{V}^h}\ \forall\ v^h \in \mathcal{V}_b^h. \quad (5.270)$$

Then, with C independent of h, and the broken Sobolev norms, $\|\cdot\|_{\mathcal{V}^h}$

$$\left\|u_0 - u_0^h\right\|_{\mathcal{V}^h} \le C\left(\inf_{u^h\in\mathcal{V}_b^h}\|u_0 - u^h\|_{\mathcal{V}^h} + \sup_{0\neq v^h\in\mathcal{V}_b^h}\frac{\left|a_h\left(u_0 - u_0^h, v^h\right)\right|}{\|v^h\|_{\mathcal{V}^h}}\right). \quad (5.271)$$

Proof. For any $u^h \in \mathcal{V}_b^h$,

$$\begin{aligned}
\left\|u_0 - u_0^h\right\|_{\mathcal{V}^h} &\le \|u_0 - u^h\|_{\mathcal{V}^h} + \left\|u^h - u_0^h\right\|_{\mathcal{V}^h} \ \text{(triangle inequality)} \\
&\le \|u_0 - u^h\|_{\mathcal{V}^h} + \frac{1}{\epsilon}\sup_{v^h\in\mathcal{V}_b^h\setminus\{0\}}\frac{\left|a_h\left(u^h - u_0^h, v^h\right)\right|}{\|v^h\|_{\mathcal{V}^h}} \ \text{in (4.170)} \\
&= \|u_0 - u^h\|_{\mathcal{V}^h} + \frac{1}{\epsilon}\sup_{v^h\in\mathcal{V}_b^h\setminus\{0\}}\frac{\left|a_h(u^h - u_0, v^h) + a_h\left(u_0 - u_0^h, v^h\right)\right|}{\|v^h\|_{\mathcal{V}^h}} \\
&\le \|u_0 - u^h\|_{\mathcal{V}^h} + \frac{1}{\epsilon}\sup_{v^h\in\mathcal{V}_b^h\setminus\{0\}}\frac{|a_h(u^h - u_0, v^h)|}{\|v^h\|_{\mathcal{V}^h}} \quad (5.272) \\
&\quad + \frac{1}{\epsilon}\sup_{v^h\in\mathcal{V}_b^h\setminus\{0\}}\frac{\left|a_h\left(u_0 - u_0^h, v^h\right)\right|}{\|v^h\|_{\mathcal{V}^h}} \ \text{(triangle inequality)} \\
&\le \|u_0 - u^h\|_{\mathcal{V}^h} + \frac{C}{\epsilon}\|u^h - u_0\|_{\mathcal{V}^h} \ \text{(continuity in } \mathcal{V}_b^h \times \mathcal{V}_b^h\text{)} \\
&\quad + \frac{1}{\epsilon}\sup_{v^h\in\mathcal{V}_b^h\setminus\{0\}}\frac{\left|a_h\left(u_0 - u_0^h, v^h\right)\right|}{\|v^h\|_{\mathcal{V}^h}} \quad (5.273) \\
&= \left(1 + \frac{C}{\epsilon}\right)\|u_0 - u^h\|_{\mathcal{V}^h} + \frac{1}{\epsilon}\sup_{v^h\in\mathcal{V}_b^h\setminus\{0\}}\frac{\left|a_h\left(u_0 - u_0^h, v^h\right)\right|}{\|v^h\|_{\mathcal{V}^h}}
\end{aligned}$$

Note that, by continuity,

$$\frac{\left|a_h\left(u_0 - u_0^h, v^h\right)\right|}{\|v^h\|_{\mathcal{V}^h}} \le C\left\|u_0 - u_0^h\right\|_{\mathcal{V}^h} \quad (5.274)$$

so that

$$\left\|u_0 - u_0^h\right\|_{\mathcal{V}^h} \ge \frac{1}{C}\sup_{v^h\in\mathcal{V}_b^h\setminus\{0\}}\frac{\left|a_h\left(u_0 - u_0^h, v^h\right)\right|}{\|v^h\|_{\mathcal{V}^h}}. \quad (5.275)$$

Combining (5.272) and (5.275), we find

$$\max\left\{\frac{1}{C}\sup_{v^h\in\mathcal{V}_b^h\setminus\{0\}}\frac{\left|a_h\left(u_0-u_0^h,v^h\right)\right|}{\|v^h\|_{\mathcal{V}^h}},\inf_{u^h\in\mathcal{V}_b^h}\|u_0-u^h\|_{\mathcal{V}^h}\right\} \tag{5.276}$$
$$\le \left\|u_0-u_0^h\right\|_{\mathcal{V}^h}$$
$$\le\left(1+\frac{C}{\epsilon}\right)\inf_{u^h\in\mathcal{V}_b^h}\|u_0-u^h\|_{\mathcal{V}^h}+\frac{1}{\epsilon}\sup_{v^h\in\mathcal{V}_b^h\setminus\{0\}}\frac{\left|a_h\left(u_0-u_0^h,v^h\right)\right|}{\|v^h\|_{\mathcal{V}^h}}.$$

□

Remark 5.53.

1. *Inequality (5.276) indicates that the sum of approximation terms* $\inf_{u^h\in\mathcal{V}_b^h}\|u_0-u^h\|_{\mathcal{V}^h}$, *and nonconformity terms* $\sup_{v^h\in\mathcal{V}_b^h\setminus\{0\}}|a_h(u_0-u_0^h,v^h)|/\ \|v^h\|_{\mathcal{V}^h}$ *correctly reflects the size of the discretization error* $\|u_0-u_0^h\|_{\mathcal{V}^h}$. *The same result for the general cases is obtained by the general discretization results in Chapter 3. Indeed by (3.33), the classical consistency error is estimated by the variational consistency and the interpolation error for the exact solution. For a* $a_h(u^h,v^h)$, *satisfying a uniform discrete inf-sup condition, or a stable* A^h *this again yields (5.271), cf.(3.33).*
2. *As a consequence of (5.266), the second term on the right-hand side of (5.271) would be zero if* $\mathcal{V}_b^h\subseteq\mathcal{V}_b$, *compare (4.156) and Theorem 4.54. Therefore, it measures the effect of nonconformity in* $\mathcal{V}_b^h\not\subseteq\mathcal{V}_b$ *or* $\mathcal{V}_b^h\not\subseteq\mathcal{V}$.

5.5.3 FEMs with crimes for linear and quasilinear problems

In Subsection 5.5.2 we have studied the impact of FEMS with crimes for a special linear equation, the Helmholtz equation. Here we generalize these results to general linear and quasilinear equations and systems of order 2 in $\mathbb{R}^2$. Since there are strong similarities we do not repeat all details for (5.251)–(5.261). The corresponding Propositions 5.46–5.47 and Lemma 5.52 will be formulated for all these cases simultaneously. Extensions to $\mathbb{R}^n, n\ge 3$ can be worthwhile, whenever the available cubature formulas for $\mathbb{R}^{n-1}, n\ge 3$, cf. Cools [208–210] are good enough for the specific problem.

Compared to (5.243)–(5.245) with the original Δu and $\partial u/\partial\nu$ we replace this special differential and boundary operator by the general forms Au and $B_a\ u$ in the following (5.278), (5.279). An essential tool is again Green's formula (2.9): for $v,w\in H^1(\Omega)$, we obtain by partial integration and with the outer normal $\nu=(\nu_1,\nu_2)^\mathsf{T}$ for $\partial\Omega$

$$\int_\Omega w\partial v/\partial x_i dx+\int_\Omega v\partial w/\partial x_i dx=\int_\Omega\partial(vw)/\partial x_i dx=\int_{\partial\Omega}vw\nu_i ds. \tag{5.277}$$

We find the bounded bilinear forms and operators for $\Omega \subset \mathbb{R}^2$

$$\begin{aligned}
a(u,v) &= \langle Au, v\rangle_{\mathcal{V}'\times\mathcal{V}} = \int_\Omega \left(\sum_{i,j=0}^{2} a_{ij}\partial^i u\ \partial^j v\right) dx \text{ with } a_{ij} \in W^{1-\delta_{j0},\infty}(\Omega), \\
&= \int_\Omega \left(\sum_{i,j=0}^{2} (-1)_{j>0}\partial^j\ (a_{ij}\partial^i u)\right)\ vdx + \int_{\partial\Omega} (B_a\ u)vds \qquad (5.278) \\
&=: \int_\Omega (A_s\ u)vdx + \int_{\partial\Omega} (B_a\ u)vds\ \forall v \in \mathcal{V} = H^1(\Omega), u \in \mathcal{V}\cap H^2(\Omega), \\
&\quad \text{with } \sum_{i,j=1}^{2} a_{ij}\xi_i\xi_j \geq \epsilon'|\xi|_2^2\ , \epsilon' > 0, \text{ for elliptic operators, and} \\
B_a\ u &= \sum_{i,j=1}^{2} \nu_j a_{ij}\partial^i\ u + \sum_{j=1}^{2} \nu_j a_{0j}\ u, \quad \text{and} \\
B_a\ u &= \partial u/\partial\nu \text{ for } a_{ij} = \delta_{i,j}\ \forall i,j = 0,1,2, \text{e.g. } A_s u = -\Delta u + cu.
\end{aligned}$$

It is possible to transform the problem such that the above $B_a u$ is replaced by $B_a\ u = \sum_{i,j=1}^{2} \nu_j a_{ij}\partial^i\ u$, always satisfied for $B_a\ u = \partial u/\partial\nu$. Then sometimes one obtains a *higher order of consistency and convergence* by the additional factor h.

Obviously, $a(u^h, v^h)$ is defined $\forall\ u^h,\ v^h \in \mathcal{V}_b^h \subset \mathcal{V}$ independent of the boundary conditions. For FEs with crimes $u^h,\ v^h \in \mathcal{V}_b^h \not\subset \mathcal{V}_b, \mathcal{V}$, we again use the extended $\langle f_h, \cdot\rangle_{W^{-k,p'}(\mathcal{T}_c^h)\times W^{k,p}(\mathcal{T}_c^h)}$, $1/p + 1/p' = 1$, $a_h(\cdot,\cdot)$, and the broken Sobolev norms $\|\cdot\|_{W^{k,p}(\mathcal{T}_c^h)}$ and obtain, cf. (5.250)–(5.252), (5.248), (5.254),

$$\begin{aligned}
a_h(u^h, v^h) &:= \langle A_h u^h, v^h\rangle_{H^{-1}(\mathcal{T}_c^h)\times H^1(\mathcal{T}_c^h)} := \sum_{T\in\mathcal{T}_c^h}\left(\int_T \sum_{i,j=0}^{2} a_{ij}\partial^i\ u^h\partial^j\ v^h\right) dx \\
&= \sum_{T\in\mathcal{T}_c^h}\int_T \left[(A_{s,h}\ u^h)v^h := \sum_{i,j=0}^{2}(-1)_{j>0}\partial^j(a_{ij}\partial^i\ u^h)\ v^h\right] dx \qquad (5.279) \\
&\quad + \sum_{e\in\mathcal{T}_c^h\setminus\partial\Omega}\int_e (\{v^h\}\ [B_a u^h] + [v^h]\{B_a u^h\})ds + \sum_{e\in\partial\Omega}\int_e v^h B_a u^h\, ds \\
&=: a_{s,h}(u^h, v^h) + \sum_{e\in\mathcal{T}_c^h\setminus\partial\Omega}\int_e (\{v^h\}\ [B_a u^h] + [v^h]\{B_a u^h\})\, ds \\
&\quad + \sum_{e\in\partial\Omega}\int_e v^h B_a u^h\, ds \quad \forall\ u^h \in (\mathcal{V}_b \cap H^2(\Omega)) + \mathcal{V}_b^h \subset H^2\left(\mathcal{T}_c^h\right),
\end{aligned}$$

$$v^h \in \mathcal{V}_b + \mathcal{V}_b^h \subset H^1\left(\mathcal{T}_c^h\right), \text{ with}$$
$$a^h(\cdot,\cdot) := a_h(\cdot,\cdot)|_{\mathcal{V}^h\times\mathcal{V}^h},\ a(\cdot,\cdot) = a_h(\cdot,\cdot)|_{\mathcal{V}\times\mathcal{V}} \text{ and}$$
$$a_s^h(\cdot,\cdot) := a_{s,h}(\cdot,\cdot)|_{H^2(\mathcal{T}_c^h)\times L^2(\Omega)}, a_s(\cdot,\cdot) = a_{s,h}(\cdot,\cdot)|_{H^2(\Omega)\times L^2(\Omega)},$$

with ellipticity as in (4.138). The exact and FE solutions u_0 and u_0^h are defined by

$$u_0 \in \mathcal{V}_b : \langle Au_0, v\rangle_{H^{-1}(\Omega)\times H^1(\Omega)} = a(u,v) = \langle f, v\rangle_{\mathcal{V}'\times\mathcal{V}}\ \forall\ v \in \mathcal{V}_b, \tag{5.280}$$
$$u_0^h \in \mathcal{V}_b^h :\ \left\langle A_h u_0^h, v^h\right\rangle_{H^{-1}(\mathcal{T}_c^h)\times H^1(\mathcal{T}_c^h)} = a^h\left(u_0^h, v^h\right) = \left\langle f^h, v^h\right\rangle_{\mathcal{V}^{h\prime}\times\mathcal{V}^h}\ \forall\ v^h \in \mathcal{V}_b^h.$$

These relations (5.278)–(5.280) are easily extendable to elliptic linear second order systems or quasilinear equations and systems. We use the notation in (4.136) and ansatz and test functions $\vec{u}^h = \left(u_1^h,\cdots,u_q^h\right),\ \vec{v}^h \in \mathcal{V}_b^h := \mathcal{V}_b^h(\mathbb{R}^q)$, now with $\mathcal{V}_b^h \not\subset \mathcal{V}_b = H_0^1(\Omega,\mathbb{R}^q)$, so violating the boundary conditions or below the continuity, $\mathcal{V}_b^h \not\subset H^1(\Omega,\mathbb{R}^q)$. We start with linear systems, and, because of the high similarity with (5.278) ff., we only formulate the FE equations: again $f, a(\cdot,\cdot), f_h, a_h(\cdot,\cdot)$ and A, A_h are bounded linear and bilinear forms and operators.

$$\langle A_h\vec{u}^h, \vec{v}^h\rangle_{H^{-1}(\mathcal{T}_c^h)\times H^1(\mathcal{T}_c^h)} = a_h(\vec{u}^h,\vec{v}^h) := \sum_{T\in\mathcal{T}_c^h}\int_T \sum_{k,l=0}^{2}(A_{kl}\partial^l\vec{u}^h, \partial^k\vec{v}^h)_q dx$$
$$\forall \vec{u}^h, \vec{v}^h \in \mathcal{V}_b + \mathcal{V}_b^h, \quad \text{with} \quad A_{kl} \in L^\infty(\Omega,\mathbb{R}^{q\times q}), \text{cf.}(5.254) \text{ for } \mathcal{V}_b^h. \tag{5.281}$$

The $\mathcal{V}_b$-coercivity of the principal part follows from the *uniform Legendre condition*, see (2.343), (2.345), (4.138), (4.139), hence with $\mathcal{W} = L^2(\Omega,\mathbb{R}^q)$, cf. Subsection 5.5.4,

$$\exists\ \lambda,\Lambda,\ \text{s.t. } 0<\lambda<\Lambda<\infty:\ \lambda|\vec{\vartheta}|_{nq}^2 \le \sum_{k,l=1}^{2}(A_{kl}(x)\vec{\vartheta}^l, \vec{\vartheta}^k)_q \le \Lambda|\vec{\vartheta}|^2 \forall x \in \Omega, \tag{5.282}$$
$$\Longrightarrow a_p(\vec{u}^h,\vec{u}^h) > \lambda^*\|\vec{u}^h\|_{\mathcal{V}}^2 \text{ and } a(\vec{u}^h,\vec{u}^h) \ge \lambda^{**}\|\vec{u}^h\|_{\mathcal{V}}^2 - C_c\|\vec{u}^h\|_{\mathcal{W}}^2\ \forall \vec{u}^h \in \mathcal{V}_b + \mathcal{V}_b^h.$$

For relating $a_h(\cdot,\cdot), A_h$ and $a_{s,h}(\cdot,\cdot), A_{s,h}$, we require $A_{kl} \in W^{1-\delta_{k0},\infty}(\Omega,\mathbb{R}^{q\times q})$:

$$a_h(\vec{u}^h,\vec{v}^h) := \langle A_h\vec{u}^h,\vec{v}^h\rangle_{H^{-1}(\mathcal{T}_c^h)\times H^1(\mathcal{T}_c^h)} := \sum_{T\in\mathcal{T}_c^h}\int_T\sum_{k,l=0}^{2}\left(A_{kl}\partial^l\ \vec{u}^h, \partial^k\ \vec{v}^h\right)_q dx$$
$$= \left[(A_{s,h}\ \vec{u}^h, \vec{v}^h)_{L^2(\Omega,\mathbb{R}^q)} := \sum_{T\in\mathcal{T}_c^h}\int_T\sum_{k,l=0}^{2}(-1)_{k>0}(\partial^k\ (A_{kl}\partial^l\ \vec{u}^h),\ \vec{v}^h)_q dx\right] \tag{5.283}$$
$$+ \sum_{e\in\mathcal{T}_c^h\setminus\partial\Omega}\int_e ((\{\vec{v}^h\}, [B_a\vec{u}^h])_q + ([\vec{v}^h], \{B_a\vec{u}^h\})_q)ds + \sum_{e\in\partial\Omega}\int_e \left(\vec{v}_l^h, B_a\vec{u}^h\right)_q ds$$

$$=: a_{s,h}(\vec{u}^h, \vec{v}^h) + \sum_{e\in\mathcal{T}_c^h\setminus\partial\Omega} \int_e \big((\{\vec{v}^h\}, [B_a\vec{u}^h])_q + ([\vec{v}^h], \{B_a\vec{u}^h\})_q\big)ds$$

$$+ \sum_{e\in\partial\Omega} \int_e (\vec{v}^h, B_a\vec{u}^h)_q\, ds \;\forall\; \vec{u}^h \in (\mathcal{V}_b \cap H^2(\mathcal{T}_c^h, \mathbb{R}^q)) + \mathcal{V}_b^h \subset H^2\left(\mathcal{T}_c^h, \mathbb{R}^q\right),$$

$$\vec{v}^h \in \mathcal{V}_b + \mathcal{V}_b^h \subset H^1\left(\mathcal{T}_c^h, \mathbb{R}^q\right) \text{ with } B_a\,\vec{u} := \sum_{k,l=1}^{2} \nu_k A_{kl}\partial^l\,\vec{u} + \sum_{l=1}^{2} \nu_k A_{k0}\,\vec{u},$$

again with $a^h(\cdot,\cdot) := a_h(\cdot,\cdot)|_{\mathcal{V}^h\times\mathcal{V}^h}$, $a(\cdot,\cdot) = a_h(\cdot,\cdot)|_{\mathcal{V}\times\mathcal{V}}$ and

$a_s^h(\cdot,\cdot) := a_{s,h}(\cdot,\cdot)|_{H^2(\mathcal{T}_c^h)\times L^2(\Omega)}, a_s(\cdot,\cdot) = a_{s,h}(\cdot,\cdot)|_{H^2(\Omega)\times L^2(\Omega)}$.

The exact and FE solutions $\vec{u}_0$ and $\vec{u}_0^h$ are defined by

$$\vec{u}_0 \in \mathcal{V}_b:\; \langle A\vec{u}_0, \vec{v}\rangle_{H^{-1}(\Omega)\times H^1(\Omega)} = a(\vec{u},\vec{v}) = \langle \vec{f}, \vec{v}\rangle_{\mathcal{V}'\times\mathcal{V}} \;\forall\; \vec{v}\in\mathcal{V}_b, \tag{5.284}$$
$$\vec{u}_0^h \in \mathcal{V}_b^h : \left\langle A_h\vec{u}_0^h, \vec{v}^h\right\rangle_{H^{-1}(\mathcal{T}_c^h)\times H^1(\mathcal{T}_c^h)} = a^h\left(\vec{u}_0^h, \vec{v}^h\right) = \langle \vec{f}^h, \vec{v}^h\rangle_{\mathcal{V}^{h\prime}\times\mathcal{V}^h} \forall \vec{v}^h \in \mathcal{V}_b^h.$$

We generalize these relation between $a_h(\cdot,\cdot)$ and $a_{s,h}(\cdot,\cdot)$. to quasilinear equations and systems. For avoiding a double formulation as for the linear case, we use the standard notation (4.129), but the *same symbols* for all functions

$$u, u_0 \in \mathcal{V}_b = W_0^{1,p}(\Omega, \mathbb{R}^q), u^h, u_0^h \in \mathcal{V}_b^h \subset W_0^{1,p}(\mathcal{T}_c^h, \mathbb{R}^q), q \geq 1, p \geq 2, \tag{5.285}$$

for equations and systems. Only $(.,.)_q$ indicates the possibility for systems, $q \geq 1$. Again we only formulate the FE equations, cf (4.173), (4.174) and use (5.277). The necessary conditions for the Nemickii operators $A_i(x, u^h, \nabla u^h)$, e.g. for bounded forms and operators, $a_h(u^h, v^h)$ and G_h, are formulated in Sections 2.5.4, 2.5.6, 4.4 in (4.174) and 4.5 in (4.193). We additionally modify them, such that G is Lipschitz-continuous. With smooth enough

$$A_i : \Omega\times\mathbb{R}^q\times\mathbb{R}^{n\times q} \to \mathbb{R}^q, \quad \text{we consider } \mathcal{V}_b^h \text{ as in (5.254), and}$$
$$\forall\; u^h \in (\mathcal{V}_b \cap W^{2,p}(\Omega, \mathbb{R}^q)) + \mathcal{V}_b^h \subset W^{2,p}\left(\mathcal{T}_c^h, \mathbb{R}^q\right),\; v^h \in \mathcal{V}_b + \mathcal{V}_b^h \subset W^{1,p}\left(\mathcal{T}_c^h, \mathbb{R}^q\right).$$

Then we define the weak and strong forms, $a_h(u^h, v^h)$ and $a_{s,h}(u^h, v^h)$, related by

$$a_h(u^h, v^h) := \langle G_h u^h, v^h\rangle_{W^{-1,p'}(\mathcal{T}_c^h)\times W^{1,p}(\mathcal{T}_c^h)} \tag{5.286}$$
$$:= \sum_{T\in\mathcal{T}_c^h} \int_T \sum_{i=0}^{2} (A_i(x, u^h, \nabla u^h), \partial^i v^h)_q dx =$$
$$= \Big[a_{s,h}(u^h, v^h) := (G_{s,h}\; u^h, v^h)_{L^2(\Omega,\mathbb{R}^q)}$$
$$:= \sum_{T\in\mathcal{T}_c^h} \int_T \sum_{i=0}^{2} (-1)_{i>0}(\partial^i A_i(x, u^h, \nabla u^h), v^h)_q dx\Big]$$

$$+ \sum_{e\in\mathcal{T}_c^h\setminus\partial\Omega} \int_e ((\{v^h\}, [B_G u^h])_q + ([v^h], \{B_G u^h\})_q) ds$$

$$+ \sum_{e\in\partial\Omega} \int_e (v^h, B_G u^h)_q \, ds$$

and $B_G\, u^h := B_G\,(x, u^h, \nabla u^h) := \sum_{i=1}^{2} \nu_i A_i(x, u^h, \nabla u^h)$ with

$$a^h(\cdot,\cdot) := a_h(\cdot,\cdot)|_{\mathcal{V}^h\times\mathcal{V}^h},\ a(\cdot,\cdot) = a_h(\cdot,\cdot)|_{\mathcal{V}\times\mathcal{V}} \text{ and}$$
$$a_s^h(\cdot,\cdot) := a_{s,h}(\cdot,\cdot)|_{H^2(\mathcal{T}_c^h,\mathbb{R}^q)\times L^2(\Omega,\mathbb{R}^q)}, a_s(\cdot,\cdot) = a_{s,h}(\cdot,\cdot)|_{H^2(\Omega,\mathbb{R}^q)\times L^2(\Omega,\mathbb{R}^q)}.$$

Again the exact and FE solutions u_0 and u_0^h are defined by

$$u_0 \in \mathcal{V}_b:\ a(u_0, v) = \langle Gu_0, v\rangle_{\mathcal{V}'\times\mathcal{V}} = 0\ \forall\ v \in \mathcal{V}_b, \tag{5.287}$$
$$u_0^h \in \mathcal{V}_b^h:\ a^h\left(u_0^h, v^h\right) = \left\langle G_h u_0^h, v^h\right\rangle_{W^{-1,p'}(\mathcal{T}_c^h,\mathbb{R}^q)\times W^{1,p}(\mathcal{T}_c^h,\mathbb{R}^q)} = 0\ \forall\ v^h \in \mathcal{V}_b^h,$$

often for $\mathcal{U}_b = \mathcal{V}_b, \mathcal{U}_b^h = \mathcal{V}_b^h$. We summarize the results for second order linear and quasilinear equations and systems and use the notation as in (5.285):

Proposition 5.54.

1. *For either Dirichlet or natural boundary conditions and the bilinear or nonlinear forms* $a_h(u^h, v^h), a_{s,h}(u^h, v^h)$ *and the operators* $A_h, A_{s,h}, G_h, G_{s,h}$ *in (5.279), (5.283) and (5.286), with the previously mentioned conditions for the* $A_i(\cdot,\cdot,\cdot)$*, are continuous. We find, cf. (5.256), with* $Bu := B_a u, p = 2$ *for linear, and* $Bu := B_G u, 2 \le p < \infty$ *for quasilinear problems with* $q \ge 1$

$$a_s(u, v) = a(u, v)\ \forall\ u \in W^{2,p}(\Omega,\mathbb{R}^q), v \in W^{1,p}(\Omega,\mathbb{R}^q), \text{and } vBu|_{\partial\Omega} = 0. \tag{5.288}$$

   Note that $\mathcal{V}_b = W_0^{1,p} \cap W^{2,p}(\Omega,\mathbb{R}^q)$ *for Dirichlet boundary conditions.*
2. *Valid for any* $u^h \in \mathcal{V}_b^h + (\mathcal{V}_b \cap H^2(\Omega,\mathbb{R}^q)) \subset W^{2,p}\left(\mathcal{T}_c^h,\mathbb{R}^q\right)$ *and* $v^h \in \mathcal{V}_b^h + \mathcal{V}_b \subset W^{1,p}\left(\mathcal{T}_c^h,\mathbb{R}^q\right)$ *we get, cf. (5.258),* $2 \le p < \infty, 1 \le q$,

$$a_h(u^h, v^h) = a_{s,h}(u^h, v^h)$$
$$+ \sum_{e\in\mathcal{T}_c^h\setminus\partial\Omega} \int_e ((\{v^h\}, [Bu^h])_q + ([v^h], \{Bu^h\})_q) ds \tag{5.289}$$
$$+ \sum_{e\in\partial\Omega} \int_e (v^h, (Bu^h))_q ds\ \forall\ u^h \in (\mathcal{V}_b \cap H^2(\Omega,\mathbb{R}^q)) + \mathcal{V}_b^h,\ v \in \mathcal{V}_b + \mathcal{V}_b^h.$$

For important special cases (5.289) simplifies, cf. (5.249), (5.263)–(5.265), as

$$a^h(u^h,v^h) = a_s^h(u^h,v^h) \text{ for } \mathcal{V}_b^h \subset W_0^{1,p}(\Omega,\mathbb{R}^q),\ \mathcal{V}_b^h \subset W^{2,p}(\Omega,\mathbb{R}^q), \tag{5.290}$$

$$a^h(u^h,v^h) = a_s^h(u^h,v^h) + \sum_{e\in\partial\Omega}\int_e v^h(Bu^h)ds \text{ for only violated boundary cond.}$$

$$\text{and } \mathcal{V}_b^h \subset C(\Omega,\mathbb{R}^q) \in W^{2,p}(\Omega,\mathbb{R}^q),\ \mathcal{V}_b^h \subset C^1(\Omega,\mathbb{R}^q) \in W^{1,p}(\Omega,\mathbb{R}^q), \tag{5.291}$$

$$a^h(u^h,v^h) = a_s^h(u^h,v^h) + \sum_{e\in\mathcal{T}_c^h\setminus\partial\Omega}\int_e v^h[Bu^h]ds \tag{5.292}$$

$$+ \sum_{e\in\partial\Omega}\int_e v^h(Bu^h)ds \text{ for } \mathcal{V}_b^h \subset C(\Omega,\mathbb{R}^q) \in W^{1,p}(\Omega,\mathbb{R}^q).$$

Simplifying and shortening these formulas we use the notation

$$\langle A_h u, v^h\rangle_{H^{-1}(\mathcal{T}_c^h,\mathbb{R}^q)\times\mathcal{V}^h} := \langle A_h u, v^h\rangle_{H^{-1}(\mathcal{T}_c^h,\mathbb{R}^q)\times H^1(\mathcal{T}_c^h,\mathbb{R}^q)} \text{ for } v^h \in \mathcal{V}^h,\ldots \tag{5.293}$$

We generalize the sequences of projectors Q'^h in (4.117). For variational crimes we have chosen for the f used in (5.244) the general form (5.250). Their obvious extensions in (5.250) are defined for $v \in \mathcal{V}$ and simultaneously for $v^h \in \mathcal{V}_b^h$. Here we explicitely reformulate Propositions 4.49, 4.50 for the $\mathcal{V}_b$ in (5.285). We only need the f, f_h in (5.250). We define the generalized projectors Q'^h by testing as

$$Q'^h \in \mathcal{L}\left(W^{-1,p'}\left(\mathcal{T}_c^h\right), \mathcal{V}_b^{h'}\right) : \langle Q'^h f_h - f_h, v^h\rangle_{W^{-1,p'}(\mathcal{T}_c^h)\times\mathcal{V}^h} = 0 \forall v^h \in \mathcal{V}_b^h. \tag{5.294}$$

Nearly analogous to (4.117) we find here with (5.296)

$$Q'^h f_h - f_h \perp \mathcal{V}_b^h,$$

with

$$\begin{aligned}
\|Q'^h f_h\|_{\mathcal{V}^{h'}} &= \sup_{v^h\in\mathcal{V}_b^h,\|v^h\|_{\mathcal{V}^h}=1} \langle f_h, v^h\rangle_{W^{-1,p'}(\mathcal{T}_c^h)\times\mathcal{V}^h} \\
&= \|f_h\|_{\mathcal{V}^{h'}} = \|f\|_{\mathcal{V}'}(1+o(h)) \\
&\Longrightarrow \lim_{h\to 0}\|Q'^h f_h\|_{\mathcal{V}^{h'}} = \lim_{h\to 0}\|f_h\|_{\mathcal{V}'} = \|f\|_{\mathcal{V}'}, \\
f^h &:= Q'^h f_h|_{\mathcal{V}^h},\quad \lim_{h\to 0}\|Q'^h\|_{\mathcal{L}(\mathcal{V}',\mathcal{V}^{h'})} = \lim_{h\to 0}\|Q'^h\|_{\mathcal{L}(\mathcal{V}^{h'},\mathcal{V}^{h'})} = 1.
\end{aligned} \tag{5.295}$$

Proposition 5.55. *For the extension* $f_h \in \mathcal{V}' + W^{-1,p'}(\mathcal{T}_c^h)$ *in (5.250), of* $f \in \mathcal{V}' \subset W^{-1,p'}(\Omega)$ *with* $W^{1,p}(\mathcal{T}_c^h) \supset \mathcal{V}^h$, $f^h = f_h|_{\mathcal{V}^h}$, *and the projectors* $Q'^h \in \mathcal{L}(W^{-1,p'}(\mathcal{T}_c^h), \mathcal{V}^{h'})$ *in (5.294), the norms are asymptotically equal* $\forall\ 1 \le p' \le \infty, 1/p + 1/p' = 1$,

$$\|f^h\|_{\mathcal{V}^{h'}} = \|f_h\|_{W^{-1,p'}(\mathcal{T}_c^h)}(1+o(1)) = \|f\|_{\mathcal{V}'}(1+o(1)),\ f^h := Q'^h f_h|_{\mathcal{V}^h},$$

$$\lim_{h\to 0}\|Q'^h f_h\|_{\mathcal{V}^{h'}} = \|f\|_{\mathcal{V}'}, \lim_{h\to 0}\|Q'^h\|_{\mathcal{L}(\mathcal{V}',\mathcal{V}^{h'})} = \lim_{h\to 0}\|Q'^h\|_{\mathcal{L}(\mathcal{V}^{h'},\mathcal{V}^{h'})} = 1. \tag{5.296}$$

Particularly interesting are the $\langle f_h, v\rangle_{W^{-1,p'}(\mathcal{T}_c^h)\times\mathcal{V}^h}$ *with* $f_h = A_h u^h$.

$\mathcal{U}$	$\overset{G}{\longrightarrow}$	$\mathcal{U}'$	$\overset{\text{tested by}}{\longleftrightarrow}$	$\mathcal{U}$
$P^h\downarrow$	$\Phi^h\downarrow$	$Q'^h\downarrow$		
$\mathcal{U}^h$	$\overset{G^h}{\longrightarrow}$	$\mathcal{U}^{h'}$	$\overset{\text{tested by}}{\longleftrightarrow}$	$\mathcal{U}^h$ with $\not\subset \mathcal{U}_b$ or $\not\subset \mathcal{U}$

Figure 5.7 Nonconforming FEMs: Spaces and operators.

We summarize: with the projectors $Q^{'h}$ in (5.294), and the operators A_h and G_h, extended to $\mathcal{V} + \mathcal{V}_b^h$ in (5.254), (5.279), (5.283), (5.286), we determine the discrete (weak) solutions $u_0^h \in \mathcal{V}_b^h$ from the following equations,

$$\left\langle A_h u_0^h - f_h, v^h \right\rangle_{H^{-1}(\mathcal{T}_c^h)\times\mathcal{V}^h} = 0 \;\forall\; v^h \in \mathcal{V}_b^h \;\Leftrightarrow\; A^h u_0^h := Q^{'h} A_h u_0^h = f^h = Q^{'h} f_h,$$

$$\text{and } \left\langle G_h u_0^h, v^h \right\rangle_{W^{-1,p'}(\mathcal{T}_c^h)\times\mathcal{V}^h} = 0 \;\forall\; v^h \in \mathcal{V}_b^h \;\Leftrightarrow\; G^h u_0^h := Q^{'h} G_h u_0^h = 0. \quad (5.297)$$

Reformulating, the discrete operators A^h and G^h have the form

$$\Phi^h A := A^h := Q^{'h} A_h|_{\mathcal{V}_b^h} = Q^{'h} A^h, \;\Phi^h G := G^h := Q^{'h} G_h|_{\mathcal{U}_b^h},$$
$$\text{with the discrete solutions } A^h u_0^h - f^h = 0, \; G^h u_0^h = 0. \quad (5.298)$$

Again we sketch the different spaces and operators constituting the method in the *diagram* in Figure 5.7.

The necessary consistency errors for the nonconforming FEMs require, cf. Lemma 5.52, estimates for the following $a_h(u - u^h, v^h)$.

Proposition 5.56.

1. *For FEs with violated boundary conditions and/or continuity, choose an arbitrary $u \in H^2(\Omega), f_u := Au \in L^2(\Omega)$, e.g. the exact solution $u_0 \in H^2(\Omega), f_{u_0} := Au_0$, with the discrete solution u_0^h, cf. (5.279), (5.283), (5.286). Then we obtain for u and the discrete solution u^h of $A^h u^h = Q^{'h} Au$*

$$a_h(u - u^h, v^h) = \langle Q^{'h} A_h u - A^h u^h, v^h \rangle_{H^{-1}(\mathcal{T}_c^h)\times\mathcal{V}^h} = \langle A_h u - Au, v^h \rangle_{\dots} \quad (5.299)$$
$$= \sum_{e\in\mathcal{T}_c^h\setminus\partial\Omega} \int_e ([v^h], \{Bu\})_q ds + \sum_{e\in\partial\Omega} \int_e (v^h, (Bu))_q ds$$
$$\forall\; u \in H^2(\Omega, \mathbb{R}^q), \; v^h \in \mathcal{V}_b^h.$$

2. *For quasilinear (5.286), the $A, A_h, H_0^{-1}\left(\mathcal{T}_c^h, \mathbb{R}^q\right)$, have to be replaced by G, G_h, $W_0^{-1,p}\left(\mathcal{T}_c^h, \mathbb{R}^q\right), p \geq 2$.*

Proof. The proof of Proposition 5.50 remains correct. □

Remark 5.57.

A solution u_0 of $Au_0 = f$ with $u_0 \in \mathcal{V} = H^1(\Omega)$ implies $f \in H^{-1}(\Omega)$. Requiring $u_0 \in H^2(\Omega)$ implies $f \in L^2(\Omega)$, so $\langle f, v\rangle_{H^{-1}(\Omega)\times H^1(\Omega)} = (f, v)_{L^2(\Omega)} = \int_\Omega f v dx$.

5.5.4 Discrete coercivity and consistency

In Proposition 5.56 we have determined, not yet estimated, the variational errors for linear and quasilinear equations and systems of order 2. As in (5.285), we use the same symbols $u, u^h, \ldots \in H^1(\Omega), \ldots$ instead of $\in H^1(\Omega, \mathbb{R}^q)$ and, as coefficients, a_{ij} instead of A_{ij}, for real and vector valued functions. In fact, the same arguments are valid for all the linear problems in the previous Subsection. For quasilinear problems slight modifications are necessary, e.g. in Lemma 5.65.

The general convergence theory in Chapter 3 requires uniform discrete inf-sup conditions, equivalently the stability, of the discrete linear and linearized operators and estimates for the variational discretization errors. These estimates will be discussed in Subsections 5.5.6, 5.5.7. Towards stability, we do need for $a(\cdot,\cdot)$, $a^h(\cdot,\cdot)$ the $\mathcal{V}_b^h$ coercivity for Dirichlet or natural boundary conditions. We prove this, as a first step, for compactly perturbed principal parts, $a_p(\cdot,\cdot) + \epsilon'(\cdot,\cdot)_{L^2(\Omega)}$, of arbitrary elliptic bilinear (weak) forms. In Subsection 5.5.8 we will extend these results to the general case of a consistent compact perturbation $A = B + C, A^h = B^h + C^h$ with boundedly invertible A, B and stable B^h; e.g. the original operator, A, induced by $a(\cdot,\cdot)$ is a compact perturbation of the operator $B := A_p + \epsilon' I$ induced by $a_p(\cdot,\cdot) + \epsilon'(\cdot,\cdot)_{L^2(\Omega)}$. Thus the results of Subsection 5.5.8 show the stability of A^h.

For the FEs with crimes in this section the standard approach requires $\mathcal{U}_b^h = \mathcal{V}_b^h$.

$$
\begin{aligned}
&\text{For } \mathcal{U}_b = H_0^1(\Omega) \text{ or } = H^1(\Omega) \subset L^2(\Omega) = L^2(\Omega)' \subset \mathcal{U}_b' = H^{-1}(\Omega)\\
&\text{we assume } \forall\ u^h \in \mathcal{U}_b^h \neq \mathcal{V}_b^h \ \ \exists v^h \in \mathcal{V}_b^h \text{ s.t } \|u^h - v^h\|_{\mathcal{V}^h} < \beta\|u^h\|_{\mathcal{V}^h} \qquad (5.300)\\
&\textbf{or} \text{ vice versa } \forall\ v^h \in \mathcal{V}_b^h \neq \mathcal{U}_b^h \ \ \exists u^h \in \mathcal{U}_b^h \text{ s.t } \|u^h - v^h\|_{\mathcal{V}^h} < \beta\|v^h\|_{\mathcal{V}^h},\\
&\text{here } \beta \text{ is an arbitrarily small constant and } \dim \mathcal{U}_b^h = \dim \mathcal{V}_b^h, \text{ with the}\\
&\text{same norms in } \mathcal{U}^h, \mathcal{V}^h, \text{ e.g. } \|u^h\|_{\mathcal{U}^h} = \|u^h\|_{\mathcal{V}^h} = \|u^h\|_{H^1(\mathcal{T}_c^h)}.
\end{aligned}
$$

Remark 5.58. *Beyond the usual case $\mathcal{U}_b^h = \mathcal{V}_b^h$, this condition is satisfied, if $\mathcal{U}_b^h$ and $\mathcal{V}_b^h$ are defined via the same interpolation or transition points along the interior and boundary edges $e \in \mathcal{T}_c^h$. Different degrees in $\mathcal{U}_b^h$ and $\mathcal{V}_b^h$ are still allowed. The nonlocal smooth approximations in Subsection 4.2.6 have this property as well, unless the application to fully nonlinear problems requires the splitting there.*

A similar condition will be satisfied for the general case of Gelfand triples: let $\mathcal{U}_b, \mathcal{U}, \mathcal{X}$ be Hilbert spaces and let $\mathcal{X} = \mathcal{X}'$ be identified such that $\mathcal{U}_b$ is dense in $\mathcal{X}$ and $I : \mathcal{U}_b \hookrightarrow \mathcal{X}$, $Iv := v \in \mathcal{X}$ for all $v \in \mathcal{U}_b$ is continuous. Then (5.300) can be generalized as $\mathcal{U}_b \subset \mathcal{X} = \mathcal{X}' \subset \mathcal{U}_b'$ and $\mathcal{U}_b \subset \mathcal{U}$ is densely and continuously embedded.

The convergence theory for mesh free methods in [120] will allow to generalize this (5.300) for FEMs as well.

Obviously the linear and bilinear forms studied in the last subsection are *uniformly continuous,* such that there exists a constant, C, independent of h, with

$$
\begin{aligned}
&\|v\|_{\mathcal{V}} = \|v\|_{H^1(\Omega)}, \|u^h\|_{\mathcal{V}^h} = \|u^h\|_{H^1(\mathcal{T}_c^h)}, \cdots\\
&|a_h(u^h, v^h)| \le C\|u^h\|_{\mathcal{V}^h} \cdot \|v^h\|_{\mathcal{V}^h} \ \forall\ u^h, v^h \in \mathcal{V} + \mathcal{V}^h, \qquad (5.301)\\
&|\langle f_h, v^h\rangle| \le C\|f_h\|_{\mathcal{V}^{h'}}\|v^h\|_{\mathcal{V}^h} \ \forall\ v^h \in \mathcal{V} + \mathcal{V}_b^h, f_h \in \mathcal{V}' + \mathcal{V}^{h'}.
\end{aligned}
$$

The following theorem applies to violated boundary conditions and continuity as discussed in Subsections 5.5.6, 5.5.7 and with the necessary machinery in Subsection 5.5.10 and Section 5.4 as well. For the general case beyond $a_p(\cdot,\cdot)+\epsilon'(\cdot,\cdot)_{L^2(\Omega)}$ the stability proof is delayed to Subsection 5.5.8.

Theorem 5.59. Stability for modified principal parts:

1. *Let this implies* A, $a(\cdot,\cdot)$, A_h, $a_h(\cdot,\cdot)$ *and* Dirichlet *and/or* natural boundary conditions *be given as in (5.278)–(5.280), (5.284) and* $\epsilon'>0$. *Let* $a_p(\cdot,\cdot)$, $a_{p,h}(\cdot,\cdot)$ *be the principal parts. Then the* $a_{p,h}(\cdot,\cdot)+\epsilon'(\cdot,\cdot)_{L^2(\Omega)}$ *is bounded, say by* $M=C$ *in (5.301), and* $\mathcal{V}_b+\mathcal{V}_b^h$-coercive, *hence a constant* $\alpha>0$ *exists such that*
$$a_{p,h}(u^h,u^h)+\epsilon'(u^h,u^h)_{L^2(\Omega)} \ge \alpha(\|u^h\|_{\mathcal{V}^h})^2 \ \forall u^h \in\ \mathcal{V}_b+\mathcal{V}_b^h. \tag{5.302}$$
2. *If for* $\mathcal{U}_b^h\neq\mathcal{V}_b^h$, *(5.300) is satisfied with* $\beta M/\epsilon'<1$, *then the uniform discrete inf-sup conditions (2.52), (2.53) are valid,*
$$\sup_{0\neq v^h\in\mathcal{V}_b+\mathcal{V}_b^h} |a_{p,h}(u^h,v^h)+\epsilon'(u^h,u^h)_{L^2(\Omega)}|/\|v^h\|_{\mathcal{V}^h}\ge\epsilon\|u^h\|_{\mathcal{U}^h}\forall u^h\in\mathcal{U}_b+\mathcal{U}_b^h,$$
3. *In particular, for Dirichlet boundary conditions* A_p *and* A_p^h *are boundedly and equiboundedly invertible, hence* A_p^h *is stable.*

Note that the inf–sup condition for $a(u,v)$ does not include the discrete inf–sup condition. This is possible under the conditions of compact perturbation results, e.g. in Theorem 5.78. Theorem 5.59 implies the unique existence of the exact solution u_0 and of the discrete solutions u_0^h for the special case of $a_p(\cdot,\cdot)+\epsilon'(\cdot,\cdot)_{L^2(\Omega)}$, if the consistency results in Subsections 5.5.6 and 5.5.7 are valid. For isoparametric FEs for curved boundaries, see Subsection 5.5.10, this result is available in Theorem 5.84.

Proof. The principal part of the above *linear elliptic operators or bilinear forms,*
$$a_p(u,v):=\int_\Omega\left(\sum_{i,j=1}^{2} a_{ij}\partial^i\ u\ \partial^j\ v\right)dx$$
satisfies, cf. (4.131) ff.,
$$\sum_{i,j=1}^{2} a_{ij}\xi_i\xi_j+\epsilon'\eta^2\ge\epsilon'\left(|\xi|_2^2+\eta^2\right)\ ,\epsilon'>0,$$
with $\xi_j,\eta\in\mathbb{R}$, and $|\xi|_2$ the Euclidean norm in $\mathbb{R}^2$. We have chosen twice the same $\epsilon=\epsilon'$ as in (4.131). Now let $\xi_j:=\partial_j u^h(x),\eta:=u^h(x)$ and obtain
$$\sum_{i,j=1}^{2} a_{ij}\partial_i u^h\partial_j u^h(x)+\epsilon'(u^h(x))^2\ge\epsilon'\left(|\nabla u^h(x)|_2^2+(u^h(x))^2\right)$$
and by integration

$$a_{p,h}(u^h,u^h)+\epsilon'(u^h,u^h)_{L^2(\Omega)} = \sum_{T\in\mathcal{T}_c^h}\int_T\left(\sum_{i,j=1}^{2} a_{ij}\partial_i u^h\partial_j u^h+\epsilon'(u^h)^2\right)dx$$
$$\geq \sum_{T\in\mathcal{T}_c^h}\int_T \epsilon'(|\nabla u^h|_2^2+(u^h)^2\,)dx \geq \epsilon'(\|u^h\|_{\mathcal{V}^h})^2$$
$$\forall u^h \in \mathcal{V}_b+\mathcal{V}_b^h.$$

For the inf-sup conditions we need (5.300). Let $\mathcal{U}_b^h \not\subset \mathcal{U}_b, \mathcal{V}_b^h \not\subset \mathcal{V}_b, \mathcal{U}_b^h \neq \mathcal{V}_b^h$. By (5.300) we know that for any $u^h \in \mathcal{V}^h$ we may choose a $u^h \in \mathcal{U}^h$ with $\|u^h - v^h\|_{\mathcal{V}^h} < \beta\|u^h\|_{\mathcal{V}^h}$. with the notation $b^h(u^h,u^h) := a_p^h(u^h,u^h)+e'(u^h,v^h)_{L^2(\Omega)}$ we find

$$\|b^h(u^h,v^h)-b^h(u^h,u^h)\| = \|b^h(u^h,u^h-v^h)\| \leq M(\|u^h\|_{\mathcal{V}^h}\|u^h-v^h\|_{\mathcal{V}^h})$$

since $b^h(\cdot,\cdot)$ is continuous. With $\|b^h(u^h,u^h)\| \geq \epsilon'\|u^h\|^2_{H^1(\Omega)}$ this implies

$$\|b^h(u^h,v^h)\| \geq \|b^h(u^h,u^h)\| - \|b^h(u^h,v^h)-b^h(u^h,u^h)\|$$
$$\geq \epsilon'(\|u^h\|_{\mathcal{V}^h})^2 - M(\|u^h\|_{\mathcal{V}^h}\|u^h-v^h\|_{\mathcal{V}^h})$$
$$\geq \epsilon'\|u^h\|_{\mathcal{V}^h}\left[\|u^h\|_{\mathcal{V}^h} - (M/\epsilon')\beta\|u^h\|_{\mathcal{V}^h}\right]$$

hence, going back to v^h and for $(M/\epsilon')\beta < 1$ the inf-sup condition is proved. For systems, with (4.131) replaced by (4.138), the proof remains nearly unchanged. □

As the next step, we prepare estimating the *variational consistency errors* as in Subsections 5.5.6 and 5.5.7. In contrast to *classical consistency errors*, we always have to assume here that the discrete solution u_0^h exists. So, we assume the existence of the exact and approximate solutions u_0 and u_0^h of

$$\exists u_0 : a(u_0,v) = (f,v)\ \forall v\in\mathcal{V}_b \text{ or } \exists u_0^h : a^h\left(u_0^h,v^h\right) = (f,v^h)^h\ \forall v^h\in\mathcal{V}_b^h. \tag{5.303}$$

For nonconforming FEs we obtain as a special case of Proposition 5.56 the following results for the *variational consistency error*.

Variational consistency error for violated boundary conditions: (5.304)

$$a_h\left(u_0-u_0^h,v^h\right) = \int_{\partial\Omega} v^h B_a u_0 ds \text{ for } u_0\in H^2(\Omega),\ \forall\ v^h\in\mathcal{V}_b^h\subset H^1(\Omega).$$

Variational consistency error for violated continuity, see (5.258): (5.305)

$$a_h\left(u_0-u_0^h,v^h\right) = \sum_{e\in\mathcal{T}_c^h}\int_e [v^h]B_a u_0 ds \text{ for } u_0\in H^2(\Omega), \forall v^h\in\mathcal{V}_b^h\not\subset H^1(\Omega).$$

Variational consistency error for general nonconforming FEs

$$a_h\left(u_0-u_0^h,v^h\right) = \sum_{e\in\mathcal{T}_c^h}\int_e [v^h]B_a u_0 ds + \int_{e\in\partial\Omega} v^h B_a u_0 ds \tag{5.306}$$

for $u_0\in H^2(\Omega),\ \forall\ v^h\in\mathcal{V}_b^h\not\subset\mathcal{V}_b, \mathcal{V}_b^h\not\subset H^1(\Omega)$.

In Subsections 5.5.6 ff. we will aim for the classical consistency errors as well. So we generalize (5.304), (5.305), (5.306) from u_0, u_0^h to general u and its corresponding discrete counterpart u^h, defined by $A^h u^h = Q'^h Au$, cf. (5.297), (5.298). Note that the classical consistency error is independent of the existence of a (unique) discrete solution, in contrast to the variational consistency error.

We require, beyond Condition 4.16, specific approximating spaces:

Condition 5.60. Conditions for nonconforming FEs: *Choose piecewise polynomial FEs locally in $\mathcal{P} = \mathcal{P}^2_{d-1}$ in $\mathbb{R}^2$. Recall that $\mathcal{P}^{d-1} = \mathcal{P}^1_{d-1}$ and $\mathcal{P}_{d-1} = \mathcal{P}^2_{d-1}$ denote the uni-and bi-variate polynomials of degree $d-1$, respectively Every $T \in \mathcal{T}_c^h$ at the boundary has at most one curved side, see Figures 4.21–4.23 above. Let $\partial\Omega$ be piecewise smooth. Nonsmooth points of $\partial\Omega$ are used as vertices of subtriangles, $T \in \mathcal{T}_c^h$, compare Condition 5.64. Let the subdivision be quasiuniform, or generalizd nondegenerate, see Definition 4.12. This condition of quasiuniformity can be relaxed in $\mathbb{R}^2$ by applying the results in Subsection 4.2.5. Extensions to $\mathbb{R}^n, n \geq 3$, are possible using results as in Cools [208–210]. We only discuss $\mathbb{R}^2$.*

5.5.5 High order quadrature on edges

These are an important tool for estimates of all types of nonconformity errors. The main idea is very simple: the variational errors $a_h\left(u_0 - u_0^h, v^h\right)$, listed in (5.304)–(5.306), are small if the $[v^h]$ disappear often enough along interior edges e and v^h along the boundary $\partial\Omega$. Then in $\mathbb{R}^2$ the Gauss-type approximations along edges e and the boundary $\partial\Omega$ vanish for the right-hand sides of (5.304)–(5.306). Consequently the corresponding right-hand side integrals, and hence the variational errors $a_h\left(u_0 - u_0^h, v^h\right)$, are small. Similarly to the techniques in Section 5.4, this is shown with inverse estimates, costing some powers of h. There quadrature is applied to the bilinear or higher nonlinear forms, compare (5.179) and (5.306), (5.330), on the $T \in \mathcal{T}_c^h$ in $\mathbb{R}^2$. Here we apply quadrature to boundary forms on edges of the triangles. Since the degree of freedom for a polynomial increases with the dimension n, cf. (4.9), this concept works well for Gauss-type formulas on edges and is still not too bad for high order cubature on triangles, but seldom for $\mathbb{R}^n, n \geq 4$. We often have to distinguish interpolation and quadrature errors and points.

Univariate Gauss–type quadrature formulas

For one-dimensional intervals or edges, we use three types of Gauss formulas, see Lemma 5.61 below. They are distinguished by the requirement that either none, one or two boundary points of the interval are included in the corresponding quadrature grid points. They are defined for Legendre and Chebyshev polynomials with respect to the weight functions $w \equiv 1$ and $w = (1-x^2)^{\frac{1}{2}}$, respectively, on the interval $[-1,1]$. We introduce three types of univariate polynomials, $Q^{\rho}_{d'} \in \mathcal{P}^{d'} := \mathcal{P}^1_{d'}$, $\rho = 0,1,2$, based on the Legendre and Chebyshev polynomials $P_{d'} \in \mathcal{P}^{d'}$ and $T_{d'} \in \mathcal{P}^{d'}$, respectively, defining quadrature formulas of the type of

$$
\begin{aligned}
&\text{Gauss:} && \rho = 0 : Q^0_{d'} := P_{d'} \text{ or } Q^0_{d'} := T_{d'} \text{ and for both cases}\\
&\text{Gauss–Radau:} && \rho = 1 : Q^1_{d'} := Q^0_{d'} + aQ^0_{d'-1}, \text{with } a\\
& && \qquad\qquad \text{such that } Q^1_{d'}(-1) = 0,\\
&\text{Gauss–Lobatto:} && \rho = 2 : Q^2_{d'} := Q^0_{d'} + aQ^0_{d'-1} + bQ^0_{d'-2}, \text{ with } a, b\\
& && \qquad\qquad \text{such that } Q^2_{d'}(\pm 1) = 0.
\end{aligned}
\tag{5.307}
$$

This implies $d' \geq \max\{1, \rho\}$. The d' quadrature points $y^\rho_j := y_j \in [-1, 1]$, $j = 0, \ldots, d'-1$ for the quadrature formulas are defined as the roots of the polynomials

$$
Q^\rho_{d'}\left(y^\rho_j\right) = 0 \text{ for } j \in \{0, \ldots, d'-1\}, \quad \rho = 0, 1, 2, \quad \max\{1, \rho\} \leq d', \tag{5.308}
$$

with $y^\rho_0 = -1$ for $\rho = 1, 2$ and $y^\rho_{d'-1} = 1$ for $\rho = 2$. The corresponding weights are then computed via the Lagrange elementary polynomials, p^ρ_j, as

$$
p^\rho_j \in \mathcal{P}^{d'} : p^\rho_j\left(y^\rho_l\right) = \delta_{j,l} \ \forall\, 0 \ \leq \ j, l \ \leq d' - 1 \text{ as}
$$

$$
w_j := w^\rho_j := \int_{-1}^{1} p^\rho_j(x) w(x) dx, \ \rho = 0, 1, 2, \ d' \ \geq \ \max\{1, \rho\},
$$

with the above weight function, w. The following proposition is proved in many textbooks, e.g. [158].

Proposition 5.61.

1. *Let $-1 \leq y_0 < \cdots < y_{d'-1} \leq +1$ be the above roots of the $Q^\rho_{d'}$, for $\rho = 0, 1, 2$, and $w_j := w^\rho_j$ the corresponding weights. Then the quadrature approximation $q^{d'}_w(u)$ is defined as*

$$
q^{d'}_w(u) := \sum_{j=0}^{d'-1} u\left(y^\rho_j\right) w^\rho_j \approx \int_{-1}^{1} u(x) w(x) dx \text{ and } q^{d'}(u) := q^{d'}_{w\equiv 1}(u),
$$

for $u \in C[-1, 1]$, $\rho = 0, 1, 2$, $\max\{1, \rho\} \leq d' \in \mathbb{N}$, and satisfies

$$
q^{d'}_w(p) = \sum_{j=0}^{d'-1} p\left(y^\rho_j\right) w^\rho_j = \int_{-1}^{1} p(x) w(x) dx, \text{ for all } p \in \mathcal{P}^{2d'-\rho-1}. \tag{5.309}
$$

2. *This implies for scalar products, $(u, v)_w$, with respect to the weight function w that*

$$
(u, v)_w = (u, v)^{d'}_w := q^{d'}_w(u \cdot v) = \sum_{j=0}^{d'-1} (u \cdot v)\left(y^\rho_j\right) w^\rho_j \ \forall u \in \mathcal{P}^{d'-\rho}, \ v \in \mathcal{P}^{d'-1}. \tag{5.310}
$$

3. *So $(u, v)^{d'}_w = q^{d'}_w(u \cdot v)$ exactly reproduces scalar products for $u, v \in \mathcal{P}^{d'-1}$ for $\rho = 0, 1$. For $\rho = 2$ we have to require at least one of the u or $v \in \mathcal{P}^{d'-2}$.*

The next proposition is applied for proving Theorems 5.66, ff.

Proposition 5.62. *For all* $d' \in \mathbb{N}$ *with* $\max\{1,\rho\} \leq d'$, *hence* $2d' - \rho \geq d' \geq 1$, *there exists* $C = C_{d'}$ *such that* $\forall 0 < h < 1, 1 \leq q \leq \infty$, $f \in W^{2d'-\rho,q}[0,h] \cap C[0,h]$,

$$\left|\int_0^h f(x)dx - h/2\sum_{j=0}^{d'-1} w_j f(h\xi_j)\right| \leq C_{d'}\ h^{2d'-\rho+(1-1/q)}|f^{(2d'-\rho)}|_{L^q(0,h)}. \quad (5.311)$$

With $\rho = 0,1,2$, *for Gauss, Gauss–Radau and Gauss–Lobatto quadrature, respectively, the points* $h\xi_j := h\xi_j^\rho := h\left(y_j^\rho + 1\right)/2$ *and* y_j^ρ *are zeros of the Legendre polynomials* $Q_{d'}^\rho$ *of degree* d', *see (5.308).*

Proof. We use, for any $T \subset \mathbb{R}^2$, the *Hölder inequality*, see (1.45),

$$\left|\int_T f(x)g(x)dx\right| \leq \left(\left|\int_T |f(x)|^p dx\right|\right)^{1/p}\left(\left|\int_T |g(x)|^q dx\right|\right)^{1/q} \quad (5.312)$$

for all $1 \leq p,q \leq \infty$ with $1/p + 1/q = 1$ and $f \in L^p(T), g \in L^q(T)$.
Now we add and subtract the Taylor polynomial $Q^{2d'-\rho}f$ for f in (5.311). Then (5.309) implies with $f \in W^{2d'-\rho,q}(0,h)$, and the Bramble–Hilbert–Lemma 4.15

$$\begin{aligned}\left|\int_0^h f(x)dx - h/2\sum_{j=0}^{d'-1} w_j f(h\xi_j)\right| & \\ \leq \left|\int_0^h f(x) - Q^{2d'-\rho}f + Q^{2d'-\rho}f dx - h/2\sum_{j=0}^{d'-1} w_j f(h\xi_j)\right| & \\ \leq \int_0^h 1\cdot|f(x) - Q^{2d'-\rho}f|dx \leq C_{d'}\ h^{1-1/q}|f - Q^{2d'-\rho}f|_{L^q(0,h)} & \\ \leq C_{d'}\ h^{2d'-\rho+1-1/q}|f^{(2d'-\rho)}|_{L^q(0,h)}. &\end{aligned}$$

□

5.5.6 Violated boundary conditions

In this Subsection we estimate the consisitency errors of continuous FEs violating Dirichlet boundary conditions. Natural boundary value problems without prescribed boundary conditions do not play any role in this context. We do need here the *classical and variational consistency error.* With the discrete solution u^h of $A^h u^h = Q'^h A_h u^h = Q'^h f_u = Q'^h Au$ they are introduced in Definition 3.14 as $Q'^h Au - A^h P^h u$ and $Q'^h A_h u - A^h u^h = Q'^h A_h u - Q'^h A^h u^h$, respectively (Note that for conforming FEMs with $A_h = A$ the variational consistency error, $A^h u^h - Q'^h Au$ vanishes!) By

Chapter 3, (3.33), they are, for a linear operator, A, related as

$$\underbrace{Q'^h A_h u - A^h u^h}_{\text{var.cons.error}} = Q'^h A_h \underbrace{(u - P^h u)}_{\text{interp.error}} + \underbrace{A^h P^h u - Q'^h Au}_{-\text{class.cons.error}} \text{ with } A^h u^h = Q'^h Au \tag{5.313}$$

For the first term, the variational consistency error, tested with v^h,

$$\langle Q'^h A_h u - A^h u^h, v^h\rangle_{\mathcal{V}^{h'}\times\mathcal{V}^h} = \langle A_h u - A_h u^h, v^h\rangle_{\mathcal{V}^{h'}\times\mathcal{V}^h} = a_h(u - u^h, v^h),$$

we have an explicit form in (5.299). With the interpolation operator, $P^h = I^h$, the middle term $Q'^h A_h(u - P^h u) = Q'^h A_h(u - I^h u)$ can be estimated by the interpolation error. So the *classical consistency error* can be estimated as

$$\begin{aligned}\|A^h P^h u - Q'^h Au\|_{\mathcal{V}^{h'}} &\le \|Q'^h A_h u - A^h u^h\|_{\mathcal{V}^{h'}} + \|Q'^h A_h(u - P^h u)\|_{\mathcal{V}^{h'}} \qquad (5.314)\\ &\le (1 + o(h))\|A\|_{\mathcal{V}'\hookrightarrow\mathcal{V}}\|P^h u - u\|_{\mathcal{V}} + \|Q'^h A_h u - A^h u^h\|_{\mathcal{V}^{h'}}.\end{aligned}$$

This (5.314) has to be slightly modified for quasilinear operators. Again, we introduce $f_u := Gu$ and u^h, the discrete solution of $G^h u^h = Q'^h f_u = Q'^h Gu$. From the possibly many solutions u^h for the nonlinear G^h we choose that with small $\|u^h - u\|$. We find with $G^h = Q'^h G_h|_{\mathcal{V}_b^h}$ in (5.287),

$$Q'^h G_h u - G^h u^h = Q'^h (G_h u - G_h P^h u) + (G^h P^h u - Q'^h Gu). \tag{5.315}$$

So we have to estimate these variational consistency errors, cf. (5.299),

$$\begin{aligned}&a_h(u - u^h, v^h) = \langle A_h u - A^h u^h, v^h\rangle_{\mathcal{V}^{h'}\times\mathcal{V}^h}\ \forall v^h \in \mathcal{V}_b^h, \text{ with } A^h u^h = Q'^h Au \text{ and}\\ &\text{for nonlinear } G : \langle G_h u - G^h u^h, v^h\rangle_{\mathcal{V}^{h'}\times\mathcal{V}^h}\ \forall v^h \in \mathcal{V}_b^h, \text{ for } G^h u^h = Q'^h Gu. \qquad (5.316)\end{aligned}$$

Explicit formulas for linear and quasilinear problems are listed in Proposition 5.56 and the G, G^h modification of (5.299). We will update it for violated boundary conditions.

Proposition 5.63. *Let Q'^h and A^h, G^h be defined in (5.294)–(5.297), and choose an arbitrary $u \in \mathcal{V}_b \cap H^2(\Omega)$ and the corresponding u^h from $A^h u^h = Q'^h Au$ or $G^h u^h = Q'^h Gu$. Then the violated boundary conditions are reflected in the* variational consistency error *$\forall\, v^h \in \mathcal{V}_b^h$ as, cf. Proposition 5.56,*

$$\langle Q'^h A_h u - A^h u^h, v^h\rangle_{\mathcal{V}^{h'}\times\mathcal{V}^h} \text{ or } \langle Q'^h G_h u - G^h u^h, v^h\rangle_{\mathcal{V}^{h'}\times\mathcal{V}^h} = \int_{\partial\Omega} v^h B u ds. \tag{5.317}$$

We start with linear problems and B_a, for nonlinear problems with $B = B_G$ cf. Theorem 5.68. According to (5.317), an estimate for the consistency error for $v^h \in \mathcal{V}_b^h \not\subset \mathcal{V}_b$, hence $v^h|_{\partial\Omega} \not\equiv 0$, depends on how well v^h approximates $v^h|_{\partial\Omega} = 0$ and how this implies small $|\int_{\partial\Omega} v^h B_a\ u_0 ds|$. The following steps involve quadrature errors and inverse estimates. First, we use Proposition 5.61 estimating $|\int_{\partial\Omega} f ds - \sum_{P_j\in\partial\Omega} w_{P_j} f(P_j)|$ for $f \in W^{d',\infty}(\Omega)$, cf. (5.308) ff. Here, $P_j := P_j^\rho \in \partial\Omega$ indicate the quadrature points along one or all edges on $\partial\Omega$, or on a subsection of $\partial\Omega$ or its approximation. The w_j indicate the corresponding quadrature weights. For

FEMs we only consider Legendre polynomials with $w \equiv 1$. The Gauss-type points are used for nonconforming FEs to specify $v^h \in \mathcal{V}_b^h$. We choose for the boundary Ω a parametrization

$$\begin{aligned}\partial\Omega = \{x_b(s) : s \in [0, s_b], s = \text{ arc length}, \; s_b = \text{ arc length of } \partial\Omega, \\ x_b = \text{ linear on straight parts of } \partial\Omega\}.\end{aligned} \tag{5.318}$$

We delay the isoparametric case to Subsection 5.5.10. There we will impose Dirichlet boundary conditions in d' points close to $\partial\Omega$. Now we proceed as in Subsubsection 4.2.7. For each edge e with endpoints $P_i, P_e \in e \cap \partial\Omega$ such that $P_i = x_b(s_e), P_e = x_b(s_e + h_e)$ we define a parametrization

$$e = \{x_e(s) : s \in [s_e, s_e + h_e]\}; \quad x_e := x_b|_{[s_e, s_e+h_e]}. \tag{5.319}$$

Any derivatives of this global and local parametrization in (5.318) and (5.319), respectively, are determined solely by $\partial\Omega$ and can be bounded independently of h. With the ξ_j in (5.311) we define the additional boundary nodes on the (curved) boundary, cf. (4.98),

$$\begin{aligned} e \ni P_j^{\rho,e} &:= P_j^e := x_e(s_e + h_e\xi_j) \in e, j = 0, 1, \ldots, d' - 1 \\ h\xi_j &:= h\left(y_j^2 + 1\right)/2 \;\; \forall \;\; e \in \mathcal{T}_c^h \text{ with } |\{\bar{e} \cap \partial\Omega\}| = 2 \\ &\text{often with Gauss–Lobatto quadrature } \rho = 2 \text{ and } P_i = P_0^e, P_e = P_{d'-1}^e, \end{aligned} \tag{5.320}$$

cf. Proposition 5.61. This allows defining the $\mathcal{V}_b^h \subset \mathcal{V}^h$, cf. (5.241), as

$$\mathcal{V}_b^h := \left\{v : \Omega \to \mathbb{R} : v \in \mathcal{V}^h, \forall e \in \mathcal{T}_c^h, v\left(P_j^e\right) = 0 \forall P_j^e \text{ in (5.320)}\right\}, \Omega = \cup_{T \in \mathcal{T}_c^h} T. \tag{5.321}$$

We apply (5.311) to the special case of $f = (v^h B_a u) \circ x_b$ with the above parametrization x_b for $\partial\Omega$. We use (1.67) for $p = \infty$, and $v^h B_a u \in W^{k+1,\infty}(\Omega)$, (5.278) and the product and chain rule estimating

$$\begin{aligned} \|((v^h B_a\, u) \circ x_b)^{(k)}\|_{L^\infty(\partial\Omega)} \le C_{x_b} \|u\|_{W^{k+1,\infty}(\Omega)} \|v^h\|_{W^{\min\{k,d-1\},\infty}(\Omega)} \\ \forall u \in W_\infty^{k+2}(\Omega), \; v^h \in \mathcal{V}^h; \end{aligned} \tag{5.322}$$

this C_{x_b} depends on powers of order $\le k$ of different derivatives $x_b^{(j)}, j = 0, \ldots, k$, for the above parametrization. Finally, inverse estimates yield for (5.311) and Condition 5.60, cf. e.g. Theorems 4.17, 4.19 and Proposition 4.18 with $n = 2$,

$$\begin{aligned} \|v^h\|_{W^{j,p}(\Omega)} \le C \;\; h^{l-j-(2/q-2/p)} \|v^h\|_{W^{l,q}(\Omega)} \text{ for FEs and} \\ 0 \le l \le j \le d-1, 1 \le q \le p \le \infty \;\; \forall v^h \in \mathcal{V}^h. \end{aligned} \tag{5.323}$$

We get estimates for the interpolation errors with Condition 5.60 and $v \in W^{d,p}(\Omega)$:

$$\begin{aligned} \|v - I^h v\|_{W^{s,p}(\Omega)} \le C \; h^{d-s} |v|_{W^{d,p}(\Omega)} \text{ and} \\ \|v - I^h v\|_{W^{s,p}(T)} \le C \; (\text{diam } T)^{d-s} |v|_{W^{d,p}(T)} \text{ for } 0 \le s \le d. \end{aligned} \tag{5.324}$$

We have to consider the consequence of (5.320), (5.321) for imposing Dirichlet boundary conditions in d' points on the edges, e. In $\mathbb{R}^2$ we require

$$\text{for } \rho = 0, 1, 2 \text{ choose } \max\{1, \rho\} \leq d' \text{ hence } 2d' - \rho \geq d' \geq 1. \tag{5.325}$$

This would still allow many different orders of convergence in the following results and rather complicated case studies yielding only few really interesting results. For avoiding these technicalities, we choose the other way round. We exclude all nonoptimal cases, and prove the results only for the optimal cases. Exotic cases violate $d - 1 \leq 2d' - \rho$. Furthermore, as already indicated in the introduction 5.5.1, we want to maintain the order of convergence, $d - 1$, of the original FEM in its violated form as well. For $d - 1 \leq 2d' - \rho$ a detailed discussion would yield as order of convergence, with $2d' - \rho - (d - 1) \geq 0$, the $\min\{2d' - \rho - (d - 1) + 1/2, d - 1\}$ aimed to be $\geq d - 1$. So we assume

$$d - 1 \leq 2d' - \rho \text{ and } \min\{2d' - \rho - (d - 1) + 1/2, d - 1\} \geq d - 1. \tag{5.326}$$

We distinguish two cases. For polygonal boundaries the trivial boundary conditions can be satisfied exactly. This is no longer correct in combination with discontinuous FEs, see Subsection 5.5.7. So we include polygonal boundaries here, too, mainly with $\rho = 0$. For curved boundaries, Dirichlet boundary conditions usually cannot be satisfied exactly by polynomial FEs.

Polygonal $\partial\Omega$ violating boundary conditions, mainly for discontinuous FEs:

The piecewise linear x_b, see (5.318), and $v^h|_T \in \mathcal{P}_{d-1}$ yield $(v^h \circ x_b)|_e \in \mathcal{P}_{d-1}$. Now $d' > d - 1$ boundary points with $v^h(P_j) = v^h(x_b(s_e + h_e\xi_j)) = 0, j = 0, \ldots, d' - 1$ imply $(v^h \circ x_b)_e \equiv 0$, hence the Dirichlet boundary conditions are satisfied exactly. So violated Dirichlet boundary conditions imply $d' \leq d - 1$. Condition 4.16, 6. assumes the same number of functionals on every edge by requiring affinely equivalent $(K, \mathcal{P}, \mathcal{N})$ and $(T, \mathcal{P}_T, \mathcal{N}_T)$ $\forall T \in \mathcal{T}_c^h$. In this sense, $\max\{1, \rho\} \leq d' \leq d - 1$ is even equivalent to violated continuity conditions, see Subsection 5.5.7. The second part of (5.326) yields, for $\rho = 2$, $d' \geq d$ and, for $\rho = 0, 1$, $d' \geq d - 1$. With the above $d' \leq d - 1$, this excludes $\rho = 2$ and allows $\rho = 0, 1$ with $d' = d - 1$ and $\max\{1, \rho\} = 1$. So we require for polygonal and allow curved $\partial\Omega$ also, so

$$\text{for polygonal and general } \partial\Omega : \rho = 0, 1, \text{ and } 1 \leq d' = d - 1. \tag{5.327}$$

Curved $\partial\Omega$ with violated boundary conditions:

In this situation a $d' > d + \tau$ and $v^h \circ x_b|_e \notin \mathcal{P}^{d+\tau}$ does not imply $v^h \circ x_b|_e \equiv 0$ as above. For τ cf. Condition 4.16. Thus, violated Dirichlet boundary conditions usually do not imply violated continuity. However, if we assume the same number of interpolation points on each e, including curved boundary edges as well, the above arguments for a polygonal boundary apply again and yield (5.327), mainly with $\rho = 0$. Another possibility, combining curved boundaries and violated continuity via $\rho = 0, 1, 2$ is discussed in the introduction of Subsection 5.5.7. Finally, (5.326) implies, for $\rho = 2$, $d' \geq d$. Since by (5.326) a $d' > d$ would not increase the convergence rate, the only interesting case is the

$$\text{standard case for curved } \partial\Omega: \ \rho = 2 \le d' = d, \mathcal{P}_{d-1} = \mathcal{P}. \tag{5.328}$$

We summarize:

Condition 5.64. Conditions for boundary and boundary quadrature points: *Let $\partial\Omega$ be a piecewise smooth, e.g. a curved boundary and introduce the points $P_j = P_j^e$ along $\partial\Omega$ as in Subsubsection 4.2.7, see (4.92) and (5.311), (5.318)–(5.320). Let the segments $e \subset \overline{\partial\Omega}$, see (5.318),(5.319), be defined by $P_i = P_j$, $P_e = P_{j+1}$. Choose a quadrature approximation according to Propositions 5.61 and 5.62, along the above closed (curved) segments e of the piecewise smooth $\partial\Omega \subset \mathbb{R}^2$. Impose Dirichlet boundary conditions in d' points $P_i^{\rho,e} \in \partial\Omega$, $i = 0,\ldots,d'-1$ on each edge, e. We require $\rho = 0, 1 \le d' = d-1$ and $\rho = 2 \le d' = d$ only for curved $\partial\Omega$, see Proposition 5.62 and (5.320),(5.321), (5.327), (5.328). In practical computations only $\rho = 0, 2$ are used.*

The $P_i^{\rho,e}$, cf. (5.319), (5.320), are defined by the Gauss, Gauss–Radau and Gauss–Lobatto points according to the chosen quadrature approximation $\rho = 0, 1, 2$ along the segment e, parametrized by x_e with respect to the arc length, see (5.319).

For the proof of Theorem 5.66 we need Lemma 5.65, discussing the estimates for violated boundary values along a single edge e. The following notation allows a simultaneous formulation for the two cases (5.327), (5.328).

$$(\nu_0, \nu^0) := \begin{cases} (d-\rho, 2d-\rho-1) & \text{for (5.327): } \rho = 0, 1 \le d' = d-1, \\ (d, 2d-1) & \text{for (5.328): } \rho = 2 \le d' = d. \end{cases} \tag{5.329}$$

Lemma 5.65. *Under Conditions 5.60 and 5.64 we choose a linear or a quasilinear differential operator, A or G, and the corresponding boundary operator $B = B_a$ or $B = B_G$, satisfying (5.335), homogeneous Dirichlet boundary conditions as in (5.278) and an arbitrary $u \in W^{\nu^0+1,\infty}(\Omega) \cap W^{\nu^0+1,\infty}(\partial\Omega)$, or $u \in H^{\nu^0+2}(\Omega)$, cf. Remark 5.67.*[39] *Furthermore, we define $\mathcal{V}_b^h$ as in (5.321) and choose $e \subset \bar{T}, T \in \mathcal{T}_c^h$. Then there exists a constant $C = C(d,\chi)$ such that*

$$\left| \int_e v^h Bu \, ds \right| \le C \, h^{\nu_0} \|u\|_{W^{\nu^0,\infty}(T)} |v^h|_{H^1(T)}, \ \forall \ v^h \in \mathcal{V}_b^h \not\subset H_0^1(\Omega). \tag{5.330}$$

Proof. The (5.311), (5.322) and (5.323) with $l = 1, j = d-1, p = \infty, q \ge 2$ are applied to an edge $e \subset \bar{T} \in \mathcal{T}_c^h$ with length h_e and $|\{\bar{e} \cap \partial\Omega\}| \ge 2$. We use the parametrization x_e for e, see (5.319). Now we combine $q^{d'}((v^h B_a u) \circ x_e) = 0 \ \forall v^h \in \mathcal{V}_b^h$, Proposition 5.62 for $q = \infty$ with (5.321), (5.322), Theorem 4.19 with (5.323), and (1.67), requiring

39 The transition from line 1 to 2 in (5.331) requires more than $u \in W^{\nu^0+1,\infty}(\Omega)$, since Theorem 1.37 is wrong for $p = \infty$. This is not surprising since $\partial\Omega$ has measure 0 in $\mathbb{R}^2$ So $u \in W^{\nu^0+1,\infty}(\Omega)$ does not imply any form of $u \in W^{\nu^0+1,\infty}(\partial\Omega)$. So, motivated by Remark 5.67 and Theorem 1.26, we impose $u \in W^{\nu^0+1,\infty}(\Omega) \cap W^{\nu^0+1,\infty}(\partial\Omega)$ or $u \in H^{\nu^0+2}(\Omega)$.

$u \in W^{\mu'+2,\infty}(\Omega)$ with $\mu' := 2d' - \rho$, for (5.331). We estimate for $B = B_a$

$$\left|\int_e v^h B_a u ds\right| = \left|\int_{s_e}^{s_e+h_e} (v^h B_a u) \circ x_e ds\right| \leq Ch_e^{\mu'+1} \|v^h B_a u\|_{W^{\mu',\infty}(e)}$$

$$\leq Ch_e^{\mu'+1} \|u\|_{W^{\mu'+1,\infty}(T)} \|v^h\|_{W^{d-1,\infty}(T)} \text{ with } \min\{\mu', d-1\} = d-1$$

$$\leq Ch^{\mu'+1} h^{1-2/q-(d-1)} \|u\|_{W^{\mu'+1,\infty}(T)} \|v^h\|_{W^1_q(T)} \text{ or} \tag{5.331}$$

$$\leq Ch^{\mu'+2-d} \|u\|_{W^{\mu'+1,\infty}(T)} \cdot \|v^h\|_{H^1(T)}, \text{ for } q = 2,$$

$$\leq Ch^{\mu'+2-d} \|u\|_{W^{\mu'+1,\infty}(T)} \cdot |v^h|_{H^1(T)},$$

where we have replaced $\|v^h\|_{H^1(T)}$ by $|v^h|_{H^1(T)}$ in (5.331). This is correct, since $v^h(P_j) = 0$ for at least one point $P_j \in e$. So we can use a *modified Poincaré–Friedrich inequality* of the form

$$\|u\|_{L^2(T)} \leq h|u|_{H^1(T)} \forall u \in H^1(T) \text{ with } u(P_j) = 0 \text{ for at least one } P_j \in e.$$

The additional changes for quasilinear problems with Lipschitz-continuous G are obvious by (5.286). □

Theorem 5.66. Consistency for FEMs violating boundary conditions and convergence under coercivity conditions:

1. *Under Conditions 5.60 and 5.64 we choose* A, $a(\cdot,\cdot)$, *homogeneous* Dirichlet boundary conditions *as in (5.278),* $\rho = 0, 2$, *and an arbitrary* $u \in W^{2d,\infty}(\Omega) \cap W^{2d,\infty}(\partial\Omega)$, *or* $u \in H^{2d+1}(\Omega)$, *cf. Remark 5.67. Furthermore, we define* $\mathcal{V}_b^h$ *as in (5.321) and choose the FEMs in (5.278)–(5.280), (5.294)–(5.298).*
2. *Then* $A^h = Q'^h A_h|_{\mathcal{V}_b^h}$ *is consistent with* A *in* u. *Let* u^h , *defined by* $A^h u^h = Q'^h Au$, *exist. Then the* variational *and* classical consistency errors *vanish,*[40] *cf. (5.329) and Remark 5.67, such that, with an h-independent* $C = C_{(d,n,\rho,x_b)}$,

$$\sup_{0 \neq v^h \in \mathcal{V}_b^h} \frac{|a_h(u - u^h, v^h)|}{\|v^h\|_{H^1(\mathcal{T}_c^h)}} \leq Ch^{\nu_0 - \frac{1}{2}} \|u\|_{W^{2d-1,\infty}(\mathcal{J}_c^h)}, \text{ and, by (5.329),}$$

$$\|A^h I^h u - Q'^h Au\|_{\mathcal{V}^{h'}} \leq Ch^{d-1} \|u\|_{W^{2d-1,\infty}(\Omega)} \text{ for } u \in W^{2d,\infty}(\Omega). \tag{5.332}$$

3. *Let* $a^h(\cdot,\cdot)$ *be* $\mathcal{V}_b^h$*-coercive, cf. (5.301) and Theorem 5.59. Then, for* $\rho = 0, 2$, *the solutions* u^h *exist uniquely and converge to* u *according to*

$$\|u^h - u\|_{H^1(\Omega)} \leq Ch^{d-1} \|u\|_{W^{2d-1,\infty}(\Omega)}. \tag{5.333}$$

4. *For* $\rho = 1$ *in (5.327) the convergence is only* $\leq Ch^{d-3/2} \|u\|_{W^{2d-2,\infty}(\Omega)}, u \in W^{2d-1,\infty}$.

[40] In this and the next subsection we often use the pair u, u^h with $A^h u^h = Q'^h Au$ instead of the exact solution u_0, u_0^h. This is motivated by the classical consistency error. It is defined not only for u_0 but for general u as well.

Remark 5.67.

1. *Here we consider Gauss, Gauss–Radau, and Gauss–Lobatto points. For violated continuity we restrict the discussion to Gauss points. Consequently, different techniques for the proofs are necessary.*
2. *An alternative result to (5.330), (5.332) is possible. Replace the* $\leq Ch_e^{\mu'+1} \|v^h B_a u\|_{W^{\mu',\infty}(e)}$ *by* $\leq Ch_e^{2(d-1)} |([v^h]B_a u)^{2(d-1)})|_{L^1(e)}$ *in the first line of (5.331), and combine it with (1.68). Then the* $\|u\|_{W^{2d-1,\infty}(\Omega)}$ *in (5.330), (5.332) can be replaced by* $\|u\|_{H^{2d}(\Omega)}$*. The analogous result to (5.350) is incorrect for curved boundaries, hence Theorems 4.60 does not hold either.*

Proof. This collects the preceding results in Proposition 5.63, Lemma 5.65 estimating (5.332). Relation (5.317) shows that we have to estimate the sum over all edges e with $|\{e \cap \partial\Omega\}| \geq 2$ along $\partial\,\Omega$,

$$
\begin{aligned}
\left|\int_{\partial\,\Omega} v^h B_a u ds\right| &\leq \sum_e \left|\int_e v^h B_a u ds\right| \text{ with } \sum_e := \sum_{e:|\{e\,\cap\partial\Omega\}|\geq 2} \\
&\leq Ch^{\mu'+1-(d-1)-1/2} \|u\|_{W^{\mu'+1,\infty}(\Omega)} \left(\sum_e h_e^{1/2} |v^h|_{H^1(T)}\right) \\
&\leq Ch^{\mu'-(d-1)+1/2} \|u\|_{W^{\mu'+1,\infty}(\Omega)} |v^h|_{H^1(\Omega)},
\end{aligned}
$$

since, by the Hölder inequality and (4.36),

$$
\begin{aligned}
\sum_e h_e^{1/2} |v^h|_{H^1(T)} &\leq \left(\sum_e h_e\right)^{1/2} \left(\sum_e |v^h|^2_{H^1(T)}\right)^{1/2} \\
&\leq \left(\sum_e h_e\right)^{1/2} |v^h|_{H^1(\Omega)} \text{ and } \sum_e h_e \leq 2 \times (\text{arc length of } \partial\Omega)
\end{aligned}
$$

for small enough h. So we obtain (5.332).

Relation (5.333) is obtained either by combining the stability result in Theorem 5.78 with the consistency estimate in (5.332) or by a combination with Lemma 5.52, see (5.271).

The last claim immediately follows, since, e.g. $H^{2d+1}(\Omega)$ is dense in $H^1(\Omega)$. So it is always possible to choose a $\hat{u} \in H^{2d+1}(\Omega)$, such that $\|\hat{u} - u\|$ is small enough. □

For quasilinear problems the previous terms in (5.330)–(5.332),

$$
\begin{aligned}
&u \in W^{2d,\infty}(\Omega), \|u\|_{W^{2d-1,\infty}(\Omega)}, \|u\|_{W^{d,\infty}(\Omega)} < \infty, \\
&u \in W^{2d'-\rho+2,\infty}(T), \|u\|_{W^{2d'-\rho+1,\infty}(T)} < \infty, \qquad (5.334)
\end{aligned}
$$

have to be replaced, with $Bu = B_G u$ in (5.286) and (5.330) in Lemma 5.65, by

$$Bu \in W^{2d-1,\infty}(\Omega), \|Bu\|_{W^{2d-2,\infty}(\Omega)}, \|Bu\|_{W^{d-1,\infty}(\Omega)} < \infty,$$
$$Bu \in W^{2d'-\rho+1,\infty}(T), \|Bu\|_{W^{2d'-\rho,\infty}(T)} < \infty. \tag{5.335}$$

These estimates for the Bu in terms of u are possible with the results in Subsections 2.5.4 and 2.5.6, based upon the second line of (5.334).

The linearized forms of these quasilinear problem are studied in Section 2.7, Subsections 2.7.3 ff., e.g. Theorem 2.122 and have to require $p \geq 2$ in (5.286). Furthermore, we already mentioned and require the necessary conditions for the $A_i(x, u^h, \nabla u^h)$, e.g. for bounded forms and operators, $a_h(u^h, v^h)$ and G_h, formulated in Sections 2.5.4, 2.5.6, 4.4 and (4.174), modified such that G is Lipschitz-continuous.

Theorem 5.68. Violated boundary conditions for quasilinear problems:

1. *Let the FEM in (5.286) for the quasilinear problem in Section 4.4, (4.174), satisfy Conditions 5.60 and 5.64, and define $\mathcal{V}_b^h$ as in (5.321). Furthermore let, for $u \in W^{\nu^0+1,\infty}(\Omega) \cap W^{\nu^0+1,\infty}(\partial\Omega)$, or $u \in H^{\nu^0+2}(\Omega)$, (5.334) imply (5.335), cf. Remark 5.67 and let the previous conditions be satisfied. Then all the consistency results in this subsection remain correct for these quasilinear problems as well.*
2. *Let the bilinear form of the linearized quasilinear problem be bounded, $\mathcal{V}_b^h$-coercive. Notice that we only consider $W^{1,p}(\Omega)$ for $2 \leq p < \infty$ and the corresponding stability and convergence results as in Theorem 4.63. with respect to the discrete $H^1(\mathcal{T}^h)$ norm. In this sense, the stability of the corresponding G^h and the existence of u_0^h are secured. Then the solutions u_0^h converge to u_0 according to (5.333).*

Proof. The claims concerning consistency are an immediate consequence of Chapter 3, Theorem 3.21, and Lemma 5.65. It remains correct, if (5.334) implies (5.335).

Finally, the linearized principal part of the linearized quasilinear operator is $H_0^1(\Omega)$-coercive for $2 \leq p < \infty$ or satisfies an inf-sup condition. If the quasilinear operator is bounded, implying its discretization to be equibounded, and consistent, cf. (5.286), we obtain the claim. □

5.5.7 Violated continuity

We consider nonconformity for piecewise polynomial $\mathcal{V}^h \not\subset \mathcal{V}$, including violated continuity and boundary conditions. We have already seen that, for the case of polygonal $\partial\Omega$ and the same number of functionals on every edge for FEs as in Definitions 4.6, 4.8, violated boundary conditions are equivalent to violated continuity. For polygonal $\partial\Omega$, violated boundary conditions are directly included in the following discussion. In contrast to [141], we allow curved boundaries and discontinuous FEs by combining the different results and techniques here with those in Subsubsection

5.5.6 for the case $\rho = 0$. Isoparametric FEs in Subsubsection 5.5.10 require other approaches.

Other combinations would be possible as well: Curved boundaries with $\rho = 2$ and discontinuous FEs with $\rho = 0$ require $\rho = 1$ for discontinuous FEs on the edges of the triangulation with exactly one boundary point. In this case the affine equivalence of the FEs, cf. Definition 4.6, would have to be replaced by three different types of FEs on triangles with none, one and more than one point on the boundary $\partial\Omega$. The whole theory would essentially remain correct. We leave its formulation with the many technical details to the interested reader.

So we only formulate the results with the same value of $\rho = 0$ for violated boundary conditions and continuity. We restrict, as in Subsubsection 5.5.6, the presentation here to the only practically relevant, *standard case,* cf. (5.327), required for the interior inner and the boundary edges,

$$\text{standard discontinuous FEs with straight edges: } 0 = \rho < d' = d - 1. \tag{5.336}$$

For FEs $\mathcal{V}_b^h \not\subset \mathcal{V}$, we have already introduced the generalized $\langle f_h, \cdot\rangle$, in (5.250), the $a_h(\cdot,\cdot)$ in (5.289) ff. and norms $\|\cdot\|_{\mathcal{V}^h}$, defined in (4.36). We study the impact of discontinuous $\mathcal{V}_b^h \not\subset \mathcal{V}$ onto the consistency estimates, cf. (5.279):

$$\begin{aligned} a_h(u^h, v^h) &= \sum_{T\in\mathcal{T}_c^h} \left(\int_T \sum_{i,j=0}^{2} a_{ij}\partial^i\, u^h \partial^j\, v^h dx \right) \quad \forall\, u,\ v \in \mathcal{V}_b + \mathcal{V}_b^h \\ &= a_{s,h}(u^h, v^h) + \sum_{e\in\mathcal{T}_c^h\setminus\partial\Omega} \int_e (\{v^h\}[B_a u^h] + [v^h]\{B_a u^h\}) ds \\ &\quad + \sum_{e\in\partial\Omega} \int_e v^h (B_a u^h) ds, \end{aligned} \tag{5.337}$$

with $u^h \in \mathcal{V}_b \cap H^2(\Omega)$ for $a_{s,h}(u^h, v^h)$ and with B_a as in (5.278). Again we use the unified notation in (5.285). The exact and discrete solutions satisfy, cf. (5.280)

$$a(u_0, v) = \langle f, v\rangle_{\mathcal{V}'\times\mathcal{V}}\ \forall v \in \mathcal{V}_b, \text{and } a^h\left(u_0^h, v^h\right) = \langle f^h, v^h\rangle_{\mathcal{V}^{h\prime}\times\mathcal{V}^h}\ \forall v^h \in \mathcal{V}_b^h. \tag{5.338}$$

In this two different approaches are possible for estimating the nonconformity errors. The μ-and σ-case are based on different proofs in Lemmas 5.65 and 5.70 The σ-case yields slightly inferior consistency orders for less smooth solutions u_0. The μ-case allows slightly higher consistency orders but often needs much higher smoothness of u_0. For example, we get for the σ-case $h^{d-1}|u|_{H^d(\Omega)}$ instead of $h^d|u|_{H^{2d}(\Omega)}$ for the μ-case.

Condition 5.69. Conditions for interpolation points on edges: *We require Condition 5.60 and modify Condition 5.64. In (5.336) we replace the Dirichlet conditions by interpolation conditions on all* straight *inner and (curved) boundary edges,* $e \in \mathcal{T}_c^h$. *The* $u^h, v^h \in \mathcal{V}^h$, *interpolate prescribed values* $v^h(P_i) = y_i, i = 0, \ldots, d' - 1, \forall e \in \mathcal{T}_c^h$

in the above $P_i = P_i^{\rho=0,e} \in e, i = 0, \ldots, d'-1$, see (5.320). This implies continuity of the FEs in the P_i, but not in e, even less in Ω.

For problems with natural boundary conditions *we do not have to impose boundary conditions. For* Dirichlet boundary conditions *we additionally impose the original Condition 5.64 along the boundary $\partial\Omega$ again with $\rho = 0$.*

Accordingly we define for $\overline{\Omega} = \cup_{T\in\mathcal{T}_c^h}\overline{T}$, see (5.321), possibly with curved $e \in \partial\Omega$,

$$\begin{aligned} \mathcal{V}_b^h &:= \{v : \Omega \to \mathbb{R} : v|_T \in \mathcal{P}_{d-1}\ \forall\ T \in \mathcal{T}_c^h,\ [v](P_i) = 0, \forall\ e \in\ \mathcal{T}_c^h \\ &\quad v(P_i) = 0, \forall\ e \in\ \partial\Omega\ \forall\ i = 0, \ldots, d'-1,\ P_i = P_i^{\rho=0,e}\ \}, \\ &\quad \|v^h\|_{\mathcal{V}^h} := \|v^h\|_{H^1(\mathcal{T}_c^h)}. \end{aligned} \tag{5.339}$$

The second condition, $v(P_i^{\rho=0,e}) = 0, e \in \partial\Omega$, only applies to Dirichlet boundary conditions; for natural boundary conditions it has to be omitted.

Except these new $\mathcal{V}_b^h$ in (5.339), instead of (5.321) the whole formal procedure in this and the preceding subsection is the same. Differences occure in the following formulas, estimates and theorems due to the discontinuity in the many interior edges here, compared to only violated boundary conditions in the few boundary edges before. Correspondingly, we only recall the important concepts, Q'^h and discrete linear and quasilinear equations and operators are, cf. (5.294) ff.:

$$Q'^h \in \mathcal{L}\left(W^{-1,p'}\left(\mathcal{T}_c^h\right), \mathcal{V}_b^{h'}\right) : \langle Q'^h f_h - f_h, v^h\rangle_{W^{-1,p'}(\mathcal{T}_c^h)\times\mathcal{V}^h} = 0 \forall v^h \in \mathcal{V}_b^h, \tag{5.340}$$

$$\begin{aligned} a^h\left(u_0^h, v^h\right) - \langle f^h, v^h\rangle_{\mathcal{V}^{h'}\times\mathcal{V}^h} &= \left\langle A_h u_0^h - f_h, v^h\right\rangle_{H^{-1}(\mathcal{T}_c^h)\times\mathcal{V}^h} = 0\ \forall\ v^h \in \mathcal{V}_b^h \\ \text{and } a^h\left(u_0^h, v^h\right) &= \left\langle G_h u_0^h, v^h\right\rangle_{W^{-1,p'}(\mathcal{T}_c^h)\times\mathcal{V}^h} = 0\ \forall\ v^h \in \mathcal{V}_b^h. \end{aligned} \tag{5.341}$$

$$A^h := Q'^h A_h|_{\mathcal{V}_b^h}, \text{and } G^h := Q'^h G_h|_{\mathcal{V}_b^h} \Rightarrow\ A^h u_0^h - f^h = 0, \text{ and } G^h u_0^h = 0. \tag{5.342}$$

The *variational discretization error* for $u \in H^2(\Omega)$ with the discrete solution u^h of the linear $A^h u^h = Q'^h Au$ and the quasilinear generalization, see (5.305), (5.299), is

$$a_h(u - u^h, v^h) = \sum_{e\in\mathcal{T}_c^h} \int_e [v^h] Buds + \sum_{e\in\partial\Omega} \int_e v^h Buds\ \forall u \in H^2(\Omega). \tag{5.343}$$

The *classical consistency errors* are, cf. (5.313), (5.339), (3.35),

$$\|A^h I^h u - Q'^h Au\|_{\mathcal{V}'},\ \|G^h I^h u - Q'^h Gu\|_{\mathcal{V}'}, \tag{5.344}$$

and we need

$$\sup_{v^h\neq 0} \frac{|a_h(u - u^h, v^h)|}{\|v^h\|_{\mathcal{V}^h}} = \sup \frac{|\sum_{e\in\mathcal{T}_c^h} \int_e [v^h] Buds + \sum_{e\in\partial\Omega} \int_e v^h Buds|}{\|v^h\|_{\mathcal{V}^h}}.$$

The next lemma could be proved by the same technique as the previous Lemma 5.65. However, we neglect the fact that along the straight edges e the parametrization is affine. This allows modified estimates compared to Lemma 5.65 and a reduced smoothness conditions for u. Since the interpolation error dominates the consistency error, cf. (5.348), we only discuss this updated form.

Lemma 5.70.

1. *Under Conditions 5.60,5.69 and with the notation in (5.258), (5.318), we choose* $A,\ a(\cdot,\cdot),\ a_h(\cdot,\cdot),\ G,\ G_h$ *as in (5.278) ff., the* $\mathcal{V}_b^h$ *as in (5.339), and* $u \in H^d(\Omega)$. *Let* $T_1, T_2 \in \mathcal{T}_c^h$ *be two neighboring or* T_1 *a boundary triangle with* diam $T_i \le h,\ i = 1,2$ *and a joint or a boundary* straight *edge and let*

$$U_e := \textit{ interior } (\overline{T_1 \cup T_2}) \textit{ for } e \subset \overline{T_1} \cap \overline{T_2} \textit{ or } U_e := T_1, \textit{ for } |\overline{T_1} \cap \partial\Omega| \ge 2.$$

Then there exists a constant $C = C(d,\chi)$ *such that, with* $|v^h|_{H_h^1(U_e)} := |v^h|_{H_h^1(T_1 \cup T_2)}$,

$$\left| \int_e [v^h - v] B_a u \, ds \right| \le C\, h^{d-1} |u|_{H^d(U_e)} \|v^h - v\|_{H_h^1(U_e)}, \ \forall v \in H^1(U_e) \qquad (5.345)$$
$$= \left| \int_e [v^h] B_a u \, ds \right| \le C\, h^{d-1} |u|_{H^d(U_e)} |v^h|_{H_h^1(U_e)} \textit{ for } v = 0.$$

2. *This remains correct for a straight boundary edge* $e \subset \partial\Omega$ *with* $[v^h - v], \ldots,$ *replaced by* $v^h - v, \ldots,$ *but not for curved boundaries.*
3. *For quasilinear problems under the condition (5.335) for Theorem 5.68, the previous estimates in (5.345) remain correct with* B_a *replaced by* $B = B_G$.

Proof. We apply a technique similar to Lemma 5.65 to our problem. For every straight edge $e \in \bar{T} \in \mathcal{T}_c^h$ with length h_e, the parameterization x_e is affine, see (5.319). This allows the following modification. We combine $q^{d'}([v^h]w) \circ x_e = 0\ \forall v^h \in \mathcal{V}_b^h$, Propositions 5.61, 5.62 for $q = 2$ in (5.309), (5.311), with (5.339) to yield

$$\int_{s_e}^{s_e+h_e} ([v^h]w) \circ x_e ds = \int_{s_e}^{s_e+h_e} ([v^h - v](w - p_2)) \circ x_e ds \qquad (5.346)$$
$$\forall v \in H^1(\Omega) \text{ or } \forall c_1 \in \mathbb{R}\ \forall\, p_2 \in \mathcal{P}^{d-2}, [c] = 0 = [v], [v^h] = [v^h - v - c_1],$$

for these c_1, v. Note $v^h p_2 \in \mathcal{P}^{2d-3}$, hence $\int_{s_e}^{s_e+h_e} ([v^h - v]p_2) \circ x_e ds = 0$. We choose in (4.34), $Q^1 v^h = c_1, Q^{d-1} B_a u = p_2$. This allows estimating

$$\left|\int_e [v^h]B_a u ds\right| = \left|\int_e [v^h - v]B_a u ds\right| \tag{5.347}$$

$$= \left|\int_{s_e}^{s_e+h_e} ([v^h - v - c_1](B_a u - p_2)) \circ x_e ds\right|$$

$$\leq C\|x_e\|_{L^\infty(e)}\|[v^h - v - c_1]\|_{L^2(e)}\|(B_a u - p_2)\|_{L^2(e)} \text{ by (1.67)}$$

$$\leq C\left(\|[v^h - v - c_1]\|_{L^2(U_e)}\|[v^h - v - c_1]\|_{H_h^1(U_e)}\right)^{1/2}$$

$$\left(\|(B_a u - p_2)\|_{L^2(U_e)}\|(B_a u - p_2)\|_{H_h^1(U_e)}\right)^{1/2}$$

$$\leq C\left(h\|v^h - v\|^2_{H^1(U_e)}\right)^{1/2} \left(h^{d-1}\|B_a u\|_{H^d(U_e)}h^{d-2}\|u\|_{H^d(U_e)}\right)^{1/2} \text{ with}$$

(4.34),(4.41) and choosing in (5.346) the infimum for c_1, p_2,

$$\leq Ch^{d-1}|v^h|_{H_h^1(U_e)}|B_a u|_{H^{d-1}(U_e)} \quad \text{for } v = 0 \text{ and}$$

$$\leq Ch^{d-1}\|v^h - v\|_{H_h^1(U_e)}|B_a u|_{H^{d-1}(U_e)} \quad \forall v \in H^1(\Omega),$$

hence, (5.345). The $v^h - v$ modification is an immediate consequence of (5.346) and the use of $\|v^h\|_{H^1(\mathcal{T}_c^h)}$ instead of $|v^h|_{H^1(\mathcal{T}_c^h)}$.

With the above change of $[v^h - v], \ldots$, into $v^h - v, \ldots$, for a boundary triangle $T_2 \cap \partial\Omega \neq \emptyset$ this result (5.345) remains valid as well.

For quasilinear problems (5.347) remains correct for the nonlinear $Bu = B_G u$. □

Theorem 5.71. Consistency for FEMs violating continuity conditions and convergence for coercive forms:

1. *Let Ω and the FEs satisfy Conditions 5.60 and 5.69. With the notation in (5.258), (5.318), we choose A, $a(\cdot,\cdot)$, A_h, A^h, $a^h(\cdot,\cdot)$, $\mathcal{V}_b^h$, Q'^h as in (5.278) ff., (5.339) ff., and* natural *or* Dirichlet boundary conditions.
2. *Then $A^h = Q'^h A_h|_{\mathcal{V}_b^h}$ and $G^h = Q'^h G_h|_{\mathcal{V}_b^h}$ are consistent with A and G in $u \in H^d(\Omega)$, respectively. The* variational *and* classical consistency errors *can be estimated for* straight *edges and polygonal $\partial\Omega$, with an h-independent $C = C_{(d,n,\rho,x_b)}$ as*

$$\sup_{0 \neq v^h \in \mathcal{V}_b^h} \frac{|a_h(u - u^h, v^h)|}{\|v^h\|_{\mathcal{V}^h}} \leq Ch^{d-1}\|u\|_{H^d(\Omega)}, \text{ and} \tag{5.348}$$

$$\|A^h I^h u - Q'^h Au\|_{\mathcal{V}^{h\prime}}, \ \|G^h I^h u - Q'^h Gu\|_{\mathcal{V}^{h\prime}} \leq Ch^{d-1}\|u\|_{H^d(\Omega)}.$$

3. *Let $a^h(\cdot,\cdot)$ be $\mathcal{V}_b^h$-coercive, e.g. the principal part, $a_p^h(\cdot,\cdot)$. Then the solutions u_0^h for $A^h u_0^h = Q'^h Au_0$ exist and converge to u_0 according to*

$$\|u_0^h - u_0\|_{H^1(\Omega)} \leq Ch^{d-1}\|u_0\|_{H^d(\Omega)}. \tag{5.349}$$

4. *For curved boundary edges in a curved $\partial\Omega$, (5.349) has to be combined with $\rho = 0$ in (5.332), and Condition 5.64 has to be imposed. Then (5.349) is replaced by* $\left\|u_0^h - u_0\right\|_{H^1(\Omega)} \le Ch^{d-1}\|u_0\|_{H^{2d}(\Omega)}$. *For the $\rho = 0, 1, 2$ combination indicated in the introduction, we only have* $\left\|u_0^h - u_0\right\|_{H^1(\Omega)} \le Ch^{d-3/2}\|u_0\|_{H^{2d}(\Omega)}$.
5. *This remains correct for quasilinear problems under the conditions previous to Theorem 5.68, if the coercivity of $a^h(\cdot,\cdot)$ is replaced by the coercivity of the bilinear form of the linearized operator $G'(u_0)$.*
6. *As in Lemma 5.70 the first estimate in (5.348) remains correct for $v^h - v$,*

$$\sup \frac{|a_h(u-u^h, v^h - v)|}{\|v^h - v\|_{\mathcal{V}^h}} \le Ch^{d-1}\|u\|_{H^d(\Omega)}, \ \forall\ v \in H^1(\Omega). \tag{5.350}$$

Proof. Similarly to Theorem 5.66 the consistency errors are immediate consequences of Lemma 5.70. We sum over all $e \in \mathcal{T}_c^h$ in Condition 5.60, where U_e is defined in Lemma 5.70. We use $|\int_e [v^h] B_a u\ ds| \le C\ h^{d-1}|u|_{H^d(U_e)}|v^h|_{H_h^1(U_e)}$, see (5.306) and (5.345) estimating[41]

$$\begin{aligned}
|a_h(u-u^h, v^h)| &= \left|\sum_{e\in\mathcal{T}_c^h}\int_e [v^h] B_a u_0 ds\right| + \left|\int_{e\in\partial\Omega} v^h B_a u_0 ds\right| \\
&\le Ch^{d-1}\sum_{e\in\mathcal{T}_c^h\cup\partial\Omega}|v^h|_{H_h^1(U_e)}|u|_{H^d(U_e)} \qquad (5.351)\\
&\le Ch^{d-1}\sum_{T\in\mathcal{T}_c^h\cup\partial\Omega}|v^h|_{H^1(T)}|u|_{H^d(T)} \text{ (by Hölder)}\\
&\le Ch^{d-1}|v^h|_{\mathcal{V}^h}\cdot|u|_{H^d(\Omega)}.
\end{aligned}$$

The analogous $\|v^h - v\|_{\mathcal{V}^h}$ extension of Lemma 5.70 is verified as there. □

Remark 5.72. *Lemma 5.70 is strictly local with the local step size h and the locally appropriate, e.g. $\|u\|_{H^d(U_e)}$ valid for all $u \in H^d(U_e)$. This can be used in two directions. Either* locally large *Bu and discretization errors are compensated by mesh refinements or the so-called hp-methods are employed. For* locally smooth *solutions u_0 and Bu_0, high orders of convergence $d-1$ and large h_e, for* locally unpleasant *u_0 and $B_a\ u_0$, small orders $d-1$ and small h_e have to be combined. These are the so-called adaptive FEMs, introduved in Chapter 6. The hp viewpoint is discussed for DCGMs in Section 7.14.*

We generalize the Aubin–Nitsche Lemma 4.59 and the corresponding convergence result in Theorem 4.60 to the case of variational crimes, with straight edges e, see, e.g. [135]. Here, we need the extension $a_h(\cdot,\cdot) : (\mathcal{V}_b + \mathcal{V}_b^h) \times (\mathcal{V}_b + \mathcal{V}_b^h) \to \mathbb{R}$ instead of $\mathcal{V}_b^h \times \mathcal{V}_b^h \to \mathbb{R}$. In addition to the original problem $a(u_0, v) = (f, v)\forall v \in \mathcal{V}_b$ we need the solution, ϕ_g, of its *dual problem,*

$$\phi_g \in \mathcal{V}_b \text{ s.t. } a(w, \phi_g) = (g, w)\forall w \in \mathcal{V}_b. \tag{5.352}$$

[41] For $0 \le a_1 \le a_2, 0 \le b_1 \le b_2$ we have $(a_1 + a_2)(b_1 + b_2) \le a_1b_1 + 3a_2b_2$ and every $T \in \mathcal{T}_c^h$ contains only a few edges, e.g. three for triangulations.

Lemma 5.73. Generalized Aubin–Nitsche lemma: *Let $a(\cdot,\cdot) : \mathcal{V}_b \times \mathcal{V}_b \to \mathbb{R}$ and its corresponding $a_h(\cdot,\cdot) : \left(\mathcal{V}_b + \mathcal{V}_b^h\right) \times \left(\mathcal{V}_b + \mathcal{V}_b^h\right) \to \mathbb{R}$ be continuous bilinear forms with $a_h(\cdot,\cdot)|_{\mathcal{V}_b\times\mathcal{V}_b} = a(\cdot,\cdot)$, see (5.279). Assume that the original and its* dual problem *satisfy the conditions for unique exact and FE solutions u_0, u_0^h and ϕ_g, ϕ_g^h. For the Banach space $\mathcal{W} \supset \mathcal{V}_b$ with inner product and norm (g,w) and $\|g\|_{\mathcal{W}}$, respectively, let the embedding $\mathcal{V}_b \hookrightarrow \mathcal{W}$ be continuous. Then the error for the FE solution can be estimated in the $\mathcal{W}$ norm as*

$$\begin{aligned}\left\|u_0 - u_0^h\right\|_{\mathcal{W}^h} \le \sup_{0\ne g\in\mathcal{W}} (1/\|g\|_{\mathcal{W}})\big(C\left\|u_0 - u_0^h\right\|_{\mathcal{V}^h}\left\|\phi_g - \phi_g^h\right\|_{\mathcal{V}^h} \\ + \left|a_h\left(u_0 - u_0^h, \phi_g\right) - \left(u_0 - u_0^h, g_h\right)\right| \\ + \left|a_h\left(u_0, \phi_g - \phi_g^h\right) - \left(f_h, \phi_g - \phi_g^h\right)\right|\big).\end{aligned} \tag{5.353}$$

Proof. We employ the dual norm of an element w of a Banach space in (1.32) and combine the solutions u_0, u_0^h and ϕ_g, ϕ_g^h of both problems to find, see [135],

$$\begin{aligned}\left(u_0 - u_0^h, g_h\right) &= a_h(u_0,\phi_g) - a^h\left(u_0^h, \phi_g^h\right) \\ &= a_h\left(u_0 - u_0^h, \phi_g - \phi_g^h\right) + a_h\left(u_0^h, \phi_g - \phi_g^h\right) + a_h\left(u_0 - u_0^h, \phi_g^h\right) \\ &= a_h\left(u_0 - u_0^h, \phi_g - \phi_g^h\right) - \left[a_h\left(u_0 - u_0^h, \phi_g\right) - \left(u_0 - u_0^h, g_h\right)\right] \\ &\quad - \left[a_h\left(u_0, \phi_g - \phi_g^h\right) - \left(f_h, \phi_g - \phi_g^h\right)\right].\end{aligned}$$

For the equality in the second and fourth lines we use the linearity of the $a_h(\cdot,\cdot)$ and split and replace the linear functionals in the [..] by the corresponding $a(\cdot,\cdot), a_h(\cdot,\cdot)$, respectively. Finally, use (1.32), and the continuity of $a_h(\cdot,\cdot)$ to obtain (5.353). □

We combine Lemma 5.73 with Theorem 5.71, but not with Theorem 5.66:

Theorem 5.74. Aubin–Nitsche theorem for nonconforming FEMs: *Under the conditions of Theorem 5.71 and Lemma 5.73 we allow standard FEs and FEs violating the continuity in (5.336) and assume $u \in H^d(\Omega)$. Then*

$$\left\|u_0 - u_0^h\right\|_{L^2(\Omega)} \le Ch^d\|u\|_{H^d(\Omega)}. \tag{5.354}$$

Proof. We apply Lemma 5.73 to our situation: we choose $\mathcal{W} = L^2(\Omega)$ and $\mathcal{V}_b = H_0^1(\Omega)$. This is justified since, by $\|u\|_{L^2(\Omega)} \le \|u\|_{H^1(\Omega)}\ \forall\, u \in H_0^1(\Omega)$, the $H_0^1(\Omega)$ is continuously embedded into $L^2(\Omega)$.

We estimate the terms in (5.353). For the first term in the second line of (5.353) we get, with the unusual defining relations for $\phi_g\ \phi_g^h$ and (5.347), (5.350),

$$\begin{aligned}&\left[a_h\left(u_0 - u_0^h, \phi_g\right) - \left(u_0 - u_0^h, g\right)\right] \\ &\qquad = \left[a(u_0,\phi_g) - (u_0, g)\right] - \left[a_h\left(u_0^h, \phi_g\right) - \left(u_0^h, g\right)\right] \\ &\qquad = -\left[a_h\left(u_0^h, \phi_g\right) - a_h\left(u_0^h, \phi_g^h\right)\right] = -\left[a_h\left(u_0^h, \phi_g - \phi_g^h\right)\right] \\ &\qquad = -\left[a_h\left(u_0 - u_0^h, \phi_g - \phi_g^h\right)\right].\end{aligned}$$

Similarly we find for the second term in the second line of (5.353), with the defining relations for u_0 and u_0^h,

$$\begin{aligned}
&[a_h\left(u_0,\phi_g-\phi_g^h\right)-\left(f,\phi_g-\phi_g^h\right)] \\
&\qquad =[a(u_0,\phi_g)-(f,\phi_g)]-\left[a_h\left(u_0,\phi_g^h\right)-\left(f,\phi_g^h\right)\right] \\
&\qquad =-\left[a_h\left(u_0,\phi_g^h\right)-a_h\left(u_0^h,\phi_g^h\right)\right] \qquad (5.355)\\
&\qquad =-\left[a_h\left(u_0-u_0^h,\phi_g^h\right)\right]=-\left[a_h\left(u_0-u_0^h,\phi_g^h-\phi_g\right)\right],
\end{aligned}$$

again with (5.347), and (5.350), now applied to the dual problem (5.352). We combine the (5.347) with the continuity of the $a_h(.,.)$ and the estimates for

$$\begin{aligned}
\left\|u_0-u_0^h\right\|_{H^1(\mathcal{T}_c^h)} &\le Ch^{d-1}\|u_0\|_{H^d(\Omega)}, \\
\left\|\phi_g-\phi_g^h\right\|_{H^1(\mathcal{T}_c^h)} &\le Ch\|\phi_g\|_{H^2(\Omega)}\le Ch\|g\|_{L^2(\Omega)}.
\end{aligned}$$

This allows, with (5.355) and by the continuity of $a_h(.,.)$, the estimates

$$\begin{aligned}
\left|a_h\left(u_0^h,\phi_g-\phi_g^h\right)\right| &= \left|a_h\left(u_0^h-u_0,\phi_g-\phi_g^h\right)\right| \\
&\le C\left\|u_0-u_0^h\right\|_{\mathcal{V}^h}\left\|\phi_g-\phi_g^h\right\|_{\mathcal{V}^h} \\
&\le Ch^d\|u_0\|_{H^d(\Omega)}\|\phi_g\|_{H^2(\Omega)} \\
&\le Ch^d\|u_0\|_{H^d(\Omega)}\|g\|_{L^2(\Omega)}.
\end{aligned}$$

In fact the second and the following lines in this inequality can be used for the term in the first line of (5.353) and by (5.355) the two last terms coincide. So estimating the three terms in (5.353) yields, with $\mathcal{W}=L^2(\Omega)$ and $\mathcal{V}=H^1(\Omega)$,

$$\begin{aligned}
\left\|u_0-u_0^h\right\|_{L^2(\Omega)} \le & \sup_{0\ne g\in L^2(\Omega)} (1/\|g\|_{L^2(\Omega)})\big(C\left\|u_0-u_0^h\right\|_{\mathcal{V}^h}\left\|\phi_g-\phi_g^h\right\|_{\mathcal{V}^h} \\
& +\left|a_h\left(u_0^h,\phi_g-\phi_g^h\right)\right|+\left|a_h\left(u_0^h-u_0,\phi_g-\phi_g^h\right)\right|\big) \\
\le & \sup_{0\ne g\in L^2(\Omega)} (1/\|g\|_{L^2(\Omega)})\left(Ch^d\|u_0\|_{H^d(\Omega)}\|g\|_{L^2(\Omega)}\right),
\end{aligned}$$

the claim. □

5.5.8 Stability for nonconforming FEMs

We turn to proving the stability for FEMs with variational crimes. Chapter 3 shows that we can confine the proof to the linear problems. By Theorem 3.23, stability is implied by the stability of the linearized A^h and some technical conditions. In Section 4.3 we could directly apply the results in Chapter 3. In Sections 5.2 and 5.4 specific ideas allowed us to overcome the difficulties with variational crimes in the fully nonlinear and quadrature approximate FEMs. Here we will procede in a similar way.

We have already proved in Theorem 5.59, assuming (5.300), see Remark 5.58, the stability of the above nonconforming FEMs for the principal parts, A_p, or specific compact perturbations, B, yielding a stable B^h. Here A was a linear operator for any of our above elliptic equations or systems of order 2. Furthermore, we have proved in Theorems 5.66 and 5.71 the consistency for general linear operators and extended it to quasilinear operators in Theorems 5.68 and 5.71, 5.

We have seen in Sections 4.3 and 5.2 that the existence of a bounded inverse of a linear A is related to, but not sufficient to ensure that the discrete A^h is stable, see [128, 567]. The next Theorem 5.78 guarantees the stability for nonconforming FEMs studied in this section for all types of elliptic equations and systems of order 2. It shows that a linear invertible B with stable B^h inherits the stability to A^h, for a linear invertible compactly perturbed A, such that A^h is consistent with A. This consistency is always valid for conforming FEs and is assumed for general discretizations in Theorem 3.29. It is not too surprising that the perturbations by the nonconformity of the FEs have to be small. Good convergence of the discrete solutions is only to be expected for consistency of higher order, so only for higher order methods and smooth enough u_0.

For the following considerations we recall the different consistency errors

$$\underbrace{Q'^h A_h u - A^h u^h}_{\text{var.cons.err.}} = Q'^h A_h \underbrace{(u - P^h u)}_{\text{interp.err.}} + \underbrace{A^h P^h u - Q'^h A u}_{\text{class.cons.err.}}, \text{ with } A^h u^h = Q'^h A u. \tag{5.356}$$

Condition 5.75. *Choose $\mathcal{V}_b^h \subset \mathcal{V}$ as in (5.321), or (5.339) and require Condition 5.60, (5.300), and $P^h = I^h$ and Q'^h as in Proposition 5.55, (5.294), (5.295). Choose the original linear problems and the FEMs as listed in (5.278)–(5.280), (5.281)–(5.284), (5.297)–(5.298).*

Lemma 5.76. *Under Condition 5.75, let $A_p : \mathcal{V}_b \to \mathcal{V}'$ be a linear principal part, cf. (5.278) or (5.281). Then $A_p u = f \in H^{-1}(\Omega)$, and for small enough $h_0 > h \in H$, $A_p^h u^h = Q'^h f$ as well, have unique solutions, and u_0^h converge to u_0 as*

$$\left\|u_0^h - u_0\right\|_{\mathcal{V}^h} \le \epsilon \left\|u\right\|_{\mathcal{V}}, \text{ for } h < h_0, \text{ hence } \lim_{h\to 0} \left\|u_0^h - u_0\right\|_{\mathcal{V}^h} = 0. \tag{5.357}$$

For general A, $u \in \mathcal{V}_b$, $P^h = I^h$, with $A^h u^h = Q'^h A u$, the preceding consistency errors converge to 0, as

$$\lim_{h\to 0} \|Q'^h A_h u - A^h u^h\|_{\mathcal{V}^{h\prime}} = 0, \quad \lim_{h\to 0} \|Q'^h A u - A^h P^h u\|_{\mathcal{V}^{h\prime}} = 0. \tag{5.358}$$

Proof. By Proposition 5.54, all the bilinear forms $a_h(\cdot,\cdot)$ are bounded and continuous. By Theorem 5.59, A_p^h is stable in $P^h u$. In Theorems 5.66, 5.71, we have formulated the smoothness of the u guaranteeing the optimal order of consistency. So we choose, e.g. $\tilde{u} \in H^d(\Omega)$, near u. Then Theorem 4.53 implies that for u and $\tilde{u}$ we have convergence, hence, e.g. for $A_p^h \tilde{u}^h = Q'^h A_p \tilde{u}$, cf. (5.357), (5.358),

$$\|\tilde{u} - \tilde{u}^h\|_{\mathcal{V}^h} \le C h^{d-1} \|\tilde{u}\|_{H^d(\Omega)} < \|u\|_{\mathcal{V}} \epsilon/3 \text{ for } h < h_0.$$

For estimating

$$\|u - u^h\|_{\mathcal{V}^h} \le \|u - \tilde{u}\|_{\mathcal{V}^h} + \|\tilde{u} - \tilde{u}^h\|_{\mathcal{V}^h} + \|u^h - \tilde{u}^h\|_{\mathcal{V}^h} < \epsilon,$$

we handle the first and the last term, $\|u-\tilde{u}\|_{\mathcal{V}^h} = \|u-\tilde{u}\|_{\mathcal{V}}$ and $\|u^h-\tilde{u}^h\|_{\mathcal{V}^h}$. We employ the definition of A_p^h, A_p cf. (5.279), and choose $\tilde{u}$ such that with the following bounded maximum,

$$\max\left\{1, \left\|\left(A_p^h\right)^{-1}\right\|_{\mathcal{V}_b^h \hookleftarrow \mathcal{V}_b'^h} \|Q'^h\|_{\mathcal{V}_b'^h \hookleftarrow \mathcal{V}_b'} \|A_p\|_{\mathcal{V}_b' \hookleftarrow \mathcal{V}_b}\right\} \|(u-\tilde{u})\|_{\mathcal{V}} < \|u\|_{\mathcal{V}}\ \epsilon/3,$$

hence

$$\left\|A_p^h(u^h-\tilde{u}^h)\right\|_{\mathcal{V}'^h} = \|Q'^h A_p(u-\tilde{u})\|_{\mathcal{V}'^h} \Longrightarrow$$

$$\left\|u^h-\tilde{u}^h\right\|_{\mathcal{V}^h} \le \left\|\left(A_p^h\right)^{-1}\right\|_{\mathcal{V}_b^h \hookleftarrow \mathcal{V}_b'^h} \|Q'^h\|_{\mathcal{V}_b'^h \hookleftarrow \mathcal{V}_b'} \|A_p\|_{\mathcal{V}_b' \hookleftarrow \mathcal{V}_b} \|(u-\tilde{u})\|_{\mathcal{V}} \le \|u\|_{\mathcal{V}}\ \epsilon/3,$$

hence (5.357). For the general A^h this argument yields (5.358). □

We combine the form of the consistency errors in Proposition 5.56, the stability of the principal parts in Theorems 5.59, and the consistency, specified in Theorems 5.66 and 5.71, to obtain the convergence of the discrete to the exact solution. This can be used defining an *anticrime transformation* from $\mathcal{V}_b^h$ to $\mathcal{V}_b$, the essential tool for proving the following stability result in Theorem 5.78, cf. Proposition 5.42.

Lemma 5.77. Anticrime transformation: *Let $A_0 \in \mathcal{L}(\mathcal{V}_b, \mathcal{V}_b')$, e.g. $A_0 = \Delta$ be boundedly invertible, $A_0^h : \mathcal{V}_b^h \to \mathcal{V}^{h'}$ be stable and A_0^h be consistent with A_0 for every $u \in \mathcal{V}_b$. For a given $u^h \in \mathcal{V}_b^h$ define $\hat{f}_h := A_0^h u^h \in H^{-1}(\mathcal{T}_c^h)$, cf. (5.250), (5.279). This $\hat{f}_h \in H^{-1}(\mathcal{T}_c^h)$ defines, as proved in Proposition 5.42, by its components f_{-1}, f_0 via (5.250) an $f_h \in H^{-1}(\Omega)$, such that $\|f_h\|_{H^{-1}(\Omega)} \le \left\|A_0^h\right|_{\mathcal{V}^{h'} \hookleftarrow \mathcal{V}_b^h} \|u^h\|_{\mathcal{V}^h}$. Define $E^h : \mathcal{V}_b^h \to \mathcal{V}_b$ by $u_h := E^h u^h := A_0^{-1} f_h \in \mathcal{V}_b$ with $\|E^h u^h\|_{H^1(\Omega)} \le C_1 \|u^h\|_{H^1(\mathcal{T}_c^h)}$. Then we obtain convergence as: $\forall \epsilon > 0\ \exists h_0 = h_0\left(\|u^h\|_{H^1(\mathcal{T}_c^h)}\right) : \ \forall h < h_0$:*

$$\|E^h u^h - u^h\|_{H^1(\mathcal{T}_c^h)} < \epsilon \|u^h\|_{H^1(\mathcal{T}_c^h)}, \text{hence, } \lim_{h\to 0} \|E^h u^h - u^h\|_{H^1(\mathcal{T}_c^h)} = 0. \quad (5.359)$$

For $f_h \in L^2(\Omega)$, we even get $\|E^h u^h - u^h\|_{\mathcal{V}^h} \le Ch\|f_h\|_{L^2(\Omega)}$ for $h \to 0$ implying convergence of order 1.

Theorem 5.78. Compactly perturbed invertible and consistent discretizations inherit stability: *Under the Condition 5.75, let $A \in \mathcal{L}(\mathcal{V}_b, \mathcal{V}_b')$ be boundedly invertible or be one of the linear(ized) operators for equations or systems, studied in this section, see (5.278), or (5.281), their extensions (5.279), or (5.283), and the corresponding FE equations (5.280), or (5.284). Finally, let $B \in \mathcal{L}(\mathcal{V}_b, \mathcal{V}_b')$, B^h be stable and A^h, B^h, hence C^h, be consistent with A, B, C, for $u \in \mathcal{V}_b$. Then, for small enough h,*

$$A^{-1} \in \mathcal{L}(\mathcal{V}', \mathcal{V}_b) \Longrightarrow A^h (\text{or its quadrature approx. } \tilde{A}^h) \text{ is stable in } P^h u. \quad (5.360)$$

Proof.

1. We employ the anticrime transformation, $E^h : \mathcal{V}_b^h \to H_0^1(\Omega)$, introduced in Lemma 5.77. We derive several relations and inequalities. The proof is similar to that of Theorem 5.9. The differences, mainly from (5.369) ff., are caused by the

necessary $P^h E^h u^h$. As discussed it suffices to show that $A^{-1} \in \mathcal{L}(\mathcal{V}', \mathcal{V}_b)$ implies the stability of A^h with

$$\begin{aligned}
&A, B, C \in \mathcal{L}(\mathcal{V}_b, \mathcal{V}'),\ C \text{ compact},\ A =: B + C, \quad \text{with } B, B^h \\
&\text{boundedly invertiblestable, and } A^h,\ B^h,\ C^h \text{ consistent with } A, B, C \\
&\text{in } u \in H^1(\Omega) \text{ or } \in H^2(\Omega).
\end{aligned} \tag{5.361}$$

In fact, for all nonconforming FEMs in this section, the consistency of the B^h, C^h is as correct as that of A^h. We use the notation in Proposition 5.55, (5.294)–(5.298),

$$A_h, B_h, C_h : (\mathcal{V}_b + \mathcal{V}_b^h) \to \mathcal{V}^{h'}, Q'^h : (\mathcal{V}' + \mathcal{V}^{h'}) \to \mathcal{V}^{h'},$$
$$A^h = Q'^h A_h|_{\mathcal{V}_b^h} : \mathcal{V}_b^h \to \mathcal{V}^{h'}, \ldots$$

with $\Phi^h A = A^h, \ldots$, cf. (5.298). The bounded invertibility and stability of B and B^h and consistency with B in u imply the unique existence of the following solutions $\hat{u}$ and $\hat{u}^h$, defined, for an arbitrary $u \in \mathcal{V}_b$ and $v' := Cu$ by

$$B\hat{u} = v' \text{ and } B^h \hat{u}^h = Q'^h v'.$$

With equicontinuous B^h, and Q'^h, and $\Phi^h(B \cdot -v') = B^h \cdot -Q'^h v'$, let

$$T := B^{-1} \text{ and } T^h := (B^h)^{-1} Q'^h \in \mathcal{L}\left(\mathcal{V}' + \mathcal{V}^{h'}, \mathcal{V}_b^h\right). \tag{5.362}$$

Then the convergence of the $\hat{u}^h$ to $\hat{u}$ can be written as

$$\forall u \in \mathcal{V}_b : \lim_{h \to 0} ||(T - T^h)Cu||_{H^1(\mathcal{T}_c^h)} = 0. \tag{5.363}$$

This implies, since C is compact and the $T - T^h$ are equibounded,

$$||(T - T^h)C||_{H^1(\mathcal{T}_c^h) \leftarrow \mathcal{V}_b} \to 0, \quad \text{for } h \to 0. \tag{5.364}$$

The limit (5.363) remains correct if $u \in \mathcal{V}_b$ is replaced by any bounded $u_h = E^h u \in \mathcal{V}_b$, depending upon h, with $||u_h||_{\mathcal{V}} = \|E^h u^h\|_{H^1(\Omega)} \le C_2 \|u^h\|_{H^1(\mathcal{T}_c^h)}$. Indeed the consistency estimate, cf. Remark 5.49, and the stability of B^h can be applied to any bounded $u_h \in \mathcal{V}_b$ as well:

$$||(T - T^h)CE^h u^h||_{H^1(\mathcal{T}_c^h) \leftarrow \mathcal{V}_b^h} \le C_3 ||(T - T^h)C||_{H^1(\mathcal{T}_c^h) \leftarrow \mathcal{V}_b} \|u^h\|_{H^1(\mathcal{T}_c^h)}. \tag{5.365}$$

We combine $u^h \in \mathcal{V}_b^h$ with E^h. With the boundedly invertible A we estimate

$$\begin{aligned}
\|u^h\|_{\mathcal{V}^h} &\le 2||E^h u^h||_{\mathcal{V}} \le 2||A^{-1}||_{\mathcal{V}_b \leftarrow \mathcal{V}'} ||AE^h u^h||_{\mathcal{V}'} \\
&= 2||A^{-1}||_{\mathcal{V}_b \leftarrow \mathcal{V}'} ||B(I + TC)E^h u^h||_{\mathcal{V}'} \\
&\le 2||A^{-1}||_{\mathcal{V}_b \leftarrow \mathcal{V}'} ||B||_{\mathcal{V}' \leftarrow \mathcal{V}_b} ||(I + TC)E^h u^h||_{\mathcal{V}},
\end{aligned} \tag{5.366}$$

hence,

$$||(I + TC)E^h u^h||_{\mathcal{V}} \ge \|u^h\|_{\mathcal{V}^h} / (2||A^{-1}||_{\mathcal{V}_b \leftarrow \mathcal{V}'} ||B||_{\mathcal{V}' \leftarrow \mathcal{V}_b}). \tag{5.367}$$

With the stability of B^h we get

$$\begin{aligned}\|(I+T^hC^h)u^h\|_{\mathcal{V}^h} &= \|(B^h)^{-1}B^h(I+T^hC^h)u^h\|_{\mathcal{V}^h} \\ &\leq \|(B^h)^{-1}\|_{\mathcal{V}_b^h \leftarrow \mathcal{V}^{h'}} \|B^h(I+T^hC^h)u^h\|_{\mathcal{V}^{h'}}.\end{aligned}$$

This implies

$$\|B^h(I+T^hC^h)u^h\|_{\mathcal{V}^{h'}} \geq \|(I+T^hC^h)u^h\|_{\mathcal{V}^h} \;/\; \|(B^h)^{-1}\|_{\mathcal{V}_b^h \leftarrow \mathcal{V}^{h'}}. \quad (5.368)$$

The linearity of FEM with variational crimes implies

$$\begin{aligned}\Phi^h(A) = A^h &= \Phi^h(B+C) = B^h + C^h \\ &= B^h(I+(B^h)^{-1}C^h) = B^h(I+T^hC^h) : \mathcal{V}_b^h \to \mathcal{V}^{h'}.\end{aligned}$$

For the following estimate we refer to (5.359), and Theorem 4.17 and obtain with (5.357),

$$\begin{aligned}\|u^h - P^hE^hu^h\|_{H^1(\mathcal{T}_c^h)} &\leq \|(E^hu^h - P^hE^hu^h)\|_{H^1(\mathcal{T}_c^h)} \quad (5.369)\\ &+ \|(u^h - E^hu^h)\|_{H^1(\mathcal{T}_c^h)} \leq 2\epsilon\|u^h\|_{\mathcal{V}^h}.\end{aligned}$$

2. For the final stability estimates we return to the $(B^h)^{-1}A^hu^h = (I+T^hC^h)u^h$ term. The identities $Q'^h|_{\mathcal{V}^{h'}} = I_{\mathcal{V}^{h'}}, T^h|_{\mathcal{V}^{h'}} = (B^h)^{-1}Q'^h|_{\mathcal{V}^{h'}} = (B^h)^{-1}$, the equi-boundedness of C^h and T^h, the consistency of C^h, and the indicated error terms, are combined with the triangle inequality estimating

$$\begin{aligned}&\|(B^h)^{-1}\|_{\mathcal{V}_b^h \leftarrow \mathcal{V}_b^{h'}} ||A^hu^h||_{\mathcal{V}^{h'}} = \|(B^h)^{-1}\|_{\mathcal{V}_b^h \leftarrow \mathcal{V}_b^{h'}} ||B^h(I+(B^h)^{-1}Q'^hC^h)u^h||_{\mathcal{V}^{h'}} \\ &\geq \|(B^h)^{-1}\|_{\mathcal{V}_b^h \leftarrow \mathcal{V}_b^{h'}} ||B^h(I+T^hC^h)u^h||_{\mathcal{V}^{h'}} \quad (5.370)\\ &\geq \Big(||(I+TC)E^hu^h||_{\mathcal{V}} - ||u^h - E^hu^h||_{\mathcal{V}^h} - ||T^hC^hu^h - TCE^hu^h||_{\mathcal{V}^h}\Big) \\ &\geq \Big(||(I+TC)E^hu^h||_{\mathcal{V}} - ||u^h - E^hu^h||_{\mathcal{V}^h} - ||(T^h - T)CE^hu^h||_{\mathcal{V}^h} \\ &\qquad - ||T^h(C^hu^h - CE^hu^h)||_{\mathcal{V}^h}\Big) \\ &\geq \Big(||(I+TC)E^hu^h||_{\mathcal{V}} - ||u^h - E^hu^h||_{\mathcal{V}^h} - ||(T^h - T)CE^hu^h||_{\mathcal{V}^h} \\ &\qquad - ||T^h(C^hP^hE^hu^h - CE^hu^h)||_{\mathcal{V}^h} - ||T^h(C^h - C^hP^hE^h)u^h||_{\mathcal{V}^h}\Big)\end{aligned}$$

$$\geq \Big(||\underbrace{(I+TC)E^h u^h}_{\text{by (5.367)}}||_{\mathcal{V}} - ||\underbrace{(u^h - E^h u^h)}_{\text{FEM error (5.369)}}||_{\mathcal{V}^h} - 2||\underbrace{(T^h - T)C}_{\text{by (5.364)}}||_{\mathcal{V}_b^h \hookleftarrow \mathcal{V}_b}||u^h||_{\mathcal{V}^h}$$

$$- ||T^h \underbrace{(C^h P^h E^h u^h - C E^h u^h)||_{\mathcal{V}^h}}_{\text{class.consistency of C,cf.Theorems 5.66, 5.71}}$$

$$- ||T^h C^h (\underbrace{(u^h - E^h u^h)}_{\text{FEM error (5.369)}} + \underbrace{(E^h u^h - P^h E^h u^h)}_{\text{interpol.error (4.41)}})||_{\mathcal{V}^h} \Big).$$

We choose h small enough, such that all the preceding errors in (5.369), (5.364), (5.358), (4.41) are $< \epsilon ||u^h||_{\mathcal{V}^h}$ with

$$\epsilon < 1/\big(4(3 + ||T^h||_{\mathcal{V}_b^h \hookleftarrow \mathcal{V}_b^{h\prime}} + 2||T^h C^h||_{\mathcal{V}_b^h \hookleftarrow \mathcal{V}_b^h})||A^{-1}||_{\mathcal{V}_b \hookleftarrow \mathcal{V}'}||B||_{\mathcal{V}' \hookleftarrow \mathcal{V}_b}\big).$$

This implies

$$||A^h u^h||_{\mathcal{V}^{h\prime}} \geq \Big(||u^h||_{\mathcal{V}^h} / (4||A^{-1}||_{\mathcal{V}_b \hookleftarrow \mathcal{V}'}||B||_{\mathcal{V}' \hookleftarrow \mathcal{V}_b}) \Big) / ||B^{h^{-1}}||_{\mathcal{V}_b^h \hookleftarrow \mathcal{V}^{h\prime}}$$
$$\geq K' ||u^h||_{\mathcal{V}^h}, \tag{5.371}$$

i.e. $(A^h)_{h \in H}$ is stable. □

Theorem 5.78 yields a criterion for the stability of discretizations of operators, which are compact perturbations of coercive operators. An important class of compactly perturbed coercive operators are the $A \in \mathcal{L}(\mathcal{V}_b, \mathcal{V}_b')$ satisfying a so-called *Gårding inequality*. Hence Theorem 3.32 can directly be generalized to our FEMs with variational crimes.

5.5.9 Convergence, quadrature and solution of FEMs with crimes

Now we have proved all necessary results guaranteeing consistency, stability, and convergence for the FEMs with variational crimes studied in this section. It would be boring to repeate all the different complex conditions for the many different cases in this section. We prefer the formulation of a summarizing theorem. It collects the different desired results.

Theorem 5.79. FEMs for elliptic problems of second order, violating homogeneous Dirichlet boundary conditions and/or continuity:

1. *These are stable under the conditions in Theorems 5.59 (coercivity), 5.78 (stability), consistent for violated Dirichlet conditions and/or continuity in Theorems 5.66 and 5.71. Hence all these methods converge with the same estimates as their consistency, multiplied by a stability constant S.*
2. *Extensions to quasilinear problems in Theorems 5.68 and 5.71, 5. imply the corresponding convergence results.*
3. *An improvement of the convergence rate for linear problems with violated continuity and with respect to the $L^2(\Omega)$ norm is formulated in Theorem 5.74.*

The quadrature approximations are illustrated in Figure 5.5. Now $\mathcal{V}_b^h$ are the FEs violating boundary conditions and/or continuity. Formulating with the quadrature projectors yields, cf. (5.183), with the $\tilde{Q}'^h$ in (5.180), and A_h in Theorem 5.66 and in (5.342),

$$\tilde{A}^h = \tilde{Q}'^h A_h|_{\mathcal{V}^h}, \tilde{A}^h \tilde{u}_0^h = \tilde{Q}'^h f. \tag{5.372}$$

The stability for $\tilde{A}^h = \tilde{Q}'^h A_h|_{\mathcal{V}^h}$ is proved in Theorem 5.30. So according to Theorems 3.21, 4.53 and the continuity of A, we only have to show the classical consistency of $\tilde{A}^h$ with A in the exact solution u_0 yielding convergence in Theorem 5.80.

The corresponding smoothness of Sobolev norms is collected in $\hat{C} = C^\sigma$ or $\hat{C} = C^\mu$. We require, for the coefficients a_{ij} and the inhomogeneity f,

$$a_{ij} \in W_\infty^k(\Omega),\ i,j = 0,\ldots,n,\ f \in H^k(\Omega). \tag{5.373}$$

We impose the condition $k > \max\{2(m+\tau-1), 2\}$ for a quadrature version of the same order of convergence as the conforming FEM. $u_0 \in H^{\hat{\mu}}(\Omega)$ is motivated by Theorem 5.71 and Remark 5.72. We have proved in Theorem 5.30 that for $k > 2(d-2)$ the $a^h(\cdot,\cdot)$ and $\tilde{a}^h(\cdot,\cdot)$ are simultaneously bounded and define stable A^h and $\tilde{A}^h$.

Theorem 5.80. Quadrature for nonconforming FEs:

1. *Assume $n = 2$, nonconforming FEs and quadrature formulas, exact for piecewise polynomials of degree $k-1$ with $k > \max\{2(d-2), 2\}$. Let (5.176),(5.373) for the a_{ij}, f, and Conditions 5.60 and 5.69 be satisfied. Let $a^h(\cdot,\cdot)$ induce a stable A^h. Then*

$$\tilde{a}^h\left(\tilde{u}_0^h, v^h\right) = \widetilde{\langle f, v^h\rangle}^h \forall v^h \in \mathcal{V}^h, \tag{5.374}$$

defines a stable and convergent discretization for $Au_0 = f$. For the other approximations, mentioned above, this result remains correct if the $\tilde{a}^h, \tilde{A}^h$ are replaced by $a^{\tilde{h}}, \tilde{a}^{\tilde{h}}, A^{\tilde{h}}, \tilde{A}^{\tilde{h}}$.

2. *For $u_0 \in H^{d-p}(\Omega), p \geq 0$, we obtain $(M := M_{a_{ij}} := \max_{i,j=0}^{2} \|a_{ij}\|_{W^{k,\infty}(\Omega_c^h)})$:*

$$\left\|u_0 - \tilde{u}_0^h\right\|_{\mathcal{V}^h} \leq Ch^{k+2-d}(\|f\|_{\mathcal{V}'} + M\|P^h u_0\|_{H^{d-1}(\mathcal{T}_c^h)}) + Ch^{d-1-p}\|u_0\|_{H^{d-p}(\Omega)}. \tag{5.375}$$

3. *For violated boundary conditions on curved boundaries, $h^{d-1}\|u_0\|_{H^d(\Omega)}$ has to be replaced by $h^{d-1}\|u_0\|_{W^{2d-1,\infty}(\Omega)}$. For isoparametric FEs the $P^h u_0$ has to be combined with $E_c u_0$.*

4. *If only approximations for $\widetilde{\langle f, v^h\rangle}^h$ and none for $\tilde{a}^h(u^h, v^h)$ are used in (5.374), then the term $M\|u_0\|_{H^{\min\{s,k+1\}}(\Omega)}$ in (5.375) is dropped.*

Proof. $\tilde{A}^h$ is stable by Theorems 5.59 and 5.30 for $k > 2(d-2)$ and the conditions in the theorem. So it suffices to estimate the classical consistency errors

$$\|\tilde{A}^h P^h u_0 - \tilde{Q}'^h A u_0\|_{\mathcal{V}^{h\prime}} \leq \|\tilde{A}^h P^h u_0 - A^h P^h u_0\|_{\mathcal{V}^{h\prime}}$$
$$+\|A^h P^h u_0 - Q'^h A u_0\|_{\mathcal{V}^{h\prime}} + \|Q'^h f - \tilde{Q}'^h f\|_{\mathcal{V}^{h\prime}}.$$

The terms on the right represent the quadrature errors for Au_0 and $Q'^h f$ and the nonconforming consistency errors of the exact FEM for A^h in u_0. Estimates are proved in Theorems 5.66, 5.71, and 5.28, cf. Remark 5.67. We get

$$\leq Ch^{k+2-d}(\|f\|_{H_k^{-1}(\mathcal{T}_c^h)} + M_{a_{ij}}\|P^h u_0\|_{H^{d-1}(\mathcal{T}_c^h)}) + Ch^{d-1}d\|u_0\|_{H^d(\Omega)}. \quad (5.376)$$

If only $u_0 \in H^{d-p}(\Omega), p \geq 0$, the consistency error is reduced to $Ch^{d-1-p}d\|u_0\|_{H^{d-p}(\Omega)}$. Now the above $k > 2(d-2)$ implies $k+2-d \geq d-1$. So the term $Ch^{d-1-p}d \|u_0\|_{H^{d-p}(\Omega)}$ dominates in (5.376) and we do not specify the other terms. □

Theorem 5.68 for violated boundary conditions and quasilinear problems in the last subsection remains nearly unchanged.

Theorem 5.81. Quadrature approximate FEMs with crimes for quasilinear problems: *For the quasilinear problem in Section 4.4, (4.174) let its FEM in (5.286) satisfy the conditions of Theorems 5.68 and 5.71. In particular we require the necessary conditions for the Nemickii operators $A_i(x, u^h, \nabla u^h)$, e.g. for bounded forms and operators, $a_h(u^h, v^h)$ and G_h, as formulated in Sections 2.5.4, 2.5.6, 4.4 and (4.174). We additionally modify them, such that G is Lipschitz-continuous. Then for a boundedly invertible linearized operator the convergence and quadrature approximation results in this subsection remain correct.*

We discuss the *solution of the nonlinear equations* here for all the previous FEMs with crimes, including quadrature approximations. These FEMs with crimes are *linear*, cf. (4.154). Hence all the quasilinear equations and their quadrature approximations can be efficiently solved with the methods in Section 3.7, combining continuation with Newton's methods and the mesh independence priciple in Theorems 5.21, 3.40. By (5.375) and for $u_0 \in H^{d-p}(\Omega, \mathbb{R}^q), d-p > 1$, we obtain convergence of order $d-p-1 > 0$. We impose the following Lipschitz-continuity for G',

$$G'(\cdot) \in C_L(B_r(u_0) \cup B_r(P^h u_0)), \quad (5.377)$$
$$G|_{\mathcal{V}_{d-p}} : \mathcal{D}(G) \cap \left(\mathcal{V}_{d-p} := H^{d-p}(\Omega) \cup H^{d-p}\left(\Omega_c^h\right)\right) \to H^{d-p-1}(\Omega) \cup H^{d-p-1}\left(\Omega_c^h\right).$$

Theorem 5.82. Newton's method for the FE equations: *In addition to the conditions in Theorems 5.80 and 5.81 we assume (5.377). Start the Newton process for G and G^h or $\tilde{G}^h$ with $u_1 \in B_r(u_0) \cap \mathcal{D}(G) \cap \mathcal{V}_{d-p}$ and $u_1^h := P^h u_1$. Then, for small enough $\|u_0 - u_1\|_{\mathcal{V}}$ and h, both methods converge quadratically and the sequences u_i, u_i^h satisfy, independend of h, for $i = 1, \ldots,$*

$$\|u_{i+1}^h - u_i^h\|_{\mathcal{V}^h} \leq C\left(\|u_i^h - u_{i-1}^h\|_{\mathcal{V}^h}\right)^2 \text{ and } \|u_i^h - P^h u_i\|_{\mathcal{V}^h} \leq Ch^{d-p-1}\|u_i\|_{\mathcal{V}_{d-p}}. \quad (5.378)$$

We have presented the proof and futher details for fully nonlinear problems in Theorems 5.21 and 4.64.

5.5.10 Isoparametric FEMs

In this subsection we generalize isoparametric FEMs from the standard second order linear elliptic equations to second order linear and quasilinear elliptic equations and systems. Even extensions to order $2m, m \geq 1$, in $\mathbb{R}^n$, would be possible, if the interpolation theory in Subsection 4.2.7 could be approporiately generalized to C^{m-1} reference elements. The essential tool is, as in the case of fully nonlinear problems, the general discretization theory in Chapter 3 and Summary 4.52.

Motivated by Proposition 5.63 we only discuss *Dirichlet boundary conditions.* For simplicity we again formulate the computations for $n = 2$ and the Laplacian, and indicate the generalizations to linear equations and systems of order 2 and the modifications for the quasilinear case and $n \geq 2$.

For the *consistency estimates* we recall the different tools introduced in Definition 4.42, (4.99), (4.100), (4.101). We recall the definition of the original and approximating polygonal domains Ω and Ω^h, the isoparametric FEs and the mappings in (4.102), (4.103), (4.100), (4.101). The isoparametric FEM uses $\mathcal{V}_b^h = \mathcal{V}_b^h$.

$$
\begin{aligned}
F^h : \Omega^h \to \Omega_c^h &:= F^h(\Omega^h),\ F_c : \Omega^h \to \Omega, \text{ and } E_c := F^h \circ F_c^{-1} : \Omega \to \Omega_c^h, \\
u^h, v^h : \Omega_c^h &= F^h(\Omega^h) \to \mathbb{R}, \hat{u}^h = (u^h \circ E_c) : \Omega \to \mathbb{R}, \\
\mathcal{V}_b^h &= \mathcal{V}_b^h,\ \widehat{\mathcal{V}}_b^h = \widehat{\mathcal{V}}_b^h := \left\{\hat{u}^h : u^h \in \mathcal{V}_b^h\right\}, \text{ and} \\
\check{f} : \Omega_c^h \to \mathbb{R},\ \check{f}(x) &:= f((E_c)^{-1}(x)) = (f \circ (E_c)^{-1})(x) \text{ with} \qquad (5.379) \\
t - E_c(t) &= \mathcal{O}(h^d) \Longrightarrow (E_c)' - Id_\Omega = \mathcal{O}(h^{d-1}),\ ((E_c)^{-1})' - Id_{\Omega_c^h} = \mathcal{O}(h^{d-1}).
\end{aligned}
$$

The isoparametric FEM with the above $\mathcal{V}_b^h$ are formulated directly for the general, including the quaislinear case. We use the notation $a_{ij}\partial^i\ u\ \partial^j\ v$, redefining them for systems and the matrices $a_{ij} = A_{ij}$ as $a_{ij}\partial^i\ u^h, \partial^j\ v^h = (A_{ij}\partial^i\ u^h, \partial^j\ v^h)_q$. For defining the extensions $a_h(\cdot,\cdot), \ldots, G_h$, as above, we

$$\text{assume } a_{ij}, f, A_i \text{ defined on } \Omega_\cup := \Omega \cup \Omega_c^h \text{ for } A_i \text{ in the sense of (5.5).} \qquad (5.380)$$

We denote the linear and quasilinear cases and the projectors as $a_h(\cdot,\cdot), a^h(\cdot,\cdot)$, A_h, A^h, G_h, G^h, and Q'^h and $Q_c'^h$., cf. (5.286), Proposition 5.55. For a unified notation we use $W^{1,p}(\Omega_\cup) = H(\Omega_\cup)$ with $A = G, p = 2$ for[42] the linear case and $\mathcal{D}(G) = \cap_{i=0}^n \mathcal{D}(A_i) \subset \mathcal{V}_b$. Then we consider

$$A : H^1(\Omega) = \mathcal{V} \to H^{-1}(\Omega) = \mathcal{V}' \text{ or } \mathcal{V}_b \to \mathcal{V}_b' \text{ and } G : \mathcal{D}(G) \subset \mathcal{V} \to \mathcal{V}' \text{ or } \subset \mathcal{V}_b \to \mathcal{V}_b'$$

[42] For simplicity often only $f \in L_2(\Omega_c^h)$. In this case the definition $(f, v^h)^h := \int_{\Omega_c^h} f(x)v^h(x)dx$ is used. $J(E_c)'(t)$ denotes the absolute value of the determinant of the Jacobian of $(E_c)'(t)$.

$$\begin{array}{ccccc}
\mathcal{U}_b & \overset{A,\,G}{\longrightarrow} & \mathcal{U}' & \overset{\text{tested by}}{\longleftrightarrow} & \mathcal{U}_b \\
P^h\big\downarrow & \Phi^h\big\downarrow & Q_c'^h\big\downarrow & & \\
\mathcal{U}_b^h & \overset{A^h,\,G^h}{\longrightarrow} & \mathcal{U}^{h'} & \overset{\text{tested by}}{\longleftrightarrow} & \mathcal{U}_b^h \not\subset \mathcal{U}
\end{array}$$

Figure 5.8 Nonconforming FEMs: Spaces and operators

$$A_h : H^1(\Omega_\cup) \to H^{-1}(\Omega_\cup) : \langle A_h u, v\rangle_{H^{-1}(\Omega_\cup)\times H^1(\Omega_\cup)} := \quad \forall u, v \in H^1(\Omega_\cup),$$

$$a_h(u,v) : = \int_{\Omega_\cup} \left(\sum_{i,j=0}^{n} a_{ij}\partial^i \; u \; \partial^j \; v \right)(x)dx, \qquad \text{and}$$

$$a_h(u^h, v^h) = \langle G_h u^h, v^h\rangle_{W^{-1,p'}\times W^{1,p}(\Omega_\cup)} = \int_{\Omega_\cup} \sum_{i=0}^{n} (A_i(x, u^h, \nabla u^h), \partial^i v^h)_q dx$$

$$\text{and} \quad a^h(\cdot,\cdot) := a_h(\cdot,\cdot)|_{W^{-1,p}(\Omega_c^h)\times \mathcal{V}_b^h}, \quad a(\cdot,\cdot) = a_h(\cdot,\cdot)|_{W^{-1,p}(\Omega)\times\mathcal{V}_b}$$

$$Q'^h : H^{-1}(\Omega) \to \mathcal{V}^{h'} : \langle Q'^h f - \check{f}, v^h\rangle_{H^{-1}(\Omega_c^h)\times\mathcal{V}_b^h} \quad \forall v^h \in \mathcal{V}_b^h$$

$$= \int_{\Omega_c^h} (Q'^h f - \check{f})(x) v^h(x) dx \quad \text{for } f \in L^2(\Omega) \tag{5.381}$$

$$= \int_{\Omega} (\widehat{Q'^h f} - f)(t)\hat{v}^h(t) J(E_c)'(t)dt = 0 \text{ and}$$

$$Q_c'^h : H^{-1}\left(\Omega_c^h\right) \to \mathcal{V}^{h'} : \langle Q_c'^h \check{f} - \check{f}, v^h\rangle_{H^{-1}(\Omega_c^h)\times\mathcal{V}_b^h} \quad \forall v^h \in \mathcal{V}_b^h$$

$$= \int_{\Omega_c^h} \left(Q_c'^h \check{f} - \check{f}\right)(x) v^h(x) dx = 0 \;\; \text{for } \check{f} \in L^2\left(\Omega_c^h\right) \qquad \text{and}$$

$$\langle f^h, v^h\rangle := \langle f, v^h\rangle^h := \langle \check{f}, v^h\rangle_{H^{-1}(\Omega_c^h)\times H_0^1(\Omega_c^h)} = \langle Q'^h f, v^h\rangle_{H^{-1}(\Omega_c^h)\times\mathcal{V}_c^h} \text{ and}$$

$$A^h := Q_c'^h A_h|_{\mathcal{V}_b^h} \text{ and } Q_c'^h G_h|_{\mathcal{V}_b^h} : \mathcal{V}_b^h \to \mathcal{V}^{h'}, \quad \text{with the}$$

$$\text{above} \;\; a^h(u^h, v^h) = \langle A^h u^h, v^h\rangle_{\mathcal{V}^{h'}\times\mathcal{V}^h} \text{ and } \langle G^h u^h, v^h\rangle_{\mathcal{V}^{h'}\times\mathcal{V}^h}.$$

Here the spaces $\mathcal{V}_b^h \not\subset \mathcal{V}$ are defined on Ω_c^h, and thus cannot satisfy the boundary conditions, cf. Figure 5.8.

Theorem 5.83. Summary for isoparametric FEMs for linear and quasilinear second order equations and systems, cf. the following proofs:

1. *Let $A \in \mathcal{L}(\mathcal{V}_b, \mathcal{V}_b')$ be a boundedly invertible linear elliptic differential operator and $f \in L^2(\Omega)$. Then the exact solution u_0 of $Au_0 = f$ uniquely exists, with $u_0 \in H^2(\Omega)$ for convex Ω. For the quasilinear G assume a unique solution $u_0 \in H^2(\Omega)$ with boundedly invertible $G'(u_0)$. Finally, let $\Omega \subset \mathbb{R}^2$ be bounded and have a piecewise smooth boundary.*

2. *Then the isoparametric FEMs with the equations* $A^h u_0^h = f^h$ *and* $G^h u_0^h = 0$*, see Theorems 5.84 ff. for the necessary conditions, are consistent and stable and (locally) unique discrete solutions* u_0^h *exist and converge to* $u_0 \in H^d(\Omega)$

$$\left\| u_0 - \hat{u}_0^h \right\|_{H^1(\Omega)} \leq C h^{d-1} \left(\|u_0\|_{H^d(\Omega)} + \|u_0\|_{W^1_\infty(\Omega)} \right).$$

3. *Quadrature approximations as in Theorems 5.80 and 5.81 yield corresponding convergence.*
4. *For isoparametric FEs the above nonlinear* E_c*, see (5.379), determines the discretization. However, since* $(E_c)' - id = \mathcal{O}(h^d)$*, see (4.101), the discrete* A^h *and* G^h *for the isoparametric case is still* k*-times* consistently differentiable *with* A *and* G *in* u_0*, respectively. This concept will play an essential role for the* mesh independence principle *and in Böhmer [120] for numerical methods in bifurcation and center manifolds.*

In a first step we prove the *discrete coercivity*, cf. Theorems 5.59 and 4.61:

Theorem 5.84. Isoparametric coercivity: *We choose the above* $\Omega \subset \mathbb{R}^2$*, the isoparametric FEs and the* $a^h(\cdot,\cdot), \langle f, \cdot\rangle^h$ *and* Q'^h*. If* $a(\cdot,\cdot)$ *is* $\mathcal{V}_b$*-coercive, see (5.302), then* $a^h(\cdot,\cdot)$ *is* $\mathcal{V}_b^h$*-coercive.*

Proof. We present it for the Laplace operator using the substitution rule in

$$a^h(u^h, v^h) = \int_{\Omega_c^h} \nabla u^h(x) \cdot \nabla v^h(x)(dx) \tag{5.382}$$

$$= \int_\Omega \nabla u^h(E_c(t)) \cdot \nabla v^h(E_c(t)) \det(E_c)'(t) dt. \tag{5.383}$$

The usual $|\det(E_c)'(t)|$ is replaced by $\det(E_c)'(t)$, since, with the identity id,

$$(E_c)'(t) = id + \mathcal{O}(h^{d-1}), \quad \text{hence, } \det(E_c)'(t) = 1 + \mathcal{O}(h^{d-1}) > 0 \tag{5.384}$$

for sufficiently small h. For the above $\hat{u}^h(t) = u^h(E_c(t))$ and $u^h(x)$ the derivatives ∂^{t_i}, ∇^t and ∂^{x_i}, ∇^x are always understood with respect to the variable t for $\hat{u}^h$ and x for u^h, respectively. Then we find with the unit vector e_i

$$\begin{aligned}
\partial^i \hat{u}^h(t) = \partial^{t_i} \hat{u}^h(t) &= \partial((u^h \circ E_c)(t))/\partial t_i = \nabla^x u^h(E_c(t)) \partial^i E_c(t) \\
&= \nabla^x u^h(x)(E_c)'(t) e_i, \text{ with } x = E_c(t) \text{ and} \\
\nabla \hat{u}^h(t) = \nabla^t \hat{u}^h(t) &= \nabla^x \left(u^h(E_c(t))\right)(E_c)'(t) = \nabla^x u^h(x)(E_c)'(t) \text{ or} \\
\partial^i u^h(x) = \partial^{x_i} u^h(x) &= \partial((\hat{u}^h \circ (E_c)^{-1})(x))/\partial x_i = \nabla^t \hat{u}^h(t) \partial^{x_i}((E_c)^{-1}(x)) \\
&= \nabla^t \hat{u}^h(t)((E_c)^{-1})'(x) e_i \text{ and} \qquad (5.385) \\
\nabla u^h(x) = \nabla^x u^h(x) &= \nabla^t \left((\hat{u}^h \circ (E_c)^{-1}(x))\right)((E_c)^{-1})'(x) \\
&= \nabla^t \hat{u}^h(t)((E_c)^{-1})'(x), \text{ with } t = (E_c)^{-1}(x).
\end{aligned}$$

Applied to the above (5.383) we find with (4.101)

$$a^h(u^h, v^h) = \int_\Omega \left((E_c)'(t)^T\right)^{-1} \nabla \hat{u}^h(t) \cdot \nabla \hat{v}^h(t)((E_c)'(t))^{-1} \det(E_c)'(t)dt$$

$$= a(\hat{u}^h, \hat{v}^h) + \mathcal{O}(h^{d-1})\|\hat{u}^h\|_{H^1(\Omega)} \cdot \|\hat{v}^h\|_{H^1(\Omega)}. \tag{5.386}$$

By (5.384), $\|\hat{v}^h\|_{H^1(\Omega)} = \|v^h\|_{\mathcal{V}^h}(1 + \mathcal{O}(h^{d-1}))$. So with (5.386) and $\hat{\mathcal{V}}_b^h \subset \mathcal{V}_b$, the $\mathcal{V}_b$-coercivity of $a(\hat{u}^h, \hat{u}^h)$ implies the $\mathcal{V}_b^h$-coercivity of $a^h(u^h, u^h)$. Hence, there exists a unique solution u_0^h for $a^h\left(u_0^h, v^h\right) = (f, v^h)^h \ \ \forall\, v^h \in \mathcal{V}_b^h$, see (5.381).

For the general case in (5.381) we find similarly, for smooth enough a_{ij}

$$\langle A^h u^h, v^h\rangle_{\mathcal{V}^{h\prime} \times \mathcal{V}_b^h} = a^h(u^h, v^h) = \int_{\Omega_c^h} \sum_{i,j=0}^{n} a_{ij}(x)\partial^i\, u^h \partial^j v^h(x))$$

$$= a(\hat{u}^h, \hat{v}^h) + \mathcal{O}(h^{d-1})\|\hat{u}^h\|_{H^1(\Omega)}\|\hat{v}^h\|_{H^1(\Omega)}. \tag{5.387}$$

An immediate consequence is

$$\sup_{0 \neq v^h \in \mathcal{V}_b^h} |a^h(u^h, v^h) - a(\hat{u}^h, \hat{v}^h)| \ / \ \|v^h\|_{H^1(\Omega)} \leq Ch^{d-1}\|u\|_{H^d(\Omega)}, \tag{5.388}$$

yielding the $\mathcal{V}_b^h$-coercivity of $a^h(u^h, v^h)$ for $u^h = v^h$. □

As a second step we consider the *consistency and convergence,* starting with linear problems. With $a^h(\cdot,\cdot), \langle f, \cdot\rangle^h, Q'^h, Q_c'^h$ as introduced in (5.381), we solve

$$u_0^h \in \mathcal{V}_b^h : \left\langle A^h u_0^h, v^h\right\rangle_{\mathcal{V}^{h\prime} \times \mathcal{V}_b^h} = a^h\left(u_0^h, v^h\right) = \langle f, v^h\rangle^h \ \ \forall\, v^h \in \mathcal{V}_b^h. \tag{5.389}$$

Theorem 5.85. Isoparametric FEMs for linear problems:

1. *For a bounded, piecewise smooth $\Omega \subset \mathbb{R}^n$, our isoparametric FEs, a boundedly invertible A and for $u \in H^d(\Omega), d \geq 1$, let the discrete solution u^h for (5.389) with $f := Au$ exist. Then A^h is consistent with A in u, and the* classical *and* variational consistency errors *for u vanish of order $d-1$. They can be estimated, with C, independent of h, and for h sufficiently small, by*

$$\sup_{0 \neq v^h \in \mathcal{V}_b^h} |a^h(u^h, v^h) - a(u, \hat{v}^h)|/\|v^h\|_{H^1(\Omega)} \leq Ch^{d-1}\|u\|_{H^1(\Omega)}, \tag{5.390}$$

$$\|A^h I^h u - Q'^h Au\|_{\mathcal{V}^{h\prime}} \leq Ch^{d-1}\|u\|_{H^d(\Omega)}, \tag{5.391}$$

$$\sup_{0 \neq v^h \in \mathcal{V}_b^h} |(f, v^h) - (\check{f}, \hat{v}^h)|/\|v^h\|_{H^1(\Omega)}$$

$$\leq Ch^{d-1}\|f\|_{H^{-1}(\Omega)} \leq Ch^{d-1}\|u\|_{H^1(\Omega)}. \tag{5.392}$$

2. *For an invertible A, and a stable A^h, let $u_0 \in W_\infty^d(\Omega)$ and u_0^h be the exact and discrete solutions of $Au_0 = f$ and (5.389), respectively. Then*

$$\left\|u_0 - \hat{u}_0^h\right\|_{H^1(\Omega)} \leq Ch^{d-1}\left(\|u_0\|_{H^d(\Omega)} + \|u_0\|_{W_\infty^1(\Omega)}\right).$$

Proof. By (5.381) the $I^h, Q'^h, Q_c'^h$ are equibounded linear operators with, e.g.

$$\lim_{h\to 0} \|Q'^h f\|_{H^{-1}(\Omega_c^h)} = \|f\|_{H^{-1}(\Omega)}.$$

Obviously (5.388) implies (5.390). The error term for f is estimated as:

$$\begin{aligned}
\left|\langle f, \hat{v}^h\rangle - \int_{\Omega_c^h} \check{f} v^h dx\right| &= \left|\int_\Omega \big(f(t)\hat{v}^h(t) - f(t)\hat{v}^h(t)dt \det((E_c)^{-1})'(t)\big)dt\right| \\
&\leq \left|\int_\Omega f(t)(1 - \det((E_c)^{-1})'(t))\hat{v}^h(t)dt\right| \\
&\leq Ch^{d-1}\|f\|_{H^{-1}(\Omega)}\|\hat{v}^h\|_{L^2(\Omega)} \qquad (5.393)\\
&\leq Ch^{d-1}\|u\|_{H^1(\Omega)}\|\hat{v}^h\|_{L^2(\Omega)} \text{ for } f = Au.
\end{aligned}$$

The proof of (5.391) is delayed to the next theorem for the quasilinear case. □

The next theorem, some conditions, and its proof are very similar to Theorem 5.4, and e.g. (5.66), however with the ℓ there replaced by d here.

Theorem 5.86. Isoparametric FEMs for quasilinear problems:

1. *We require the conditions for Theorems 4.43, 4.44, 5.84, 5.85, a Lipschitz-continuous quasilinear operator G with a global constant L,*

$$G, A_i \in C_L(\mathcal{D}(G)), i = 0, \dots n. \qquad (5.394)$$

2. *Then for $u \in H^d(\Omega) \cap \mathcal{D}(G), d > 1$, G^h is consistent with G in $I^h u$ and*

$$\|G^h I^h u - Q'^h Gu\|_{\mathcal{V}_\Pi^{h'}} = \left\|Q_c'^h G_h I^h u - Q'^h Gu\right\|_{\mathcal{V}_\Pi^{h'}} \leq CLh^{d-1}\|u\|_{H^d(\Omega)}. \qquad (5.395)$$

3. *If $G(u) = 0$ posesses a unique solution $u_0 \in H^d(\Omega)$ with boundedly invertible $G'(u_0)$, then the $G^h\left(u_0^h\right) = 0$ are locally uniqely solvable as well and converge according to*

$$\left\|u_0 - \hat{u}_0^h\right\|_{H^1(\Omega)} \leq CLh^{d-1}\|u_0\|_{H^d(\Omega)}.$$

Proof. We estimate the left-hand side with (5.381), (4.105), and L in (5.394):

$$\begin{aligned}
&\left\|Q_c'^h G_h I^h u - Q'^h Gu\right\|_{\mathcal{V}^{h'}} - \left\|Q_c'^h G_h E_c u - Q'^h Gu\right\|_{\mathcal{V}^{h'}} \qquad (5.396)\\
&\leq \left\|Q_c'^h G_h I^h u - Q_c'^h G_h E_c u\right\|_{\mathcal{V}^{h'}} \leq CL\|I^h u - E_c u\|_{\mathcal{V}^h} \leq CLh^{d-1}\|u\|_{H^d(\Omega)}.
\end{aligned}$$

We test $Q_c'^h G_h E_c u - Q'^h Gu$ with v^h, and combine it with (5.379), (5.380), (5.381). For the estimate we need the set of triangles with at most one boundary point,

$$\mathcal{T}_i^h := \left\{T \in \mathcal{T}_c^h \text{ in Definition 4.42 1.}\right\} : \Omega_b := \Omega \setminus \left(\cup_{T \in \mathcal{T}_i^h} T\right),$$

$$\Omega_b^h := \Omega_c^h \setminus \left(\cup_{T\in\mathcal{T}_i^h} T\right).$$

Then we obtain with $G_h(E_c u)v^h - Gu(v^h \circ E_c) \equiv 0$ in $\Omega_i^h := \Omega_c^h \cap \left(\cup_{T \in \mathcal{T}_i^h} T\right) = \Omega_i$

$$\begin{aligned}\left|\left\langle Q_c'^h G_h E_c u - Q'^h Gu, v^h \right\rangle_{\mathcal{V}^{h'} \times \mathcal{V}^h}\right| &= \left|\int_{\Omega_c^h} G_h E_c u v^h dx - \int_\Omega Gu\hat{v}^h dx\right| \; \forall v^h \in \mathcal{V}_b^h \\ &\le \left|\int_{\Omega_b^h} G_h(E_c u) v^h dx - \int_{\Omega_b} Gu(v^h \circ E_c) dx\right|. \end{aligned} \tag{5.397}$$

In $\Omega_b^h \cap \Omega_b$ the integrands are $G_h(E_c u)v^h - Gu(v^h \circ E_c) = \mathcal{O}(h^d)$ by (4.101) and (5.394). Furthermore, the remaining strips $\Omega_b^h \setminus \Omega_b$ and $\Omega_b \setminus \Omega_b^h$ have a width $\mathcal{O}(h^d)$ and the integrands are equibounded. So (5.394), (4.100), (4.101) imply (5.395). The convergence claim is, for a coercive $G'(u_0)$, a consequence of Theorem 5.84 and (5.395). □

6

Adaptive finite element methods, by W. Dörfler

6.1 Introduction

The finite element method (FEM) has become a flexible tool to solve partial differential equations numerically. It is especially useful in program packages where the user typically adds his geometry. This is comfortable even for the nonexpert. However, such use can be dangerous if no controls exist. For example, in 1991, the offshore oil platform *Sleipner* broke down and this failure could be traced back to an underestimation of the shear stresses in some part of the structure because of insufficient grid resolution [51], Ch. 1.2. Thus the remedy would be to set up an algorithm that provides sufficient refinement in an efficient way and that provides a measure for the quality of the solution. The task of this chapter is to present techniques to overcome these deficiencies by a posteriori error estimation and local mesh refinement.

The theory of a posteriori error estimation for partial differential equations came up about 1980 [49,50]. Combined with methods of local mesh refinement (or coarsening) it is called the *adaptive finite element method* (AFEM). Its applicability has now been demonstrated in many circumstances and can be viewed as a standard tool nowadays.

These techniques are now widespread, concerning the type of error estimation and mesh refinement techniques. In this short overview we will present the main ideas only and we will further restrict to the Poisson problem with Dirichlet boundary conditions (6.1), (6.2). This equation can be seen as a simplified model of linear elasticity which played a role in the example mentioned above (Section 6.1.1). We explain that solutions of such equations may exhibit singular behavior and the consequences for the finite element method (Section 6.1.2). A heuristic argument shows that local mesh refinement can overcome this problem (Section 6.1.4). We develop the theory of a posteriori error estimation with the residual error estimator (Section 6.2). This states that the adaptive algorithm can be designed to converge as an iterative method (Section 6.2.6) and that it leads in fact to the optimal grid in a certain sense (Section 6.2.7).

Only briefly will we mention some alternatives to the residual error estimator for estimating the energy norm in Section 6.2.8 and refer to the literature. In the final Section 6.3 we report on the methodology of estimating errors in quantities of interest.

Finally we mention a few books and overview articles on all these methods such as, for example, [6,51,77,499,655].

6.1.1 The model problem

Let $\Omega \subset \mathbb{R}^3$ be the domain occupied by an elastic body. The continuum mechanical model yields the equations

$$-\nabla\cdot\left(\frac{1}{2}C(\nabla u + (\nabla u)^d)\right) = f \quad \text{in } \Omega$$

for the deformation field $u : \Omega \to \mathbb{R}^3$ and the elasticity tensor C. For simplification we study the equation

$$-\nabla\cdot(a\nabla u) = f \quad \text{in } \Omega \tag{6.1}$$

to obtain $u : \bar{\Omega} \to \mathbb{R}$ for given $a : \bar{\Omega} \to \mathbb{R}_+$ and $f : \bar{\Omega} \to \mathbb{R}$. To state the boundary conditions we assume a disjoint decomposition of $\partial\Omega$

$$\partial\Omega = \partial_{\mathrm{D}}\Omega \cup \partial_{\mathrm{N}}\Omega$$

and we require the solution of (6.1) to satisfy

$$u\big|_{\partial_{\mathrm{D}}\Omega} = g^{\mathrm{D}}, \quad a\nabla u \cdot n\big|_{\partial_{\mathrm{N}}\Omega} = g^{\mathrm{N}}. \tag{6.2}$$

6.1.2 Singular solutions

Let us consider the boundary value problem (6.1), (6.2). Our question is: are solutions of this equation in $C^2(\bar{\Omega})$ as the formulation would require? In general the answer is no, if $\partial\Omega$ is not smooth.

***Corner singularities in* 2D**

A point x at the boundary $\partial\Omega$ is called a *corner point* if the boundary arc is continuous and one-sided differentiable in x, but not differentiable. It can be shown that such a situation can be locally described by that of a solution in a suitable angular domain with angle θ and tip at 0. Special solutions for the case $a = 1$, $f = 0$ are of the form

$$u(r, \phi) = r^\gamma \sin\left(m\pi\frac{\phi}{\theta}\right), \quad r := |x|, \ \phi = \arg(x),$$

in the case of homogeneous Dirichlet boundary values for $\phi = 0$ and $\phi = \theta$. Insertion into $-\Delta u = 0$ yields

$$\left(-\gamma^2 + m^2\frac{\pi^2}{\theta^2}\right) r^{\gamma-2} \sin\left(m\pi\frac{\phi}{\theta}\right) = 0$$

for all $r \in (0, \infty)$ and $\phi \in (0, \theta)$. This equation for γ is solved by

$$\gamma^2 = m^2\frac{\pi^2}{\theta^2}.$$

The smallest positive solution for γ is obtained for $m = 1$:

$$u(r, \phi) = r^{\pi/\theta} \sin\left(\pi\frac{\phi}{\theta}\right). \tag{6.3}$$

For $\gamma < 0$, the solution would not satisfy our least requirement that u has finite energy, i.e. $\int_\Omega |\nabla u|^2$ exists. In the case of homogeneous Neumann conditions at $\{\phi = 0\}$ or $\{\phi = 0, \pi\}$ we have to take

$$u(r,\phi) = r^\gamma \sin\left(\frac{\pi}{2}\frac{\phi}{\theta}\right) u(r,\phi) = r^\gamma \cos\left(\pi\frac{\phi}{\theta}\right),$$

respectively. Note that we would get in the first case

$$u(r,\phi) = r^{\pi/(2\theta)} \sin\left(\frac{\pi}{2}\frac{\phi}{\theta}\right).$$

For the slit domain ($\theta = 2\pi$), this yields the worst cases

$$u(r,\phi) = r^{1/2} \quad \text{or} \quad u(r,\phi) = r^{1/4}.$$

We will study the regularity of these solutions in *Sobolev function spaces*. Recall the definitions in Section 1.4.3:

$$W^{m,p}(\Omega) := \left\{v : \Omega \to \mathbb{R} \, : \, \|v\|_{W^{m,p}(\Omega)} := \Big(\sum_{|\alpha|\le m} \int_\Omega |\nabla^\alpha v|^p\Big)^{1/p} < \infty\right\},$$

$$H^m(\Omega) := W^{m,2}(\Omega).$$

Moreover, recall that this definition can extended to $m \in \mathbb{R}$, cf. Theorem 1.30 and [668]. We may consider u as above on the sector $\Omega = \{(r,\phi) \, : \, r \in (0,1), \, \phi \in (0,\theta)\}$. For u that behaves like r^γ near 0 we find that

$$\|u\|^p_{W^{m,p}(\Omega)} \le C \int_0^1 |r^{\gamma - m}|^p r \, dr = C \int_0^1 r^{(\gamma-m)p+1} \, dr \overset{!}{<} \infty$$

holds if $(\gamma - m)p + 1 > -1 \Leftrightarrow \gamma > m - \frac{2}{p}$. For $u \in H^m(\Omega)$ we need the bound $m < 1 + \gamma = 1 + \pi/\theta$, respectively $1 + \pi/(2\theta)$. Solutions of Dirichlet boundary value problems are thus at least in $H^{3/2}(\Omega)$; for mixed Dirichlet and Neumann conditions at least in $H^{5/4}(\Omega)$. If we consider the existence in $W^{2,p}$ spaces, this allows arbitrary $p \in [1,\infty]$ for $\gamma \ge 2$ but $p < \frac{2}{2-\gamma}$ for $\gamma \in (0,2)$. For $\gamma < 1$ we will have $W^{2,p}$ regularity for some $p \in [1,2)$. Especially, $u \in W^{2,1}(\Omega)$ holds in general. Only for $\gamma \ge 1$ do we have H^2 regularity.

The *general structure of singular solutions* u of elliptic partial differential equations is as follows. Given $m \ge 2$, there exists $n_m \in \mathbb{N}$, smooth functions w_j that are locally supported near the singular point, e.g. 0, of $\partial\Omega$, and singular functions η_j (often $\eta_j(x) = |x|^{\alpha_j}$ for some $\alpha_j \in \mathbb{R}$) such that

$$u(x) = \sum_{j=1}^{n_m} w_j(x)\eta_j(x) + w(x)$$

for $w \in W^{m,2}(\Omega)$ [255, 372, 446, 455].

***Corner and edge singularities in* 3D**

A well studied example of a corner point in three dimensions is the *Fichera corner* in 0 for $\Omega := (-1,1)^3 \setminus [0,1)^3$. The solution of $\Delta u = 0$ near 0 behaves like $x \mapsto |x|^\gamma$ for $\gamma = 0.45\ldots$. The problem of finding γ leads to a *quadratic eigenvalue problem* $(\lambda^2 A + \lambda B + C)u = 0$ for $\lambda \in \mathbb{R}$, $u \neq 0$ [35]. Here A, B, C are suitable operators. Besides the corners, there are additional singular solutions along the edges [212].

Problems with jumping coefficients

We consider Equation (6.1) for a that is piecewise constant in the four quadrants of the coordinate system with values $a_1, \ldots, a_4$. This problem could appear for flow in porous media where large jumps in the permeability a may in fact occur. Again we try an ansatz of the form $u(r,\phi) = r^\gamma s(\phi)$ to get in each quadrant

$$a(s'' - \gamma^2 s) r^{\gamma-2} \stackrel{!}{=} 0,$$

that is, the following eigenvalue problem for γ

$$s'' = \gamma^2 s \quad \text{in } \left(j\frac{\pi}{4}, (j+1)\frac{\pi}{4}\right) \quad \text{for } j = 0,1,2,3.$$

The solution must have a continuous flux $(a\nabla u)\cdot n$ along the normal n of the interfaces, thus we can require periodicity of s and $a_j s'\left(j\frac{\pi}{4}-\right) = a_{j+1} s'\left(j\frac{\pi}{4}+\right)$ for $j = 0,1,2,3$ [533]. In the case of $a_1 = a_3 = 1$ and $a_2 = a_4 \approx 100$ one gets a value $\gamma = 0.1$. Thus $u(x)$ behaves like $\sqrt[10]{|x|}$ near 0! Moreover, $u \in H^{1.1}(\Omega)$ [444, 498].

Boundary layers

We consider the case of, cf. [559]

$$\begin{aligned} -\varepsilon\Delta u + b\cdot\nabla u + cu &= f \quad \text{in } \Omega, \\ u &= 0 \quad \text{on } \partial\Omega, \end{aligned}$$

with constant coefficients ε, b, c, where $b \neq 0$ or $c \neq 0$ and $0 < \varepsilon \ll |b| + |c|$. For example, the typical behavior of solutions near *outflow points* of $\partial\Omega$ (i.e. where $b \cdot n > 0$) is $|\nabla^l u| \sim 1/\varepsilon^l$.[43] At other parts of the boundary, a weaker dependence on $1/\varepsilon$ will occur. Although u might be smooth, it appears as a nonsmooth function unless the discretization method resolves the scale ε, which will lead to prohibitive work on uniform meshes.

6.1.3 A priori error bounds

In the following, we will simplify (6.1) further by assuming $a = 1$. The *weak form* of

$$\begin{aligned} -\nabla\cdot\nabla u_0 = -\Delta u_0 &= f \quad \text{in } \Omega, \\ u &= 0 \quad \text{on } \partial\Omega \end{aligned}$$

[43] $a \sim b :\Leftrightarrow \exists c_1, c_2 : 0 < c_1 < c_2$ such that $c_1 a \leq b \leq c_2$.

is given by

$$\int_\Omega \nabla u_0 \cdot \nabla v = \int_\Omega f v \qquad \text{for all } v \in C_0^\infty(\Omega). \tag{6.4}$$

As previously in this book, we denote solutions of specific problems, e.g. of (6.4) and (6.5), as u_0 and u_0^h, respectively For given $f \in L^2(\Omega)$, there exists a solution $u_0 \in H_0^1(\Omega)$ of this problem (cf. Theorem 2.2 and Theorem 2.43 in Chapter 2). By density of $C_0^\infty(\Omega)$ in $H_0^1(\Omega)$ we can equally well state "for all $v \in H_0^1(\Omega)$" in (6.4).

Let $\mathcal{V}^h$ be a finite dimensional subspace of $\mathcal{V} := H_0^1(\Omega)$. Then the *Galerkin approximation* $u_0^h \in \mathcal{V}^h$ is defined by

$$\int_\Omega \nabla u_0^h \cdot \nabla v^h = \int_\Omega f v^h \quad \text{for all } v^h \in \mathcal{V}^h. \tag{6.5}$$

It is well known that both problems are uniquely solvable and that (Cea's lemma, cf. Lemma 4.48 in Section 4.3)

$$\|\nabla\left(u_0 - u_0^h\right)\|_{L^2(\Omega)} \le C \inf_{v^h \in \mathcal{V}^h} \|\nabla(u_0 - v^h)\|_{L^2(\Omega)}. \tag{6.6}$$

If we have a continuous mapping $P^h : H_0^1(\Omega) \to \mathcal{V}^h$, we can further estimate the right-hand side of (6.6) to get

$$\|\nabla\left(u_0 - u_0^h\right)\|_{L^2(\Omega)} \le C \|\nabla\left(u_0 - P^h u_0\right)\|_{L^2(\Omega)}.$$

Let $\mathcal{V}^h$ be a finite element space that at least contains the linear finite elements. Then there exists a continuous *interpolation operator* $P^h : H_0^1(\Omega) \longrightarrow \mathcal{V}^h$ with the property

$$\|\nabla(v - P^h v)\|_{L^2(\Omega)} \le C h_{\max} \|\nabla^2 v\|_{L^2(\Omega)}$$

for all $v \in H^2(\Omega)$. Here, $h_{\max}$ is the maximal diameter of the mesh cells in the underlying discretization. This is a special case of Theorem 4.17 in Section 4.2 of Chapter 4. However, $u \in H^2(\Omega)$ is not always true as we have seen in Section 6.1.2. In fact, we get $u \in H^{1+s}(\Omega)$ for $s > 0$ only.

Theorem 6.1. Interpolation operator: *Let $\mathcal{V}^h$ be a finite element space that at least contains the linear finite elements. Then there exists a continuous* interpolation operator *$P^h : H_0^1(\Omega) \longrightarrow \mathcal{V}^h$ with the property*

$$\|\nabla(v - P^h v)\|_{L^2(\Omega)} \le C h_{\max}^s \|\nabla v\|_{H^s(\Omega)} \quad \textit{for all } v \in H^{1+s}(\Omega)$$

and suitable $s \in (0, 1]$. Here, $h_{\max}$ is the maximal diameter of the mesh cells in the underlying discretization and C depends on properties of the mesh [141], Ch. 14.3.

This leads to the following error bound.

Theorem 6.2. A priori error bound: *The error for the finite element solution of* (6.5) *on $\mathcal{V}^h$ as in Theorem 6.1 obeys the* a priori error bound

$$\|\nabla\left(u_0 - u_0^h\right)\|_{L^2(\Omega)} \le C h_{\max}^s \|\nabla u_0\|_{H^s(\Omega)}$$

for some $s \in (0,1]$. Here, $h_{\max}$ is the maximal diameter of the mesh cells in the underlying discretization and C depends on properties of the mesh (cf. Section 6.1.7).

The meaning of this result is that a lack of regularity (smaller s) leads to a low convergence rate for the error in terms of the maximal mesh size in the underlying discretization. Or, in other words, for a fixed mesh, there is a significant loss in accuracy if one approximates less regular functions.

6.1.4 Necessity of nonuniform mesh refinement

As we have observed in Section 6.1.3, there are cases where solutions of (6.4) are in $H^{1+s}(\Omega)$ for $s \in (0,1]$. By Theorem 6.2 the H^1 error behaves like h^s on, say, a uniform mesh of mesh size h. In this case we have $N^h = O(1/h^n)$ unknowns for linear finite elements in a bounded domain in $\mathbb{R}^n$. In terms of N^h the H^1 error thus behaves like $h^s \sim (N^h)^{-s/n}$. We discuss this, as an example, in the case $n = 2$.

- $\Omega \subset \mathbb{R}^2$ and $s = 1$ (which is the full regular case). Now the error behaves like $(N^h)^{-1/2}$ and to achieve a prescribed tolerance TOL, one needs $(N^h)^{-1/2} \sim$ TOL, or, $N^h \sim \mathrm{TOL}^{-2}$. Thus, if we start with an error of size 1 and if we want to reduce the error by a factor of 100, we have to increase N^h by a factor of 10^4.
- $\Omega \subset \mathbb{R}^2$ and $s = 1/2$ (the case of the "slitted domain" $\Omega = B_1(0) \setminus \{x \in \mathbb{R}^2 : x_1 < 0,\ x_2 = 0\}$). Now the error decreases like $(N^h)^{-1/4}$ and to achieve a prescribed tolerance TOL we need $N^h \sim \mathrm{TOL}^{-4}$ unknowns. That means, in the same situation as above we now need about 10^8 unknowns!

6.1.5 Optimal meshes – A heuristic argument

(See [78].) Let $\Omega \subset \mathbb{R}^n$ be bounded and let Ω be decomposed into simplices or rectangles. We assume that we are given a *mesh-size function* $h : \Omega \longrightarrow \mathbb{R}_+$ with $h(x)$ describing the local mesh size near x (i.e. $x \in T \Rightarrow h(x) \approx \mathrm{diam}(T)$). Let the function σ denote a geometry dependent function,

$$\sigma : \Omega \longrightarrow \mathbb{R}_+ \quad \text{with} \quad \sigma(x_T) = \frac{h(x_T)^n}{\mathrm{vol}(T)}.$$

For linear finite elements we have approximately one unknown per simplex. Therefore the total *number of unknowns* $\mathcal{N}(h)$ for the linear finite element space is approximately given by

$$\mathcal{N}(h) \approx \sum_T 1 = \sum_{T \in \mathcal{T}^h} \frac{\mathrm{vol}(T)}{\mathrm{vol}(T)} = \sum_{T \in \mathcal{T}^h} \frac{\sigma(x_T)}{h(x_T)^n} \mathrm{vol}(T) \approx \int_\Omega \frac{\sigma(x)}{h(x)^n}\, dx. \tag{6.7}$$

Thus we define for a given mesh-size function h the number of unknowns as

$$\mathcal{N}(h) := \int_\Omega \frac{\sigma}{h^n}.$$

Assume that we have a representation of the local error in the form

$$\left|\nabla \left(u_0 - u_0^h\right)(x)\right| \approx h(x)^\alpha E(x), \tag{6.8}$$

where α is the convergence order and E does not depend on h. For linear finite elements and $u_0 \in H^2(\Omega)$ we have for example $\alpha = 1$, $E(x) = |\nabla^2 u_0(x)|$. If the solution u_0, we are trying to approximate, is in $H^2_{\text{loc}}(\Omega)$, then the global error $\mathcal{E}(h)$ is

$$\mathcal{E}(h) := \Big(\int_\Omega (h^\alpha E)^2\Big)^{\frac{1}{2}}.$$

We further assume that the total work to establish the approximate solution is proportional to $\mathcal{N}(h)$ (e.g. use a multigrid method cf. [141]). Our aim is to find a mesh-size function h that gets an error of TOL with lowest possible number of unknowns. In other words, we have to find a mesh-size function h such that

$$\mathcal{N}(h) \text{ is minimal while } \mathcal{E}(h) = \text{TOL}.$$

This is a constraint optimization problem! To solve it we introduce the Lagrange function

$$\mathcal{L}(h,\lambda) = \mathcal{N}(h) - \lambda\big(\text{TOL}^2 - \mathcal{E}(h)^2\big) \tag{6.9}$$

$$= \int_\Omega \frac{\sigma}{h^n} - \lambda\Big(\text{TOL}^2 - \int_\Omega h^{2\alpha}E^2\Big). \tag{6.10}$$

The point (h,λ) we are looking for is a stationary point of $\mathcal{L}$. Thus for all $\phi \in C^\infty(\Omega)$ we get

$$0 = \partial_1 \mathcal{L}(h,\lambda)[\phi] = \int_\Omega \Big\{-\frac{n\sigma}{h^{n+1}} + 2\alpha\lambda h^{2\alpha-1}E^2\Big\}\phi.$$

This requires the pointwise condition $-n\sigma/h^{n+1} + 2\alpha\lambda h^{2\alpha-1}E^2 = 0$ and leads to

$$h^{2\alpha}E^2 = \frac{n}{2\alpha\lambda}\frac{\sigma}{h^n} \quad \text{in } \Omega.$$

To satisfy the constraint

$$\text{TOL}^2 = \mathcal{E}(h)^2 = \int_\Omega h^{2\alpha}E^2 = \int_\Omega \frac{n}{2\alpha\lambda}\frac{\sigma}{h^n} = \frac{n}{2\alpha\lambda}\mathcal{N}(h)$$

λ should be taken as

$$\lambda = \frac{n}{2\alpha}\frac{\mathcal{N}(h)}{\text{TOL}^2}.$$

Insertion into the equation for h yields

$$h^{2\alpha}E^2 = \frac{\text{TOL}^2}{\mathcal{N}(h)}\frac{\sigma}{h^n} \quad \text{or} \quad \frac{1}{\sigma}h^{n+2\alpha}E^2 = \frac{\text{TOL}^2}{\mathcal{N}(h)}. \tag{6.11}$$

The left-hand side depends on x, but the right-hand side is independent of x. Hence our result is to distribute $h^{n+2\alpha}E^2/\sigma$ equally on the mesh. For any T in the discretization, let $\mathcal{E}_T^2(h) := \int_T h^{2\alpha}E^2$ be the local error on T and $\mathcal{N}_T(h) := \int_T \sigma/h^n$ be the local number of degrees of freedom in T. Then we have by (6.11)

$$\mathcal{E}_T^2(h) = \int_T h^{2\alpha}E^2 = \frac{\text{TOL}^2}{\mathcal{N}(h)}\int_T \frac{\sigma}{h^n} = \frac{\text{TOL}^2}{\mathcal{N}(h)}\mathcal{N}_T(h).$$

In our setting, so far, $\mathcal{N}_T$ is a constant depending on the finite element method.

Thus our result is that we have to equidistribute the error over the decomposition of Ω.

From (6.11) we get the optimal mesh size h_* and if $\int_\Omega |E|^{2n/(n+2\alpha)} < \infty$, we obtain from this

$$\mathcal{N}(h_*) \sim \mathrm{TOL}^{-n/\alpha}.$$

This recovers the result we would get for the finite element method for the full regular problem (see Section 6.1.4)!

6.1.6 Optimal meshes for 2D corner singularities

We apply the previous result to the case of a typical corner singularity in two space dimensions for $-\Delta u = 0$ as in Section 6.1.2. For simplicity, we assume that the corner point is located at 0 and that Ω is near 0 a sector of the unit circle with angle $\theta = \beta\pi$, $\beta \in (1,2)$. Then the solution (6.3) behaves like $|\nabla^l u_0(x)| \sim r^{1/\beta - l}$ for $r = |x| \in (0,1)$, $l \geq 0$.

Let us assume a discretization with a uniform mesh size $h(x) := h_0$ for all $x \in \Omega$. In $\Omega \cap B^h(0)$ the error is given by

$$\|\nabla\left(u_0 - u_0^h\right)\|^2_{L^2(B_{h_0}(0))} \approx \|\nabla u_0\|^2_{L^2(B_{h_0}(0))} \sim \int_0^{h_0} r^{\frac{2}{\beta}-2} r\, dr \sim h_0^{\frac{2}{\beta}},$$

while in the remaining domain, by (6.8),

$$\|\nabla\left(u_0 - u_0^h\right)\|^2_{L^2(\Omega\setminus B_{h_0}(0))} \sim \|h_0 \nabla^2 u_0\|^2_{L^2(\Omega\setminus B_{h_0}(0))} \sim h_0^2 \int_{h_0}^1 r^{\frac{2}{\beta}-3}\, dr \sim h_0^2 h_0^{\frac{2}{\beta}-2} \sim h_0^{\frac{2}{\beta}}.$$

Thus $\|\nabla\left(u_0 - u_0^h\right)\|_{L^2(\Omega)} \sim h_0^{\frac{1}{\beta}}$ and the requirement

$$\|\nabla\left(u_0 - u_0^h\right)\|_{L^2(\Omega)} \sim \mathcal{E}(h_0) \stackrel{!}{=} \mathrm{TOL}$$

leads to the relation, cf. (6.7),

$$N(h_0) \sim h_0^{-2} = \left(h_0^{\frac{1}{\beta}}\right)^{-2\beta} = \mathrm{TOL}^{-2\beta}.$$

We already noticed this relation for $\beta = 2$ in Section 6.1.4. But, what does the optimal mesh-size distribution h look like, and what can we achieve with it? Let us assume that we established the optimal mesh-size distribution. We observed in (6.11) that $h^{n+2\alpha}|\nabla^2 u_0|^2/\sigma = \mathrm{TOL}^2/\mathcal{N}(h) =: K$. For a radial symmetric mesh-size distribution it follows from this that (recall $n = 2$, $\alpha = 1$)

$$h(r)^4 r^{\frac{2}{\beta}-4} \sim K \quad \text{or} \quad h(r) \sim K^{\frac{1}{4}} r^{1-\frac{1}{2\beta}}. \tag{6.12}$$

The constants in this estimate depend on σ, but we assume $\sigma \sim 1$. So the error for such a choice of h will be

$$\mathrm{TOL}^2 = \int_\Omega h^2 |\nabla^2 u|^2 \sim \int_0^1 K^{\frac{1}{2}} r^{2-\frac{1}{\beta}} r^{\frac{2}{\beta}-4}\, r dr = K^{\frac{1}{2}} \int_0^1 r^{\frac{1}{\beta}-1}\, dr \tag{6.13}$$

$$\sim K^{\frac{1}{2}} = \frac{\mathrm{TOL}}{\mathcal{N}(h)^{\frac{1}{2}}}, \tag{6.14}$$

and so we finally achieve

$$\mathcal{N}(h) \sim \mathrm{TOL}^{-2},$$

independent of β. *We thus recover the complexity for the full regular case as stated in Section* 6.1.4 (these statements are only valid in the sense that they show the same asymptotic behavior for large $\mathcal{N}(h)$; the constants have not been considered).

Compared with the second example in Section 6.1.4, we notice that the number of unknowns on the optimal mesh is a factor TOL^2 less than on a uniform mesh, that is a factor 10^{-4} for $\mathrm{TOL} = 0.01$ for this two-dimensional problem.

6.1.7 The finite element method–Notation and requirements

For the reader's convenience we summarize some definitions for the conforming finite element method for Equation (6.4) for simplicial discretizations, cf. Section 4.2.

Let $\Omega \subset \mathbb{R}^n$, an open bounded set and polygonally bounded, be decomposed into a set of simplices $\mathcal{T}^h$, so that

$$\overline{\Omega} = \bigcup_{T \in \mathcal{T}^h} T.$$

In this chapter, simplices are assumed to be closed sets, $T^\circ \cap D^\circ = \emptyset$ for all $T, D \in \mathcal{T}^h$, $T \neq D$ and each vertex of T is also a vertex of a neighbor of T. $\mathcal{T}^h$ is then called a *regular discretization* of Ω (or *triangulation* if $n = 2$). For $T \in \mathcal{T}^h$ let

$$h_T := \mathrm{diam}(T)$$

and let $h : \Omega \to \mathbb{R}_+$ defined by $h\big|_T = h_T$ be the mesh-size function. Furthermore let ρ_T be the radius of the largest inscribed ball in T, σ^h the piecewise constant function with $\sigma_T := h_T/\rho_T \geq 1$, the *shape factor* or *chunkiness parameter* of T. Let $\mathcal{N}^h$ be the set of all vertices, $\mathcal{E}^h$ the set of all edges. Boundary vertices and edges are $\mathcal{N}^{h,\partial}$ and $\mathcal{E}^{h,\partial}$, respectively, and for the interior sets we define $\mathcal{E}^{h,\circ} := \mathcal{E}^h \setminus \mathcal{E}^{h,\partial}$ and $\mathcal{N}^{h,\circ} := \mathcal{N}^h \setminus \mathcal{N}^{h,\partial}$. The space of piecewise polynomials of degree $p \in \mathbb{N}$ is

$$\mathbb{P}_p(\mathcal{T}^h) := \left\{ v : \overline{\Omega} \to \mathbb{R} \,:\, v\big|_T \in \mathbb{P}_p \text{ for all } T \in \mathcal{T}^h \right\}.$$

The space of *conforming finite elements* of degree p is

$$\mathcal{S}^h_p := \mathbb{P}_p(\mathcal{T}^h) \cap C^0(\overline{\Omega}) \subset H^1(\Omega).$$

In the following we will work with $\mathcal{V} := H_0^1(\Omega)$ and its subspace

$$\mathcal{V}^h := \mathcal{S}_p^h \cap H_0^1(\Omega) = \{v \in \mathcal{S}_p^h : v\big|_{\partial\Omega} = 0\}. \tag{6.15}$$

For each $z \in \mathcal{N}^{h,\circ}$, there is a unique function $\phi_z \in \mathcal{S}_1^h$ such that

$$\phi_z(z') = \delta_{zz'} \quad \text{for all } z \in \mathcal{N}^{h,\circ},\ z' \in \mathcal{N}^h.$$

Each $v^h \in \mathcal{S}_1^h$, the space of *linear conforming finite elements*, can be written as

$$v^h = \sum_{z \in \mathcal{N}^{h,\circ}} \alpha_z \phi_z.$$

Inserting such an ansatz into the equation

$$\int_\Omega \nabla u_0^h \cdot \nabla v^h = \int_\Omega f^h v^h \quad \text{for all } v^h \in \mathcal{V}^h$$

yields a linear system of equations

$$A\vec{u} = \vec{b}$$

with $A_{zz'} := \int_\Omega \nabla\phi_{z'} \cdot \nabla\phi_z$, $b_z := \int_\Omega f^h \phi_z$ for $z, z' \in \mathcal{N}^{h,\circ}$ and the vector $\vec{u} = [\alpha_z]_{z\in\mathcal{N}^{h,\circ}}$ of the representation

$$u_0^h = \sum_{z \in \mathcal{N}^{h,\circ}} \alpha_z \phi_z.$$

In contrast to the interpolation estimates cited before in Theorem 6.1 the following are stated for nonuniform meshes.

Theorem 6.3. Interpolation operator: *There exists a continuous* interpolation operator

$$P^h : L^2(\Omega) \to \mathcal{S}_p^h$$

that has for all $v \in H^m(\Omega)$ with $m = 0, 1, \ldots, p+1$ and $l = 0, 1, \ldots, \min(p, m)$ the following properties:

$$\|\nabla^l P^h v\|_{L^2(\Omega)} \le C\|\nabla^l v\|_{L^2(\Omega)} \tag{6.16}$$

and

$$\|\nabla^l (v - P^h v)\|_{L^2(T)} \le C h_T^{m-l} \|\nabla^m v\|_{L^2(S_T)} \quad \textit{for all } T \in \mathcal{T}^h, \tag{6.17}$$

$$\|v - P^h v\|_{L^2(E)} \le C h_E^{l-\frac{1}{2}} \|\nabla^l v\|_{L^2(S_E)} \quad \textit{for all } E \in \mathcal{E}^h, \tag{6.18}$$

where S_T or S_E are the domains composed of all triangles that touch $T \in \mathcal{T}^h$ or $E \in \mathcal{E}^h$, respectively. If $v \in H_0^1(\Omega)$, then $P^h v \in H_0^1(\Omega)$. The constants only depend on m, p and $\sigma_{T'}$ for $T' \subset S_T$, respectively $T' \subset S_E$.

Proof. See [141], Ch. 4.8 or Theorem 4.17 in Section 4.2 of Chapter 4 and 7.7, Corollary 7.8 in Section 7.7 of Chapter 7. □

6.2 The residual error estimator for the Poisson problem

6.2.1 Upper a posteriori bound

Definition 6.4. Jumps and residuals:

1. *Let* $E \in \mathcal{E}^{h,\circ}$. *The* edge jump $[v]_E$ *of a piecewise continuous function* v *on* $\mathcal{T}^h$ *along the edge* E *is defined by*

$$[v]_E(x) := \lim_{s \searrow 0} \{v(x + sn_E) - v(x - sn_E)\} \quad \textit{for } x \in E,$$

where n_E *is a fixed unit normal vector on* E.

2. *Let* u_0^h *be the solution of the discrete equation* (6.5) *in* $\mathcal{V}^h$ *as in* (6.15). *Define the* element residual

$$R_T := \Delta u_0^h + f^h \quad \textit{on } T \in \mathcal{T}^h$$

and the edge residual

$$R_E := [\partial_n u_0^h]_E \quad \textit{on } E \subset \mathcal{E}^{h,\circ}.$$

Note that this definition does not depend on the choice of n_E.

Theorem 6.5. Upper a posteriori error estimate: *Let* u_0, u_0^h *be the solutions of the continuous equation* (6.4) *in* $\mathcal{V} = H_0^1(\Omega)$ *and the discrete equation* (6.5) *in* $\mathcal{V}^h$ *as in* (6.15), *respectively. Then there are sets of positive constants* $\{c_T\}_{T\in\mathcal{T}^h}$, $\{c_E\}_{E\in\mathcal{E}^{h,\circ}}$ *such that*

$$\begin{aligned}\|\nabla\left(u_0 - u_0^h\right)\|^2_{L^2(\Omega)} \leq \sum_{T\in\mathcal{T}^h} c_T h_T^2 \|R_T\|^2_{L^2(T)} + \sum_{E\in\mathcal{E}^{h,\circ}} c_E h_E \|R_E\|^2_{L^2(E)} \\ + C_P(\Omega)\|f - f^h\|^2_{L^2(\Omega)}.\end{aligned}$$

c_T, c_E *depend on* σ^h *(the shape constant) and* p *(the polynomial degree of the finite element space).* $C_P(\Omega)$ *is Poincaré's constant for* Ω *(cf.* (1.57) *and the residuals* R_T, R_E *have been defined in Definition 6.4.*

Proof. Let u_0, u_0^h be the respective solutions and let $v \in \mathcal{V}$, $v^h \in \mathcal{V}^h$ be arbitrary. Then by (6.4) and (6.5) and partial integration

$$\begin{aligned}
&\int_\Omega \nabla\left(u_0 - u_0^h\right)\cdot\nabla v \\
&= \int_\Omega \{fv - \nabla u_0^h \cdot \nabla(v - v^h) - f^h v^h\} \\
&= \int_\Omega \{(f - f^h)v + f^h(v - v^h) - \nabla u_0^h \cdot \nabla(v - v^h)\} \\
&= \int_\Omega (f - f^h)v + \sum_{T\in\mathcal{T}^h}\left\{\int_T \left(f^h + \Delta u_0^h\right)(v - v^h) - \int_{\partial T} \partial^h u_0^h (v - v^h)\right\} \\
&= \int_\Omega (f - f^h)v + \sum_{T\in\mathcal{T}^h}\int_T \left(f^h + \Delta u_0^h\right)(v - v^h) - \sum_{E\in\mathcal{E}^{h,\circ}}\int_E [\partial^h u_0^h]_E (v - v^h). \qquad (6.19)
\end{aligned}$$

Equation (6.19) is called the *error representation formula*. Now we choose $v^h = P^h v$ with P^h as in Theorem 6.3 and use the interpolation estimates to get (note that c_T, c_E will change from estimate to estimate)

$$\begin{aligned}\left|\int_\Omega \nabla\left(u_0 - u_0^h\right)\cdot\nabla v\right| &\le \sum_{T\in\mathcal{T}^h} c_T h_T \|R_T\|_{L^2(T)}\|\nabla v\|_{L^2(S_T)} \\ &\quad + \sum_{E\in\mathcal{E}^{h,\circ}} c_E h_E^{1/2}\|R_E\|_{L^2(E)}\|\nabla v\|_{L^2(S_E)} + \|f - f^h\|_{L^2(\Omega)}\|v\|_{L^2(\Omega)} \\ &\le \left(\sum_{T\in\mathcal{T}^h} c_T^2 h_T^2 \|R_T\|_{L^2(T)}^2 + \sum_{E\in\mathcal{E}^{h,\circ}} c_E^2 h_E \|R_E\|_{L^2(E)}^2\right)^{1/2} \|\nabla v\|_{L^2(\Omega)} \\ &\quad + C_P(\Omega)\|f - f^h\|_{L^2(\Omega)}\|\nabla v\|_{L^2(\Omega)}.\end{aligned}$$

Here we used the Cauchy–Schwarz inequality for sums and that $\cup_{E\in\mathcal{E}^{h,\circ}} S_E$ and $\cup_{T\in\mathcal{T}^h} S_T$ give a finite coverage of Ω with a constant that only depends on σ^h. Since v is arbitrary, the bound will follow dividing by $\|\nabla v\|_{L^2(\Omega)}$ and taking the supremum over all $v \in \mathcal{V}$. □

Remark 6.6.

1. *This estimate in Theorem 6.5 is in principle computable. However, the constants c_T, c_E have to be estimated. There are some attempts to get explicit or tight bounds, e.g. [164].*
2. *The estimate in Theorem 6.5 shows the expected dependence on h for smooth u_0, since*

$$\left(\sum_{T\in\mathcal{T}^h} h_T^2 \|R_T\|_{L^2(T)}^2\right)^{1/2} = O(h_{\max})$$

and

$$\left(\sum_{E\in\mathcal{E}^{h,\circ}} h_E \|R_E\|_{L^2(E)}^2\right)^{1/2} = O(h_{\max})$$

because of $[\partial_n u_0^h]_E \approx \partial_n^2 u_0\, h_E$ and

$$h_E \|\left[\partial_n u_0^h\right]_E\|_{L^2(E)}^2 \sim h_E^4 \left|\partial_n^2 u_0\right|^2 \sim h_T^2 \|\nabla^2 u_0\|_{L^2(T)}^2$$

for T with $E \subset \partial T \setminus \partial\Omega$. Note that $\sum_{E\in\mathcal{E}^{h,\circ}} h_T^2 \sim \sum_{T\in\mathcal{T}^h} |T| = |\Omega|$.
3. *Note that the estimate in Theorem 6.5 does not require any higher regularity than $u_0 \in \mathcal{V}$. Thus it covers also the cases when $u_0 \notin H^2(\Omega)$, like in Section 6.1.2.*

6.2.2 Lower a posteriori bound

Definition 6.7. Bubble functions: *If ϕ_1, ϕ_2, ϕ_3 are the linear basis functions at the vertices of a triangle T, then $\beta_T := \phi_1\phi_2\phi_3/h_T^3$ defines a cubic function on T called* an element bubble. *Note that β_T can be extended by 0 to a function in $\mathcal{V}$ with support in T. For each edge $E = T_1 \cap T_2 \neq \emptyset$ $(T_1, T_2 \in \mathcal{T}^h)$, say $E = \overline{z_1 z_2}$ for some $z_1, z_2 \in \mathcal{N}^h$, we define a piecewise quadratic polynomial $\theta_E := \phi_{z_1}\phi_{z_2}/h_E^2$ on $\omega_E := T_1 \cup T_2$ called* an edge bubble. *Note that θ_E is continuous on E and it can be extended by 0 to a function in $\mathcal{V}$ with support in ω_E.*

The following theorem gives a lower bound for the actual error of u_0^h.

Theorem 6.8. Lower a posteriori error estimate: *Let u_0, u_0^h be as in Theorem 6.5. Then there is a constant c that depends only on σ^h and p, such that for each $T \in \mathcal{T}^h$ and for R_T, R_E and ω_E as in Definitions 6.4 and 6.7, respectively,*

$$c\left(c_E h_E \|R_E\|^2_{L^2(E)} + \sum_{T\subset\omega_E} c_T h_T^2 \|R_T\|^2_{L^2(E)}\right)$$
$$\leq \|\nabla\left(u_0 - u_0^h\right)\|^2_{L^2(\omega_T)} + \|h(f - f^h)\|^2_{L^2(\omega_T)}.$$

The constants c_T, c_E are as in Theorem 6.5.

Proof. Let $E \subset \mathcal{E}^{h,\circ}$ and T be any of the adjacent triangles, and let $v := \beta_T R_T$. Thus $v \in \mathcal{V}$ and the error representation formula (6.19) specializes to

$$\int_T \nabla\left(u_0 - u_0^h\right) \cdot \nabla(\beta_T R_T) = \textstyle\int_T \left\{\beta_T R_T^2 + (f - f^h)\beta_T R_T\right\}.$$

Since R_T and $\beta_T R_T$ are polynomials, there is a constant c, depending on σ_T and p, such that

$$\|R_T\|_{L^2(T)} \leq C\|\beta_T^{1/2} R_T\|_{L^2(T)}$$

and furthermore we have the *inverse estimate*

$$\|\nabla(\beta_T R_T)\|_{L^2(T)} \leq C h_T^{-1}\|\beta_T R_T\|_{L^2(T)} \leq C h_T^{-1}\|R_T\|_{L^2(T)}.$$

Hence we can derive the bound

$$\begin{aligned}\|R_T\|^2_{L^2(T)} &\leq C\int_T \beta_T R_T^2 = C\int_T \{\nabla\left(u_0 - u_0^h\right) \cdot \nabla(\beta_T R_T) + (f - f^h)\beta_T R_T\}\\ &\leq C\left(\|\nabla\left(u_0 - u_0^h\right)\|_{L^2(T)}\|\nabla(\beta_T R_T)\|_{L^2(T)} + \|f - f^h\|_{L^2(T)}\|R_T\|_{L^2(T)}\right)\\ &\leq C\left(h_T^{-1}\|\nabla\left(u_0 - u_0^h\right)\|_{L^2(T)} + \|f - f^h\|_{L^2(T)}\right)\|R_T\|_{L^2(T)}\end{aligned}$$

so that

$$h_T\|R_T\|_{L^2(T)} \leq C\left(\|\nabla\left(u_0 - u_0^h\right)\|_{L^2(T)} + \|h(f - f^h)\|_{L^2(T)}\right).$$

For the edge term there is a similar trick: we extend R_E onto $\mathbb{R}^2$ by a constant extension normal to E [654] and define as a test function $v := \theta_E R_E$. Clearly, $v \in \mathcal{V}$ and the error representation yields

$$\int_{\omega_E} \nabla\left(u_0 - u_0^h\right) \cdot \nabla(\theta_E R_E) = \int_E \theta_E R_E^2 + \sum_{T \subset \omega_E} \int_T R_T \theta_E R_E + \int_{\omega_E} (f - f^h)\theta_E R_E.$$

Similar to the above, the inequality

$$\begin{aligned}
\int_E |R_E|^2 &\le C \int_E \theta_E R_E^2 \\
&\le C\Bigg(\|\nabla\left(u_0 - u_0^h\right)\|_{L^2(\omega_E)} \|\nabla(R_E\theta_E)\|_{L^2(\omega_E)} \\
&\quad + \left(\sum_{T \subset \omega_E} \|R_T\|^2_{L^2(T)} \right)^{1/2} \|\theta_E R_E\|_{L^2(\omega_E)} + \|f - f^h\|_{L^2(\omega_E)} \|\theta_E R_E\|_{L^2(\omega_E)} \Bigg) \\
&\le C\Bigg(h_E^{-1} \|\nabla\left(u_0 - u_0^h\right)\|_{L^2(\omega_E)} + \left(\sum_{T \subset \omega_E} \|R_T\|^2_{L^2(T)} \right)^{1/2} \\
&\quad + \|f - f^h\|_{L^2(\omega_E)} \Bigg) \|R_E\|_{L^2(\omega_E)}
\end{aligned}$$

leads, with $\|R_E\|_{L^2(\omega_E)} \le C h_E^{1/2} \|R_E\|_{L^2(E)}$ (recall the definition of R_E on ω_E), to

$$\begin{aligned}
& h_E^{1/2} \left(\int_E |R_E|^2 \right)^{1/2} \\
&\quad \le C \left(\|\nabla\left(u_0 - u_0^h\right)\|_{L^2(\omega_E)} + \left(\sum_{T \subset \omega_E} h_T^2 \|R_T\|^2_{L^2(T)} \right)^{1/2} + \|h(f - f^h)\|_{L^2(\omega_E)} \right).
\end{aligned}$$

Now observe that the second term on the right has already been estimated in the first step of the proof in terms of the other two terms. This yields the result. □

6.2.3 The a posteriori error estimate

We summarize the results of the last two sections in the following theorem.

Theorem 6.9. A posteriori error estimate: *Let u_0, u_0^h be the solutions of the continuous equation (6.4) in $\mathcal{V} = H_0^1(\Omega)$ and the discrete equation (6.5) in $\mathcal{V}^h$ as in (6.15), respectively. For each $E \in \mathcal{E}^{h,\circ}$ we define the* local error indicator

$$\eta_E^2 := c_E h_E \|R_E\|^2_{L^2(E)} + \sum_{T \subset \omega_E} c_T h_T^2 \|R_T\|^2_{L^2(E)} \tag{6.20}$$

with constants c_T, c_E from Theorem 6.5, and residuals *R_T, R_E from Definition 6.4. Then there are constants c, C, both depending only on σ^h and the polynomial degree p, such that*

$$c_{\text{eff}} \left(\sum_{E \in \mathcal{E}^{h,\circ}} \eta_E^2 \right)^{1/2} \leq \|\nabla \left(u_0 - u_0^h\right) \|_{L^2(\Omega)} + \|h(f - f^h)\|_{L^2(\Omega)}$$

and

$$\|\nabla \left(u_0 - u_0^h\right) \|_{L^2(\Omega)} \leq \left(\sum_{E \in \mathcal{E}^{h,\circ}} \eta_E^2 \right)^{1/2} + C_P(\Omega)\|f - f^h\|_{L^2(\Omega)}.$$

Proof. Use Theorem 6.5 and Theorem 6.8 (sum over $E \in \mathcal{E}^{h,\circ}$). □

Remark 6.10. A posteriori error estimate:

1. *The upper bounds were first proved in [49, 50], the lower bound first appeared in [653].*
2. *Usually, f^h is not given explicitly, but is implicitly defined by using quadrature rules for the integrals $\int_T f\varphi$, where φ is a basis function. In general, this quadrature rule should be as cheap as possible but at least converge as fast as the error or one degree higher (e.g. for linear finite elements choose a quadrature rule of order* 1 *or* 2*). This same approximation f^h to f should also be taken into account when we compute R_T and $\|f - f^h\|_{L^2(T)}$.*
3. *The edge residuals R_E can be avoided in one dimension since $v - v^h$ in the proof of Theorem 6.5 can be chosen to vanish in the vertices. The jump residuals R_E are unavoidable and even dominant in more than one dimension [165]. This is, for example, easy to see if $p = 1$ and $f = 0$ (for nonhomogeneous Dirichlet data). Note that globally Δu_0^h is not a function but a distribution. Its regular part is zero, its singular part is $[\partial_n u_0^h]_E$. For $\varphi \in C_0^\infty(\omega_E)$ and for some $E \in \mathcal{E}^{h,\circ}$ we obtain*
$$-\Delta u_0^h[\varphi] = \int_{\omega_E} \nabla u_0^h \cdot \nabla\varphi = \int_E [\partial_n u_0^h]_E \varphi - \int_{T_1 \cup T_2} \Delta u_0^h \varphi = \int_E R_E \varphi.$$
4. *The upper bound Theorem 6.5 is global while the lower bound Theorem 6.8 is local. The upper bound cannot be local since a local error source will have global influence. Only a part of the error can be judged by a local interpolation error and local data error, while other parts of the error have its source somewhere else. This part is also called* pollution error.
5. *Such residual error estimates can also be derived for nonlinear equations. For this see [655].*

6.2.4 The adaptive finite element method

The macro discretization

We need a first decomposition of Ω into simplices that resolves the geometric details. This is often a very difficult problem, especially in 3D. A major difficulty is to avoid simplices T with large shape constant σ_T.

Initial discretization

The macro-discretization may not resolve the data required by the problem. In order to judge and improve data resolution we define a data error estimator (take $\|f - f^h\|_{L^2(\Omega)}$ and perform mesh refinement until a certain coarse tolerance is met), e.g.

$$\delta_{\mathcal{T}^h} := \left(\sum_{T \in \mathcal{T}^h} \delta_T^2 \right)^{1/2} \quad \text{with} \quad \delta_T^2 := \|f - f^h\|_{L^2(T)}^2.$$

The adaptive iteration

1. *Solve* for u_0^h (e.g. by a multigrid method). We assume in the following that u_0^h is the exact finite element solution.
2. *Estimate*: Compute η_E as in (6.20) for all $E \in \mathcal{E}^{h,\circ}$. Stop the iteration if

$$\eta_{\mathcal{T}^h} := \left(\sum_{E \in \mathcal{E}^{h,\circ}} \eta_E^2 \right)^{1/2} \leq \text{TOL}$$

   is satisfied (or use TOL $\|\nabla u_0^h\|_{L^2(\Omega)}$ on the right-hand side in the case of relative errors).
3. *Mark* a set of edges $\mathcal{F}^h \subset \mathcal{E}^{h,\circ}$ for refinement. Possible strategies are:
   - *Maximum strategy.* Let

$$E \in \mathcal{F}^h \iff \eta_E \geq (1 - \theta)\eta_{\max}$$

     with $\eta_{\max} := \max_{E \in \mathcal{E}^{h,\circ}} \eta_E$ and some $\theta \in [0, 1]$. Note that we get $\mathcal{F}^h \approx \emptyset$ for $\theta \approx 0$ and $\mathcal{F}^h \approx \mathcal{E}^{h,\circ}$ for $\theta \approx 1$.
   - *Fixed energy fraction strategy.* For $\theta \in [0, 1]$ seek a minimal set $\mathcal{F}^h \subset \mathcal{E}^{h,\circ}$ such that

$$\sum_{E \in \mathcal{F}^h} \eta_E^2 \geq \theta^2 \eta_{\mathcal{T}^h}^2. \tag{6.21}$$

     Note that we get $\mathcal{F}^h \approx \emptyset$ for $\theta \approx 0$ and $\mathcal{F}^h \approx \mathcal{E}^{h,\circ}$ for $\theta \approx 1$.
4. Mesh refinement. Aim: create a new regular mesh from $\mathcal{T}^h$ and the *set of marked triangles*

$$\mathcal{A}^h := \{T \in \mathcal{T}^h \, : T \subset \omega_E, \, E \in \mathcal{F}^h\}. \tag{6.22}$$

   - 1D problems. Interval bisection: bisect $T \in \mathcal{A}^h$ in its center.
   - 2D problems. A refinement of one triangle introduces a ***hanging node*** (a vertex that is not shared with a neighboring triangle [655], Sect. 4.1) and enforces the refinement of neighboring triangles if we have to work on regular meshes. On the other hand, our refining algorithm has to stay local. Arbitrary bisection of triangles may lead to very small angles over several refinement steps and as a consequence the shape constant will grow. We will study this topic in Section 6.2.5.

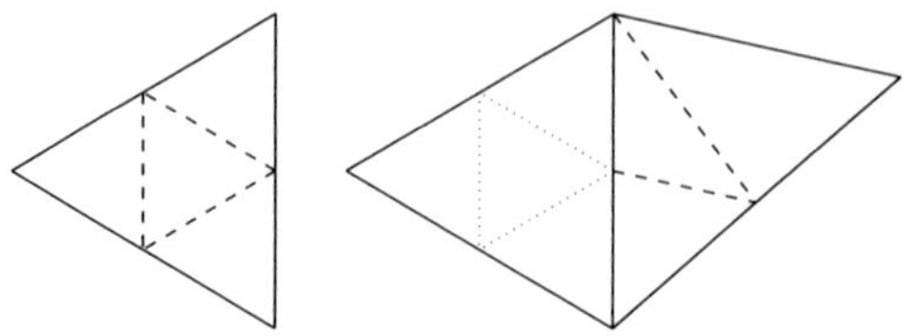

Figure 6.1 Left: Red refinement. Right: Blue refinement of the right triangle.

Remark: If we are bound to rectangular decompositions we cannot avoid working with hanging nodes.

- 3D problems. The principal problems are the same as in 2D but solutions to it are more involved.

Now go back to 1.

6.2.5 Stable refinement methods for triangulations in $\mathbb{R}^2$

The aim of this section is to describe some methods that return a (refined) regular mesh for a given regular mesh $\mathcal{T}^h$ and a set $\mathcal{A}^h$ of marked triangles to be refined. Such a method is called *geometrically stable* if repeated application of the method leads to a sequence of meshes such that the corresponding sequence of shape constants is uniformly bounded. The following methods will fulfill this requirement.

Red–green (–blue) refinement [655], Sect. 4.1
In case $T \in \mathcal{A}^h$ we will use *red refinement* as shown in Figure 6.1, left. Note that all new triangles have the same angles as their original triangle. For the neighbors that are not in $\mathcal{A}^h$, we enforce bisection to remove the hanging node, regardless of the size of the new angles. These so-called *green refinements*, Figure 6.1, right, will be coarsened before such a triangle will be refined in a subsequent step. If the opposite angle that is divided by bisection is very small, one might use bisection methods that avoid these (like the one in the next subsection). In this context, this refinement is called *blue refinement*. Triangles refined in this way need not be removed in later steps.

Newest node bisection [67]
We store the nodes P_1, P_2, P_3 of triangle T in a counterclockwise sense: NODES $[\mathrm{T},:] = [\mathrm{P}_1,\mathrm{P}_2,\mathrm{P}_3]$. If T is marked, we plan to bisect T from P_1 towards the opposite edge as in Figure 6.2. The two new triangles must again have oriented NODES – entries with P_{new} in the first place. However, the neighboring triangle has to be added to $\mathcal{A}^h$ if P_{new} is a hanging node. In a practical implementation one actually only refines simple situations:

- two triangle share the marked edge;
- the marked edge is a boundary edge;

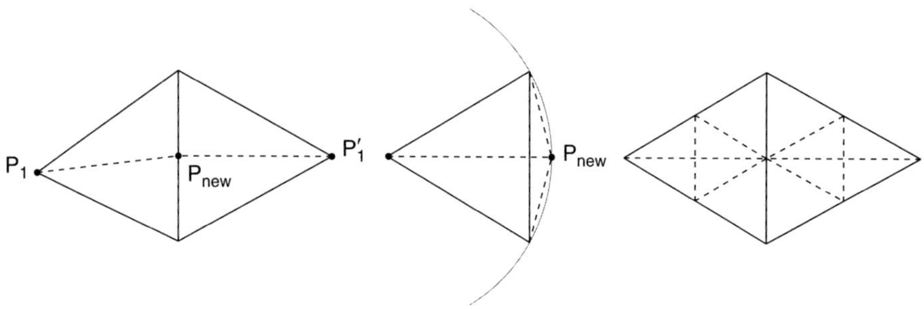

Figure 6.2 Left: Bisection in the interior. Middle: Bisection near the boundary. Right: Refinement with interior node property.

as seen in Figure 6.2. It can be shown that Algorithm 6.1, which is based on these ideas, stops after finitely many steps and results in a locally refined mesh. The refinement method is geometrically stable if the macro-triangulation $\mathcal{T}_0$ has a "correct" distribution of marked edges. We only have to avoid all edges emanating from one vertex being marked edges.

A newest node bisection with m-times repeated bisection is as follows. The marking procedure assigns to each T a number $m(T) \in \mathbb{N}$. Then $\mathcal{A}^h := \{T \in \mathcal{T}^h : m(T) > 0\}$. After each bisection, we decrease $m(T)$ until we obtain 0, see Algorithm 6.1.

Algorithm 6.1.
Repeated newest node bisection

```
while any m(T) > 0
    for T with m(T) > 0
        identify neighbor T' opposite P1
        situation Figure 6.2 left/right?;
        Yes :
            bisect T (and T');
            m(T) := max{m(T) − 1, 0};
            (respectively m(T') := max{m(T') − 1, 0});
        No :
            m(T') := max{m(T'), 1}
```

Longest node bisection

Always refine towards the longest edge in T and only in one of the situations as in Figure 6.2. For that, work on a copy of the coordinate field with a random distortion (to assure that the longest edge in T is unique almost surely). The important question is now whether this ends after finitely many steps in one of the mentioned situations? The answer is affirmative since following these choices the length of the largest edge increases and thus will end at an edge of locally maximal length or at the boundary. The algorithm is like Algorithm 6.1, but a special ordering in the NODES field is not necessary.

6.2.6 Convergence of the adaptive finite element method

The following description is very special to the case of linear finite elements. Nevertheless, the method works for arbitrary polynomial degree with some modifications, see Remark 6.15.

Definition 6.11. Interior node property: *A refinement method has the* interior node property, *if all marked triangles of the coarse mesh are refined in such a way that all their edge centers are vertices of the refined mesh and that each such triangle contains in its interior a vertex of the refined mesh.*

An example for such a method is the repeated newest node bisection with $m(T) = 3$, Algorithm 6.11.

Theorem 6.12. Computable a posteriori estimates: *Let $\mathcal{T}^H$ be a triangulation, $\mathcal{V}^H$ the linear finite element space over $\mathcal{T}^H$ and $u_0^H \in \mathcal{V}^H$ the finite element solution to the discrete Poisson problem (6.5) in $\mathcal{V}^H$. Let $\mathcal{T}^h$ be a refinement of $\mathcal{T}^H$, $\mathcal{V}^H \subset \mathcal{V}^h$ the refined finite element space and u_0^h the discrete solution in $\mathcal{V}^h$. Assume that both approximations f^H and f^h of f are piecewise constant. If $\mathcal{F}^H \subset \mathcal{E}^{H,\circ}$ is any set of refined edges and if all triangles in $\mathcal{A}^H$ as in (6.22) have been refined by a refinement method with the interior node property (see Figure 6.2), then*

$$c'_{\text{eff}} \sum_{E \subset \mathcal{F}^H} \eta_E^2 \left(u_0^H\right) \leq \|\nabla \left(u_0^h - u_0^H\right)\|_{L^2(\Omega)}^2 + \|H(f^h - f^H)\|_{L^2(\Omega)}^2$$

with some constant c'_{eff} depending only on σ^H, σ^h. Here, H is the mesh-size function for $\mathcal{T}^H$ and $\eta_E \left(u_0^H\right)$ is the error indicator (6.20) with respect to u_0^H.

Proof. [499], Lemma 4.2. We can assume the following situation. Let $E \in \mathcal{E}^{H,\circ}$ and $T_1, T_2 \in \mathcal{T}^H$ with $E = T_1 \cap T_2$. Both T_1 and T_2 have been refined and $\mathcal{T}^h$ has vertices x_E (the center of E) and x_i in the interior of T_i for $i \in \{1, 2\}$. Let φ_E, φ_i be the (maximally coarse) basis functions related to these vertices in $\mathcal{V}^h$ with $\omega_i := \operatorname{supp}(\varphi_i) \subseteq T_i$ and $\omega_E := \operatorname{supp}(\varphi_E) \subseteq T_1 \cup T_2$. For all φ with $\omega_\varphi := \operatorname{supp}(\varphi) \subseteq T_1 \cup T_2$

$$\begin{aligned}\int_{\omega_\varphi} \nabla \left(u_0^h - u_0^H\right) \cdot \nabla \varphi &= \int_{\omega_\varphi} f^h \varphi - \int_{\omega_\varphi \cap E} \left[\partial_n u_0^H\right]_E \varphi \\ &= \int_{\omega_\varphi} (f^h - f^H)\varphi + \int_{\omega_\varphi} f^H \varphi - \int_{\omega_\varphi} \left[\partial_n u_0^H\right]_E \varphi.\end{aligned}$$

Let $\varphi = \varphi_i$ for $i \in \{1, 2\}$. Since f^H is constant and $\varphi_i\big|_E = 0$ we have

$$\begin{aligned}|f^H|^2 \int_{\omega_i} \varphi_i = \int_{\omega_i} f^H f^H \varphi_i &= \int_{\omega_i} \nabla \left(u_0^h - u_0^H\right) \cdot \nabla (f^H \varphi_i) - \int_{\omega_i} (f^h - f^H) f^H \varphi_i \\ &\leq \|\nabla \left(u_0^h - u_0^H\right)\|_{L^2(\omega_i)} \|\nabla (f^H \varphi_i)\|_{L^2(\omega_i)} + \|f^h - f^H\|_{L^2(\omega_i)} \|f^H \varphi_i\|_{L^2(\omega_i)}.\end{aligned}$$

Using inverse estimates on φ_i (see Theorem 4.19) and considering the nodal scaling $\varphi_i(x_i) = 1$ yields

$$H_i \|f^H\|_{L^2(\omega_i)}^2 \leq C \left(H_i^{-1} \|\nabla \left(u_0^h - u_0^H\right)\|_{L^2(\omega_i)} + \|f^h - f^H\|_{L^2(\omega_i)}\right) \|f^H\|_{L^2(\omega_i)}$$

with a constant C that depends on σ_{T_i}. Similar, we derive the bound for the constant edge residual taking the test function $\varphi = R_E\varphi_E$

$$
\begin{aligned}
\|R_E\|^2_{L^2(E)} &\le C\int_E |R_E|^2\varphi_E = C\int_E R_E(R_E\varphi_E)\\
&= C\left(\int_{\omega_E}\nabla\left(u_0^h-u_0^H\right)\cdot\nabla(R_E\varphi_E) - \int_{\omega_E} f^h R_E\varphi_E\right)\\
&\le C\Big(\|\nabla\left(u_0^h-u_0^H\right)\|_{L^2(\omega_E)} h_E^{-1/2}\|R_E\|_{L^2(E)}\\
&\quad + \left(\|f^H\|_{L^2(\omega_E)} + \|f^h - f^H\|_{L^2(\omega_E)}\right) h_E^{1/2}\|R_E\|_{L^2(\omega_E)}\Big).
\end{aligned}
$$

$\|f^H\|_{L^2(\omega_E)}$ can be estimated with the first step of the proof and our stated result follows readily. □

Definition 6.13. Saturated data approximation: *Let $\mathcal{T}^H$ be a triangulation and $\mathcal{T}^h$ a refinement of $\mathcal{T}^H$, $u_0^H \in \mathcal{V}^H$ the solution of the discrete Poisson problem* (6.5) *in $\mathcal{V}^H$. We say that the* data approximation is saturated with a factor $\mu > 0$, *if*

$$
\begin{aligned}
\delta_{Hh} &:= \max\left\{C_P(\Omega)\|f-f^H\|_{L^2(\Omega)}, C_P(\Omega)\|f-f^h\|_{L^2(\Omega)}, \|H(f^H-f^h)\|_{L^2(\Omega)}\right\}\\
&\le \mu\eta_{\mathcal{T}^H}\left(u_0^H\right).
\end{aligned}
\tag{6.23}
$$

δ_{Hh} is called the data approximation error. *Note that this bound is quite natural for small h, since generally δ_{Hh} will be of lower order compared with $\eta_{\mathcal{T}^H}(u_0^H)$ (for example, if f is piecewise smooth, then $\|f-f^H\|_{L^2(\Omega)} = O(h^2_{\max})$. However, the following results will depend on the quantitative bound* (6.23) *and not on a vage higher order argument.*

Theorem 6.14. Convergence of adaptive finite element method: *Let $\mathcal{T}^H, \mathcal{V}^H, u_0^H, u_0^h$, and f^H, f^h be as in Theorem 6.12. Assume that we define $\mathcal{F}^H$ by the fixed energy fraction strategy* (6.21) *for some $\theta \in (0,1)$ and that we refine $\mathcal{T}^H$ to get $\mathcal{T}^h$ as required in Theorem 6.12. Assume that*

$$\delta_{Hh} \le \mu\eta_{\mathcal{T}^H}$$

holds for a sufficiently small $\mu > 0$, only depending on σ^H, σ^h and θ. Then there is a constant c that again depends only on σ^H, σ^h, such that

$$\|\nabla\left(u_0 - u_0^h\right)\|_{L^2(\Omega)} \le \sqrt{1-c\theta^2}\,\|\nabla\left(u_0-u_0^H\right)\|_{L^2(\Omega)}$$

as long as $\eta_{\mathcal{T}^H} > \mathrm{TOL}$. Thus, for a given tolerance $\mathrm{TOL} > 0$ and a geometrically stable refinement algorithm, this method reaches $\eta_{\mathcal{T}^H} < \mathrm{TOL}$ after finitely many steps.

Proof. We can achieve $\delta_{Hh} \le \mu\eta_{\mathcal{T}^H}$ by refining the mesh. We could even establish $\delta_{Hh} \le \mu\,\mathrm{TOL}$ for the given μ, so our requirement would hold as long as $\eta_{\mathcal{T}^H} > \mathrm{TOL}$. Since the finite element approximations to f converge on refined meshes, this can be achieved in finitely many steps. (See [499], where δ_{Hh} and $\eta_{\mathcal{T}^H}$ are treated within one iteration.)

We get for the lower bound from Theorem 6.12 with the definition of $\mathcal{F}^H$

$$c'_{\text{eff}}\eta_{\mathcal{F}^H} \le \|\nabla\left(u_0^h - u_0^H\right)\|_{L^2(\Omega)} + \mu\eta_{\mathcal{T}^H}$$

which leads with the marking strategy (6.21) to

$$(c'_{\text{eff}}\theta - \mu)\eta_{\mathcal{T}^H} \le c'_{\text{eff}}\eta_{\mathcal{F}^H} - \mu\eta_{\mathcal{T}^H} \le \|\nabla\left(u_0^h - u_0^H\right)\|_{L^2(\Omega)}.$$

We use the approximate *Galerkin orthogonality* (since $\mathcal{V}^H \subset \mathcal{V}^h$!)

$$\int_\Omega \nabla\left(u_0 - u_0^h\right) \cdot \nabla\left(u_0^h - u_0^H\right) = \int_\Omega (f - f^h)\left(u_0^h - u_0^H\right)$$

and conclude from this

$$\begin{aligned}
2\left|\int_\Omega \nabla\left(u_0 - u_0^h\right) \cdot \nabla\left(u_0^h - u_0^H\right)\right| &\ge -2\|f - f^h\|_{L^2(\Omega)}\|u_0^h - u_0^H\|_{L^2(\Omega)} \\
&\ge -2C_P(\Omega)\|f - f^h\|_{L^2(\Omega)}\|\nabla\left(u_0^h - u_0^H\right)\|_{L^2(\Omega)} \\
&\ge -2\mu\eta_{\mathcal{T}^H}\|\nabla\left(u_0^h - u_0^H\right)\|_{L^2(\Omega)} \\
&\ge -2\mu^2\eta_{\mathcal{T}^H}^2 - \frac{1}{2}\|\nabla\left(u_0^h - u_0^H\right)\|_{L^2(\Omega)}^2.
\end{aligned}$$

Using this inequality and Theorem 6.12, we get

$$\begin{aligned}
\|\nabla\left(u_0 - u_0^H\right)\|_{L^2(\Omega)}^2 &= \|\nabla\left(u_0 - u_0^h + u_0^h - u_0^H\right)\|_{L^2(\Omega)}^2 \\
&= \|\nabla\left(u_0 - u_0^h\right)\|_{L^2(\Omega)}^2 + \|\nabla\left(u_0^h - u_0^H\right)\|_{L^2(\Omega)}^2 + 2\int_\Omega \nabla\left(u_0 - u_0^h\right) \cdot \nabla\left(u_0^h - u_0^H\right) \\
&\ge \|\nabla\left(u_0 - u_0^h\right)\|_{L^2(\Omega)}^2 + \frac{1}{2}\|\nabla(u_0^h - u_0^H)\|_{L^2(\Omega)}^2 - 2\mu^2\eta_{\mathcal{T}^H}^2 \\
&\ge \|\nabla(u_0 - u_0^h)\|_{L^2(\Omega)}^2 + \left(\frac{1}{2}(c'_{\text{eff}}\theta - \mu)^2 - 2\mu^2\right)\eta_{\mathcal{T}^H}^2 \\
&\ge \|\nabla(u_0 - u_0^h)\|_{L^2(\Omega)}^2 + c\theta^2\|\nabla(u_0 - u_0^H)\|_{L^2(\Omega)}^2
\end{aligned}$$

for a sufficiently small μ, depending on c'_{eff} and θ, and this yields the assertion

$$\|\nabla(u_0 - u_0^h)\|_{L^2(\Omega)} \le \sqrt{1 - c\theta^2}\,\|\nabla(u_0 - u_0^H)\|_{L^2(\Omega)}.$$

□

Remark 6.15. *Convergence of adaptive finite element method:*

1. *The first convergence proof was in [54] ($n = 1$ with no rate). The method of Theorem 6.14 was developed in [300] and [498]. A more operator oriented approach was [96].*

2. *For $p > 1$ the error indicator has the element residual $R_T\left(u_0^H\right) = \Delta u_0^H + f^H$. The first step towards convergence is to prove (compare Theorem 6.12)*

$$\sum_{E\in\mathcal{E}^{H,\circ}} \eta_E^2\left(u_0^H\right)$$
$$\leq C\left(\|\nabla(u_0^h - u_0^H)\|^2_{L^2(\Omega)} + \|H(f-f^H)\|^2_{L^2(\Omega)} + \|h(f-f^h)\|^2_{L^2(\Omega)}\right),$$

where one has to assume that marked edges and triangles in $\mathcal{A}^H$ have been refined in a suitable way. More generally, one has to prove the inequality

$$ch_T\|R_T(u_0^H)\|_{L^2(T)} \leq \sup_{v^h\in X_T} \frac{\int_T R_T(u_0^H)v^h}{\|v^h\|_{L^2(T)}}$$

for some constant c depending on σ^h and p. Here X_T is the surplus space *on T: $X_T = \mathcal{V}^h\big|_T \backslash \mathcal{V}^H\big|_T$. A similar inequality has to hold for R_E [302].*

3. *The error decreases geometrically for fixed θ. If we miss the tolerance* TOL *slightly in one step, we have to perform an additional step and we will very likely end up with an iterate that is more accurate than necessary. Note that the last step determines the total work if the number of elements increases geometrically. However, we can use the error decay to lower θ near the tolerance level. Assume that the error satisfies*

$$E_k^2 = (1 - c\theta^2)E_{k-1}^2.$$

From this we may eliminate the unknown constant c by

$$c = \tfrac{1}{\theta^2}\left(1 - \left(\tfrac{E_k}{E_{k-1}}\right)^2\right) \approx \tfrac{1}{\theta^2}\left(1 - \left(\tfrac{\eta_k}{\eta_{k-1}}\right)^2\right),$$

where the error E_k is approximated by our guess $\eta_k := \eta_{\mathcal{T}_k}$. When the next error is presumably below the tolerance, that is if

$$(1 - c\theta^2)E_k^2 < \mathrm{TOL}^2,$$

we use a safety factor 0.9 and define a new value θ_{new} such that

$$(0.9\,\mathrm{TOL})^2 = \left(1 - c\theta^2_{\mathrm{new}}\right)E_k^2 \approx \left(1 - c\theta^2_{\mathrm{new}}\right)\eta_k^2.$$

This yields the explicit formula

$$\theta^2_{\mathrm{new}} = \theta^2\,\frac{1 - (0.9\,\mathrm{TOL})^2/\eta_k^2}{1 - \eta_k^2/\eta_{k-1}^2}$$

under the assumption that both nominator and denominator are positive [300].

4. *How do we compute $\mathcal{F}^H$? We may sort $\{\eta_E\}_{E\in\mathcal{E}^{H,\circ}}$ by size and then put $E \in \mathcal{F}^H$ beginning from the largest value until $\eta_{\mathcal{F}^H}$ meets the requirement* (6.21). *The complexity of this algorithm is $N_\mathcal{E}\log(N_\mathcal{E})$ in the number $N_\mathcal{E}$ of edges in $\mathcal{T}^H$. Alternatively, we may do only a rough sort and keep the linear complexity in $N_\mathcal{E}$, Algorithm 6.2 [300].*

Algorithm 6.2.
Fixed energy fraction marking
sum := 0; $\tau := 1$; $\nu = 0.1$; $\mathcal{F}^H := \emptyset$;
while sum $< \theta^2 \eta^2_{\mathcal{T}^H}$
    $\tau := \tau - \nu$;
    for all $E \in \mathcal{E}^{H,\circ}$
        if $E \notin \mathcal{F}^H$;
            if $\eta_E^2 > \tau^2 \eta^2_{\max}$
                $\mathcal{F}^H := \mathcal{F}^H \cup \{E\}$;
                sum := sum $+ \eta_E^2$

6.2.7 Optimality

Although the convergence of the algorithm is an important result, it does not guarantee that we found the most efficient method. In this section we introduce optimal approximation spaces and show that the proposed algorithm does indeed produce meshes of the optimal asymptotic complexity. This description follows [600].

If $u_0 \in H^2(\Omega)$ and if we choose a linear finite element method on a uniform mesh ($h_T \equiv h$ for all $T \in \mathcal{T}^h$), we know that

$$\|\nabla(u_0 - u_0^h)\|_{L^2(\Omega)} \sim h$$

is optimal with respect to the exponent in h. For $\Omega \subset \mathbb{R}^2$ we have the relation

$$h \sim N^{-1/2}$$

with the number of elements (and thus number of unknowns) N. To achieve

$$\|\nabla(u_0 - u_0^h)\| \sim \mathrm{TOL},$$

we need $N \sim \mathrm{TOL}^{-2}$ degrees of freedom and this is again optimal with respect to the exponent, cf. Section 6.1.4. If $u_0 \in H^{1+r}(\Omega)$ for some $r \in (0,1]$, then we know

$$\|\nabla(u_0 - u_0^h)\|_{L^2(\Omega)} \sim h^r$$

and this yields that $N \sim \mathrm{TOL}^{-\frac{2}{r}}$. We say that $u_0 \in \mathcal{B}^s$, for some $s > 0$, if the best approximation error of u_0 in an N-dimensional linear finite element space behaves asymptotically as N^{-s}. Thus $N = O(\mathrm{TOL}^{-\frac{1}{s}})$ is the necessary dimensionality to achieve the approximation error TOL for u_0. We now define the function space

$$\mathcal{B}^s := \left\{ v \in H_0^1(\Omega) \,:\, \sup_{\varepsilon>0} \left\{ \varepsilon \inf_{\mathcal{T}^c \,:\, \inf_{v_{\mathcal{T}^c} \in \mathcal{V}_{\mathcal{T}^c}} \|\nabla(v - v_{\mathcal{T}^c})\|_{L^2(\Omega)} \le \varepsilon} \{(\sharp\mathcal{T}^c - \sharp\mathcal{T}_0^c)^s\} < \infty \right\} \right\},$$

where $\sharp\mathcal{T}$ denotes the number of triangles in a triangulation $\mathcal{T}$. The infimum is taken over all conforming triangulations $\mathcal{T}^c$ that can be achieved from the given initial mesh $\mathcal{T}_0^c$ by newest node bisection, Section 6.2.5. $\mathcal{V}_{\mathcal{T}^c}$ is the linear finite element space over $\mathcal{T}^c$. The space $\mathcal{B}^s$ is equipped with the norm

$$\|v\|_{\mathcal{B}^s} := \|\nabla v\|_{L^2(\Omega)} + [v]_{\mathcal{B}^s}.$$

By construction, $v \in \mathcal{B}^s$ guarantees that there is a refinement $\mathcal{T}^c$ of $\mathcal{T}_0^c$ with

$$\sharp\mathcal{T}^c - \sharp\mathcal{T}_0^c \leq O(\varepsilon^{-\frac{1}{s}})$$

such that

$$\inf_{v_{\mathcal{T}^c} \in \mathcal{V}_{\mathcal{T}^c}} \|\nabla(v - v_{\mathcal{T}^c})\|_{L^2(\Omega)} \leq \varepsilon.$$

As seen above (for $r = 2s$), a function in $H^{1+2s}(\Omega)$ fulfills this requirement, hence $H^{1+2s}(\Omega) \subset \mathcal{B}^s$, but $\mathcal{B}^s$ is definitely a larger set. This kind of space is related to Besov spaces for which a detailed analysis is well known [97]. Based on this definition we will formulate the following notion of optimality.

Definition 6.16. *An adaptive method is of* optimal complexity *if it realizes a mesh $\mathcal{T}^c$ and a numerical approximation $u_{\mathcal{T}^c}$ of u_0 such that*[44]*.*

$$\|\nabla(u_0 - u_{\mathcal{T}^c})\|_{L^2(\Omega)} \leq \varepsilon$$

with a number of $O(\varepsilon^{-\frac{1}{s}})$ operations for every solution $u_0 \in \mathcal{B}^s$ of (6.4)*.*

The adaptive algorithm, for which we sketch the optimality proof, is the one we proposed before with fixed energy fraction marking (6.21) for some $\theta > 0$ and a refinement as required in Theorem 6.12. It is important to control the amount of the actual refinement that is caused by our marking strategy. For this, the refinement step is split into two steps, REFINE and MAKECONFORM.

Theorem 6.17. Complexity of MAKECONFORM: *The refinement edges on $\mathcal{T}_0^c$ can be arranged so that the following holds true. Construct a sequence of meshes $\{\mathcal{T}_k^c\}_{k\geq 1}$ by*

% Refine marked triangles only, $\mathcal{T}_{k+1}$ is nonconforming
$\mathcal{T}_{k+1}$ = REFINE $(\mathcal{T}_k^c, f_k, u_k)$;
% Refine further to get a conforming mesh
$\mathcal{T}_{k+1}^c$ = MAKECONFORM$(\mathcal{T}_{k+1})$;

Here, f_k is the representation of f and u_k the discrete solution of (6.5) *on $\mathcal{T}_k^c$. If REFINE refines according to the requirements in Theorem 6.12 and if MAKECONFORM uses the nodes of newest node bisection, then for $n \in \mathbb{N}$*

$$\sharp\mathcal{T}_n^c - \sharp\mathcal{T}_0^c \leq C \sum_{k=1}^{n} \left\{\sharp\mathcal{T}_k - \sharp\mathcal{T}_{k-1}^c\right\}, \tag{6.24}$$

where C only depends on $\mathcal{T}_0^c$.

Proof. [96], Lemma 2.1, Theorem 2.4. □

Let $\mathcal{T}$ be a triangulation that is a refinement of $\mathcal{T}_0^c$ and $\mathcal{E}_{\mathcal{T}}^\circ$ the set of interior edges in $\mathcal{T}$. For the following arguments we will require that f is piecewise constant on $\mathcal{T}_0^c$, hence we can take $f = f_{\mathcal{T}}$ for any refinement $\mathcal{T}$ of $\mathcal{T}_0^c$. For $v_{\mathcal{T}} \in \mathcal{V}_{\mathcal{T}}$ and $E \in \mathcal{E}_{\mathcal{T}}^\circ$ we

[44] To avoid over-indexing, we abbreviate our usual $u_{0,\mathcal{T}^c}^h$ to $u_{\mathcal{T}^c}$, and below $u_{0,\tilde{\mathcal{T}}}^h, u_{0,k}^h$ to $u_{\tilde{\mathcal{T}}}, u_k$.

Now we can assume that $\check{\mathcal{T}}$ is the smallest common refinement of $\mathcal{T}^c$ and $\widehat{\mathcal{T}}$. Thus we have

$$\sharp\check{\mathcal{T}} - \sharp\mathcal{T}^c \leq \sharp\widehat{\mathcal{T}} - \sharp\mathcal{T}_0^c$$

and we conclude

$$\sharp\mathcal{T} - \sharp\mathcal{T}^c \leq C(\sharp\check{\mathcal{T}} - \sharp\mathcal{T}^c) \leq C\left(\sharp\widehat{\mathcal{T}} - \sharp\mathcal{T}_0^c\right) \leq C\|\nabla(u_0 - u_{\mathcal{T}^c})\|_{L^2(\Omega)}^{-\frac{1}{s}}[u_0]_{\mathcal{B}^s}^{\frac{1}{s}}$$

with C depending on $\lambda^{-1/s}$ hence on θ and any lower bound $s_0 > 0$ for s. □

Algorithm 6.3.
Algorithm SOLVE, with $f \in \mathbb{P}_0(\mathcal{T}_0^c)$ assumed!
$[\mathcal{T}_k^c, u_{\mathcal{T}_k^c}] = \mathrm{SOLVE}\,(\mathcal{T}_0^c, f, \varepsilon)$
  Solve for $u_{\mathcal{T}_0^c}$; $k := 0$;
  while $c_1\eta\left(\mathcal{T}_k^c, f, u_{\mathcal{T}_k^c}\right) \geq \varepsilon$
    $\mathcal{T}_{k+1} := \mathit{REFINE}\left(\mathcal{T}_k^c, f, u_{\mathcal{T}_k^c}\right)$;
    $\mathcal{T}_{k+1}^c := \mathit{MAKECONFORM}(\mathcal{T}_{k+1})$;
    Solve for $u_{\mathcal{T}_{k+1}^c}$;
    $k = k + 1$

We summarize the method in the routine SOLVE in Algorithm 6.3. The next theorem proves the optimality of this procedure according to Definition 6.16.

Theorem 6.20. Complexity of SOLVE: *$\mathcal{T}_0^c$ macro-triangulation, $f \in \mathbb{P}_0\left(\mathcal{T}_0^c\right)$, $u_0 \in \mathcal{B}^s$ solution of the continuous problem (6.4) for some $s > 0$ and $u_{\mathcal{T}^c}$ the solution of the discrete problem (6.5) on $\mathcal{T}^c$. Then the adaptive algorithm $(\mathcal{T}^c, u_{\mathcal{T}^c}) = \mathit{SOLVE}\,(\mathcal{T}_0^c, f, \varepsilon)$ (see Algorithm 6.3) terminates after finitely many steps with*

$$\|\nabla(u_0 - u_{\mathcal{T}^c})\|_{L^2(\Omega)} \leq \varepsilon$$

and

$$\sharp\mathcal{T}^c - \sharp\mathcal{T}_0^c \leq C\varepsilon^{-\frac{1}{s}}[u_0]_{\mathcal{B}^s}.$$

Proof. The *Galerkin orthogonality* yields

$$\|\nabla\left(u_0 - u_{\mathcal{T}_k^c}\right)\|_{L^2(\Omega)}^2 = \|\nabla\left(u_0 - u_{\mathcal{T}_{k+1}^c}\right)\|_{L^2(\Omega)}^2 + \|\nabla\left(u_{\mathcal{T}_{k+1}^c} - u_{\mathcal{T}_k^c}\right)\|_{L^2(\Omega)}^2.$$

By the previous estimates (6.28) and (6.25)

$$\begin{aligned}\|\nabla\left(u_{\mathcal{T}_{k+1}^c} - u_{\mathcal{T}_k^c}\right)\|_{L^2(\Omega)}^2 &\geq (c_2\theta)^2\eta\left(\mathcal{T}_k^c, f, u_{\mathcal{T}_k^c}\right)^2 \\ &\geq \left(\frac{c_2}{c_1}\theta\right)^2\|\nabla\left(u_0 - u_{\mathcal{T}_k^c}\right)\|_{L^2(\Omega)}^2.\end{aligned}$$

and so with

$$\delta := \left(1 - \left(\frac{c_2}{c_1}\theta\right)^2\right)^{1/2} < 1$$

we arrive at

$$\|\nabla\left(u_0 - u_{\mathcal{T}_{k+1}^c}\right)\|_{L^2(\Omega)} \le \delta \|\nabla(u_0 - u_{\mathcal{T}_k^c})\|_{L^2(\Omega)}$$

(this is the simplified version of Theorem 6.14). From this it follows that the algorithm SOLVE terminates after finitely many steps. It remains to show optimality. Let n be the index with

$$c_1 \eta\left(\mathcal{T}_{n-1}^c, f, u_{\mathcal{T}_{n-1}^c}\right) \ge \varepsilon$$

and

$$c_1 \eta\left(\mathcal{T}_n^c, f, u_{\mathcal{T}_n^c}\right) < \varepsilon.$$

With Lemma 6.19 we get

$$\sharp\mathcal{T}_{k+1} - \sharp\mathcal{T}_k^c \le C\|\nabla(u_0 - u_{\mathcal{T}_k^c})\|_{L^2(\Omega)}^{-\frac{1}{s}} [u_0]_{\mathcal{B}^s}^{\frac{1}{s}}$$

for $k = 0, \ldots, n-1$. From our results on newest node bisection, Lemma 6.19,

$$\begin{aligned}
\sharp\mathcal{T}_n^c - \sharp\mathcal{T}_0^c &\le C\sum_{k=0}^{n-1}\{\sharp\mathcal{T}_{k+1} - \sharp\mathcal{T}_k^c\} \le C\sum_{k=0}^{n-1}\|\nabla(u_0 - u_{\mathcal{T}_k^c})\|_{L^2(\Omega)}^{-\frac{1}{s}}[u_0]_{\mathcal{B}^s}^{\frac{1}{s}} \\
&\le C\sum_{k=0}^{n-1}(\delta^{n-1-k})^{\frac{1}{s}}\|\nabla(u_0 - u_{\mathcal{T}_{n-1}^c})\|_{L^2(\Omega)}^{-\frac{1}{s}}[u_0]_{\mathcal{B}^s}^{\frac{1}{s}} \\
&\le C\|\nabla(u_0 - u_{\mathcal{T}_{n-1}^c})\|_{L^2(\Omega)}^{-\frac{1}{s}}[u_0]_{\mathcal{B}^s}^{\frac{1}{s}} \le C\eta\left(\mathcal{T}_{n-1}^c, f, u_{\mathcal{T}_{n-1}^c}\right)^{-\frac{1}{s}}[u_0]_{\mathcal{B}^s}^{\frac{1}{s}} \\
&\le C\varepsilon^{-\frac{1}{s}}[u_0]_{\mathcal{B}^s}^{\frac{1}{s}}.
\end{aligned}$$

□

This is the first part of the proof in [600]. It continues with the general case of variable f and $u_{\mathcal{T}_k^c}$ being replaced by an approximate solution of the discrete system. Using interior loops, one establishes a right-hand side $f_{\mathcal{T}^c}$ and an approximate discrete solution $w_{\mathcal{T}^c}$ so that

$$\|f - f_{\mathcal{T}^c}\|_{H^{-1}(\Omega)} + \|\nabla(u_{\mathcal{T}^c} - w_{\mathcal{T}^c})\|_{L^2(\Omega)} \le \omega\eta(\mathcal{T}^c, f_{\mathcal{T}^c}, w_{\mathcal{T}^c})$$

(for ω sufficiently small, not depending on the mesh size and the data). If the discrete approximate solution can be established with optimal complexity, then this adaptive algorithm is of optimal complexity with regard to the regularity properties of u_0 and f.

6.2.8 Other types of estimators

There are several possibilities to obtain alternative a posteriori error estimates in the energy norm. The objective of these variants is to get formulas that are easy to implement or that allow us to better quantify the constants (c_T, c_E in (6.20)). However, from the mathematical viewpoint it is decisive to have the upper and lower bounds as in Theorem 6.9 that would allow us to prove convergence of the adaptive method and optimality. Actually, these estimates mean that these error indicators are equivalent to the residual error estimator [655], Sect. 1.3, and these extensions are straightforward (besides the technical details).

Subdomain residual method

One (approximately) solves a local error equation

$$\int_{\omega_z} \nabla\chi_z \cdot \nabla w = \textstyle\int_{\omega_z} \{f^h w - \nabla u_0^h \cdot \nabla w\} \quad \text{for all } w \in H_0^1(\omega_z), \tag{6.29}$$

where ω_z is a patch of triangles around the node z and $\mathcal{V}_z$ is an enhanced finite element space with support in ω_z. The local error estimators, given by $\|\nabla\chi_z\|_{L^2(\omega_z)}$, are equivalent to the residual error estimator [6], Ch. 3.2, [655], Sect. 1.3.

Gradient recovery

The idea is to replace $\|\nabla(u_0 - u_0^h)\|_{L^2(\Omega)}$ by $\|G(u_0^h) - \nabla u_0^h\|_{L^2(\Omega)}$, where $G\left(u_0^h\right)$ is a smoothed gradient of u_0^h. For example, if u_0^h is piecewise linear, ∇u_0^h is piecewise constant and $G\left(u_0^h\right)$ is defined as a projection of ∇u_0^h into the space of linear finite elements. The method is easy and works well in practise [6], Ch. 4.

Hierarchical error estimate

The idea is to define an enhanced discrete space $(\mathcal{V}^h)^*$ and write the equation on $(\mathcal{V}^h)^*$ as a system with respect to the decomposition $(\mathcal{V}^h)^* = \mathcal{V}^h \oplus W^h$. W^h is called the surplus space. With some simplifications one can approximate the error on $\mathcal{V}^h$ by an equation on W^h that is easy to solve. For this method to work it is decisive that the data error is well resolved [6], Ch. 5.

Equilibrated residual method

In order to work more locally than the previous method it is necessary to solve local Neumann problems. To obtain such bounds one needs to carefully choose the fluxes that define these local boundary problems. There are several approaches to constructing such fluxes [6], Ch. 6.

6.2.9 *hp* finite element method

The method so far will work for mesh refinement (h method) for fixed polynomial degree. If one is interested in variable polynomial degrees, one faces the problem of deciding whether to refine the mesh (for fixed p) or increase the polynomial degree for fixed h. The optimal choice of h and p will lead to exponential (instead of algebraic) decay of the error with the number of unknowns. For the error estimation one may also use the residual error estimator, however, the explicit dependence of the error indicator from the polynomial degree is important [575]. A convergence proof for such a method is only known in one space dimension [301]. Many numerical issues can be found in [280, 281].

6.3 Estimation of quantities of interest

6.3.1 Quantities of interest

In the following considerations the functions mostly, but not always, are exact and discrete solutions. So we use the notation u and u^h instead of the previous u_0 and u_0^h.

So far we have been interested in discrete solutions u^h of $-\Delta u = f$ and looked for estimates for $\|\nabla(u-u^h)\|_2$. In practical problems, however, one has often more specific questions. For example, we could ask

$$u(x_0) = ?\,, \quad \partial_1 u(x_0) = ? \quad \text{for some } x_0 \in \Omega,$$

$$\int_\Gamma \partial_n u = ? \text{ for some curve } \Gamma \subset \overline{\Omega}, \quad \|u-u^h\|_{\Omega'} \text{ for } \Omega' \subset\subset \Omega.$$

A mesh to answer these specific questions might look very different from those we considered before. Therefore we will pose the generalized question

$$Q(u) = ?\,,$$

where $Q : \mathcal{V} \to \mathbb{R}$ is a continuous operator. However, the examples $Q(u) := u(x_0)$, $Q(u) := \partial_1 u(x_0)$ or $Q(u) := \int_\Gamma \partial_n u$ do not satisfy this if $u \in \mathcal{V} = H^1(\Omega)$ is assumed! Hence we have to regularize these operators. Note that solutions are usually more regular than just in $\mathcal{V}$, so that Q will have some meaning unless very special situations occur (e.g. $\partial_1 u(x_0)$ and x_0 is a corner point). For example, we let

$$Q_\epsilon(u) := \tfrac{1}{|B_\epsilon(x_0)|}\int_{B_\epsilon(x_0)} u \quad \text{for } x_0 \in \Omega \text{ and sufficiently small } \epsilon > 0.$$

Since in this case $Q \equiv Q_\epsilon$ is linear, we can write

$$Q(u) - Q(u^h) = Q(u-u^h) = Q(e).$$

Our aim is to define a computable quantity $\eta(u^h)$ for which one can show $c\eta(u^h) \le |Q(e)| \le C\eta(u^h)$ for positive constants c, C.

6.3.2 Error estimates for point errors

Again we need a reasonable error representation. For this we use the *duality argument*, to represent $Q(\,.\,)$ as a scalar product in $\mathcal{V}$. We choose $u^* \in \mathcal{V}$ to be the solution of

$$\int_\Omega \nabla v \cdot \nabla u^* = Q(v) = \fint_{B_\epsilon(x_0)} v \approx v(x_0) = \delta_{x_0}[v] \quad \text{for all } v \in \mathcal{V},$$

where δ_{x_0} is the *Dirac measure* in x_0. This means that u^* is an approximate *Green's function*

$$u^*(x) \approx G(x; x_0),$$

that is, $G(\,.\,; x_0)$ is the solution of $-\Delta G(\,.\,, x_0) = \delta_{x_0}$, $G(\,.\,, x_0)\big|_{\partial\Omega} = 0$. Now we can localize the error using

$$e(x_0) = \delta_{x_0}[e] \approx Q(e) = \int_\Omega \nabla e^* \cdot \nabla u^* = \int_\Omega \nabla e \cdot \nabla(u^* - w^h).$$

The last equality uses the definitions of u and u^h. Now we can also solve the discrete dual problem

$$\int_\Omega \nabla v^h \cdot \nabla (u^h)^* = Q(v^h) \quad \text{for all } v^h \in \mathcal{V}^h$$

and by taking $v^h = (u^h)^*$ we obtain

$$Q(e) = \int_\Omega \nabla(u - u^h) \cdot \nabla(u^* - (u^h)^*) = \int_\Omega \nabla e \cdot \nabla e^*. \tag{6.30}$$

Thus, the dual error $\nabla(u^* - (u^h)^*)$ can be viewed as a weight for the error $\nabla(u - u^h)$ of the primal problem. To proceed we transform the first integral in (6.30) as in the well-known error representation (6.19) to get

$$Q(e) = \sum_{T \in \mathcal{T}^h} \int_T R_T(u^* - (u^h)^*) + \sum_{E \in \mathcal{E}^{h,\circ}} \int_E R_E(u^* - (u^h)^*),$$

with $R_T \equiv R_T(u^h)$ and $R_E \equiv R_E(u^h)$ as in Definition 6.4. Proceeding as in Theorem 6.5 leads to

$$|Q(e)| \leq C \sum_{T \in \mathcal{T}^h} \left(h^2 \|R_T\|_{L^2(T)} + \sum_{E \subset \partial T \setminus \partial\Omega} h_E^{3/2} \|R_E\|_{L^2(E)} \right) \zeta_T$$

with $\zeta_T := \max\left\{h_T^{-2}\|u^* - (u^h)^*\|_{L^2(T)}, h_T^{-3/2}\|u^* - (u^h)^*\|_{L^2(\partial T)}\right\}$ and some positive constant C that depends only on the shape regularity σ^h and the polynomial degree p. Unfortunately, we do not know u^* and it is additional work to compute $(u^h)^*$.

Heuristic approach
In $\mathbb{R}^2$, we know that u^* has the form $u^*(x) = \frac{1}{2\pi}\log(|x - x_0|) + w(x)$ where w is smooth. Therefore we take

$$\zeta_T \approx \frac{h_T}{|x - x_0|^2 + \epsilon^2}.$$

Defining $\eta_T := h_T\|f\|_{2;T} + \sum_{E \subset \partial T \setminus \partial\Omega} h_E^{3/2}\|R_E\|_{2;E}$, we obtain with $\epsilon = h_T$

$$|Q(e)| \leq C \sum_{T \in \mathcal{T}^h} \frac{h_T^2}{|x_T - x_0|^2 + h_T^2} \eta_T,$$

where x_T denotes the *center of gravity* of T and C is as above.

Thus we have derived a computable estimate for $|Q(e)|$ in terms of the local contribution and we can now proceed with the marking and refinement strategies of Section 6.2.4.

Estimates using the discrete dual solution
One computes the dual solution $(u^h)^*$ on the same mesh for simplicity. Using difference quotients or gradient recovery (see Section 6.2.8), one can replace $\|u^* - (u^h)^*\|_{L^2(T)}$ by $h_T^2\|\nabla(\tilde{u}^h)^*\|_{2;T}$ or by $\|u^*_{h,\text{recover}} - (u^h)^*\|_{L^2(T)}$. Another possibility is to compute dual

solutions on a fine (h) and a coarse (H) mesh and let $\|u^* - (u^h)^*\|_{L^2(T)} \approx \|(u^H)^* - (u^h)^*\|_{L^2(T)}$.

6.3.3 Optimal meshes–A heuristic argument

First recall Section 6.1.5, where we defined an optimal mesh-size distribution as the minimizer of a constraint optimization problem. As before, we let $\mathcal{N}(h) := \int_\Omega \sigma/h^n$ and $Q(e) = \mathcal{E}(h) = \int_\Omega h^2 E^2$, but now with $E^2(x) = |\nabla^2 u(x)|\,|\nabla^2 u^*(x)|$, motivated by the previous section. Observe that our error criterion $\mathcal{E}(h) \stackrel{!}{=} \mathrm{TOL}$ leads to the *Lagrange function*

$$\mathcal{L}(h,\lambda) = \mathcal{N}(h) - \lambda\big(\mathrm{TOL} - \mathcal{E}(h)\big) = \int_\Omega \frac{\sigma}{h^n} - \lambda\big(\mathrm{TOL} - \int_\Omega h^2 E^2\big),$$

which differs from the former definition (6.9) by the exponent of TOL and $\mathcal{E}(h)$, and by the formula for E^2. Therefore we will conclude, instead of (6.11), that

$$h^2 E^2 = \frac{\mathrm{TOL}}{N(h)} \frac{\sigma}{h^n}.$$

Extracting h and inserting this into the definition of $\mathcal{N}(h)$ we find the relation

$$\mathcal{N}(h) \sim \mathrm{TOL}^{-n/(2\alpha)}$$

if the integral $\int_\Omega E^{2n/(n+2\alpha)}$ exists. To compare this to our previous situation we consider first a smooth solution on a uniform mesh. Since for linear finite elements $\|u - u^h\|_{L^2(\Omega)} = O\left(h_0^2\right)$, that is $\alpha = 1$, the requirement $h_0^2 \sim \mathrm{TOL}$ gives $\mathcal{N}(h_0) \sim \mathrm{TOL}^{-n/2}$ as above. If we establish a solution by the adaptive finite element method using the energy estimator (6.20) with tolerance TOL, we need $\mathcal{N}(h) \sim \mathrm{TOL}^{-n}$ (cf. Section 6.1.4). However, these results do not come with a reliable a posteriori error estimate for the point values. A rigorous a posteriori error bound, that would very likely give the optimal complexity, has been given by [513].

A theory of convergence and optimality is the subject of current research. A recent result is [495].

6.3.4 The general approach

Now we assume that the problem is given by:

$$\text{seek } u \in \mathcal{V} \text{ such that } A(u)[\phi] := \langle A(u), \phi\rangle_{\mathcal{V}'\times\mathcal{V}} = 0 \quad \text{for all } \phi \in \mathcal{V}.$$

Here, $\mathcal{V}$ is some Hilbert space and $A : \mathcal{V} \to \mathcal{V}'$ has two derivatives A' and A''. For example, the nonlinear Poisson problem $-\Delta u = f(u)$ in Ω with homogeneous Dirichlet boundary conditions leads to

$$A(u)[\phi] := \int_\Omega \{\nabla u \cdot \nabla\phi - f(u)\phi\}, \quad A'(u)[\phi,\psi] = \int_\Omega \{\nabla\psi \cdot \nabla\phi - f'(u)\psi\phi\}$$

$$A''(u)[\phi,\psi,\chi] = \int_\Omega \{-f''(u)\psi\phi\chi\}.$$

Now let $\mathcal{V}^h \subset \mathcal{V}$ be a finite element space and assume that integration is performed exactly. Then the discrete problem reads:

$$\text{seek } u^h \in \mathcal{V}^h \text{ such that } A(u^h)[\phi^h] = 0 \quad \text{for all } \phi^h \in \mathcal{V}^h.$$

Our quantity of interest may be given by some nonlinear, twice differentiable functional $Q : \mathcal{V} \to \mathbb{R}$. We define the error functional by

$$\triangle Q(u^h) := Q(u) - Q(u^h).$$

As before, we seek a representation formula for $\triangle Q$. We start with a Taylor series (or twice applying the Mean Value Theorem 1.43): with $e := u - u^h$ we get

$$\begin{aligned} A(u)[\,.\,] &= A(u^h)[\,.\,] + \int_0^1 A'(u^h + se)[e,\,.\,](s-1)\,ds \\ &= A(u^h)[\,.\,] + A'(u^h)[e,\,.\,] - \int_0^1 A''(u^h + se)[e,e,\,.\,](s-1)\,ds \end{aligned} \tag{6.31}$$

and

$$Q(u) = Q(u^h) + Q'(u^h)[e] - \int_0^1 Q''(u^h + se)[e,e](s-1)\,ds$$

which gives

$$\triangle Q(u^h) = Q'(u^h)[e] - \int_0^1 Q''(u^h + se)[e](s-1)\,ds. \tag{6.32}$$

Now let $u^* \in \mathcal{V}$ be the solution of the *dual problem*

$$A'(u^h)[\phi, u^*] = Q'(u^h)[\phi] \quad \text{for all } \phi \in \mathcal{V}$$

and $(u^h)^* \in \mathcal{V}^h$ be the solution of its discrete version

$$A'(u^h)[\phi^h, (u^h)^*] = Q'(u^h)[\phi^h] \quad \text{for all } \phi^h \in \mathcal{V}^h.$$

Taking $\phi = e$ we obtain by definition of u and u^h

$$\begin{aligned} \triangle Q(u^h) &= A'(u^h)[e, u^*] - \int_0^1 Q''(u^h + se)[e,e](s-1)\,ds \\ &= A(u)[u^*] - A(u^h)[u^*] \\ &\quad + \int_0^1 \Big(A''(u^h + se)[e,e,u^*] - Q''(u^h + se)[e,e] \Big)(s-1)\,ds \\ &\equiv A(u)[u^*] - A(u^h)[u^*] + R(u^h, e, u^*) \\ &= -A(u^h)[u^* - \phi^h] + R(u^h, e, u^*) \end{aligned}$$

for all $\phi^h \in \mathcal{V}^h$. Hence

$$|\triangle Q(u^h)| \le \inf_{\phi^h \in \mathcal{V}^h} |A(u^h)[u^* - \phi^h]| + O(\|e\|_{\mathcal{V}}^2).$$

This generalizes Equation (6.30). In the linear case the higher order term R vanishes. In the example from the beginning of this section, the nonlinear Poisson problem, and with a linear Q, we find the dual problem

$$\int_\Omega \left\{ \nabla \phi^h \cdot \nabla u^* - f'(u^h)\phi u^* \right\} = Q[\phi] \quad \text{for all } \phi \in \mathcal{V},$$

whose strong form is $-\Delta u^* - f'(u^h)u^* = Q[\,.\,]$ in Ω, $u^*\big|_{\partial\Omega} = 0$. Since $\triangle Q(u^h) = Q[e]$,

$$\big|Q[e]\big| \le \inf_{\phi^h \in \mathcal{V}^h} \left| \int_\Omega \nabla u^h \cdot \nabla(u^* - \phi^h) - f(u^h)(u^* - \phi^h) \right| + \left| \int_0^1 \int_\Omega f''(u^h + se)e^2 u^* \right|.$$

By partial integration in the first term we obtain

$$\sum_{T \in T^h} \left\{ \int_T f(u^h)(u^* - \phi^h) + \sum_{E \subset \partial T \setminus \partial\Omega} \left[\partial_n u^h\right]_E (u^* - \phi^h) \right\}.$$

The development of the a posteriori error estimate and the practical aspects are as before in Section 6.3.2.

The special case of A=L' and Q=L

This holds in many physical applications where L is an "energy" and the weak equation is the extremal condition $L'(u)[v] \equiv A(u)[v] \equiv 0$ for all $v \in \mathcal{V}$. Our quantity of interest may be the value of the energy $Q(u) := L(u)$. Hence $\triangle Q(u^h) := Q(u) - Q(u^h) = L(u) - L(u^h)$. Then (6.32) translates to

$$\begin{aligned} L'(u)[\,.\,] &= L'(u^h)[\,.\,] + \int_0^1 L''(u^h + se)[e, e, \,.\,](s-1)\, ds \\ &= L'(u^h)[\,.\,] + L''(u^h)[e, \,.\,] - \int_0^1 L'''(u^h + se)[e, e, \,.\,](s-1)\, ds \end{aligned}$$

and instead of (6.32) we have the expansion

$$\begin{aligned} \triangle Q(u^h) &= \int_0^1 L'(u^h + se)[e]\, ds = sL'(u^h + se)[e]\Big|_0^1 - \int_0^1 sL''(u^h + se)[e, e]\, ds \\ &= -\frac{1}{2}(s^2 - 1)L''(u^h + se)[e, e]\Big|_0^1 + \int_0^1 \frac{1}{2}(s^2 - 1)L'''(u^h + se)[e, e, e]\, ds \\ &= -\frac{1}{2}L''(u^h)[e, e] + \frac{1}{2}\int_0^1 (s^2 - 1)L'''(u^h + se)[e, e, e]\, ds. \end{aligned}$$

We now observe that we do not need to solve a dual problem since it now suffices to use $-1/2\, e$ as a test function in the expansion of $L'(u)[\,.\,]$ to eliminate $-\frac{1}{2}L''(u^h)[e, e]$

and so we get

$$\begin{aligned}\triangle Q(u^h) &= -\frac{1}{2}L''(u^h)[e,e] + \frac{1}{2}\int_0^1 (s^2-1)L'''(u^h+se)[e,e,e]\,ds \\ &= \frac{1}{2}L'(u^h)[e] - \frac{1}{2}\int_0^1 s(1-s)L'''(u^h+se)[e,e,e]\,ds \\ &= \frac{1}{2}L'(u^h)[e-\phi^h] + O\big(\|e\|_{\mathcal{V}}^3\big) \\ &= \frac{1}{2}L'(u^h)[u-\phi^h] + O\big(\|e\|_{\mathcal{V}}^3\big)\end{aligned}$$

and thus

$$\triangle Q(u^h) = \inf_{\phi^h \in \mathcal{V}^h} \left|\frac{1}{2}L'(u^h)[u-\phi^h]\right| + O\big(\|e\|_{\mathcal{V}}^3\big).$$

We repeat that this holds without a dual solution and that the second term vanishes if L is quadratic! This is especially useful if we consider eigenvalue problems [398].

7
Discontinuous Galerkin methods (DCGMs), with V. Dolejší

7.1 Introduction

Discontinuous Galerkin methods (DCGMs) represent a class of FEMs. In contrast to the usual FEMs, piecewise polynomials of varying degrees in different elements K are used. Continuity between neighboring elements and boundary conditions are not required. These violations are punished by a penalty function, see (7.41), (7.61), increasingly unpleasant for finer mesh sizes h. In some sense, DCGMs are therefore nonconforming FEMs.

So, many of the remarks in the introduction to Chapter 4 remain valid for DCGMs as well. Again we use the general discretization theory in Chapter 3. The proofs for convergence via stability and consistency techniques are again based upon linearization.

The original DCGM was introduced by Reed and Hill [546] for the solution of the neutron transport equation. It was analyzed for the first time by Le Saint and Raviart [472]; later improvements were achieved by Wheeler [665] and Johnson and Pitkäranta [426]. These methods were also applied to purely elliptic problems. Examples are the original method of Bassi and Rebay [75] the stabilized version of the original Bassi–Rebay method studied by Brezzi *et al.* [146], and its generalization, called the local discontinuous Galerkin methods (LDGMs), introduced by Cockburn and Shu [194], and further studied by Castillo *et al.* [166, 190]. In the 1970s, Galerkin methods for elliptic and parabolic equations using discontinuous finite elements were independently proposed and a number of variants introduced and studied, cf. Arnold [39], Baker [56] and Wheeler [665]. These DCGMs were called interior penalty (IP) methods, and their development remained independent of the development of the DCGMs for hyperbolic equations. A unified analysis and comparison of various discontinuous Galerkin techniques for elliptic problems was developed by Arnold *et al.* [40]. For a survey see, e. g. Cockburn *et al.* [182, 192], summarizing the development of DCGMs for elliptic, parabolic and hyperbolic problems.

Finally, we mention the so-called *hybridized discontinuous Galerkin methods* proposed and developed by Cockburn and co-workers in [163, 183–189]. This approach belongs among the mixed techniques where the hybridization reduces the globally coupled unknowns.

Here and throughout this book, we restrict the discussion to elliptic problems. In this chapter, we present a detailed study of a class of DG methods for second-order elliptic problems which includes most of the above-mentioned methods.

Many papers (e.g. [166, 297, 319, 517]) consider the linear diffusive operator $-\Delta u$ or $\nabla \cdot (A\nabla u)$, where A is a symmetric matrix independent of u. In addition, lower order terms are allowed, cf. (7.13). Several generalizations to nonlinear diffusion operators were studied. Rivière and Wheeler [552] consider the nonlinear diffusion operator

$$-\nabla \cdot (a(x,u)\nabla u), \quad a : \Omega \times \mathbb{R} \to \mathbb{R}, \tag{7.1}$$

where the function $a(x,u)$ is Lipschitz-continuous with respect to its second variable and there exist constants γ and $\gamma\prime$ such that

$$0 < \gamma \leq a(x,u) \leq \gamma\prime \qquad \forall (x,u) \in \Omega \times \mathbb{R}. \tag{7.2}$$

Extensions and improved energy estimates with applications to single phase flow in porous media are described by Wheeler *et al.* [554, 555].

Houston, Robson and Süli considered in [407], and extended in Houston, Süli and Wihler [412], the quasilinear diffusive operator in the form

$$-\nabla \cdot (\mu(x, |\nabla u|)\nabla u)\,, \text{with} |\nabla u| = |\nabla u|_n, \tag{7.3}$$

where $\mu \in C(\overline{\Omega} \times [0,\infty))$ is a function satisfying the assumption

$$m_\mu(t-s) \leq \mu(x,t)t - \mu(x,s)s \leq M_\mu(t-s), \qquad t \geq s \geq 0,\ x \in \overline{\Omega}, \tag{7.4}$$

with positive constants m_μ and M_μ.

Here we start with $\varepsilon\Delta u$ plus a low order flux term in (7.13) and proceed to more general cases. For quasilinear systems, e.g. the diffusion term takes the form

$$-\nabla \cdot \vec{R}(x, \vec{u}, \nabla \vec{u}), \quad \vec{R} : \Omega \times \mathbb{R}^q \times \mathbb{R}^{n\times q} \to \mathbb{R}^q. \tag{7.5}$$

Among several discontinuous Galerkin techniques two approaches, SIPG (symmetric interior penalty Galerkin) and NIPG (nonsymmetric interior penalty Galerkin), are very popular. The main idea, applied to a linear diffusion problem, here to $-\Delta u$, is the following. By applying Green's theorem, we obtain, for every inner face e and its unit normal vector ν, an integral

$$\int_e (\{\nabla u\}, \nu)_n [v]\, dS, \tag{7.6}$$

appropriately modified for boundary faces. The symbols $\{\cdot\}$ and $[\cdot]$ denote an average value and jump of a function beyond the face e, cf. (7.32), (7.33). Then we add, by formally exchanging the regular function u and the test function v, an integral in the form

$$\theta \int_e (\{\nabla v\}, \nu)_n [u]\, dS, \tag{7.7}$$

where $\theta = 1$ for SIPG and $\theta = -1$ for NIPG. Obviously, the integrals (7.7) vanish for a regular function $u \in H^1(\Omega)$. Moreover, the so-called *interior penalty,*

$$\int_e \sigma[u][v]\, dS, \qquad \sigma > 0 \tag{7.8}$$

again vanishing for a regular function, is included in the interior penalty variants of DCGMs. In Antonietti *et al.* [33] a different type of stabilization of DCGMs is presented. Moreover, Burman and Stamm [153] propose a method which requires stabilization only on a part of the skeleton.

The antisymmetry of the sum of terms in (7.6) and (7.7) (with $\theta = -1$) implies the coercivity of an extended bilinear form plus the penalty term, cf. (7.57), (7.161), of the NIPG technique for any penalty parameter $\sigma > 0$. On the other hand, for ensuring the coercivity of this bilinear form for the other DCGMs it is necessary to choose the parameter σ sufficiently large. However, this pays for the SIPG type discretizations. It yields a symmetric discrete problem. This is important for numerical methods and allows optimal error estimates in the L^2 norm.

The SIPG and NIPG techniques can be directly extended to nonlinear diffusion terms as in (7.1)–(7.3). The face integrals corresponding to (7.6) and (7.7) are

$$\int_e (\{a(u)\nabla u\}, \nu)_n [v]\, dS, \qquad \int_e (\{\mu(x, |\nabla u|)\nabla u\}, \nu)_n [v]\, dS \tag{7.9}$$

and

$$\theta \int_e (\{a(u)\nabla v\}, \nu)_n [u]\, dS, \qquad \theta \int_e (\{\mu(x, |\nabla u|)\nabla v\}, \nu)_n [u]\, dS, \tag{7.10}$$

respectively. The incomplete replacement of u by v from (7.9) to (7.10) maintains the essential linearity with respect to v. The linearity with respect to u is lost anyway.

This technique is no longer possible for general quasilinear diffusion terms of the form (7.5). Hence, SIPG as well as NIPG methods are not suitable for the discretization of quasilinear terms $\nabla \cdot \vec{R}(x, \vec{u}, \nabla \vec{u})$, unless $\vec{R}(x, \vec{u}, \nabla \vec{u})$ is linear in $\nabla \vec{u}$, or specific conditions as for (7.3) are imposed. Applying Green's theorem, the face integral of type (7.6) and/or (7.9) yields, differently from before,

$$-\int_e (\{\vec{R}(x, u, \nabla u)\}, \nu)_n [v]\, dS. \tag{7.11}$$

Exchanging the arguments u and v as in (7.7) would cause the resulting integral to become nonlinear with respect to v. Therefore we employ the so-called *incomplete interior penalty Galerkin* (IIPG) method studied by Dawson *et al.* [269,615,616]. Our method here for elliptic problems has been afterwards, without reference, modified and generalized to the parabolic case in [295]. We obtain a nonsymmetric diffusive form and have to choose the penalty parameter sufficiently large.

As for FEMs with variational crimes, the interplay between the strong and the weak form of the problem plays an essential role. So the original strong problem is transformed into its weak form. This excludes, for our approach, general fully nonlinear problems. For smooth enough functions, the strong forms of our operators can be easily determined, cf. (7.1)–(7.5). The corresponding weak forms we systematically define via testing in the appropriate spaces, cf. e.g. (7.19)–(7.22).

We start discussing these ideas for the important model problem (7.13)–(7.15). For $u \in H^2(\Omega) \cap L^\infty(\Omega)$, (7.13) is transformed into its weak form (7.23), valid for $u \in H^1(\Omega) \cap L^\infty(\Omega)$. These *convection–diffusion problems* are among the most interesting applications of the DCGMs. Here the diffusion term is multiplied by a small diffusion coefficient $\varepsilon > 0$.

In most of the chapter, we deal with convection–diffusion problems. For a reaction term $f(u)$ instead of $\sum_{s=1}^n \partial f_s(u)/\partial x_s$, (7.13) represents a *reaction–diffusion problem.* Sometimes a reaction term is added to the equation, e.g. for obtaining better properties. However, it is not present in the equations of fluid dynamics.

The terms in (7.6)–(7.11) are combined with the piecewise or broken Sobolev norms in (7.29)–(7.31) on the triangulation $\mathcal{T}^h$ for $u \in H^k(\mathcal{T}^h)$ for all $k \geq 0$ for convergence results. Penalty terms, e.g. (7.8), measure the transition errors along the interior and boundary edges for broken bilinear and nonlinear forms and penalty terms, cf. (7.41). Thus the original problem, e.g. (7.23), with its boundary condition, is translated into the discrete (7.70) or (7.71), cf. (7.57)–(7.61). For the strong form, (7.13), with $u \in H^2(\Omega), v \in L^2(\Omega)$, this often requires $u \in L^\infty(\Omega)$ for the nonlinearity. For the transition to the weak form, and the DCGMs, we have to impose conditions $u \in H^1(\mathcal{T}^h)$ or $u, v \in H^{3/2+\epsilon}(\mathcal{T}^h)$. This up and down between $u, v \in H^2(\mathcal{T}^h)$, $u, v \in H^{3/2+\epsilon}(\mathcal{T}^h)$, and $u, v \in H^1(\mathcal{T}^h)$, all $u \in L^\infty(\Omega)$, sometimes complicates the presentation, cf. Remarks 7.1 and 2.1, (2.156), and Theorem 7.5.

As indicated, we extend the DCGMs from the semilinear model problem (7.13) to more general quasilinear systems with diffusive terms in (7.5). Furthermore, we study, besides the standard numerical flux for convection diffusion equations, other types of fluxes as well, cf. (7.45)–(7.50). As nonconforming FEMs, the DCGMs are possible for linear to quasilinear, but not for fully nonlinear problems. The modification from our second order equations to the different types of semilinear or quasilinear systems is totally analogous to that for the previous FEMs. So we do not formulate the DCGMs for all these types of elliptic problems, but only for some order 2 including general quasilinear cases.

The results are presented in the *Hilbert space setting,* although extensions to the Banach space setting would be essentially possible.

Extensions to order $2m$ are problematic. The main tool for second order problems, the appropriate relation between the strong an weak forms of the problem, would become highly technical. So we omit problems of order $2m, m > 1$.

Recall that the following problems are only discussed for FEMs: General convergence theory for monotone operators, cf. Section 4.5, variational methods for eigenvalue problems, cf. Section 4.7, methods for nonlinear boundary conditions, cf. Section 5.3, quadrature approximate methods, cf. Section 5.4. They remain valid for DCGMs as well. Our techniques for fully nonlinear elliptic problems, cf. Section 5.2, are not possible for DCGMs. However, the monotone operator approach, cf. Section 4.5, allows DCGMs for problems of order $2m, m \geq 1$.

7.2 The model problem

Let $\Omega \subset \mathbb{R}^n$ (here $n = 2$ or 3) be a bounded *polyhedral domain* with closure $\overline{\Omega}$ and boundary $\partial\Omega = \partial\Omega_D \cup \partial\Omega_N$, $\partial\Omega_D \cap \partial\Omega_N = \emptyset$. We consider the following *boundary value problem,* for the exact solution u_0 in Ω :

$$u_0 : \Omega \to \mathbb{R}, u_0 \in H^1(\Omega) \cap L^\infty(\Omega), \Omega \ \subset \mathbb{R}^n, \Omega \in C^{0,1}, \text{ open bounded,} \tag{7.12}$$

$$G(u_0) := \sum_{s=1}^{n} \frac{\partial f_s(u_0)}{\partial x_s} - \varepsilon \Delta u_0 - g = 0 \quad \text{in } \Omega, \tag{7.13}$$

$$u_0 \,|_{\partial\Omega_D} \ = u_D \text{ on } \partial\Omega_D, \tag{7.14}$$

$$\frac{\partial u_0}{\partial \nu} \,|_{\partial\Omega_N} \ = g_N \text{ on } \partial\Omega_N, \tag{7.15}$$

cf. Remark 7.1, with an appropriate *flux term.* We require the following conditions:

$$\text{(a)} \ \ f_s \in C^1(\mathbb{R}), \ f_s(0) = 0, \ s = 1, \ldots, n; \tag{7.16}$$

(b) $\varepsilon > 0$;

(c) $g \in L^2(\Omega)$, or below $g \in H^{-1}(\Omega)$ with $\langle g, v \rangle := \langle g, v \rangle_{H^{-1}(\Omega) \times H^1(\Omega)}$;

(d) u_D is the trace of some $u^* \in H^1(\Omega) \cap L^\infty(\Omega)$ on $\partial\Omega_D$;

(e) $g_N \in L^2(\partial\Omega_N)$.

This $u^* \in H^1(\Omega) \cap L^\infty(\Omega)$ allows trivial boundary conditions by substituting $u := u - u^*$ and updating g. We avoid that here to demonstrate the role of the violated boundary conditions in DCGMs more clearly, cf. (7.41).

Equation (7.13) shows that the assumption $f_s(0) = 0$, $s = 1, \ldots, n$, does not cause any loss of generality. The functions f_s, called fluxes, represent *convective terms,* $\varepsilon > 0$ is the *diffusion coefficient.* The diffusion term can be more complicated, in some cases even nonlinear, see Section 7.5. For our model problem in (7.13)–(7.15), we consider mixed Dirichlet–Neumann boundary conditions. The same technique is possible for the following general cases as well. There we only present the Dirichlet condition on the whole boundary.

A sufficiently regular function, $u_0 \in C^2(\overline{\Omega})$, satisfying (7.13) – (7.15) pointwise is called a *classical solution.* Let

$$\mathcal{V}_b = \{v \in H^1(\Omega), \ v|_{\partial\Omega_D} = 0\}. \tag{7.17}$$

Multiplying (7.13) with $v \in \mathcal{V}_b$ and partial integration yields, via the *first Green's formula,* (2.9), (2.10), and with (7.15), and the Euclidean product $(.,.)_n$ in $\mathbb{R}^n$,

$$\int\limits_\Omega (-\Delta \, u) \ v dx = \int\limits_\Omega (\nabla \, u, \nabla \, v)_n dx - \int_{\partial\Omega_D} \frac{\partial u}{\partial \nu} \ v dS - \int_{\partial\Omega_N} g_N \, v dS. \tag{7.18}$$

Now we have to discuss, e.g. the $\int_{\partial\Omega_D} \partial u/\partial\nu\, v dS$, vanishing for $v \in \mathcal{V}_b$. The restriction $v|_{\partial\Omega}$ for $v \in H^1(\Omega)$ yields $v|_{\partial\Omega} \in H^{1/2}(\partial\Omega)$ in the trace sense. This is, for our Ω in (7.12), an immediate consequence of Theorems 1.37, 1.39.

Remark 7.1. *For Neumann conditions (7.15), and for the evaluation of the $\partial u/\partial\nu|_{\partial\Omega_D}$ in (7.18) and the formulation of the DCGMs, the situation is more complicated. An arbitrary $u \in H^1(\Omega)$ does not allow a meaningful restriction to $\partial\Omega$ or to e for $\partial u/\partial\nu \in L^2(\partial\Omega)$ or $\partial u/\partial\nu \in L^2(e)$.*

There are two ways out of this dilemma. Either Theorems 1.37, 1.39 are used for replacing the above $u \in H^1(\Omega)$ by $u \in H^{3/2+\epsilon}(\Omega)$ with any $\epsilon > 0$. Then $\partial u/\partial\nu|_{\partial\Omega} \in H^\epsilon(\partial\Omega) \subset L^2(\partial\Omega)$ or $\partial u/\partial\nu|_e \in L^2(e)$ is valid. Hence, the $\int_{\partial\Omega} v\nabla u dS$, and $\int_e v\nabla u dS$, are well defined, cf. (7.42). Or $u \in H^1(\Omega)$ is a function in the domain, e.g. of the Laplacian, then again $\int_e v\nabla u dS$ are well defined. This problem of boundary traces has attracted some attention recently, cf. Taylor [618], Chapter 4, Proposition 4.5, and Jonsson and Wallin [425], Jerison and Kenig [419], Schwab [575] and Grisvard [373].

Readers not so familiar with the $H^{3/2+\epsilon}(\Omega)$ results can in this chapter nearly everywhere replace $H^{3/2+\epsilon}(\Omega)$ by $H^2(\Omega)$, with $\partial u/\partial\nu|_{\partial\Omega} \in L^2(\partial\Omega)$ and $\partial u/\partial\nu|_e \in L^2(e)$, thus requiring a bit more than necessary.

So we assume $u \in H^{3/2+\epsilon}(\Omega)$, hence $\int_{\partial\Omega_D} \partial u/\partial\nu\ v ds$ is well defined, and $\int_{\partial\Omega_D} \partial u/\partial\nu\ v dS = 0$ for our $v \in \mathcal{V}_b$. This allows the transition, motivated by (7.18), to the following well-defined forms in their appropriate spaces:

$$a(u,v) = \varepsilon \int_\Omega (\nabla u, \nabla v)_n\, dx, \quad u \in \mathcal{U} := H^1(\Omega) \cap L^\infty(\Omega),\ v \in \mathcal{V}_b \text{ cf. (7.17):} \tag{7.19}$$

$$b(u,v) = \int_\Omega \sum_{s=1}^n \frac{\partial f_s(u)}{\partial x_s}\, v\, dx, \quad u \in \mathcal{U}, v \in L^2(\Omega), \text{ cf. (7.16) (a)}, \tag{7.20}$$

$$\ell(v) = (g,v)_{L^2(\Omega)} - (g_N, v)_{L^2(\partial\Omega_N)}, \quad v \in L^2(\Omega), v|_{\partial\Omega} \in L^2(\partial\Omega). \tag{7.21}$$

These $a(u,v), b(u,v), \ell(v)$ allow introducing the nonlinear operator G :

$$G : u \in \mathcal{D}(G) \subset \mathcal{U} \to \mathcal{U}' : \langle G(u), v\rangle - b(u,v) - a(u,v) + \ell(v) = 0\ \forall\ v \in \mathcal{V}_b. \tag{7.22}$$

A *weak solution*, u_0, of (7.13)–(7.15) satisfies the weak equation

$$u_0 \in \mathcal{U} :\ \langle G(u_0), v\rangle = 0 \qquad \forall\ v \in \mathcal{V}_b \text{ and } u_0\,|_{\partial\Omega_D} = u_D \text{ on } \partial\Omega_D \tag{7.23}$$
$$\frac{\partial u_0}{\partial\nu}\,|_{\partial\Omega_N} = g_N \text{ on } \partial\Omega_N.$$

A regular solution $u_0 \in H^2(\Omega)$ of (7.23) solves (7.13)–(7.15) as well.

In Chapter 2, we discussed existence and uniqueness results for a large class of problems, generalizing (7.13), (7.14), (7.23). A classical solutions solves (7.23) for $\forall\ v \in \mathcal{V}_b$. Vice versa, if (7.23) has smooth input data and a smooth enough solution, $u_0 \in H^2(\Omega)$, then its solves (7.13),(7.14). So we assume a unique solution u_0 of (7.23) with boundedly invertible $G'(u_0)$. A similar discussion applies to the more general cases, (7.74), (7.78), (7.79) and (7.90), (7.91).

7.3 Discretization of the problem

7.3.1 Triangulations

We formulate the DCGM on a polyhedral domain

$$\Omega \subset \mathbb{R}^n, n \geq 2, \text{ bounded } \textit{polyhedral} \text{ with boundary } \partial\Omega \in C^{0,1}. \tag{7.24}$$

Generalizations to curved domains Ω are possible, admitting

$$\Omega \subset \mathbb{R}^n, n \geq 2, \text{ bounded } \textit{curved domain} \text{ with boundary } \partial\Omega \in C^{0,1}, \tag{7.25}$$

if we require that, except at the edges of $\partial\Omega$, the solution $u_0 \in H^k(\Omega)$ can be extended into $u_0 \in H^k(\Omega_e)$ with an $\Omega_e \supset \Omega$. Under this condition, we can employ the technique in Section 5.2, Theorem 5.4, for *extending DCGMs on polyhedral to curved domains.* We can obtain at most second order accuracy, if we choose a

$$\text{sequence of approximating polygonal } \Omega^h \text{ for } \Omega, \text{ s.t. } \operatorname{dist}(\partial\Omega, \partial\Omega^h) \leq Ch^2. \tag{7.26}$$

Higher order versions would require Theorems 7.6 and 7.5 also for curved elements. Theorem 7.6 is proved for this case; for Theorem 7.5 we did not succeed.

So we restrict the formulation of the following *DCGMs to polyhedral domains.* Let $\mathcal{T}^h$ $(h > 0)$ be a *partition of the domain* Ω into a finite number of open n-dimensional (convex or nonconvex) mutually disjoint polyhedra K. In contrast to triangles $T \in \mathcal{T}^h$, for the standard FEMs, we denote the often more general elements here as K. In 2D problems we usually choose $K \in \mathcal{T}^h$ as triangles or quadrilaterals, or as in Figure 7.1. In 3D, $K \in \mathcal{T}^h$ can be, e. g. tetrahedra, pyramids or hexahedra. But we can construct even more general elements K, as dual finite volumes from [314]. In the analysis carried out in Section 7.7 it is important to assume star-shaped elements $K \in \mathcal{T}^h$. Then the standard error and inverse estimates in Theorems 7.6, 7.7 are valid, cf. Theorems 4.17, 4.19, and [141], p. 104. The exact assumptions on $\mathcal{T}^h$ will be given in Section 7.7.1. We modify the *notation* for a *triangulation,* $\mathcal{T}^h$, of Ω as

$$\forall K \in \mathcal{T}^h : K \text{ is open}, \forall K_1 \neq K_2 \in \mathcal{T}^h : K_1 \cap K_2 = \emptyset,\ \overline{\Omega} = \cup_{K \in \mathcal{T}^h} \overline{K}, \tag{7.27}$$
$$\mathcal{T}^h = \{K\}_{K \in \mathcal{T}^h} \text{ with } h_K = \operatorname{diam}(K), h = \max_{K \in \mathcal{T}^h} h_K \text{ and } \{e\}_{e \in \mathcal{T}^h},$$

the *set of all faces* of $\mathcal{T}^h$. Then for any $e \in \mathcal{T}^h$ there are two possibilities:

(i) $e \in \mathcal{T}^h \setminus \partial\Omega$ is an *inner face,* so there exist two elements K_l, $K_r \in \mathcal{T}^h$, $K_l \neq K_r$ sharing e, i.e. $e \subset \overline{K}_l \cap \overline{K}_r$.[45] If $\overline{K}_l \cap \overline{K}_r$ is not a straight face we split it into a minimal possible number of straight faces, see Figure 7.1, showing a possible 2D situation. It is arbitrary which one of the elements sharing the inner face is denoted as K_l and as K_r, but this choice should be kept.

(ii) $e \subset \partial\Omega$, often denoted as $e \in \partial\Omega$, is a boundary face. Then there exists an element $K_l \in \mathcal{T}_h$ such that $e \subset \partial K_l$.

By $\nu_e := \nu := (\nu_1, \ldots, \nu_n)$ we denote the unit normal on the face e, which is the outer normal to the element K_l sharing e (see Figure 7.1). Moreover, the symbols

[45] Subscripts $_l$ and $_r$ indicate "left" and "right" elements sharing e.

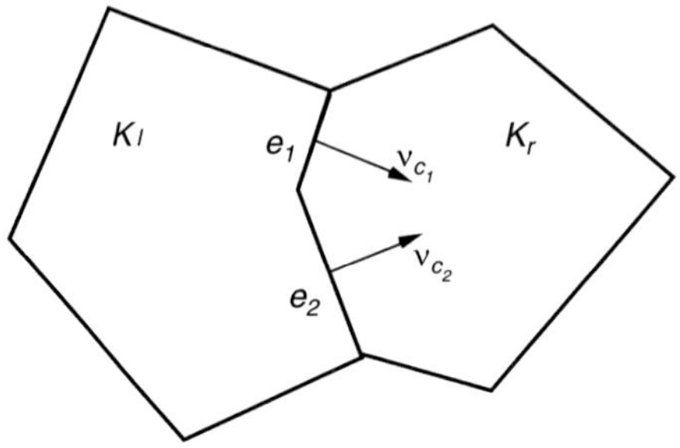

Figure 7.1 Neighboring elements K_l, K_r sharing two faces $e_1, e_2 \in \mathcal{T}^h$.

$|K|$ and $|e|$ mean the n-, and $(n-1)$-dimensional Lebesgue measure of K, and of e, respectively, (i. e. the length and the area of e, if $n = 2$ and $n = 3$), and diam(e) its diameter. It is obvious that for $K \in \mathcal{T}^h$ and its face $e \subset \overline{K}$ we have

$$|K| \leq C_1 h_K^n \leq C_1 h^n, \ |e| \leq C_1 h_K^{n-1} \leq C_1 h^{n-1}, \ d(e) := \text{diam}(e) \leq h_K \leq C_1 h. \quad (7.28)$$

7.3.2 Broken Sobolev spaces

We recapitulate the *broken Sobolev space, norms and seminorms* over the triangulation $\mathcal{T}^h$ only for the case of Hilbert spaces, see (4.35), (4.36).

$$H^k(\mathcal{T}^h) = \{v : \ v|_K \in H^k(K) \ \forall \, K \in \mathcal{T}^h\}, \quad (7.29)$$

$$\|v\|_{H^k(\mathcal{T}^h)} = \left(\sum_{K \in \mathcal{T}^h} \|v\|^2_{H^k(K)} \right)^{1/2}, \quad (7.30)$$

$$|v|_{H^k(\mathcal{T}^h)} = \left(\sum_{K \in \mathcal{T}^h} |v|^2_{H^k(K)} \right)^{1/2}. \quad (7.31)$$

For $v \in H^1(\mathcal{T}^h)$, we recall jumps and means, cf (5.82): for a face $e \in \mathcal{T}^h \cap \overline{K}$ we extend v to $\overline{K}$; for a boundary face $e \in \partial\Omega$, or an inner face with $e = \overline{K}_l \cap \overline{K}_r$, let

$$v_l = v|_{\overline{K}_l}, \ v_r = v|_{\overline{K}_r}, \ [v] = [v]_e = v_l|_e - v_r|_e, \ \{v\} = \{v\}_e = (v_l|_e + v_r|_e)/2, \quad (7.32)$$

$$[v] = [v]_e = \{v\} = \{v\}_e = v|_{\partial\Omega^h} \text{ along } e \in \partial\Omega^h, \ v \text{ arbitrary in } \mathbb{R}^n \backslash \Omega^h. \quad (7.33)$$

Obviously, the value $[v]_e$ depends upon the setting of which of the sharing element is K_l and which is K_r. On the other hand, the interesting term $\nu_e [v]_e$ is independent of the choice of K_l and K_r. In the following, omitting the index, $_e$, we agree

$$\sum_{e \in \mathcal{T}^h} \int_e [v](\{\nabla u\}, \nu)_n \text{ means } \sum_{e \in \mathcal{T}^h} \int_e [v]_e(\{\nabla u\}_e, \nu_e)_n \text{ etc.}$$

7.3.3 Extended variational formulation of the problem

In order to derive the discrete problem, we start from the strong solution $u_0 \in H^2(\Omega) \cap L^\infty(\Omega)$ of (7.13). We consider (7.13) on each K, multiply it by an arbitrary $v \in H^1(\mathcal{T}^h)$, integrate over each $K \in \mathcal{T}^h$, apply Green's theorem and summarize over $K \in \mathcal{T}^h$. Note that we no longer require $\in H_0^1(\mathcal{T}^h)$. Whenever defined, we obtain the identity

$$\sum_{K\in\mathcal{T}^h} \Big(- \int_K (\vec{f}(u_0), \nabla v)_n \, dx + \int_{\partial K} (\vec{f}(u_0), \nu)_n \, v \, dS \tag{7.34}$$
$$+ \varepsilon \int_K (\nabla u_0, \; \nabla v)_n \, dx - \varepsilon \int_{\partial K} (\nabla u_0, \nu)_n \, v \, dS - \int_K g \, v \, dx \Big) = 0 \;\; \forall v \in H^1(\mathcal{T}^h)$$

with $(\vec{f}(u), \nu)_n = \sum_{s=1}^{n} f_s(u) \, \nu_s$, and $(\vec{f}(u), \nabla v)_n = \sum_{s=1}^{n} f_s(u) \, \partial_s v.$

We combine (7.32), (7.33) with (5.259), obtaining for the sum of the second boundary integrals in (7.34), whenever defined,

$$-\varepsilon \sum_{K\in\mathcal{T}^h} \Big(\int_{\partial K} (\nabla u, \nu)_n \, v \, dS \Big) = \tag{7.35}$$
$$-\varepsilon \sum_{e\in\mathcal{T}^h\backslash\partial\Omega} \int_e \Big([v](\{\nabla u\}, \nu)_n + \{v\}([\nabla u], \nu)_n \Big) \, dS - \varepsilon \sum_{e\in\partial\Omega} \int_e v(\nabla u, \nu)_n dS.$$

Similarly, we obtain as in (5.259), (5.260),

$$\sum_{K\in\mathcal{T}^h} \int_{\partial K} (\vec{f}(u), \nu)_n \, v \, dS \tag{7.36}$$
$$= \sum_{e\in\mathcal{T}^h\backslash\partial\Omega} \int_e \Big([v](\{\vec{f}(u)\}, \nu)_n + \{v\}([\vec{f}(u)], \nu)_n \Big) \, dS + \sum_{e\in\partial\Omega} \int_e v(\vec{f}(u), \nu)_n dS.$$

We have to discuss the question of when the terms in (7.34), (7.35), (7.36) are defined. The extended variational form (7.34) is certainly not defined for the original $u_0 \in \mathcal{U}$ as in (7.19). Instead (7.34) motivates spaces $u_0, u \in \mathcal{U}_e^o = H^{3/2+\epsilon}(\Omega)$ for the original problem, or even an extension to $\mathcal{T}^h$, allowing $\partial u/\partial\nu|_e, \partial v/\partial\nu|_e \in L^2(e)$. Similarly we will choose $\mathcal{V}_e = \mathcal{U}_e$ for (7.42) ff. We use $\mathcal{V}_b$ in (7.17) and introduce, always restricting $\mathcal{U}$, and for simplicity the $\mathcal{V}$ as well, to $\cap L^\infty(\Omega)$,

$$\begin{aligned} \mathcal{U} &= \mathcal{V} = H^1(\Omega) \cap L^\infty(\Omega), \; \mathcal{U}' \subset H^{-1}(\Omega), \; \mathcal{V}_b, \mathcal{U}_b = H_0^1(\Omega) \cap L^\infty(\Omega), \\ \mathcal{U}_e^o &:= \mathcal{V}_e^o := H^{3/2+\epsilon}(\Omega) \cap L^\infty(\Omega), \mathcal{U}_e := \mathcal{V}_e := H^{3/2+\epsilon}(\mathcal{T}^h) \cap L^\infty(\mathcal{T}^h), \dots, \end{aligned} \tag{7.37}$$
and for the strong forms $\mathcal{U}_s = H^2(\Omega) \cap L^\infty(\Omega), \;\; \mathcal{U}_{s,e} := H^2(\mathcal{T}^h) \cap L^\infty(\Omega).$

We are going to introduce DCGMs. They are defined on $S^h \subset H^{3/2+\epsilon}(\mathcal{T}^h) \cap \mathcal{U}_e$. So a combination now of the exact solution $u_0 \in \mathcal{U}_e$, with $v \in \mathcal{U}_e$, the Neumann condition

(7.15), and (7.34), (7.35), (7.36), yields the following equation:

$$u_0 \in \mathcal{U}_e^o \subset \mathcal{U}_e : - \sum_{K\in\mathcal{T}^h} \int_K (\vec{f}(u_0), \nabla v)_n dx \tag{7.38}$$

$$+ \sum_{e\in\mathcal{T}^h\setminus\partial\Omega} \int_e ([v](\{\vec{f}(u_0)\}, \nu)_n + \{v\}([\vec{f}(u_0)], \nu)_n) dS + \sum_{e\in\partial\Omega} \int_e v(\vec{f}(u_0), \nu)_n dS$$

$$+\varepsilon \sum_{K\in\mathcal{T}_h} \int_K (\nabla u_0,\ \nabla v)_n\, dx - \varepsilon \sum_{e\in\mathcal{T}^h\setminus\partial\Omega} \int_e \Big([v](\{\nabla u_0\}, \nu)_n + \{v\}([\nabla u_0], \nu)_n\Big)\, dS$$

$$-\varepsilon \sum_{e\in\partial\Omega_D} \int_e v(\nabla u_0, \nu)_n\, dS - \int_\Omega g\, v\, dx - \varepsilon \int_{\partial\Omega_N} g_N\, v\, dS = 0 \quad \forall v \in \mathcal{U}.$$

There are different ways for further modifying the diffusion terms: The unchanged $\varepsilon \sum_{e\in\mathcal{T}^h} \int_e [v](\{\nabla u\}, \nu)_n dS$, based upon the vanishing $([\nabla u_0], \nu)_n$, is used by Wheeler [665], p. 153, extended to the case of nonlinear diffusion in (7.1).

Here we use the fact that for the exact solution $u_0 \in H^{3/2+\epsilon}(\Omega) \subset \mathcal{U}_e$ the terms $[\vec{f}(u_0)]$, $[\nabla u_0]$ and $(u_0 - u_D)|_{\partial\Omega_D}$ in (7.38) vanish. Furthermore, preparing the DCGMs, we first omit the terms $\{v\}([\vec{f}(u_0)], \nu)_n$, $\{v\}([\nabla u_0], \nu)_n$.

Then we add to (7.38), for $e \in \mathcal{T}^h \setminus \partial\Omega$, $0 = \theta\varepsilon(\{\nabla v\}, \nu)_n[u_0]$, and, for $e \in \partial\Omega_D$, $0 = \theta\varepsilon(\nabla v, \nu)_n(u_0 - u_D)$. This is used in the integration over the edges e in (7.38). The standard choices for θ, are, see below,

$$\theta = -1, \text{ or } \theta = 1, \text{ or } \theta = 0, \text{ sometimes even } \theta \in [-1, 1].$$

For the DCGMs we need the function σ, defined by, cf. (7.28),

$$\sigma : \bigcup_{e\in\mathcal{T}^h} e \to \mathbb{R} : \ \sigma|_e = \frac{c_w}{d(e)}, \text{ with } d(e) = \operatorname{diam}(e), \quad c_w > 0. \tag{7.39}$$

Another possibility is to define σ, e.g. as

$$\sigma|_e = \frac{c_w}{\max\{\operatorname{diam}(K_{e_1}), \operatorname{diam}(K_{e_2})\}} \quad \text{or} \quad \sigma|_e = \frac{c_w \min\{|K_{e_1}|, |K_{e_2}|\}}{d(e)}, \tag{7.40}$$

where K_{e_1} and K_{e_2} are neighboring elements sharing face e, and $K_{e_1} = K_{e_2}$ for $e \subset \partial\Omega$. These choices of interior penalties lead to equivalent results. Moreover, we should note that c_w may depend on the degree of the polynomial approximation. Finally, we add the following *penalty terms* to the last two lines of (7.38).

$$\varepsilon \sum_{e\in\mathcal{T}^h\setminus\partial\Omega} \int_e \sigma\, [u]\, [v]\, dS, \ \ \varepsilon \sum_{e\in\partial\Omega_D} \int_e \sigma\, (u - u_D)\, v\, dS. \tag{7.41}$$

This allows the desired coercivity results for DC piecewise polynomial test functions $v^h \in S^h$, cf. (7.63), Theorems 7.16–7.19. They are based upon the penalty norm

$\|v\|^J_{H^1(\mathcal{T}^h)} = \|v\|$, see (7.153). But the original $\|v\|_{H^1(\mathcal{T}^h)}$ and the new $\|v\|^J_{H^1(\mathcal{T}^h)}$ are nonequivalent norms on $H^1(\mathcal{T}^h)$ for $h \to 0$, see Theorem 7.12. We will formulate the corresponding consistency results in Theorems 7.21–7.33, tested by $v^h \in S^h$, linear and bounded with respect to $\|v\|^J_{H^1(\mathcal{T}^h)}$.

All the terms in (7.38), (7.41), including those evaluating $u, \nabla u_0, v, \nabla v$ along edges, e, are well defined $\forall u, v \in \mathcal{U}_e$, $e \in \mathcal{T}^h$, and many of them vanish or equal u_D for $e \in \mathcal{T}^h$, for the exact solution $u_0 \in H^{3/2+\epsilon}(\Omega)$, cf. Theorem 1.37 on trace operators. This would be incorrect for a $u_0 \in \mathcal{U}$.

We use (7.32), (7.33), (7.35), and collect $\sum_{e\in\mathcal{T}^h\setminus\partial\Omega}$ and $\sum_{e\in\partial\Omega_D}$ into $\sum_{e\in\mathcal{T}^h\setminus\partial\Omega_N}$, now for $u_0, v \in H^{3/2+\epsilon}(\Omega) = \mathcal{U}^o_e$. Then we obtain

$$
\begin{aligned}
u_0 \in \mathcal{U}^o_e : \quad & - \sum_{K\in\mathcal{T}^h} \int_K (\vec{f}(u_0),\, \nabla v)_n \, dx + \sum_{e\in\mathcal{T}^h} \int_e [v](\{\vec{f}(u_0)\}, \nu)_n dS \\
& + \varepsilon \sum_{K\in\mathcal{T}_h} \int_K (\nabla u_0,\, \nabla v)_n \, dx - \varepsilon \sum_{e\in\mathcal{T}^h\setminus\partial\Omega_N} \int_e \Big([v](\{\nabla u_0\}, \nu)_n + \theta(\{\nabla v\}, \nu)_n [u_0] \Big) \, dS, \\
& + \varepsilon \left(\sum_{e\in\mathcal{T}^h\setminus\partial\Omega_N} \int_e \sigma [u_0] \, [v] \, dS - \sum_{e\in\partial\Omega_D} \int_e \sigma u_D \, v \, dS \right) \\
& - \left(\int_\Omega g \, v \, dx + \varepsilon \int_{\partial\Omega_N} g_N \, v \, dS - \varepsilon\theta \sum_{e\in\partial\Omega_D} \int_e (\nabla v, \nu)_n \, u_D \, dS \right) = 0 \quad \forall v \in \mathcal{U}_e.
\end{aligned}
\tag{7.42}
$$

For the exact solution $u = u_0$ of (7.23), satisfying the boundary conditions (7.14), and the test functions $v \in H^{3/2+\epsilon}(\Omega) \subset \mathcal{U}_e$ we obtain

$$
[u_0] = [u_0]_e = 0, \ \{\nabla u_0\} = \{\nabla u_0\}_e = \nabla u_0|_e, \qquad \text{and similarly for } v, \tag{7.43}
$$

thus we get the weak form (7.23), and the strong form (7.13) for $u = u_0 \in \mathcal{U}_s$.

For the "convective" face integrals

$$
\sum_{e\in\mathcal{T}^h} \int_e [v](\{\vec{f}(u_0)\}, \nu)_n dS, \tag{7.44}
$$

the more sophisticated concept of a *numerical flux* in DCGMs is employed. This is motivated, e.g. by finite volume techniques for conservation laws. There the use of the numerical flux ensures the stability of numerical schemes using explicit time discretization, see, e.g. [313, 315]. For the different choices of numerical fluxes the present type of problems plays a crucial role. For example, the center numerical flux $(\{\vec{f}(u_0)\}, \nu)_n$ is unconditionally unstable for explicit discretization of time dependent problems. However, for implicit time discretization we have no stability restriction, but we have to solve a system of nonlinear algebraic equations at each time step. Our results below show stability and convergence for elliptic problems.

In our context the numerical flux emphasizes the evaluation of the term $(\vec{f}(u), \nu)_n$ appearing as the integrand in the first line of (7.34). We consider the edge, e, again as the face of K_l and K_r, $e \subset \overline{K}_l \cap \overline{K}_r$ given by (7.27) ff., (7.32) with the outer normal, ν. The numerical flux $H(u_l|_e, u_r|_e, \nu)$ approximates the evaluation along e as the face of K_l in the first argument of $H(\cdot,\cdot,\cdot)$ and indicates the impact of e as the face of K_r in its second argument.

A famous example is the *Lax–Friedrich numerical flux* in the form

$$H(u_l|_e, u_r|_e, \nu) = \frac{1}{2}\Big(P(u_l|_e, \nu) + P(u_r|_e, \nu) - \lambda(u_r|_e - u_l|_e)\Big), \quad \lambda > 0, \tag{7.45}$$

$$(\text{e.g.}) = \frac{1}{2}\Big((\vec{f}(u_l|_e), \nu)_n + (\vec{f}(u_r|_e), \nu)_n - \lambda(u_r|_e - u_l|_e)\Big) \tag{7.46}$$

$$= \Big((\{\vec{f}(u|_e)\}, \nu)_n - \frac{1}{2}\lambda[u|_e]\Big) \tag{7.47}$$

hence, $P(u, \nu) := (\vec{f}(u), \nu)_n$, for (7.46).

Another example is the upwind scheme given by

$$H(u_l|_e, u_r|_e, \nu) = \begin{cases} (\vec{f}(u_l|_e), \nu)_n = \sum_{s=1}^n f_s(u_l|_e)\nu_s, & \text{if } A > 0 \\ (\vec{f}(u_r|_e), \nu)_n = \sum_{s=1}^n f_s(u_r|_e)\nu_s, & \text{if } A \le 0, \end{cases} \tag{7.48}$$

where

$$A = \sum_{s=1}^{n} f_s'(\{u\}_e)\nu_s = (\vec{f}'(\{u\}_e), \nu)_n. \tag{7.49}$$

The face integrals (7.44) correspond to the central numerical flux

$$H(u_l|_e, u_r|_e, \nu_e) = (\{\vec{f}(u|_e)\}, \nu_e)_n \quad \forall e \in \mathcal{T}^h. \tag{7.50}$$

For further examples of numerical fluxes we refer to, e.g. [313].

We shall assume that the numerical flux has the following properties:

Assumptions (H):

1. $H(u, v, \nu)$ is defined in $\mathbb{R}^2 \times B_1$, where $B_1 = \{\nu \in \mathbb{R}^n; |\nu| = 1\}$, and *Lipschitz-continuous* with respect to u, v:

$$|H(u, v, \nu) - H(u^*, v^*, \nu)| \le C_2(|u - u^*| + |v - v^*|), \tag{7.51}$$
$$u,\ v,\ u^*,\ v^* \in \mathbb{R},\ \nu \in B_1.$$

2. $H(u, v, \nu)$ is *flux-consistent*:

$$H(u, u, \nu) = (\vec{f}(u), \nu)_n, \quad u \in \mathbb{R},\ \nu \in B_1. \tag{7.52}$$

3. $H(u, v, \nu)$ is *conservative*:

$$H(u, v, \nu) = -H(v, u, -\nu), \quad u,\ v \in \mathbb{R},\ \nu \in B_1. \tag{7.53}$$

Moreover, (7.51) and (7.52) imply that the function $\vec{f}$ is Lipschitz-continuous with constant $L_f = 2C_2$ and from (7.16) (a) and (7.52) we see that

$$H(0,0,\nu) = 0 \quad \forall \nu \in B_1. \tag{7.54}$$

Obviously, numerical fluxes given by (7.45), (7.48) and (7.50) satisfy Assumptions **(H)**.

So we obtain in the first line of (7.34), (7.42) with (7.33) the terms

$$\begin{aligned}
&\sum_{K\in\mathcal{T}^h} \left(\int_{\partial K} (\vec{f}(u),\, \nu)_n \, v \, dS \right) \\
&\quad = \sum_{e\in\mathcal{T}_h\setminus\partial\Omega} \int_e \big(H(u_l|_e, u_r|_e, \nu_e) v_l|_e + H(u_r|_e, u_l|_e, -\nu_e) v_r|_e\big) dS \\
&\quad + \sum_{e\in\partial\Omega} \int_e \big(H(u_l|_e, u_r|_e, \nu_e) v_l|_e\big) dS \qquad\qquad (7.55) \\
&\quad = \sum_{e\in\mathcal{T}_h\setminus\partial\Omega} \int_e H(u_l|_e, u_r|_e, \nu_e)(v_l|_e - v_r|_e)\, dS + \sum_{e\in\partial\Omega} \int_e \big(H(u_l|_e, u_r|_e, \nu_e) v_l|_e\big) dS \\
&\quad = \sum_{e\in\mathcal{T}_h} \int_e H(u_l|_e, u_r|_e, \nu_e)[v]_e \, dS, \quad \forall v \in H^1(\mathcal{T}^h).
\end{aligned}$$

Consequently (7.42) has the form

$$\begin{aligned}
&u_0 \in \mathcal{U}_e : -\sum_{K\in\mathcal{T}^h} \int_K (\vec{f}(u_0),\, \nabla v)_n \, dx + \sum_{e\in\mathcal{T}_h} \int_e H(u_{0,l}|_e, u_{0,r}|_e, \nu_e)[v]_e \, dS \qquad (7.56) \\
&+\varepsilon \sum_{K\in\mathcal{T}_h} \int_K (\nabla u_0,\, \nabla v)_n \, dx - \varepsilon \sum_{e\in\mathcal{T}^h\setminus\partial\Omega_N} \int_e \Big([v](\{\nabla u_0\}, \nu)_n + \theta(\{\nabla v\}, \nu)_n [u_0] \Big) dS \\
&+\varepsilon \sum_{e\in\mathcal{T}^h\setminus\partial\Omega_N} \int_e \sigma[u_0]\,[v]\, dS - \int_\Omega g\, v\, dx - \int_{\partial\Omega_N} g_N \, v\, dS \\
&- \left(\varepsilon \sum_{e\in\partial\Omega_D} \int_e \sigma u_D \, v\, dS - \varepsilon\theta \sum_{e\in\partial\Omega_D} \int_e (\nabla v, \nu)_n \, u_D \, dS \right) = 0 \quad \forall v \in \mathcal{U}_e.
\end{aligned}$$

On the basis of equations (7.42) and (7.56) we introduce:

$$\begin{aligned}
\forall u, v \in \mathcal{U}_e : a_h(u,v) &:= \varepsilon \sum_{K\in\mathcal{T}_h} \int_K (\nabla u,\, \nabla v)_n \, dx \qquad\qquad (7.57) \\
&- \varepsilon \sum_{e\in\mathcal{T}^h\setminus\partial\Omega_N} \int_e \Big((\{\nabla u\}, \nu)_n \, [v] + \theta(\{\nabla v\}, \nu)_n [u] \Big) dS
\end{aligned}$$

$$b_h(u,v) = -\sum_{K\in\mathcal{T}^h}\int_K (\{\vec{f}(u)\},\nabla v)_n\,dx + \sum_{e\in\mathcal{T}^h}\int_e (\{\vec{f}(u)\},\nu)_n[v]dS \tag{7.58}$$

$$c_h(u,v) := -\sum_{K\in\mathcal{T}^h}\int_K (\{\vec{f}(u)\},\nabla v)_n\,dx + \sum_{e\in\mathcal{T}_h}\int_e H(u_l,u_r,\nu)[v]\,dS \tag{7.59}$$

$$\ell_h(v) := \langle g,v\rangle + (g_N,v)_{L^2(\partial\Omega_N)} - \varepsilon\theta\sum_{e\in\partial\Omega_D}\int_e (\nabla v,\nu)_n\,u_D\,dS \tag{7.60}$$

$$+\varepsilon\sum_{e\in\partial\Omega_D}\int_e \sigma u_D\,v\,dS \quad \text{with}$$

$$J_h^\sigma(u,v) := \sum_{e\in\mathcal{T}^h\setminus\partial\Omega_N}\int_e \sigma[u]\,[v]\,dS, \text{ and } J_h^1(u,v) := \sum_{e\in\mathcal{T}^h\setminus\partial\Omega_N}\int_e [u]\,[v]\,dS. \tag{7.61}$$

In the discrete problem the form J_h^σ represents a *penalty function for interior discontinuities* along the edges e. The *boundary errors* are collected in the terms $\varepsilon\sum_{e\in\partial\Omega}\int_e \sigma(u_D\,v\; - \;u\,v)dS$ obtained from $\varepsilon J_h^\sigma(u,v) - \ell_h(v)$. These terms replace the requirement of continuity and boundary conditions of the approximate solution in conforming FE methods. Note that for increasingly finer $\mathcal{T}^h$ the σ in (7.39) increases. Thus violated conditions are punished more strongly. For some of the later discussions we mainly consider $u^h, v^h \in S^h$, with bounded $J_h^\sigma(v,v)$, e.g. those well approximating $u,v\in\mathcal{U}$. This is achieved by introducing the so-called *penalty norms,* cf. Definition 7.11.

So we have modified the above (7.38). These (7.57)–(7.61) allow formulating different DCGMs. As mentioned above, the standard choices for θ, are

$$\theta = -1, \text{ or } \theta = 1, \text{ or } \theta = 0, \text{ sometimes even } \theta\in[-1,1], \tag{7.62}$$

with $\theta = -1$ for the so-called *nonsymmetric form of the interior penalty Galerkin method, NIPG* cf. Rivière *et al.* [552, 553, 555], $\theta = 1$ for the *symmetric form, SIPG,* cf. Arnold [39], $\theta = 0$ for the *incomplete form, IIPG,* cf. Dawson *et al.* [269, 615, 616], and $-1 \le \theta \le 1$ for the *general form for the general penalty Galerkin method, GIPG,* cf. Houston *et al.* [407]. Hence $a_h(u,v) \ne a_h(v,u)$ or, for $\theta = 1$, the $a_h(u,v) = a_h(v,u)$ is obtained.

Our analysis includes all four cases. Standard numerical methods essentially use $\theta = -1, 0, 1$. For $\theta = -1$, we only have to impose $c_w > 0$. For $\theta = 0, 1$, and $\theta\in(-1,1]$ we have to observe lower bounds for $c_w > 0$, cf. Theorem 7.16 ff. There we will come back to the discussion of appropriate choices of the $c_w > 0$.

These different choices have "pros and cons": The symmetric form maintains the symmetry of the previous and later $a(u,v) = a(v,u)$ with desirable consequences for numerical methods. However, the c_w in (7.39) has to be chosen, for the symmetric form, the incomplete form and for $\theta\in(-1,1]$ according to (7.164), (7.180), (7.194). This implies coercive principal parts of the corresponding linearized discrete operators. This is valid for the unsymmetric form for any $c_w > 0$.

The symmetric bilinear form $a_h(\cdot,\cdot)$ allows deriving *optimal a priori error estimates* in the L^2 norm using the well-known *Aubin–Nitsche trick,* cf. Lemma 5.73 and Theorem 5.74 for FEMs and [40] for DCGMs and the references therein. Nevertheless, the methods with $\theta = 1$ give in some situations optimal experimental order of convergence, namely for odd degrees of polynomial approximations when (almost) equidistant partitions are employed. On the other hand, for ensuring the coercivity for the symmetric form it is necessary to choose the penalty parameter σ, in particular the c_w, more carefully than in the nonsymmetric case, cf. Theorems 7.16, 7.18, 7.19. For general quasilinear problems we only formulate this last type of IIPGM below.

7.3.4 Discretization

Now we introduce the discrete problem approximating (7.56) via the DCGM. We define the space of discontinuous piecewise polynomial functions

$$S^h := S^{d-1,-1}(\mathcal{T}^h) := \{v : \mathcal{T}^h \to \mathbb{R};\ v|_K \in \mathcal{P}_{d-1}(K)\ \forall\, K \in \mathcal{T}^h\}, d \geq 2, \qquad (7.63)$$

where $\mathcal{P}_{d-1}(K)$ denotes the space of all polynomials on K of degree $\leq d-1 \in \mathbb{N}$.

We derive the discrete problem in such a way that u_0 in (7.56) is approximated by $u_0^h \in S^h$ and tested by $v^h \in S^h$. In the open $K \in \mathcal{T}^h$, $\nabla u_0^h, \nabla v^h$ and their restrictions to $e \subset \overline{K}$ are well defined for $u^h, v^h \in S^h \subset H^{3/2+\epsilon}(\mathcal{T}^h) \subset H^1(\mathcal{T}^h)$. Hence the forms $a_h(u,v)$, $c_h(u,v)$, $b_h(u,v), \ell_h(v), J_h^\sigma(u,v)$, cf. (7.57)–(7.61) make sense for $u, v \in \mathcal{U}_e$ and $u = u^h,\ v = v^h \in S^h$.

We have seen that extended analytical problems require $u, v \in \mathcal{U}_e$. So it is surprising and advantageous that for the DCGMs a combination of $J_h^\sigma(v,v)$ and $|v|^2_{H^1(\mathcal{T}^h)}$ is sufficient for proving the boundedness, and consistency of the previous $a^h(\cdot,\cdot), b^h(\cdot,\cdot)$, $c^h(\cdot,\cdot)$, $\ell^h(\cdot)$, and the coercivity of the principal part of and the stability for $a^h(\cdot,\cdot)$ with respect to such a norm. This so-called *penalty norm,* cf. (7.61), and Definition 7.11, is

$$\|v\| := \|v\|_{\mathcal{V}^h} := \|v\|^J_{H^1(\mathcal{T}^h)} := \left(J_h^\sigma(v,v) + |v|^2_{H^1(\mathcal{T}^h)}\right)^{1/2}. \qquad (7.64)$$

This norm is chosen for $\mathcal{U}^h = \mathcal{V}^h = S^h$, cf (7.153)

$$u^h, v^h \in \mathcal{U}^h = \mathcal{V}^h = S^h \text{ with } \|v^h\|_{\mathcal{U}^h} = \|v^h\|_{\mathcal{V}^h} := \|v^h\|^J_{H^1(\mathcal{T}^h)} = \|v\|. \qquad (7.65)$$

It is not equivalent to the broken Sobolev norm for $h \to 0$, cf. Theorem 7.12.

Before we formulate the DCGMs, we discuss the relations between the original and discrete operators, G and G^h. This is essentially the same idea for all the following problems. So we demonstrate it for our model problem. With the $a(u,v), \ldots$ in (7.19)–(7.21), the $a_h(u,v), \ldots, J_h^\sigma(u,v)$ in (7.57)–(7.61), and, e.g. the $\langle G_h(u), v\rangle = \langle G_h(u), v\rangle_{\mathcal{U}_e' \times \mathcal{U}_e}$, we define, for a given fixed u or u^h, the weak operators G, G_h, G^h, by testing with v or v^h. We indicate this in (7.66), e.g. by $u, \forall v \in \mathcal{U}$. For G we use the $b(u,v)$ in (7.20), and in G_h we allow replacing $c_h(u,v)$ by $b_h(u,v)$ in (7.66). For G we have two possibilities: either we impose the Dirichlet condition directly as in

(7.66) (b), or we enforce them for $h \to 0$ as in (7.66) (c) by the penalty terms.

$$\begin{aligned}
&\text{(a)}\ \ G : \mathcal{D}(G) \subset \mathcal{U} \to \mathcal{U}', G_h : \mathcal{D}(G_h) \subset \mathcal{U}_e \to \mathcal{U}_e', G^h : \mathcal{D}(G^h) \subset \mathcal{U}^h \to \mathcal{U}'^h, && (7.66)\\
&\text{(b)}\ \ \langle G(u), v\rangle - a(u,v) - b(u,v) + \ell(v) = 0\ \forall v \in \mathcal{V}_b, (u - u_D)|_{\partial\Omega_D} = 0, u \in \mathcal{U},\\
&\text{(c)}\ \ \langle G(u), v\rangle - a_h(u,v) - b_h(u,v) - \varepsilon J_h^\sigma(u,v) + \ell_h(v) = 0,\ u, \forall v \in \mathcal{U}_e^o,\\
&\text{(d)}\ \ \langle G_h(u), v\rangle - a_h(u,v) - b_h(u,v) - \varepsilon J_h^\sigma(u,v) + \ell_h(v) = 0,\ u, \forall v \in \mathcal{U}_e,\\
&\text{(e)}\ \ \langle G^h(u^h), v^h\rangle - a^h(u^h,v^h) - b^h(u^h,v^h) - \varepsilon J_h^\sigma(u^h,v^h) + \ell^h(v^h) = 0, u^h, \forall v^h\\
&\qquad \in S^h, a^h(u^h,v^h), b^h(u^h,v^h) := a_h(u^h,v^h), b_h(u^h,v^h)|_{S^h\times S^h}, \ell^h(v^h) := \ell_h(v^h)|_{S^h},
\end{aligned}$$

or we may replace in (c)–(e) the b_h by c_h. In (c) we either have to consider $u, \forall v \in \mathcal{U}_e^o$, or $u, \forall v \in \mathcal{U}$, with $[u] = [v] = 0$ for interior faces e and such that $u|_{\partial\Omega}, v|_{\partial\Omega} \in L^2(\partial\Omega)$.

For relating G, G_h, and G^h we use the previous $\mathcal{U}_e^o = H^{3/2+\epsilon}(\Omega) \cap L^\infty(\Omega)$, cf. (7.37). For all $u, v \in \mathcal{U}_e^o$, many terms in G_h cancel:

$$u, v \in \mathcal{U}_e^o \text{ and } [u]_e = [v]_e = 0, \text{ for all interior faces } e \in \mathcal{T}^h, \text{ imply}$$

$$a_h(u,v) := \varepsilon \sum_{K\in\mathcal{T}_h} \int_K (\nabla u,\ \nabla v)_n\, dx - \varepsilon \sum_{e\in\partial\Omega_D} \int_e \Big((\nabla u, \nu)_n v + \theta(\nabla v, \nu)_n u\Big)\, dS$$

$$b_h(u,v) = - \sum_{K\in\mathcal{T}^h} \int_K (\vec{f}(u), \nabla v)_n\, dx + \sum_{e\in\partial\Omega} \int_e (\vec{f}(u), \nu)_n v dS$$

$$c_h(u,v) := - \sum_{K\in\mathcal{T}^h} \int_K (\vec{f}(u), \nabla v)_n\, dx + \sum_{e\in\partial\Omega} \int_e H(u_l, u_r, \nu) v\, dS$$

$$\ell_h(v) := \langle g, v\rangle + (g_N, v)_{L^2(\partial\Omega_N)} - \varepsilon\theta \sum_{e\in\partial\Omega_D} \int_e (\nabla v, \nu)_n\, u_D\, dS + \varepsilon \sum_{e\in\partial\Omega_D} \int_e \sigma u_D\, v\, dS$$

$$J_h^\sigma(u,v) := \sum_{e\in\mathcal{T}^h} \int_e \sigma[u]\,[v]\, dS.$$

We collect these terms in $\langle G_h(u), v\rangle$ and find

$$\begin{aligned}
\forall u,\ v\ \in \mathcal{U}_e^o : \langle G_h(u), v\rangle &= a_h(u,v) + b_h(u,v) + \varepsilon J_h^\sigma(u,v) - \ell_h(v)\\
&= \langle G(u), v\rangle + \sum_{e\in\partial\Omega} \int_e (\vec{f}(u), \nu)_n v dS \qquad (7.67)\\
&\quad - \varepsilon \sum_{e\in\partial\Omega_D} \int_e \Big((\nabla u, \nu)_n v + \theta(\nabla v, \nu)_n u\Big)\, dS + \varepsilon \sum_{e\in\partial\Omega_D} \int_e \sigma u\, v\, dS\\
&\quad + \varepsilon\theta \sum_{e\in\partial\Omega_D} \int_e (\nabla v, \nu)_n\, u_D\, dS - \varepsilon \sum_{e\in\partial\Omega_D} \int_e \sigma u_D\, v\, dS\\
&= \langle G(u), v\rangle \forall\ u \in \mathcal{U}_e^o, (u - u_D)|_{\partial\Omega_D} = 0\ \forall v \in \mathcal{U}_b.
\end{aligned}$$

Note that we used for the last equality in (7.67) the

$$b(u,v) = b_h(u,v) = -\sum_{K\in\mathcal{T}^h}\int_K(\{\vec{f}(u)\},\nabla dxv)_n$$
$$+\sum_{e\in\partial\Omega_N}\int_e(\vec{f}(u),\nu)_n v dS\ \forall\ v\in\mathcal{U}_b. \tag{7.68}$$

This allows summarizing these relations as

$$\begin{aligned}\langle G_h(u),v\rangle &= a_h(u,v)+b_h(u,v)+\varepsilon J_h^\sigma(u,v)-\ell_h(v)\forall\, u,v\in\mathcal{U}_e^o, \qquad (7.69)\\ &= \langle G(u),v\rangle = a(u,v)+b(u,v)-\ell(v)\ \forall\, u:(u-u_D)|_{\partial\Omega_D}=0\ \forall\, v\in\mathcal{U}_b,\\ &= \langle G(u),v\rangle\ \forall\, u\in\mathcal{U}:(u-u_D)|_{\partial\Omega_D}=0,\ \forall\, v\in\mathcal{U}_b,\ \text{so, with}\\ \mathcal{U}_{e,b}^o &:= \{u\in\mathcal{U}_e^o,(u-u_D)|_{\partial\Omega_D}=0\},\ \ \mathcal{U}_b:=\{u\in\mathcal{U},(u-u_D)|_{\partial\Omega_D}=0\}:\\ G^h &= G_h|_{S^h},\ \text{and}\ G_h|_{\mathcal{U}_{e,b}^o}=G|_{\mathcal{U}_{e,b}^o}=G\uparrow_{\mathcal{U}_b}.\end{aligned}$$

Remark 7.2. *This last equality in (7.69), $G_h|_{\mathcal{U}_{e,b}^o}=G|_{\mathcal{U}_{e,b}^o}=G\uparrow_{\mathcal{U}_b}$, indicates the restriction from G_h, defined in $\mathcal{U}_e^o$ to $\mathcal{U}_{e,b}^o$, and the extension of G, defined in $\mathcal{U}_{e,b}^o$, to G defined in the original $\mathcal{U}_b\supset\mathcal{U}_{e,b}^o$. Note that we have to test $G^h=G_h|_{S^h}$ by $v\in S^h$, $G_h u$ by $v\in\mathcal{U}_e$, $G_h|_{\mathcal{U}_e^o}=G|_{\mathcal{U}_e^o}$ by $v\in\mathcal{U}_e^o$, and $G\uparrow_{\mathcal{U}_b}$ by $v\in\mathcal{U}_b$. We observe an analogous situation for all the following problems as well.*

We define the *nonstandard DCGM* and its *discrete approximate solution* as a function $u_0^h:\mathcal{T}^h\to\mathbb{R}$ satisfying the conditions, cf. (7.66), (7.69),

$$u_0^h\in S^h;\quad \left\langle G^h\left(u_0^h\right),v^h\right\rangle=0\ \forall\, v^h\in S^h. \tag{7.70}$$

Similarly, we define the discrete approximate solution of a *standard DCGM*, by replacing b_h in (7.58), (7.69) by c_h in (7.59), for a function $u_0^h:\Omega\to\mathbb{R}$ satisfying

$$\begin{aligned}&\text{(a)}\ \ u_0^h\in S^h,\ \text{s.t.}\ \ \forall\, v^h\in S^h, \qquad (7.71)\\ &\text{(b)}\ \ \left\langle G^h\left(u_0^h\right),v^h\right\rangle := a^h\left(u_0^h,v^h\right)+c^h\left(u_0^h,v^h\right)+\varepsilon J_h^\sigma\left(u_0^h,v^h\right)-\ell^h(v^h)=0.\end{aligned}$$

Many terms in (7.71) cancel, in particular $J_h^\sigma(u,v)=0$, and we have obtained $G_h|_{\mathcal{U}_e^o}=G|_{\mathcal{U}_e^o}=G\uparrow_{\mathcal{U}}$, in (7.66). Proceeding as in (7.67), (7.68), (7.69), the exact solution $u_0\in\mathcal{U}$, obviously satisfies the original (7.23) as

$$\langle G(u_0),v\rangle=\big(a_h(u_0,v)+c_h(u_0,v)+\varepsilon J_h^\sigma(u_0,v)\big)-\ell_h(v)=0\ \forall\, v\in\mathcal{V}_b. \tag{7.72}$$

With this regular exact solution u_0 we find the identity

$$\langle G_h(u_0),v^h\rangle := a_h(u_0,v^h)+c_h(u_0,v^h)+\varepsilon J_h^\sigma(u_0,v^h)-\ell_h(v^h)=0\ \forall v^h\in S^h \tag{7.73}$$

by (7.57). This implies the so-called *Galerkin orthogonality property* of the error. This means that, for this and the following cases,

$$\langle G^h(u_0^h), v^h\rangle = 0 = \langle G_h(u_0), v^h\rangle \ \forall v^h \in S^h \Rightarrow \langle G^h(u_0^h) - G_h(u_0), v^h\rangle = 0 \ \forall v^h \in S^h.$$

7.4 General linear elliptic problems

Now we are ready to formulate more general problems. For simplicity, we consider only the Dirichlet boundary condition on the whole boundary of Ω in this and the following sections, i.e. $\partial\Omega = \partial\Omega_D$. A possible generalization to a case with Neumann boundary conditions on a part of the boundary can be carried out simply as in the previous section.

We have fully formulated FEMs for elliptic linear or semi- and quasilinear second order equations and systems. So we restrict the presentation here to linear equations and to semilinear equations and quasilinear systems in Section 7.5. The *linear boundary value problem* and its DCGMs are discussed briefly. We give more details for the following semilinear second order equations. Since there are strong similarities we do not repeat all details for extending (7.13), (7.14) ff. to general linear equations. Compared to (7.13), we replace the original $\Delta u, \partial u/\partial\nu$ by the general $Au, (B_a\nabla u, \nu)_n$ or $Au, (B_a\nabla^0 u, \nu)_n$ in the following (7.74), (7.75) and omit the flux terms, and the restriction to $L^\infty(\Omega)$ in $\mathcal{U}, \ldots,$ in (7.37). As usual, multiplying the strong operator, A_s, with $v \in \mathcal{U}$, we obtain by partial integration and Green's formula, the relation to the weak operator, A, cf. (5.278): Compute[46]

$$u_0 \in \mathcal{U} := H_0^1(\Omega) : a(u_0, v) = \langle Au_0, v\rangle_{\mathcal{V}'\times\mathcal{V}} = \langle f, v\rangle_{\mathcal{V}'\times\mathcal{V}} \ \forall v \in \mathcal{U}, \text{ with} \tag{7.74}$$

$$\begin{aligned}
a(u,v) &= \langle Au, v\rangle_{\mathcal{V}'\times\mathcal{V}} = \int_\Omega \left(\sum_{i,j=0}^{n} a_{ij}\partial^i u \, \partial^j v\right) dx \text{ with } \max_{x\in\overline{\Omega}}\{|a_{ij}(x)|\} \le C_0, \\
&= \int_\Omega \left(\sum_{i,j=0}^{n} (-1)_{j>0}\partial^j \left(a_{ij}\partial^i u\right)\right) v dx + \int_{\partial\Omega} (B_a\nabla^0 u, \nu)_n v ds \\
&=: (A_s u, v)_{L^2(\Omega)} + \int_{\partial\Omega} (B_a\nabla^0 u, \nu)_n v ds \\
&\quad \forall v \in \mathcal{U}, u \in \mathcal{U}_s = H^2(\Omega), \text{ if } a_{ij} \in W^{1-\delta_{j0},\infty}(\Omega), \\
&\quad \text{with } (B_a\nabla^0 u, \nu)_n := \sum_{i=0,\ldots,n}^{j=1,\ldots,n} \nu_j a_{ij}\partial^i u, \text{ and} \\
&\quad \sum_{i,j=1}^{n} a_{ij}(\cdot)\xi_i\xi_j \ge \lambda|\xi|_n^2, \lambda > 0, \text{ for our elliptic operators.}
\end{aligned}$$

46 We require $\max_{x\in\overline{\Omega}}\{|a_{ij}(x)|\} \le C_0$ instead of the usual $a_{ij} \in L^\infty(\Omega)$, since we need bounds for $\|a_{ij}\|_{L^\infty(e)}$ along $e \in \mathcal{T}^h$ as well, not available for $a_{ij} \in L^\infty(\Omega)$.

This $(B_a\nabla^0 u,\nu)_n$ depends upon $\nabla^0 u=(\partial^i u)_{i=0}^n$, e.g. $A_s u=-\Delta u$ yields $(B_a\nabla^0 u,\nu)_n=(B_a\nabla u,\nu)_n=\partial u/\partial\nu=(\nabla u,\nu)_n$.

Again we replace $H^k(\Omega)$ by $H^k(\mathcal{T}^h)$ and obtain with Green's formula as for (7.34) ff., with $\mathcal{U}^h,\mathcal{V}^h$ in (7.63), (7.65), and with (7.32), (7.33),

$$
\begin{aligned}
&\langle A_h u,v\rangle_{H^{-1}(\mathcal{T}^h)\times H^1(\mathcal{T}^h)}\\
&:=\sum_{K\in\mathcal{T}^h}\left(\int_K\sum_{i,j=0}^n a_{ij}\partial^i\, u\partial^j\, v\right)dx\quad\forall u,v\in\mathcal{U}_e\supset\mathcal{U}^h\\
&=\sum_{K\in\mathcal{T}^h}\int_K\left[(A_{s,h}\,u)v:=\sum_{i,j=0}^n(-1)_{j>0}\partial^j(a_{ij}\partial^i\,u)\,v\right]dx \qquad (7.75)\\
&\quad+\sum_{e\in\mathcal{T}^h\setminus\partial\Omega}\int_e\left([v]\{(B_a\nabla^0 u,\nu)_n\}+\{v\}\left[(B_a\nabla^0 u,\nu)_n\right]\right)ds\\
&\quad+\sum_{e\in\partial\Omega}\int_e v(B_a\nabla^0 u,\nu)_n\,ds\\
&=:a_{s,h}(u,v)+\sum_{e\in\mathcal{T}^h}\int_e\left([v]\{(B_a\nabla^0 u,\nu)_n\}+\{v\}\left[(B_a\nabla^0 u,\nu)_n\right]\right)ds.
\end{aligned}
$$

As above, for preparing the DCGMs, we consider several terms, which vanish for the exact solution $u_0=u\in\mathcal{U}_e$, and $v^h=v\in\mathcal{V}^h$. We omit, for all $e\in\mathcal{T}^h$, the $\{v^h\}[(B_a\nabla^0 u,\nu)_n]$. We add, for all $e\in\mathcal{T}^h\setminus\partial\Omega$, $0=\theta\{(B_a\nabla^0 v,\nu)_n\}[u]$, and for $e\in\partial\Omega$, $0=\theta(B_a\nabla^0 v,\nu)_n(u-u_D)$. Again the *nonsymmetric*, *symmetric*, *incomplete* and *general form* are defined by $\theta=-1$, $\theta=1$, $\theta=0$ and $\theta\in[-1,1]$, respectively. For all $u,v\in\mathcal{U}_e$ we use the $J_h^\sigma(u,v)$ in (7.61), and introduce, cf Remark 7.2:

$$
\begin{aligned}
a_h(u,v)&:=\sum_{K\in\mathcal{T}_h}\left(\int_K\sum_{i,j=0}^n a_{ij}\partial^i u\,\partial^j v\right)dx-\sum_{e\in\mathcal{T}^h}\int_e\Big(\{(B_a\nabla^0 u,\nu)_n\}[v] \qquad (7.76)\\
&\qquad+\theta\{(B_a\nabla^0 v,\nu)_n\}[u]\Big)\,dS\quad\text{for}\quad u,v\in\mathcal{U}_e\\
\ell_h(v)&:=\langle g,v\rangle-\theta\sum_{e\in\partial\Omega}\int_e((B_a\nabla^0 v,\nu)_n)\,u_D\,dS+\sum_{e\in\partial\Omega}\int_e\sigma u_D\,v\,dS\ \text{and}\\
\langle A_h u,v\rangle&-a_h(u,v)-J_h^\sigma(u,v)+\ell_h(v)=0\ \text{ for a fixed } u\in\mathcal{U}_e,\forall\, v\in\mathcal{U}_e,\\
&\text{with } a^h(\cdot,\cdot):=a_h(\cdot,\cdot)|_{\mathcal{U}^h\times\mathcal{V}^h},\ a(\cdot,\cdot)=a_h(\cdot,\cdot)|_{\mathcal{U}\times\mathcal{V}},\\
&\qquad a_s^h(\cdot,\cdot):=a_{s,h}(\cdot,\cdot)|_{\mathcal{U}^h\times L^2(\Omega)},a_s(\cdot,\cdot)=a_{s,h}(\cdot,\cdot)|_{\mathcal{U}_s\times L^2(\Omega)}\\
&\qquad\ell^h(\cdot):=\ell_h(\cdot)|_{\mathcal{V}^h},\ \ell(\cdot)=\ell_h(\cdot)|_{\mathcal{U}},\ \text{and again,}\\
\langle A_h u,v\rangle&=\langle Au,v\rangle\ \forall u,v\in\mathcal{U}_e\ \text{and}\ =\langle Au,v\rangle\ \forall u,v\in\mathcal{U}.
\end{aligned}
$$

This yields for linear elliptic equations the *DCGM* and its *discrete approximate solution* as a function $u_0^h \in \mathcal{S}^h$ satisfying the conditions

$$u_0^h \in S^h : \left\langle A^h \left(u_0^h\right), v^h \right\rangle = 0 \; \forall v^h \in S^h = \mathcal{V}^h. \tag{7.77}$$

The Galerkin orthogonality results in (7.73) remain correct with the obvious modifications.

Another interesting type, the advection–diffusion–reaction problem, is studied by Antonietti and Houston [34]

7.5 Semilinear and quasilinear elliptic problems

7.5.1 Semilinear elliptic problems

Next we discuss *semilinear boundary value problems,* from now on only with Dirichlet boundary conditions. Determine the exact solution $u_0 : \Omega \to \mathbb{R}$ such that

$$G(u_0) := \sum_{s=1}^{n} \partial^s f_s(u_0) - \varepsilon \sum_{i,j=1}^{n} \left(\partial^j \left(a_{ij}(x, u_0)\partial^i u_0\right) - g(x, u_0) = 0 \quad \text{in } \Omega, \tag{7.78}$$

$$u_0 \,|_{\partial\Omega} = u_D \text{ on } \partial\Omega. \tag{7.79}$$

We choose, similarly to (7.13), $\sum_{i,j=1}^{n}$ instead of an equally possible $\sum_{i,j=0}^{n}$, and $\partial\Omega_D = \partial\Omega$. We do not formulate the obvious modifications for $\sum_{i,j=0}^{n}$. This generalizes (7.13), (7.14). Its linearized second order principal part of the form $\sum_{i,j=1}^{n} a_{ij}(x, u_0(x))\partial^j \partial^i u)$ is a special case of (7.75). The special case

$$\nabla \cdot a(x, u)\nabla u = \sum_{i=1}^{n} \partial^i \left(a(x, u)\right)\partial^i u$$

of (7.78) has been discussed by Rivière and Wheeler [552] for parabolic problems. The ellipticity of (7.78) is defined by its linearized principal part, cf. (7.80) (e). We assume the following conditions in a form that is simply to handle, e.g. (c):

(a) $f_s \in C^1(\mathbb{R}),\ f_s(0) = 0,\ s = 1, \ldots, n,$ (7.80)

(b) $\varepsilon > 0,$

(c) $a_{ij} : \Omega \times \mathbb{R} \to \mathbb{R},\ i, j = 1, \ldots, n,$ are bounded by $|a_{ij}(x, u)| \le \Lambda_a,$

(d) a_{ij} are Lipschitz-continuous with respect to u, i.e. $\exists L > 0\ \forall x \in \Omega,\ u_1, u_2 \in \mathbb{R}$:

$$|a_{ij}(x, u_1) - a_{ij}(x, u_2)| \le L|u_1 - u_2|, \quad i, j = 1, \ldots, n,$$

(e) Ellipticity: For $u \in \mathcal{U}$, with $\|u - u_0\|_{\mathcal{U}} \le \rho$ there exist $\lambda, \Lambda \in \mathbb{R}_+$ such that

$$0 < \lambda|\vartheta|^2 \le \sum_{i,j=1}^{n} a_{ij}(x, u(x))\vartheta^j \vartheta^i \le \Lambda|\vartheta|^2 \quad \forall x \in \Omega, \vartheta = (\vartheta^1, \ldots, \vartheta^n) \in \mathbb{R}^n,$$

(f) $a_{ij}(\cdot, u(\cdot)) \in C^1(\overline{\Omega})$ for $u \in C^2(\Omega)$ in (7.78), and

$\in L^\infty(\Omega)$ for $u \in \mathcal{U}$ in (7.81),

(g) $g : \Omega \times \mathbb{R} \to \mathbb{R}$, $g(\cdot, u(\cdot)) \in L^2(\Omega)$ or $\in \mathcal{U}'$ for $u \in \mathcal{U}$,

(h) $g \in \mathcal{U}'$ is Lipschitz-continuous with respect to u, i.e.

$\exists L' > 0 \;\; \forall x \in \Omega \; \forall u_1, u_2 \in \mathbb{R} : |g(x, u_1) - g(x, u_2)| \leq L'|u_1 - u_2|$,

(i) u_D is the trace of some $u^* \in \mathcal{U}$ on $\partial\Omega$.

A *classical solution* $u_0 \in \mathcal{U}_s$ satisfies (7.78)–(7.79) pointwise. For the *weak solutions*, $u, v \in \mathcal{U}$, we use the *notation*, cf (7.19), (7.20), (7.74):

$$a(u,v) = \varepsilon \int_\Omega \sum_{i,j=1}^n a_{ij}(x,u)\partial^i u \partial^j v \, dx = \varepsilon \int_\Omega (B_a(u)\nabla u, \nabla v)_n dx, \text{ with} \tag{7.81}$$

$$B_a(u)\nabla u := \left(\sum_{i=1}^n (a_{ij}(x,u)\partial^i u \right)_{j=1}^n,$$

$$b(u,v) = \int_\Omega \sum_{s=1}^n \partial^s f_s(u) \, v \, dx, \text{ with boundary condition (7.79) } u, v \in \mathcal{U}. \tag{7.82}$$

Thus we update the above ∇u, and $a(u,v)$ in (7.19), by $B_a(u)\nabla u$, and $a(u,v)$ in (7.81). A *weak solution* of (7.78) – (7.79) is defined by the updated (7.23).

7.5.2 Variational formulation and discretization of the problem

In analogy to (7.34), we obtain $\forall K \in \mathcal{T}^h$, by replacing ∇u by $B_a(u)\nabla u$,

$$\sum_{K \in \mathcal{T}^h} \left(- \int_K (\vec{f}(u), \nabla v)_n \, dx + \int_{\partial K} (\vec{f}(u), \nu)_n \, v \, dS + \varepsilon \int_K (B_a(u)\nabla u, \nabla v)_n dx \right. \tag{7.83}$$

$$\left. -\varepsilon \int_{\partial K} (B_a(u)\nabla u, \nu)_n v \, dS - \int_K g(x,u) \, v \, dx \right) = 0 \;\; \forall v \in H^1(\mathcal{T}^h),$$

or we replace $\int_K g(x,u) \, v \, dx$ by $\langle g(x,u), \, v \rangle_{H^{-1}(K) \times H^1(K)}$. Further considerations follow (7.38), (7.41)–(7.42): Here, we again omit the terms $\{v\}([\vec{f}(u_0)], \nu)_n$, $\{v\}([B_a(u_0)\nabla u_0], \nu)_n$ in (7.83) and add

$$\varepsilon \int_e \sigma \, [u] \, [v] \, dS, \;\; \theta\varepsilon \int_e (\{B_a(u)\nabla v\}, \nu)_n [u] \, dS, \;\; \text{for } e \in \mathcal{T}^h \setminus \partial\Omega \;\; \text{and} \tag{7.84}$$

$$\varepsilon \int_e \sigma \, (u - u_D) v \, dS, \;\; \theta\varepsilon \int_e (B_a(u)\nabla v, \nu)_n (u - u_D) \, dS, \;\; \text{for } e \in \partial\Omega,$$

with $\theta = -1$ $\theta = 1$, $\theta = 0$, $\theta \in [-1, 1]$ for the NIPG, SIPG, IIPG, GIPG methods, respectively.

So we obtain, instead of (7.42), again with (7.32), (7.33), and u_0 replacing u,

$$u_0 \in \mathcal{U}_e^o = H^{3/2+\epsilon}(\Omega) : -\sum_{K\in\mathcal{T}^h}\int_K (\vec{f}(u_0),\ \nabla v)_n\,dx + \sum_{e\in\mathcal{T}^h}\int_e [v](\{\vec{f}(u_0)\},\nu)_n dS$$

$$+\,\varepsilon\sum_{K\in\mathcal{T}_h}\int_K (B_a(u_0)\nabla u_0,\ \nabla v)_n\,dx - \varepsilon\sum_{e\in\mathcal{T}^h}\int_e ((\{B_a(u_0)\nabla u_0\},\nu)_n[v] \tag{7.85}$$

$$+\,\theta(\{B_a(u_0)\nabla v\},\nu)_n[u_0])\,dS + \varepsilon\sum_{e\in\mathcal{T}^h}\int_e \sigma[u_0]\,[v]\,dS$$

$$-\int_\Omega g\,v\,dx - \varepsilon\sum_{e\in\partial\Omega}\left(\int_e \sigma u_D\,v - \theta(B_a(u_0)\nabla v,\nu)_n u_D\right)dS = 0\ \forall v\in\mathcal{U}_e^o,$$

cf. (7.37). Now we return to the exact solution $u = u_0 \in \mathcal{U}_e$, of (7.78), satisfying the boundary conditions, (7.79), and the test functions $v \in \mathcal{U}_e^o \cap \mathcal{U}_b$. Along the edges, e, we have

$$[u_0]_e = 0, \{B_a(u_0)\nabla u_0\}_e = B_a(u_0)\nabla u_0|_e\ \forall e\in\mathcal{T}^h\setminus\partial\Omega \text{ and}$$
$$\{B_a(u_0)\nabla u_0\}_e = B_a(u_0)\nabla u_0|_e,\ [u_0]_e = u_0|_e = u_D\ \forall e\in\partial\Omega;$$

similarly for v, we end up again with the weak form, the modified (7.23).

On the basis of the above considerations and notation, we introduce, similarly to (7.42), (7.56), the strong and a provisional and the final weak discrete semilinear forms, $a_{s,h}(\cdot,\cdot)$, $\hat{a}_h(\cdot,\cdot)$, and $a_h(\cdot,\cdot)$, by including the transition terms, as

$$a_{s,h}(u,v) := -\varepsilon\sum_{K\in\mathcal{T}^h}\int_K \sum_{i,j=1}^{n} (\partial^j(a_{ij}(x,u)\partial^i u)v dx, \tag{7.86}$$

$$\hat{a}_h(u,v) := \varepsilon\sum_{K\in\mathcal{T}^h}\int_K \Big[(B_a(u)\nabla u,\ \nabla v)_n = \sum_{i,j=1}^{n}\big(a_{ij}(x,u)\partial^i u\,\partial^j v\big)\Big]\,dx$$

$$a_h(u,v) := \varepsilon\sum_{K\in\mathcal{T}^h}\int_K (B_a(u)\nabla u,\ \nabla v)_n\,dx \tag{7.87}$$

$$-\,\varepsilon\sum_{e\in\mathcal{T}^h}\int_e ((\{B_a(u)\nabla u\},\nu)_n\,[v] + \theta(\{B_a(u)\nabla v\},\nu)_n\,[u])\,dS,$$

$$\ell_h(u,v) := (g(x,u),v)_{L^2(\Omega)} + \varepsilon\sum_{e\in\partial\Omega}\int_e \big(\sigma u_D\,v\ - \theta(B_a(u)\nabla v,\nu)_n\,u\big)\,dS \tag{7.88}$$

$\forall\ u,v\ \in\ \mathcal{U}_e \supset S^h$, and $c_h(u,v), J_h^\sigma(u,v)$ as in (7.59), (7.61).

The $a_{s,h}(u,v)$ and $\hat{a}_h(u,v)$ even require $u\in H^2(\mathcal{T}^h), v\in L^2(\mathcal{T}^h)$, and $u,v\in H^1(\mathcal{T}^h)$. Or $(g(x,u),v)$ may be replaced by $\langle g(x,u),\ v\,\rangle = \langle g(x,u),\ v\,\rangle_{H^{-1}(\mathcal{T}^h)\times H^1(\mathcal{T}^h)}$.

This yields

$$a_{s,h}(u,v) = a_s(u,v) \forall u \in \mathcal{U}_s, v \in L^2(\Omega).$$

Totally analogously to (7.66) we define the operators G, G_h, G^h, by replacing those $a_h(u,v), \ell_h(v)$ in (7.58), (7.59), (7.61), by the new terms in (7.87), (7.88). We do not explicitly update (7.66), cf (7.89) for G^h.

We define the *standard DCGM* in NIPG, SIPG, IIPG, and GIPG forms and its *discrete approximate solution* as a function $u_0^h : \Omega \to \mathbb{R}$ satisfying the conditions

$$\begin{aligned} &\text{(a)}\ \ u_0^h \in S^h, \text{ s.t. } \forall v^h \in S^h && (7.89)\\ &\text{(b)}\ \ \langle G^h(u_0^h), v^h\rangle := a^h(u_0^h, v^h) + c^h(u_0^h, v^h) + \varepsilon J_h^\sigma(u_0^h, v^h) - \ell^h(u_0^h, v^h) = 0 \text{ with}\\ &\qquad a^h(u^h,v^h), c^h(u^h,v^h), \ell^h(u^h,v^h) := a_h(u^h,v^h), c_h(u^h,v^h), \ell_h(u^h,v^h)|_{S^h\times S^h}. \end{aligned}$$

The *nonstandard DCGM* in NIPG, SIPG, IIPG, and GIPG form is again defined by replacing in the last equation $c^h(u_0^h, v^h)$ by $b^h(u_0^h, v^h)$.

A particularly interesting problem is the stationary incompressible Navier-Stokes equation. This has to be discretized by *mixed formulations* of DCGMs, namely by the so-called LDG, the *local discontinuous Galerkin method.* It could be treated similarly to Section 4.6 on the basis of the *Ladyženskaja–Brezzi–Babuska condition,* cf. (4.214), for the Stokes operator and for DCGMs, e.g. [191, 434].

7.5.3 Quasilinear elliptic systems

Previously we discussed equations. The following quasilinear problems are formulated as systems of q equations of second order. Again we start with the standard divergent quasilinear systems in strong form in the Hilbert space setting. We distinguish the direct generalization of (7.78) and (7.79) in (7.90), (7.91) and its further generalization in (7.93). We determine the solution $\vec{u}_0 \in \mathcal{D}(G) \subset H^2(\Omega, \mathbb{R}^q) \cap L^\infty(\Omega, \mathbb{R}^q)$ of the problem in its strong form, such that,

$$G(\vec{u}_0) := \sum_{s=1}^n \partial^s f_s(\vec{u}_0) - \varepsilon \sum_{k=1}^n \partial^k A_k(\cdot, \vec{u}_0, \nabla\vec{u}_0) - g(\cdot, \vec{u}_0) = 0 \in \mathbb{R}^q \text{ in } \Omega, \quad (7.90)$$

$$\vec{u}_0\,|_{\partial\Omega} = \vec{u}_D \text{ on } \partial\Omega, \text{ with smooth enough } A_k : \Omega \times \mathbb{R}^q \times \mathbb{R}^{n\times q} \to \mathbb{R}^q. \quad (7.91)$$

In this case we maintain the conditions for $\vec{f}, \varepsilon, g$ in (7.80). The precise condition for the A_k will be discussed below. For our quasilinear problem we introduce, with an appropriate $\mathcal{D}(G)$, the spaces, cf. (7.37),

$$\begin{aligned} &\mathcal{U} = H^1(\Omega,\mathbb{R}^q) \cap \mathcal{D}(G), \quad \mathcal{V}_b = H_0^1(\Omega,\mathbb{R}^q), \mathcal{U}_e := \mathcal{V}_e := H^{3/2+\epsilon}(\mathcal{T}^h,\mathbb{R}^q) \cap \mathcal{D}(G),\\ &\quad \mathcal{U}_b = H_0^1(\Omega,\mathbb{R}^q) \cap \mathcal{D}(G), \ \mathcal{U}' \subset H^{-1}(\Omega,\mathbb{R}^q), \ldots, \quad \text{, and for the} \qquad (7.92)\\ &\quad \text{strong forms } \mathcal{U}_s = H^2(\Omega,\mathbb{R}^q) \cap \mathcal{D}(G), \ \ \mathcal{U}_{s,e} := H^2(\mathcal{T}^h,\mathbb{R}^q) \cap \mathcal{D}(G). \end{aligned}$$

The general quasilinear weak form is

$$\vec{u}_0 \in \mathcal{U}: \langle G(\vec{u}_0), \vec{v}\rangle := \sum_{K\in\mathcal{T}^h} \int_K \sum_{k=0}^{n} \left(A_k(x, \vec{u}_0, \nabla\vec{u}_0), \partial^k \vec{v}\right)_q dx = 0 \ \forall \vec{v} \in \mathcal{V}_b,$$
$$\vec{u}_0 \mid_{\partial\Omega} = \vec{u}_D \text{ on } \partial\Omega. \tag{7.93}$$

For equations we admitted $u_0 \in \mathcal{D}(G) \subset W^{2,\alpha}(\Omega)$ in Chapter 2, Section 4.4 and the following results would have to be modified accordingly. The necessary conditions for the Nemickii operators $A_i(x, u^h, \nabla u^h)$, e.g. for bounded forms and operators, $a_h(u^h, v^h)$ and G_h, are formulated in Sections 2.5, 2.6, 4.4 and (4.174). We have studied growth conditions for systems for $A_k(x, \vec{u}, \nabla\vec{u}) = \left(a_k^i(x, \vec{u}, \nabla\vec{u})\right)_{i=1}^q$ in Subsection 2.6.4, cf. e.g. (2.361) ff., (2.368) ff. and (2.372). They allow for the existence, uniqueness and regularity results for the solutions of our quasilinear systems, e.g., Theorems 2.89, 2.96–2.101. Their bounded linearizations with the Legendre condition and the coercive principal part are discussed in Section 2.7.4, cf. Theorem 2.125.

For the A_k, we modify (7.80) into (7.94) on the basis of these results. As an example we consider the so-called *natural growth conditions,* cf. (2.380). The *controllable growth conditions* in (2.379) can be treated in a similar way.

We summarize one type of these natural growth conditions. The ellipticity of (7.93) is defined by the ellipticity of its linearized principal part. This can be guaranteed by (7.94) (e). With $\vec{\mathbf{u}} = (\mathbf{u}^1, \ldots, \mathbf{u}^n)^T \in \mathbb{R}^{n\times q}$, we need the partials of $A_k(x, \vec{u}, \vec{\mathbf{u}})$ with respect to $x \in \mathbb{R}^n, \vec{u}, \mathbf{u}^l \in \mathbb{R}^q$, evaluated in $(x, \vec{u}, \nabla\vec{u})$ and, e.g. applied to $\vec{\vartheta}^k, \vec{\vartheta}^l \in \mathbb{R}^q$, so the $(\partial A_k/\partial \mathbf{u}^l)(x, \vec{u}, \nabla\vec{u})\vec{\vartheta}^l \in \mathbb{R}^q$ or $\left((\partial A_k/\partial \mathbf{u}^l)(x, \vec{u}, \nabla\vec{u})\vec{\vartheta}^k, \vec{\vartheta}^l\right)_q \in \mathbb{R}$, or $\partial A_k/\partial \mathbf{u} = (\partial A_k/\partial \mathbf{u}^1, \ldots, \partial A_k/\partial \mathbf{u}^n)$. We do not distinguish $A_k : \Omega \times \mathbb{R}^q \times \mathbb{R}^{n\times q} \to \mathbb{R}^q$, and $A_k : H^1(\Omega, \mathbb{R}^q) \to H^{-1}(\Omega, \mathbb{R}^q)$, and omit the details in the norms, e.g. write $\|\partial A_k/\partial \mathbf{u}^l\|$, instead of $\|\partial A_k/\partial \mathbf{u}^l\|_{\mathbb{R}^q \leftarrow \mathbb{R}^{q^2}}$.

$$\begin{aligned}
&\text{(a)}\ \ A_k : w := (x, \vec{u}, \vec{\mathbf{u}}) \in \Omega \times \mathbb{R}^q \times \mathbb{R}^{n\times q} \to \mathbb{R}^q,\ k = 0, 1, \ldots, n, && (7.94)\\
&\text{(b)}\ \ A_k : \vec{u} \in \mathcal{D}(G) \subset \mathcal{D}(A_k) \subset H^1(\Omega, \mathbb{R}^q) \to A_k(\vec{u}) := A_k(\cdot, \vec{u}, \nabla\vec{u}) \in H^{-1}(\Omega, \mathbb{R}^q),\\
&\text{(c)}\ \ \|A_k\|, \|\partial A_k/\partial x\|, \|\partial A_k/\partial \vec{u}\| \le L(1 + \|\vec{\mathbf{u}}\|), \|\partial A_k/\partial \vec{\mathbf{u}}\| \le L\ \forall k > 0,\\
&\text{(d)}\ \ \|A_0\|, \|\partial A_0/\partial x\|, \|\partial A_0/\partial \vec{u}\| \le L(1 + \|\vec{\mathbf{u}}\|^2),\ \ \|\partial A_0/\partial \vec{\mathbf{u}}\| \le L(1 + \|\vec{\mathbf{u}}\|),\\
&\qquad \forall \|\vec{u}\| \le M,\ \text{all terms in c), d) are evaluated in } w = (x, \vec{u}, \vec{\mathbf{u}}),\\
&\text{(e)}\ \ \text{Ellipticity: For } \vec{u} \in H^1(\Omega, \mathbb{R}^q), \text{ with } \|\vec{u} - \vec{u}_0\|_{H^1(\Omega, \mathbb{R}^q)} \le \rho\ \ \exists \lambda, \Lambda \in \mathbb{R}\ \forall x \in \Omega,
\end{aligned}$$

$$0 \ < \lambda\|\vec{\vartheta}\|^2 \le \sum_{k,l=1}^{n} \left(\frac{\partial A_k}{\partial \mathbf{u}^l}(x, \vec{u}, \nabla\vec{u})\vec{\vartheta}^k, \vec{\vartheta}^l\right)_q < \Lambda\|\vec{\vartheta}\|^2 \forall 0 \ne \vec{\vartheta} = (\vec{\vartheta}^1, \ldots, \vec{\vartheta}^n) \in \mathbb{R}^{n\times q},$$

(f) $\vec{u}_D$ is the trace of some $\vec{u}^* \in \mathcal{U}$ on $\partial\Omega$.

7.5.4 Discretization of the quasilinear systems

Now we start formulating the DCGMs for these divergent quasilinear equations of second order in (7.93), cf. the results for conforming FEMs in Theorem 5.81. We choose $d \geq 2$, generalizing (7.63),

$$\mathcal{U}^h = \mathcal{V}^h = S^h := S_q^{d-1,-1}(\mathcal{T}^h) := \{v : \mathcal{T}^h \to \mathbb{R}^q; v|_K \in \mathcal{P}_{d-1}(K)\ \forall K \in \mathcal{T}^h\}. \quad (7.95)$$

By the standard multiplication of (7.93) with $\vec{v}^h$, partial integration over K, and summation, we relate the strong and weak discrete quasilinear forms $a_{s,h}(\cdot,\cdot)$, and $\hat{a}_h(\cdot,\cdot)$ and the final weak discrete form, $a_h(\cdot,\cdot)$, in (7.99). We define these strong and weak forms for

$$\forall \vec{u}^h, \vec{v}^h \in \mathcal{U}_e \text{ with } \vec{u}^h \in \mathcal{U}_{s,e}, \vec{v}^h \in L^2(\Omega,\mathbb{R}^q) \text{ for } a_{s,h}(\vec{u}^h,\vec{v}^h), \quad (7.96)$$

$$a_{s,h}(\vec{u}^h,\vec{v}^h) := (G_{s,h}(\vec{u}^h),\vec{v}^h) := \sum_{K\in\mathcal{T}^h}\int_K \sum_{k=0}^n \big((-1)_{k>0}\partial^k A_k(x,\vec{u}^h,\nabla\vec{u}^h),\vec{v}^h\big)_q dx,$$

$$\hat{a}_h(\vec{u}^h,\vec{v}^h) := \langle G_h(\vec{u}^h),\vec{v}^h\rangle := \sum_{K\in\mathcal{T}^h}\int_K \sum_{k=0}^n (A_k(x,\vec{u}^h,\nabla\vec{u}^h),\partial^k\vec{v}^h)_q dx$$

where we use the abbreviations $(G_{s,h}(\vec{u}^h),\vec{v}^h) = (G_{s,h}(\vec{u}^h),\vec{v}^h)_{L^2(\Omega,\mathbb{R}^q)}$, $\langle G_h(\vec{u}^h),\vec{v}^h\rangle = \langle G_h(\vec{u}^h),\vec{v}^h\rangle_{\mathcal{V}_e'\times\mathcal{V}_e}$. Since the summation in (7.96) starts with $k=0$ in $\sum_{K\in\mathcal{T}^h}\int_K$, we choose the above notation for $\hat{a}_h(\vec{u}^h,\vec{v}^h)$ in contrast to the semilinear case. This yields the relation for

$$\forall \vec{u}^h \in \mathcal{U}_{s,e}, \forall \vec{v}^h \in \mathcal{U}_e \text{ with } B_A(\vec{u}^h,\nabla\vec{u}^h) := \big(A_k(x,\vec{u}^h,\nabla\vec{u}^h)\big)_{k=1}^n \in \mathbb{R}^{n\times q}, \quad (7.97)$$

$$a_{s,h}(\vec{u}^h,\vec{v}^h) = \hat{a}_h(\vec{u}^h,\vec{v}^h) - \sum_{e\in\mathcal{T}^h\setminus\partial\Omega}\int_e \Big(\big((\{B_A(\vec{u}^h,\nabla\vec{u}^h)\},\nu)_n,[\vec{v}^h]\big)_q$$

$$+\big(([B_A(\vec{u}^h,\nabla\vec{u}^h)],\nu)_n,\{\vec{v}^h\}\big)_q\Big)dS - \sum_{e\in\partial\Omega}\int_e \big((B_A(\vec{u}^h,\nabla\vec{u}^h),\nu)_n,\vec{v}^h\big)_q dS.$$

cf. (5.286). All the previous face integrals $\int_e$ disappear for $\vec{u}^h = \vec{u} \in H^{3/2+\epsilon}(\Omega,\mathbb{R}^q)$, $\vec{v}^h = \vec{v} \in H_0^1(\Omega,\mathbb{R}^q)$ implying for the exact solution $G_{s,h}(\vec{u}_0) = 0$ and $G_h(\vec{u}_0) = 0$.

We proceed as from (7.83)–(7.88): we replace $B_a(u)\nabla u$ by $B_A(\vec{u},\nabla\vec{u})$, $(B_a(u^h)\nabla u^h,\nu)_n v^h$ by $\big((B_A(\vec{u}^h,\nabla\vec{u}^h),\nu)_n,\vec{v}^h\big)_q$ etc. We omit the flux terms, and $\int_e \big(\{\vec{v}\},([B_A(\vec{u}_0,\nabla\vec{u}_0)],\nu)_n\big)_q$, vanishing in (7.97) for $\vec{u}^h = \vec{u}_0$, and add the penalty terms in (7.84). We only consider the IIPG with $\theta = 0$, otherwise $\theta\big((\{B_A(\vec{v}^h,\nabla\vec{v}^h)\},\nu)_n,[\vec{u}^h]\big)_q$ would cause a nonlinearity with respect to $\vec{v}^h$. Furthermore, we collect $-\sum_{e\in\mathcal{T}^h\setminus\partial\Omega}$ and $-\sum_{e\in\partial\Omega}$ into $-\sum_{e\in\mathcal{T}^h}\int_e$. So (7.87), (7.88), is generalized, with $\vec{u}^h,\vec{v}^h$ as in (7.96), as

$$\sum_{K\in\mathcal{T}_h}\int_K \sum_{k=0}^n (A_k(x,\vec{u}^h,\nabla\vec{u}^h),\partial^k\vec{v}^h)_q\,dx - \sum_{e\in\mathcal{T}^h}\int_e \big((\{B_A(\vec{u}^h,\nabla\vec{u}^h)\},\nu)_n,[\vec{v}^h]\big)_q dS$$

$$+\left(\sum_{e\in\mathcal{T}^h}\int_e \sigma([\vec{u}],[\vec{v}])_q dS - \sum_{e\in\partial\Omega}\int_e \sigma(\vec{u}_D,\vec{v})_q dS\right) = 0. \quad (7.98)$$

Again the exact solution $\vec{u} = \vec{u}_0 \in H^{3/2+\epsilon}(\Omega, \mathbb{R}^q)$, of (7.93), (7.91), with the test functions $\vec{v} \in H_0^1(\Omega, \mathbb{R}^q)$ satisfy (7.98).

With this notation, we introduce, similarly to (7.57)–(7.61), the quasilinear, linear and bilinear forms, $a_h(\cdot,\cdot)$, $\ell_h(\cdot)$, and $J_h^\sigma(\vec{u},\vec{v})$, cf. (7.97), and rewrite (7.98), cf. (7.96), as

$$a_h(\vec{u}^h, \vec{v}^h) := \hat{a}_h(\vec{u}^h, \vec{v}^h) - \sum_{e\in\mathcal{T}^h} \int_e \left((\{B_A(\vec{u}^h, \nabla\vec{u}^h)\}, \nu)_n, [\vec{v}^h]\right)_q dS, \tag{7.99}$$

$$\ell_h(\vec{u}) := \sum_{e\in\partial\Omega} \int_e \sigma(\vec{u}_D, \vec{v})_q \, dS \text{ and } J_h^\sigma(\vec{u},\vec{v}) := \sum_{e\in\mathcal{T}^h} \int_e \sigma([\vec{u}], [\vec{v}])_q \, dS. \tag{7.100}$$

Again analogously to (7.66) we define G, G_h, G^h, by replacing those $a_h(u,v)$, $c_h(u,v), b_h(u,v), \ell_h(v), J_h^\sigma(u,v)$ in (7.58), (7.59), (7.61), by the new terms

$$a_h(\vec{u},\vec{v}), \ell_h(\vec{v}), J_h^\sigma(\vec{u},\vec{v}), c_h(\vec{u},\vec{v}) = b_h(\vec{u},\vec{v}) = 0,$$

in (7.99), (7.100). Instead of explicitly updating (7.66), we refer to (7.89) for G^h.

We only consider the *IIPG form, so* $\theta = 0$, *of the DCGM for the quasilinear problem (7.93), (7.91).* Its *discrete approximate solution* $\vec{u}_0^h : \Omega \to \mathbb{R}^q$ is defined as

$$\begin{aligned}
&\text{(a)}\ \ \vec{u}_0^h \in S^h, \text{ cf. (7.95)}, \qquad\qquad\qquad\qquad (7.101)\\
&\text{(b)}\ \ \left\langle G^h\left(\vec{u}_0^h\right), \vec{v}^h\right\rangle := a^h\left(\vec{u}_0^h, \vec{v}^h\right) + J_h^\sigma\left(\vec{u}_0^h, \vec{v}^h\right) - \ell^h(\vec{u}^h, \vec{v}^h) = 0 \ \ \forall \vec{v}^h \in S^h\\
&\qquad \text{with } a^h(\cdot,\cdot) := a_h(\cdot,\cdot)|_{S^h\times S^h}, \ \text{ and } \ell^h(\cdot) := \ell_h(\cdot)|_{S^h}.
\end{aligned}$$

A direct generalization of the *NIPG, SIPG, GIPG form of the DCGM* would yield problematic forms, nonlinear in v. Possible linearizations seem to be pretty artificial.

The quasilinear problem and their DCGMs, studied by Houston *et al.* [407] somehow represent a combination of our approaches for semilinear and quasilinear problems. As for semilinear problems they consider all types of DCGM with $\theta = -1, = 0, = 1, \in [-1,1]$, but for a quasilinear equation of the special form

$$G(u) := -\nabla\cdot(\mu(x,|\nabla u|)\nabla u) := -\sum_{i=1}^n \partial^i\left(\mu(x,|\nabla u|)\partial^i u\right) = g(x,u). \tag{7.102}$$

Here $\mu \in C(\overline{\Omega}\times[0,\infty))$ is a real function satisfying the assumption, with $m_\mu < M_\mu \in \mathbb{R}_+$,

$$\begin{aligned}
m_\mu(t-s) \le \mu(x,t)t - \mu(x,s)s \le M_\mu(t-s), \ t \ge s \ge 0, \ x \in \overline{\Omega} \qquad (7.103)\\
\text{and for } t > s = 0 \Longrightarrow 0 < m_\mu \le \mu(x,t) \le M_\mu.
\end{aligned}$$

In addition, we obtain $\mu(\cdot, |\nabla u(\cdot)|) \in L^\infty(\Omega)$ for $u \in W^{1,\infty}(\Omega)$. Garret and Liu [338] proved that (7.103) implies, for symmetric matrices, $X, Y = (Y_{ij})_{i,j=1}^n \in \mathbb{R}^{n^2}$, or for $x, y \in \mathbb{R}$,

$$\begin{aligned}
&|\mu(x,|Y|)Y - \mu(x,|X|)X| \le C_1|Y-X|, \text{ and} \qquad (7.104)\\
&C_2|Y-X|^2 \le \left(\mu(x,|Y|)Y - \mu(x,|X|)X, (Y-X)\right)_{n^2},
\end{aligned}$$

with $|Y|$ the Euclidean (Frobenius) scalar product and norm, hence $|Y| = \big(\sum_{i,j=1}^n = Y_{ij}^2\big)^{1/2}$. This includes, via diagonal matrices, $Y = (Y_1, \ldots, Y_n) \in \mathbb{R}^n$.

This can be used in two ways: linearization of the weak form of $G(u)$ yields

$$(G'(u_0)w, v) = \sum_{i=1}^n \int_\Omega \mu(x, |\nabla u_0|)\partial^i w \partial^i v dx \tag{7.105}$$

$$+ \sum_{i,j=1}^n \int_\Omega \mu_t(x, |\nabla u_0|)\partial^j u_0 |\nabla u_0|^{-1/2} \partial^j w \partial^i v, \ \textit{with}\ \mu_t(x,t) := (\partial\mu/\partial t)(x,t),$$

so for $|\mu_t(x, |\nabla u_0|)||\nabla u_0|^{1/2}$ *small compared to* m_μ, *a coercive principal part* of the linearized operator and hence stability of the DCGM as in Theorem 7.18.

Or Houston *et al.* [407] use (7.103) for the discrete form of the weak G^h, cf. (7.109). They prove that G^h is Lipschitz-continuous and strongly monotone with respect to u^h.

Compared to (7.78), here we have chosen $\varepsilon = 1$. Then the lower order flux terms $\sum_{s=1}^n \partial^s f_s(u_0)$ are no longer dominant. Hence we omit them here. For $g(x, u)$ we impose the conditions in (7.80) (g), (h).

In this *special quasilinear equation* in (7.102), cf. [407], in (7.78), the second order term is replaced by $\nabla \cdot (\mu(x, |\nabla u|)\nabla u)$. Hence we obtain, cf. (7.3), (7.23),

$$a(u,v) = \int_\Omega \sum_{i=1}^n \mu(x, |\nabla u|)\partial^i u \partial^i v \, dx = \int_\Omega (B_a(u)\nabla u, \nabla v)_n dx = \int_\Omega g(x,u) v dx \tag{7.106}$$

$$B_a(u)\nabla u := \mu(x, |\nabla u|)\nabla u, \ (B_a(u)\nabla u, \nabla v)_n = \mu(x, |\nabla u|) \sum_{i,=1}^n \left(\partial^i u \, \partial^i v\right) \ \forall u, v \in \mathcal{U},$$

and determine the weak solution $u_0 \in \mathcal{U} : a(u_0, v) = \int_\Omega g(x, u_0) v dx \ \forall v \in \mathcal{U}$.

The DCGMs for (7.102) is similar to (7.89). We choose $\varepsilon = 1$ and replace the $a^h(u^h, v^h)$ there by (7.106). We avoided, in (7.99), the θ terms

$$\theta \sum_{e \in \mathcal{T}^h} \int_e (\{B_a(u)\nabla v\}, \nu)_n \, [u] \, dS = \theta \sum_{e \in \mathcal{T}^h} \int_e (\{\mu(x, |\nabla u|)\nabla v\}, \nu)_n \, [u] \, dS$$

by choosing $\theta = 0$. To obtain a monotone operator, Houstan *et al.* [407] employ instead the form

$$\theta \sum_{e \in \mathcal{T}^h \setminus \partial\Omega} \int_e (\{\mu(x, h^{-1}|[u]|)\nabla v\}, \nu)_n \, [u] dS$$

$$+\theta \sum_{e \in \partial\Omega} \int_e (\{\mu(x, h^{-1}|u - u_D|)\nabla v\}, \nu)_n \, (u - u_D) dS.$$

Correspondingly, we reformulate $a_h(u,v)$, cf. (7.87), for $-1 \le \theta \le 1$ as

$$a_h(u,v) := \sum_{K\in\mathcal{T}^h} \int_K (B_a(u)\nabla u,\ \nabla v)_n\, dx - \sum_{e\in\mathcal{T}^h} \int_e ((\{B_a(u)\nabla u\},\nu)_n\,[v])\, dS$$
$$+\theta \sum_{e\in\mathcal{T}^h\setminus\partial\Omega} \int_e ((\{\mu(\cdot,h^{-1}|[u]|)\nabla v\},\nu)_n\,[u])\, dS. \tag{7.107}$$

Similarly, we modify ℓ_h (7.60) and use the J_h^σ (7.61):

$$\ell_h(u,v) := (g(\cdot,u),v)_{L^2(\Omega)} - \theta \sum_{e\in\partial\Omega} \int_e (\{\mu(x,h^{-1}|u-u_D|)\nabla v\},\nu)_n\,(u-u_D) dS$$
$$+ \sum_{e\in\partial\Omega_D} \int_e \sigma u_D\, v\, dS \tag{7.108}$$

Finally, we determine the discrete solution $u_0^h \in S^h$, such that

$$\left\langle G^h\left(u_0^h\right), v^h\right\rangle := a^h\left(u_0^h, v^h\right) + J_h^\sigma\left(u_0^h, v^h\right) - \ell^h\left(u_0^h, v^h\right) = 0\ \forall v^h \in S^h. \tag{7.109}$$

As for equations the *generalization to systems* is considered by Houston *et al.* [407]. Again either the monotony or the stability of the linearization can be used. Then the $u_0^h, v^h, \mu(x,|\nabla u|)$ in (7.109) or additionally in (7.105) with $\vec{v} = (v_1,\dots,v_n)$ are replaced *for systems* by

$$\vec{u}_0^h, \vec{v}^h, \mu(x,|e(\vec{u})|), e(\vec{u}) := \frac{1}{2}\left(\nabla\vec{u} + (\nabla\vec{u})^{\mathrm{T}}\right) \text{ or } e(\vec{u}) := (\partial_1 v_1,\dots,\partial_n v_n) \tag{7.110}$$

with $B_a(\vec{v})e(\vec{v}) := \mu(x,|e(\vec{u})|)e(\vec{v})$ and finally,

$$\left\langle G^h\left(\vec{u}_0^h\right), \vec{v}^h\right\rangle := a^h\left(\vec{u}_0^h, \vec{v}^h\right) + J_h^\sigma\left(\vec{u}_0^h, \vec{v}^h\right) - \ell^h\left(\vec{u}_0^h, \vec{v}^h\right) = 0\ \forall \vec{v}^h \in S^h. \tag{7.111}$$

Again (7.103), (7.104) implies, cf. Houston *et al.* [407], Lemmas 2.2 and 2.3, that G^h is Lipschitz-continuous and strongly monotone with respect to $\vec{u}^h$.

7.6 DCGMs are general discretization methods

For these methods, stability and consistency implying convergence play a crucial role. For realizing this interpretation, we relate the original and discrete function spaces with the corresponding operators. We only distinguish the spaces $H^{3/2+\epsilon}(\Omega)$ and $H^{3/2+\epsilon}(\Omega,\mathbb{R}^q)$, ..., where it is necessary, e.g. in the consistency results for $a^h(\vec{u}^h,\vec{v}^h)$ below, in Theorem 7.33. Most of the other results are valid componentwise, with obvious modifications, for both cases, e.g. Theorems 7.5–7.19.

Directly comparing G and G^h, similarly A and A^h, required a detour via the G_h. This has been shown for the model problem in (7.13)–(7.15), (7.66), and indicated for the other problems as well. We recall the situation here for nonlinear operators. Previously we studied $\mathcal{U} = \mathcal{V}, \mathcal{U}^h = \mathcal{V}^h$ but will allow, for possible generalizations, in this section, $\mathcal{U} \neq \mathcal{V}$, or $\mathcal{U}^h \neq \mathcal{V}^h$, as well. So we start with the original

$$\begin{aligned} G\ &: \mathcal{D}(G) \subset \mathcal{U},\ \text{e.g.}, \mathcal{U} = H^1(\Omega) \to \mathcal{V}',\ \text{e.g. } \mathcal{V}' = H^{-1}(\Omega),\ \text{extended as} \quad (7.112)\\ G_h &: \mathcal{D}(G_h) \subset \mathcal{U}_e^o,\ \text{e.g.}, \ = H^{3/2+\varepsilon}(\Omega) \to (\mathcal{V}_e^o)',\ \text{e.g. } = H^{-3/2-\varepsilon}(\Omega),\ \text{extended as}\\ G_h &: \mathcal{D}(G_h) \subset \mathcal{U}_e,\ \text{e.g.}, \ = H^{3/2+\varepsilon}(\mathcal{T}^h) \to \mathcal{V}_e', \text{e.g.} = H^{-3/2-\varepsilon}(\mathcal{T}^h),\ \text{restricted as}\\ G^h &: \mathcal{D}(G^h) \subset \mathcal{U}^h,\ \text{e.g.}, \mathcal{U}^h = S^h \to \mathcal{V}'^h,\ \text{e.g. } \mathcal{V}'^h = S'^h,\ \text{in (7.63), (7.95).} \end{aligned}$$

For a nonlinear G the restriction to $\mathcal{D}(G) \subset \mathcal{U}$ is important. This $\mathcal{D}(G)$ has to be chosen such that $G(u)$ is well defined and has the desired properties, e.g. listed in (7.16), (7.51) ff. for (7.13), (7.14), or (7.80) for (7.78), (7.79), or (7.94) for (7.90), (7.91). Other possibilities are the conditions imposed for nonlinear operators and their linearization discussed in Chapter 2, here Sections 2.5.3 ff. and 2.7.2.

We combine $\mathcal{U}, \mathcal{V}$ with the discrete spaces $\mathcal{U}^h, \mathcal{V}^h$, e.g. S^h via the operators $P^h := I^h \in \mathcal{L}(\mathcal{U}, \mathcal{U}^h)$, and $Q'^h \in \mathcal{L}(\mathcal{V}', \mathcal{V}'^h)$. We may choose the P^h relatively arbitrarily, here the approximation operator P^h introduced in Theorem 7.7. The P^h, Q'^h have to be combined such that we obtain optimal consistency order. In particular, the Q'^h has to be chosen to fit the above variational method. Similarly to (7.112), we need local pairs of $P_e^h, Q_e'^h$. The difference is the piecewise definition, with P_e^h the local form on the $K \in \mathcal{T}^h$ in Theorem 7.7. For $Q_e'^h$, we have to include the integrals along edges and boundaries. Correspondingly, see (5.294),(5.295), we define $Q'^h, Q_e'^h$. We start with the local version $Q_e'^h$ and return to Q'^h in (7.115).

$$Q_e'^h \in \mathcal{L}(\mathcal{V}_e', \mathcal{V}'^h) : Q_e'^h f_h - f_h \perp \mathcal{V}^h \Leftrightarrow \langle Q_e'^h f_h - f_h, v^h \rangle = 0\ \forall v^h \in \mathcal{V}^h \quad (7.113)$$

with $\langle \cdot,\cdot \rangle = \langle \cdot,\cdot \rangle_{\mathcal{V}_e' \times \mathcal{V}_e}$. The general form of these f_h is, cf. (2.107):

$$\text{for } f_{-1} \in L^2(\Omega, \mathbb{R}^n), f_0 \in L^2(\Omega, \mathbb{R}), f_{-1}^e, f_0^e \in L^2(\partial\mathcal{T}^h, \mathbb{R}),\ \text{define } \langle f_h, v \rangle_{\mathcal{V}_e' \times \mathcal{V}_e} :=$$

$$\langle f_h, v \rangle := \sum_{K \in \mathcal{T}^h} \int_K ((f_{-1}, \nabla v)_n + f_0 v)\, dx + \sum_{e \in \mathcal{T}^h} \int_e \left(f_{-1}^e (\nabla v, \nu)_n + f_0^e v\right) dS, \quad (7.114)$$

$$f_h \in \mathcal{V}_e',\quad \forall v \in \mathcal{V}_e \supset \mathcal{V} + \mathcal{V}_e^o + \mathcal{V}^h,\ \text{with } \partial\mathcal{T}^h := \{e \in \mathcal{T}^h\},\quad f = f_h|_{\mathcal{V}}, f^h := f_h|_{\mathcal{V}^h}.$$

Examples are the previous cases, $f_h := A_h u^h, u^h \in \mathcal{U}_e = H^{3/2}(\mathcal{T}^h)$, and $f_h := A_h u, u \in \mathcal{U}_e^o$, induced by the $a_h(u^h, v^h)$ in (7.57), (7.76), (7.87), (7.99). In contrast to these $a_h(\cdot,\cdot)$, in the original $a(\cdot,\cdot) = a_h(\cdot,\cdot)|_{\mathcal{U}\times\mathcal{V}}$, in (7.19), (7.74), (7.81), the $\int_e$ are missing. The DCGMs for the previous problems have shown that it is impossible: neither evaluating Au^h for a $u^h \in \mathcal{U}^h$, nor testing it nor the corresponding $f := Au, u \in \mathcal{U}$, with $v^h \in \mathcal{V}^h \subset \mathcal{V}_e$. So we define, for A and G,

$$\begin{aligned} (Q'^h A)\Big|_{\mathcal{U}^h} &:= Q_e'^h A_h\Big|_{\mathcal{U}^h}, \quad \Phi^h A := (Q'^h A)\Big|_{\mathcal{U}^h} := Q_e'^h A_h\Big|_{\mathcal{U}^h}, \quad \text{and} \quad (7.115)\\ (Q'^h G)\Big|_{\mathcal{U}^h} &:= Q_e'^h G_h\Big|_{\mathcal{U}^h}, \quad \Phi^h G := (Q'^h G)\Big|_{\mathcal{U}^h} := Q_e'^h G_h\Big|_{\mathcal{U}^h}, \end{aligned}$$

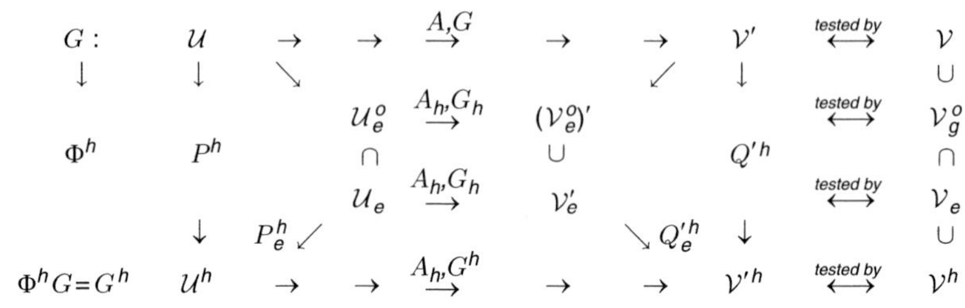

Figure 7.2 Original and discrete spaces and operators in DCGMs.

cf. Figure 7.2. The combination of $P_e^h \swarrow$ and $Q_e^{'h} \searrow$ in this figure indicate the ranges $P_e^h \mathcal{U}_e = \mathcal{U}^h, Q_e^{'h}\mathcal{V}_e' = \mathcal{V}'^h$. Note that by (7.115), the $Q^{'h} \in \mathcal{L}(\mathcal{V}', \mathcal{V}'^h)$ are only defined via $Q_e^{'h} \in \mathcal{L}(\mathcal{V}_e', \mathcal{V}'^h)$ in (7.113),(7.115), including the $\int_e$.

The uniformly bounded $P^h, Q^{'h}$ and $P_e^h, Q_e^{'h}$ have the properties required for the general discretization theory, cf. Definition 3.12, (3.24), Theorems 7.7, Corollary 7.8:

$$\text{for } d \geq 2: \lim_{h\to 0} P^h u = u \;\forall\; u \;\in\; \mathcal{U}, \;\; \lim_{h\to 0} \|Q^{'h} f\|_{\mathcal{V}'^h}^h = \|f\|_{\mathcal{V}'} \forall f \in \mathcal{V}',$$
$$\text{and }: \lim_{h\to 0} P_e^h u_h = u_h \;\forall\; u_h \;\in\; \mathcal{U}_e, \;\; \lim_{h\to 0} \left\|Q_e^{'h} f_h\right\|_{\mathcal{V}'^h}^h = \|f_h\|_{\mathcal{V}_e'} \forall f_h \in \mathcal{V}_e', \tag{7.116}$$

with the local degree $d-1$, cf. (7.63), (7.95). Thus, the $\mathcal{U}^h$ are *approximating spaces* for $\mathcal{U}, \mathcal{U}_e$, and their subspaces, similarly for $\mathcal{V}^h, \mathcal{V}, \mathcal{V}_e$.

Summarizing, we have assigned to $\mathcal{T}^h$ with $h,\; 0 < h \in H \subset \mathbb{R}, \inf_{h\in H} h = 0$, the sequence of spaces $\{\mathcal{U}^h, \mathcal{V}^h\}_{h,\; 0<h\in H}$ of the same finite dimension. Often we use the short notation $\mathcal{U}^h, \mathcal{V}^h$. Thus we have defined by (7.70), (7.71) and the corresponding generalizations (7.77), (7.89), (7.101), an operator Φ^h transforming the original G into the discrete $G^h = \Phi^h(G)$, similarly the linear $A^h = \Phi^h(A)$, cf. (7.115),

$$\Phi^h G := G^h := (Q^{'h} G)|_{\mathcal{U}^h} = \left(Q_e^{'h} G_h\right)|_{\mathcal{U}^h}, \text{ and } \Phi^h A := A^h := Q_e^{'h} A_h|_{\mathcal{U}^h}. \tag{7.117}$$

In contrast to (5.2), we enforce the continuity and boundary conditions in $\mathcal{U}_b$ via penalty terms and *penalty norms* for $\mathcal{U}^h$, cf. Theorems 7.12, 7.13.

This procedure defines a mapping, Φ^h, applied to linear or nonlinear operators, Φ^h : $A \in \mathcal{L}(\mathcal{U}, \mathcal{V}') \to \mathcal{L}(\mathcal{U}^h, \mathcal{V}'^h)$ or $\Phi^h : G \in \mathcal{NL}(\mathcal{D}(G), \mathcal{V}') \to \mathcal{NL}(\mathcal{D}(G^h), \mathcal{V}'^h)$, according to the chosen DCGM. It allows reinterpreting (7.70), (7.71), (7.89), (7.101):

We determine the approximate solution $u_0^h \in \mathcal{V}^h$ as

$$G^h u_0^h = Q_e^{'h} G_h u_0^h = 0 \Leftrightarrow \left\langle G_h u_0^h, v^h \right\rangle = 0 \;\forall v^h \in \mathcal{V}^h \Leftrightarrow G^h \left(u_0^h\right) \perp \mathcal{V}^h. \tag{7.118}$$

Next, we study the relations between operators and their discretizations. Here the penalty norms for $\mathcal{U}_e$, mentioned before, will play an essential role. In particular, we will prove stability and consistency.

Stability is a kind of perturbation insensitivity, cf. Definition 3.19. For a nonlinear operator $G^h : \mathcal{D}(G^h) \subseteq \mathcal{U}^h \to \mathcal{V}'^h$, and $u^h \in \mathcal{D}(G^h)$ we assume

$$B_r(u^h) = \{v^h \in \mathcal{U}^h : \|v^h - u^h\|_{\mathcal{U}^h} < r\} \subset \mathcal{D}(G^h)\ \forall h < h_0, h \in H.$$

Furthermore, let $h_0, r, S \in \mathbb{R}_+$ be fixed constants, such that, *uniformly in* $h \in H$, $h < h_0$:

$$u_i^h \in B_r(u^h), i = 1, 2, \Longrightarrow \|u_1^h - u_2^h\|_{\mathcal{U}^h} \leq S\|G^h\left(u_1^h\right) - G^h\left(u_2^h\right)\|_{\mathcal{V}'^h}. \quad (7.119)$$

Then G^h are called *stable in* u^h. For a linear A^h or for $A^h - f^h$ this reduces to $(A^h)^{-1} \in \mathcal{L}(\mathcal{V}'^h, \mathcal{U}^h)$ exists and $\|(A^h)^{-1}\|_{\mathcal{U}^h \leftarrow \mathcal{V}'^h} \leq S$.

Stability is proved using Theorems 3.23 and 3.29 in Chapter 3. It suffices to study stability for the linearized operator. It is here always a compact perturbation of a coercive operator, the latter defining a stable discretization. We assumed the linearized operator to be boundedly invertible. Then the stability of the linearized operator follows as in Section 5.5 by proving the coercivity for a perturbed principal part. This follows with the penalty norm from our Theorems 7.16–7.19. Combined with a nearly verbatim copy of the arguments in Lemma 5.77 and Theorem 3.29, it yields stability for $(G'(u_0))^h$ with respect to the penalty norm $\|v\|$, cf. Theorem 7.12, (7.64), (7.153), for general boundedly invertible linearizations, $G'(u_0)$. This is left as an exercise for the interested reader. We will summarize the exact conditions and results in Theorem 7.40.

The *classical consistency error* has to converge to 0. It has, for a linear or nonlinear operator, A or G, the following general form, cf. (5.313), (3.34), Chapter 3, (3.33). These $G^h P^h u - (Q'^h G)u$, cf. (7.117), have to be tested by $v^h \in S^h$:

$$A^h P^h u - (Q'^h A)u \text{ and } G^h P^h u - (Q'^h G)u \text{ tested by } 0 \neq v^h \in S^h \text{ yield} \quad (7.120)$$
$$\frac{|\langle G^h P^h u - (Q'^h G)u, v^h\rangle|}{\|v^h\|} \to 0 \text{ or } \leq C(u)h^k \text{ with } C(u) \in \mathbb{R}_+, k = k(u) \in \mathbb{N};$$

these $C(u), k = k(u)$ depend on the regularity of u and the degree $d-1$ of the chosen polynomials in $\mathcal{V}^h$, cf Theorem 7.40.

The classical consistency error can be estimated by combining (7.71) or the corresponding generalizations (7.77), (7.89), (7.101), with the Lipschitz-continuity of G and G_h and the penalty norm, cf. Theorem 7.21 ff. The combination of (7.71) with Q'^h in (7.113) yields for our first example (7.23) for any $u \in \mathcal{U}$ and $v^h \in S^h$, cf. (7.118),

$$\begin{aligned}\langle G^h P^h u - G_h u, v^h\rangle = \big(&a^h(P^h u, v^h) + b^h(P^h u, v^h) + \varepsilon J_h^\sigma(P^h u, v^h) \quad (7.121)\\ &-\ell^h(P^h u, v^h)\big) - \big(a_h(u, v^h) + b_h(u, v^h) + \varepsilon J_h^\sigma(u, v^h)\\ &-\ell_h(u, v^h)\big)\forall v^h \in S^h.\end{aligned}$$

For our next examples (7.74), (7.78), and (7.93) we simply replace $a_h(P^h u, v^h)$, $\ell_h(P^h u, v^h)$ in (7.57), (7.60) by the updated forms in (7.76), (7.87), (7.88), and in (7.99), (7.100). The $b^h(P^h u, v^h)$ is skipped in (7.74), (7.93), (7.99). All these differences of corresponding pairs can be estimated by combining the interpolation errors in Theorem 7.7 and Corollary 7.8 on the K and ∂K with specific properties of the $a_h(P^h u, v^h), \dots, \ell_h(P^h u, v^h)$ in Theorems 7.21–7.33.

Continuity, consistency plus stability imply the unique existence of the discrete solution and its convergence. For the reader's convenience we reformulate the essential Theorem 3.21 in the following theorem. It applies to the original problems, and their discretizations, (7.13), (7.14), (7.70), (7.71), and (7.74), (7.77); (7.78), (7.79), and (7.89); (7.93), and (7.101).

Theorem 7.3. Unique existence and convergence of discrete solutions: *Let the original problem have the exact solution u_0. Let $G^h : \mathcal{D}(G^h) \subset \mathcal{U}^h \to \mathcal{V}'^h$ be its discretization, cf. (7.118), and Definitions 3.5, 3.6, 3.12 (here nonlinear Φ^h are allowed) and satisfy the following conditions*

1. $G^h = \Phi^h(G) : \mathcal{D}(G^h) \subset \mathcal{U}^h \to \mathcal{V}'^h$ *is defined and continuous in* $B_r(P^h u_0) \subset \mathcal{D}(G^h)$ *with* $r > 0$ *independent of* h;
2. G^h *is (classically) consistent with* G *in* $P^h u_0$, *in the sense of (7.120);*
3. G^h *is stable for* $P^h u_0$ *with respect to* $\| \cdot \|$.

Then the discrete problem $G^h(u^h) = 0$ possesses a unique solution $u_0^h \in \mathcal{D}(G^h) \subset \mathcal{U}^h$ near u_0 for all sufficiently small $h \in H$ and u_0^h converges to u_0. If G^h is consistent and consistent of order p, then u_0^h converges and converges of order p, hence

$$\|u_0^h - u_0\| \le S \|G^h(P^h u_0)\|_{\mathcal{V}'^h} \text{ and } \le \mathcal{O}(h^p), \text{ respectively} \tag{7.122}$$

For all the cases considered here, G_h and G^h are differentiable simultaneously with G. They even satisfy *consistent differentiability*, cf. (5.165), so $\forall u, v \in B_r(u_0) \cap \mathcal{U}_s$:

$$\|(Q^h G'(u))v - Q^h G'(P^h u) P^h v\|_{\mathcal{V}^{h\prime}} \le C_2 h^{p'} (1 + \|u\|_{\mathcal{U}_s}) \|v\|_{\mathcal{U}_s}, \tag{7.123}$$

with a "smooth" norm $\|v\|_{\mathcal{U}_s}$. This is important for the *mesh independence principle.* It guarantees the quadratic convergence of the Newton method for solving the corresponding discrete nonlinear systems, essentially independent of the mesh size, cf. Section 7.13.

So the program for the next sections is the proof for *stability, consistency, and consistent differentiability* for our preceding problems.

7.7 Geometry of the mesh, error and inverse estimates

In this section we shall be concerned with the necessary approximation results. The advantage of the approach in this section compared to the usual FE approach is the avoided condition of quasiuniform subdivisions. For FEMs this is possible by modifying the results for boundary integral methods of Graham *et al.* [362], cf. Chapter 4. For DCGMs some unusual conditions upon the geometry of the subdivision have to be imposed. Then a multiplicative trace inequality is combined with inverse estimates.

7.7.1 Geometry of the mesh

Let us consider a system $\{\mathcal{T}^h\}_{h\in(0,h_0)}$, $h_0 > 0$, of partitions of the domain Ω, i.e. $\mathcal{T}^h = \{K \in \mathcal{T}^h\}$. *We assume throughout the chapter that the system* $\{\mathcal{T}^h\}_{h\in(0,h_0)}$ *has the following properties:*

(A0) $\mathcal{T}^h$ is a covering partition of the domain $\overline{\Omega} = \cup_{K\in\mathcal{T}^h}\overline{K}$.

(A1) Each element $K \in \mathcal{T}^h$, $h \in (0, h_0)$, is a *star-shaped* domain with respect to at least one interior point $x_K = (x_{K_1}, \dots, x_{K_n})$ of K. We assume that

(i) there exists a constant $\kappa > 0$ independent of K and h such that

$$\frac{\max_{x\in\partial K}|x - x_K|}{\min_{x\in\partial K}|x - x_K|} \le \kappa \quad \forall K \in \mathcal{T}^h,\ h \in (0, h_0); \tag{7.124}$$

(ii) each K can be divided into a finite number of closed simplices:

$$\overline{K} = \bigcup_{S\in\mathcal{S}(K)} S\ ; \tag{7.125}$$

there exist positive constants $C_3, \tilde{\kappa}$ independent of K, S and h such that

$$\frac{h_S}{\rho_S} \le C_3 \quad \forall S \in \mathcal{S}(K) \qquad \text{(shape regularity)} \tag{7.126}$$

where h_S is the diameter of S, ρ_S is the radius of the largest n-dimensional ball inscribed into S and, moreover,

$$1 \le \frac{h_K}{h_S} \le \tilde{\kappa} < \infty \quad \forall S \in \mathcal{S}(K). \tag{7.127}$$

(A2) There exists a constant $C_4 > 0$ such that

$$h_K \le C_4\, d(e) \quad \forall K \in \mathcal{T}^h, e \subset \partial K,\ h \in (0, h_0). \tag{7.128}$$

Remark 7.4. *The conditions* (A1) *and* (A2) *imply a bound on the maximum number of edges, i.e. there exists a constant* c_e *such that* $\#(e \in K) \le c_e$ *for all* $K \in \mathcal{T}^h$. *Moreover, it follows that the number of hanging nodes for each element is bounded.*

Let us note that these properties can be verified, e.g. in the case of dual finite volumes constructed over a regular simplicial mesh. Furthermore, nondegenerate subdivisions in FEMs satisfy these conditions as well, see [314]. Figure 7.1 satisfies (A0)–(A2), it is star-shaped and the K are nonconvex. Figure 7.3 shows a type of mesh used in [314]. For practical computation these grids are not very suitable.

We cite some technical results formulated as Theorems 7.5–7.7.

7.7.2 Inverse and interpolation error estimates

Under the above assumptions, (A1), (A2) the following results can be established.

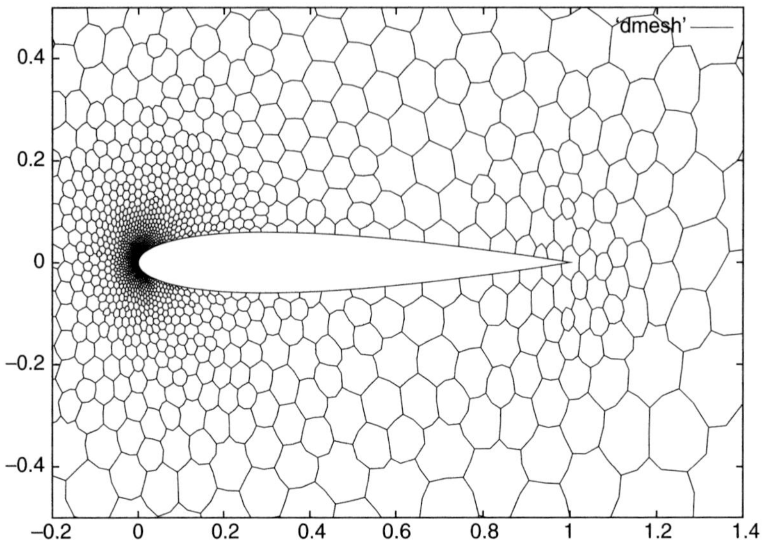

Figure 7.3 Dual mesh.

Theorem 7.5. Multiplicative trace inequality: *Assume (A1), (A2). Then there exists a constant* $C_5 > 0$ *independent of* v, h *and* K *such that*

$$\|v\|^2_{L^2(\partial K)} \leq C_5\big(\|v\|_{L^2(K)}\,|v|_{H^1(K)} + h_K^{-1}\|v\|^2_{L^2(K)}\big), \tag{7.129}$$
$$K \in \mathcal{T}^h, \ v \in H^1(K), \ h \in (0, h_0).$$

Proof. Using the approach from [296] and [297], we prove the theorem in two steps.

(i) Let $K \in \mathcal{T}_h$ and $S \in \mathcal{S}(K)$ be arbitrary but fixed. We denote by x_S the center of the largest d-dimensional ball inscribed into S. Without loss of generality we suppose that x_S is the origin of the coordinate system. We start from the following relation, a consequence, e.g. of (2.9),

$$\int_{\partial S} v^2 x \cdot \nu \,\mathrm{d}S = \int_S \nabla \cdot (v^2 x)\,\mathrm{d}x, \qquad v \in H^1(S). \tag{7.130}$$

Let ν_e be the unit outer normal to S on the face $e \subset \partial S$. Then

$$x \cdot \nu_e = |x||\nu_e| \cos\alpha = |x|\cos\alpha = \rho_S \quad x \in e, \tag{7.131}$$

since ρ_S is the distance of x_S from the face e, see Figure 7.4.

From (7.131) we have

$$\int_{\partial S} v^2 x \cdot \nu \,\mathrm{d}S = \sum_{e \subset \partial S} \int_e v^2 x \cdot \nu_{ij} \ \mathrm{d}S \tag{7.132}$$
$$= \sum_{e \subset \partial S} t_e \int_e v^2 \,\mathrm{d}S \geq \rho_S \sum_{e \subset \partial S} \int_e v^2 \ \mathrm{d}S = \rho_S \|v\|^2_{L^2(\partial S)}.$$

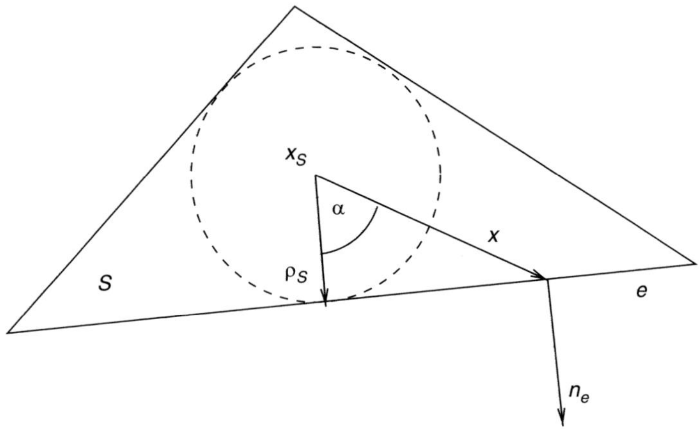

Figure 7.4 Simplex S with its face e and the distances t_e.

Moreover,

$$\int_S \nabla \cdot (v^2 x)\ \mathrm{d}x = \int_S \left(v^2 \nabla \cdot x + x \cdot \nabla v^2\right) \mathrm{d}x \tag{7.133}$$

$$= d \int_S v^2\ \mathrm{d}x + 2 \int_S vx \cdot \nabla v\ \mathrm{d}x \le d \|v\|^2_{L^2(S)} + 2 \int_S |vx \cdot \nabla v|\ \mathrm{d}x.$$

With the aid of the Cauchy inequality the second term of (7.133) is estimated as

$$2 \int_S |vx \cdot \nabla v|\ \mathrm{d}x \le 2 \sup_{x \in S} |x| \int_S |v||\nabla v|\ \mathrm{d}x \le 2 h_S \|v\|_{L^2(S)} |v|_{H^1(S)}. \tag{7.134}$$

Then (7.130), (7.132), (7.133) and (7.134) give

$$\|v\|^2_{L^2(\partial S)} \le \frac{1}{\rho_S} \left[2 h_S \|v\|_{L^2(S)} |v|_{H^1(S)} + d \|v\|^2_{L^2(S)}\right] \tag{7.135}$$

$$\le C_3 \left[2 \|v\|_{L^2(S)} |v|_{H^1(S)} + \frac{d}{h_S} \|v\|^2_{L^2(S)}\right],$$

where C_3 is the constant given by (7.126) and is independent of S, K, v and h.

(ii) Let $K \in \mathcal{T}_h$ and $S \in \mathcal{S}(K)$. Then (7.135) yields

$$\|v\|^2_{L^2(\partial S)} \le C \left(\|v\|_{L^2(S)} |v|_{H^1(S)} + h_S^{-1} \|v\|^2_{L^2(S)}\right), \tag{7.136}$$

$$S \in \mathcal{S}(K),\ v \in H^1(K),$$

where $C = C_3 \max\{2, d\}$. Using the inclusion $\partial K \subset \bigcup_{S\in\mathcal{S}(K)} \partial S$, relations (7.136) and (7.127) and the Cauchy inequality, we find that

$$\begin{aligned}
\|v\|^2_{L^2(\partial K)} &\le \sum_{S\in\mathcal{S}(K)} \|v\|^2_{L^2(\partial S)} \\
&\le C\left\{ \sum_{S\in\mathcal{S}(K)} \left(\|v\|_{L^2(S)}\, |v|_{H^1(S)} + h_S^{-1} \|v\|^2_{L^2(S)} \right) \right\} \qquad (7.137) \\
&\le C\left\{ \left(\sum_{S\in\mathcal{S}(K)} \|v\|^2_{L^2(S)} \right)^{1/2} \left(\sum_{S\in\mathcal{S}(K)} |v|^2_{H^1(S)} \right)^{1/2} + \tilde{\kappa} h_K^{-1} \|v\|^2_{L^2(K)} \right\} \\
&= C_M \left(\|v\|_{L^2(K)}\, |v|_{H^1(K)} + h_K^{-1} \|v\|^2_{L^2(K)} \right),
\end{aligned}$$

where $C_M = \tilde{\kappa} C = \tilde{\kappa} C_3 \max\{2, d\}$. □

Theorem 7.6. Inverse inequality: *Under the conditions (A1), (A2), there exists a constant $C_6 = C_6(d-1) > 0$ independent of v, h and K such that*

$$|v|_{H^1(K)} \le C_6 h_K^{-1} \|v\|_{L^2(K)}, \quad v \in \mathcal{P}_{d-1}(K),\ K \in \mathcal{T}^h,\ h \in (0, h_0). \qquad (7.138)$$

The proof of Theorem 4.35 remains valid for Theorem 7.6 as well. According to Schwab [575], the constant C_6 is $\mathcal{O}((d-1)^2)$.

Important are the approximation properties of the space S^h constructed over the mesh $\mathcal{T}^h$. The following theorem is for the FE situation, a consequence of Definition 4.8 and Theorem 4.17; for the DCG situation it is proved by Verfürth [656].

Theorem 7.7. Error estimates: *Assume (A1), (A2). Then there exists a constant $C_7 > 0$ independent of v and h and a mapping $P^h : H^1(K) \to \mathcal{P}_{d-1}(K)$ such that $\forall\ k = 0, 1, 2,\ v \in H^s(\mathcal{T}^h),\ K \in \mathcal{T}^h,\ h \in (0, h_0)$,*

$$\begin{aligned}
|P^h v - v|_{H^k(K)} &\le C_7 h_K^{\min\{d,s\}-k} |v|_{H^{\min\{d,s\}}(K)}, \quad \textit{with } |\cdot|_{H^0(K)} = |\cdot|_{L^2(K)} \\
|P^h v - v|_{H^k(\mathcal{T}^h)} &\le C_7 \left(\sum_{K\in\mathcal{T}^h} \left(h_K^{\min\{d,s\}-k} |v|_{H^{\min\{d,s\}}(K)} \right)^2 \right)^{1/2} \qquad (7.139)
\end{aligned}$$

By Veeser and Verfürth [652], and Thrun [628], (7.139) remain correct, if the $H^k, H^{\min\{d,s\}}$ are replaced by $W^{k,\infty}, W^{\min\{d,s\},\infty}$.

If in K, polynomials of degree d_K are employed we replace in Theorem 7.21 ff.

$$h^{\min\{d,s\}} |u|_{H^{\min\{d,s\}}(\Omega)} \text{ by } h^{\{\min d, \min s\}} \left(\sum_{K\in\mathcal{T}^h} h_K^{2(d_K-d)} |u|^2_{H^{\min\{d_K, s_K\}}(K)} \right)^{1/2}. \qquad (7.140)$$

The precise conditions for these modified results, used in adaptive and *hp*- DCGMs, are described in Section 7.14.

Proof. According to [656], we denote by

$$\pi_K \varphi \equiv \frac{1}{|K|} \int_K \varphi \, dx \tag{7.141}$$

the mean value of any integrable function φ on K. For any $v \in H^s(K)$ we recursively define polynomials $p_{d-1,K}(v), p_{d-2,K}(v), \ldots, p_{0,K}(v)$ in $\mathcal{P}_{d-1}(K)$ by

$$p_{d-1,K}(v) \equiv \sum_{\substack{\alpha \in N^{d-1} \\ |\alpha| = d-1}} \frac{1}{(d-1)!} x^\alpha \pi_K(D^\alpha v) \tag{7.142}$$

and, for $k = d-1, d-2, \ldots, 1$

$$p_{k-1,K}(v) \equiv p_{k,K}(v) + \sum_{\substack{\alpha \in N^{d-1} \\ |\alpha| = k-1}} \frac{1}{(d-1)!} x^\alpha \pi_K \left(D^\alpha (v - p_{k,K}(u)) \right). \tag{7.143}$$

Finally, we set

$$P^h v \equiv p_{0,K}(v). \tag{7.144}$$

For the proof of the approximation properties (7.139) we refer to [656]. □

As a consequence of the Trace Theorem 1.38 we obtain

Corollary 7.8. *There exist a constant $C_8 > 0$ independent of v and h_0 such that*

$$|P^h v - v|_{H^k(\partial K)} \le \|P^h v - v\|_{H^k(\partial K)} \le C_7 h_K^{\min\{d,s\}-k-1/2} |v|_{H^{\min\{d,s\}}(K)}, \tag{7.145}$$

for all $v \in H^s(K)$, $K \in \mathcal{T}^h$, $0 \le k \le s$, and $h \in (0, h_0), h_0 < 1$.

7.8 Penalty norms and consistency of the J_h^σ

The need for the *penalty norms* has been mentioned several times before. Their form here differs from the energy norms in Chapters 2 and 4. So we maintain the notation of penalty norms for the DCGMs here.

The coercivity, boundedness and even more so the consistency properties in the next two sections are not possible with respect to broken Sobolev norms, but only with respect to these penalty norms. For the following considerations we assume, cf. (7.39),

$$\sigma_{\min} := \min_{e \in \mathcal{T}^h} \{\sigma|_e = 1/d(e)\} \ge 1. \tag{7.146}$$

The consistency equation (7.121) and the estimates in the following theorems show that a successful convergence approach requires carefully monitoring the $J_h^\sigma(u^h, v^h)$. Beginning with Theorem 7.16, a naive approach to the following coercivity and consistency estimates would yield unbounded results with respect to broken Sobolev norms, $\|v\|_{H^1(\mathcal{T}_h)}$. This is caused by the $J_h^\sigma(v^h, v^h)$, strongly growing for arbitrary v^h and increasingly finer meshes, $\mathcal{T}^h$. We compare, e.g. (7.121) with the consistency results for FEMs with boundary crimes in (5.348). This motivates, e.g. for the $a^h(u^h, v^h)$ in (7.57), nonlinear in u^h, but linear and bounded in v^h, a condition

$$|a^h(P^h u, v^h) - a_h(u, v^h)| \leq C(u)\text{dist}(u, S^h)\|v^h\|_?, \tag{7.147}$$

with appropriate norms in $\text{dist}(u, S^h)$ and $\|v^h\|_?$ and $C(u)$ usually depending upon u. $\text{dist}(u, S^h)$ is well estimated by the interpolation errors in Theorem 7.7, Corollary 7.8. However it neglects the violated continuity and boundary conditions. For $\mathcal{V}^h = S^h$ the $\|v^h\|_?$ will be chosen as a combination of the standard broken Sobolev norm and $J_h^\sigma(v^h, v^h)$, allowing the necessary (7.147).

In our general discretization approach we are mainly interested in those $u^h, v^h \in \mathcal{U}_e$, approximating the elements u, v in a bounded subset of $H^{3/2+\varepsilon}(\Omega)$. By Proposition 7.9 and Theorem 7.12, for smooth v and the interesting $v^h \approx P^h v \in S^h$ or $\mathcal{U}_e$ the above $\|v^h\|_?$ is very close to a standard broken Sobolev norm.

Proposition 7.9. Essential properties of J_h^σ : *Choose a bounded* $\mathcal{B} \subset H^s(\Omega)$, *with* $s \geq 1$, *and the local polynomial degree,* $d - 1 > 0$. *Then the* $J_h^\sigma(\cdot,\cdot) : \mathcal{B} \times \mathcal{B} \to \mathbb{R}$ *is continuous. With* $e_{\mathcal{T}^h} = \max_{K \in \mathcal{T}^h}\{\#(e \in K)\}$, $d = \min_{K\in\mathcal{T}^h}\{d_K\}$, *a constant* $C_{14} := 2^2 e_{\mathcal{T}^h} c_w C_4 C_7^2 > 0$ *exists, such that* $\forall u, v \in \mathcal{B}$

$$\left|J_h^\sigma(P^h u - u, P^h v - v)\right| \leq C_{14} h^{2\min\{d,s\}-2} |u|_{H^{\min\{d,s\}}(\Omega)} |v|_{H^{\min\{d,s\}}(\Omega)}. \tag{7.148}$$

Proof. Corollary 7.8, the Cauchy–Schwarz inequality, and (7.128) imply for $e \subset \partial K$

$$\int_e \sigma([u - P^h u]\,[v - P^h v] dS \leq \sigma|_e 2^2 \|u - P^h u\|_{L^2(\partial K)} \|v - P^h v\|_{L^2(\partial K)} \tag{7.149}$$

$$\leq \sigma|_e \left(2C_7 h_K^{\min\{d,s\}-1/2}\right)^2 |u|_{H^{\min\{d,s\}}(K)} |v|_{H^{\min\{d,s\}}(K)}$$

$$\leq c_w C_4 2^2 C_7^2 h_K^{2\min\{d,s\}-2} |u|_{H^{\min\{d,s\}}(K)} |v|_{H^{\min\{d,s\}}(K)}.$$

The final $\sum_{e\in\mathcal{T}^h}$, the discrete Cauchy–Schwarz inequality, and a local degree $d_K = d$ show, with $C_{14} = c_w C_4 2^2 C_7^2 \max_{K\in\mathcal{T}^h}\{\#(e \in K)\}$,

$$\left|J_h^\sigma(P^h u - u, P^h v - v)\right| \leq C_{14} \sum_{K\in\mathcal{T}^h} h_K^{2\min\{d_K,s\}-2} |u|_{H^{\min\{d_K,s\}}(K)} |v|_{H^{\min\{d_K,s\}}(K)}$$

$$\leq h^{2\min\{d,s\}-2} C_{14} \sum_{K\in\mathcal{T}^h} |u|_{H^{\min\{d,s\}}(K)} |v|_{H^{\min\{d,s\}}(K)}$$

$$\text{with } d = \min_{K\in\mathcal{T}^h}\{d_K\} \tag{7.150}$$

hence (7.148) with (7.150). □

Theorem 7.10. Consistency estimates for the J_h^σ: *We obtain for* $u \in H^s(\Omega), v^h \in H^s(\mathcal{T}^h)$, *the local degree*, $d-1>0$, *and with* $C_{15} := 2\sqrt{C_5 c_w} C_7$.

$$\left|J_h^\sigma(P^h u, v^h) - J_h^\sigma(u, v^h)\right| \le h^{\min\{d,s\}-1} C_{15} |u|_{H^{\min\{d,s\}}(\Omega)} \|v^h\| \forall h \in (0, h_0). \quad (7.151)$$

where $P^h u$ *is the previous* S^h*-interpolant of* u.

Proof. From (7.61), the Cauchy inequality, (7.149)–(7.150), the multiplicative trace inequality (7.129) and (7.139) (but (7.317) for hp methods) we have

$$\begin{aligned}
&\left|J_h^\sigma(u, v^h) - J_h^\sigma(P^h u, v^h)\right| && (7.152)\\
&\le \sum_{e \in \mathcal{T}^h} \int_e \sigma |[u - P^h u]| \, |[v^h]| \, dS \\
&\le J_h^\sigma(v^h, v^h)^{1/2} \left(\sum_{e \in \mathcal{T}^h} \int_e \sigma [u - P^h u]^2 \, dS \right)^{1/2} \\
&\le \|v^h\| \sum_{K \in \mathcal{T}^h} \left(c_w h_K^{-1} \|u - P^h u\|_{L^2(\partial K)}^2 \right)^{1/2} \\
&\le \sqrt{C_5 c_w} \|v^h\| \left(\sum_{K \in \mathcal{T}^h} h_K^{-1} \|u - P^h u\|_{L^2(K)} |u - P^h u|_{H^1(K)} + h_K^{-2} \|u - P^h u\|_{L^2(K)}^2 \right)^{1/2} \\
&\le \sqrt{2 C_5 c_w} \|v^h\| C_A^2 C_7^2 \\
&\quad \times \left(\sum_{K \in \mathcal{T}_h} \left(\frac{1}{h_K} h_K^{\min\{d,s\}} h_K^{\min\{d,s\}-1} + \frac{1}{h_K^2} h_K^{2\min\{d,s\}} \right) |u|_{H^{\min\{d,s\}}(K)}^2 \right)^{1/2} \\
&\le 2\sqrt{C_5 c_w} C_7 h_K^{2\min\{d,s\}-1} \left(\sum_{K \in \mathcal{T}_h} |u|_{H^{\min\{d,s\}}(K)}^2 \right)^{1/2} \|v^h\|,
\end{aligned}$$

thus (7.151) with $C_{15} := 2\sqrt{C_5 c_w} C_7$. □

Before we discuss DCGMs, we introduce the announced penalty norm. It penalizes elements $v \in H^s(\mathcal{T}^h)$, violating continuity and boundary conditions.

Definition 7.11. *For* σ *in (7.39), (7.146), we introduce the* penalty norms *for* $1 \le s \in \mathbb{R}$,

$$\|v\|_{H^s(\mathcal{T}^h)}^J := \left(J_h^\sigma(v, v) + |v|_{H^s(\mathcal{T}^h)}^2 \right)^{1/2}, \forall v \in H^s(\mathcal{T}^h), \text{ and for} \quad (7.153)$$

$$s = 1: \ \|v\| := \|v\|_{H^1(\mathcal{T}^h)}^J \text{ used as } \|v^h\|_{\mathcal{V}^h} := \|v^h\| \text{ in } S^h, \text{ and finally}$$

$$\|v\|_{\|H^s(\mathcal{T}^h)\|}^J := \left(J_h^\sigma(v, v) + \|v\|_{H^s(\mathcal{T}^h)}^2 \right)^{1/2}, \text{ used e.g. in subsection 7.11.4.}$$

Theorem 7.12. Banach spaces with respect to $\|v\|$: *For $s \geq 1$ and σ in (7.146), we obtain*

$$\exists c_e > 1: \ (c_e)^{-1}\|v\|^J_{\|H^1(\mathcal{T}^h)\|} \leq \|v\|^J_{H^1(\mathcal{T}^h)} \leq c_e\|v\|^J_{\|H^1(\mathcal{T}^h)\|} \ \forall v \in H^1(\mathcal{T}^h), \quad (7.154)$$

hence, equivalent norms in $H^s(\mathcal{T}^h)$, a Banach-space with respect to both norms. In particular, we obtain for $s = 1$, $\|\cdot\| = \|v\|^J_{H^1(\mathcal{T}^h)}$. The original $\|\cdot\|_{H^s(\mathcal{T}^h)}$ and the new $\|\cdot\|^J_{H^s(\mathcal{T}^h)}$ are nonequivalent.

Proof. Obviously $\|\cdot\|^J_{H^s(\mathcal{T}^h)}$ are norms for $H^s(\mathcal{T}^h)$, nonequivalent with the original norm. A Cauchy sequence $\{v_\nu\}_{\nu=1}^\infty$ with respect to $\|\cdot\|^J_{H^s(\mathcal{T}^h)}$ is Cauchy sequence with respect to $\|\cdot\|_{H^s(\mathcal{T}^h)}$ as well, hence has a limit element $v_0 \in H^s(\mathcal{T}^h)$ with respect to $\|\cdot\|_{H^s(\mathcal{T}^h)}$. Since J_h^σ is continuous in $H^s(\mathcal{T}^h)$, this v_0 is limit element with respect to $\|\cdot\|^J_{H^s(\mathcal{T}^h)}$ as well, hence $H^s(\mathcal{T}^h)$ is a Banach space with respect to both norms. The equivalence of both norms $\|v\|^J_{H^s(\mathcal{T}^h)}$ and $\|v\|^J_{\|H^s(\mathcal{T}^h)\|}$ is an immediate consequence of Proposition 1.25. □

Theorem 7.13. Interpolation errors with respect to $\|\cdot\|^J_{H^k(\mathcal{T}^h)}$ and the equivalence of these norms:

1. *There exists a constant $C_{16} > 0$ independent of v and h, such that*

$$\|P^h v - v\|^J_{H^k(\mathcal{T}^h)} \leq C_{16} h^{\min\{d,s\}-(1/2+k)} |v|_{H^{\min\{d,s\}}(\mathcal{T}^h)}, \ k = 0, 1, 2 \leq s \quad (7.155)$$

   for all $v \in H^s(\mathcal{T}^h)$ or $v \in H^s(\Omega)$ and $h \in (0, h_0)$.

2. *For a given $u \in H^s(\Omega)$, $d \geq s \geq 1$, define $\mathcal{B}_u^h \in S^h$ as*

$$\mathcal{B}_u^h := \{v^h \in H^s(\mathcal{T}^h) \cap S^h : \|u - v^h\|^J_{H^k(\mathcal{T}^h)} \leq 1/(4|u|_{H^s(\mathcal{T}^h)}), k = 0, 1, 2 \leq s\}$$

   with small enough h. Then all the norms $\|\cdot\|^J_{H^s(\mathcal{T}^h)}$, and $\|\cdot\|_{H^s(\mathcal{T}^h)}$ are equivalent in $\mathcal{B}_u^h$, and $\lim_{h\to 0} \|P^h u\|^J_{H^s(\mathcal{T}^h)} = \|u\|_{H^s(\Omega)}$.

Proof. Inequality (7.155) is a consequence of Theorem 7.7, and Corollary 7.8. Claim 2 follows from Proposition 7.9 and (7.155). □

7.9 Coercive linearized principal parts

7.9.1 Coercivity of the original linearized principal parts

Existence and uniqueness results for the original previous problems are available by the Fredholm alternative, if the linearized operators are compact perturbations of bounded coercive operators, cf. Chapter 2. This is essential for the proof of stability as well. So we summarize these results.

The bilinear form for the Laplacian is, cf. (7.19),

$$c(u, v) := a(u, v) = \varepsilon \int_\Omega (\nabla u, \ \nabla v)_n \, dx \quad \text{for} \quad u, v \in \mathcal{U}, \quad (7.156)$$

or $a(u,v)$ in (7.74) for the general bilinear form. For the more general forms of $a(u,v)$, nonlinear in u, in (7.81), and (7.96) ff., the *linearized principal parts,* the $\partial a_p/\partial u$, evaluated in u in the neighborhood of u_0 and applied to w, v, are

$$c(w,v) := \left(\frac{\partial a_p}{\partial u}(u)\right)(w,v) = \left(G_p'(u)w, v\right) = \varepsilon \int_\Omega \sum_{i,j=1}^n a_{ij}(x,u)\partial^i w\, \partial^j v\, dx \text{ and} \tag{7.157}$$

$$c(\vec{w},\vec{v}) := \left(\frac{\partial a_p}{\partial \vec{u}}(\vec{u})\right)(\vec{w},\vec{v}) = \left(G_p'(\vec{u})\vec{w}, \vec{v}\right)_q = \int_\Omega \sum_{k,l=1}^n \left(\frac{\partial A_k}{\partial \mathbf{u}^l}(x,\vec{u},\nabla\vec{u})\partial^l \vec{w}, \partial^k \vec{v}\right)_q \tag{7.158}$$

for $u, w, v \in \mathcal{U}$ and $\vec{u}, \vec{w}, \vec{v} \in H^1(\Omega, \mathbb{R}^q)$. The corresponding conditions are listed in (7.80) and (7.94), in particular the ellipticity conditions (e).

Theorem 7.14. Coercivity of the original (linearized) principal parts: *For a polygonal Ω the bilinear principal forms $c(\cdot,\cdot)$ in (7.156)–(7.158), and its generalization to (7.74), are bounded and coercive with respect to the energy (semi-) norm. For the nonlinear problems in (7.81) and (7.96) ff., we have to impose the additional conditions (7.80) and (7.94):*

$$\begin{aligned} &c(v,v) \geq C\varepsilon |v|_{\mathcal{U}}^2, \quad |c(u,v)| \leq C'\varepsilon |u|_{\mathcal{U}} |v|_{\mathcal{U}} \ \forall u, v \in \mathcal{U} = H^1(\Omega) \\ &c(\vec{v},\vec{v}) \geq C|\vec{v}|_{\mathcal{U}}^2, \quad |c(\vec{u},\vec{v})| \leq C' |\vec{u}|_{\mathcal{U}} |\vec{v}|_{\mathcal{U}} \ \forall \vec{u}, \vec{v} \in \mathcal{U} = H^1(\Omega, \mathbb{R}^q). \end{aligned} \tag{7.159}$$

For $u, v \in H_0^1(\Omega)$ and $\vec{u}, \vec{v} \in H_0^1(\Omega, \mathbb{R}^q)$ this bounded coercivity (7.159) holds for the norm $\|\cdot\|_{H_0^1(\Omega, R^q)}, q \geq 1$, the equivalent standard Sobolev norm.

7.9.2 Coercivity and boundedness in $\mathcal{V}^h$ for the Laplacian

Stability for the previous discrete problems can be guaranteed by two properties. We need the bounded invertibility of the corresponding linearized operator and the bounded coercivity in $\mathcal{V}^h$ of its linearized principal part $A_h(u,v) + J_h^\sigma(u,v)$ or $A_h(u;w,v) := \big(\partial a_h/\partial u(u)\big)(w,v) + J_h^\sigma(w,v)$ in the following subsections.

Remark 7.15. *Boundedness and coercivity in $\mathcal{V}^h$ always means equiboundedness and equicoercivity with respect to h.*

For the above different principal parts $c(u,v)$, we define the corresponding $c_h(u,v)$, and

$$A_h(u,v) := c_h(u,v) + \varepsilon J_h^\sigma(u,v) \text{ with } J_h^\sigma(\cdot,\cdot) \text{ in (7.61).} \tag{7.160}$$

For the nonlinear cases, let $c_h(w,v), c_h(\vec{w},\vec{v})$ indicate the corresponding discrete linearizations analogous to $c(w,v), c(\vec{w},\vec{v})$ in (7.157), (7.158). Then, cf. (7.99),

$$A_h(u;w,v) := c_h(w,v) + \varepsilon J_h^\sigma(w,v) \text{ and } A_h(\vec{u};\vec{w},\vec{v}) := c_h(\vec{w},\vec{v}) + J_h^\sigma(\vec{w},\vec{v}). \tag{7.161}$$

As above, we discuss the three types of DCGMs, the nonsymmetric interior, NIPG, with $\theta = -1$, the symmetric interior, SIPG, with $\theta = 1$, and the incomplete interior penalty Galerkin, IIPG, methods with $\theta = 0$. For completeness, we extend it to the general interior penalty Galerkin, GIPG, method with $-1 \le \theta \le 1$, cf. Houston *et al.* [407]. We prove the coercivity and boundedness of these $A_h(\cdot,\cdot)$ with respect to $\|\cdot\|$

For our model problem, these bilinear forms can be written in the form

$$a_h(u,v) = \varepsilon \sum_{K\in\mathcal{T}_h} \int_K (\nabla u,\ \nabla v)_n \, dx - \varepsilon \sum_{e\in\mathcal{T}^h\setminus\partial\Omega_N} \int_e \big(\{\nabla u\},\nu)_n[v] \quad (7.162)$$

$$+\theta(\{\nabla v\},\nu)_n[u]\big)\, dS \quad \forall u,v\in\mathcal{U}_e \text{ for } -1\le\theta\le 1.$$

The multiplicative trace and the inverse inequality (7.129) and (7.138) yield

$$\|v\|^2_{L^2(\partial K)} \le C_5(1+C_6)h_K^{-1}\|v\|^2_{L^2(K)} \ \forall v\in\mathcal{V}^h. \quad (7.163)$$

Theorem 7.16. Bounded and coercive $a_h(u,v)$ for the Laplacian: *Choose an arbitrary constant $c_w > 0$ in (7.39) for the NIPG method. For the SIPG, IIPG, and GIPG methods (with $\theta = 1, 0, \in [-1,1]$) choose*

$$c_w \ge C_5(1+C_6)(1+\theta)^2 \quad (7.164)$$

where C_5 and C_6 are given by (7.138), respectively. Then the corresponding bilinear forms $A_h(\cdot,\cdot)$ given by (7.160), (7.162), are bounded and coercive with respect to the norm, $\|v\|$, such that

$$A_h(v,v) \ge c_c\varepsilon\|v\|^2, \quad (7.165)$$

$$|A_h(u,v)| \le c_b\varepsilon\|u\|\ \|v\| \ \forall u,v\in\mathcal{V}^h. \quad (7.166)$$

Remark 7.17. *Theoretically it would be possible to determine the constants, defining c_w in (7.164): Before starting computations, check all elements $K\in\mathcal{T}_h$ and find the maximal values of κ, C_3, $\tilde{\kappa}$, C_4 appearing in (7.124)–(7.128). This is combined with the C_5, C_6 in (7.129), (7.138). However, we doubt that anybody uses this approach. However, Epshteyn and Rivière [309] present computable lower bounds of the penalty parameters in the SIPG method. In particular, they derive the explicit dependence of the coercivity constants with respect to the polynomial degree and the angles of the mesh elements.*

Usually, a suitable value of c_w is sought empirically. Furthermore, a too high value of c_w leads to bad computational properties of the corresponding discrete problem. Some numerical experiences for the choice of c_w for the system of the compressible Navier-Stokes equations, cf. (7.343), are given in [294]. Houston et al. use a similar approach, see their [407], text between relations (2.4) and (2.5). They have an explicit relation containing constants from a multiplicative trace inequality (C_d) and the inverse inequality (C_3).

Proof.

1. *Coercivity of the NIPG method*: Inequality (7.165) for A_h immediately follows from the definition (7.160), (7.162) with $\theta = -1, c_c = 1$.
2. *Coercivity of the SIPG, IIPG, and GIPG methods*: By (7.160) and (7.162) we have

$$A_h(v,v) = \varepsilon |v|^2_{H^1(\mathcal{T}^h)} - \varepsilon(1+\theta) \sum_{e \in \mathcal{T}^h \setminus \partial\Omega_N} \int_e (\{\nabla v\}, \nu)_n [v] \, dS + \varepsilon J_h^\sigma(v,v). \tag{7.167}$$

This, the Cauchy inequality and Young's inequality in the form,

$$ab \le ca^2 + b^2/(4c) \text{ for } a,b,c \in \mathbb{R}, \; c > 0, \tag{7.168}$$

imply that for $c = 1$ and any $\delta > 0$,

$$\begin{aligned} A_h(v,v) &\ge \varepsilon |v|^2_{H^1(\mathcal{T}^h)} + \varepsilon J_h^\sigma(v,v) - \varepsilon(1+\theta) \left(\frac{1}{\delta} \sum_{e \in \mathcal{T}^h \setminus \partial\Omega_N} \int_e d(e) \{\nabla v\}^2 dS \right)^{1/2} \\ &\quad \times \left(\delta \sum_{e \in \mathcal{T}^h \setminus \partial\Omega_N} \int_e \frac{1}{d(e)} [v]^2 dS \right)^{1/2} \\ &\ge \varepsilon \|v\|^2 - \varepsilon\omega - \varepsilon \frac{\delta}{c_w} J_h^\sigma(v,v), \end{aligned} \tag{7.169}$$

where

$$\omega = \frac{(1+\theta)^2}{4\delta} \sum_{e \in \mathcal{T}^h \setminus \partial\Omega_N} \int_e d(e) |\{\nabla v\}|^2 dS. \tag{7.170}$$

Further, from (7.129), (7.163), we get for $v \in \mathcal{V}^h$

$$\begin{aligned} \sum_{K \in \mathcal{T}_h} h_K \|\nabla v\|^2_{L^2(\partial K)} &\le C_5 \sum_{K \in \mathcal{T}_h} h_K \left(|v|_{H^1(K)} |\nabla v|_{H^1(K)} + h_K^{-1} |v|^2_{H^1(K)} \right) \\ &\le C_5 (1 + C_6) |v|^2_{H^1(\mathcal{T}^h)} \end{aligned} \tag{7.171}$$

and thus

$$\omega \le \frac{(1+\theta)^2}{4\delta} \sum_{K \in \mathcal{T}_h} h_K \|\nabla v\|^2_{L^2(\partial K)} \le \frac{C_5(1+C_6)(1+\theta)^2}{4\delta} |v|^2_{H^1(\mathcal{T}^h)}. \tag{7.172}$$

Now let us choose

$$\delta = C_5(1+C_6)(1+\theta)^2/2. \tag{7.173}$$

Then it follows from (7.164), (7.167) and (7.169)–(7.173) that

$$A_h(v,v) \geq \varepsilon\|v\|^2 - \frac{\varepsilon}{2}|v|^2_{H^1(\mathcal{T}^h)} - \frac{\varepsilon\, C_5(1+C_6)(1+\theta)^2}{2c_w} J_h^\sigma(v,v) \tag{7.174}$$
$$\geq \varepsilon\|v\|^2 - \frac{1}{2}\varepsilon\left(|v|^2_{H^1(\mathcal{T}^h)} + J_h^\sigma(v,v)\right) = \frac{\varepsilon}{2}\|v\|^2,$$

which proves (7.165) with $c_c = 1/2$.

3. *Boundedness*: In view of the definition of the form A_h, we can write

$$A_h(u,v) = \vartheta_1 + \vartheta_2 + \vartheta_3, \tag{7.175}$$
$$\vartheta_1 = \varepsilon \sum_{K\in\mathcal{T}_h} \int_K (\nabla u,\ \nabla v)_n\, dx$$
$$\vartheta_2 = -\varepsilon \sum_{e\in\mathcal{T}^h\setminus\partial\Omega_N} \int_e \left(\{\nabla u\},\nu)_n[v] + \theta(\{\nabla v\},\nu)_n[u]\right) dS,$$
$$\vartheta_3 = \varepsilon J_h^\sigma(u,v),$$

where $\theta = -1$ for NIPG, $\theta = 1$ for SIPG, $\theta = 0$ for IIPG, and $\theta \in [-1,1]$ for GIPG methods.

4. Using the Cauchy inequality, we get the bound for ϑ_1 and ϑ_3

$$|\vartheta_1| \leq \varepsilon|u|_{H^1(\mathcal{T}^h)}|v|_{H^1(\mathcal{T}^h)} \leq \varepsilon\,\|u\|\,\|v\|, \text{ and} \tag{7.176}$$
$$|\vartheta_3| \leq \varepsilon\left(J_h^\sigma(u,u)\right)^{1/2}\left(J_h^\sigma(v,v)\right)^{1/2} \leq \varepsilon\,\|u\|\,\|v\|.$$

For estimating ϑ_2, we apply the Cauchy inequality, with $|\theta| \leq 1$ yielding

$$|\vartheta_2| \leq \varepsilon\left(\sum_{e\in\mathcal{T}^h\setminus\partial\Omega_N} \frac{d(e)}{c_w}\|\nabla u\|^2_{L^2(e)}\right)^{1/2}\left(\sum_{e\in\mathcal{T}^h\setminus\partial\Omega_N} \frac{c_w}{d(e)}\|[v]\|^2_{L^2(e)}\right)^{1/2} \tag{7.177}$$
$$+\varepsilon\left(\sum_{e\in\mathcal{T}^h\setminus\partial\Omega_N} \frac{d(e)}{c_w}\|\nabla v\|^2_{L^2(e)}\right)^{1/2}\left(\sum_{e\in\mathcal{T}^h\setminus\partial\Omega_N} \frac{c_w}{d(e)}\|[u]\|^2_{L^2(e)}\right)^{1/2}$$
$$\leq \frac{\varepsilon}{\sqrt{c_w}}\left(\sum_{K\in\mathcal{T}_h} h_K\|\nabla u\|^2_{L^2(\partial K)}\right)^{1/2} J_h^\sigma(v,v)^{1/2}$$
$$+\frac{\varepsilon}{\sqrt{c_w}}\left(\sum_{K\in\mathcal{T}_h} h_K\|\nabla v\|^2_{L^2(\partial K)}\right)^{1/2} J_h^\sigma(u,u)^{1/2}.$$

We combine (7.171), (7.177), and the discrete Cauchy–Schwarz inequality

$$s_1r_2 + s_2r_1 \leq \left(s_1^2+s_2^2\right)^{1/2}\left(r_1^2+r_2^2\right)^{1/2}, \quad s_1,\ s_2,\ r_1,\ r_2 \in \mathbb{R}$$

to obtain

$$|\vartheta_2| \leq \varepsilon \sqrt{\frac{C_5(1+C_6)}{c_w}} \times \left(|u|_{H^1(\mathcal{T}^h)} J_h^\sigma(v,v)^{1/2} + |v|_{H^1(\mathcal{T}^h)} J_h^\sigma(u,u)^{1/2}\right)$$

$$\leq \varepsilon \sqrt{\frac{C_5(1+C_6)}{c_w}} \times \left(|u|^2_{H^1(\mathcal{T}^h)} + J_h^\sigma(u,u)\right)^{1/2} \left(|v|^2_{H^1(\mathcal{T}^h)} + J_h^\sigma(v,v)\right)^{1/2}$$

$$\leq \varepsilon \sqrt{\frac{C_5(1+C_6)}{c_w}} \, \|u\| \, \|v\|. \tag{7.178}$$

Now from (7.175), (7.176), (7.178) we immediately get (7.166) with $c_b = 2 + \varepsilon \sqrt{C_5(1+C_6)/c_w}$. □

7.9.3 Coercivity and boundedness in $\mathcal{V}^h$ for the general linear and the semilinear case

As a consequence of (7.89), we omit the lower order compact perturbations terms b^h, c^h, ℓ^h and confine the discussion here to the principal part in (7.179). Let, for $u, w, v \in \mathcal{U}_e$, and Dirirchlet boundary conditions on Ω,

$$A_h(u; w, v) = \varepsilon J_h^\sigma(w, v) + \varepsilon \sum_{K \in \mathcal{T}_h} \int_K \sum_{i,j=1}^n a_{ij}(x,u)\, \partial^i w\, \partial^j v\, dx \tag{7.179}$$

$$- \varepsilon \sum_{e \in \mathcal{T}^h} \int_e \sum_{i,j=1}^n \left(\{a_{ij}(x,u)\, \partial^i w\, \nu^j\}\, [v] + \theta \{a_{ij}(x,u)\, \partial^i v\, \nu^j\}\, [w]\right) dS$$

with $\theta = -1, 1, 0, \in [-1, 1]$ for the NIPG, SIPG, IIPG, and GIPG methods and J_h^σ in (7.61). For general linear problems, we replace the $a_{ij}(x,u)$ by $a_{ij}(x)$ in (7.179).

Theorem 7.18. Bounded and coercive $a_h(u,v)$ for general linear and semilinear problems: *For the Λ_a, λ in (7.74), (7.80), and $\overline{\lambda} = \min(1, \lambda)$, we choose c_w arbitrarily positive for the NIPG and, for the SIPG, IIPG or GIPG methods, as*

$$c_w \geq \frac{C_5(1+C_6)(1+\theta)^2\, \Lambda_a^2 n^2}{\overline{\lambda}^2} \tag{7.180}$$

with C_5 and C_6 in (7.129) and (7.138), respectively. Then the corresponding bilinear forms $A_h(\cdot,\cdot)$, given by (7.179) or by replacing in (7.179) the $a_{ij}(x,u)$ by $a_{ij}(x)$, are bounded and coercive with respect to the norm, $\|v\|$, such that

$$A_h(u, v, v) \geq c_c \varepsilon \|v\|^2, \tag{7.181}$$

$$|A_h(u, w, v)| \leq c_b \varepsilon \|w\| \, \|v\| \quad \forall u, w, v \in \mathcal{V}^h. \tag{7.182}$$

Proof. We formulate the proof only for semilinear problems.

1. *The coercivity of the NIPG method* again is a consequence of (7.179) with $\theta = -1$ and (7.80) (e):

$$A_h(u, v, v) = \varepsilon J_h^\sigma(v, v) + \varepsilon \sum_{K \in \mathcal{T}_h} \int_K \sum_{i,j=1}^n a_{ij}(x, u)\, \partial^i v\, \partial^j v\, dx \qquad (7.183)$$

$$\geq \varepsilon \left(J_h^\sigma(v, v) + \lambda |v|^2_{H^1(\mathcal{T}^h)} \right),$$

which proves (7.181) with $c_c := \min(1, \lambda)$.

2. *Coercivity of SIPG, IIPG, and GIPG methods*: Let $\rho = (\rho_1, \ldots, \rho_n), \vartheta = (\vartheta_1, \ldots, \vartheta_n) \in \mathbb{R}^n$ and $a_{ij}(\cdot, \cdot)$ satisfy assumption (7.80) (c). Then using an appropriately reordered discrete Chebyshev inequality (1.43), and the Cauchy–Schwarz inequality (1.44), for $p = q = 2$, we obtain

$$\left| \sum_{i,j=1}^n f_i g_j \right| = \left| \sum_{i=1}^n f_i \sum_{j=1}^n g_j \right| \leq n \left(\sum_{j=1}^n f_j^2 \right)^{1/2} \left(\sum_{j=1}^n g_j^2 \right)^{1/2} = n|f|_n |g|_n = n|f||g|. \qquad (7.184)$$

This yields

$$\sum_{i,j=1}^n a_{ij}(\cdot, \cdot) \rho_i \vartheta_j \leq \left| \sum_{i,j=1}^n a_{ij}(\cdot, \cdot) \rho_i \vartheta_j \right| \leq \Lambda_a \sum_{i,j=1}^n |\rho_i|\, |\vartheta_j| \qquad (7.185)$$

$$\leq n\Lambda_a \left(\sum_{i=1}^n \rho_i^2 \right)^{1/2} \left(\sum_{i=1}^n \theta_i^2 \right)^{1/2} = n\Lambda_a\, |\rho|\, |\vartheta|.$$

From (7.179), (7.80) (e) and (7.185) we get, with $\theta \in (-1, 1]$,

$$A_h(u, v, v) = \varepsilon J_h^\sigma(v, v) + \varepsilon \sum_{K \in \mathcal{T}_h} \int_K \sum_{i,j=1}^n a_{ij}(x, u)\, \partial^i v\, \partial^j v\, dx$$

$$- (1 + \theta)\varepsilon \sum_{e \in \mathcal{T}^h} \int_e \sum_{i,j=1}^n \{a_{ij}(x, u)\, \partial^i v\, \nu^j\}\, [v]\, dS$$

$$\geq \varepsilon J_h^\sigma(v, v) + \varepsilon\, \lambda |v|^2_{H^1(\mathcal{T}^h)} - n(1 + \theta)\varepsilon\, \Lambda_a \sum_{e \in \mathcal{T}^h} \int_e |\nabla v||[v]|\, dS.$$

Applying a similar technique as in the proof of Theorem 7.16, relations (7.167) – (7.174) yield

$$A_h(v, v) \geq \overline{\lambda}\varepsilon \|v\|^2 - \varepsilon\omega - \varepsilon \frac{\delta}{c_w} J_h^\sigma(v, v),$$

where for $v \in \mathcal{V}^h$

$$\omega = \frac{(1+\theta)^2\Lambda_a^2 n^2}{4\delta} \sum_{e\in\mathcal{T}^h} \int_e d(e)|\{\nabla v\}|^2 dS \tag{7.186}$$

$$\le \frac{C_5(1+C_6)(1+\theta)^2\Lambda_a^2 n^2}{4\delta} |v|^2_{H^1(\mathcal{T}^h)}.$$

Now, putting

$$\delta = \frac{C_5(1+C_6)(1+\theta)^2\Lambda_a^2 n^2}{2\overline{\lambda}} \tag{7.187}$$

then (7.167) and (7.169)–(7.173) imply that

$$A_h(v,v) \ge \varepsilon\,\overline{\lambda}\,\|v\|^2 - \frac{\varepsilon\overline{\lambda}}{2}|v|^2_{H^1(\mathcal{T}^h)} - \frac{\varepsilon\, C_5(1+C_6)(1+\theta)^2\Lambda_a^2 n^2}{2\overline{\lambda}\, c_w} J_h^\sigma(v,v)$$

$$\ge \varepsilon\overline{\lambda}\|v\|^2 - \frac{\overline{\lambda}\varepsilon}{2}\left(|v|^2_{H^1(\mathcal{T}^h)} + J_h^\sigma(v,v)\right) = \frac{\varepsilon\overline{\lambda}}{2}\|v\|^2, \tag{7.188}$$

which proves (7.182) with $c_c = \varepsilon\overline{\lambda}/2$.

3. *Boundedness*: In view of the definition of the form A_h (7.179), we can write

$$A_h(u,w,v) = \vartheta_1 + \vartheta_2 + \vartheta_3, \tag{7.189}$$

$$\vartheta_1 = \varepsilon \sum_{K\in\mathcal{T}_h} \int_K \sum_{i,j=1}^n a_{ij}(x,u)\,\partial^i w\,\partial^j v\,dx$$

$$\vartheta_2 = -\varepsilon \sum_{e\in\mathcal{T}^h} \int_e \sum_{i,j=1}^n$$

$$\times \left(\{a_{ij}(x,u)\,\partial^i w\,\nu^j\}\,[v] + \theta\{a_{ij}(x,u)\,\partial^i v\,\nu^j\}\,[w]\right) dS,$$

$$\vartheta_3 = \varepsilon J_h^\sigma(w,v).$$

Using relation (7.185), we get

$$|\vartheta_1| \le \varepsilon\,\Lambda_a\, n|w|_{H^1(\mathcal{T}^h)}|v|_{H^1(\mathcal{T}^h)}. \tag{7.190}$$

In order to estimate ϑ_2, we again use (7.185) and obtain

$$|\vartheta_2| \le \varepsilon\,\Lambda_a\, n \sum_{e\in\mathcal{T}^h} \int_e \left(|\{\nabla w\}|\,[v] + \theta|\{\nabla v\}|\,[w]\right) dS. \tag{7.191}$$

With the same technique as in the proof of Theorem 7.16, (7.177)–(7.178), we find

$$|\vartheta_2| \le \varepsilon\,\Lambda_a\, n\,\sqrt{C_5(1+C_6)/c_w}\;\|w\|\,\|v\|. \tag{7.192}$$

Now from (7.189), (7.190), (7.192) and (7.176) we immediately get (7.182) with $c_b = \Lambda_a\, n(1+\sqrt{C_5(1+C_6)/c_w}) + 1$. □

7.9.4 $\mathcal{V}^h$-coercivity and boundedness for quasilinear problems

We have to prove the coercivity and boundedness of the linearized principal part for the IIPG method, so the summation starts with $k, l = 1$, instead of $k, l = 0$ above

$$A_h(\vec{u}; \vec{w}, \vec{v}) := J_h^\sigma(\vec{w}, \vec{v}) + \sum_{K\in\mathcal{T}_h} \int_K \sum_{k,l=1}^{n} \left(\frac{\partial A_k}{\partial \mathbf{u}^l}(x, \vec{u}, \nabla\vec{u}) \partial^l \vec{w}, \partial^k \vec{v} \right)_q \tag{7.193}$$

$$- \sum_{e\in\mathcal{T}^h} \int_e \sum_{k,l=1}^{n} \left(\left\{ \frac{\partial A_k}{\partial \mathbf{u}^l}(x, \vec{u}, \nabla\vec{u}) \partial^l \vec{w} \cdot \nu_k \right\}, [\vec{v}] \right)_q dS, \text{ for } \vec{u}, \vec{w}, \vec{v} \in \mathcal{U}_e.$$

Theorem 7.19. Bounded and coercive $a_h(u, v)$ for quasilinear systems: *With the constant λ from the assumption (7.94) (e), let $\overline{\lambda} = \min(1, \lambda)$. We choose the IIPG method ($\theta = 0$), and a constant c_w in (7.39) as*

$$c_w \geq \frac{C_5(1 + C_6)\,(nqL)^2}{\overline{\lambda}^2}, \tag{7.194}$$

where C_5, C_6 and L are given by (7.129), (7.138) and (7.94) (c), respectively. Then the corresponding bilinear form $A_h(\cdot,\cdot)$ given by (7.193) is bounded and coercive with respect to $\vec{w}, \vec{v}$ and the norm, $\|\vec{v}\|$, such that

$$A_h(\vec{u}, \vec{v}, \vec{v}) \geq c_c \|\vec{v}\|^2, \tag{7.195}$$

$$|A_h(\vec{u}, \vec{w}, \vec{v})| \leq c_b \|\vec{w}\|\;\|\vec{v}\| \quad \forall \vec{u}, \vec{v}, \vec{w} \in H^{3/2+\epsilon}(\mathcal{T}^h, \mathbb{R}^q). \tag{7.196}$$

Proof. *Coercivity*: Similarly as in (7.185) we have with L in (7.94) (c)

$$\sum_{k,l=1}^{n} \left(\frac{\partial A_k}{\partial \mathbf{u}^l}(\cdot,\cdot,\cdot) \partial^l \vec{v}, \nu_k [\vec{v}] \right)_q \leq \left| \sum_{k,l=1}^{n} \left(\frac{\partial A_k}{\partial \mathbf{u}^l}(\cdot,\cdot,\cdot) \partial^l \vec{v}, \nu_k [\vec{v}] \right)_q \right| \tag{7.197}$$

$$\leq nqL \sum_{k,l=1}^{n} (|\partial^l \vec{v},\; |\nu_k[\vec{v}]|)_q \leq nqL \sum_{k,l=1}^{n} (|\partial^l \vec{v}|,\; |[\vec{v}]|)_q \leq nqL |\nabla \vec{v}|_{q\times n}\; |[\vec{v}]|_q,$$

for the n components of the $\partial^l \vec{v}, \nu_k[\vec{v}] \in \mathbb{R}^q$, their absolute values, $|\partial^l \vec{v}|, |\nu_k[\vec{v}]| \in \mathbb{R}^q$ and their norms, $|\partial^l \vec{v}|_q, |\nabla \vec{v}|_{q\times n}, |\nu_k[\vec{v}]|_q \in \mathbb{R}$. Relations (7.94), (7.193) and (7.197), imply

$$A_h(\vec{u}, \vec{v}, \vec{v}) = J_h^\sigma(\vec{v}, \vec{v}) + \sum_{K\in\mathcal{T}_h} \int_K \sum_{k,l=1}^{n} \left(\frac{\partial A_k}{\partial \mathbf{u}^l}(x, \vec{u}, \nabla\vec{u}) \partial^l \vec{v}, \partial^k \vec{v} \right)_q$$

$$- \sum_{e\in\mathcal{T}^h} \int_e \sum_{k,l=1}^{n} \left(\left\{ \frac{\partial A_k}{\partial \mathbf{u}^l}(x, \vec{u}, \nabla\vec{u}) \partial^l \vec{v} \cdot \nu_k \right\}, [\vec{v}] \right)_q dS,$$

$$\geq J_h^\sigma(v, v) + \lambda |\vec{v}|^2_{H^1(\mathcal{T}^h)} - nqL \sum_{e\in\mathcal{T}^h} \int_e |\{\nabla \vec{v}\}|_{q\times n}\; |[\vec{v}]|_q \, dS.$$

With very minor changes the remaining proof of coercivity as well as boundness for this theorem follows that of Theorem 7.18. □

7.10 Consistency results for the c^h, b^h, ℓ^h

For the general discretization theory we have to prove the consistency of the different terms, the a^h, c^h, b^h, ℓ^h. We formulate the estimates with respect to global norms and step sizes. Local versions are possible as well, cf. Section 7.14.

Depending upon the different choices of c^h or b^h we obtain the standard or nonstandard version, for $\theta = -1$ or $\theta = 1$ or $\theta = 0$ or $\theta \in [-1, 1]$ in the NIPG or SIPG or IIPG or GIPG version of the DCGMs. So the following results again cover all these cases.

Since the consistency for the c^h, b^h, ℓ^h is much easier, at least for the case $\partial\Omega_D = \partial\Omega$, we delay the consistency for the $a^h, \overline{a}^h$ to the next section. The complementary case $\partial\Omega_N \neq 0$, requires for c^h, b^h a much more complicated proof, cf. [394]. The exact result is presented in Theorem 7.22. The forms c^h, b^h are independent of θ. ℓ^h depends on θ. So the results in Section 7.10.2 are valid for $\theta \in [-1, 1]$.

7.10.1 Consistency of the c_h and b_h

We start with the standard numerical flux c_h for $\partial\Omega_D = \partial\Omega$. The results for the more general c_h in (7.59) are valid for its special case b_h in (7.58) as well. Let (7.16) (a), (A1), (A2) in Subsection 7.7.1 and, for c_h, Assumptions (H) in Subsection 7.3.3 be satisfied. We shall be concerned with the Lipschitz-continuity of the form c_h. The reason for requiring $\partial\Omega_D = \partial\Omega$ is the last inequality in (7.202). J_h^σ is defined in (7.61) as a sum over interior and Dirichlet boundary edges but not the Neumann edges, i.e.

$$\sum_{K\in\mathcal{T}^h}\sum_{e\subset\partial K}\int_e \frac{[v]^2}{d(e)}\,dS \le 2\sum_{e\in\mathcal{T}^h}\int_e \frac{[v]^2}{d(e)}\,dS \not\le 2J_h^\sigma(v,v) = 2\sum_{e\in\mathcal{T}^h\setminus\partial\Omega_N}\int_e \frac{[v]^2}{d(e)}\,dS.$$

Proposition 7.20. *Under* **Assumption (H)** *and for $\partial\Omega_D = \partial\Omega$, there exist constants $C_8 > 0,\ C_9 > 0$ such that for c_h, and b_h,*

$$|c_h(u,v) - c_h(\overline{u},v)| \le C_8\|v\|\left(\|u-\overline{u}\|_{L^2(\Omega)} + \left(\sum_{K\in\mathcal{T}_h} h_K\|u-\overline{u}\|^2_{L^2(\partial K)}\right)^{1/2}\right),$$

$$\forall\ u,\overline{u}\in\mathcal{U}_e,\ v\in H^1(\mathcal{T}^h),\ h\in(0,h_0). \tag{7.198}$$

For $\forall u^h, \overline{u}^h, v^h \in S^h$, the last term in (7.198) is replaced by

$$\sum_{K\in\mathcal{T}^h}\left(h_K\|u-\overline{u}\|_{L^2(K)}\|u-\overline{u}\|_{H^1(K)}\right)^{1/2}. \tag{7.199}$$

These estimates remain valid for $|b_h(u,v) - b_h(\overline{u},v)|$, $|b_h(u^h,v^h) - b_h(\overline{u}^h,v^h)|$ as well, if f is Lipschitz-continuous.

Proof. By (7.59), for $u,\ \overline{u},\ v \in H^1(\mathcal{T}^h)$,

$$c_h(u,v) - c_h(\overline{u},v) = \underbrace{- \sum_{K\in\mathcal{T}^h} \int_K \big(\vec{f}(u) - \vec{f}(\overline{u}), \nabla v\big)\, dx}_{:=\sigma_1} \tag{7.200}$$

$$+ \underbrace{\sum_{e\in\mathcal{T}^h} \int_e \big(H(u_l,u_r,\nu) - H(\overline{u}_l,\overline{u}_r,\nu)\big)[v]\, dS}_{:=\sigma_2}.$$

From the Lipschitz-continuity of the function $\vec{f}$, we have

$$|\sigma_1| \le L_f \sum_{K\in\mathcal{T}^h} \int_K |u-\overline{u}|\ |\nabla v|\ dx \le L_f \|u-\overline{u}\|_{L^2(\Omega)} |v|_{H^1(\mathcal{T}^h)}. \tag{7.201}$$

The Lipschitz-continuity (7.51) of H, the Cauchy inequality, $d(e)$ in (7.28) imply

$$|\sigma_2| \le C_2 \sum_{e\in\mathcal{T}^h} \left\langle \int_e |[v]|\,\big(|u_l-\overline{u}_l| + |u_r-\overline{u}_r|\big)\, dS \right\rangle \tag{7.202}$$

$$\le C_2 \sum_{K\in\mathcal{T}^h} \left\langle \sum_{e\subset\partial K} \int_e |[v]|\,|u-\overline{u}|\, dS \right\rangle$$

$$\le 2C_2 \sum_{K\in\mathcal{T}^h} \left\langle \left(\sum_{e\subset\partial K} \int_e \frac{[v]^2}{d(e)}\, dS \right)^{1/2} \left(\sum_{e\subset\partial K} d(e) \int_e (u-\overline{u})^2\, dS \right)^{1/2} \right\rangle$$

$$\le 4C_2\, J_h^\sigma(v,v)^{1/2} \left(\sum_{K\in\mathcal{T}^h} h_K \|u-\overline{u}\|^2_{L^2(\partial K)} \right)^{1/2}.$$

Now from (7.201) and (7.202) we get (7.198) with $C_8 = \max(L_f, 4C_2)$.

We want to prove (7.199). In (7.198) we set $v := v^h \in S^h$. Using the multiplicative trace and the inverse inequalities (7.129) and (7.138),

$$|\sigma_2| \le C J_h^\sigma(v^h,v^h)^{1/2} \sum_{K\in\mathcal{T}^h} \big(h_K \|u-\overline{u}\|^2_{L^2(K)} + \|u-\overline{u}\|^2_{H^1(K)}\big)^{1/2}. \tag{7.203}$$

From (7.200), (7.201) and (7.203) we get the final result (7.198).

In this proof we only have to replace C_2 by C_f for obtaining the estimates for $|b_h(u,v^h) - b_h(\overline{u},v^h)|$. We do need the Lipschitz-continuity for f. □

Theorem 7.21. Consistency of the c_h, and b_h: *Under* **Assumption (H)** *and for* $\partial\Omega_D = \partial\Omega$, *there exists a constant* $C_{10} > 0$ *such that for* c_h, *and* b_h,

$$|c_h(u,v^h) - c_h(P^h u, v^h)| \le C_{10} h^{\min\{d,s\}} \|v^h\| |u|_{H^{\min\{d,s\}}(\Omega)},$$

$$u \in H^s(\Omega),\ v^h \in S^h,\ h \in (0,h_0), \tag{7.204}$$

or (7.140) with $d_K \geq d$ and where $P^h u$ is the S^h-interpolant of u from Theorem 7.7. This estimate remains valid for $|b_h(u, v^h) - b_h(P^h u, v^h)|$, as well.

Proof. We start from (7.198) with $u \in H^{p+1}(\Omega)$, $\overline{u} := P^h u$ and $v := v^h \in S^h$. Using the multiplicative trace inequality (7.129), the discrete Cauchy inequality and the approximating properties (7.139), we obtain

$$\begin{aligned} \sum_{K \in \mathcal{T}^h} h_K \|u - P^h u\|^2_{L^2(\partial K)} & \qquad (7.205) \\ \leq \sum_{K \in \mathcal{T}^h} C_5 h_K \|u - P^h u\|_{L^2(K)} \|u - P^h u\|_{H^1(K_i)} \\ \leq \sum_{K \in \mathcal{T}^h} C_5 C_7^2 h_K h_K^{\min\{d_K, s\}} h_K^{\min\{d_K, s\}-1} |u|^2_{H^{\min\{d_K,s\}}(K)} \\ \leq C_5 C_7^2 h^{2\min\{d,s\}} \left(\sum_{K \in \mathcal{T}^h} h_K^{2(d_K - d)} |u|^2_{H^{\min\{d_K,s\}}(K)} \right)^{1/2} \\ \leq 2 C_5 C_7^2 h^{2\min\{d,s\}} |u|^2_{H^{\min\{d,s\}}(\Omega)}. \end{aligned}$$

This estimate, (7.139) and (7.198) yield (7.204) with $C_{10} = C_7 C_8 (\sqrt{2C_5} + 1)$. □

In the following theorem, we recall results presented in [394], where the consistency of the forms c_h and b_h, both are independent of θ, is proved in the case when $\partial\Omega_N \neq 0$.

Theorem 7.22. Consistency of the c_h, and b_h: *Under* **Assumption (H)** *and for $\partial\Omega_N \neq 0$, there exists a constant $C'_{10} > 0$ such that for c_h, and b_h,*

$$\begin{aligned} &|c_h(u, v^h) - c_h(P^h u, v^h)| \leq C'_{10} h^{\min\{d,s\}-1/2} \|v^h\|_{L^2(\Omega)} (1 + \|v^h\|) |u|_{H^{\min\{d,s\}}(\Omega)}, \\ &u \in H^s(\Omega), \ v^h \in S^h, \ h \in (0, h_0), \end{aligned} \qquad (7.206)$$

or (7.140) with $d_K \geq d$ and where $P^h u$ is the S^h-interpolant of u from Theorem 7.7. This estimate remains valid for $|b_h(u, v^h) - b_h(P^h u, v^h)|$, as well.

Proof. See [394], Lemma 3. □

7.10.2 Consistency of the ℓ^h

For the model problem (7.13), (7.14), $\ell_h(v)$ in (7.60) is independent of u and has the form, with $\theta = 1, = -1, 0$, and $\theta \in [-1, 1]$, for the SIPG, NIPG, IIPG, and GIPG methods, respectively

$$\begin{aligned} \ell_h(v) &= \langle g, v \rangle + (g_N, v)_{L^2(\partial\Omega_N)} - \varepsilon \sum_{e \in \partial\Omega_D} \int_e u_D \left(\theta(\nabla v, \nu)_n + \sigma v \right) dS, \\ \ell^h(v^h) &= \ell_h(v^h). \end{aligned} \qquad (7.207)$$

For all the other problems (7.77), (7.78), (7.79), (7.100), we have $\partial\Omega_D = \partial\Omega$, and omit the $(g_N, v)_{L^2(\partial\Omega_N)}$ term and, for (7.78), we replace $g(\cdot)$ by $g(\cdot, u)$:

$$\ell_h(u,v) = \langle g(\cdot,u), v(\cdot)\rangle - \varepsilon \sum_{e\in\partial\Omega_D} \int_e \theta(B_a(u)\nabla v, \nu)_n u + \sigma u_D\, v\, dS. \quad (7.208)$$

The $\ell_h(\vec{u},\vec{v})$ in (7.100) is

$$\ell^h(\vec{v}^h) = \sum_{e\in\partial\Omega} \int_e \sigma(\vec{u}_D, \vec{v}^h)_q\, dS, \quad \ell^h(\vec{v}^h) = \ell_h(\vec{v}^h). \quad (7.209)$$

These $\ell^h(v^h) = \ell_h(v^h)$, $\ell^h(\vec{v}^h) = \ell_h(\vec{v}^h)$ show by (7.121), the following:

Theorem 7.23. Consistency of the ℓ_h: *For the problems (7.13), (7.14), (7.77), with $\ell^h(v) = \ell_h(v)$, in (7.207), and the quasilinear problem in (7.90), (7.100) with $\ell^h(\vec{v}) = \ell_h(\vec{v})$ in (7.209), the consistency errors vanish for the SIPG, NIPG, IIPG, and GIPG methods.*

However for the more general semilinear problem (7.78), (7.79) we have to consider the $\ell_h(u,v)$ in (7.208), with $(B_a(u)\nabla v, \nu)_n$ as in (7.81) and $\langle g(x,u), v\rangle = \langle g(x,u), v\rangle_{H^{-1}\times H^1(\mathcal{T}^h)}$:

$$\ell_h(u,v) = \varepsilon \sum_{e\in\partial\Omega} \int_e \sigma u_D\, v - \theta(B_a(u)\nabla v, \nu)_n u dS + \langle g(x,u), v\rangle. \quad (7.210)$$

Theorem 7.24. Consistency for the semilinear ℓ_h: *For the $\ell^h(u, v^h)$ in (7.88), (7.208), with properties (7.80), a constant, $C_{13} = (n+1)\max\{L, \Lambda_a\}C_7(1 + \sqrt{2C_5(C_6+1)}) > 0$, exists such that $\forall h \in (0, h_0), h_0 < 1$,*

$$|\ell^h(P^h u, v^h) - \ell_h(u, v^h)| \le C_{13} \left(\sum_{K\in\mathcal{T}^h} \left(h_k^{\min\{d,s\}-1} |u|_{H^{\min\{d,s\}}(K)} \right)^2 \right)^{1/2} \quad (7.211)$$

$$(1 + \|u\|_{L^\infty(\partial\Omega)})\|v^h\|_{H^1(\mathcal{T}^h)} \ \forall u \in H^s(\mathcal{T}^h), v^h, P^h u \in S^h.$$

Proof. We combine the Lipschitz-continuity of the $g_j(x,u), j = 0,\dots,n$, and $a_{ij}(x,u)$, $i,j = 1,\dots,n$, with

$$\mp a_{ij}(x, P^h u)u = -a_{ij}(x,P^h u)u + a_{ij}(x,P^h u)u = 0 = |\ell^h(P^h u, v^h) - \ell_h(u, v^h)|$$

$$\le \underbrace{\left| \sum_{K\in\mathcal{T}^h} \int_K \sum_{j=0}^n (g_j(x, P^h u) - g_j(x,u))\, \partial^j v^h\, dx \right|}_{:=\sigma_9}$$

$$+\varepsilon \underbrace{\left| \sum_{e\in\partial\Omega} \int_e \sum_{i,j=1}^n \Big(a_{ij}(x,P^h u)P^h u \mp a_{ij}(x,P^h u)u - a_{ij}(x,u)u \Big) \partial^i v^h\, \nu_j dS \right|}_{:=\sigma_{10}}.$$

Obviously we obtain with the Lipschitz constant, L, for the g_j

$$\begin{aligned}|\sigma_9| &\leq (n+1)L\|P^h u - u\|_{L^2(\mathcal{T}^h)}\|v^h\|_{H^1(\mathcal{T}^h)}\\ &\leq (n+1)LC_7\left(\sum_{K\in\mathcal{T}^h}\left(h_k^{\min\{d,s\}}|u|_{H^{\min\{d,s\}}(K)}\right)^2\right)^{1/2}\|v^h\|_{H^1(\mathcal{T}^h)}.\end{aligned}$$

We estimate by combining (7.129) and (7.138)

$$\begin{aligned}h_K\|\nabla v^h\|^2_{L^2(\partial K)} &\leq C_5 h_K\|\nabla v^h\|_{L^2(K)}|\nabla v^h|_{H^1(K)} + C_5\|\nabla v^h\|^2_{L^2(K)}\\ &\leq 2C_5(C_6+1)\|\nabla v^h\|^2_{L^2(K)}\end{aligned}\tag{7.212}$$

hence,

$$\frac{1}{C_6+1}\sum_{K\in\mathcal{T}^h} h_K\|\nabla v^h\|^2_{L^2(\partial K)} \leq 2C_5\sum_{K\in\mathcal{T}^h}\|\nabla v^h\|^2_{L^2(K)} = 2C_5|v^h|^2_{H^1(\mathcal{T}^h)}.\tag{7.213}$$

Then the Lipschitz constant, L, and the bound Λ_a for the a_{ij}, yield

$$\begin{aligned}&|\sigma_{10}|/\big(n(\Lambda_a + L\|u\|_{L^\infty(\partial\Omega)})\big)\\ &\leq \sum_{e\in\partial\Omega}\int_e |\nabla v^h|\,|\,[u-P^h u]\,|\,dS\\ &\leq \left(\sum_{e\in\mathcal{T}^h}\int_e h_K|\nabla v^h|^2\,dS\right)^{1/2}\left(\sum_{e\in\mathcal{T}^h}\int_e h_K^{-1}([u-P^h u])^2\,dS\right)^{1/2}\\ &\leq \left(\left(\sum_{K\in\mathcal{T}^h} h_K\|\nabla v^h\|^2_{L^2(\partial K)}\right)^{1/2}\left(\sum_{K\in\mathcal{T}^h} h_K^{-1}\|(u-P^h u)\|^2_{L^2(\partial K)}\right)^{1/2}\right)\\ &\leq \sqrt{2C_5(C_6+1)}C_7|v^h|_{H^1(\mathcal{T}^h)}\left(\sum_{K\in\mathcal{T}^h}\left(h_k^{\min\{d,s\}-1}|u|_{H^{\min\{d,s\}}(K)}\right)^2\right)^{1/2}.\end{aligned}\tag{7.214}$$

Adding the estimates for σ_9 and σ_{10} proves the claimed (7.211). □

7.11 Consistency properties of the a_h

7.11.1 Consistency of the a_h for the Laplacian

We prove, for $\partial\Omega_D \cap \partial\Omega_N = \partial\Omega$, the consistency for a^h in (7.215). The necessary changes for semilinear and quasilinear a^h in (7.87) and (7.99) are discussed in

Proposition 7.29 and Proposition 7.32. The a^h for the model problem have the form

$$a_h(u,v) := \varepsilon \sum_{K\in\mathcal{T}^h} \int_K (\nabla u,\ \nabla v)_n \, dx \qquad (7.215)$$
$$-\varepsilon \sum_{e\in\mathcal{T}^h\setminus\partial\Omega_N} \int_e [v](\{\nabla u\},\nu)_n + \theta[u](\{\nabla v\},\nu)_n \, dS,$$

where $\theta = -1$ for NIPG, $\theta = 1$ for SIPG, $\theta = 0$ for IIPG, and $\theta \in [-1,1]$ for the GIPG methods.

Proposition 7.25. *These $a_h(u,v)$ are bounded with respect to u, v, such that u, v, $\overline{u} \in \mathcal{U}_e$*

$$|a_h(u,v) - a_h(\overline{u},v)| \le \varepsilon \ |\overline{u}-u|_{H^1(\mathcal{T}^h)}|v|_{H^1(\mathcal{T}^h)} \qquad (7.216)$$
$$+ \varepsilon \ 2n \left(\left(\frac{1}{c_w}\right)^{1/2} (J_h^\sigma(v,v))^{1/2} \left(\sum_{K\in\mathcal{T}^h} h_K |u-\overline{u}|^2_{H^1(\partial K)}\right)^{1/2}\right.$$
$$\left. + \left(\sum_{K\in\mathcal{T}^h} h_K \|\nabla v\|^2_{L^2(\partial K)}\right)^{1/2} \left(\sum_{K\in\mathcal{T}^h} h_K^{-1} \|(u-\overline{u})\|^2_{L^2(\partial K)}\right)^{1/2}\right).$$

The last line vanishes for the IIPG method.

Proof. We consider $u,\ \overline{u},\ v \in \mathcal{U}_e$ as arguments of $a_h(\cdot,\cdot)$ and estimate the different terms σ_j in (7.217) ff.:

$$|a_h(\overline{u},v) - a_h(u,v)| \le \varepsilon \underbrace{\left| \sum_{K\in\mathcal{T}^h} \int_K (\nabla(\overline{u}-u), \nabla v)_n \, dx \right|}_{:=\sigma_1} \qquad (7.217)$$
$$+\varepsilon \underbrace{\left| \sum_{e\in\mathcal{T}^h\setminus\partial\Omega_N} \int_e ((\{\nabla(\overline{u}-u)\},\nu)_n[v]) \, dS \right|}_{:=\sigma_2}$$
$$+\varepsilon \underbrace{\left| \theta \sum_{e\in\mathcal{T}^h\setminus\partial\Omega_N} \int_e ((\{\nabla v\},\nu)_n[(\overline{u}-u)]) \, dS \right|}_{:=\sigma_3}.$$

σ_1 is estimated with both Cauchy–Schwarz inequalities (1.44), for $p=q=2$, (1.45),

$$|\sigma_1| \le \sum_{K\in\mathcal{T}^h} |\overline{u}-u|_{H^1(K)}|v|_{H^1(K)} \le |\overline{u}-u|_{H^1(\mathcal{T}^h)}|v|_{H^1(\mathcal{T}^h)}. \qquad (7.218)$$

For σ_2 the same inequalities (1.44), (1.45), and $d(e)$ in (7.28) yield

$$|\sigma_2| \leq n \sum_{e\in\mathcal{T}^h\setminus\partial\Omega_N} \int_e |[v]|\,|\nabla u - \nabla\overline{u}|\,dS \tag{7.219}$$

$$\leq n\left(\sum_{e\in\mathcal{T}^h\setminus\partial\Omega_N} \int_e \frac{[v]^2}{d(e)/c_w}\,dS\right)^{1/2} \left(\sum_{e\in\mathcal{T}^h\setminus\partial\Omega_N} \frac{d(e)}{c_w}\int_e |\nabla u - \nabla\overline{u}|^2\,dS\right)^{1/2}$$

$$\leq 2n\left(\frac{1}{c_w}\right)^{1/2} (J_h^\sigma(v,v))^{1/2}\left(\sum_{K\in\mathcal{T}^h} h_K|u-\overline{u}|^2_{H^1(\partial K)}\right)^{1/2}.$$

σ_3 vanishes for the IIPG method. Otherwise, again (1.44), (1.45) yield

$$|\sigma_3| \leq n \sum_{e\in\mathcal{T}^h\setminus\partial\Omega_N} \int_e |\nabla v|\,|\,[u-\overline{u}]\,|\,dS \tag{7.220}$$

$$\leq n\left(\sum_{e\in\mathcal{T}^h\setminus\partial\Omega_N} \int_e h_K|\nabla v|^2\,dS\right)^{1/2} \left(\sum_{e\in\mathcal{T}^h\setminus\partial\Omega_N} \int_e h_K^{-1}([u-\overline{u}])^2\,dS\right)^{1/2}$$

$$\leq 2n\left(\sum_{K\in\mathcal{T}^h} h_K\|\nabla v\|^2_{L^2(\partial K)}\right)^{1/2} \left(\sum_{K\in\mathcal{T}^h} h_K^{-1}\|(u-\overline{u})\|^2_{L^2(\partial K)}\right)^{1/2}.$$

Adding the estimates for σ_1 to σ_3 in (7.218)–(7.220) proves (7.216). □

Proposition 7.25 implies with $\overline{u} = P^h u$, and for $\forall v^h \in S^h = \mathcal{V}^h$ as in Theorem 7.21

Theorem 7.26. Consistency of the Laplacian a_h: *The $a^h(P^h u, v^h)$ in (7.87) are consistent with $a(u,v)$ in (7.81) $\forall\ u \in \mathcal{U}_e$, $\forall v^h \in S^h = \mathcal{V}^h$, so a constant, $C_{17} := \max\{C_7, 2n\sqrt{2C_5C_7}((1/c_w)^{1/2} + 2C_5\sqrt{C_7(C_6+1)}\} > 0$, exists, such that for the S^h-interpolant $P^h u$ of u in Theorem 7.7,*

$$|a_h(u,v^h) - a^h(P^h u, v^h)| \leq \varepsilon\left(\sum_{K\in\mathcal{T}^h}\left(h_k^{\min\{d,s\}-1}|u|_{H^{\min\{d,s\}}(K)}\right)^2\right)^{1/2} \tag{7.221}$$

$$\left(C_7|v^h|_{H^1(\mathcal{T}^h)} + 2n\left(c_w^{-1/2}\sqrt{2C_5C_7/c_w}\,(J_h^\sigma(v^h,v^h))^{1/2}\right.\right.$$

$$\left.\left. +\ 2C_5\sqrt{C_7(C_6+1)}|v^h|_{H^1(\mathcal{T}^h)}\right)\right)$$

$$\leq c_e\ C_{17}h^{\min\{d,s\}-1}|u|_{H^{\min\{d,s\}}(\mathcal{T}^h)}\|v^h\|.$$

The $2C_5\sqrt{C_7(C_6+1)}|v^h|_{H^1(\mathcal{T}^h)}$ can be deleted for the IIPG method.

Proof. For (7.221) and by (7.216) we get by (7.129)–(7.139). The following first and third lines are straightforward

$$|P^h u - u|_{H^1(\mathcal{T}^h)} \leq C_7 \left(\sum_{K\in\mathcal{T}^h} \left(h_k^{\min\{d,s\}-1} |u|_{H^{\min\{d,s\}}(K)} \right)^2 \right)^{1/2}, \text{ by (7.139)}, \tag{7.222}$$

$$\left(\sum_{K\in\mathcal{T}^h} h_K |u - P^h u|^2_{H^1(\partial K)} \right)^{1/2} \leq \sqrt{2C_5C_7} \left(\sum_{K\in\mathcal{T}^h} \left(h_k^{\min\{d,s\}-1} |u|_{H^{\min\{d,s\}}(K)} \right)^2 \right)^{1/2},$$

$$\left(\sum_{K\in\mathcal{T}^h} h_K \|\nabla v^h\|^2_{L^2(\partial K)} \right)^{1/2} \leq \sqrt{2C_5(C_6+1)}\ |v^h|_{H^1(\mathcal{T}^h)}, \text{ by (7.129), (7.138)},$$

$$\left(\sum_{K\in\mathcal{T}^h} h_K^{-1} \|u - P^h u\|^2_{L^2(\partial K)} \right)^{1/2} \leq \sqrt{2C_5C_7} \left(\sum_{K\in\mathcal{T}^h} \left(h_k^{\min\{d,s\}-1} |u|_{H^{\min\{d,s\}}(K)} \right)^2 \right)^{1/2}.$$

For the term in the second line of (7.222) we get, by (7.129), (7.139),

$$\begin{aligned} &h_K |u - P^h u|^2_{H^1(\partial K)} \\ &\quad \leq C_5 h_K |u - P^h u|_{H^1(K)} |u - P^h u|_{H^2(K)} + C_5 |u - P^h u|^2_{H^1(K)} \\ &\quad \leq C_5C_7 (|u|_{H^{\min\{d,s\}}(K)})^2 \left(h_K h_K^{\min\{d,s\}-1} h_K^{\min\{d,s\}-2} + h_K^{2(\min\{d,s\}-1)} \right) \\ &\quad \leq 2C_5C_7 \left(|u|_{H^{\min\{d,s\}}(K)} h_K^{(\min\{d,s\}-1)} \right)^2 \end{aligned}$$

implying the second line in (7.222). The third line coincides with (7.213).

For the last line in (7.222) we need, with (7.129), (7.138), and (7.139),

$$\begin{aligned} &h_K^{-1} \|u - P^h u\|^2_{L^2(\partial K)} \\ &\quad \leq C_5 h_K^{-1} \|u - P^h u\|_{L^2(K)} |u - P^h u|_{H^1(K)} + C_5 h_K^{-2} \|u - P^h u\|^2_{L^2(K)} \\ &\quad \leq C_5C_7 (|u|_{H^{\min\{d,s\}}(K)})^2 \left(h_K^{-1} h_K^{\min\{d,s\}} h_K^{\min\{d,s\}-1} + h_K^{-2} h_K^{2(\min\{d,s\})} \right) \\ &\quad \leq 2C_5C_7 \left(|u|_{H^{\min\{d,s\}}(K)} h_K^{(\min\{d,s\}-1)} \right)^2. \end{aligned} \tag{7.223}$$

Relation (7.222) is applied in this order to the different error terms $u - P^h u$ in (7.216):

$$|a_h(u, v^h) - a_h(P^h u, v^h)| \leq \varepsilon \left(\sum_{K\in\mathcal{T}^h} \left(h_k^{\min\{d,s\}-1} |u|_{H^{\min\{d,s\}}(K)} \right)^2 \right)^{1/2}$$

$$\left(C_7 |v^h|_{H^1(\mathcal{T}^h)} + 2n (c_w^{-1/(2} \sqrt{2C_5C_7}\ (J_h^\sigma(v^h, v^h))^{1/2} + 2C_5 \sqrt{C_7(C_6+1)} |v^h|_{H^1(\mathcal{T}^h)}) \right).$$

This shows, with (7.154) and Proposition 7.9, the estimate (7.221). Finally, Theorems 7.12 and 7.13 imply the last estimate in (7.221). □

7.11.2 Consistency of the a_h for general linear problems

In Section 7.4 we have introduced for general linear elliptic problems the bilinear form, $a_h(u,v)$, cf (7.74),(7.76), for $\theta=-1, \theta=1, \theta=0, \theta\in[-1,1]$,

$$a_h(u,v) := \sum_{K\in\mathcal{T}_h}\left(\int_K \sum_{i,j=0}^{n} a_{ij}\partial^i\, u\partial^j\, v\right)dx - \sum_{e\in\mathcal{T}^h}\int_e ((\{B_a\nabla^0 u\},\nu)_n^0[v] \tag{7.224}$$
$$+\theta\,(\{B_a\nabla^0 v\},\nu)_n[u])\;dS \quad \text{for} \quad u,v\in\mathcal{U}_e$$

with

$$(B_a\nabla^0 u,\nabla^0 v)_n^0 := \sum_{i,j=0}^{n} a_{ij}\partial^i\, u\partial^j\, v,\; (B_a\nabla^0 u,\nu)_n = \sum_{i=0,\dots,n}^{j=1,\dots,n} \nu_j a_{ij}\partial^i\, u.$$

We prove its consistency.

Proposition 7.27. *These bilinear $a_h(u,v)$ are bounded with respect to u,v, such that $u,v,\overline{u}\in\mathcal{U}_e$*

$$|a_h(u,v)-a_h(\overline{u},v)| \le C_0\;\;\|\overline{u}-u\|_{H^1(\mathcal{T}^h)}\|v\|_{H^1(\mathcal{T}^h)} \tag{7.225}$$
$$+\,2nC_0\left(\left(\frac{1}{c_w}\right)^{1/2}(J_h^\sigma(v,v))^{1/2}\left(\sum_{K\in\mathcal{T}^h} h_K|u-\overline{u}|^2_{H^1(\partial K)}+h_K\|u-\overline{u}\|^2_{L^2(\partial K)}\right)^{1/2}\right.$$
$$\left.+\left(\sum_{K\in\mathcal{T}^h} h_K\left(\|\nabla v\|^2_{L^2(\partial K)}+\|v\|^2_{L^2(\partial K)}\right)\right)^{1/2}\left(\sum_{K\in\mathcal{T}^h} h_K^{-1}\|(u-\overline{u})\|^2_{L^2(\partial K)}\right)^{1/2}\right).$$

The last line vanishes for the IIPG method.

Proof. The boundedness follows from (7.225) for $\overline{u}=0$. We consider $u,\ \overline{u},\ v\in\mathcal{U}_e$ as arguments of $a_h(\cdot,\cdot)$ and use, cf. (7.74), the bound $|a_{i,j}(x)|\le C_0$. We estimate the different terms σ_j in (7.226) ff.:

$$|a_h(\overline{u},v)-a_h(u,v)| \le \varepsilon\;\underbrace{\left|\sum_{K\in\mathcal{T}^h}\int_K (B_a\nabla^0(\overline{u}-u),\nabla^0 v)_n^0\,dx\right|}_{:=\sigma_1} \tag{7.226}$$
$$+\varepsilon\;\underbrace{\left|\sum_{e\in\mathcal{T}^h}\int_e ((\{B_a\nabla^0(\overline{u}-u)\},\nu)_n[v])\,dS\right|}_{:=\sigma_2}$$

$$+\varepsilon \underbrace{\left|\theta \sum_{e\in\mathcal{T}^h}\int_e ((\{B_a\nabla^0 v\},\nu)_n[(\overline{u}-u)])\,dS\right|}_{:=\sigma_3}.$$

σ_1 is estimated with both Cauchy–Schwarz inequalities (1.44), for $p=q=2$, (1.45),

$$|\sigma_1| \le C_0 \sum_{K\in\mathcal{T}^h} \|\overline{u}-u\|_{H^1(K)}\|v\|_{H^1(K)} \le C_0\|\overline{u}-u\|_{H^1(\mathcal{T}^h)}\|v\|_{H^1(\mathcal{T}^h)}. \quad (7.227)$$

For σ_2 the same inequalities (1.44), (1.45), and $d(e)$ in (7.39) yield

$$\begin{aligned}
|\sigma_2| &\le n \sum_{e\in\mathcal{T}^h}\int_e |[v]|\,|\,B_a\nabla^0 u - B_a\nabla^0\overline{u}|\,dS \qquad (7.228)\\
&\le n\left(\sum_{e\in\mathcal{T}^h}\int_e \frac{[v]^2}{d(e)/c_w}\,dS\right)^{1/2}\left(\sum_{e\in\mathcal{T}^h}\frac{d(e)}{c_w}\int_e |\,B_a\nabla^0 u - B_a\nabla^0\overline{u}|^2\,dS\right)^{1/2}\\
&\le 2nC_0\left(\frac{1}{c_w}\right)^{1/2}\\
&\quad\times\ (J_h^\sigma(v,v))^{1/2}\left(\sum_{K\in\mathcal{T}^h}\left(h_K|u-\overline{u}|^2_{H^1(\partial K)} + h_K\|u-\overline{u}\|^2_{L^2(\partial K)}\right)^{1/2}\right).
\end{aligned}$$

σ_3 vanishes for the IIPG method. Otherwise, again (1.44), (1.45) yield

$$\begin{aligned}
|\sigma_3| &\le n\sum_{e\in\mathcal{T}^h}\int_e |B_a\nabla^0 v|\,|\,[u-\overline{u}]\,|\,dS \qquad (7.229)\\
&\le nC_0\left(\sum_{e\in\mathcal{T}^h}\int_e h_K(|\nabla v|^2+|v|^2)\,dS\right)^{1/2}\left(\sum_{e\in\mathcal{T}^h}\int_e h_K^{-1}([u-\overline{u}])^2\,dS\right)^{1/2}\\
&\le 2nC_0\left(\sum_{K\in\mathcal{T}^h} h_K\left(\|\nabla v\|^2_{L^2(\partial K)}+\|v\|^2_{L^2(\partial K)}\right)\right)^{1/2}\\
&\quad\times\left(\sum_{K\in\mathcal{T}^h} h_K^{-1}\|(u-\overline{u})\|^2_{L^2(\partial K)}\right)^{1/2}
\end{aligned}$$

Adding the estimates for σ_1 to σ_3 in (7.227)–(7.229) proves (7.225). □

Proposition 7.27 implies with $\overline{u}=P^h u$, as in Theorem 7.21:

Theorem 7.28. Consistency for the general linear a_h: *The $a^h(P^h u, v^h)$ in (7.76) are consistent with $a(u,v)$ in (7.74) $\forall\, u \in \mathcal{U}_e$, $\forall v^h \in S^h = \mathcal{V}^h$. So a constant, $C_{18} := 2\max\left\{C_7, 2n\sqrt{2C_5C_7}(2(1/c_w)^{1/2} + C_5\sqrt{C_7(C_6+1)}\right\} > 0$, exists, such that for the S^h-interpolant $P^h u$ of u in Theorem 7.7,*

$$|a_h(u,v^h) - a^h(P^h u, v^h)| \le C_0 \left(\sum_{K\in\mathcal{T}^h}\left(h_k^{\min\{d,s\}-1}|u|_{H^{\min\{d,s\}}(K)}\right)^2\right)^{1/2} \tag{7.230}$$

$$2\Big(C_7\|v^h\|_{H^1(\mathcal{T}^h)} + 2n\left(c_w^{-1/2}\sqrt{2C_5C_7/c_w}\left(J_h^\sigma(v^h,v^h)\right)\right)^{1/2}$$

$$+ C_5\sqrt{C_7(C_6+1)}\|v^h\|_{H^1(\mathcal{T}^h)}\Big)$$

$$\le \varepsilon\; C_{18} h^{\min\{d,s\}-1}|u|_{H^{\min\{d,s\}}(\mathcal{T}^h)}\|v^h\|.$$

The $C_5\sqrt{C_7(C_6+1)}\|v^h\|_{H^1(\mathcal{T}^h)}$ term can be deleted for the IIPG method.

Proof. The proof is nearly identical with that of Theorem 7.26. For (7.230) and by (7.225) we again use (7.231), however replacing the first two lines by

$$\|P^h u - u\|_{H^1(\mathcal{T}^h)} \le 2C_7\left(\sum_{K\in\mathcal{T}^h}\left(h_k^{\min\{d,s\}-1}|u|_{H^{\min\{d,s\}}(K)}\right)^2\right)^{1/2}, \text{ by (7.139)},$$

$$\times\left(\sum_{K\in\mathcal{T}^h} h_K\|u - P^h u\|^2_{H^1(\partial K)}\right)^{1/2} \tag{7.231}$$

$$\le 2\sqrt{2C_5C_7}\left(\sum_{K\in\mathcal{T}^h}\left(h_k^{\min\{d,s\}-1}|u|_{H^{\min\{d,s\}}(K)}\right)^2\right)^{1/2},$$

(7.222), (7.231) are applied to the different error terms $u - P^h u$ in (7.225):

$$|a_h(u,v^h) - a_h(P^h u, v^h)| \le C_0\left(\sum_{K\in\mathcal{T}^h}\left(h_k^{\min\{d,s\}-1}|u|_{H^{\min\{d,s\}}(K)}\right)^2\right)^{1/2}$$

$$\Big(2C_7\|v^h\|_{H^1(\mathcal{T}^h)} + 4n(c_w^{-1/(2}\sqrt{2C_5C_7}\left(J_h^\sigma(v^h,v^h)\right))^{1/2}$$

$$+ 2C_5\sqrt{C_7(C_6+1)}\|v^h\|_{H^1(\mathcal{T}^h)}\Big).$$

This shows, with (7.225) and Proposition 7.9, the estimate (7.230). Finally, Theorems 7.12 and 7.13 imply the last estimate in (7.230). □

7.11.3 Consistency of the semilinear a_h

We turn to the consistency for the semilinear a^h in (7.87) where $\partial\Omega_N = \emptyset$. Here it is worthwhile introducing new $\overline{a}_h(\cdot;\cdot,\cdot)$, based upon $B_a(u)\nabla w$ instead of the $B_a(u)\nabla u$ in (7.87).

For $u, v, w \in \mathcal{U}_e \supset S^h$, define (7.232)

$$B_a(u)\nabla w := \left(\sum_{i=1}^{n} (a_{ij}(x,u)\partial^i w\right)_{j=1}^{n}, \text{ and } (B_a(u)\nabla w, \nabla v)_n, (B_a(u)\nabla w, \nu)_n,$$

$$\overline{a}_h(u; w, v) := \varepsilon \sum_{K\in\mathcal{T}^h} \int_K (B_a(u)\nabla w,\ \nabla v)_n\, dx$$

$$-\varepsilon \sum_{e\in\mathcal{T}^h} \int_e ((\{B_a(u)\nabla w\},\nu)_n\, [v] + \theta(\{B_a(u)\nabla v\},\nu)_n\, [w])\, dS.$$

Proposition 7.29. *The $a_h(u,v) = \overline{a}_h(u;u,v)$ in (7.232) with properties (7.80) are modified Lipschitz-continuous with respect to u and linear and bounded with respect to v, such that $\overline{a}_h(u;u,v)$ satisfies $\forall\ u,v \in \mathcal{U}_e\ \forall \overline{u} \in W^{2,\infty}(\mathcal{T}^h)$*

$$|a_h(u,v) - a_h(\overline{u},v)|$$

$$= |\overline{a}_h(u;u,v) - \overline{a}_h(\overline{u};\overline{u},v)| \le \varepsilon\ \ n\Lambda_a|\overline{u}-u|_{H^1(\mathcal{T}^h)}|v|_{H^1(\mathcal{T}^h)}$$

$$+\varepsilon\ \ 2n\Lambda_a/(c_w)^{1/2}\ (J_h^\sigma(v,v))^{1/2}\left(\sum_{K\in\mathcal{T}^h} h_K|u-\overline{u}|^2_{H^1(\partial K)}\right)^{1/2} \quad (7.233)$$

$$+\varepsilon\ \ 2n\Lambda_a\left(\sum_{K\in\mathcal{T}^h} h_K\|\nabla v\|^2_{L^2(\partial K)}\right)^{1/2}\left(\sum_{K\in\mathcal{T}^h} h_K^{-1}\|(u-\overline{u})\|^2_{L^2(\partial K)}\right)^{1/2}$$

$$+\varepsilon\ \ nL\|\overline{u}-u\|_{L^2(\mathcal{T}^h)}\|\nabla\overline{u}\|_{L^\infty(\Omega)}\|v^h\|_{H^1(\mathcal{T}^h)}$$

$$+\varepsilon\ \ 2nL/(c_w)^{1/2}\|\overline{u}\|_{W^{1,\infty}(\mathcal{T}^h)}\ (J_h^\sigma(v,v))^{1/2}\left(\sum_{K\in\mathcal{T}^h} h_K\|u-\overline{u}\|^2_{L^2(\partial K)}\right)^{1/2}$$

$$+\varepsilon\ \ 2nL\|\overline{u}\|_{L^\infty(\Omega)}\left(\sum_{K\in\mathcal{T}^h} h_K\|\nabla v\|^2_{L^2(\partial K)}\right)^{1/2}\left(\sum_{K\in\mathcal{T}^h} h_K^{-1}\|(u-\overline{u})\|^2_{L^2(\partial K)}\right)^{1/2}.$$

For the special case $a(u,v)$ in (7.19) the last three lines drop out and $\overline{u} \in W^{2,\infty}(\mathcal{T}^h)$ is replaced by $\overline{u} \in \mathcal{U}_e$.

Remark 7.30. *The* modified Lipschitz-continuity *with respect to u indicates the dependency upon $\|\overline{u}\|_{W^{1,\infty}(\mathcal{T}^h)}$. The condition $\overline{u} \in W^{2,\infty}(\mathcal{T}^h)$ is imposed for allowing*

$\|\nabla\overline{u}\|_{L^\infty(e)} \le \|\nabla\overline{u}\|_{L^\infty(K)}$, *cf. (7.202), (7.242). We will need that for* $\overline{u} = P^h u$. *By the* $W^{k,\infty}$ *form of (7.139),* $u \in W^{2,\infty}(\mathcal{T}^h)$ *implies* $P^h u \in W^{2,\infty}(\mathcal{T}^h)$. *Moreover,*

$$\|P^h u\|_{W^{2,\infty}(\mathcal{T}^h)} \le \|P^h u - u\|_{W^{2,\infty}(\mathcal{T}^h)} + \|u\|_{W^{2,\infty}(\mathcal{T}^h)} \le C\|u\|_{W^{2,\infty}(\mathcal{T}^h)} \tag{7.234}$$

hence $\|\nabla P^h u\|_{L^\infty(e)} \le \|\nabla P^h u\|_{L^\infty(K)} \le C\|\nabla u\|_{L^\infty(K)}$.

For good convergence we will impose later on $u_0 \in H^s(\mathcal{T}^h), s \ge 2$, *for the exact solution anyway. For* $s - 2 \ge n/2$ *this implies* $u_0 \in W^{2,\infty}(\mathcal{T}^h)$, *cf. Theorem 1.26, otherwise we have to assume it additionally. For the* $u \in S^h$ *with* $S^h \subset W^{2,\infty}(\mathcal{T}^h)$ *it is automatically satisfied.*

Proof. We consider $u,\ \overline{u},\ v \in \mathcal{U}_e$ as arguments of $\overline{a}_h(\cdot;\cdot,\cdot)$:

$$\begin{aligned} &|\overline{a}_h(\overline{u};\overline{u},v) - \overline{a}_h(u;u,v)| \\ &\qquad \le |\overline{a}_h(u;\overline{u},v) - \overline{a}_h(u;u,v)| + |\overline{a}_h(\overline{u};\overline{u},v) - \overline{a}_h(u;\overline{u},v)|. \end{aligned} \tag{7.235}$$

and estimate the different terms σ_j in (7.236) ff. For the first term we obtain with (7.232)

$$|\overline{a}_h(u;\overline{u},v) - \overline{a}_h(u;u,v)| \le \varepsilon \underbrace{\left|\sum_{K\in\mathcal{T}^h}\int_K (B_a(u)\nabla(\overline{u}-u),\nabla v)_n\,dx\right|}_{:=\sigma_4} \tag{7.236}$$

$$+\,\varepsilon \underbrace{\left|\sum_{e\in\mathcal{T}^h}\int_e \left((\{B_a(u)\nabla(\overline{u}-u)\},\nu)_n\,[v] + \theta(\{B_a(u)\nabla v\},\nu)_n\,[(\overline{u}-u)]\right)dS\right|}_{:=\sigma_5+\theta\overline{\sigma}_5}. \tag{7.237}$$

σ_4 is estimated with the bounded $a_{ij}(x,u)$, cf.(7.80) (c), and an appropriately reordered discrete Chebyshev inequality (1.43), and the Cauchy–Schwarz inequality, (1.44), for $p = q = 2$, cf. (7.185),

$$\begin{aligned} |\sigma_4| &\le \sum_{K\in\mathcal{T}^h}\int_K |(B_a(u)\nabla(\overline{u}-u),\nabla v)_n|\,dx \\ &= \sum_{K\in\mathcal{T}^h}\int_K \sum_{i,j=1}^n |a_{ij}(x,u)\partial^i(\overline{u}-u)\,\partial^j v|\,dx \\ &\le \Lambda_a \sum_{K\in\mathcal{T}^h}\int_K \sum_{i,j=1}^n |\partial^i(\overline{u}-u)\,\partial^j v|\,dx \\ &\le n\Lambda_a \sum_{K\in\mathcal{T}^h} |\overline{u}-u|_{H^1(K)}|v|_{H^1(K)} \\ &\le n\Lambda_a |\overline{u}-u|_{H^1(\mathcal{T}^h)}|v|_{H^1(\mathcal{T}^h)}. \end{aligned} \tag{7.238}$$

Similarly as in (7.202), we obtain for σ_5 with the Chebyshev and the Cauchy–Schwarz inequality (1.43), (1.44), for $p = q = 2$, and $d(e)$ in (7.28),

$$
\begin{aligned}
|\sigma_5| &\leq n\Lambda_a \sum_{K\in\mathcal{T}^h} \sum_{e\subset\partial K} \int_e |[v]|\,|\nabla(u-\overline{u})|\,dS \qquad (7.239)\\
&\leq n\Lambda_a \sum_{K\in\mathcal{T}^h} \left(\sum_{e\subset\partial K} \int_e \frac{[v]^2}{d(e)/c_w}\,dS\right)^{1/2} \left(\sum_{e\subset\partial K} \frac{d(e)}{c_w} \int_e |\nabla(u-\overline{u})|^2\,dS\right)^{1/2}\\
&\leq 2n\Lambda_a/(c_w)^{1/2}\,\left(J_h^\sigma(v,v)\right)^{1/2} \left(\sum_{K\in\mathcal{T}^h} h_K|u-\overline{u}|^2_{H^1(\partial K)}\right)^{1/2}.
\end{aligned}
$$

The reordered discrete Chebyshev inequality (1.43), and the Hölder inequality, (1.44) for $p = q = 2$, yield for $\overline{\sigma}_5$, cf. (7.220),

$$
\begin{aligned}
|\overline{\sigma}_5| &\leq n\Lambda_a \sum_{e\in\mathcal{T}^h} \int_e |\nabla v|\,|\,[u-\overline{u}]\,|\,dS \qquad (7.240)\\
&\leq n\Lambda_a \left(\sum_{e\in\mathcal{T}^h} \int_e h_K|\nabla v|^2\,dS\right)^{1/2} \left(\sum_{e\in\mathcal{T}^h} \int_e h_K^{-1}([u-\overline{u}])^2\,dS\right)^{1/2}\\
&\leq 2n\Lambda_a \left(\sum_{K\in\mathcal{T}^h} h_K\|\nabla v\|^2_{L^2(\partial K)}\right)^{1/2} \left(\sum_{K\in\mathcal{T}^h} h_K^{-1}\|(u-\overline{u})\|^2_{L^2(\partial K)}\right)^{1/2}.
\end{aligned}
$$

For the second term in (7.235) we combine the Lipschitz-continuity of the functions $a_{ij},\ i,j = 1,\ldots,n$ or B_a with (7.232), getting

$$
\begin{aligned}
&|\overline{a}_h(\overline{u};\overline{u},v) - \overline{a}_h(u;\overline{u},v)|\\
&\leq \varepsilon \underbrace{\left|\sum_{K\in\mathcal{T}^h} \int_K \left((B_a(\overline{u}) - B_a(u))\nabla\overline{u}, \nabla v\right)_n dx\right|}_{:=\sigma_6} \qquad (7.241)\\
&+\varepsilon \underbrace{\left|\sum_{e\in\mathcal{T}^h} \int_e \Big((\{(B_a(\overline{u}) - B_a(u))\nabla\overline{u}\},\nu)_n\,[v] + \theta(\{(B_a(\overline{u}) - B_a(u))\nabla v\},\nu)_n\,[\overline{u}]\Big)\,dS\right|}_{:=\sigma_7+\theta\overline{\sigma}_7}.
\end{aligned}
$$

Obviously we obtain with the Lipschitz constant, L, for the a_{ij} or the B_a with (1.43),

$$
|\sigma_6| \leq nL\|\overline{u}-u\|_{L^2(\mathcal{T}^h)}\|\nabla\overline{u}\|_{L^\infty(\Omega)}\|v^h\|_{H^1(\mathcal{T}^h)}. \qquad (7.242)
$$

A combination with the ideas for, e.g. (7.202)–(7.203) allows the estimation of σ_7 : here we need $\|\nabla\overline{u}\|_{L^\infty(e)} \le \|\nabla\overline{u}\|_{L^\infty(K)}$ requiring $\overline{u} \in W^{2,\infty}(\mathcal{T}^h)$ by Theorem 1.38.

$$|\sigma_7| \le 2nL \left(\sum_{e\in\mathcal{T}^h} \|\nabla\overline{u}\|^2_{L^\infty(e)} \int_e \frac{[v]^2}{d(e)/c_w}\, dS\right)^{1/2} \left(\sum_{e\in\mathcal{T}^h} \frac{d(e)}{c_w} \int_e \sum_{i=1}^n (u-\overline{u})^2\, dS\right)^{1/2}$$

$$\le 2nL/(c_w)^{1/2}\|\overline{u}\|_{W^{1,\infty}(\mathcal{T}^h)}\, (J_h^\sigma(v,v))^{1/2} \left(\sum_{K\in\mathcal{T}^h} h_K\|u-\overline{u}\|^2_{L^2(\partial K)}\right)^{1/2}. \quad (7.243)$$

For $\overline{\sigma}_7$ we obtain, as in (7.220),

$$|\overline{\sigma}_7| \le 2nL\|\overline{u}\|_{L^\infty(\Omega)} \sum_{e\in\mathcal{T}^h} \int_e |\nabla v|\; |(u-\overline{u})|\, dS \quad (7.244)$$

$$\le 2nL\|\overline{u}\|_{L^\infty(\Omega)} \left(\sum_{K\in\mathcal{T}^h} h_K\|\nabla v\|^2_{L^2(\partial K)}\right)^{1/2} \left(\sum_{K\in\mathcal{T}^h} h_K^{-1}\|(u-\overline{u})\|^2_{L^2(\partial K)}\right)^{1/2}.$$

Adding the estimates for σ_4 to $\overline{\sigma}_7$ in (7.238)–(7.244) proves the claimed (7.233). □

As consequence of Proposition 7.29 we obtain with $\overline{u} = P^h u$, the S^h-interpolant from Theorem 7.7, but now only for $\forall v^h \in S^h = \mathcal{V}^h$ as in Theorem 7.21, compare Remark 7.30.

Theorem 7.31. Consistency for the semilinear a_h: *The $a_h(u,v^h)$ in (7.87) with properties (7.80) are Lipschitz-continuous with respect to u and linear and bounded with respect to v^h. Furthermore the $a^h(P^h u, v^h)$ are consistent with the $a_h(u,v^h)$. So constants C_1^*, C_2^* exist, cf. (7.247) such that, $\forall\, h<1,\ u \in W^{2,\infty}(\mathcal{T}^h)$, $\forall v^h \in S^h = \mathcal{V}^h$, and the S^h-interpolant, $P^h u \in W^{2,\infty}(\mathcal{T}^h)$, with $d_K \ge d$,*

$$|a_h(u,v^h) - a_h(P^h u, v^h)| \quad (7.245)$$

$$\le \varepsilon \left(\sum_{K\in\mathcal{T}^h} \left(h_k^{\min\{d_K,s\}-1}|u|_{H^{\min\{d_K,s\}}(K)}\right)^2\right)^{1/2} \left(C_1^* + C_2^*\|P^h u\|_{W^{1,\infty}(\mathcal{T}^h)}\right)\|v^h\|$$

$$\le \varepsilon\, h^{\min\{d,s\}-1}|u|_{H^{\min\{d,s\}}(\mathcal{T}^h)} \left(C_1^* + 2C_2^*\|u\|_{W^{1,\infty}(\mathcal{T}^h)}\right)\|v^h\|.$$

The term $+2C_2^*\|u\|_{W^{1,\infty}(\mathcal{T}^h)}$ drops out for $a(u,v)$ in (7.19) and $u \in W^{2,\infty}(\mathcal{T}^h)$ is replaced by $u \in \mathcal{U}_e$.

Proof. It proceeds totally analogously to that of Theorem 7.26. We essentially apply the estimates in (7.222) to the terms in (7.246) and obtain, with the equivalence of the penalty norms in (7.154), (1.43),

$$|\overline{a}_h(u;u,v^h) - \overline{a}_h(P^hu;P^hu,v^h)|$$

$$\leq \varepsilon \left(\sum_{K\in\mathcal{T}^h} \left(h_k^{\min\{d_K,s\}-1} |u|_{H^{\min\{d_K,s\}}(K)} \right)^2 \right)^{1/2}$$

$$\times \Big(n\Lambda_a C_7 |v^h|_{H^1(\mathcal{T}^h)} + nLC_7 h \|v^h\|_{H^1(\mathcal{T}^h)} \|\nabla P^h u\|_{L^\infty(\mathcal{T}^h)}$$

$$+ 2n\sqrt{2C_5C_7/c_w}(\Lambda_a + hL\|P^hu\|_{W^{1,\infty}(\mathcal{T}^h)}) \left(J_h^\sigma(v^h,v^h)\right)^{1/2}$$

$$+ 4nC_5\sqrt{C_7(C_6+1)}(\Lambda_a + L\|P^hu\|_{L^\infty(\mathcal{T}^h)})|v^h|_{H^1(\mathcal{T}^h)} \Big)$$

$$\leq \varepsilon \left(\sum_{K\in\mathcal{T}^h} \left(h_k^{\min\{d_K,s\}-1} |u|_{H^{\min\{d_K,s\}}(K)} \right)^2 \right)^{1/2} \tag{7.246}$$

$$\|v^h\| \Big(n\Lambda_a \sqrt{[(C_7 + 4nC_5\sqrt{C_7(C_6+1)})^2 + 8nC_5C_7/c_w]}$$

$$+ \|P^hu\|_{W^{1,\infty}(\mathcal{T}^h)} (1+c_e) nL\sqrt{C_7}\sqrt{h^2C_7 + 8h^2C_5/c_w \ + 16C_5^2(C_6+1)} \Big)$$

and with the following C_1^*, C_2^*,

$$\leq \varepsilon \left(\sum_{K\in\mathcal{T}^h} \left(h_k^{\min\{d_K,s\}-1} |u|_{H^{\min\{d_K,s\}}(K)} \right)^2 \right)^{1/2} \left(C_1^* + C_2^*\|P^hu\|_{W^{1,\infty}(\mathcal{T}^h)}\right) \|v^h\|.$$

with

$$C_1^* := n\Lambda_a \sqrt{[(C_7 + 4nC_5\sqrt{C_7(C_6+1)})^2 + 8nC_5C_7/c_w]} \quad \text{and} \tag{7.247}$$

$$C_2^* := (1+c_e)\, nL\sqrt{C_7}\sqrt{h^2C_7 + 8h^2C_5/c_w \ + 16C_5^2(C_6+1)} \Big).$$

Thus we have proved the estimate (7.245). □

7.11.4 Consistency of the quasilinear a_h for systems

Again we introduce a new $\overline{a}^h(\cdot;\cdot,\cdot)$, based upon $B_A(\vec{u},\nabla\vec{w})$ instead of the $B_A(\vec{u},\nabla\vec{u})$ in (7.97), (7.99). We update the notation in (7.92), with $(\mathcal{T}^h,\mathbb{R}^q)$ replacing $(\mathcal{T}^h)$, and define, e.g.

$$B_A(\vec{u}^h,\nabla\vec{w}^h) := \left(A_k(x,\vec{u}^h,\nabla\vec{w}^h)\right)_{k=1}^n \in \mathbb{R}^{n\times q}, (B_A(\vec{u}^h,\nabla\vec{w}^h),\nu)_n \in \mathbb{R}^q \tag{7.248}$$

yielding $\forall \vec{u}^h, \vec{w}^h \in \mathcal{U}_e, \forall \vec{v}^h \in \mathcal{U}_{1,e} := H^1(\mathcal{T}^h, \mathbb{R}^q)$:

$$\overline{a}_h(\vec{u}^h; \nabla \vec{w}^h, \vec{v}^h) = \sum_{K \in \mathcal{T}^h} \int_K \sum_{k=0}^{n} (A_k(x, \vec{u}^h, \nabla \vec{w}^h), \partial^k \vec{v}^h)_q dx$$
$$- \sum_{e \in \mathcal{T}^h} \int_e \big((\{B_A(\vec{u}^h, \nabla \vec{w}^h)\}, \nu)_n, [\vec{v}^h]\big)_q dS.$$

Proposition 7.32. *The $\overline{a}_h(\vec{u}; \vec{u}, \vec{v})$ in (7.248) with properties (7.94) are modified Lipschitz-continuous with respect to $\vec{u}$ and linear in $\vec{v}$, and bounded with respect to $\|\vec{v}\|$. So $\forall\ \vec{u}, \vec{v} \in \mathcal{V}_e$, $\overline{\vec{u}} \in W^{2,\infty}(\mathcal{T}^h, \mathbb{R}^q)$ and $\|\overline{\vec{u}}\|_{\mathcal{U}_{1,e}}, \|\vec{u}\|_{\mathcal{U}_{1,e}} \le M$ this $\overline{a}_h(\vec{u}; \vec{u}, \vec{v})$ satisfies*

$$\begin{aligned} |a_h(\vec{u}, \vec{v}) - a_h(\overline{\vec{u}}, \vec{v})| &= |\overline{a}_h(\vec{u}; \vec{u}, \vec{v}) - \overline{a}_h(\overline{\vec{u}}; \overline{\vec{u}}, \vec{v})| \le L|\overline{\vec{u}} - \vec{u}|_{\mathcal{U}_{1,e}} |\vec{v}|_{\mathcal{U}_{1,e}} \\ &+ L(1 + |\vec{u}|_{W^{1,\infty}(\mathcal{T}^h, \mathbb{R}^q)}) |\overline{\vec{u}} - \vec{u}|_{\mathcal{U}_{1,e}} |\vec{v}|_{L^2(\mathcal{T}^h, \mathbb{R}^q)} \\ &+ nL/(c_w)^{1/2}\ (J_h^\sigma(\vec{v}, \vec{v}))^{1/2} \left(\sum_{K \in \mathcal{T}^h} h_K |\vec{u} - \overline{\vec{u}}|^2_{H^1(\partial K)} \right)^{1/2} \\ &+ L\|\overline{\vec{u}} - \vec{u}\|_{L^2(\mathcal{T}^h, \mathbb{R}^q)} \big(1 + \|\vec{u}\|_{W^{1,\infty}(\mathcal{T}^h, \mathbb{R}^q)} \\ &+ \|\vec{u}\|^2_{W^{1,\infty}(\mathcal{T}^h, \mathbb{R}^q)}\big) \|\vec{v}^h\|_{\mathcal{U}_{1,e}} \\ &+ nL/(c_w)^{1/2} \left(\sum_{K \in \mathcal{T}^h} h_K \|\vec{u} - \overline{\vec{u}}\|^2_{L^2(\partial K)} \right)^{1/2} \\ &\times J_h^\sigma(\vec{v}, \vec{v}))^{1/2} (1 + \|\vec{u}\|_{W^{1,\infty}(\mathcal{T}^h, \mathbb{R}^q)}). \end{aligned} \tag{7.249}$$

Proof. We consider $\vec{u}$, $\overline{\vec{u}}$, $\vec{w}, \vec{v} \in \mathcal{U}_e$ as arguments of $\overline{a}_h(\cdot; \cdot, \cdot)$ and estimate

$$\begin{aligned} |\overline{a}_h(\overline{\vec{u}}; \overline{\vec{u}}, \vec{v}) - \overline{a}_h(\vec{u}; \vec{u}, \vec{v})| &\le |\overline{a}_h(\overline{\vec{u}}; \overline{\vec{u}}, \vec{v}) - \overline{a}_h(\vec{u}; \overline{\vec{u}}, \vec{v})| \\ &+ |\overline{a}_h(\vec{u}; \overline{\vec{u}}, \vec{v})| - \overline{a}_h(\vec{u}; \vec{u}, \vec{v})|. \end{aligned} \tag{7.250}$$

For the second term we obtain with (7.248)

$$\begin{aligned} &|\overline{a}_h(\vec{u}; \overline{\vec{u}}, \vec{v}) - \overline{a}_h(\vec{u}; \vec{u}, \vec{v})| \\ &\le \underbrace{\left| \sum_{K \in \mathcal{T}^h} \int_K \sum_{k=0}^{n} \Big((A_k(x, \vec{u}, \nabla \overline{\vec{u}}) - A_k(x, \vec{u}, \nabla \vec{u})), \partial^k \vec{v} \Big)_q\ dx \right|}_{:= \sigma_9 = \sigma_{9_{k>0}} + \sigma_{9_{k=0}}} \\ &+ \underbrace{\left| \sum_{e \in \mathcal{T}^h} \int_e ([\vec{v}], (\{B_A(\vec{u}, \nabla \overline{\vec{u}})\}, \nu)_n - (\{B_A(\vec{u}, \nabla \vec{u})\}, \nu)_n)_q\, dS \right|}_{:= \sigma_{10}}. \end{aligned} \tag{7.251}$$

Since the conditions for the partials with respect to $\vec{u}$ of the A_k in (7.94) for $k=0$ and $k>0$ are different, σ_9 and σ_{11} below are split. $\sigma_{9_{k>0}}$, obtained by $\sum_{k=1}^n$, is estimated with the Lipschitz-continuous $A_k(x,\vec{w},\nabla\vec{u})$, cf. (7.94) (c), and the Cauchy–Schwarz inequalities (1.44) for $p=q=2$, (1.45),

$$
\begin{aligned}
|\sigma_{9_{k>0}}| &\le \sum_{K\in\mathcal{T}^h}\int_K \sum_{k=1}^n \Big(\big(A_k(x,\vec{u},\nabla\overline{\vec{u}}) - A_k(x,\vec{u},\nabla\vec{u})\big), \partial^k\vec{v}\Big)_q\, dx \qquad (7.252)\\
&\le L\sum_{K\in\mathcal{T}^h}\int_K \|\nabla\overline{\vec{u}}-\nabla\vec{u}\|_{n\times q}\|\nabla\vec{v}\|_{n\times q}\, dx\\
&\le L\sum_{K\in\mathcal{T}^h} |\overline{\vec{u}}-\vec{u}|_{H^1(K)}|\vec{v}|_{H^1(K)}\\
&\le L|\overline{\vec{u}}-\vec{u}|_{\mathcal{U}_{1,e}}|\vec{v}|_{\mathcal{U}_{1,e}}.
\end{aligned}
$$

For $\sigma_{9_{k=0}}$, we combine (7.94) (d), with (1.44), (1.45),

$$
\begin{aligned}
|\sigma_{9_{k=0}}| &\le \sum_{K\in\mathcal{T}^h}\int_K \Big(\big(A_0(x,\vec{u},\nabla\overline{\vec{u}}) - A_0(x,\vec{u},\nabla\vec{u})\big), \vec{v}\Big)_q\, dx\\
&\le L\sum_{K\in\mathcal{T}^h}\int_K (1+\|\nabla\vec{u}\|_{n\times q})\|\nabla\overline{\vec{u}}-\nabla\vec{u}\|_{n\times q}\|\vec{v}\|_q\, dx\\
&\le L\sum_{K\in\mathcal{T}^h} (1+|\vec{u}|_{W^{1,\infty}(K)})|\overline{\vec{u}}-\vec{u}|_{H^1(K)}|\vec{v}|_{L^2(K)}\\
&\le L(1+|\vec{u}|_{W^{1,\infty}(\mathcal{T}^h,\mathbb{R}^q)})|\overline{\vec{u}}-\vec{u}|_{\mathcal{U}_{1,e}}|\vec{v}|_{L^2(\mathcal{T}^h,\mathbb{R}^q)}. \qquad (7.253)
\end{aligned}
$$

Since $B_A(\vec{u},\nabla\overline{\vec{u}})$ is independent of A_0 we do not need a $\sigma_{10_{k=0}}$, but only σ_{10}. Similarly as in (7.238), (7.252), we obtain with $d(e)$ in (7.28),

$$
\begin{aligned}
|\sigma_{10}| &\le nL\sum_{e\in\mathcal{T}^h}\int_e \big(\|\,[\vec{v}]\,\|_q\|\nabla\vec{u}-\nabla\overline{\vec{u}}\|_{n\times q}\big)dS\\
&\le nL\left(\sum_{e\subset\mathcal{T}^h}\int_e \frac{\|\,[\vec{v}]\,\|_q^2}{d(e)/c_w}\, dS\right)^{1/2}\left(\sum_{e\subset\mathcal{T}^h}\frac{d(e)}{c_w}\int_e (\|\nabla\vec{u}-\nabla\overline{\vec{u}}\|_{n\times q})^2\, dS\right)^{1/2}\\
&\le nL/(c_w)^{1/2}\,(J_h^\sigma(\vec{v},\vec{v}))^{1/2}\left(\sum_{K\in\mathcal{T}^h} h_K|\vec{u}-\overline{\vec{u}}|^2_{H^1(\partial K)}\right)^{1/2}.
\end{aligned}
$$

For the first term in (7.250) we combine the bounds for the partials of the functions $A_k,\ k = 1, \dots, n$ in (7.94) with (7.248), hence σ_{11} is split:

$$
\begin{aligned}
&|\bar{a}_h(\overline{\vec{u}}; \overline{\vec{u}}, \vec{v})| - \bar{a}_h(\vec{u}; \overline{\vec{u}}, \vec{v})| \qquad (7.254)\\
&\le \underbrace{\left| \sum_{K\in\mathcal{T}^h} \int_K \sum_{k=0}^{n} \Big((A_k(x, \overline{\vec{u}}, \nabla\overline{\vec{u}}) - A_k(x, \vec{u}, \nabla\overline{\vec{u}})), \partial^k \vec{v} \Big)_q \, dx \right|}_{:=\sigma_{11}=:\sigma_{11_{k>0}}+\sigma_{11_{k=0}}} \\
&\quad + \underbrace{\left| \sum_{e\in\mathcal{T}^h} \int_e ([\vec{v}], (\{B_A(\overline{\vec{u}}, \nabla\overline{\vec{u}})\} - \{B_A(\vec{u}, \nabla\overline{\vec{u}})\}, \nu)_n\,)_q)\, dS \right|}_{:=\sigma_{12}} .
\end{aligned}
$$

We obtain with the different bounds for the partials of the functions A_k for $k = 0$ and $k = 1, \dots, n$, and for $\vec{u} \in W^{1,\infty}(\mathcal{T}^h, \mathbb{R}^q)$ for the $\sigma_{11_{k>0}}$ and $\sigma_{11_{k=0}}$

$$
\begin{aligned}
|\sigma_{11_{k>0}}| &\le L \|\overline{\vec{u}} - \vec{u}\|_{L^2(\mathcal{T}^h,\mathbb{R}^q)} (1 + \|\vec{u}\|_{W^{1,\infty}(\mathcal{T}^h,\mathbb{R}^q)}) \|\vec{v}^h\|_{\mathcal{U}_{1,e}} \qquad (7.255)\\
|\sigma_{11_{k=0}}| &\le L \|\overline{\vec{u}} - \vec{u}\|_{L^2(\mathcal{T}^h,\mathbb{R}^q)} (1 + \|\vec{u}\|^2_{W^{1,\infty}(\mathcal{T}^h,\mathbb{R}^q)}) \|\vec{v}^h\|_{L^2(\mathcal{T}^h,\mathbb{R}^q)} .
\end{aligned}
$$

A combination of (7.94) (c) with the ideas in (7.202)–(7.203) allows the estimation of σ_{12}. In (7.256), we need $\|\nabla\overline{\vec{u}}\|_{L^\infty(e)} \le \|\nabla\overline{\vec{u}}\|_{L^\infty(K)}$ valid for $\overline{\vec{u}} \in W^{2,\infty}(\mathcal{T}^h, \mathbb{R}^q)$, cf. Theorem 1.38.

$$
\begin{aligned}
|\sigma_{12}| &\le nL \sum_{K\in\mathcal{T}^h} \left(\sum_{e\subset\partial K} \int_e \frac{\|[\vec{v}]\ \|_q^2}{d(e)/c_w}\, dS \right)^{1/2} \qquad (7.256)\\
&\quad \times \left(\sum_{e\subset\partial K} \frac{d(e)}{c_w} \int_e \|(\vec{u} - \overline{\vec{u}})\ \|_q^2\, dS \right)^{1/2} (1 + \|\nabla\overline{\vec{u}}\|_{L^\infty(e)}) \\
&\le nL/(c_w)^{1/2} \left(J_h^\sigma(\vec{v}, \vec{v})\right)^{1/2} \left(\sum_{K\in\mathcal{T}^h} h_K \|\vec{u} - \overline{\vec{u}}\|^2_{L^2(\partial K)} \right)^{1/2} (1 + |\overline{\vec{u}}|_{W^{1,\infty}(\mathcal{T}^h,\mathbb{R}^q)}).
\end{aligned}
$$

Adding the estimates for σ_{11} to σ_{12} in (7.252)–(7.256), proves the claim. □

As consequence of Proposition 7.32 we obtain with $\overline{\vec{u}} = P^h u$ as in Theorem 7.31

Theorem 7.33. Consistency for the quasilinear a_h: *The $a_h(\vec{u}, \vec{v}^h)$ in (7.248) with properties (7.94) are modified Lipschitz-continuous with respect to $\vec{u}$, and linear and bounded with respect to $\vec{v}^h$. Furthermore the $a^h(P^h\vec{u}, \vec{v}^h)$ are consistent with the $a_h(\vec{u}, \vec{v}^h)$. So for sufficiently small $h < 1$, and $\|\overline{\vec{u}}\|_{\mathcal{U}_{1,e}}, \|\vec{u}\|_{\mathcal{U}_{1,e}} \le M$, a constant $C_{16} := L \max\{2 + 2n^2 C_5 C_7 (1 + h^2)/c_w, c_e^2\}^{1/2}$ exists for $a_h(\vec{u}, \vec{v}^h)$, s.t. $\forall\ \vec{v}^h \in S_q^h$ and*

$\forall\, \vec{u} \in W^{2,\infty}(\mathcal{T}^h, \mathbb{R}^q)$, *the* S^h*-interpolant,* $P^h\vec{u}$ *from Theorem 7.7, cf. (7.234):*

$$|a_h(\vec{u}, \vec{v}^h) - a^h(P^h\vec{u}, \vec{v}^h)| \leq C_{16} h^{\min\{d,s\}-1} |\vec{u}|_{H^{\min\{d,s\}}(\mathcal{T}^h,\mathbb{R}^q)} \times \left(1 + \|\vec{u}\|_{W^{1,\infty}(\mathcal{T}^h,\mathbb{R}^q)} + \|\vec{u}\|^2_{W^{1,\infty}(\mathcal{T}^h,\mathbb{R}^q)}\right) \|\vec{v}^h\|. \tag{7.257}$$

Proof. Again we apply the slightly modified error estimates in (7.222):

$$\|P^h\vec{u} - \vec{u}\|_{\mathcal{U}_{1,e}} \leq C_7 \left(\sum_{K\in\mathcal{T}^h} \left(h_k^{\min\{d,s\}-1} |\vec{u}|_{H^{\min\{d,s\}}(K,\mathbb{R}^q)} \right)^2 \right)^{1/2} \times \left(\sum_{K\in\mathcal{T}^h} h_K |\vec{u} - P^h\vec{u}|^2_{H^1(\partial K)} \right)^{1/2}$$
$$\leq \sqrt{2C_5C_7} \left(\sum_{K\in\mathcal{T}^h} \left(h_k^{\min\{d,s\}-1} |\vec{u}|_{H^{\min\{d,s\}}(K,\mathbb{R}^q)} \right)^2 \right)^{1/2}$$

$$\|P^h\vec{u} - \vec{u}\|_{L^2(\mathcal{T}^h,\mathbb{R}^q)} \leq C_7 \left(\sum_{K\in\mathcal{T}^h} \left(h_k^{\min\{d,s\}} |\vec{u}|_{H^{\min\{d,s\}}(K,\mathbb{R}^q)} \right)^2 \right)^{1/2} \times \left(\sum_{K\in\mathcal{T}^h} h_K \|\vec{u} - P^h\vec{u}\|^2_{L^2(\partial K)} \right)^{1/2}$$
$$\leq \sqrt{2C_5C_7} \left(\sum_{K\in\mathcal{T}^h} \left(h_k^{\min\{d,s\}} |\vec{u}|_{H^{\min\{d,s\}}(K,\mathbb{R}^q)} \right)^2 \right)^{1/2}.$$

These estimates are applied to the error terms in (7.249) to get

$$\begin{aligned}
&|a_h(\vec{u}, \vec{v}^h) - a_h(P^h\vec{u}, \vec{v}^h)| \\
&\quad \leq \max\{C_7, \sqrt{2C_5C_7}\} h^{\min\{d,s\}-1} |\vec{u}|_{H^{\min\{d,s\}}(\mathcal{T}^h,\mathbb{R}^q)} \\
&\qquad \times \Big(L|\vec{v}^h|_{\mathcal{U}_{1,e}} + L(1 + |\vec{u}|_{W^{1,\infty}(\mathcal{T}^h,\mathbb{R}^q)}) |\vec{v}^h|_{L^2(\mathcal{T}^h,\mathbb{R}^q)} \\
&\qquad + nL/(c_w)^{1/2} \sqrt{2C_5C_7}\, \left(J_h^\sigma(\vec{v}^h, \vec{v}^h)\right)^{1/2} \\
&\qquad + L\big(1 + \|\vec{u}\|_{W^{1,\infty}(\mathcal{T}^h,\mathbb{R}^q)} + \|\vec{u}\|^2_{W^{1,\infty}(\mathcal{T}^h,\mathbb{R}^q)}\big) \|\vec{v}^h\|_{\mathcal{U}_{1,e}} \\
&\qquad + nLh/(c_w)^{1/2} \sqrt{2C_5C_7} \big(J_h^\sigma(\vec{v}^h, \vec{v}^h)\big)^{1/2} (1 + \|\vec{u}\|_{W^{1,\infty}(\mathcal{T}^h,\mathbb{R}^q)})\Big).
\end{aligned}$$

This implies with $P^h\vec{u} \in W^{2,\infty}(\Omega, \mathbb{R}^q)$ and the equivalence of $\|\vec{v}\|$ and $\|\vec{v}\|^J_{\|H^1(\mathcal{T}^h,\mathbb{R}^q)\|}$ with c_e in (7.154)

$$\begin{aligned}
|a_h(\vec{u}, \vec{v}^h) - a_h(P^h\vec{u}, \vec{v}^h)| &\le \max\left\{C_7, \sqrt{2C_5C_7}\right\} h^{\min\{d,s\}-1}|\vec{u}|_{H^{\min\{d,s\}}(\mathcal{T}^h,\mathbb{R}^q)} \\
&\|v^h\| \left(L\left(2 + |\vec{u}|_{W^{1,\infty}(\mathcal{T}^h,\mathbb{R}^q)}\right) + nL/(c_w)^{1/2}\sqrt{2C_5C_7} + Lc_e\big(1 + \|\vec{u}\|_{W^{1,\infty}(\mathcal{T}^h,\mathbb{R}^q)}\right. \\
&\qquad \left. + \|\vec{u}\|^2_{W^{1,\infty}(\mathcal{T}^h,\mathbb{R}^q)}\big) + nLh/(c_w)^{1/2}\sqrt{2C_5C_7}\big(1 + \|\vec{u}\|^J_{W^{1,\infty}(\mathcal{T}^h,\mathbb{R}^q)}\big)\right) \\
&\le L\max\left\{2 + 2n^2/c_wC_5C_7(1+h^2), c_e^2\right\}^{1/2} h^{\min\{d,s\}-1}|\vec{u}|_{H^{\min\{d,s\}}(\mathcal{T}^h,\mathbb{R}^q)}\|v^h\| \\
&\quad \left(1 + \|\vec{u}\|_{W^{1,\infty}(\mathcal{T}^h,\mathbb{R}^q)} + \|\vec{u}\|^2_{W^{1,\infty}(\mathcal{T}^h,\mathbb{R}^q)}\right).
\end{aligned}$$

This shows the claim (7.257). □

7.11.5 Consistency of the quasilinear a_h for the equations of Houston, Robson, Süli, and for systems

We turn to consistency for the DCGMs for the quasilinear problem $G(u) = 0$ in Houston *et al.* [407], again only for the principal part. All other terms are equal to those for the semilinear problem. As an essential tool for consistency, we refer to Barett and Liu [68], Lemma 2.1: From (7.103)–(7.106) Houston *et al.* [407] obtain for $-1 \le \theta \le 1$

$$|(B_a(u)\nabla u - B_a(v)\nabla v)(x)| \le C|(\nabla u - \nabla v)(x)| \quad \forall x \in \Omega, \forall u, v \in \mathcal{U}_e. \tag{7.258}$$

We have

Proposition 7.34. *The $a_h(u, v)$ in (7.107) with properties (7.103), and C in (7.258), and $C^* := 4M_\mu C_5(1 + C_6)$, are modified Lipschitz-continuous with respect to u and linear and bounded with respect to v, such that $a_h(u, v)$ satisfies $\forall\ u, \overline{u}, v \in \mathcal{U}_e$,*

$$\begin{aligned}
|a_h(u, v) - a_h(\overline{u}, v)| &\le nC|\overline{u} - u|_{H^1(\mathcal{T}^h)}|v|_{H^1(\mathcal{T}^h)} \qquad (7.259)\\
&\quad + 2nC/(c_w)^{1/2}\left(J_h^\sigma(v, v)\right)^{1/2}\left(\sum_{K\in\mathcal{T}^h} h_K|u - \overline{u}|^2_{H^1(\partial K)}\right)^{1/2} \\
&\quad + C^*|v|_{H^1(\mathcal{T}^h)}\left(\sum_{K\in\mathcal{T}^h} h_K^{-1}|u - \overline{u}|^2_{L^2(\partial K)}\right)^{1/2}.
\end{aligned}$$

Remark 7.35. *In (7.107) we have replaced our usual*

$$\theta\sum_{e\in\mathcal{T}^h}\int_e(\{B_a(u)\nabla v\}, \nu)_n\,[u]\,dS = \theta\sum_{e\in\mathcal{T}^h}\int_e(\{\mu(x, |\nabla u|)\nabla v\}, \nu)_n\,[u]\,dS \tag{7.260}$$

by the form [407]

$$\theta \sum_{e\in\mathcal{T}^h\setminus\partial\Omega} \int_e (\{\mu(x,h^{-1}|[u]|)\nabla v\},\nu)_n\,[u]\,dS + \theta \sum_{e\in\partial\Omega} \int_e (\{\mu(x,h^{-1}|u-u_D|)\nabla v\}\ldots.$$

Our form (7.260) yields the term $\left(\sum_{K\in\mathcal{T}^h} h_K^{-1}|u-\overline{u}|^2_{H^1(\partial K)}\right)^{1/2}$ *in the last line of (7.259). This would reduce the consistency order by 1, unless we restrict the method to $\theta = 0$. The above form [407] yields* $\left(\sum_{K\in\mathcal{T}^h} h_K^{-1}|u-\overline{u}|^2_{L^2(\partial K)}\right)^{1/2}$, *and thus the desired order of consistency.*

Proof.

1. We estimate the different terms σ_j in (7.261) ff. and obtain with (7.107)

$$\begin{aligned} a_h(\overline{u},v) - a_h(u,v)| &\le \underbrace{\left|\sum_{K\in\mathcal{T}^h}\int_K (B_a(\overline{u})\nabla\overline{u} - B_a(u)\nabla u,\nabla v)_n\,dx\right|}_{:=\sigma_4} \qquad (7.261)\\ &+ \underbrace{\left|\sum_{e\in\mathcal{T}^h}\int_e ((\{B_a(\overline{u})\nabla\overline{u} - B_a(u)\nabla u\},\nu)_n\,[v])\,dS\right|}_{:=\sigma_5}\\ &+ \theta\underbrace{\left|\sum_{e\in\mathcal{T}^h\setminus\partial\Omega}\int_e \left(\mu(x,h^{-1}|[u]|)\,[u] - \mu(x,h^{-1}|[\overline{u}]|)\,[\overline{u}]\right)(\{\nabla v\},\nu)_n\,dS + \sum_{e\in\partial\Omega}\cdots\right|}_{:=\theta\overline{\sigma}_5}. \end{aligned}$$

2. σ_4 and σ_5 are estimated, similarly to the previous cases. With (7.258), an appropriately reordered discrete Chebyshev inequality (1.43), and the Cauchy–Schwarz inequality, (1.44), for $p = q = 2$, cf. (7.185), we obtain

$$\begin{aligned} |\sigma_4| &\le \sum_{K\in\mathcal{T}^h}\int_K |(B_a(\overline{u})\nabla\overline{u} - B_a(u)\nabla u,\nabla v)_n|\,dx \qquad (7.262)\\ &\le C\sum_{K\in\mathcal{T}^h}\int_K |\nabla\overline{u} - \nabla u\,||\nabla v|\,dx\\ &\le nC\sum_{K\in\mathcal{T}^h} |\overline{u}-u|_{H^1(K)}|v|_{H^1(K)}\\ &\le nC|\overline{u}-u|_{H^1(\mathcal{T}^h)}|v|_{H^1(\mathcal{T}^h)}. \end{aligned}$$

Similarly as in (7.202), we obtain for σ_5 with (7.258), the Chebyshev and the Cauchy–Schwarz inequalities (1.43), (1.44), for $p = q = 2$, and $d(e)$ in (7.28),

$$
\begin{aligned}
|\sigma_5| &\le nC \sum_{K\in\mathcal{T}^h} \sum_{e\subset\partial K} \int_e |[v]|\, |\nabla(u-\overline{u})|\, dS \qquad (7.263)\\
&\le nC \sum_{K\in\mathcal{T}^h} \left(\sum_{e\subset\partial K} \int_e \frac{[v]^2}{d(e)/c_w}\, dS \right)^{1/2} \left(\sum_{e\subset\partial K} \frac{d(e)}{c_w} \int_e |\nabla(u-\overline{u})|^2\, dS \right)^{1/2}\\
&\le 2nC/(c_w)^{1/2} \left(J_h^\sigma(v,v)\right)^{1/2} \left(\sum_{K\in\mathcal{T}^h} h_K |u-\overline{u}|^2_{H^1(\partial K)} \right)^{1/2}.
\end{aligned}
$$

3. We start estimating the integrals $\int_e$ in σ_5. Then (7.103) implies

$$
\begin{aligned}
&\left| \int_e \left(\mu(x,h^{-1}|[u]|)\,[u] - \mu(x,h^{-1}|[\overline{u}]|)\,[\overline{u}]\right) (\{\nabla v\},\nu)_n\, dS \right|\\
&\quad\le \int_e \left| \mu(x,h^{-1}|[u]|)\,[u] - \mu(x,h^{-1}|[\overline{u}]|)\,[\overline{u}] \right| |\{\nabla v\}|\, dS\\
&\quad\le \int_e h \left| \mu(x,h^{-1}|[u]|)\,h^{-1}[u] - \mu(x,h^{-1}|[\overline{u}]|)\,h^{-1}[\overline{u}] \right| |\{\nabla v\}|\, dS\\
&\quad\le \int_e h M_\mu h^{-1} |[u]-[\overline{u}]| |\{\nabla v\}|\, dS = \int_e M_\mu |[u-\overline{u}]| |\{\nabla v\}|\, dS.
\end{aligned}
$$

Summation yields

$$
\begin{aligned}
|\overline{\sigma}_5| &\le \sum_{e\in\mathcal{T}^h\setminus\partial\Omega} \int_e M_\mu |[u-\overline{u}]| |\{\nabla v\}|\, dS\\
&\le 4M_\mu \sum_{e\in\mathcal{T}^h\setminus\partial\Omega} \int_e |u-\overline{u}| |\nabla v|\, dS\\
&\le 4M_\mu \left(\sum_{e\in\mathcal{T}^h\setminus\partial\Omega} \int_e h_K |\nabla v|^2\, dS \right)^{1/2} \left(\sum_{e\in\mathcal{T}^h\setminus\partial\Omega} \int_e h_K^{-1} |u-\overline{u}|^2\, dS \right)^{1/2}\\
&\le 4M_\mu C_5(1+C_6)|v|_{H^1(\mathcal{T}^h)} \left(\sum_{K\in\mathcal{T}^h} h_K^{-1} |u-\overline{u}|^2_{L^2(\partial K)} \right)^{1/2}. \qquad (7.264)
\end{aligned}
$$

For the last estimate $\left(\sum_{e\in\mathcal{T}^h\setminus\partial\Omega} \int_e h_K |\nabla v|^2\, dS\right)^{1/2} \le C_5(1+C_6)|v|_{H^1(\mathcal{T}^h)}$ we have combined (7.129), (7.138). Similarly for $\theta \sum_{e\in\partial\Omega} \int_e (\{\mu(x,h^{-1}|u - u_D|)\nabla v\} \dots$.

Adding the estimates for σ_4, σ_5 and $\overline{\sigma}_5$ proves the claimed (7.259). □

As a consequence of Proposition 7.34 we obtain the consistency for the quasilinear equations. These results can be directly extended to the system of the form, cf (7.102),

$$-\nabla \cdot (\mu(x, |\nabla \vec{u}|)\nabla \vec{u}) = -\sum_{k=1}^{n} \partial^k \left(\mu(x, |\nabla \vec{u}|)\partial^k \vec{u}\right) = \vec{g}, \tag{7.265}$$

where $\vec{u} = (u_1, \ldots, u_q) : \Omega \to \mathbb{R}^q$, $\vec{g} \in L^2(\Omega)^q$ and the previous $\mu \in C(\overline{\Omega} \times [0, \infty))$ is a real function satisfying the assumption (7.4). Here we put $|\nabla \vec{u}|^2 = \sum_{i=1}^{q} |\nabla u_i|^2$. Using Barett and Liu [68], Lemma 2.1, and (7.102)–(7.106) Houston *et al.* [407] obtain for $-1 \leq \theta \leq 1$ similarly as in (7.258)

$$|(B_a(\vec{u})\nabla \vec{u} - B_a(\vec{v})\nabla \vec{v})(x)| \leq C|(\nabla \vec{u} - \nabla \vec{v})(x)| \quad \forall x \in \Omega, \forall \vec{u}, \vec{v} \in \mathcal{U}_e, \tag{7.266}$$

where $\mathcal{U}_e$ is given by (7.92) and

$$B_a(\vec{v})\nabla \vec{v} = \mu(x, |\nabla \vec{v}|)\nabla \vec{v}. \tag{7.267}$$

Therefore, replacing (7.258) by (7.266) we immediately obtain the consistency results represented by Proposition 7.34 and Theorem 7.36 also for system (7.265).

Theorem 7.36. Consistency for the quasilinear a_h in [407] and for systems:

1. *The $a_h(u^h, v^h)$ in (7.106) with properties (7.103) are Lipschitz-continuous with respect to u and linear and bounded with respect to v^h. Furthermore the $a^h(P^h u, v^h) = a_h(P^h u, v^h)$ are consistent with the $a_h(u, v^h)$. So a constant C_1^* exists, cf. (7.269) such that, $\forall\ h < 1$, $\forall\ u \in \mathcal{U}_e$, $v^h \in S^h = \mathcal{V}^h$,*

$$|a_h(u, v^h) - a_h(P^h u, v^h)| \leq C_1^* h^{\min\{d,s\}-1} |u|_{H^{\min\{d,s\}}(\mathcal{T}^h)} \|v^h\|. \tag{7.268}$$

2. *These results remain correct, if (7.102) is replaced by the system in (7.265) ff., and the $u^h, v^h, B_a(v)v$ in $a_h(u, v^h)$ by $\vec{u}^h, \vec{v}^h, B_a(\vec{v})\vec{v}$ to obtain $a_h(\vec{u}, \vec{v}^h)$.*

Proof. The proof proceeds totally analogously to that of Theorem 7.31. With $\overline{u} = P^h u$ we obtain:

$$|a_h(u, v^h) - a_h(P^h u, v^h)| \leq \left(\sum_{K \in \mathcal{T}^h} (h_k^{\min\{d_K,s\}-1} |u|_{H^{\min\{d_K,s\}}(K)})^2 \right)^{1/2} \tag{7.269}$$

$$\Big(nCC_7 |v^h|_{H^1(\mathcal{T}^h)} + 2nC\sqrt{2C_5C_7/c_w}\, \left(J_h^\sigma(v^h, v^h)\right)^{1/2}$$

$$+ 4nM_\mu C_5(1 + C_6)\sqrt{2C_5C_7} |v^h|_{H^1(\mathcal{T}^h)}\Big).$$

$$\leq C_1^* \left(\sum_{K\in\mathcal{T}^h} \left(h_k^{\min\{d_K,s\}-1} |u|_{H^{\min\{d_K,s\}}(K)} \right)^2 \right)^{1/2} \|v^h\|$$

$$\text{with } C_1^* := n\sqrt{\left[(C^2C_7^2 + 8C^2C_5C_7/c_w + 32M_\mu^2C_5^3C_7(1+C_6)^2\right]}.$$

Thus we have proved the estimate (7.268). □

7.12 Convergence for DCGMs

We achieve convergence in the following way. The $\mathcal{V}^h$-coercivity and boundedness of the discrete principal parts $(G^h)'_p(u^h)$ of the linearized operators $G'(u)$ plus the penalty term with respect to the penalty norm $\|v^h\|$, implies its stability, cf. Theorems 7.16–7.19. Its combination with consistency allows the proof of the stability of the discrete derivatives $(G^h)'(u^h)$, for boundedly invertible $G'(u_0)$, and hence of the discrete nonlinear G^h. With the consistency and some technical conditions this yields the desired convergence, cf. Theorem 3.21, 3.23, 3.29.

For consistency, we tested all the $a_h(u^h,v^h)$, $b_h(u^h,v^h)$, $c_h(u^h,v^h), \ell_h(u^h,v^h)$, $J_h^\sigma(u^h,v^h)$, by v^h. They are linear and bounded in v^h with respect to the $\|\cdot\|$, but some are nonlinear in u^h. The orders of consistency are different for these forms and the constants in the estimates depend nonlinearly upon u, cf. Theorems 7.21–7.33.

The proof that a boundedly invertible compact perturbation of the previous principal parts generates a stable discretization is pretty technical. However, we apply exactly the same technique and ideas for DCGMs as for FEMs with variational crimes, only indicated here. We use the notation $\mathcal{U}$ for $H^1(\Omega,\mathbb{R}^q), q\geq 1$, $\mathcal{U}_e$ for $H^{3/2+\epsilon}(\mathcal{T}^h,\mathbb{R}^q)$, and $\mathcal{U}_e^o$ for $H^{3/2+\epsilon}(\Omega,\mathbb{R}^q)$ here and below.

In a first step towards general stability we define, similarly to nonconforming FEs, an *anticrime transformation* from $u^h\in\mathcal{U}^h$ to $u_h = E^hu^h\in\mathcal{U}$. In Lemma 5.77 we had combined the $a_h(\cdot,\cdot)$-coercivity for specific $a(\cdot,\cdot)$, implying the stability, with its consistency, hence the convergence, however here in the opposite direction of the exact solution $u_h = E^hu^h$ to the chosen discrete $u^h\in\mathcal{U}^h$. For nonconforming FEs we reduced continuity and boundary conditions along the full edges only to a few appropriate points along these edges. For the DCG elements we instead consider the penalty terms.

Lemma 7.37. Anticrime transformation for DCGMs: *We choose the boundedly invertible Laplacian, with the inhomogeneous $A = \Delta - g : \mathcal{U}\to\mathcal{U}'$, and Dirichlet conditions on $\partial\Omega$. Then its IIPG form with $\theta = 0$, the $A^h:\mathcal{U}^h\to\mathcal{U}'^h$ is stable and consistent with A for every $u\in\mathcal{U}$.*

For a given $u^h\in\mathcal{U}^h$ define $\hat{f}^h := A^hu^h\in\mathcal{U}_e$ with $\|f_h\|_{\mathcal{U}_e'}\leq C\|u^h\|$, cf. (7.114). Define $E^h:\mathcal{U}^h\to\mathcal{U}$ by $\langle AE^hu^h,v\rangle = \langle f_h,v\rangle\forall v\in\mathcal{V}_b$ and the boundary condition $(u_h - u_D)|_{\partial\Omega} = 0$. Then we obtain convergence as $\lim_{h\to 0}\|E^hu^h - u^h\| = 0$.

Proof. In contrast to the previous nonlinear problems, we could omit here the $\cap L^\infty(\mathcal{T}^h)$ in the definition of $\mathcal{U}_e = H^{3/2+\epsilon}(\mathcal{T}^h) \cap L^\infty(\Omega)$, etc. That the IIPG form $A^h : \mathcal{U}^h \to \mathcal{U}'^h$ is stable and consistent with A is proved above.

With the $a_h(u,v), \ldots, J_h^\sigma(u,v)$ in (7.57)–(7.61), $\partial\Omega_N = \emptyset$, and our notation, e.g. $\langle A_h u, v\rangle = \langle A_h u, v\rangle_{\mathcal{U}_e' \times \mathcal{U}_e}$, we define the weak operators A, A_h, A^h, as

$$
\begin{aligned}
&\text{(a)}\ \ A : \mathcal{U} \to \mathcal{U}' \text{ with } (u - u_D)|_{\partial\Omega} = 0,\ A_h : \mathcal{U}_e \to \mathcal{U}_e',\ A^h : \mathcal{U}^h = S^h \to \mathcal{U}'^h, \qquad (7.270)\\
&\text{(b)}\ \ \langle Au, v\rangle - a(u,v) + \ell(v) = 0,\quad (u - u_D)|_{\partial\Omega} = 0 \text{ for fixed } u \in \mathcal{U},\ \forall v \in \mathcal{V}_b,\\
&\text{(c)}\ \ \langle A_h u, v\rangle - a_h(u,v) - \varepsilon J_h^\sigma(u,v) + \ell_h(v) = 0,\ \text{fixed } u,\ \forall v \in \mathcal{U}_e,\\
&\text{(d)}\ \ \langle A^h u^h, v^h\rangle - a^h(u^h, v^h) - \varepsilon J_h^\sigma(u^h, v^h) + \ell^h(v^h) = 0,\ \text{fixed } u^h, \forall v^h \in S^h,\\
&\qquad\qquad a^h(u^h, v^h) := a_h(u^h, v^h)|_{S^h \times S^h}, \ell^h(v^h) := \ell_h(v^h)|_{S^h}.
\end{aligned}
$$

Similarly as for nonconforming FEMs, we define, for $u^h \in S^h$ the $\hat{f}^h \in S'^h$ by

$$
u^h : \langle \hat{f}^h, v^h\rangle := \langle A^h u^h, v^h\rangle = a^h(u^h, v^h) + \varepsilon J_h^\sigma(u^h, v^h) - \ell^h(v^h) \forall v^h \in S^h. \quad (7.271)
$$

In the next step we determine $u_h := E^h u^h \in \mathcal{U}$ from $Au_h = \hat{f}_h$. We start with $u_h, v \in \mathcal{U}_e \supset \mathcal{U}_e^o$, such that the $a_h(u_h, v), \varepsilon J_h^\sigma(u_h, v), \ell_h(v), \hat{f}_h$, are well defined. By $u_h \in \mathcal{U}_e^o$, the boundary condition and $v \in \mathcal{V}_b \cap \mathcal{U}_e^o$, the terms with jumps, $[u] = [v] = 0$, and along the boundary vanish, and we end up with $u_h \in \mathcal{U}$, and $v \in \mathcal{V}_b$. So we determine

$$
\begin{aligned}
E^h u^h = u_h \in \mathcal{U} \qquad & \text{with boundary conditions } (u_h - u_D)|_{\partial\Omega} = 0 \text{ implying} \qquad (7.272)\\
\langle Au_h, v\rangle &= \langle \hat{f}_h, v\rangle = a_h(u_h, v) + \varepsilon J_h^\sigma(u_h, v) - \ell_h(v) \quad \forall v \in \mathcal{V}_b\\
&= \varepsilon \sum_{K \in \mathcal{T}_h} \int_K (\nabla u_h,\ \nabla v)_n\, dx\\
&\quad - \varepsilon \sum_{e \in \partial\Omega} \int_e \Big((\nabla u_h, \nu)_n v + (\nabla v, \nu)_n u_h \Big)\, dS + \varepsilon \sum_{e \in \partial\Omega} \int_e \sigma u_h\, v\, dS\\
&\quad - \langle g, v\rangle - \varepsilon \sum_{e \in \partial\Omega} \int_e \big(\ \sigma u_D\, v - (\nabla v, \nu)_n\, u_D\ \big) dS\\
&= \varepsilon \sum_{K \in \mathcal{T}_h} \int_K (\nabla u_h,\ \nabla v)_n\, dx - \langle g, v\rangle = a(u_h, v) - \ell(v)\ \forall v \in \mathcal{V}_b.
\end{aligned}
$$

For this $u_h \in \mathcal{U}_e$ we determine the discrete solution $u_h^h \in S^h$ solving, cf. (7.270),

$$
\langle A u_h^h, v^h\rangle = a^h\left(u_h^h, v^h\right) + \varepsilon J_h^\sigma\left(u_h^h, v^h\right) - \ell^h(v^h) \quad \forall v^h \in S^h.
$$

This is Equation (7.271) with u_h^h replaced by u^h. Then the coercivity of $a^h\left(u_h^h, v^h\right) + \varepsilon J_h^\sigma\left(u_h^h, v^h\right)$ implies $u^h = u_h^h$ and, with the consistency, the convergence $\lim_{h\to 0} \|u^h - u_h^h\| = 0$ to the chosen discrete $u^h \in \mathcal{U}^h$. $\square$

This allows the strategy, starting with Lemma 5.77, for proving the stability of A^h for boundedly invertible linear A as in Theorem 3.29. We reformulate this theorem for our DCGMs.

Theorem 7.38. Stability criterion for DCGMs: *Choose $\mathcal{U}^h = \mathcal{V}^h \not\subset \mathcal{U}$ as in (7.65),(7.95). Choose $P^h = I^h$ and Q'^h as in Theorem 7.7, Corollary 7.8, and in (7.113), (7.115), and one of the DCGMs as formulated above. Let $A : \mathcal{U}_b \to \mathcal{V}' = \mathcal{V}'$ be boundedly invertible and be one of the linear(ized) operators for equations or systems, studied in this chapter. These DCGMs for A, the A^h, are consistent with A in a smooth enough u. Then*

$$A^{-1} \in \mathcal{L}(\mathcal{V}', \mathcal{U}_b) \Longrightarrow A^h \quad \textit{is stable in} \quad P^h u. \tag{7.273}$$

Under these conditions the linear A^h is stable.

By Theorem 3.23 this stability is inherited for boundedly invertible $G'(u_0)$ to $\Phi^h G$. With our above consistency results, Theorem 4.53 implies the unique existence of the discrete approximation u_0^h and its convergence to the exact solution u_0.

We summarize, for the different problems and their discrete operators.

Condition 7.39. Conditions for stability and consistency for DCGMs; summary of these results for the different previous problems:

1. Choose $\mathcal{U}^h = \mathcal{V}^h = S^h \not\subset \mathcal{U}$ as in (7.63), (7.95) with the penalty norm $\|u^h\|_{\mathcal{V}^h} = \|u^h\|$ in (7.153) and the parameter c_w as in (7.39), (7.164). For $b(\cdot,\cdot)$ we impose **Assumption (H)**, cf. (7.51)–(7.53), and for our DCG elements S^h of local degree $d-1$ the conditions (A_0)–(A_2), (7.124)–(7.128), (7.146). In particular, the S^h are approximating spaces for the function spaces considered here. We always denote the S^h-interpolant for u from Theorem 7.7 as $P^h u$.
2. We consider the *nonsymmetric*, the *symmetric* or the *incomplete or general interior penalty Galerkin method,* the NIPG, SIPG, IIPG or GIPG method, respectively, with $\theta = -1, 1, 0$ or $\theta \in [-1, 1]$.
3. For all problems, the operator G and its linearization in u_0, the $G'(u_0)u = f \in \mathcal{U}'$, has to be combined with the boundary condition (7.14), so $u\,|_{\partial\Omega_D} = u_D$ on $\partial\Omega_D$, and $\partial u/\partial\nu\,|_{\partial\Omega_N} = \phi$ on $\partial\Omega_N$. In particular, we require for all cases the bounded invertibility of $G'(u_0)$ with the Dirichlet and Neumann or only Dirichlet boundary conditions, e.g. (7.14), cf. Theorem 7.14.
4. If in the consistency results a $|u|_{H^{\min\{d,s\}}(\Omega)}$ is shown, we always require $u \in H^{\min\{d,s\}}(\Omega)$; for the hp-methods in Section 7.14 we replace this by $|u|_{H^{\min\{d_K,s\}}(K)}$ for $u \in H^{\min\{d_K,s\}}(K)$, respectively Throughout this section we use the abbreviation

$$C(h,u) := h^{\min\{d,s\}-1} |u|_{H^{\min\{d,s\}}(\Omega)}. \tag{7.274}$$

5. For the special convection–diffusion problem (7.13), (7.14) we impose (7.16), and choose a numerical flux satisfying **Assumption (H)**: (7.51)–(7.53). Define its *standard DCGM* as in (7.70). Then coercivity and stability is guaranteed by Theorem 7.16, (7.165), (7.166). The following consistency estimates are valid, cf.

Theorems 7.10, 7.21, 7.22, 7.23, 7.26,

$$|c_h(u,v^h) - c^h(P^h u, v^h)| \leq C_{10}C(h,u)\|v^h\| \textit{ for Dirichlet boundary cd.,} \tag{7.275}$$

$$|c_h(u,v^h) - c^h(P^h u, v^h)| \leq C_{10}C(h,u)(1+\|v^h\|) \textit{ for Dir. and Neumann b.cs.}$$

$$|a_h(u,v^h) - a^h(P^h u, v^h)| \ \leq \varepsilon\, C_{17}C(h,u)\|v^h\|,$$

$$\left|\varepsilon\ J_h^\sigma(u,v^h) - \varepsilon\ J_g^\sigma(P^h u, v^h)\right| \ \leq \varepsilon\, C_{15}C(h,u)\|v^h\|$$

$$\text{and } \ell^h(v^h) - \ell_h(v^h) = 0\ \forall v^h \in \mathcal{V}^h = S^h.$$

6. *The nonstandard DCGM* for this problem (7.13), (7.14), (7.16) is defined in (7.71). The coercivity and consistency estimates for the a_h, b_h, ℓ_h are the same as for the previous a_h, c_h, ℓ_h.
7. For the linear operator (7.74), (7.75) with its conditions we define its DCGM in (7.77). Then coercivity (and stability) and the consistency estimates are guaranteed by obvious modifications of Theorems 7.18, and 7.10, 7.23, 7.31. Modifying the previous consistency estimates requires changing the constant in (7.275), (7.276).
8. For the semilinear convection–diffusion operator (7.78), (7.79) we impose (7.80), and define its DCGM as in (7.89).Then coercivity (and stability) is guaranteed by Theorems 7.10, 7.18, (7.181), (7.182) for c_w in (7.180). For $u \in W^{2,\infty}(\mathcal{T}^h)$, we obtain the following consistency estimates, cf. Theorems 7.10, 7.21, 7.24, 7.31, here (7.204), (7.211), (7.245),

$$|c_h(u,v^h) - c_h(P^h u, v^h)| \leq C_{10}C(h,u)\|v^h\| \textit{ for Dirichlet boundary cd.,} \tag{7.276}$$

$$|c_h(u,v^h) - c^h(P^h u, v^h)| \leq C_{10}C(h,u)(1+\|v^h\|) \textit{ for Dir. and Neumann b.cs.}$$

$$|a_h(u,v^h) - a_h(P^h u, v^h)| \leq \varepsilon C(h,u)\big(C_1^* + 2C_2^*\|u\|_{W^{1,\infty}(\mathcal{T}^h)}\big)\|v^h\|, \tag{7.277}$$

$$\left|\varepsilon\ J_h^\sigma(u,v^h) - \varepsilon\ J_g^\sigma(P^h u, v^h)\right| \ \leq \varepsilon\, C_{15}C(h,u)\|v^h\|$$

$$|\ell_h(P^h u, v^h) - \overline{\ell}_h(u,v^h)| \leq C_{13}C(h,u)\ \big(1+\|u\|_{L^\infty(\partial\Omega)}\big)\|v^h\|_{H^1(\mathcal{T}^h)}\forall v^h \in \mathcal{V}^h.$$

9. For the quasilinear system (7.93), we impose (7.94), and define its DCGM as in (7.101). Then coercivity (and stability) is guaranteed by Theorem 7.19, (7.195), (7.196) for c_w in (7.194). For $u \in W^{2,\infty}(\mathcal{T}^h, \mathbb{R}^q)$, we get the following consistency estimates, cf. Theorems 7.10, 7.23, 7.33, here (7.257),

$$|a_h(\vec{u},\vec{v}^h) - a_h(P^h\vec{u},\vec{v}^h)| \leq C_{16}C(h,u) \tag{7.278}$$

$$\Big(1+\|\vec{u}\|_{W^{1,\infty}(\mathcal{T}^h,\mathbb{R}^q)} + \|\vec{u}\|^2_{W^{1,\infty}(\mathcal{T}^h,\mathbb{R}^q)}\Big)\|\vec{v}^h\|,$$

$$\left|J_h^\sigma(u,v^h) - J_g^\sigma(P^h u, v^h)\right| \ \leq C_{15}C(h,u)\|v^h\|$$

$$\ell^h(v^h) - \ell_h(v^h) = 0\ \forall v^h \in \mathcal{V}^h = S^h.$$

10. For the special quasilinear equation in (7.102), Houston *et al.* [407] formulate their DCGM according to (7.109); for systems they provide all necessary tools, but do not formulate (7.110). Under the conditions (7.103)–(7.105), the coercivity (and stability) is guaranteed by Theorem 7.19, (7.195), (7.196) for c_w in (7.194). The following consistency estimates are valid, cf. Theorem 7.10, 7.23, 7.36. So instead of $a_h(\cdot,\cdot),\ldots,$ in (7.278) we now formulate for the G_h, analogous to the G^h in (7.109), (7.110), cf. (7.268),

$$|G_h(u,v^h) - G_h(P^h u, v^h)| \le C_1^* C(h,u)\|v^h\| \ \ \forall \theta \in [-1,1]. \qquad (7.279)$$

We summarize the convergence results for all DCGMs in Theorem 7.40. In particular, we have the same order of convergence for the DCGMs as for the classical FEMs. It is straightforward along the lines of the above proofs, to prove consistent differentiability as well. This is the condition for the mesh independence principle in Theorem 3.40.

Theorem 7.40. Convergence for the DCGMs for the above cases 5–10: *Under the previous conditions and notation (1)–(4) and for inhomogeneous* Dirichlet and/or Neumann boundary conditions *choose, for the problems in (5)–(10), the quoted DCGMs with the above coercivity. Hence stability results for a boundedly invertible $G'(u_0)$, and the consistency estimates (7.274)–(7.279) are granted.*

Then for the $G(u_0) = 0$, the application of these DCGMs define the $G^h\left(u_0^h\right) = 0$. Then for sufficiently small h, a unique solution $u_0^h \in S^h$ exists near u_0 and

$$\left\|u_0^h - u_0\right\|_{H^1(\mathcal{T}^h)}^J \le CC(h,u_0), \ \ \text{with } C(h,u_0) := h^{\min\{d,s\}-1}|u_0|_{H^{\min\{d,s\}}(\Omega)} \qquad (7.280)$$

with C depending upon u_0 for problems (8)–(10). There, we have to impose additionally $u_0 \in W^{2,\infty}(\mathcal{T}^h)$.

For the special quasilinear equation (system) in [407] we formulate:

Theorem 7.41. Convergence via monotony for the DCGMs for case 10: *We assume the previous conditions and notation (1), (2), (4) and (7.103), consequently (7.105). Then for inhomogeneous* Dirichlet and/or Neumann boundary conditions *choose the DCGMs in (7.109) and (7.110) for equations and systems, respectively, with the consistency estimates (7.279). Then there exist unique solutions u_0 and $u_0^h \in S^h$ for $G(u_0) = f$ and $G^h\left(u_0^h\right) = Q^h f$, respectively These u_0^h converge to u_0 according to*

$$\left\|u_0^h - u_0\right\|_{H^1(\mathcal{T}^h)}^J \le CC(h,u_0), \ \ \text{with } C(h,u_0) \text{ in (7.280)}. \qquad (7.281)$$

Proof. Under the above conditions, Lemmas 2.2 and 2.3 in [407] show that G^h is Lipschitz-continuous and monotone with respect to u^h in $W^{1,p}(\mathcal{T}^h)$. Therefore, Theorem 2.68 implies the unique existence of the solutions, and Theorem 4.67 *(5)*, and Theorems 7.5 ff., and Remark 7.4, show the convergence according to (7.281). □

Theorem 7.42. Discontinuous Galerkin methods for monotone operators, eigenvalue problems, nonlinear boundary operators, quadrature approximations: *Under the conditions of Theorem 7.40, discontinuous Galerkin methods applied to the following* *problems yield unique converging solutions for variational methods for eigenvalue problems, cf. Section 4.7, methods for nonlinear boundary operators, cf. Section 5.3, quadrature approximate methods, cf. Section 5.4. For all these problems (7.280) remains valid.*

Remark 7.43. *By the appropriate modifications, cf. Section 7.14, these three theorems remain valid for hp-methods.*

7.13 Solving nonlinear equations in DCGMs

7.13.1 Introduction

For solving the nonlinear DCG equations, we restrict ourselves to presenting a discrete Newton method based upon the mesh independence principle (MIP). It is formulated and proved only for quasilinear systems. Except for the modified Lipschitz condition for the derivatives, the results are the same for semilinear equations as well.

For this MIP we have to determine the linearized form of the quasilinear system and its discretization, based upon the same S^h as before. This derivative has to satisfy the following property, cf. (3.83). A general discretization method is *differentiably consistent* and *of order* p, if it is classically consistent and if for a Frechet differentiable nonlinear operator G,

$$\|(G^h)'(P^h u)P^h v - Q'^h(G'(u)v)\|_{\mathcal{V}'^h} \to 0 \text{ and} \tag{7.282}$$
$$\|(G^h)'(P^h u)P^h v - Q'^h(G'(u)v)\|_{\mathcal{V}'^h} \le Ch^p(1 + \|u\|_{\mathcal{U}_s})\|v\|_{\mathcal{U}_s} \text{ for } h \to 0.$$

These $\|v\|_{\mathcal{U}_s}$ are usually $\|v\|_{W^{k,q}(\Omega)}$ and p, k, q appropriately chosen. For our most complicated DCGM, essentially including the previous problems as special cases, we prove this property.

7.13.2 Discretized linearized quasilinear system and differentiable consistency

Hence we have to determine the Frechet derivatives of the $a_h(\cdot,\cdot)$ in (7.93), (7.285). With the $A_k(\cdot, \vec{u}_1, \nabla\vec{u}_1)$ in (7.93), we define these weak forms, cf.(7.94), (7.93),

$$\begin{aligned}
\langle G'(\vec{u}_1)\vec{u}, \vec{v}\rangle &:= \sum_{K\in\mathcal{T}^h}\int_K \sum_{k=0}^{n}\left(\frac{\partial A_k}{\partial\vec{u}}(\cdot, \vec{u}_1, \nabla\vec{u}_1)\vec{u} + \frac{\partial A_k}{\partial\mathbf{u}}(\cdot, \vec{u}_1, \nabla\vec{u}_1)\nabla\vec{u}, \partial^k\vec{v}\right)_q dx \\
&= \sum_{K\in\mathcal{T}^h}\int_K \sum_{k=0}^{n}\left(A'_k(\cdot, \vec{u}_1, \nabla\vec{u}_1)\nabla^0\vec{u}, \partial^k\vec{v}\right)_q dx \\
&= \sum_{K\in\mathcal{T}}\int_K \left(B'_A(\vec{u}_1, \nabla\vec{u}_1)\nabla^0\vec{u}, \nabla^0\vec{v}\right)_q^0 dx,
\end{aligned} \tag{7.283}$$

where we used and will use the notation

$$A'_k(\cdot,\vec{u}_1,\nabla\vec{u}_1)\nabla^0\vec{u} := \frac{\partial A_k}{\partial \vec{u}}(\cdot,\vec{u}_1(\cdot),\nabla\vec{u}_1(\cdot))\vec{u}(\cdot) + \frac{\partial A_k}{\partial \vec{\mathbf{u}}}(\cdot,\vec{u}_1,\nabla\vec{u}_1)\nabla\vec{u} \in \mathbb{R}^q \,\mathrm{in}\, x\in\Omega$$

$$\left(B'_A(\vec{u}_1,\nabla\vec{u}_1)\nabla^0\vec{u},\nabla^0\vec{v}\right)^0_q := \sum_{k=0}^n \left(A'_k(\cdot,\vec{u}_1,\nabla\vec{u}_1)\nabla^0\vec{u},\partial^k\vec{v}\right)_q \quad \in \mathbb{R} \text{ in } x\in\Omega, \tag{7.284}$$

and

$$\left(B'_A(\vec{u}_1,\nabla\vec{u}_1)\nabla^0\vec{u},\nu\right)_n := \sum_{k=1}^n A'_k(\cdot,\vec{u}_1,\nabla\vec{u}_1)\nabla^0\vec{u}\nu_k \quad \in \mathbb{R}^q \text{ in } x\in\Omega,$$

Again we omitted, in the nonlinear and consequently in the linearized form of the DCGM, the $\left(\left(\left[B'_A(\vec{u}_1,\nabla\vec{u}_1)\nabla^0\vec{u}\right],\nu\right)_n,\{\vec{v}\}\right)_q$, vanishing for $\vec{u}=\vec{u}_0$ and only consider the IIPG for $\theta=0$. We collect the $\sum_{e\in\mathcal{T}^h\setminus\partial\Omega}$ and $\sum_{e\in\partial\Omega}$ into $\sum_{e\in\mathcal{T}^h}$. Compared to (7.76), we replace the $(B_a\nabla^0 u,\nu)_n$ by $\left(B'_A(\vec{u}_1,\nabla\vec{u}_1)\nabla^0\vec{u},\nu\right)_n$ etc. With this notation, we introduce, similarly to (7.57)–(7.61), the linearized quasilinear, and linear forms, $a'_h(\vec{u}_1;\vec{u},\vec{v})$, $\ell_h(\vec{v})$, and use $J^\sigma_h(\vec{u},\vec{v})$,

$$\begin{aligned} a'_h(\vec{u}_1;\vec{u},\vec{v}) &:= \sum_{K\in\mathcal{T}^h}\int_K \left(B'_A(\vec{u}_1,\nabla\vec{u}_1)\nabla^0\vec{u},\nabla^0\vec{v}\right)^0_q dx \\ &\quad - \sum_{e\in\mathcal{T}^h}\int_e \left(\left(\left\{B'_A(\vec{u}_1,\nabla\vec{u}_1)\nabla^0\vec{u}\right\},\nu\right)_n,[\vec{v}]\right)_q dS \\ \ell_h(\vec{v}) &:= \langle g,\vec{v}\rangle + \sum_{e\in\partial\Omega}\int_e \sigma(\vec{u}_D,\vec{v})_q\, dS. \end{aligned} \tag{7.285}$$

For this discrete *IIPG method* ($\theta=0$) for the linearized quasilinear problem (7.93), (7.91) we determine the *discrete approximate solution* $\vec{u}_2^h$ such that

$$\begin{aligned} &\text{(a)}\quad \vec{u}_2^h\in S^h, \text{ cf. (7.95)}, \\ &\text{(b)}\quad \left\langle (G^h)'\left(\vec{u}_1^h\right)\vec{u}_2^h,\vec{v}^h\right\rangle := a'_h\left(\vec{u}_1^h;\vec{u}_2^h,\vec{v}^h\right) + J^\sigma_h\left(\vec{u}_2^h,\vec{v}^h\right) - \ell^h(\vec{v}^h) = 0 \quad \forall\vec{v}^h\in S^h. \end{aligned} \tag{7.286}$$

For quadratic convergence of Newton's method, Lipschitz-continuous derivatives of the operator are necessary. The same holds for its discrete counterpart. We assume for semilinear problems, in addition to (7.80),

(a) For (b) –(d) we assume $|u|\le R_0$ and an $L'=L'(R_0)$ such that (7.287)

(b) $f_s\in C^1_{L'}(\mathbb{R})$ are Lipschitz-continuous with the constant L',

(c) $(\partial a_{ij}/\partial u)(\cdot,u)$ are Lipschitz-continuous with respect to u with the constant L',

(d) $(\partial g/\partial u)(\cdot,u)$ are Lipschitz-continuous with respect to u with the constant L'.

For quasilinear problems, we avoid a formulation analogous to (7.94), by imposing conditions, similar to (7.287), in the form

$$\begin{aligned}&\text{(a) We assume the } |\vec{u}| \le R_0 \text{ and an } L' = L'(R_0) \text{ such that} \\ &\text{(b) } (\partial A_k/\partial \vec{u})(\vec{u}, \mathbf{\vec{u}}),\ (\partial A_k/\partial \mathbf{\vec{u}})(\vec{u}, \mathbf{\vec{u}}), (\partial A_0/\partial \vec{u})(\vec{u}, \mathbf{\vec{u}}), (\partial A_0/\partial \mathbf{\vec{u}})(\vec{u}, \mathbf{\vec{u}}) \\ &\quad \text{are Lipschitz-continuous with respect to } \vec{u}, \mathbf{\vec{u}}, \text{ with the constant } L' \forall |\vec{u}| \le R_0.\end{aligned} \tag{7.288}$$

We turn to our main goal, the consistency of the terms in (7.286). This has been proved for the $J_h^\sigma(\vec{u}^h, \vec{v}^h)$ and $\ell^h(\vec{v}^h)$ in the previous sections; only $a_h'\left(\vec{u}_1^h; \vec{u}^h, \vec{v}^h\right)$ is missing. We proceed as above in two steps.

Proposition 7.44. *Under the condition (7.288), and for all $\|\vec{u}_1\|_{H^1(\mathcal{T}^h)} \le C'$, the $a_h'(\vec{u}_1; \vec{u}, \vec{v})$ are nonlinear and bounded with respect to $\vec{u}_1$, and bilinear and bounded with respect to $\vec{u}, \vec{v}$ and satisfy the estimate for all $\vec{u}, \vec{v}, \overline{\vec{u}}, \vec{u}_1 \in \mathcal{U}_e$, $\vec{u} \in W^{2,\infty}(\mathcal{T}^h)$, and with $C_0 := L(1 + (C')^2), L'$ cf. (7.94), (7.288), $H^1(\mathcal{T}^h) = H^1(\mathcal{T}^h, \mathbb{R}^q), \ldots$,*

$$\begin{aligned}\left|a_h'(\vec{u}_1; \vec{u}, \vec{v}) - a_h'(\overline{\vec{u}}_1; \overline{\vec{u}}, \vec{v})\right| &\le C_0 \|\overline{\vec{u}} - \vec{u}\|_{H^1(\mathcal{T}^h)} \|\vec{v}\|_{H^1(\mathcal{T}^h)} \\ &\quad + 2nC_0 \left(\frac{1}{c_w}\right)^{1/2} \left(J_h^\sigma(\vec{v}, \vec{v})\right)^{1/2} \\ &\quad \times \left(\sum_{K \in \mathcal{T}^h} h_K |\vec{u} - \overline{\vec{u}}|^2_{H^1(\partial K)} + h_K \|\vec{u} - \overline{\vec{u}}\|^2_{L^2(\partial K)}\right)^{1/2} \\ &\quad + nL' \|\overline{\vec{u}}_1 - \vec{u}_1\|_{H^1(\mathcal{T}^h)} \|\vec{u}\|_{W^{1,\infty}(\Omega)} \|\vec{v}\|_{H^1(\mathcal{T}^h)} \\ &\quad + 2nL'/(c_w)^{1/2} \|\vec{u}\|_{W^{1,\infty}(\mathcal{T}^h)} \left(J_h^\sigma(\vec{v}, \vec{v})\right)^{1/2} \\ &\quad \times \left(\sum_{K \in \mathcal{T}^h} h_K \|\vec{u} - \overline{\vec{u}}\|^2_{L^2(\partial K)}\right)^{1/2}.\end{aligned} \tag{7.289}$$

Proof. The boundedness follows from (7.289) for $\overline{\vec{u}} = 0$ and use, cf. (7.94), of the bounds for $|A_k, A_k'| \le C_0$. We consider the arguments of $a_h'(\vec{u}_1; \vec{u}, \vec{v})$ as $\vec{u}, \overline{\vec{u}}, \vec{u}_1, \overline{\vec{u}}_1, \vec{v} \in \mathcal{U}_e$. Then

$$\begin{aligned}|a_h'(\overline{\vec{u}}_1; \overline{\vec{u}}, \vec{v}) - a_h'(\vec{u}_1; \vec{u}, \vec{v})| &\le \left|a_h'(\overline{\vec{u}}_1; \overline{\vec{u}}, \vec{v}) - a_h'(\overline{\vec{u}}_1; \vec{u}, \vec{v})\right| \\ &\quad + \left|a_h'(\overline{\vec{u}}_1; \vec{u}, \vec{v}) - \overline{a}_h(\vec{u}_1; \vec{u}, \vec{v})\right|.\end{aligned} \tag{7.290}$$

The first term is estimate via the σ_{19}, σ_{20} as

$$|a_h'(\overline{\vec{u}}_1;\overline{\vec{u}},\vec{v}) - a_h'(\overline{\vec{u}}_1;\vec{u},\vec{v})| \leq \underbrace{\left|\sum_{K\in\mathcal{T}^h}\int_K \left(B_A'(\overline{\vec{u}}_1,\nabla\overline{\vec{u}}_1)\nabla^0(\overline{\vec{u}}-\vec{u}),\nabla^0\vec{v}\right)_q^0 dx\right|}_{:=\sigma_{19}}$$

$$+\underbrace{\left|\sum_{e\in\mathcal{T}^h}\int_e \left(\left(\{B_A'(\overline{\vec{u}}_1,\nabla\overline{\vec{u}}_1)\nabla^0(\overline{\vec{u}}-\vec{u})\},\nu\right)_n [\vec{v}]\right)_q dS\right|}_{:=\sigma_{20}}.$$

σ_{19} is estimated with both Cauchy–Schwarz inequalities (1.44), for $p = q = 2$, (1.45),

$$|\sigma_{19}| \leq C_0 \sum_{K\in\mathcal{T}^h} \|\overline{\vec{u}}-\vec{u}\|_{H^1(K)}\|\vec{v}\|_{H^1(K)} \leq C_0\|\overline{\vec{u}}-\vec{u}\|_{H^1(\mathcal{T}^h)}\|\vec{v}\|_{H^1(\mathcal{T}^h)}.$$

For σ_{20} the same inequalities (1.44), (1.45), and $d(e)$ in (7.39) yield

$$\begin{aligned}
|\sigma_{20}| &\leq nC_0 \sum_{e\in\mathcal{T}^h}\int_e |[\vec{v}]|_q \,|\, \nabla^0\vec{u}-\nabla^0\overline{\vec{u}}|_q\, dS \\
&\leq nC_0\left(\left(\sum_{e\in\mathcal{T}^h}\int_e \frac{([\vec{v}],[\vec{v}])_q}{d(e)/c_w}\,dS\right)^{1/2}\left(\sum_{e\in\mathcal{T}^h}\frac{d(e)}{c_w}\int_e |\,\nabla^0\vec{u}-\nabla^0\overline{\vec{u}}|_q^2\, dS\right)^{1/2}\right) \\
&\leq 2nC_0\left(\frac{1}{c_w}\right)^{1/2} \\
&\qquad\times\ (J_h^\sigma(\vec{v},\vec{v}))^{1/2}\left(\sum_{K\in\mathcal{T}^h}\left(h_K|\vec{u}-\overline{\vec{u}}|_{H^1(\partial K)}^2 + h_K\|\vec{u}-\overline{\vec{u}}\|_{L^2(\partial K)}^2\right)\right)^{1/2}.
\end{aligned}$$

For the second term in (7.290) we introduce σ_{21}, σ_{22} as

$$\begin{aligned}
&|a_h'(\overline{\vec{u}}_1;\vec{u},\vec{v}) - a_h'(\vec{u}_1;\vec{u},\vec{v})| \\
&\quad\leq \underbrace{\left|\sum_{K\in\mathcal{T}^h}\int_K \left(\left(B_A'(\overline{\vec{u}}_1,\nabla\overline{\vec{u}}_1)-B_A'(\vec{u}_1,\nabla\vec{u}_1)\right)\nabla^0\vec{u},\nabla^0\vec{v}\right)_q^0 dx\right|}_{:=\sigma_{21}} \\
&\quad+\underbrace{\left|\sum_{e\in\mathcal{T}^h}\int_e \left(\left(\{(B_A'(\overline{\vec{u}}_1,\nabla\overline{\vec{u}}_1)-B_A'(\vec{u}_1,\nabla\vec{u}_1))\nabla^0\vec{u}\},\nu\right)_n [\vec{v}]\right)_q dS\right|}_{:=\sigma_{22}}.
\end{aligned}$$

Obviously we obtain with the Lipschitz constant, L', for the B_A with (1.43),

$$|\sigma_{21}| \leq nL'\|\overline{\vec{u}}_1 - \vec{u}_1\|_{H^1(\mathcal{T}^h)}\|\vec{u}\|_{W^{1,\infty}(\Omega)}\|\vec{v}\|_{H^1(\mathcal{T}^h)}.$$

A combination with the ideas in (7.212) ff., allows the estimation of σ_{22}. Here we need $\|\nabla^0\vec{u}\|_{L^\infty(e)} \leq \|\vec{u}\|_{W^{1,\infty}(K)}$, valid for $\vec{u} \in W^{2,\infty}(\mathcal{T}^h)$ by Theorem 1.38.

$$|\sigma_{22}| \leq 2nL' \left(\sum_{e\in\mathcal{T}^h} \|\nabla^0\vec{u}\|^2_{L^\infty(e)} \int_e \frac{([\vec{v}],[\vec{v}])_q}{d(e)/c_w}\, dS \right)^{1/2}$$

$$\left(\sum_{e\in\mathcal{T}^h} \frac{d(e)}{c_w} \int_e \sum_{i=1}^{n} (\vec{u} - \overline{\vec{u}}, \vec{u} - \overline{\vec{u}})_q\, dS \right)^{1/2}$$

$$\leq 2nL'/(c_w)^{1/2}\|\vec{u}\|_{W^{1,\infty}(\mathcal{T}^h)}\, (J_h^\sigma(\vec{v},\vec{v}))^{1/2} \left(\sum_{K\in\mathcal{T}^h} h_K\|\vec{u} - \overline{\vec{u}}\|^2_{L^2(\partial K)} \right)^{1/2}.$$

Adding the estimates for σ_{19}–σ_{22} proves (7.289). □

Proposition 7.44 implies with $\overline{\vec{u}} = P^h\vec{u}$, as in Theorem 7.21

Theorem 7.45. Differentiable consistency for quasilinear DCGMs: *We assume $\vec{u}, \vec{u}_1 \in \mathcal{U}_e \subset \mathcal{U}$, $\vec{u} \in W^{2,\infty}(\mathcal{T}^h)$, $\vec{u}_1 \in \mathcal{D}(G)$, with $\left\|\vec{u}_1^h\right\|_{H^1(\mathcal{T}^h)} \leq C'$, $\left\|\vec{u}_1^h\right\|_{L^\infty(\mathcal{T}^h)} \leq R_0$ and $C_0 := L(1+(C')^2), L' = L'(R_0)$, cf. (7.94), (7.288) and these latter conditions. Then $(G^h)'(P^hu_1)P^hu$ are consistent with $G'(u_1)u$. More precisely, $a'_h\left(\vec{u}_1^h;\vec{u}^h,\vec{v}^h\right)$ are bounded, nonlinear and bilinear with respect to $\vec{u}_1^h$, and $\vec{u}^h,\vec{v}^h$. A constant, $C_{18} := C_0c_e(2+nL'\|\vec{u}\|_{W^{1,\infty}(\Omega)}) \max\{C_7, 2n\sqrt{2C_5C_7/c_w}\}$ exists, such that for the S^h-interpolants $P^h\vec{u}, P^h\vec{u}_1$ of $\vec{u}, \vec{u}_1$ in Theorem 7.7,*

$$|a'_h(\vec{u}_1;\vec{u},\vec{v}^h) - a'_h(P^h\vec{u}_1;P^h\vec{u},\vec{v}^h)| \tag{7.291}$$

$$\leq C_{18} \left(\sum_{K\in\mathcal{T}^h} \left(h_k^{\min\{d,s\}-1} |\vec{u}|_{H^{\min\{d,s\}}(K)} \right)^2 \right)^{1/2} \|\vec{v}^h\|.$$

With the classical consistency in Theorems 7.10, 7.33, this DCGM for the quasilinear system is differentiably consistent.

Proof. The proof is nearly identical to that of Theorem 7.26. For (7.291) and by (7.289) we again use (7.222), however replacing the first two lines by

$$\|P^h\vec{u} - \vec{u}\|_{H^1(\mathcal{T}^h)} \leq 2C_7 \left(\sum_{K\in\mathcal{T}^h} \left(h_k^{\min\{d,s\}-1} |\vec{u}|_{H^{\min\{d,s\}}(K)} \right)^2 \right)^{1/2}, \text{ by (7.139)},$$

$$\times \left(\sum_{K\in\mathcal{T}^h} h_K \|\vec{u} - P^h\vec{u}\|^2_{H^1(\partial K)} \right)^{1/2} \tag{7.292}$$

$$\leq 2\sqrt{2C_5C_7} \left(\sum_{K\in\mathcal{T}^h} \left(h_k^{\min\{d,s\}-1} |\vec{u}|_{H^{\min\{d,s\}}(K)} \right)^2 \right)^{1/2},$$

(7.222), (7.292) are applied to the different error terms $\vec{u} - P^h\vec{u}$ in (7.289):

$$\begin{aligned}
&\left|a_h'(\vec{u}_1; \vec{u}, \vec{v}^h) - a_h'(P^h\vec{u}_1; P^h\vec{u}, \vec{v}^h)\right| \\
&\leq C_0 \left(\sum_{K\in\mathcal{T}^h} \left(h_k^{\min\{d,s\}-1} |\vec{u}|_{H^{\min\{d,s\}}(K)} \right)^2 \right)^{1/2} \\
&\times \Big(2C_7 \|\vec{v}^h\|_{H^1(\mathcal{T}^h)} + 4n\sqrt{2C_5C_7/c_w}\, \left(J_h^\sigma(\vec{v}^h, \vec{v}^h)\right)^{1/2} + nL'C_7 \|\vec{u}\|_{W^{1,\infty}(\Omega)} \|\vec{v}^h\|_{H^1(\mathcal{T}^h)} \\
&\qquad + 2nL'\sqrt{2C_5C_7/c_w}\, h\|\vec{u}\|_{W^{1,\infty}(\mathcal{T}^h)} \left(J_h^\sigma(\vec{v}^h, \vec{v}^h)\right)^{1/2} \Big) \\
&\qquad \leq C_0 c_e (2 + nL'\|\vec{u}\|_{W^{1,\infty}(\Omega)}) \max\left\{C_7, 2n\sqrt{2C_5C_7/c_w}\right\} \\
&\qquad \left(\sum_{K\in\mathcal{T}^h} \left(h_k^{\min\{d,s\}-1} |\vec{u}|_{H^{\min\{d,s\}}(K)} \right)^2 \right)^{1/2} \|\vec{v}^h\|.
\end{aligned}$$

This shows, with (7.225) and Proposition 7.9, the estimate (7.291). Finally, Theorems 7.12 and 7.13 imply the last estimate in (7.291). □

We modify Theorem 7.45 for the semilinear, in particular, the model problem. We only have to replace (7.94), (7.288) by (7.80), (7.287), and C_0 by L'.

Theorem 7.46. Differentiable consistency for semilinear and the special quasilinear DCGMs: *We assume $u, u_1 \in \mathcal{U}_e \subset \mathcal{U}$, $u \in W^{2,\infty}(\mathcal{T}^h)$, $u_1 \in \mathcal{D}(G)$, and, the model problem, (7.13), the conditions (7.80), (7.287) (a), (b), for the semilinear problem, (7.78), the conditions (7.80), (7.287), and finally for the special quasilinear problem, (7.102), the conditions (7.288). Then the $(G^h)'(P^hu_1)P^hu$ are consistent with the $G'(u_1)u$.*

A $C'_{18} := c_e L'(2 + nL' \; \|u\|_{W^{1,\infty}(\Omega)}) \; \max\left\{C_7, 2n\sqrt{2C_5C_7/c_w}\right\}$ *exists, such that*

$$|a'_h(u_1; u, v^h) - a'_h(P^h u_1; P^h u, v^h)| \tag{7.293}$$

$$\leq C_{18} \left(\sum_{K \in \mathcal{T}^h} \left(h_k^{\min\{d,s\}-1} |u|_{H^{\min\{d,s\}}(K)} \right)^2 \right)^{1/2} \|v^h\|.$$

With the classical consistency in Theorems 7.21–7.31, this DCGM for the semilinear systems, (7.13), (7.78), are differentiably consistent. For the special quasilinear problem, (7.102), slight modifications of the constant C'_{18} *are necessary.*

We summarize the results of Theorem 3.40 and Corollary 3.41 for our DCGMs

Theorem 7.47. Newton's method for DCGMs G^h: *Let* $u_0 \in H^s(\Omega)$ *be an isolated solution of* $G(u_0) = 0$*, such that* $G'(u_0) : \mathcal{U} \to \mathcal{V}'$ *is boundedly invertible and let* G^h *indicate one of the DCGMs for nonlinear problems in Theorems 7.45 and 7.46 with the corresponding conditions. Start the Newton process with* $u_1 \in H^s(\Omega)$*, and the discrete Newton process for* $i = 1, \ldots,$ *with* $u_1^h := P^h u_1$*, for small enough* $\|u_0 - u_1\|_{\mathcal{U}}$ *and* h*. That yields*

$$u_{i+1}^h := u_i^h - \left(G^h \left(u_i^h \right)' \right)^{-1} G^h \left(u_i^h \right), \tag{7.294}$$

cf. (3.84). The u_{i+1}^h *in (7.294) uniquely exist and converge quadratically to* u_0^h *and to* u_{i+1} *of order* $\min\{d,s\} - 1$*, such that*

$$\left\|u_{i+1}^h - u_i^h\right\|_{\mathcal{U}^h} \leq C \left\|u_i^h - u_{i-1}^h\right\|_{\mathcal{U}^h}^2 \quad \text{and}$$
$$\left\|u_i^h - P^h u_i\right\|_{\mathcal{U}^h} \leq C h^{\min\{d,s\}-1} \|u_i\|_{\mathcal{U}_s}, \; i = 1, \ldots.$$

The following relation between the numbers of iterations of the exact and the discrete method is the reason for calling this a "mesh independence principle", MIP. For small enough η *:*

$$\left| \min\{i \geq 1 : \left\|u_i^h - P^h u_0\right\|_{\mathcal{U}^h} < \eta\} - \min\{i \geq 1 : \|u_i - u_0\|_{\mathcal{U}} < \eta\} \right| \leq 1. \tag{7.295}$$

7.14 *hp*-variants of DCGM

For many important problems the local smoothness of the solution strongly varies in Ω. It is appropriate to choose the local degrees according to this smoothness. DCGMs are based upon a discontinuous piecewise polynomial approximation. So there is no principal obstacle for using these different degrees of polynomial approximation on different elements.

The corners of the domain and singularities of the data are particularly interesting. The real power of *hp*-methods, e.g. the *exponential convergence*, strongly depends upon a careful combination of the analytic structure of the singularities of the solution with mesh refinement strategies. In Chapter 6 we have discussed these problems for FEMs. Here we only formulate the necessary tools for *hp*-methods.

We list a few from the many papers published for *hp* DCGMs, by B. Cockburn, P. Houston, M. Luskin, I. Perugia, D. Schötzau, Ch. Schwab, C-W. Shu, E. Süli, and T.P. Wihler [193, 405, 408, 410–412].

7.14.1 *hp*-finite element spaces

To each $K \in \mathcal{T}^h$, we assign a positive integer s_K (local Sobolev index) and a positive integer p_K (local polynomial degree). Then we define a new type of vector

$$\mathsf{s} \equiv \{s_K, K \in \mathcal{T}^h\}, \qquad \mathsf{p} \equiv \{p_K, K \in \mathcal{T}^h\}. \tag{7.296}$$

We generalize the above *broken Sobolev space* to the vector s and triangulation $\mathcal{T}^h$

$$H^{\mathsf{s}}(\mathcal{T}^h) \equiv \{v: \ v|_K \in H^{s_K}(K) \ \forall\, K \in \mathcal{T}^h\} \tag{7.297}$$

with the norm

$$\|v\|_{H^{\mathsf{s}}(\mathcal{T}^h)} \equiv \left(\sum_{K \in \mathcal{T}^h} \|v\|^2_{H^{s_K}(K)} \right)^{1/2} \tag{7.298}$$

and the seminorm

$$|v|_{H^{\mathsf{s}}(\mathcal{T}^h)} \equiv \left(\sum_{K \in \mathcal{T}^h} |v|^2_{H^{s_K}(K)} \right)^{1/2} \tag{7.299}$$

for the Sobolev space $H^{s_K}(K) \equiv W^{s_K,2}(K)$. If, as in (7.63), (7.95), $s_K = d-1 \ \forall K \in \mathcal{T}^h$, $d-1 \in \mathbb{N}$ then we use the notation $H^{d-1}(\mathcal{T}^h) = H^{\mathsf{s}}(\mathcal{T}^h)$. Obviously,

$$H^{\overline{s}}(\mathcal{T}^h) \subset H^{\mathsf{s}}(\mathcal{T}^h) \subset H^{\underline{s}}(\mathcal{T}^h), \tag{7.300}$$

where $\overline{s} = \max\{s_K, \ s_K \in \mathsf{s}\}$ and $\underline{s} = \min\{s_K, \ s_K \in \mathsf{s}\}$.

Furthermore, we define the space of discontinuous piecewise polynomial functions associated with the vector p by

$$S^{h\mathsf{p}} \equiv \{v: \ v \in L^2(\Omega), \ v|_K \in \mathcal{P}_{p_K-1}(K) \ \forall K \in \mathcal{T}^h\}, \tag{7.301}$$

where $\mathcal{P}_{p_K-1}(K)$ denotes the space of all polynomials on K of degree $\le p_K - 1$, $K \in \mathcal{T}^h$. For deriving a priori *hp* error estimates we assume that there exists a constant $C_P \ge 1$ such that

$$\frac{p_K}{p_{K'}} \le C_P \quad \forall K, K' \in \mathcal{T}^h \text{ such that } K, K' \text{ have a common face.} \tag{7.302}$$

The assumption (7.302) may seem to be rather restrictive. However, we suppose that an application of *hp*-methods to practical problems is efficient and accurate when the polynomial degrees of approximation on neighboring elements do not differ too much. Moreover, to each $K \in \mathcal{T}^h$ we define the parameter

$$d_{h\mathsf{p}}(K) := \frac{h_K}{p_K^2}, \qquad K \in \mathcal{T}^h. \tag{7.303}$$

Another approach was presented by Dryja *et al.* [304] where the harmonic average of h_K and p_K of neighboring elemenents was used. Let $e \in \mathcal{T}^h$ be an inner face shared by two elements K_l^e and K_r^e of $\mathcal{T}^h$. Then we put

$$d_{h\mathsf{p}}(e) := \min\left(d_{h\mathsf{p}}\left(K_l^e\right), d_{h\mathsf{p}}\left(K_r^e\right)\right). \tag{7.304}$$

For a boundary face $e \in \partial\Omega \cap \overline{K_l^e}$ we put

$$d_{h\mathsf{p}}(e) := d_{h\mathsf{p}}\left(K_l^e\right). \tag{7.305}$$

7.14.2 *hp*-DCGMs

Since the essential differences for the *hp*-DCGMs are well demonstrated for the model problem (7.13)–(7.15), we restrict the discussion to this case. Then many of the ideas for adaptive FEMs in Chapter 6 can be applied to DCGMs as well. Similarly as in (7.70) we define the *standard hp-DCGM* and its *discrete approximate solution* as a function $u_0^h : \mathcal{T}^h \to \mathbb{R}$ satisfying the conditions

(a) $u_0^h \in S^{h\mathsf{p}}$, (7.306)

(b) $\left(G^h\left(u_0^h\right), v^h\right) := a^h\left(u_0^h, v^h\right) + c^h\left(u_0^h, v^h\right) + \varepsilon J_h^\sigma\left(u_0^h, v^h\right) - \ell^h(v^h) = 0 \;\forall\, v^h \in S^{h\mathsf{p}}$,

with $a^h(u^h, v^h) := a_h(u^h, v^h)|_{S^{h\mathsf{p}} \times S^{h\mathsf{p}}}$, $c^h(u^h, v^h) := c_h(u^h, v^h)|_{S^{h\mathsf{p}} \times S^{h\mathsf{p}}}$,

and $\ell^h(v^h) := \ell_h(v^h)|_{S^{h\mathsf{p}}}$

and the forms a_h, c_h, ℓ_h and J_h^σ are defined by (7.57)–(7.61). The only difference occurs in the definition of the penalty parameter σ: here (7.39) is replaced by

$$\sigma|_e = \frac{c_w}{d_{h\mathsf{p}}(e)} \qquad e \in \mathcal{T}^h \tag{7.307}$$

where $d_{h\mathsf{p}}(e)$ is given by (7.304) or (7.305) and $c_w > 0$.

Similarly, the *hp*-DCGMs are defined by a formal exchange of S^h by $S^{h\mathsf{p}}$ for the general linear (7.77), semilinear (7.89) and quasilinear (7.286) problems considered within this chapter. This is a great advantage for DCGM.

7.14.3 *hp*-inverse and approximation error estimates

For analysing the *hp*-variant of DCGM we have to restrict the discussion to elements, K, which are simplices and/or parallelograms. We allow $(n-1)$- dimensional faces e to be shared by more than two elements, cf. (7.304). We still do not require the conforming properties, usually imposed upon FEMs, but the elements cannot be as general as in (7.27) ff. In particular, a *triangulation*, $\mathcal{T}^h = \{K\}_{K \in \mathcal{T}^h}$ $(h > 0)$, is a partition of Ω into a finite number of open n-dimensional mutually disjoint simplexes and/or parallelograms K, hence star-shaped elements. We assume that $\mathcal{T}^h$ is

- *locally quasiuniform*, i.e. there exists a constant $C_Q > 0$ such that

$$h_K \le C_Q\, h_{K'} \quad \forall K, K' \in \mathcal{T}^h \text{ sharing a face } e \in \mathcal{T}^h \setminus \partial\Omega; \tag{7.308}$$

- *shape-regular*, i.e. there exists a constant $C_S > 0$ such that, cf. (7.126),

$$h_K \le C_S\, \rho_K \quad \forall K \in \mathcal{T}^h, \tag{7.309}$$

where ρ_K is the radius of the largest n-dimensional ball inscribed into K.

The assumptions (7.308) and (7.309) mean that the diameters of two neighboring elements do not differ too much and that elements do not degenerate, respectively.

The definitions (7.304)–(7.305) and the assumptions (7.302) and (7.308) imply that

$$d_{hp}(e) \le d_{hp}(K) \quad \text{and} \quad \frac{1}{d_{hp}(e)} \le C_Q C_P^2 \frac{1}{d_{hp}(K)} \quad \forall e \subset \partial K,\ K \in \mathcal{T}^h. \tag{7.310}$$

The multiplicative trace inequality (7.129) is still valid for this section. On the other hand, we need *hp*-variants of the inverse inequality and approximation error estimates.

Theorem 7.48. Inverse inequality: *There exists a constant $C_I > 0$ independent of v, h and K such that*

$$|v|_{H^1(K)} \le C_I \frac{p_K^2}{h_K} \|v\|_{L^2(K)}, \quad v \in \mathcal{P}_{p_K}(K),\ K \in \mathcal{T}^h,\ h \in (0, h_0). \tag{7.311}$$

Proof. Let $\hat{K}$ be a reference triangle or square for $K \in \mathcal{T}^h$, and $F_K : \hat{K} \to K$ be an affine mapping such that $F_K(\hat{K}) = K$. From Proposition 5.41 it follows that if the function $v \in H^m(K)$, $\mathbb{N} \ni m \ge 0$, then $\hat{v}(\hat{x}) = v(F_K(\hat{x})) \in H^m(\hat{K})$ and

$$|v|_{H^m(K)} \le c_c h_K^{n/2-m} |\hat{v}|_{H^m(\hat{K})}, \tag{7.312}$$

$$|\hat{v}|_{H^m(\hat{K})} \le c_c h_K^{m-n/2} |v|_{H^m(K)}, \tag{7.313}$$

where $c_c > 0$ depends on C_S but not on K and v. From [575], Theorem 4.76 we have

$$|\hat{v}|_{H^1(\hat{K})} \le c_s p_K^2 \|\hat{v}\|_{L^2(\hat{K})}, \quad v \in \mathcal{P}_{p_K}(\hat{K}), \tag{7.314}$$

where $c_s > 0$ depends on n but not on $\hat{v}$ and p_K. A simple combination of (7.312)–(7.314) proves (7.311) with $C_I = c_s c_c^2$. □

For FE interpolation, Condition 4.16 is essential: for the local results in Theorem 4.17 we know that (i): our K are star-shaped, (ii) we choose $\tau = -1$ and any interpolation condition according to (iii) or (iv). The above shape regularity condition (7.309) implies the nondegeneracity of the $\mathcal{T}^h$, cf. Condition 4.16 (v), and (4.30). Note that the constants appearing in Theorem 4.17 are not indepenent of p. Then the local DCG interpolation operator, $I^h = P^h$ yields:

Theorem 7.49. Local approximation properties: *There exists a constant $C_A > 0$, independent of v and h, and a mapping $\pi_p^K : H^s(K) \to \mathcal{P}_p(K)$, $s \ge 1$, e.g. the previous $I^h = P^h$, such that $\forall v \in H^s(K)$, $K \in \mathcal{T}^h$, $h \in (0, h_0)$ the following inequality is valid*

$$\left\|\pi_p^K v - v\right\|_{H^j(K)} \le C_A \frac{h_K^{\min(p,s)-j}}{p^{s-j}} \|v\|_{H^{\min(p,s)}(K)}, \tag{7.315}$$

where $0 \le j \le s$.

Proof. See Lemma 4.5 in [53] for the case $n = 2$. The $n = 3$ arguments are completely analogous. □

Definition 7.50. *Let* s *and* p *be the vectors introduced by (7.296). We define the mapping* $\Pi^{h\mathsf{p}} : H^{\mathsf{s}}(\mathcal{T}^h) \to S^{h\mathsf{p}}$ *by*

$$(\Pi^{h\mathsf{p}} u)\,|_K := \pi^K_{p_K}(u|_K) \quad \forall K \in \mathcal{T}^h, \tag{7.316}$$

where $\pi^K_{p_K} : H^{s_K} \to P_{p_K}(K)$ *is the mapping introduced in Theorem 7.49.*

Theorem 7.51. Global approximation properties: *Let* $\Pi^{h\mathsf{p}} : H^{\mathsf{s}}(\mathcal{T}^h) \to S^{h\mathsf{p}}$ *be the mapping introduced by (7.316), and* $v \in H^{\mathsf{s}}(\mathcal{T}^h)$. *Then*

$$\left\|\Pi^{h\mathsf{p}} v - v\right\|^2_{H^q(\mathcal{T}^h)} \le C_A^2 \sum_{K\in\mathcal{T}^h} \frac{h_K^{2\min(p_K,s_K)-2q}}{p_K^{2s_K-2q}} \|v\|^2_{H^{\min(p_K,s_K)}(K)}, \tag{7.317}$$

where $0 \le q \le \min_{s_K\in\mathsf{s}} s_K$.

Proof. Using definition (7.316), the Cauchy inequality, and the approximation properties (7.315) we obtain (7.317) and C_A is given by Theorem 7.49. □

7.14.4 Consistency and convergence of *hp*-DCGMs

Here we analyze the *hp*–DCGM (7.306) for our model problem. We consider for simplicity $\partial\Omega_D = \partial\Omega$. More general linear, semilinear, and quasilinear cases can be analyzed analogously. The proofs are very similar to *h*-variant (but not identical), since we have to use *hp*-inverse inequality or *hp*-interpolation error estimates. The coercivity as well as boundedness of the form $A_h(u,v) := a_h(u,v) + \varepsilon J_h^\sigma(u,v)$, proved in Theorem 7.16, remain valid for $u, v \in S_{h\mathsf{p}}$. Let us analyze the consistency of forms b_h, a_h and J_h^σ. ℓ_h is independent of u and therefore its consistency immediately follows from Theorem 7.23.

Theorem 7.52. Consistency for the c_h and b_h: *For* $\partial\Omega = \partial\Omega_D$, *there exists a constant* $C_{10} > 0$ *such that for* c_h *and* b_h

$$|c_h(u,v^h) - c^h(\Pi^{h\mathsf{p}}u, v^h)| \le C_{10}\|v^h\| \left(\sum_{K\in\mathcal{T}^h} \frac{h_K^{2\min(p_K,s_K)}}{p_K^{2s_K}} \|u\|^2_{H^{\min(p_K,s_K)}(K)} \right)^{1/2},$$
$$u \in H^{\mathsf{s}}(\mathcal{T}^h),\ v^h \in S^h,\ h \in (0,h_0), \tag{7.318}$$

where $\Pi^{h\mathsf{p}}u$ *is the* $S^{h\mathsf{p}}$*-interpolant of* u *from Theorem 7.51. This estimate remains valid for* $|b_h(u,v^h) - b_h(\Pi^{h\mathsf{p}}u, v^h)|$, *as well.*

Proof. The first part of the proof of Proposition 7.20 remains unchanged, except replacing $d(e), h_K$ in (7.202) by $d_{h\mathsf{p}}(e), d_{h\mathsf{p}}(K)$ here. After (7.202) the proof continues as:

Now, we insert $\overline{u} := \Pi^{h\mathsf{p}}u$ and $v := v^h \in S^{h\mathsf{p}}$ into (7.200)–(7.202) in the modified form. Using the multiplicative trace inequality (7.129), the discrete Cauchy inequality and the approximating properties (7.317), we obtain

$$
\begin{aligned}
&\sum_{K\in\mathcal{T}^h} d_{h\mathsf{p}}(K)\|u-\Pi^{h\mathsf{p}}u\|^2_{L^2(\partial K)} \qquad (7.319)\\
&\quad\le \sum_{K\in\mathcal{T}^h} C_5 d_{h\mathsf{p}}(K)\|u-\Pi^{h\mathsf{p}}u\|_{L^2(K)}\|u-\Pi^{h\mathsf{p}}u\|_{H^1(K)}\\
&\quad\le C_5C_A^2\sum_{K\in\mathcal{T}^h}\frac{h_K}{p_K^2}\frac{h_K^{\min(p_K,s_K)}}{p_K^{s_K}}\frac{h_K^{\min(p_K,s_K)-1}}{p_K^{s_K-1}}\|u\|^2_{H^{\min(p_K,s_K)}(K)}\\
&\quad\le C_5C_A^2\sum_{K\in\mathcal{T}^h}\frac{h_K^{2\min(p_K,s_K)}}{p_K^{2s_K+1}}\|u\|^2_{H^{\min(p_K,s_K)}(K)}.
\end{aligned}
$$

This estimate, (7.317) and (7.198), yield

$$
\begin{aligned}
&|c_h(u,v)-c_h(\Pi^{h\mathsf{p}}u,v)| \qquad (7.320)\\
&\quad\le C_8\|v\|\left(\|u-\Pi^{h\mathsf{p}}u\|_{L^2(\Omega)}+\left(\sum_{K\in\mathcal{T}^h} h_K\|u-\Pi^{h\mathsf{p}}u\|^2_{L^2(\partial K)}\right)^{1/2}\right)\\
&\quad\le C_8C_A\|v\|\left(\left(\sum_{K\in\mathcal{T}^h}\frac{h_K^{2\min(p_K,s_K)}}{p_K^{2s_K}}\|u\|^2_{H^{\min(p_K,s_K)}(K)}\right)^{1/2}\right.\\
&\qquad\left.+\left(C_5\sum_{K\in\mathcal{T}^h}\frac{h_K^{2\min(p_K,s_K)}}{p_K^{2s_K+1}}\|u\|^2_{H^{\min(p_K,s_K)}(K)}\right)^{1/2}\right),
\end{aligned}
$$

which proves (7.318) with $C_{10}=C_AC_8(\sqrt{C_5}+1)$ since $1/p_K\le 1$. □

The consistency of the form a_h is given by

Theorem 7.53. Consistency for the a_h: *The $a^h(\Pi^{h\mathsf{p}}u,v^h)$ in (7.57) are consistent with $a(u,v)$ in (7.19) $\forall\ u\in\mathcal{U}_e$, $\forall v^h\in S^h=\mathcal{V}^h$, so a constant, $C_{14}^\star:=C_A+2nC_A\sqrt{C_5}(1/c_w^{-1/2}+\sqrt{C_5(1+C_6)})$ exists, such that for the $S^{h\mathsf{p}}$-interpolant $\Pi^{h\mathsf{p}}u$ of u in Theorem 7.51,*

$$
\begin{aligned}
&|a_h(u,v^h)-a^h(\Pi^{h\mathsf{p}}u,v^h)| \qquad (7.321)\\
&\quad\le \varepsilon\ C_{14}^\star\|v^h\|\left(\sum_{K\in\mathcal{T}^h}\frac{h_K^{2\min(p_K,s_K)-2}}{p_K^{2s_K-3}}\|u\|^2_{H^{\min(p_K,s_K)}(K)}\right)^{1/2}.
\end{aligned}
$$

Proof. In Proposition 7.25, we consider $u,\ \overline{u},\ v\in\mathcal{U}_e$ as arguments of $a_h(\cdot,\cdot)$. Again, for the generalization to *hp*-methods, the proof remains unchanged, except replacing $d(e),h_K$ there by $d_{h\mathsf{p}}(e),d_{h\mathsf{p}}(K)$ here. After that the proof of Theorem 7.26 is modified as:

Putting $\overline{u} := \Pi^{h\mathsf{p}} u$ and $v := v^h$ we have from (7.217)–(7.220)

$$|a_h(u, v^h) - a_h(\Pi^{h\mathsf{p}} u, v^h)| \le \varepsilon \ \ |\Pi^{h\mathsf{p}} u - u|_{H^1(\mathcal{T}^h)} |v^h|_{H^1(\mathcal{T}^h)} \tag{7.322}$$
$$+ \varepsilon \ 2n \left(\left(\frac{1}{c_w}\right)^{1/2} \left(J_h^\sigma(v^h, v^h)\right)^{1/2} \left(\sum_{K\in\mathcal{T}^h} d_{h\mathsf{p}}(K) |u - \Pi^{h\mathsf{p}} u|^2_{H^1(\partial K)} \right)^{1/2} \right.$$
$$\left. + \left(\sum_{K\in\mathcal{T}^h} d_{h\mathsf{p}}(K) \|\nabla v^h\|^2_{L^2(\partial K)} \right)^{1/2} \left(\sum_{K\in\mathcal{T}^h} d_{h\mathsf{p}}(K)^{-1} \|(u - \Pi^{h\mathsf{p}} u)\|^2_{L^2(\partial K)} \right)^{1/2} \right).$$

Similarly as in (7.319), it is possible to show that

$$\sum_{K\in\mathcal{T}^h} d_{h\mathsf{p}}(K)^{-1} |u - \Pi^{h\mathsf{p}} u|^2_{L^2(\partial K)} \le C_5 C_A^2 \sum_{K\in\mathcal{T}^h} \frac{h_K^{2\min(p_K, s_K) - 2}}{p_K^{2s_K - 3}} \|u\|^2_{H^{\min(p_K, s_K)}(K)}. \tag{7.323}$$

and

$$\sum_{K\in\mathcal{T}^h} d_{h\mathsf{p}}(K) |u - \Pi^{h\mathsf{p}} u|^2_{H^1(\partial K)} \le C_5 C_A^2 \sum_{K\in\mathcal{T}^h} \frac{h_K^{2\min(p_K, s_K) - 2}}{p_K^{2s_K - 3}} \|u\|^2_{H^{\min(p_K, s_K)}(K)} \tag{7.324}$$

These formulas yield, with (7.64), (7.171), (7.317), (7.319), (7.322), (7.323) and (7.324),

$$|a_h(u, v^h) - a_h(\Pi^{h\mathsf{p}} u, v^h)| \tag{7.325}$$
$$\le \varepsilon C_A \|v^h\| \left(\sum_{K\in\mathcal{T}^h} \frac{h_K^{2\min(p_K, s_K) - 2}}{p_K^{2s_K - 2}} \|v\|^2_{H^{\min(p_K, s_K)}(K)} \right)^{1/2}$$
$$+ \frac{2n\varepsilon}{\sqrt{c_w}} \|v^h\| \left(C_5 C_A^2 \sum_{K\in\mathcal{T}^h} \frac{h_K^{2\min(p_K, s_K) - 2}}{p_K^{2s_K - 3}} \|u\|^2_{H^{\min(p_K, s_K)}(K)} \right)^{1/2}$$
$$+ 2n\varepsilon \sqrt{C_5(1 + C_6)} \|v^h\| \left(C_5 C_A^2 \sum_{K\in\mathcal{T}^h} \frac{h_K^{2\min(p_K, s_K) - 2}}{p_K^{2s_K - 3}} \|u\|^2_{H^{\min(p_K, s_K)}(K)} \right)^{1/2}$$
$$\le \varepsilon C_{14}^\star \|v^h\| \left(\sum_{K\in\mathcal{T}^h} \frac{h_K^{2\min(p_K, s_K) - 2}}{p_K^{2s_K - 3}} \|u\|^2_{H^{\min(p_K, s_K)}(K)} \right)^{1/2},$$

$C_{14}^\star := C_A + 2nC_A\sqrt{C_5}(1/c_w^{-1/2} + \sqrt{C_5(1 + C_6)})$. □

Finally, we show the consistency of the penalty form J_h^σ.

Theorem 7.54. Consistency for the J_h^σ: *There exists a constant* $C_{15}^\star > 0$ *such that*

$$\left|J_h^\sigma(u, v^h) - J_h^\sigma(\Pi^{h\mathsf{p}}u, v^h)\right| \leq C_{15}^\star \|v^h\| \left(\sum_{K\in\mathcal{T}^h} \frac{h_K^{2\min(p_K,s_K)-2}}{p_K^{2s_K-1}} \|u\|^2_{H^{\min(p_K,s_K)}(K)}\right)^{\frac{1}{2}},$$

$$u \in H^{\mathsf{s}}(\mathcal{T}^h),\ v^h \in S^h,\ h \in (0, h_0), \tag{7.326}$$

where $\Pi^{h\mathsf{p}}u$ *is the* $S^{h\mathsf{p}}$*-interpolant of* u *from Theorem 7.51.*

Proof. From (7.61), the Cauchy inequality, (7.149), and the multiplicative trace inequality (7.129) and (7.317) we have

$$\left|J_h^\sigma(u, v^h) - J_h^\sigma(\Pi^{h\mathsf{p}}u, v^h)\right| \tag{7.327}$$

$$\leq \sum_{e\in\mathcal{T}^h} \int_e \sigma|[u - \Pi^{h\mathsf{p}}u]|\,|[v^h]|\,dS$$

$$\leq J_h^\sigma(v^h, v^h)^{1/2} \left(\sum_{e\in\mathcal{T}^h} \int_e \sigma[u - \Pi^{h\mathsf{p}}u]^2\,dS\right)^{1/2}$$

$$\leq \|v^h\| \sum_{K\in\mathcal{T}^h} \left(c_w d_{h\mathsf{p}}(K)^{-1} \|u - \Pi^{h\mathsf{p}}u\|^2_{L^2(\partial K)}\right)^{1/2}$$

$$\leq \sqrt{C_5 c_w}\|v^h\| \Bigg(\sum_{K\in\mathcal{T}^h} d_{h\mathsf{p}}(K)^{-1} \left(\|u - \Pi^{h\mathsf{p}}u\|_{L^2(K)} |u - \Pi^{h\mathsf{p}}u|_{H^1(K)}\right.$$

$$\left.\left. + d_{h\mathsf{p}}(K)^{-1} \|u - \Pi^{h\mathsf{p}}u\|^2_{L^2(K)}\right)\right)^{1/2}$$

$$\leq \sqrt{2C_5 c_w}\|v^h\| C_A^2 \Bigg(\sum_{K\in\mathcal{T}^h} \Bigg(\frac{1}{h_K} \frac{h_K^{\min(p_K,s_K)}}{p_K^{s_K}} \frac{h_K^{\min(p_K,s_K)-1}}{p_K^{s_K-1}}$$

$$\left.\left. + \frac{1}{h_K^2} \frac{h_K^{2\min(p_K,s_K)}}{(p_K^{s_K})^2}\right) \|u\|^2_{H^{\min(p_K,s_K)}(K)}\right)^{1/2}$$

$$\leq 2\sqrt{C_5 c_w}\|v^h\| C_A^2 \left(\sum_{K\in\mathcal{T}^h} \frac{h_K^{2\min(p_K,s_K)-2}}{p_K^{2s_K-1}} \|u\|^2_{H^{\min(p_K,s_K)}(K)}\right)^{1/2},$$

which proves Theorem 7.54 with $C_{15}^\star := 2\sqrt{C_5 c_w} C_A^2$. □

For our previous problems we impose the conditions listed in Section 7.12 for the *standard DCGM* and the new conditions in Subsections 7.14.1–7.14.3. We define the

hp-DCGMs by replacing the previous S^h by the S^{hp} and (7.274) by

$$C(h^{\mathrm{s}}, u) := \left(\sum_{K \in \mathcal{T}^h} \frac{h_K^{2\min(p_K, s_K)-2}}{p_K^{2s_K-1}} \|u\|^2_{H^{\min(p_K,s_K)}(K)} \right)^{\frac{1}{2}} \quad \forall u \in H^{\mathrm{s}}(\mathcal{T}^h), h \in (0, h_0). \tag{7.328}$$

Previously we have proved the consistency for our model problem. For the other cases the proofs follow similar modifications. The necessary stability with respect to the $\|\cdot\|$ is available via the $\mathcal{U}$- and $\mathcal{U}^h$-coercivity as in Section 7.9. Then our hp-updated consistency results can be applied for proving the general stability and convergence as in Section 7.12. We summarize

Theorem 7.55. Convergence for the hp- DCGMs for the problem 5.–10. in Condition 7.39. *Under these conditions 1.–4. and for inhomogeneous* Dirichlet boundary conditions *choose, for the problems in 5.–10. the quoted hp-DCGMs. This implies coercivity, hence stability results for a boundedly invertible $G'(u_0)$, and the necessary consistency estimates in Theorems 7.53,7.54 and the corresponding modifications for the other methods.*

Then for $G(u_0) = 0$, the application of these hp-DCGMs define the $G^h\left(u_0^h\right) = 0$ with a uniquely existing solution u_0^h near u_0 converging according to

$$\|u^h - u\|^J_{H^{\mathrm{s}}(\mathcal{T}^h)} \le C(h^{\mathrm{s}}, u). \tag{7.329}$$

Again a numerical solution is efficiently computed via Newton's method under the conditions of and as in Theorem 7.47. This again implies quadratic convergence essentially independent of h.

7.15 Numerical experiences

7.15.1 Scalar quasilinear equation

In this section we verify theoretical results presented in the previous sections by numerical experiments. Similarly as in [407], we consider the nonlinear diffusion equation

$$-\sum_{s=1}^{2} \frac{\partial}{\partial x_s} \left(\nu(|\nabla u|) \frac{\partial}{\partial x_s} u \right) = g, \; u : \Omega \to \mathbb{R}, \tag{7.330}$$

where $\nu(w) : (0, \infty) \to \mathbb{R}$ is chosen in the form

$$\nu(w) = \nu_\infty + \frac{\nu_0 - \nu_\infty}{(1+w)^\gamma}, \quad \gamma > 0. \tag{7.331}$$

The corresponding quasilinear weak form is

$$\langle G(u), v\rangle = \int_\Omega \sum_{k=0}^{n} (A_k(x,u,\nabla u), \partial^k v)\,dx \tag{7.332}$$
$$:= \int_\Omega \sum_{k=1}^{2} \left(\nu(|\nabla u|)\frac{\partial}{\partial x_k}u\right)\frac{\partial}{\partial x_k}v dx - \int_\Omega g v dx.$$

These $A_k(x,u,\nabla u), k = 1,\ldots,n = 2$, satisfy the conditions (7.94) (a)–(c) and $A_0 \equiv 0$, cf. (2.284), (2.285), (2.286), (2.293).

We set $\nu_0 = 0.15$, $\nu_\infty = 0.1$, $\gamma = 1/2$, $\Omega = (0,1)^2$, and define the function g and the boundary conditions in such a way that the exact solution has the form

$$u(x_1,x_2) = 2r^\alpha x_1 x_2(1-x_1)(1-x_2) \tag{7.333}$$
$$= r^{\alpha+2}\sin(2\varphi)(1-x_1)(1-x_2),$$

where $(r,\varphi)(r := (x_1^2+x_2^2)^{1/2})$ are the polar coordinates and $\alpha \in \mathbb{R}$ is a constant. The function u is equal to zero on $\partial\Omega$ and its regularity depends on the value of α, namely (cf. [52])

$$u \in H^\beta(\Omega) \quad \forall \beta \in (0,\ \alpha+3), \tag{7.334}$$

where $H^\beta(\Omega)$ denotes (in general) the Sobolev–Slobodetskii space of functions with "noninteger derivatives".

In the presented numerical tests we use the values $\alpha = 2$ and $\alpha = -3/2$. The value $\alpha = 2$ gives a function u sufficiently regular ($\in H^\beta(\Omega)$ for $\beta < 5$), whereas the value $\alpha = -3/2$ gives $u \in H^\beta(\Omega),\ \beta < 3/2$. Figure 7.5 shows functions u for both values of α.

The above problem is discretized by the IIPG method (7.286). The resulting system of nonlinear algebraic equations by a *pseudo-time* discretization, i.e. we solve a time dependent problem

$$\left(\frac{\partial w_0^h(t)}{\partial t}, v^h\right) + \left(G^h\left(w_0^h(t)\right), v^h\right) = 0 \qquad \forall\, v^h \in S^h, \tag{7.335}$$

where $w_0^h(t,x) : (0,T)\times S^h \to \mathbb{R}$, $T > 0$ and G^h is an operator introduced by (7.286). We seek a steady state solution, i.e.

$$\frac{\partial w_0^h(t,x)}{\partial t} = 0 \qquad \text{for } T \to \infty$$

and put

$$u_0^h(x) := \lim_{T\to\infty} w_0^h(T,x), \qquad x \in \Omega.$$

The system of ordinary differential equations (7.335) is solved by a three step backward differential formula, which is formally of third order, see [293]. The following numerical

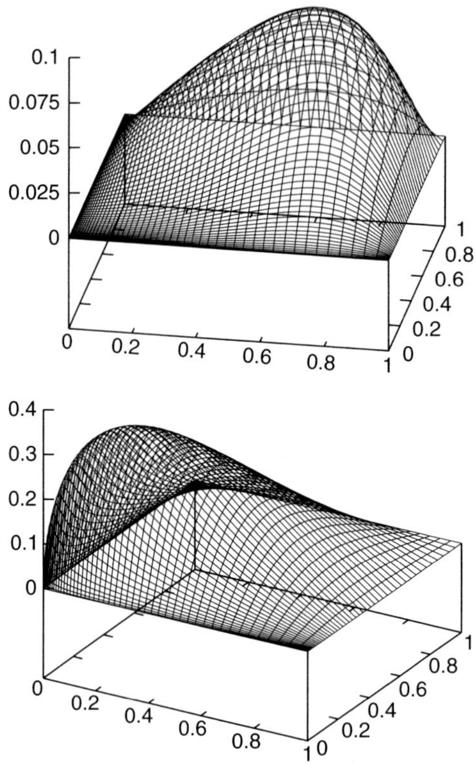

Figure 7.5 The exact solution (7.333) for $\alpha = 2$ (above) and $\alpha = -3/2$ (below).

flux is used for the discretization of the convective term

$$H(u_1, u_2, \boldsymbol{n}) = \begin{cases} \displaystyle\sum_{s=1}^{2} f_s(u_1) n_s & \text{if } A > 0, \\ \displaystyle\sum_{s=1}^{2} f_s(u_2) n_s & \text{if } A \leq 0, \end{cases} \tag{7.336}$$

where

$$A = \sum_{s=1}^{2} f\prime_s(\overline{u}) n_s, \quad \overline{u} = \frac{1}{2}(u_1 + u_2) \text{ and } \boldsymbol{n} = (n_1, n_2). \tag{7.337}$$

Numerical experiments are carried out with the use of piecewise linear ($\mathcal{P}^1$), quadratic ($\mathcal{P}^2$), and cubic ($\mathcal{P}^3$) elements on six triangular meshes $\mathcal{T}_{h_l}$, $l = 1, \ldots, 6$ having 128, 288, 512, 1152, 2048 and 4608 elements. Figure 7.6 shows the coarsest mesh and the finest one. We investigate the experimental orders of convergence (EOC) for grid

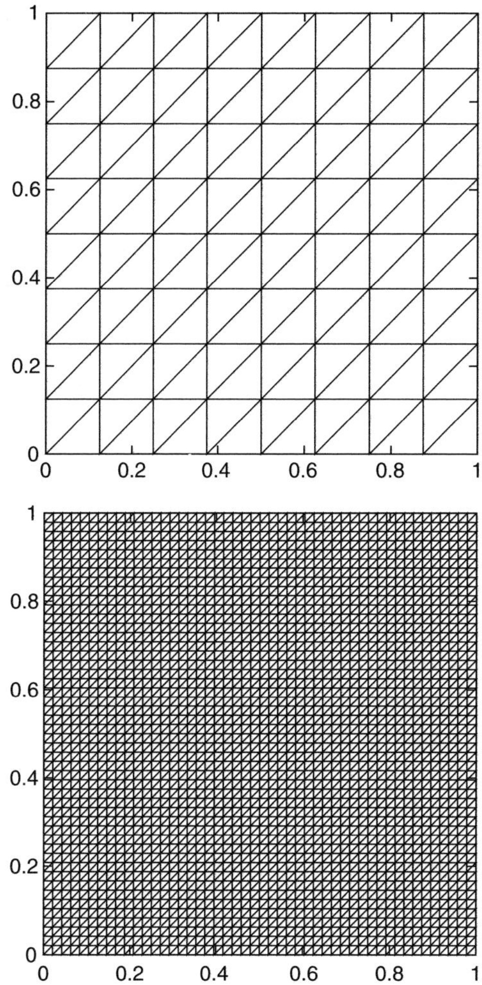

Figure 7.6 The coarsest and the finest grids used for the computations.

pairs $\mathcal{T}_{h_l}$ and $\mathcal{T}_{h_{l-1}}$, $l = 2, \ldots 6$, defined by

$$\mathrm{EOC}_l = \frac{\log(e_{h_l}/e_{h_{l-1}})}{\log(h_l/h_{l-1})}, \qquad l = 2, \ldots, 6, \tag{7.338}$$

where e_{h_l} is the computational error obtained on mesh $\mathcal{T}_{h_l}$.

Tables 7.1–7.4 show computational errors in the $L^2(\Omega)$ norm and in the $H^1(\Omega)$ seminorm and the corresponding experimental orders of convergence (EOC) for $\alpha = 2$ and $\alpha = -3/2$. These values together with the marked EOC are shown in Figures 7.7 and 7.8. We observe the following:

Table 7.1: Computational errors and the corresponding experimental orders of convergence (EOC) of the $\mathcal{P}^1$, $\mathcal{P}^2$ and $\mathcal{P}^3$ approximations for $\alpha = 2$ in the L^2 norm, purely diffusive case; *notice the definition of* α *in (7.333) ff.*

Grid	h	$\mathcal{P}_1$		$\mathcal{P}_2$		$\mathcal{P}_3$	
		$\|e_h\|_{L^2(\Omega)}$	EOC	$\|e_h\|_{L^2(\Omega)}$	EOC	$\|e_h\|_{L^2(\Omega)}$	EOC
1	0.177E+00	0.1809E−02	—	0.9906E−04	—	0.4470E−05	—
2	0.118E+00	0.8373E−03	1.90	0.3102E−04	2.86	0.9031E−06	3.94
3	0.884E−01	0.4791E−03	1.94	0.1354E−04	2.88	0.2895E−06	3.95
4	0.589E−01	0.2162E−03	1.96	0.4202E−05	2.89	0.5811E−07	3.96
5	0.442E−01	0.1224E−03	1.98	0.1839E−05	2.87	0.1856E−07	3.97
6	0.295E−01	0.5477E−04	1.98	0.5832E−06	2.83	0.3703E−08	3.98

Table 7.2: Computational errors and the corresponding experimental orders of convergence (EOC) of the $\mathcal{P}^1$, $\mathcal{P}^2$ and $\mathcal{P}^3$ approximations for $\alpha = 2$ in the H^1 seminorm, purely diffusive case.

Grid	h	$\mathcal{P}_1$		$\mathcal{P}_2$		$\mathcal{P}_3$	
		$\|e_h\|_{H^1(\Omega)}$	EOC	$\|e_h\|_{H^1(\Omega)}$	EOC	$\|e_h\|_{H^1(\Omega)}$	EOC
1	0.177E+00	0.5869E−01	—	0.6045E−02	—	0.3884E−03	—
2	0.118E+00	0.3994E−01	0.95	0.2756E−02	1.94	0.1157E−03	2.99
3	0.884E−01	0.3026E−01	0.97	0.1570E−02	1.96	0.4886E−04	3.00
4	0.589E−01	0.2037E−01	0.98	0.7073E−03	1.97	0.1448E−04	3.00
5	0.442E−01	0.1535E−01	0.98	0.4006E−03	1.98	0.6107E−05	3.00
6	0.295E−01	0.1029E−01	0.99	0.1793E−03	1.98	0.1808E−05	3.00

Table 7.3: Computational errors and the corresponding experimental orders of convergence (EOC) of the $\mathcal{P}^1$, $\mathcal{P}^2$ and $\mathcal{P}^3$ approximations for $\alpha = -3/2$ in the L^2 norm, purely diffusive case.

Grid	h	$\mathcal{P}_1$		$\mathcal{P}_2$		$\mathcal{P}_3$	
		$\|e_h\|_{L^2(\Omega)}$	EOC	$\|e_h\|_{L^2(\Omega)}$	EOC	$\|e_h\|_{L^2(\Omega)}$	EOC
1	0.177E+00	0.1580E−01	—	0.5292E−02	—	0.3121E−02	—
2	0.118E+00	0.8664E−02	1.48	0.2898E−02	1.49	0.1518E−02	1.78
3	0.884E−01	0.5659E−02	1.48	0.1888E−02	1.49	0.9201E−03	1.74
4	0.589E−01	0.3050E−02	1.52	0.1033E−02	1.49	0.4623E−03	1.70
5	0.442E−01	0.1950E−02	1.56	0.6725E−03	1.49	0.2866E−03	1.66
6	0.295E−01	0.1032E−02	1.57	0.3662E−03	1.50	0.1481E−03	1.63

Table 7.4: Computational errors and the corresponding experimental orders of convergence (EOC) of the $\mathcal{P}^1$, $\mathcal{P}^2$ and $\mathcal{P}^3$ approximations for $\alpha = -3/2$ in the H^1 seminorm, purely diffusive case.

Grid	h	$\mathcal{P}_1$		$\mathcal{P}_2$		$\mathcal{P}_3$	
		$\lvert e_h\rvert_{H^1(\Omega)}$	EOC	$\lvert e_h\rvert_{H^1(\Omega)}$	EOC	$\lvert e_h\rvert_{H^1(\Omega)}$	EOC
1	0.177E+00	0.4512E+00	—	0.2098E+00	—	0.1548E+00	—
2	0.118E+00	0.3730E+00	0.47	0.1662E+00	0.57	0.1129E+00	0.78
3	0.884E−01	0.3251E+00	0.48	0.1415E+00	0.56	0.9082E−01	0.76
4	0.589E−01	0.2668E+00	0.49	0.1133E+00	0.55	0.6779E−01	0.72
5	0.442E−01	0.2314E+00	0.49	0.9707E−01	0.54	0.5576E−01	0.68
6	0.295E−01	0.1890E+00	0.50	0.7822E−01	0.53	0.4305E−01	0.64

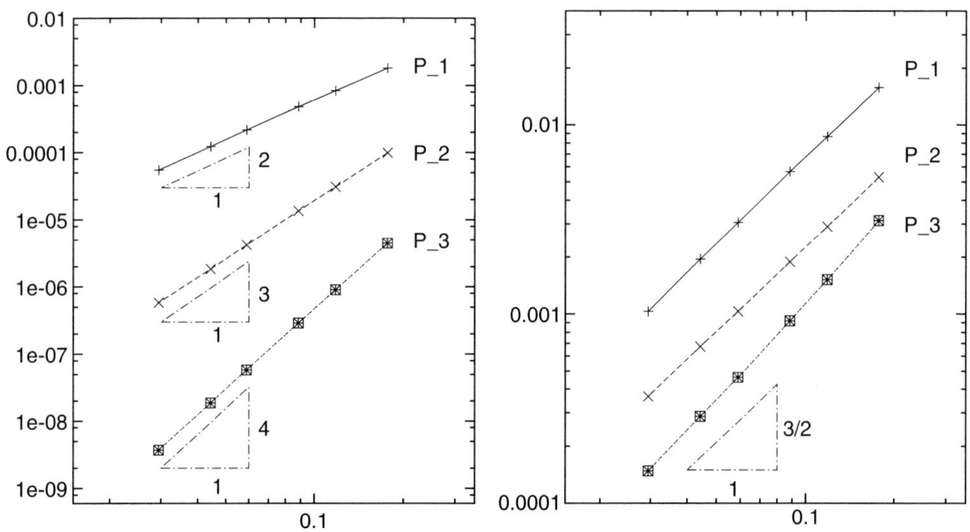

Figure 7.7 Computational errors and the corresponding experimental orders of convergence of the $\mathcal{P}^1$, $\mathcal{P}^2$ and $\mathcal{P}^3$ approximations for $\alpha = 2$ (left) and $\alpha = -3/2$ (right), purely diffusive case.

(i) For a sufficiently regular exact solution (case $\alpha = 2$) we observe the optimal order of convergence $O(h^{p+1})$ for $p = 1, 2, 3$ in the L^2 norm (a small decrease of the experimental order of convergence for the $\mathcal{P}^3$ approximation on the finest grid is caused by rounding off errors, so that we obtain a better EOC than the theoretical results). It is a rather surprising fact that the IIPG variant gives (almost) optimal order of convergence also for an even degree polynomial approximation, i.e. $\mathcal{P}^2$, cf. Table 7.1. On the other hand, Guzmán and Rivière [380] show that the NIPG method gives only suboptimal EOC also for odd polynomial degrees when special nonuniform grids are used.

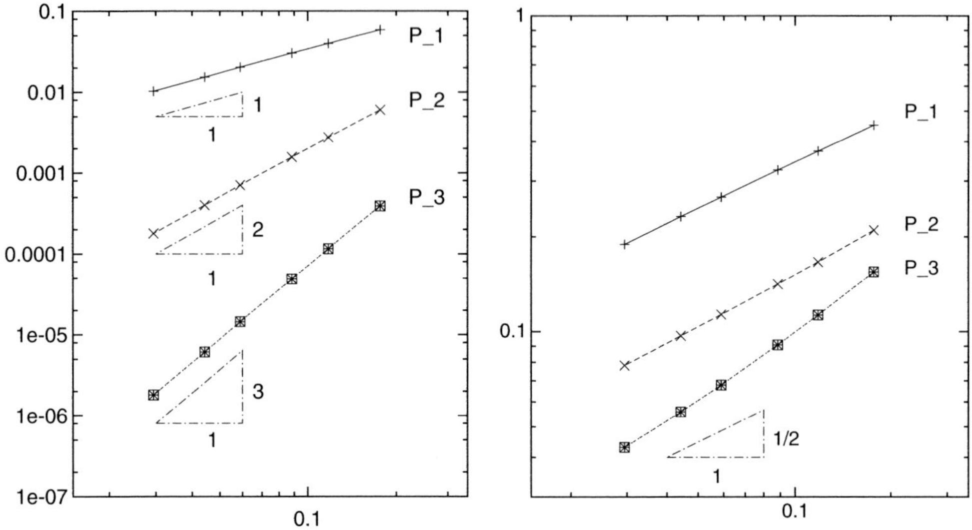

Figure 7.8 Computational errors and the corresponding experimental orders of convergence of the $\mathcal{P}^1$, $\mathcal{P}^2$ and $\mathcal{P}^3$ approximations for $\alpha = 2$ (left) and $\alpha = -1/2$ (right), purely diffusive case.

(ii) For a sufficiently regular exact solution (case $\alpha = 2$) we observe the optimal order of convergence $O(h^p)$ for $p = 1, 2, 3$ in the H^1 seminorm which is in good agreement with the theoretical results, cf. Table 7.2.

(iii) For the case $\alpha = -3/2$, we obtain an EOC equal to 3/2 in the L^2 norm and an EOC equal to 1/2 in the H^1 seminorm for $p = 1, 2, 3$. Using the result from [312], for any $\beta \in (1, 3/2)$ we get, cf. Table 7.3,

$$\|v - I_h v\|_{L^2(\Omega)} \leq C(\beta) h^{\mu} \|v\|_{H^\beta(\Omega)}, \quad v \in H^\beta(\Omega), \tag{7.339}$$

where $I_h v$ is a piecewise polynomial Lagrange interpolation to v of degree $\leq p$, $\mu = \min(p+1, \beta)$ and $C(\beta)$ is a constant independent of h and v. The exact approximation corresponding precisely to our experimental results can be obtained with the use of the interpolation in so-called Bessov spaces. See [51], Section 3.3 and the references therein.

Moreover, we consider a generalization of (7.330) given by

$$\sum_{s=1}^{2} u \frac{\partial u}{\partial x_s} - \sum_{s=1}^{2} \frac{\partial}{\partial x_s} \left(\nu(|\nabla u|) \frac{\partial}{\partial x_s} u \right) = g, u : \Omega \to \mathbb{R}, \tag{7.340}$$

where $\nu(w) : (0, \infty) \to \mathbb{R}$ is given by (7.331). This problem represents the *convection–diffusion* equation with nonlinear convective as well as diffusive terms. The corresponding quasilinear weak form is given by (7.332) but with $A_0 \not\equiv 0$. Although this form violates condition (7.94)(d) it is possible to solve it numerically. The convective term is discretized in the same way as in (7.89) where the discrete convective form c_h

is defined by (7.59). There the following numerical flux is used

$$H(u_1, u_2, \boldsymbol{n}) = \begin{cases} \sum_{s=1}^{2} f_s(u_1) n_s & \text{if } A > 0, \\ \sum_{s=1}^{2} f_s(u_2) n_s & \text{if } A \leq 0, \end{cases} \tag{7.341}$$

where

$$A = \sum_{s=1}^{2} f_s'(\overline{u}) n_s, \quad \overline{u} = \frac{1}{2}(u_1 + u_2) \text{ and } \boldsymbol{n} = (n_1, n_2). \tag{7.342}$$

Tables 7.5–7.8 show computational errors in the $L^2(\Omega)$ norm and in the $H^1(\Omega)$ seminorm and the corresponding experimental orders of convergence (EOC) for $\alpha = 2$ and $\alpha = -3/2$. These values together with the marked EOC are shown in Figures 7.9 and 7.10. We observe the completely same results as for problem (7.330).

Table 7.5: Computational errors and the corresponding experimental orders of convergence (EOC) of the $\mathcal{P}^1$, $\mathcal{P}^2$ and $\mathcal{P}^3$ approximations for $\alpha = 2$ in the L^2 norm, convection–diffusion case.

Grid	h	$\mathcal{P}_1$		$\mathcal{P}_2$		$\mathcal{P}_3$	
		$\lVert e_h \rVert_{L^2(\Omega)}$	EOC	$\lVert e_h \rVert_{L^2(\Omega)}$	EOC	$\lVert e_h \rVert_{L^2(\Omega)}$	EOC
1	0.177E+00	0.1036E−02	—	0.4985E−04	—	0.2300E−05	—
2	0.118E+00	0.4846E−03	1.87	0.1552E−04	2.88	0.4681E−06	3.93
3	0.884E−01	0.2785E−03	1.93	0.6724E−05	2.91	0.1508E−06	3.94
4	0.589E−01	0.1261E−03	1.95	0.2052E−05	2.93	0.3042E−07	3.95
5	0.442E−01	0.7155E−04	1.97	0.8767E−06	2.96	0.9763E−08	3.95
6	0.295E−01	0.3205E−04	1.98	0.2669E−06	2.93	0.2003E−08	3.91

Table 7.6: Computational errors and the corresponding experimental orders of convergence (EOC) of the $\mathcal{P}^1$, $\mathcal{P}^2$ and $\mathcal{P}^3$ approximations for $\alpha = 2$ in the H^1 seminorm, convection–diffusion case.

Grid	h	$\mathcal{P}_1$		$\mathcal{P}_2$		$\mathcal{P}_3$	
		$\lvert e_h \rvert_{H^1(\Omega)}$	EOC	$\lvert e_h \rvert_{H^1(\Omega)}$	EOC	$\lvert e_h \rvert_{H^1(\Omega)}$	EOC
1	0.177E+00	0.3024E−01	—	0.3098E−02	—	0.1962E−03	—
2	0.118E+00	0.2061E−01	0.95	0.1415E−02	1.93	0.5843E−04	2.99
3	0.884E−01	0.1561E−01	0.96	0.8074E−03	1.95	0.2468E−04	3.00
4	0.589E−01	0.1051E−01	0.98	0.3640E−03	1.96	0.7314E−05	3.00
5	0.442E−01	0.7925E−02	0.98	0.2063E−03	1.97	0.3084E−05	3.00
6	0.295E−01	0.5309E−02	0.99	0.9239E−04	1.98	0.9133E−06	3.00

Table 7.7: Computational errors and the corresponding experimental orders of convergence (EOC) of the $\mathcal{P}^1$, $\mathcal{P}^2$ and $\mathcal{P}^3$ approximations for $\alpha = -3/2$ in the L^2 norm, convection–diffusion case.

		$\mathcal{P}_1$		$\mathcal{P}_2$		$\mathcal{P}_3$	
Grid	h	$\|e_h\|_{L^2(\Omega)}$	EOC	$\|e_h\|_{L^2(\Omega)}$	EOC	$\|e_h\|_{L^2(\Omega)}$	EOC
1	0.177E+00	0.3916E−02	—	0.1413E−02	—	0.6157E−03	—
2	0.118E+00	0.2161E−02	1.47	0.7447E−03	1.58	0.3530E−03	1.37
3	0.884E−01	0.1408E−02	1.49	0.4765E−03	1.55	0.2352E−03	1.41
4	0.589E−01	0.7654E−03	1.50	0.2577E−03	1.52	0.1315E−03	1.43
5	0.442E−01	0.4943E−03	1.52	0.1682E−03	1.48	0.8659E−04	1.45
6	0.295E−01	0.2651E−03	1.54	0.9297E−04	1.46	0.4772E−04	1.47

Table 7.8: Computational errors and the corresponding experimental orders of convergence (EOC) of the $\mathcal{P}^1$, $\mathcal{P}^2$ and $\mathcal{P}^3$ approximations for $\alpha = -3/2$ in the H^1 seminorm, convection–diffusion case.

		$\mathcal{P}_1$		$\mathcal{P}_2$		$\mathcal{P}_3$	
Grid	h	$\|e_h\|_{H^1(\Omega)}$	EOC	$\|e_h\|_{H^1(\Omega)}$	EOC	$\|e_h\|_{H^1(\Omega)}$	EOC
1	0.177E+00	0.2068E+00	—	0.7954E−01	—	0.3507E−01	—
2	0.118E+00	0.1699E+00	0.48	0.6488E−01	0.50	0.2988E−01	0.40
3	0.884E−01	0.1475E+00	0.49	0.5622E−01	0.50	0.2650E−01	0.42
4	0.589E−01	0.1206E+00	0.50	0.4598E−01	0.50	0.2220E−01	0.44
5	0.442E−01	0.1045E+00	0.50	0.3987E−01	0.50	0.1947E−01	0.46
6	0.295E−01	0.8534E−01	0.50	0.3262E−01	0.50	0.1607E−01	0.47

7.15.2 System of the steady compressible Navier–Stokes equations

Let $\Omega \subset \mathbb{R}^n$, $n = 2, 3$, be a bounded domain, and by $\partial\Omega$ we denote the boundary of Ω which consists of several disjoint parts. We distinguish inlet $\partial\Omega_i$, outlet $\partial\Omega_o$ and impermeable walls $\partial\Omega_w$, i.e. $\partial\Omega = \partial\Omega_i \cup \partial\Omega_o \cup \partial\Omega_w$. We use the following notation: ρ – density, p – pressure, e – total energy, $\boldsymbol{v} = (v_1, \ldots, v_n)$ – velocity, θ – temperature, γ – Poisson adiabatic constant, Re – Reynolds number, Pr – Prandtl number.

The system of Navier-Stokes equations describing steady state viscous compressible fluids can be written in dimensionless form

$$\nabla \cdot \vec{\boldsymbol{f}}(\boldsymbol{w}) - \nabla \cdot \vec{\mathbf{R}}(\boldsymbol{w}, \nabla \boldsymbol{w}) = 0 \quad \text{in } \Omega, \tag{7.343}$$

where

$$\boldsymbol{w} = (w_1, \ldots, w_{n+2})^{\mathrm{T}} = (\rho,\ \rho v_1, \ldots, \rho v_n,\ e)^{\mathrm{T}} \tag{7.344}$$

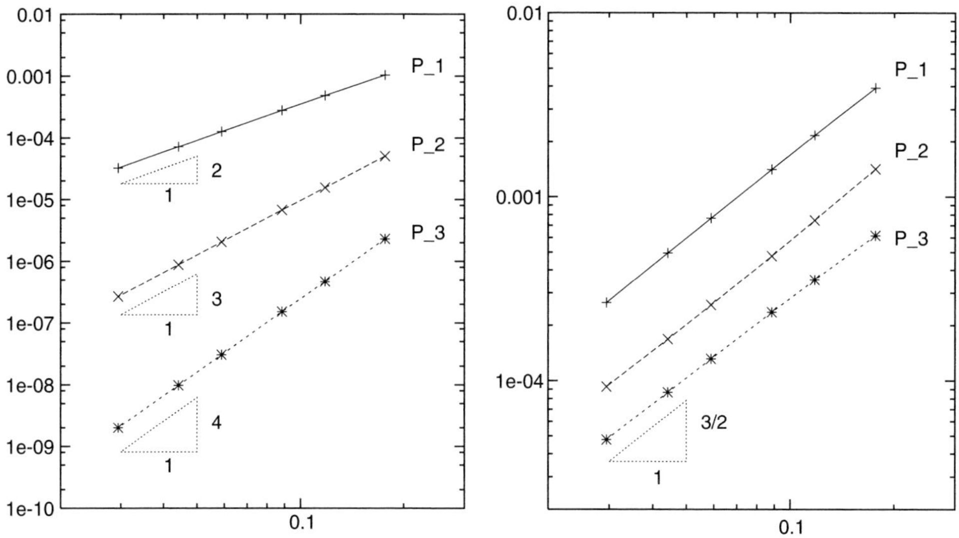

Figure 7.9 Computational errors and the corresponding experimental orders of convergence of the $\mathcal{P}^1$, $\mathcal{P}^2$ and $\mathcal{P}^3$ approximations for $\alpha = 2$ (left) and $\alpha = -3/2$ (right), convection–diffusion case.

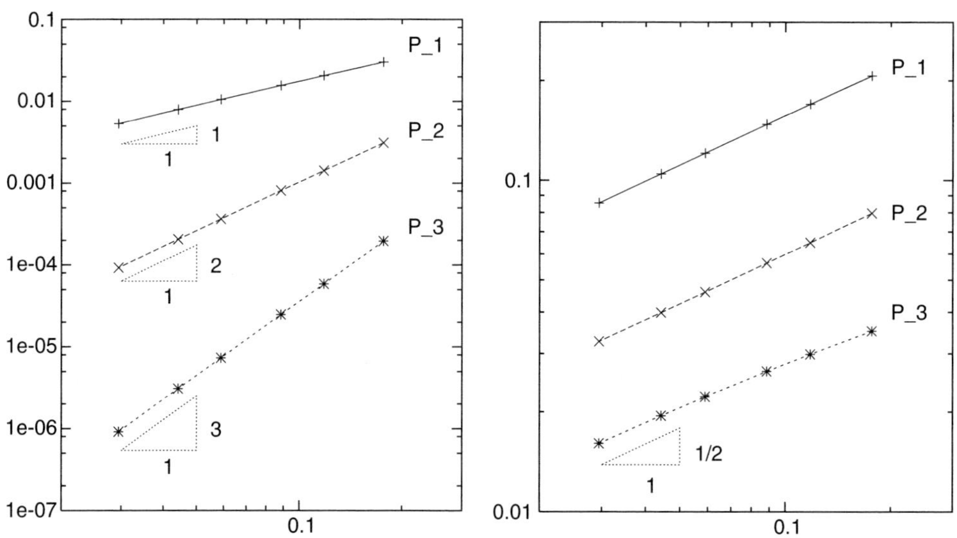

Figure 7.10 Computational errors and the corresponding experimental orders of convergence of the $\mathcal{P}^1$, $\mathcal{P}^2$ and $\mathcal{P}^3$ approximations for $\alpha = 2$ (left) and $\alpha = -3/2$ (right), convection–diffusion case.

is the so-called *state vector*, and

$$\vec{\boldsymbol{f}}(\boldsymbol{w}) = (\boldsymbol{f}_1(\boldsymbol{w}), \ldots, \boldsymbol{f}_n(\boldsymbol{w})), \tag{7.345}$$

$$\text{with} \quad \boldsymbol{f}_s(\boldsymbol{w}) = \left(f_s^{(1)}(\boldsymbol{w}), \ldots, f_s^{(n+2)}(\boldsymbol{w})\right)^{\mathrm{T}}$$

$$= (\rho v_s,\ \rho v_s v_1 + \delta_{s1} p,\ \ldots,\ \rho v_s v_n + \delta_n p,\ (e+p)\, v_s)^{\mathrm{T}}, \quad s = 1, \ldots, n$$

the so-called *inviscid (Euler) fluxes*. Finally

$$\vec{\mathbf{R}}(\boldsymbol{w}, \nabla \boldsymbol{w}) = (\mathbf{R}_1(\boldsymbol{w}, \nabla \boldsymbol{w}), \ldots, \mathbf{R}_n(\boldsymbol{w}, \nabla \boldsymbol{w})), \tag{7.346}$$

$$\text{with} \quad \mathbf{R}_s(\boldsymbol{w}, \nabla \boldsymbol{w}) = \left(R_s^{(1)}(\boldsymbol{w}, \nabla \boldsymbol{w}), \ldots, R_s^{(n+2)}(\boldsymbol{w}, \nabla \boldsymbol{w})\right)^{\mathrm{T}}$$

$$= \left(0,\ \tau_{s1},\ \ldots,\ \tau_{sn},\ \sum_{k=1}^{n} \tau_{sk} v_k + \frac{\gamma}{\mathrm{Re}\ \mathrm{Pr}} \frac{\partial \theta}{\partial x_s}\right)^{\mathrm{T}}, \quad s = 1, \ldots, n$$

are the so-called *viscous fluxes*. The symbols ∇ and $\nabla\cdot$ mean the gradient and divergence operators, i.e.

$$\nabla \boldsymbol{w} := \left(\frac{\partial \boldsymbol{w}}{\partial x_1}, \ldots, \frac{\partial \boldsymbol{w}}{\partial x_n}\right) \in \mathbb{R}^{n+2} \times \cdots \times \mathbb{R}^{n+2} \tag{7.347}$$

and

$$\nabla \cdot \vec{\boldsymbol{f}}(\boldsymbol{w}) := \sum_{s=1}^{n} \frac{\partial \boldsymbol{f}_s(\boldsymbol{w})}{\partial x_s} \in \mathbb{R}^{n+2}, \tag{7.348}$$

respectively.

Again, we can, but do not explicitly, formulate this as the quasilinear weak form

$$\langle G(\vec{u}), \vec{v}\rangle = \int_\Omega \sum_{k=0}^{n} \left(A_k(x, \vec{u}, \nabla \vec{u}), \partial^k \vec{v}\right) dx. \tag{7.349}$$

These $A_k(x, u, \nabla u), k = 1, \ldots, n$, satisfy the conditions (7.94) (a)–(c); $A_0(x, u, \nabla u)$, violates (d). If we omit $A_0(x, u, \nabla u)$, we would fully satisfy (7.94) (a)–(c), the (d) would be dropped, and would have an example similar to [407]. There probably exist some physical models whose mathematical description completely satisfies our assumption for a quasilinear model, but is not the case of this viscous compressible flow. In our context it cerainly would not make any physical sense avoiding this term, e.g.

$$A_0(x, u, \nabla u) = \nabla \cdot \vec{\boldsymbol{f}}(\boldsymbol{w}),$$

where $\vec{\boldsymbol{f}}$ is the convective flux. This would mean considering a flow with only with diffusion and not convection.

We consider a Newtonian type of fluid, i.e. the viscous part of the stress tensor has the form

$$\tau_{sk} = \frac{1}{\mathrm{Re}} \left[\left(\frac{\partial v_s}{\partial x_k} + \frac{\partial v_k}{\partial x_s} \right) - \frac{2}{3} \sum_{i=1}^{n} \frac{\partial v_i}{\partial x_i} \, \delta_{sk} \right], \quad s, k = 1, \ldots, n. \tag{7.350}$$

In order to close the system, we consider the state equation for a perfect gas and the definition of the total energy

$$p = (\gamma - 1)\,(e - \rho|\boldsymbol{v}|^2/2), \quad e = c_V \rho\theta + \rho|\boldsymbol{v}|^2/2, \tag{7.351}$$

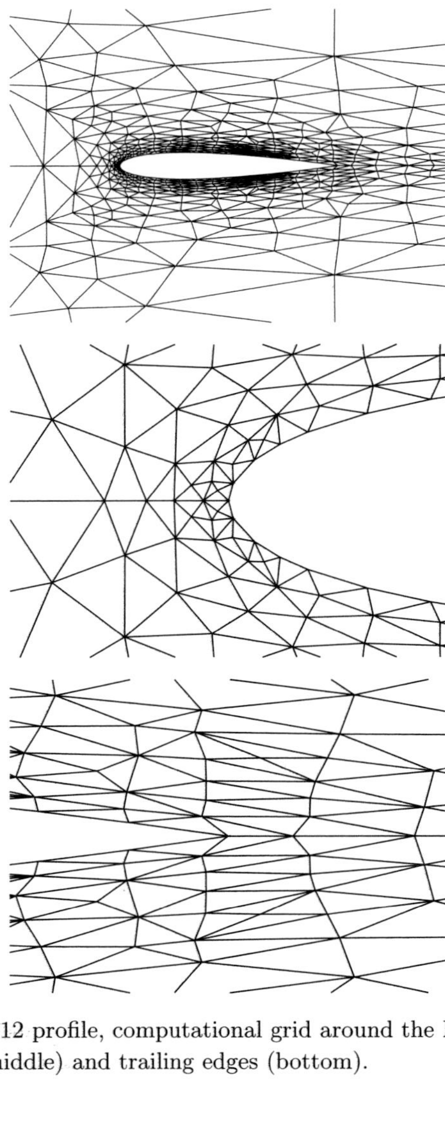

Figure 7.11 NACA 0012 profile, computational grid around the NACA0012 profiles (top), details of the leading (middle) and trailing edges (bottom).

where c_V is the specific heat at constant volume which we assume to be equal to 1 in the dimensionless case. The system (7.343)–(7.351) is equipped with the following set of boundary conditions on appropriate parts of the boundary:

$$\text{(a) } \rho = \rho_D, \quad \boldsymbol{v} = \boldsymbol{v}_D, \quad \sum_{k=1}^{n} \left(\sum_{l=1}^{n} \tau_{lk} n_l \right) v_k + \frac{\gamma}{\text{Re Pr}} \frac{\partial \theta}{\partial \boldsymbol{n}} = 0 \quad \text{on } \partial\Omega_i, \tag{7.352}$$

$$\text{(b) } \sum_{k=1}^{n} \tau_{sk} n_k = 0, \ s = 1, \dots, n, \quad \frac{\partial \theta}{\partial \boldsymbol{n}} = 0 \quad \text{on } \partial\Omega_o,$$

$$\text{(c) } \boldsymbol{v} = 0, \ \frac{\partial \theta}{\partial \boldsymbol{n}} = 0 \quad \text{on } \partial\Omega_w,$$

where ρ_D and $\boldsymbol{v}_D$ are given functions and $\boldsymbol{n} = (n_1, \dots, n_n)$ is a unit outer normal to $\partial\Omega$. Another possibility is to replace the adiabatic boundary condition (7.352) (c) by

$$(c') \ \boldsymbol{v} = 0, \quad \theta = \theta_D \quad \text{on } \partial\Omega_w. \tag{7.353}$$

Although the existence of the solution of system (7.343)–(7.352) is still an open problem we solve it numerically. The question of the existence and convergence of

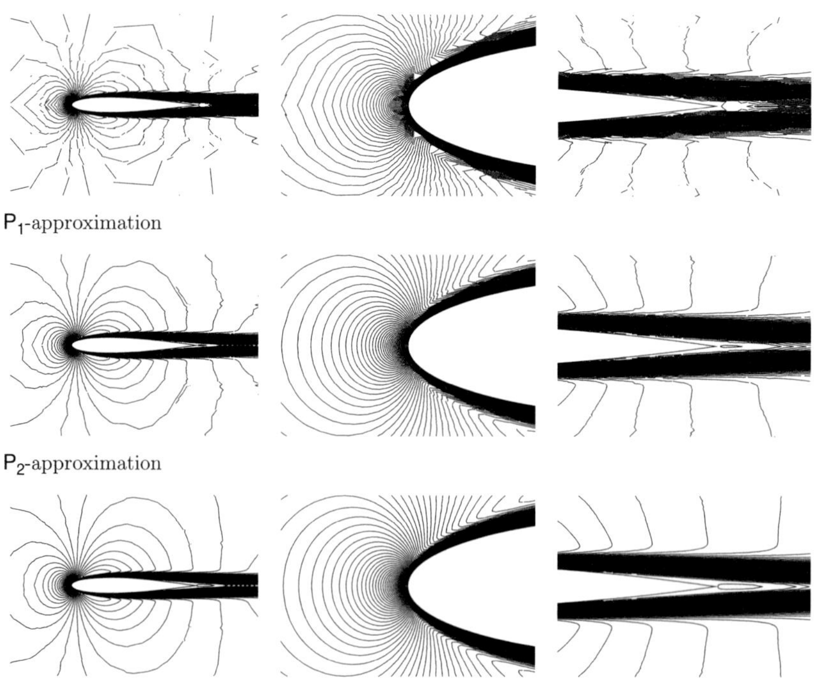

Figure 7.12 Flow around NACA0012 profiles, isolines of the Mach number (left) with details around the leading (middle) and trailing edges (right) computed by $\mathcal{P}_1$, $\mathcal{P}_2$ and $\mathcal{P}_3$ approximations.

discrete solutions is equally open. However, the formulation (7.343) of the problem is pretty close to the quasilinear problem (7.90). We employ the IIPG variant of the DCGM (7.286) to the solution of the system of the compressible Navier-Stokes equations. We consider a flow around the NACA0012 profile characterized by the inlet Mach number $M_{\text{in}} = 0.5$ $(= |\boldsymbol{v}|/\sqrt{\gamma p/\rho}$ at infinity), angle of attack $\overline{\alpha} = 0$ $(= \text{arctg}(v_2/v_1)$ at infinity) and Reynolds number $Re = 5\,000$. The Reynolds number is near to the upper limit for steady laminar flow. A characteristic feature of this flow problem is the separation of the flow occurring near to the trailing edge. The computations were carried out on a relatively coarse triangular grid having 1 360 elements which was obtained by the anisotropic mesh adaptation method (cf. [292, 298]), see Figure 7.11. We employ $\mathcal{P}_1$, $\mathcal{P}_2$ and $\mathcal{P}_3$ polynomial approximation and the corresponding isolines of the Mach number are shown in Figure 7.12. We observe a significantly smooth resolution for the $\mathcal{P}_3$ polynomial approximation although the corresponding grid is coarse.

Other interesting papers, applying DCGMs to the Stokes problem and to two-dimensional incompressible and to compressible Navier-Stokes equations, are due to Hartmann *et al.* [390–392, 409, 505].

8
Finite difference methods

8.1 Introduction

In our synopsis it is impossible to present all methods in the same generality as the dominant FEMs. Our choice for the results in finite difference methods is motivated by the following aspects: The most popular approach between 1960 and 1990, based upon M matrices, allows error estimates, e.g. for the Laplacian, of the form $\sup_{x\in\Omega^h}\left\{|u_0(x) - u_0^h(x)|\right\} \leq Ch^2\|u_0\|_{C^4(\overline{\Omega})}$. Here u_0 and u_0^h are the exact and the discrete solutions of the strong equation $\Delta u_0 = f \in C(\Omega)$ and the corresponding strong symmetric difference equations on a grid $\Omega^h \subset \Omega$ with step size h. We apply our general discretization theory in Chapter 3, thus avoiding the "external approximation approach". This motivates considering finite difference methods as variational methods similar to FEMs, cf. earlier approaches, discussed below. The above $|u_0(x) - u_0^h(x)|$ can then be estimated by a stronger discrete Sobolev energy norm on Ω^h. More generally, convergence results and error estimates can be proved here for similarly general problems as for FEMs. However, handling boundary conditions and adaptivity is strongly restricted. These difficulties are discussed by Heinrich [395] with appropriate combinations of finite difference methods and finite volume techniques, on irregular grids, Ω^h. For intended applications to bifurcation, finite difference methods are often used on cuboidal domains $\Omega = \Pi_{i=1}^n(a_i, .b_i)$. So this case will play an essential role. Finally we only discuss equidistant grids Ω^h. All our results are based upon the analytical results for general elliptic problems, summarized in Chapter 2.

Section 8.2 is devoted to simple examples and basic tools. We will discuss the changes from *unsymmetric to symmetric difference equations.* Symmetric difference equations remain unchanged, if h is replaced by $-h$, and thus they automatically are of order two, cf. Lemmas 8.1–8.2 and Grossmann and Roos [374–376]. The corresponding convergence theory is nearly the same as for unsymmetric forms. Section 8.3 presents the facts for discrete Sobolev spaces.

Section 8.4 elaborates general elliptic problems with Dirichlet boundary conditions, and their difference methods, Section 8.5 the unique existence of the discrete solutions and their convergence.

For general problems, we consider unsymmetric difference methods first and generalize them to curved boundaries. This yields methods converging of order 1. A modification to symmetric difference methods converges of order 2. We consider for all cases the quartets of strong and weak differential and difference equations. These weak difference forms go back to the Russian literature, cf. Samarskii [564, 565], and are used, e.g. by Temam [621] and Grossmann and Roos [375]. By combining the ∂_-^j and $\partial_+^i u^h$, we define an *unsymmetric divergence form* as, e.g. in (8.44). We proceed from unsymmetric to symmetric forms in Subsection 8.4.2. The transformation of the strong forms $A_s^h u^h = f^h$, essentially the classical difference equation, into the weak forms $A^h u_0^h = f^h$, and the discussion of solving $A^h u_0^h = f^h$, proceeds along the same lines for all cases. So we give the details only for second order linear equations in Subsections 8.4.2, 8.4.3. For the other cases we restrict ourselves to listing the equations in Subsections 8.4.4 ff. For nonlinear operators the domains are subsets of the corresponding Sobolev spaces in (8.43), (8.47). Conditions for the coefficients are listed in Summary 8.6, and Remark 8.7. The above transformation from strong to weak forms becomes more complicated for natural boundary conditions in Section 8.6, and is not applicable in Section 8.7.

With one exception, the proof for stability and convergence follows "the same general procedure as" for all other discretization methods: We combine discrete coercivity, cf. (8.120), compact perturbation of a boundedly invertible linear operator and consistency. This will yield the existence of a unique solution for the difference equation, and its convergence to the exact solution u_0, and the corresponding generalizations. For equations of higher order and general domains violated boundary conditions are treatable as for FEMs. We do not reformulate these rather technical considerations. Convergence for monotone and quasilinear operators follows the lines of Section 4.5 and again yields Theorem 4.67.

Except for Section 8.6, we restrict the discussion to homogeneous Dirichlet conditions. In Lemma 2.26 we have formulated conditions allowing the transformation of a general Dirichlet boundary value problem into a homogeneous form. For differential and difference equations the transformation of strong into weak forms is based upon partial integration and summation. Here the vanishing Dirichlet boundary conditions at least for the test functions, often for the solutions as well, are crucial, cf. e.g. Proposition 8.8. For nonlinear discrete problems the *mesh independence principle,* cf. Theorem 8.33, implies the quadratic convergence of Newton's method for solving the difference equations, essentially independent of the step size.

In Section 8.6 we formulate, for problems with natural boundary conditions, difference methods and prove existence and convergence of the discrete solutions by modifying the previous technique.

In Section 8.7, we discuss different methods of orders ≥ 1 on curved domains for special differential equations. This will require distinguishing points more or less near the boundary. These results are only summarized, but not proved.

We have confined the discussion to difference methods of first or second order accuracy. By our convergence results with respect to discrete Sobolev norms, the extrapolation or defect correction methods, based upon asymptotic expansions, presented in Section 8.8, allow a very efficient formulation of higher order methods anyway. In particular, our symmetric forms yield second order accuracy and the advantageous asymptotic expansions in powers of h^2.

The last Section 8.9 applies Richardson extrapolation in an academic example to the von Kármán equations.

Notice that the following problems are only discussed for FEMs. General convergence theory for monotone operators, cf. Section 4.5, variational methods for eigenvalue problems, cf. Section 4.7, methods for fully nonlinear elliptic problems, cf. Section 5.2, methods for nonlinear boundary conditions, cf. Section 5.3, and quadrature approximate methods, cf. Section 5.4. All the results remain correct for difference methods. For appropriate updates see Remark 8.27.

Many difference methods for many problems are considered in the literature. From the huge number of publications, we only list a very small choice of books and surveys. Papers are cited in the corresponding context. Ansorge [31], Cheng [171], Ciarlet and Lions (ed.) [177] Collatz [206], Cuvelier *et al.* [231], Forsythe and Wasow [319], Grossmann *et al.* [375,376], Hackbusch [387], Heinrich [395], Isaacson and Keller [414], Marchuk [483], Quarteroni [536], Samarskii [564, 565], Sloan *et al.* [592], Strikwerda [604, 606], Temam [624], Thomas [626], Vainberg [643], Vainikko [644]. Xu [670] and Zeidler [677, 678].

8.2 Difference methods for simple examples, notation

We start with the simple boundary value problem

$$-u''(x) = f(x)\ \forall\ x\ \in \Omega = (0,1), u(0) = u_a, u(1) = u_e, \tag{8.1}$$

hence $-u'' = -\Delta u$ in $\mathbb{R}^1 = \mathbb{R}$. To determine an approximate solution we replace the second derivative u'' by divided differences in the grid points. We choose a *step size* h, and introduce an equidistant one-dimensional grid as, cf. Figure 8.1,

$$\begin{aligned} \mathcal{G}_1^h &:= \mathcal{G}^h := \{x_i \in \mathbb{R} : x_i = ih, h := 1/k, k, i \in \mathbb{Z}, k > 0\}, \\ \Omega_0^h &:= \{x_i \in \mathbb{R} : x_i = ih, 0 < i < k\} \subset \mathcal{G}^h,\ \ \partial\Omega^h := \{x_i = ih, i = 0, k\}, \\ \Omega^h &:= \Omega_0^h \cup \partial\Omega^h = \{x_i \in \mathbb{R} : x_i = ih, 0 \le i \le k\} \subset \mathcal{G}^h. \end{aligned} \tag{8.2}$$

We define for a function v: $\mathbb{R} \to \mathbb{R}$. For $v : \Omega^h \to \mathbb{R}$, we restrict x to $x \in \Omega^h$ or $x \in \mathcal{G}^h$:

$$\begin{aligned} &\partial_+ v(x) := \partial_+^h v(x) := \partial_+^{(1)} v(x) := [v(x+h) - v(x)]/h,\ 0 \le x \le 1-h, \\ &\partial_- v(x) := \partial_+^{-h} v(x) := \partial_-^{(1)} v(x) := [v(x) - v(x-h)]/h,\ h \le x \le 1,\ \text{recursively}, \\ &\partial_+^{(j)} v(x) := \partial_{+,h}^{(j)} v(x) := \partial_+ \left(\partial_+^{(j-1)} v(x)\right),\ \ \partial_-^{(j)} v(x) := \partial_{-,h}^{(j)} v(x) := \partial_- \left(\partial_-^{(j-1)} v(x)\right), \\ &\quad \text{e.g.}\quad \partial_+^{(2)} v(x) = [v(x+2h) - 2v(x+h) + v(x)]/h^2,\ 0 \le x \le 1-2h. \end{aligned} \tag{8.3}$$

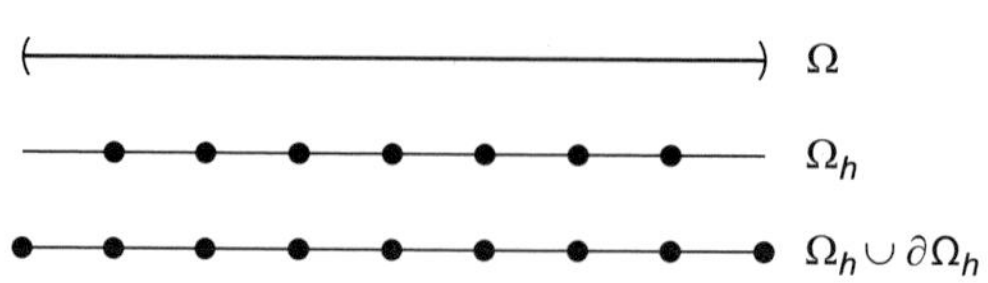

Figure 8.1 One-dimensional grid.

The recursive definition of *symmetric formulas* is a bit more complicated than for (8.3), cf. Collatz [206], pp. 538 ff. In difference equations, these (8.4) can be directly used for constant coefficients, but will be modified below for nonconstant coefficients:

$$\begin{aligned}
\partial_h^{(0)} v(x) &:= [v(x+h) + v(x-h)]/2, \qquad (8.4)\\
\partial_h^{(1)} v(x) &:= [v(x+h) - v(x-h)]/(2h) = \left(\partial_+^h + \partial_-^h v(x)\right)/2,\\
\partial_h^{(2)} v(x) &:= \partial_+^h \left(\partial_-^h v(x)\right) = [v(x+h) - 2v(x) + v(x-h)]/h^2\\
&= \left(\partial_-^h \left(\partial_+^h v(x)\right) + \partial_+^h \left(\partial_-^h v(x)\right)\right)/2\\
\partial_h^{(3)} v(x) &:= \partial_h^{(1)} \left(\partial_h^{(2)} v(x)\right) = [v(x+2h) - 2v(x+h) + 2v(x-h) - v(x-2h)]/(2h^3)\\
&\qquad \text{with } \partial_h^{(1)} \left(\partial_h^{(2)} v(x)\right) = \partial_h^{(2)} \left(\partial_h^{(1)} v(x)\right),\\
\partial_h^{(4)} v(x) &:= \partial_h^{(2)} \left(\partial_h^{(2)} v(x)\right)\\
&= [v(x+2h) - 4v(x+h) + 6v(x) - 4v(x-h) + v(x-2h)]/(h^4), \text{ etc.}
\end{aligned}$$

The formulas in (8.4) have the *invariance property* with respect to replacing h by $-h$, so $\partial_h^{(i)} v(x) = \partial_{-h}^{(i)} v(x)$.

Compared to the *nonsymmetric divided differences* in (8.3), the corresponding *symmetric formulas* in (8.4) approximate the function values and derivatives in x of order 2, instead of the previous order 1. This implies an asymptotic expansion in powers of h^2 instead of h, cf. the following Lemma 8.1, cf. [387], Lemma 4.1.1, and [103]. These advantages are inherited by the difference methods.

Lemma 8.1. *The* mean value *and* difference operators $\partial_h^{(0)}$, *and* ∂_+, ∂_-, $\partial_\pm^{(i)}, \partial_h^{(i)}$ *satisfy the following relations, estimates, and asymptotic expansions. For* $i = 0, 1, \ldots$, *and with* $c_{i,j} \in \mathbb{R}$, *independent of* u, h, *we obtain*

$$\partial_h^{(0)} = (\partial_+ + \partial_-)/2, \partial_h^{(2)} = (\partial_+ - \partial_-)/h = \partial_+ \circ \partial_- = \partial_- \circ \partial_+, \text{ and} \qquad (8.5)$$

$$\left|\partial_\pm^{(i)} u(x) - u^{(i)}(x)\right| \le Lh \text{ for } u^{(i)} \in C_L(\Omega), \text{ and for symmetric forms}$$

$$\left|\partial_h^{(i)} u(x) - u^{(i)}(x)\right| \le L\frac{h^2}{i+1} \text{ for } u^{(i+1)} \in C_L(\Omega) \text{ or asymptotic expansions}$$

$$\left|\partial_+ u(x) - u'(x) - \sum_{i=1}^{m} h^i \frac{u^{(i+1)}(x)}{(i+1)!}\right| \le L\frac{h^{m+1}}{(m+1)!} u^{(m+1)} \in C_L(\Omega), \qquad (8.6)$$

$$\left|\partial_h^{(i)} u(x) - u^{(i)}(x) - \sum_{j=1}^{q} h^{2j} c_{i,j} \frac{u^{(2j+i)}(x)}{(2j+i)!}\right| \le L\frac{h^{2q+1}}{(2q)!}, \text{ for } u^{(2q+i)} \in C_L(\Omega).$$

These results remain valid, with obvious modifications, for the multidimensional generalizations in (8.12) ff., cf. Lemma 8.2.

Replacing the $u''(x)$ in (8.1) by the $\partial_h^{(2)}u(x)$ for $x = x_i \in \Omega_0^h$ in (8.2) yields

$$-\partial_h^{(2)}u(x_i) = f(x_i) + \mathcal{O}(h^2), 0 < i < k, h := \frac{1}{k}, x_i = ih, u(x_0) = u_a, u(x_k) = u_b.$$

Omitting the $\mathcal{O}(h^2)$ term we end up with the (symmetric) difference equation for a discrete function $u^h : \Omega^h \to \mathbb{R}$ of the form

$$-\partial_h^{(2)}u^h(x_i) = -[u^h(x_i - h) - 2u^h(x_i) + u^h(x_i + h)]/h^2 = f(x_i),$$
$$0 < i < k, u_0^h = u^h(x_0) = u_a, u_k^h = u^h(x_k) = u_b. \quad (8.7)$$

As for all the previous discretization methods, the deciding question is: Does an approximate solution $u_0^h(x_i), 0 \le i \le k$, uniquely exist, and do these values converge to the values of the exact solution $u_0(x_i)$? Do small perturbations in the equation only slightly perturb the solution $u_0^h(x_i)$? This has to be generalized, and discussed in this chapter.

For motivating difference methods for PDEs, we generalize our above model problem (8.1) as

$$A\,u = -\Delta u = f \in C(\Omega), (u - \phi)|_{\partial\Omega} = 0. \quad (8.8)$$

We only consider finite difference methods with equal step sizes, h, in all directions. Different h_i can be treated with essentially the same techniques.

For cuboidal Ω, *Dirichlet boundary conditions* can be exactly discretized:

$$\Omega := \prod_{i=1}^{n}(a_i, b_i) \subset \mathbb{R}^n \text{ s.t. } \exists h > 0 \text{ with } a_i/h,\ b_i/h\ \in \mathbb{Z}, i = 1, \cdots, n. \quad (8.9)$$

The following approach works as well for unions of cuboids in (8.9), connected along $(n-1)$-dimensional faces, perpendicular to the coordinate axes. We extend it to curved boundaries below.

For $\Omega \subset \mathbb{R}^n$ in (8.9) or in (8.18) below we introduce step sizes, h, and grids:

$$H := \{h \text{ as in (8.9) or } h \in \mathbb{R}_+ \text{ for (8.18) below}\} : \quad (8.10)$$
$$\mathcal{G}^h := \mathcal{G}_D^h := \{x = (x_1, \ldots, x_n)^T \in \mathbb{R}^n : x_i/h\ \in \mathbb{Z}\},$$
$$\mathcal{G}_l^h := \{\forall j = 1, \ldots, n : x = (x_1, \ldots, x_n)^T \in \mathbb{R}^n : \forall i \ne j : x_i/h\ \in \mathbb{Z}, x_j/h\ \in \mathbb{R}\}.$$

For Ω in (8.9) we introduce

$$\Omega_0^h := \Omega \cap \mathcal{G}^h, \Omega^h := \overline{\Omega} \cap \mathcal{G}^h,\ \partial\Omega^h := \partial\Omega \cap \mathcal{G}^h = \Omega^h \setminus \Omega_0^h \quad (8.11)$$

but we have to modify it for Ω in (8.18).

We consider grid functions v^h, defined on Ω^h, and define their *multidimensional forward, backward, and normal (divided) differences* for $x \in \mathcal{G}^h$, generalizing

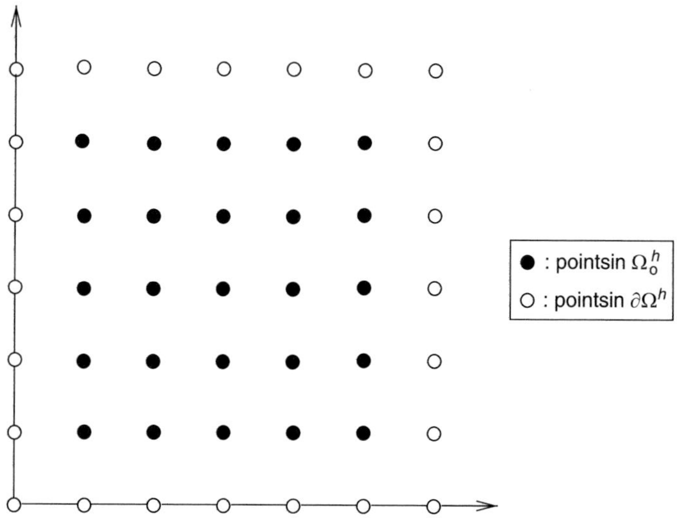

Figure 8.2 Two-dimensional grid for Dirichlet boundary conditions.

(8.3), (8.4), as, cf. Figure 8.2,

$$\begin{aligned}
&\partial^i_+ v^h(x) := \partial^i_{+,h} v(x) := [v^h(x+he^i) - v^h(x)]/h, \; e^i = i\text{-th unit vector} \qquad (8.12)\\
&\partial^i_- v^h(x) := [v^h(x) - v^h(x-he^i)]/h = \partial^i_{-,h} v(x), i = 1\dots n, \forall x \in \mathcal{G}^h,\\
&\partial^\alpha_\pm v^h(x) := \Pi^n_{i=1} \left(\partial^i_\pm\right)^{\alpha_i} v^h(x), \text{ and } \partial^\nu_\pm v^h(x) := [v^h(x \pm h\nu) - v^h(x)]/h,
\end{aligned}$$

with the *normal differences*, where, for $x \in \partial\Omega$, $\Omega \subset \mathbb{R}^n$ in (8.9), e^i is replaced by the outer normal ν or by $-\nu$. The *symmetric divided differences* are

$$\begin{aligned}
&\partial^i_h v^h(x) := \partial^i_{-h} v(x) := [v^h(x+he^i) - v^h(x-he^i)]/(2h), \qquad (8.13)\\
&\partial^{(2,0)}_h v^h(x) := \partial^1_- \left(\partial^1_+ v^h(x)\right) = \partial^1_+ \left(\partial^1_-\right) v^h(x)), \partial^{(0,2)}_h v^h(x) := \partial^2_- \left(\partial^2_+ v^h(x)\right), \text{ and}\\
&\partial^{(1,1)}_h v^h(x) := \partial^1_h \left(\partial^2_h v^h(x)\right) \text{ or } \partial^{(1,1)}_h v^h(x) := \left(\partial^1_- \partial^2_+ + \partial^1_+ \partial^2_-\right) v^h(x),\\
&\quad \partial^{(3,0)}_h := \partial^1_h \partial^{(2,0)}_h, \partial^{(2,1)}_h := \partial^2_h \partial^{(2,0)}_h \text{ or } \partial^1_h \partial^{(1,1)}_h, \partial^{(1,2)}_h := \partial^2_h \partial^{(1,1)}_h \text{ or } \partial^1_h \partial^{(0,2)}_h,\\
&\quad \partial^{(4,0)}_h := \partial^{(2,0)}_h \partial^{(2,0)}_h, \partial^{(3,1)}_h := \partial^1_h \partial^2_h \partial^{(2,0)}_h, \dots, \text{ and } \partial^{(j)}_{\nu,h} v^h(x) := \left(\partial^\nu_h\right)^j v^h(x).
\end{aligned}$$

These symmetric formulas are invariant with respect to replacing h by $-h$. The problematic

$$\partial^{(2,0)}_{h,u} v^h(x) := \left(\partial^1_h \left(\partial^1_h v^h(x)\right)\right) = [v^h(x+2he^1) - 2v(x) + v^h(x-2he^1)]/(2h)^2 \quad (8.14)$$

is avoided. It would yield in (8.16) a five-point star of double step size $2h$. Therefore, the previous higher order $\partial^{(i,j)}_h v^h(x)$ in (8.13) are chosen to include only points as

Table 8.1: Stencil for the five-point star around $x \in \Omega^h \subset \mathbb{R}^2$ for $-h^2\Delta_h$.

0	$-u^h(x+he_2)$	0		0	−1	0
$-u^h(x-he_1)$	$4u^h(x)$	$-u^h(x+he_1)$	abbreviated as	−1	4	−1
0	$-u^h(x-he_2)$	0		0	−1	0

close as possible to x. So we allow only points $x \pm \nu he_1 + \mu he_2$ in $\partial_h^{(i,j)} v^h(x)$ show that in

$$\partial_h^{(i,j)} v^h(x) \text{ with} 2(m-1) \leq i+j \leq 2m, \text{only } x \pm \nu he_1 + \mu he_2 \text{with} |\nu|, |\mu| \leq m. \quad (8.15)$$

For the two-dimensional Laplacian in (8.8) and $\Omega = (0,1) \times (0,1)$ we formulate its difference approximation analogously to (8.1) and (8.7). We replace the partial $-\Delta u(x) = -(\partial^{(2,0)} + \partial^{(0,2)})u(x)$ in (8.8) by the difference approximations $-(\partial_h^{(2,0)} + \partial_h^{(0,2)})u^h(x)$, and only admit $x = (x_i, y_j) \in \Omega_0^h$. Omitting the $\mathcal{O}(h^2)$ term, the difference equation for $u_0^h : \Omega^h \to \mathbb{R}$ has the form of a *five-star approximation*

$$\begin{aligned} f(x) = A^h u_0^h(x) &:= -\Delta_h u_0^h(x) := -\left(\partial_h^{(2,0)} + \partial_h^{(0,2)}\right) u_0^h(x) = -\left[u_0^h(x+he_1)\right. \quad (8.16) \\ &+ u_0^h(x-he_1) + u_0^h(x+he_2) + u_0^h(x-he_2) - 4u_0^h(x)\left.\right]/h^2, \forall x = (x_i, y_j) \in \Omega_0^h, \\ h = 1/k, \forall 0 < i,j < k,\ &u_0^h(x) = \phi(x) \forall x \in \partial\Omega^h \Leftrightarrow \forall 0 \leq i,j \leq k,\ i \cdot j = 0. \end{aligned}$$

Sometimes we approximate the $f(x)$ as in Proposition 8.5. Equation (8.16) is visualized in Table 8.1. We will use the same standard abbreviation, showing only the factors of the $u^h(x \pm he_i), u^h(x)$ in the appropriate positions, in the later tables.

Equation (8.16), and the following symmetric, and unsymmetric finite difference methods, essentially maintain the order 2 and 1, and the asymptotic expansion in powers of h^2 and h, in Lemma 8.2 for the difference between the exact solution, u_0, e.g. of the two-dimensional Laplacian and its discrete approximation, u_0^h, e.g. of (8.16), evaluated in the grid points, cf. Theorem 8.31 in Section 8.5, and Theorem 8.43 in Section 8.8.

8.3 Discrete Sobolev spaces

Essential for our stability and consistency results is the modification of continuous to discrete Sobolev spaces. These are spaces of grid functions only defined on Ω^h or $\mathcal{G}^h$. Correspondingly, we introduce these concepts, based upon *partial differences*. For the intended convergence results, we have to consider sequences of discrete Sobolev spaces, defined on increasingly finer grids, indicated by $h \in H$, cf. (8.10).

8.3.1 Notation and definitions

Similarly to the above, we define the higher *partial differences* with the usual multi-indices $\alpha := (\alpha_1, \ldots, \alpha_n) \in \mathbf{N}_0^n$, and the corresponding reals, and vectors in (8.17).

They are based upon the above $\partial_+^{(i)}u^h, \partial_-^{(i)}u^h, \partial_h^{(i)}u^h$ in (8.3), (8.4), generalized in (8.12)–(8.15). We formulate the ∂_+^α, version, and indicate the $\partial_-^\alpha, \partial_h^\alpha$ forms. The notation $\partial_\pm^\alpha$ indicates ∂_+^α or ∂_-^α or their combination, applied recursively.

$$\text{With } \partial_+^i u,\ \ \partial_-^i u,\ \ \partial_h^i u \ \text{ as in (8.12), (8.13) let} \tag{8.17}$$

$$\partial_+^\alpha u := \partial_+^{\alpha_1}\dots\partial_+^{\alpha_n}u := \left(\partial_+^1\right)^{\alpha_1}\cdots\left(\partial_+^n\right)^{\alpha_n}u, \nabla_+^k u := \left(\partial_+^\alpha u\right)_{|\alpha|=k},$$

$$\nabla_+^{\le k}u := \left(\partial_+^\alpha u\right)_{|\alpha|\le k},\ \nabla_+^0 u := u,\ \partial_+ u := \nabla_+ u := \nabla_+^1 u,$$

and correspondingly $\partial_-^i u, \partial_-^\alpha u, \nabla_-^k u, \partial_- u$, and $\partial_h^i u, \partial_h^\alpha u, \nabla_h u, \nabla_h^k u, \partial_h u$,

with, e.g., $u(x), \partial_+^i u(x), \partial_+^\alpha u(x) \in \mathbb{R}, \partial_+ u(x) \in \mathbb{R}^n, n_k, N_k$ s.t. $\nabla_+^k u(x) \in \mathbb{R}^{n_k}$,

$$\nabla_+^{\le k}u(x) \in \mathbb{R}^{N_k},\ \ \vartheta^i, \vartheta^\alpha := (\vartheta^1)^{\alpha_1}\cdots(\vartheta^n)^{\alpha_n} \in \mathbb{R}, \vartheta := (\vartheta^1,\dots,\vartheta^n) \in \mathbb{R}^n,$$

$$\Theta^k := (\vartheta^\alpha)_{|\alpha|=k} \in \mathbb{R}^{n_k},\ \vartheta^\alpha := \vartheta^i := 1 \text{ for } |\alpha| = i = 0, \Theta^0 \in \mathbb{R}, \Theta := \Theta^1.$$

For the elliptic equations below, the boundary conditions, cf. (8.8), are transformed into trivial Dirichlet conditions, see (8.62). For the *unsymmetric and symmetric difference methods*, this is translated into vanishing $u^h(x)$ for grid points x near the boundary, cf. (8.19), (8.20). Then the $\partial_\pm^\alpha u^h$ in the difference equations are defined in the standard way for all points, in particular for those near to, and at the boundary. The *symmetric difference methods,* cf. Subsection 8.4.2, (8.60), ff., and Section 8.7, are the motivation for choosing different forms for the relevant grid points, cf. (8.20), compared to (8.19), e.g. [678].

For relating the strong forms in (8.7), (8.16), and (8.44) ff. below to the corresponding generalized weak forms, we split the set of grid points into *all relevant, interior, boundary and near boundary grid points,* $\Omega^h, \Omega_0^h, \partial\Omega^h, \dots, \partial\Omega_m^h$ or $\Omega^h \setminus \Omega_0^h$, generalizing (8.11). We allow curved boundaries, such that

$$\Omega \subset \mathbb{R}^n \ \text{ is open, bounded, nonempty, and } \partial\Omega \text{ is Lipschitz-continuous.} \tag{8.18}$$

We distinguish two cases for defining the relevant grid points. Often $u_0 \in H^k(\Omega), k < m$, so it cannot be boundedly extended to an $E_c u_0 \in H^m(\mathbb{R}^n)$. Or the solution $u_0 \in H^k(\Omega), k \ge m$ with (8.18) allows the bounded extension $E_c u_0 \in H^k(\mathbb{R}^n)$, cf. Theorem 4.37.

The Ω-relevant grid points $x \in \mathcal{G}^h$, cf. (8.10), are defined via half open cubes $c^h(x)$. In both cases, (8.19) and (8.20), the strong and weak difference equations will be formulated on Ω_0^h and Ω^h, respectively, and the boundary conditions on $\Omega^h \setminus \Omega_0^h$.

For *both cases* and for the *first case,* $u_0 \in H^k(\Omega), k < m$, we define

$$c^h(x) := \prod_{i=1}^n [x - he_i/2, x + he_i/2),\ c^h(\Omega^h) := \bigcup_{x\in\Omega^h} \overline{c^h(x)}, \text{ and for } k < m \tag{8.19}$$

$$\Omega^h := \{x \in \mathcal{G}^h : \overline{c^h(x)} \subset \overline{\Omega}\},\ \ \partial\Omega_1^h := \{x \in \mathcal{G}^h \setminus \Omega^h : \overline{c^h(x)} \cap \partial\Omega \ne \emptyset\},$$

$$\Omega_0^h := \left\{x \in \Omega^h : \operatorname{dist}\left(x, \partial\Omega_1^h\right) \ge mh\right\}, \text{ and } \partial\Omega^h := \partial\Omega_m^h := \Omega^h \setminus \Omega_0^h,$$

of *interior* and *boundary knots*. In the second case, we define the interior knots as Ω_0^h and $\partial\Omega^h$, respectively.

$$\Omega_0^h := \{x \in \mathcal{G}^h : \overline{c^h(x)} \subset \overline{\Omega}\}, \text{and define boundary knots recursively: For} \tag{8.20}$$

$$m = 1 : \partial\Omega^h := \partial\Omega_1^h := \Big\{x \in \mathcal{G}^h \setminus \Omega_0^h : \overline{c^h(x)} \cap \partial\Omega \neq \emptyset\Big\},\ \Omega^h := \Omega_1^h := \Omega_0^h \cup \partial\Omega_1^h,$$

$$m \geq 2 : \Omega^h := \Omega_m^h := \Omega_0^h \cup \partial\Omega_1^h \cup \ldots \partial\Omega_m^h,\quad \partial\Omega^h := \Omega^h \setminus \Omega_0^h = \partial\Omega_1^h \cup \ldots \partial\Omega_m^h \text{ with}$$

$$\partial\Omega_m^h := \Big\{x \in \mathcal{G}^h \setminus \big(\Omega_0^h \cup \partial\Omega_1^h \cup \ldots \partial\Omega_{m-1}^h\big) : \overline{c^h(x)} \cap \big(\overline{\cup_{x' \in \Omega_0^h .. \cup \partial\Omega_{m-1}^h} c^h(x')}\big) \neq \emptyset\Big\}.$$

The half open cubes $c^h(x)$ yield a disjoint covering of $\mathbb{R}^n$, and Ω :

$$\mathbb{R}^n = \bigcup_{x \in \mathcal{G}^h} c^h(x),\ \Omega \subset \bigcup_{x \in \Omega^h \cup \partial\Omega_1^h} c^h(x) \text{ for (8.19)), and} \tag{8.21}$$

$$\Omega \subset \bigcup_{x \in \Omega_0^h \cup \partial\Omega_1^h} c^h(x) \text{ for (8.20)), with } c^h(x) \cap c^h(x') = \emptyset \forall x' \neq x, x' \in \mathcal{G}^h.$$

For Ω in (8.18), the above $\partial\Omega_1^h$ in (8.19) or (8.20) is equivalently defined by $\partial\Omega_1^h = \Big\{x \in \mathcal{G}^h \setminus \Omega^h \text{ or } \setminus \Omega_0^h : \overline{c^h(x)} \cap \overline{\Omega} \neq \emptyset\Big\}$. We have defined the $\Omega_0^h, \Omega^h, \partial\Omega_m^h$, in (8.20) differently from other authors, e.g. Temam [624]. This allows the presentation of symmetric difference methods of order 2 instead of the nonsymmetric order 1.

For the consistency of these methods we need

Lemma 8.2. *The* difference operators $\partial_\pm^{(\alpha)}, \partial_h^{(\alpha)}$ *satisfy the following estimates, and asymptotic expansions: With* $\mathbf{1} := (1, \ldots, 1)$, $c_{\alpha,j}, c'_{\alpha,j} \in \mathbb{R}$, *independent of* h, *and* $c_{\alpha,j} = c_{\alpha,j}\big(u^{(\alpha+\beta)}(x), |\beta| = j\big)$, $c'_{\alpha,j} = c'_{\alpha,j}\big(u^{(\alpha+\beta)}(x), |\beta| = 2j, \beta \text{ even }\big)$, *we get*

$$\big|\partial_\pm^{(\alpha)} u(x) - u^{(\alpha)}(x)\big| \leq c_{\alpha,1} h \text{ for } u^{(\alpha)} \in C_L(\Omega) \text{ for unsymmetric forms} \tag{8.22}$$

$$\big|\partial_h^{(\alpha)} u(x) - u^{(\alpha)}(x)\big| \leq c'_{\alpha,1} h^2 \text{ for } u^{(\alpha+\mathbf{1})} \in C_L(\Omega) \text{ for symmetric forms,}$$

with $\partial_h^{(0,i)} u(x) = \partial_h^{(\alpha=0e_i)} u(x) = (u(x+e_i) + u(x-e_i))/2$, *and asymptotic expansions*

$$\Big|\partial_\pm^{(\alpha)} u(x) - u^{(\alpha)}(x) - \sum_{j=1}^m h^j c_{\alpha,j}\Big| \leq L h^{m+1} c_{\alpha,m+1} \text{ for } u \in C_L^{\alpha+m}(\Omega), \tag{8.23}$$

$$\Big|\partial_h^{(\alpha)} u(x) - u^{(\alpha)}(x) - \sum_{j=1}^m h^{2j} c'_{\alpha,j}\Big| \leq L h^{2m+1} c'_{\alpha+2m+1}, \text{ for } u \in C_L^{(\alpha+2m)}(\Omega);$$

$u \in C_L^{\alpha+m}(\Omega)$ *indicates derivatives of* u *only in directions* e_i *with* $\alpha_i \neq 0$ *or* $0e_i$.

8.3.2 Discrete Sobolev spaces

Convergent difference methods require regularity assumptions for the boundary. For our chosen $\Omega \in C_L$ and the definition of Ω^h_m the following, usually imposed conditions are satisfied by Theorem 1.26 and by construction. In different papers the following conditions are imposed:

1. Assume $\partial\Omega$ smooth enough, such that the embedding $W^{1,p}(\Omega) \hookrightarrow L^p(\partial\Omega)$ is continuous, cf. Nečas [510].
2. The boundary strip $\Omega^h_m := \cup_{x\in\Omega^h} c^h(x) \setminus \Omega$ is uniformly small. Hence there exist $0 < d, h_0 \in \mathbb{R}$ such that for all $0 < h < h_0, h \in H$, for all $x \in \Omega^h_m$, there exists an α with $|\alpha| \le m$ such that $\exists x + d\alpha h$ or $\exists x - d\alpha h \notin \Omega$, cf. Raviart [545], p. 22.

The continuous embedding $W^{1,p}(\Omega) \hookrightarrow L^p(\partial\Omega)$ is a consequence of (8.9), (8.18), the uniformly small boundary strip of our construction.

We consider the following grid functions, u^h, and test functions $v^h : \mathcal{G}^h \to \mathbb{R}$ or $\mathbb{R}^q$, always vanishing outside of Ω^h. The *imposed vanishing Dirichlet boundary conditions* (of higher order) imply test functions, $v^h(x)$, vanishing on $\Omega^h \setminus \Omega^h_0$,

$$u^h, v^h : \mathcal{G}^h \to \mathbb{R} \text{ or } \to \mathbb{R}^q, \text{ with } u^h(x),\ v^h(x) = 0 \ \ \forall x \in \mathcal{G}^h \setminus \Omega^h, \text{ and} \tag{8.24}$$

$$\text{for test functions } v^h \in W^{m,p}_{0,+}(\Omega^h): \quad v^h(x) := 0 \ \ \forall x \in \Omega^h \setminus \Omega^h_0, \text{ cf. (8.19), (8.20).}$$

In Section 8.6, we will discuss natural boundary conditions, and modify the previous, and following definitions.

We define, for the grid functions, u^h, and their differences in (8.17)–(8.20), the *discrete integral,* the (scaled) *discrete Lebesgue, and Sobolev scalar products, their norms, seminorms, and spaces,* (of grid functions of finite norms $\|u^h\|_{L^2_h(\Omega^h)}$ or $\|u^h\|_{W^{m,p}_+(\Omega^h)}$), cf. e.g. Hackbusch [387] and Zeidler [678]:

$$\int_{\Omega^h} u^h dx^h := h^n \sum_{x\in\Omega^h} u^h(x), \int_{\Omega^h_0} u^h dx^h := h^n \sum_{x\in\Omega^h_0} u^h(x), \tag{8.25}$$

$$(v^h, w^h)_{L^2_h(\Omega^h)} := \int_{\Omega^h} v^h w^h dx^h, \quad \|v^h\|_{L^2_h(\Omega^h)} := (v^h, v^h)^{1/2}_{L^2_h(\Omega^h)} \text{ and}$$

$$1 \le p \le \infty : \|v^h\|_{L^p_h(\Omega^h)} := \left(\int_{\Omega^h} |v^h(x)|^p dx^h\right)^{1/p} = \left(h^n \sum_{x\in\Omega^h} |v^h(x)|^p\right)^{1/p}, \tag{8.26}$$

$$\text{with } \|v^h\|_{L^\infty_h(\Omega^h)} := \max_{x\in\Omega^h}\{|v^h(x)|\}, \text{ and } L^p_h(\Omega^h) := \left\{v^h : \|v^h\|_{L^p_h(\Omega^h)} < \infty\right\}.$$

We turn to *discrete Sobolev spaces* and the *dual norms* for bounded linear forms. Depending upon the different types of divided differences, mainly the ∂_+, we introduce the higher order Sobolev spaces of grid functions, denoted as $W^{m,p}_+(\Omega^h)$ or $W^{m,p}_-(\Omega^h)$, $W^{m,p}_\pm(\Omega^h)$, $W^{m,p}_h(\Omega^h)$. Compared to $W^{m,p}_+(\Omega^h)$ with u^h of finite norms, we replace in $W_-^{m,p}(\Omega^h)$, or the symmetric forms, $W_h^{m,p}(\Omega^h)$, the $\partial_+^\alpha v^h$ by the corresponding $\partial_-^\alpha v^h$, and appropriate symmetric $\partial_h^\alpha v^h$, cf. (8.12)–(8.17). The notation $L_h^p(\Omega^h)$, with $_h$ indicating the discretization parameter, is not appropriate for the following $W_+^{m,p}(\Omega^h),\ldots,$ since the $_+$, $_-$, $_h$ have to indicate the chosen type of divided differences, assuming, e.g. well-defined $\partial_+^\alpha u^h(x)\forall x\in\Omega^h$, $u^h:\mathcal{G}^h\to\mathbb{R}$.

$$(u^h,v^h)_{H_+^m(\Omega^h)} := \sum_{|\alpha|\le m}\left(\partial_+^\alpha u^h,\partial_+^\alpha v^h\right)_{L_h^2(\Omega^h)}, \text{ and} \tag{8.27}$$

$$|v^h|_{W_+^{k,p}(\Omega^h)} := \left(\sum_{|\alpha|=k}\|\partial_+^\alpha v^h\|_{L_h^p(\Omega^h)}^p\right)^{1/p}, \|v^h\|_{W_+^{m,p}(\Omega^h)} := \left(\sum_{0\le k\le m}|v^h|_{W_+^{k,p}(\Omega^h)}^p\right)^{1/p},$$

$$W_{0,+}^{m,p}(\Omega^h) := \left\{u^h\in W_+^{m,p}(\Omega^h) : u^h(x)=0\ \forall x\notin\Omega_0^h\right\} =: W_{0,-}^{m,p}(\Omega^h) =: W_{0,h}^{m,p}, \tag{8.28}$$

$$H_+^m(\Omega^h) := W_+^{m,2}(\Omega^h),\ldots,W_+^{-m,p'}(\Omega^h) := \mathcal{L}\left(W_+^{m,p}(\Omega^h),\mathbb{R}\right), 1/p+1/p'=1, \tag{8.29}$$

$$\langle f^h,v^h\rangle_{W_+^{-m,p'}(\Omega^h)\times W_+^{m,p}(\Omega^h)} = \sum_{|\alpha|\le m}\int_{\Omega^h} f_\alpha^h\partial_+^\alpha v^h dx^h, 1<p<\infty, \text{ cf. (2.107), and}$$

$$\|f^h\|_{W_+^{-m,p'}(\Omega^h)} := \sup_{\|v^h\|_{W_+^{m,p}(\Omega^h)}=1}\{|\langle f^h,v^h\rangle|\} = \left(\sum_{|\alpha|\le m}\|f^\alpha\|_{L_h^{p'}(\Omega^h)}^{p'}\right)^{1/p'}. \tag{8.30}$$

By (8.24), (8.29), $\|\cdot\|_{W_+^{m,p}(\Omega^h)}$, and $\|\cdot\|_{W_h^{m,p}(\Omega^h)}$ are equivalent. For systems we modify these definitions, corresponding to (8.69), as

$$(\vec{u}^h,\vec{v}^h)_{H_+^m(\Omega^h,\mathbb{R}^q)} := \sum_{i=1}^q\sum_{|\alpha|\le m}\left(\partial_+^\alpha u_i^h,\partial_+^\alpha v_i^h\right)_{L_h^2(\Omega^h)}, \quad \vec{u}^h=\left(u_1^h,\ldots,u_q^h\right), \tag{8.31}$$

$$|\vec{u}^h|_{W_+^{m,p}(\Omega^h,\mathbb{R}^q)}, \|\vec{u}^h\|_{W_+^{m,p}(\Omega^h,\mathbb{R}^q)} := \left(\sum_{i=1}^q\sum_{|\alpha|\le m}\|\partial_+^\alpha u_i^h\|_{L_h^p(\Omega^h)}^p\right)^{1/p}, \text{ cf. (2.107), with}$$

$$\|\vec{f}^h\|_{W_+^{-m,p'}(\Omega^h,\mathbb{R}^q)} := \left(\sum_{i=1}^q\|f_i^h\|_{W_+^{-m,p'}(\Omega^h)}^{p'}\right)^{1/p'}, \text{ for } \vec{f}^h=\left(f_1^h,\cdots,f_q^h\right).$$

For strong difference equations, and in Proposition 8.4, we additionally need $W_+^{k,p}(\Omega^h)$ with $u^h(x)=0\forall x\notin\Omega^h$ for $k=2m$ or $0\le k<m$.

Proposition 8.3. *All these spaces of discrete functions with any, but fixed choices of ∂_+^α or ∂_-^α or ∂_h^α, are* discrete Lebesgue and Sobolev spaces, *in particular Banach spaces, with respect to the scalar products, and norms defined in (8.25)–(8.31).*

As for standard Sobolev spaces, discrete Sobolev spaces satisfy (8.32), sometimes called the *Sobolev inequality,* cf. Yulin [673], Theorems 4, 9. The norm and seminorm in $W^{k,p}_{0,+}(\Omega^h, \mathbb{R}^q)$ in (8.33) are equivalent, cf. Ladyženskaja [462], Schumann and Zeidler, [574], Lemma 4, Temam [624], a consequence of his Proposition 3.3. Its generalization to $m > 1$ is straightforward.

Proposition 8.4. *For any discrete function* $\vec{u}^h \in W^{m,p}_+(\Omega^h)$, *and for any given* $\epsilon > 0$ *there exists a constant* $K(\epsilon, m, q)$ $\forall 0 \le k < m, 2 \le p \le \infty$, *independent of* h, *and* u^h, *such that the following inequality is valid* $\forall u^h \in W^{k,p}_+(\Omega^h)$:

$$|\vec{u}^h|_{W^{k,p}_+(\Omega^h)} \le \epsilon |\vec{u}^h|_{H^m_+(\Omega^h)} + K(\epsilon, m, q)|\vec{u}^h|_{L^2_h(\Omega^h)}. \tag{8.32}$$

$\|\vec{u}^h\|_{W^{k,p}_+(\Omega^h)}$ *and* $|\vec{u}^h|_{W^{k,p}_+(\Omega^h)}$ *are equivalent norms in* $W^{k,p}_{0,+}(\Omega^h)$, *hence there exists* $C = C(m, q) \in \mathbb{R}_+$, *independent of* h, *such that*

$$\|\vec{u}^h\|_{W^{k,p}_+(\Omega^h)} \le C|\vec{u}^h|_{W^{k,p}_+(\Omega^h)} \le C\|\vec{u}^h\|_{W^{k,p}_+(\Omega^h)} \ \forall u^h \in W^{k,p}_{0,+}(\Omega^h). \tag{8.33}$$

This allows a modification of (8.32) with $\|\vec{u}^h\|_{W^{k,p}_+(\Omega^h)}, \|\vec{u}^h\|_{W^{m,p}_+(\Omega^h)}$ *in* $W^{k,p}_{0,+}(\Omega^h)$.

In the following difference equations we sometimes employ the following mean value function. We will come back to it in our later approximations. For (8.20), we assume u or f to be extended to an $\Omega_1 \supset \Omega^h$ or to $\mathbb{R}^n$ and define:

$$\bar{f}^h(x) := \frac{1}{h^n} \int_{c^h(x)} f(\bar{x}) d\bar{x}. \tag{8.34}$$

Proposition 8.5. *This* integral mean value $\bar{f}^h(x)$ *approximates* $f(x) \in C^1_L\left(\Omega^h_e\right)$:

$$\bar{f}^h(x) := \frac{1}{h^n} \int_{c^h(x)} f(\bar{x}) d\bar{x}, \quad |\bar{f}^h(x) - f(x)| \le Lh^2. \tag{8.35}$$

For functions only defined in $\Omega \subset \Omega^h_e$, *extensions beyond* Ω *as (8.24) or in Gilbarg and Tru-dinger [346], Lemma 6.37 have to be used.*

With $\sum_{|\alpha|=2}^{2m',ev}$ *indicating even multi-indices and* $\alpha + \mathbf{1}$, $\mathbf{1} := (1, \ldots, 1)$, *we obtain the asymptotic expansion for* $f \in C^{2m'}_L\left(\Omega^h_e\right)$

$$\left| \bar{f}^h(x) - \left(f(x) + \sum_{|\alpha|=2}^{2m',ev} \frac{(h/2)^\alpha}{(\alpha + \mathbf{1})!} \partial^\alpha f(x) \right) \right| \le Ch^{2m'+1}. \tag{8.36}$$

The analogous result holds for $f \in W^{2m'+1,p}(\Omega^h_e)$ *and* $\sum_{|\alpha|=2}^{2m',ev} \cdots$ *and* $|f|_{C^{2m'}_L(\Omega^h_e)}$ *replaced by the averaged Taylor polynomial* $Q^{2m'+1,ev}f$ *in (4.33) and by* $|f|_{W^{2m'+1,p}(\Omega^h_e)}$.

Proof. We use the Taylor expansion:

$$f(x+te_j) = f(x) + \sum_{i=1}^{m} (t)^i \frac{\partial_{x_j}^i f(x)}{i!} + R_1 \text{ with } \partial_{x_j}^i f = \partial^i f / \left(\partial x_j{}^i\right) \text{ and}$$

$$R_1^m := R_1 := \int_0^t \frac{(t-\sigma)^{m-1}}{(m-1)!} \left(\partial_{x_j}^m f(x+\sigma) - \partial_{x_j}^m f(x)\right) d\sigma \text{ with} \tag{8.37}$$

$$|R_1| \leq L \frac{(t)^{m+1}}{m!} \text{ for } \partial_{x_j}^m f \in C_L(\Omega).$$

This is inserted into

$$\int_{c^h(x)} f(x)dx = \int_{-h/2}^{h/2} \cdots \int_{-h/2}^{h/2} f\left(x + \sum_{j=1}^{m} t_j e_j\right) dt_1 \cdots dt_n$$

$$= \int_{-h/2}^{h/2} \cdots \left(\int_{-h/2}^{h/2} f\left(x + t_1 e_1 + \sum_{j=2}^{m} t_j e_j\right) dt_1\right) \cdots dt_n \text{ and with (8.37)}$$

$$= \int_{-h/2}^{h/2} \cdots \left(\int_{-h/2}^{h/2} \left(f\left(x + \sum_{j=2}^{m} t_j e_j\right)\right.\right.$$

$$\left.\left. + \sum_{i=1}^{m,ev} (t_1)^i \frac{\partial_{x_1}^i f\left(x + \sum_{j=2}^m t_j e_j\right)}{i!} + R_1\right) dt_1\right) \cdots dt_n$$

and, with $m = 2m'$ and $x_+ := x + \sum_{j=2}^{m,ev} t_j e_j$

$$= \int_{-h/2}^{h/2} \cdots h \int_{-h/2}^{h/2} \left(f(x_+) + \sum_{i=2}^{m,ev} (h/2)^i \frac{\partial_{x_1}^i f(x_+)}{(i+1)!} + R_1\right) dt_2 \cdots dt_n.$$

By induction we obtain the claim. □

8.4 General elliptic problems with Dirichlet conditions, and their difference methods

8.4.1 General elliptic problems

In Section 8.1 we have already described the unsymmetric and the symmetric types of difference methods. The previous definitions allow a generalization of these difference equations to all types of elliptic differential equations studied in Chapter 2. To avoid obscuring the basic ideas we start, similarly to the above two examples in (8.1) and (8.8) with the case that Ω has the form (8.9). In contrast to (8.1) and (8.8) we consider unsymmetric difference methods first and generalize them to curved boundaries. This yields methods converging of order 1. We list for all cases the quartets of strong and weak differential and difference equations, e.g. A_s, A_s^h, A, A^h. We have seen in (8.12)–(8.15) that for difference equations, replacing $\partial_+ u^h$, $\partial_- u^h$ by $\partial_h u^h$ does not make sense, cf. (8.14). Nevertheless, we will formulate *symmetric divergence forms* for all our problems. This modification to symmetric difference methods converges of order 2.

We start, summarizing for the different elliptic problems studied here, with the definitions of ellipticity, and the conditions for the coefficients, implying the boundedness, and $H_0^m(\Omega, \mathbb{R}^q)$-or $W_0^{m,p}(\Omega, \mathbb{R}^q)$-coercivity of the weak forms of the principal parts for $m, q \geq 1$. All equations have to satisfy a strong Legendre, or in the case of higher order systems, a strong Legendre–Hadamard condition. We use the unified notation $a_{\alpha,\beta}$ for equations, $q = 1$, and $A_{\alpha,\beta}(x) = \left(a_{\alpha,\beta}^{ij}(x)\right)_{i,j=1}^q \in \mathbb{R}^{q\times q}$ for systems, $q > 1$, with $a_{\alpha,\beta} = a_{ij}$, and $A_{\alpha,\beta} = A_{kl}$ for $m = 1$, and $\forall |\alpha|, |\beta| \leq 1$. We include the conditions for nonlinear operators G implying a bounded derivative $G'(u_0)$ with a coercive principal part. Difference equations will require appropriate modifications.

Summary 8.6. *Let the bilinear form, $a(\cdot,\cdot)$, be induced by the (weak) linear operator A or the linearized $G'(u_0)$. We assume the following conditions 1.–4. for boundedness, and (strong, and uniform) ellipticity with $\vartheta = (\vartheta^1, \ldots \vartheta^n) \in \mathbb{R}^n$:*

1. *For linear equations of second order, cf. (8.46), (8.47), (8.60), we require $\forall 0 \leq i, j \leq n$, for almost every (a.e.) $x \in \Omega$, cf. Chapter 2, (2.136),*

$$a_{ij} \in L^\infty(\Omega),\ \exists \lambda > 0, a.e.\ \forall x \in \Omega:\ \lambda|\vartheta|_n^2 \leq \sum_{i,j=1}^n a_{ij}(x)\vartheta^i\vartheta^j\ \forall \vartheta \in \mathbb{R}^n, \tag{8.38}$$

where $a_{ij} \in L^\infty(\Omega)$ implies a.e. $\sum_{i,j=1}^n a_{ij}(x)\vartheta^i\vartheta^j < \Lambda|\vartheta|_n^2$ for $\Lambda > \lambda > 0$.

2. *For linear equations of order $2m, m > 1$, in (8.63), cf. Theorems 2.36, (2.79), (2.106), let*

$$\forall |\alpha|, |\beta| \leq m:\ a_{\alpha,\beta} \in L^\infty(\Omega),\ \text{and for } \forall |\alpha|, |\beta| = m:\ a_{\alpha,\beta} \in C(\overline{\Omega}), \tag{8.39}$$
$$\exists \lambda > 0,\ a.e.\ \forall x \in \Omega:\ \lambda|\vartheta|_n^{2m} \leq \sum_{|\alpha|=|\beta|=m} a_{\alpha,\beta}(x)\vartheta^\alpha\vartheta^\beta\ \forall \vartheta \in \mathbb{R}^n,\ x \in \Omega.$$

This implies a.e. $\sum_{|\alpha|=|\beta|=m} a_{\alpha,\beta}(x)\vartheta^\alpha\vartheta^\beta \leq \Lambda|\vartheta|_n^{2m}\ \ \forall \vartheta \in \mathbb{R}^n,\ x \in \Omega$.

3. *For linear systems of order $2m, m \geq 1, q > 1$, in (8.69), (8.70), we combine for $m = 1$ the strong Legendre condition (8.40), cf. Theorems 2.89, (2.343), and for $m > 1$, the strong Legendre–Hadamard condition (8.41), cf. Theorems 2.104, (2.394). Assume $A_{kl}, A_{\alpha,\beta} \in L^\infty(\Omega, \mathbb{R}^{q\times q})$ for $m \geq 1$. For $m > 1$, and $\forall |\alpha|, |\beta| = m$ let $A_{\alpha,\beta} \in C(\overline{\Omega}, \mathbb{R}^{q\times q})$. With $\vec{\vartheta} = (\vec{\vartheta}^1, \ldots, \vec{\vartheta}^n) \in \mathbb{R}^{n\times q}$, $\vartheta \in \mathbb{C}^n, \eta \in \mathbb{C}^q$, we assume for (8.70) that $\exists \lambda > 0$ such that $\forall \vec{\vartheta}, \vartheta, \eta$*

$$\text{for } m = 1:\ \ \lambda|\vec{\vartheta}|_{nq}^2 \leq \sum_{k,l=1}^n \left(A_{kl}(x)\vec{\vartheta}^l, \vec{\vartheta}^k\right)_q \leq \Lambda|\vec{\vartheta}|_{nq}^2,\ \ a.e. x \in \Omega, \tag{8.40}$$

$$\text{for } m > 1:\ \lambda|\vartheta|^{2m}|\eta|^2 \leq \sum_{|\alpha|=|\beta|=m}^{i,j=1,\ldots,q} a_{\alpha\beta}^{ij}(x)\vartheta^\beta\eta_j\overline{\vartheta}^\alpha\overline{\eta}_i\ \ \forall x \in \Omega. \tag{8.41}$$

For $m > 1$ *this*[47] *implies* $\sum_{|\alpha|=|\beta|=m}^{i,j=1,\dots,q} a_{\alpha\beta}^{ij}(x)\vartheta^\beta\eta_j\overline{\vartheta}^\alpha\overline{\eta}_i \leq \Lambda|\vartheta|^{2m}|\eta|^2 \; \forall x \in \Omega.$

4. *A nonlinear problem,* G, *as in Subsections 8.4.5–8.4.7, is called elliptic if its derivative,* $G'(u_0)$, *satisfies the appropriate condition 1.–3., for* $m, q \geq 1$.
5. *Then* $a(\cdot,\cdot)$, *induced by the previous linear or linearized operators, is bounded, and its principal part,* $a_p(\cdot,\cdot)$, *is* $H_0^m(\Omega,\mathbb{R}^q)$*-coercive. This* H_0^m*-coercivity remains correct for bilinear forms, induced by* $G'(u_0)$ *of nonlinear* G, *for* $m, q \geq 1$, *and in* $W^{m,p}(\Omega,\mathbb{R}^q)$ *for* $2 \leq p \leq \infty$, *cf. Section 2.7, and Theorems 2.118, 2.119, 2.122, 2.124, 2.125, 2.126, 2.127.*
6. *For a constant* $a_0 > 0$, *the extended principal part,* $a_p^e(u,v) := a_p(u,v) + a_0 \cdot (u,v)_{L^2(\Omega,\mathbb{R}^q)}$, *is also* $H^m(\Omega,\mathbb{R}^q)$*- coercive, cf. Theorem 1.24.*
7. *The strong differential operators,* A_s *or* $(G'(u_0))_s$, *are bounded if some coefficients* $a_{\alpha,\beta}$, *and* $A_{\alpha,\beta}$,*respectively, additionally are boundedly differentiable, more precisely:*

$$a_{\alpha,\beta} \in W^{|\beta|,\infty}(\Omega), \text{ and } A_{\alpha,\beta} \in W^{|\beta|,\infty}(\Omega,\mathbb{R}^{q\times q}) \; \forall|\alpha|,|\beta| \leq m. \tag{8.42}$$

8. *We consider difference equations in the weak forms (8.47), (8.63), (8.65), (8.67), (8.71)–(8.73), and in the corresponding strong forms, essentially the classical FDEs, (8.44), (8.62), (8.64), (8.66), (8.76). These* difference equations *are well defined, if the previous* $L^\infty(\Omega), W^{|\beta|,\infty}(\Omega), \dots$, *are replaced by* $C(\overline{\Omega}), C^{|\beta|}(\overline{\Omega}), \dots$, *or by bounds on* $\mathcal{G}^h$ *or* $a_{ij} \in H^1(\Omega), \dots$, *by* $\bar{a}_{ij}(x)$, *cf. (8.34), and Remark 8.7.*
9. *The* discrete principal parts, $a_p^h(\cdot,\cdot)$, *are defined as the sum of the highest order difference terms, corresponding to the coefficients, e.g. in (8.38)–(8.41).*

The discrete forms of these linear operators, the A^h, or the linearizations of the discrete operator, G^h, for $m, q \geq 1$, and the coercivity of the principal parts are essential for stability.

In Subsections 8.4.2 and 8.4.3, we give the details for second order linear equations for unsymmetric and symmetric forms. The other cases we restrict to listing the equations in Subsections 8.4.4 ff. The domains, and ranges, e.g. the $\mathcal{D}\left(A_s^h\right) = H_+^2(\Omega^h) \cap H_{0,+}^1(\Omega^h)$, $\mathcal{R}\left(A_s^h\right) = L_h^2\left(\Omega_0^h\right)$ of the A_s, A_s^h in (8.43), and the A^h in (8.47) have to be obviously modified. For nonlinear operators the domains are subsets of the corresponding Sobolev spaces in (8.43), (8.47). Conditions for the coefficients are listed in Remark 8.7 and Summary 8.6.

For general domains as in (8.18) with boundary grid points in (8.19) not in the boundary of Ω, specific difference methods require a separate discretization of the differential equation, and the boundary conditions. This would be necessary for fully nonlinear problems in Subsection 8.4.7, if we were to discuss the details. We indicate some other examples in Section 8.7.

8.4.2 Second order linear elliptic difference equations

We present the detailed relations between the strong and weak differential and difference equations, A_s, A_s^h, A, A^h. We start assuming Ω *in the form* (8.9), and

[47] For difference methods several definitions for regular elliptic systems are used in the literature. They are all more or less equivalent, cf. [151]. Here we have enforced in (8.41), towards the needs for difference methods, the condition $\vartheta \in \mathbb{R}^n, \eta \in \mathbb{C}^q, \vartheta^\alpha$ for elliptic systems into $\vartheta \in \mathbb{C}^n, \eta \in \mathbb{C}^q, \overline{\vartheta}^\alpha$.

consider the *strong unsymmetric form* of difference equations in *divergence form.* Combining the ∂_-^j, $\partial_+^i u^h(x)$, yields one of the *classical or natural difference equations.* The unsymmetric methods of order 1 are modified in Subsection 8.4.3 on rectangular domains to symmetric methods of order 2.

We apply partial summation (8.49) in Proposition 8.8 to difference equations in the strong form to get the weak forms. The following A_s is well defined for $u \in H^2(\Omega), a_{ij} \in H^1(\Omega)$, cf. (2.16), and A_s^h for $u^h \in H_+^2(\Omega^h)$. We use $(-1)_{j>0} = 1$ for $j = 0$, else $= -1$, and sometimes replace $a_{ij}(x), f(x)$ by $\bar{a}_{ij}^h(x), \bar{f}^h(x)$ in (8.34).

$$A_s : H^2(\Omega) \cap H_0^1(\Omega) \to L^2(\Omega), A_s^h : H_+^2(\Omega^h) \cap H_{0,+}^1(\Omega^h) \to L_h^2\left(\Omega_0^h\right),$$

$$A_s u_0 = \sum_{i,j=0}^{n} (-1)_{j>0} \partial^j (a_{ij} \partial^i \ u_0) = f \text{ in } L^2(\Omega), \text{ cf. (8.42)}, \tag{8.43}$$

$$A_s^h u_0^h(x) = \sum_{i,j=0}^{n} (-1)_{j>0} \partial_-^j \left(a_{ij}(x) \partial_+^i u_0^h(x)\right) = f^h(x) \ \ \forall x \in \Omega_0^h, \text{ with} \tag{8.44}$$

$$\text{boundary conditions, cf.(8.24), (2.16), } u_0 \in H_0^1(\Omega), u_0^h(x) \in H_{0,+}^1(\Omega^h). \tag{8.45}$$

Similarly to Table 8.1, we visualize the unsymmetric (8.44) for $x \in \Omega_0^h \subset \mathbb{R}^2$. We use the same abbreviation as in Table 8.1, by listing the coefficients of, but omitting the unknowns $u^h(x \pm he_i)$. We include, here and below, only the highest order terms. The three-point stars in the direction of $e_i, i = 1, 2$, for the $-h^2 \partial_-^i a_{ii} \partial_+^i u^h(x)$ are listed in Table 8.2. They have to be combined with the seven-point stars for the $-h^2 \left(\partial_-^2 a_{12} \partial_+^1 + \partial_-^1 a_{21} \partial_+^2\right) u^h(x)$ in Table 8.3. Summation of the numbers in Table 8.2, $i = 1, 2$, and in Table 8.3 yields the seven-point star for the full (8.44) in Table 8.4. Obviously Table 8.4 reduces to Table 8.1 for $a_{11} = a_{22} = 1$, $a_{12} = a_{21} = 0$.

The following dilemma applies to all our elliptic difference equations. For (8.43) we only need $a_{ij} \in H^1(\Omega)$. This is certainly problematic for (8.44).

Remark 8.7.

1. $a_{ij} \in H^1(\Omega)$ *in (8.43) is sufficient for analytical results: We require, for simplicity, and fitting to the standard situation in difference methods,* $a_{ij} \in C(\overline{\Omega})$, *with well-defined* $a_{ij}(x) \forall x \in \Omega^h$ *in (8.44). For interesting cases* $a_{ij} \in H^1(\Omega), \ldots,$ *we might replace the* $a_{ij}(x) \forall x \in \Omega^h$ *by* $\bar{a}_{ij}(x)$*, as defined in (8.34). This is unproblematic for linear(ized) problems. With minor modifications all results remain correct. The proof for this claim is left as an exercise for the interested reader.*
2. *For semi-, quasi-, and fully nonlinear problems below, this averaging, necessarily defined with respect to* x*, but for fixed values of* $u^h(x)$*, e.g., (8.64),(8.65), would be problematic.*
3. *The conditions for boundedness and ellipticity of linear(ized) operators in Summary 8.6 may be appropriately modified for difference equations, and systems for higher order problems as well. In particular, generalizations to the*

Table 8.2: Three-point star for (8.44), $x \in \mathcal{G}^h \subset \mathbb{R}^2$ for $-h^2(\partial_-^i a_{ii}\partial_+^i)u^h(x), i = 1, 2$.

$-a_{ii}(x - he_i)$	$a_{ii}(x) + a_{ii}(x - he_i)$	$-a_{ii}(x)$

Table 8.3: Seven-point star for (8.44), $x \in \mathbb{R}^2$, and $-h^2(\partial_-^2 a_{12}\partial_+^1 + \partial_-^1 a_{21}\partial_+^2)u^h(x)$.

$a_{21}(x - he_1)$	$-a_{21}(x)$	0
$-a_{21}(x - he_1)$	$(a_{12} + a_{21})(x)$	$-a_{12}(x)$
0	$-a_{12}(x - he_2)$	$a_{12}(x - he_2)$

Table 8.4: Seven-point star for the complete $-h^2\times$ (8.43) and $x \in \mathbb{R}^2$.

$a_{21}(x - he_1)$	$-a_{22}(x) - a_{21}(x)$	0
$-(a_{11} + a_{21})(x - he_1)$	$a_{11}(x - he_1) + a_{22}(x - he_2)$	$-(a_{11} + a_{12})(x)$
0	$+(a_{11} + a_{22} + a_{12} + a_{21})(x)$	$a_{12}(x - he_2)$
	$-(a_{22} + a_{12})(x - he_2)$	

$W_0^{1,p}(\Omega), W_0^{m,p}(\Omega)$ situation and the appropriate conditions for this and the following subsections are discussed there.

The usual multiplication of $A_s u$ with $v \in H_0^1(\Omega)$, integration over Ω, and partial integration yields the (standard) *weak bilinear form* $a(\cdot,\cdot)$, and the induced A. Both are well defined for $u, v \in H^1(\Omega)$ and $a_{ij} \in L^\infty(\Omega)$, see (2.18). With

$$\langle Au, v\rangle_{H^{-1}(\Omega)\times H_0^1(\Omega)} = (A_s u, v)_{L^2(\Omega)} \forall u \in H^2(\Omega), v \in H_0^1(\Omega), \tag{8.46}$$

$$a_{ij} \in H^1(\Omega) \forall j > 0,$$

$$\text{compute } u_0 \in H_0^1(\Omega) : a(u_0, v) := \sum_{i,j=0}^n \int_\Omega a_{ij}\partial^i u_0 \partial^j v dx = \langle Au_0, v\rangle_{H^{-1}(\Omega)\times H_0^1(\Omega)}$$

$$= \langle f, v\rangle_{H_0^{-1}(\Omega)\times H_0^1(\Omega)} \forall v \in H_0^1(\Omega) \text{ with } f \in H_0^{-1}(\Omega).$$

Similarly to the previous strong form, A_s^h, we introduce the corresponding *weak operator*, A^h, and *weak bilinear form*, $a^h(\cdot,\cdot)$. Both are well defined for $u^h, u_0^h, v^h \in H_+^1(\Omega^h)$ and $a_{ij} \in C(\overline{\Omega})$. Choosing u^h, v^h as in (8.24), we determine u_0^h from

$$a^h : H_{0,+}^1(\Omega^h) \times H_{0,+}^1(\Omega^h) \to \mathbb{R}, A^h : \mathcal{V}_b^h = H_{0,+}^1(\Omega^h) \to H_{0,+}^{-1}(\Omega^h) = \mathcal{V}_b^{\prime h}, u_0^h \in \mathcal{V}_b^h$$

$$a^h\left(u_0^h, v^h\right) := \sum_{i,j=0}^n \int_{\Omega^h} a_{ij}\partial_+^i u_0^h \partial_+^j v^h dx^h = \sum_{i,j=0}^n h^n \sum_{x\in\Omega^h} \left(a_{ij}\partial_+^i u_0^h \partial_+^j v^h\right)(x)$$

$$=: \left\langle A^h u_0^h, v^h\right\rangle_{\mathcal{V}_b^{\prime h}\times\mathcal{V}_b^h} = \left\langle f^h, v^h\right\rangle_{\mathcal{V}_b^{\prime h}\times\mathcal{V}_b^h} \forall v^h \in \mathcal{V}_b^h = H_{0,+}^1(\Omega^h). \tag{8.47}$$

Depending upon the structure of f (and of Au) we employ different kinds of approximations. The $(f, v^h)_{L^2_h(\Omega^h)}$ or $\langle f, v^h \rangle_{\mathcal{V}_b'^h \times \mathcal{V}_b^h}$ can be evaluated exactly or appropriate quadrature approximations may be used, cf. Section 5.4. For nonsmooth f, we may replace $f(x)$ by $f^h(x) := \bar{f}^h(x)$ in (8.34). We denote all these approximations as $\langle f^h, v^h \rangle_{\mathcal{V}_b'^h \times \mathcal{V}_b^h}$. For other strategies, cf., e.g. Thomée and Westergren [627], Heinrich [395], Bube *et al.* [151, 605] and Grossmann *et al.* [374–376].

Similarly to partial integration transforming differential equations in *strong form* into the *weak form*, we apply *partial summation.* For avoiding unnecessary repetition, we formulate Proposition 8.8 for linear and quasilinear elliptic equations of orders $2m, m \geq 1$, in (8.62), (8.66), with obvious extensions to systems. We require for $v^h \in H^m_{0,+}(\Omega^h)$ the extension (8.24), but not for all u^h considered here.

We simplify the notation for the proof, and the following results by introducing the grid points, and their function values in Ω^h in the e_j direction as, cf. (8.16).

$$x_i := x_0 + ihe_j, i \in \mathbb{Z}, x_i \in \Omega^h_0 \text{ for } i = 1, \ldots, k-1, x_i \in \partial\Omega^h \text{ for } i \leq 0, i \geq k,$$

$$u_i := u^h(x_i), \text{ and the } \Omega^j := \{x_i, i = 1, \ldots, k-1\}, \overline{\Omega}^j := \{x_i, i = 0, \ldots, k\},$$

$$\overline{\Omega}^{h,j}_m := \{x_i, i = -m+1, \ldots, k+m-1\}, \overline{\Omega}^j := \overline{\Omega}^{h,j}_1, \text{ with sums, e.g.,}$$

$$\sum_{\overline{\Omega}^j} v^h \partial^j_- u^h := \sum_{i=0}^{k} \left(v^h \partial^j_- u^h \right)(x_i), \text{and obvious modifications for } \Omega^j, \overline{\Omega}^{h,j}_m, \Omega^h, \ldots.$$

Proposition 8.8. *For* $u^h \in H^{2m}_+(\Omega^h), v^h \in H^m_{0,+}(\Omega^h), m \geq 1$, *(8.48)–(8.52) are valid, with obvious* $(v^h, \cdots)_{L^2_h(\Omega^h_0)} = (v^h, \cdots)_{L^2_h(\Omega^h)}, \ldots$ *in (8.49)– (8.52), hence*

$$\textit{for } j > 0: -\sum_{\Omega^j} v^h \partial^j_- u^h = -\sum_{\overline{\Omega}^j} v^h \partial^j_- u^h = \sum_{\overline{\Omega}^j} u^h \partial^j_+ v^h \Rightarrow \quad \textit{for } m = 1, \tag{8.48}$$

$$\left(\sum_{i,j=0}^{n} v^h, (-1)_{j>0} \partial^j_- \left(a_{ij} \partial^i_+ u^h_0 \right) \right)_{L^2_h(\Omega^h_0) = L^2_h(\Omega^h)} = \left(\sum_{i,j=0}^{n} a_{ij} \partial^i_+ u^h_0, \partial^j_+ v^h \right)_{L^2_h(\Omega^h)} \tag{8.49}$$

$$\textit{for } m > 1, |\beta| \leq m: \; (-1)^{|\beta|} \left(v^h, \partial^\beta_- u^h \right)_{L^2_h(\Omega^h_0) = L^2_h(\Omega^h)} = \left(u^h, \partial^\beta_+ v^h \right)_{L^2_h(\Omega^h)}, \tag{8.50}$$

$$(-1)^{|\beta|} \left(\partial^\beta_- \left(A_\beta \left(\cdot, u^h, \nabla^m_+ u^h \right) \right), v^h \right)_{L^2_h(\Omega^h_0) = \ldots} = \left(A_\beta \left(\cdot, u^h, \nabla^m_+ u^h \right), \partial^\beta_+ v^h \right)_{L^2_h(\Omega^h)} \Rightarrow$$

$$\left(\sum_{|\alpha|,|\beta| \leq m} v^h, (-1)^{|\beta|} \partial^\beta_- \left(a_{\alpha\beta} \partial^\alpha_+ u^h_0 \right) \right)_{L^2_h(\Omega^h_0) = ..} = \left(\sum_{|\alpha|,|\beta| \leq m} a_{\alpha\beta} \partial^\alpha_+ u^h_0, \partial^\beta_+ v^h \right)_{L^2_h(\Omega^h)} \tag{8.51}$$

$$\left(\sum_{|\beta|\le m} v^h, (-1)^{|\beta|}\partial_-^\beta A_\beta\left(\cdot, u_0^h, \cdots, \nabla_+^m u_0^h\right)\right)_{L_h^2(\Omega_0^h)=L_h^2(\Omega^h)}$$

$$= \left(\sum_{|\beta|\le m} A_\beta\left(\cdot, u_0^h, \cdots, \nabla_+^m u_0^h\right), \partial_+^\beta v^h\right)_{L_h^2(\Omega^h)}. \tag{8.52}$$

Equations (8.48) ff. remain correct if ∂_+, and ∂_- are interchanged.

Remark 8.9.

1. *According to (8.48)–(8.52), the following weak forms of the difference methods use $(\cdot,\cdot)_{L_h^2(\Omega^h)}$. The unequal previous condition $u^h \in H_+^m(\Omega^h), m \ge 1$, $v^h \in H_{0,+}^m(\Omega^h)$ for u^h, v^h is important, since we apply (8.48)–(8.52) in the difference methods to very different terms u^h, cf. (8.44), Proposition 8.10, and (8.56),(8.65).*
2. *These results remain correct, if we allow $u^h \in W_+^{m,p}(\Omega^h), 2 \le p < \infty$, and for the linear case v^h and the quasilinear case $\forall u^h : A_\beta\left(\cdot, u^h, \cdots, \nabla_+^m u^h\right) \in W_{0,+}^{-m,p'}(\Omega^h), 1/p + 1/p' = 1$, respectively*

Proof. We start with the first order result (8.48) for $j > 0$. Since the proofs for $\partial_+^j u^h$, and $\partial_-^j u^h$ are very similar, we only consider $\partial_-^j u^h$. We use $H_{0,+}^1(\Omega^h) \ni v^h$, $v^h(x) = 0 \forall x \in \mathcal{G}^h \setminus \Omega_0^h$, only for v^h, but not for u^h.

$$j > 0 : -\sum_{\Omega^j} v^h \partial_-^j u^h = -\sum_{\overline{\Omega}^j} v^h \partial_-^j u^h \tag{8.53}$$

$$= -\sum_{i=0}^{k} v_i(u_i - u_{i-1})/h = -\left(\sum_{i=0}^{k} v_i u_i - \sum_{i'=-1}^{k-1} v_{i'+1} u_{i'}\right)/h$$

$$\overset{v_0=v_k=v_{k+1}=0}{=} \left(\sum_{i=0}^{k} (v_{i+1} - v_i) u_i\right)/h = \sum_{\overline{\Omega}^j} u^h \partial_+^j v^h.$$

Replacing the previous u^h by $w^h := \left(a_{\ell j}\partial_+^\ell u^h\right)$ we obtain with $v_0 = v_k = v_{k+1} = 0$

$$j > 0 : \quad -\sum_{\Omega^j} v^h \partial_-^j \left(a_{\ell j}\partial_+^\ell u^h\right) = \sum_{\overline{\Omega}^j} \left(a_{\ell j}\partial_+^\ell u^h\right) \partial_+^j v^h. \tag{8.54}$$

For (8.49), we multiply (8.54) with h^n and sum over all $0 \le i = \ell, j \le n$.

The first line of (8.50) with $1 < m \ge |\beta| \ge 1$ is proved inductively. Equation (8.54) with $m > 1 \ge |\alpha|, |\beta| = 1$ represents the beginning of the induction. So we continue with considering the components $\beta_j \ge 2$, and the sum over the lines Ω^j. In (8.53), we replace, for the next step of the induction, the v^h, and Ω^j, and $\overline{\Omega}^j$ by, e.g. $\partial_+^\ell v^h$, and

$\overline{\Omega}^j$, and $\overline{\Omega}_2^{h,j}$, respectively. We obtain with $0 = \partial_+^\ell v_{-1} = \partial_+^\ell v_{k+1}$:

$$\partial_-^{\beta_j - 1} u_i := \partial_-^{\beta_j-1} u(x_i) : -\sum_{\overline{\Omega}^j} \partial_+^\ell v^h \partial_-^{\beta_j} u^h = -\sum_{\overline{\Omega}_2^{h,j}} \partial_+^\ell v^h \partial_-^j \partial_-^{\beta_j-1} u^h \tag{8.55}$$

$$= -\sum_{i=-1}^{k+1} \partial_+^\ell v_i \partial_-^j \partial_-^{\beta_j-1} u_i$$

$$= -\left(\sum_{i=-1}^{k+1} \partial_+^\ell v_i \partial_-^{\beta_j-1} u_i - \sum_{i=-1}^{k+1} \partial_+^\ell v_i \partial_-^{\beta_j-1} u_{i-1}\right)/h$$

$$\overset{0=\partial_+^\ell v_{-1}=\partial_+^j v_{k+2}}{=} \left(\sum_{i=-1}^{k+1} \left(\partial_+^\ell v_{i+1} - \partial_+^\ell v_i\right) \partial_-^{\beta_j-1} u_i\right)/h$$

$$= \sum_{\overline{\Omega}_2^{h,j}} \left(\partial_+^\ell \partial_+^j\right) v^h \partial_-^{\beta_j-1} u^h.$$

This allows reducing inductively each component $\beta_j > 0$ in (8.55) to 0, and proceeding as above. Then $v^h \in H_{0,+}^m$ is sufficient for the first line of (8.50).

For the second line of (8.50), the quasilinear equations of order $2m$ in (8.66) below, we obtain with

$$\beta^j := \beta - e_j, \partial_-^\beta = \partial_-^j \partial_-^{\beta^j},\ w^h := (-1)^{|\beta^j|} \partial_-^{\beta^j} A_\beta\left(\cdot, u^h(\cdot), \ldots, \nabla_+^m u^h(\cdot)\right) :$$

$$\sum_{\overline{\Omega}^j} v^h \left((-1)^{|\beta|} \partial_-^\beta A_\beta\left(\cdot, u^h(\cdot), \ldots, \nabla_+^m u^h(\cdot)\right)\right)$$

$$= -\sum_{\overline{\Omega}^j} v^h \partial_-^j \left((-1)^{|\beta^j|} \partial_-^{\beta^j} A_\beta\left(\cdot, u^h(\cdot), \ldots, \nabla_+^m u^h(\cdot)\right)\right)$$

$$= -\sum_{\Omega^j} v^h \partial_-^j w^h = -\sum_{\overline{\Omega}^j} v^h \partial_-^j w^h = -\sum_{i=0}^{k} v_i (w_i - w_{i-1})/h$$

$$= -\left(\sum_{i=0}^{k} v_i w_i - \sum_{i=0}^{k} v_i w_{i-1}\right)/h \overset{0=v_0=v_{k+1}}{=} \left(\sum_{i=0}^{k} w_i (v_{i+1} - v_i)/h\right)$$

$$= \sum_{\overline{\Omega}^j} w^h \partial_+^j v^h = \sum_{\overline{\Omega}^j} \left((-1)^{|\beta^j|} \partial_-^{\beta^j} A_\beta\left(\cdot, u^h(\cdot), \ldots, \nabla_+^m u^h(\cdot)\right)\right) \partial_+^j v^h.$$

Again, we reduce $|\beta|$ to 0. Now multiplying the first and second line of (8.50) with h^n, and summation over all directions yields (8.51) and (8.52). □

We reformulate the results in Proposition 8.8, (8.49), (8.51) now for the weak and strong difference operators, A_s^h, and A^h, and extend them to systems.

Proposition 8.10. *For the difference operators A_s^h, A^h in (8.44), (8.47) for $m = 1$ and (8.62), (8.63), (8.71) for $m > 1, q \geq 1$ we obtain $\forall v^h \in \mathcal{V}_b^h = H_{0,+}^m(\Omega^h, \mathbb{R}^q)$:*

$$\left\langle A_s^h u^h, v^h \right\rangle_{L^2\left(\Omega_0^h, \mathbb{R}^q\right)} = \left\langle A_s^h u^h, v^h \right\rangle_{L^2(\Omega^h, \mathbb{R}^q)} = \left\langle A^h u^h, v^h \right\rangle_{\mathcal{V}_b'^h \times \mathcal{V}_b^h}. \tag{8.56}$$

Equation (8.56) remains correct for the quasilinear problems G_s^h, and G^h in (8.65), (8.67), and linear and quasilinear systems (8.71) and (8.72), (8.73), cf. Remark 8.9

Exchanging the two difference operators ∂_+^h, and ∂_-^h in the previous (8.44) (8.47), and all the following cases, allows the same convergence results. In the symmetric form below we will use the mean value of these two equations.

Equation (8.44) is a classical system of difference equations for (8.43), and has to be solved. It is, possibly with slight modifications, included in the usual software packages. For the corresponding weak difference form (8.47) of (8.44), and its generalizations below, a direct stability, and convergence proof with respect to the $\|\cdot\|_{H_+^1(\Omega^h)}$ will be given in Subsections 8.5.4 ff. This implies the results for (8.44) as well. In fact, a strong solution of (8.44) satisfies, by (8.56), the weak (8.47) as well, and vice versa:

$$u_0^h \in H_+^2(\Omega^h) \cap \left(\mathcal{V}_b^h = H_{0,+}^1(\Omega^h)\right) : \ \left(A_s^h u_0^h, v^h\right)_{L_h^2\left(\Omega_0^h\right)} = \left(f^h, v^h\right)_{L_h^2\left(\Omega_0^h\right)} \forall v^h \in \mathcal{V}_b^h$$

$$f \in L^2(\Omega) \implies u_0^h \in \mathcal{V}_b^h : \ \left\langle A^h u_0^h, v^h \right\rangle_{\mathcal{V}_b'^h \times \mathcal{V}_b^h} = (f^h, v^h)_{L_h^2(\Omega^h)} \forall v^h \in \mathcal{V}_b^h. \tag{8.57}$$

Hence we solve (8.44) by standard software packages. This solution inherits the convergence properties of the weak solution, u_0^h, of (8.47), cf. Theorem 8.11.

If a discrete solution $u_0^h(x) \forall x \in \Omega^h$ for (8.44) or (8.47) with the boundary conditions (8.45) is computed, it is straightforward to determine its first and second divided differences, $\partial_+^i u_0^h(x), \partial_+^j \partial_+^i u_0^h(x)$, yielding the desired $u_0^h \in \mathcal{V}_b^h$ or $u_0^h \in H_+^2(\Omega^h) \cap \mathcal{V}_b^h$. This remains correct for order $2m$, and the solutions $u_0^h \in \mathcal{V}_b^h = H_{0,+}^m(\Omega^h, \mathbb{R}^q)$, and $u_0^h \in H_+^{2m}(\Omega^h, \mathbb{R}^q) \cap \mathcal{V}_b^h$, $m, q \geq 1$, of the weak, and strong difference problems in the following subsections.

We use the same approach as in the previous chapters, related to Stetter [596]. This contrasts to, e.g. Raviart [545], Temam [621] and Schumann and Zeidler [574, 678]. They apply external approximations.

We want to outline the differences in alternative approaches. Let $\breve{u}_0^h$ and $\hat{u}_0^h$ indicate the solutions of the strong and the weak problem, respectively. For the strong $A_s^h \breve{u}_0^h = f^h (= f)$ with special A, often stability results are obtained via *M-matrices*, see, e.g. Hackbusch [387], Sections 4.2 ff. This yields estimates of the form

$$A_s^h \breve{u}_0^h(x) = f(x) \forall x \in \Omega_0^h, \ \breve{u}_0^h(x) = 0 \forall x \in \partial\Omega^h \tag{8.58}$$

$$\implies \sup_{x \in \Omega^h} \left|\breve{u}_0^h(x)\right| = \left\|\breve{u}_0^h\right\|_{L_h^\infty(\Omega^h)} \leq C\|f\|_{L_h^\infty(\Omega^h)}.$$

The approach, based upon the weak $Au_0 = f$, starts with the coercive bilinear form, $a_p(\cdot,\cdot)$, of the principal part, A_p, of the above general A. We will show that its discrete counterpart $a_p^h(\cdot,\cdot)$, cf. Summary 8.6, is again coercive, so stable as well. Thus the consistency, and Theorem 3.29, imply, for a boundedly invertible A, the stability of

A^h, hence the existence of a weak solution $\hat{u}_0^h = \breve{u}_0^h$ by (8.57), such that

$$\text{for } f^h \in L_h^\infty(\Omega^h), \text{ and for } \hat{u}_0^h(x) = 0 \ \forall\ x \in \partial\Omega^h : \quad A_s^h \hat{u}_0^h = f^h \iff \tag{8.59}$$
$$\langle A^h \hat{u}_0^h, v^h \rangle_{\mathcal{V}_b'^h \times \mathcal{V}_b^h} = a^h\left(\breve{u}_0^h, v^h\right) = \langle f^h, v^h \rangle_{\mathcal{V}_b'^h \times \mathcal{V}_b^h} \forall v^h \in \mathcal{V}_b^h = H_{0,+}^1(\Omega^h), \text{ and}$$
$$\|\hat{u}_0^h\|_{H_+^1(\Omega^h)} = \|\hat{u}_0^h\|_{\mathcal{V}_b^h} \le C\|f^h\|_{\mathcal{V}_b'^h} \implies \|\breve{u}_0^h\|_{\mathcal{V}_b^h} \le C'\|f^h\|_{L_h^\infty(\Omega^h)} \le C\|f^h\|_{\mathcal{V}_b'^h},$$

since $\|f^h\|_{L_h^\infty(\Omega^h)}$ is dominated by $\|f^h\|_{\mathcal{V}_b'^h}$. This argument from (8.57) to (8.59) remains correct for the following linear cases as well, if only the $\|\breve{u}_0^h\|_{H_+^1(\Omega^h)}$ are replaced by $\|\breve{u}_0^h\|_{H_+^m(\Omega^h,\mathbb{R}^q)}, m, q \ge 1$. For quasilinear problems corresponding stability–consistency arguments apply. Fully nonlinear problems require another treatment.

Theorem 8.11. H_+^m convergence for classical (strong) difference methods: *We assume a stability estimate (8.59) for the weak* $A^h\hat{u}_0^h = f^h \in H_{0,+}^{-m}(\Omega^h, \mathbb{R}^q), m, q \ge 1$, *in the form* $\|\hat{u}_0^h\|_{H_+^m(\Omega^h,\mathbb{R}^q)} \le C\|f^h\|_{H_{0,+}^{-m}(\Omega^h,\mathbb{R}^q)}$. *Let* $A : H_0^m(\Omega, \mathbb{R}^q) \to H_0^{-m}(\Omega, \mathbb{R}^q)$, *and* $A_s : H^{2m}(\Omega, \mathbb{R}^q) \cap H_0^m(\Omega, \mathbb{R}^q) \to L^2(\Omega, \mathbb{R}^q) \supset L^\infty(\Omega, \mathbb{R}^q)$ *be boundedly invertible. This implies* $\|\breve{u}_0^h\|_{H_+^m(\Omega^h,\mathbb{R}^q)} \le C\|f\|_{L_h^\infty(\Omega^h,\mathbb{R}^q)}$ *for the solution of* $A_s^h\breve{u}_0^h = f^h \in L_h^\infty(\Omega^h, \mathbb{R}^q)$ *as well, strongly improving (8.58).*

This $\|\breve{u}_0^h\|_{H_+^m(\Omega^h,\mathbb{R}^q)} \le C\|f\|_{L_h^\infty(\Omega^h,\mathbb{R}^q)}$ is a special case of regularity results for the solutions of difference equations in the style of, e.g. Bube *et al.* [151, 605], Fössmeier [320], Geissert [339], Hackbusch [381, 383, 384, 387] and Thomée and Westergren [627].

Other concepts of generalized weak solutions for difference equations are studied by Jovanovich *et al.* [416,430–432,613], and the viscosity solutions for nonlinear problems by Crandall and Lions [217] ff., cf. Subsection 8.4.7

8.4.3 Symmetric difference methods

After our excursion to linear and quasilinear equations, and systems of orders 2 and $2m$, we return to linear equations of order 2, and their *symmetric difference equations.* The symmetric forms are particularly interesting, since the discretization error for the difference equations is $\mathcal{O}(h^2)$, compared to $\mathcal{O}(h)$ for unsymmetric forms. The above discussion has shown that replacing the $\partial_\pm^i$ by ∂_h^i does not yield usable symmetric forms. However, for the previous unsymmetric forms a simple modification yields the desired *symmetric forms.* This idea is identically applicable to all the following cases. Only the boundary conditions for higher order equations have to be modified, cf. Subsection 8.4.4. We define the *strong and the corresponding weak symmetric discrete operators* A_{ss}^h, and A_{ws}^h, and the bilinear form $a_{ws}^h(\cdot,\cdot)$ as the mean of two difference equations, exchanging ∂_+^i, and ∂_-^i :

$$A_{ss}^h u^h(x) := \frac{1}{2}\sum_{i,j=0}^{n}(-1)_{j>0}\left(\partial_-^j(a_{ij}(x)\partial_+^i u^h(x)) + \partial_+^j(a_{ij}(x)\partial_-^i u^h(x))\right), \tag{8.60}$$

$$a_{ws}^h(u^h,v^h) := \frac{1}{2}\sum_{i,j=0}^{n}\int_{\Omega^h} a_{ij}(\partial_+^i u^h\partial_+^j v^h + \partial_-^i u^h\partial_-^j v^h)dx^h =: \langle A_{ws}^h u^h, v^h\rangle_{\mathcal{V}_b'^h\times\mathcal{V}_b^h}.$$

We turn to the boundary conditions. The error for the difference equations, of $\mathcal{O}(h^2)$, only makes sense if the boundary conditions are discretized equally accurately or better. For the cuboidal domains Ω in (8.9) this is straightforward: we choose $\Omega^h, \Omega_0^h, \partial\Omega^h$ as in (8.11). Then the boundary points $x \in \partial\Omega^h$ satisfy $x \in \partial\Omega$, hence, allow an exact discretization $u^h(x) = 0\ \forall x \in \partial\Omega^h$. For curved domains the techniques in Section 8.7 apply.

So, based upon Proposition 8.10, and this boundary condition we determine

$$\begin{aligned} u_0^h \in \mathcal{V}_b^h : A_{ss}^h u_0^h(x) = f(x)\forall x \in \Omega_0^h \iff \\ \langle A_{ws}^h u_0^h, v^h\rangle_{\mathcal{V}_b'^h\times\mathcal{V}_b^h} = \langle f, v^h\rangle_{\mathcal{V}_b'^h\times\mathcal{V}_b^h}\forall v^h \in \mathcal{V}_b^h. \end{aligned} \tag{8.61}$$

Only the highest order divided differences in (8.60), so $i, j = 1, \ldots, n$, are required in this special form. For the lower order terms with $i \cdot j = 0$ we may choose any of the symmetric formulas, e.g. as in (8.13). In fact, they do satisfy the conditions in Proposition 8.4, thus the $\mathcal{V}_b^h$-coercivity of the principal part is maintained.

We visualize the symmetric A_{ss}^h in (8.60), starting with the three-point stars in Table 8.5, the seven-point stars in Tables 8.6, 8.7, and the final seven-point star in Table 8.8. Note that the factor 1/2 in front of every table represents the factor 1/2 in (8.60). Furthermore we define, in Tables 8.5–8.7, several $a_{ij}^{\pm}$ used in Table 8.8.

Table 8.5: Three-point star for (8.60) for $-h^2\left(\partial_-^i a_{ii}\partial_+^i + \partial_+^i a_{ii}\partial_-^i\right)u^h(x)$.

1/2	$-a_{ii}^- := -a_{ii}(x) - a_{ii}(x - he_i)$	$a_{ii}^- + a_{ii}^+$	$-a_{ii}^+ := -a_{ii}(x) - a_{ii}(x + he_i)$

Table 8.6: Seven-point star for (8.60), and $-h^2(\partial_-^2 a_{12}\partial_+^1 + \partial_+^2 a_{12}\partial_-^1)u^h(x)$.

	$a_{12}(x + he_2)$	$-a_{12}^+ := -a_{12}(x + he_2)$	0
1/2	$-a_{12}(x)$	$2a_{12} := 2a_{12}(x)$	$-a_{12}(x)$
	0	$-a_{12}^- := -a_{12}(x - he_2)$	$a_{12}(x - he_2)$

Table 8.7: Seven-point star for (8.60), and $-h^2(\partial_-^1 a_{21}\partial_+^2 + \partial_+^1 a_{21}\partial_-^2)u^h(x)$.

	$a_{21}^- := a_{21}(x - he_1)$	$-a_{21}(x)$	0
1/2	$-a_{21}(x - he_1)$	$2a_{21} := 2a_{21}(x)$	$-a_{21}(x + he_1)$
	0	$-a_{21}(x)$	$a_{21}^+ := a_{21}(x + he_1)$

Table 8.8: Seven-point star for the complete $-h^2\times$ (8.60) and $x \in \Omega_0^h \subset \mathbb{R}^2$.

	$a_{12}^+ + a_{21}^-$	$-a_{22}^+ - a_{12}^+ - a_{21}$	0
1/2	$-a_{11}^- - a_{12} - a_{21}^-$	$a_{11}^+ + a_{11}^- + a_{22}^+ + a_{22}^-$ $+2a_{12} + 2a_{21}$	$-a_{11}^+ - a_{12} - a_{21}^+$
	0	$-a_{22}^- - a_{12}^- - a_{21}$	$a_{12}^- + a_{21}^+$

8.4.4 Linear equations of order $2m$

For these equations we need the higher order Dirichlet boundary conditions mentioned above. So we assume the discrete boundary values in the form $u^h \in \mathcal{V}_b^h = H_{0,+}^m(\Omega^h)$ as in (8.28), and the same for $H_{0,-}^m(\Omega^h)$. Symmetric formulas only are worthwhile for cuboidal domains or for the special methods in Section 8.7.

We formulate the equations of order $2m$: for $a_{\alpha\beta} \in W^{|\beta|,\infty}(\Omega), f \in L^2(\Omega)$, and $a_{\alpha\beta} \in C^{|\beta|}(\overline{\Omega}), f \in C(\overline{\Omega})$, cf. Summary 8.6, and Remark 8.7, we determine u_0, and u_0^h, respectively:

$$u_0 \in H^{2m}(\Omega) \cap (\mathcal{V}_b := H_0^m(\Omega)), \text{ and } u_0^h \in H_+^{2m}(\Omega^h) \cap \left(\mathcal{V}_b^h := H_{0,+}^m(\Omega^h)\right) \text{ s.t.}$$

$$A_s u_0 = \sum_{|\alpha|,|\beta|\le m} (-1)^{|\beta|}\partial^\beta(a_{\alpha\beta}\partial^\alpha u_0) = f \in L^2(\Omega), \text{ and}$$

$$A_s^h u_0^h(x) = \sum_{|\alpha|,|\beta|\le m} (-1)^{|\beta|}\partial_-^\beta\left(a_{\alpha\beta}(x)\partial_+^\alpha u_0^h(x)\right) = f(x)\ \forall x \in \Omega_0^h. \tag{8.62}$$

Similarly as above we formulate, for $a_{\alpha\beta} \in L^\infty(\Omega)$, the *weak forms* $a(u,v), Au$, and, for $a_{\alpha\beta} \in C(\overline{\Omega})$, the $a^h(u^h,v^h), A^h u^h$, cf. (8.46), (8.47), and compute $u_0 \in \mathcal{V}_b$, and $u_0^h \in \mathcal{V}_b^h$ for $f \in \mathcal{V}_b'$ from

$$a(u_0,v) := \langle Au_0, v\rangle_{\mathcal{V}_b'\times\mathcal{V}_b} := \int_\Omega \sum_{|\alpha|,|\beta|\le m} a_{\alpha\beta}\partial^\alpha u_0 \partial^\beta v dx\ = \langle f,v\rangle_{\mathcal{V}_b'\times\mathcal{V}_b}\ \forall v \in \mathcal{V}_b,$$

$$\begin{aligned} u_0^h \in \mathcal{V}_b^h : a^h\left(u_0^h, v^h\right) &:= h^n \sum_{x\in\Omega^h}\sum_{|\alpha|,|\beta|\le m}\left(a_{\alpha\beta}\partial_+^\alpha u_0^h \partial_+^\beta v^h\right)(x) \\ &=: \langle A^h u_0^h, v^h\rangle_{\mathcal{V}_b'^h\times\mathcal{V}_b^h} \\ &= \langle f^h, v^h\rangle_{\mathcal{V}_b'^h\times\mathcal{V}_b^h} \\ &= h^n \sum_{x\in\Omega^h}\sum_{|\beta|\le m}\left(f_\beta^h\partial_+^\beta v^h\right)(x)\forall v^h \in \mathcal{V}_b^h, \end{aligned} \tag{8.63}$$

cf (2.107). The form for f in (8.63) is unusual in classical difference methods, cf. (2.110). However it makes sense for the weak form, and the later error estimates.

Generalizations to Banach spaces with $a_{\alpha\beta} \in W^{\infty}(\Omega)$, $u, u_0, v \in W^{m,p}(\Omega), 2 \le p < \infty$ or $a_{\alpha\beta} \in W^{p_1}(\Omega)$, $u, u_0 \in W^{m,p_2}(\Omega), v \in W^{m,p_3}(\Omega)$, with $1/p_1 + 1/p_2 + 1/p_3 = 1$ and the corresponding discrete spaces $W^{m,p}_{0,+}(\Omega^h)$ are possible, similarly to Section 2.3 and Subsection 2.4.4, considering the corresponding coercivity problems. This allows results for the discrete $H^m_+(\Omega^h)$ norm.

As in (8.60) *symmetric difference equations* can be defined, again only for cuboidal domains Ω, cf. (8.9), and $\forall x \in \Omega^h_0$ or the methods in Section 8.7.

8.4.5 Quasilinear elliptic equations of orders 2, and $2m$

According to the theoretical results we assume homogeneous Dirichlet boundary conditions of the form (8.45) or (8.62). Even more general Dirichlet systems, see (2.82), Lemma 2.26, (2.158), (2.159), and Theorem 2.50 can be transformed into the standard homogeneous Dirichlet boundary conditions. Furthermore, the first types of nonlinear elliptic equations in Section 2.5, Subsection 2.5.3, are either special quasilinear equations, or simply the term $a_{00}u$ in (8.62) has to be replaced by $f(u)$. So we only give the difference equations corresponding to a general quasilinear form, with the boundary conditions in (8.45) or (8.62), compare (2.270), (2.306). The inhomogeneous function f is included in the nonlinear A_j and A_α. Appropriate conditions for the unique existence of solutions are listed in Subsections 2.5.4 and 2.5.6.

These strong quasilinear equations are defined for $u, u_0 \in W^{2,p}(\Omega) \cap W^{1,p}_0(\Omega)$, and $u^h, u^h_0 \in W^{2,p}_+(\Omega^h) \cap \left(\mathcal{V}^h_b := W^{1,p}_{0,+}(\Omega^h)\right)$. Applying (8.49) to (8.64) we get the *weak form* G^h, cf. (8.65), defined for $u^h, u^h_0, v^h \in \mathcal{V}^h_b$, and $1/p + 1/p' = 1$:

$$G_s u_0 = \sum_{j=0}^{n} (-1)_{j>0} \partial^j A_j(\cdot, u_0, \nabla u_0) = 0 \ \in L^{1/p'}(\Omega), \tag{8.64}$$

$$u_0 \in W^{2,p} \cap W^{1,p}_0(\Omega),$$

$$G^h_s u^h_0(x) = \sum_{j=0}^{n} (-1)_{j>0} \partial^j_- A_j\left(\cdot, u^h_0, \nabla_+ u^h_0\right)(x) = 0 \forall x \in \Omega^h_0, \text{and} \tag{8.65}$$

$$\begin{aligned}\langle G^h u^h_0, v^h \rangle_{\mathcal{V}'^h_b \times \mathcal{V}^h_b} &:= h^n \sum_{x \in \Omega^h} \sum_{j=0}^{n} A_j\left(\cdot, u^h_0, \nabla_+ u^h_0\right)(x) \partial^j_+ v^h(x) \\ &:= a^h\left(u^h_0, v^h\right) = 0 \ \forall \ v^h \in \mathcal{V}^h_b = W^{1,p}_{0,+}(\Omega^h) \text{ with } 2 \le p < \infty \\ &\qquad u^h_0 \in \mathcal{V}^h_b, \forall u^h \in \mathcal{V}^h_b \cap \mathcal{D}(A_j) : A_j(\cdot, u^h, \nabla_+ u^h) \\ &\qquad \in \mathcal{V}'^h_b = W^{-1,p'}_{0,+}(\Omega^h)\end{aligned}$$

for the weak form. Notice that $a^h(u^h, v^h)$ here, and for the following nonlinear problems, is linear and bounded in v^h, but not linear in u^h.

For the analysis of linearized quasilinear problems we will need general Sobolev spaces $W^{2m,p}(\Omega) \cap W_0^{m,p}(\Omega)$ with $1 \leq p \leq \infty$, and their discrete counterparts. According to Theorem 2.122 this linearization essentially requires the $H_0^m(\Omega)$-coercivity of the principal part, cf. the end of Subsection 8.4.4. For our discretization approach via linearization, we again restrict p from $1 < p < \infty$ to $2 \leq p < \infty$ in $v^h \in \mathcal{V}_b^h = W_{0,+}^{1,p}(\Omega^h)$ and $A_j(\cdot, u^h, \nabla_+ u^h) \in \mathcal{V}_b^{\prime h} = W_{0,+}^{-1,p'}(\Omega^h)$ or $v^h \in W_{0,+}^{m,p}(\Omega^h)$. If we employ the monotone approach, $1 < p < \infty$ is still possible, cf. Section 4.5 and Theorem 8.32.

The following strong equations G_s^h of order $2m$ are transformed, by iteratively applying (8.49), into the *weak form* G^h. Again we consider trivial Dirichlet conditions, and require $u, u_0 \in W^{2m,p}(\Omega) \cap W_0^{m,p}(\Omega)$, and $u^h, u_0^h \in W_+^{2m,p}(\Omega^h) \cap \left(\mathcal{V}_b^h := W_{0,+}^{m,p}(\Omega^h)\right)$ for the strong, and $u^h, u_0^h, v^h \in \mathcal{V}_b^h$ for the weak form, cf. (8.17) for $\nabla_+^m u$, then

$$G_s u_0 = \sum_{|\beta| \leq m} (-1)^{|\beta|} \partial^\beta A_\beta(\cdot, u_0, \cdots, \nabla^m u_0) = 0 \in L^2(\Omega), \tag{8.66}$$

$$G_s^h u_0^h(x) = \sum_{|\beta| \leq m} (-1)^{|\beta|} \partial_-^\beta A_\beta\left(\cdot, u_0^h, \cdots, \nabla_+^m u_0^h\right)(x) = 0 \ \forall x \in \Omega_0^h, \text{ and}$$

$$\left\langle G^h u_0^h, v^h \right\rangle_{\mathcal{V}_b^{\prime h} \times \mathcal{V}_b^h} = h^n \sum_{x \in \Omega^h} \sum_{|\beta| \leq m} A_\beta\left(\cdot, u_0^h, \cdots, \nabla_+^m u_0^h\right)(x) \partial_+^\beta v^h(x) \tag{8.67}$$

$$:= a^h\left(u_0^h, v^h\right) \ := 0 \ \forall \ v^h \in \mathcal{V}_b^h = W_{0,+}^{m,p}(\Omega^h) \text{ with } 2 \leq p < \infty$$

$$u_0^h \in \mathcal{V}_b^h, \forall u^h \in \mathcal{V}_b^h \cap \mathcal{D}(A_j) : A_j\left(\cdot, u^h, \ldots, \nabla_+^m u^h\right) \in \mathcal{V}_b^{\prime h} = W_{0,+}^{-m,p'}(\Omega^h).$$

Schumann and Zeidler [574, 678] prove for these equations convergence of order 1 via “external approximation schemes”, so totally differently from our approach.

We update Proposition 8.10 with Proposition 8.11, to the quasilinear case:

Proposition 8.12. *For the difference operators G_s^h, and G^h in (8.65)–(8.67) we obtain $\forall v^h \in \mathcal{V}_b^h = W_{0,+}^{m,p}(\Omega^h)$, hence $v^h \in W^{2m,p}(\Omega^h)$ as well, cf. Remark 8.9:*

$$\begin{aligned}\int_{\Omega_0^h} v^h\left(G_s^h u^h\right) dx^h &= \int_{\Omega^h} v^h \left(\sum_{|\beta| \leq m} (-1)^{|\beta|} \partial_-^\beta A_\beta\left(\cdot, u^h(\cdot), \ldots, \nabla_+^m u^h(\cdot)\right) \right) dx^h \\ &= \int_{\Omega^h} \sum_{|\beta| \leq m} A_\beta\left(\cdot, u^h(\cdot), \ldots, \nabla_+^m u^h(\cdot)\right) \partial_+^\beta v^h dx^h \\ &= \left\langle G^h u^h, v^h \right\rangle_{\mathcal{V}_b^{\prime h} \times \mathcal{V}_b^h}.\end{aligned} \tag{8.68}$$

For the weak form we require $u^h, v^h \in W_+^{m,p}(\Omega^h), 2 \leq p < \infty$, and

$$\forall u^h : A_\beta\left(\cdot, u_0^h, \cdots, \nabla_+^m u_0^h\right) \in W_{0,+}^{-m,p'}(\Omega^h), 1/p + 1/p' = 1.$$

For the symmetric difference equations *we proceed as in (8.68).*

8.4.6 Systems of linear and quasilinear elliptic equations

We formulate the differential and difference systems only for order $2m$, cf. (2.391), (2.306). The transition from order $2m$ back to second order is obvious from (8.65)–(8.67). The main reason is that for systems of order $2m$ the proof for coercivity requires a modification, based upon the linearized forms. Preparing the quasilinear problems, we formulate the linear problems in the Sobolev setting $W_+^{2m,p}(\Omega, \mathbb{R}^q)$. We maintain the notation in Chapter 2, e.g. (2.389), and formulate the necessary difference modifications compared to (8.17), e.g. as

$$\vec{u}^h = \left(u_1^h, \cdots, u_q^h\right), \partial_+^l \vec{u}^h := \left(\partial_+^l u_1^h, \cdots, \partial_+^l u_q^h\right), \partial_+^\alpha \vec{u}^h := \left(\partial_+^\alpha u_1^h, \cdots, \partial_+^\alpha u_q^h\right), \quad (8.69)$$

$$\nabla_+^{\leq k} \vec{u}^h := \left\{\partial_+^\alpha \vec{u}^h \ \forall |\alpha| \leq k\right\}, \nabla_+^k \vec{u}^h := \left\{\partial_+^\alpha \vec{u}^h \ \forall |\alpha| = k\right\},$$

with $\nabla_+ \vec{u}^h := \nabla_+^1 \vec{u}^h, \quad \nabla_+^k \vec{u}^h := \partial_+^l \vec{u}^h := \partial_+^\alpha \vec{u}^h := \vec{u}^h$ for $k = l = |\alpha| = 0$, and

$\vec{u}^h(x), \partial_+^l \vec{u}^h(x) \in \mathbb{R}^q, \nabla_+ \vec{u}^h(x) \in \mathbb{R}^{q\times n}, \nabla_+^k \vec{u}^h(x) \in \mathbb{R}^{q\times n_k}$, and

$\vec{\vartheta}_j = \left(\vartheta_j^1, \ldots, \vartheta_j^n\right) \in \mathbb{R}^n, j = 1, \ldots, q, \quad \vec{\vartheta}^l = \left(\vartheta_1^l, \ldots, \vartheta_q^l\right) \in \mathbb{R}^q, l = 1, \ldots, n,$ with

$$\vec{\vartheta} := \left(\vec{\vartheta}^1, \cdots, \vec{\vartheta}^n\right) = (\vec{\vartheta}_1, \cdots, \vec{\vartheta}_q) \in \mathbb{R}^{n\times q}, \quad \text{and} |\vec{\vartheta}| = |\vec{\vartheta}|_{nq} \in \mathbb{R},$$

$$\vec{\vartheta}^\alpha := \left(\vartheta_1^\alpha, \cdots, \vartheta_q^\alpha\right) \in \mathbb{R}^q, \text{ and } \vec{\Theta}^k := (\vec{\vartheta}^\alpha)_{|\alpha|=k} \in \mathbb{R}^{q\times n_k}, \quad \vec{\Theta}^{\leq k} := (\vec{\vartheta}^\alpha)_{|\alpha|\leq k}.$$

Correspondingly we define, e.g. the $\nabla_-^k \vec{u}^h, \nabla_h^k \vec{u}^h$. Then for the strong linear differential and difference systems, and the weak difference systems we determine the solutions $\vec{u}_0$, and $\vec{u}_0^h$ from (8.70), (8.71). We choose $\vec{u}_0 \in W^{2m,p}(\Omega, \mathbb{R}^q) \cap W_0^{m,p}(\Omega, \mathbb{R}^q)$, $\vec{u}_0^h \in W_+^{2m,p}(\Omega^h, \mathbb{R}^q) \cap \left(\mathcal{V}_b^h := W_{0,+}^{m,p}(\Omega^h, \mathbb{R}^q)\right), \vec{v}^h \in \mathcal{V}_b^h$, for the strong, and delete the $W_+^{2m,p}(\Omega, \mathbb{R}^q)$ for the weak problems. The conditions for the *Nemyckii operators,* cf. Subsection 2.5.5, $A_{\alpha\beta}$ in (8.70), and A_β in (8.72), and their second order form, A_{kl}, and A_k, are discussed in Subsections 2.6.3–2.6.6, and are summarized for difference methods in Summary 8.6.

$$\vec{f} = A_s \vec{u}_0 = \sum_{|\beta|,|\alpha|\leq m} (-1)^{|\beta|} \partial^\beta (A_{\alpha\beta}\, \partial^\alpha \vec{u}_0) \in L^{p'}(\Omega, \mathbb{R}^q), A_{\alpha\beta}(x) \in \mathbb{R}^{q\times q}, \quad (8.70)$$

$$1/p + 1/p' = 1, \quad \vec{u}_0^h \in \mathcal{V}_b^h \cap W_+^{2m,p}(\Omega^h, \mathbb{R}^q), \vec{f}^h \in L_+^{p'}(\Omega^h, \mathbb{R}^q), \text{ and}$$

$$\vec{f}^h(x) = A_s^h \vec{u}_0^h(x) = \sum_{|\beta|,|\alpha|\leq m} (-1)^{|\beta|} \partial_-^\beta \left(A_{\alpha\beta}(x)\, \partial_+^\alpha \vec{u}_0^h(x)\right) \forall\, x \in \Omega_0^h \quad (8.71)$$

$$\langle \vec{f}^h, \vec{v}^h \rangle_{\mathcal{V}_b'^h \times \mathcal{V}_b^h} = h^n \sum_{x\in\Omega^h} \sum_{|\alpha|\leq m} \left(\vec{f}_\alpha^h, \partial_+^\alpha \vec{v}^h\right)_q (x) = \langle A^h \vec{u}_0^h, \vec{v}^h \rangle_{\mathcal{V}_b'^h \times \mathcal{V}_b^h} :=$$

$$a^h\left(\vec{u}_0^h, \vec{v}^h\right) := \int_{\Omega^h} \sum_{|\beta|,|\alpha|\leq m} \left(A_{\alpha\beta}\, \partial_+^\alpha \vec{u}_0^h, \partial_+^\beta \vec{v}^h\right)_q dx^h \ \forall \vec{v}^h \in \mathcal{V}_b^h, \vec{f}^h \in W_{0,+}^{-m,p'}(\Omega^h, \mathbb{R}^q).$$

The quasilinear versions with $\vec{f}$ included in the A_β are defined, and the solutions, $\vec{u}_0, \vec{u}_0^h$, are computed in $\mathcal{D}(G), \mathcal{D}(G^h)$, subsets of the same spaces as in (8.70), (8.71), see (2.424), (2.425), so

$$G_s\vec{u}_0 := \sum_{|\alpha|\le m}(-1)^{|\beta|}\partial^\beta A_\beta(\cdot,\nabla^{\le m}\vec{u}_0) = 0 \in L^{p'}(\Omega,\mathbb{R}^q), \quad \text{and} \tag{8.72}$$

$$G_s^h\vec{u}_0^h(x) := \sum_{|\beta|\le m}(-1)^{|\beta|}\partial_-^\beta A_\beta\left(\cdot,\nabla_+^{\le m}\vec{u}_0^h\right)(x) = 0,\ \forall x\in\Omega_0^h,$$

$$a^h\left(\vec{u}_0^h,\vec{v}^h\right) := \langle G^h\vec{u}_0^h,\vec{v}^h\rangle_{\mathcal{V}_b'^h\times\mathcal{V}_b^h} \text{ with } \mathcal{V}_b^h := W_{0,+}^{m,p}(\Omega^h,\mathbb{R}^q) \tag{8.73}$$

$$:= \int_{\Omega^h}\sum_{|\beta|\le m}\left(A_\beta\left(\cdot,\nabla_+^{\le m}\vec{u}_0^h\right),\partial_+^\beta\vec{v}^h\right)_q dx^h = 0\ \forall\vec{v}^h\in\mathcal{V}_b^h.$$

Again we proceed as in (8.60) for *symmetric difference equations.*

8.4.7 Fully nonlinear elliptic equations and systems

For the fully nonlinear equation (8.74) in general no weak formulations are possible, cf. Subsections 2.5.7, 2.6.8, and Section 5.2. So, as for FEMs in (5.2) ff., we formulate only the strong form of the difference methods. As for FEMs, stability is proved via the detour to the linearized operator $A = G'(u_0)$. It requires regularity results of the form $Au = f \in L^2(\Omega) \Longrightarrow u \in H^{2m}(\Omega) =: \mathcal{U}$. These results are available for equations of order $2m$ for boundaries $\partial\Omega\in C^{2m}$ or $\partial\Omega\in C^\infty$, cf. Theorems 2.79 ff. For equations and systems of order 2 this is known for convex polygonal boundaries, cf. Kozlov *et al.* [452]. We consider essentially unsymmetric difference methods for curved boundaries of second order equations, cf. (8.76). For special cuboidal domains as in (8.9), *symmetric methods* are formulated at the end of this subsection. For G in (8.76) we find its derivative in the form (8.43). We restrict the discussion to order $2m = 2$.

Crandall *et al.*, e.g. [216, 217] discuss the concept of *viscosity solutions* for fully nonlinear uniformly elliptic, e.g. [216], and parabolic equations of second order using specific monotonicity properties of the operator. They extend these concepts to difference methods on equidistant grids and their convergence to the exact solutions. Kuo and Trudinger [456] consider discrete versions of maximum principles, Hölder estimates and Harnack inequalities for linear elliptic difference operators of positive type and generalize them to fully nonlinear problems on equidistant and moderately nonequidistant grids. Relations between discrete and continuous elliptic equations, but no convergence results, are discussed.

We compute $u_0(x), u_0^h(x)$ for (8.74), (8.76), cf. (8.45),

$$G_s : \mathcal{D}(G)\subset\mathcal{U}_b := H^{2m}(\Omega)\cap H_0^m(\Omega)\to\mathcal{V} := L^2(\Omega), \tag{8.74}$$

$$G_s u_0 := G(\cdot,u_0,\nabla u_0,\nabla^2 u_0) = 0\ \in L^2(\Omega),\ \text{with } u_0(x) = 0 \text{ on } \partial\Omega, \text{ and} \tag{8.75}$$

$$G_s^h : \mathcal{D}(G_s^h)\subset\mathcal{U}_b^h := H_{0,+}^{2m}(\Omega^h)\to\mathcal{V}^h := L^2(\Omega^h), G_s^h u_0^h := G(.,u_0^h,\nabla_+u_0^h,\nabla_+^2u_0^h)|_{\Omega_0^h} = 0,$$

$$u_0^h(x)|_{\partial\Omega^h} = 0 \tag{8.76}$$

$$\Longleftrightarrow \left(G_s^h u_0^h, v^h\right)_{L^2(\Omega^h)} = 0\,\forall v^h\in\mathcal{V}^h, u_0^h|_{\partial\Omega^h} = 0,\ \text{hence } u_0(x)\in\mathcal{U}_b, u_0^h(x)\in\mathcal{U}_b^h.$$

We do not reformulate the highly technical proof for stability, consistency and convergence for fully nonlinear problems for the finite difference form (8.76). We only indicate the necessary steps, strongly analogous to the FEM approach:

For special polyhedral, e.g. cuboidal, domains in (8.9) the boundary conditions are satisfied exactly, otherwise they are violated. Correspondingly, (8.74) is discretized as (8.76) for (8.9). The general discretization approach in Subsection 5.2.4 is then reduced to elaborating the difference operator, our G^h, the $F_1^h = G^h$ in (5.41). Consequently, we only need projectors for the G component and no extension operators for the consistency. This strongly simplifies the analogue of the proof for Theorem 5.4. For the other domains we again need the full machinery of discretization errors for differential and boundary operators G^h, B^h as for FEMs in Section 5.2. In the proof of stability via the linearized operator, $G'(u_0)$, we use the equality of weak and strong discrete bilinear forms induced by $G'(u_0)$, cf. Proposition 8.10, (8.56). We obtain the regularity of the difference solutions, as in Lemma 5.10, and Hackbusch [387], Theorem 9.2.26. The corresponding convergence results are summarized in Theorem 8.31 below.

This difference method avoids the main disadvantage of the preceding FEMs, the need of C^1 FEs. These require necessarily piecewise polynomials of higher degree, e.g. ≥ 5 for $n = 2$ and ≥ 9 for $n = 3$.

For *symmetric difference equations*, we slightly modify (8.60). The *boundary conditions* are formulated in (8.19), (8.20). So we require:

$$G_{ss}^h u_0^h(x) := \frac{1}{2}\left(G\left(x, u_0^h, \nabla_+ u_0^h, \nabla_+^2 u_0^h\right) + G\left(x, u_0^h, \nabla_- u_0^h, \nabla_-^2 u_0^h\right)\right)(x) = 0 \ \forall x \in \Omega_0^h.$$

Extensions of these results to equations and systems of order $2m$, and to curved boundaries $\partial\Omega \in C^{2m}$ can be formulated as in the preceding subsections.

A totally different approach is due to Crandall and Lions. They developed, in a series of papers, the theoretical tools and full convergence proofs for difference methods for quasi- and fully nonlinear elliptic and parabolic equations [217, 226, 227], and with Kocan and Swiech in [216]. The basis is a concept of generalized weak solutions, the viscosity solutions; for a user's guide see Crandall *et al.* [215].

For other types of results Kuo and Trudinger combine nonlinear elliptic and parabolic difference equations of positive type with their concepts of weak solutions, local, and Schauder, Harnack and Hölder estimates, and a discrete local maximum principle, see [456–460].

8.5 Convergence for difference methods

Here we apply our general discretization theory from Chapter 3 to the difference methods formulated in the last chapter. As mentioned, we have only studied methods of orders 1 and 2. We formulate the Richardson extrapolation and the defect correction methods, presented in Section 8.8, as an appropriate strategy for developing high order methods as well. The obtained convergence results with respect to the $H_+^m(\Omega^h)$ norms is a special case of the regularity results for difference methods mentioned at the end of Subsection 8.4.2. Important in this context are exact conditions for the linearized operators, collected in Summary 8.6 and in Remark 8.7. From the many papers for general difference equations we only mention Borzì *et al.* [134] and Vainikko with Tamme [646, 646–649].

Although $W^{-m,p'}(\Omega) \neq W_0^{-m,p'}(\Omega)$ and $W_+^{-m,p'}(\Omega^h) \neq W_{0,+}^{-m,p'}(\Omega^h)$, we often do not distinguish them. The restriction, e.g. of $f \in W^{-m,p'}(\Omega)$ yields an $f|_{W_0^{-m,p'}(\Omega)} \in W_0^{-m,p'}(\Omega)$.

8.5.1 Discretization concepts in discrete Sobolev spaces

We briefly summarize the general discretization theory in Chapter 3, cf. Summary 4.52 as well. Our approach is much more similar to the FEM or DCGM form than to the earlier combination of strong forms of the difference equations with M-matrices. In contrast, we discuss here stability and convergence for the weak forms, cf. Proposition 8.11. These *variational difference methods* yield results that are usually stronger than M-matrix results. These ideas have been applied to several problems, e.g. by Samarskii [564, 565] and Temam [621].

However, we simplify the major part of the following proof for convergence. We restrict the full presentation to the case of discretely exactly reproduced boundary conditions. This is satisfied for the cuboidal domains Ω in (8.9), and second order elliptic problems, hence $m = 1, q \geq 1$. The extension to general Ω in (8.18), and $m > 1, q \geq 1$, can be proved by modifying the FE approach for fully nonlinear problems in Section 5.2, and Subsection 5.2.4. We would have to replace the differential operator A or G by the two-component operator, F, including the boundary operator, B, thus $F = (A, B)$ or $F = (G, B)$, e.g. $F = (A, B) : H_0^1(\Omega) \to H_0^{-1}(\Omega) \times H^{1/2}(\partial\Omega)$. This allows verifying the consistency of the "nonconforming finite difference method" as in Subsection 5.2.4, and the stability by combining the techniques in Subsections 8.5.5 and 5.2.7.

This technique will be applied to the natural boundary conditions in Section 8.6 as well. For the difference methods for curved domains in Section 8.7 totally different techniques are necessary. These proofs will be omitted there.

In the last section we have transformed the strong (classical) form of difference equations into the corresponding weak form except for fully nonlinear problems. For the case of trivial Dirichlet conditions we need lines 1, 2 of (8.76), for the strong and weak form the next two lines for their weak discretization.

$$\begin{aligned} \mathcal{U} &= H^{2m}(\Omega), \text{ or } W^{2m,p}(\Omega), \text{ or } H^{2m}(\Omega, \mathbb{R}^q), \textit{ and } \mathcal{V}' = L^{p'}(\Omega), \ldots, \quad (8.77)\\ \mathcal{U}_b &= \mathcal{U} \cap H_0^m(\Omega), W_0^{m,p}(\Omega), \mathcal{V}_b' = L^{p'}(\Omega), \text{ for } m, q \geq 1, 1 \leq p \leq \infty, \end{aligned}$$

$$\begin{aligned} \mathcal{U}^h &= \mathcal{V}^h = H_+^m(\Omega^h), W_+^{m,p}(\Omega^h), \quad \mathcal{V}'^h = \mathcal{U}'^h = W_+^{-m,p'}(\Omega^h),\\ \mathcal{V}_b^h &= \mathcal{U}_b^h = H_{0,+}^m(\Omega^h), W_{0,+}^{m,p}(\Omega^h), \quad \mathcal{V}_b'^h = \mathcal{U}_b'^h = W_{+,0}^{-m,p'}(\Omega^h). \end{aligned}$$

or one can use the or the $W_-^{m,p}(\Omega^h), \ldots,$ or the $W_{0,-}^{m,p}(\Omega^h)$ forms.

For nonlinear G, G^h, the domains $\mathcal{D}(G), \mathcal{D}(G^h)$ are subsets of $\mathcal{U}, \mathcal{U}^h$. The A, G, will usually be defined on $\mathcal{U} \to \mathcal{V}'$, and the A^h, G^h on $\mathcal{U}^h \to \mathcal{V}'^h$, cf. Figure 8.3. Uniqueness

$$\begin{array}{ccccccc}
\mathcal{U}_b \subset & \mathcal{U} & \supset D(G) & \overset{A,\,G}{\longrightarrow} & \mathcal{V}' & \overset{\text{tested by}}{\longleftrightarrow} & \mathcal{V}_b \\
 & P^h\big\downarrow & & \Phi^h\big\downarrow & Q'^h\big\downarrow & & \\
\mathcal{U}_b^h \subset & \mathcal{U}^h & \supset D(G^h) & \overset{A^h,\,G^h}{\longrightarrow} & \mathcal{V}'^h & \overset{\text{tested by}}{\longleftrightarrow} & \mathcal{V}_b^h
\end{array}$$

Figure 8.3 Difference methods: Spaces, and operators, usually with $Au, Gu \in \mathcal{V}_b' \forall u \in \mathcal{U}_b$ and $A^h u^h, G^h u^h \in \mathcal{V}_b'^h \forall u^h \in \mathcal{U}_b^h$.

requires boundary conditions $\mathcal{U}_b, \mathcal{V}_b$ and $\mathcal{U}_b^h, \mathcal{V}_b^h$, tested according to $Au, Gu \in \mathcal{V}_b'$ and $A^h u^h, G^h u^h \in \mathcal{V}_b'^h$.

For applying the general discretization theory to A, G, we need restriction, and projection operators, P^h, and Q'^h, relating the original and the discrete spaces.

$$P^h : \mathcal{U} \to \mathcal{U}^h, Q'^h : \mathcal{V}_b' \to \mathcal{V}_b'^h, \text{ or only } P^h : \mathcal{U}_b \to \mathcal{U}_b^h. \tag{8.78}$$

Then $A^h = \Phi^h A$ and $G^h = \Phi^h G$ are defined by the constructions in Section 8.4, cf. Figure 8.3. We formulate the $W_+^{k,p}(\Omega^h)$ case. The modification to $W_-^{k,p}(\Omega^h)$ is obvious.

For our difference method, applicable to the problem $Au_0 = f$ or $Gu_0 = 0$, see Definition 3.12, we thus have the *sequence of quintuples*

$$\left\{\mathcal{U}_b^h, \mathcal{V}_b^h, P^h, Q'^h, G^h\right\}_{h\in H}, \text{with } \inf\{0 < h \in H\} = 0, \dim \mathcal{U}_b^h = \dim \mathcal{V}_b^h < \infty, \tag{8.79}$$

and $h < h_0$, thus $\lim_{h\to 0} := \lim_{h\in H, h\to 0}$ always makes sense.

According to Section 3.2, cf. (3.16), (3.24), we need bounded, not necessarily linear P^h, Q'^h in (8.78), such that $P^h 0 = 0, Q'^h 0 = 0$, and

$$\lim_{h\to 0} \|P^h u\|_{\mathcal{U}^h} = \|u\|_{\mathcal{U}} \forall u \in \mathcal{U}, \quad \lim_{h\to 0} \|Q'^h f\|_{\mathcal{V}_b'}^h = \|f\|_{\mathcal{V}_b'} \forall f \in \mathcal{V}_b'. \tag{8.80}$$

We employ the claim of Theorem 3.21 in Chapter 3: consistency and stability essentially imply uniquely existing discrete solutions, u_0^h, and their convergence to the exact solution u_0. The *(classical) consistency error, and consistency* for differential operators in u, are defined as, cf. Definition 3.14,

$$\begin{gathered}(A^h P^h u - f^h) - Q'^h(Au - f) \in \mathcal{V}_b'^h, \text{ or } G^h P^h u - Q'^h G u \in \mathcal{V}_b'^h, \text{ and} \\ \lim_{h\to 0} \|(A^h P^h u - f^h) - Q'^h(Au - f)\|_{\mathcal{V}_b'^h} = 0 \text{ or } \lim_{h\to 0} \|G^h P^h u - Q'^h G u\|_{\mathcal{V}_b'^h} = 0.\end{gathered} \tag{8.81}$$

G^h are called *stable in* u^h, if $\exists h_0, r, S \in \mathbb{R}_+$ fixed, such that for all $h \in H, h < h_0$:

$$u_1^h, u_2^h \in B_r(u^h) \cap \mathcal{U}_b \Rightarrow \|u_1^h - u_2^h\|_{\mathcal{U}^h} \le S \left\|G^h\left(u_1^h\right) - G^h\left(u_2^h\right)\right\|_{\mathcal{V}_b'^h}. \tag{8.82}$$

For A^h or $A^h - f^h$ this reduces to (for any r, and u^h, but for $h < h_0$)

$$(A^h)^{-1} \in \mathcal{L}\left(\mathcal{V}_b'^h, \mathcal{U}_b^h\right) \text{ exists and } \|(A^h)^{-1}\|_{\mathcal{U}_b^h \leftarrow \mathcal{V}_b'^h} \le S.$$

G^h is stable if its derivative $(G'(u_0))^h$ is stable, cf. Theorem 3.23.

So we have to introduce P^h, Q'^h, and prove consistency and stability, via coercivity or an inf-sup condition of its principal part. Alternatively the monotony arguments in Section 4.5 are applicable.

8.5.2 The operators P^h, Q'^h

We do not repeat the basic proofs for well-defined, linear and bounded P^h and Q'^h, cf. e.g. Raviart [545], Schumann and Zeidler [574, 678] and Temam [621]. But we prove, in a slightly nonstandard way, consistency and stability.

Again we consider the two cases (8.19) and (8.20), with the two different definitions of the $\Omega_0^h \subset \Omega^h, \partial\Omega^h$ in parallel. Either $\Omega^h \subset \Omega$ for the original $u \in W^{k,p}(\Omega)$ or we extend u to $u \in W^{k,p}(\Omega_1)$ with $\Omega^h \subset \Omega_1$. This is possible for Ω in (8.9) by Theorem 4.37. The same holds for functionals.

The definition of the $P^h : \mathcal{U} \to \mathcal{U}^h$, and $Q'^h : \mathcal{V}' \to \mathcal{V}'^h$ is nearly the same. But the trivial boundary conditions in $P^h : \mathcal{U}_b \to \mathcal{U}_b^h$ require modifications. This is achieved by defining $P^h u(x)$ either $\forall x \in \Omega^h$ or $\forall x \in \Omega_0^h$. We discuss two different possibilities. $P^h = P_1^h$ is studied, e.g. in Raviart [545], Schumann and Zeidler [574, 678], and $P^h = P_2^h$, e.g. in Temam [621]. We mainly consider $P^h = P_2^h$, and only sometimes indicate P_1^h. Note that in (8.84) supp u is closed, and Ω is open.

Definition 8.13. *For functions, $u \in \mathcal{U}$, defined in Ω, or extended to Ω_1 as $E_c u$, and with $c^h(x) \subset \Omega^h \subset \Omega$ or $\Omega^h \subset \Omega_1$, for $E_c, \Omega^h, \Omega_0^h$ cf. (8.19)–(8.20), define $P^h = P_1^h$ by the* integral mean value *$\bar{u}^h(x)$ of u or $E_c u$ as*

$$P^h = P_1^h : \mathcal{U} \to \mathcal{U}^h : \quad \forall x \in \Omega^h : \; P_1^h u(x) := \bar{u}^h(x) := \frac{1}{h^n} \int_{c^h(x)} u(\bar{x}) d\bar{x}, \text{ and}$$

$$P^h = P_1^h : \mathcal{U}_b \to \mathcal{U}_b^h : \; \forall x \in \Omega_0^h : \; P_1^h u(x) := \bar{u}^h(x) \text{ otherwise } = 0. \tag{8.83}$$

For $P^h = P_2^h$ we modify (8.83) by replacing $P_1^h u(x)$ by $P_2^h u(x) \forall x \in \Omega^h$ or Ω_0^h: For

$$u \in \mathcal{C}(\Omega) := \{v \in C(\Omega) \text{ or } C(\Omega_1, \mathbb{R}) : \text{carr } v \subset \Omega \text{ or} \subset \Omega_1\} : P_2^h u(x) := u(x) \tag{8.84}$$

and uniquely extend P_2^h from $\mathcal{C}(\Omega)$ to $\mathcal{U} = W^{k,p}(\Omega)$ defining $P^h = P_2^h : \mathcal{U} \to \mathcal{U}^h$.

Obviously these P^h and the Q'^h below can be trivially extended as

$$P^h : \mathcal{U} \cup \mathcal{U}^h \to \mathcal{U}^h : P^h|_{\mathcal{U}} = P^h \text{ in (8.83) and } P^h|_{\mathcal{U}^h} = I, \text{ identity on } \mathcal{U}^h. \tag{8.85}$$

Parallel to the standard grid functions, u^h, we employ the *step functions, u_e^h*, with $\chi_{c^h(x)}$, the characteristic functions of the $c^h(x)$, cf. (8.19),

$$E^h : \mathcal{U}^h \to \mathcal{U}_e^h := \left\{ u_e^h := E^h u^h := \sum_{x \in \Omega^h} u^h(x) \chi_{c^h(x)} \right\}. \tag{8.86}$$

We find, possibly except for a subset of measure 0 on the boundaries of the $c^h(x)$, that

$$\partial_+^\alpha E^h u^h = E^h \partial_+^\alpha u^h. \tag{8.87}$$

This implies that

$$\|\mathcal{U}^h\|_{\mathcal{U}^h} := \|u^h\|_{W_+^{k,p}(\Omega^h)} := [[E^h u^h]]_{W_+^{k,p}(\Omega^h)} := \left(\sum_{|\alpha|\le k} \|\partial_+^\alpha E^h u^h\|^p_{L^p(\Omega^h)}\right)^{1/p}. \quad (8.88)$$

Remark 8.14. *It is important to realize that the results in this and the next subsection are measured with respect to this specific norm $\|u^h\|_{\mathcal{U}^h} = \|u^h\|_{W_+^{k,p}(\Omega^h)}$, unless stated otherwise. This holds, in particular, for the consistency estimates.*

The following proposition is an obvious extension of Temam [621], Lemma 3.1 on p. 51 for $L^p(\Omega)$, to higher order derivatives, and to both our cases (8.19), (8.20). His Proposition 3.1, p. 42, shows that Proposition 8.15 had only to be proved for functions u in a dense subspace of $W^{k,p}(\Omega)$. This remains correct for Theorem 8.21, (8.104) as well. We do not repeat the argument in [621], Proposition 3.1, since we use it in part (b) of the proof of the following Theorem 8.21. The E_c in Theorem 4.37 combined with this result yields, cf. (8.19)–(8.20):

Proposition 8.15. *P^h, E^h, cf. (8.84), (8.86), satisfy $\forall u \in W^{k,p}(\Omega), |\alpha| \le m$*

$$\text{for (8.19): } \lim_{h\to 0} \|\partial_+^\alpha E^h P^h u - \partial^\alpha u\|_{L_h^p(c^h(\Omega^h))} = 0 \quad (8.89)$$

$$\text{for (8.20): } \lim_{h\to 0} \|\partial_+^\alpha E^h P^h u - E_c \partial^\alpha u\|_{L_h^p(c^h(\Omega^h))} = 0, \text{ this implies}$$

$$\lim_{h\to 0} \|P^h u\|_{W_+^{k,p}(\Omega^h)} = \lim_{h\to 0} [[E^h P^h u]]_{W_+^{k,p}(\Omega^h)} = \|u\|_{W^{k,p}(\Omega)},$$

cf. (8.80). Similarly, we obtain

$$\text{for (8.19): } \lim_{h\to 0} \|\partial_+^\alpha P^h u - P^h(\partial^\alpha u)\|_{L_h^p(\Omega^h)} = 0 \quad (8.90)$$

$$\text{for (8.20): } \lim_{h\to 0} \|\partial_+^\alpha P^h u - P^h(\partial^\alpha E_c u)\|_{L_h^p(\Omega^h)} = 0.$$

Remark 8.16. *A direct comparison between the original functions and grid functions is not possible. So the concept of* external approximation schemes *was introduced, cf. Petryshyn [528–530, 532], Temam [621] and Zeidler [677]. To this end a mapping is defined for $\mathcal{V} = W^{m,p}(\Omega)$ as*

$$\omega : \mathcal{V} \to \mathcal{W} := \Pi_{|\alpha|\le m} L^p(\Omega), \omega u := (\partial^\alpha u)_{|\alpha|\le m}, \|\omega u\|_{\mathcal{W}} := \max_{|\alpha|\le m} \|\partial^\alpha u\|_{L^p(\Omega)}.$$

ω is bounded, since all norms in $\Pi_{|\alpha|\le m}\mathbb{R}$ are equivalent. Then, e.g. the first line of (8.89) can be reformulated as

$$\lim_{h\to 0} \| \left(\partial_+^\alpha E^h P^h u\right)_{|\alpha|\le m} - \omega u\|_{\mathcal{W}} = 0. \quad (8.91)$$

In this generalized sense, grid functions approximate functions in Sobolev spaces.

By (8.30), cf. Proposition 2.34, (2.107), (2.108), for $1 < p < \infty, 1/p + 1/p' = 1$, and functionals, $f \in \mathcal{V}' = W^{-m,p'}(\Omega)$, unique $f_\beta \in L^{p'}(\Omega)$ exist, such that,

$$\langle f, v\rangle_{\mathcal{V}'\times\mathcal{V}} := \int_\Omega \sum_{|\beta|\le m} f_\beta \partial^\beta v dx \ \forall\ v \in \mathcal{V}, \text{ with } \|f\|_{\mathcal{V}'} = \left(\sum_{|\beta|\le m} \|f_\beta\|^{p'}_{L^{p'}(\Omega)}\right)^{1/p'}, \quad (8.92)$$

cf. Proposition 2.34. For $1 = p$ or $p = \infty$ the $f_\beta \in L^{p'}(\Omega)$ are no longer unique, and we obtain $\|f\|_{\mathcal{V}'} = \min_{f_\beta}\{\sum_{|\beta|\le m} \|f_\beta\|_{L^{p'}(\Omega)}\}$ over all possible f_β.

Two different possibilities for Q'^h correspond to the above $P^h = P^h_1$ or $= P^h_2$.

Definition 8.17. *For functionals, $f \in \mathcal{V}' = W^{-m,p'}(\Omega)$, tested by $v \in \mathcal{V} = W^{m,p}$, we define the $Q'^h : \mathcal{V}' \to \mathcal{V}'^h$ componentwise for the f_β in (8.92). Using the first line of (8.83) or (8.84), and testing $\forall\ v^h \in \mathcal{V}^h = W^{m,p}_+$ yields*

$$Q'^h : \mathcal{V}' \to \mathcal{V}'^h : Q'^h f := \left(f^h_\beta(x) := P^h f_\beta(x) \forall x \in \Omega^h\right)_{|\beta|\le m}, f^h_\beta \in L^{p'}_h(\Omega^h), \quad (8.93)$$

$$\Longrightarrow \langle Q'^h f, v^h\rangle_{\mathcal{V}'^h\times\mathcal{V}^h} := \int_{\Omega^h} \sum_{|\beta|\le m} f^h_\beta \partial^\beta_+ v^h dx^h = \sum_{x\in\Omega^h} h^n \sum_{|\beta|\le m} \partial^\beta_+ v^h(x)(P^h f_\beta)(x).$$

In contrast to Definition 8.13, usually $P^h f_\beta(x) \neq 0$ for $x \in \Omega^h \setminus \Omega^h_0$.

Proposition 8.18. *The previous $P^h : \mathcal{V} \to \mathcal{V}^h$ in (8.83), (8.84) and $Q'^h : \mathcal{V}' \to \mathcal{V}'^h$ in (8.93), are well-defined linear bounded operators, satisfying (8.80), (8.89).*

Proof. We start with the properties of the P^h. Obviously the P^h_1 are well defined. By Raviart [545], pp. 22–25, and Schumann and Zeidler [574, 678], Lemma 6, $P^h = P^h_1$ are linear bounded operators. For $P^h = P^h_2$ this is a consequence of Temam [621], Proposition 3.1, and Lemma 3.1, cf. its proof. In particular, the extension of P^h_2 from $\mathcal{C}(\Omega)$ to $\mathcal{U} = W^{k,p}(\Omega)$ is possible, and yields this result. Relation (8.80) is implied, and obtained for P^h_2 by combining [621], Proposition 3.1 with the proof of Lemma 3.1.

Obviously Q'^h is well defined and linear. The boundedness follows from (8.93), and twice the discrete Hölder inequality (1.44) with $1/p + 1/p' = 1$:

$$\begin{aligned} \langle Q'^h f, v^h\rangle_{\mathcal{V}'^h\times\mathcal{V}^h} &= \sum_{x\in\Omega^h} h^n \sum_{|\beta|\le m} \partial^\beta_+ v^h(x)(P^h f_\beta)(x) \qquad (8.94)\\ &\le \sum_{|\beta|\le m} \left(\sum_{x\in\Omega^h} h^n \left(\partial^\beta_+ v^h(x)(x)\right)^p\right)^{1/p} \left(\sum_{x\in\Omega^h} h^n (P^h f_\beta(x))^{p'}\right)^{1/p'}\\ &= \sum_{|\beta|\le m} \|\partial^\beta_+ v^h)\|_{L^p_h(\Omega^h)} \|P^h f_\beta\|_{L^{p'}_h(\Omega^h)}\\ &\le \|v^h\|_{W^{m,p}_h(\Omega^h)} \|Q'^h f\|_{W^{-m,p'}_h(\Omega^h)}. \end{aligned}$$

This shows that Q'^h is bounded. A combination with (8.80) for the P^h, and (8.92), (8.35) with the minimal f_β, implies (8.80). □

8.5.3 Consistency for difference equations

For the previous examples consistency does not require $u \in \mathcal{U}_b, v^h \in \mathcal{V}_b^h$, but is valid for $u \in \mathcal{U}, v^h \in \mathcal{V}^h$ as well. In fact, the boundary conditions are exactly satisfied. Since here the norms in $\mathcal{V}$ and $\mathcal{V}_b$ are equivalent, we often do not distinguish them. The previously introduced P^h, Q'^h, A^h, G^h allow the proof of this consistency.

We only consider finite difference methods with equal step sizes, h, in all directions. For special Ω, *Dirichlet boundary conditions* can be exactly reproduced:

$$\Omega := \prod_{i=1}^{n}(a_i, b_i) \subset \mathbb{R}^n \text{ s.t. } \exists h > 0 \text{ with } \; a_i/h, \; b_i/h \; \in \mathbb{Z}, i = 1, \cdots, n. \tag{8.95}$$

For $\Omega \subset \mathbb{R}^n$ in (8.95) or in (8.98) below we introduce step sizes, h, and grids:

$$\begin{aligned} H &:= \{h \text{ as in (8.95) or } h \in \mathbb{R}_+ \text{ for (8.98)}\} : \\ \mathcal{G}^h &:= \mathcal{G}_D^h := \{x = (x_1, \ldots, x_n)^T \in \mathbb{R}^n : x_i/h \; \in \mathbb{Z}\}, \\ \mathcal{G}_l^h &:= \{\forall j = 1, \ldots, n : x = (x_1, \ldots, x_n)^T \in \mathbb{R}^n : \forall i \neq j : x_i/h \; \in \mathbb{Z}, x_j/h \; \in \mathbb{R}\}. \end{aligned} \tag{8.96}$$

For Ω in (8.95) we introduce, cf. Figures 8.4,

$$\Omega_0^h := \Omega \cap \mathcal{G}^h, \Omega^h := \overline{\Omega} \cap \mathcal{G}^h, \; \partial\Omega^h := \partial\Omega \cap \mathcal{G}^h = \Omega^h \setminus \Omega_0^h. \tag{8.97}$$

This has to be modified as in (8.20) for

$$\Omega \subset \mathbb{R}^n \; \text{ is open, bounded, nonempty, and } \; \partial\Omega \text{ is Lipschitz-continuous.} \tag{8.98}$$

Let, with the norm for $\mathcal{U}^h$ in (8.87),

$$P^h = P_1^h : \mathcal{U}_b \to \mathcal{U}_b^h : \; \forall x \in \Omega_0^h : \; P_1^h u(x) := \bar{u}^h(x) \text{ otherwise } = 0. \tag{8.99}$$

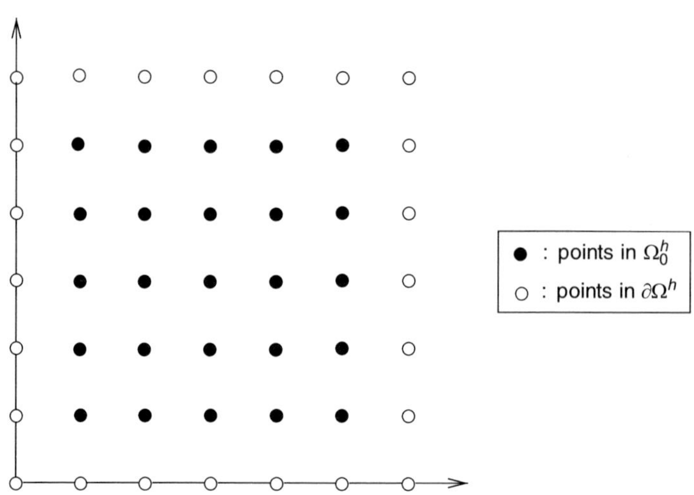

Figure 8.4 Two-dimensional grid for Dirichlet boundary conditions.

For proving consistency of orders 1 or 2, for the difference methods, we have to approximate the $u \in \mathcal{V} = H^{m+\mu}(\Omega)$. A very up-to-date choice, and very appropriate for our bounded Ω, and equidistant Ω^h, are *radial basis functions.* Since they are not quite as well known as other approximations, we give a very short introduction, and refer for all the specific conditions to Buhmann [152] and Böhmer [120]. We start with ϕ, defined by a positive completely monotonic g

$$\phi : \mathbb{R}_+ \to \mathbb{R}_+, \quad \phi(r) := g(r^2), \quad g \in C(\mathbb{R}_+), \quad \text{e.g. } g(r^2) = \sqrt{r^2 + c^2} \tag{8.100}$$

$$\text{and define } \psi(x) := \sum_{k \in \mathbb{Z}^n, |k| < N} c_k \phi(\|x - k\|) c_k \in \mathbb{R}, x \in \overline{\Omega} \subset \mathbb{R}^n,$$

with just one ψ, a finite sum, and appropriate c_k. We introduce

$$\mathcal{S}^h := \left\{ s^h(x) := \sum_{j \in \mathbb{Z}^n} c_j \phi(\|x/h - j\|) \right\}, \text{e.g. } s^h(x) := \sum_{j \in \mathbb{Z}^n} f(jh) \psi(\|x/h - j\|), \tag{8.101}$$

the latter interpolating f in the jh, if ψ is appropriately chosen. The following proposition requires a slightly modified semi-inner product and norm in $H^k(\Omega)$:

$$(u, v)_{H_*^k(\Omega)} := \sum_{|\alpha| = k} \frac{k!}{\alpha!} (\partial^\alpha u, \partial^\alpha v)_{L_h^2(\Omega^h)}, \|u\|_{H_*^k(\Omega)} := ((u, u)_{H_*^k(\Omega)})^{1/2}, \tag{8.102}$$

$$\text{and } \|p\|_{H_*^k(\Omega)} = 0 \forall p \in \mathcal{P}_{k-1}^n.$$

If Ω is not too narrow and h is small enough, Ω^h contains a unisolvent subset Ω_u^h for $\mathcal{P}_{k-1}^n$, the set of polynomials of degree $\leq k - 1$, in $\mathbb{R}^n$, such that

$$p \in \mathcal{P}_{k-1}^n : \forall x \in \Omega_u^h : p(x) = 0 \quad \Rightarrow \quad p \equiv 0.$$

We have mentioned that the second s^h in (8.101) interpolates f in Ω^h, however not uniquely in $\mathcal{S}^h$. This is possible in a modified way by, cf. [152], Proposition 5.2.

Proposition 8.19. *Assume Ω^h contains a unisolvent subset for $\mathcal{P}_{k-1}^n$, for $u \in H^k(\Omega)$, the function values $u(x) \forall x \in \Omega^h$ are well defined, cf. Theorem 1.26 or [152]. Then there exists a unique minimal norm interpolant $s^h \in \mathcal{S}^h$ interpolating $s^h(x) = u(x) \forall x \in \Omega^h$. Compared to any other interpolating $g \in \mathcal{S}^h$, this s^h has minimal seminorm, hence $\|s^h\|_{H_*^k(\Omega)} \leq \|g\|_{H_*^k(\Omega)}$.*

For these interpolants Buhmann [152] proves, cf. Theorem 5.5 for $m = 0$ and Wendland [662], for $m \geq 0$:

Theorem 8.20. Convergence for radial basis functions: *For $u \in \mathcal{V} = W^{m,p}(\Omega)$, and small enough h, the conditions of Proposition 8.19 are satisfied. They guarantee a unique minimal norm interpolant $s_u^h \in \mathcal{S}^h$. For $\mu = 1, 2$ we obtain convergence and* *bounds for*

$$u \in \mathcal{Z} := W^{m+\mu,p}(\Omega), \mu = 1, 2, \quad \Longrightarrow \|u - s_u^h\|_{\mathcal{V}} < Ch^{\mu}\|u\|_{\mathcal{Z}} \text{ and } \tag{8.103}$$

$$\Longrightarrow \|\omega u - \omega s_u^h\|_{\mathcal{W}} < 2h^{\mu}\|\omega\|C\|u\|_{\mathcal{Z}} \text{ with } \|\omega s_u^h\|_{\mathcal{W}} < 2\|\omega\|\|u\|_{\mathcal{V}}.$$

Theorem 8.21. Consistency of linear operators with respect to $\|\cdot\|_{\mathcal{V}^h}$ in (8.87):

1. *With P^h, Q'^h in (8.99), (8.84), (8.93), and the conditions for the coefficients of A in Summary 8.6 and Remark 8.7, we obtain for $Au = f \in \mathcal{V}', u \in \mathcal{U}_b = W_0^{m,p}(\Omega), A \in \mathcal{L}(\mathcal{U}, \mathcal{V}')$ and $f^h = Q'^h f$, consistent unsymmetric and symmetric approximations A^h, A^h_{sw}, cf. (8.60), for linear operators, e.g.*

$$\forall v^h \in \mathcal{V}^h, v^h \neq 0 : \lim_{h\to 0} \langle A^h P^h u - Q'^h Au, v^h\rangle_{\mathcal{V}'^h \times \mathcal{V}^h} / ||v^h||_{\mathcal{V}^h} = 0, \text{ or } \tag{8.104}$$

$$\lim_{h\to 0} ||A^h P^h u - Q'^h Au||_{\mathcal{V}'^h} = \lim_{h\to 0} ||A^h_{sw} P^h u - Q'^h Au||_{\mathcal{V}'^h} = 0 \ \forall u \in W^{m,p}(\Omega).$$

2. *More precisely, we obtain, for $u \in W^{m+\mu,p}(\Omega)$, $A \in \mathcal{L}(W^{m+\mu,p}(\Omega), W^{-m+\mu,p'}(\Omega))$ consitency of order $\mu = 1$ for unsymmetric, and $\mu = 2$ for symmetric differences:*

$$||A^h P^h u - Q'^h Au||_{\mathcal{V}'^h} \leq Ch\|u\|_{W^{m+1,p}(\Omega)} \ \forall u \in W^{m+1,p}(\Omega), \text{ or } \tag{8.105}$$

$$||A^h_{sw} P^h u - Q'^h Au||_{\mathcal{V}'^h} \leq Ch^2\|u\|_{W^{m+2,p}(\Omega)} \forall u \in W^{m+2,p}(\Omega). \tag{8.106}$$

3. *For possible choices of $f^h \neq Q'^h f$, $\lim_{h\to 0} ||f^h - Q'^h f||_{\mathcal{V}'^h} = 0$ has to be proved.*

Proof. We prove the result only for $m \geq q = 1$; the generalization to $q \geq 1$ is straightforward. The proof requires three steps:

(a) We estimate $\langle A^h P^h u - Q'^h Au, v^h\rangle_{\mathcal{V}'^h \times \mathcal{V}^h}$ for $u \in C^m_L(\Omega)$, and $\langle A^h_{sw} P^h u - Q'^h Au, v^h\rangle_{\mathcal{V}'^h \times \mathcal{V}^h}$ for $u \in C^{m+1}_L(\Omega)$, thus (8.105), (8.106), cf. (8.108)
(b) The dense $C^m_L(\Omega) \subset W^{m,p}(\Omega)$ implies consisteny in (8.104) for $u \in W^{m,p}(\Omega)$.
(c) For the stronger forms with order 1 in (8.105), and 2 in (8.106) we need interpolation with radial basis functions with the error estimates in Theorem 8.20 on the rectangular grids in (8.95), (8.98).

(a) For $a_{\alpha\beta} \notin C(\overline{\Omega})$ the $a_{\alpha\beta}(x)$ have to be replaced by $P^h a_{\alpha\beta}(x)$. We obtain, cf. (8.47), for $\forall u \in \mathcal{V} = W^{m,p}(\Omega)$, $\forall v^h \in \mathcal{V}^h = W^{m,p}_+(\Omega^h)$, $a^h : \mathcal{V}^h \times \mathcal{V}^h \to \mathbb{R}$:

$$\begin{aligned} &a^h(P^h u, v^h) - \langle Q'^h Au, v^h\rangle_{\mathcal{V}'^h \times \mathcal{V}^h} \qquad (8.107)\\ &= \langle A^h P^h u - Q'^h Au, v^h\rangle_{\mathcal{V}'^h \times \mathcal{V}^h}\\ &= \int_{\Omega^h} \sum_{|\alpha|,|\beta|\leq m} \left(a_{\alpha\beta}\partial^\alpha_+ P^h u - P^h\left(a_{\alpha\beta}\partial^\alpha u\right)\right) \partial^\beta_+ v^h dx^h\\ &= h^n \sum_{x\in\Omega^h} \sum_{|\alpha|,|\beta|\leq m} \Big((a_{\alpha\beta}\partial^\alpha_+ P^h u)(x) - P^h\left(a_{\alpha\beta}\partial^\alpha u\right)(x)\Big)\partial^\beta_+ v^h(x)\\ &= h^n \sum_{x\in\Omega^h} \sum_{|\alpha|,|\beta|\leq m} a_{\alpha\beta}(x)\Big(\left(\partial^\alpha_+ P^h u\right)(x) - P^h\left(\partial^\alpha u\right)(x)\Big)\partial^\beta_+ v^h(x). \end{aligned}$$

In (8.107), only the terms with $|\alpha| > 0$ are different. We use Lemma 8.2. For the unsymmetric and symmetric divided difference, respectively, This yields with (8.22) for $u \in C_L^m(\Omega)$ and $u \in C_L^{m+1}(\Omega)$, and with (8.99), (8.84),

$$\left|\partial_+^\alpha P^h u(x) - P^h\big(\partial^\alpha u\big)(x)\right| \le c_{\alpha,1} h \ \ \forall x \in \Omega^h \text{ for } u \in C_L^m(\Omega) \text{ and} \tag{8.108}$$
$$\left|\partial_h^\alpha P^h u(x) - P^h\big(\partial^\alpha u\big)(x)\right| \le c'_{\alpha,1} h^2 \ \ \forall x \in \Omega^h \text{ for } u \in C_L^{m+1}(\Omega).$$

(b) This part is an update of Temam's proof of his Proposition 3.1 [621]. $C_L^m(\Omega)$ is a dense subset of $\mathcal{V} = W^{m,p}(\Omega)$. So we approximate the $u \in W^{m,p}(\Omega)$ by appropriate $u_n \in C_L^m(\Omega)$, and employ the approximation results in (8.91), (8.108). For each $n \in \mathbb{N}$ a $u_n \in C_L^m(\Omega)$ exists such that

$$\|u - u_n\|_{\mathcal{V}} < \frac{1}{n} \quad \Rightarrow \quad \|\omega u - \omega u_n\|_{\mathcal{W}} < \frac{\|\omega\|}{n}. \tag{8.109}$$

We fix this u_n. Since $\|\partial_+^\alpha E^h P^h u - \partial^\alpha u\|_{L_h^p(c^h(\Omega^h))} \to 0$ by (8.89), (8.91), there exists an $\eta_n > 0$, such that for $0 < h < \eta_n$,

$$\|\left(\partial_+^\alpha E^h P^h u_n\right)_{|\alpha|\le m} - \omega u_n\|_{\mathcal{W}} < \frac{1}{n}. \tag{8.110}$$

We may additionally choose $\eta_n < 1/n$, thus $\{\eta_n\}$ monotonely converge to 0. For

$$P_+^h u := P^h u_n \text{ for } \eta_{n+1} < h < \eta_n$$

we obtain that

$$\begin{aligned}
&\|\omega u - \left(\partial_+^\alpha E^h P_+^h u\right)_{|\alpha|\le m}\|_{\mathcal{W}} \\
&\quad \le \|\omega u - \omega u_n\|_{\mathcal{W}} + \|\omega u_n - \left(\partial_+^\alpha E^h P^h u_n\right)_{|\alpha|\le m}\|_{\mathcal{W}} \\
&\qquad + \|\left(\partial_+^\alpha E^h P^h u_n\right)_{|\alpha|\le m} - \left(\partial_+^\alpha E^h P_+^h u\right)_{|\alpha|\le m}\|_{\mathcal{W}} \\
&\quad \le \frac{1 + \|\omega\|}{n}.
\end{aligned}$$

This $\|\omega u - \left(\partial_+^\alpha E^h P_+^h u\right)_{|\alpha|\le m}\|_{\mathcal{W}} \le (1 + \|\omega\|)/n$ for $h < \eta_n$ with (8.91), (8.107), (8.108), proves the claimed consistency for $u \in W^{m,p}(\Omega)$.

(c) A slight modification of the proof in (b) yields the stronger results (8.105), (8.106). Only for this case do we need the approximation properties in (8.103). It implies with $P^h u := P^h s_u^h$ that

$$\begin{aligned}
&\|\omega_e E_c u - \left(\partial_+^\alpha E^h P^h u\right)_{|\alpha|\le m}\|_{\mathcal{W}_e} \\
&\le \|\omega_e E_c u - \omega_e s_u^h\|_{\mathcal{W}_e} + \|\omega_e s_u^h - \left(\partial_+^\alpha E^h P^h s_u^h\right)_{|\alpha|\le m}\|_{\mathcal{W}_e} \\
&\quad + \|\left(\partial_+^\alpha E^h P^h s_u^h\right)_{|\alpha|\le m} - \left(\partial_+^\alpha E^h P^h u\right)_{|\alpha|\le m}\|_{\mathcal{W}_e} \\
&\le h(C' + \|\omega\|)\|u\|_{\mathcal{Z}} \text{ and} \\
&\le h^2(C' + \|\omega\|)\|u\|_{\mathcal{Z}}.
\end{aligned}$$

This $\|\omega_e E_c u - \left(\partial_+^\alpha E^h P^h u\right)_{|\alpha|\le m}\|_{\mathcal{W}_e} \le h^\mu(C' + \|\omega_e\|)\|u\|_{\mathcal{Z}}$ proves the claimed consistency of orders $\mu = 1, 2$, for $u \in W^{m+\mu,p}(\Omega)$ for unsymmetric and symmetric differences.

□

In Subsections 8.4.5 and 8.4.6 we introduced the difference equations for the quasilinear equations of orders 2 and $2m$, cf. (8.65) and (8.67), and for linear and quasilinear systems in (8.71) and (8.72). The symmetric forms only were described analogously to the linear problems. For the different types of quasilinear problems different conditions are discussed for unique existence of solutions and their regularity, cf. Subsections 2.5.4, 2.5.6, 2.6.4, 2.6.6 and bounded Frechet derivatives, cf. Subsections 2.7.2, 2.7.3, 2.7.5. To simplify the following consistency result for quasilinear problems, we assume a unifying condition. It is oriented towards that for bounded Frechet derivatives, needed anyway for the proof of the coercivity, stability and convergence in Subsections 8.5.4 and 8.5.5. For minimal conditions the reader has to consult Subsections 2.5.4, 2.5.6, 2.6.4, 2.6.6, 2.7.2, 2.7.3, 2.7.5. We do not fully formulate the next condition, and the theorem for systems, since the generalization is obvious.

We assume all the A_j in (8.65) and A_α in (8.67) to be equicontinuous in a "neighborhood" of u_0, more precisely, with N_m in (8.17),

$$\begin{aligned} &A_j, A_\alpha \text{ equicontinuous in } \Omega \times N_{\nabla^m} \subset \Omega \times \mathbb{R}^{N_m} \text{ with bounded } N_{\nabla^m}, \text{ s.t.} \\ &\forall u \in \mathcal{U}_b : \|u - u_0\|_{\mathcal{V}} \le \rho, \nabla^{\le m} u(x) \in N_{\nabla^m} \forall x \in \Omega, \nabla_{\mp}^{\le m} u(x) \in N_{\nabla^m} \forall x \in \Omega_0^h. \end{aligned} \tag{8.111}$$

Theorem 8.22. Consistency of quasilinear equations and systems with $\|\cdot\|_{\mathcal{V}^h}$ in (8.87)

1. *Choose the P^h, Q'^h in (8.84), (8.99), (8.93), and let A_j, A_α in (8.67) and u satisfy (8.111). Choose G^h, similarly G^h_{sw}, as in (8.65), (8.67), and (8.71), (8.72). Then we obtain for $Gu \in W^{-m,p'}(\Omega)$ consistent unsymmetric and symmetric approximations G^h, G^h_{sw}, hence, e.g.*

$$\forall v^h \in \mathcal{V}^h, v^h \ne 0 : \lim_{h\to 0} \langle G^h P^h u - Q'^h G u, v^h\rangle_{\mathcal{V}'^h \times \mathcal{V}^h} / ||v^h||_{\mathcal{V}^h} = 0, \tag{8.112}$$

thus

$$\lim_{h\to 0} ||G^h P^h u - Q'^h G u||_{\mathcal{V}'^h} = \lim_{h\to 0} ||G^h_{sw} P^h u - Q'^h G u||_{\mathcal{V}'^h} = 0 \ \forall u \in W^{m,p}(\Omega).$$

2. *For $u \in W^{m+\mu,p}(\Omega)$, $A_\alpha : W^{m+\mu,p}(\Omega) \to W^{-m+\mu,p'}(\Omega))$ with Lipschitz-continuous A_j, A_α with respect to the variables in N_{∇^m}, we obtain consitency of order $\mu = 1$ for unsymmetric, and $\mu = 2$ for symmetric differences:*

$$||G^h P^h u - Q'^h G u||_{\mathcal{V}'^h} \le Ch\|u\|_{W^{m+1,p}(\Omega)} \ \forall u \in W^{m+1,p}(\Omega), \tag{8.113}$$

or

$$||G^h_{sw} P^h u - Q'^h G u||_{\mathcal{V}'^h} \le Ch^2\|u\|_{W^{m+2,p}(\Omega)} \forall u \in W^{m+2,p}(\Omega). \tag{8.114}$$

Proof. The proof is very similar to the proof of Theorem 8.21, here restricted to $m = q = 1$. The generalization to $m, q \geq 1$ is straightforward. From the three steps, we only formulate (a) and only indicate (b) and (c)

(a) Again, we formulate this proof for $u, \nabla u \in C(\overline{\Omega})$, $A_j \in C(\overline{\Omega}, \mathbb{R}, \mathbb{R}^n)$. Otherwise these $u, \nabla u, A_j$ have to be replaced by $P^h u, P^h \nabla u, P^h A_j$. We obtain, cf. (8.65), for $\forall u \in \mathcal{V} = W^{1,p}(\Omega)$, $\forall v^h \in \mathcal{V}^h = W_+^{1,p}(\Omega^h)$,

$$\begin{aligned} a^h(P^h u, v^h) - \langle Q'^h G u, v^h\rangle_{\mathcal{V}'^h \times \mathcal{V}^h} &= \langle G^h P^h u - Q'^h G u, v^h\rangle_{\mathcal{V}'^h \times \mathcal{V}^h} \quad (8.115)\\ &= \int_{\Omega^h} \sum_{j=0}^{n} \big(A_j(\cdot, P^h u, \nabla_+ P^h u) \\ &\quad -P^h A_j(\cdot, u, \nabla u)\big)\partial_+^j v^h dx^h \\ &= h^n \sum_{x\in\Omega^h} \sum_{j=0}^{n} \Big(A_j(\cdot, P^h u, \nabla_+(P^h u)) \\ &\quad -A_j(\cdot, u, \nabla u)\Big)(x)\partial_+^j v^h(x). \end{aligned}$$

Lemma 8.2 yields with (8.22) for $u \in C_L^1(\overline{\Omega})$, $u \in C_L^2(\overline{\Omega})$, but now for $A_j \in C_L(\overline{\Omega}, \mathbb{R}, \mathbb{R}^n)$,

$$\begin{aligned} \left|\Big(A_j(\cdot, (P^h u), (D_+(P^h u))) - A_j(\cdot, u, \nabla u)\Big)(x)\right| &\leq c_{1,1} h \text{ for } u \in C_L^1(\overline{\Omega}) \quad (8.116)\\ \left|\Big(A_j(\cdot, (P^h u), (D_+(P^h u))) - A_j(\cdot, u, \nabla u)\Big)(x)\right| &\leq c'_{1,1} h^2 \text{ for } u \in C_L^2(\overline{\Omega}). \end{aligned}$$

(b) We approximate the $u \in W^{1,p}(\Omega)$ by $u_k \in C_L^1(\overline{\Omega})$. Then a combination of the equicontinuity in (8.111), (8.109), (8.110), implies the consistency (8.112).

(c) The final consistency of orders 1 and 2 follows as in the proof of Theorem 8.21, now with the Lipschitz-continuity of the A_j. □

Finally, we introduced in Subsection 8.4.7 the difference equations for the fully nonlinear equations restricted to order 2, cf. (8.76) and (8.67). The symmetric form is listed at the end of Subsection 8.4.7. Again we restrict the discussion to the Hilbert space setting $\mathcal{U} := H^2(\Omega) \to \mathcal{V} := L^2(\Omega)$. For fully nonlinear equations different conditions for unique existence of solutions and their regularity are listed in Subsection 2.5.7 and the bounded Frechet derivatives in Subsections 2.7.3 and 2.7.5. For systems nearly nothing is known, cf. Section 5.2.8.

Similarly to (8.111), we assume the fully nonlinear G to be equicontinuous in a "neighborhood" of u_0, more precisely, with N_m in (8.17),

$$\begin{aligned} &G \text{ equicontinuous in } \Omega \times N_{\nabla^2} \subset \Omega \times \mathbb{R}^{N_2} \text{ with bounded } N_{\nabla^2}, \text{ s.t.} \quad (8.117)\\ &\forall u \in \mathcal{V} : \|u - u_0\|_{\mathcal{V}} \leq \rho, \forall x \in \Omega \nabla^{\leq 2} u(x) \in N_{\nabla^2} \forall x \in \Omega_0^h . \nabla_{\mp}^{\leq 2} u(x) \in N_{\nabla^2}. \end{aligned}$$

Theorem 8.23. Consistency for fully nonlinear problems with $\|\cdot\|_{\mathcal{V}^h} = \|\cdot\|_{L_h^2(\Omega^h)}$:

1. *Choose the* P^h, Q'^h *in (8.99), (8.84), (8.93), and let* G *now in (8.74) satisfy (8.117). Then we obtain for* $G : \mathcal{D}(G) \subset \mathcal{U} := H^{2m}(\Omega) \to \mathcal{V} := L^2(\Omega)$ *and the grid functions,* $\mathcal{U}^h, \mathcal{V}^h$, *a consistent strong approximation,* $G^h = G^h_s$, *and the strong symmetric approximation,* G^h_{ss}, *hence we find for* $\|u - u_0\|_{\mathcal{U}} \le \rho$,

$$\forall v^h \in \mathcal{V}^h, v^h \neq 0 : \lim_{h\to 0} (G^h P^h u - Q'^h Gu, v^h)_{\mathcal{V}^h} / ||v^h||_{\mathcal{V}^h} = 0, \qquad (8.118)$$

thus

$$\lim_{h\to 0} ||G^h P^h u - Q'^h Gu||_{\mathcal{V}^h} = \lim_{h\to 0} ||G^h_{ss} P^h u - Q'^h Gu||_{\mathcal{V}^h} = 0 \ \forall u \in \mathcal{U}.$$

2. *For* $u \in H^{2+\mu}(\Omega)$, $G : \mathcal{D}(G) \subset H^{2+\mu}(\Omega) \to H^\mu(\Omega))$ *with Lipschitz-continuous* G *with respect to the variables in* N_{∇^2}, *we obtain consitency of order* $\mu = 1$ *for unsymmetric, and* $\mu = 2$ *for symmetric differences:*

$$||G^h P^h u - Q'^h Gu||_{\mathcal{V}'^h} \le Ch^\mu \|u\|_{H^{2+\mu}(\Omega)} \ \forall u \in H^{2+\mu}(\Omega). \qquad (8.119)$$

Proof. This is an obvious modification of the last proofs. □

8.5.4 $\mathcal{V}^h_b$-coercivity for linear(ized) elliptic difference equations

In contrast to the previous consistency results, the boundary conditions are crucial for coercivity and the following stability of elliptic difference operators. $\mathcal{V}^h_b$-coercivity is, by Theorems 3.23, 2.12 ff., and 8.29, the essential step towards stability of our difference methods. In fact, let the principal part of a linear, or linearized, elliptic difference operator be $\mathcal{V}^h_b := H^m_{0,+}(\overline{\Omega}, \mathbb{R}^q)$-coercive, see below, and the method be consistent, cf. Subsection 8.5.3. Assume a bounded and boundedly invertible Frechet derivative $G'(u_0)$ of the original problem in the exact solution, u_0. Then the difference equation, $G^h(u^h) = 0$, similarly $A^h(u^h) = 0$, is stable, by Theorem 8.29. With its consistency, this implies its uniquely existing solution, u^h_0, converging to u_0 with respect to $\|\cdot\|_{H^m_+(\overline{\Omega}, \mathbb{R}^q)}$. We summarize these results in Subsection 8.5.5.

We start with coercivity results for the original linear operators. In Chapter 2, Theorems 2.43, 2.89, and 2.104, we have seen that coercive bilinear forms induce boundedly invertible linear operators. For the different cases, we have formulated in Summary 8.6 the definitions of ellipticity and the conditions for the coefficients, implying $H^m_0(\Omega, \mathbb{R}^q)$-or $W^{m,p}_0(\Omega, \mathbb{R}^q)$-coercivity, and boundedness of the weak (and strong) forms of the principal parts of the original operators for $m, q \ge 1$. As we have previously discussed, cf. Chapters 2–5, we will usually only have $H^m_0(\Omega, \mathbb{R}^q)$-coercivity, even for (nonlinear) problems defined in $W^{m,p}_0(\Omega, \mathbb{R}^q)$.

As indicated in Remark 8.7 with its pros and cons, the conditions in Summary 8.6 for the coefficients of the linear difference equations, (8.38)–(8.41), could be considerably reduced towards those of differential equations, by replacing the $C(\Omega, \mathbb{R}^q), C^{|\beta|}(\Omega, \mathbb{R}^q)$ by $L^\infty(\Omega, \mathbb{R}^q), W^{|\beta|,\infty}(\Omega, \mathbb{R}^q)$. The nonlinear problems in Subsections 8.4.5 and 8.4.6 are included via linearizations of their discrete operators, for $m, q \ge 1$. They are bounded and have $H^m_0(\Omega, \mathbb{R}^q)$-coercive principal parts, for $2 \le p < \infty$, under appropriate conditions, cf. (8.38)–(8.41), by Theorems 2.124, 2.125, 2.126, 2.127. Thus we obtain stability for $2 \le p < \infty$, with respect to the $H_+^m(\Omega^h, \mathbb{R}^q)$ norm. The case $1 \le p < \infty$, with respect to the $W_+^{m,p}(\Omega^h, \mathbb{R}^q)$ norm can be studied via monotone operators.

We memorize the discrete $\mathcal{V}_b^h$-coercivity and ellipticity. We choose the discrete spaces $\mathcal{V}_b^h = H_{0,+}^m(\Omega^h, \mathbb{R}^q)$, or $\mathcal{V}_b^h = W_{0,+}^{m,p}(\Omega^h, \mathbb{R}^q)$, $m, q \ge 1, 1 \le p \le \infty$ for the homogeneous Dirichlet boundary value problem. Let $a^h(u^h, v^h)$, cf. e.g. (8.47), (8.63), (8.71) for $m \ge 1$, be a *bounded* bilinear form, defined for $\mathcal{V}^h$, and $A^h : \mathcal{V}^h \to \mathcal{V}'^h$ the induced linear operator. Then $a^h(u^h, v^h)$ are called $\mathcal{V}_b^h$*-coercive*, and $\mathcal{V}_b^h$*-elliptic,* if h-independent constants $\alpha \in \mathbb{R}_+, C_c \in \mathbb{R}$ exist such that

$$a^h(u^h, u^h) > \alpha \|u^h\|^2_{\mathcal{V}^h}, \text{ and } \ge \alpha \|u^h\|^2_{\mathcal{V}^h} - C_c \|u^h\|^2_{\mathcal{W}^h} \; \forall u^h \in \mathcal{V}_b^h \subset \mathcal{W}^h, \tag{8.120}$$

$$\text{respectively, e.g. } \mathcal{V}_b^h = H_{0,+}^m(\Omega^h, \mathbb{R}^q) \text{ or } W_{0,+}^{m,p}(\Omega^h, \mathbb{R}^q) \subset \mathcal{W}^h = L_h^2(\Omega^h, \mathbb{R}^q),$$
$$m, q \ge 1.$$

If additionally $C \in \mathbb{R}_+$ exist such that

$$A^h : \mathcal{V}_b^h \to \mathcal{V}_b'^h \text{ is boundedly invertible with } \|(A^h)^{-1}\|_{\mathcal{V}_b'^h \to \mathcal{V}_b^h} \le C \;\; \forall h > 0, \tag{8.121}$$

then A^h is called $\mathcal{V}_b^h$*-stable* or $\mathcal{V}_b^h$*-regular,* e.g. by Thomée and Westergren [627], Hackbusch [381, 383, 387] and Bube and Strikwerda [151].

The proofs for the $\mathcal{V}_b$-coercivity in Chapter 2, and the $\mathcal{V}_b^h$-coercivity here are very similar. So we only indicate the proof for one equation of order 2, but present it for a system of order $2m$. This case shows the strongest differences between the $\mathcal{V}_b$-, and the $\mathcal{V}_b^h$-situation.[48]

[48] We recapitulate the discrete Fourier transform, cf. Theorem 2.105, Thomée and Westergren [627], Stoer and Bulirsch [603] and Bube and Strikwerda [151], pp. 660 ff., and Bube *et al.* [605]. In standard applications to periodic functions it yields (the coefficients of) the trigonometric polynomials interpolating discrete function values $u^h : \mathcal{G}_n^h = \mathcal{G}^h \to \mathbb{R}$, see (8.96), e.g. via FFT, but not the Fourier series. This method will be extensively studied in spectral methods, see [120]; see also any standard textbook on numerical methods, cf. [151, 603]. We only summarize the results needed for the following coercivity proof in this special case. We assume Ω in (8.95) or (8.98). For (8.98) we extend Ω to a cuboidal Ω_e with $u^h(x) = 0 \forall x \notin \Omega_0^h$. Then we transform it, such that $\Omega = (-\pi, \pi)^n$, and

$$\Omega^h := \{x_\nu := h\nu\} \subseteq [-\pi, \pi]^n \text{ with } h := \pi/N, \; N \in \mathbb{N}, \tag{8.122}$$
$$\nu = (\nu^1, \cdots, \nu^n) \in \mathbb{Z}^n, \; \forall j = 1, \cdots, n : \; |\nu^j| \le N.$$

We periodically extend the grid functions, $u^h \in H_{0,+}^m(\Omega^h)$, to $\mathcal{G}^h := \{h\nu : \; \nu = (\nu^1, \cdots, \nu^n) \in \mathbb{Z}^n\}$, since u^h has identical, here even vanishing, values on opposite faces of $[-\pi, \pi]^n$. We extend scalar products, and norms, see (8.25), to $u^h, v^h : \Omega^h \to \mathbb{C}$ with $u^h(x)^* = \overline{u^h(x)}$ as

$$(u^h, v^h)_{L_h^2(\Omega^h)} := h^n \sum_{x \in \Omega^h} u^h(x)^* v^h(x), \|v^h\|_{H_+^m(\Omega^h)} := \left[\sum_{|\alpha| \le m} h^n \sum_{x \in \Omega^h} |\partial_+^\alpha v^h(x)|^2 \right]^{1/2}.$$

We define the discrete Fourier transform with the imaginary unit, ι, $\iota^2 = -1$, and

$$\langle \xi, x \rangle_n = \sum_{k=1}^n \xi_k x_k \; \forall \, x \in \Omega^h, \; \forall \, \xi \in \Gamma^h := \{\xi = (\xi_1, \cdots, \xi_n) \in \mathbb{Z}^n : |\xi_j| \le N\} \tag{8.123}$$

For order $2m$, discrete Fourier transforms are employed. They have been used in the literature to prove regularity results for the solutions of difference equations, similar to those of differential equations, and the convergence to exact solutions, cf. Agmon *et al.* [2, 3], Thomée and Westergren [627], Bube and Strikwerda [151], Strikwerda *et al.* [605] and Hackbusch [387]. But the author did not find any application towards discrete coercivity as in this section.

Theorem 8.25. $\mathcal{V}_b^h$-coercive principal parts: *Assume the conditions in Summary 8.6, and Remark 8.7, let* $\mathcal{V}_b^h := H_{0,+}^m(\Omega^h, \mathbb{R}^q)$, $\mathcal{U}_b^h := H_+^{2m}(\Omega^h, \mathbb{R}^q) \cap \mathcal{V}_b^h$, $\mathcal{V}_b^h = \mathcal{V}_b'^h := L_h^2(\Omega^h, \mathbb{R}^q)$, *and let* $f^h \in \mathcal{V}_b'^h$ *be bounded so* $p = 2$ *compared to Subsections 8.5.2, 8.5.3.*

1. *Then the weak bilinear forms* $a^h(u^h, v^h)$ *are bounded with respect to* $\mathcal{V}^h$, *the principal parts* $a_p^h(u^h, v^h)$, *and* $a^h(u^h, v^h)$ *are* $\mathcal{V}_b^h$-coercive, and $\mathcal{V}_b^h$-elliptic, *respectively. The* $a_p^h(u^h, v^h)$ *induce a stable* A_p^h, *which is consistent with* A_p.
2. *The strong difference operators,* A_s^h, *or* $(G'(u_0))_s^h$, *are also bounded with respect to* $\mathcal{V}_s^h$.

Proof.

(1) The claim that the strong operators, A_s^h, $(G'(u_0))_s^h$ and $a^h(u^h, v^h)$ are bounded with respect to $\mathcal{V}_s^h$, and $\mathcal{V}^h$, respectively, is nearly obvious.
(2) Since the $a^h(u^h, v^h)$ are compact perturbations of the principal parts $a_p^h(u^h, v^h)$ the $\mathcal{V}_b^h$-*ellipticity* of the $a^h(u^h, v^h)$ can be shown, via the detour

as

$$(F_h u^h)(\xi) := \hat{u}^h(\xi) := h^n \sum_{x \in \Omega^h} e^{-\iota \langle \xi, x \rangle_n} u^h(x) = (e^{\iota \langle \xi, \cdot \rangle_n}, u^h(\cdot))_{L_h^2(\Omega^h)} \quad \forall u^h \in H_{0,+}^m(\Omega^h).$$

Obviously this $\hat{u}^h(\xi)$ is $2N$ periodic in each ξ_j for $\xi \in \mathbb{Z}^n$. The fact that $u^h(x)$ and $\hat{u}^h(\xi)$ are defined on different grids Ω^h and Γ^h or their extensions, is reflected in a certain asymmetry of the inverse discrete Fourier transform, in contrast to (2.397). The following properties are proved in standard textbooks, and, e.g. Bube and Strikwerda [151].

Theorem 8.24. *For the discrete Forier transform its inverse is defined as*

$$\left(F_h^{-1} \hat{u}^h\right)(x) := \sum_{\xi \in \Gamma^h} e^{\iota \langle \xi, x \rangle_n} \hat{u}^h(\xi), \text{ it yields } u^h(x) = \left(F_h^{-1} \hat{u}^h\right)(x), \text{ and} \tag{8.124}$$

$$(u^h, v^h)_{L_h^2(\Omega^h)} = h^n \sum_{x \in \Omega^h} u^h(x)^* v^h(x) = \sum_{\xi \in \Gamma^h} \hat{u}^h(\xi)^* \hat{v}^h(\xi) = (\hat{u}^h, \hat{v}^h)_{L_h^2(\Gamma^h)} \forall u^h, v^h \in L^2(\Omega^h),$$

the so-called Parseval formula, and $F_h\left(\partial_+^\beta u^h\right)(\xi) = \widehat{\partial_+^\beta u^h}(\xi) = \xi^{h,\beta} \hat{u}^h(\xi)$, *with*

$$\xi^{h,\beta} := \prod_{\ell=1}^n \frac{(exp(\iota h \xi_\ell) - 1)^{\beta_\ell}}{h^{\beta_\ell}}, \xi^{h,\beta} = \prod_{\ell=1}^n \frac{2(exp(\iota h \xi_\ell/2)^{\beta_\ell} \sin(h\xi_\ell/2))^{\beta_\ell}}{h^{\beta_\ell}}, |\xi^{h,\beta}| \approx \pm \prod_{\ell=1}^n 2(\xi_\ell/2)^{\beta_\ell}$$

for small h. *The squares of the norm and seminorm,* $\|u^h\|_{H_+^k(\Omega^h)}^2$, *and* $|u^h|_{H_+^k(\Omega^h)}^2$, *are related as*

$$\left\|u^h\right\|_{H_+^k(\Omega^h)}^2 = \sum_{\xi \in \Gamma^h} \sum_{|\alpha| \le k} \left|\xi^{h,\alpha}\right|^2 \hat{u}^h(\xi)^* \hat{u}^h(\xi), \left|u^h\right|_{H_+^k(\Omega^h)}^2 = \sum_{\xi \in \Gamma^h} \sum_{|\alpha| = k} \left|\xi^{h,\alpha}\right|^2 \hat{u}^h(\xi)^* \hat{u}^h(\xi).$$

$a_p(u,v)$, $a(u,v)$ as in Chapter 2 by results similar to Theorem 1.26 and Proposition 2.24.

(3) So we turn to the $\mathcal{V}_b^h$-*coercivity* of the $a_p^h(u^h,v^h)$, only indicating the proof for $m=q=1$. In fact for (8.38) we only have to replace ϑ^i by $\partial_+^i u(x)$, and sum over $x\in\Omega^h$ to get the claim. This is analogous to (8.38) where $\vartheta^i=\partial^i u(x)$ is combined with integrating over Ω.

(4) For $m,q>1$, so for systems (8.71) of order $2m$, we impose the above (8.41). For equations and systems of order $2m$, similarly to the situation in Theorem 2.36, we cannot repeat the above argumentation in (3) based upon (8.38). The reason is again the inequality $\left(\partial_+^i u\right)^{\alpha_i}\neq\partial_+^{\alpha_i}u$ for $\alpha_i>1$. So we apply instead the discrete Fourier transform to each component $\partial_+^\beta u_j$ and $\partial_+^\alpha u_j$. So we prove:

(4) With Theorem 8.24 for the discrete Fourier transform, we prove the $\mathcal{V}_b^h=H_{0,+}^m(\Omega^h)$-coercivity of the principal part. We only discuss constant coefficients $A_{\alpha\beta},|\alpha|,|\beta|=m$. The reduction of $A_{\alpha\beta}\in C^m(\overline{\Omega})$ to constant coefficients by the technique of frozen (constant) coefficients is presented in the proof of Theorem 2.43. It can be applied to difference methods as well.

We combine Theorem 8.24 with the Legendre–Hadamard condition (8.41). With the notation in (8.123), (8.124) and

$$\xi^{h,\beta}=\prod_{\ell=1}^n\frac{(\exp(\iota h\xi_\ell)-1)^{\beta_\ell}}{h^{\beta_\ell}}=\prod_{\ell=1}^n\frac{2(\exp(\iota h\xi_\ell/2)^{\beta_\ell}\sin(h\xi_\ell/2))^{\beta_\ell}}{h^{\beta_\ell}},\tag{8.125}$$

we obtain with $\xi^{h,\beta}=\prod_{\ell=1}^n\left(\xi_\ell^h\right)^{\beta_\ell}$, $\vartheta^\ell:=\xi_\ell^h$,

$$\begin{aligned}
a_p^h(\vec{u}^h,\vec{u}^h)&=\left(\sum_{j,k=1}^q\sum_{|\alpha|=|\beta|=m}h^n\sum_{x\in\Omega^h}\left(a_{\alpha\beta}^{jk}\partial_+^\beta u_k^h(x)\right)\partial_+^\alpha u_j^h(x)\right)\\
&=\left(\sum_{j,k=1}^q\sum_{|\alpha|=|\beta|=m}a_{\alpha\beta}^{jk}\left(h^n\sum_{x\in\Omega^h}\partial_+^\beta u_k^h(x)\partial_+^\alpha u_j^h(x)\right)\right)\\
&=\left(\sum_{j,k=1}^q\sum_{|\alpha|,|\beta|=m}a_{\alpha\beta}^{jk}\left(\partial_+^\beta u_k^h,\partial_+^\alpha u_j^h\right)_{L^2(\Omega^h)}\right)\\
&=\left(\sum_{j,k=1}^q\sum_{|\alpha|,|\beta|=m}a_{\alpha\beta}^{jk}\left(F_h\left(\partial_+^\beta u_k^h\right),F_h\left(\partial_+^\alpha u_j^h\right)\right)_{L^2(\Gamma^h)}\right)\\
&=\sum_{j,k=1}^q\sum_{|\alpha|=|\beta|=m}a_{\alpha\beta}^{jk}\sum_{\xi\in\Gamma^h}\left(\left[\xi^{h,\beta}\hat{u}_k(\xi)\right]\overline{\left[\xi^{h,\alpha}\hat{u}_j(\xi)\right]}\right),\ \text{by (8.41),(8.125)}\\
&\ge\lambda\sum_{\xi\in\Gamma^h}\left(\sum_{\ell=1}^n\left|\xi_\ell^h\right|^2\right)^m\sum_{j=1}^q\left|\hat{u}_j^h(\xi)\right|^2.
\end{aligned}$$

For a fixed $m > 1$ there exists an $\epsilon > 0$, such that $\left(\sum_{\ell=1}^{n} \left|\xi_\ell^h\right|^2\right)^m > \epsilon \sum_{|\alpha|=m} |\xi^{h,\alpha}|^2$. This is verified by the polynomial formula. So we find with (8.124)

$$a_p^h(\vec{u}^h, \vec{u}^h) > \lambda\epsilon \sum_{\xi\in\Gamma^h} \sum_{i=1}^{q} \left(\sum_{|\alpha|=m} (\xi^{h,\alpha})^2 \right) \left|\hat{u}_i^h(\xi)\right|^2 = \lambda\epsilon |\vec{u}^h|^2_{H_+^m(\Omega^h,\mathbb{R}^q)}.$$

With the equivalent norms $|\vec{u}^h|_{H_+^m(\Omega^h,\mathbb{R}^q)}$ and $\|\vec{u}^h\|_{H_+^m(\Omega^h,\mathbb{R}^q)}$ on $\mathcal{V}_b^h = H_{0,+}^m(\Omega^h, \mathbb{R}^q)$, the $\mathcal{V}_b^h$-coercivity of $a_p^h(\vec{u}^h, \vec{v}^h)$ is proved for constant coefficients. The general case is proved as in Theorem 2.42. □

8.5.5 Stability and convergence for general elliptic difference equations

In Subsection 8.5.3 we proved the consistency of our discrete operator with the original operators. Subsection 8.5.4 showed the $\mathcal{V}_b$-coercivity of the principal parts of the linear(ized) operators, and hence its stability. This implies, by Theorem 8.21, the convergence of the discrete to the exact solutions of these elliptic problems for the principal part. The corresponding Lemma 8.28 is combined in Theorem 8.29 for proving the stability for boundedly invertible linear elliptic operators.

Condition 8.26. *Choose $\mathcal{U}_b^h = \mathcal{V}_b^h$ and their duals as $H_+^m(\Omega^h, \mathbb{R}^q)$, $\mathcal{V}'^h = H_+^{-m}(\Omega^h, \mathbb{R}^q)$, cf. (8.27)–(8.31), (8.77), and $P^h = I^h$ and Q'^h as in Subsection 8.5.2 for $p = 2$. The original linear and nonlinear problems and the difference methods are listed in Section 8.4. For the consistency, and its order 1 or 2, we refer to Theorems 8.21, 8.22, and 8.23; for the coercivity of the principal parts we refer to Theorem 8.25.*

Remark 8.27. *Notice that we have chosen the $\mathcal{V}^h = H_+^m(\Omega^h, \mathbb{R}^q)$, in contrast to $\mathcal{V} = W^{m,p}(\Omega, \mathbb{R}^q)$. This allows our linearization approach based upon coercive principal parts, as used in this section. The direct discrete counterpart $\mathcal{V}^h = W_+^{m,p}(\Omega^h, \mathbb{R}^q)$ for $\mathcal{V} = W^{m,p}(\Omega, \mathbb{R}^q)$ is possible for quasilinear problems and their difference methods as well. Then we have to use the monotony technique in Section 4.5, and Theorem 4.67 remains valid for $1 < p < \infty$ and e.g. cf. (4.197), (8.146)*

$$\left\|P^h u_0 - u_0^h\right\|_{\mathcal{U}_b^h} \leq \left(h^\mu L \|u_0\|_{W^{m+\mu}(\Omega)}\right)^{1/(p-1)}, \tag{8.126}$$

for $\mu = 1$ and $= 2$ for unsymmetric $\mathcal{V}^h = H_+^m$ and symmetric difference methods.

Even better results are obtainable by linearization techniques. They require $2 \leq p < \infty$ and yield stability and convergence with respect to the discrete $H_+^m(\Omega^h)$ norm, cf. the discussion in Subsection 8.4.5, following (8.65).

The following lemma enforces the consequences of Theorem 8.21. The different errors are related with $A^h u^h = Q'^h A u$ as:

$$\underbrace{Q'^h A_h u - A^h u^h}_{\text{var.cons.error}} = Q'^h A_h \underbrace{(u - P^h u)}_{\text{interp.error}} + \underbrace{A^h P^h u - Q'^h A u}_{\text{class.cons.err.}};\ A^h u^h = Q'^h A u. \tag{8.127}$$

Lemma 8.28. *Under Condition 8.26, let $A_p : \mathcal{U}_b \to \mathcal{V}'$ be a coercive principal part, cf. (5.278) or (5.281). Then $A_p u = f \in W^{-m,p'}(\Omega)$, and for small enough $h_0 > h \in H$, $A_p^h u^h = Q'^h f$ as well, have unique solutions, and u^h converge to u as*

$$\|u^h - P^h u\|_{\mathcal{U}^h} \le \epsilon \, \|u\|_{\mathcal{U}}, \text{ for } h < h_0, \text{ hence } \lim_{h\to 0} \|u^h - P^h u\|_{\mathcal{U}^h} = 0. \quad (8.128)$$

For general A, and $u \in \mathcal{U}_b$, the preceding consistency errors satisfy, for $h < h_0$,

$$\|Q'^h A u - A^h P^h u\|_{\mathcal{V}^{h'}} \le \epsilon \, \|u\|_{\mathcal{U}}, \quad \text{hence } \lim_{h\to 0} \|Q'^h A u - A^h P^h u\|_{\mathcal{V}^{h'}} = 0. \quad (8.129)$$

Proof. We can essentially dublicate the proof of Lemma 5.76, if we replace $\|u - u^h\|_{\mathcal{U}^h}$ there by $\|P^h u - u^h\|_{\mathcal{U}^h}$ here. □

Theorem 8.29. Stability inherited to invertible compact perturbations. *Under Condition 8.26, let $A, B \in \mathcal{L}(\mathcal{U}_b, \mathcal{V}_b')$, $\mathcal{U}_b = W_0^{m,p}(\Omega), \mathcal{V}_b' = W_0^{-m,p'}(\Omega)$, be boundedly invertible, and $B^h \in \mathcal{L}\left(\mathcal{U}_b^h, \mathcal{V}_b'^h\right)$ be stable, e.g. $B = A_p$. Let $A := B + C$, with $C \in \mathcal{C}(\mathcal{U}_b, \mathcal{V}_b')$, the set of compact operators from $\mathcal{U}_b \to \mathcal{V}_b'$, and let A^h, B^h, hence C^h, be consistent with A, B, C. Then, for small enough $h \in H$,*

$$A^{-1} \in \mathcal{L}(\mathcal{V}', \mathcal{U}_b) \Longrightarrow A^h \quad \text{is stable in} \quad P^h u \quad \text{hence}$$
$$(A^h)^{-1} \in \mathcal{L}\left(\mathcal{V}_b'^h, \mathcal{U}_b^h\right), \quad \|(A^h)^{-1}\|_{\mathcal{U}_b^h \leftarrow \mathcal{V}_b'^h} \le C.$$

Remark 8.30. *We can choose $B = \Delta^m$. Then, for $\mathcal{U}_b^h = W_{0,+}^{m,p}(\Omega^h)$ with $2 < p \le \infty$ only $H_{0,+}^m(\Omega^h)$-coercivity, stability and convergence is obtained, cf. Summary 8.6, Remarks 8.7, 8.14 and Theorems 8.31, 8.32.*

Proof. This proof essentially modifies that for FEMs with variational crimes. We consider the appropriate extensions E^h, and E_h, cf. (8.86), (8.131).

(1) Again we define a kind of *anticrime transformation* from $\mathcal{U}_b^h$ to $\mathcal{U}_b = W_0^{m,p}(\Omega) \subset \mathcal{U}$, cf. Lemma 5.77. It allows a direct comparison between the elements in, e.g. $\mathcal{U}_b^h = W_{0,+}^{m,p}\left(\mathcal{T}_c^h\right)$ and $\mathcal{U}_b = W_0^{m,p}(\Omega), \mathcal{V}_b' = W_0^{-m,p'}(\Omega)$. We again choose the boundedly invertible $B : \mathcal{U}_b \to \mathcal{V}_b'$, here $B = \Delta^m$. We have proved $B^h : \mathcal{U}_b^h \to \mathcal{V}_b^{h'}$ to be stable and consistent with B for every $u \in \mathcal{U}_b$, with a consistency order $p > 0$ for smoother u. In the following procedure we combine Proposition 8.18 and Theorem 8.21 with the results in Theorems 8.25 and 3.21:

$$\text{For any given } u^h \in \mathcal{U}_b^h \text{ define } \hat{f}^h := B^h u^h \in \mathcal{V}^{h'}, \text{ and } \hat{f}_h := E^h \hat{f}^h \in \mathcal{V}',$$
$$\text{with } \|\hat{f}_h\|_{\mathcal{V}'} \le \|B^h\|_{\mathcal{V}^{h'} \leftarrow \mathcal{U}_b^h} \|u^h\|_{\mathcal{U}^h}(1 + o(1)), \text{ cf. (8.30).} \quad (8.130)$$

We define the bounded

$$E_h : \mathcal{U}_b^h \to \mathcal{U}_b : \; u_h := u_{0,h} := E_h u^h := (B)^{-1}\hat{f}_h \in \mathcal{U}_b = W_0^{m,p}(\Omega) \text{ yielding}$$
$$\|E_h u^h\|_{\mathcal{U}} \le C_1 \|u^h\|_{\mathcal{U}^h}, \quad \text{and } \forall \epsilon > 0\; \exists h_0 = h_0\left(\left\|u^h\right\|_{H^1(\mathcal{T}_c^h)}\right) : \; \forall h < h_0 : \tag{8.131}$$
$$\left\|P^h E_h u^h - u^h\right\|_{\mathcal{U}_b^h} < \epsilon \|u^h\|_{\mathcal{U}_b^h}, \text{ hence, } \lim_{h\to 0} \|P^h E_h u^h - u^h\|_{\mathcal{U}_b^h} = 0.$$

(2) We derive several relations and inequalities. For an arbitrary $u \in \mathcal{U}_b$, and $v' := Cu \in \mathcal{V}_b'$, we determine the exact, and discrete solutions, $\hat{u}$, and $\hat{u}^h$, of

$$B\hat{u} = v' \text{ and } B^h \hat{u}^h = Q'^h Cu = Q'^h v' \in \mathcal{V}_b'^h.$$

By assumption they uniquely exist and the $\hat{u}^h$ converge to $\hat{u}$ in the sense that

$$\forall \epsilon > 0\; \exists h_0 = h_0(\|\hat{u}\|_{\mathcal{U}}) : \; \forall h < h_0 : \;\; \left\|P^h\hat{u} - \hat{u}^h\right\|_{\mathcal{U}^h} < \epsilon \left\|\hat{u}\right\|_{\mathcal{U}}. \tag{8.132}$$

With the operators $B^{-1}, (B^h)^{-1}Q'^h$, and the notation

$$T := B^{-1} \in \mathcal{L}\left(\mathcal{V}_b', \mathcal{U}_b\right), \text{ and}$$
$$T^h := (B^h)^{-1}Q'^h \in \mathcal{L}\left(\mathcal{V}_b' \cup \mathcal{V}_b'^h, \mathcal{U}_b^h\right), Q'^h|_{\mathcal{V}_b'^h} = Id,$$

we relate this result to the stability and consistency of B^h, as

$$B^h\hat{u}^h - B^hP^h\hat{u} = Q'^hCu - Q'^hB\hat{u} + Q'^hB\hat{u} - B^hP^h\hat{u} = Q'^hB\hat{u} - B^hP^h\hat{u} \Rightarrow$$
$$\|P^h\hat{u} - \hat{u}^h\|_{\mathcal{U}^h} \le \|(B^h)^{-1}\|_{\mathcal{U}_b^h \leftrightarrow \mathcal{V}_b'^h} \|Q'^hB\hat{u} - B^hP^h\hat{u}\|_{\mathcal{V}_b'^h} \to 0 \text{ for } h \to 0.$$

Applying $B\hat{u} = Cu$ and the equibounded $(B^h)^{-1}$ to the terms in the last norm, we find

$$\|(B^h)^{-1}Q'^hCu - P^hB^{-1}Cu\|_{\mathcal{U}^h} = \|(T^h - P^hT)Cu\|_{\mathcal{U}^h} \to 0$$
$$\text{for } h \to 0 \;\forall\; u \in \mathcal{U}_b.$$

Since C is compact, and $\|(T^h - P^hT)\|_{\mathcal{U}_b^h \leftarrow \mathcal{V}_b'}$ are equibounded, this implies

$$\left\|(T^h - P^hT)C\right\|_{\mathcal{U}_b^h \leftarrow \mathcal{U}_b} \to 0 \quad \text{for } h \to 0. \tag{8.133}$$

We combine $u^h \in \mathcal{U}_b^h$ with E_h. With the boundedly invertible A we estimate

$$\begin{aligned}\|u^h\|_{\mathcal{U}^h} &\le 2\left\|E_hu^h\right\|_{\mathcal{U}} \le 2\left\|A^{-1}\right\|_{\mathcal{U}_b \leftrightarrow \mathcal{V}'} \left\|AE_hu^h\right\|_{\mathcal{V}'} \\ &= 2\left\|A^{-1}\right\|_{\mathcal{U}_b \leftrightarrow \mathcal{V}'} \left\|B(I + TC)E_hu^h\right\|_{\mathcal{V}'} \\ &\le 2\left\|A^{-1}\right\|_{\mathcal{U}_b \leftrightarrow \mathcal{V}'} \|B\|_{\mathcal{V}' \leftrightarrow \mathcal{U}_b} \left\|(I + TC)E_hu^h\right\|_{\mathcal{U}},\end{aligned} \tag{8.134}$$

hence,

$$\left\|(I + TC)E_hu^h\right\|_{\mathcal{U}} \ge \left\|u^h\right\|_{\mathcal{U}^h} / (2\left\|A^{-1}\right\|_{\mathcal{U}_b \leftrightarrow \mathcal{V}'} \|B\|_{\mathcal{V}' \leftrightarrow \mathcal{U}_b}). \tag{8.135}$$

With the stability of B^h we get

$$\left\|(I+T^hC^h)u^h\right\|_{\mathcal{U}^h} = \left\|(B^h)^{-1}B^h(I+T^hC^h)u^h\right\|_{\mathcal{U}^h}$$
$$\le \left\|(B^h)^{-1}\right\|_{\mathcal{U}_b^h \leftarrow \mathcal{V}^{h'}} \left\|B^h(I+T^hC^h)u^h\right\|_{\mathcal{V}^{h'}}.$$

This implies

$$\left\|B^h(I+T^hC^h)u^h\right\|_{\mathcal{V}^{h'}} \ge \left\|(I+T^hC^h)u^h\right\|_{\mathcal{U}^h} \ / \ \left\|(B^h)^{-1}\right\|_{\mathcal{U}_b^h \leftarrow \mathcal{V}^{h'}}. \quad (8.136)$$

The linearity of difference methods implies

$$\Phi^h(A) = A^h = \Phi^h(B+C) = B^h + C^h \quad (8.137)$$
$$= B^h(I+(B^h)^{-1}C^h) = B^h(I+T^hC^h) : \mathcal{U}_b^h \to \mathcal{V}^{h'}.$$

(3) For the final stability estimates we return to (8.136) with $(B^h)^{-1}A^hu^h = (I+T^hC^h)u^h$, and use (8.137), cf. Remark 8.14. We extend $Q'^h : \mathcal{V}' \to \mathcal{V}^{h'}$ as $Q'^h : \mathcal{V}' \cup \mathcal{V}^{h'} \to \mathcal{V}^{h'}$ by defining for the new component $\mathcal{V}^{h'}$, the $Q'^h|_{\mathcal{V}^{h'}} = I_{\mathcal{V}^{h'}}$.

This is combined with $T^h|_{\mathcal{V}^{h'}} = (B^h)^{-1}Q'^h|_{\mathcal{V}^{h'}} = (B^h)^{-1}$, the equiboundedness of C^h and T^h, the consistency of C^h, the indicated error terms, and the triangle inequality estimating

$$\|(B^h)^{-1}\|_{\mathcal{U}_b^h \leftarrow \mathcal{V}_b^{h'}} ||A^hu^h||_{\mathcal{V}^{h'}} \quad (8.138)$$
$$= \|(B^h)^{-1}\|_{\mathcal{U}_b^h \leftarrow \mathcal{V}_b^{h'}} ||B^h(I+(B^h)^{-1}Q'^hC^h)u^h||_{\mathcal{V}^{h'}}$$
$$\ge \|(B^h)^{-1}\|_{\mathcal{U}_b^h \leftarrow \mathcal{V}_b^{h'}} ||B^h(I+T^hC^h)u^h||_{\mathcal{V}^{h'}}$$
$$\ge \Big(||P^h(I+TC)E_hu^h||_{\mathcal{U}} - ||u^h - P^hE_hu^h||_{\mathcal{U}^h} - ||T^hC^hu^h - P^hTCE_hu^h||_{\mathcal{U}^h}\Big)$$
$$\ge \Big(P^h||(I+TC)E_hu^h||_{\mathcal{U}} - ||u^h - P^hE_hu^h||_{\mathcal{U}^h}$$
$$- ||(T^h - P^hT)CE_hu^h||_{\mathcal{U}^h} - ||T^h(C^hu^h - CE_hu^h)||_{\mathcal{U}^h}\Big)$$
$$\ge \Big(||(I+TC)E_hu^h||_{\mathcal{U}}/2 - ||u^h - P^hE_hu^h||_{\mathcal{U}^h}$$
$$- ||(T^h - P^hT)CE_hu^h||_{\mathcal{U}^h} - ||T^h(C^hP^hE_hu^h - CE_hu^h)||_{\mathcal{U}^h}$$
$$- ||T^h(C^h - C^hP^hE_h)u^h||_{\mathcal{U}^h}\Big)$$
$$\ge \left(||\underbrace{(I+TC)E_hu^h}_{\text{by (8.135)}}||_{\mathcal{U}}/2 - ||\underbrace{u^h - P^hE_hu^h}_{\text{diff. err. (8.131)}}||_{\mathcal{U}^h} \right.$$
$$- 2||\underbrace{(T^h - P^hT)C}_{by\ (8.133)}||_{\mathcal{U}_b^h \leftarrow \mathcal{U}_b}||u^h||_{\mathcal{U}^h}$$
$$\left. - ||T^h\underbrace{(C^hP^hE_hu^h - CE_hu^h)}_{\text{class.consist. } C,(8.129)}||_{\mathcal{U}^h} - ||T^hC^h\underbrace{(u^h - P^hE_hu^h)}_{\text{diff. error (8.131)}}||_{\mathcal{U}^h} \right).$$

We choose h small enough, such that all the preceding errors in (8.131), (8.133), (8.129), (8.89), (8.91) are $< \epsilon ||u^h||_{\mathcal{U}^h}$ with

$$\epsilon < 1/\Big(8\left(2 + ||T^h||_{\mathcal{U}_b^h \hookrightarrow \mathcal{V}_b^{h'}} + ||T^h C^h||_{\mathcal{U}_b^h \hookrightarrow \mathcal{U}_b^h}\right) ||A^{-1}||_{\mathcal{U}_b \hookrightarrow \mathcal{V}'} ||B||_{\mathcal{V}' \hookrightarrow \mathcal{U}_b}\Big).$$

This implies

$$\begin{aligned} ||A^h u^h||_{\mathcal{V}^{h'}} &\geq \Big(||u^h||_{\mathcal{U}^h} / (8||A^{-1}||_{\mathcal{U}_b \hookrightarrow \mathcal{V}'} ||B||_{\mathcal{V}' \hookrightarrow \mathcal{U}_b})\Big) / ||B^{h^{-1}}||_{\mathcal{U}_b^h \hookrightarrow \mathcal{V}^{h'}} \qquad (8.139) \\ &\geq K' ||u^h||_{\mathcal{U}^h}, \end{aligned}$$

i.e. $(A^h)_{h \in H}$ is stable. □

For difference methods, Theorem 8.29, similarly to Theorem 5.78 for FEMs, yields a criterion for the stability of discretizations of operators, which are compact perturbations of coercive operators. An important class are $A \in \mathcal{L}(\mathcal{U}_b, \mathcal{U}_b')$ satisfying a so-called *Gårding inequality.* Hence Theorem 3.32 can directly be generalized to our difference methods.

We combine the consistency and coercivity/stability results in Theorems 8.21–8.23, 8.29 with Condition 8.26 for obtaining the following theorem. We only formulate the convergence results of orders μ for $u \in W^{m+\mu,p}(\Omega)$ and $\mu = 1, 2$. For the change from $H^m(\Omega)$ to $W^{m+\mu,p}(\Omega)$, compare the remarks at the end of Subsection 8.4.4. For convergence for $u \in W^{m,p}(\Omega)$ we refer directly to Theorems 8.21–8.23. We start listing the conditions for linear, quasilinear and fully nonlinear problems.

1. We use in Theorem 8.31 the following spaces, cf. (8.77), often we omit $\mathbb{R}^q$ and set $p = p' = 2$

$$\begin{aligned} &\mathcal{U}_b = \mathcal{V}_b = W_0^{m,p}(\Omega, \mathbb{R}^q),\ \mathcal{V}_b' = W_0^{-m,p'}(\Omega, \mathbb{R}^q), \quad \text{for } m, q \geq 1, 1 \leq p \leq \infty, \\ &\mathcal{U}_b^h = \mathcal{V}_b^h = W_{0,+}^{m,p}(\Omega^h, \mathbb{R}^q),\ \mathcal{V}_b'^h = \mathcal{U}_b'^h = W_{+,0}^{-m,p'}(\Omega^h, \mathbb{R}^q), \qquad (8.140) \\ &\text{and for strong equations } \mathcal{U}_s^h = W_+^{2m,p}(\Omega^h), \mathcal{U}_{s,b}^h = \mathcal{U}_s^h \cap W_{0,+}^{m,p}(\Omega^h), \\ &\mathcal{V}_s^h = \mathcal{V}_s'^h = L_+^2(\Omega^h). \end{aligned}$$

2. For $\mu = 1, 2$, and for linear problems we assume, cf. Remark 8.7,

$$a_{\alpha,\beta} \in C(\overline{\Omega}) \cap W^{\mu,\infty}(\Omega), u \in W^{m+\mu,p}(\Omega), A : W^{m+\mu,p}(\Omega) \to W^{-m+\mu,p'}. \qquad (8.141)$$

3. For quasilinear problems we assume, with N_m in (8.17),

$$\begin{aligned} &A_j, A_\alpha \text{ equicontinuous in } \Omega \times N_{\nabla^m} \subset \Omega \times \mathbb{R}^{N_m},\ N_{\nabla^m} \text{ bounded, s.t.} \qquad (8.142) \\ &\forall u : \|u - u_0\|_{\mathcal{V}} \leq \rho, \mathcal{V}, \nabla^{\leq m} u(x) \in N_{\nabla^m} \forall x \in \Omega, \nabla_{\mp}^{\leq m} u(x) \in N_{\nabla^m} \forall x \in \Omega_0^h, \end{aligned}$$

with Lipschitz-continuous A_j, A_α with respect to the variables in N_{∇^m}. Finally let for

$$u \in W^{m+\mu,p}(\Omega), A_\alpha : \mathcal{D}(A_\alpha) \cap W^{m+\mu,p}(\Omega) \to W^{-m+\mu,p'}(\Omega)) \qquad (8.143)$$

4. For fully nonlinear problems, here only formulated for order $2m = 2$, let

$$G \text{ be equicontinuous in } \Omega \times N_{\nabla^2} \subset \Omega \times \mathbb{R}^{N_2} \text{ with bounded } N_{\nabla^2}, \text{ s.t.} \quad (8.144)$$
$$\forall u : \|u - u_0\|_{\mathcal{V}} \leq \rho, \mathcal{V}, \nabla^{\leq 2} u(x) \in N_{\nabla^2} \forall x \in \Omega, \nabla_{\mp}^{\leq 2} u(x) \in N_{\nabla^2} \forall x \in \Omega_0^h,$$

with Lipschitz-continuous G with respect to the variables in N_{∇^2}. Furthermore require

$$\text{for } u \in H^{2+\mu}(\Omega), G : \mathcal{D}(G) \subset H^{2+\mu}(\Omega) \to H^{\mu}(\Omega)). \quad (8.145)$$

We obtain consitency of order 1 and 2, respectively.

Theorem 8.31. Convergence for symmetric and unsymmetric finite difference methods: *Let $\mathcal{U}_b = \mathcal{V}_b \subset \mathcal{V}$ and $\mathcal{U}_b^h = \mathcal{V}_b^h$ be the spaces and their duals in (8.140), cf. (8.27)–(8.31), (8.77), and $P^h = I^h$ and Q'^h the projectors in Subsection 8.5.2. For the original linear and nonlinear elliptic problems, $Au = f \in W^{-m,p'}(\Omega, \mathbb{R}^q)$ and $G(u) = 0 \in W^{-m,p'}(\Omega, \mathbb{R}^q)$ define the difference methods as listed in Section 8.4, and impose, for the different cases, the conditions (8.141)–(8.145). Finally, let A or $G'(u_0) : W^{m,p}(\Omega, \mathbb{R}^q) \to W^{-m,p'}(\Omega, \mathbb{R}^q)$ be boundedly invertible.*

1. *Then under the conditions in Chapter 2, a (locally) unique exact solution u_0 and for small enough h, a unique discrete solution u_0^h for these problems exist.*
2. *Stability, and hence convergence, follows for $2 \leq p < \infty$ from the $H_{0,+}^m(\Omega^h)$ coercivity of the principal part $a_p^h(.,.)$ in the $\|\cdot\|_{H_+^m}(\Omega^h)$ norm.*
3. *If for $\mathcal{U}_0^h = \mathcal{V}_b^h = \mathcal{W}_+^{m,p}(\Omega^h)$ with $2 \leq p < \infty$ an inf-sup condition for $a_p^h(.,.)$ is valid, the u_0^h converge in $\|\cdot\|_{W_+^{m,p}(\Omega^h)}$.*
4. *For strong difference equations Theorem 8.21 can be updated. More precisely, we obtain convergence of order μ, with $\mu = 1$, and 2, for unsymmetric and symmetric difference methods, cf. (8.105), (8.106), (8.103), for the linear and quasilinear problems. Symmetric methods, with $\mu = 2$, are worthwhile in cuboidal domains, cf. (8.95).*

$$\|P^h u_0 - u_0^h\|_{\mathcal{U}^h} \leq Ch^{\mu}\|u_0\|_{W^{m+\mu,p}(\Omega)} \text{ for } \mu = 1, 2, \quad u_0 \in W^{m+\mu,p}(\Omega), \quad (8.146)$$

with $p = 2$ and $p > 2$ for the two previous cases in 2. and 3., respectively. For fully nonlinear second order problems we find

$$\|P^h u_0 - u_0^h\|_{\mathcal{U}^h} \leq Ch^{\mu}\|u_0\|_{H^{2+\mu}(\Omega)} \quad \text{for } u_0 \in H^{2+\mu}(\Omega). \quad (8.147)$$

Theorem 8.32. Difference methods for monotone operators, eigenvalue problems, nonlinear boundary operators, quadrature approximations: *Under the conditions of Theorems 8.29 and 8.31, difference methods applied to the following problems yield unique converging solutions. For monotone operators the convergence satisfies (4.196)–(4.198). For the other problems (8.146) and (8.147) are valid: variational methods for eigenvalue problems, cf. Section 4.7, for nonlinear boundary operators, cf. Section 5.3, quadrature approximations, cf. Section 5.4.*

Solving nonlinear equations in difference methods

As for FEMs and DCGMs we use a discrete Newton's method based upon the mesh independence principle (MIP). For the MIP we have to determine the linearized form of the nonlinear system and its discretization, cf. Subsections 8.4.6, 8.4.7. The derivatives have to be *consistently differentiable* and *of order* μ, hence, additionally to the standard consistency,

$$\|(G^h)'(P^h u)P^h v - Q'^h(G'(u)v)\|_{\mathcal{V}'^h} \to 0 \text{ for small } \|u - u_0\|_{\mathcal{V}} \text{ and, } \mu = 1, 2,$$

$$\|(G^h)'(P^h u)P^h v - Q'^h(G'(u)v)\|_{\mathcal{V}'^h} \le Ch^\mu(1 + \|u\|_{\mathcal{U}_s})\|v\|_{\mathcal{U}_s} \text{ for } h \to 0. \quad (8.148)$$

These $\|v\|_{\mathcal{U}_s}$ are usually $\|v\|_{W^{m+\mu,p}(\Omega)}$ and the above m, μ, p. The proof of this property follows essentially the lines of FEMs and DCGMs, so we do not repeat it here, cf. Theorems 5.21, 7.46, 7.47. We transform these results to difference equations:

Theorem 8.33. Discrete Newton method for difference equations G^h: *Let $u_0 \in H^s(\Omega)$ be an isolated solution of $G(u_0) = 0$, such that $G'(u_0) : \mathcal{U} \to \mathcal{V}'$ is boundedly invertible and let $G^h\left(u_0^h\right) = 0$ indicate one of the difference equations for the nonlinear problems in Subsections 8.4.6, 8.4.7. Assume G and $G'(u)$ to be Lipschitz-continuous with respect to the variables u, for $\|u - u_0\|_{\mathcal{V}}$ small enough. Let the conditions in Theorems 8.21–8.23 be satisfied for G and G'. Then the corresponding difference equation G^h for G is consistently differentiable.*

Start the original Newton process with $u_1 \in W^{m+\mu,p}(\Omega), \mu = 1, 2,$ and the discrete Newton process for $i = 1, \ldots,$ with $u_1^h := P^h u_1$, for small enough $\|u_0 - u_1\|_{\mathcal{U}}$ and h. Then in the original Newton method and in

$$u_{i+1}^h := u_i^h - \left(G^h\left(u_i^h\right)'\right)^{-1} G^h\left(u_i^h\right). \quad (8.149)$$

the u_{i+1} and the u_{i+1}^h in (3.84) uniquely exist and converge quadratically to u_0 and to u_0^h and u_i^h to u_{i+1} of order μ, such that

$$\left\|u_{i+1}^h - u_i^h\right\|_{\mathcal{U}^h} \le C\left\|u_i^h - u_{i-1}^h\right\|_{\mathcal{U}^h}^2 \text{ and}$$

$$\left\|u_i^h - P^h u_i\right\|_{\mathcal{U}^h} \le Ch^\mu\|u_i\|_{\mathcal{U}_s}, \ i = 1, \ldots.$$

8.6 Natural boundary value problems of order 2

We turn to *natural boundary, e.g. Neumann conditions,* instead of the previous Dirichlet case. We only consider second order equations, and their symmetric difference methods on cuboidal domains, cf. (8.95). Generalizations to order $2m$ are possible, but require many technicalities, cf. the discussion in Subsection 2.4.2 (2.92) ff. and Subsection 2.4.4, Lions and Magenes [478], p. 120, Oden and Reddy [518], p. 290 ff. Other boundary conditions are discussed, e.g. by Collatz [206], Forsythe and Wasow [319] and Hackbusch [387].

Standard natural boundary conditions as in (8.152) for linear problems (8.151) are generalized to quasilinear problems in (8.160). Mixed boundary value problems can be included by generalized test functions. As in the previous sections, there is nearly as close a relation between the weak and strong differential equations as in (8.151)–(8.152) for natural boundary conditions, cf. Proposition 8.35. This allows us to prove unique existence of discrete solutions, u_0^h, and their convergence of order 1 very similarly to the previous sections. For the proof of the full order 2 of convergence of this difference method, the technique in Section 5.2 for fully nonlinear elliptic problems with FEMs would have to be employed. We omit that here.

8.6.1 Analysis for natural boundary value problems

As mentioned already, we only study second order elliptic problems in

$$\mathcal{U} := \mathcal{V} := H^1(\Omega), \mathcal{U}' := H^{-1}(\Omega), \text{ with } \mathcal{U}_b := \mathcal{V}_b := H_0^1(\Omega), \mathcal{U}_b' := H_0^{-1}(\Omega). \quad (8.150)$$

We give a short summary of the analytic results for natural boundary value problems of second order, cf. Chapter 2. They again allow the transformation from strong to weak forms, cf. (2.24), and Hackbusch [387], Section 7.4.

The A_s, A and $a(\cdot,\cdot)$ are maintained and the boundary conditions in (8.43)–(8.45) are modified. The conditions for the coefficients of A are formulated in Summary 8.6, and, for difference methods, in Remark 8.7. We solve $Au_0 = f \in \mathcal{U}'$, $u_0 \in \mathcal{U}$ (or for $u_0 \in H^2(\Omega)$: $A_s u_0 = g \in L^2(\Omega), B_a u_0|_{\partial\Omega} \in L^2(\partial\Omega)$)

$$\begin{aligned} a(u_0, v) &= \sum_{i,j=0}^{n} \int_\Omega a_{ij}\partial^i u_0 \partial^j v dx = \langle Au_0, v\rangle_{\mathcal{U}'\times\mathcal{U}} = \langle f, v\rangle_{\mathcal{U}'\times\mathcal{U}} \quad (8.151)\\ &= (A_s u_0, v)_{L^2(\Omega)} + \int_{\partial\Omega} (B_a\ u_0) v ds \end{aligned}$$

with the *natural boundary operator*

$$B_a\ u := \sum_{i=0,\dots,n}^{j=1,\dots,n} \nu^j a_{ij}\partial^i u \text{ cf. (2.25).} \quad (8.152)$$

We do not consider the general $f \in \mathcal{U}'$, e.g. in (8.71), but restrict it in (8.151), (8.156), to the form

$$\begin{aligned} f(v) &= \langle f, v\rangle_{\mathcal{U}'\times\mathcal{U}} = \int_\Omega \hat{f} v dx + \int_{\partial\Omega} \varphi v ds, \forall v \in \mathcal{U} \quad \text{with} \quad (8.153)\\ \hat{f} &\in L^2(\Omega), \varphi \in H^{-1/2}(\partial\Omega), \quad \|f\|_{\mathcal{U}'} = \left(\|\hat{f}\|^2_{\mathcal{U}_b'} + \|\varphi\|^2_{H^{-1/2}(\partial\Omega)}\right)^{1/2}. \end{aligned}$$

For $c_0 > 0$, the extended principal part of $a(u, v)$, in (8.151),

$$a_{p,e}(u, v) := a_p(u, v) + c_0(u, v)_{L^2(\Omega)} : \mathcal{U}\times\mathcal{U} \to \mathbb{R}, \quad c_0 > 0, \quad (8.154)$$

is $\mathcal{U}$-coercive, cf. Theorems 2.43, 2.89, Corollary 2.44.

Let A, and $A_{p,e} : \mathcal{U} \to \mathcal{U}'$, be the bounded linear operators induced by $a(u, v)$, and $a_{p,e}(u, v)$. Hence Theorem 2.12 implies a unique solution for $f \in \mathcal{U}'$,

$$\forall f \in \mathcal{U}' : \exists_1 u_1 \in \mathcal{U} : \ a_{p,e}(u_1, v) = \langle f, v\rangle_{\mathcal{U}'\times\mathcal{U}} \forall v \in \mathcal{U}, \quad (8.155)$$

and for

$$A^{-1} \in \mathcal{L}(H^{-1}(\Omega), \mathcal{U}) \Rightarrow \exists_1 u_0 \in \mathcal{U} : \ a(u_0, v) = \langle f, v \rangle_{\mathcal{U}' \times \mathcal{U}} \forall v \in \mathcal{U}. \tag{8.156}$$

$A_{p,e}$ is coercive and its difference form $A^h_{p,e}$ will turn out to be consistent with $A_{p,e}$. So Theorem 3.29 implies the stability of a boundedly invertible A, as a compact perturbation of $A_{p,e}$.

The most important special case, showing some difficulties, is the Laplacian

$$A_s u = -\Delta u, A u_0 = \hat{f}, a(u,v) = \sum_{i=1}^{n} \int_\Omega \partial^i u \partial^i v = (A_s u, v)_{L^2(\Omega)} + \int_{\partial\Omega} v \partial u / \partial \nu ds$$

$$\textit{Neumann boundary condition } B_a u = \partial u/\partial \nu = \varphi, \hat{f} \in L^2(\Omega), \ \varphi \in L^2(\partial\Omega). \tag{8.157}$$

Obviously, for any constant $c \in \mathbb{R}$ we find $-\Delta c = 0$, $\partial c/\partial \nu = 0$, so (8.157) is not uniquely solvable. Hence the kernel, $\ker F'(u_0) \cong \mathbb{R}$ of the operator $Fu := (-\Delta u, \partial u/\partial \nu)$ is not empty. By the Fredholm alternative, cf. Theorem 2.21, a unique solution exists only for the modified problem

$$\exists_1 u_0 \perp \ker F'(u_0) \Leftrightarrow f \perp \ker F'(u_0) \Leftrightarrow \langle f, 1 \rangle_{\mathcal{U}' \times \mathcal{U}} = \int_\Omega \hat{f}_0 dx + \int_{\partial\Omega} \varphi ds = 0.$$

This is a special form of conditions relevant in bifurcation. A direct approach is discussed in [387]. We will study these problems in [120]. The *numerical Liapunov–Schmidt method* systematically proves the convergence for these cases. For avoiding these difficulties, we assume here a *boundedly invertible linear operator, A.*

Then we solve (8.156), and summarize consequences for the differential equation and the boundary condition, cf. Remark 2.6. We will partially prove again the results in Theorem 2.43 and Corollary 2.44. The transition in (8.159) from the weak $Au_0 - f = 0$ to the strong form is only possible for special $f \in H^{-1}(\Omega)$, cf. (8.153), so we assume a Lipschitz-continuous $\partial\Omega$, and

$$Au_0 - f = 0 \Leftrightarrow A_s u_0 = \hat{f} = \hat{f}_0 \in L^2(\Omega), B_a u_0 = \varphi \in H^{-1/2}(\partial\Omega). \tag{8.158}$$

Theorem 8.34. Analytical results for natural boundary conditions:

1. *The $a_{p,e}(\cdot,\cdot)$ in (8.154) is $\mathcal{V}$-coercive, the original in (8.156) is $\mathcal{V}$-elliptic, and satisfies the Fredholm alternative. Regularity results are available in Theorems 2.45 and 2.47. The latter yields $u_0 \in H^2(\Omega)$ for convex Ω and coefficients satisfying (2.149), (2.47), and $\hat{f} \in L^2(\Omega), \varphi \in H^{1/2}(\partial\Omega)$ in (8.158).*
2. *The solution, $u_0 \in H^1(\Omega)$, of (8.156), (8.158), with f in (8.153), satisfies*

$$Au_0 = \hat{f} \in L^2(\Omega), B_a u_0 = \varphi \in H^{-1/2}(\partial\Omega) \textit{ or, cf (8.158), for smooth data}$$

$$u_0 \in H^2(\Omega), A_s u_0 = \hat{f} \in L^2(\Omega), B_a u_0 = \varphi \in H^{1/2}(\partial\Omega). \tag{8.159}$$

3. *An analogous result holds for the quasilinear problems in (8.160), (8.161).*

Proof. We start testing (8.156) with $\forall v \in \mathcal{U}_b$. This eliminates the boundary term, $\int_{\partial\Omega} \varphi v ds = \langle \varphi, v \rangle_{H^{-1/2}(\partial\Omega) \times H^{1/2}(\partial\Omega)} = 0$, by $0 = v|_{\partial\Omega} \in H^{1/2}(\partial\Omega)$, and yields $a(u_0, v) = (\hat{f}, v)_{L^2(\Omega)} \forall v \in \mathcal{U}_b$, hence $Au_0 - \hat{f} \in L^2(\Omega)$, by (8.156), (8.151), and since

$L^2(\Omega)$ is dense in $\mathcal{V}$. We even get $u_0 \in H^2(\Omega)$, for $\hat{f} \in L^2(\Omega)$, $\varphi \in H^{1/2}(\partial\Omega)$, smooth enough $\partial\Omega$ and coefficients, or convex $\partial\Omega$, hence $A_s u_0 = \hat{f}$ cf. Theorem 2.45. Next, testing (8.156) with general $\forall v \in \mathcal{U}$ implies $\int_{\partial\Omega}(B_a\, u - \varphi)v ds = 0\ \forall v|_{\partial\Omega} \in H^{1/2}(\partial\Omega)$. □

For the general quasilinear systems of order 2 we obtain

$$\begin{aligned} G_s(\vec{u}) &:= \sum_{k=0}^{n} \left((-1)_{k>0}\partial^k A_k(x, \vec{u}, \nabla\vec{u})\right) \qquad (8.160)\\ \forall \vec{u} \in \mathcal{V} &= H^2(\Omega, \mathbb{R}^q), \vec{v} \in H^1(\Omega, \mathbb{R}^q)\\ \langle Gu, v\rangle_{\mathcal{V}'\times\mathcal{V}} &= a(\vec{u}, \vec{v}) = \int \sum_{k=0}^{n} (A_k(x, \vec{u}, \nabla\vec{u}), \partial^k v)_q dx\\ &\quad - \int_{\partial\Omega} \left(B_A(\vec{u}, \nabla\vec{u}), \vec{v}\right)_q ds,\\ \text{where } B_A(\vec{u}, \nabla\vec{u}) &:= \sum_{k=1}^{n} \nu^k A_k(\cdot, \vec{u}, \nabla\vec{u}), \end{aligned}$$

For these general quasilinear systems of order 2 we obtain, with the precise conditions for the A_k discussed in Chapter 2, for difference equations in Remark 8.7, (8.111), and with $A_k(\cdot, \vec{u}, \nabla\vec{u})(x) \in \mathbb{R}^q$, $B_A(\vec{u}, \nabla\vec{u})(x) \in \mathbb{R}^q$:

$$G_s(\vec{u}) := \sum_{k=0}^{n}(-1)_{k>0}\partial^k A_k(x, \vec{u}, \nabla\vec{u}) \in L^2(\Omega, \mathbb{R}^q),\ \ \forall \vec{u} \in H^2(\Omega, \mathbb{R}^q), \vec{v} \in H^1(\Omega, \mathbb{R}^q)$$

$$\begin{aligned} (G_s(\vec{u}), \vec{v})_{L^2(\Omega,\mathbb{R}^q)} + \sum_{e\in\partial\Omega} \int_e \left(B_A(\vec{u}, \nabla\vec{u}), \vec{v}\right)_q ds &= \int_\Omega \sum_{k=0}^{n} (A_k(x, \vec{u}, \nabla\vec{u}), \partial^k v)_q dx\\ &= \langle G(\vec{u}), \vec{v}\rangle_{H^{-1}(\Omega,\mathbb{R}^q)\times H^1(\Omega,\mathbb{R}^q)},\\ &\quad \text{and } \vec{u} \in H^1(\Omega, \mathbb{R}^q). \qquad (8.161) \end{aligned}$$

8.6.2 Difference methods for natural boundary value problems

For these natural boundary value problems we are going to define the corresponding difference methods. As in Section 8.4 we could formulate the first order unsymmetric forms for domains Ω in (8.98). This would yield similar equations as for Dirichlet conditions, however with a modified discrete natural boundary operator B^h_a. Or we assume (8.95) and restrict the presentation to *symmetric difference approximations*, cf. Hackbusch [387], Subsections 4.7.2, 4.7.3 and Section 7.4. We only present this second choice in two different variants. We start by extending the original Ω^h by a layer of exterior points. Another possibility, a shifted grid, will only be indicated.

As in Subsection 8.4.3, we have to avoid the $\partial^j_h u^h$, and instead use arithmetic means, $\left(\partial^j_- + \partial^j_+\right) u^h/2$. This complicates the presentation. It requires modifying the previous

Ω^h in (8.96), (8.97). For

$$\Omega := \prod_{i=1}^{n} (a_i, b_i), \quad \Omega^h := \overline{\Omega} \cap \mathcal{G}^h, \tag{8.162}$$

we add to Ω^h left and right outer boundary grid points, $x \in \partial\Omega_l^h$ and $x \in \partial\Omega_r^h$ both $\subset \mathcal{G}^h$ in the distance h from Ω^h, more precisely

$$\begin{aligned}
\partial\Omega_l^h &:= \{x = (x_1, \ldots, x_n) \in \mathcal{G}^h : \exists j : x_j = a_j - h, \forall i \neq j : a_i \le x_i \le b_i\}, \quad (8.163)\\
\partial\Omega_r^h &:= \{x = (x_1, \ldots, x_n) \in \mathcal{G}^h : \exists j : x_j = b_j - h, \forall i \neq j : a_i \le x_i \le b_i\} \text{ and}\\
\Omega_r^h &:= \Omega^h \cup \partial\Omega_r^h, \quad \Omega_l^h := \Omega^h \cup \partial\Omega_l^h, \Omega_e^h := \Omega_r^h \cup \partial\Omega_l^h, \partial\Omega_e^h := \partial\Omega_r^h \cup \partial\Omega_l^h.
\end{aligned}$$

The results for unsymmetric difference equations are obtained essentially by choosing one of the unsymmetric versions used in the following formulas for the averaging. Straightforward extensions to systems with $\vec{u}(x) \in \mathbb{R}^q$ are omitted here.

In this section the outer normal vectors ν in boundary grid points, $x \in \partial\Omega^h$, play a crucial role. Sometimes even the number, $\#\nu(x) \ge 1$, of these unit normals in x, or their parerallel shifts to $\partial\Omega_e^h$, are relevant:

$$\text{for } x \in \partial\Omega^h, x \in \text{interior of a face of } \partial\Omega : \#\nu(x) = 1, \ \text{otherwise } \#\nu(x) > 1. \tag{8.164}$$

In particular, we find for $n = 2$ two normals in the vertices, and for $n = 3$ two normals in the edges, and three in the vertices. For an interior point x of a face of $\partial\Omega$, hence $\#\nu(x) = 1$, this $\nu = \pm e^j$ for one j. For an x in an edge or a vertex x, we need $\#\nu(x)$ unit vectors e^j.

For every $j = 1, \ldots, n$, there are two boundary points $x \in \partial\Omega_e^j \subset \partial\Omega^h$, cf. (8.176), with the outer normal $\nu = -e^j$ in $\partial\Omega_l^j$ and $\nu = +e^j$ in $\partial\Omega_r^j$, $\nu^j = \mp 1$, cf. (8.163). Consequently, we use in the following formulas the notation $\pm \left(a_{\ell,j}\partial_\pm^\ell u^h\right)(\cdot \mp he^j) = \nu^j \left(a_{\ell,j}\partial_\pm^\ell u^h\right)(\cdot \mp h\nu)$.

Accordingly, we modify, for our symmetric forms, the discrete functions, u^h, the *Sobolev spaces*, the *inner product, seminorms, norms.* This yields, with the $\int_{\Omega^h}$, $\int_{\Omega_l^h}$, $\int_{\Omega_r^h}$, cf. (8.25), (8.163), and a new boundary integral, e.g.

$$u^h, v^h : \Omega_e^h \to \mathbb{R}, \int_{\Omega^h} u^h dx^h := h^n \sum_{x \in \Omega^h} u^h(x), \int_{\partial\Omega_e^h} u^h ds^h := h^{n-1} \sum_{x \in \partial\Omega_e^h} u^h(x),$$

$$(u^h, v^h)_{H_\pm^1(\Omega_e^h)} := \frac{1}{2}\Big(\int_{\Omega_l^h} \sum_{j}^{n} \partial_+^j u^h \partial_+^j v^h dx^h + \int_{\Omega_r^h} \sum_{j=0}^{n} \partial_-^j u_0^h \partial_-^j v^h dx^h\Big), \tag{8.165}$$

and

$$\mathcal{V}^h = H_\pm^1\left(\Omega_e^h\right) := \left\{u^h : \Omega_e^h \to \mathbb{R}, \|u^h\|_{H_\pm^1(\Omega_e^h)} < \infty\right\}. \tag{8.166}$$

The relation between the weak and the strong linear difference operators is based upon the *natural discrete boundary operator* $B_a^h u^h$ and $\int_{\partial\Omega_e^h} v^h B_a^h u^h ds^h$, cf. (8.174). Accordingly, we modify the previous A_s^h, A^h, $a^h(\cdot,\cdot)$, $\hat{f}^h(\cdot)$, into the form for natural boundary conditions motivated by (8.151)–(8.152), (8.153), and Proposition 8.35. We maintain the notation $A^h, A^h_s, \dots$, instead of $A^h_{ws}, A^h_{ss}, \dots$, for the weak and strong form of *symmetric difference equations* in Subsection 8.4.3:

$$A^h_s : H^2_\pm\left(\Omega^h_e\right) \to L^2(\Omega^h), A^h : \mathcal{V}^h := H^1_\pm\left(\Omega^h_e\right) \to \mathcal{V}'^h, B^h_a : H^1_\pm\left(\partial\Omega^h_e\right) \to L^2\left(\partial\Omega^h_e\right),$$

$$f^h : \mathcal{V}'^h \to \mathbb{R}, u^h, u^h_0, v^h \in \mathcal{V}^h, \text{ cf. (8.166):}$$

$$A^h_s u^h(x) := \frac{1}{2}\sum_{j,\ell=0}^{n} (-1)_{j>0}\Big(\partial^j_-\left(a_{\ell,j}\partial^\ell_+ u^h\right) + \partial^j_+\left(a_{\ell j}\partial^\ell_- u^h\right)\Big)(x) \tag{8.167}$$

$$a^h(u^h, v^h) := \langle A^h u^h, v^h\rangle_{\mathcal{V}'^h\times\mathcal{V}^h} \tag{8.168}$$

$$:= \frac{1}{2}\left(\int_{\Omega^h_l}\sum_{j,\ell=0}^{n} a_{\ell,j}\partial^\ell_+ u^h_0 \partial^j_+ v^h dx^h + \int_{\Omega^h_r}\sum_{j,\ell=0}^{n} a_{\ell j}\partial^\ell_- u^h_0 \partial^j_- v^h dx^h\right),$$

$$B^h_a u^h|_{\partial\Omega^h_l} := B^h_{a,l} u^h := \frac{1}{2}\left(\sum_{\ell=0,\dots,n}^{j=1\dots,n} \nu^j\Big(a_{\ell,j}\partial^\ell_+ u^h + \left(a_{\ell j}\partial^\ell_- u^h\right)(\cdot + he^j)\Big)\right), \text{ in } x \in \partial\Omega^h_l$$

$$B^h_a u^h|_{\partial\Omega^h_r} := B^h_{a,r} u^h := \frac{1}{2}\left(\sum_{\ell=0,\dots,n}^{j=1\dots,n} \nu^j\Big((a_{\ell,j}\partial^\ell_+ u^h)(\cdot - he^j) + a_{\ell j}\partial^\ell_- u^h\Big)\right) \text{ in } x \in \partial\Omega^h_r,$$

$$\int_{\partial\Omega^h_e} v^h B^h_a u^h ds^h := \int_{\partial\Omega^h_l} v^h B^h_{a,l} u^h ds^h + \int_{\partial\Omega^h_r} v^h B^h_{a,r} u^h ds^h, \text{ and} \tag{8.169}$$

$$\langle f^h, v^h\rangle^h := \int_{\Omega^h} \hat{f}^h v^h dx^h + \int_{\partial\Omega^h_e} v^h \varphi^h(\cdot - h\nu/2) ds^h, \text{cf.(8.153)}. \tag{8.170}$$

Quasilinear difference systems satisfy, with $\nabla^h_\pm = \left(\partial^1_\pm, \dots, \partial^n_\pm\right)$ and $\mathcal{V}^h = H^1_\pm(\Omega^h, \mathbb{R}^q)$,

$$\langle G^h\vec{u}^h, \vec{v}^h\rangle_{\mathcal{V}'^h\times\mathcal{V}^h} = a^h(\vec{u}^h, \vec{v}^h) \tag{8.171}$$

$$:= \frac{1}{2}\left(\int_{\Omega^h_l}\sum_{j=0}^{n} A_j\left(\cdot, \vec{u}^h, \nabla^h_+\vec{u}^h\right)\partial^j_+ v^h dx^h + \int_{\Omega^h_r}\sum_{j=0}^{n} A_j\left(\cdot, \vec{u}^h, \nabla^h_-\vec{u}^h\right)\partial^j_- v^h dx^h\right)$$

$$= \int_{\Omega^h}\frac{1}{2}\sum_{j=0}^{n}\Big(\vec{v}^h, (-1)_{j>0}\Big(\partial^j_- A_j\left(\cdot, \vec{u}^h, \nabla^h_+\vec{u}^h\right) + \partial^j_+ A_j\left(\cdot, \vec{u}^h, \nabla^h_-\vec{u}^h\right)\Big)\Big)_q dx^h \tag{8.172}$$

$$+ \left(\int_{\partial\Omega^h_l} v^h B^h_{a,l} u^h ds^h + \int_{\partial\Omega^h_r} v^h B^h_{a,r} u^h ds^h\right)$$

$$=: \left(G^h_s\vec{u}^h, \vec{v}^h\right)_{L^2(\Omega^h)} + \int_{\partial\Omega^h_e} v^h B^h_a u^h ds^h, \text{ with} \tag{8.173}$$

$$B_a^h u^h|_{\partial\Omega_l^h} := B_{a,l}^h u^h := \frac{1}{2}\left(\sum_{j=1}^n \nu^j \Big(A_j\left(\cdot, \vec{u}^h, \nabla_+^h \vec{u}^h\right) + A_j\left(\cdot, \vec{u}^h, \nabla_-^h \vec{u}^h\right)(\cdot + he^j)\Big)\right),$$

$$B_a^h u^h|_{\partial\Omega_r^h} := B_{a,r}^h u^h := \frac{1}{2}\left(\sum_{j=1}^n \nu^j \Big(A_j\left(\cdot, \vec{u}^h, \nabla_+^h \vec{u}^h\right)(\cdot - he^j) + A_j\left(\cdot, \vec{u}^h, \nabla_-^h \vec{u}^h\right)\Big)\right)$$

Proposition 8.35. *For natural boundary conditions, the linear symmetric weak and strong difference operators, $A^h u^h, A_s^h u^h$, $a^h(u^h, v^h)$, and the discrete boundary operators, $B_a^h u^h$, in (8.167)–(8.169), are related by*

$$a^h(u^h, v^h) = \langle A^h u^h, v^h\rangle_{\mathcal{V}'^h \times \mathcal{V}^h} = \left(A_s^h u^h, v^h\right)_{L^2(\Omega^h)} + \int_{\partial\Omega_e^h} v^h B_a^h u^h ds^h. \quad (8.174)$$

Quasilinear difference systems satisfy, with $\nabla_\pm^h = \left(\partial_\pm^1, \ldots, \partial_\pm^n\right)$ and $\mathcal{V}^h = H_\pm^1(\Omega^h, \mathbb{R}^q)$, cf. (8.171)–(8.173)

$$\langle G^h \vec{u}^h, \vec{v}^h\rangle_{\mathcal{V}'^h \times \mathcal{V}^h} = a^h(\vec{u}^h, \vec{v}^h) = \left(G_s^h \vec{u}^h, \vec{v}^h\right)_{L^2(\Omega^h)} + \int_{\partial\Omega_e^h} v^h B_a^h u^h ds^h. \quad (8.175)$$

Proof. Again, cf. (8.53), we start with the first order result for $j > 0$ and the simplified notation for grid points and function values in Ω^h in the e_j direction:

$$\begin{aligned}
\Omega^j &:= \{x_i := x_0 + ihe_j, i = 0, \ldots, k\}, &(8.176)\\
\Omega_r^j &:= \{x_i := x_0 + ihe_j, i = 0, \ldots, k+1\},\\
\Omega_l^j &:= \{x_i := x_0 + ihe_j, i = -1, \ldots, k\}, \subset \mathcal{G}^h \text{ cf. (8.163)}.
\end{aligned}$$

Then we obtain for $j > 0$:

$$\begin{aligned}
-\sum_{\Omega^j} v^h \partial_-^j u^h &= -\left(\sum_{i=0}^k v_i u_i - \sum_{i'=-1}^{k-1} v_{i'+1} u_{i'}\right)/h\\
&= -\left(\sum_{i=0}^k (v_i u_i - v_{i+1} u_i) - v_0 u_{-1} + v_{k+1} u_k\right)/h\\
&= \sum_{i=0}^k u_i \partial_+^j v_i + (v_0 u_{-1} - v_{-1} u_{-1} + v_{-1} u_{-1} - v_{k+1} u_k)/h\\
&= \sum_{\Omega_l^j} u^h \partial_+^j v^h + (v_{-1} u_{-1} - v_{k+1} u_k)/h\\
&= -\sum_{\Omega^j} v^h \partial_-^j u^h &(8.177)
\end{aligned}$$

and

$$-\sum_{\Omega^j} v^h \partial_+^j u^h = \sum_{i=0}^{k} u_i \partial_-^j v_i + \big(v_{-1}u_0 - v_k u_{k+1} + v_{k+1}u_{k+1} - v_{k+1}u_{k+1}\big)/h$$

$$= \sum_{\Omega_r^j} u^h \partial_-^j v^h + \big(v_{-1}u_0 - v_{k+1}u_{k+1}\big)/h = -\sum_{\Omega^j} v^h \partial_+^j u^h. \tag{8.178}$$

With $\partial_\pm^0 u^h = u^h$, we insert for u^h in (8.177) the $w^h := \big(\sum_{\ell=0}^n a_{\ell,j}\partial_+^\ell u^h\big)$, and in (8.178) the $w^h := \big(\sum_{\ell=0}^n a_{\ell j}\partial_-^\ell u^h\big)$, multiply with h^n, and sum for $j \geq 0$:

$$\int_{\Omega^h} v^h \sum_{j,\ell=0}^{n} (-1)_{j>0} \partial_-^j \big(a_{\ell,j}\partial_+^\ell u^h\big) dx^h = \int_{\Omega_l^h} \sum_{j,\ell=0}^{n} a_{\ell,j}\partial_+^\ell u^h \partial_+^j v^h dx^h \tag{8.179}$$

$$+ \int_{\partial\Omega_l^h} v^h \left(\sum_{\ell=0,\dots,n}^{j=1\dots,n} a_{\ell,j}\partial_+^\ell u^h \right) ds^h$$

$$- \int_{\partial\Omega_r^h} v^h \left(\sum_{\ell=0,\dots,n}^{j=1\dots,n} a_{\ell,j}\partial_+^\ell u^h \right) (\cdot - he^j) ds^h$$

and

$$\sum_{\Omega^h} v^h \sum_{j,\ell=0}^{n} (-1)_{j>0} \partial_+^j \big(a_{\ell j}\partial_-^\ell u^h\big) dx^h = \int_{\Omega_r^h} \sum_{j,\ell=0}^{n} a_{\ell j}\partial_-^\ell u^h \partial_-^j v^h dx^h \tag{8.180}$$

$$+ \int_{\partial\Omega_l^h} v^h \left(\sum_{\ell=0,\dots,n}^{j=1\dots,n} a_{\ell j}\partial_-^\ell u^h \right) (\cdot + he^j) ds^h$$

$$- \int_{\partial\Omega_r^h} v^h \left(\sum_{\ell=0,\dots,n}^{j=1\dots,n} a_{\ell j}\partial_-^\ell u^h \right) ds^h.$$

We take the arithmetic mean of the last two equations and find (8.174).
Similarly, applying these results to $u^h = A_j(\cdot, \vec{u}^h, \nabla\vec{u}^h)$, yields (8.175). □

For the second variant we consider the shifted grid, shown for $n = 2$ in Figure 8.5, and we modify (8.163). To the inner grid points $\Omega^h \subset \Omega$ of the shifted grid $\mathcal{G}^h$, we add those grid points in $\mathcal{G}^h$ with distance $\leq h$ to Ω^h, cf. (8.95)–(8.97):

$$\text{For } \Omega^h \text{ let } \Omega^h := \Omega^h \cup \{\circ \text{ in Figure 8.5}\} \text{ and } \partial\Omega^h := \{\times \text{ in Figure 8.5}\}. \tag{8.181}$$

Compared to the first variant, we mainly have to modify the boundary conditions, but do not give the details. For, e.g. *Neumann conditions,* choose for each $x \in \partial\Omega^h$, never a vertex of Ω, the next neighbor $x^h = x \mp h\nu/2 \in \Omega^h$, and the corresponding outer next neighbor $x_{\pm h}^h = x \pm h\nu/2 \in \{\circ \text{ in Figure 8.5}\} \subset \mathcal{G}_n^h \setminus \overline{\Omega}$. Then we require,

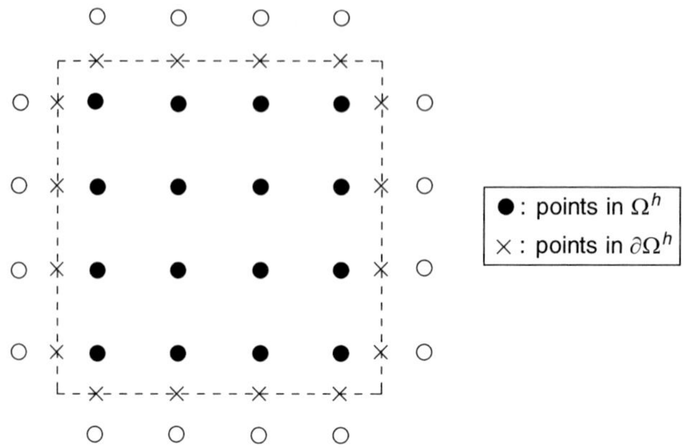

Figure 8.5 Two-dimensional grid for a square, and Neumann conditions.

cf. (8.174),

$$\varphi(x) = \partial^{\nu}_{h/2} u^h(x) \approx \frac{\partial}{\partial \nu} u(x) \forall x \in \partial\Omega^h. \tag{8.182}$$

We might want to avoid outer points $x^h_{\pm h} \in \{\circ \text{ in Figure 8.5}\}$ outside Ω^h. To this end we combine, in the five-point star, the above $x^h = x \mp h\nu/2 \in \Omega^h$ with the corresponding $x^h_{\pm h} \notin \Omega^h$. We replace the value for the $u^h\left(x^h_{\pm h}\right)$ by a linear combination of $\varphi(x)$, and $u^h(x^h)$, computed from (8.182). For an $x \in \partial\Omega^h$, next to an edge for $n > 2$ or a vertex of Ω^h, for $n \geq 2$, e.g. the point • next to $(0,0)$ in Figure 8.5, there are at least two directions with neigboring points $x^h_{\pm h}$ for x to apply this replacement. For Neumann boundary conditions we are done.

Now we have to solve first the linear then the quasilinear problems.

$$A^h_s u^h_0(x) = \hat{f}^h(x) \quad \forall x \in \Omega^h, \hat{f}^h \in H^{-1}_{\pm}(\Omega^h), \tag{8.183}$$

$$u^h_0 \in \mathcal{V}^h : \quad a^h\left(u^h_0, v^h\right) = \langle A^h u^h_0, v^h \rangle_{\mathcal{V}'^h \times \mathcal{V}^h} = \langle f^h, v^h \rangle^h, \forall v^h \in \mathcal{V}^h, \tag{8.184}$$

$$B^h_a u^h_0(x) = \varphi^h(x - h\nu/2) \forall x \in \partial\Omega^h_e,$$

or

$$\int_{\partial\Omega^h_e} v^h B^h_a u^h_0 ds^h := \int_{\partial\Omega^h_e} v^h \varphi^h(\cdot - h\nu/2) ds^h,$$

$$a^h\left(u^h_0, v^h\right) = \left(A^h_s u^h_0, v^h\right)_{L^2_h(\Omega^h)} + \int_{\partial\Omega^h_e} v^h B^h_a u^h_0 ds^h = \langle f^h, v^h \rangle^h \forall v^h \in \mathcal{V}^h. \tag{8.185}$$

Either $\hat{f}, \varphi$ are continuous or we use appropriate approximations as in Proposition 8.5. The original f and φ define the corresponding $\hat{f}^h := Q'^h \hat{f}$ and $\varphi^h := Q'^h \varphi$ in (8.170) and (8.191). The quasilinear difference systems has the form

$$\langle G^h \vec{u}^h, \vec{v}^h \rangle_{\mathcal{V}'^h \times \mathcal{V}^h} = a^h(\vec{u}^h, \vec{v}^h) = \left(G^h_s \vec{u}^h, \vec{v}^h\right)_{L^2(\Omega^h)} + \int_{\partial\Omega^h_e} v^h B^h_a u^h ds^h = 0. \tag{8.186}$$

As in Subsection 8.6.1 we have the two choices of either the weak or the strong formulation. The latter essentially yields the classical difference equations and boundary conditions.

The weak form allows using our variational techniques. For this we prove stability, unique discrete solutions and their convergence for our difference method. The necessary relation between the weak A^h, the strong A_s^h, the boundary condition, B_a^h, the coercivity and consistency are collected in Propositions 8.36–8.38. They guarantee the claim in Theorem 8.39.

So we contrast the weak solution, u_0^h, cf. Proposition 8.35, similarly to Theorem 8.34, defined by

$$a^h\left(u_0^h, v^h\right) = \left\langle A^h u_0^h, v^h\right\rangle_{\mathcal{V}'^h \times \mathcal{V}^h} = \langle f^h, v^h\rangle \forall v \in \mathcal{V}^h = H_\pm^1\left(\Omega_e^h\right). \tag{8.187}$$

The strong solution solves the difference equation and the discrete natural boundary conditions:

$$\left(A_s^h u_0^h - \hat{f}^h\right)(x) = 0 \forall x \in \Omega^h \text{ and } B_a^h u_0^h(x) - \varphi^h = 0 \forall x \in \partial\Omega_e^h. \tag{8.188}$$

Proposition 8.36. *Every solution, $u_0^h \in H_\pm^2\left(\Omega_e^h\right)$, of (8.188) satisfies (8.187) as well, and vice versa. An analogous result holds for quasilinear problems.*

Proof. We modify the proof of Theorem 8.34 according to (8.169)–(8.174). With $\mathcal{V}_0^h := \left\{v^h \in \mathcal{V}^h : v^h|_{\partial\Omega_e^h} = 0\right\}$ we obtain

$$a^h\left(u_0^h, v^h\right) = \left\langle A^h u_0^h, v^h\right\rangle_{\mathcal{V}'^h \times \mathcal{V}^h} = \left(A_s^h u_0^h, v^h\right)_{L_h^2(\Omega^h)} = \left(\hat{f}^h, v^h\right)_{L_h^2(\Omega^h)} \forall v^h \mathcal{V}_0^h,$$

hence the strong difference equation in (8.188). Next we extend testing (8.169) with $v^h \in \mathcal{V}^h$. By the previous equation, we are left with: for all $v^h \in \mathcal{V}^h$:

$$\int_{\partial\Omega_e^h}\left(B_a^h u_0^h - \varphi^h\left(\cdot - \frac{h\nu}{2}\right)\right) v^h ds^h = 0 \Rightarrow B_a^h\, u_0^h(x) - \varphi^h\left(x - \frac{h\nu}{2}\right) \forall x \in \partial\Omega_e^h, \tag{8.189}$$

so a *natural boundary condition.*

Vice versa, a solution $u_0^h \in H_\pm^1\left(\Omega_e^h\right)$, satisfying the strong difference equation, and natural boundary conditions in (8.188), satisfies the weak problem (8.187) as well. The same strategy applies to quasilinear problems. □

By an obvious modification of part (3) of the proof of Theorem 8.25 we get

Proposition 8.37. *Under the condition (8.38), and with $c_0 > 0$, the extended principal part*

$$a_{p,e}^h(u^h, v^h) := a_p^h(u^h, v^h) + c_0(u^h, v^h)_{L^2(\Omega^h)} : \mathcal{V}^h \times \mathcal{V}^h \to \mathbb{R} \tag{8.190}$$

is $\mathcal{V}^h$-coercive, hence stable; the original $a^h(\cdot,\cdot)$ in (8.169) is $\mathcal{V}^h$-elliptic.

We finally need the consistency of our approximations. To that end, we use the obvious extensions of the P^h, Q'^h in Subsection 8.5.2 to the $\Omega_e^h, \partial\Omega_e^h$ in this section.

The only not totally obvious modification concerns the boundary integral terms. We come back to the boundary term of the f^h in (8.170): for

$$\langle \varphi^h, v^h \rangle^h := \int_{\partial\Omega_e^h} v^h \varphi^h(\cdot - h\nu/2) ds^h = h^{n-1} \sum_{x\in\partial\Omega_e^h} (v^h(x)\varphi^h(x - h\nu/2), \text{ let } \quad (8.191)$$

$$Q_b^{'h} : L^2(\partial\Omega) \to \mathcal{V}^{'h} : Q_b^{'h}\varphi(x) := \varphi^h(x - h\nu/2) := P^h(E_c\varphi)(x - h\nu/2) \forall x \in \partial\Omega_e^h,$$

$$\Rightarrow \left\langle Q_b^{'h}\varphi, v^h \right\rangle_{\mathcal{V}^{'h}\times\mathcal{V}^h} := \int_{\partial\Omega_e^h} v^h Q_b^{'h}\varphi ds^h = \sum_{x\in\partial\Omega_e^h} h^{n-1} v^h(x)(P^h E_c\varphi)(x - h\nu/2).$$

Proposition 8.38. *With the extended* $P^h, Q^{'h}$, *as in (8.99), (8.84), (8.93),* $Q_b^{'h}$ *in (8.191), and the conditions for the coefficients of* A *in Summary 8.6 and Remark 8.7, we obtain for*

$$Au = f \in \mathcal{V}', A \in \mathcal{L}(\mathcal{V}, \mathcal{V}'), A_s \in \mathcal{L}(H^2(\Omega), L^2(\Omega)), \qquad (8.192)$$

$$\hat{f}^h = Q^{'h}\hat{f}_0, \varphi^h = Q_b^{'h}\varphi, \quad u, v \in \mathcal{V} = H^1(\Omega), \mathcal{V}^h = H^1_{\pm}(\Omega_e^h), \qquad (8.193)$$

consistent symmetric approximations A^h, A_s^h, B_a^h *cf. (8.60), (8.194), for linear operators, hence,*

$$\lim_{h\to 0} ||A^h P^h u - Q^{'h} Au||_{\mathcal{V}^{'h}} = \lim_{h\to 0} ||A_s^h P^h u - Q^{'h} Au||_{\mathcal{V}^{'h}} = 0 \quad \forall u \in H^1(\Omega),$$

$$\lim_{h\to 0} ||B_a^h P^h u - Q_b^{'h} B_a u||_{\mathcal{V}_b^{'h}} = 0 \forall u \in H^2(\Omega). \qquad (8.194)$$

More precisely, we obtain, for the weak form and $u \in H^2(\Omega), A \in \mathcal{L}(H^2(\Omega), L^2(\Omega))$, $B_a \in \mathcal{L}(H^2(\Omega), L^2(\partial\Omega))$, *consistency of order* 1 *for* A^h, *and for* B_a^h,

$$||A^h P^h u - Q^{'h} Au||_{\mathcal{V}^{'h}} \le Ch\|u\|_{H^2(\Omega)}, \quad ||B_a^h P^h u - Q_b^{'h} B_a u||_{\mathcal{V}_b^{'h}} \le Ch\|u\|_{H^2(\Omega)}.$$

For $u \in H^3(\Omega)$ *with* $A \in \mathcal{L}(H^3(\Omega), H^1(\Omega))$, *and* $B_a \in \mathcal{L}(H^3(\Omega), H^1(\partial\Omega))$, *we obtain consistency of orders* 2 *and* 1 *for* A_s^h, *and for* B_a^h,

$$||A_s^h P^h u - Q^{'h} Au||_{\mathcal{V}^{'h}} \le Ch^2\|u\|_{H^3(\Omega)}, ||B_a^h P^h u - Q_b^{'h} B_a u||_{\mathcal{V}_b^{'h}} \le Ch\|u\|_{H^3(\Omega)}$$

With the technique in Section 5.2 we obtain consistency of order 2 *for* B_a^h *as well.*

For possible choices of $f^h \neq Q^{'h} f$, *and* $\varphi^h \neq Q_b^{'h}\varphi$, *the* $\lim_{h\to 0} ||f^h - Q^{'h} f||_{\mathcal{V}^{'h}} = 0$ *and* $\lim_{h\to 0} ||\varphi^h - Q_b^{'h}\varphi||_{\mathcal{V}_b^{'h}} = 0$ *have to be proved.*

Analogous results hold for the quasilinear problems.

Proof. Except for $||B_a^h P^h u - Q_b^{'h} B_a u||_{\mathcal{V}_b^{'h}}$ we can argue for the strong form as in Theorem 8.21. In $\int_{\partial\Omega_e^h} v^h B_a^h u^h ds^h$ we had to shift from $x \in \partial\Omega^h$ to $\infty - h\nu/2$. This reduces the consistency to order 1. Order 2 would be obtained if we could compare B_a and B_a^h exactly along $\partial\Omega^h$. The technique in Section 5.2, splitting into differential and boundary ooperators, allows this exact comparison along $\partial\Omega^h$, and thus yields order 2. We do not repeat that proof. □

We summarize the necessary conditions for our difference methods for natural boundary conditions, essentially restricted to convergence of order 2.

1. We use the following spaces, cf. (8.77), often we omit the $\mathbb{R}^q$:

$$\mathcal{V} = H^1(\Omega, \mathbb{R}^q), \quad \mathcal{V}' = H^{-1}(\Omega, \mathbb{R}^q), \text{ for } q \geq 1,$$
$$\mathcal{V}^h = H^1_{\pm}(\Omega^h, \mathbb{R}^q), \ \mathcal{V}^h_b = H^1_{0,\pm}(\partial\Omega^h, \mathbb{R}^q), \ \mathcal{V}'^h = H^{-1}_{\mp}(\Omega^h, \mathbb{R}^q). \quad (8.195)$$

2. For linear problems we assume, cf. Remark 8.7,

$$a_{ij} \in C(\overline{\Omega}), \text{ and } A : H^3(\Omega) \to H^1(\Omega), B_a : H^3(\Omega) \to H^1(\partial\Omega). \quad (8.196)$$

3. For quasilinear problems we assume, with N_1 in (8.17),

$$A_j \text{ equicontinuous in } \Omega \times N_{\nabla^1} \subset \Omega \times \mathbb{R}^{N_1}, \ N_{\nabla^1} \text{ bounded, s.t.} \quad (8.197)$$
$$\forall u \in \mathcal{V} : \|u - u_0\|_{\mathcal{V}} \leq \rho, \nabla^{\leq 1} u(x) \in N_{\nabla^1} \forall x \in \Omega, \nabla_h^{\leq 1} u(x) \in N_{\nabla^1} \forall x \in \Omega^h,$$

with Lipschitz-continuous A_j with respect to the variables in N_{∇^1}. Finally we assume

$$A_j : \mathcal{D}(A_j) \cap H^3(\Omega) \to H^1(\Omega)). \quad (8.198)$$

Theorem 8.39. Difference methods for natural boundary conditions:

1. *Let A or $G'(u_0) : H^1(\Omega, \mathbb{R}^q) \to H^{-1}(\Omega, \mathbb{R}^q)$ be boundedly invertible, for a (locally) unique exact solutions u_0 for the exact equations, cf. the conditions in Chapter 2. Under the conditions (8.195)–(8.197) we extend $P^h = I^h$ and Q'^h, from Subsection 8.5.2, to Ω^h_e and define Q'^h_b as in (8.191). For the exact linear and nonlinear problems, $Au = f \in H^{-1}(\Omega, \mathbb{R}^q)$ and $G(u) = 0 \in H^{-1}(\Omega, \mathbb{R}^q)$ define the symmetric difference methods as in (8.167)–(8.175), defined on cuboidal domains, cf. (8.95).*
2. *A unique discrete solution u_0^h, simultaneously for the weak and strong discrete linear and nonlinear problems exist near u_0 for small enough h. It converges of order 1. With the f in (8.153), and the corresponding f^h in (8.170) with $\hat{f}^h = Q'^h \hat{f}_0, \varphi^h = Q'^h_b \varphi$, in (8.192) we find*

$$A^h u_0^h = f^h \Rightarrow \big||P^h u_0 - u_0^h|\big|_{\mathcal{V}^h} \leq Ch\|u_0\|_{H^2(\Omega)} \text{ for } u_0 \in H^2(\Omega). \quad (8.199)$$

3. *This u_0^h satisfies the classical difference and a modified boundary equation:*

$$A^h u_0^h = f^h \Rightarrow \left(A_s^h u_0^h - \hat{f}^h\right)(x) = 0 \forall x \in \Omega^h, \text{ and} \quad (8.200)$$
$$B_a^h u_0^h(x) = \varphi^h(x - h\nu/2) \forall x \in \partial\Omega_e^h.$$

4. *If we maintain, in (8.200), $(A_s^h u_0^h - \hat{f}^h)(x) = 0$, and replace the modified boundary condition by the classical discrete boundary condition, then this implies*

convergence of order 2:

$$\left(A_s^h u_0^h - \hat{f}^h\right)(x) = 0 \forall x \in \Omega^h, B_a^h u_0^h(x) = \varphi^h(x) \forall x \in \partial\Omega^h \tag{8.201}$$
$$\Longrightarrow \left||P^h u_0 - u_0^h\right||_{\mathcal{V}^h} \leq Ch^2 \|u_0\|_{H^3(\Omega)} \text{ for } u_0 \in H^2(\Omega).$$

5. *We replace the above (8.167) ff., (8.174) for linear problems by (8.171) ff., (8.175) for quasilinear problems. Then we obtain the corresponding results.*

Proof. We only prove first order convergence, and indicate second order. We have shown that $A^h, A_{p,e}^h : H_\pm^1\left(\Omega_e^h\right) \to H_\pm^{-1}\left(\Omega_e^h\right)$, induced by $a^h(u^h, v^h)$, $a_{p,e}^h(u^h, v^h)$, and $A, A_{p,e}$ induced by $a(u,v), a_{p,e}(u,v)$, are bounded linear operators. By Proposition 8.37, the $a_{p,e}^h(u^h, v^h)$, and $a_{p,e}(u,v)$ are $\mathcal{V}^h$-, and $\mathcal{V}$-coercive. This A is a compact perturbation of $A_{p,e}$ with stable $A_{p,e}^h$. Furthermore, the $A_{p,e}^h, A^h$ are consistent with the $A_{p,e}, A$, by Proposition 8.38.

Hence for a boundedly invertible A, the $H_\pm^1\left(\Omega_e^h\right)$-stability of $A_{p,e}^h$, and the consistencies in Proposition 8.38 and Theorem 3.29 imply the stability of A^h. With $A^{-1} \in \mathcal{L}(H^{-1}(\Omega), H^1(\Omega))$, and the corresponding $f \in H^{-1}(\Omega)$, $f^h = Q'^h f \in H_\pm^{-1}\left(\Omega_e^h\right)$, as in (8.170), we obtain

$$\exists_1 u_0^h \in H_h^1(\Omega^h) : a^h\left(u_0^h, v^h\right) = \left\langle A^h u_0^h, v^h\right\rangle = \langle f^h, v^h\rangle \forall v^h \in H_h^1(\Omega^h), \tag{8.202}$$

with the convergence. By the equivalence in (8.174), we verify the strong difference equation, and the modified discrete natural boundary condition. This is shown as in the previous sections.

For second order convergence, we have to replace the above modified boundary condition in (8.200) by classical discrete natural boundary conditions in (8.201), and similarly for the quasilinear case. Then the full operator is split into differential and boundary operators, (A_s, B_a). The corresponding classical pair of discrete difference and boundary operators, $\left(A_s^h, B_a^h\right)$, is consistent of order 2. This allows the full proof by modifying the technique in Section 5.2. We omit that here. □

8.7 Other difference methods on curved boundaries

In Section 8.4 we discussed nonsymmetric first order methods for domains with curved boundaries for general elliptic problems. Here we modify these methods for the following special cases, allowing, by appropriate treatment of the boundary, methods of higher order. Points more or less near the boundary will have to be distinguished. We summarize the results for problems with Dirichlet boundary conditions, but do not prove them for several reasons:

For FEMs, an extended theory essentially for all types of elliptic equations and systems is available, and is preferably used and presented in Chapter 4. For the mentioned special cases, the proofs for stability, consistency and convergence use techniques totally different from those in Sections 8.5, 8.6. They do not seem to allow generalizing and unifying them towards the problems in Section 8.4.

The results in Subsections 8.7.1 and 8.8.1 are only known for special cases. Hackbusch [387] formulates his results only for the Laplacian, but his proofs apply (as he mentions) to most of the problems in Subsection 8.7.1. A coercivity proof as in the last sections is not known for the Shortley-Weller method. We refer, cf. (8.210), to the $\mathcal{V}^h = H_h^1(\Omega^h)$-regularity result in [387], Theorems 9.2.3, and 9.2.8 for the principal part. Our techniques imply the stability of (8.205) for a boundedly invertible operator. For the polynomial interpolation case, even asymptotic expansions for the discretization error are available, allowing higher order methods via extrapolation or defect corrections. Proving the convergence in the following subsections would probably be possible for special cases by applying the techniques in Section 5.2, splitting the discretization error into errors for the differential and the boundary operator. However this is beyond the scope of this chapter.

Finally, Heinrich [395] discusses with appropriate machinery finite difference methods on irregular networks.

8.7.1 The Shortley–Weller–Collatz method for linear equations

We assume Ω as in (8.98), and have to modify the $\Omega^h, \overline{\Omega}^h$ and $\partial\Omega^h$. The original grid $\mathcal{G}^h$, the j-th grid line $\mathcal{G}_j^h$, and different types of grid points, Ω^h, Ω_i^h, $\overline{\Omega^h}$ are distinguished, cf. Figure 8.6. With e_i, the i-th unit vector, $i = 1, \cdots, n$, let

$$\begin{aligned}
\mathcal{G}^h &:= \{x = (x_1, \ldots, x_n)^T : x_i/h \in \mathbb{Z} \text{ for } i = 1, \cdots, n\}, \qquad (8.203)\\
\mathcal{G}_j^h &:= \{x = (x_1, \ldots, x_n)^T \in \mathbb{R}^n : x_j \in \mathbb{R},\ x_i/h \in \mathbb{Z} \text{ for } j \neq i = 1, \cdots, n\},\\
\Omega^h &:= \{x \in \Omega \cap \mathcal{G}^h : x \pm h e_i \in \Omega \text{ for } i = 1, \cdots, n\},\quad \Omega_i^h := (\Omega \cap \mathcal{G}^h) \setminus \Omega^h,\\
\partial\Omega^h &:= \left\{P \in \partial\Omega \cap \mathcal{G}_j^h \text{ for } j = 1, \cdots, n\right\},\ \overline{\Omega}^h := \Omega^h \cup \Omega_i^h \cup \partial\Omega^h, \Omega_e^h := \Omega^h \cup \Omega_i^h.
\end{aligned}$$

We denote Ω^h and Ω_i^h, and $\partial\Omega^h$ as the set of regular, Ω_i^h as the irregular, $\partial\Omega^h$ as the boundary, $\overline{\Omega}^h$ as all mesh points in Ω, and Ω_e^h as the essential mesh points in Ω. In Figure 8.6 we have marked these points by •, by ■, and by □

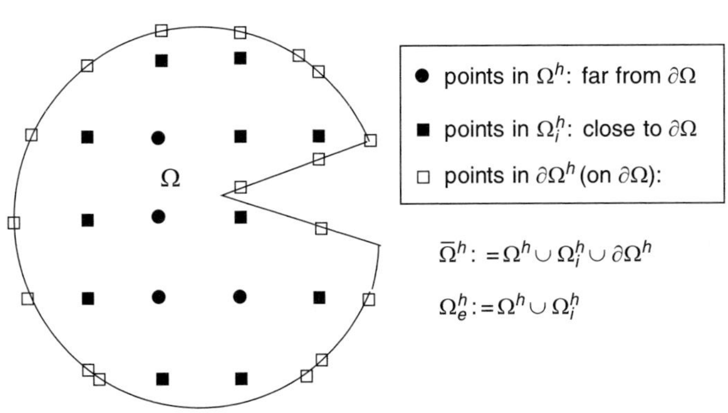

Figure 8.6 Grid for curved $\partial\Omega$.

We only consider the following special second order equations, and formulate their classical strong difference equation, cf. (8.43)–(8.47). Note that we exclude mixed derivatives, and use symmetric formulas

$$A_s u(x) = \sum_{j=1}^{n} \big(- a_j(x)(\partial^j)^2 u(x) + b_j(x)\partial^j u(x)\big) + c(x)u(x) = f(x), \tag{8.204}$$

and with $\partial_h^j := \left(\partial_-^j + \partial_+^j\right)/2$, and $\partial_h^{2e_j} := \partial_-^j \partial_+^j$, cf. (8.12), (8.13), define

$$A_s^h u^h(x) = \sum_{j=1}^{n} \big(- a_j(x)\partial_h^{2e_j} u^h(x) + b_j(x)\partial_h^j u^h(x)\big) + c(x)u^h(x) = f(x) \tag{8.205}$$

$$\forall x \in \Omega^h,\ u^h : \overline{\Omega}^h \to \mathbb{R}, \text{ with boundary conditions } u^h(x) = 0 \forall x \in \partial\Omega^h, \tag{8.206}$$

with bounded and continuous $a_j, b_j, j = 1, \ldots, n$, cf. Remark 8.7.

Then $A_s^h u^h$ is well defined for $\forall x \in \Omega^h$, but not for $x \in \Omega_i^h$. The ellipticity condition (8.38) for this A_s has the form, see (2.16), (2.20),

$$\sum_{j=1}^{n} a_j(x)(\vartheta^j)^2 \geq \epsilon|\vartheta|^2 \text{ or } a_j(x) \geq \epsilon > 0, \forall x \in \Omega. \tag{8.207}$$

There is a long history of dealing with the missing $x \in \Omega_i^h$ in (8.205). We discuss only two different approaches.

We choose an $x \in \Omega_i^h$ with at least one neighbor outside of $\overline{\Omega}$, hence

$$\text{for } x \in \Omega_i^h \text{ choose the smallest possible } 0 < s_j^+, s_j^- \leq 1,\ j = 1, \cdots, n, \tag{8.208}$$

$$\text{with } \min_{j=1,\cdots,n} \left\{s_j^+, s_j^-\right\} < 1 \text{ s.t. } \ x + s_j^+ he_j, x - s_j^- he_j \in \overline{\Omega}^h.$$

Shortley and Weller [588] use, instead of the symmetric differences, $\partial_h^{2e_j} u^h(x)$, in (8.205), the unsymmetric differences with respect to $x' = x - s_j^- he_j < x < x'' = x + s_j^+ he_j$, with $u(x') = 0$ or $u(x'') = 0$ for x' or $x'' \in \partial\Omega^h$, [588] replace, cf. [387], (4.8.5), (4.8.6),

$$\partial_h^{2e_j} u^h(x) \text{ by } (\partial^j)_h^2 u(x) := 2h^{-2}\Big(\frac{u(x'') - u(x)}{s_j^+} - \frac{u(x) - u(x')}{s_j^-}\Big)/\left(s_j^+ + s_j^-\right)$$

$$\text{with } \left|u''(x) - (\partial^j)_h^2 u(x)\right| \leq h \max\left\{s_j^+, s_j^-\right\} \|u'''\|_{C[x',x'']}/3 \tag{8.209}$$

$$\text{and } \partial_h^j u^h(x) \text{ by } (\partial^j)^h u(x) := h^{-1}\frac{u(x'') - u(x)}{s_j^+}$$

with $|u'(x) - (\partial^j)^h u(x)| \leq hs_j^+ \|u''\|_{C[x',x'']}$.

The previous discrete Sobolev spaces have to be modified accordingly. The special differences in (8.205), corresponding to (8.209), are indicated by special multi-indices, α, so let $\alpha^c := je_i, j = 0, 1, 2, i = 1, \ldots, n$. Finite differences, $\partial_h^{\alpha^c}$, are chosen as in (8.205), (8.209). With these divided differences, $\partial_h^{\alpha^c} u^h$, the *discrete Sobolev inner products, norms, and spaces* are defined, for $m = 0, 1, 2$, as

$$\left(u^h, v^h\right)_{H_h^m(\overline{\Omega}^h)} := \left(u^h, v^h\right)_{L_h^2(\overline{\Omega}^h)} + \sum_{0<|\alpha^c|\le m} \left(\partial_h^{\alpha^c} u^h, \partial_h^{\alpha^c} v^h\right)_{L_h^2(\Omega_e^h)}, \text{ and} \tag{8.210}$$

$$|v^h|^2_{H_h^k(\overline{\Omega}^h)} := \sum_{0<|\alpha^c|=k} \left\|\partial_h^{\alpha^c} v^h\right\|^2_{L_h^2(\overline{\Omega}^h))}, \|v^h\|_{H_h^m(\overline{\Omega}^h)} := \left(\sum_{0\le k\le m} |v^h|_{H_h^k(\overline{\Omega}^h)}\right)^{1/2}.$$

Applying (8.209) for (8.208) instead of (8.205), we obtain the *Shortley–Weller approximation* for an $x \in \Omega_i^h$, with $\min_{j=1,\cdots,n}\left\{s_j^+, s_j^-\right\} < 1$, usually for one j :

$$\begin{aligned} A_s^h u^h(x) := \sum_{j=1}^n \Bigg[& a_j(x)h^{-2}\left(-\frac{2}{s_j^+\left(s_j^+ + s_j^-\right)}u^h\left(x + s_j^+ h e_j\right) + \frac{2}{s_j^+ s_j^-}u^h(x)\right. \\ & \left. - \frac{2}{s_j^-\left(s_j^+ + s_j^-\right)}u^h\left(x - s_j^- h e_j\right)\right) + b_j(x)\frac{u^h\left(x + s_j^+ h e_j\right) - u^h(x)}{s_j^+ h}\Bigg] \\ & + c(x)u^h(x). \end{aligned} \tag{8.211}$$

Reinterpreted, this is essentially Collatz's method as well [205, 206], see below. For computations we multiply $A_{SW}^h = A_s^h$ in (8.205), (8.211) by the diagonal matrix D^h obtaining, cf. [387], Corollary 9.2.10, with $s_j^- s_j^+ := 1$ for $x \in \Omega^h$,

$$D^h := (\text{diag } d(x))_{x\in\Omega_e^h}, d(x) := \min_{j=1\dots,n}\left\{1, 2s_j^- s_j^+\right\}, \text{ and } A_{SW}^{\prime h} := D^h A_{SW}^h. \tag{8.212}$$

Then instead of $A_{SW}^h u_0^h = f^h$ we have to solve $A_{SW}^{\prime h} u_0^h = D^h f^h$. Hackbusch's [387] Theorems 4.3.16, 4.8.4 and 9.2.8 are formulated for $A_s u = -\Delta u$. His proofs for consistency in (4.8.11) or Theorem 9.2.14 including his regularity in Theorems 9.2.3, 9.2.8 apply to $b_j = c = 0$ in (8.205), (8.211). The extension to boundedly invertible operators $A : H_0^1(\Omega) \to H^{-1}(\Omega)$ follows with the previous consistency and regularity of the principal part by Theorem 3.29.

The preceding Shortly–Weller method can be generalized to special *strong quasilinear equations* of the form, cf. (8.64), yielding a linearized form (8.204),

$$G : \mathcal{D}(G) \subset H^1(\Omega) \to H^{-1}(\Omega), G^h : \mathcal{D}(G^h) \subset H_h^1\left(\Omega_e^h\right) \to H_h^{-1}\left(\Omega_e^h\right),$$

$$G_s u_0 = \sum_{j=0}^n (-1)_{j>0}\partial^j A_j(\cdot, u_0, \partial^j u_0) = 0 \in L^2(\Omega), \; u_0 \in H^2 \cap H_0^1(\Omega), \tag{8.213}$$

$$G_s^h u_0^h(x) = \sum_{j=0}^n (-1)_{j>0}\left(\partial_-^j A_j\left(\cdot, u_0^h, \partial_+^j u_0^h\right) + \partial_+^j A_j\left(\cdot, u_0^h, \partial_-^j u_0^h\right)\right)(x)/2 = 0$$

$$\forall x \in \Omega^h, \text{ and } u_0^h \in \mathcal{V}_b^h. \tag{8.214}$$

For the proof, we combine, as usual, Theorems 3.23, 8.40, 3.29, 3.21 with the above consistency. We summarize these results in

Theorem 8.40. Convergence of the Shortly–Weller method:

1. *Let Ω be bounded, and in C^2 or convex, and $A : H_0^1(\Omega) \to H_0^{-1}(\Omega)$ in (8.204) be boundedly invertible. Then A'^h_{SW}, the matrix defined by the Shortley–Weller method (8.205), (8.211), (8.212) as $A'^h_{SW} : H_h^1(\Omega_e^h) \to H_h^{-1}(\Omega_e^h)$, cf. (8.210), is stable and consistent. The Shortley–Weller equations have, for small enough h, a unique solution u_0^h, converging to the exact solution u_0 of $Au_0 = f$:*

$$\|u_0^h - P^h u_0\|_{H_h^1(\Omega_e^h)} \le Ch\|u_0\|_{H^2(\Omega)}. \tag{8.215}$$

2. *Whenever the conditions for an M matrix are satisfied, cf. [387], Theorems 4.8.4, 4.8.6, then*

$$\max_{x\in\Omega_e^h} |u_0^h(x) - u_0(x)\| \le Ch^2\|u_0\|_{C_L^3(\overline{\Omega})}.$$

3. *Whenever the conditions for well-posed quasilinear equations of the form (8.213) and for a locally unique solution u_0 with boundedly invertible $G'(u_0)$ are satisfied, cf. Chapter 2, the Shortly–Weller equations (8.214) have, for a small enough step size h, a unique solution u_0^h, converging as in (8.215).*

8.8 Asymptotic expansions, extrapolation, and defect corrections

This area of research for ordinary and partial differential equations florished between 1960 and 1990. Since the interest in these results is pretty low these days, we will only list a choice of relevant results and apply them to extrapolation methods.

The following authors have contributed to this area: textbooks are, e.g. Conte and de Boor, Deuflhard and Hohmann, and Fox, Isaacson and Keller, and Stetter [207, 284,324,414,596]. Papers are due to, e.g. Axelsson and Layton, Böhmer, with Hemker, Gross, Schmitt, R. Schwarz, and with Stetter, Chibi and Moore, Doedel with Reddien, Fox, Frank and Ueberhuber, Gragg, Hackbusch, Keller and Pereyra, Stetter, Monnet, Pereyra *et al.*, Reinhardt, and Thoma [44,101,103,106,107,110,125,126,130,172,288–291,321–323,325,326,358,381–384,442,496,526,527,548,549,596,597,625]. Extensions to asymptotic expansions for FEMs are due to Lin, Liu, Rannacher, Zou, [476,477,541].

In particular, Hackbusch's regularity results applied to defect corrections [382–384] showed that defect corrections yielded the same orders of acccuracy as the extrapolation methods. The approach presented in this chapter, with convergence results with respect to the discrete Sobolev norms, $\|\cdot\|_{H_\mp^m(\Omega^h)}$, is a special regularity result as well, now proved for all the previous problems. So the result [384] is valid for our methods and problems as well.

Here we combine extrapolation techniques or defect corrections with asymptotic expansions for the discretization error. They are available for cuboidal and curved domains, cf (8.95) and (8.98). Similarly to extrapolation for numerical integration, cf. e.g. [414, 602, 603], they allow, by linear combinations of numerical results for step sizes, say $h, h/2, h/4$, higher order methods via extrapolation. Defect corrections are strongly related to discrete Newton methods. Deferred corrections are another choice, but are more difficult to explain. So we omit them here, cf. the above papers.

8.8.1 A difference method based on polynomial interpolation for linear, and semilinear equations

The methods discussed now, and the necessary tools for their proofs, fit even less than in Subsection 8.7.1, to the previous approach. We include them here, since their discretization error admits an asymptotic expansion.

For smooth Ω, and small enough h we find for $x \in \Omega_i^h$, cf. (8.203), e.g.

$$x + he_j \notin \overline{\Omega}, \text{ the } x + s_j he_j \in \partial\Omega, 0 < s_j < 1, \qquad (8.216)$$
$$\text{but } x - (\nu - 1)he_j \in \overline{\Omega}, \nu = 1, \cdots, k \geq 1.$$

Again we choose the smallest $0 < s_j < 1$, such that $x + s_j he_j \in \partial\Omega$ and the full segment $\{x - t\ h\ e_j\} \subset \overline{\Omega}, -s_j \leq t \leq k - 1$.

All the following methods essentially replace $u^h(x + he_j)$ in (8.216) by the value of a polynomial interpolating the boundary function $\varphi(x + s_j he_j)$ in $x + s_j he_j \in \partial\Omega$, and the $u^h(x)$ at the other points in (8.216): Gerschgorin [340] replaced $u^h(x + he_j)$ by a nontrivial $\varphi(x + s_j he_j)$. Pereyra *et al.* [527], referring to unpublished notes of H.-O. Kreiss, use the following polynomial P_j^k for $k \geq 2$. It is a univariate k-th order interpolating polynomial

$$P_{j+}^{xk} := P_j^k \in \Pi_k \text{ interpolates: } P_j^k(x + s_j he_j) = \varphi(x + s_j he_j),$$
$$\text{and } P_j^k(x - (\nu - 1)he_j) = u^h(x - (\nu - 1)he_j), \nu = 1, \cdots, k. \qquad (8.217)$$

Then they replace $u^h(x + he_j)$ by $P_j^k(x + he_j)$. If in (8.205), $b_j(x) \equiv 0$, the above Shortley–Weller method is obtained for $k = 2$.

Pereyra *et al.* [527] conjectured an asymptotic expansion for the discretization error of order k based upon Bramble–Hubbard techniques [139]. Böhmer [106] proved this expansion $\|u(x) - u^h(x) - h^2 e_2(x) - h^4 e_4(x)\|_\infty \leq \mathcal{O}(h^{k+1})$ for $k \leq 4$. Since large values for k impose restrictive conditions for Ω, and h, $k \leq 4$ is not a really severe restriction. [106, 139] essentially work with M-matrix arguments, not applicable to our previous technique.

Differently, Collatz [205,206] replaced $u^h(x)$, not $u^h(x + he_j)$, by the value of the linear polynomial interpolating in $(x + s_j he_j, \varphi(x + s_j he_j))$, and $(x - he_j, u^h(x - he_j))$. Then again (8.211) results. In fact, rescale (4.8.14a) and (4.8.14b) in [387], by the equibounded $2/(s_\ell + s_r)$, and $2/(s_u + s_o)$ and then proceed to the modified (4.8.16), yielding (4.8.7). So [205, 206] is included in the last subsection.

For $x + he_j \notin \overline{\Omega}$, see (8.216) [106, 527], we compute the approximate,

$$u^h(x + he_j) := P_j^k(x + he_j) = \left(\varphi(x + s_j he_j) - \sum_{\nu=1}^{k} c_j^\nu u^h(x - (\nu - 1)he_j)\right) / c_j^0, \qquad (8.218)$$

with

$$c_j^\nu := \prod_{\nu \neq \ell = 0}^{k} \frac{(1 - s_j) - \ell}{\nu - \ell},$$

and with our $\varphi(x + s_j h e_j) = 0$.

The situation $x - he_j \notin \overline{\Omega}$, and $x - s_j^- he_j \in \partial\Omega, 0 < s_j^- < 1$, in (8.216)–(8.218), is handled analogously by replacing h by $-h$.

Summarizing, we replace in (8.205), e.g. the $\partial_+^j u^h(x)$ by $\big(P_j^k(x+he_j) - u^h(x - he_j)\big)/h$ and obtain for an $x \in \Omega_i^h, x + he_j \notin \overline{\Omega}$

$$\begin{aligned} f(x) =\ & A_s^h u^h(x) \\ =\ & \sum_{j\neq i=1}^{n} \left[-a_j(x)\partial_h^{2e_j} u^h(x) + b_j(x)\partial_h^j u^h(x)\right] + c(x)u^h(x) \\ & - a_j(x)h^{-2}\left(\left[0 - \sum_{\nu=1}^{k} c_j^\nu u^h(x - (\nu-1)he_j)\right]\Big/ c_j^0 - 2u^h(x) + u^h(x - he_j)\right) \\ & + b_j(x)h^{-1}\left(\left[0 - \sum_{\nu=1}^{k} c_j^\nu u^h(x - (\nu-1)he_j)\right]\Big/ c_j^0 - u^h(x - he_j)\right)/2. \end{aligned} \tag{8.219}$$

By (8.216), the situation $x \in \Omega_i^h$, $x + he_j \notin \overline{\Omega}$, and $x - he_{j_1} \notin \overline{\Omega}$ for $j_1 \neq j$ or even $x \pm he_j \notin \overline{\Omega}$ for $k = 1$ is not excluded. Then in (8.219) the first sum over $j \neq i = 1, \ldots, n$ has to be modified into $i = 1, \ldots, n, x \pm he_i \in \overline{\Omega}$, and the missing i have to be collected in the last two lines of (8.219). We avoid the messy formulation.

For (8.219) we need a connectedness condition for the grid Ω^h. We define, for a fixed $x \in \Omega^h \cup \Omega_i^h$, the set of neighbors as

$$N(x) := \left\{\bar{x} \in \Omega^h \cup \Omega_i^h \text{ occurring in (8.205) or (8.219)}\right\}.$$

Then we reformulate (8.205), and (8.219) with $\sigma(x,\bar{x}) \neq 0$ as

$$x \in \Omega^h \cup \Omega_i^h : \ A_s^h u^h(x) - f(x) =: \sum_{\bar{x}\in N(x)} \sigma(x,\bar{x})u^h(\bar{x}) + B^h(-a_j, b_i, f, 0 = \varphi, x).$$

Then the grid $\Omega^h \cup \Omega_i^h$ is called *connected,* if for every proper subset, S^h,

$$\forall\ S^h \subset\subset \Omega^h \cup \Omega_i^h : \ \left(\bigcup_{x\in S^h} N(x)\right) \cap \left(\left(\Omega^h \cup \Omega_i^h\right) \setminus S^h\right) \neq \emptyset. \tag{8.220}$$

Furthermore we assume for small enough h the existence of an $\Omega^* \supset \Omega$ such that

$$h \le \min\left\{\min_{x\in\overline{\Omega}}\left(\frac{2\sum_{j=1}^n a_j(x)}{|c(x)|}\right)^{1/2}, \min_{j=1}^{n}\min_{x\in\overline{\Omega}} \frac{a_j(x)}{|b_j(x)|}\right\} \tag{8.221}$$

and that

$$\Omega \subset \Omega^* \supset \left\{x \pm he_j : j = 1, \cdots, n,\ x \in \Omega^h \cup \Omega_i^h\right\}. \tag{8.222}$$

We summarize parts of Propositions 4.1, 4.3, and Theorems 4.4, 4.11 in [106]:

Theorem 8.41. Convergence for a Collatz–Pereyra method: *For Ω in (8.98) let $\Omega^h \cup \Omega_i^h$ be connected, let $A_s : C^2(\Omega) \cap C(\overline{\Omega}) \to C(\Omega)$ in (8.204) satisfy (8.207), and* *be boundedly invertible. Let* A_k^h *be the matrix defined by the preceding method, see (8.216)–(8.219), and let* h *satisfy (8.221). Then* $A_k^h : L^\infty(\overline{\Omega}^h) \to L^\infty(\overline{\Omega}^h)$ *is invertible with bounded* $\left\| \left(A_k^h\right)^{-1} \right\|_\infty \leq C$*, hence* A_k^h *is stable in the sense of (8.58). If additionally for* Ω^* *in (8.222) an extension* $u_0^* \in C^{q+\alpha}(\Omega^*), 2 \leq q \leq 6, 0 < \alpha \leq 1$ *exists, then the discrete solution* u_0^h *converges according to*

$$\left\| u_0^h - u_0 \right\|_{L_h^\infty(\overline{\Omega}^h)} \leq C h^{\min\{2, k+1, q-2+\alpha\}}. \tag{8.223}$$

In addition let

$$-a_j, b_j, c, f \in C^{k,\alpha}(\overline{\Omega}), \varphi \in C^{k+2,\alpha}(\partial\Omega), \partial\Omega \in C^{k+2,\alpha}, 0 \leq k \leq 4.$$

Then, by Theorems 2.38 and 2.39, the exact solution satisfies $u_0 \in C^{k+2,\alpha}(\overline{\Omega})$*, and there exist* $f_2 \in C^{k+\alpha}(\overline{\Omega})$*,* $f_4 \in C^{k-2+\alpha}(\overline{\Omega})$*, independent of* h*, such that the following asymptotic expansion is valid*

$$\left\| u_0^h - (u_0 + h^2 f_2 + h^4 f_4) \right\|_{L_h^\infty(\overline{\Omega}^h)} \leq C h^{\min\{4+\alpha, k+1\}}. \tag{8.224}$$

In Böhmer [104] we have extended results in Theorem 8.41 to *quasilinear equations*

$$G_s u(x) = \sum_{j=1}^{n} \left(-a_j \left(x, u(x), \partial^j u(x)\right) (\partial^j)^2 u(x)\right) - f(x, u(x), \nabla u(x)) = 0, \tag{8.225}$$

$$G_s^h u^h(x) = \sum_{j=1}^{n} \left(-a_j \left(x, u(x), \partial_h^j u(x)\right) \partial_h^{2e_j} u^h(x)\right) - f(x, u^h(x), \nabla_h u^h(x)) = 0$$

$\forall x \in \Omega^h$ with boundary conditions $u(x) = 0 \forall x \in \partial\Omega$, and $u^h(x) = 0 \forall x \in \partial\Omega^h$,

with bounded and continuous $f, a_j, j = 1, \ldots, n$.

Then G_s^h is well defined for $\forall x \in \Omega^h$, for $x \in \Omega_i^h$ the modification as in (8.219) has to be employed. For the stability we have to consider the derivative $G'(u_0)$ of (8.225), evaluated in the exact solution, u_0. This again has the form (8.204). We impose condition (8.226) for stability, and (8.227) for consistency, and asymptotic expansions, basic for Theorems 4.5, and 5.3 in [104]:

$$\begin{aligned}
&\text{Let } u_0 \text{ be the unique solution of (8.225), and} \\
&0 < a \leq a_j(\cdot, u_0(\cdot), \partial^j u_0(\cdot)) \text{ in } \Omega, \forall j = 1, \ldots, n \\
&0 < Q_* \leq \left(\frac{\partial f}{\partial u}(\cdot, u_0(\cdot), \nabla u_0(\cdot)) - \sum_{j=1}^{n} \frac{\partial a_j}{\partial u}(\cdot, u_0(\cdot), \partial^j u_0(\cdot))(\partial^j)^2 u_0(\cdot) \right) \\
&\quad \times \left(\sum_{j=1}^{n} a_j(\cdot, u_0(\cdot), \partial^j u_0(\cdot)) \right)^{-1} \leq Q^* \text{ in } \Omega, \text{ and as the } a_j,\ b_j
\end{aligned} \tag{8.226}$$

for $G'(u_0)$ in (8.204) we get $a_j(\cdot) := a_j(\cdot, u_0(\cdot), \partial^j u_0(\cdot)), \forall j = 1, \ldots, n$

$$b_j(\cdot) := -\frac{\partial a_j}{\partial(\partial^j u)}(\cdot, u_0(\cdot), \partial^j u_0(\cdot))(\partial^j)^2 u_0(\cdot) - \frac{\partial f}{\partial(\partial^j u)}(\cdot, u_0(\cdot), \nabla u_0(\cdot)) \text{ in } \Omega,$$

and assume $\|b_j(\cdot)/a_j(\cdot)\|_{L^\infty(\Omega)} \le P_j^*$ hence for our case

$$\left|\left(\frac{\partial a_j}{\partial(\partial^j u)}(\cdot, u_0(\cdot), \partial^j u_0(\cdot))(\partial^j)^2 u_0(\cdot) + \frac{\partial f}{\partial(\partial^j u)}(\cdot, u_0(\cdot), \nabla u_0(\cdot))\right)\right.$$
$$\left. \times \left(a_j(\cdot, u_0(\cdot), \partial^j u_0(\cdot))\right)^{-1}\right| \le P_j^*, \text{ for stability, with}$$

$$a_j, \frac{\partial f}{\partial u}, \frac{\partial a_j}{\partial u}, \frac{\partial a_j}{\partial(\partial^j u)}, \frac{\partial f}{\partial(\partial^j u)}(\cdot, u_0(\cdot), \nabla u_0(\cdot)) \in C(\overline{\Omega}), \text{ and, for asymptotics,}$$

$$u_0 \in C(\overline{\Omega}) \cap C^{2q+2+\alpha}(\Omega), a_j \in C^{2q+\alpha}(\overline{\Omega} \times B),\ f \in C^{2q+\alpha}(\overline{\Omega} \times B_n),\ \partial\Omega \in C^{2q+2+\alpha},$$
$$\text{with closed } B \supset \{(x, u_0(x), \partial^j u_0(x)) \forall x \in \Omega, j = 1, \ldots, n\} \tag{8.227}$$
$$B_n \supset \{(x, u_0(x), \nabla u_0(x)) \forall x \in \Omega\}, \text{ and } 0 < \alpha < 1.$$

Theorem 8.42. Asymptotic expansions for quasilinear problems*: For Ω in (8.98) let $\Omega^h \cup \Omega_i^h$ be connected, and $G_s : \mathcal{D}(G_s) \subset C^2(\Omega) \cap C(\overline{\Omega}) \to C(\Omega)$ in (8.225) satisfy (8.226). Then $G^h(\cdot)$ in (8.225) is stable in $P^h u_0$.*

Additionally let (8.216), $k - 1 \le \min\{3, 2q + \alpha\}$, (8.227), and the more or less geometric type conditions *for these types of problems in Agmon et al. [2], Chapter 7, be satisfied, implying $u_0 \in C^{2q+2+\alpha}(\overline{\Omega})$. Then we obtain the asymptotic expansion*

$$\left\| u_0^h - R^h u_0 - \sum_{i=1}^{\tilde{q}} h^{2i} e_{2i} \right\|_{L^\infty(\overline{\Omega}^h)} \le C h^{k-1} \|u\|_{C^{2q+2+\alpha}(\Omega)}, \tag{8.228}$$

with $\tilde{q} := \max\{m \in \mathbb{N} | 2m < k - 1, m \le q\}$. *These e_{2i} are independent of h, and*

$$e_{2i} \in C^{2q+2-2i+\alpha}(\overline{\Omega}), \text{ and } e_{2i}|_{\partial\Omega} = 0, i = 1, \ldots, \tilde{q}.$$

8.8.2 Asymptotic expansions for other methods

Results based upon the earlier sections in this chapter are summarized. Basic is Proposition 8.5, where we have proved the asymptotic expansion for

$$\bar{f}^h(x) := \frac{1}{h^n} \int_{c^h(x)} f(\bar{x}) d\bar{x}, \quad |\bar{f}^h(x) - f(x)| \le L h^2 \text{ as} \tag{8.229}$$

$$\left| \bar{f}^h(x) - \left(f(x) + \sum_{|\alpha|=2}^{2m', ev} \frac{(h/2)^\alpha}{(\alpha + \vec{1})!} \partial^\alpha f(x) \right) \right| \le C h^{2m'+1}. \tag{8.230}$$

In Subsection 8.8.1, we have already listed results on asymptotic expansions of the discretization errors for those special methods. Here we extend, but do not prove, these results to the methods in Sections 8.4–8.6. We obtain asymptotic expansions in powers of h and h^2 for unsymmetric and symmetric difference methods for the Dirichlet problem on curved and cuboidal domains, respectively. The asymptotic expansions in powers of h^2 are much more interesting for extrapolation methods.

Theorem 8.43. Asymptotic expansion for the difference methods in Sections 8.4–8.6: *For unsymmetric and symmetric difference methods on curved and cuboidal domains, cf. (8.95), with $\mu = 1$ and $\mu = 2$, respectively, we assume the conditions of Theorems 8.31, 8.39. Furthermore, assume $u_0 \in W^{\mu(r+1),p}(\Omega)$, and Lipschitz-continuous $G \in C_L^{\mu(r+1)}$ with respect to the variables in N_{∇^m}. This essentially implies the additional technical conditions, e.g. the $(\mu r, \mu)$-smooth asymptotic expansion of the local discretization error, of the $A^h P^h u - Q'^h Au$, cf. (8.105), (8.106), and $G^h P^h u - Q'^h Gu$, cf. (8.113), (8.114), (8.119), cf. Stetter [596] and Böhmer [103]. Then for small enough h, unique discrete solutions u_0^h for the linear and nonlinear problems in Sections 8.4, 8.5 exist, converge, and admit an asymptotic expansion of the discretization error. More precisely, there exist $e_{\mu k} \in W^{\mu(r+1-k),p}(\Omega), 1 \leq k \leq r$, independent of h, such that*

$$\left\| P^h\left(u_0 - \sum_{k=1}^{r} h^{\mu k} e_{\mu k}\right) - u_0^h \right\|_{\mathcal{V}^h} \leq C h^{\mu r+1} \|u_0\|_{W^{\mu(r+1),p}(\Omega)} \text{ for } u_0 \in W^{\mu(r+1),p}(\Omega). \tag{8.231}$$

We combine asymptotic expansions with Richardson extrapolation: a few step sizes, $h = h_0 > h_1 > h_2 > h_3$, often $h = h_0, h/2, h/4, h/8$, are chosen with $\Omega^h \subset \Omega^{h_1} \subset \Omega^{h_2} \subset \Omega^{h_3}$. The corresponding $G^{h_i}\left(u_0^{h_i}\right) = 0, i = 0, 1, \ldots, t$, defined on Ω^{h_i} are computed. Appropriate linear combinations of these u^{h_i} require the restriction of the u^{h_i} to Ω^h. So define,

$$\text{for } \Omega^h \subset \Omega^{h_1} \subset \Omega^{h_2} \ldots : u_i : \Omega^h \to \mathbb{R}^q : u_i(x) := (R_h u^{h_i})(x) := (u^{h_i})(x) \forall x \in \Omega^h; \tag{8.232}$$

next define the $P_{i,0} := u_i, i = 0, 1, \ldots, t$, and for $i = 0, 1, \ldots, t-j, j = 1, \ldots, t < q$, finally $P_{i,j} := P_{i+1,j-1} + \frac{P_{i+1,j-1} - P_{i,j-1}}{(h_i/h_{i+j})^{\mu} - 1}$.

Theorem 8.44. High order difference methods by Richardson extrapolation: *Let a problem $G(u_0) = 0$ and its stable, consistent discretization $G^h\left(u_0^h\right) = 0$ define solutions u_0 and u_0^h with an asymptotic expansion (8.231), and let h be small enough. Compute the u_i as in (8.232). Then we obtain for $u_0 \in W^{\mu(r+1),p}(\Omega)$ and a constant $\gamma := h_i/h_{i-1} > 1$:*

$$\left\|P^h\left(u_0 - h_i^{\mu} \ldots h_{i+j}^{\mu}(-1)^j e_{\mu(j+1)}\right) - P_{i,j}\right\|_{\mathcal{V}^h} = \mathcal{O}\left(h_i^{\mu} h_i^{\mu} \ldots h_{i+j}^{\mu}\right), t < q, \tag{8.233}$$

$$\textit{for fixed } \gamma : \left\|P^h\left(u_0 - h_{i+j}^{\mu(j+1)}(-1)^j \gamma^{-j(j-1)/2} e_{\mu(j+1)}\right) - P_{i,j}\right\|_{\mathcal{V}^h} = \mathcal{O}\left(h_{i+j}^{\mu(j+1)}\right).$$

Defect and deferred correction methods have strong advantages compared to Richardson extrapolation. As indicated we restrict the discussion to defect corrections, and here to elliptic problems. The basic version has to solve the nonlinear system $G^h\left(u_0^h\right) = 0$ only once and only for one Ω^h. A combination with mesh refinement strategies is possible by computations on different Ω^{h_i}, cf. Böhmer, *et al.* [11, 12, 101, 102, 104, 107–109, 112, 125, 126]. The basic idea of this method is a modification of Newton's method. This has to be combined with the asymptotic expansions in (8.231). We start with the original discrete problem

$$\text{define } u_0^h \text{ by } G^h\left(u_0^h\right) = f^h := Q'^h f, \text{ or } A^h\left(u_0^h\right) = f^h. \tag{8.234}$$

For our linear difference methods, we can update this u_0^h iteratively by the defect correction process. Note that (8.234) is usually a nonlinear system, whereas (8.235) is a sequence of linear problems, always with the same linear $\left(G^h\left(u_0^h\right)\right)'$. The nonlinearity is reduced to computing the nonlinear defect $Q'^h G\left(P_h u_i^h\right)$:

$$\left(G^h\left(u_0^h\right)\right)'\left(u_{i+1}^h - u_i^h\right) = -Q'^h\left(G\left(P_h u_i^h\right) - f\right),\ i = 1, \ldots. \tag{8.235}$$

Here, it is important that the asymptotic expansion of the u_i^h is reproduced by P_h, cf. (8.231), hence with $e_{\mu k,i}, g_{\mu k,i}, \overline{g}_{\mu k,i}$ independent of h,

$$\left\| P^h u_0 - u_i^h - \sum_{k=i+1}^{r} h^{\mu k} e_{\mu k,i} \right\|_{\mathcal{V}^h} \le C h^{\mu r+1} \|u_0\|_{W^{\mu(r+1),p}(\Omega)} \text{ with} \tag{8.236}$$

$$e_{\mu k,i} \in W^{\mu(r+1-k),p}(\Omega), g_{\mu k,i}, \overline{g}_{\mu k,i} \in W^{\mu(r+1-k)-2m,p}(\Omega), 1 \le k \le r,$$

has to imply

$$\left\| P_h u_i^h - (u_0 - \sum_{k=i+1}^{r} h^{\mu k} e_{\mu k,i}) \right\|_{\mathcal{V}^h} \le C h^{\mu r+1} \|u_0\|_{W^{\mu(r+1),p}(\Omega)} \Longrightarrow$$

$$\left\| Q'^h G\left(P_h u_i^h\right) - (Q'^h (G(u_0) - \sum_{k=i+1}^{r} h^{\mu k} g_{\mu k,i}) \right\|_{\mathcal{V}'^h} \le C h^{\mu r+1} \|u_0\|_{W^{\mu(r+1),p}(\Omega)}$$

or

$$\left\| G_{i+1}^h\left(u_i^h\right) - (Q'^h (G(u_0) - \sum_{k=i+1}^{r} h^{\mu k} \overline{g}_{\mu k,i}) \right\|_{\mathcal{V}'^h} \le C h^{\mu r+1} \|u_0\|_{W^{\mu(r+1),p}(\Omega)},$$

where this last $G_{i+1}^h\left(u_i^h\right)$ indicates a difference approximation for $Q'^h G\left(P_h u_i^h\right)$ of order $\mu(i+1)$ again with an asymptotic expansion. The P_h can be defined as above, for $G_{i+1}^h\left(u_i^h\right)$ see e.g. [101, 102, 104, 107].

Theorem 8.45. Defect correction method for G^h: *Under the conditions of Theorem 8.44, and for our linear difference methods the defect correction method (8.235) for G^h in (8.234) yields for small enough h, high order approximations u_i^h of the order $\mu(i+1)$, hence*

$$\left\| P_h u_i^h - \left(u_0 - \sum_{k=i+1}^{r} h^{\mu k} e_{\mu k,i}\right) \right\|_{\mathcal{V}^h} \le C h^{\mu r+1} \|u_0\|_{W^{\mu(r+1),p}(\Omega)} \text{ for } i = 0, \ldots, r-1,$$

$$\text{and with } e_{\mu k,i} \in W^{\mu(r+1-k),p}(\Omega), 1 \le k \le r \text{ independent of } h. \tag{8.237}$$

8.9 Numerical experiments for the von Kármán equations, with C.S. Chien

The strong form of the von Kármán equations is, cf. (2.403)

$$G_s(u,w) := \begin{pmatrix} \Delta^2 u & -[u,w] \\ \\ \Delta^2 w & +\frac{1}{2}[u,u] \end{pmatrix} = \begin{pmatrix} f_1 \\ \\ f_2 \end{pmatrix} \tag{8.238}$$

in $\Omega = [0,1]^2 \subset \mathbb{R}^2$, with $\Delta^2 = \Delta\Delta$, boundary conditions

$$u = \Delta u = 0, \quad w = \Delta w = 0 \text{ on } \partial\Omega,$$

and the bracket operator $[u,v] := u_{xx}v_{yy} - 2u_{xy}v_{xy} + u_{yy}v_{xx} = [v,u]$.

For demonstrating the power of extrapolation, we choose

$$f_1(x,y) := \pi^4[4\sin\pi x \sin\pi y - 5\sin\pi x \sin 2\pi x \sin^2\pi y + 4\cos\pi x \cos 2\pi x \cos^2\pi y]$$

and

$$f_2(x,y) := \pi^4[25\sin 2\pi x \sin\pi y + \sin^2\pi x \sin^2\pi y - \cos^2\pi x \cos^2\pi y].$$

Then the exact solution is

$$u(x,y) = \sin\pi x \sin\pi y, \quad w(x,y) = \sin 2\pi x \sin\pi y.$$

We employ the symmetric difference method for fourth order systems on the square $\Omega = [0,1]^2$ with Ω^h as in (8.95)–(8.97). With step sizes $h = 1/8, 1/16,\ 1/32,\ 1/64$, and $1/128$, we obtain coefficient matrices of order $98 \times 98, 450 \times 450, 1922 \times 1922, 7938 \times 7938$, and 32258×32258. The nonlinear difference systems corresponding to (8.238) are solved with Newton's method. In this academic example we have guaranteed all necessary smoothness conditions.

In Table 8.9 we list the computed $e_u^h := \|P^h u_0 - u_0^h\|_{L^\infty(\Omega^h)}$, $e_w^h := \|P^h w_0 - w_0^h\|_{L^\infty(\Omega^h)}$ and determine the squares of the corresponding experimental orders of convergence, $(\mathrm{EOC})^2$, from the quotients $e_u^h/e_u^{h/2}$, $e_w^h/e_w^{h/2}$, corresponding to the expected $h^2/(h/2)^2 = 4$, cf. (8.231)

In Table 8.10 we compare the errors in the midpoint of the square, in $X := (1/2, 1/2)$ $e_u^h(X) := \left|\left(P^h u_0 - u_0^h\right)(X)\right|$ $e_w^h(X) := \left|\left(P^h w_0 - w_0^h\right)(X)\right|$. Again, we determine the

Table 8.9: Computational errors and the corresponding experimental orders of convergence $(\mathrm{EOC})^2$ of u^h, w^h approximations in the discrete L^∞ norm.

Grid	h	$\lVert e_u^h \rVert_{L^\infty(\Omega^h)}$	$(\mathrm{EOC})^2$	$\lVert e_w^h \rVert_{L^\infty(\Omega^h)}$	$(\mathrm{EOC})^2$
1	1/8	3.4731045E − 02	-	9.6361989E − 02	-
2	1/16	8.156673E − 03	4.2600	2.3262711E − 02	4.1423
3	1/32	2.0079870E − 03	4.0601	5.7658634E − 03	4.0346
4	1/64	5.0015205E − 04	4.0148	1.4383812E − 03	4.0086
5	1/128	1.2568853E − 04	3.9793	3.5913250E − 04	4.0051

Table 8.10: Computational errors and the corresponding experimental orders of convergence, $(EOC)^2$ of the $u^h(X), w^h(X)$ approximations at the midpoint $X = (1/2, 1/2)$.

Grid	h	$\lvert(e_u^h)(X)\rvert$	$(\mathrm{EOC})^2$	$\lvert(e_w^h)(X)\rvert$	$(\mathrm{EOC})^2$
1	1/8	3.0417972E − 02	–	8.2382692E − 02	–
2	1/16	7.2228714E − 03	4.2113	1.9354054E − 02	4.2566
3	1/32	1.7835678E − 03	4.0497	4.7722116E − 04	4.0556
4	1/64	4.4453207E − 04	4.0122	1.1890687E − 04	4.0134
5	1/128	1.1192101E − 04	3.9718	2.9399067E − 05	4.04455

Table 8.11: Computational errors and the corresponding experimental orders of convergence, $(EOC)^2$ of the $\overline{u}^h(X)$ approximations at the midpoint X.

Grid	h	$u^h(X)$	$\overline{u}^h(X)$	$\lvert u^h(X) - \overline{u}^h(X)\rvert$	$(\mathrm{EOC})^2$
1	1/8	1.0304179722	0.999491170	5.0883E − 04	–
2	1/16	1.0072228714	0.999970465	2.9535E − 05	17.2280
3	1/32	1.0017835678	0.999998187	1.8130E − 06	16.2906
4	1/64	1.0004445321	1.000001051	1.0510E − 06	1.7250
5	1/128	1.0001119210			

squares, $(\mathrm{EOC})^2$, corresponding to the expected $h^2/(h/2)^2 = 4$, now for the $e_u^{X,h}/e_u^{X,h/2}$, and $e_w^{X,h}/e_w^{X,h/2}$.

In Table 8.11, we compare the errors in the midpoint X, but now for the first extrapolated approximation. Instead of the $P_{i,j}$ in (8.232), (8.233), with $\gamma = 2$, we use the notation $\overline{u}^h(X) := (4u^{h/2}(X) - u^h(X))/3$. We list the $u^h(X), \overline{u}^h(X), |u(X) - \overline{u}^h(X)|$ and determine the squares, $(\mathrm{EOC})^2$, corresponding to the expected $h^4/(h/2)^4 = 16$ from the $|u(X) - \overline{u}^h(X)|/|u(X) - \overline{u}^{h/2}(X)|$. Similar results hold for $w^h(X)$, $\overline{w}^h(X)$, $|w(X) - \overline{w}^h(X)|$.

9

Variational methods for wavelets, with S. Dahlke

9.1 Introduction

In the previous variational approaches we have used FEs, DCGEs, and grid functions in this book, and will use radial and spectral functions in Böhmer [120]. In this chapter, the spaces of ansatz and test functions are defined by wavelets. Despite the unified convergence theory, cf. Chapter 3, Petryshyn [528, 532], Stetter [596], Stummel [607–609] Vainikko [644] and Zeidler [676,677], these different methods require us to carefully distinguish their different approximation characteristics.

Until now, only *conforming wavelet Galerkin methods* have been applied. We present them here for general elliptic, saddle point, and Navier–Stokes equations. This extends some recent results for wavelet methods. Again, we employ the basic concepts of invertibility, compact perturbation, approximation and general discretization theory. First results, for nonsymmetric, noncoercive, boundedly invertible linear operators have been derived by Böhmer and Dahlke [122]. Here nonlinear problems and bifurcation are indicated, however not explicitely formulated. Böhmer and Dahlke [122] constitute the initial form of this chapter. Our standard result for elliptic problems specified for conforming wavelet Galerkin schemes is applied. Let B be a boundedly invertible operator with stable discretization B^h and $A = B + C$ a compact perturbation of B, such that A is boundedly invertible as well. Then A^h is also stable. In fact, the additional condition of consistent A^h, B^h, C^h is always satisfied for conforming methods. In Chapter 3, cf. [122], we have additionally shown that, under appropriate technical conditions, the stability for the nonlinear problem is guaranteed if it is correct for the linearized operator, cf. Stetter [596] and Böhmer [116], for variational methods. This applies to wavelet methods, if the exact evaluation of all nonlinear terms in the operator is possible. For wavelet methods this is still a problem for future research, in particular since in the wavelet setting, exact evalutions of nonlinear operators are hard to achieve, see e.g. [69, 251] for details. Therefore we restrict the discussion to nonlinear problems, e.g. in (9.13), where the nonlinear terms can be evaluated for wavelet arguments, see Subsection 9.3.5.

The linearized Navier–Stokes equations can be interpreted as compact perturbations of the Stokes equations, at least for moderate Reynolds numbers. In this sense, we obtain convergence results for wavelet applied to the Navier–Stokes equations on the same level as the finite element methods. These problems have been carefully studied by a different approach in, e.g. Girault and Raviart [348] and Temam [622].

For our conforming wavelet methods, convergence is, for linear coercive problems, an immediate consequence of Cea's Lemma 4.48. If the solution u_0 is well approximated, here $\inf_{v_n \in S_n} \|u_0 - v_n\|_{L^2(\Omega) \to 0}$, with the stability implied by the coercivity, we obtain "optimal" convergence.

During the last few years, wavelets have shown some potential for the numerical treatment of partial differential and integral equations, c.f. e.g. Barinka *et al.* [70], Cohen *et al.* [197, 198], Dahlke *et al.* [236, 237, 241] and Dahmen [243].

The *advantages of wavelet methods* can be described as follows. It turns out that a simple diagonal scaling applied to stiffness matrices relative to wavelet bases suffices to produce *uniformly bounded condition numbers.* Moreover, for a wide class of integral or pseudo-differential operators the stiffness matrix relative to wavelet bases can be shown to be sufficiently close to sparse matrices so that efficient sparse solvers can be applied and yield powerful stable and convergent Galerkin schemes. The most far-reaching results were obtained for linear self-adjoint and saddle point problems. For these problems, it has even been possible to derive optimal convergent adaptive wavelet schemes, see Cohen *et al.* [197] and Dahlke *et al.* [236,237]. Numerical tests for self-adjoint and saddle point problems, e.g. the Stokes problem, with wavelet methods are documented in Barinka *et al.* [70] and Dahlke *et al.* [237].

Adaptive wavelet techniques require highly technical proofs, but do have a lot of advantages whenever they are needed. They can be applied only to boundedly invertible (linearized) operators A. This is the general assumption in the cited papers: Barinka *et al.* [70], Cohen *et al.* [197, 198], Dahlke *et al.* [236] and Dahmen [243]. For a nonsymmetric, boundedly invertible A, for these techniques the use of the symmetric coercive A^*A is unavoidable, but expensive, cf. e.g. Sections 3 and 7 in Cohen *et al.* [198] for further details. Here A^* denotes the adjoint of A. This excludes bifurcation problems. There is another difficulty related to A^*A: It is well-known that cond $(A^*A) = (\text{cond}(A))^2$ explodes for increasing $\text{cond}(A)$. First results for nonsymmetric, boundedly invertible problems that avoid these difficulties and which are based on perturbation arguments have been derived by Gantumur [332].

The adaptive treatment of problems where the underlying operator is not boundedly invertible, i.e. of ill-posed problems, is a very challenging task that has recently been intensively studied. We refer to Bonesky *et al.* [131,132] for further details. First results for nonlinear equations have been developed by Cohen *et al.* [199].

This chapter is organized as follows. In Section 9.2, we briefly discuss the scope of problems we shall be concerned with. Especially, we introduce the setting of elliptic problems. In Section 9.3 we present the basic concepts of wavelet analysis, i.e. the discrete wavelet transform, biorthogonal bases, approximations of functions by wavelets, construction of wavelets on domains and manifolds, and the evaluation of nonlinear functionals applied to wavelet expansions. Section 9.4 discusses the first steps, definitions and basic tools, e.g. stability and preconditioning for wavelet methods. Section 9.5 is devoted to wavelet methods for general linear and nonlinear elliptic problems. We explain how suitable Galerkin schemes based on wavelets can be constructed and discuss their stability and consistency properties. Then, in Section 9.6, we apply these results via its linearization to the Navier–Stokes equations. Finally, in Section 9.7, we explain how adaptive numerical methods based upon wavelets can be designed.

The following problems, only presented for FEMs, can be solved and yield the corresponding results for wavelet methods as well: general convergence theory for monotone operators, cf. Section 4.5, variational methods for eigenvalue problems, cf. Section 4.7, methods for nonlinear boundary operators which can be evaluated for wavelet functions, cf. Section 5.3, quadrature approximate methods, cf. Section 5.4.

All this is limited to problems where the nonlinearities can be evaluated for wavelets, cf. Theorem 9.13. For all the nonlinear problems, considered here, the corresponding nonlinear equations can be solved with Newton's method, satisfying the mesh independence principle. Methods for fully nonlinear elliptic problems are impossible with our technique.

In this chapter we use the notation $\mathbb{R}^d, \mathbb{C}^d, \ldots$, instead of the previous $\mathbb{R}^n, \mathbb{C}^n, \ldots$, and restrict the presentation to a Hilbert space setting.

9.2 The scope of problems

In this section, we shall briefly explain the scope of problems we shall be concerned with. The goal is to derive a stable and convergent wavelet scheme for problems with a linearization, that can be interpreted as boundedly invertible compact perturbations of an operator equation with stable discretization. Especially, we are interested in problems related to well-known elliptic and saddle point problems. The Stokes operator and, for moderate Reynolds numbers, its compact perturbation, the (linearized) Navier–Stokes operator, fall into this category. Therefore we first recall the general setting of operator equations.

Suppose that $\mathcal{U}$ is a Hilbert space with norm $\|\cdot\|_{\mathcal{U}}$ induced by the inner product $\langle\cdot,\cdot\rangle$. Let $A : \mathcal{U} \longrightarrow \mathcal{U}'$ denote a linear operator into the normed dual $\mathcal{U}'$ of $\mathcal{U}$. In contrast to the other methods, we assume throughout that in this chapter $\mathcal{U}$ and the approximating subspaces, $S_\Lambda := \{\psi_\lambda : \lambda \in \Lambda\} \subset \mathcal{U}$, *include the appropriate boundary conditions.* We shall mainly discuss the case that A can be written as

$$A = B + C, \tag{9.1}$$

with a boundedly invertible operator, $B \in \mathcal{L}(\mathcal{U},\mathcal{U}')$, with stable discretization, and a compact $C \in \mathcal{C}(\mathcal{U},\mathcal{U}')$. We start with a short presentation of general elliptic equations. It is well known that, given a Hilbert space $\mathcal{U}$, its dual $\mathcal{U}'$ and the dual pairing $\langle\cdot,\cdot\rangle_{\mathcal{U}'\times\mathcal{U}}$, a general elliptic operator $A : \mathcal{U} \to \mathcal{U}'$ induces a continuous and *elliptic* bilinear form, see Sections 2.3 and 2.4,

$$\begin{aligned} a \,:\, \mathcal{U}\times\mathcal{U} \to \mathbb{R},\ a(u,v) := \langle Au, v\rangle_{\mathcal{U}'\times\mathcal{U}}\ \forall\, u,v \in \mathcal{U} \text{ and} \\ a(v,v) \geq \alpha\|v\|_{\mathcal{U}}^2 - \gamma\|v\|_{\mathcal{W}_{\mathcal{U}}}^2,\ \alpha > 0, \text{ for Hilbert spaces } \mathcal{U} \hookrightarrow \mathcal{W}_{\mathcal{U}} = \mathcal{W}_{\mathcal{U}}' \hookrightarrow \mathcal{U}', \end{aligned} \tag{9.2}$$

usually a *Gelfand triple* $\mathcal{U} \hookrightarrow \mathcal{W}_{\mathcal{U}} = \mathcal{W}_{\mathcal{U}}' \hookrightarrow \mathcal{U}'$ with dense and continuous embedding $\mathcal{U} \hookrightarrow \mathcal{W}_{\mathcal{U}}$. Furthermore, in $A = B + C$ we usually assume B, inducing a *coercive* bilinear form $b(\cdot,\cdot)$, i.e.

$$\langle Bv, v\rangle_{\mathcal{U}'\times\mathcal{U}} = b(v,v) \geq \alpha\|v\|_{\mathcal{U}}^2, \qquad \alpha > 0, \qquad \text{for all } v \in \mathcal{U}. \tag{9.3}$$

It is known that, for an invertible A, in particular for a B in (9.3), the equation

$$a(u_0, v) = \langle f, v\rangle_{\mathcal{U}'\times\mathcal{U}} \quad \text{for all } v \in \mathcal{U} \tag{9.4}$$

is uniquely solvable.

Consequently, typical examples for A and B are given by general elliptic partial differential operators and their invertible special cases satisfying (9.3). For example, the operator induced by the Poisson equation[49]

$$\begin{aligned} -\triangle u &= f, \quad \text{in } \Omega \subset \mathbb{R}^d \\ u &= 0, \quad \text{on } \partial\Omega, \end{aligned} \tag{9.5}$$

would play the role of B. This $B = -\triangle$ is a bounded and boundedly invertible mapping of $H_0^1(\Omega)$ onto its dual $H_0^{-1}(\Omega)$, i.e. $\|Bu\|_{\mathcal{U}'} \sim \|u\|_{\mathcal{U}}$. Here "$a \sim b$" means that both quantities can be uniformly bounded by some constant multiple of each other. Likewise, "$\lesssim$" indicates inequality up to constant factors.

We generalize the Laplacian to elliptic operators of order $2m$. The generalization to systems follows exactly Section 4.3.2, cf. (9.13).

For the reader's convenience we repeat the relevant notation and equations: for the *multi-indices* α, we defined, see (2.73), partials, vectors and reals as

$$\partial^i u = \frac{\partial u}{\partial x_i}, \nabla u = (\partial^1 u, \ldots, \partial^d u) \text{ and } \partial^\alpha u = \partial_1^{\alpha_1} \ldots \partial_d^{\alpha_d} u = \frac{\partial^{|\alpha|}}{\partial x_1^{\alpha_1} \ldots \partial x_d{}^{\alpha_d}} u,$$
$$\vartheta = (\vartheta^1, \ldots, \vartheta^d) \in \mathbb{R}^d, \vartheta^\alpha = (\vartheta^1)^{\alpha_1} \cdots (\vartheta^d)^{\alpha_d}, \vartheta^i, \partial^i u(x), \partial^\alpha u(x) \in \mathbb{R}.$$

A linear second order elliptic equation, see (2.160), determines the solution u_0, such that

$$u_0 \in \mathcal{U}: \quad a(u_0, v) = \langle Au_0, v\rangle_{\mathcal{U}'\times\mathcal{U}} = \int_\Omega \sum_{i,j=0}^{d} a_{ij}\partial^i u_0 \partial^j v dx = \langle f, v\rangle_{\mathcal{U}'\times\mathcal{U}}$$
$$\forall\, v \in \mathcal{U} = H_0^1(\Omega), \text{ with } \; a_{ij} \in L^\infty(\Omega). \tag{9.6}$$

The linear form f and our elliptic bilinear form $a(u,v)$ are bounded, so

$$|\langle f, v\rangle_{\mathcal{U}'\times\mathcal{U}}| \leq C'\|v\|_{\mathcal{U}}, \quad |a(u,v)| \leq C\|u\|_{\mathcal{U}}\|v\|_{\mathcal{U}}. \tag{9.7}$$

The principal part, $a_p(u,v)$, satisfies for all $\vartheta \in \mathbb{R}^d$

$$\lambda|\vartheta|_d^2 \leq \lambda(x)|\vartheta|_d^2 \leq \sum_{i,j=1}^{d} a_{ij}(x)\vartheta^i\vartheta^j \leq \Lambda|\vartheta|_d^2 \;\; \forall\, x \in \Omega \text{ with } \lambda > 0. \tag{9.8}$$

[49] Note that we use in this chapter $\Omega \subset \mathbb{R}^d$ instead of $\Omega \subset \mathbb{R}^n$ in the previous chapters.

By (2.22), (2.79), $a_p(u,v)$ is bounded, and $\mathcal{U} = H_0^1(\Omega)$-coercive, hence there exists $\lambda^* > 0$, such that

$$a_p(u,u) = \int_\Omega \sum_{i,j=1}^{d} a_{ij}\partial^i u \partial^j u dx \geq \lambda^* \|u\|_{\mathcal{U}}^2, \forall u \in \mathcal{U}, \text{ i.e. } |a_p(u,u)| \sim \|u\|_{\mathcal{U}}^2. \tag{9.9}$$

Elliptic equations of order $2m$ require $\mathcal{U} = H_0^m(\Omega)$. For linear A, determine

$$u_0 \in \mathcal{U} = H_0^m(\Omega): \quad a(u_0,v) := \int_\Omega \sum_{|\alpha|,|\beta|\leq m} a_{\alpha\beta}\partial^\alpha u_0 \partial^\beta v dx = \langle f, v\rangle_{\mathcal{U}'\times\mathcal{U}} \tag{9.10}$$

$$\forall v \in \mathcal{U} \text{ with } a_{\alpha\beta} \in L^\infty(\Omega), \text{ and for } 1 < m = |\alpha| = |\beta|: \ a_{\alpha\beta} \in C(\bar{\Omega}).$$

The previous $a(u,v)$ is bounded and has to be elliptic, hence the principal part, $a_p(u,v)$, is $\mathcal{U}$-coercive, satisfying with $\lambda, \lambda^* > 0$ and $\forall \vartheta \in \mathbb{R}^d$, see (2.79), (4.132),

$$\lambda|\vartheta|_d^{2m} \leq \lambda(x)|\vartheta|_d^{2m} \leq (-1)^m \sum_{|\alpha|=|\beta|=m} a_{\alpha\beta}(x)\vartheta^\alpha\vartheta^\beta \leq \Lambda|\vartheta|_d^{2m} \ \forall\, x \in \Omega, \tag{9.11}$$

$$\forall u \in \mathcal{U}: a_p(u,u) = \int_\Omega \sum_{|\alpha|=|\beta|=m} a_{\alpha\beta}\partial^\alpha u_0 \partial^\beta u dx \geq \lambda^* \|u\|_{\mathcal{U}}^2, \text{ i.e. } |a_p(u,u)| \sim \|u\|_{\mathcal{U}}^2.$$

Quasilinear elliptic systems of order $2m$ extend and modify the above equations and conditions, cf. Subsection 2.6.3 ff. In the introduction we have already mentioned the difficulties in evaluating nonlinear functionals with wavelet arguments, cf. Subsection 9.3.5. Therfore, we impose the following condition: we admit only coefficients such that

$$A_\alpha \textit{ in (9.13)}, \ A_\alpha(x, \vec{u}, \ldots, \nabla^m \vec{u}), \textit{ can be evaluated for wavelet arguments.} \tag{9.12}$$

Under this condition we consider for $\mathcal{U} = H_0^m(\Omega, \mathbb{R}^q), q \geq 1$, the weak form, and determine $\vec{u}_0 \in \mathcal{D}(G) \subset \mathcal{U}$

$$\begin{aligned} &\vec{u}_0 \in \mathcal{D}(G) \text{ s.t.} \\ &a(\vec{u}_0, \vec{v}) := \langle G\vec{u}_0, \vec{v}\rangle_{\mathcal{U}'\times\mathcal{U}} := \textstyle\int_\Omega \sum_{|\alpha|\leq m}(A_\alpha(x, \vec{u}_0, \ldots, \nabla^m \vec{u}_0), \partial^\alpha \vec{v})_q dx \\ &\qquad = \langle \vec{f}, \vec{v}\rangle_{\mathcal{U}'\times\mathcal{U}} = \textstyle\sum_{|\alpha|\leq m}(\vec{f}_\alpha \partial^\alpha \vec{v})_q dx \ \forall\, \vec{v} \in \mathcal{U} = H_0^m(\Omega, \mathbb{R}^q), \\ &\qquad \vec{f}_\alpha \in L^2(\Omega, \mathbb{R}^q), q \geq 1, \end{aligned} \tag{9.13}$$

cf. Proposition 2.34. The appropriate conditions are listed in Sections 2.6.4, 2.6.6. For the linearized principle part, (9.11) has to be replaced by (2.394).

Applications to saddle point and Navier–Stokes equations will be discussed in Section 9.6.

9.3 Wavelet analysis

Our goal is conforming Galerkin methods for the approximate solution of $Au = f$ or $Gu = f$ for operators A, G as in Section 9.2. However, in contrast to conventional finite element discretizations we will work with trial spaces that not only exhibit the usual approximation properties and good localization but in addition lead to *expansions* of any element in the underlying Hilbert spaces in terms of *multiscale* or *wavelet* bases with certain stability properties. This fact can be used to derive, e.g. very efficient preconditioning strategies as well as powerful adaptive algorithms, see Section 9.4, Theorem 9.6, and Section 9.7. To understand these relationships, we shall first of all briefly recall the basic setting of wavelet analysis as far as it is needed for our purposes. In Subsection 9.3.1, we want to collect some facts concerning the discrete wavelet transform. Then, in Subsection 9.3.2, we discuss the biorthogonal wavelet approach. It is one of the most important properties of wavelets that they form stable bases for scales of smoothness spaces such as Sobolev and Besov spaces. These aspects will be discussed in Section 9.3.3. Then, in Subsection 9.3.4, we briefly recall the construction of wavelet bases on bounded domains. Finally, in Subsection 9.3.5, we discuss how nonlinear functionals applied to wavelet expansions can be evaluated.

9.3.1 The discrete wavelet transform

In general, a function ψ is called a *wavelet* if all its scaled, dilated, and integer-translated versions

$$\psi_{j,k}(x) := 2^{j/2}\psi(2^j x - k), \quad j,k \in \mathbb{Z}, \tag{9.14}$$

form a (Riesz) basis of $L^2(\mathbb{R})$. Usually, they are constructed by means of a *multiresolution analysis* introduced by Mallat [482]:

Definition 9.1. *A sequence* $\{\mathcal{V}_j\}_{j\in\mathbb{Z}}$ *of closed subspaces of* $L^2(\mathbb{R})$ *is called a* multiresolution analysis (MRA) *of* $L^2(\mathbb{R})$ *if*

$$\cdots \subset \mathcal{V}_{j-1} \subset \mathcal{V}_j \subset \mathcal{V}_{j+1} \subset \ldots; \tag{9.15}$$

$$\overline{\bigcup_{j=-\infty}^{\infty} \mathcal{V}_j} = L^2(\mathbb{R}); \tag{9.16}$$

$$\bigcap_{j=-\infty}^{\infty} \mathcal{V}_j = \{0\}; \tag{9.17}$$

$$f(\cdot) \in \mathcal{V}_j \Longleftrightarrow f(2\cdot) \in \mathcal{V}_{j+1}; \tag{9.18}$$

$$f(\cdot) \in \mathcal{V}_0 \Longleftrightarrow f(\cdot - k) \in \mathcal{V}_0 \qquad \text{for all} \quad k \in \mathbb{Z}. \tag{9.19}$$

Moreover, we assume that there exists a function $\varphi \in \mathcal{V}_0$ *such that*

$$\mathcal{V}_0 := \overline{\operatorname{span}\{\varphi(\cdot - k),\, k \in \mathbb{Z}\}} \tag{9.20}$$

and that φ *has* stable integer translates, *i.e., with*

$$\ell_2(\mathbb{Z}) := \left\{ \mathbf{c} := \{c_k\}_{k\in\mathbb{Z}} : \|\mathbf{c}\|_{\ell_2} := \left(\sum_{k\in\mathbb{Z}} c_k^2\right)^{1/2} < \infty \right\},$$

$$C_1\|\mathbf{c}\|_{\ell_2} \le \|\sum_{k\in\mathbb{Z}} c_k\,\varphi(\cdot - k)\|_{L^2} \le C_2\|\mathbf{c}\|_{\ell_2}, \qquad \mathbf{c} := \{c_k\}_{k\in\mathbb{Z}} \in \ell_2(\mathbb{Z}). \quad (9.21)$$

The function φ *is called the* generator *of the multiresolution analysis.*

The properties (9.15), (9.18), (9.20), and (9.21) immediately imply that φ is *refinable*, i.e. it satisfies a *two-scale relation*

$$\varphi(x) = \sum_{k\in\mathbb{Z}} a_k\varphi(2x-k), \qquad (9.22)$$

with the *mask* $\mathbf{a} = \{a_k\}_{k\in\mathbb{Z}} \in \ell_2(\mathbb{Z})$. A function satisfying a relation of the form (9.22) is sometimes also called a *scaling function.* Because the union of the spaces $\{\mathcal{V}_j\}_{j\in\mathbb{Z}}$ is dense, and their intersection is $\{0\}$, it is easy to see that the construction of a wavelet basis reduces to finding a function whose translates span a complement space $\mathcal{W}_0$ of $\mathcal{V}_0$ in $\mathcal{V}_1$,

$$\mathcal{V}_1 = \mathcal{V}_0 \oplus \mathcal{W}_0, \qquad \mathcal{W}_0 = \overline{\operatorname{span}\{\psi(\cdot - k) \mid k \in \mathbb{Z}\}}. \qquad (9.23)$$

Indeed, if we define

$$\mathcal{W}_j := \{f(\cdot) \in L^2(\mathbb{R}) \mid f(2^{-j}\cdot) \in \mathcal{W}_0\}, \qquad (9.24)$$

it follows from (9.16), (9.17) and (9.18) that

$$L^2(\mathbb{R}) = \bigoplus_{j=-\infty}^{\infty} \mathcal{W}_j, \qquad (9.25)$$

so that

$$\psi_{j,k}(\cdot) = 2^{j/2}\psi(2^j\cdot - k), \qquad j,k\in\mathbb{Z} \qquad (9.26)$$

forms a wavelet basis of $L^2(\mathbb{R})$.

Obviously, (9.18) and (9.20) imply that the wavelet ψ can be found by means of a functional equation of the form

$$\psi(x) = \sum_{k\in\mathbb{Z}} b_k\varphi(2x-k), \qquad (9.27)$$

where the sequence $\mathbf{b} := \{b_k\}_{k\in\mathbb{Z}}$ has to be judiciously chosen; see, e.g. [173, 254, 490] for details.

The construction outlined above is quite general. In many applications, it is convenient to impose some more conditions, i.e. to require that functions on different scales are orthogonal with respect to the usual L^2 inner product, i.e.

$$\langle \psi(2^j\bullet - k), \psi(2^{j'}\bullet - k')\rangle = 0, \qquad \text{if} \quad j \ne j'. \qquad (9.28)$$

This can be achieved if the translates of ψ not only span an (algebraic) complement but the orthogonal complement,

$$\mathcal{V}_0 \perp \mathcal{W}_0, \qquad \mathcal{W}_0 = \overline{\mathrm{span}\{\psi(\bullet - k) \mid k \in \mathbb{Z}\}}. \tag{9.29}$$

The resulting functions are sometimes called *pre-wavelets.* The basic properties of scaling functions and (pre-) wavelets can be summarized as follows:

- Reproduction of polynomials. If φ is contained in $C_0^r(\mathbb{R}) := \{g \mid g \in C^r(\mathbb{R}) \text{ and } \operatorname{supp} g \text{ compact}\}$, then every monomial x^α has an expansion of the form

$$x^\alpha = \sum_{k\in\mathbb{Z}} c_k^\alpha \varphi(x-k), \quad \alpha \leq r. \tag{9.30}$$

- Oscillations. If the generator φ is contained in $C_0^r(\mathbb{R})$, then the associated wavelet ψ has vanishing moments up to order r, i.e.

$$\int_{\mathbb{R}} x^\alpha \psi(x)dx = 0 \qquad \text{for all} \quad 0 \leq \alpha \leq r. \tag{9.31}$$

- Approximation. If $\varphi \in C_0^r(\mathbb{R})$ and $f \in H^r(\mathbb{R})$, then the following Jackson-type inequality holds, cf. Lemma 9.7:

$$\inf_{g\in\mathcal{V}_j} \|f-g\|_{L^2(\mathbb{R})} \leq C2^{-jr}|f|_{H^r}. \tag{9.32}$$

In practice, it is clearly desirable to work with an *orthonormal* wavelet basis. This can be realized as follows. Given an ℓ_2-stable generator in the sense of (9.21), one may define another generator ϕ by

$$\hat{\phi}(\xi) := \frac{\hat{\varphi}(\xi)}{(\sum_{k\in\mathbb{Z}} |\hat{\varphi}(\xi+2\pi k)|^2)^{1/2}}. \tag{9.33}$$

It can be checked directly that the translates of ϕ are orthonormal and span the same space $\mathcal{V}_0$. The generator ϕ is also refinable,

$$\phi(x) = \sum_{k\in\mathbb{Z}} \hat{a}_k \phi(2x-k) \tag{9.34}$$

and it can be shown that the function

$$\psi(x) = \sum_{k\in\mathbb{Z}} b_k \phi(2x-k), \qquad b_k := (-1)^k \hat{a}_{1-k} \tag{9.35}$$

is an orthonormal wavelet with the same regularity properties as the original generator φ. However, this approach has a serious disadvantage. If the generator φ is compactly supported, this property will in general not carry over to the resulting wavelet since it gets lost during the orthonormalization procedure (9.33). Therefore the compact support will only be preserved if we can dispense with the orthonormalization procedure, i.e. if the translates of φ are already orthonormal. This observation was the starting point for the investigations of I. Daubechies who constructed a family ϕ^N, $N \in \mathbb{N}$, of generators with the following properties [253, 254].

Theorem 9.2. Good wavelet generators: *There exists a constant* $\beta > 0$ *and a family* ϕ^N *of generators satisfying* $\phi^N \in C^{\beta N}(\mathbb{R})$, supp $\phi^N = [0, 2N-1]$, *and*

$$\langle \phi^N(\cdot), \phi^N(\cdot - k)\rangle = \delta_{0,k}, \qquad \phi^N(x) = \sum_{k=N_1}^{N_2} a_k \phi^N(2x-k). \tag{9.36}$$

Obviously, (9.36) and (9.35) imply that the associated wavelet ψ^N is also compactly supported with the same regularity properties as ϕ^N.

Given such an orthonormal wavelet basis, any function $v \in \mathcal{V}_j$ has two equivalent representations, the *single-scale representation* with respect to the functions $\phi_{j,k}(x) = 2^{j/2}\phi(2^j x - k)$ and the *multiscale representation* which is based on the functions $\phi_{0,k}, \psi_{l,m}, k, m \in \mathbb{Z}, 0 \le l < j, \psi_{l,m} := 2^{j/2}\psi(2^j x - m)$. From the coefficients of v in the single-scale representation, the coefficients in the multiscale representation can easily be obtained by some kind of filtering, and vice versa. Indeed, given

$$v = \sum_{k \in \mathbb{Z}} \lambda_k^j \phi_{j,k}$$

and using the refinement equation (9.34) and the functional equation (9.35), it turns out that

$$v = \sum_{l \in \mathbb{Z}} 2^{-1/2} \left(\sum_{k \in \mathbb{Z}} \bar{a}_{k-2l} \lambda_k^j \right) \phi_{j-1,l} + \sum_{m \in \mathbb{Z}} 2^{-1/2} \left(\sum_{k \in \mathbb{Z}} \bar{b}_{k-2l} \lambda_k^j \right) \psi_{j-1,m}. \tag{9.37}$$

From (9.37) we observe that the coefficient sequence $\lambda^{j-1} = \{\lambda_k^{j-1}\}_{k \in \mathbb{Z}}$ which describes the information corresponding to $\mathcal{V}_{j-1}$ can be obtained by applying the low-pass filter H induced by $\mathbf{a}$ to λ^j,

$$\lambda^{j-1} = H\lambda^j, \qquad \lambda_l^{j-1} = \sum_{k \in \mathbb{Z}} 2^{-1/2} \bar{a}_{k-2l} \lambda_k^j. \tag{9.38}$$

The wavelet space $\mathcal{W}_{j-1}$ describes the detailed information added to $\mathcal{V}_{j-1}$. From (9.37), we can conclude that this information can be obtained by applying the high-pass filter D induced by $\mathbf{b}$ to λ^j:

$$c^{j-1} = D\lambda^j, \qquad c_l^{j-1} = 2^{-1/2} \sum_{k \in \mathbb{Z}} \bar{b}_{k-2l} \lambda_k^j. \tag{9.39}$$

By iterating this decomposition method, we obtain a pyramid algorithm, the so-called *fast wavelet transform*:

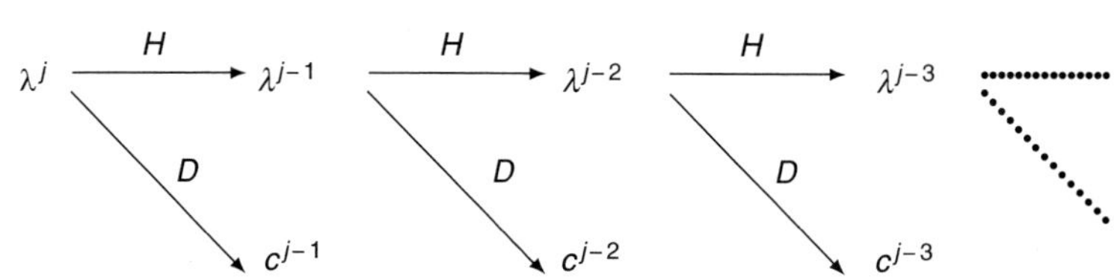

A reconstruction algorithm can be obtained in a similar fashion. Indeed, a straightforward computation shows

$$\sum_{l\in\mathbb{Z}} \lambda_l^j \phi_{j,l} = \frac{1}{\sqrt{2}} \sum_{l\in\mathbb{Z}} \left(\sum_{k\in\mathbb{Z}} a_{l-2k}\lambda_k^{j-1} + \sum_{n\in\mathbb{Z}} b_{l-2n}c_n^{j-1} \right) \phi_{j,l}$$

so that

$$\lambda_l^j = 2^{-1/2} \sum_{k\in\mathbb{Z}} a_{l-2k}\lambda_k^{j-1} + 2^{-1/2} \sum_{n\in\mathbb{Z}} b_{l-2n}c_n^{j-1}. \tag{9.40}$$

Similar decomposition and reconstruction schemes also exist for the pre-wavelet case.

There are several methods to generalize this concept to higher dimensions. The simplest way is to use *tensor products*. Given a univariate multiresolution analysis with generator φ, it turns out that

$$\phi(x_1, \ldots, x_d) := \varphi(x_1) \cdots \varphi(x_d) \tag{9.41}$$

generates a multiresolution analysis of $L^2(\mathbb{R}^d)$. Let E denote the vertices of the unit cube in $\mathbb{R}^d$. Defining $\psi^0 := \varphi$ and $\psi^1 := \psi$, it can be shown that the set Ψ of $2^d - 1$ functions

$$\psi^e(x_1, \ldots, x_d) := \prod_{l=1}^{d} \psi^{e_l}(x_l) \qquad e = (e_1, \ldots, e_d) \in E \backslash \{0\}, \tag{9.42}$$

generates by shifts and dilates a basis of $L^2(\mathbb{R}^d)$. There also exist multivariate wavelet constructions with respect to *nonseparable* refinable functions ϕ satisfying

$$\phi(x) = \sum_{k\in\mathbb{Z}^d} a_k \phi(2x - k), \qquad \{a_k\}_{k\in\mathbb{Z}^d} \in \ell_2(\mathbb{Z}^d), \tag{9.43}$$

see, e.g. [420] for details. Analogously to the tensor product case, a family ψ^i, $i = 1, \ldots, 2^d - 1$, of wavelets is needed. Each ψ^i satisfies a functional equation similar to (9.27):

$$\psi^i(x) = \sum_{k\in\mathbb{Z}^d} b_k^i \phi(2x - k). \tag{9.44}$$

The basic properties of wavelets and scaling functions (approximation, oscillation, etc.) carry over to the multivariate case in the usual way.

9.3.2 Biorthogonal bases

Given an orthonormal wavelet basis, the basic calculations are usually quite simple. For instance, the wavelet expansion of a function $v \in L^2(\mathbb{R})$ can be computed as

$$v = \sum_{j,k\in\mathbb{Z}} \langle v, \psi_{j,k} \rangle \psi_{j,k}, \qquad \psi_{j,k}(x) := 2^{j/2}\psi(2^j x - k). \tag{9.45}$$

However, requiring smoothness and orthonormality is quite restrictive, and consequently, as we have already seen above, the resulting wavelets are usually not compactly supported. It is one of the advantages of the pre-wavelet setting that the compact support property of the generator can be preserved. Moreover, since we have to deal with weaker conditions, the pre-wavelet approach provides us with much more flexibility. Therefore, given a generator φ, many different families of pre-wavelets adapted to a specific application can be constructed. Nevertheless, since orthonormality is lost, one is still interested in finding suitable alternatives which in some sense provide a compromise between both concepts. This can be performed by using the *biorthogonal* approach. For a given (univariate) wavelet basis $\{\psi_{j,k},\ j,k \in \mathbb{Z}\}$, $\psi_{j,k}(x) := 2^{j/2}\psi(2^j x - k)$, one is interested in finding a second system $\{\tilde{\psi}_{j,k},\ j,k \in \mathbb{Z}\}$ satisfying

$$\langle \psi_{j,k}(\cdot), \tilde{\psi}_{j',k'}(\cdot)\rangle = \delta_{j,j'}\delta_{k,k'}, \quad j,j',k,k' \in \mathbb{Z}. \tag{9.46}$$

Then all the computations are as simple as in the orthonormal case, i.e.

$$v = \sum_{j,k\in\mathbb{Z}} \langle v, \tilde{\psi}_{j,k}\rangle \psi_{j,k} = \sum_{j',k'\in\mathbb{Z}} \langle v, \psi_{j',k'}\rangle \tilde{\psi}_{j',k'}. \tag{9.47}$$

To construct such a biorthogonal system, one needs *two* sequences of approximation spaces $\{\mathcal{V}_j\}_{j\in\mathbb{Z}}$ and $\{\tilde{\mathcal{V}}_j\}_{j\in\mathbb{Z}}$. As for the orthonormal case, one has to find bases for certain algebraic complement spaces $\mathcal{W}_0$ and $\tilde{\mathcal{W}}_0$ satisfying the biorthogonality conditions

$$\mathcal{V}_0 \perp \tilde{\mathcal{W}}_0, \quad \tilde{\mathcal{V}}_0 \perp \mathcal{W}_0, \quad \mathcal{V}_0 \oplus \mathcal{W}_0 = \mathcal{V}_1, \quad \tilde{\mathcal{V}}_0 \oplus \tilde{\mathcal{W}}_0 = \tilde{\mathcal{V}}_1. \tag{9.48}$$

This is quite easy if the two generators ϕ and $\tilde{\phi}$ form a *dual pair*,

$$\langle \phi(\cdot), \tilde{\phi}(\cdot - k)\rangle = \delta_{0,k}. \tag{9.49}$$

Indeed, then two biorthogonal wavelets ψ and $\tilde{\psi}$ can be constructed as

$$\psi(x) = \sum_{k\in\mathbb{Z}} (-1)^k d_{1-k}\phi(2x-k), \qquad \tilde{\psi}(x) = \sum_{k\in\mathbb{Z}} (-1)^k a_{1-k}\tilde{\phi}(2x-k) \tag{9.50}$$

where

$$\phi(x) = \sum_{k\in\mathbb{Z}} a_k\phi(2x-k), \qquad \tilde{\phi}(x) = \sum_{k\in\mathbb{Z}} d_k\tilde{\phi}(2x-k). \tag{9.51}$$

Therefore, given a *primal* generator ϕ, one has to find a smooth and compactly supported *dual* generator $\tilde{\phi}$ satisfying (9.49) which is much less restrictive than the orthonormal setting. Elegant constructions can be found, e.g. in [202]. There, especially, the important case where the primal generator is a cardinal B-spline, $\phi = N_m,\ m \geq 1$, is discussed in detail. Generalizations to higher dimensions also exist [201].

For further information on wavelet analysis, the reader is referred to one of the excellent textbooks on wavelets which have appeared quite recently [173, 254, 433, 490, 669].

9.3.3 Wavelets and function spaces

It is one of the most important properties of wavelets that they give rise to stable bases in scales of function spaces. For wavelet applications, the *Besov* spaces are the most important smoothness spaces. Therefore in this section we introduce the Besov spaces and recall their characterization by means of wavelet expansions. For simplicity, we restrict ourselves to the biorthogonal setting and the univariate case. Similar results also hold for the multivariate and the pre-wavelet setting. We refer to [329] and [490] for details.

Let us start by recalling the definition of Besov spaces. The *modulus of smoothness* $\omega_r(v,t)_{L_p(\mathbb{R})}$ of a function $v \in L_p(\mathbb{R})$, $0 < p \leq \infty$, is defined by

$$\omega_r(v,t)_{L_p(\mathbb{R})} := \sup_{|h|\leq t} \|\Delta_h^r(v,\cdot)\|_{L_p(\mathbb{R})}, \quad t > 0,$$

with Δ_h^r the r-th difference with step h. For $s > 0$ and $0 < q, p \leq \infty$, the *Besov space* $B_q^s(L_p(\mathbb{R}))$ is defined as the space of all functions v for which

$$|v|_{B_q^s(L_p(\mathbb{R}))} := \begin{cases} \left(\int_0^\infty [t^{-s}\omega_r(v,t)_{L_p(\mathbb{R})}]^q dt/t\right)^{1/q}, & 0 < q < \infty, \\ \sup_{t\geq 0} t^{-s}\omega_r(v,t)_{L_p(\mathbb{R})}, & q = \infty\ , \end{cases} \tag{9.52}$$

is finite with $r := [s] + 1$. Then, (9.52) is a (quasi-)seminorm for $B_q^s(L_p(\mathbb{R}))$. If we add $\|v\|_{L_p(\mathbb{R})}$ to (9.52), we obtain a (quasi-)norm for $B_q^s(L_p(\mathbb{R}))$. Let P_0 be the biorthogonal projector which maps $L^2(\mathbb{R})$ onto $\mathcal{V}_0$, i.e.

$$P_0(v) := \sum_{k\in\mathbb{Z}} \langle v, \tilde{\phi}(\cdot - k)\rangle \phi.$$

Then, P_0 has an extension as a projector to $L_p(\mathbb{R})$, $1 \leq p \leq \infty$. For each $v \in L_p(\mathbb{R})$, we have

$$v = P_0(v) + \sum_{j\geq 0, k\in\mathbb{Z}} \langle v, \tilde{\psi}_{j,k}\rangle \psi_{j,k}. \tag{9.53}$$

The Besov spaces $B_q^s(L_p(\mathbb{R}))$ can be characterized by wavelet coefficients provided the parameters s, p, q satisfy certain restrictions. For simplicity, we shall only discuss the case $q = p$. Then, the following characterization holds:

Proposition 9.3. *Let ϕ and $\tilde{\phi}$ be in $C^r(\mathbb{R})$. If $0 < p \leq \infty$ and $r > s > (1/p - 1)$, then a function v is in the Besov space $B_p^s(L_p(\mathbb{R}))$, if and only if,*

$$v = P_0(v) + \sum_{j\geq 0, k\in Z} \langle v, \tilde{\psi}_{j,k}\rangle \psi_{j,k} \tag{9.54}$$

with

$$\|P_0(v)\|_{L_p(\mathbb{R})} + \left(\sum_{j\geq 0, k\in\mathbb{Z}} 2^{j(s+(\frac{1}{2}-\frac{1}{p}))p} |\langle v, \tilde{\psi}_{j,k}\rangle|^p \right)^{1/p} < \infty \tag{9.55}$$

and (9.55) provides an equivalent (quasi-)norm for $B_p^s(L_p(\mathbb{R}))$.

In the case $p \geq 1$, this is a standard result and can be found for example in Meyer [490] (Section 10 of Chapter 6). For the general case of p, this can be deduced from general results in Littlewood–Paley theory (see e.g. Section 4 of Frazier and Jawerth [329]) or proved directly. We also refer to [636] and [195] for further information. The condition that $s > (1/p - 1)$ implies that the Besov space $B_p^s(L_p(\mathbb{R}))$ is embedded in $L_{\tilde{p}}(\mathbb{R})$ for some $\tilde{p} > 1$ so that the wavelet decomposition of v is defined. Also, with this restriction on s, the Besov space $B_p^s(L_p(\mathbb{R}))$ is equivalent to the nonhomogeneous Besov spaces $B_{p,p}^s$ defined via Fourier transforms and Littlewood–Paley theory.

Another important scale of function spaces we shall be concerned with are the *Sobolev spaces* $H^s(\mathbb{R})$. Usually, these spaces are defined by means of weak derivatives or by Fourier transforms, see, e.g. [1] for details. However, it can be shown that for $p = q = 2$, Sobolev and Besov spaces coincide:

$$H^s(\mathbb{R}) = B_2^s(L_2(\mathbb{R})), \qquad s > 0.$$

Therefore Proposition 9.3 immediately provides us with a characterization of Sobolev spaces.

Proposition 9.4. *Let ϕ and ψ be in $C^r(\mathbb{R}), r > s$. Then a function v is in the Sobolev space $H^s(\mathbb{R})$, if and only if,*

$$\|P_0(v)\|_{L_p(\mathbb{R})} + \left(\sum_{j \geq 0, k \in \mathbb{Z}} 2^{2js} |\langle v, \tilde{\psi}_{j,k} \rangle|^2 \right)^{1/2} < \infty \tag{9.56}$$

and (9.56) *provides an equivalent (quasi-)norm for $H^s(\mathbb{R})$.*

9.3.4 Wavelets on domains

Our aim is to employ wavelet bases for the design of efficient numerical schemes. However, usually the operator under consideration is defined on a bounded domain $\Omega \subset \mathbb{R}^d$, and therefore we are faced with the problem of designing a wavelet basis adapted to this domain. We cannot simply cut all elements of a wavelet basis on the whole Euclidean plane, for this would destroy almost all of the important properties described above. Nevertheless, it is by now well known how to construct suitable *biorthogonal* wavelet bases for many cases of interest, see [159, 160, 204, 250]. The first step is always to construct nested sequences $\mathcal{S} = \{S_j\}_{j \geq j_0}, \tilde{\mathcal{S}} = \{\tilde{S}_j\}_{j \geq j_0}$ whose unions are dense in $L^2(\Omega)$. Then, one has to find suitable bases

$$\Psi_j = \{\psi_{j,k} : k \in \nabla_j\}, \qquad \tilde{\Psi}_j = \{\tilde{\psi}_{j,k} : k \in \nabla_j\},$$

for some complements $\mathcal{W}_j$ of S_j in S_{j+1} and $\tilde{W}_j$ of $\tilde{S}_j$ in $\tilde{S}_{j+1}$, such that the biorthogonality condition

$$\langle \psi_{j,k}, \tilde{\psi}_{j',k'} \rangle = \delta_{j,j'} \delta_{k,k'} \tag{9.57}$$

holds. Moreover, one has to ensure that all the convenient properties of wavelets, especially the characterization of Sobolev spaces according to Propositon 9.4, can still be saved. What turns out to matter is that

$$\Psi = \bigcup_{j \geq j_0} \Psi_j$$

forms a *Riesz basis* of $L^2(\Omega)$, i.e. every $v \in L^2(\Omega)$ has a unique expansion

$$v = \sum_{j=j_0}^{\infty} \sum_{k \in J_j} \langle v, \tilde{\psi}_{j,k} \rangle \psi_{j,k} \tag{9.58}$$

such that

$$\|v\|_{L^2(\Omega)} \sim \left(\sum_{j=j_0}^{\infty} \sum_{k \in \nabla_j} |\langle v, \tilde{\psi}_{j,k} \rangle|^2 \right)^{\frac{1}{2}}, \quad v \in L^2(\Omega), \tag{9.59}$$

and that both $\mathcal{S}$ and $\tilde{\mathcal{S}}$ should have some approximation and regularity properties which can be stated in terms of the following pair of estimates. There exists some $\rho > 0$ such that the inverse estimate

$$\|v_n\|_{H^s(\Omega)} \lesssim 2^{ns} \|v_n\|_{L^2(\Omega)}, \quad v_n \in S_n, \tag{9.60}$$

holds for $s < \rho$. Moreover, there exists some $m \geq \rho$ such that the direct estimate

$$\inf_{v_n \in S_n} \|v - v_n\|_{L^2(\Omega)} \lesssim 2^{-sn} \|v\|_{H^s(\Omega)}, \quad v \in H^s(\Omega), \tag{9.61}$$

holds for $s \leq m$, compare with (9.32). Such estimates are known to hold for every finite element or spline space. For instance, for piecewise linear finite elements one has $\rho = 3/2, m = 2$.

It will be convenient to introduce the following notation. Let

$$J := \{\lambda = (j,k) : k \in \nabla_j, j \geq j_0\} = \bigcup_{j=j_0}^{\infty} (\{j\} \times \nabla_j),$$

and define

$$|\lambda| := j \quad \text{if} \quad \lambda = (j,k) \in J, \quad \text{that is,} \quad k \in \nabla_j.$$

Then the following result holds, [242].

Theorem 9.5. Equivalent norms: *Suppose that* $\Psi = \{\psi_\lambda : \lambda \in J\}$ *and* $\tilde{\Psi} = \{\tilde{\psi}_\lambda : \lambda \in J\}$ *are biorthogonal collections in* $L_2(\Omega)$ *satisfying* (9.59). *If both* $\mathcal{S}$ *and* $\tilde{\mathcal{S}}$ *satisfy* (9.60) *and* (9.61) *relative to some* $\rho, \rho' > 0$, $\rho \leq m$, $\rho' \leq m'$, *then*

$$\|v\|_{H^s(\Omega)} \sim \left(\sum_{\lambda \in J} 2^{2|\lambda| s} |\langle v, \tilde{\psi}_\lambda \rangle|^2 \right)^{\frac{1}{2}}, \quad s \in (-\rho', \rho), \tag{9.62}$$

$$\sim \left(\sum_{\lambda \in J} 2^{2|\lambda| s} |\langle v, \psi_\lambda \rangle|^2 \right)^{\frac{1}{2}}, \quad s \in (-\rho, \rho'), \quad v \in H^s(\Omega).$$

It will also be convenient to introduce the diagonal matrix

$$(\mathbf{D^s})_{\lambda,\lambda'} := 2^{s|\lambda|}\delta_{\lambda,\lambda'}. \tag{9.63}$$

By using (9.63), the norm equivalence (9.62) can be written as

$$\|v\|_{H^s} \sim \|\mathbf{D}^s\langle v, \tilde{\Psi}\rangle^T\|_{\ell_2(J)}, \qquad s \in (-\rho', \rho). \tag{9.64}$$

All the construction in [159, 160, 204, 250] induce isomorphisms of the form (9.62) at least for $-1/2 < s < 3/2$. For special domains such as rectangular domain even much sharper results are available.

To explain the basic idea, let us briefly recall the construction from [250]. It consists of three steps:

(a) Construct dual pairs of generator bases

$$\Phi_j = \{\phi_{j,k},: \ k \in \triangle_j\}, \quad \tilde{\Phi}_j = \{\tilde{\phi}_{j,k} : \ k \in \triangle_j\},$$

i.e.

$$\langle \Phi_j, \tilde{\Phi}_j\rangle := \Big(\langle \phi_{j,k}, \tilde{\phi}_{j,l}\rangle\Big)_{k,l\in\triangle_j} = \mathbf{I}, \tag{9.65}$$

whose elements have local support,

$$\mathrm{diam}(\mathrm{supp}\phi_{j,k}) \sim 2^{-j}, \quad \mathrm{diam}(\mathrm{supp}\tilde{\phi}_{j,k}) \sim 2^{-j}, \tag{9.66}$$

and such that their spans

$$S_j = S(\Phi_j) := \mathrm{span}\ \Phi_j, \qquad \tilde{S}_j = S(\tilde{\Phi}_j) := \mathrm{span}\ \tilde{\Phi}_j,$$

are nested,

$$S(\Phi_j) \subset S(\Phi_{j+1}), \qquad S(\tilde{\Phi}_j) \subset S(\tilde{\Phi}_{j+1}). \tag{9.67}$$

$S(\Phi_j), S(\tilde{\Phi}_j)$ are referred to as the *primal* and *dual* multiresolution spaces.

(b) Find a stable basis $\breve{\Psi}_j$ of *some* complement $S(\breve{\Psi}_j)$ of $S(\Phi_j)$ in $S(\Phi_{j+1})$, i.e.

$$S(\Phi_{j+1}) = S(\breve{\Psi}_j) \oplus S(\Phi_j), \quad j = j_0, \ldots. \tag{9.68}$$

(c) Given such an initial decomposition *project* the initial basis $\breve{\Psi}_j$ into a basis Ψ_j which is perpendicular to $S(\tilde{\Phi}_j)$. The union $\Psi := \Phi_{j_0} \cup \bigcup_{j=j_0}^{\infty} \Psi_j$ will be the final primal wavelet basis satisfying our requirements.

Note that (9.67) states that each coarse scaling function can be written as a linear combination of fine scale basis functions. Viewing the bases as vectors whose components are the individual scaling functions, (9.67) is equivalent to saying that there must be $\#\triangle_{j+1} \times \#\triangle_j$ matrices $\mathbf{M}_{j,0}, \tilde{\mathbf{M}}_{j,0}$ such that

$$\Phi_j^T = \Phi_{j+1}^T \mathbf{M}_{j,0}, \qquad \tilde{\Phi}_j^T = \tilde{\Phi}_{j+1}^T \tilde{\mathbf{M}}_{j,0}. \tag{9.69}$$

This program can be carried out for various types of domains. As an example, let us consider the interval $\Omega = (0, 1)$. The common strategy is to start with a biorthogonal multiresolution analysis on $\mathbb{R}$ as explained in Subsection 9.3.2. Specifically one often chooses the biorthogonal system from [202] where the primal scaling functions consist of cardinal B-splines. For $j \geq j_0$ where j_0 is fixed and sufficiently large to disentangle endpoint effects, one builds Φ_j by keeping those translates $2^{j/2}\phi(2^j \cdot -k)$, $k \in \mathbb{Z}$, that are fully supported in $[0,1]$. These will be refered to as *interior basis functions*. For B-splines of order m one adds at each end of the interval m fixed linear combinations of the $2^{j/2}\phi(2^j \cdot -k)$ in such a way that the resulting collection Φ_j spans all polynomials of order m on $(0,1)$, compare with (9.30). One proceeds in the same way with the dual scaling functions restoring the original order of polynomial exactness while keeping $\#\Phi_j = \#\tilde{\Phi}_j$. At this point one can verify (9.67), (9.66) and (9.69), i.e. nestedness, refinability and locality. However, only the interior basis functions inherit the biorthogonality from the line whereas the boundary modifications have perturbed biorthogonality. One can show that in this spline family of dual multiresolution sequences one can *always* biorthogonalize [247], ending up with pairs of generator bases $\Phi_j, \tilde{\Phi}_j$ satisfying all the properties mentionend above. In fact, there are additonal noteworthy features:

(a) Exploiting symmetry properties of the original scaling functions, one can arrange the bases to be invariant under the transformation $x \to 1 - x$.
(b) One can arrange that only a single primal and dual basis function differs from zero at the endpoints of the interval.
(c) According to the above comments, the bases $\Phi_j, \tilde{\Phi}_j$ always consist of three parts signified by the index sets $\triangle_j^L, \triangle_j^I, \triangle_j^R$, identifying the left boundary, interior and right boundary basis functions. Only the size of the interior sets $\triangle_j^I$ depends on j. The number of boundary functions always stays the same. Moreover, at each endpoint one has a fixed number of scaling relations which can be computed a priori and stored. The interior basis functions satisfy the classical refinement rule from the real line.
(d) The matrices $\mathbf{M}_{j,0}, \tilde{\mathbf{M}}_{j,0}$ are invariant under reversing the order of rows and columns.
(e) The refinement matrices have fixed upper left and lower right blocks (in the leftmost and rightmost column) with the above symmetry properties. Only the interior block (column) changes its size with growing level j, see Figure 9.1.

This construction can also be modified in order to incorporate homogeneous Dirichlet boundary conditions. There are two ways that suggest themselves:

(I) In view of property (b) above one can simply remove those basis functions from the generator bases $\Phi_j, \tilde{\Phi}_j$ that do not vanish at the endpoints of the interval. Obviously, the resulting collections are still biorthogonal and span appropriate subspaces of $H_0^1(0,1)$, see [160, 204, 250].
(II) Following [250], remove the two endpoint boundary functions from Φ_j that do not vanish at 0 and 1 but discard two *interior* basis functions from the dual collection $\tilde{\Phi}_j$. It can be shown that the resulting sets can again be biorthogonalized retaining all the above properties [250]. Note that now only $\Phi_j \subset H_0^1(0,1)$ while $S(\tilde{\Phi}_j)$ still contains all polynomials up to a certain order.

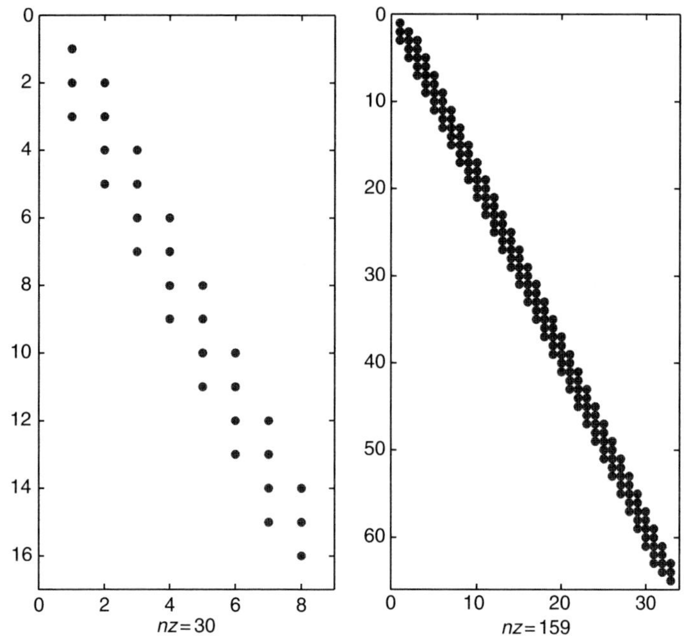

Figure 9.1 Nonzero elements (nz) of the matrix, $\mathbf{M}_{j,0}$, for a quadratic and cubic wavelet basis with three and six vanishing moments, respectively.

While (I) is most simple and immediate, option (II) appears to be conceptually preferable for the following reasons. Quite in line with the structure of the dual $H^{-1}(0,1)$, the functional on $H_0^1(0,1)$ should not be constrained at the endpoints. This is supported by the following observation. *Biorthogonality of the wavelets combined with the fact that in (II) the dual system retains full polynomial exactness immediately ensures that the wavelets have vanishing moments of a certain order, cf. (9.31).*

This approach also works for more complicated domains such as polygonal or polyhedral domains in $\mathbb{R}^d$. In this case, one can use a domain decomposition technique. The domain Ω of interest is devided into nonoverlapping subdomains Ω_i,

$$\bar{\Omega} = \bigcup_{i=1}^{N} \bar{\Omega}_i, \qquad \Omega_i \cap \Omega_j = \emptyset, i \neq j. \tag{9.70}$$

Here each Ω_i is a smooth parametric image $\Omega_i = \kappa_i(\Box)$ of the unit cube. In order to construct wavelet bases on Ω we follow the recipe from above, i.e. we have to construct first dual pairs of biorthogonal bases on the composite domain Ω. This in turn is fairly easy be stitching together parametric liftings of generator bases on the unit cube. This is essentially due to the following facts. Firstly, the boundary properties (b) of the univariate ingredients confine the gluing process to very few functions associated with the domain. Secondly, the symmetric properties (a) leaves convenient flexibility concerning invariance under rotations in the parametric mappings κ_i. On the reference domain, we use tensor products of the scaling functions constructed for the interval, i.e.

$$\varphi_{j,k}^{\square} := \varphi_{j,k_1} \oplus \cdots \oplus \varphi_{j,k_j}, \qquad k \in I_1 \times \cdots \times I_d, \tag{9.71}$$

see also (9.41). Then we define a generator bases on Ω by defining

$$\varphi_{j,k}^{i}(x) := \varphi_{j,k}^{\square}\left(\kappa_i^{-1}(x)\right), \qquad x \in \Omega \tag{9.72}$$

and gluing together those scaling functions across interfaces of subdomains which do not vanish at the interfaces. The analogous construction is applied to the dual system. The result is a global system $\Phi_j^{\Omega}, \tilde{\Phi}_j^{\Omega}$ which is globally continuous and biorthogonal to the modified inner product

$$(v, w) := \sum_{i=1}^{N} (v \circ \kappa_i, w \circ \kappa_i)$$

which is equivalent to the canonical inner product on Ω. So far, we have accomplished step (a) from above. Step (b) consists of identifying suitable bases for the complement spaces. There are several possibilities described in [250] which work for generator bases of any order. The result is the so-called *composite wavelet basis.* Once again, homogeneous Dirichlet boundary conditions can be incorporated. As before, we have the options (I) and (II).

9.3.5 Evaluation of nonlinear functionals

If we want to discretize *nonlinear* operator equations by means of wavelets, then sooner or later the problem of evaluating a nonlinear functional applied to a wavelet expansion occurs. That is, given an expansion $u_\Lambda = \sum_{\lambda \in \Lambda} d_\lambda \psi_\lambda$, one is interested in determining the coefficients $\langle f(u_\Lambda), \psi_\lambda \rangle$. Usually, this is a nontrivial problem and many different solutions have been suggested. In this chapter, we shall only discuss the approach described in [251], which, in our opinion, leads to the most far-reaching results.

Let $u \in B_q^s(L_\tau) \cap L_\infty, \tau > (s/d + 1/p)^{-1}$, for some $q \le p$ with $q \in (0, \infty]$ and suppose that $\|u\|_{L_\infty} \le \varrho < \infty$. Furthermore, let a function $f \in C^m[-\varrho, \varrho]$, $m > s$ with $f(0) = 0$ be given. Suppose that we have found an approximation in terms of a wavelet basis

$$u_\Lambda = \sum_{\lambda \in \Lambda} d_\lambda^u \psi_\lambda \quad \text{with} \quad \|u - u_\Lambda\| < \epsilon. \tag{9.73}$$

Then one is interested in computing $\langle f(u_\Lambda), \psi_\lambda \rangle$ with an accuracy of order ϵ. Expanding $f(u_\Lambda)$ with respect to the dual system $\tilde{\Psi}$ leads to

$$f(u_\Lambda) = \sum_{\lambda \in J} \tilde{d}_\lambda^{f(u_\Lambda)} \tilde{\psi}_\lambda, \qquad \tilde{d}_\lambda^{f(u_\Lambda)} = \langle f(u_\Lambda), \psi_\lambda \rangle,$$

i.e. the quantities $\langle f(u_\Lambda), \psi_\lambda \rangle$ are just the expansion coefficients of $f(u_\Lambda)$ in terms of the *dual* wavelet bases. Let $g_{\hat{\Lambda}} = \sum_{\lambda \in \hat{\Lambda}} \hat{d}_\lambda \tilde{\psi}_\lambda$ be some approximation of $f(u_\Lambda)$ in the system $\tilde{\Psi}$ satisfying

$$\|g_{\hat{\Lambda}} - f(u_\Lambda)\|_{L^2} \lesssim \epsilon. \tag{9.74}$$

Using the norm equivalences as stated in Theorem 9.5 (for $s = 0$), we obtain

$$\|\hat{\mathbf{d}}_{\hat{\Lambda}} - \langle f(u_\Lambda), \Psi_{\hat{\Lambda}}\rangle\|_{\ell_2} \leq \|\hat{\mathbf{d}}_{\hat{\Lambda}} - \langle f(u_\Lambda), \Psi\rangle\|_{\ell_2} \lesssim \|g_{\hat{\Lambda}} - f(u_\Lambda)\|_{L^2} \lesssim \epsilon.$$

This states that a good approximation of $f(u_\Lambda)$ in terms of the dual wavelet basis automatically yields a good approximation of the desired coefficients. The algorithm in [251] focuses on constructing such an approximation. Roughly speaking, this is done by first enlarging the set Λ to $\tilde{\Lambda}$ in such a way that the corresponding space $\tilde{S}_{\hat{\Lambda}}$ contains a function satisfying (9.74). Under the above assumptions, there exists a constant $C(f, \|u\|_{L_\infty})$ such that

$$\|f(u)\|_{B^s_q(L_p)} \leq C(f, \|u\|_{L_\infty})\|u\|_{B^s_q(L_p)}, \tag{9.75}$$

and it can be shown that this guarantees that only a finite enlargement has to be done. In a second step one now constructs the functions $g_{\hat{\Lambda}}$. This is done by performing a local transformation to scaling functions $\tilde{\phi}_\lambda$ for all indices in $\tilde{\Lambda}$, then approximating $\langle f(u), \phi_\lambda\rangle$ (e.g. by quadrature via point evaluations), and decomposing this in an appropriate way.

9.4 Stable discretizations and preconditioning

Our goal is to develop a suitable Galerkin scheme for approximating the solution of $Au = f$ for an A as in (9.1) which is based on a wavelet basis as introduced in Section 9.3. Therefore we consider subspaces of the form

$$S_\Lambda := \{\psi_\lambda : \lambda \in \Lambda\} \quad \subset \mathcal{U}, \qquad \Lambda \subset J. \tag{9.76}$$

We project our problem onto these spaces, i.e. the Galerkin approximation $u_\Lambda \in S_\Lambda$ is defined by

$$\langle Au_\Lambda, v\rangle = \langle f, v\rangle \qquad \text{for all} \quad v \in S_\Lambda. \tag{9.77}$$

With the pair of biorthogonal bases Ψ and $\tilde{\Psi}$ we can associate canonical projectors Q_Λ and Q'_Λ by

$$Q_\Lambda(v) := \sum_{\lambda\in\Lambda} \langle v, \tilde{\psi}_\lambda\rangle\psi_\lambda, \qquad Q'_\Lambda(v) = \sum_{\lambda\in\Lambda} \langle v, \psi_\lambda\rangle\tilde{\psi}_\lambda. \tag{9.78}$$

By means of these projectors, the Galerkin scheme (9.77) may very conveniently be written as

$$Q'_\Lambda A Q_\Lambda u_\Lambda = Q'_\Lambda f. \tag{9.79}$$

Obviously, (9.79) is equivalent to a linear system with the stiffness matrix

$$A_\Lambda := \langle A\Psi_\Lambda, \Psi_\Lambda\rangle^T. \tag{9.80}$$

The most far-reaching results were obtained for linear operators $A : \mathcal{U} \to \mathcal{U}'$, which are boundedly invertible,

$$\|Au\|_{\mathcal{U}'} \sim \|u\|_{\mathcal{U}}, \quad u \in \mathcal{U}. \tag{9.81}$$

In this case, the equation

$$Au = f \tag{9.82}$$

clearly has a unique solution, compare again with Section 9.2. Typical examples are second order coercive boundary value problems with Dirichlet boundary conditions on some open domain $\Omega \subset \mathbb{R}^d$ such as the Poisson equation (9.5). In this case $\mathcal{U} = H_0^1(\Omega)$ and $\mathcal{U}' = H_0^{-1}(\Omega)$. Other examples are obtained by turning an exterior boundary value problem into a singular integral operator on the boundary Γ of the domain. For a formulation in terms of the single layer potential operator one obtains for instance $\mathcal{U} = H^{-1/2}(\Gamma)$ and $\mathcal{U}' = H^{1/2}(\Gamma)$. Thus $\mathcal{U}$ typically is a Sobolev space and

$$\mathcal{U} \subset L^2 \subset \mathcal{U}' \quad \text{or} \quad \mathcal{U}' \subset L^2 \subset \mathcal{U}.$$

Unless A is a differential operator, one further property of A will matter. Whenever A is an operator with *global Schwartz kernel*

$$(Av)(x) = \int K(x, y) v(y)\, dy,$$

we will assume in addition that when (9.81) holds for $\mathcal{U} = H^t$ then

$$\left|\partial_x^\alpha \partial_y^\beta K(x, y)\right| \lesssim \operatorname{dist}(x, y)^{-\left(d+t+|\alpha|+|\beta|\right)}. \tag{9.83}$$

This assumption covers a wide range of cases, including *Calderón–Zygmund operators,* cf. [248].

In any case, to obtain an applicable numerical algorithm, it is essential that the Galerkin scheme has some basic *stability* properties. By using the projectors Q_Λ, Q'_Λ this requirement can be formulated as

$$\|Q'_\Lambda A u_\Lambda\|_{\mathcal{U}'} = \|Q'_\Lambda A Q_\Lambda u_\Lambda\|_{\mathcal{U}'} \sim \|u_\Lambda\|_{\mathcal{U}}, \qquad u_\Lambda \in S_\Lambda. \tag{9.84}$$

Before we study this issue in detail, let us discuss one of its consequences. It is one of the main advantages of wavelet Galerkin schemes that they give rise to very simple preconditioning strategies. Essentially, this is a consequence of the norm equivalences stated in Theorem 9.5. Indeed, if the Galerkin scheme is stable in the sense of (9.84), then a simple diagonal preconditioning involving the matrices introduced in (9.63) gives rise to uniformly bounded condition numbers of the stiffness matrices (9.80). As we shall see below, (9.84) holds under natural conditions for all the problems considered in this chapter. Therefore the following result is very important:

Theorem 9.6. Simple diagonal preconditioning for all problems in this chapter: *Suppose that the Galerkin scheme* (9.79) *for* $A : H^t \to H^{-t}$ *is stable* (9.84). *Furthermore, suppose that the parameters* ρ, ρ' *in* (9.62) *satisfy*

$$|t| < \rho, \rho'. \tag{9.85}$$

Let $\mathbf{D}^s_\Lambda$ be the diagonal matrix defined in (9.63). Then the matrices

$$\mathbf{B}_\Lambda := \mathbf{D}_\Lambda^{-t}\mathbf{A}_\Lambda\mathbf{D}_\Lambda^{-t} \tag{9.86}$$

have uniformly bounded spectral condition numbers,

$$\|\mathbf{B}_\Lambda\|\,\|\mathbf{B}_\Lambda^{-1}\| = \mathcal{O}(1), \qquad \Lambda \subset J. \tag{9.87}$$

Proof. The first step is to introduce the projectors

$$Q_j(v) := \sum_{k\in\triangle_j} \langle v, \tilde{\phi}_{j,k}\rangle \phi_{j,k}, \qquad Q_j'(v) := \sum_{k\in\triangle_j} \langle v, \phi_{j,k}\rangle \tilde{\phi}_{j,k}. \tag{9.88}$$

With the aid of these projections, we can define the operators

$$\Sigma_s v := \sum_{j=j_0}^{\infty} 2^{js}(Q_j - Q_{j-1})v, \qquad \text{where} \qquad Q_{j_0-1} := 0, \tag{9.89}$$

which, due to the norm equivalences (9.62), act as shifts in the Sobolev scale,

$$\|\Sigma_s v\|_{L^2(\Omega)} \sim \|v\|_{H^s(\Omega)}, \qquad s \in (-\rho', \rho). \tag{9.90}$$

It can be shown that

$$\Sigma_s^{-1} = \Sigma_{-s}, \qquad \Sigma_s' = \sum_{j=j_0}^{\infty} 2^{js}\left(Q_j' - Q_{j-1}'\right). \tag{9.91}$$

Now consider any $v \in S_\Lambda$ and set $w := \Sigma_t v$. Thus, by (9.85), (9.90) and (9.84), we obtain

$$\|w\|_{L^2} = \|\Sigma_t v\|_{L^2} \sim \|v\|_{H^t} \sim \|Q_\Lambda' A Q_\Lambda v\|_{H^{-t}}.$$

Employing the norm equivalence (9.90), now relative to the dual basis, and bearing (9.91) in mind, yields

$$\|w\|_{L^2} \sim \left\|\Sigma_{-t}' Q_\Lambda' A Q_\Lambda \Sigma_{-t} w\right\|_{L^2}.$$

This means that the operators

$$A_{t,\Lambda} := \Sigma_{-t}' Q_\Lambda' A Q_\Lambda \Sigma_{-t} : \; S_\Lambda \to \tilde{S}_\Lambda$$

are uniformly boundedly invertible, that is,

$$\|A_{t,\Lambda}\|\left\|A_{t,\Lambda}^{-1}\right\| = \mathcal{O}(1), \qquad \#\Lambda \to \infty. \tag{9.92}$$

It is now a matter of straightforward calculation to verify that the matrix representation of $A_{t,\Lambda}$ relative to Ψ_Λ is

$$\langle A_{t,\Lambda}\Psi_\Lambda, \Psi_\Lambda\rangle^T = \mathbf{D}_\Lambda^{-t}\mathbf{A}_\Lambda\mathbf{D}_\Lambda^{-t}, \tag{9.93}$$

proving the claim. □

Sending $\#\Lambda$ to infinity, the original equation $Au = f$ can be viewed as an infinite discrete system

$$\mathbf{D}^{-t}\mathbf{A}\mathbf{D}^{-t}\mathbf{d} = \mathbf{D}^{-t}\mathbf{f}, \tag{9.94}$$

where $\mathbf{D}^{-t}$ and $\mathbf{A}$ are the infinite counterparts of $\mathbf{D}_\Lambda^{-t}, \mathbf{A}_\Lambda$, respectively, and $\mathbf{f} := \langle f, \Psi\rangle^T$ is the coefficient sequence of f expanded relative to the dual basis. The sequence $\mathbf{d}$ then consists of the wavelet coefficients of the solution,

$$u = \mathbf{d}^T\Psi.$$

The infinite matrix $\mathbf{D}^{-t}\mathbf{A}\mathbf{D}^{-t}$ is, on account of Theorem 9.6, a boundedly invertible mapping from $\ell_2(J)$ onto $\ell_2(J)$.

In the remaining part of this chapter, we shall mainly discuss Galerkin schemes based on sets of the form, cf. (9.88),

$$\Lambda^n := \{\lambda \in J : \ |\lambda| \leq n\}, \qquad \text{i.e.} \qquad Q_{\Lambda^n} = Q_n, \ S_{\Lambda^n} = S_n. \tag{9.95}$$

Therefore, in this case the approximation spaces consist of the spaces of the underlying multiresolution analysis. We project our problem onto these spaces, i.e. the Galerkin approximation $u^n := u_{\Lambda^n} \in S_{\Lambda^n}$ for $Au_0 = f, \mathcal{U} = H_0^t(\Omega)$ is defined by

$$u^n = u_{\Lambda^n} \in S_n = S_{\Lambda^n} : \quad \langle Au^n, v\rangle = \langle f, v\rangle \quad \text{for all} \quad v \in S_n = S_{\Lambda^n}. \tag{9.96}$$

These methods correspond to *uniform mesh refinements.* Later on, in Subsection 9.7, we shall also discuss adaptive numerical schemes based on wavelets, and then we will clearly dispense with these assumptions.

In terms of the adjoint projectors Q_n and Q_n',

$$Q_n(v) = \sum_{\lambda\in\Lambda^n} \langle v, \tilde{\psi}_\lambda\rangle\psi_\lambda, \qquad Q_n'(v) = \sum_{\lambda\in\Lambda^n} \langle v, \psi_\lambda\rangle\tilde{\psi}_\lambda, \tag{9.97}$$

the Galerkin scheme (9.96) now reads as

$$Q_n'Au^n = Q_n'AQ_nu^n = Q_n'f. \tag{9.98}$$

(We sometimes use the redundant notation $Q_n'AQ_n$ to indicate self-adjointness.)

For later use, we summarize a generalized version of Propositions 4.50 for the wavelet setting which implies that the operators Q_n and Q_n' are projection and linear approximation operators.

Lemma 9.7. Q_n, Q_n' are linear approximation and projection operators: *The previous projectors $Q_n \in \mathcal{L}(\mathcal{U}, S_n), Q_n' \in \mathcal{L}(\mathcal{U}', \tilde{S}_n)$, satisfy*

$$\forall v \in \mathcal{U}, f \in \mathcal{U}' : \quad \lim_{n\to\infty} \|Q_nv - v\|_{\mathcal{U}} = 0 \ \ \textit{and} \ \lim_{n\to\infty} \|Q_n'f - f\|_{\mathcal{U}'} = 0. \tag{9.99}$$

More precisely we obtain for $\mathcal{U} = H^m(\Omega)$, cf. (9.32),

$$\mathcal{U} = H^m(\Omega) : \forall v \in H^{m+s}(\Omega) : \quad \|Q_nv - v\|_{\mathcal{U}} \lesssim C2^{-ns}\|v\|_{H^{m+s}(\Omega)}. \tag{9.100}$$

Proof. By using the norm equivalences (9.62), we obtain

$$\begin{aligned}
\|Q_n v - v\|_{H^m}^2 &\lesssim \sum_{|\lambda|>n} 2^{2|\lambda|m} |\langle v, \tilde{\psi}_\lambda \rangle|^2 \\
&\lesssim \sum_{|\lambda|>n} 2^{2|\lambda|(m+s)} 2^{-2ns} |\langle v, \tilde{\psi}_\lambda \rangle|^2 \\
&\lesssim 2^{-2ns} \sum_{|\lambda|\in J} 2^{2|\lambda|(m+s)} |\langle v, \tilde{\psi}_\lambda \rangle|^2 \\
&\lesssim 2^{-2ns} \|v\|_{H^{m+s}}^2 .
\end{aligned}$$

□

Let us now come back to the stability requirement as stated in (9.84). For uniform wavelet schemes, this property can be formulated as

$$\|Q_n' A u^n\|_{\mathcal{U}'} \sim \|u^n\|_{\mathcal{U}}, \qquad u^n \in S_n. \tag{9.101}$$

When A is positive definite and self-adjoint, then the property (9.101) holds for *any* trial subspace of $\mathcal{U}$. The same is correct, if A induces a coercive bilinear form as B in (9.3). Moreover, in the framework of pseudo-differential operators, sufficient conditions have been derived by Dahmen *et al.* [248]. Let us briefly recall the definition of pseudo-differential operators. To do this consider the Fourier inversion formula, cf. Theorem 2.105, and notice that here $i \in \mathcal{C}$ with $i^2 = -1$,

$$u(x) = (2\pi)^{-d} \int_{\mathbb{R}^d} e^{ix\cdot\xi} \hat{u}(\xi) d\xi \ , \qquad u \in C_0^\infty(\mathbb{R}^d). \tag{9.102}$$

Differentiating this expression and employing here the notation $\partial^j = \left(\frac{\partial}{\partial x_j}\right)/i$ yields

$$D^\beta u(x) = (2\pi)^{-d} \int_{\mathbb{R}^d} \xi^\beta e^{ix\cdot\xi} \hat{u}(\xi) d\xi \ . \tag{9.103}$$

Let $P = p(x, D) = \sum_{|\beta|\le k} a_\beta(x) D^\beta$ be a differential operator defined on a domain $\Omega \subset \mathbb{R}^d$. Then

$$(Pu)(x) = (2\pi)^{-d} \int_{\mathbb{R}^d} p(x,\xi) e^{ix\cdot\xi} \hat{u}(\xi) d\xi \ . \tag{9.104}$$

Instead of the polynomial $p(x,\xi)$ one can take a function $\sigma(x,\xi)$ belonging to a more general class of functions. Then the operator A defined by

$$(Au)(x) := \int_{\mathbb{R}^d} \sigma(x,\xi) e^{ix\cdot\xi} \hat{u}(\xi) d\xi \tag{9.105}$$

is called a *(linear) pseudo-differential operator* with *symbol* $\sigma(x,\xi)$. (For details we refer to the standard literature concerning pseudo-differential operators, e.g. Taylor [617].) It turns out that injectivity of A and a *Gårding inequality* also imply stability. More precisely, it turns out that the uniform Galerkin scheme based on the projectors Q_n, Q'_n will be stable and convergent if the following three conditions are satisfied [248]:

- A is a pseudo-differential operator with symbol σ and is in the class $S^m_{1,0}$, the subclass of the Hörmander class with the property that
$$\left|D_x^\beta D_\xi^\alpha \sigma(x,\xi)\right| \le c_{\alpha,\beta}(1+|\xi|)^{(m-|\alpha|)}. \tag{9.106}$$
The symbols in $S^m_{1,0}$ can be split as a sum of the principal part σ_0 representing an operator of order m and a second part σ_1 which represents an operator of order $m' < m$.
- A satisfies a Gårding inequality, i.e.
$$\Re\sigma_0(x,\xi) \ge c|\xi|^m, \qquad \xi \in \mathbb{R}^d, \tag{9.107}$$
holds.
- A is injective,
$$\text{Ker } A = \{0\}. \tag{9.108}$$

In summary, it is by now possible to construct stable wavelet Galerkin schemes for a large class of problems. One of the aims of this chapter is to investigate to what extent these stability properties are preserved under perturbations. The relations are clarified in the following theorem. This is, for our Hilbert space setting, a known result, see e.g. Zeidler [677], extended to our situation in Chapter 3, and Böhmer *et al.* [129,567]. Nevertheless, we want to repeat the simple and straightforward proof for Theorem 9.8 formulated for wavelets.

Theorem 9.8. Stability inherited to compact perturbations: *Let $B \subset \mathcal{L}(\mathcal{U},\mathcal{U}')$ and suppose that the biorthogonal wavelet Galerkin scheme $B^n := Q'_n B Q_n$ is stable. Let $A := B + C$ with $C \in \mathcal{C}(\mathcal{U},\mathcal{U}')$, the* set of compact operators from $\mathcal{U} \to \mathcal{U}'$. *Then*

$$A^{-1} \in \mathcal{L}(\mathcal{U}',\mathcal{U}) \implies A^n := Q'_n A Q_n \quad \textit{is stable.}$$

Proof. See Chapter 3, and e.g. Böhmer *et al.* [129, 567]. We determine for an arbitrary $u \in \mathcal{U}$ and $v' := Cu$ the unique exact and discrete solutions, $\hat{u}$ and $\hat{u}_n$, of the equations $B\hat{u} = v'$ and $B^n\hat{u}_n = Q'_n v' = Q'_n B\hat{u}$. Since B^n is assumed to be stable, the corresponding Galerkin scheme converges, hence for any $u \in \mathcal{U}$ we obtain, with the notation $T = B^{-1}$, $T_n = (B^n)^{-1}Q'_n$,

$$\lim_{n\to\infty} \left\|(B^n)^{-1}Q'_n Cu - B^{-1}Cu\right\|_{\mathcal{U}} = \lim_{n\to\infty} \|(T_n - T)Cu\|_{\mathcal{U}} = 0.$$

C is compact, so we get

$$\lim_{n\to\infty} \|(T - T_n)C\| = 0. \tag{9.109}$$

Now let $u^n \in S_n$. Because A is boundedly invertible, we can estimate

$$\begin{aligned}\|u^n\|_{\mathcal{U}} &\le \|A^{-1}Au^n\|_{\mathcal{U}} \le \|A^{-1}\|\|Au^n\|_{\mathcal{U}'} \le \|A^{-1}\|\|B(I + B^{-1}C)u^n\|_{\mathcal{U}'} \\ &\le \|A^{-1}\|\|B\|\|(I + TC)u^n\|_{\mathcal{U}}.\end{aligned} \tag{9.110}$$

Hence, we obtain

$$
\begin{aligned}
\|A^n u^n\|_{\mathcal{U}'} &= \|Q_n'(B+C)Q_n u^n\|_{\mathcal{U}'} \\
&= \left\|B^n \left(I + (B^n)^{-1}Q_n' C Q_n\right) u^n\right\|_{\mathcal{U}'} \\
&\geq 1/\|B^n\|^{-1} \left\|\left(I + (B^n)^{-1}Q_n' C Q_n\right) u_n\right\|_{\mathcal{U}} \\
&= 1/\|B^n\|^{-1}\|(I + T_n C Q_n)u^n\|_{\mathcal{U}} \\
&\geq 1/\|B^n\|^{-1}(\|(I+TC)Q_n u^n\|_{\mathcal{U}} - \|(T-T_n)CQ_n u^n\|_{\mathcal{U}}).
\end{aligned}
$$

By using the fact that $A = B + C$ implies $I + B^{-1}C = B^{-1}A$, this reduces to

$$\|A^n u^n\|_{\mathcal{U}'} \geq 1/\|B^n\|^{-1}\left(1/\|A^{-1}B\| - \|(T-T_n)C\|\right)\|u^n\|_{\mathcal{U}}.$$

Because of (9.109) and the stability of B^n there exists a positive constant K, independent of n, such that for all $n \geq n_0$ the following holds:

$$\|A^n u^n\|_{\mathcal{U}'} \geq K\|u^n\|_{\mathcal{U}} \quad \text{for all} \quad u^n \in S_n,$$

hence A^n is stable. □

9.5 Applications to elliptic equations

Based upon the previous results, in this section we will employ our general discretization approach for general convergence results for general linear and semilinear elliptic problems. Quasilinear problems are still excluded, since they cannot yet be evaluated for wavelet arguments. Combining the general stability theory with the conforming wavelets in Theorem 9.11, we obtain our goal, the convergence of (uniform) wavelet methods. The convergence analysis of nonuniform methods, i.e. of adaptive wavelet schemes, will be carried out in Section 9.7.

The situation here is very similar to conforming FEMs in Subsection 4.3.3. In fact, only *conforming wavelets* are employed for our *Galerkin methods.* For the reader's convenience, we update the general approach in Subsection 4.3.3 to the wavelet situation.

For the following it is important that any elliptic operator A induces an elliptic bilinear form $a(\cdot,\cdot)$. This can be split into the sum of a coercive bilinear form $b(\cdot,\cdot)$ and its complement $c(\cdot,\cdot)$ such that the induced operators B, C satisfy $A = B + C$ with a compact perturbation, C. For our $\mathcal{U} = H_0^m(\Omega), m \geq 1$, we use for $b(\cdot,\cdot)$ the principal part $a_p(\cdot,\cdot)$.

Wavelet methods for linear elliptic equation of order $2m$ require $\mathcal{U} = H_0^m(\Omega)$. For linear A, determine $u_0^n \in S_n \subset \mathcal{U} = H_0^m(\Omega)$ such that

$$a(u_0^n, v^n) = \langle Au_0^n, v^n\rangle_{\mathcal{U}'\times\mathcal{U}} = \int_\Omega \sum_{|\alpha|,|\beta|\leq m} a_{\alpha\beta}\partial^\alpha u_0^n \partial^\beta v^n dx = \langle f, v^n\rangle_{\mathcal{U}'\times\mathcal{U}} \quad (9.111)$$

$$\forall\, v^n \in S_n \text{ with } a_{\alpha\beta} \in L^\infty(\Omega), \text{ and for } 1 < m = |\alpha| = |\beta| : \; a_{\alpha\beta} \in C(\bar{\Omega}).$$

$$
\begin{array}{ccccccc}
\mathcal{U} & \xrightarrow{A,\,G} & & \mathcal{U}' & \overset{\text{tested by}}{\longleftrightarrow} & \mathcal{U} \\
Q_n\big\downarrow & & \Phi^n\big\downarrow & Q_n'\big\downarrow & & \\
S_n & \xrightarrow{A^n,\,G^n} & & \tilde{S}_n & \overset{\text{tested by}}{\longleftrightarrow} & S_n
\end{array}
$$

Figure 9.2 Conforming wavelet methods: Spaces and operators.

(Of course, for preconditioning reasons, we shall always employ the multiscale wavelet bases instead of the single-scale bases, compare with Theorem 9.6.) Equation (9.111) defines a bounded, elliptic $a(u^n, v^n)$, with $\mathcal{U}$-coercive principle part, cf. (9.11).

Linear pseudo-differential operators: As already stated in Subsection 9.4, a uniform wavelet Galerkin scheme for operator equations induced by a linear pseudo-differential operator, B, in the Hörmander class is stable if the fundamental conditions (9.107) and (9.108) are satisfied.

Wavelet methods for quasilinear elliptic equation and systems of order $2m$*:* We may use again the spaces S_n of our multiresolution analysis as test and approximation spaces, i.e. we require $S_n \subset \mathcal{U} = H_0^m(\Omega, \mathbb{R}^q), q \geq 1$. In fact, we use here and in the following Summary the

$$
\begin{aligned}
&\text{same symbols for } u_0 \in \mathcal{U} = H_0^m(\Omega, \mathbb{R}^q), u_0^n \in S_n \subset H_0^m(\Omega, \mathbb{R}^q), q \geq 1, \\
&a\left(u_0^n, v^n\right) = \langle A u_0^n, v^n \rangle_{\mathcal{U}' \times \mathcal{U}} = \langle A^n u_0^n, v^n \rangle_{\mathcal{U}' \times \mathcal{U}} = \langle f, v^n \rangle_{\mathcal{U}' \times \mathcal{U}} \qquad (9.112) \\
&\forall\, v^n \in S^n, \text{hence, } A^n := Q_n' A|_{S^n} = Q_n' A Q_n|_{S_n}, \text{ similarly } G^n := Q_n' G|_{S^n}.
\end{aligned}
$$

We recall and impose the condition (9.12) for the A_α in (9.113): *the* $A_\alpha(x, \vec{u}, \ldots, \nabla^m \vec{u})$ *can be evaluated for wavelet arguments.* Then we determine

$$
\begin{aligned}
u_0^n \in S_n \text{ s.t. } a\left(u_0^n, v^n\right) = {} & \langle G u_0^n, v^n \rangle_{\mathcal{U}' \times \mathcal{U}} := \int_\Omega \sum_{|\alpha| \leq m} \left(A_\alpha\left(\cdot, u_0^n, \ldots, \nabla^m u_0^n\right), \partial^\alpha v^n\right)_q dx \\
= {} & \langle f, v^n \rangle_{\mathcal{U}' \times \mathcal{U}} = \sum_{|\alpha| \leq m} (f_\alpha, \partial^\alpha v^n)_q dx \\
& \forall\, v^n \in S_n \subset \mathcal{U} = H_0^m(\Omega, \mathbb{R}^q), q \geq 1. \qquad (9.113)
\end{aligned}
$$

All these methods are special cases of the general discretization methods in Chapter 3. For the proof of the convergence we have to verify the conditions in Subsections 3.3–3.4 for the different spaces and operators constituting these methods. Their relations are sketched in the *diagram* in Figure 9.2. We summarize the necessary conditions and definitions. As mentioned, the wavelet methods in this chapter are conforming variational methods.

We *discuss wavelet methods for quasilinear problems* only under the condition (9.12), but *do not discuss wavelet methods for fully nonlinear problems* at all. There are still too many problems open in this context.

Summary 9.9. Wavelet methods are special discretization methods:

1. *The wavelet method has to be applicable to the problem* $Au_0 = f$ *or* $Gu_0 = 0$, *cf. Definition 3.12. This requires, throughout a* sequence *of approximating spaces* S_n *for* $\mathcal{U}$, *projectors* Q_n, Q'_n, *cf. Lemma 9.7, and the discrete problems* A^n, G^n, *cf. (9.112),*
$$\{S_n, Q_n, Q'_n, G^n\}_{n\in N}, \text{ with } \dim S_n < \infty. \tag{9.114}$$
This is provided by the spaces of the underlying multiresolution analysis together with the associated biorthogonal projectors. The wavelet spaces S_n *are approximating spaces, compare with (9.100):*
$$\forall v \in \mathcal{U} : \lim_{n\to\infty} \text{dist}\,(v, S_n) = 0, \text{ e.g. } \lim_{n\to\infty} \|v - Q_n v\|_{\mathcal{U}} = 0. \tag{9.115}$$
2. *For semi- and quasilinear problems, we require that the* $A_\alpha(x, u, \ldots, \nabla^m u)$ *in (9.113) can be evaluated for wavelet arguments.*
3. $Q_n \in \mathcal{L}(\mathcal{U}, S_n), Q'_n \in \mathcal{L}(\mathcal{U}', \tilde{S}_n)$, *are linear approximation operators, such that*
$$\forall v \in \mathcal{U}, f \in \mathcal{U}' : \lim_{n\to\infty} \|Q_n v\|_{\mathcal{U}} = \|v\|_{\mathcal{U}}, \lim_{n\to\infty} \|Q'_n f\|_{\mathcal{U}'} = \|f\|_{\mathcal{U}'}. \tag{9.116}$$
4. *Wavelet methods, cf. (9.112), define a mapping* $\Phi^n : \mathcal{D}(\Phi^n) \subset (\mathcal{D}(G) \subset \mathcal{U} \to \mathcal{U}') \to (S_n \to \tilde{S}_n)$, *transforming* A, f, G *into* A^n, f^n, G^n *as*
$$A^n := \Phi^n A := Q'_n A|_{S_n}, f^n := \Phi^n f := Q'_n f|_{S_n}, G^n := \Phi^n G := Q'_n G|_{S^n}. \tag{9.117}$$
5. *According to Definition 3.14, we define* u^n, *for a chosen* u, *as* $A^n u^n = Q'_n Au$ *and* $G^n u^n = Q'_n Gu$. *Then*
$$Q'_n Au - A^n u^n = Q'_n A(u - u^n) \text{ and } Q'_n Gu - G^n u^n = Q'_n (Gu - Gu^n) \tag{9.118}$$
are the so-called (variational) consistency errors, *cf. Theorem 9.11.*
6. *The* (classical) consistency error or local discretization error *in* u *is*
$$AQ_n u - Q'_n Au \text{ or } GQ_n u - Q'_n Gu. \tag{9.119}$$
7. *The* G^n *are called* stable in u^n, *and* A^n stable, *if for fixed* $n_0, r, S \in \mathbb{R}_+$,
$$u_i^n \in B_r(u^n), i = 1, 2, \Rightarrow \|u_1^n - u_2^n\|_{\mathcal{U}} \le S\,\|G^n(u_1^n) - G^n(u_2^n)\|_{\tilde{S}_n} \quad \forall n > n_0 \tag{9.120}$$
For A^n *or* $A^n \cdot - f^n$ *this reduces to (for any* r *and* u^n, *but for* $n > n_0$*)*
$$(A^n)^{-1} \in \mathcal{L}(\tilde{S}_n, S_n) \text{ exists and } \|(A^n)^{-1}\|_{S_n \leftarrow \tilde{S}_n} \le S.$$
The stability of G^n *is essentially implied by the stability of its derivative* $(G'(u_0))^n$, *cf. Theorems 9.8, 3.23.*
8. *For possible quadrature approximations, the* Q'_n, A^n, G^n *have to be replaced by appropriate* $\tilde{Q}'_n, \tilde{A}^n, \tilde{G}^n$.
9. *All our wavelet methods are so-called* linear methods *in the sense*
$$\Phi^n(B + C) = \Phi^n B + \Phi^n C, \text{ and } \Phi^n(G + f) = \Phi^n G + \Phi^n f. \tag{9.121}$$

Theorem 9.10. Existence and convergence for wavelet methods, cf. Theorem 3.21: *Let the original problem $Gu = 0$ have the exact solution $u_0 \in \mathcal{U}$. Let its discretization $G^n = Q_n' G\,|_{S^n} = Q_n' G Q_n|_{S^n} : S^n \to (S^n)'$ in (9.117) satisfy*

1. *$G^n : S^n \to (S^n)'$ is defined and continuous in $B_r(Q_n u_0)$, $r > 0, n$-independent;*
2. *G^n is consistent with G in $Q_n u_0$, hence $\|Q_n' G^n(Q_n u_0)\|_{\mathcal{U}'} \to 0$ for $n \to \infty$;*
3. *G^n is stable for $Q_n u_0$.*

Then the discrete problem $G^n(u^n) = 0$ possesses a unique solution $u_0^n \in S^n$ near u_0 for all sufficiently large n and u_0^n converge to u_0 as

$$\|u_0^n - u_0\|_{\mathcal{U}} \lesssim \|Q_n' G^n(Q_n u_0)\|_{\mathcal{U}'} \to 0. \tag{9.122}$$

Hence, good convergence, essentially determined by the consistency, is usually only satisfied if u_0 is smoother than $u_0 \in \mathcal{U}$.

For our claim of a uniquely determined discrete solution, u_0^n, we have to verify the previous conditions 1., 2., 3. in Theorem 9.10. The continuity of G implies that of G^n. So consistency and stability have to be proved.

Theorem 9.11. Variational and classical consistency errors vanish and go to 0, respectively:

1. Wavelet methods *applied to linear problems, $A \in \mathcal{L}(\mathcal{U},\mathcal{U}')$, yield vanishing variational consistency errors $Q_n' A u_0 - A^n u_0^n \equiv 0$.*
2. *For $A \in \mathcal{L}(\mathcal{U},\mathcal{U}')$, the* classical consistency or local discretization error *is*

$$\|A^n(Q_n u) - Q_n' A u\|_{\mathcal{U}'} = \|Q_n' A(Q_n u - u)\|_{\mathcal{U}'} \lesssim \|Q_n u - u\|_{\mathcal{U}}. \tag{9.123}$$

   This remains correct for G if it is Lipschitz-continuous and (9.12) is satisfied.

3. *By Lemmas 9.7, the classical consistency error tends to 0, according to $\|Q_n' A(Q_n u - u)\|_{\mathcal{U}'} \to 0$ and $\|Q_n'(G Q_n u - G u)\|_{\mathcal{U}'} \to 0$ for $u \in \mathcal{U}$ or for*

$$\mathcal{U} = H^m(\Omega), m \geq 1 : u \in H^{m+s}(\Omega) : \quad \|Q_n u - u\|_{\mathcal{U}} \leq C 2^{-ns} \|u\|_{H^{m+s}(\Omega)}.$$

4. *The above (9.116) is a consequence of (9.115).*

Proof. The *variational consistency errors* vanish by (9.117) for the exact solution u_0. This is a simple consequence of $S_n \subset \mathcal{U}$, e.g. (9.112) implies for A, similarly for G,

$$\langle A u_0 - A^n u_0^n, v^n \rangle_{\mathcal{U}' \times \mathcal{U}} = a\,(u_0 - u_0^n, v^n) = \langle f, v^n \rangle - \langle f, v^n \rangle = 0 \quad \forall\, v^n \in S_n. \tag{9.124}$$

For the classical consistency error we use $S_n \subset \mathcal{U}$, and $(\Phi^n A) u^n = A^n u^n = Q_n' A u^n$. Then we find $\|A^n Q_n u - Q_n' A u\|_{\mathcal{U}'} \leq C \|A\|_{\mathcal{U}' \hookleftarrow \mathcal{U}} \|Q_n u - u\|_{\mathcal{U}} \to 0$, by Lemma 9.7. □

Finally, the stability of the nonlinear G^n is an immediate consequence of the stability of its linearized $(G^n)'(u_0^n)$ or the invertibility of $G'(u_0)$, cf. Theorem 3.23.

Theorem 9.12. Convergence for the wavelet method applied to the above cases: *Let $Au = f$ and let $G(u) = 0$ have (locally) unique solutions u_0. Hence $A : \mathcal{U} \to \mathcal{U}'$ or, for the nonlinear operator $G : \mathcal{D}(G) \subset \mathcal{U} \to \mathcal{U}'$, the linearization $G'(u_0) : \mathcal{U} \to \mathcal{U}'$ be boundedly invertible. For nonlinar G in (9.13), (9.113), we impose the condition (9.12), so the $A_\alpha(x, \vec{u}, \ldots, \nabla^m \vec{u})$ can be evaluated for wavelet arguments. For the* *problems (9.111), (9.13) choose the wavelet methods according to (9.111)–(9.113). They are stable and consistent. Hence for $n > n_0$ the wavelet equations have a (locally) unique solution u_0^n as well near the exact u_0. If this exact solution satisfies $u_0 \in H^{m+s}(\Omega)$, the $u_0^n \in S^n$ converge according to*

$$\|u_0^n - u_0\|_{\mathcal{U}} \lesssim 2^{-sn} \|u_0\|_{H^{m+s}(\Omega)}. \tag{9.125}$$

As already indicated in Summary 9.9 7., these results remain correct, if the exact evaluation of the integrals in (9.111), (9.113) is replaced by applying quadrature formulas. The following results are obtained by a straightforward modification of the results in Section 5.4.

$$\begin{aligned}\langle \tilde{A}^n u^n,\ v^n\rangle|_{S'^n \times S^n} &:= \tilde{a}^n(u^n, v^n) := \textit{quadr. formula applied to } a(u^n, v^n)|_{S^n \times S^n},\\ \langle \tilde{f}^n, v^n\rangle|_{S'^n \times S^n} &:= \textit{quadr. formula applied to } := \langle f, v^n\rangle|_{S'^n \times S^n}, \qquad (9.126)\\ \langle \tilde{G}^n u^n,\ v^n\rangle|_{S'^n \times S^n} &:= \tilde{a}^n(u^n, v^n) := \textit{quadr. formula applied to } a(u^n, v^n)|_{S^n \times S^n}.\end{aligned}$$

We require good enough quadrature formulas, such that for smooth enough functions $u^n, v^n \in H^{m+s}(\Omega)$ and coefficients, $a_{ij} \in H^{m+s}(\Omega)$, terms $A_\alpha(u^n) \in H^{-(m+s)}(\Omega)$ for $u^n \in H^{m+s}(\Omega)$, and $f \in H^{-(m+s)}(\Omega)$ the following quadrature errors hold:

$$\begin{aligned}|\langle f, v^n\rangle - \langle \tilde{f}^n, v^n\rangle| &\lesssim 2^{-sn} \|f\|_{H^{-(m+s)}(\Omega)} \cdot \|v^n\|_{H^{m+s}(\Omega)}, \qquad (9.127)\\ |a^n(u^n, v^n) - \tilde{a}^n(u^n, v^n)| &\lesssim 2^{-sn} \cdot \|u^n\|_{H^{m+s}(\Omega)} \|v^n\|_{H^{m+s}(\Omega)}.\end{aligned}$$

We formulate these quadrature approximate results with other generalizations:

Theorem 9.13. Wavelet methods for monotone operators, eigenvalue problems, nonlinear boundary operators, quadrature approximations: *Under the conditions of Theorem 9.12, wavelet methods applied to the following problems yield unique converging solutions. For monotone operators the convergence has to be modified as in Section 4.5. For the remaining problems (9.125) is valid: variational methods for eigenvalue problems, cf. Section 4.7, and methods for nonlinear boundary operators, cf. Section 5.3. Additionally, let the quadrature approximate methods in (9.126) satisfy (9.127), cf. Section 5.4. All these general convergence properties are only valid for problems where the nonlinearities can be evaluated for wavelets. Finally, all the nonlinear equations for all problems, considered in this chapter, can be solved with Newton's method, satisfying the mesh independence principle.*

We aim for a better convergence with respect to the L^2 norm for second order equations, cf. Theorem 4.60. The solution, ϕ_g, of the *dual problem,*

$$\phi_g \in \mathcal{U} \text{ s.t. } a(w, \phi_g) = (g, w) \forall w \in \mathcal{U}$$

can be combined with the Aubin–Nitsche Lemma 5.73 for modifying Theorem 5.74:

$$\|u_0 - u_0^n\|_{\mathcal{U}} \lesssim \|u_0 - u_0^n\|_{S^n} \sup_{0 \neq g \in L^2(\Omega)} \|\phi_g - \phi_g^n\|_{S^n} / \|g\|_{L^2(\Omega)} \tag{9.128}$$

to obtain

Theorem 9.14. Aubin–Nitsche result: *Under the conditions of Theorem 9.10 and for an exact solution $u_0 \in H^{1+s}(\Omega)$, the solution of the wavelet method equation satisfies*

$$\|u_0 - u_0^n\|_{L^2(\Omega)} \lesssim C 2^{-(s+1)n} \|u_0\|_{H^{1+s}(\Omega)}. \tag{9.129}$$

9.6 Saddle point and (Navier–)Stokes equations

We start with saddle point equations as a generalization of Stokes equations, a special case of the Navier–Stokes equations. For the reader's convenience, we partially repeat Section 2.8.

9.6.1 Saddle point equations

Saddle point problems cf. Subsection 2.8.2 and, for the name, Theorem 2.129, require *two pairs* of Hilbert spaces $\mathcal{U} \subseteq \mathcal{W}_{\mathcal{U}}$, $\mathcal{M} \subseteq \mathcal{W}_{\mathcal{M}}$, with continuous bilinear forms

$$a: \mathcal{U} \times \mathcal{U} \to \mathbb{R},\ b: \mathcal{U} \times \mathcal{M} \to \mathbb{R} \text{ with } \mathcal{U} \hookrightarrow \mathcal{W}_{\mathcal{U}} \hookrightarrow \mathcal{U}', \mathcal{M} \hookrightarrow \mathcal{W}_{\mathcal{M}} \hookrightarrow \mathcal{M}'. \tag{9.130}$$

Usually, $\mathcal{U} \hookrightarrow \mathcal{W}_{\mathcal{U}} \hookrightarrow \mathcal{U}'$ and $\mathcal{M} \hookrightarrow \mathcal{W}_{\mathcal{M}} \hookrightarrow \mathcal{M}'$ are Gelfand triples with continuous and dense embeddings $\mathcal{U} \hookrightarrow \mathcal{W}_{\mathcal{U}}$, $\mathcal{M} \hookrightarrow \mathcal{W}_{\mathcal{M}}$.

For given $(f, g) \in \mathcal{U}' \times \mathcal{M}'$ one has to determine a pair $(u_0, p_0) \in \mathcal{U} \times \mathcal{M}$ such that

$$\begin{aligned} a(u_0, v) + b(v, p_0) &= \langle f, v\rangle_{\mathcal{U}' \times \mathcal{U}} \quad \text{for all } v \in \mathcal{U}, \\ b(u_0, q) &= \langle g, q\rangle_{\mathcal{M}' \times \mathcal{M}} \text{ for all } q \in \mathcal{M}. \end{aligned} \tag{9.131}$$

In general, we assume the bilinear form $a(\cdot,\cdot)$ to be coercive on the subspace

$$\mathcal{U}_0 := \{v \in \mathcal{U} : b(v, q) = 0\ \forall\ q \in \mathcal{M}\},\ \text{hence, } a(v, v) \geq \alpha \|v\|_{\mathcal{U}}^2\ \forall v \in \mathcal{U}_0,\ \alpha > 0, \tag{9.132}$$

compare with (9.3). To ensure that the problem (9.131) is uniquely solvable, we also have to assume that $\mathcal{U}$ and $\mathcal{M}$ fulfill the *inf–sup condition*:

$$\inf_{0 \neq q \in \mathcal{M}} \sup_{0 \neq v \in \mathcal{U}} \frac{b(v, q)}{\|v\|_{\mathcal{U}}\, \|q\|_{\mathcal{M}}} \geq \beta \tag{9.133}$$

for some constant $\beta > 0$. For details, we refer to Subsection 2.8.2, and e.g. to Hackbusch [387]. The following equivalent formulation will be very useful in the sequel. Defining the operators

$$\begin{aligned} A: \mathcal{U} \to \mathcal{U}',\ &\langle Au, v\rangle_{\mathcal{U}' \times \mathcal{U}} := a(u, v),\ v \in \mathcal{U}, \\ B: \mathcal{M} \to \mathcal{U}',\ &\langle Bp, v\rangle_{\mathcal{U}' \times \mathcal{U}} := b(v, p),\ v \in \mathcal{U}, \\ B': \mathcal{U} \to \mathcal{M}',\ &\langle B'u, q\rangle_{\mathcal{M}' \times \mathcal{M}} := b(u, q),\ q \in \mathcal{M}, \end{aligned} \tag{9.134}$$

the problem (9.131) is equivalent to: find $(u_0, p_0) \in \mathcal{U} \times \mathcal{M}$ such that

$$\begin{aligned} Au_0 + Bp_0 &= f \text{ in } \mathcal{U}', \\ B'u_0 &= g \text{ in } \mathcal{M}'. \end{aligned} \tag{9.135}$$

If (9.131) is well posed, the operator

$$\mathcal{B} := \begin{pmatrix} A & B \\ B' & 0 \end{pmatrix} \tag{9.136}$$

is bounded and boundedly invertible with respect to the usual graph norm, see again Subsection 2.8.2, and Hackbusch [387] for details,

$$\|(u,p)\|_{\mathcal{U}\times\mathcal{M}} \sim \|\mathcal{B}(u,p)\|_{\mathcal{U}'\times\mathcal{M}'} \text{ where } \|(u,p)\|^2_{\mathcal{U}\times\mathcal{M}} := \|u\|^2_{\mathcal{U}} + \|p\|^2_{\mathcal{M}}. \tag{9.137}$$

As saddle point problems are defined on product spaces of the form $\mathcal{U}\times\mathcal{M}$, we need *two* pairs of biorthogonal wavelet bases $\Psi = \{\psi_\lambda : \lambda \in J^{\mathcal{U}}\}$, $\tilde{\Psi} = \{\tilde{\psi}_\lambda : \lambda \in J^{\mathcal{U}}\}$ and $\Theta = \{\vartheta_\mu : \mu \in J^{\mathcal{M}}\}$, $\tilde{\Theta} = \{\tilde{\vartheta}_\mu : \mu \in J^{\mathcal{M}}\}$ forming Riesz bases for $\mathcal{W}_{\mathcal{U}}$ and $\mathcal{W}_{\mathcal{M}}$, respectively. The second pair of biorthogonal bases Θ and $\tilde{\Theta}$ also induces a pair of projectors in the sense of (9.79), (9.88):

$$P_n(q) := \sum_{j=0}^{n} \sum_{k\in J_j^{\mathcal{M}}} \langle q, \tilde{\vartheta}_{j,k}\rangle \vartheta_{j,k}, \quad P_n'(q) := \sum_{j=0}^{n} \sum_{k\in J_j^{\mathcal{M}}} \langle q, \vartheta_{j,k}\rangle \tilde{\vartheta}_{j,k}. \tag{9.138}$$

(We use the notation P_n here since in our setting the projectiors Q_n are related with the space $\mathcal{U}$.)

In our applications, $\mathcal{U}, \mathcal{M}_{\mathcal{U}}$ and $\mathcal{M}, \mathcal{W}_{\mathcal{M}}$ are mainly Hilbertian Sobolev spaces on suitable domains or manifolds $\Omega_1 \subset \mathbb{R}^d$, $\Omega_2 \subset \mathbb{R}^{d'}$, i.e.

$$\mathcal{U} = H^t(\Omega_1),\ \mathcal{W}_{\mathcal{U}} = L^2(\Omega_1), \quad \mathcal{M} = H^s(\Omega_2),\ \mathcal{W}_{\mathcal{U}} = L_0^2(\Omega_2), \tag{9.139}$$

where

$$L_0^2(\Omega) := \left\{ q \in L^2(\Omega) \ : \ \int_\Omega q(x)dx = 0 \right\}. \tag{9.140}$$

Then, we assume that the norm equivalences of the form (9.62) hold for both spaces, cf. Theorem 9.5,

$$\|v\|_{H^\tau(\Omega_1)} \sim \left(\sum_{\lambda\in J^{\mathcal{U}}} 2^{2|\lambda|\tau} |\langle v, \tilde{\psi}_\lambda\rangle|^2 \right)^{1/2} \sim \left(\sum_{\lambda\in J^{\mathcal{U}}} 2^{2|\lambda|\tau} |\langle v, \psi_\lambda\rangle|^2 \right)^{1/2}, \tau \in [-t,t],$$

$$\|q\|_{H^\varsigma(\Omega_2)} \sim \left(\sum_{\lambda\in J^{\mathcal{U}}} 2^{2|\mu|\varsigma} |\langle q, \tilde{\vartheta}_\mu\rangle|^2 \right)^{1/2} \sim \left(\sum_{\lambda\in J^{\mathcal{U}}} 2^{2|\lambda|\tau} |\langle v, \vartheta_\lambda\rangle|^2 \right)^{1/2}, \varsigma \in [-s,s].$$

So we choose for (9.135), the trial spaces $(\mathcal{U}_\Lambda, \mathcal{M}_\Lambda) \subset (\mathcal{U}, \mathcal{M})$ with the previous pair of index sets, $(J^{\mathcal{U}}, J^{\mathcal{M}})$,

$$\Lambda := (\Lambda_{\mathcal{U}}, \Lambda_{\mathcal{M}}) \subset (J^{\mathcal{U}}, J^{\mathcal{M}}). \tag{9.141}$$

It is well known [144] that stability of the discretization of (9.135) is ensured if the *Ladyshenskaja–Babuska–Brezzi (LBB) condition*

$$\inf_{q_\lambda\in\mathcal{M}_\Lambda} \sup_{v_\lambda\in\mathcal{U}_\Lambda} \frac{b(v_\lambda, q_\lambda)}{\|v_\lambda\|_{\mathcal{U}}\, \|q_\lambda\|_{\mathcal{M}}} \geq \beta \tag{9.142}$$

is satisfied. Quite recently, explicit conditions to check (9.142) in the wavelet context have been derived by Dahlke *et al.* [241], cf. also Dahmen *et al.* [246], and the conditions for FEMs in Subsection 4.6.2. For any subset $\bar{\mathcal{U}} \subseteq \mathcal{U}$ we will use the notation

$$\bar{\mathcal{U}}^{\perp_b} := \{q \in \mathcal{M} : b(v,q) = 0 \quad \text{for all } v \in \bar{\mathcal{U}}\}, \tag{9.143}$$

and similarly for $\bar{\mathcal{M}} \subseteq \mathcal{M}$

$$\bar{\mathcal{M}}^{\perp_b} := \{v \in \mathcal{U} : b(v,q) = 0 \quad \text{for all } q \in \bar{\mathcal{M}}\}. \tag{9.144}$$

In terms of these sets, the fundamental result from Dahlke *et al.* [241] reads as follows.

Theorem 9.15. Ladyshenskaja–Babuska–Brezzi (LBB) condition: *The multiscale spaces $\mathcal{U}_\Lambda$, $\mathcal{M}_\Lambda$ defined above fulfill the LBB condition (9.142) provided that one of the following equivalent conditions holds (for B c.f.* (9.134)*):*

1. $\mathcal{M}_\Lambda \subseteq (\mathcal{U} \ominus \mathcal{U}_\Lambda)^{\perp_b}$;
2. $B(\mathcal{M}_\Lambda) \subseteq \tilde{\mathcal{U}}_\Lambda = \ \text{span}\{\tilde{\psi}_\lambda, \lambda \in \Lambda\}$;
3. $B'(\mathcal{U} \ominus \mathcal{U}_\Lambda) \subseteq \mathcal{M}' \ominus \tilde{\mathcal{M}}_\Lambda$ *and* $\tilde{\mathcal{M}}_\Lambda$ *is defined analogously to* $\tilde{\mathcal{U}}_\Lambda$.

For simplicity, we treat (9.135) again by a uniform method, i.e. we consider multiscale spaces of the form $\mathcal{V}^n := \mathcal{V}_{J_n^{\mathcal{V}}}$, $J_n^{\mathcal{V}} = \{\lambda \in J^{\mathcal{V}},\ |\lambda| \le n\}$, $\mathcal{M}_{n'} := \mathcal{M}_{J_{n'}^{\mathcal{M}}}$, $J_{n'}^{\mathcal{M}} = \{\mu \in J^{\mathcal{M}},\ |\mu| \le n'\}$. Let Q_n and $P_{n'}$ denote the associated biorthogonal projectors as defined in (9.79), (9.88), and (9.138), respectively. Then the resulting Galerkin scheme for (9.135) is given by: find a pair $(u_0^n, p_0^{n'})$

$$u_0^n \in \mathcal{V}^n := \mathcal{V}_{J_n^{\mathcal{V}}},\ \ p_0^{n'} \in \mathcal{M}^{n'} := \mathcal{M}_{J_{n'}^{\mathcal{M}}} : \ \nu Q_n' A u_0^n + Q_n' B p_0^{n'} = Q_n' f,$$
$$P_{n'}' B' u_0^n = P_{n'}' g. \tag{9.145}$$

Consequently we obtain for saddle point equations:

Theorem 9.16. Stability and convergence for the saddle point wavelet methods:

1. *Assume (9.133), and choose the multiscale spaces $\mathcal{V}^n$ and $\mathcal{M}^{n'}$ in such a way that one of the conditions in Theorem 9.15 is satisfied.*
2. *Then (9.145), the saddle point wavelet method, is stable and consistent for (9.131) with unique and converging solutions, cf. (9.158).*

Adaptive numerical schemes for saddle point problems have been derived in [237], see also Section 9.7.

9.6.2 Navier–Stokes equations

For the *Stokes problem* let Ω be a bounded, simply connected domain in $\mathbb{R}^d$. Then, given a vector field $f \in H^{-1}(\Omega, \mathbb{R}^d)$ and a function $g \in L_0^2(\Omega)$ one has to determine the velocity $u_0 \in H_0^1(\Omega, \mathbb{R}^d)$ and the pressure $p_0 \in L_0^2(\Omega)$ such that, in the strong formulation,

$$-\triangle u_0 + \nabla p_0 = f \quad \text{in } \Omega, \qquad (9.146)$$
$$-\nabla \cdot u_0 = g \quad \text{in } \Omega.$$

In the *mixed formulation* this problem reads as follows: find a solution $(u_0, p_0) \in \big(\mathcal{U} := H_0^1(\Omega, \mathbb{R}^d) \times \big(\mathcal{M} := L_0^2(\Omega)\big)$ for (9.131),with

$$a(u,v) := \langle \nabla u, \nabla v \rangle = \sum_{i,j=1}^{d} \int_\Omega \frac{\partial u_i}{\partial x_j}(x) \frac{\partial v_i}{\partial x_j}(x) dx \text{ for all } u, v \in H_0^1(\Omega, \mathbb{R}^d),$$

$$b(v,q) := -\langle \nabla \cdot v, q \rangle = -\sum_{i=1}^{d} \int_\Omega q(x) \frac{\partial}{\partial x_i} v_i(x) dx \text{ for all } q \in L_0^2(\Omega).$$

For further information concerning the theory and the numerical treatment of the Stokes equations, the reader is referred to Section 4.6, and, e.g. to Hackbusch [387] and Temam [622]. Our goals is to present a stable and convergent wavelet scheme for the stationary (nonlinear) Navier–Stokes equation. It has the form

$$G(u,p) : = \begin{pmatrix} -\nu\Delta u + \sum_{i=1}^{d} u_i \partial^i u + \nabla\, p \\ \text{div } u \end{pmatrix}, \; G(u_0, p_0) = \begin{pmatrix} f \\ 0 \end{pmatrix} \text{ in } \Omega,$$

$$u_0 = 0 \text{ on } \partial\Omega, \qquad \int_\Omega p_0 dx = 0. \qquad (9.147)$$

After multiplying with test functions in the usual way, this problem fits into our setting as follows: find a pair $(u_0, p_0) \in H_0^1(\Omega, \mathbb{R}^d) \times L_0^2(\Omega)$ such that

$$\begin{aligned} \nu a(u_0, v) + d(u_0, u_0, v) + b(v, p_0) &= \langle f, v \rangle \quad \text{for all } v \in H_0^1(\Omega, \mathbb{R}^d), \\ b(u_0, q) &= \langle g, q \rangle \quad \text{for all } q \in L_0^2(\Omega), \end{aligned} \qquad (9.148)$$

where

$$d(u,v,w) := \sum_{i,j=1}^{d} \int_\Omega u_i (\partial^i v_j) w_j dx. \qquad (9.149)$$

For bounded Ω, and $d \leq 4$, see Temam [622], Lemma 1.2, Ch. II, U 1,

$$d(u,v,w) \text{ is a bounded trilinear form on } H_0^1(\Omega, \mathbb{R}^d) \times H_0^1(\Omega, \mathbb{R}^d) \times H_0^1(\Omega, \mathbb{R}^d).$$

To discuss wavelet methods, in particular the stability, for (9.148), we consider its linearized form. We obtain the derivative $G'(u,p)$ evaluated in (u,p) and applied to (w,r) as

$$G'(u,p)(w,r) = \begin{pmatrix} -\nu\Delta w + \sum_{i=1}^{d} (w_i \partial^i u + u_i \partial^i w) + \nabla\, r \\ \text{div } w \end{pmatrix} \qquad (9.150)$$

and, with the $a(\cdot,\cdot), b(\cdot,\cdot), d(\cdot,\cdot)$ in (9.148), and the pair of test functions (v,q)

$$\langle G'(u,p)(w,r),(v,q)\rangle_2 := \begin{pmatrix} \nu a(w,v) + d(u,w,v) + d(w,u,v) + b(v,r) \\ b(w,q) \end{pmatrix}. \tag{9.151}$$

In fact, the term $\sum_{i=1}^{d}(w_i\partial^i u + u_i\partial^i w)$ in (9.150) of lower order derivatives represents, for a moderate Reynolds number, a compact perturbation of the Stokes operator. This can immediately be seen as in Section 4.6.

We shall derive a stable numerical scheme for the treatment of (9.150), (9.151). Some preparations are necessary. We want to solve the problem

$$\langle G'(u,p)(w,r),(v,q)\rangle_2 = \begin{pmatrix} \langle f_1, v\rangle \\ \langle f_2, q\rangle \end{pmatrix} = F(v,q). \tag{9.152}$$

For fixed u, the continuous bilinear forms $d(u,\cdot,w)$ and $d(\cdot,u,v)$ define elements in $H^{-1}(\Omega)$, hence they induce linear continuous operators, D_1, D_2,

$$d_1(v) := d(u,\cdot,v) \text{ and } d_2(v) := d(\cdot,u,v) \text{ induce } D_1, D_2 \in \mathcal{L}(\mathcal{U},\mathcal{U}'). \tag{9.153}$$

Therefore we observe that (9.151) can be written as

$$\mathcal{A}(w,r) = \mathcal{B}(w,r) + \mathcal{C}(w,r) = F \tag{9.154}$$

where, slightly different from (9.136), cf. (9.134), (9.135),

$$\mathcal{B} = \begin{pmatrix} \nu A & B \\ B' & 0 \end{pmatrix}, \quad \mathcal{C} = \begin{pmatrix} D_1 + D_2 & 0 \\ 0 & 0 \end{pmatrix}. \tag{9.155}$$

We treat this problem again by a uniform method, i.e. we consider the same multicale spaces as in Section 9.6.1. Then the resulting Galerkin scheme for the linear (9.154) requires: find a pair $(u_0^n, p^{n'})$

$$u_0^n \in \mathcal{V}^n := \mathcal{V}_{J_n^{\mathcal{V}}}, \quad p_0^{n'} \in \mathcal{M}^{n'} := \mathcal{M}_{J_{n'}^{\mathcal{M}}} : \tag{9.156}$$

$$\nu Q_n' A u_0^n + Q_n' B p_0^{n'} + Q_n'(D_1 + D_2) u_0^n = Q_n' f_1,$$
$$P_{n'}' B' u_0^n = P_{n'}' f_2.$$

The stationary (nonlinear) Navier-Stokes equation has the form (9.148), (9.147). So the wavelet method requires: find a pair $(u_0^n, p_0^{n'}) \in \mathcal{V}^n \times \mathcal{M}^{n'}$ such that

$$\begin{aligned} \nu a(u_0^n, v^n) + d(u_0^n, u_0^n, v^n) + b\left(v^n, p_0^{n'}\right) &= \langle f, v^n\rangle \quad \text{for all } v^n \in \mathcal{V}^n, \\ b\left(u_0^n, q^{n'}\right) &= \langle g, q^{n'}\rangle \quad \text{for all } q^{n'} \in \mathcal{M}^{n'}. \end{aligned} \tag{9.157}$$

Consequently we obtain for the (linearized) Navier–Stokes equations, cf. Subsection 4.6.2 and Böhmer *et al.* [122, 129, 567].

Theorem 9.17. Stability and convergence for the Navier–Stokes wavelet methods:

1. *Choose the multiscale spaces* $\mathcal{V}^n$ *and* $\mathcal{M}^{n'}$ *such that one of the conditions in Theorem 9.15 is satisfied for the Stokes equations.*
2. *Then the linearized form of the Navier–Stokes operator* $\mathcal{A}$ *in* (9.154) *represents, for sufficiently large* ν, *a compact perturbation of the Stokes operator* $\mathcal{B}$ *in* (9.155). *For boundedly invertible* $\mathcal{A}$, *in particular for sufficiently large* ν, *the Galerkin scheme* (9.156) *yields stable and consistent* $\mathcal{B}^n$ *and* $\mathcal{A}^n$.
3. *The wavelet Navier–Stokes problems (9.156) and (9.157) for the linearized and the nonlinear equations, possess, for large enough* n, *unique discrete solutions,* $u_0^n, p_0^{n'}$, *near the exact solution,*u_0, p_0. *For the nonlinear equations the nonlinear terms have to be evaluatible for wavelet arguments.*
4. *If the exact solution satisfies* $u_0 \in H^{1+s}(\Omega), p_0 \in H^{1+s}(\Omega)$, *the* $u_0^n, p_0^{n'}$ *converge according to, cf. [246],*

$$\|u_0^n - u_0\|_{\mathcal{U}} + \|p_0 - p_0^{n'}\|_{\mathcal{M}} \lesssim 2^{-sn}\|u_0\|_{H^{1+s}(\Omega)} + 2^{-sn'}\|p_0\|_{H^{1+s}(\Omega)}. \quad (9.158)$$

Proof. Compare Subsection 4.6.2 and Böhmer *et al.* [122, 129, 567]. All the operators are continuously differentiable. So we show that our wavelet method for both problems satisfies the conditions of Theorem 3.21. The method is stable by assumption or by the compact perturbation property of the Navier–Stokes equation, by Theorem 9.8 for large enough parameters ν. Finally the conformity of our methods implies their consistency. This finishes the proof. □

9.7 Adaptive wavelet methods, by T. Raasch

For the efficient numerical simulation of realistic problems coming from technical applications, *adaptive* approximation methods with a *highly nonuniform spatial discretization* are of ultimate importance. The core ingredient of any adaptive algorithm is an appropriate coupling of *a posteriori error estimators* and *adaptive space refinement.* By following this general strategy, reliable approximations of the unknown solution can be provided within a prescribed error tolerance and with a reasonable amount of computational work.

For more than 25 years, adaptive finite element methods have been successfully used in practical applications. However, a full theoretical comprehension of their convergence and complexity properties, even for second order elliptic problems, could be acquired only recently. We refer to Morin *et al.* [499], Stevenson [600] and to Chapter 6 for an overview. Since the late 1990s, particular interest has also been drawn to the analysis of adaptive discretization schemes based on *wavelets.* Contrary to adaptive finite element methods, the convergence and complexity properties of adaptive wavelet algorithms could be analyzed already at an early stage of their development. This is in particular due to the strong analytic properties of wavelet systems. In the sequel, we will give a brief overview of the current status in adaptive wavelet methods; see also Dahmen [243] and Cohen [195].

Unless otherwise stated, we shall assume in this section that the underlying wavelet system $\Psi = \{\psi_\lambda\}_{\lambda \in J}$ is a *Riesz basis* for the solution space $\mathcal{U}$. By this we mean that any $u \in \mathcal{U}$ can be expanded with respect to Ψ into a series $u = \sum_{\lambda \in J} u_\lambda \psi_\lambda$ with a unique coefficient array $\mathbf{u} := (u_\lambda)_{\lambda \in J}$, and with the stability condition

$$c_R \|\mathbf{u}\|_{\ell^2(J)} \le \|u\|_{\mathcal{U}} \le C_R \|\mathbf{u}\|_{\ell^2(J)}, \tag{9.159}$$

where $c_R, C_R > 0$ are suitable Riesz constants. Therefore, in contrast to the previous sections, the wavelet system is normalized in $\mathcal{U}$, i.e. we have $\|\psi_\lambda\|_{\mathcal{U}} \eqsim 1$ for all $\lambda \in J$. From the density of span Ψ in $\mathcal{U}$, it follows that the variational problem $a(u, v) = \langle f, v\rangle_{\mathcal{U}' \times \mathcal{U}}$ for all $v \in \mathcal{U}$ is equivalent to the infinite system

$$\mathbf{A}\mathbf{u} = \mathbf{f}, \tag{9.160}$$

where we define

$$\mathbf{A} := \big(a(\psi_\mu, \psi_\lambda)\big)_{\lambda,\mu \in J}, \quad \mathbf{f} = \big(\langle f, \psi_\lambda\rangle_{\mathcal{U}' \times \mathcal{U}}\big)_{\lambda \in J}.$$

The following observation is of fundamental importance to any adaptive wavelet algorithm:

Theorem 9.18. Bounded bilinear form and $\mathbf{A}$: *If the bilinear form $a : \mathcal{U} \times \mathcal{U} \to \mathbb{R}$ is continuous, $a(v, w) \le \beta \|v\|_{\mathcal{U}} \|w\|_{\mathcal{U}}$ for all $v, w \in \mathcal{U}$ and some $\beta > 0$, then $\mathbf{A}$ is bounded on $\ell^2(J)$ with $\|\mathbf{A}\| \le \beta C_R^2$.*

Moreover, if $a = b + c$ decomposes into bilinear forms b and c, where b is coercive, $b(v, v) \ge \alpha \|v\|_{\mathcal{U}}^2$, and c is bounded, $c(u, v) \le \gamma \|u\|_{\mathcal{U}} \|v\|_{\mathcal{U}}$ with $\gamma < \alpha$, then $\mathbf{A}$ is boundedly invertible on $\ell^2(J)$ with $\|\mathbf{A}^{-1}\| \le (\alpha - \gamma)^{-1} c_R^{-2}$.

Proof. The claim immediately follows by inserting the respective properties of the bilinear forms and (9.159) into the identity

$$\|\mathbf{A}\mathbf{v}\|_{\ell^2(J)} = \sup_{0 \ne \mathbf{w} \in \ell^2(J)} \frac{\left|a\left(\sum_\mu v_\mu \psi_\mu, \sum_\lambda w_\lambda \psi_\lambda\right)\right|}{\|\mathbf{w}\|_{\ell^2(J)}}.$$

□

In the mathematical analysis of adaptive wavelet methods, typically the following three major issues are addressed.

(i) First of all, one is interested in establishing *convergence* of the adaptive refinement strategy. By this we mean that, given a target accuracy $\epsilon > 0$ as input argument, the algorithm $(\mathbf{u}, \epsilon) \mapsto \mathbf{u}_\epsilon$ under consideration terminates after finitely many steps and outputs a finitely supported coefficient sequence $\mathbf{u}_\epsilon$, such that

$$\|\mathbf{u} - \mathbf{u}_\epsilon\|_{\ell^2(J)} \le \epsilon.$$

Note that via the Riesz stability (9.159), such convergence estimates are equivalent to controlling the error in the $\mathcal{U}$ norm.

(ii) As a second issue, the *convergence rate* of the adaptive method is usually analyzed in detail, i.e. the ratio between the number of active degrees of freedom and the corresponding discretization error. Results on attainable convergence rates of adaptive approximation schemes and on the conditions to guarantee them can be used to quantify whether, for a given operator equation, the additional computational work of adaptive algorithms asymptotically pays off at all, compared to nonadaptive strategies.

In order to analyze convergence rates of adaptive approximation schemes in detail, certain tools from nonlinear approximation theory are necessary. We will summarize them in Subsection 9.7.1.

(iii) The third issue in the mathematical analysis of an adaptive refinement strategy is related with its *computational complexity.* In order to scale in an optimal way when applied to a realistic problem, the algorithm should work in linear complexity, i.e. the number of floating point and storage operations needed to compute the output $\mathbf{u}_\epsilon$ should stay proportional to the number of degrees of freedom $\#\mathrm{supp}\mathbf{u}_\epsilon$. In that case, a reasonable response time of the adaptive algorithm can be expected for small target accuracies ϵ also.

Implementable algorithms which feature one or more of these key properties typically have an iterative structure and make use of a posteriori error estimators. Choosing the infinite dimensional system (9.160) as a starting point, the following two major strategies for the development of adaptive wavelet schemes for stationary problems have been pursued in the literature.

(i) A first natural strategy is to consider adaptive wavelet–*Galerkin* methods. They select certain finite subsets $\Lambda \subset J$ and compute the solutions $\mathbf{u}_\Lambda \in \ell^2(\Lambda)$ to the associated Galerkin systems

$$\mathbf{P}_\Lambda \mathbf{A} \mathbf{I}_\Lambda \mathbf{u}_\Lambda = \mathbf{P}_\Lambda \mathbf{f},$$

where $\mathbf{P}_\Lambda : \ell^2(J) \to \ell^2(\Lambda)$ denotes the restriction operator and $\mathbf{I}_\Lambda = \mathbf{P}_\Lambda^* : \ell^2(\Lambda) \to \ell^2(J)$ is the trivial inclusion. The particular choice of the active index sets Λ is steered by the biggest coefficients in the current discrete residual

$$\mathbf{r}_\Lambda = \mathbf{r}(\mathbf{I}_\Lambda \mathbf{u}_\Lambda) = \mathbf{f} - \mathbf{A} \mathbf{I}_\Lambda \mathbf{u}_\Lambda \in \ell^2(J)$$

or in an approximation thereof. This strategy leads to the adaptive wavelet–Galerkin algorithms considered in [70,197,203,236,333], and we note that it is similar to analogous residual-based refinement strategies in a finite element setting [499,600]. In any of these variants, the actual computation of approximate residuals is made feasible by exploiting certain matrix compression properties of wavelet systems, discussed in Subsection 9.7.2. However, we remark that for wavelet–Galerkin methods, it is mostly assumed that the bilinear form a and hence $\mathbf{A}$ are symmetric. This is due to the fact that convergence is usually proven in the energy norm, see Subsection 9.7.3 for details. Only for a compact antisymmetric part of a, has convergence of adaptive wavelet–Galerkin methods recently been analyzed in [332].

(ii) A second approach, initially propagated in [198], is focused on the approximate application of well-known *iterative methods* for finite dimensional systems to infinite dimensional systems like (9.160). For linear iterative schemes, like the damped Richardson iteration and the steepest descent method, converge can be verified also in an infinite-dimensional setting [161,198] by straightforward arguments. Moreover, by replacing the evaluations of the infinitely supported right-hand side $\mathbf{f}$ and of biinfinite matrix-vector products $\mathbf{Av}$ by inexact computable versions thereof, inexact linear iterations were derived in [161,198] and in subsequent related papers. We refer to Subsection 9.7.4 for an overview.

For several reasons, adaptive descent iterations are more flexible than adaptive Galerkin methods. On the one hand, linear descent iterations of the aforementioned type can be readily applied also to more general problems than those with a linear symmetric operator $\mathbf{A}$. Generalizations towards nonsymmetric elliptic equations have been discussed in [198]. By using adaptive variants of the classical Uzawa iteration, also saddle point problems can be handled [237]. Moreover, the analysis of best tree approximation methods in [69, 196, 200] has enabled the development of adaptive wavelet algorithms for semilinear stationary equations, based on an adaptive Newton-type iteration [199]. We refer to Subsection 9.7.5 for details.

In order to improve the linear convergence properties of iterations like the Richardson and the steepest descent method, one may also consider adaptive variants of more complicated Krylov iterations. As an example, inexact conjugate gradient methods have been analyzed in [252], although their numerical performance usually does not outperform the simpler steepest descent scheme.

Another important feature is that, when choosing approximate iterative schemes, one has the possibility of replacing the wavelet basis Ψ by a *redundant* system, e.g. by a wavelet frame. Frames are usually easier to construct than bases, in particular on nontrivially shaped bounded domains. Moreover, the matrix compression properties of wavelet bases can be preserved when using suitable frame systems. It must be noted that in a frame discretization, the discrete system (9.160) may become singular and the coefficient representation of a given function u will no longer be unique. However, these potential drawbacks can be handled in practice, e.g. for Richardson iterations [239,538,598] and for the steepest descent method [240]. Moreover, domain decomposition methods in $\ell^2(J)$ can be used to improve the quantitative performance of wavelet frame discretizations considerably, see [601,663] for details.

9.7.1 Nonlinear approximation with wavelet systems

The underlying principle of adaptive wavelet methods consists in approximating the unknown solution u in the norm of $\mathcal{U}$ by finite linear combinations from the ansatz system Ψ, such that the number of nonzero coefficients is as small as possible. By using the Riesz stability (9.159), we may confine the discussion to the sequence space regime, approximating the unknown coefficient array $\mathbf{u}$ in $\ell^2(J)$ by sparsely populated sequences.

As a benchmark, the most economical approximations of a given $\mathbf{v} \in \ell^2(J)$ are the *best N-term approximations* $\mathbf{v}_N$, defined by discarding in $\mathbf{v}$ all but the N largest coefficients in absolute value. The error of best N-term approximation of $\mathbf{v}$ is then defined as

$$\sigma_N(\mathbf{v}) := \|\mathbf{v} - \mathbf{v}_N\|_{\ell^2(\mathcal{I})}.$$

Obviously, an upper bound for the convergence rate of an adaptive discretization scheme is given by the largest value $s > 0$, such that the best N-term approximation error of the unknown coefficient array $\mathbf{u}$ decays like $\sigma_N(\mathbf{u}) \lesssim N^{-s}$, as N tends to infinity. All sequences with this property are usually collected in the nonlinear approximation space

$$\mathcal{A}^s := \{\mathbf{v} \in \ell^2(J) : \|\mathbf{v}\|_{\mathcal{A}^s} := \|\mathbf{v}\|_{\ell^2(J)} + \sup_{N\in\mathbb{N}} N^s \sigma_N(\mathbf{v}) < \infty\}.$$

Note that $\|\cdot\|_{\mathcal{A}^s}$ is in general only a *quasinorm.*[50] It is well known that for $\epsilon > 0$, $0 < p \leq 2 - \epsilon$ and $s = \frac{1}{p} - \frac{1}{2} > 0$, we have the continuous and dense embeddings

$$\ell^p(J) \hookrightarrow \mathcal{A}^s \hookrightarrow \ell^{p+\epsilon}(J), \tag{9.161}$$

see [286, 331] for a proof. Therefore, $\mathcal{A}^s$ is also frequently referred to as the *weak* ℓ^p space $\ell^p_w(J)$, and $\mathcal{A}^s$ may be endowed with several equivalent quasinorms; we refer to [286] for details.

In order to ensure that the unknown coefficient array $\mathbf{u}$ is contained in $\mathcal{A}^s$, certain nonclassical regularity assumptions on the solution u play a crucial role. Wavelet theory tells us, see [195], that if Ψ is a wavelet Riesz basis for $\mathcal{U} = H^t(\Omega)$ which is sufficiently smooth and has polynomial order r, then for all $0 < s < (r-t)/d$ we have

$$\mathbf{v} \in \ell^p(J) \quad \text{if and only if} \quad v = \sum_{\lambda \in J} v_\lambda \psi_\lambda \in B^{sd+t}_p(L^p(\Omega)), \tag{9.162}$$

where $p = \left(s + \frac{1}{2}\right)^{-1}$ and d is the spatial dimension of the domain Ω. Therefore, combining (9.162) with the embedding (9.161), we see that whenever the unknown solution u is contained in the *Besov space* $B^{sd+t}_p(L^p(\Omega))$, then the nonlinear H^t-approximation rate of u with respect to the wavelet system Ψ is at least s, $\|\mathbf{u}\|_{\mathcal{A}^s} \lesssim \|u\|_{B^{sd+t}_p(L^p(\Omega))}$. The maximum nonlinear H^t-convergence rate that can be expected by only assuming appropriate smoothness conditions on the unknown solution u is $(r-t)/d$.

Remark 9.19.

1. *In order to obtain an H^t-approximation rate $s > 0$ for* linear *approximation methods with* uniform *grid refinement, it would be necessary that u belongs to the smaller space $H^{sd+t}(\Omega)$. However, recent results from regularity theory of PDEs show that there exist several relevant classes of operator equations whose solutions are much smoother in the scale $B^{sd+t}_p(L^p(\Omega))$ than in the Sobolev scale, see [233, 235, 238] and the references therein. For example, corner singularities of polygonal domains in $\mathbb{R}^2$ have infinitely high regularity in the Besov scale $B^{2s+t}_p(L^p(\Omega))$, whereas their Sobolev regularity is strongly limited [234]. For problems with those kinds of solutions, nonlinear approximation methods will*

[50] Here the triangle inequality is replaced by $\exists C > 0.\ \|x + y\| \leq C(\|x\| + \|y\|) \forall x, y \in \mathcal{A}^s$

therefore asymptotically outperform linear approximation strategies with uniform refinement.

2. *Note that we will not address the approximation of u in weaker norms than that of $\mathcal{U}$. This is due to the fact that, to the best of our knowledge, a counterpart of Nitsche duality arguments is not known in the case of nonlinear approximation methods.*

In practical computations, where the target quantity $\mathbf{v}$ is only implicitly given, we will not be able to compute a best N-term approximation $\mathbf{v}_N$ exactly. Instead, it is easier to aim at the computation of an *approximate* or *near* best N-term approximation $\tilde{\mathbf{v}}_N$, by which we mean that $\#\mathrm{supp}\tilde{\mathbf{v}}_N \leq N$ and

$$\|\mathbf{v} - \tilde{\mathbf{v}}_N\|_{\ell^2} \leq C\sigma_N(\mathbf{v}),$$

with $C \geq 1$ being a uniform constant which only depends on the particular algorithm. Due to the fact that the best N-term approximation of an element $\mathbf{v} \in \mathcal{A}^s$ up to accuracy ϵ needs at most $\epsilon^{-1/s}\|\mathbf{v}\|_{\mathcal{A}^s}^{1/s}$ entries, a convergent algorithm $(\mathbf{u}, \epsilon) \mapsto \mathbf{u}_\epsilon$ is called *asymptotically optimal* if, for any $\mathbf{u}$ from the unit ball of $\mathcal{A}^s$, the output $\mathbf{u}_\epsilon$ fulfills

$$\#\mathrm{supp}\mathbf{u}_\epsilon \lesssim \epsilon^{-1/s},$$

as ϵ tends to 0. The algorithm has *asymptotically optimal complexity* if, moreover, the number of floating point and storage operations to compute $\mathbf{u}_\epsilon$ are bounded by a constant multiple of $\#\mathrm{supp}\mathbf{u}_\epsilon$. The algorithm is called s^**-optimal* and of s^**-optimal complexity*, respectively, if these implications hold true for all $0 < s < s^*$.

In many adaptive schemes, computational optimality is ensured by inserting certain thresholding operations into the algorithm. The rationale is to thin out the current iterate in such a way that the perturbed algorithm still converges, but it additionally features an optimal work/accuracy balance. It is therefore of crucial importance to perform thresholding operations onto a given finitely supported vector $\mathbf{v}$ in a computationally optimal way. The naive approach of sorting all entries of $\mathbf{v}$ in modulus and then collecting the most important ones has the disadvantage of log-linear complexity. For this reason, sorting operations were not taken into account in the initial complexity analysis of adaptive wavelet methods [197]. However, it has been shown in [69, 598] that quasisorting is sufficient to compute an approximate best N-term approximation $\tilde{\mathbf{v}}_N$ in linear complexity. The following binning variant of hard thresholding is therefore used in current adaptive wavelet schemes.

Algorithm 9.1.
COARSE$[\mathbf{v}, \epsilon] \to \mathbf{v}_\epsilon$
$q := \lceil \log_2((\#\mathrm{supp}\mathbf{v})^{1/2}\|\mathbf{v}\|_{\ell^2}/\epsilon) \rceil$
sort nonzero entries of $\mathbf{v}$ *into bins:*
for all $\lambda \in \mathrm{supp}\mathbf{v}$ do
    if $|v_\lambda| \leq 2^{-q}\|\mathbf{v}\|_{\ell^2}$ then
        put (λ, v_λ) into the bin V_q
    else
        put (λ, v_λ) into that bin V_i with $2^{-(i+1)}\|\mathbf{v}\|_{\ell^2} < |v_\lambda| \leq 2^{-i}\|\mathbf{v}\|_{\ell^2}$

```
    end if
end do
aggregate large entries until tolerance is met:
v_ε := 0; err := ||v||²_ℓ²
for i = 0 to q do
    for all (λ, v_λ) ∈ V_i do
        if err ≤ ε² then return v_ε
        (v_ε)_λ := v_λ; err := err − |v_λ|²
    end for
end for
```

The computational optimality of the subroutine COARSE has been analyzed in [69, 598].

Theorem 9.20. $\mathbf{v}_\epsilon$ has few nonzero entries: *Let $\mathbf{v} \in \ell^2(J)$ be finitely supported and $\epsilon > 0$. Then for the output $\mathbf{v}_\epsilon := \mathrm{COARSE}[\mathbf{v}, \epsilon]$, it holds that $\|\mathbf{v} - \mathbf{v}_\epsilon\|_{\ell^2} \leq \epsilon$ and $\mathbf{v}_\epsilon$ has at most $\#\mathrm{supp}\mathbf{v}_\epsilon \lesssim \inf\{N : \sigma_N(\mathbf{v}) \leq \epsilon\}$ nonzero entries. Moreover, the number of arithmetic operations and storage locations needed to compute $\mathbf{v}_\epsilon$ is bounded by a constant multiple of $\#\mathrm{supp}\mathbf{v} + \max\{\log(\epsilon^{-1}\|\mathbf{v}\|_{\ell^2}), 1\} \lesssim \epsilon^{-1/s}\|\mathbf{v}\|_{\mathcal{A}^s}^{1/s}$.*

Moreover, finitely supported approximations to a given vector $\mathbf{v} \in \mathcal{A}^s$ can be thinned out efficiently by an application of the subroutine COARSE, at the price of a slightly enlarged error, see [197, 598].

Theorem 9.21. Good approximation by COARSE with few nonzero entries: *Let $\theta < 1/3$ be fixed and $s > 0$. Then, for any $\epsilon > 0$, $\mathbf{v} \in \mathcal{A}^s$, and a finitely supported $\mathbf{w} \in \ell^2(J)$ with $\|\mathbf{v} - \mathbf{w}\|_{\ell^2} \leq \theta\epsilon$, the output $\bar{\mathbf{w}} := \mathrm{COARSE}[\mathbf{w}, (1-\theta)\epsilon]$ fulfills $\|\mathbf{v} - \bar{\mathbf{w}}\|_{\ell^2} \leq \epsilon$ and the number of nonzero entries in $\bar{\mathbf{w}}$ is bounded by*

$$\#\mathrm{supp}\bar{\mathbf{w}} \lesssim \epsilon^{-1/s}\|\mathbf{v}\|_{\mathcal{A}^s}^{1/s}.$$

Consequently, there is a constant $C > 0$ which only depends on $s > 0$, such that

$$\|\bar{\mathbf{w}}\|_{\mathcal{A}^s} \leq C\|\mathbf{v}\|_{\mathcal{A}^s}.$$

However, we note that thresholding routines like COARSE are not a necessary ingredient of any adaptive scheme. Recently, it has been proved that certain variants of adaptive wavelet–Galerkin methods attain optimal convergence rates also without intermediate coarsening of the iterands, see [333].

9.7.2 Wavelet matrix compression

In almost all variants of adaptive wavelet methods, it is exploited that the biinfinite system matrix $\mathbf{A}$ is not arbitrarily structured but it often features certain *compressibility properties*. By this we mean that $\mathbf{A}$ can be approximated well by sparse matrices with a finite number of entries per row and column.

Such compressibility properties can indeed be observed and verified, see [91, 197, 245, 571, 599], whenever the underlying operator is *local* or at least *pseudo-local*, i.e. for differential equations and certain classes of integral equations. In both cases, using the locality and the vanishing moment properties of wavelet systems, *off-diagonal decay estimates* of the following type can be shown for individual matrix entries, see [243, 571, 599]:

$$\left|a(\psi_{\lambda'}, \psi_\lambda)\right| \lesssim 2^{-||\lambda|-|\lambda'||\sigma}\left(1+\delta(\lambda,\lambda')\right)^{-\beta}, \quad \text{for all } \lambda, \lambda' \in J. \tag{9.163}$$

Here $\sigma > d/2$ and $\beta > d$ depend on the particular wavelet system Ψ and on the operator, and δ is given by

$$\delta(\lambda,\lambda') := 2^{\min\{|\lambda|,|\lambda'|\}}\text{dist}(\text{supp}\psi_\lambda, \text{supp}\psi_{\lambda'}), \quad \text{for all } \lambda, \lambda' \in J.$$

In the special case of a differential operator, the decay parameter $\beta > 0$ in (9.163) may be chosen arbitrarily large, i.e. the off-diagonal decay is essentially exponential in the level distance $||\lambda| - |\lambda'||$,

$$\left|a(\psi_{\lambda'}, \psi_\lambda)\right| \lesssim 2^{-||\lambda|-|\lambda'||\sigma}, \quad \text{for all } \lambda, \lambda' \in J. \tag{9.164}$$

Moreover, it is important here that the wavelets ψ_λ are normalized in the solution space $\mathcal{U} = H^t(\Omega)$. Otherwise, in the case of an L^2 normalization, the left-hand sides of (9.163) and of (9.164) would entail also the normalization factors $\|\psi_\lambda\|^{-1}_{H^t(\Omega)} \eqsim 2^{-|\lambda|t}$.

For an appropriate formalization of compressibility from the viewpoint of the full system matrix, we recall the following definition from [197] and related works. A bounded operator $\mathbf{A} : \ell^2(J) \to \ell^2(J)$ is called *s^*-compressible*, when for each $j \in \mathbb{N}$ there exists an infinite matrix $\mathbf{A}_j$ with at most $\alpha_j 2^j$ nonzero entries per row and column and $\sum_{j=0}^{\infty} \alpha_j < \infty$, such that $\sum_{j=0}^{\infty} 2^{js}\|\mathbf{A} - \mathbf{A}_j\| < \infty$ holds for any $0 < s < s^*$. Equivalently, for any $0 < s < s^*$ and any $N \in \mathbb{N}$, there exists an infinite matrix with at most N nonzero entries per row and column and with distance to $\mathbf{A}$ of order N^{-s}.

Remark 9.22.

1. *We do not assume that the nonzero entries of $\mathbf{A}_j$ coincide with those of $\mathbf{A}$, i.e. quadrature errors are allowed.*
2. *Any s^*-compressible $\mathbf{A}$ is bounded from $\mathcal{A}^s$ to $\mathcal{A}^s$, $0 < s < s^*$ [331].*

In general, the particular compression rules mostly work by retaining in $\mathbf{A}_j$ only those entries $\mathbf{A}_{\lambda,\lambda'}$ (or approximations thereof), for which the level difference is below the threshold $||\lambda| - |\lambda'|| \leq j/d$. For differential operators this is already sufficient, at least for a certain range of the compressibility exponent s^*. In the case of singular integral operators, additional thresholding techniques have to be applied per column; we refer to [243, 571, 599] for details.

In order to estimate the associated computational work, an s^*-compressible matrix $\mathbf{A}$ is called *s^*-computable*, if the computation of any column of $\mathbf{A}_j$ takes at most $\mathcal{O}(2^j)$ floating point operations. An interpretation is that the *average* computational cost of each individual entry of $\mathbf{A}_j$ should be constant. This notion of computability was introduced in [334, 335] in order to verify s^*-optimal complexity of adaptive wavelet schemes for differential and singular integral operators, also taking the quadrature subproblems into account.

Within an adaptive numerical scheme, the matrices $\mathbf{A}_j$ will be used to realize the multiplication of $\mathbf{A}$ with a given finitely supported vector $\mathbf{v}$, up to a prescribed error tolerance ϵ. In particular, the estimate $\|\mathbf{A} - \mathbf{A}_j\| \leq C_j 2^{-js}$ for $0 < s < s^*$ can be used to construct a numerical routine APPLY for approximate matrix-vector multiplication. The initial definition of APPLY goes back to [197]. However, analogous to the COARSE Algorithm 9.1 from Subsection 9.7.1, also the APPLY subroutine has undergone a certain evolution in the literature. In Algorithm 9.2 below we present the most advanced binning variant from [598], for which also optimal complexity estimates could be established.

Algorithm 9.2.
APPLY$[\mathbf{A}, \mathbf{v}, \epsilon] \rightarrow \mathbf{w}_\epsilon$
$q := \lceil \log_2((\#\mathrm{supp}\mathbf{v})^{1/2}\|\mathbf{v}\|_{\ell^2}\|\mathbf{A}\|2/\epsilon)\rceil$
sort nonzero entries of $\mathbf{v}$ *into bins:*
for all $\lambda \in \mathrm{supp}\mathbf{v}$ do
    if $|v_\lambda| \leq 2^{-q}\|\mathbf{v}\|_{\ell^2}$ then
        put (λ, v_λ) into the bin V_q
    else
        put (λ, v_λ) into that bin V_i with $2^{-(i+1)}\|\mathbf{v}\|_{\ell^2} < |v_\lambda| \leq 2^{-i}\|\mathbf{v}\|_{\ell^2}$
    end if
end do
regroup large nonzero entries of $\mathbf{v}$ *into segments* $\mathbf{v}_{[k]}$*:*
$k := 0$; $\mathbf{v}_{[0]} := \mathbf{0}$; $err := \|\mathbf{v}\|_{\ell^2}^2$; $i := 0$
while $err > \epsilon^2/(4\|\mathbf{A}\|^2)$ do
    if $V_i = \emptyset$ then $i := i + 1$
    if $\#\mathrm{supp}\mathbf{v}_{[k]} > 2^k - \lfloor 2^{k-1} \rfloor$ then $k := k + 1$; $\mathbf{v}_{[k]} := \mathbf{0}$
    select $(\lambda, v_\lambda) \in V_i$ and remove it from bin V_i
    $(\mathbf{v}_{[k]})_\lambda := v_\lambda$; $err := err - |v_\lambda|^2$
end do
$l := k$
compute approximation to $\mathbf{Av}$*:*
compute smallest $j \geq l$ with $\sum_{k=0}^{l} C_{j-k} 2^{-(j-k)s}\|\mathbf{v}_{[k]}\|_{\ell^2} \leq \epsilon/2$
$\mathbf{w}_\epsilon := \sum_{k=0}^{l} \mathbf{A}_{j-k}\mathbf{v}_{[k]}$

Theorem 9.23. Bounded operations, storage, nonzero entries by APPLY: *Let* $\mathbf{A}$ *be* s^**-compressible and* $\mathbf{v}$ *be finitely supported. Then, for any* $\epsilon > 0$*, the output* $\mathbf{w}_\epsilon :=$ APPLY$[\mathbf{A}, \mathbf{v}, \epsilon]$ *fulfills* $\|\mathbf{Av} - \mathbf{w}_\epsilon\|_{\ell^2} \leq \epsilon$ *and* $\mathbf{w}_\epsilon$ *has at most* $\#\mathrm{supp}\mathbf{w}_\epsilon \lesssim \epsilon^{-1/s}\|\mathbf{v}\|_{\mathcal{A}^s}^{1/s}$ *nonzero entries. Moreover, the number of arithmetic operations and storage locations needed to compute* $\mathbf{w}_\epsilon$ *is bounded by a constant multiple of* $\epsilon^{-1/s}\|\mathbf{v}\|_{\mathcal{A}^s}^{1/s} + \#\mathrm{supp}\mathbf{v}$.

It remains to specify the concrete value of s^* for which compressibility and computability of $\mathbf{A}$ can be expected, being of crucial importance to the convergence properties of an adaptive wavelet scheme. In order not to spoil the theoretically optimal value $(r - t)/d$, it is desirable that the system matrix $\mathbf{A}$ is s^*-computable for values of s^* strictly greater than $(r - t)/d$.

The first results on the compressibility of operators in wavelet coordinates can be found in [197, 243], assuming only generic properties of the wavelet system, like smoothness and vanishing moments, and certain properties of the operator. However, the results from [197, 243], on the actual range of s^* were still suboptimal. It turned out shortly afterwards that the restriction to spline wavelet systems can substantially improve the compressibility properties of the system matrix. This is mostly due to the fact that spline wavelets are locally smooth, away from their lower-dimensional singular support. Therefore, in the case of spline wavelets more entries of $\mathbf{A}$ can be expected to have small modulus, compared to the case of generic wavelet systems. Considering the Laplace operator, compressibility of order $s^* = (\gamma - t)/d$ has been proved in [70, 237], where

$$\gamma = \sup\{\nu : \psi_\lambda \in H^\nu(\Omega) \text{ for all } \lambda \in J\}$$

is the critical Sobolev smoothness of Ψ. The most far-reaching compressibility results, however, have been obtained in [599]. For spline wavelet discretizations of differential operators and of integral operators with *Schwartz kernel*, the results from [599] show that $\mathbf{A}$ is indeed s^*-compressible for $s^* > (r - t)/d$. In particular, there exist compression rules $\mathbf{A} \mapsto \mathbf{A}_j$, such that $\|\mathbf{A} - \mathbf{A}_j\| \lesssim 2^{-js}$ for s larger than any value for which $\mathbf{u} \in \mathcal{A}^s$ can be expected. Therefore, the theoretically optimal convergence rate will not be limited by the compressibility of the system matrix.

Concerning computability, in the early results on the convergence rates of adaptive wavelet schemes [197], it was assumed that the entries of $\mathbf{A}$ are available exactly at uniformly bounded computational cost. However, using spline wavelets, this is only possible for differential operators with constant coefficients. The computational complexity of the adaptive matrix-vector multiplication APPLY in the case of general differential operators and of integral operators with Schwartz kernel was discussed recently in [335] and [334], respectively. Computing the nonzero entries of $\mathbf{A}_j$ by suitable quadrature rules with an error proportional to the compression error of merely dropping entries from $\mathbf{A}$, the computational optimality of APPLY could be verified for a large range of compressible operators as well.

9.7.3 Adaptive wavelet–Galerkin methods

In the history of adaptive wavelet methods, convergence was first established for certain Galerkin strategies [197, 236] Due to the fact that the respective proofs are using the energy norm, we will assume in this subsection that the bilinear form $a : \mathcal{U} \times \mathcal{U} \to \mathbb{R}$ is coercive and symmetric. Moreover, we denote with $|||\mathbf{v}||| := \langle \mathbf{A}\mathbf{v}, \mathbf{v}\rangle^{1/2}$ the discrete energy norm of a given coefficient sequence $\mathbf{v} \in \ell^2(J)$. The following norm equivalence is well known:

$$\|\mathbf{A}^{-1}\|^{-1/2}\|\mathbf{v}\|_{\ell^2(J)} \le |||\mathbf{v}||| \le \|\mathbf{A}\|^{1/2}\|\mathbf{v}\|_{\ell^2(J)}, \quad \text{for all } \mathbf{v} \in \ell^2(J).$$

Therefore, convergence of the coefficient arrays in the discrete energy norm is equivalent to convergence in $\ell^2(J)$ and, by the Riesz basis property (9.159), to $\mathcal{U}$-convergence of the expansions.

Adaptive Galerkin methods typically have an iterative structure. Starting with an initial Galerkin set Λ_0, they compute a sequence of nested index sets Λ_j, $j \in \mathbb{N}_0$.

The iteration stops as soon as the associated Galerkin solution $u_{\Lambda_j} \in \mathcal{U}$ is considered sufficiently good by an embedded a posteriori error estimator. During the iteration, successive Galerkin approximations u_{Λ_j} are chosen to realize a certain *saturation property*, i.e. the new iterate $u_{\Lambda_{j+1}}$ and the corresponding coefficient sequence $\mathbf{u}_{\Lambda_{j+1}}$ are strictly better than the previous ones:

$$|||\mathbf{u} - \mathbf{u}_{\Lambda_{j+1}}||| \le \rho|||\mathbf{u} - \mathbf{u}_{\Lambda_j}|||, \quad \rho \in (0,1). \tag{9.165}$$

In order to predict a new Galerkin set Λ_{j+1} with a prescribed linear error reduction (9.165), the current discrete residual $\mathbf{r}_{\Lambda_j} = \mathbf{f} - \mathbf{A}\mathbf{I}_{\Lambda_j}\mathbf{u}_{\Lambda_j}$ plays a key role. The following lemma points out that we have to determine the positions of the biggest entries of $\mathbf{r}_{\Lambda_j}$. The result is implicitly used in [197, 236], and the proof can be found in [331, 333]:

Lemma 9.24. *Let $0 < \mu \le 1$ and $\mathbf{w} \in \ell^2(J)$ with finite support* $\operatorname{supp}\mathbf{w} \subset \Lambda \subset J$ *be such that*

$$\|\mathbf{P}_\Lambda \mathbf{r}(\mathbf{w})\|_{\ell^2(J)} \ge \mu \|\mathbf{r}(\mathbf{w})\|_{\ell^2(J)}. \tag{9.166}$$

Then the Galerkin solution $\mathbf{u}_\Lambda$ associated with Λ fulfills

$$|||\mathbf{u} - \mathbf{I}_\Lambda \mathbf{u}_\Lambda||| \le \Big(1 - \frac{\mu^2}{\kappa(\mathbf{A})}\Big)^{1/2} |||\mathbf{u} - \mathbf{w}|||,$$

with $\kappa(\mathbf{A}) = \|\mathbf{A}\|\|\mathbf{A}^{-1}\|$.

In fact, if $\mathbf{w} = \mathbf{I}_{\Lambda_j}\mathbf{u}_{\Lambda_j}$ is the current iterate, an application of Lemma 9.24 shows that the new Galerkin set $\Lambda_{j+1} \supset \Lambda_j$ should fulfill (9.166) with

$$\mu = \big(\kappa(\mathbf{A})(1 - \rho^2)\big)^{1/2}.$$

Note that due to the constraint $\mu \le 1$ in Lemma 9.24, the attainable error reduction per iteration is bounded from below by

$$\rho \ge \Big(1 - \frac{1}{\kappa(\mathbf{A})}\Big)^{1/2}.$$

However, it is still a nontrivial task to detect the biggest entries in the discrete residual since $\mathbf{r}_{\Lambda_j}$ will almost always be an ℓ^2 sequence with infinite support. Moreover, in order to end up with an optimal computational complexity, the new index set Λ_{j+1} should ideally be the minimal superset of Λ_j with the prescribed saturation property. Both tasks can in fact be accomplished in practice, by the development of a suitable numerical subroutine GROW for the extension process. In [236], approximations of the discrete residual $\mathbf{r}_{\Lambda_j}$ were constructed by the aforementioned adaptive matrix-vector multiplication and by inexact evaluations of the right-hand side $\mathbf{f}$, ending up with a convergent Galerkin scheme. In [197], the problem of computational optimality was addressed. By interleaving the extension steps with certain thresholding operations, it was shown that the resulting sequence of Galerkin approximations shows an optimal convergence rate. Moreover, the corresponding algorithm was numerically tested and compared with finite element schemes in [70]. The experiments revealed that in many cases and for judicious parameter choice, the intermediate thresholding steps could even be avoided. This observation was analyzed and justified for a specific variant of GROW in [333]; we cite the corresponding main result.

Theorem 9.25. Bounds for operations and storage by GROW: *There exists an implementable algorithm* GROW$[\mathbf{w}, \bar{\nu}, \epsilon] \to (\Lambda, \nu)$ *with the following properties:*

1. *For finitely supported* $\mathbf{w} \in \ell^2(J)$, $\bar{\nu} \geq \|\mathbf{r}(\mathbf{w})\|_{\ell^2(J)}$ *and* $\epsilon > 0$, GROW *terminates with* $\nu \geq \|\mathbf{r}(\mathbf{w})\|_{\ell^2(J)}$ *and* $\nu \gtrsim \min\{\nu, \epsilon\}$. *If* $\mathbf{u} \in \mathcal{A}^s$ *for some* $s < s^*$, *then the number of floating point and storage operations for the call is bounded by a constant multiple of* $\min\{\nu, \bar{\nu}\}^{-1/s}(\|\mathbf{w}\|_{\mathcal{A}^s} + \|\mathbf{u}\|_{\mathcal{A}^s} + \bar{\nu}^{1/2}(\#\mathrm{supp}\mathbf{w} + 1))$.
2. *If* GROW *terminates with* $\nu > \epsilon$, *then*

$$\left\|\mathbf{P}_\Lambda \mathbf{r}(\mathbf{w})\right\|_{\ell^2(J)} \geq \frac{\alpha - \omega}{1 + \omega} \nu$$

and

$$\#(\Lambda \setminus \mathrm{supp}\mathbf{w}) \lesssim \nu^{-1/s} \|\mathbf{u}\|_{\mathcal{A}^s}^{1/s}.$$

Here $0 < \omega < \alpha$ *are internal parameters of* GROW *with* $(\alpha + \omega)/(1 - \omega) < \kappa(\mathbf{A})^{-1/2}$.

Therefore, unless the GROW subroutine outputs a new Galerkin set with residual error below the target accuracy ϵ, the new iterate $u_{\Lambda_{j+1}}$ fulfills (9.166) for $\mu = \frac{\alpha - \omega}{1 + \omega}$. In consequence, the error reduction rate of the global iteration is at least $\rho = \left(1 - (\alpha - \omega)^2/((1 + \omega)^2 \kappa(\mathbf{A}))\right)^{1/2}$.

9.7.4 Adaptive descent iterations

A conceptually different approach to the discretization of stationary elliptic equations has been proposed in [198]. Instead of cutting out finite portions of the equivalent ℓ^2 problem (9.160) and then solving them with standard algorithms from numerical linear algebra, the strategy advocated in [198] proceeds the other way around. The idea is to directly apply well-known iterative schemes in the infinite dimensional setting, preferably with fixed error reduction per iteration step. Only after that, are all infinite dimensional quantities replaced by finitely supported and computable ones. In particular, one has to work with an inexact right-hand side $\bar{\mathbf{f}} \approx \mathbf{f}$, approximate matrix-vector operations, and appropriately matched tolerances for subroutines like APPLY or COARSE.

Among the relevant iterative schemes are, e.g. *Richardson iteration*

$$\mathbf{u}^{(n+1)} = \mathbf{u}^{(n)} + \alpha \mathbf{r}^{(n)}, \quad \mathbf{r}^{(n)} = \mathbf{f} - \mathbf{A}\mathbf{u}^{(n)}, \quad n = 0, 1, \ldots \tag{9.167}$$

with fixed relaxation parameter $\alpha > 0$, and the closely related *steepest descent method*

$$\mathbf{u}^{(n+1)} = \mathbf{u}^{(n)} + \alpha_n \mathbf{r}^{(n)}, \quad \alpha_n = \frac{\langle \mathbf{r}^{(n)}, \mathbf{r}^{(n)} \rangle}{\langle \mathbf{A}\mathbf{r}^{(n)}, \mathbf{r}^{(n)} \rangle}, \quad n = 0, 1, \ldots \tag{9.168}$$

with a posteriori choice of the descent parameter α_n. Both iterative schemes can be shown to converge also in an ℓ^2 setting. For instance, convergence of the exact

Richardson iteration (9.167) follows from the contractivity

$$\rho = \rho(\alpha) = \|\mathbf{I} - \alpha\mathbf{A}\| < 1,$$

which can be achieved whenever $0 < \alpha < 2/\|\mathbf{A}\|$. In that case, we have linear error reduction per iteration,

$$\|\mathbf{u} - \mathbf{u}^{(n+1)}\|_{\ell^2(J)} \leq \rho\|\mathbf{u} - \mathbf{u}^{(n)}\|_{\ell^2(J)},$$

and the exact solution $\mathbf{u}$ has the series representation

$$\mathbf{u} = \mathbf{A}^{-1}\mathbf{f} = \alpha\sum_{n=0}^{\infty}(\mathbf{I} - \alpha\mathbf{A})^n\mathbf{f}.$$

If $\mathbf{A}$ is symmetric, then the optimal relaxation parameter $\hat{\alpha} = \arg\min_\alpha \rho(\alpha)$ can be computed as

$$\hat{\alpha} = \frac{2}{\|\mathbf{A}\| + \|\mathbf{A}^{-1}\|^{-1}},$$

with

$$\rho(\hat{\alpha}) = \frac{\kappa(\mathbf{A}) - 1}{\kappa(\mathbf{A}) + 1}.$$

In practical applications, however, we will almost never be able to determine $\hat{\alpha}$ exactly. Implementations of the Richardson iteration (9.167) will therefore often suffer from suboptimal performance. In contrast, the steepest descent iteration (9.168) automatically adapts the descent parameter to the current iterate, and it usually outperforms the Richardson iteration exactly for this reason. We refer to [240, 252] for illustrative numerical experiments.

The convergence and complexity analysis of the resulting inexact iterations follows general principles. Usually, it is exploited that the exact iteration

$$\mathbf{u}^{(n+1)} = \mathbf{G}(\mathbf{u}^{(n)}), \quad n = 0, 1, \ldots \tag{9.169}$$

has at least linear error reduction

$$\|\mathbf{u} - \mathbf{u}^{(n+1)}\| \leq \rho\|\mathbf{u} - \mathbf{u}^{(n)}\|, \quad n = 0, 1, \ldots \tag{9.170}$$

for some $0 < \rho < 1$, and $\|\cdot\|$ is either the ℓ^2 norm or the discrete energy norm. Many important schemes have this property: we mention Richardson iteration (9.167), the steepest descent method (9.168) and other Krylov methods, like Chebyshev iteration [331] or the conjugate gradient method [252]. In practice, the exact iteration mapping $\mathbf{G}$ is replaced by an inexact algorithm $\text{ITER}[\mathbf{w}, \epsilon] \to \mathbf{v}_\epsilon$ whose output $\mathbf{v}_\epsilon$ fulfills

$$\left\|\mathbf{G}(\mathbf{w}) - \mathbf{v}_\epsilon\right\| \leq \epsilon, \quad \text{for all } \epsilon > 0.$$

Linear convergence of the perturbed but implementable iteration

$$\mathbf{v}^{(n+1)} = \text{ITER}[\mathbf{v}^{(n)}, \epsilon_n], \quad n = 0, 1, \ldots \tag{9.171}$$

with a slightly larger error reduction constant can easily be shown.

Lemma 9.26. *Assume that the exact iteration* (9.169) *has linear error reduction* (9.170) *with* $\rho < \bar{\rho} < 1$. *Then for all tolerances* $\epsilon_n \le \|\mathbf{u} - \mathbf{v}^{(0)}\|(\bar{\rho} - \rho)\bar{\rho}^n$, *the iterates from* (9.171) *fulfill the error estimate*

$$\|\mathbf{u} - \mathbf{v}^{(n)}\| \le \|\mathbf{u} - \mathbf{v}^{(0)}\|\bar{\rho}^n, \quad n = 0, 1, \ldots$$

Proof. The claim immediately follows from the induction

$$\begin{aligned}\|\mathbf{u} - \mathbf{v}^{(n+1)}\| &\le \left\|\mathbf{u} - \mathbf{G}(\mathbf{v}^{(n)})\right\| + \left\|\mathbf{G}(\mathbf{v}^{(n)}) - \mathrm{ITER}[\mathbf{v}^{(n)}, \epsilon_n]\right\| \\ &\le \rho\|\mathbf{u} - \mathbf{v}^{(n)}\| + \epsilon_n \\ &\le \rho\|\mathbf{u} - \mathbf{v}^{(0)}\|\bar{\rho}^n + (\bar{\rho} - \rho)\|\mathbf{u} - \mathbf{v}^{(0)}\|\bar{\rho}^n \\ &= \|\mathbf{u} - \mathbf{v}^{(0)}\|\bar{\rho}^{n+1}.\end{aligned}$$

□

However, with the simple iteration (9.171) it is in general impossible to verify either convergence rates or linear computational complexity of the overall algorithm. In order to end up with an optimal convergence rate, one may consider intermediate thresholding operations with the COARSE subroutine. Note that Theorems 9.20 and 9.21 have analogs when the ℓ^2 norm is replaced by the discrete energy norm. Due to the fact that coarsening of a given iterate slightly enlarges the approximation error, the tolerances have to be adjusted carefully to preserve convergence. We therefore consider the following generic SOLVE Algorithm 9.3, see also [198, 331].

Algorithm 9.3.
SOLVE$[\mathbf{v}^{(0)}, \epsilon] \to \mathbf{v}_\epsilon$
let $\theta < 1/3$ be fixed and $K \in \mathbb{N}$ such that $\rho^K + (\bar{\rho} - \rho)\sum_{i=0}^{K-1} \rho^i \le \theta\bar{\rho}$
$n := 0$
$\mathbf{w}^{(0,0)} := \mathbf{v}^{(0)}$
$\epsilon_0 := \|\mathbf{A}^{-1}\|\|\mathbf{f} - \mathbf{A}\mathbf{v}^{(0)}\|$
while $\epsilon_n > \epsilon$ do
  for $i = 1$ to K do
    $\mathbf{w}^{(n,i)} := \mathrm{ITER}\left[\mathbf{w}^{(n,i-1)}, (\bar{\rho} - \rho)\epsilon_n\right]$
  end do
  $\mathbf{v}^{(n+1)} := \mathrm{COARSE}\left[\mathbf{w}^{(n,K)}, (1 - \theta)\bar{\rho}\epsilon_n\right]$
  $\epsilon_{n+1} := \bar{\rho}\epsilon_n$
  $n := n + 1$
end do
$\mathbf{v}_\epsilon := \mathbf{v}^{(n)}$

Theorem 9.27. Estimates for errors and support in SOLVE: *Assume that the exact iteration* (9.169) *fulfills* (9.170) *with* $\rho < \bar{\rho} < 1$. *Then* SOLVE$[\mathbf{v}^{(0)}, \epsilon] \to \mathbf{v}_\epsilon$ *terminates with* $\|\mathbf{u} - \mathbf{v}_\epsilon\| \le \epsilon$. *If, moreover,* $\mathbf{u} \in \mathcal{A}^s$ *for some* $s > 0$, *then* $\#\mathrm{supp}\mathbf{v}_\epsilon \lesssim \epsilon^{-1/s}\|\mathbf{u}\|_{\mathcal{A}^s}^{1/s}$.

Proof. Concerning convergence, it is sufficient to verify that

$$\|\mathbf{u} - \mathbf{v}^{(n)}\| \le \epsilon_n = \epsilon_0\bar{\rho}^n = \|\mathbf{A}^{-1}\|\|\mathbf{f} - \mathbf{A}\mathbf{v}^{(0)}\|\bar{\rho}^n. \tag{9.172}$$

For $n = 0$, this is trivial. Otherwise, after the K iterations in the while loop, then with an analogous argument as in Lemma 9.26,

$$\begin{aligned}\|\mathbf{u} - \mathbf{w}^{(n,K)}\| &\leq \rho\|\mathbf{u} - \mathbf{w}^{(n,K-1)}\| + (\bar{\rho} - \rho)\epsilon_n \\ &\leq \rho^K \|\mathbf{u} - \mathbf{v}^{(n)}\| + (\bar{\rho} - \rho)\epsilon_n \textstyle\sum_{i=0}^{K-1} \rho^i \\ &\leq \big(\rho^K + (\bar{\rho} - \rho) \textstyle\sum_{i=0}^{K-1} \rho^i\big)\epsilon_n \leq \theta\bar{\rho}\epsilon_n.\end{aligned}$$

By the properties of COARSE, see Theorem 9.20, we can deduce (9.172),

$$\|\mathbf{u} - \mathbf{v}^{(n+1)}\| \leq \theta\bar{\rho}\epsilon_n + (1-\theta)\bar{\rho}\epsilon_n = \epsilon_{n+1}.$$

As an immediate consequence of (9.172) and Theorem 9.21, we finally estimate

$$\#\mathrm{supp}\mathbf{v}_\epsilon \lesssim (\bar{\rho}\epsilon_n)^{-1/s}\|\mathbf{u}\|_{\mathcal{A}^s}^{1/s} \lesssim \epsilon^{-1/s}\|\mathbf{u}\|_{\mathcal{A}^s}^{1/s}.$$

□

In order to obtain an algorithm with linear computational complexity, further assumptions on the subroutine ITER are necessary. The following theorem is adapted from [331].

Theorem 9.28. Estimates for operations, storage and support in SOLVE: *In the situation of Theorem 9.27, assume that* $\mathbf{w} = \mathrm{ITER}[\mathbf{v}, \epsilon]$ *fulfills*

$$\#\mathrm{supp}\mathbf{w} \lesssim \#\mathrm{supp}\mathbf{v} + \epsilon^{-1/s}\|\mathbf{u}\|_{\mathcal{A}^s}^{1/s}, \qquad \|\mathbf{w}\|_{\mathcal{A}^s} \lesssim (\#\mathrm{supp}\mathbf{v})^{1/s}\epsilon + \|\mathbf{u}\|_{\mathcal{A}^s},$$

where the number of floating point and storage operations needed to compute $\mathbf{w}$ *is bounded by a constant multiple of* $\epsilon^{-1/s}\|\mathbf{u}\|_{\mathcal{A}^s}^{1/s} + \#\mathrm{supp}\mathbf{v} + 1$. *Then the output* $\mathbf{v}_\epsilon = \mathrm{SOLVE}[\mathbf{v}^{(0)}, \epsilon]$ *can be computed with at most an absolute multiple of* $\epsilon^{-1/s}\|\mathbf{u}\|_{\mathcal{A}^s}^{1/s}$ *floating point and storage operations, as* ϵ *tends to zero.*

Remark 9.29. *The generic algorithm* SOLVE *requires knowledge of the error reduction constant* $\rho < 1$ *or at least an approximation thereof. For a particular iterative scheme like Richardson iteration or the steepest descent method, such estimates are available whenever the spectral bounds* $\|\mathbf{A}\|$ *and* $\|\mathbf{A}^{-1}\|$ *can be approximated.*

9.7.5 Nonlinear stationary problems

The paradigm which has been outlined in the previous subsection, namely the treatment of the original operator equation in the sequence space regime by infinite-dimensional iterations and by a subsequent discretization of the single iteration step, may be readily applied to problems with nonlinearities also. Starting out from [199], the adaptive discretization of certain nonlinear operator equations has been discussed recently in the literature.

Assume that we are looking for a solution $u \in \mathcal{U}$ to

$$F(u) = 0, \tag{9.173}$$

where $F : \mathcal{U} \to \mathcal{U}'$ has a continuous Fréchet derivative $F' : \mathcal{U} \to \mathcal{U}'$ which is invertible in a neighborhood of u,

$$\|F'(v)w\|_{\mathcal{U}'} \eqsim \|w\|_{\mathcal{U}}, \quad \text{for all } v \approx u, w \in \mathcal{U}.$$

Using a wavelet Riesz basis $\Psi = \{\psi_\lambda\}_{\lambda \in J}$ for $\mathcal{U}$, (9.173) has an equivalent reformulation as the infinite nonlinear system

$$\mathbf{F}(\mathbf{u}) = \mathbf{0},$$

where

$$\mathbf{F}(\mathbf{v}) = \big(\langle F(v), \psi_\lambda \rangle_{\mathcal{U}' \times \mathcal{U}}\big)_{\lambda \in J}, \quad v = \sum_{\lambda \in J} v_\lambda \psi_\lambda.$$

Such systems of equations can be handled by iterative schemes of the form

$$\mathbf{u}^{(n+1)} = \mathbf{u}^{(n)} - \mathbf{R}_n \mathbf{F}(\mathbf{u}^{(n)}), \quad n = 0, 1, \ldots \tag{9.174}$$

Here the $\mathbf{R}_n : \mathcal{U} \to \mathcal{U}$ are suitable bounded linear operators. As typical examples, we mention Richardson iteration with stationary choice $\mathbf{R}_n = \alpha \mathbf{I}$, or the Newton method with nonstationary $\mathbf{R}_n = \mathbf{F}'(\mathbf{u}^{(n)})^{-1}$. Note that the discrete mapping $\mathbf{F}$ has the Fréchet derivative

$$\mathbf{F}'(\mathbf{v}) = \big(\langle F'(v)\psi_\lambda, \psi_\mu \rangle_{\mathcal{U}' \times \mathcal{U}}\big)_{\mu, \lambda \in J}.$$

The convergence properties of the exact method (9.174) have been analyzed in [199], for several classes of nonlinearities and iterative strategies. In case we are dealing with semilinear operator equations

$$Au + g(u) = f, \tag{9.175}$$

where $A : \mathcal{U} \to \mathcal{U}'$ is boundedly invertible and $g : \mathcal{U} \to \mathcal{U}'$ is monotone,

$$\langle v - w, g(v) - g(w) \rangle_{\mathcal{U} \times \mathcal{U}'} \geq 0, \quad \text{for all } v, w \in \mathcal{U},$$

global convergence of the exact discrete Richardson iteration

$$\mathbf{u}^{(n+1)} = \mathbf{u}^{(n)} - \alpha\big(\mathbf{A}\mathbf{u}^{(n)} + \mathbf{g}(\mathbf{u}^{(n)}) - \mathbf{f}\big), \quad n = 0, 1, \ldots$$

can be verified for sufficiently small values of $\alpha > 0$. For general nonlinear problems like (9.173), the simplified Newton iteration $\mathbf{R}_n = \mathbf{F}'(\mathbf{u}^{(0)})^{-1}$ has been discussed in [199]. As a first-order scheme, it at least shows linear error reduction (9.170) with rate $\rho < 1$ in the vicinity of the exact solution u.

When it comes to concrete numerical implementations, the question arises of how to efficiently realize the application of the nonlinearity in (9.174). As already mentioned in Subsection 9.3.5, *tree approximation techniques* are an appropriate tool for this kind of problem. By imposing a tree structure on their output arguments, numerical routines for thresholding and for the adaptive application of nonlinear mappings to finitely supported sequences have been developed in [199], based on the findings from [251]. We cite the corresponding main results in the following theorems.

Thresholding of finitely supported vectors can be performed with a subroutine TCOARSE, i.e. tree coarsening.

Theorem 9.30. Quasi-optimal tree in TCOARSE: *There exists an implementable algorithm* TCOARSE$[\mathbf{v}, \epsilon] \to \mathbf{v}_\epsilon$ *with the following properties:*

1. *For finitely supported* $\mathbf{v} \in \ell^2(J)$ *and* $\epsilon > 0$, TCOARSE *computes a finitely supported* $\mathbf{v}_\epsilon \in \ell^2(J)$, *such that* $\|\mathbf{v} - \mathbf{v}_\epsilon\|_{\ell^2(J)} \leq \epsilon$ *and* $\mathrm{supp}\mathbf{v}_\epsilon$ *is a quasioptimal tree, i.e.* $\mathrm{supp}\mathbf{v}_\epsilon$ *is a tree with*
$$\#\mathrm{supp}\mathbf{v}_\epsilon \leq C \inf\left\{\#\mathcal{T} : \mathcal{T} \subset J \textit{ is a tree and } \|\mathbf{v} - \mathbf{I}_\mathcal{T}\mathbf{P}_\mathcal{T}\mathbf{v}\|_{\ell^2(J)} \leq \epsilon\right\}.$$
2. *The number of floating point and storage operations to compute the output* $\mathbf{v}_\epsilon$ *stays proportional to*
$$\inf\left\{\#\mathcal{T} : \mathcal{T} \textit{ is a tree with } \mathrm{supp}\mathbf{v} \subset \mathcal{T} \subset J\right\}.$$

Due to its similarity with the COARSE subroutine from Subsection 9.7.1, results like Theorems 9.20 and 9.21 carry over to the case of tree coarsening with TCOARSE.

The inexact evaluation of the nonlinearity is done with a subroutine EVAL. As outlined in Subsection 9.3.5, the algorithmic flow of EVAL decomposes into two steps. The first step is the recovery step, where an estimate for the support of the output is computed. The concrete entries indexed by the recovery set are computed in a second step. The basic properties of EVAL read as follows, see [199].

Theorem 9.31. Approximation of nonlinear $\mathbf{F}(\mathbf{v})$ by $\mathbf{w}_\epsilon$: *There exists an implementable algorithm* EVAL$[\mathbf{v}, \epsilon] \to \mathbf{w}_\epsilon$ *with the following properties:*

1. *For finitely supported* $\mathbf{v} \in \ell^2(J)$ *and* $\epsilon > 0$, *EVAL computes a finitely supported* $\mathbf{w}_\epsilon \in \ell^2(J)$, *such that* $\|\mathbf{F}(\mathbf{v}) - \mathbf{w}_\epsilon\|_{\ell^2(J)} \leq \epsilon$ *and* $\mathrm{supp}\mathbf{w}_\epsilon$ *is a tree with*
$$\#\mathrm{supp}\mathbf{w}_\epsilon \leq C \inf\left\{\#\mathrm{supp}\mathbf{w} : \mathrm{supp}\mathbf{w} \textit{ is a tree and } \|\mathbf{F}(\mathbf{v}) - \mathbf{w}\|_{\ell^2(J)} \leq \epsilon\right\}.$$
2. *The number of floating point and storage operations to compute the output* $\mathbf{w}_\epsilon$ *stays proportional to* $\#\mathrm{supp}\mathbf{w}_\epsilon + \#\mathrm{supp}\ \mathbf{v} + 1$

By combining the subroutines TCOARSE, EVAL and APPLY, an implementale adaptive iteration has been designed in [199]. At least in the case of semilinear problems (9.175), the overall algorithm converges at an optimal rate and with linear computational complexity. In [69, 72], the recovery algorithm is explained in more detail, with an explicit analysis of all computational ingredients and with extensive numerical experiments for nonlinearities depending on function values only. As an interesting variant, we mention that the recovery algorithm can also be applied to adaptive quadrature schemes for the stiffness matrix $\mathbf{A}$ of linear problems; see [71] for details.

Moreover, in [435], it has been shown that also in the case of redundant ansatz systems, tree approximation techniques can be used to obtain adaptive recovery algorithms with optimal computational complexity.

Bibliography

[1] R. Adams. *Sobolev Spaces.* Academic Press, New York, 1978.

[2] S. Agmon, A. Douglis, and L. Nirenberg. Estimates near the boundary for solutions of elliptic partial differential equations, i, ii. *Comm. Pure Appl. Math.*, 12:623–727, 1959.

[3] S. Agmon, A. Douglis, and L. Nirenberg. Estimates near the boundary for solutions of elliptic partial differential equations, i, ii. *Comm. Pure Appl. Math.*, 17:35–92, 1964.

[4] N. E. Aguilera and L. A. Caffarelli. Regularity results for discrete solutions of second order elliptic problems in the finite element method. *Calcolo*, 23:327–353, 1986.

[5] M. Ahues. A note on finite element error bounds for multiple defective eigenvalues and maximal invariant subspaces. *Appl. Math. Lett.*, 4(3):65–68, 1991.

[6] M. Ainsworth and J. T. Oden. *A Posteriori Error Estimation in Finite Element Analysis.* John Wiley, New York, 2000.

[7] M. A. Alimov and A. K. Pulatov. Eigenvalues of a non-selfadjoint operator. In *Mathematical Analysis, Algebra and Probability Theory, Collect. Sci. Works, Tashkent, 8–11*, 1987.

[8] E. L. Allgower, P. Ashwin, K. Böhmer, and Zhen Mei. Liapunov-Schmidt reduction for a bifurcation problem with periodic boundary conditions on a square domain. In K. Georg, E. L. Allgower, and R. Miranda, editors, *Exploiting Symmetry in Applied and Numerical Analysis*, Lectures in Applied Mathematics 29, 11–22. AMS, Providence, 1993.

[9] E. L. Allgower and K. Böhmer. A mesh independence principle for operator equations and their discretization. Preprint, Dept. of Mathematics, Colorado State University, Fort Collins, 1984. Revision: Arbeitspapier der GMD 129, Bonn.

[10] E. L. Allgower and K. Böhmer. Application of the mesh independence principle to mesh refinement strategies. *SIAM J. Numer. Anal.*, 24:1335–1351, 1987.

[11] E. L. Allgower, K. Böhmer, and S. F. McCormick. Discrete correction methods for operator equations. In K. Glashoff, E. Allgower, and H.-O. Peitgen, editors, *Numerical Solution of Nonlinear Equations: Simplicial and Classical Methods*, Springer Lecture Notes in Mathematics 878, 30–97. Springer, Vienna, New York, 1981.

[12] E. L. Allgower, K. Böhmer, and S. F. McCormick. Discrete defect corrections: The basic ideas. *ZAMM*, 62:371–377, 1982.

[13] E. L. Allgower, K. Böhmer, and Z. Mei. On a problem decomposition for semi-linear nearly symmetric elliptic problems. In W. Hackbusch, editor, *Parallel Algorithms for PDE's*, Notes on Numerical Fluid Mechanics 31, 1–17. Vieweg Verlag, Braunschweig, Fed. Rep. Germany, 1991.

[14] E. L. Allgower, K. Böhmer, and Z. Mei. On new bifurcation results for semi-linear elliptic equations with symmetries. In J. Whiteman, editor, *The Mathematics of Finite Elements and Applications*, volume VII, 487–494. Academic Press, Brunel, UK, 1991.

[15] E. L. Allgower, K. Böhmer, and Z. Mei. A complete bifurcation scenario for the 2d-nonlinear Laplacian with Neumann boundary conditions on the unit square. In R. Seydel, F. W. Schneider, T. Küpper, and H. Troger, editors, *Bifurcation and Chaos: Analysis, Algorithms, Applications*, volume 97 of *International Series of Numerical Mathematics*, 1–18. Birkhäuser Verlag, Basel, 1991.

[16] E. L. Allgower, K. Böhmer, and Z. Mei. Exploiting equivariance in the reduced bifurcation equations. In E. L. Allgower, K. Böhmer, and M. Golubitsky, editors, *Bifurcation and Symmetry: Cross Influence and Application*, volume 104 of *ISNM*, 1–10. Birkhäuser Verlag, Basel, 1992.

[17] E. L. Allgower, K. Böhmer, F. A. Potra, and W. C. Rheinboldt. A mesh independence principle for operator equations and their discretization. *SIAM J. Numer. Anal.*, 23:160–169, 1986.

[18] E. L. Allgower, S-G. Cruceanu, and S. Tavener. Application of numerical continuation to compute all solutions of semilinear elliptic equations. *Advanced Geometry*, 9(3):377–400, 2009.

[19] E. L. Allgower, S-G. Cruceanu, and S. Tavener. Turning points and bifurcation for homotopies of analytic maps. Contemperary Mathematics, 496, 1–10, 2009.

[20] E. L. Allgower and K. Georg. *Numerical Continuation Methods: An Introduction*, Series in Computational Mathematics 13 Springer, Berlin, Heidelberg, New York, 1990.

[21] E. L. Allgower and K. Georg. Numerical path following. In P. G. Ciarlet and J. L. Lions, editors, *Handbook of Numerical Analysis*, volume 5, 3–207. North-Holland, Amsterdam, 1997.

[22] E. L. Allgower and M. M. Jeppson. The approximation of solutions of nonlinear elliptic boundary value problems having several solutions. In R. Ansorge and W. T. Törnig, editors, *Numerische, insbesondere approximationstheoretische Behandlung von Funktionalgleichungen*, Lecture Notes in Mathematics 33, 1–20. Oberwolfach, Springer Verlag, New York, 1973.

[23] H. W. Alt and S. Luckhaus. Quasilinear elliptic-parabolic differential equations. *Math. Z.*, 183:311–341, 1983.

[24] M. Altman. *A Unified Theory of Nonlinear Operator and Evolution Equations with Applications. A New Approach to Nonlinear Partial Differential Equations.* Pure and Applied Mathematics 103. Marcel Dekker, Inc., New York, Basel, 1986.

[25] H. Amann and P. Quittner. Elliptic boundary value problems involving measures: Existence, regularity, and multiplicity. *Adv. Differ. Equ.*, 3(6):753–813, 1998.

[26] W. F. Ames. *Nonlinear partial differential equations in engineering.* Academic Press, New York, San Francisco, London, 1965.

[27] W. F. Ames. *Nonlinear partial differential equations in engineering, Vol. II.* Academic Press, New York, San Francisco, London, 1972.

[28] P. M. Anselone and R. Ansorge. Compactness principles in nonlinear operator approximation theory. *Numer. Funct. Anal. and Optimiz.*, 1:589–618, 1979.

[29] P. M. Anselone and R. Ansorge. A unified framework for the discretization of nonlinear operator equations. *Numer. Funct. Anal. and Optimiz.*, 4:61–99, 1981.

[30] P. M. Anselone and R. Ansorge. Discrete closure and asymptotic (quasi-) regularity in discretization algorithms. *IMA J. Num. Anal.*, 7:431–448, 1987.

[31] R. Ansorge. Konvergenz von Differenzenverfahren für quasilineare Anfangswertaufgaben. *Numer. Math.*, 13:217–225, 1969.

[32] R. Ansorge. Iterated discretization. *Numer. Funct. Anal. and Otimiz.*, 17:691–701, 1996.

[33] P. F. Antonietti, F. Brezzi, and L. D. Marini. Bubble stabilization of discontinuous Galerkin method. *Comput. Methods Appl. Mech. Engrg.*, 198: 1651–1659, 2009.

[34] P. F. Antonietti and P. Houston. A pre-processing moving mesh method for discontinuous Galerkin approximations of advection-diffusion-reaction problems. *Int. J. Num. Anal. and Model.*, 5(4):704–728, 2008.

[35] T. Apel, A.-M. Sändig, and S. I. Solov'ev. Computation of 3D vertex singularities for linear elasticity: Error estimates for a finite element method on graded meshes. *Math. Model. Numer. Anal.*, 36:1043–1070, 2002.

[36] I. K. Argyros. A mesh-independence principle for nonlinear operator equations and their discretizations under mild differentiability conditions. *Computing*, 45(3):265–268, 1990.

[37] I. K. Argyros. Newton methods on Banach spaces with a convergence structure and applications. *Comput. Math. Appl.*, 40(1):37–48, 2000.

[38] M. Arias. Nonselfadjoint boundary value problems at resonance with nonlinearities which may grow linearly. *Nonlin. Anal., Theor. Meth. Appl.*, 15(2):155–163, 1990.

[39] D. N. Arnold. An interior penalty finite element method with discontinuous elements. *SIAM J. Numer. Anal.*, 19(4):742–760, 1982.

[40] D. N. Arnold, F. Brezzi, B. Cockburn, and L. D. Marini. Unified analysis of discontinuous Galerkin methods for elliptic problems. *SIAM J. Numer. Anal.*, 39(5):1749–1779, 2002.

[41] P. Ashwin, K. Böhmer, and Zhen Mei. A numerical Liapunov–Schmidt method for finitely determined problems. In K. Georg, E. L. Allgower, and R. Miranda, editors, *Exploiting Symmetry in Applied and Numerical Analysis*, Lectures in Applied Mathematics 29, 49–69, AMS, Providence, 1993.

[42] P. Ashwin, K. Böhmer, and Zhen Mei. A numerical Liapunov–Schmidt method with applications to hopf bifurcation on a square. *Math.Comp.*, 64:649–670 and S19–S22, 1995.

[43] G. I. Aslanov. The distribution of the eigenvalues of elliptic non-selfadjoint operators of second order in unbounded n-dimensional domains. *Tr. Inst. Mat. Mekh., Baku*, 2:30–44, 1981.

[44] O. Axelsson and W. Layton. Iteration method as discretization procedures. In *Preconditioned Conjugate Gradient Methods*, Proc. Conf., Nijmegen/Neth. 1989, 174–193. Lecture Notes in Mathemetics 1457, Springer, Berlin, 1990.

[45] I. Babuška and A. Aziz. Survey lectures on the mathematical foundations of the finite element method. In A. K. Aziz, editor, *The Mathematical Foundations of the Finite Element Method with Applications to Partial Differential Equations*, Academic Press, New York, 1972.

[46] I. Babuška and J. E. Osborn. Estimates for the errors in eigenvalue and eigenvector approximation by Galerkin methods, with particular attention to the case of the multiple eigenvalues. *SIAM J. Numer. Anal.*, 24:1249–1276, 1987.

[47] I. Babuška and J. E. Osborn. Finite element-Galerkin approximation of the eigenvalues and eigenvectors of selfadjoint problems. *Math. Comput.*, 52(186):275–297, 1989.

[48] I. Babuška and J. E. Osborn. Regularity and numerical solution of eigenvalue problems with piecewise analytic data. *SIAM J. Numer. Anal.*, 26:1534–1560, 1989.

[49] I. Babuška and W. C. Rheinboldt. Error estimators for adaptive finite element computations. *SIAM J. Numer. Anal.*, 15:736–754, 1978.

[50] I. Babuška and W. C. Rheinboldt. A posteriori error estimates for the finite element method. *Internat. J. Numer. Meth. Engrg.*, 12:1597–1615, 1978.

[51] I. Babuška and T. Strouboulis. *The Finite Element Method and its Reliability.* Oxford Science Publications, Oxford, 2001.

[52] I. Babuška and M. Suri. The *p*- and *h-p* versions of the finite element method. an overview. *Comput. Meth. Appl. Mech. Eng.*, 80:5–26, 1990.

[53] I. Babuška and M. Suri. The h-p version of the finite element method with quasiuniform meshes. Math. Model. Numer. Anal. 21, 199–238, 1987.

[54] I. Babuška and M. Vogelius. Feedback and adaptive finite element solution of one-dimensional boundary value problems. *Numer. Math.*, 44:75–102, 1984.

[55] I. Ya. Bakelman. *Geometric Methods of Solutions of Elliptic Equations.* Nauka, Moscow, 1965.

[56] C. A. Baker. Finite element methods for elliptic equations using nonconforming elements. *Math. Comp.*, 31:45–59, 1977.

[57] U. Banerjee and M. Suri. Analysis of numerical integration in *p*-version finite element eigenvalue approximation. *Numer. Meth. Part. Differ. Eq.*, 8(4):381–394, 1992.

[58] R. E. Bank. *PLTMG: A Software Package for Solving Elliptic Partial Differential Equations. User's Guide 6.0.* SIAM, Philadelphia, 1990.

[59] R. E. Bank, T. F. Dupont, and H. Yserentant. The hierarchical basis multigrid method. *Numer. Math.*, 52:427–458, 1988.

[60] R. E. Bank. Hierarchical bases and the finite element method. In A. Iserles, editor, Acta Numerica 5, 1–43. Cambridge, Cambridge University Press, 1996.

[61] R. E. Bank. *PLTMG: A Software Package for Solving Elliptic Partial Differential Equations. User's Guide 8.0.* Software – Environments – Tools. 5. SIAM, Philadelphia, 1998.

[62] R. E. Bank. Private communication about lectures, 2000.

[63] R. E. Bank and T. F. Chan. PLTMGC: A multi-grid continuation program for parameterized nonlinear elliptic systems. *SIAM J. Sci. Stat. Comput.*, 7:540–559, 1986.

[64] R. E. Bank and T. F. Chan. An analysis of the composite step biconjugate gradient method. *Numer. Math.*, 66(3):295–319, 1993.

[65] R. E. Bank and R. K. Smith. A posteriori error estimates based on hierarchical bases. *SIAM J. Numer. Anal.*, 30(4):921–935, 1993.

[66] R. E. Bank and J. Xu. Asymptotically exact a posteriori error estimators. I: Grids with superconvergence. *SIAM J. Numer. Anal.*, 41(6):2294–2312, 2003.

[67] E. Bänsch. Local mesh refinement in 2 and 3 dimensions. *Impact Comput. Sci. Engrg.*, 3:181–191, 1991.

[68] J. W. Barett and W. Liu. Quasi-error bounds for the finite element approximation of a non-Newtonian flow. *Numer. Math.*, 68:437–456, 1994.

[69] A. Barinka. *Fast computation tools for adaptive wavelet schemes.* PhD thesis, RWTH Aachen, 2005.

[70] A. Barinka, T. Barsch, P. Charton, A. Cohen, S. Dahlke, W. Dahmen, and K. Urban. Adaptive wavelet schemes for elliptic problems: Implementation and numerical experiments. *SIAM J. Sci. Comput.*, 23(3):910–939, 2001.

[71] A. Barinka, S. Dahlke, and W. Dahmen. Adaptive application of operators in standard representation. *Adv. Comp. Math.*, 24(1–4):5–34, 2006.

[72] A. Barinka, W. Dahmen, and R. Schneider. Fast computation of adaptive wavelet expansions. *Num. Math.*, 105(4):549–589, 2007.

[73] G. Barles and P. E. Souganidis. Convergence of approximation schemes for fully nonlinear second order equations. *Asymptotic Anal.*, 4:271–283, 1991.

[74] M. Barner and F. Flohr. *Analysis II.* De Gruyter, Berlin, New York, 1983.

[75] F. Bassi and S. Rebay. A high-order accurate discontinuous finite element method for the numerical solution of the compressible Navier–Stokes equations. *J. Comput. Phys.*, 131:267–279, 1997.

[76] C. Beattie. Galerkin eigenvector approximations. *Math. Comput.*, 69(232):1409–1434, 2000.

[77] R. Becker and R. Rannacher. A feed-back approach to error control in finite element methods: Basic analysis and examples. *East-West J. Numer. Math.*, 4:237–264, 1996.

[78] R. Becker and R. Rannacher. An optimal control approach to a posteriori error estimation in finite element methods. *Acta Numerica*, 10:1–102, 2001.

[79] Y. I. Belopol'skaya. A priori estimates for solutions of the first initial boundary-value problem for systems of fully nonlinear partial differential equations. *Ukr. Math. J.*, 49(3):373–402, 1997.

[80] Y. I. Belopol'skaya. A priori estimates for solutions of the first initial boundary-value problem for systems of fully nonlinear partial differential equations. *Ukr. Mat. J.*, 49(3):338–363, 1997.

[81] A. Bensoussan and J. Frehse. *Regularity Results for Nonlinear Elliptic Systems and Applications.* Applied Mathematical Sciences 151. Springer, Berlin, Heidelberg, 2002.

[82] M. Berger. *Nonlinearity and Functional Analysis.* Academic Press, New York, 1977.

[83] M. S. Berger. On Von Karman's equations and the buckling of a thin elastic plate, i. The clamped plate. *Comm. Pure Appl. Math.*, 20:687–719, 1967.

[84] M. S. Berger and P. C. Five. On Von Karman's equations and the buckling of a thin elastic plate, ii. Plate with general edge conditions. *Comm. Pure Appl. Math.*, 21:227–241, 1968.

[85] M. Bernadou. $\mathcal{C}^1$-curved finite elements with numerical integration for thin plate and thin shell problems. I: Construction and interpolation properties of curved $\mathcal{C}^1$ finite elements. *Comput. Meth. Appl. Mech. Eng.*, 102(2):255–289, 1993.

[86] M. Bernadou. $\mathcal{C}^1$-curved finite elements with numerical integration for thin plate and thin shell problems. II: Approximation of thin plate and thin shell problems. *Comput. Meth. Appl. Mech. Eng.*, 102(3):389–421, 1993.

[87] M. Bernadou. $\mathcal{C}^1$-curved finite elements and applications to plate and shell problems. *J. Comput. Appl. Math.*, 50(1–3):133–144, 1994.

[88] M. Bernadou. Straight and curved finite elements of class C^1 and some applications to thin shell problems. In M. Krizek *et al.*, editors, *Finite Element Methods. 50 Years of the Courant Element.* Conference held at the Univ. of Jyväskylä, Finland, 1993. Lect. Notes Pure Appl. Math. 164, 63–77. Marcel Dekker, New York, 1994.

[89] M. Bernadou and J.-M. Boisserie. Curved finite elements of class C^1: Implementation and numerical experiments. I: Construction and numerical tests of the interpolation properties. *Comput. Meth. Appl. Mech. Eng.*, 106(1–2):229–269, 1993.

[90] J. Bey. *Finite-Volumen- und Mehrgitterverfahren für elliptische Randwertprobleme.* Teubner, Stuttgart, 1998.

[91] G. Beylkin, R. Coifman, and V. Rokhlin. Fast wavelet transforms and numerical algorithms I. *Commun. Pure Appl. Math.*, 44:141–183, 1991.

[92] W.-J. Beyn. Zur numerischen Berechnung mehrfacher Verzweigungspunkte. *ZAMM*, 65:370–371, 1985.

[93] W.-J. Beyn and W. Kless. Numerical Taylor expansions of invariant manifolds in large dynamical systems. *Numer. Math.*, 80(1):1–38, 1998.

[94] W.-J. Beyn, W. Kless, and Vera Thümmler. Continuation of low-dimensional invariant subspaces in dynamical systems of large dimension. In Bernold Fiedler, editor, *Ergodic Theory, Analysis, and Efficient Simulation of Dynamical Systems*, 47–72. Springer, Berlin, 2001.

[95] B. Bialecki and G. Fairweather. Orthogonal spline collocation methods for partial differential equations. *J. Comput. Appl. Math.*, 128:55–82, 2001.

[96] P. Binev, W. Dahmen, and R. DeVore. Adaptive finite element methods with convergence rates. *Numer. Math.*, 97:219–268, 2004.

[97] P. Binev, W. Dahmen, R. DeVore, and P. Petruchev. Approximation classes for adaptive methods. *Serdica Math. J.*, 28:391–416, 2002.

[98] H. Blum and R. Rannacher. Finite element eigenvalue computation on domains with reentrant corners using Richardson extrapolation. *J. Comput. Math.*, 8(4):321–332, 1990.

[99] A. Boccuto and E. Mitidieri. Regularity results for positive solutions of a semilinear elliptic system. *Ann. Mat. Pura Appl., IV. Ser.*, 179:125–147, 2001.

[100] I. Bock. On nonstationary von Kármán equations. *Z. Angew. Math. Mech.*, 76(10):559–571, 1996.

[101] K. Böhmer. A defect correction method for functional equations. In R. Schaback and K. Scherer, editor, *Approximation Theory*, 16–29, Lecture Notes in Mathematics 556. Springer-Verlag, Berlin, 1976.

[102] K. Böhmer. Defect correction via neighbouring problems. i. General theory. MRC Report N0. 1750, University of Wisconsin-Madison, 1977.

[103] K. Böhmer. Fehlerasymptotik bei Diskretisierungsverfahren und ihre numerische Anwendung. Vorlesungsmanuskript vom WS 1976/77, Universität Karlsruhe, 1977.

[104] K. Böhmer. Defekt-Korrektur-Methoden für das zentrale Euler-Verfahren bei Randwertproblemen gewöhnlicher Differentialgleichungen. In L. Collatz, J. Albrecht, and G. Hämmerlin, editors, *Numerische Behandlung von Differentialgleichungen mit besonderer Berücksichtigung freier Randwertaufgaben*, International Series of Numerical Mathematics, Vol 39, Birkhäuser, Basel, Sluttgart, 42–61, 1978.

[105] K. Böhmer. Error asymptotics for linear elliptic boundary value problems on general domains. In L. Collatz and G. Meinardus, editors, *Approximation Theory III*, 237–242. Birkhäuser, Boston, 1980.

[106] K. Böhmer. Asymptotic expansion for the discretization error in linear ellliptic boundary value problems on general regions. *Math. Z.*, 177:235–255, 1981.

[107] K. Böhmer. Discrete Newton methods and iterated defect corrections. *Numer. Math.*, 37:167–192, 1981.

[108] K. Böhmer. Defect corrections for Hartree-Fock methods in the Schrödinger equation. *In Proceedings of the Symposium on Differential Geometry and Differential Equations*, 1–30, Series Press, Beijing, 1984.

[109] K. Böhmer. Defect corrections and grid strategy for the Kreiss' method for stiff boundary value problems. *ZAMM*, 67:459–460, 1987.

[110] K. Böhmer. Difference solutions for linear least squares boundary value problems with asymptotic expansions. In I. Marek, editor, *Proceedings of the Second International Symposium on Numerical Analysis*, 120–124. Teubner, Stuttgart, 1987.

[111] K. Böhmer. On a numerical Liapunov–Schmidt method for operator equations. Preprint 2, Philipps-Universität Marburg, Fachbereich Mathematik, 1989.

[112] K. Böhmer. Defect correction method and mesh independence principle for operator equations and their discretizations. In E. Allgower and K. Georg, editors, *Computational Solutions of Nonlinear Systems of Equations*, Lectures in Applied Mathematics 26, 61–78. AMS, Providence, 1990.

[113] K. Böhmer. On a numerical Liapunov–Schmidt method for operator equations. *Computing*, 51:237–269, 1993.

[114] K. Böhmer. On numerical bifurcation studies for general operator equations. In J. Sprekels, B. Fiedler, and K. Gröger, editors, *International Conference on Differential Equations*, Proceedings, Equadiff '99, Berlin, Germany, 1–7 August 1999, volume 2, 877–883. World Scientific, Singapore, 2000.

[115] K. Böhmer. On hybrid methods for bifurcation studies for general operator equations. In B. Fiedler, editor, *Ergodic Theory, Analysis, and Efficient Simulation of Dynamical Systems*, 73–107. Springer, Berlin, Heidelberg, New York, 2001.

[116] K. Böhmer. Finite elements methods, variational crimes and collocation. Bericht 95, Fachbereich Mathematik und Informatik / Mathematik, Philipps–Universität Marburg, 2002.

[117] K. Böhmer. Elements of linear and nonlinear elliptic differential operators. Bericht 9, Fachbereich Mathematik und Informatik / Mathematik, Philipps–Universität Marburg, 2003.

[118] K. Böhmer. On finite element methods for fully nonlinear elliptic equations and systems of orders 2 and $2m$. Bericht 98, Fachbereich Mathematik und Informatik / Mathematik, Philipps–Universität Marburg, 2008.

[119] K. Böhmer. On finite element methods for fully nonlinear elliptic equations of second order. *SIAM J. Numer. Anal.*, 46(3):1212–1249, 2008.

[120] K. Böhmer. *Numerical Methods for Bifurcation and Center Manifolds in Nonlinear Elliptic and Parabolic Differential Equations.* Oxford University Press, Oxford, 2011.

[121] K. Böhmer and S. Dahlke. On the stability of multiscale wavelet methods with applicvations to Navier–Stokes equations. Bericht 5, Fachbereich Mathematik und Informatik / Mathematik, Philipps–Universität Marburg, 2002.

[122] K. Böhmer and S. Dahlke. Stability and convergence for discretization methods with applications to wavelet Galerkin schemes. *Inter. J. of Pure Appl. Math.*, 6(4):465–489, 2003.

[123] K. Böhmer, C. Geiger, and J. Rodriguez. On a numerical Liapunov–Schmidt spectral method, part I: Review of spectral methods. Technical report, SP-Report, 1996.

[124] K. Böhmer, C. Geiger, and J. Rodriguez. On a numerical Liapunov–Schmidt spectral method and applications to bio logical pattern formation. *SIAM J. Numer. Anal.*, 40:683–701, 2002.

[125] K. Böhmer, W. Gross, B. Schmitt, and R. Schwarz. Defect corrections and Hartree–Fock method. *Computing Supplementum*, 5:193–210, 1984.

[126] K. Böhmer, P. Hemker, and H. J. Stetter. The defect correct approach. In K. Böhmer and H. J. Stetter, editors, *Defect Correct Methods: Theory and Applications*, 1–32. Springer, Vienna, New York, 1984.

[127] K. Böhmer and Z. Mei. On a numerical Liapunov–Schmidt method. In E. L. Allgower and K. Georg, editors, *Computational Solutions of Nonlinear Systems of Equations*, Lectures in Applied Mathematics 26, 79–98. AMS, Providence, 1990.

[128] K. Böhmer and N. Sassmannshausen. Numerical Liapunov–Schmidt spectral method for k-determined problems. *Comput. Meth. Appl. Mech. and Engng.*, 170:277–312, 1999.

[129] K. Böhmer and N. Sassmannshausen. Stability for generalized Petrov-Galerkin methods applied to bifurcation, 2001. included in [[120]].

[130] K. Böhmer and H. J. Stetter (eds.), *Defect Correct Methods: Theory and Applications.* Springer, Vienna, New York, 1984.

[131] T. Bonesky, S. Dahlke, P. Maass, and T. Raasch. Adaptive wavelet Galerkin methods and sparse reconstruction of inverse heat conduction problems. Accepted for publication in Appl. Anal.

[132] T. Bonesky and P. Maass. Iterated soft shrinkage with adaptive operator evaluations. *Inverse and Ill-Posed Problems*, 17(4):337–358, 2009.

[133] J. Bošek. *Continuation of invariant subspaces in bifurcation problems.* PhD thesis, Philipps-Universität Marburg, 2002.

[134] A. Borzì, K. Kunisch, and D. Y. Kwak. Accuracy and convergence properties of the finite difference multigrid solution of an optimal control optimality system. *SIAM J. Cont. Optimiz.*, 41(5):1477–1497, 2003.

[135] D. Braess. *Finite Elemente*, 2nd edn. Springer, Berlin, 1997.

[136] D. Braess. *Finite Elements. Theory, Fast Solvers and Applications in Solid Mechanics*, 3rd edn. Cambridge University Press, Cambridge, 1997.

[137] D. Braess. A posteriori error estimators for obstacle problems – another look. *Numer. Math.*, 101:415–421, 2005.

[138] D. Braess, C. Carstensen, and B. D. Reddy. Uniform convergence and a posteriori error estimators for the enhanced strain finite element method. *Numer. Math.*, 96:461–479, 2004.

[139] J. H. Bramble and B. E. Hubbard. On the formulation of finite difference analogues of the Dirichlet problem for Poisson's equation. *Numer. Math.*, 4:313–327, 1962.

[140] J. H. Bramble and J. E. Osborn. Rate of convergence estimates for non selfadjoint eigenvalue approximations. *Math. Comput.*, 27:525–549, 1973.

[141] S. C. Brenner and L. R. Scott. *The Mathematical Theory of Finite Element Methods.* Springer, New York, 1996.

[142] S. C. Brenner and L. R. Scott. *The Mathematical Theory of Finite Element Methods*, 2nd edn. Springer, New York, 2002.

[143] H. Brézis and F. Browder. Partial differential equations in the 20th century. *Adv. Math.*, 135:76–144, 1998.

[144] F. Brezzi and M. Fortin. *Mixed and Hybrid Finite Element Methods.* Springer, Berlin, 1991.

[145] F. Brezzi, T. J. R. Hughes, D. Marini, A. Russo, and E. Süli. A priori error analysis of a finite element method with residual-free bubbles for advection dominated equations. *SIAM J. Numer. Anal.*, 36(6):1933–1948, 1999.

[146] F. Brezzi, G. Manzini, D. Marini, P. Pietra, and A. Russo. Discontinuous Galerkin approximations for elliptic problems. *Numer. Meth. Part. Differ. Eq.*, 16:365–378, 2000.

[147] F. Brezzi, J. Rappaz, and P. A. Raviart. Finite-dimensional approximation of problems, I. Branches of nonsingular solutions. *Numer. Math.*, 36:1–25, 1980.

[148] F. Brezzi, J. Rappaz, and P. A. Raviart. Finite-dimensional approximation of nonlinear problems, II. Limit points. *Numer. Math.*, 37:1–28, 1981.

[149] F. Brezzi, J. Rappaz, and P. A. Raviart. Finite-dimensional approximation of nonlinear problems, III. Simple bifurcation points. *Numer. Math.*, 38:1–30, 1981.

[150] F. Browder. Nonlinear elliptic boundary value problems. *Bull. Amer. Math. Soc.*, 69:862–874, 1963.

[151] K. P. Bube and J. C. Strikwerda. Interior regularity estimates for elliptic systems of difference equations. *SIAM J. Numer. Anal.*, 4:653–670, 1983.

[152] M. D. Buhmann. *Radial-Basis Functions*, Cambridge Monographs on Applied and Computational Mathematics 12. Cambridge University Press, Cambridge, 2003.

[153] E. Burman and B. Stamm. Local discontinuous Galerkin method with reduced stabilization for diffusion equations. *Commun. Comput. Phys.*, 5(2–4):498–514, 2009.

[154] L.A. Caffarelli and R. Glowinski. Numerical solution of the Dirichlet problem for a Pucci equation in dimension two. Application to homogenization. *J. Numer. Math.*, 16(3):185–216, 2008.

[155] X. Cai, H. P. Langtangen, B. F. Nielsen, and A. Tveito. A finite element method for fully nonlinear water waves. *J. Comput. Phys.*, 143(2):544–568, 1998.

[156] G. Caloz and J. Rappaz. Numerical analysis for nonlinear and bifurcation problems. In P. G. Ciarlet and J. L. Lions, editors, *Handbook of Numerical Analysis*, volume V of *Techniques of Scientific Computing (Part 2)*. Elsevier Science, Amsterdam, 1998.

[157] M. Cannone and G. Karch. Smooth or singular solutions to the Navier–Stokes system? *J. Differ. Eq.*, 197(2):247–274, 2004.

[158] C. Canuto, M. Y. Hussaini, A. Quarteroni, and T. A. Zang. *Spectral Methods in Fluid Dynamics*, 3rd edn. Springer, Berlin, 1997.

[159] C. Canuto, A. Tabacco, and K. Urban. The wavelet element method, part I: Construction and analysis. *Appl. Comp. Harm. Anal.*, 6:1–52, 1999.

[160] C. Canuto, A. Tabacco, and K. Urban. The wavelet element method, part II: Realization and additional features in 2D and 3D. *Appl. Comp. Harm. Anal.*, 8:123–165, 2000.

[161] C. Canuto and K. Urban. Adaptive optimization in convex Banach spaces. SIAM J. Numer. Anal. 42(5):2043–2075, 2005.

[162] S. Carl and S. Heikkilä. *Nonlinear Differential Equations In Ordered Spaces.* Chapman and Hall/CRC, London, 2000.

[163] J. Carrero, B. Cockburn, and D. Schötzau. Hybridized, globally divergence-free LDG methods. Part I: The Stokes problem. *Math. Comp.*, 75:533–563, 2006.

[164] C. Carstensen and S. A. Funken. Fully realiable localised error control in the FEM. *SIAM J. Sci. Comput.*, 21:1465–1484, 2000.

[165] C. Carstensen and R. Verfürth. Edge residuals dominate a posteriori error estimates for low order finite element methods. *SIAM J. Numer. Anal.*, 36:1571–1587, 1999.

[166] P. Castillo, B. Cockburn, I. Perugia, and D. Schötzau. An a priori error analysis of the Local Discontinuous Galerkin method for elliptic problems. *SIAM J. Numer. Anal.*, 38(5):1676–1706, 2000.

[167] F. Chatelin. *Spectral Approximation of Linear Operators.* Academic Press, New York, 1983.

[168] F. Chatelin. *Eigenvalues of Matrices.* Wiley, New York, 1993.

[169] J.-Y. Chemin and C.-J. Xu. Remarque sur la régularité de solutions faibles d'équations elliptiques semi-linéaires. (Remarks on the regularity of weak solutions of some semilinear elliptic equations). Private communication, 1997.

[170] Y. Z. Chen and L. C. Wu. *Second Order Elliptic Equations and Elliptic Systems*, trans. B. Hu. Translations of Mathematical Monographs. AMS, Providence, 1998.

[171] S. S. Cheng. *Partial Difference Equations.* Taylor and Francis, London, 2003.

[172] A.-S. Chibi and G. Moore. Defect correction and Galerkin's method for second-order boundary value problems. *Numer. Meth. Part. Differ. Eq.*, 8(5):469–491, 1992.

[173] C. K. Chui. *An Introduction to Wavelets.* Academic Press, Boston, 1992.

[174] P.-G. Ciarlet. *The Finite Element Method for Elliptic Problems.* North-Holland, Amsterdam, New York, 1978.

[175] P.-G. Ciarlet. A justification of the Von Karman equations. *Arch. Rat. Mech. and Anal.*, 73:349–389, 1980.

[176] P.-G. Ciarlet and P. A. Raviart. Conforming and nonconforming finite element methods for solving the stationary stokes problem. *RAIRO Anal. Numer.* 3:33–75, 1973.

[177] P. G. Ciarlet and J. L. Lions. editors, *Handbook of Numerical Analysis*, Volume I. *Finite Difference Methods (Part 1). Solution of equations in* R^n *(Part 1).* North-Holland, Amsterdam, 1990.

[178] P. Clément. Approximation by finite element functions using local regularization. *RAIRO Anal. Numer.*, 9:77–84, 1975.

[179] K. A. Cliffe, T. J. Garratt, and A. Spence. Eigenvalues of the discretized Navier–Stokes equation with application to the detection of Hopf bifurcations. *Adv. Comput. Math.*, 1(3–4):337–356, 1993.

[180] K. A. Cliffe, A. Spence, and S. J. Tavener. The numerical analysis of bifurcation problems with application to fluid mechanics. *Acta Numerica*, 9:39–131, 2000.

[181] C. Coatmélec. Approximation et interpolation des fonctions différentiables de plusieurs variables. *Ann. Sci. Ecole Norm. Sup.*, 83:271–341, 1966.

[182] B. Cockburn. Discontinuous Galerkin methods for convection dominated problems. In T. J. Barth and H. Deconinck, editors, *High–Order Methods for Computational Physics*, Lecture Notes in Computational Science and Engineering 9, 69–224. Springer, Berlin, 1999.

[183] B. Cockburn, B. Dong, and J. Guzmáan. A superconvergent LDG-hybridizable Galerkin method for second-order elliptic problems. *Math. Comp.*, 77:1887–1916, 2008.

[184] B. Cockburn and J. Gopalakrishnan. A characterization of hybridized mixed methods for second order elliptic problems. *SIAM J. Numer. Anal.*, 42:283–301, 2004.

[185] B. Cockburn and J. Gopalakrishnan. Error analysis of variable degree mixed methods for elliptic problems via hybridization. *Math. Comp.*, 74:1653–1677, 2005.

[186] B. Cockburn and J. Gopalakrishnan. Incompressible finite elements via hybridization. Part I: The Stokes system in two space dimensions. *SIAM J. Numer. Anal.*, 43:1627–1650, 2005.

[187] B. Cockburn and J. Gopalakrishnan. Incompressible finite elements via hybridization. Part II: The Stokes system in three space dimensions. *SIAM J. Numer. Anal.*, 43:1651–1672, 2005.

[188] B. Cockburn and J. Gopalakrishnan. New hybridization techniques. *GAMM-Mitt.*, 2:154–193, 2005.

[189] B. Cockburn, J. Guzmán, and H. Wang. Superconvergent discontinuous Galerkin methods for second-order elliptic problems. *Math. Comp.*, 78:1–24, 2009.

[190] B. Cockburn, G. Kanschat, I. Perugia, and D. Schötzau. Superconvergence of the local discontinuous Galerkin method for elliptic problems on Cartesian grids. *SIAM J. Numer. Anal.*, 39:264–285, 2001.

[191] B. Cockburn, G. Kanschat, and D. Schötzau. A locally conservative ldg method for the incompressible Navier-Stokes equations. *Math. Comput.*, 74(251):1067–1095, 2005.

[192] B. Cockburn, G. E. Karniadakis, and C.-W. Shu, editors, *Discontinuous Galerkin Methods: Theory, Computation and Applications*, Lecture Notes in Computational Science and Engineering 11. Springer, Berlin, 2000.

[193] B. Cockburn, M. Luskin, C.-W. Shu, and E. Süli. Postprocessing of the discontinuous Galerkin finite element method. *Math. Comput.*, 72:577–606, 2003.

[194] B. Cockburn and C.-W. Shu. The local discontinuous Galerkin method for time-dependent convection-diffusion systems. *SIAM J. Numer. Anal.*, 35:2440–2463, 1998.

[195] A. Cohen. *Numerical Analysis of Wavelet Methods*, Studies in Mathematics and its Applications 32. North-Holland, Amsterdam, 2003.

[196] A. Cohen, W. Dahmen, I. Daubechies, and R. DeVore. Tree approximation and optimal encoding. *Appl. Comp. Harmonic Anal.*, 11(2):192–226, 2001.

[197] A. Cohen, W. Dahmen, and R. DeVore. Adaptive wavelet methods for elliptic operator equations – convergence rates. *Math. Comp.*, 70(233):22–75, 2001.

[198] A. Cohen, W. Dahmen, and R. DeVore. Adaptive wavelet methods ii: Beyond the elliptic case. *Found. Comput. Math.*, 2(3):203–245, 2002.

[199] A. Cohen, W. Dahmen, and R. DeVore. Adaptive wavelet schemes for nonlinear variational problems. *SIAM J. Numer. Anal.*, 41(5):1785–1823, 2003.

[200] A. Cohen, W. Dahmen, and R. DeVore. Sparse evaluation of compositions of functions using multiscale expansions. *SIAM J. Math. Anal.*, 35(2):279–303, 2003.

[201] A. Cohen and I. Daubechies. Non–separable bidimensional wavelet bases. *Rev. Mat. Iberoamericana*, 9:51–137, 1993.

[202] A. Cohen, I. Daubechies, and C. Feauveau. Biorthogonal bases of compactly supported wavelets. *Commun. Pure Appl. Math.*, 45:485–560, 1992.

[203] A. Cohen and R. Masson. Wavelet methods for second-order elliptic problems, preconditioning, and adaptivity. *SIAM J. Sci. Comp.*, 21(3):1006–1026, 1999.

[204] A. Cohen and R. Masson. Wavelet adaptive methods for second order elliptic problems, boundary conditions and domains decompositions. *Numer. Math.*, 86(2):193–238, 2000.

[205] L. Collatz. Bemerkungen zur Fehlerabschätzung für das Differenzenverfahren bei partiellen Differentialgleichungen. *Z. Angew. Math. Mech.*, 13:56–57, 1933.

[206] L. Collatz. *The Numerical Treatment of Differential Equations*, 3rd edn. Berlin, Heidelberg, New York, 1960.

[207] S. D. Conte and C. de Boor. *Elementary Numerical Analysis*. McGraw-Hill, New York, London, Sydney, 1972.

[208] R. Cools. Monomial cubature rules since "stroud": a compilation – part 1. *J. of Comp. and Appl. Math.*, 112:21–27, 1999.

[209] R. Cools. Monomial cubature rules since "stroud": a compilation – part 2. *J. of Comp. and Appl. Math.*, 149:1–12, 2002.

[210] R. Cools. An encyclopedia of cubature formulas. *Journal of Complexity*, 19:445–453, 2003.

[211] M. Costabel. Boundary integral operators on Lipschitz domains: Elementary results. *SIAM J. Numer. Math.*, 19(3):613–626, 1988.

[212] M. Costabel and M. Dauge. General edge asymptotics of solutions of second-order elliptic boundary value problems. I-II. *Proc. R. Soc. Edinb., Sect. A*, 123:109–184, 1993.

[213] R. Courant, K. Friedrichs, and H. Lewy. Über die partiellen Differenzengleichungen der mathematischen Physik. *Math. Ann.*, 100:32–74, 1928.

[214] R. Courant and D. Hilbert. *Methods of Mathematical Physics*, Volume II: *Partial Differential Equations*. Trans. and rev. from the German original. Reprint of the First English edn 1962. Wiley Classics, John Wiley and Sons/Interscience Publishers, New York, 1989.

[215] M. G. Crandall, H. Ishii, and P.-L. Lions. User's guide to viscosity solutions of second order partial differential equations. *Bull. Am. Math. Soc., New Ser.*, 27(1):1–67, 1992.

[216] M. G. Crandall, M. Kocan, P. L. Lions, and A. Swiech. Existence results for boundary problems for uniformly elliptic and parabolic fully nonlinear equations. *Electr. J. Diff. Eq.* 1999(24) 1–20, 1999.

[217] M. G. Crandall and P.-L. Lions. Hamilton–Jacobi equations in infinite dimensions. I: Uniqueness of viscosity solutions. *J. Funct. Anal.*, 62:379–396, 1985.

[218] M. G. Crandall and P.-L. Lions. Hamilton–Jacobi equations in infinite dimensions. II: Existence of viscosity solutions. *J. Funct. Anal.*, 65:368–405, 1986.

[219] M. G. Crandall and P.-L. Lions. Hamilton–Jacobi equations in infinite dimensions. III. *J. Funct. Anal.*, 68(2):214–247, 1986.

[220] M. G. Crandall and P.-L. Lions. On existence and uniqueness of solutions of Hamilton–Jacobi equations. *Nonlin. Anal., Theor. Meth. Appl.*, 10:353–370, 1986.

[221] M. G. Crandall and P.-L. Lions. Remarks on the existence and uniqueness of unbounded viscosity solutions of Hamilton–Jacobi equations. *Ill. J. Math.*, 31:665–688, 1987.

[222] M. G. Crandall and P. L. Lions. Quadratic growth of solutions of fully nonlinear second order equations in $\mathbb{R}^n$. *Differ. Integral Eq.*, 3(4):601–616, 1990.

[223] M. G. Crandall and P. L. Lions. Viscosity solutions of Hamilton–Jacobi equations in infinite dimensions. IV: Hamiltonians with unbounded linear terms. *J. Funct. Anal.*, 90(2):237–283, 1990.

[224] M. G. Crandall and P. L. Lions. Viscosity solutions of Hamilton–Jacobi equations in infinite dimensions. V: Unbounded linear terms and B-continuous solutions. *J. Funct. Anal.*, 97(2):417–465, 1991.

[225] M. G. Crandall and P. L. Lions. Hamilton–Jacobi equations in infinite dimensions. VI: Nonlinear A and Tataru's method refined. In Philippe Clément *et al.*, editors, *Evolution Equations, Control Theory, and Biomathematics*, Third International Workshop-Conference, held at the Han-sur-Lesse Conference Center of the Belgian Ministry of Education, Proceedings, 155, 51–89. Lect. Notes Pure Appl. Math. 155, Dekker, Basel, 1993.

[226] M. G. Crandall and P. L. Lions. Viscosity solutions of Hamilton–Jacobi equations in infinite dimensions. VII: The HJB equation is not always satisfied. *J. Funct. Anal.*, 125(1):111–148, 1994.

[227] M. G. Crandall and P.-L. Lions. Convergent difference schemes for nonlinear parabolic equations and mean curvature motion. *Numer. Math.*, 75(1):17–41, 1996.

[228] M. G. Crandall, P.-L. Lions, and P. E. Souganidis. Maximal solutions and universal bounds for some partial differential equations of evolution. *Arch. Ration. Mech. Anal.*, 105(2):163–190, 1989.

[229] M. Crouzeix and J. Rappaz. *On Numerical Approximation in Bifurcation Theory.* Masson, Paris and Springer, Berlin, 1990.

[230] M. Cui. Fully discrete finite element schemes for nonlinear parabolic equation in a three-dimensional variable domain. *Numer. Math., Nanjing*, 19(4):312–323, 1997.

[231] C. Cuvelier, J. Descloux, J. Rappaz, C. Stuart, and B. Zwahlen. *Partial Differential Equations for Engineers. Theory and Numerical Methods*, Vols 1 and 2. (*Éléments d'équations aux dérivées partielles pour ingénieurs. Théorie et méthodes numériques.* Vol. 1 et 2.). Cahiers Mathématiques de l'École Polytechnique Fédérale de Lausanne. Presses Polytechniques Romandes, Lausanne, 1988.

[232] S. Dachkovski. *The Q-method and regularity problems of some integral equations and PDE's.* PhD thesis, Univ. Jena, Fakultät für Mathematik und Informatik, 2001.

[233] S. Dahlke. *Wavelets: Construction Principles and Applications to the Numerical Treatment of Operator Equations.* Shaker, Aachen, 1997.

[234] S. Dahlke. Besov regularity for elliptic boundary value problems on polygonal domains. *Appl. Math. Lett.*, 12:31–38, 1999.

[235] S. Dahlke, W. Dahmen, and R. DeVore. Nonlinear approximation and adaptive techniques for solving elliptic operator equations. In W. Dahmen, A. Kurdila, and P. Oswald, editors, *Multiscale Wavelet Methods for Partial Differential Equations*, 237–283. Academic Press, San Diego, 1997.

[236] S. Dahlke, W. Dahmen, R. Hochmuth, and R. Schneider. Stable multiscale bases and local error estimation for elliptic problems. *Appl. Numer. Math.*, 23(1):21–48, 1997.

[237] S. Dahlke, W. Dahmen, and K. Urban. Adaptive wavelet methods for saddle point problems – optimal convergence rates. *SIAM J. Numer. Anal.*, 40(4):1230–1226, 2002.

[238] S. Dahlke and R. DeVore. Besov regularity for elliptic boundary value problems. *Commun. Part. Differ. Eq.*, 22(1&2):1–16, 1997.

[239] S. Dahlke, M. Fornasier, and T. Raasch. Adaptive frame methods for elliptic operator equations. *Adv. Comput. Math.*, 27(1):27–63, 2007.

[240] S. Dahlke, M. Fornasier, T. Raasch, R. Stevenson, and M. Werner. Adaptive frame methods for elliptic operator equations: The steepest descent approach. *IMA J. Numer. Anal.*, 27(4):717–740, 2007.

[241] S. Dahlke, R. Hochmuth, and K. Urban. Adaptive wavelet methods for saddle point problems. *M2AN Math. Model. Numer. Anal.*, 34:1003–1022, 2000.

[242] W. Dahmen. Stability of multiscale transformations. *J. Fourier Anal. Appl.*, 2:341–361, 1996.

[243] W. Dahmen. Wavelet and multiscale methods for operator equations. *Acta Numerica*, 6:55–228, 1997.

[244] W. Dahmen, B. Faermann, I. G. Graham, W Hackbusch, and S. A. Sauter. Inverse inequalities on non-quasiuniform meshes and application to the mortar element method. *Math. Comp.*, 73:1107–1138, 2004.

[245] W. Dahmen, H. Harbrecht, and R. Schneider. Compression techniques for boundary integral equations – asymptotically optimal complexity estimates. *SIAM J. Numer. Anal.*, 43(6):2251–2271, 2006.

[246] W. Dahmen and W. Kunoth. Appending boundary conditions by Lagrange multipliers: Analysis of the lbb condition. *Numer. Math.*, 88:9–42, 2001.

[247] W. Dahmen, W. Kunoth, and K. Urban. Biorthogonal spline-wavelets on the interval – stability and moment conditions. *Appl. Comp. Harm. Anal.*, 6:132–196, 1999.

[248] W. Dahmen, S. Prössdorf, and R. Schneider. in: "Wavelets: Theory, Calperthuns and Applications" (C. K. Chai, L. Montefusco and L. Puccio, Eds.) Multiscale methods for pseudo-differential equations on smooth manifolds. *Academic Press*, 385–424, 1994.

[249] W. Dahmen and R. Schneider. In: "Wavelets: Theory, Calperthuns and Applications" (C. K. Chai, L. Montefusco and L. Puccio, Eds.) Wavelets on maniferds I: Construction and domain decomposition, *SIAM J. Math. Anal.* 31, 184–230, 1999.

[250] W. Dahmen and R. Schneider. Composite wavelet bases for operator equations. *Math. Comput.*, 68:1533–1567, 1999.

[251] W. Dahmen, R. Schneider, and Y. Xu. Nonlinear functionals of wavelet expansions – Adaptive reconstruction and fast evaluation. *Numer. Math.*, 86(1):49–101, 2000.

[252] W. Dahmen, J. Vorloeper, and K. Urban. Adaptive wavelet methods – basic concepts and applications to the Stokes problem. In D.-X. Zhou, editor, *Proceedings of the International Conference of Computational Harmonic Analysis*, 39–80. World Scientific, Singapore, 2002.

[253] I. Daubechies. Orthonormal bases of compactly supported wavelets. *Commun. Pure Appl. Math.*, 41(7):909–996, 1988.

[254] I. Daubechies. *Ten Lectures on Wavelets*, CBMS–NSF Regional Conference Series in Applied Math. 61. SIAM, Philadelphia, 1992.

[255] M. Dauge. *Elliptic Boundary Value Problems on Corner Domains. Smoothness and Asymptotics of Solutions*, Lecture Notes in Mathematics 1341. Springer, Berlin, 1988.

[256] R. Dautray and J.-L. Lions. *Mathematical Analysis and Numerical Methods for Science and Technology, Vol. 1: Physical Origins and Potential Energy*, 2000. Springer, Berlin, 1992.

[257] R. Dautray and J. L. Lions. *Mathematical Analysis and Numerical Methods for Science and Technology. Vol. 6, Evolution Problems II: The Navier–Stokes and Transport Equation in Numerical Methods.* Springer, Berlin, Heidelberg, New York, 2000.

[258] R. Dautray and J. L. Lions. *Mathematical Analysis and Numerical Methods for Science and Technology. Vol. 4, Integral Equations and Numerical Methods.* Springer, Berlin, Heidelberg, New York, 2000.

[259] R. Dautray and J. L. Lions. *Mathematical analysis and numerical methods for science and technology. Vol. 3: Spectral Theory and Applications.* Berlin Heidelberg New York, Springer, 2000.

[260] R. Dautray and J. L. Lions. *Mathematical Analysis and Numerical Methods for Science and Technology*, Vol. 2: Functional and Variational Methods. Springer, Berlin, Heidelberg, New York, 2000.

[261] R. Dautray and J. L. Lions. *Mathematical Analysis and Numerical Methods for Science and Technology. Vol. 5: Evolution Problems I.* Springer, Berlin, Heidelberg, New York, 2000.

[262] E. B. Davies. *Spectral Theory and Differential Operators.* Cambridge Studies in Advanced Mathematics 42. Cambridge Cambridge, University Press, 1996.

[263] O. Davydov. Stable local bases for multivariate spline spaces. *J. Approx. Theory*, 111:267–297, 2001.

[264] O. Davydov. Smooth finite elements and stable splitting.Technical Report 4, University of Marburg, Fachbereich Mathematik und Informatik, 2007.

[265] O. Davydov and P. Petrushev. Nonlinear approximation from differentiable piecewise polynomials. *SIAM J. Math. Anal.*, 35:708–758, 2004.

[266] O. Davydov and L. L. Schumaker. Stable local nodal bases for c^1 bivariate polynomial splines. In C. Rabut, A. Cohen, and L. L. Schumaker, editors, *Curve and Surface Fitting: Saint-Malo 1999*, 171–180. Vanderbilt University Press, Nashville, TN, 2000.

[267] O. Davydov and L. L. Schumaker. On stable local bases for bivariate polynomial spline spaces. *Constr. Approx.*, 18:87–116, 2002.

[268] O. Davydov and R. Stevenson. Hierarchical Riesz bases for $h^s(\omega)$, $1 < s < \frac{5}{2}$. *Constr. Approx.*, 43:365–394, 2005.

[269] C. N. Dawson, S. Sun, and M. F. Wheeler. Compatible algorithms for coupled flow and transport. *Comput. Meth. Appl. Mech. Engng.*, 193:2565–2580, 2004.

[270] C. de Boor. *A Practical Guide to Splines.* Springer, New York, Heidelberg, Berlin, 1978.

[271] G. De Donno. Interior regularity for semilinear partial differential equations in anisotropic Sobolev spaces. *Nonlin. Anal., Theor. Meth. Appl.*, 47(4):2829–2834, 2001.

[272] E. J. Dean and R. Glowinski. Numerical solution of the two-dimensional elliptic Monge–Ampère equation with Dirichlet boundary conditions: An augmented Lagrangian approach. *C. R. Math., Acad. Sci. Paris*, 336(9):779–784, 2003.

[273] E. J. Dean and R. Glowinski. Numerical solution of the two-dimensional elliptic Monge–Ampère equation with Dirichlet boundary conditions: a least-squares approach. *C. R., Math., Acad. Sci. Paris, Série I*, 339(12):887–892, 2004.

[274] E. J. Dean and R. Glowinski. On the numerical solution of a two-dimensional Pucci's equation with Dirichlet boundary conditions: a least squares approach. *C. R., Math., Acad. Sci. Paris, Série I*, 341(6):375–380, 2005.

[275] E. J. Dean and R. Glowinski. Numerical methods for fully nonlinear elliptic equations of the Monge-Ampère type. *Comput. Methods Appl. Mech. Eng.*, 195(13-16):1344–1386, 2006.

[276] E. J. Dean and R. Glowinski. An augmented Lagrangian approach to the numerical solution of the Dirichlet problem for the elliptic Monge–Ampère equation in two dimensions. *ETNA, Electron. Trans. Numer. Anal.*, 22:71–96, 2006.

[277] L. Debnath. *Nonlinear Partial Differential Equations for Scientists and Engineers.* Boston, Birkhäuser, 1997.

[278] D. W. Decker and H. B. Keller. Path following near bifurcation. *Commun. Pure Appl. Math.*, XXXIV:149–175, 1981.

[279] G. Delic, E. J. Janse van Rensburg, and G. Welke. A non-selfadjoint general matrix eigenvalue problem. *J. Comput. Phys.*, 69:325–340, 1987.

[280] L. F. Demkowicz. *Computing with hp-Adaptive Finite Elements*, volume 1: *One- and Two-Dimensional Elliptic and Maxwell Problems.* Chapman and Hall/CRC, Boca Raton, 2007.

[281] L. F. Demkowicz, J. Kurtz, D. Pardo, M. Paszyński, W. Rachowicz, and A. Zdunek. *Computing with hp-Adaptive Finite Elements*, volume 2: *Frontiers: Three-dimensional Elliptic and Maxwell Problems with Applications.* Chapman and Hall/CRC, Boca Raton, 2008.

[282] N. Dencker, J. Sjstrand, and M. Zworski. Pseudospectra of semiclassical (pseudo-) differential operators. *Commun. Pure Appl. Math.*, 57(3):384–415, 2004.

[283] Y. Deng. On the regularity and existence of solutions for semilinear elliptic equations of second order. *Acta Math. Sci.*, 7:217–228, 1987.

[284] P. Deuflhard and A. Hohmann. *Numerische Mathematik I.* De Gruyter, Berlin, 1993.

[285] P. Deuflhard and F. A. Potra. Asymptotic mesh independence of Newton Galerkin methods via a refined Mysovskii theorem. *SIAM J. Numer. Anal.*, 29:1395–1412, 1992.

[286] R. DeVore. Nonlinear approximation. *Acta Numerica*, 7:51–150, 1998.

[287] L. Dieci and M. J. Friedman. Continuation of invariant subspaces. *Numer. Linear Algebra Appl.*, 8(5):317–327, 2001.

[288] E. J. Doedel. On the existence of extraneous solutions to discretizations of boundary value problems. In *Numerical Mathematics and Computing*, Proc. 9th Manitoba Conf., 1979, 173–185. 1980.

[289] E. J. Doedel. Some stability theorems for finite difference collocation methods on nonuniform meshes. *BIT, Nord. Tidskr. Inf.-behandl.*, 2:58–66, 1980.

[290] E. J. Doedel and G. W. Reddien. Computation of bifurcation branches using projection methods. In *Numerical Analysis and Computers*, Proc. Army Conf., Moffett Field, CA 1980, 223–231, 1980.

[291] E. J. Doedel and G. W. Reddien. Finite-difference collocation methods for singular boundary value problems. In *Approximation Theory III*, Proc. Conf. Hon. G. G. Lorentz, Austin, TX 1980, 349–354, 1980.

[292] V. Dolejší. Anisotropic mesh adaptation for finite volume and finite element methods on triangular meshes. *Comput. Vis. Sci.*, 1(3):165–178, 1998.

[293] V. Dolejší. Higher order semi-implicit discontinuous Galerkin finite element schemes for nonlinear convection-diffusion problems. In A. Bermúdez de Castro, D. Gómez, P. Quintela, and P. Salgado, editors, *Numerical Mathematics and Advanced Applications, ENUMATH 2005*, 432–439. Springer-Verlag Berlin, Heidelberg, 2006.

[294] V. Dolejší. Semi-implicit interior penalty discontinuous Galerkin methods for viscous compressible flows. *Commun. Comput. Phys.*, 4(2):231–274, 2008.

[295] V. Dolejší. Analysis and application of IIPG method to quasilinear nonstationary convection-diffusion problems. *J. Comp. Appl. Math.*, published online doi:10.1016/j.cam.2007.10.055, 2007.

[296] V. Dolejší, M. Feistauer, and C. Schwab. A finite volume discontinuous Galerkin scheme for nonlinear convection–diffusion problems. *Calcolo*, 39:1–40, 2002.

[297] V. Dolejší, M. Feistauer, and V. Sobotíková. A discontinuous Galerkin method for nonlinear convection–diffusion problems. *Comput. Meth. Appl. Mech. Engng.*, 194:2709–2733, 2005.

[298] V. Dolejší and J. Felcman. Anisotropic mesh adaptation and its application for scalar diffusion equations. *Numer. Meth. Part. Differ. Eq.*, 20:576–608, 2004.

[299] G.-C. Dong. *Nonlinear Partial Differential Equations of Second Order.* Trans. Kai Seng Chou. Translations of Mathematical Monographs. AMS, Providence, 1991.

[300] W. Dörfler. A convergent adaptive algorithm for Poisson's equation. *SIAM J. Numer. Anal.*, 33:1106–1124, 1996.

[301] W. Dörfler and V. Heuveline. Convergence of an adaptive *hp* finite element strategy in one space dimension. *Appl. Numer. Math.*, 57:1108–1124, 2007.

[302] W. Dörfler and O. Wilderotter. An adaptive finite element method for a linear elliptic equation with variable coefficients. *Z. Angew. Math. Mech.*, 80:481–491, 2000.

[303] J. Douglas and T. Dupont. *Interior Penalty Procedures for Elliptic and Parabolic Galerkin Methods*, Lecture Notes in Physics, volume 58, 207–216. Springer, Berlin, 1976.

[304] M. Dryja, J. Galvis, and M. Sarkis. DDC methods for discontinuous Galerkin discretization of elliptic problems. *J. Complexity*, 23(4-6):715–739, 2007.

[305] P. DuChateau and D. Zachmann. *Partial Differential Equations.* Schaum's Outline Series. McGraw-Hill, New York, 1986.

[306] P. DuChateau and D. Zachmann. *Applied Partial Differential Equations.* Harper and Row, New York, 1989.

[307] P. DuChateau and D. Zachmann. *Applied Partial Differential Equations.* Harper and Row, New York, 1996.

[308] Le Dung. Boundary $C^{1,\alpha}$ regularity of singular semilinear elliptic equations. *Nonlin. Anal., Theor. Meth. Appl.*, 26(1):67–80, 1996.

[309] Y. Epshteyn and B. Rivière. Estimation of penalty parameters for symmetric interior penalty Galerkin methods. *J. Comput. Appl. Math.*, 206(2):843–872, 2007.

[310] L. C. Evans. *Partial Differential Equations.* Graduate Studies in Mathematics 19. AMS, Providence, 1998.

[311] M. S. Fanciullo. $\mathcal{L}^{2,\lambda}$ regularity for second order nonlinear non variational elliptic systems. *Matematiche*, 50(1):163–172, 1995.

[312] M. Feistauer. On the finite element approximation of functions with noninteger derivatives. *Numer. Funct. Anal. and Optimiz.*, 10(91-110), 1989.

[313] M. Feistauer. *Mathematical Methods in Fluid Dynamics.* Longman Scientific and Technical, Harlow, 1993.

[314] M. Feistauer, J. Felcman, M. Lukáčová-Medvidová, and G. Warnecke. Error estimates of a combined finite volume – finite element method for nonlinear convection – diffusion problems. *SIAM J. Numer. Anal.*, 36(5):1528–1548, 1999.

[315] M. Feistauer, J. Felcman, and I. Străskraba. *Mathematical and Computational Methods for Compressible Flow.* Series in Numerical Mathematics and Scientific Computation. Oxford University Press, Oxford, 2003.

[316] M. Feistauer and A. Zenísek. Compactness method in the finite element theory of nonlinear elliptic problems. *Numer. Math.*, 52(2):147–163, 1988.

[317] M. Feistauer and A. Zenísek. Finite element variational crimes in nonlinear elliptic problems. In *Numerical Analysis*, Proc. 2nd Int. Symp., Prague, Czechoslovakia, 1987, 28–35. Texte Math. 107. Teubner, Leipzig, 1988.

[318] X. Feng and M. Neilan. Analysis of Galerkin methods for the fully nonlinear Monge–Ampère type. Technical report, 2008.

[319] G. W. Forsythe and W. R. Wasow. *Finite-Difference Methods for Partial Differential Equations.* Applied Mathematics Series. John Wiley and Sons, New York, 1960.

[320] R. Fössmeier. *Differenzenverfahren höherer Ordnung für elliptische Randwertprobleme mit gekrümmten Rändern.* PhD thesis, Technische Universität München, 1984.

[321] L. Fox. Some improvments in the use of relaxation methods for the solution of ordinary and partial differential equations. *Proc. Roy. Soc. Ser. A*, 190:31–59, 1947.

[322] L. Fox. The solution by relaxation methods of ordinary and partial differential equations. *Proc. Phil. Soc.*, 45:50–68, 1949.

[323] L. Fox. The numerical solution of elliptic differential equations when the boundary conditions involve a derivative. *Phil. Trans. R. Soc. Lond. A.*, 242: 345–378, 1950.

[324] L. Fox, editor. *Numerical Solution of Ordinary and Partial Differential Equations*. Oxford University Press, Oxford, 1962.

[325] R. Frank. The method of iterated defect-correction and its application to boundary value problems. *Numer. Math.*, 25:407–420, 1977.

[326] R. Frank and R. Ueberhuber. Iterated defect-correction for the efficient solution of stiff systems of ordinary differential equations. *BIT*, 17:146–159, 1977.

[327] C. Franke and R. Schaback. Convergence order estimates of meshless collocation methods using radial basis functions. *Adv. Comput. Math.* 8(4):381–399, 1998.

[328] C. Franke and R. Schaback. Solving partial differential equations by collocation using radial basis functions. *Appl. Math. Comp.*, 93(1):73–91, 1998.

[329] M. Frazier and B. Jawerth. A discrete transform and decompositions of distribution spaces. *J. Funct. Anal.*, 93(1):34–170, 1990.

[330] S. R. Fulton. Multigrid solution of non-linear balance equations in meteorology. *Commun. Appl. Numer. Meth.*, 8(10):695–706, 1992.

[331] T. Gantumur. *Adaptive wavelet algorithms for solving operator equations*. PhD thesis, Utrecht University, 2006.

[332] T. Gantumur. An optimal adaptive wavelet method for nonsymmetric and indefinite elliptic problems. *J. Comput. Appl. Math.*, 211(1):90–102, 2008.

[333] T. Gantumur, H. Harbrecht, and R. Stevenson. An optimal adaptive wavelet method without coarsening of the iterands. *Math. Comp.* 76(258):615–629, 2007.

[334] T. Gantumur and R. Stevenson. Computation of differential operators in wavelet coordinates. *Math. Comp.* 75(254):697–709, 2006.

[335] T. Gantumur and R. Stevenson. Computation of singular integral operators in wavelet coordinates. *Computing*, 76(1–2):77–107, 2006.

[336] P. R. Garabedian. *Partial Differential Equations*. John Wiley and Sons, New York, London, Sydney, 1964.

[337] G. Garello. Generalized Sobolev algebras and regularity for solutions of multi-quasi-elliptic semilinear equations. *Commun. Appl. Anal.*, 3(4):563–574, 1999.

[338] J. W. Garret and W. Liu. Quasi-error bounds for the finite element approximation of a non-Newtonian flow. *Numer. Math.*, 68:437–456, 1994.

[339] M. Geissert. *Maximal lp-lq regularity for solutions of difference equations.* Dissertation at Department of Mathematics, University of Darmstadt, 1992.

[340] S. Gerschgorin. Fehlerabschätzung für das Differenzenverfahren zur Lösung partieller Differentialgleichungen. *Z. Angew. Math. Mech*, 10:373–382, 1930.

[341] M. Giaquinta and S. Hildebrandt. *Calculus of Variation 1. The Lagrangian Formalism.* Springer, Berlin, 1996.

[342] M. Giaquinta and S. Hildebrandt. *Calculus of Variation 2. The Hamiltonian Formalism.* Springer, Berlin, 1996.

[343] M. Giaquinta, Jindrich Necas, Oldrich John, and Jana Stará. On the regularity up to the boundary for second order nonlinear elliptic systems. *Pac. J. Math.*, 99:1–17, 1982.

[344] D. Gilbarg and N. S. Trudinger. *Elliptic Partial Differential Equations of Second Order*, 2nd edn. Grundlehren der Mathematischen Wissenschaften 224. Springer, Berlin, 1983.

[345] D. Gilbarg and N. S. Trudinger. *Elliptic Partial Differential Equations of Second Order*, 3rd edn. Springer, Berlin, 1998.

[346] D. Gilbarg and N. S. Trudinger. *Elliptic Partial Differential Equations of Second Order.* Springer, Berlin, 2001. Reprint of the 1998 edition.

[347] M. Girardi, L. Mastroeni, and M. Matzeu. Existence and regularity results for non-negative solutions of some semilinear elliptic variational inequalities via mountain pass techniques. *Z. Anal. Anwend.*, 20(4):845–857, 2001.

[348] V. Girault and P. A. Raviart. *Finite Element Approximation of the Navier–Stokes Equations.* Lecture Notes in Math. 749. Springer, Berlin, New York, 1979.

[349] V. Girault and P. A. Raviart. *Finite Element Methods for Navier–Stokes Equations*, 2nd edn. Springer Series in Computational Mathematics 5. Springer, Berlin, New York, 1986.

[350] R. Glowinski. *Numerical methods for nonlinear variational problems.* Springer Series in Computational Physics. New York etc.: Springer-Verlag, 1984.

[351] R. Glowinski. Finite element methods for incompressible viscous flow. In P.G. Ciarlet and J.L. Lions, editors, *Handbook of Numerical Analysis*, volume IX of *Techniques of Scientific Computing*, 3–1176. Elsevier Science B.V., 2003.

[352] R. Glowinski, E.J. Dean, G. Guidoboni, L.H. Juárez, and T.-W. Pan. Applications of operator-splitting methods to the direct numerical simulation of particulate and free-surface flows and to the numerical solution of the two-dimensional elliptic Monge-Ampère equation. *Japan J. Ind. Appl. Math.*, 25(1):1–63, 2008.

[353] R. Glowinski and J.L. Lions. Exact and approximate controllability for distributed parameter systems. Iserles, A. (ed.), Acta Numerica 1994. Cambridge: Cambridge University Press. 269–378, 1994.

[354] R. Glowinski, J.L. Lions, and R. Tremolieres. *Numerical Analysis of Variational Inequalities.* Dunod, Paris, 1981.

[355] R. Glowinski and J.L. Lions (eds.). *Computing methods in applied sciences and engineering. Part 1. International symposium, Versailles, December 17-21, 1973.* Lecture Notes in Computer Science. 10. Berlin-Heidelberg-New York: Springer-Verlag. X, 1974.

[356] R. Glowinski and P. Neittaanmäki (eds.). *Partial differential equations. Modelling and numerical simulation. Some papers based on the presentations at the international conference, Helsinki, Finland, fall 2005.* Computational Methods in Applied Sciences 16. Dordrecht: Springer. xvi, 2008.

[357] W. Govaerts. *Numerical Methods for Bifurcations of Dynamical Equlibria.* SIAM, Philadelphia, 2000.

[358] W. B. Gragg. On extrapolation alorithms for ordinary initial value problems. *SIAM J. Numer. Anal.*, 2:384–403, 1965.

[359] I. G. Graham, L. Grasedyck, W. Hackbusch, and S. A. Sauter. Optimal panel-clustering in the presence of anisotropic mesh refinement. *SIAM J. Numer. Anal.*, 46:517–543, 2008.

[360] I. G. Graham, W. Hackbusch, and S. A. Sauter. Fast integration in 3d boundary elements. In C.-H. Lai, P. E. Bjørstad, M. Cross, and O. B. Widlund, editors, *Domain Decomposition Methods in Science and Engineering*, 37–48. Domain Decomposition Press, Bergen, 1999.

[361] I. G. Graham, W. Hackbusch, and S. A. Sauter. Discrete boundary element methods on general meshes in 3d. *Numer. Math.*, 86:103–137, 2000.

[362] I. G. Graham, W. Hackbusch, and S. A. Sauter. Hybrid boundary elements: theory and implementation. *Numer. Math*, 86:139–172, 2000.

[363] I. G. Graham, W. Hackbusch, and S. A. Sauter. Hybrid Galerkin boundary elements on degenerate meshes. In M. Bonnet, A.-M. Sändig, and W. L. Wendland, editors, *Mathematical Aspects of Boundary Element Methods.* CRC Press, Boca Raton, 2000.

[364] I. G. Graham, W. Hackbusch, and S. A. Sauter. Finite elements on degenerate meshes: Inverse-type inequalities and applications. *IMA J. Numer. Anal.*, 25:379–407, 2005.

[365] I. G. Graham, A. Spence, and E. Vainikko. Parallel iterative methods for Navier–Stokes equations and application to eigenvalue computation. *Concurrency Comput. Pract. Exp.*, 15(11-12):1151–1168, 2003.

[366] A. Griewank and G. W. Reddien. Characterization and computation of generalized turning points. *SIAM J. Numer. Anal.*, 21:176–185, 1984.

[367] A. Griewank and G. W. Reddien. The approximate solution of defining equations for generalized turning points. *SIAM J. Numer. Anal.*, 33:1912–1920, 1996.

[368] A. Griewank and G. W. Reddien. The approximation of generalized turning points by projection methods with superconvergence to the critical parameter. *Numer. Math.*, 48:591–606, 1984.

[369] A. Griewank and G. W. Reddien. Characterization and computation of generalized turning points. *SIAM J.Numer. Anal.*, 21:176–185, 1984.

[370] A. Griewank and G. W. Reddien. Computation of cusp singularities for operator equations and their discretization. *J. Comput. Appl. Math.*, 26:133–153, 1989.

[371] R. D. Grigorieff. Zur Theorie linearer approximationsregulärer Operatoren, i und ii. *Math. Nachr.*, 55:233–249 and 251–263, 1973.

[372] P. Grisvard. *Elliptic Problems in Nonsmooth Domains*. Pitman Publishing, Boston, 1985.

[373] P. Grisvard. *Singularities in Boundary Value Problems*. Springer, Masson, Paris, 1992.

[374] C. Grossmann and H.-G. Roos. *Numerik partieller Differentialgleichungen*. Teubner Studienbücher Mathematik. Teubner, Stuttgart, 1992.

[375] C. Grossmann and H.-G. Roos. *Numerische Behandlung partieller Differentialgleichungen, 3. Auflage*. Teubner Studienbücher Mathematik. Teubner, Wiesbaden, 2005.

[376] C. Grossmann, H.-G. Roos, and M. Stynes. *Numerical Treatment of Partial Differential Equations*. Springer, Berlin, 2007.

[377] H.-C. Grunau and G. Sweers. Classical solutions for some higher order semilinear elliptic equations under weak growth conditions. *Nonlin. Anal., Theor. Meth. Appl.*, 28(5):799–807, 1997.

[378] H.-C. Grunau and W. von Wahl. Regularity considerations for semilinear parabolic systems. Rend. Jst. Mat. Univ. Trieste 28, Suppt., 221–233(1996).

[379] Z. Guo and J. Li. Partial regularity for weak solutions of semilinear elliptic equations with supercritical exponents. *Pac. J. Math.*, 214(1):89–107, 2004.

[380] J. Guzmán and B. Rivière. Sub-optimal convergence of non-symmetric discontinuous Galerkin method for odd polynomial approximations. *J. Sci. Comput.*, 40(1–3):273–280, 2009.

[381] W. Hackbusch. On the regularity of difference schemes. *Ark. Mat.*, 19:71–95, 1981.

[382] W. Hackbusch. On multi-grid iterations with defect correction. In *Multigrid Methods*, Proc. Conf., Köln-Porz 1981, Lect. Notes in Maths. Springer-Verlag, 960, 461–473. Berlin, 1982.

[383] W. Hackbusch. On the regularity of difference schemes-part II. *Ark. Mat.*, 21:3–28, 1982.

[384] W. Hackbusch. Local defect correction method and domain decomposition techniques. In *Defect Correction Methods. Theory and Applications, Comput. Suppl.*, 5: 89–113, 1984.

[385] W. Hackbusch. *Multi-Grid Methods and Applications.* Springer, Berlin Heidelberg, New York, 1985.

[386] W. Hackbusch. *Theorie und Numerik elliptischer Differentialgleichungen.* Teubner, Stuttgart, 1986.

[387] W. Hackbusch. *Elliptic Differential Equations: Theory and Numerical Treatment.* Springer, Berlin, 1992.

[388] W. Hackbusch. *Iterative Lösung grosser schwach besetzter Gleichungssysteme.* Teubner, Mannheim, 1993.

[389] G. J. Haltiner. *Numerical Weather Prediction.* Wiley and Sons, New York, 1971.

[390] R. Hartmann and P. Houston. Symmetric interior penalty DG methods for the compressible Navier–Stokes equations I: method formulation. *Inter. J. Num. Anal. Model.*, 3(1):1–20, 2006.

[391] R. Hartmann and P. Houston. Symmetric interior penalty DG methods for the compressible Navier–Stokes equations II: goal-oriented a posteriori error estimation. *Inter. J. Num. Anal. Model.*, 3(2):141–162, 2006.

[392] R. Hartmann and P. Houston. An optimal order interior penalty discontinuous Galerkin discretization of the compressible Navier–Stokes equations. *J. Comput. Phys.*, 227:9670–9685, 2008.

[393] J. Haslinger, I. Hlavácek, and J. Nečas. Numerical methods for unilateral problems in solid mechanics. In P. G. Ciarlet and J. L. Lions, editors, *Handbook of Numerical Analysis*, volume IV: *Finite Element Methods* of *Techniques of Scientific Computing*, 313–485. North-Holland, Amsterdam, 1995.

[394] O. Havle, V. Dolejší, and M. Feistauer. DGFEM for nonlinear convection-diffusion problems with mixed Dirichlet–Neumann boundary conditions. *Applications of Mathematics*, (in press).

[395] B. Heinrich. *Finite Difference Methods on Irregular Networks. A Generalized Approach to Second Order Elliptic Problems*, Mathematical Research. Mathematische Forschung 33. Akademie-Verlag, Berlin, 1987.

[396] H. Heuser. *Lehrbuch der Analysis, Teil 1.* Mathematische Leitfäden. Teubner, Stuttgart, 1981.

[397] H. Heuser. *Lehrbuch der Analysis, Teil 2.* Mathematische Leitfäden. Teubner, Stuttgart, 1991.

[398] V. Heuveline and R. Rannacher. Duality-based adaptivity in the *hp*-finite element method. *J. Numer. Math.*, 11:95–113, 2003.

[399] I. Hlavácek and J. Naumann. Inhomogeneous boundary value problems for von Kármán equations. *Apl. Mat.*, 19:253–269, 1974.

[400] K. Höllig, U. Reif, and J. Wipper. Weighted extended B-spline approximation of Dirichlet problems. *SIAM J. Numer. Anal.*, 39(2):442–462, 2001.

[401] Y. C. Hon and R. Schaback. An efficient numerical scheme for Burger's equation. *Appl. Math. Comp.*, 95:37–50, 1998.

[402] L. Hörmander. *Linear Partial Differential Operators.* Springer, Berlin, 1963.

[403] L. Hörmander. *The Analysis of Linear Partial Differential Operators*, volume III. Springer, Berlin, 1985.

[404] M. A. Horn. A trace regularity result for solutions to plate equations with simply supported boundary conditions. In Guillermo Ferreyra *et al.*, editors, *Evolution Equations.* International Conference, Louisiana State Univ., Baton Rouge, LA, 7–11 January 1992. Lect. Notes Pure Appl. Math. 168, 245–256. Marcel Dekker, New York, 1994.

[405] P. Houston, I. Perugia, and D. Schötzau. A posteriori error estimation for discontinuous Galerkin discretizations of H(curl)-elliptic partial differential equations. *IMA J. Numer. Anal.*, 27:122–150, 2007.

[406] P. Houston, R. Rannacher, and E. Süli. A posteriori error analysis for stabilised finite element approximations of transport problems. *Comp. Meth. Appl. Mech. Engnn.*, 190(11-12):1483–1508, 2000.

[407] P. Houston, J. Robson, and E. Süli. Discontinuous Galerkin finite element approximation of quasilinear elliptic boundary value problems I: The scalar case. *IMA J. Numer. Anal.*, 25(4):726–749, 2005.

[408] P. Houston, D. Schötzau, and T. P. Wihler. Energy norm a posteriori error estimation of *hp*-adaptive discontinuous Galerkin methods for elliptic problems. *Math. Models Methods Appl. Sci.*, 17(1):33–62, 2007.

[409] P. Houston, D. Schötzau, and T. P. Wihler. Energy norm a posteriori error estimation for mixed discontinuous Galerkin approximations of the Stokes problem. *J. Sci. Comput.*, 22(1):357–380, 2005.

[410] P. Houston, C. Schwab, and E. Süli. Discontinuous *hp*-finite element methods for advection-diffusion problems. *SIAM J. Numer. Anal.*, 39(6):2133–2163, 2002.

[411] P. Houston and E. Süli. A note on the design of hp-adaptive finite element methods for elliptic partial differential equations. Comput. Methods Appl. Mech. Engry, 194: 229–243, 2005.

[412] P. Houston, E. Süli, and T. P. Wihler. A posteriori error analysis of hp-version discontinuous Galerkin finite element methods for second-order quasilinear elliptic problems. *IMA J. Numer. Anal.*, 28:245–273, 2008.

[413] G. C. Hsiao and W. L. Wendland. Boundary element methods. In *Encyclopedia of Computational Mechanics*. John Wiley and Sons, New York, 2004.

[414] E. Isaacson and H. B. Keller. *Analysis of Numerical Methods*. John Wiley and Sons Inc., New York, London, Sydney, 1969.

[415] A. V. Ivanov. *Quasilinear Degenerate and Nonuniform Elliptic and Parabolic Equations of Second Order*. AMS, Providence, RI, 1984.

[416] L. D. Ivanovich, B. S. Jovanovich, and E. Süli. Convergence of difference schemes for the biharmonic equation. *Zhurnal Vychislitelnoi Matematiki i Matematicheskoi Fiziki*, 26(5):776–799, 1986.

[417] B. Jiang, Z.-C. Shi, and W. Xue. A new superconvergence property of Wilson nonconforming finite element. *Numer. Mathmatik*, 78:259–268, 1997.

[418] A. D. Jepson and A. Spence. On a reduction process for nonlinear equations. *SIAM J. Math. Anal.*, 20:39–56, 1989.

[419] D. Jerison and C. E. Kenig. The inhomogeneous Dirichlet problem in Lipschitz domains,. *J. Funct. Anal.*, 130:161–219, 1995.

[420] R. Q. Jia and C. A. Micchelli. Using the refinement equations for the construction of prewavelets II: Powers of two. In "Curves and surfaces" (P. J. Laurent, A. Le'Mehaute, and L. L. Schumaker, Eds.), Academic Press, New York, 209–246 (1991), 1991.

[421] K. Johannsen, W. Kinzelbach, S. Oswald, and G. Wittum. The saltpool benchmark problem - numerical simulation of saltwater upconing in a porous medium. *Advances in Water Resources*, 25(3):335–348, 2002.

[422] O. John and J. Naumann. On regularity of variational solutions of the von Kármán equations. *Math. Nachr.*, 71:23–36, 1976.

[423] O. John and J. Necas. On the solvability of von Kármán equations. *Apl. Mat.*, 20:48–62, 1975.

[424] J. Johnsen. Regularity properties of semilinear boundary problems in Besov and Triebel–Lizorkin spaces. Math. Scand. 79, No 1, 25–85 (1996).

[425] J. Johnson and H. Wallin. *Function Spaces on Subsets of* $\mathbb{R}^n$. Harwood Academic Publishers, Chur, 1984.

[426] C. Johnson and J. Pitkäranta. An analysis of the discontinuous Galerkin method for a scalar hyperbolic equation. *Math. Comp.*, 46:1–26, 1986.

[427] K. Jörgens. *Spectral Theory of Second-Order Ordinary Differential Operators*. Lecture Notes Series 2. Matematisk Institut, Aarhus Universitet, 1964.

[428] D. D. Joseph and T. S. Lundgren. Quasilinear Dirichlet problems driven by positive sources. *Arch. Rational Mech. Anal.*, 49:241–269, 1973.

[429] A. D. Joshi. Dual algebra techniques and the invariant subspace problem. In *H. P. Dikshit et al., editors, Analysis and Applications. Outcome of the Conference Dedicated to Professor Tribikram Pati on the Occasion of his 70th Birthday,* Ujjain, India, 1999, 101–120. Narosa Publishing House, New Delhi; Chapman and Hall/CRC, Boca Raton, 2002.

[430] B. S. Jovanovic, L. D. Ivanovic, and E. Süli. Convergence of finite-difference schemes for elliptic equations with variable coefficients. *IMA J. Numer. Anal.*, 7(3):301–305, 1987.

[431] B. S. Jovanovic, E. Süli, and L. D. Ivanovic. On finite difference schemes of high order accuracy for elliptic equations with mixed derivatives. *Matematicki Vesnik*, 38(2):131–136, 1986.

[432] B. S. Jovanovich, L. D. Ivanovich, and E. Süli. Convergence of difference schemes for the equation $-\Delta u + cu = f$ on generalised solutions from $W^s_{2,*}; (-\infty < s < +\infty)$. *Mathematique. Publications. Nouvelle Serie Publ. Inst. Math. (Beograd)*, 37(51):129–138, 1985.

[433] J.-P. Kahane and P.-G. Lemarié-Rieusset. *Fourier Series and Wavelets.* Breach Science Publishers, Luxembourg, 1995.

[434] G. Kanschat. Block preconditioners for LDG discretizations of linear incompressible flow problems. *J. Sci. Comput.*, 22–23:371–384, 2005.

[435] J. Kappei. Adaptive frame methods for nonlinear elliptic problems. Bericht 2009–3, Fachbereich Mathematik und Informatik, Philipps-Universität Marburg, 2009.

[436] B. Karaivanov and P. Petrushev. Nonlinear piecewise polynomial approximation beyond Besov spaces. *Appl. Comput. Harmon. Anal.*, 15:177–223, 2003.

[437] A. Kasahara. Significance of non-elliptic regions in balanced flows of the tropical atmosphere. Preprint, National Center for Atmospheric Research, University of Boulder, Colorado, 1981.

[438] B. Kawohl, J. Stara, and G. Wittum. Analysis and numerical studies of a problem of shape design. *Arch. Ration. Mech. Anal.*, 114(4):349–363, 1991.

[439] H. B. Keller. The bordering algorithm and path following near singular points of higher nullity. *SIAM J. Sci. Statist. Comput.*, 4:573–582, 1983.

[440] H. B. Keller. *Lectures on Numerical Methods in Bifurcation Problems.* Springer, Berlin, Heidelberg, New York, 1987.

[441] H. B. Keller. Approximation methods for nonlinear problems with application to two-point boundary value problems. *Math. Comput.*, 29:464–474, 1975.

[442] H. B. Keller and V. Pereyra. Difference methods and deferred corrections for ordinary boundary value problems. *SIAM J. Numer. Anal.*, 16:241–259, 1979.

[443] R. B. Kellogg and J. E. Osborn. A regularity result for the Stokes problem in a convex polygon. *J. Funct. Anal.*, 21:379–431, 1976.

[444] R. B. Kellogg. On the Poisson equation with intersecting interfaces. *Applicable Analysis*, 4:101–129, 1975.

[445] P. Knabner and L. Angermann. *Numerical Methods for Elliptic and Parabolic Differential Equations*, Texts in Applied Mathematics 44. Springer, Berlin, 2003.

[446] V. A. Kondrat'ev. Boundary problems for elliptic equations in domains with conical or angular points. *Trans. Moscow Math. Soc.*, 16:227–313, 1967.

[447] V. A. Kondrat'ev and O. A. Olejnik. Unimprovable estimates in Hölder spaces for generalized solutions of the biharmonic equation, the Navier–Stokes system of equations, and the von Kármán system in nonsmooth two-dimensional domains. *Mosc. Univ. Math. Bull.*, 38(6):24–43, 1983.

[448] V. A. Kondrat'ev and O. A. Olejnik. Unimprovable estimates in Hölder spaces for generalized solutions of the biharmonic equation, the system of Navier–Stokes equations, and the von Kármán system in non-smooth domains. *Vestn. Mosk. Univ., Ser. I*, 1983(6):22–39, 1983.

[449] A. Koshelev. *Regularity Problems for Quasilinear Elliptic and Parabolic Systems.* Springer, Berlin, 1995.

[450] A. Koshelev and S. Chelkak. *Regularity of Solutions of Quasilinear Elliptic Systems.* Teubner, Leipzig, 1985.

[451] A. Koshelev and S. Chelkak. Regularity and weak regularity of solutions of quasilinear elliptic systems. In O. Ladyženskaja, editor, *Proceedings of the St. Petersburg Mathematical Society*, Vol. II, 163–185, AMS, Providence, 1994.

[452] V. A. Kozlov, V. G. Maz'ya, and J. Rossmann. *Elliptic Boundary Value Problems in Domains with Point Singularities.* Mathematical Surveys and Monographs 52. AMS, Providence, 1997.

[453] M. Kötter. A free boundary value ptoblem for the stationary Navier–Stokes equation. Technical report, Universität Bremen, 1988.

[454] H.-O. Kreiss and J. Lorenz. *Initial-Boundary Value Problems and the Navier–Stokes Equations.* Reprint of the 1989 edition. Classics in Applied Mathematics 47. SIAM, Princeton, NJ, 2004.

[455] A. Kufner and A.-M. Sändig. *Some Applications of Weighted Sobolev Spaces.* Teubner, Leipzig, 1987.

[456] H.-J. Kuo and N. S. Trudinger. Linear elliptic difference inequalities with random coefficients. *Math. Comput.*, 55(191):37–53, 1990.

[457] H.-J. Kuo and N. S. Trudinger. Discrete methods for fully nonlinear elliptic equations. *SIAM J. Numer. Anal.*, 29(1):123–135, 1992.

[458] H.-J. Kuo and N. S. Trudinger. Local estimates for parabolic difference operators. *J. Differ. Eq.*, 122(2):398–413, 1995.

[459] H.-J. Kuo and N. S. Trudinger. Evolving monotone difference operators on general space-time meshes. *Duke Math. J.*, 91(3):587–607, 1998.

[460] H.-J. Kuo and N. S. Trudinger. Schauder estimates for fully nonlinear elliptic difference operators. *Proc. R. Soc. Edinb., Sect. A, Math.*, 132(6):1395–1406, 2002.

[461] O. Ladyženskaja. *Funktionalanalytische Untersuchungen der Navier–Stokesschen Gleichungen.* Akademie-Verlag, Berlin, 1965.

[462] O. Ladyženskaja. *The Boundary Value Problems of Mathematical Physics.* Springer, Berlin, Heidelberg, New York, 1985.

[463] O. Ladyženskaja and N. Uralceva. *Linear and Quasilinear Elliptic Equations* (in Russian). Nauka, Moscow, 1964.

[464] O. Ladyženskaja and N. Uralceva. *Linear and Quasilinear Elliptic Equations.* Academic Press, New York, 1968.

[465] O. A. Ladyzhenskaya. The sixth millennium problem: Navier–Stokes equations, existence and smoothness. *Russ. Math. Surv.*, 58(2):251–286, 2003.

[466] O. A. Ladyzhenskaya and N. N. Ural'tseva. Estimates on the boundary of a domain for the first derivatives of functions satisfying an elliptic or parabolic inequality. *Proc. Steklov Inst. Math.*, 179:109–135, 1989.

[467] M.-J. Lai and L. L. Schumaker. *Spline Functions on Triangulations.* Cambridge University Press, Cambridge, 2007.

[468] O. Lakkis. A short proof of regularity for solutions to semilinear elliptic problems with exponential critical growth. *Rend. Ist. Mat. Univ. Trieste*, 30(1–2):181–183, 1998.

[469] S. Larsson and V. Thomée. *Partial Differential Equations with Numerical Methods.* Texts in Applied Mathematics 45. Springer, Berlin, 2003.

[470] L. Lassoued. Existence of regular solutions to $-\Delta u = V(x)e^u$ in bounded domains of $\mathcal{R}^n$, $n \geq 1$. *Panam. Math. J.*, 6(3):1–23, 1996.

[471] M. Lenoir. Optimal isoparametric finite elements and error estimates for domains involving curved boundaries. *SIAM J. Numer. Anal.*, 23:562–580, 1986.

[472] P. Le Saint and P.-A. Raviart. On a finite element method for solving the neutron transport equation. In C. de Boor, editor, *Mathematical Aspects of Finite Elements in Partial Differential Equations*, 89–145. Academic Press, New York, 1974.

[473] E. W. Leung. *Systems of Nonlinear Partial Differential Equations.* Graduate Studies in Mathematics. Kluwer Academic Publishers, Dordrecht, 1989.

[474] S. Z. Levendorskij. Asymptotics of the spectrum of operators elliptic in the sense of Douglis-Nirenberg. 1989.

[475] S. Li and X. Liu. Viscosity solutions for the weakly coupled system of second-order fully nonlinear elliptic equation with obstacle. *Chin. Ann. Math., Ser. A*, 20(5):605–616, 1999.

[476] Q. Lin and J. Liu. Defect corrections of finite element approximation for a singular problem. In *Finite Element Methods*, Proc. China-France Symp., Beijing, China, 1982, 361–368. 1983.

[477] Q. Lin and A. Zhou. Defect correction for finite element gradient. *Syst. Sci. Math. Sci.*, 5(3):287–288, 1992.

[478] J. L. Lions and E. Magenes. *Non-Homogeneous Boundary Value Problems and Applications*, volume I. Springer, Berlin, Heidelberg, New York, 1972.

[479] J. L. Lions and E. Magenes. *Non-Homogeneous Boundary Value Problems and Applications*, volume II. Springer, Berlin, Heidelberg, New York, 1972.

[480] J. L. Lions and E. Magenes. *Non-Homogeneous Boundary Value Problems and Applications*, volume III. Springer, Berlin, Heidelberg, New York, 1973.

[481] S. Luckhaus. Existence and regularity of weak solutions to the Dirichlet problem for semilinear elliptic systems of higher order. *J. Reine Angew. Math.*, 306:192–207, 1979.

[482] S. Mallat. Multiresolution approximation and wavelet orthonormal bases of $L_2(\mathbb{R})$. *Trans. Amer. Math. Soc.*, 315:69–87, 1989.

[483] G. I. Marchuk. *Methods of Numerical Mathematics*, 2nd edn, trans. Arthur A. Brown. Applications of Mathematics 2. Springer, New York, Heidelberg, Berlin, 1982.

[484] A. Le Méhauté. Unisolvent interpolation in $\mathbb{R}^n$ and the simplicial polynomial finite element method. In L. L. Schumaker, C. K. Chui, and F. Utreras, editors, *Topics in Multivariate Approximation*, Academic Press, New York, 1987.

[485] J. M. Melenk. *hp-Finite Element Methods for Singular Perturbations*, Lecture Notes in Mathematics 1796. Springer, Berlin, 2002.

[486] J. M. Melenk and C. Schwab. *hp* FEM for reaction-diffusion equations. I: Robust exponential convergence. *SIAM J. Numer. Anal.*, 35(4):1520–1557, 1998.

[487] J. M. Melenk and C. Schwab. Analytic regularity for a singularly perturbed problem. *SIAM J. Math. Anal.*, 30(2):379–400, 1999.

[488] U. Mertins. Zur Konvergenz des Rayleigh–Ritz-Verfahrens bei Eigenwertaufgaben. (Convergence of the Rayleigh–Ritz method for eigenvalue problems). *Numer. Math.*, 59(7):667–682, 1991.

[489] U. Mertins. Asymptotische Fehlerschranken für Rayleigh–Ritz-Approximationen selbstadjungierter Eigenwertaufgaben. (Asymptotic error estimates for Rayleigh–Ritz-approximations of selfadjoint eigenvalue problems). *Numer. Math.*, 63(2):227–241, 1992.

[490] Y. Meyer. *Wavelets and Operators.* Cambridge Studies in Advanced Mathematics 37. Cambridge University Press, Cambridge, 1992.

[491] G. Minty. On a monotonicity method for the solution of non-linear equations in Banach spaces. *Proc. Nat. Acad. Sci. U.S.A.*, 50:1038–1041, 1963.

[492] H. D. Mittelmann. Multilevel continuation techniques for nonlinear boundary value problems with parameter dependence. *Appl. Math. and Comp.*, 19:265–282, 1986.

[493] H. D. Mittelmann and H. Weber. *Bifurcation Problems and their Numerical Solutions.* ISNM54. Birkhäuser, Basel, 1980.

[494] H. D. Mittelmann and H. Weber. Multigrid solution of bifurcation problems. *SIAM J. Sci. Stat. Comput.*, 6:49–60, 1985.

[495] M. S. Mommer and R. Stevenson. A goal-oriented adaptive finite element method with convergence rates. *SIAM J. Numer. Anal.*, 47:861–886, 2009.

[496] J. P. Monnet. A defect correction method for the effective computation of high-degree finite element solutions. *Syst. Sci. Math. Sci.*, 1(1):72–93, 1988.

[497] J. Morgan and R. Scott. A nodal basis for c^1 piecewise polynomials of degree $n \geq 5$,. *Math. Comp.*, 29:736–740, 1975.

[498] P. Morin, R. H. Nochetto, and K. G. Siebert. Data oscillation and convergence of adaptive FEM. *SIAM J. Numer. Anal.*, 38:466–488, 2000.

[499] P. Morin, R. H. Nochetto, and K. G. Siebert. Convergence of adaptive finite element methods. *SIAM Review*, 44:631–658, 2002.

[500] C. Morrey. *Multiple Integrals and the Calculus of Variations.* Springer, Berlin, 1966.

[501] C. B. Morrey. Partial regularity results for non-linear elliptic systems. *J. Math. Meth.*, 17:649–670, 1968.

[502] C. B. Morrey. Differentiability theorems for weak solutions of non-linear elliptic differential equations. *Bull. Am. Math. Soc.*, 75:684–705, 1969.

[503] C. B. Morrey. Differentiability theorems for non-linear elliptic equations. In *Actes Congr. Internat. Math*, volume 2, 859–866, 1971.

[504] C. B. Morrey. Differentiability of weak solutions of elliptic systems. In Proc. Int. Conf. Sus. Funct. Anal. Rel. Topics, Tokyo, 1969, 131–141 (1970).

[505] I. Mozolevski, E. Süli, and P. Bösing. Discontinuous Galerkin finite element approximation of the two-dimensional Navier–Stokes equations in stream-function formulation. *Commun. Numer. Meth. Engng.*, 23(6):447–459, 2007.

[506] F. Müller. Analyticity of solutions for semilinear elliptic systems of second order. *Calc. Var. Part. Differ. Eq.*, 15(2):257–288, 2002.

[507] Z. Naniewicz and P. D. Panagiotopoulos. On regularity results for variational-hemivariational inequalities. In P. Robert Gilbert, *et al.*, editor, *Direct and Inverse Problems of Mathematical Physics*, Int. Soc. Anal. Appl. Comput. 5, 301–322. Kluwer Academic Publishers, Dordrecht, 2000.

[508] J. Naumann. An existence theorem for von Kármán equations under the condition of free boundary. *Apl. Mat.*, 19:17–27, 1974.

[509] J. Nečas. *Équations aux dérivées partielle.* Presse de l'Université de Montréal, Montréal, 1965.

[510] J. Nečas. *Introduction to the Theory of Nonlinear Elliptic Equations.* Teubner, Leipzig, 1983.

[511] J. Nečas, J. Oldrich, and J. Stará. Counterexample to the regularity of weak solution of elliptic systems. *Commentat. Math. Univ. Carol.*, 21:145–154, 1980.

[512] G. Nedev. Regularity of the extremal solution of semilinear elliptic equations. *Comptes lend. Acad. Sci.* 330(11): 997–1002, 2000.

[513] R. H. Nochetto. Pointwise a posteriori error estimates for elliptic problems on highly graded meshes. *Math. Comput.*, 64:1–22, 1995.

[514] B. Nour-Omid and B. N. Parlett. Lanczos versus subspace iteration for solution of eigenvalue problems. *Inter. J. Num. Meth. Eng.*, 19:859–871, 1983.

[515] A. Novotný and I. Straškraba. *Introduction to the Mathematical Theory of Compressible flow.* Oxford Lecture Series in Mathematics and its Applications 27. Oxford University Press, Oxford, 2004.

[516] G. Nürnberger, L. L. Schumaker, and F. Zeilfelder. Lagrange interpolation by c^1 cubic splines on triangulated quadrangulations. *Advances in Comp. Math.*, 21:357–380, 2004.

[517] T. J. Oden, I. Babuska, and C. E. Baumann. A discontinuous *hp* finite element method for diffusion problems. *J. Comput. Phys.*, 146(2):491–519, 1998.

[518] J. T. Oden and J. N. Reddy. *An Introduction to the Mathematical Theory of Finite Elements.* Wiley and Sons, New York, 1976.

[519] V. I. Oliker and L. D. Prussner. On the numerical solution of the equation $\frac{\partial^2 z \partial^2 z}{\partial x^2 \partial y^2} - \left(\frac{\partial^2 z}{\partial x \partial y}\right)^2 = f$ and its discretizations. I. *Numer. Math.*, 54(3):271–293, 1988.

[520] J. M. Ortega and W. C. Rheinboldt. *Iterative Solution of Nonlinear Equations in Several Variables.* Academic Press, New York, 1970.

[521] J. E. Osborn. Spectral approximation for compact operators. *Math. Comput.*, 29(131):712–725, 1975.

[522] F. Pacard. A regularity criterion for positive weak solutions of $-\Delta u = u^{\alpha}$. *Comment. Math. Helv.*, 68(1):73–84, 1993.

[523] B. N. Parlett. *The Symmetric Eigenvalue Problem.* Prentice Hall Series in Computational Mathematics. Prentice Hall, Upper Saddle River, NJ, 1980.

[524] B. N. Parlett, D. R. Taylor, and Z. A. Liu. A look-ahead Lanczos algorithm for unsymmetric matrices. *Math. Comp.*, 44:105–124, 1985.

[525] H. O. Peitgen, D. Saupe, and K. Schmitt. Nonlinear elliptic boundary value problems versus their finite difference approximations: numerically irrelevant solutions. *J. Reine Angew. Math.*, 322:74–117, 1981.

[526] V. Pereyra. Iterated deferred corrections for nonlinear operator equations. *Numer. Math.*, 10:316–323, 1967.

[527] V. Pereyra, W. Proskurowsky, and O. Widlund. High order fast Laplace solvers for the Dirichlet problem on general regions. *Math. Comp.*, 31:1–16, 1977.

[528] W. Petryshyn. On the approximation-solvability of nonlinear equations. *Math. Ann.*, 177:156–164, 1968.

[529] W. Petryshyn. On the aproximation-solvability of equations involving a-proper and pseudo-a-proper mappings. *Bull. Amer. Math. Soc.*, 81:223–312, 1975.

[530] W. Petryshyn. Solvability of linear and quasilinear elliptic boundary value problems via the a-proper mapping theory. *Numer. Funct. Anal. Optim.*, 2:591–635, 1980.

[531] W. Petryshyn. Variational solvability of quasilinear elliptic boundary value problems at resonance. *Nonlin. Anal.*, 5:1095–1108, 1981.

[532] W. Petryshyn. Approximation-solvability of periodic boundary value problems via the a-proper mapping theory. In *Nonlinear Functional Analysis and its Applications*, Proc. Sympos. Pure Math., volume 45, 261–282. AMS, Providence, 1986.

[533] M. Petzoldt. A posteriori error estimators for elliptic equations with discontinuous diffusion coefficients. *Adv. Comput. Math.*, 16:47–75, 2002.

[534] J. Pousin and J. Rappaz. Consistency, stability, a priori and a posteriori errors for Petrov- Galerkin methods applied to nonlinear problems. *Numer. Math.*, 69(2):213–231, 1994.

[535] A. D. Poyanin and V. F. Zaitsev. *Handbook of Nonlinear Partial Differential Equations*. Chapman and Hall/CRC Press, Boca Raton, 2004.

[536] A. Quarteroni. *Numerical Modelling for Differential Problems (Modellistica numerica per problemi differenziali)*, 2nd edn. Springer, Milan, 2003.

[537] A. Quarteroni and A. Valli. *Numerical Approximation of Partial Differential Equations*, 2nd edn. Springer, Berlin, 1997.

[538] T. Raasch. *Adaptive wavelet and frame schemes for elliptic and parabolic Equations*. Dissertation, Philipps-Universität Marburg, 2007.

[539] T. Raasch. Private communication about lectures, 2008.

[540] H. H. Rachford and M. F. Wheeler. An H^{-1} Galerkin procedure for the two-point boundary value problem. In C. de Boor, editor, *Mathematical Aspects of Finite Element Methods in Partial Differential Equations*, 353–382. Academic Press, New York, 1975.

[541] R. Rannacher. Defect correction techniques in the finite element method. In M. Chipot *et al.*, editors, *Progress in Partial Differential Equations: The Metz surveys*. Papers from the "Metz days" conferences, held in Metz, France, during 1989–1990. Pitman Res. Notes Math. Ser. 249, 184–200. Longman, Harlow, 1991.

[542] J. Rappaz and G. Raugel. Finite-dimensional approximation of bifurcation problems at a multiple eigenvalue. Rapport n° 71, Centre de Mathématiques appliquées, École Polytechnique Palaiseau, 1981.

[543] J. Rappaz and G. Raugel. Approximations of double bifurcation points for nonlinear eigenvalue problems. In J. Whiteman, editor, *Mafelap 1981*, 453–461. Academic Press, New York, 1982.

[544] G. Raugel. Approximation numérique de problemes nonlinéaires, Universitée de Rennes I, France, Thesis, 1985.

[545] P. A. Raviart. Sur l'approximation de certaines équations d'evolution linéaires et nonlinéaires. *J. Math. Pures Appl.*, 46:11–107, 1967.

[546] W. H. Reed and T. R. Hill. Triangular mesh methods for the neutron transport equation. Technical Report LA–UR–73–479, Los Alamos Scientific Laboratory, 1973.

[547] H. J. Reinhardt. *Analysis of Approximation Methods for Differential and Integral Equations*. Springer, Berlin, Heidelberg, Tokyo, New York, 1985.

[548] H.-J. Reinhardt. Defect corrections with finite element methods for singular perturbation problems. In *Computational and Asymptotic Methods for Boundary and Interior Layers*, Proc. BAIL II Conf., Dublin 1982, Boole Press Conf. Ser. 4, 348–356, 1982.

[549] H.-J. Reinhardt. On a principle of direct defect correction based on a-posteriori error estimates. In *Defect Correction Methods. Theory and Applications, Comput. Suppl.* 5:43–66, 1984.

[550] W. C. Rheinboldt. *Numerical Analysis of Parameterized Nonlinear Equations.* John Wiley and Sons, New York, 1986.

[551] W. C. Rheinboldt and J. V. Burkardt. A locally parameterized continuation process. *ACM Trans. Software*, 9:215–235, 1983.

[552] B. Rivière and M. F. Wheeler. A discontinuous Galerkin method applied to nonlinear parabolic equations. In B. Cockburn, G. E. Karniadakis, and C.-W. Schu, editors, *Discontinuous Galerkin Methods. Theory, Computation and Applications.*, Lect. Notes Comput. Sci. Eng., volume 11, 231–244. Springer, Berlin, 2000.

[553] B. Rivière, M. F. Wheeler, and V. Girault. Part i: Improved energy estimates for interior penalty, constrained and discontinuous Galerkin methods for elliptic problems. *Comput. Geosci.*, 3:337–360, 1999.

[554] B. Rivière, M. F. Wheeler, and V. Girault. Part ii: Discontinuous Galerkin method aplied to a single phase flow in porous media. *Computational Geosciences*, 4(4):337–349, 2000.

[555] B. Rivière, M. F. Wheeler, and V. Girault. A priori error estimates for finite element methods based bases on discontinuous approximation spaces for elliptic problems. *SIAM J. Numer. Anal.*, 39:902–931, 2001.

[556] S. Rjasanov and O. Steinbach. The fast solution of boundary integral equations. Mathematical and Analytical Techniques with Applications to Engeneering, Springer, Berlin, 2007.

[557] C. Rogers and W. F. Ames. *Nonlinear Boundary Value Problems in Science and Engineering.* Academic Press, New York, San Francisco, London, 1989.

[558] J. Rossmann. The asymptotics of the solutions of linear elliptic variational problems in domains with edges. *Z. Anal. Anwend.*, 9(6):565–578, 1990.

[559] H.-G. Roos, M. Stynes, and L. Tobiska. *Numerical Methods for Singularly Perturbed Differential Equations.* Springer, Berlin, New York, 1996.

[560] W. Rudin. *Principles of Mathematical Analysis.* International Series in Pure and Applied Mathematics. McGraw-Hill, New York, 1976.

[561] T. Runst and W. Sickel *Sobolev Spaces of Fractional Order, Nemitskij Operators and Nonlinear Partial Differential Equations.* Walter de Gruyter, Berlin, New York, 1996.

[562] Y. Saad. *Numerical Methods for Large Eigenvalue Problems.* Algorithms and Architectures for Advanced Scientific Computing. Manchester University Press, Manchester, 1992.

[563] Y. Saad. Variants of Krylov subspace methods for eigenvalue problems, IMA 97 Minnesota, Workshop on Large-Scale Dynamical Systems, 1997.

[564] A. A. Samarskii. *Theorie der Differenzen Verfahren.* Geest und Portig, Leipzig, 1984.

[565] A. A. Samarskii. *The Theory of Difference Schemes.* Pure and Applied Mathematics 24. Marcel Dekker, New York, 2001.

[566] F. C. G. Santos. The linearization method for the numerical analysis of finite element solutions to quasi-linear elliptic partial differential equations. *SIAM J. Numer. Anal.*, 38:227–266, 2000.

[567] N. Sassmannshausen and K. Böhmer. Petrov–Galerkin methods for projected linear operator equation, stability and convergence. *DFG-Schwerpunktprogramm Danse*, 1998. Preprint.

[568] R. Schaback. Unsymmetric Meshless Methods for Operator Equations, *Numer. Math.*, 114, 629-651, 2010.

[569] R. Schaback and H. Wendland. Inverse and saturation theorems for radial basis functions interpolation. *Math. Comp.* 71, 669–681, 2001.

[570] M. Schechter. Semilinear boundary value problems of the strong resonance type. *Topol. Meth. Nonlin. Anal.*, 2(2):189–202, 1993.

[571] R. Schneider. *Multiskalen- und Wavelet-Matrixkompression. Analysisbasierte Methoden zur effizienten Lösung großer vollbesetzter Gleichungssysteme.* Habilitationsschrift, TH Darmstadt, 1995.

[572] H. M. Schultz. *Spline Analysis.* Prentice Hall, Upper Saddle River, NJ, 1973.

[573] L. L. Schumaker. On super splines and finite elements. *SIAM J. Numer. Anal.*, 26:997–1005, 1989.

[574] R. Schumann and E. Zeidler. The finite difference method for quasilinear elliptic equations of order 2m. *Numer. Funct. Anal. Optimiz.*, 1:161–194, 1979.

[575] Ch. Schwab. *p– and hp– Finite Element Methods.* Clarendon Press, Oxford, 1998.

[576] H. R. Schwarz. *Methode der finiten Elemente.* Teubner, Stuttgart, 1980.

[577] A. Schwarzer. *Skalierungstechniken für k-bestimmte Verzweigungsprobleme und Iterationsmethoden für grosse, dünn besetzte Eigenwertprobleme.* PhD thesis, Philipps-Universität Marburg, 1997.

[578] L. R. Scott and S. Zhang. Finite element interpolation of nonsmooth functions satisfying boundary conditions. *Math. Comp.*, 54:483–493, 1990.

[579] R. Sebastian. *Anwendung von Krylov–Verfahren auf Verzweigungs- und Fortsetzungsprobleme.* Dissertation am Fachbereich Mathematik, Universität Marburg, 1995.

[580] R. Seydel. Numerical computation of branch points in ordinary differential equations. *Numer. Math.*, 32:51–68, 1979.

[581] R. Seydel. *From Equilibrium to Chaos, Practikal Bifurcation and Stability Analysis.* Elsevier Science Publishing Co., Inc., New York, Amsterdam, London, 1988.

[582] H. Shahgholian. $C^{1,1}$ regularity in semilinear elliptic problems. *Commun. Pure Appl. Math.*, 56(2):278–281, 2003.

[583] Z.-C. Shi. The F-E-M test for nonconforming finite elements. *Math. Comp*, 49:391–405, 1987.

[584] Z.-C. Shi. On the nine degree quasi conforming finite elements. *Numerica Mathematica Sinica*, 10:100–106, 1988.

[585] Z.-C. Shi. On Stummel's examples to the patch test. *Computational Mechanics*, 5:81–87, 1989.

[586] Z.-C. Shi and T. Ushijima (eds.). *Numerical mathematics. Proceedings of the 1st China-Japan joint seminar, August 24–29, 1992, Beijing, P. R. China.* Series on Applied Mathematics. 5. Singapore: World Scientific. viii, 1993.

[587] Z.-C. Shi and M. Wang. Mathematical theory of some non-standard finite element methods. *Contemporary Mathematics*, 163:111–125, 1994.

[588] G. Shortley and R. Weller. On the accuracy of thr numerical solution of Laplace's equation. *J. Appl. Phys.*, 9:334–348, 1938.

[589] R. E. Showalter. *Monotone Operators in Banach Spaces and Nonlinear Partial Differential Equations.* AMS Providence, 1997.

[590] I. Skrypnik. *Nonlinear Elliptic Equations of Higher Order in* Russian. Naukova Dumka, Kiev, 1973.

[591] I. Skrypnik. *Nonlinear Elliptic Boundary Value Problems.* Teubner, Leipzig, 1986.

[592] D. Sloan, Endre Süli, and Stefan Vandewalle, editors. *Numerical Analysis 2000*, volume 7: *Partial Differential Equations.* Elsevier, Amsterdam, 2001.

[593] P. Šolín, K. Segeth, and I. Doležel. *High-Order Finite Element Methods.* Chapman and Hall/CRC Press, Boca Raton, 2003.

[594] E. Stein and G. Sagar. Theory and finite element computation of cyclic martensitic phase transformation at finite strain. *Int.J.Numer.Meth.Engng*, 74(1):1–31, 2008.

[595] O. Steinbach. *Numerische Näherungsverfahren für elliptische Randwertprobleme. Finite Elemente und Randelemente.* Advances in Numerical Mathematics. Teubner, Stuttgart, 2003.

[596] H. Stetter. *Analysis of Discretization Methods for Ordinary Differential Equations.* Springer, Berlin, Heidelberg, New York, 1973.

[597] H. J. Stetter. The defect correction principle and discretization methods. *Numer. Math.*, 29:425–443, 1978.

[598] R. Stevenson. Adaptive solution of operator equations using wavelet frames. SIAM J. Numer. Anal. 41(3):1074–1100, 2003.

[599] R. Stevenson. On the compressibility of operators in wavelet coordinates. SIAM J. Numer. Anal. 35(5):1110–1132, 2004.

[600] R. Stevenson. Optimality of a standard adaptive finite element method. *Found. Comput. Math.*, 7(2):245–269, 2007.

[601] R. Stevenson and M. Werner. A multiplicative Schwarz adaptive wavelet method for elliptic boundary value problems. Math. Comput. 78(266):619–644, 2009.

[602] J. Stoer. *Numerische Mathematik I.* Springer, Heidelberg, 1976.

[603] J. Stoer and R. Bulirsch. *Introduction to Numerical Analysis*, 2nd edn. Springer, Berlin, Heidelberg, 1992.

[604] J. C. Strikwerda. *Finite Difference Schemes and Partial Differential Equations*, 2nd edn. SIAM, Philadelphia, 2004.

[605] J. C. Strikwerda, B. A. Wade, and K. P. Bube. Interior regularity estimates up to the boundary for elliptic systems of difference equations. *SIAM J. Numer. Anal.*, 27:292–322, 1990.

[606] J. C. Strikwerda. *Finite Difference Schemes and Partial Differential Equations.* Wadsworth & Brooks/Cole Mathematics Series. Pacific Grove, CA, 1989.

[607] F. Stummel. Diskrete Konvergenz linearer Operatoren I. *Math. Ann.*, 190:45–92, 1970.

[608] F. Stummel. Diskrete Konvergenz linearer Operatoren II. *Math. Z.*, 120:231–264, 1971.

[609] F. Stummel. Diskrete Konvergenz linearer Operatoren III. In *Linear Operators and Approximation (Proc. Conf. Oberwolfach, 1971)*, 196–216, Basel, 1971. Birkhäuser. Proc. Oberwolfach, ISNM 20.

[610] F. Stummel. Stability and discrete convergence of differentiable mappings. *Rev. Roum. Math. Pures e. Appl.*, 21:63–96, 1976.

[611] F. Stummel. The generalized patch test. *SIAM J. Numer. Anal.*, 16:449–471, 1979.

[612] E. Süli. Convergence and nonlinear stability of the Lagrange–Galerkin method for the Navier–Stokes equations. *Numer. Math.*, 53:459–483, 1988.

[613] E. Süli, B. S. Jovanovic, and L. D. Ivanovic. Finite difference approximations of generalised solutions. *Mathematics of Computation*, 45(172):319–327, 1985.

[614] J. Sun. Numerical computation of the sensitivities of multiple eigenvalues. *Math. Numer. Sin.*, 14(1):10–19, 1992.

[615] S. Sun. *Discontinuous Galerkin methods for reactive transport in porous media.* PhD thesis, University of Texas, Austin, 2003.

[616] S. Sun and M. F. Wheeler. Symmetric and nonsymmetric discontinuous Galerkin methods for reactive transport in porous media. *SIAM J. Numer. Anal.*, 43(1):195–219, 2005.

[617] M. E. Taylor. *Pseudo Differential Operators.* Princeton University Press, Princeton, NJ, 1981.

[618] M. E. Taylor. *Partial Differential Equations I: Basic Theory.* Springer, New York, Berlin, Heidelberg, 1996.

[619] M. E. Taylor. *Partial Differential Equations II: Qualitative Studies of Linear Equations.* Springer, New York, Berlin, Heidelberg, 1996.

[620] M. E. Taylor. *Partial Differential Equations III: Nonlinear Equations.* Springer, New York, Berlin, Heidelberg, 1996.

[621] R. Temam. *Navier–Stokes Equations, Theory and Numerical Analysis.* Studies in Mathematics and its Applications 2. North-Holland, Amsterdam, New York, Oxford, 1977.

[622] R. Temam. *Navier–Stokes Equations, Theory and Numerical Analysis*, 2nd edn. Studies in Mathematics and its Applications. North-Holland, Amsterdam, 1979.

[623] R. Temam. Navier–Stokes equations and nonlinear functional analysis. CBMS-NSF-Regional Conference Series in Applied Mathematics. SIAM, Philadelphia, 1983.

[624] R. Temam. *Navier–Stokes Equations, Theory and Numerical Analysis*, 3rd rev. edn, Studies in Mathematics and its Applications 2. North-Holland, Amsterdam, 1984.

[625] M. Thoma. *Konvergenzverbesserung durch Defektkorrektur in der Methode der finiten Elemente für elliptische Randwertaufgaben. (Convergence acceleration by defect correction in the finite element method for elliptic boundary value problems).* PhD thesis, München: Univ. München, Fak. f. Math., Diss. 100 S., 1991.

[626] J. W. Thomas. *Numerical Partial Differential Equations. Conservation Laws and Elliptic Equations.* Texts in Applied Mathematics 33. Springer, New York., 1999.

[627] V. Thomée and B. Westergren. Elliptic difference equations and interior regularity. *Numer. Math.*, 11:196–210, 1968.

[628] A. Thrun. Über die Konstanten in Poincaréschen ungleichungen. Technical report, Fakultät für Mathemetik, Ruhr Universität Bochum, Germany, 2003.

[629] T. Tiihonen. Shape calculus and finite element method in smooth domains. *Math. Comput.*, 70(233):1–15, 2001.

[630] K. Toyonaga, M. T. Nakao, and Y. Watanabe. Verified numerical computations for multiple and nearly multiple eigenvalues of elliptic operators. *J. Comput. Appl. Math.*, 147(1):175–190, 2002.

[631] L. N. Trefethen. *Spectral Methods in MATLAB.* SIAM, Philadelphia, 2000.

[632] F. Treves. *Basic Linear Partial Differential Equations.* Academic Press, London, 1975.

[633] N. M. Tri. On the Gevrey regularity of solutions of a class of semilinear elliptic degenerate equations on the plane. *J. Math. Sci., Tokyo*, 9(2):217–255, 2002.

[634] H. Triebel. *Theory of Function Spaces.* Geest and Portig, Leipzig, 1983.

[635] H. Triebel. *Theory of Function Spaces, II.* Birkhäuser, Basel, 1992.

[636] H. Triebel. A note on wavelet bases in function spaces. In Zbigniew Ciesielski *et al.*, editors, Orlicz centenary volume. Proceedings of the conferences "The Władysław Orlicz centenary conference" and "Function spaces VII", Poznań, Poland, 21–25 July 2003. Volume I: *Plenary Lectures.* Banach Center Publications 64, 193–206. Polish Academy of Sciences, Institute of Mathematics, Warsaw, 2004.

[637] H. Troger and A. Steindl. *Nonlinear Stability and Bifurcation Theory: An Introduction for Engineers and Applied Scientists.* Springer, Vienna, New York, 1991.

[638] M.-U. Tyn and L. Debnath. *Partial Differential Equations for Scientists and Engineers*, 3rd edn. Elsevier Science Publishing, Amsterdam, 1987.

[639] M. Uchida. Regularity of weak solutions of semilinear elliptic differential equations. *J. Math. Sci., Tokyo*, 8(2):357–363, 2001.

[640] K. Urban. *Wavelets in Numerical Simulation: Problem Adapted Construction and Applications*, Springer Lecture Notes in Computational Science and Engineering 22. Springer, Berlin, 2002.

[641] K. Urban. *Wavelet Methods for Elliptic Partial Differential Equations.* Oxford University Press, Oxford, 2008.

[642] T Ushijima, Z.-C. Shi, and T. Kako (eds.). *Advances in numerical mathematics. Proceedings of the 2nd Japan-China seminar on numerical mathematics, held in August 22-26, 1994 at the University of Electro-Communications in Tokyo, Japan.* Lecture Notes in Numerical and Applied Analysis. 14. Tokyo: Kinokuniya. viii, 278 p. , 1995.

[643] M. M. Vainberg. *Variational Method and Method of Monotone Operators in the Theory of Nonlinear Equations, trans. A. Libin, trans. ed. D. Louvish.* A Halsted Press Book. New York-Toronto: John Wiley & Sons; Jerusalem-London: Israel Program for Scientific Translations. XI, 1973.

[644] G. Vainikko. *Funktionsanalysis der Diskretisierungsmethoden.* Teubner Texte zur Mathematik. Teubner, Leipzig, 1976.

[645] G. M. Vainikko. Galerkin's perturbation method and the general theory of approximate methods for non–linear equations. *Zh. Vychisl. Mat. Mat. Fiz.*, 7:723–751, 1967.

[646] G. M. Vainikko. Discrete compact sequences. *Zh. Vychisl. Mat. Mat. Fiz.*, 14:572–583, 1974.

[647] G. M. Vainikko. Über die Konvergenz der Differenzenmethode im Problem der periodischen Lösungen gewöhnlicher Differentialgleichungen. *Zh. Vychisl. Mat. Mat. Fiz.*, 15:87–100, 1975.

[648] G. M. Vainikko and E. E. Tamme. Konvergenz der Differenzenmethode im Problem der periodischen Lösungen von Gleichungen elliptischen Typs. *Zh. Vychisl. Mat. Mat. Fiz.*, 16:652–664, 1976.

[649] G. M. Vainikko and E. E. Tamme. Convergence of a difference method in the problem concerning the periodic solutions of an elliptic equation. 1977.

[650] M. Vanmaele and Roger Van Keer. An operator method for a numerical quadrature finite element approximation for a class of second-order elliptic eigenvalue problems in composite structures. *Math. Model. Numer. Anal.* 29(3), 1995.

[651] M. Vanmaele and A. Zenísek. External finite-element approximations of eigenfunctions in the case of multiple eigenvalues. *J. Comput. Appl. Math.*, 50(1-3):51–66, 1994.

[652] A. Veeser and R. Verfürth. Poincaré constants for finite element stars. Technical report, Fakultät f ür Mathemetik, Ruhr Universität Bochum, Germany, 2008.

[653] R. Verfürth. A posteriori error estimators for the Stokes equations. *Numer. Math.*, 55:309–325, 1989.

[654] R. Verfürth. A posteriori error estimation and adaptive mesh-refinement techniques. *J. Comput. Appl. Math.*, 50:67–83, 1994.

[655] R. Verfürth. *A Review of a Posteriori Error Estimation and Adaptive Mesh-Refinement Techniques.* Wiley-Teubner, Chichester, 1996.

[656] R. Verfürth. A note on polynomial approximation in Sobolev spaces. *RAIRO Model Numer. Anal.*, 33:715–719, 1999.

[657] R. Verfürth. A posteriori error estimates for nonlinear problems. $L^r(0,T;L^\rho(\Omega))$-error estimates for finite element discretizations of parabolic equations. *Math. Comput.*, 67(224):1335–1360, 1998.

[658] W. von Wahl. Über semilineare elliptische Gleichungen mit monotoner Nichtlinearität. *Nachr. Akad. Wissensch. Göttingen*, 3:27–33, 1975.

[659] W. von Wahl. Die stationären Gleichungen von Navier–Stokes und semilineare elliptische Systeme. *Am. J. Math.*, 100:1173–1184, 1978.

[660] H. Weber. A singular multi-grid iteration method for bifurcation problems. In T. Küpper, H. D. Mittelmann, and H. Weber, editors, *Numerical Methods for Bifurcation Problems*, volume 70 of *ISNM*, 547–561, Birkhäuser, Basel, 1984.

[661] H. Weber. Muligrid bifurcation iteration. *SIAM J. Numer. Anal*, 22:262–279, 1985.

[662] H. Wendland. *Scattered Data Approximation*, Cambridge Monographs on Applied and Computational Mathematics 17. Cambridge University Press, Cambridge, 2005.

[663] M. Werner. *Adaptive wavelet frame domain decomposition methods for elliptic operators.* Dissertation, Philipps-Universität Marburg, 2009.

[664] B. S. Westcott. *Shaped Reflector Antenna Design.* Research Studies Press, Letchworth, Herdfordshire, 1983.

[665] M. F. Wheeler. An elliptic collocation–finite element method with interior penalties. *SIAM J. Numer. Anal.*, 15:152–161, 1978.

[666] J. Wloka. *Funktionalanalysis und Anwendungen.* De Gruyter, Berlin, New York, 1971.

[667] J. Wloka. *Partielle Differentialgleichungen.* Teubner, Stuttgart, 1982.

[668] J. Wloka. *Partial Differential Equations,* trans. C. B. and M. J. Thomas. Cambridge University Press, Cambridge, 1987.

[669] P. Wojtaszczyk. *A Mathematical Introduction to Wavelets.* Cambridge University Press, Cambridge, 1997.

[670] C. Xu. Identity features of difference scheme and finite element scheme on rectangular grid. *J., Huazhong (Cent. China) Univ. Sci. Technol.*, 17(4):111–119, 1989.

[671] D. Yang. Mixed methods with dynamic finite-element spaces for miscible displacement in porous media. *J. Comput. Appl. Math.*, 30(3):313–328, 1990.

[672] X. Yang. The existence and regularity of solutions of a semilinear obstacle problem of higher order. *Math. Appl.*, 5(1):67–73, 1992.

[673] Z. Yulin. *Application of Discrete Functional Ananlysis to the Finite Difference Method.* International Academic Publishers, A Pergamon-CNPIEC Joint Venture, 1991.

[674] E. Zarantonello. Solving functional equations by contractive averaging. MRC Report No. 160, University of Wisconsin/Madison, 1960.

[675] E. Zeidler. *Nonlinear Functional Analysis and its Applications*, volume III: *Variational Methods and Optimization.* Springer, New York, Berlin, Heidelberg, London, Paris, Tokyo, 1985.

[676] E. Zeidler. *Nonlinear Functional Analysis and its Applications*, volume I: *Fixed-point Theorems.* Springer, New York, Berlin, Heidelberg, London, Paris, Tokyo, 1986; 2nd edn 1992.

[677] E. Zeidler. *Nonlinear Functional Analysis and its Applications*, volume II/A: *Linear Monotone Operators.* Springer, New York, Berlin, Heidelberg, London, Paris, Tokyo, 1990.

[678] E. Zeidler. *Nonlinear Functional Analysis and its Applications*, volume II/B: *Nonlinear Monotone Operators.* Springer, New York, Berlin, Heidelberg, London, Paris, Tokyo, 1990.

[679] A. Zenísek. Polynomial approximation on tetrahedrons in the finite element method. *J. Approx. Theory*, 7:334–351, 1973.

[680] A. Zenísek. Curved triangular C^m–elements and their applications. In Plzen Proceedings of the Fourth Symposium on *Basic Problems of Numerical Mathematics, CSSR 1978, pp. 183–194*, 1978.

[681] A. Zenísek. *Nonlinear Elliptic and Evolution Problems and their Finite Element Approximations.* Academic Press, London, 1990.

[682] A. Zhang. Regularity for weak solutions of semilinear elliptic systems with critical exponents. *Appl. Math., Ser. A (Chin. Ed.)*, 11(1):21–26, 1996.

[683] O. Zubelevich. On regularity of stationary solutions to the Navier–Stokes equation in 3-D torus. *Lobachevskii J. Math.*, 17:257–261, 2005.

Index

$(-1)_{j>0} := 1 \text{ for } j = 0 \text{ else } := -1$, 40
$\alpha = (\alpha_1, \ldots, \alpha_n) \in \mathbb{N}_0^n$, 54
$\nabla^0 u = u$, 55
$\nabla u = \nabla^1 u$, 54
$\vartheta = (\vartheta^1, \ldots, \vartheta^n) \in \mathbb{R}^n$, 54
(A_i, B_i), 68
$(\cdot, \cdot)_n$, 37
$(\cdot, \cdot)_n$ Euclidean scalar product, 456
$(u, v)_{H^1(\Omega)}$, 36
$(u, v)_{H^2(\Omega)}$, 36
$(u, v)_{H^m(\Omega, \mathbb{R}^q)}$, 119
$\| u \|_{C^k(\Omega)}$, 58
$\| u \|_{C^{k,\gamma}(\Omega)}$, 58
$\| u \|_{H^i(\Omega)}$, 36
$\|u\|_{W^{m,p}(\Omega, \mathbb{R}^q)}$, 119
$\| v \|_{H^i(\Omega)}, i = 0, 1, \ldots$, 36
$\| v \|_{W^{m,p}(\Omega, \mathbb{R}^q)}$, 119
A, 37
A^*, 47
A^d, 47
A_c, 37
A_p, 40
A_s, 37, 40
B, 41
B_a, 41
$C^{2m+s,\gamma}(\Omega)$, 58
$D^{\leq k}$, 55
D^k, 55
h, 293
$H^1(\Omega)$, 36
$H^2(\Omega)$, 36
$H^m(\Omega, \mathbb{R}^q)$, 119
$H_k^{-1}(\mathcal{T}_c^h)$, 351
$H^{-m}(\Omega) =??H_0^{-m}(\Omega)???$, 39
P^h, Q'^h, 591
$W^{2m+s,p}(\Omega)$, 58
$W^{m,p}(\Omega, \mathbb{R}^q)$, 119
$W^{s,p}(\partial\Omega)$ Sobolev space, 25
% Euclidean product, 43
$\mathbb{C}$: complex numbers, 36
Δu, 36, 119
$\mathbb{N}$: natural numbers $\{1, 2, \ldots\}$, 36
$\mathbb{N}_0$: $\{0\} \cup \mathbb{N} = \{0, 1, 2, \ldots\}$, 36
Ω, 36, 119
$\Omega_0 \in C^{0,1}$, 75
$\Omega_0 \subset\subset \Omega_1$ for open Ω_0, Ω_1 means $\bar{\Omega_0} \subset \Omega_1$, 75
$\mathbb{Q}_+$: positive rational numbers, 36
$\mathbb{Q}$: rational numbers, 36
$\mathbb{R}_{+,0} = \mathbb{R}_+ \cup \{0\}$, 140
$\mathbb{R}$: real numbers, 36
$\mathbb{R}^{N_k}$, 119
$\mathbb{R}^{n_k}$, 119
$\mathbb{R}_+$: positive real numbers, 36
$\mathbb{Z}$: integer numbers, 36
α, 55
$\bar{\Omega}$, 19
$\mathcal{L}(\mathcal{U}, \mathcal{V}')$, 46
$\mathcal{N}(A)$, 46
$\mathcal{NL}(\mathcal{U}_b, \mathcal{V}_b')$, 184
$\mathcal{O}(x^\alpha)$, 5
$\mathcal{O}(h^p)$, 229
$\mathcal{R}(A)$, 46
$\mathcal{U}_{\mathcal{T}^h} = W^{k,p}(\mathcal{T}^h)$, 231
$\mathcal{V} = H_0^m(\Omega, \mathbb{R}^q)$−coercivity and ellipticity, 121
$\mathcal{Q}$ set of all vertices $\mathcal{T}$, 235
$\delta_{ij} = 1 \text{for} i = j$, else$= 0$, 213
$\lesssim$, 233
$\mathbf{v}_\epsilon$ with few nonzero entries, 675
div $\vec{w}$, 114
index(A), 60
$\nabla^k, \Theta^k \in \mathbb{R}^{n_k}$, 54
$\nabla^{\leq k}, \Theta^{\leq k} \in \mathbb{R}^{N_k}$, 55
∂u, 54
$\partial\Omega$, 36, 119
  curved, 395
  polygonal, 395
$\partial^\alpha u$, 54
$\partial^i \partial^j u$, 36, 119
$\partial^i u$, 36, 119
$\vartheta = (\vartheta^1, \ldots, \vartheta^n)^\mathsf{T} \in \mathbb{R}^n$, 40
ϑ^α, 54
ϑ^i, 54
$| \cdot |_n$, 37
$|\vartheta|_n = |\vartheta|$, 56
$|\alpha|$, 54
$| \cdot |_n$ Euclidean norm, 456
$a \lesssim b$, 638
$a \sim b$, 638
$a(\cdot, \cdot)$, 37
$a := b$: a is defined by b, 36
$a =: b$: b is defined by a, 36
$a \sim b$, 423, 638
$a^d(u, v)$, 47
$a_p(\cdot, \cdot)$, 40
$a_s(\cdot, \cdot)$, 37, 40
e^i, i-th unit vector, 30
$f \in W_0^{-m,p'}(\Omega)$, 63
$f \in H_k^{-1}(\Omega)$, 351
$f \in W_0^{-m,p'}(\Omega)$, 63

$\vec{f} \in W_0^{-1,p'}(\Omega, \mathbb{R}^q)$, 119
$\langle \vec{f}, \vec{v} \rangle_{\mathcal{V}' \times \mathcal{V}}$, 119
$o(x)$, 5
$o(h^p)$, 229
$v^h \notin C(\Omega) \Rightarrow v^h \notin H^1(\Omega)$, 372
$v_l, v_r, [v], \{v\}$, 462
$\mathcal{X} \hookrightarrow \mathcal{Y}$, 21
h.o.t., 5
$\mathcal{T}^h$ non–degenerate, 229

a posteriori error estimator, 669
a priori error bound, 424
a.e. =almost everywhere, 20, 40
adaptive approximation method, 669
  convergence, 670
  convergence rate, 670
adaptive Finite Element Method, 420
adaptive wavelet-technique, 636
advantage
  wavelet methods, 636
algorithm
  APPLY, 677
  COARSE, 674
  SOLVE, 682
almost everywhere (a.e.), 20, 40
analysis
  natural boundary conditions, 611, 612
anti-crime transformation, 190, 527
approximating spaces
  admissible Galerkin, 272
  bi-dual pair, 181
approximation
  cubature, 346
  edge singularities,boundary layers, 234
  external, 180
  five star, 566
  good by COARSE, few nonzero entries, 675
  internal, 180
  nonlinear $\mathbf{F}(\mathbf{v})$ by $\mathbf{w}_\epsilon$, 685
  quadrature, 305, 346
  saturated data approximation, 439
approximation projector, P^h, Q'^h, 184
approximation scheme
  external, 592
asymptotic expansion
  extrapolation, defect correction, 626
asymptotic optimality, 674
atlas, 23, 25
Aubin–Nitsche result, 664

Banach space
  approximate, 334
  reflexive, $\mathcal{V} = \mathcal{V}''$, 15, 52
basis
  biorthogonal, 644
  interpolation, 223, 367
  nodal, 213
  stable and local, 241
  stable local, 249
basis functions
  interior, 650
Besov space, 646, 673
best N-term approximation, 672
bifurcation
  multiple, 11
bilinear form, 157
  $\mathcal{V}_b^h$-coercive, 601
  $\mathcal{V}_b^h$-elliptic, 601
  approximate, 346
  bounded, 45
  coercive, 45, 50, 70, 121, 321, 637
  coercivity implies invertibility, 49
  coinciding strong and weak, 338
  continuous, 45
  difference weak, 576
  dual, 45
  elliptic, 50, 70, 121, 637
  induced linear operator, 45
  strong, 39, 66, 159, 160, 320, 337
  strong and weak form, 36
  symmetric, 46
  weak, 40, 66, 160, 320, 321, 337, 576
biorthogonality of wavelets + full polynomial exactness imply vanishing moments, 651
blue refinement, 436
bound
  upper a priori, 430
  lower a priori, 431
  a preori, 423
boundary
  curved, 250
    isoparametric, 253
    polynomial interpolation, 252
  parametrization, 24
boundary derivative
  complementary tangential, 75
boundary condition, 5, 8, 588
  (2.177) - (2.179), 83
  Dirichlet, 42, 56, 70, 101, 132, 145, 388, 397, 403, 414, 531, 546, 564, 594
  Dirichlet and/or Neumann, 531
  excluded tangential derivative, 42
  general, 56, 70, 75
  homogeneous Dirichlet, 258
  included, 637
  incorporated, 47
  modified, 75
  natural, 8, 41, 42, 71, 74, 388, 403, 614, 617
  natural, Neumann, 610
  Neumann, 8, 56, 70, 612
  results for $2m = 2$ and B_i in (2.185), (2.186), 83
  transformation into homogeneous, 57
  vanishing Dirichlet, 569
boundary layers, 423
boundary operator
  Dirichlet B_D, 56
  natural, 41

natural discrete, 614
principal part, 75
boundary value problem, 5, 8, 459
Dirichlet, 42, 56, 66, 86
existence, uniqueness, regularity, 67
first, 66
generalized Neumann, 66
generalized Robin, 66
linear, 472
natural, 44, 66
normal, 56, 59, 61
second, 67
semilinear, 474
third, 66
bounded operations, storage, nonzero entries by APPLY, 677
bounded bilinear form and **A**, 670
boundedness, 498
for the Laplacian, 495
SIPG, NIPG, IIPG, GIPG, 501
bounds for operations, storage by GROW, 680
bubble
edge, 432
element, 432

carr(σ), 25
carr u = supp u, 591
case
μ, 400
σ, 400
Cea's Lemma, 424
center of gravity, 450
Chladny sound figures, 10, 290
chunkiness parameter, 230, 428
co-ordinate system, 25
coercivity, 49, 94, 99, 387
$H_0^1(\Omega)$, 41
$H^m(\Omega)$, 71
$H_0^m(\Omega)$, 71
$H_0^m(\Omega) \Rightarrow W_0^{m,p}(\Omega), p < 2$, 155, 191
$H_0^m(\Omega) \not\Rightarrow W_0^{m,p}(\Omega), 2 < p$, 155, 191
$W_0^{m,p}(\Omega)$, 155, 191
$\mathcal{V}$, 121
$\mathcal{V}^h$ quasi-linear, 502
$\mathcal{V}^h$ for the Laplacian, 495
$\mathcal{V}_b^h$–, 603
discrete, 416
discrete, relations, 387
in $\mathcal{V}^h$ general linear, semi-linear case, 499
isoparametric, 416
modified, 105
NIPG, SIPG, IIPGMs, 497, 500
nonlinear, 103, 143
coercivity implies invertibility, 49
collar
Dirichlet BVP, 137
estimate, 137
norm, 137
compact, 14
pre-, 14
relatively, 14
complement
generalized orthogonal, 16
complexity of MAKECONFORM, 443
complexity of SOLVE, 446
compressibility, 675
computational complexity, 671
condition
uniform Legendre, 160
boundary and quadrature, 396
boundedness, 94
Brezzi-Babuska, 47
Carathéodory, 98, 103, 131, 141, 142, 144
Carathéodory type, 128
Carathéodory, growth, 145
complementary, 59, 82
controllable growth, 128, 478
ellipticity, 94
exterior sphere, 110
for linearization, 151
growth, 98, 103, 110, 141–144
controllable, 129
first, 94
fourth, 94
natural, 129
second, 94
third, 94
H, 304,307,311,317,341
Hm, 335,340
inf-sup-, 48
Ladyshenskaja–Babuska–Brezzi, 166, 283, 477
Ladyshenskaja–Babuska–Brezzi (LBB), 665
Legendre, 120
Legendre–Hadamard, 121, 128
modified strong Legendre, 136, 339
modified Brezzi–Babuska, 166
modified ellipticity, 157
monotone strict, 103, 143
monotone strict uniform, 103
monotone uniform, 103, 143
natural boundary, 66
natural growth, 128, 478
stability, consistency for DCGMs, 529
strong Legendre, 128, 131, 144
strong Legendre–Hadamard, 122, 133, 266
uniform Legendre–Hadamard, 161
violated boundary, 378
violated continuity, 378
condition numbers
uniformly bounded, 636
conforming finite elements, 428
conjecture, smooth splitted splines, 250
connected, 628
conservative
flux, 466
consistency, 174
J_h^σ, 491
ℓ^h, 505
ℓ_h, 506

consistency (*cont.*)
a_h for general linear, 511
a_h for the Laplacian, 507
c_h and b_h, 503
classical, 185, 187, 590
difference methods
natural boundary condition, 620
differentiable
quasilinear DCGMs, 536
differentiable, semilinear DCGMs, 537
for violated boundary conditions, 392
for violated Dirichlet conditions, 393
fully nonlinear, 599
general linear a_h,, 513
linear operator, 596
linearized quasilinear, 532
of order p in u, 187
quadrature for quasilinear problems, 363
quasilinear a_h in [409], 523
quasilinear equations, systems, 598
quasilinear system a_h, 518
quasilinear systems, 526
semilinear a_h, 513
variational, 185, 188
consistency error
variational, 392
consistent
flux-, 466
consistent differentiability, 416, 486, 610
constant
depending upon parameters, 212
Poincaré's, 21, 430
shape regularity, 239
continuation
embedding method, 206
continuity
Hölder, 20
Lipschitz, 20
modified Lipschitz, 514
uniform, 387
controllable and natural growth conditions, 128
convergence, 190, 268
adaptive FEM, 440
order p, 190, 268
radial basis function, 595
symmetric, unsymmetric finite difference method, 609
weak, 15
convergence theory
general for monotone operators, 277, 279
general for quasilinear operators, 277, 279
Navier–Stokes wavelet method, 669
saddle point wavelet method, 666
solution for monotone operators, 279
solution for quasilinear operators, 279
corner point, 421
corner singularity, 421
corner stone of modern analysis, 37
covering
open, 14
cubature formula
bivariate, 390
cuboid, 214

data approximation error, 439
DCGM
hp, 539, 540
$hp-$, convergence, 546
$hp-$variants, 538
coercive linearized principal part, 494
consistency, 503, 507, 511, 531
convergence, 527, 531
convergence of hp, 542
convergence via monotony, 531
estimate, inverse and interpolation error, 487
general form, 468, 473
generalization to systems, 482
geometry of the mesh, 486
IIPG form, 480
incomplete form, 468, 473
inverse estimate, 486
linearized consistency, 534
linearized IIPG form, 533
NIPG or SIPG or GIPG, 480
non symmetric form, 468, 473
nonstandard, 471, 530
numerical experiences, 546
standard, 471, 477, 529
symmetric form, 468, 473
derivative, 32
directional, 28
Frechét, 28
Gateaux, 28
inverse function, 31
normal, 10
partial, 36, 54, 230, 264, 638
relation:Frechét/Gateaux-, 29
weak, 20
diagram, 267
diffeomorphism
C^1, 31
local C^1, 31
difference
backward, 562, 564
forward, 562, 564
nonsymmetric, 562
normal, 564, 565
operator, 563, 568
partial, 566
symmetric, 563, 565
difference approximation
symmetric, 613
difference equation, 574
classical form, 575
coercivity for elliptic, 600
consistency, 593
divergence form, 574, 575

elliptic 2. order, 574
natural form, 575
partial summation, 575, 578
symmetric, 581, 615
symmetric form, 560, 573
unsymmetric form, 561
difference method
asymptotic expansion, 631
convergence, 588, 609
symmetric, unsymmetric method, 609
strong (classical) form, 609
curved boundaries, 622
elliptic
nonlinear order $2m$, 584
elliptic linear order 2
Dirichlet, 583
fully nonlinear order 2, 587
natural boundary conditions, 611, 613, 621
nonlinear order $2m$, 586
polynomial interpolation, 626
Shortley–Weller–Collatz, 623
simple examples, 562
solving nonlinear equation, 610
strong form, 609
symmetric, 581
unsymmetric/symmetric, 567
variational, 589
differentiability
partial, 30
differentiable
Frechét, 28
Gateaux, 28
consistent, 207, 344, 532
partially, 30
differential operator
linear
linear of order $2m$, 56, 62
linear system
order $2m$, 133
order $2m$, 70
dilemma the G, G^h, 196
Dirac measure, 449
Dirichlet system, 75
discontinuous Galerkin methods (DCGMs), 455
Discrete Brezzi–Babuska condition:, 272
discrete solution
quasilinear, unique, convergence, 275
unique existence,convergence, 269
discrete space
Lebesgue, 569
discretization
DCGM, 460, 469
quasilinear system, 478
discretization and preconditioning, 653
discretization errors
difference, 189
discretization method
linear, 269, 661
wavelets, 661
discretization schemes
inner and outer, 185
distance
A, B, 254
domain
curved, 461
Lipschitz, 23
partition of, 461
polygonal, 251
polyhedral, 459, 461
reference, 213
star-shaped, 229, 487
dual pair, 645
dual problem, 452
dualgenerator, 645
duality argument, 449

edge jump, 430
edge residual, 430
eigen
function, 290
space, 290
value problem linear operator, 289
eigenvalue, 11, 51, 60, 64, 73, 81, 83, 89, 169, 290
variational method, xxiii
eigenvalues, 109
element
C^r, 223
reference, 213
element residual, 430
elliptic, 94, 96
$H^m(\Omega)$, 71
$H_0^m(\Omega)$, 71
$\mathcal{V}$, 121
$\mathcal{V}_b^h$, 601
linear system order 2, 117
strongly, 56
strongly $2m-$, 136
uniformly, 56, 89
elliptic equation
quasilinear, nonlinear coercive, 106
elliptic difference equation
stability,convergence, 604
elliptic differential operator
order 2, 35
elliptic equation
existence quasilinear, 91
fully nonlinear order 2, 108
linear, order $2m$, 639
linear, order 2, 638
nonlinear, 77
nonuniformly, 91
quasilinear
divergent, 88, 100
non divergent, 88
order 2, 88
special quasilinear, 149
special semilinear, 149
uniformly, 91
weak, order $2m$, 69

elliptic operator
  characteristic polynomial, 56
  definition nonlinear, 79
  fully nonlinear, 81
  linear
    Fredholm alternative, 54
    order 2, 64
    order $2m$, 58
    regular solution, 54
  nonlinear, linearization, 147
  order $2m$, m, 80
  quasilinear, 80
  semilinear, 78, 80
  symbol, 56
elliptic problem
  general linear DCGM, 472
  general, Dirichlet condition, 572
  semilinear and quasilinear, 474
elliptic system, 114
  characteristic polynomial, 115
  divergent quasilinear order $2m$, 137
  fully nonlinear order $2m$, 146
  linear, nonlinear, 112
  local solvability, nonlinear, 147
  nonlinear order $2m$, 586
  order, 115
  quasilinear, 477
  quasilinear of order $2m$, 639
  quasilinear, variational method, 125
  symbol, 115
ellipticity
  $\mathcal{V}_b^h-$, 602
  fully nonlinear, 112
embedding, 21
  compact, 21, 50
  continuous, 21, 50
  dense, 50
equation
  Bratu, 78, 87
  differential, 5
  elliptic
    divergent quasilinear, 273
    fully nonlinear form, 306
    general quasilinear, 306
  fully nonlinear, xxiii, 151, 156
  Kármán, 137
  Lamé, 117
  linear order $2m$, 83
  minimal surface, 12, 103
  Monge-Ampère, 13, 109, 307
    regular solution, 111
  monotone convergence, xx
  Navier–Stokes, 13, 163, 211
  Navier–Stokes, steady compressible, 554
  non coercive, 148
  non divergent, xx
  non divergent quasilinear, 156
  nonuniformly elliptic, 35, 57, 90
  order $2m$, 336
  quasilinear
    divergent, 151
  reaction-diffusion, 12
  saddle point, 664
  Stokes, 114
  surface with Gauss curvature, 13, 110, 307
  uniformly elliptic, 35, 89
  von Kármán, 116
  von Kármán, numerical experiments, 632
equilibrium, 4
error
  (classical) consistency, 187, 268, 270, 590
  (classical) consistency, nonconf. FEMs, 270
  (global) discretization, 190
  (variational) consistency, 268
  boundary, 468
  classical consistency, 179, 189, 296, 373, 389, 401, 485, 662
  classical consistency=local discretization, 275
  classical discretization, 412
  conforming variational, 188, 189
  consistency, 187, 207, 417
  FE interpolation
    global estimates, 232
    local estimates, 232
    tensor FEs, 232
  interpolation, 189, 229, 241, 424
  local discretization, 187, 268, 270, 661, 662
  local discretization, nonconf. FEMs, 270
  nonconforming variational, 189
  nonconforming, variational,classical consistency, 270
  quantities of interest, 449
  variational consistency, 179, 186, 188, 269, 296, 389, 393, 401, 661, 662
  variational discretization, 412
  variational,classical consistency,conforming FEM, 269
error bound
  a priori, 423
  a preori, 423
error estimate
  a posteriori, 433
  other types, 447
  quantities of interest, 449
error representation formula, 431
errors
  wavelet consistency, vanish, 662
estimate
  *hp*inverse and error estimate, 540
  computable a posteriori, 438
  consistency, 414
  errors and support in **SOLVE**, 682
  interior, 65, 136
  inverse, 229, 232, 237, 240, 257
  inverse, nonquasiuniform triangulation, 233
  lower a posteriori error, 432
  operations, storage, support in **SOLVE**, 683
  point error, 449
  quantities of interest, 449
  upper a posteriori error, 430
estimator

other type, 447
eststimation
quantities of interest, 449
Euclidean product
in $\mathbb{R}^n$, 43
Euclidean scalar product and norm in $\mathbb{R}^n$, 37
Euler system
strong form, 127
weak form, 127
examples nonlinear PDEs, 10
excluded tangential derivative, 43, 75
existence, 145
interior, 65
expansion, 640
extension
bilinear form, 373
energy, 37
Friedrichs, 36
from Ω to Ω_0^h, 311
extremal problem, 49

FE, 213
$\mathcal{P}_{d-1}^n(\mathcal{T}^h)$, 223
affine equivalent, 221
approximating spaces, 222
approximation theory, 212
Argyris, 218
Bell, 218
bilinear, biquadratic, 219
condition
interpolation points on edges, 400
conforming, 218, 258
cubic Hermite, 227
discontinuous Crouzeix–Raviart, 177
equivalent, 221
affine, 221
example, 221, 258
Hermite, 215
interpolation, 221
interpolation error, 256
isoparametric, 251, 254
isoparametric polynomial, 253
Langrange, 215
linear Lagrange, 226
Mortar, 234
non–degenerate, 229
nonconforming, 189, 218, 368
nonconforming, conditions, 390
of degree, 223
order $2m$, 271
polynomial, 214
at least degree, 223
degree or order, 223
quadrilateral, hexahedral, 235
rectangular K, 214
simple computation, 219
smooth
curved domain, 238
polyhedral domain, 238
space, 221
spaces, updated, 237
standard case, 390
curved boundary, 396
discontinuous FEs, 400
subdivision, 213
tensor product, 232
triangular, 215
triangular K, 214
uniformly star-shaped, 229
violated boundary condition, 371, 392, 397
violated continuity, 399
violating continuity condition, 372
violating continuity, consistency, 403
FEM
hp, 448
adaptive, 434
bubble function, 284
conforming, 209, 210, 257, 274, 275, 282
conforming generalized Petrov–Galerkin, 199
conforming,basic idea, 258
consistency, 336
convergence, 336
$\| u_0 - u_0^h \|_{L^2(\Omega)}$, 272
linear, quasilinear problem, 271
convergence of adaptive, 439
convergence, adaptive, 438
convergence, quadrature with crimes, 411
crimes
linear problems, 380
quasilinear problems, 380
crimes, quasilinear, quadrature, 413
discrete Newton method, MIP, 276
equation
order $2m$, 332
general convergence theory, 266
general linear equations, 264
general linear systems, 264
impacts of variational crimes, 373
isoparametric, 414, 417
quasilinear, 418
linear, 276
linear + quadratic, 284
linear elliptic system
order 2, 266
order $2m$, 266
linear elliptic equation
order 2, 264
order $2m$, 265
linear, sparse equation, 342
main idea, 258
mixed, 211, 282
Mixed for Navier–Stokes operator, 286
mixed, nonlinear Navier–Stokes,convergence, 288
nonconforming, 210, 283, 297, 371
nonlinear boundary conditions, 345
equation, 345
system, 345

FEM (*cont.*)
notation, requirement, 428
optimal mesh, 451
quadrature
second order linear, 350
second order nonlinear, 357
systems order $2m$, 361
quadrature for nonconforming, 412
reducing costs, 262
second order, violating boundary condition and/or continuity, 411
semi-conforming, 308
semi-conforming, fully nonlinear, 309
solution of nonlinear equations, 413
sparse linear equations, 342
sparse nonlinear system, 342
stability, 336
stability nonconforming, 406
stable
linear, quasilinear problem, 271
stable refinement, triangulations in $\mathbb{R}^2$, 436
summary isoparametric second order problems, 415
system
fully nonlinear, 297
general quasilinear, 297
order $2m$, 332
variational crimes, 297, 368
consistency, 368
convergence, 368
example, 370
stability, 368
violating boundary conditions, 371
Fichera corner, 423
finite element, xv
flux
inviscid (Euler), 556
numerical, 465
viscous, 556
flux term, 459
form
bilinear, 41, 45
bounded bilinear, 41
bounded linear, bilinear, 14
continuous, 63
divergence, 63, 66, 86
linear
special representation, 63
strong, 36
strong quasilinear, 80
strong quasilinear difference, 625
weak, 37, 76, 86, 101, 132, 583
weak difference, 584, 585
formula
fundamental Green's, 37, 101
generalized Green's, 41
Green's, 37, 77, 459
Frechét derivative, 27
Fredholm alternative, 51, 52, 59, 66, 67, 73, 134, 151, 155, 157, 290
for order $2m$, 137
divergence form, 64
Fredholm-index, 18
fully nonlinear problem
existence, uniqueness, 110
order 2 and 2m, 108
function
cut-off, 230
Dirac delta, 213
mesh-size, 425
shape, 213
step, 591
functional
regular integral, 126

Galerkin approximation, 424
Galerkin method, 659
adaptive, 671
Gelfand triple, 50, 637
general approach, 451
general discretization method
DCGM, 482
geometric problem
quasilinear, fully nonlinear, 12
geometrically stable, 436
geometry of the mesh, 487
GIPG, 468
global Schwartz kernel, 654
Gram determinant, 235
green refinement, 436
Green's function
approximate, 449
Gårding inequality, 658

hanging node, 435
Hilbert space setting, 458
hyper plane
non–degenerate, 215

IIPG, 457, 468, 529
inclusion, 21
index, 52
index A, 82
inequality
Cauchy–Schwarz, 19
discrete Chebyshev, 18
discrete Hölder, 19
Gårding, 50, 70, 411, 608
Hölder, 19, 231, 349, 392
modified Poincaré–Friedrich, 397
multivariate Markov, 240
Sobolev, 571
inf–sup condition (BB), 664
integral
discrete, 569, 601
integral mean value, 571, 591
interior node property, 438
interior regularity, 124
interpolant

FE, simple inductive computation, 219
interpolation
error, 232
interpolation operator, 424, 429
invariance property, 563
inverse estimate, 432
invertibility, equibounded, xvi
isomorphism, 16
iteration
adaptive descent, 680
iterative method, 671

Jacobian, 235
jumping coefficients, 423

Ladyshenskaja–Babuska–Brezzi (LBB) condition, 666
Lagrange function, 451
Laplacian, 10
LDC, 477
Lebesgue-/Sobolev space
discrete, 570
lemma
Aubin–Nitsche, 272
Bramble-Hilbert, 230, 349
Cea, 259
generalized Aubin–Nitsche, 404
nonconforming Aubin–Nitsche, 405
Strang, 378
Liapunov-Schmidt method
numerical, 612
linear approximation method, 673
linear complexity, 674
linear conforming finite elements, 429
linear form
approximate, 346
special representation, 63, 260
linear operator
$\mathcal{V}_b^h$-regular, 601
$\mathcal{V}_b^h$-stable, 601
linearization
bounded, 155, 191
fully nonlinear systems, 161
linearized problem, 32
local error indicator, 433
local solvability and regularity, 112
locality, 239
longest node bisection, 437
lower bound
a posteriori, 431

mapping
isoparametric, 254
mapping properties, 235
mask, 641
matrix
M, 179
mass, 257, 262
real symmetric, 109
stiffness, 257, 262
symmetric, 81
weakly diagonal dominant, 179
matrix compression, 676
mean value
symmetric, 563
measure
positive, 71
mesh
anisotropically graded, 236
optimal, 425
mesh function, 237
mesh independence principle, xviii, 205, 561
mesh parameter, 234
mesh refinement
necessary nonuniform, 425
optimal, 425
uniform, 656
mesh sequence
shape-regular, 234
method
(linear) discretization, 183
adaptive wavelet, 678
optimal FEM, 442, 451
hp FEM, 448
all discretization
quasilinear equations+systems, monotone operators, 279
applicable discretization, 184, 316
applicable Galerkin, 272
approximation
conforming, 180
Galerkin, 180
generalized Petrov–Galerkin, 180
nonconforming, 180
Petrov–Galerkin, 180
conforming wavelet Galerkin, 635
DCGMs, 538
defect correction, 631, 632
deferred correction, 631
difference, xxiv
difference equations, 610
difference for eigenvalue problems, 609
difference for monotone operators, 609
difference for nonlinear boundary operators, 609
difference for quadrature approximations, 609
discontinuous Galerkin (DCGM), xvii, xxiv
discontinuous Galerkin (DCGMs), 212, 297, 455
discrete Newton, 207
eigenvalue problems, 531, 663
existence,convergence, 662
finite difference, xvii
finite element, xvii
approximation theory, 212
general interior penalty Galerkin (GIPG), 468
generalized gradient, 280

method (*cont.*)
incomplete interior penalty Galerkin (IIPG), 468
linear discretization, 184
local discontinuous Galerkin, 477
mesh free or radial basis, xvii
monotone operators, 531, 663
Newton–Kantorovich, 207
non symmetric interior penalty Galerkin (NIPG), 468, 529
nonconforming approximation Galerkin, 311
nonlinear boundary operators, 531, 663
other, nonlinear boundary conditions, 345
Petrov–Galerkin
generalized, 211
quadrature approximations, 531, 663
quasilinear problems, 660
Runge–Kutta, 185
space discretization, xxi, xxv
spectral, xvii
symmetric, 587
symmetric interior penalty Galerkin (SIPG), 468
wavelet, xvii, xxiv
minimal surface equation, 153
minimum
potential, 4
MIP, 205
mixed formulation, 667
model
nonlinear, 3
model problem for DCGMs, 459
model problem, 421
modulus of smoothness, 646
Monge -Ampere operator
$G(u^h)$ defined and continuous, 197
monotone, 99
strictly, 99
uniform, 103, 143, 274
multi-indices, 54, 230, 264, 638
multiresolution analysis, 640
multiresolution spaces
dual, 649
primal, 649
multiscale, 640

Navier–Stokes operator
linearization, 286
near best N-term approximation, 674
Nemyckii operator, 586
newest node bisection, 436
NIPG, 468, 529
coercivity linear, semilinear, 499, 500
node hanging, 436
nondegenerate triangulations, 328
nonuniqueness
interior, 65
norm, 69
broken Sobolev, 462
discrete dual, 569
discrete Sobolev, 569
discrete Sobolev,$\|u^h\|_{H^1_\pm(\Omega^h_e)}$, 614
dual, 15
dual Au, 48
energy, 41, 46
eqivalent to Sobolev, 21
equivalent, 41, 46
Lebesgue discrete, 569
penalty, 468, 469, 484, 493
scaled Sobolev, 367
semi, 230
Sobolev, 36, 55, 119, 230
Sobolev–Slobodeckij, 23
norms
equivalent, 648
notation
often used, 259
number of unknowns, $\mathcal{N}(h)$, 425

operator
P^h, Q'^h, 591
Q_n, Q'_n linear approximation, 656
ν–Stokes, 167, 169
adjoint, 17, 47
approximate, 346
bijective, 16
bounded, 97, 278
bounded linear, 13
Calderón–Zygmund, 654
classical, strong and weak form, 36
coercive, 97, 278
compact, 13, 14
continuous, 97, 278
continuous linear, 13
difference weak, 576
dual, 16, 47
elliptic, 40, 56, 81, 388
elliptic nonlinear, 80
estimates for nonlinear, 99, 141
existence for the quasilinear, 91
extension, 16, 22
extension on Ω, 246
Fredholm, 18, 51
Fredholm-index, 18
global interpolation, 222
hemi continuous, 97, 278
induced, 46
induced boundary, 42, 66
induced linear, 45
injective, 16
interpolation, 424
isoparametric interpolation, 254, 256
kernel, 14
linear, 41
linear, continuous, bounded, 14, 41
linearization, quasilinear, order $2m$, coercive, 158
linearized Navier–Stokes, 167, 286

linearized Navier-Stokes, 114, 167
Lipschitz-continuous, 97, 278
local interpolation, 213, 222
monotone, 34, 97, 278
natural boundary, 42, 66
Navier-Stokes, 114, 167
Nemyckii, 96, 98, 141
Nemyckii for systems, 140
nonlinear, 96
nonlinear system, 140
nonlinearcoercive, 97, 278
norm, 14
null space, 14
properties for nonlinear, 98, 140
properties of Nemyckii, 100
properties of the trace, 26
properties,compact, 15
pseudo–differential, 657
quasilinear, 81, 150
H_0^m- coercive linearization, 158
quasilinear order 2, 85
range, 14
saddle point problems, 163
self-adjoint, 17
semilinear, 81, 105
semilinear in nonlinear spaces, 86, 150
semilinear non autonomous, 150
semilinear Order $2m$, 149
special linear, 14
stable, 97, 278
stationary Naviér–Stokes, 281
Stokes, 116, 163
strictly monotone, 97, 278
strongly continuous, 97, 278
strongly elliptic, 40, 56, 81
strongly monotone, 97, 278
surjective, 16
symmetric, 17
trace or restriction, 26
uniformly elliptic, 56, 81, 109, 307, 320
uniformly monotone, 97, 278
weak, 160
optimal complexity, 443
optimal mesh
for 2D corner singularity, 427
optimality, 442
order, 115, 133
orthogonality
Galerkin, 440
outflow boundary point, 423

paralellotope, 214
parameter
chunkiness, 230
parametrization of $\partial\Omega$, 25
partition of unity, 25
PDE
divergence form, 34
divergent, 34
penalty
interior, 457
penalty function
interior discontinuities, 468
penalty norm, 491
plate
buckling, 11
pollution error, 434
polynomial
$\mathcal{P}^d = \mathcal{P}^1_d$, 215
$\mathcal{P}^n_d$, 215
$\mathcal{P}_d = \mathcal{P}^2_d$, 215
characteristic, 40
average Taylor, 230, 243
bilinear,biquadratic, 219
degree d, 214
positive, 99
asymptotic, 99
pre–wavelets, 642
primal generator, 645
principal part, 40, 64, 67, 73, 76, 80, 81, 106, 122, 134, 151, 158, 160
$\mathcal{V}^h_b-$coercive, 602
discrete, 574
elliptic operator, 56, 70, 133
symbol, 115
system, 115, 120, 133, 266
principal symbol
system, 120
principle
mesh independence, 486
problem
and Stokes, 282
convection-diffusion, 458
dual, 272, 404, 663
existence and uniqueness for semilinear, 106
extremal, 49
linearization, 151
nonlinear stationary, 683
reaction-diffusion, 458
saddle point, 163, 165, 166, 282, 664
unique solution, 166
scope of, 637
unique discrete solution++, 283
variational, 49, 64, 67, 126
product
tensor, 219
projector
$Q'^h = Q^{h'}$, 260, 270, 385, 401
approximation, $P^h, Q'^h = Q^{h'}$, 184
orthogonal, 17
property
k-linear algebraic, 342
Galerkin orthogonality, 472
proposition
useful, 366

quadratic eigenvalue problem, 423
quadrature formula
  Gauss, 390
  Gauss–Lobatto, 390
  Gauss–Radau, 390
  Gauss–type, 390
  high order, 390
quasi uniform, 229, 233
quasi-interpolant, 241
quasi-norm, 673
quasi-optimal tree in **TCOARSE**, 684
quasilinear
  problem violated boundary conditions, 399
quasilinear equation, 629
  special DCGM, 481
quasilinear operator
  special, 150
quasilinear system
  order 2, $2m$, 157
quasiuniform
  locally, 234

radial basis functions, 595
red refinement, 436
redundant system, 672
refinable, 641
refinement, red-green, 436
  blue, 436
  newest node bisection, 436
  longest node bisection, 437
regular discretisation, 428
regularity, 146
  in $\bar{\Omega}$, 65
  bootstrap arguments, 84
  in Morrey spaces, 125
  inherited, 84
  inherited to perturbations, 85
  results of Ladyženskaja/Uralceva, 93
  solutions of FEMs, 327
regularity for order $2m$, 137
relation
  two–scale, 641
residual error
  estimator, Poisson problem, 429
residuals, 433
Reynolds number, 164, 281
Richardson extrapolation
  high order difference, 631
Richardson iteration, 680
Riesz basis, 648, 669
rod
  axial load, 7
  axial loaded, 15
  bending, 3
  clamped end, 3
  loaded, 27
  perp load, 3
  simply supported end, 7
  stationary position, minimum, 8

saturation property, 679
scalar product, 69
  discrete Sobolev, 569, 614
  Lebesgue discrete, 569
  Sobolev, 36, 55, 119
scale representation
  multi, 643
  single, 643
scaling function, 641
Schwartz kernel, 678
semi norm, 55, 231
  broken Sobolev, 462
  discrete, 569, 614
  Sobolev, 36, 119
sequence
  approximating spaces, 267, 661
set
  admissible, 126
  dense, compact, 14
set of marked triangles $\mathcal{A}_h$, 435
shape factor, 428
Shortley–Weller approximation, 625
Shortly–Weller Method
  convergence, 625
singular solutions
  general structure, 422
singularities
  corner, 421
  edge, 423
SIPG, 468, 529
smooth manifold, 25
Sobolev are Banach spaces, 20
Sobolev space, 20, 422, 647
  broken, 231
  discrete, 566, 568
  discretization in discrete, 589
  fractional order, 23
  trace, fractional order, 27
solution
  Ω convex, 74
  hp-DCG approximate, 540
  classical, 38, 459, 475
  discrete approximate, 471, 474
  existence, 106, 110
  existence and uniqueness, 86, 87
  existence and uniqueness semilinear, 84
  existence/regularity, 96
  for convex domain, 74, 124
  for fully nonlinear, 111
  for quasilinear, 110
  generalized, 104, 121, 134, 143
    weak Euler system, 129
  interior regularity, 75
  linear and semilinear, 88
  regular, 60, 65, 73, 74, 95, 112, 123, 130, 139, 146

singular, 421
smooth data imply smooth solution, 170
spurious, 87
strong, 38, 42
strong and weak, 133
unique, 93
uniqueness, 106, 110
viscosity, 296, 299, 587
weak, 38, 42, 121, 134, 460, 475
weak discrete, 374
weak=strong, 42
space
admissible approximating, 180
approximating, 484
bi–dual, 180
conforming, 180
Galerkin, 180
generalized Petrov–Galerkin, 180
nonconforming, 180
Petrov–Galerkin, 180
Banach, 14
Banach, $\mathcal{L}(\mathcal{X},\mathcal{Y})$, 18
Banach, dense, 14
broken Sobolev, 462
discrete Sobolev, 569, 624
discrete Sobolev,$H^1_{\pm}(\Omega^h_e)$, 614
dual, 14
Hölder, 58
Hölder/Sobolev, 58
Morrey, 125
nonconforming approximating, 311
reflexive Banach, 47, 49
Sobolev, 36, 69, 119
surplus, 441
space refinement
adaptive, 669
uniform, 673
spectrum
bilinear form, 291
operator, 51, 52
spline
super spline subspace, 247
splitting
stable, 250
stability, 174, 190, 191, 239, 268, 485, 661
bound, 191
compactly perturbed discretization inherits, 408
discrete $W^{m,p}(\Omega)$ norm, 155
elliptic operators, 406
in u^h, 191, 268, 485, 590
proofs, 194
threshold, 191
stability and consistency yield convergence, 189
stability implies existing A^{-1}, 204
stability inherited to compact perturbations, 658
star of a vertex, 240
star-shaped, 230
state vector, 556
steepest descent method, 680
step size, 562
maximal, 229
Stokes problem
unique discrete solutions,convergence, 285
unique, solution, regularity, 166
strategy
fixed energy fraction, 435
maximum, 435
subdivision, 229, 231
cf. triangulation, 214
isoparametric, 254
$supp(\sigma)$, 25
symbol, 657
elliptic operator, 70
symmetric formula, 563
system
bounded nonlinearity, 145
Dirichlet, 56, 62, 132, 145
elliptic, 115
Euler, 34, 126
fully nonlinear, 158
quasilinear divergent, 140
linear elliptic
order $2m$, 132
nonlinear
order m, 112, 147
order $2m$, 115
nonlinear elliptic, 115
normal, 56
order $2m$, 336
quasilinear, 144
divergent, 158
quasilinear elliptic, 157
strongly elliptic, 120
uniform Legendre condition, 382
uniformly elliptic, 120, 266, 382
weak Euler, 129

tensor products, 644
term
penalty, 464
tetrahedra/bricks, 233
The Stokes problem, 666
theorem
(local) inverse function, 31
Aubin–Nitsche, 272
Banach, open mapping, 16
closed range, 18
extension of linear forms, operators, 16
Gauss integral, 27
Gårding, 71
Hahn–Banach Extension, 16
implicit function, 31
mean value, 30
modified Sobolev–Stein extension, 246
nonconforming Aubin–Nitsche, 405

theorem (*cont.*)
  Riesz–representation, 17
  Riesz–Schauder, 51
  Sobolev embedding, 22, 350
  Sobolev–Stein extension, 22
  trace, 26
  transformation, 367
theory
  Riesz–Schauder, 51
trace from Ω to $\partial\Omega$, 75
transform
  discrete Fourier, 603
  Fourier, 134
  inverse discrete Fourier, 602, 603
transformation
  anti crime, xvi, xxiii, 293, 408, 605
tree approximation, 684
triangles/quadrilaterals
  curvilinear, 233
triangulation, 214, 251, 254, 428, 461
  admissible, 214
  locally quasi-uniform, 540
  non–degenerate, 229, 231
  quasi-uniform, 229, 239
  shape-regular, 540
Tychonov regularization, 57

unique existence, 169
uniqueness
  in the small, 95
  interior, 65
uniqueness results, 91
unisolvent, 213
unit ball, 23
upper bound
  a posteriori, 430

variables
  nodal, 213
variational crimes, 175
variational formulation, 462
  discretization, 475
variational problem
  regular, 126
vertex
  interior, 247
vertices, 214

wavelet, 640
  analysis, 639
  bases, 647
  biorthogonal, 647
  composite, 652
  dual, 652
  evaluation of nonlinear functionals, 652
  generator, 643
  good, 643
  nonlinear approximation, 672
  orthonormal, 642
  sdiagonal preconditioning for all problems, 654
wavelet method
  conforming, 660
  convergence, 662
  linear elliptic equation
    order $2m$, 659
  not for fully nonlinear problems, 660
wavelet transform
  discrete, 640
  fast, 643
wavelets, 669
  conforming, 659
  elliptic equations, 659
  function spaces, 645
  on domains, 647
weak $\ell^p_w(J)$ space, 673
weak form, 423
welldefined G, G^h for the Monge-Ampere operator, 196